HANDBOOK
OF X-RAY ANALYSIS OF POLYCRYSTALLINE MATERIALS

SPRAVOCHNIK
PO RENTGENOSTRUKTURNOMU ANALIZU POLIKRISTALLOV

СПРАВОЧНИК
ПО РЕНТГЕНОСТРУКТУРНОМУ АНАЛИЗУ ПОЛИКРИСТАЛЛОВ

Handbook of
X-Ray Analysis
of Polycrystalline Materials

by

Lev Iosifovich Mirkin

Authorized translation from the Russian by
J. E. S. Bradley, B.Sc., Ph.D.

CONSULTANTS BUREAU
NEW YORK
1964

Acknowledgments

The publishers and the author take this opportunity to express their appreciation to Dr. Konrad Sagel, who has provided for this edition a number of new tabulations, and has considerably expanded several tables which first appeared in: K. Sagel, *Tabellen zur Roentgenstrukturanalyse*, Springer Verlag, 1958. *(Anleitungen für die chemische Laboratoriumspraxis*, Vol. VIII.)

We are also very grateful to Dr. Ludo K. Frevel for permission to reprint the tables which appear on pages 426 to 497 of this edition, which originally appeared in the periodical literature.

The material herein credited to the following sources:

Azaroff, L. V., and Buerger, M. J., *The Powder Method in X-Ray Crystallography*, 1958, McGraw-Hill Book Company

Buerger, M.J., *X-Ray Crystallography*, John Wiley and Sons, New York, 1942

Klug, H. P., and Alexander, L. E., *X-Ray Diffraction Procedures*, John Wiley and Sons, New York, 1954

Parrish, W., Eckstein, M. G., and Irwin, B. W., *Data for X-Ray Analysis*, Vol. II, Philips Technical Library

Peiser, H. S., *X-Ray Diffraction by Polycrystalline Materials*, Chapman & Hall, London; Reinhold, New York; 1960

is published with the kind permission of the authors and/or publishers, and we take this opportunity to express our appreciation.

Both Dr. Mirkin and Consultants Bureau extend their sincere apologies to any and all authors whose data have been used but inadvertently credited to secondary sources.

ISBN-13: 978-1-4684-6062-9   e-ISBN-13: 978-1-4684-6060-5
DOI: 10.1007/978-1-4684-6060-5

The original Russian text was published by Fizmatgiz, the State Press for Physics and Mathematics Literature, in Moscow in 1961, under the editorship of Prof. Ya. S. Umanskii.

A tabulation of interplanar distances and line strengths (pages 437 to 563 in the original text) has been omitted from the English edition.

Лев Иосифович Миркин.

Справочник по рентгеноструктурному анализу поликристаллов.

Library of Congress Catalog Card Number 64-23250

This translation is dedicated
to the memory of

Prof. Isidor Fankuchen,
Fan,

whose interest, advice, and constant encouragement
stimulated and sustained the Consultants Bureau program
of translations in crystallography.

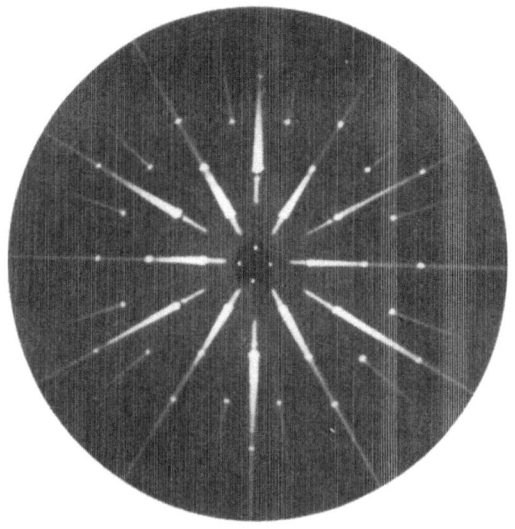

"His mind was an X-ray, his heart was a crystal."

H. F. Mark

The illustration on this page is an X-ray diffraction diagram of a single ice crystal made by
Professor Fankuchen at Polytechnic Institute of Brooklyn.

FOREWORD TO THE ENGLISH EDITION

The importance of x-ray structure analysis in physics, chemistry, metallurgy, and other branches of science has increased immeasurably since the discovery of the diffraction of x-rays by crystals 50 years ago. X-ray structure analysis has become divided into independent branches which now have little in common.

Nevertheless, the distinctive feature of almost all uses of x-ray methods is the need for extensive and laborious calculations, which require much tabulated data derived by experiment or calculation. Very often, much of the investigator's time is spent in producing these data or in extracting them from the literature (the annual output of papers on x-ray structure analysis and its applications now amounts to several thousand).

Another difficulty is the enormous variety of methods of examination and calculation; it is no easy matter to carry through the experiments and calculations and then to evaluate the results merely with the aid of the theoretical principles expounded in works on structure analysis.

No suitable work of reference for this purpose was available in the USSR or elsewhere, which led me to think of compiling the present book. In 1960 the author (jointly with N. N. Kachanov) published "X-Ray Analysis of Polycrystals: A Practical Manual" (Mashgiz, Moscow), which dealt with some problems in structure analysis, particularly those encountered in industrial research institutes and laboratories, for which detailed descriptions of methods were given. In 1961 the present book was published.

Specialization in x-ray structure analysis methods is now so great, and the amount of source material is now so vast, that it is practically impossible to include in one book all the information needed to solve all problems in structure analysis.

Some typographical errors and incorrect formulations have been pointed out since the book was published; most of these errors have been corrected in the translation.

I appreciate that all defects and errors cannot yet have been corrected, and I wish to thank in advance users of the book for their comments.

Institute of Mechanics　　　　　　　　　　　　　　L. Mirkin
Lomonosov University　　　　　　　　　　　　　　Moscow, December 1963
Moscow

FOREWORD FROM THE EDITOR

X-ray structure studies on crystalline materials have an important place in the testing and examination of materials generally. Metals, alloys, chemical compounds, and other crystalline materials are used in machines, metallurgy, construction, electronics, agriculture, and so on. Most technical materials are polycrystalline, so x-ray methods are of major importance to industry in the analysis of processes in solids.

A special feature of x-ray analysis is the extensive use of constants previously calculated or determined by experiment. The work also involves very laborious calculations, which can be much reduced by the use of suitable tables and nomograms.

Very often, the lack of suitable data in reference works forces one to carry out additional experiments and calculations, which can be very laborious.

Only one general reference work on x-ray analysis has been published in the USSR; this was compiled under the direction of A. I. Kitaigorodskii 20 years ago and is now a bibliographic rarity. It dealt reasonably fully with methods for monocrystals and polycrystals, and it gave the necessary tables, many of which have not yet lost their value; but, as a whole, it it out of date. Mikheev's recent "X-Ray Tables for the Identification of Minerals" are of great value, but they have only a narrow range of application.

Of foreign works, we may mention the "International Tables," which form a fundamental source (mainly for monocrystals), and also several works devoted to particular methods. Sagel's "Tabellen zur Röntgenstrukturanalyse" (Berlin, 1958) contains some novel and valuable tables and nomograms, but it is directed only to chemists and does not contain the information needed in the x-ray laboratory of a plant or industrial institute (e.g., in the determination of residual stresses); it even fails to give tables of interplanar distances for compounds.

The present book represents a first attempt to compile a reasonably complete collection of data for the main branches of structure analysis.

Use has been made of results from the many studies here and elsewhere that have been published up to 1959 or 1960; some of the tables have not previously been published.

The book is not intended to present the type of material to be found in textbooks; it gives only brief statements as to the use of the tables and nomograms, and only seldom are there deductions of formulas or descriptions of methods.

The book makes available a large assortment of numerical constants and accessory materials needed in the use of various methods.

The first part includes the data needed in the first stages of an analysis and is sufficient for most purposes. The material is presented in the sequence in which the analysis is usually performed.

The second part gives data needed in detailed applications; some of these (e.g., small-angle scattering, diffuse scattering, detailed crystal structure) have arisen only recently and are of great scientific and technical importance, but they do not yet have firmly established or classical methods. In such cases the data are those for the latest or most widely used methods.

Professor Ya. S. Umanskii

PREFACE

The book is best used in the following sequence. (1) The radiation and method of recording are selected in accordance with the data of Chapters 1 and 2; the detailed parameters for the recording are defined. (2) The patterns are indexed with the assistance of the graphs and tables of Chapter 3. (3) The measured intensities are compared with the values found from the tables of Chapter 4. (4) The particular problem at hand (determination of stresses, phase analysis, and so on) is solved with the aid of the tables and nomograms given in the second part of the book. The nomograms can be enlarged for use if necessary.

This is not the only mode of use; in particular, the material in the appropriate chapter may be sufficient for a particular type of routine analysis.

I have had the benefit of valuable advice from workers in various laboratories (Moscow State University, Moscow Steel Institute, the Institute of Crystallography, the Central Research Institute for Ferrous Metallurgy, the Technological Research Institute of the Automobile Industry, the Karpov Institute of Physical Chemistry, the All-Union Hard-Alloys Research Institute, and so on).

In addition, I am deeply indebted for much assistance to Professor Ya. S. Umanskii (scientific editor), Professor V. I. Iveronova, Professor A. I. Kitaigorodskii, G. A. Gol'der, and V. I. Rydnik.

I recognize that this work cannot be free from deficiencies, and I should like to thank in advance workers in x-ray laboratories who may offer criticisms.

Institute of Mechanics
Moscow State University

L. I. Mirkin

CONTENTS

Part 1

GENERAL METHODS OF X-RAY STRUCTURE ANALYSIS

Part 2

SOME SPECIAL PROBLEMS AND METHODS IN X-RAY STRUCTURE ANALYSIS

Part 1
General Methods of X-Ray Structure Analysis

CHAPTER 1

INTERACTION OF X-RAYS WITH MATTER.
X-RAY SPECTRA

This chapter deals with the processes that occur when x-rays interact with various materials.

See [1, 244] for detailed discussions of interaction processes.

1-1. Characteristic X-Rays

1-1a. Wavelengths of the K-Series X-Rays

The table below gives the wavelengths of the α_2, α_1, β_3, β_1, β_5, and β_2 lines of the K-series x-rays (in angstroms) [102]; the excitation potentials, absorption edges, and line intensities are given, as well as the initial orbits from which transitions to the K-shell give rise to the characteristic radiation. Elements shown with asterisks have been given the wavelengths of [2] multiplied by 1.00202; the others have been derived in the same way from other data [3].

z	Element	V_e, kV	$L_{II} \overset{\alpha_2}{\to} K$ strong	$L_{III} \overset{\alpha_1}{\to} K$ v. strong	$M_{II} \overset{\beta_3}{\to} K$ v. weak	$M_{III} \overset{\beta_1}{\to} K$ mod.	$N_{II} \overset{\beta_5}{\to} K$ v. weak	$N_{III} \overset{\beta_2}{\to} K$ weak	Abs. edge
3	Li	0,040	228,5	—	—	—	—	—	—
4	Be*	0,093	113,43		—	—	—	—	—
5	B*	0,166	67,64		—	—	—	—	—
6	C*	0,252	44,59		—	—	—	—	43,58
7	N*	0,372	31,634		—	—	—	—	31,168
8	O*	0,507	23,658		—	—	—	—	23,55
9	F*	0,664	18,370		—	—	—	—	—
10	Ne*	0,835	14,830		—	—	—	—	—
11	Na*	1,07	11,9090		—	11,6174	- -	—	—
12	Mg*	1,30	9,88894		—	9,55827	—		9,5115
13	Al*	1,55	8,33681		—	7,98109	—	—	7,9516
14	Si*	1,83	7,12536		—	6,76814	—	—	6,7446
15	P*	2,14	6,15441		—	5,80380	—	—	5,7866

z	Element	V_e, kV	$L_{II} \to K$ α_2 strong	$L_{III} \to K$ α_1 v. strong	$M_{II} \to K$ β_3 v. weak	$M_{III} \to K$ β_1 mod.	$N_{II} \to K$ β_5 v. weak	$N_{III} \to K$ β_2 weak	Abs. edge
16	S	2,46	5,37472	5,37196	—	5,0317	—	—	5,0184
17	Cl	2,82	4,73050	4,72760	—	4,4031	—	—	4,3969
18	Ar	—	4,18610	4,18317	—	3,8779	—	—	—
19	K	3,59	3,74462	3,74122	—	3,4538	3,44144	—	3,4365
20	Ca	4,00	3,26160	3,35825	—	3,0896	3,07420	—	3,0702
21	Sc	4,49	3,03452	3,03114	—	2,7800	2,76357	—	2,7578
22	Ti	4,95	2,75207	2,74841	—	2,51381	2,4987	—	2,4973
23	V	5,45	2,50729	2,50348	—	2,28434	2,2701	—	2,2690
24	Cr	5,98	2,29351	2,28962	—	2,08480	2,0709	—	2,0701
25	Mn	6,54	2,10568	2,10175	—	1,91015	1,89709	—	1,8964
26	Fe	7,10	1,93991	1,93597	—	1,75653	1,74432	—	1,7433
27	Co	7,71	1,79278	1,78892	—	1,62075	1,60896	—	1,6081
28	Ni	8,29	1,66169	1,65784	—	1,50010	—	1,48861	1,4880
29	Cu	8,86	1,54433	1,54050	—	1,39217	—	1,38102	1,3804
30	Zn	9,65	1,43894	1,43511	—	1,29522	1,2845	1,28366	1,2833
31	Ga	10,4	1,34394	1,34003	—	1,20785	1,1983	1,1962	1,1957
32	Ge	11,1	1,25797	1,25401	—	1,12890	1,1197	1,11684	1,1165
33	As	11,9	1,17981	1,17581	—	1,05726	1,0487	1,04492	1,0450
34	Se	12,7	1,10876	1,10471	—	0,99212	0,9843	0,97989	0,9798
35	Br	13,5	1,04376	1,03969	—	0,93273	0,9255	0,92039	0,9199
36	Kr	—	0,982	0,97810	—	0,87679	—	0,86480	—
37	Rb	15,2	0,92963	0,92551	0,82916	0,82863	0,8221	0,81641	0,8155
38	Sr	16,1	0,87938	0,87521	0,78334	0,78288	0,7769	0,77076	0,7697
39	Y	17,0	0,83300	0,82879	0,74112	0,74068	0,7347	0,72860	0,7276
40	Zr	18,0	0,79010	0,78588	0,70225	0,70169	—	0,68989	0,6888
41	Nb	19,0	0,75040	0,74615	0,66630	0,66572	—	0,65412	0,6529
42	Mo	20,0	0,71354	0,70926	0,632819	0,63225	—	0,62035	0,6198
43	Tc	—	0,6778	0,6735	—	0,6014	—	0,5899	—
44	Ru	22,1	0,64736	0,64304	0,57309	0,57246	—	0,56164	0,5605
45	Rh	23,2	0,61761	0,61325	0,54619	0,54559	—	0,53504	0,5338
46	Pd	24,4	0,58980	0,58542	0,52114	0,52052	—	0,51021	0,5092
47	Ag	25,5	0,56378	0,55936	0,49765	0,49701	—	0,48701	0,4858
48	Cd	26,7	0,53941	0,53498	0,47567	0,47506	—	0,46514	0,4641
49	In	27,9	0,51652	0,51209	0,45515	0,45451	—	0,44498	0,4439
50	Sn	29,1	0,49502	0,49056	0,43583	—	—	0,42585	0,4248

z	Element	V_e, kV	α_2 $L_{II}\to K$ strong	α_1 $L_{III}\to K$ v. strong	β_3 $M_{II}\to K$ v. weak	β_1 $M_{III}\to K$ mod.	β_5 $N_{II}\to K$ v. weak	β_2 $N_{III}\to K$ weak	Abs. edge
51	Sb	30,4	0,47479	0,47032	0,41707		—	—	0,4066
52	Te	31,8	0,45575	0,45126	0,40007		—	—	0,3897
53	I	33,2	0,43781	0,43329	0,38470	—	—	0,37547	0,3738
54	Xe	—	0,41960	0,41510	—	0,36770	—	0,35920	—
55	Cs	35,9	0,40481	0,40027	0,35508	—	—	0,34586	0,3447
56	Ba	37,4	0,38965	0,38509	0,34158	—	—	0,33289	0,3314
57	La*	38,7	0,37542	0,36996	0,32875	—	—	0,32031	—
58	Ce*	40,3	0,36494	0,35637	0,31636	—	—	0,30832	—
59	Pr*	41,9	0,34811	0,34399	0,30500	—	—	0,29685	—
60	Nd*	43,6	0,33582	0,33120	0,29410	—	—	0,28631	
61	Pm*	—	—	—	—	—	—	—	
62	Sm*	46,8	0,31325	0,30859	0,27380	—	—	0,26629	—
63	Eu*	48,6	0,30274	0,29805	0,26439	—	—	0,25697	—
64	Gd*	50,3	0,29246	0,28778	0,25522	—	—	0,24812	—
65	Tb*	52,0	0,28299	0,27825	0,24679	—	—	0,23960	—
66	Dy*	53,8	0,27369	0,26892	0,23835	—	—	0,23175	—
67	Ho*	55,8	0,26504	0,26036	—	—	—	—	—
68	Er*	57,5	0,25677	0,25204	0,22345	—	—	0,21715	—
69	Tm*	59,5	0,24865	0,24392	0,21602	—	—	—	—
70	Yb*	61,4	0,24103	0,23622	0,20958	—	—	0,20363	—
71	Lu*	63,4	0,23358	0,22887	0,20293	—	—	0,19689	—
72	Hf*	65,4	0,22658	0,22177	0,19623	—	—	0,19080	—
73	Ta	67,4	0,22029	0,215484	0,19029		—	0,18489	0,1839
74	W	69,3	0,21881	0,208992	0,18459		—	0,17934	0,1784
75	Re	—	0,20723	0,20241	—		—	—	—
76	Os	73,8	0,201626	0,196783	0,173607		—	0,16909	0,1678
77	Ir	76,0	0,195889	0,191033	0,16853		—	0,16409	0,1631
78	Pt	78,1	0,190372	0,185504	0,16366		—	0,15919	0,1581
79	Au	80,5	0,185064	0,180185	0,15897		—	0,15457	0,1534
80	Hg	82,9	0,17956	0,17473	—	0,15412	—	—	—
81	Tl	85,2	0,175038	0,170131	—	—	—	0,14568	0,1447
82	Pb	87,6	0,170285	0,165364	—	—	—	0,14154	0,1408
83	Bi	90,1	0,165704	0,160777	—	—	—	0,13649	0,1371
90	Th	109	0,137820	0,132806	—	—	—	0,1136	0,1129
92	U	115	0,130962	0,125940	—	—	—	0,10864	0,1068

1-1b. Wavelengths of the L-Series X-Rays

The table below gives the wavelengths of the L-series x-rays (in angstroms). The excitation potentials, absorption edges, and line intensities are given, as well as the initial orbits from which transitions originate.

The wavelengths have been calculated by multiplying Siegbahn's values [2] by 1.00202.

z	Element	V_e, kV	β_4 $M_{II} \to L_I$ mod.	β_3 $M_{III} \to L_I$ mod.	γ_2 $N_{II} \to L_I$ weak	γ_3 $N_{III} \to L_I$ weak	γ_4 $O_{III} \to L_{II}$ v. weak	η $M_I \to L_{II}$ weak	γ_5 $N_I \to L_{II}$ v. weak	β_1 $M_{IV} \to L_{II}$ strong
22	Ti	–	–	–	–	–	–	30,880	–	27,020
23	V	–	–	–	–	–	–	27,320	–	23,850
24	Cr	–	–	–	–	–	–	24,290	–	21,280
25	Mn	–	17,540	–	–	–	–	21,820	–	19,120
26	Fe	—	15,6415	15,6415	—	—	—	19,690	—	17,255
27	Co	—	—	—	—	—	—	17,806	—	15,652
28	Ni	—	13,1665	13,1665	—	—	—	16,203	—	14,269
29	Cu	—	12,1244	12,1244	—	—	—	14,860	—	13,056
30	Zn	1,20	11,1825	11,1825	—	—	—	13,637	—	11,984
31	Ga	1,31	—	—	—	—	—	12,585	—	11,032
32	Ge	1,41	—	—	--	—	—	11,610	—	10,174
33	As	1,52	8,9300	8,9300	—	—	—	10,733	—	9,414
34	Se	1,64	—	—	—	—	—	9,959	—	8,736
35	Br	1,77	—	—	—	—	—	9,254	—	8,125
37	Rb	2,05	6,8146	6,7831	6,0482	6,0482	—	—	—	—
38	Sr	2,19	6,4047	6,3710	5,6488	5,6488	—	7,521	6,2923	6,6234
39	Y	2,36	6,0198	5,9862	5,2809	5,2809	—	7,0442	—	6,2164
40	Zr	2,51	5,6631	5,6299	4,9508	4,9508	—	6,6072	5,4929	5,8354
41	Nb	2,68	5,3413	5,3078	4,6561	4,6561	—	6,209	—	5,4914
42	Mo	2,87	5,0512	5,0148	4,3783	4,3783	—	5,849	4,8410	5,1769
44	Ru	3,24	4,5217	4,4854	3,8958	3,8958	—	—	4,2852	4,6203
45	Rh	3,43	4,2888	4,2533	3,6889	3,6889	—	4,9211	4,0434	4,3728
46	Pd	3,64	4,0708	4,0338	3,4879	3,4879	—	4,6596	3,8193	4,1457
47	Ag	3,79	3,8689	3,8322	3,3065	3,3065	—	4,4190	3,6146	3,9345
48	Cd	4,07	3,6818	3,6437	3,1379	3,1379	—	4,1960	3,4250	3,7376
49	In	4,28	3,5061	3,4689	2,9796	2,9796	2,9250	3,9841	3,2483	3,5550
50	Sn	4,49	3,3430	3,3056	2,8358	2,8358	2,7769	3,7894	3,0836	3,3847
51	Sb	4,69	3,1907	3,1515	2,7007	2,7007	2,6389	3,6069	2,9315	3,2249
52	Te	4,93	3,0461	3,0074	2,5701	2,5701	2,5108	—	—	3,0766
53	I	5,18	2,9118	2,8740	2,4470	2,4470	2,3910	—	—	2,9368

γ_1 $N_{IV} \to L_{II}$ mod.	l $M_I \to L_{III}$ mod.	β_6 $N_I \to L_{III}$ weak	α_2 $M_{IV} \to L_{III}$ mod.	α_1 $M_V \to L_{III}$ v. strong	β_2 $N_V \to L_{III}$ v. weak	Abs. edge			Z	Element
						L_I	L_{II}	L_{III}		
—	31,360	—	27,39	—	—	—	—	—	22	Ti
—	27,770	—	24,26	—	—	—	—	—	23	V
—	24,790	—	21,67	—	—	—	—	—	24	Cr
—	22,270	—	19,45	—	—	—	—	—	25	Mn
—	20,161	—	17,616	17,616	—	—	—	—	26	Fe
—	18,237	—	15,972	15,972	—	—	—	—	27	Co
—	16,583	—	14,559	14,559	—	—	—	—	28	Ni
—	15,221	—	13,333	13,333	—	—	—	—	29	Cu
—	13,978	—	12,255	12,255	—	—	—	—	30	Zn
—	12,916	—	11,293	11,293	—	—	—	—	31	Ga
—	11,946	—	10,436	10,436	—	—	—	—	32	Ge
—	11,070	—	9,671	9,671	—	—	—	—	33	As
—	10,293	—	8,990	8,990	—	—	—	—	34	Se
—	9,583	—	8,375	8,375	—	—	—	—	35	Br
—	—	6,9822	—	—	—	5,9975	—	6,8551	37	Rb
—	7,838	6,5212	6,8624	6,8624	—	5,5826	6,1745	6,3749	38	Sr
—	—	6,0980	6,4487	6,4487	—	5,2321	5,7489	5,9564	39	Y
5,3847	6,913	5,7042	6,0689	6,0689	5,5855	4,8672	5,3767	5,5722	40	Zr
5,0350	6,523	5,3579	5,730	5,7235	5,2366	4,5809	—	5,2226	41	Nb
—	—	—	5,412	5,4059	4,9199	4,2984	4,7215	4,9141	42	Mo
4,1812	5,4975	4,4865	4,8535	4,8455	4,3707	—	4,1777	4,3692	44	Ru
3,9437	5,2175	4,2414	4,6049	4,5971	4,1304	3,6259	3,9419	4,1295	45	Rh
3,7239	4,9496	4,0151	4,3754	4,3673	3,9086	3,4275	3,7227	3,9118	46	Pd
3,5220	4,7071	3,8063	4,1622	4,1540	3,7013	3,2540	3,5138	3,6983	47	Ag
3,3347	4,4803	3,6146	3,9644	3,9558	3,5135	3,0835	3,3259	3,5034	48	Cd
3,1617	4,2679	3,4349	3,7800	3,7713	3,3379	2,9253	3,1458	3,3222	49	In
3,0009	4,0715	3,2688	3,6084	3,5995	3,1743	2,7752	2,9783	3,1557	50	Sn
2,8508	3,8881	3,1134	3,4478	3,4387	3,0227	2,6370	2,8276	2,9967	51	Sb
2,7120	3,7176	2,9704	3,2976	3,2886	2,8819	2,5090	2,6847	2,8514	52	Te
2,5827	3,5569	2,8362	3,1573	3,1480	2,7516	2,3887	2,5526	2,7194	53	I

z	Element	V_e, kV	β_4 $M_{II} \to L_I$ mod.	β_3 $M_{III} \to L_I$ mod.	γ_2 $N_{II} \to L_I$ weak	γ_3 $N_{III} \to L_I$ weak	γ_4 $O_{III} \to L_{II}$ v. weak	η $M_I \to L_{II}$ weak	γ_6 $N_I \to L_{II}$ v. weak	β_1 $M_{IV} \to L_{II}$ strong	γ_1 $N_{IV} \to L_{II}$ mod.	l $M_I \to L_{III}$ mod.
55	Cs	5,71	2,6659	2,6282	2,2367	2,2315	2,1735	2,9893	2,4160	2,6832	2,3472	3,2662
56	Ba	5,99	2,5550	2,5161	2,1383	2,1338	2,0757	2,8629	2,3070	2,5674	2,2411	3,1350
57	La	6,26	2,4487	2,4102	2,0447	2,0407	1,9827	2,740	2,2052	2,4583	2,1415	3,0061
58	Ce	6,54	2,3489	2,3106	1,9599	1,9548	1,8990	2,6200	2,1099	2,3557	2,0484	2,8915
59	Pr	6,83	2,2546	2,2169	1,8788	1,8737	1,8190	2,512	2,0202	2,2585	1,9608	2,7837
60	Nd	7,12	2,1666	2,1265	1,8010	1,7961	1,7443	2,4091	1,9352	2,1666	1,8776	2,6757
62	Sm	7,73	2,0004	1,9620	1,6592	1,6550	1,6065	2,218	1,7787	1,9976	1,7266	2,482
63	Eu	8,04	1,9260	1,8865	1,5971	1,5909	1,5438	—	1,708	1,9202	1,6576	2,3951
64	Gd	8,37	1,8530	1,8146	1,5341	1,5290	1,4848	—	1,6409	1,8462	1,5918	2,3118
65	Tb	8,70	1,7850	1,7460	1,4768	1,4713	1,4268	—	1,5774	1,7763	1,5297	2,2335
66	Dy	9,03	1,7202	1,6811	1,4232	1,4168	1,3742	1,8960	1,5183	1,7100	1,4727	2,1584
67	Ho	9,38	1,6586	1,6193	1,3705	1,3640	1,3224	1,8257	1,462	1,6468	1,4171	2,0863
68	Er	9,73	1,5996	1,5610	1,3211	1,3144	1,2758	1,7583	1,406	1,5866	1,3651	2,0192
69	Tu	10,1	1,5443	1,5053	1,2738	1,2679	1,2289	1,6957	1,3550	1,5299	1,3154	1,9550
70	Yb	10,5	1,4912	1,4523	1,2281	1,2223	1,1844	1,634	1,306	1,4755	1,2674	1,894
71	Lu	10,9	1,4401	1,4010	1,1856	1,1799	1,1434	1,5770	1,259	1,4236	1,2228	1,8355
72	Hf	11,3	1,3921	1,3524	1,1436	1,1379	1,1023	1,5228	1,2145	1,3739	1,1789	1,7810
73	Ta	11,7	1,34578	1,30672	1,1052	1,09930	1,0648	1,4709	1,1732	1,32690	1,13787	1,7284
74	W	12,1	1,30141	1,26247	1,06803	1,06201	1,0279	1,4210	1,1321	1,28175	1,09851	1,6784
75	Re	12,5	1,2588	1,2201	1,0320	1,0257	0,9930	1,3734	1,0934	1,23853	1,0608	1,6306
77	Ir	13,4	1,17953	1,14077	0,96527	0,95906	0,9276	1,2843	1,0216	1,15773	0,99076	—
78	Pt	13,9	1,14216	1,10388	0,93421	0,92786	0,8970	1,2428	0,9877	1,11984	0,95792	1,4994
79	Au	14,4	1,10645	1,06765	0,90430	0,89762	0,8672	1,2027	0,9555	1,08346	0,92648	1,4598
80	Hg	14,8	1,0714	1,03254	0,8742	0,8679	0,8378	1,1639	0,9248	1,04863	0,8964	1,42128
81	Tl	15,3	1,03908	1,00052	0,84742	0,84104	0,8117	1,1277	0,8947	1,01504	0,86746	1,3847
82	Pb	15,8	1,00766	0,96916	0,82082	0,81457	0,7859	1,0922	0,8664	0,98281	0,83970	1,3501
83	Bi	16,4	0,97698	0,93855	0,79560	0,79105	0,7608	1,0586	0,8394	0,95194	0,81307	1,3164
90	Th	20,5	0,79352	0,75476	0,64208	0,63541	0,6107	0,8545	0,6748	0,76510	0,65308	1,1150
91	Pa	—	0,7699	0,7322	0,6239	0,6168	0,5937	0,8295	0,6549	0,7422	0,6338	1,0907
92	U	21,7	0,7479	0,71022	0,60508	0,59832	0,5748	0,8051	0,6355	0,71996	0,61483	1,0671

β_6 $N_I \to L_{III}$ weak	β_7 $O_I \to L_{III}$ v. weak	α_2 $M_{IV} \to L_{III}$ mod.	α_1 $M_V \to L_{III}$ v. strong	β_2 $N_V \to L_{III}$ mod.	β_5 $O_{IV} \to L_{III}$ v. weak	β_{10} $M_{IV} \to L_I$ v. weak	Abs. edge			z	Element
							L_I	L_{II}	L_{III}		
2,5927	–	2,9014	2,8919	2,5115	–	–	2,1649	2,3122	2,4724	55	Cs
2,4822	–	2,7846	2,7752	2,4041	–	–	2,0662	2,2037	2,3616	56	Ba
2,3787		2,6743	2,6651	2,3026	–	–	1,9729	2,1031	2,2583	57	La
2,2815	2,1807	2,5703	2,5612	2,2086	–	2,1960	1,8894	2,0108	2,1641	58	Ce
2,1903	2,0916	2,4726	2,4627	2,1191	–	2,1067	1,8108	1,9240	2,0770	59	Pr
2,1035	2,0083	2,3804	2,3701	2,0355	–	2,0234	1,7352	1,8428	1,9947	60	Nd
1,9461	1,8560	2,2102	2,1994	1,8819	–	1,8695	1,5986	1,7025	1,8445	62	Sm
1,8743	1,788	2,1316	2,1206	1,8119	–	1,800	1,5364	1,6261	1,7753	63	Eu
1,8067	1,7231	2,0567	2,0460	1,7454	–	1,7316	1,4770	1,5618	1,7096	64	Gd
1,7410	1,6591	1,9863	1,9755	1,6824	–	1,667	1,4210	1,5011	1,6486	65	Tb
1,6811	1,5989	1,9195	1,9084	1,6231	–	–	1,3676	1,4443	1,5902	66	Dy
1,6221	–	1,8558	1,8447	1,5669	–	–	1,3173	1,3897	1,5353	67	Ho
1,5668	1,4922	1,7950	1,7840	1,5137	–	–	1,2681	1,33830	1,48218	68	Er
1,5146	–	1,7374	1,7263	1,4631	–	–	1,2221	1,2875	1,4328	69	Tu
1,4657	–	1,6823	1,6712	1,4157	–	–	1,1788	1,24064	1,38543	70	Yb
1,4172	1,3486	1,62965	1,61877	1,3700	1,3425	1,3425	1,13851	1,1964	1,3402	71	Lu
1,3739	1,3061	1,58023	1,56923	1,3262	1,2993	1,2993	1,0975	1,1538	1,2956	72	Hf
1,3311	1,2638	1,53287	1,52192	1,28449	1,2557	1,2539	1,059	1,1124	1,2542	73	Ta
1,2896	1,2242	1,48738	1,47634	1,24454	1,2154	1,2120	1,0256	1,0739	1,2154	74	W
1,2506	1,1857	1,4439	1,43286	1,2065	1,1771	1,1722	0,9893	1,0375	1,1779	75	Re
1,17782	1,1148	1,3625	1,35119	1,13526	1,10580	1,0970	0,9242	0,9674	1,1060	77	Ir
1,14330	1,0816	1,32422	1,31298	1,10196	1,07237	–	0,8932	0,9340	1,0732	78	Pt
1,11087	1,0519	1,28762	1,27634	1,07017	1,0401	1,0281	0,8639	0,9027	1,0403	79	Au
1,0790	1,0176	1,25203	1,24113	1,03980	1,0087	0,9956	0,8359	0,8726	1,0095	80	Hg
1,04960	1,0000	1,21872	1,20736	1,01026	0,98047	0,9635	0,8088	0,8436	0,9798	81	Tl
1,02112	0,9622	1,18647	1,17495	0,98281	0,95269	0,9342	0,7828	0,8159	0,9511	82	Pb
0,99331	0,9349	1,15534	1,14381	0,95517	0,92552	0,9053	0,7574	0,7894	0,9240	83	Bi
0,82813	0,7744	0,96780	0,95598	0,79352	0,76510	0,7301	0,6051	0,6306	0,7615	90	Th
0,8078	0,7545	0,9446	0,9328	0,7737	0,7452	0,7087	—	—	—	91	Pa
0,78838	0,7361	0,92248	0,91058	0,75459	0,72631	0,6878	0,5691	0,5925	0,7223	92	U

1-1c. Relative Intensities of the K Lines in the Characteristic Spectrum

The following table gives the relative spectral intensities [1], α_1 being taken as 100. The values for α_1, α_2, and β_1 (strong lines) are accompanied by ones for β_2 and β_5 (weaker), because the latter can sometimes be recorded with modern ionization equipment when monochromatic radiation or discrimination systems are used.

Z	Element	α_1	α_2	β_1	β_2	β_5	Z	Element	α_1	α_2	β_1	β_2	β_5
23	V	100	52,1	20,5	—	0,48	39	Y	100	50,0	23,3	3,19	—
24	Cr	100	50,6	21,0	—	0,66	40	Zr	100	49,1	21,9	3,28	—
25	Mn	100	54,9	22,4	—	0,34	41	Nb	100	49,7	21,4	3,32	—
26	Fe	100	49,1	18,2	—	0,26	42	Mo	100	50,6	23,3	3,48	—
27	Co	100	53,2	19,1	—	0,23	44	Ru	100	51,1	23,3	3,96	—
28	Ni	100	47,6	17,1	—	0,20	45	Rh	100	51,2	25,3	3,97	—
29	Cu	100	46,0	15,8	—	0,15	46	Pd	100	52,3	24,8	4,14	—
30	Zn	100	48,9	18,5	0,19	—	47	Ag	100	51,7	24,0	4,22	—
31	Ga	100	50,6	21,6	—	—	48	Cd	100	53,8	26,1	4,18	—
32	Ge	100	50,7	22,8	0,46	—	49	In	100	51,8	21,7	3,65	—
33	As	100	49,2	21,7	0,69	—	50	Sn	100	49,9	29,6	7,02	—
34	Se	100	50,3	21,0	1,07	—	51	Sb	100	50,3	31,0	7,08	—
35	Br	100	50,9	22,2	1,73	—	52	Te	100	49,7	30,6	7,35	—
37	Rb	100	49,3	23,0	2,62	—	74	W	100	47,0	18,1	—	0,25
38	Sr	100	48,6	21,8	2,72	—	78	Pt	100	52,0	20,0	—	0,38

1-1d. Widths of the Lines of the Characteristic Spectrum

The table gives the true (spectral) widths $\Delta\lambda$ for patterns free from distortion caused by structure defects in the specimen, recording conditions, and other factors [1].

The widths for α_1 and α_2 (K series) have been measured in X units by spectral methods from the ordinates of the peaks in the characteristic spectrum.

The ΔE are the line widths in electron volts.

A point to be borne in mind here is that these values relate to the anode material, not to the specimen.

Element	α_1		α_2		Element	α_1		α_2	
	$\Delta\lambda$	ΔE	$\Delta\lambda$	ΔE		$\Delta\lambda$	ΔE	$\Delta\lambda$	ΔE
20 Ca	1,60	1,76	1,50	1,69	32 Ge	0,43	3,4	0,46	3,7
22 Ti	1,22	2,00	1,43	2,33	38 Sr	0,35	5,7	0,36	5,8
23 V	1,15	2,26	1,35	2,67	40 Zr	0,33	6,6	0,35	6,9
24 Cr	1,03	2,22	1,23	2,89	41 Nb	0,33	7,4	0,31	6,9
25 Mn	1,10	3,09	1,22	3,43	42 Mo	0,29	7,2	0,32	7,7
26 Fe	1,01	3,34	1,12	3,60	44 Ru	0,29	8,7	0,29	8,6
27 Co	0,81	3,1	0,95	3,7	45 Rh	0,29	9,5	0,29	9,4
28 Ni	0,68	3,07	0,85	3,78	46 Pd	0,28	10,1	0,29	10,3
29 Cu	0,58	3,0	0,77	4,0	47 Ag	0,28	11,1	0,29	11,3
30 Zn	0,51	3,1	0,58	3,5	74 W	0,152	43,0	0,153	43,3

1-1e. Asymmetry Indices for Lines of the Characteristic Spectrum

The asymmetry indices given below for the K series have been derived by spectral methods [1].

The indices are deduced by drawing a straight line parallel to the background through the middle of the ordinate representing the peak, the long-wave (α) and short-wave (β) intercepts being measured. The index is

$$a = \frac{\alpha}{\beta} \cdot$$

Values are given for α_1 and α_2; they refer to the anode, not to the specimen.

Element	$K\alpha_1$	$K\alpha_2$	Element	$K\alpha_1$	$K\alpha_2$
Sc	1,0	1,0	Cu	1,2	1,3
	1,2	1,0	Zn	1,1	1,3
V	1,2	1,1	Ga	1,0	1,2
Cr	1,4	1,0	Ge	1,0	1,1
Mn	1,5	1,3	Sr	1,0	1,1
Fe	1,6	1,3	Zr	1,0	1,1
Co	1,4	1,3	Mo	1,0	1,0
Ni	1,2	1,3			

1-2. Conversion of kX Units to Angstroms

The table is used to convert interplanar distances d in kX units to angstroms [247]; d_0 in $\overset{\circ}{A}$ is related to d in kX by $d_0 = 1.00202d$.

The middle column gives D, the difference between d_0 and d; the following example illustrates its use: for d (or a period a) between 2.7970 and 2.8465 kX, the correction required for the conversion to $\overset{\circ}{A}$ is the addition of D = 56 · 10^{-4} or 0.0056.

See [245, 246] for questions concerning the relations between the units of wavelength.

kX	$D \cdot 10^4$	$\overset{\circ}{A}$	kX	$D \cdot 10^4$	$\overset{\circ}{A}$	kX	$D \cdot 10^4$	$\overset{\circ}{A}$	kX	$D \cdot 10^4$	$\overset{\circ}{A}$
2,2525	45	2,2571	2,9950	60	3,0010	3,7376	75	3,7451	4,4802	90	4,4893
2,3020	46	2,3067	3,0446	61	3,0507	3,7871	76	3,7947	4,5297	91	4,5388
2,3515	47	2,3563	3,0941	62	3,1004	3,8366	77	3,8443	4,5792	92	4,5884
2,4010	48	2,4059	3,1436	63	3,1500	3,8861	78	3,8939	4,6287	93	4,6380
2,4505	49	2,4555	3,1931	64	3,1996	3,9356	79	3,9435	4,6782	94	4,6876
2,5000	50	2,5050	3,2426	65	3,2492	3,9852	80	3,9932	4,7277	95	4,7372
2,5495	51	2,5546	3,2921	66	3,2988	4,0347	81	4,0429	4,7772	96	4,7868
2,5990	52	2,6042	3,3416	67	3,3484	4,0842	82	4,0925	4,8267	97	4,8364
2,6485	53	2,6538	3,3911	68	3,3980	4,1337	83	4,1421	4,8762	98	4,8860
2,6980	54	2,7034	3,4406	69	3,4476	4,1832	84	4,1917	4,9257	99	4,9356
2,7475	55	2,7530	3,4901	70	3,4972	4,2327	85	4,2413	5,0248	101	5,0350
2,7970	56	2,8026	3,5396	71	3,5467	4,2822	86	4,2909	5,0743	102	5,0846
2,8465	57	2,8522	3,5891	72	3,5963	4,3317	87	4,3405	5,1238	103	5,1342
2,8960	58	2,9018	3,6386	73	3,6459	4,3812	88	4,3901	5,1733	104	5,1838
2,9455	59	2,9514	3,6881	74	3,6955	4,4307	89	4,4397	5,2228	105	5,2336

kX	$D \cdot 10^4$	Å	kX	$D \cdot 10^4$	Å	kX	$D \cdot 10^4$	Å	kX	$D \cdot 10^4$	Å
5,2723	106	5,2830	6,4604	130	6,4735	7,6485	154	7,6639	8,8366	178	8,8544
5,3218	107	5,3326	6,5099	131	6,5230	7,6980	155	7,7135	8,8861	179	8,9040
5,3713	108	5,3822	6,5594	132	6,5726	7,7475	156	7,7631	8,9357	180	8,9538
5,4208	109	5,4318	6,6089	133	6,6222	7,7970	157	7,8127	8,9852	181	9,0034
5,4703	110	5,4814	6,6584	134	6,6718	7,8465	158	7,8623	9,0347	182	9,0530
5,5198	111	5,5309	6,7079	135	6,7214	7,8960	159	7,9119	9,0842	183	9,1026
5,5963	112	5,5805	6,7574	136	6,7710	7,9456	160	7,9617	9,1337	184	9,1522
5,6188	113	5,6301	6,8069	137	6,8206	7,9951	161	8,0113	9,1832	185	9,2018
5,6683	114	5,6797	6,8564	138	6,8702	8,0446	162	8,0609	9,2327	186	9,2514
5,7178	115	5,7293	6,9059	139	6,9198	8,0941	163	8,1105	9,2822	187	9,3010
5,7673	116	5,7789	6,9555	140	6,9696	8,1436	164	8,1601	9,3317	188	9,3506
5,8168	117	5,8285	7,0050	141	7,0192	8,1931	165	8,2097	9,3812	189	9,4002
5,8663	118	5,8781	7,0545	142	7,0688	8,2426	166	8,2593	9,4307	190	9,4498
5,9158	119	5,9277	7,1040	143	7,1184	8,2921	167	8,3089	9,4802	191	9,4994
5,9654	120	5,9775	7,1535	144	7,1680	8,3416	168	8,3585	9,5297	192	9,5489
6,0149	121	6,0271	7,2030	145	7,2176	8,3911	169	8,4081	9,5792	193	9,5985
6,0644	122	6,0767	7,2525	146	7,2672	8,4406	170	8,4577	9,6287	194	9,6481
6,1139	123	6,1263	7,3020	147	7,3168	8,4901	171	8,5073	9,6782	195	9,6977
6,1634	124	6,1759	7,31515	148	7,3664	8,5396	172	8,5568	9,7277	196	9,7473
6,2129	125	6,2255	7,4010	149	7,4160	8,5891	173	8,6064	9,7772	197	9,7969
6,2624	126	6,2751	7,4505	150	7,4656	8,6386	174	8,6560	9,8267	198	9,8465
6,3119	127	6,3247	7,5000	151	7,5152	8,6881	175	8,7056	9,8762	199	9,8961
6,3614	128	6,3743	7,5495	152	7,5647	8,7376	176	8,7552	9,9258	200	9,9459
6,4109	129	6,4239	7,5990	153	7,6143	8,7871	177	8,8048			

1-3. Relations between Units for Absorption Coefficients

Quantities concerned with the interaction of x-rays with matter may be measured in linear, mass, or atomic units.

The scattering, absorption, or attenuation coefficients (cross sections) have the dimensions cm^2/g or barn (10^{-24} cm^2/atom); the linear coefficients have the dimensions of length.

The following table gives factors for converting from barns to cm^2/g; a quantity in barns is converted to one in cm^2/g by multiplying by k [4].

Material	k	Material	k	Material	k
H	0,5997	P	0,01945	I	0,004747
Be	0,06684	S	0,01879	W	0,003276
C	0,05016	Ar	0,01508	Pt	0,003086
N	0,04301	K	0,01541	Tl	0,002948
O	0,03765	Ca	0,01503	Pb	0,002908
Na	0,02620	Fe	0,01079	U	0,002531
Mg	0,02477	Cu	0,009482	H_2O	0,03344
Al	0,02233	Mo	0,006279	NaI	0,004019
Si	0,02145	Sn	0,005076	$CaPO_3$	0,001942

1-4. Scattering of X-Rays

1-4a. Scattering of X-Rays by Electron Shells and Nuclei

Total scattering cross sections are given for Pb, I, Cu, and C; these include the cross sections for coherent and incoherent scattering and also pair formation (top line) [4]; the lower line gives the scattering cross section for the nuclear photoeffect. These cross sections are in barns.

Element \ Energy, MeV	10	12	13,2	14	15,2	16	18	19,2
Pb	16,8 0,2	17,7 0,5	18,3 0,84	18,6 0,8		19,5 0,4	20,3 0,2	
I		8,66 0,16		9,02 0,36	9,24 0,47	9,37 0,41	9,68 0,31	
Cu				3,32 0,05		3,38 0,09	3,45 0,11	3,49 0,12
C								

Element \ Energy, MeV	20	21	22	23	24	25	26
Pb							
I	9,96 0,19						
Cu	3,52 0,11		3,59 0,03				
C		0,303 0,002	0,300 0,009	0,297 0,013	0,294 0,009	0,292 0,005	0,289 0,003

1-4b. Scattering of X-Rays in Gases

Scattering coefficients σ are given for certain gases; these are useful in comparing patterns recorded in air, in vacuum, and in special atmospheres [109].

Characteristic radiation	Wave-length, X	Air 0°C; 760 mm Hg	SO_2 0°C; 760 mm Hg	C_2H_5Br 0°C; 116 mm Hg	CH_3I 0°C; 135 mm Hg
Al	8360	1,484	—	—	—
Fe	1930	0,0254	0,24	0,112	0,384
Cu	1541	0,0130	0,134	0,057	0,273
Se	1107	0,0044	0,0546	0,024	0,11
Sr	873	0,0026	0,0281	0,071	0,035
Ag	565	—	0,0079	0,0236	0,020

1-4c. Scattering Coefficients for X-Rays

The total scattering coefficient σ is made up of the true one, σ_s, and of σ_r, the incoherent scattering coefficient; values are given for the mass scattering coefficient σ/ρ (cm^2/g) for various elements and wavelengths [111].

Element \ λ, kX	0,13—2,5	0,1—1,0	0,12	0,25—0,98	0,16—0,8	0,10—0,17	0,1—0,3	0,12—0,39	0,13—1,05
H	—	—	—	0,37	—	—	—	—	0,309
Li	—	—	—	—	—	—	—	—	0,157
C	0,16	0,145	0,145	0,18	—	—	—	—	0,175
N	—	—	—	—	—	—	—	—	0,168
O	—	—	0,16	—	—	—	—	—	0,165
Mg	—	—	0,144	—	—	—	—	—	—
Al	0,14	0,147	0,144	—	0,13	0,147	—	0,12	0,173
S	0,14	—	0,15	—	—	—	—	—	—
Fe	—	0,17	0,16	—	—	—	0,18	—	0,18
Co	—	—	—	—	0,27	—	0,18	—	—
Ni	—	—	0,17	—	—	—	0,20	—	—
Cu	—	0,19	0,18	—	0,29	0,13	0,20	0,12	—
Zn	—	—	0,20	—	0,30	—	—	—	—
Mo	—	0,56	0,28	—	—	—	—	—	—
Pd	—	—	0,32	—	—	—	—	—	—
Ag	—	0,68	0,35	—	0,47	—	—	—	—
Sn	—	0,70	0,40	—	0,50	—	—	—	—
W	—	0,65	0,50	—	—	—	—	—	—
Pt	—	0,58	0,58	—	—	—	—	—	—
Au	—	0,73	0,60	—	—	—	—	—	—
Pb	—	0,69	0,67	—	0,82	—	—	—	—
Bi	—	—	0,70	—	—	—	—	—	—

1-4d. Mass Scattering Coefficients σ_s/ρ

Experimental values are given for σ_s/ρ (cm²/g) for various elements and wavelengths [111].

λ, kX \ Substance	H	Li	B	C	Air	H_2O	Na	Mg	Al	S	Cu	Ag
0,0199	—	—	—	0,080	—	—	—	—	0,077	—	—	—
0,0208	—	—	—	0,082	—	—	—	—	0,079	—	—	—
0,0248	—	—	—	0,086	—	—	—	—	0,083	—	—	—
0,0256	—	—	—	0,087	—	—	—	—	0,084	—	—	—
0,0363	—	—	—	0,100	—	—	—	—	0,097	—	—	—
0,0504	—	—	—	0,113	—	—	—	—	0,109	—	—	—
0,082	—	—	—	—	—	—	—	—	0,130	—	—	—
0,10	0,285	—	—	0,121	—	—	0,124	0,126	0,121	0,118	—	—
0,14	0,303	—	—	0,129	—	—	0,140	0,141	0,134	0,126	—	—
0,161	—	—	—	—	—	0,185	—	—	—	—	—	—
0,173	—	—	—	—	—	—	—	—	0,161	—	—	—
0,20	0,316	—	—	0,135	—	—	0,146	0,148	0,139	0,132	—	—
0,225	—	—	—	—	—	—	—	—	0,147	—	—	—
0,240	—	—	—	0,138	—	0,206	0,148	0,159	0,146	0,142	—	—
0,28	0,333	—	—	0,140	—	—	0,153	0,177	0,154	0,183	—	—
0,285	—	—	—	—	—	0,170	—	—	—	—	—	—
0,32	—	0,133	0,154	0,166	—	0,198	0,173	—	—	—	—	—
0,373	—	—	—	—	—	—	—	—	0,156	—	—	—
0,43	—	0,165	0,162	0,182	—	0,206	0,191	—	—	—	—	—
0,458	—	—	—	—	—	—	—	—	0,260	—	—	—
0,501	—	—	—	—	—	0,201	—	—	—	—	—	—
0,54	—	0,157	0,169	0,194	—	0,210	0,248	—	—	—	—	—
0,56	—	—	—	0,2	0,2	—	—	—	0,2	—	0,4	—
0,66	—	0,169	0,165	0,214	—	0,216	—	—	—	—	—	—
0,71	0,46	0,168	—	0,20	—	—	—	—	—	—	—	—
0,79	—	0,200	0,179	0,234	—	0,228	—	—	—	—	—	—
2,28	—	—	—	0,2	0,2	—	—	—	0,2	—	0,4	1,5

1-4e. Scattering Coefficients σ_e

The scattering coefficient per electron is calculated as

$$\sigma_e = \frac{\sigma}{\varrho}\,\frac{A}{ZN}\,,$$

where A is the atomic weight, Z is the atomic number, and N is Avogadro's number. The values given are $\sigma_e \cdot 10^{27}$ cm² for various elements and wavelengths [111].

λ, kX Element	0,005	0,010	0,020	0,025	0,036	0,050	0,064	0,098	0,130	0,200	0,417
H	128	195	—	270	—	380	410	460	530	620	640
Li	—	—	—	—	—	—	410	460	—	570	620
Be	—	—	—	—	—	—	—	—	—	580	600
B	—	—	—	—	—	—	440	480	—	570	590
C	128	197	276	270	331	390	420	460	500	550	570
N	—	—	—	—	—	—	—	—	—	540	—
O	—	195	—	—	—	—	420	460	—	520	540
Ne	—	—	—	—	—	—	—	—	—	530	—
Na	—	—	—	—	—	—	430	490	—	540	550
Mg	—	—	—	—	—	—	410	470	—	540	540
Al	129	197	263	270	324	380	420	490	510	530	550
S	128	198	—	270	—	—	420	460	—	600	600
Cl	—	—	—	—	—	—	430	500	—	660	680
Ar	—	—	—	—	—	—	—	—	—	700	—
Ca	—	—	—	—	—	—	430	510	—	720	820
Fe	133	203	—	280	—	430	450	575	670	900	1000
Cu	134	206	—	290	—	450	470	700	—	1400	1500
Mo	—	—	—	—	—	—	1000	1600	—	—	—
Sn	145	213	—	400	—	900	1300	2400	3200	5000	5500
Pb	172	300	—	950	—	2700	3500	4500	4100	—	—

1-4f. Incoherent Scattering Cross Sections for X-Rays

The Compton scattering coefficient for "free" electrons is given by the Klein-Nishina formula [248]; the differential scattering cross section for a photon of frequency ν having a deviation ϑ within a solid angle $d\Omega$ is given [4] by

$$d\sigma(\vartheta) = \frac{r_0^2}{2}\,\frac{1}{[1+\alpha(1-\cos\vartheta)]^2}\left\{1+\cos^2\vartheta+\frac{\alpha^2(1-\cos\vartheta)^2}{1+\alpha(1-\cos\vartheta)}\right\}d\Omega,$$

where

$$r_0^2 = \left(\frac{e^2}{mc^2}\right)^2 = 7.94\cdot10^{-26}\text{ cm}^2,\quad \alpha=\frac{h\nu}{mc^2}\,.$$

The integral scattering cross section is

$$\sigma_n = 2\pi r_0^2\left\{\frac{1+\alpha}{\alpha^3}\left[\frac{2\alpha(1+\alpha)}{1+2\alpha}-\ln(1+2\alpha)\right]+\frac{\ln(1+2\alpha)}{2\alpha}-\frac{1+3\alpha}{(1+2\alpha)^2}\right\}.$$

Values of σ_n are given for various energies [4].

Energy, MeV	σ, cm^2/electron $\times 10^{24}$	Energy, MeV	σ, cm^2/electron $\times 10^{24}$
0,010	0,640	1,0	0,2112
.015	.629	1,5	.1716
.020	.618	2,0	.1464
.030	.597	3,0	.1151
.040	.578	4,0	.0960
.050	.561	5,0	.0828
.060	.546	6,0	.0732
.080	.517	8,0	.0599
.100	.4929	10,0	.05100
.150	.4436	15,0	.03773
.200	.4066	20,0	.03024
.300	.3535	30,0	.02199
.400	.3167	40,0	.01746
.500	.2892	50,0	.01456
.600	.2675	60,0	.01254
.800	.2350	80,0	.00988
		100,0	.00820

1-5. Absorption of X-Rays

1-5a. Absorption Edges for Some Elements

The discontinuity is given as

$$S_q = \frac{\tau(\lambda_q)}{\tau'(\lambda_q)} ,$$

where q is the level, τ is the photoelectric absorption coefficient for wavelengths shorter than λ_q, and τ' is the absorption coefficient for wavelengths slightly greater than λ_q.

Values are given for S_K for some elements as excited by the K-series characteristic radiation [1].

Element	S_K	Element	S_K	Element	S_K	Element	S_K
Al	12,6	Cu	8,3	Ag	7,0	Pt	6,0
S	11,0	Zn	7,9	Sn	6,6	Au	5,65
Cl	10,4	Br	7,3	I	5,5	Pb	5,4
Ar	10,0	Sr	7,4	Ba	5,2	U	2,9
Fe	8,8	Mo	7,5	Ta	4,2		
Ni	8,3	Pd	6,8	W	5,65		

1-5b. Calculation of Absorption Coefficients

The relation between the true absorption coefficients is

$$(\tau_e)_K = \frac{\tau_a}{Z} = \frac{A}{NZ}\tau, \quad \tau/\varrho = \frac{A}{NZ\varrho}\tau, \tag{1}$$

where τ_a is the atomic coefficient, τ/ρ is the mass coefficient, τ is the linear coefficient, and $(\tau_e)_K$ is the electronic coefficient; Z is the atomic number, N is Avogadro's number, A is the atomic weight, and ρ is the density.

The table gives $N(\tau_e)_K$ as a function of $Z\lambda$, where λ is the wavelength; $N(\tau_e)_K$ is taken from the table and used in (1) to give the desired coefficient [1].

$$8,0 \leqslant Z\lambda \leqslant 29,9$$

$Z\lambda$	0,0	0,1	0,2	0,3	0,4	0,5	0,6	0,7	0,8	0,9
8	7,20	7,46	7,73	8,00	8,27	8,54	8,81	9,09	9,39	9,70
9	10,0	10,3	10,6	11,0	11,3	11,7	12,1	12,8	12,8	13,1
10	13,4	13,8	14,2	14,6	15,0	15,4	15,9	16,4	16,9	17,4
11	17,9	18,4	18,9	19,4	19,9	20,5	21,1	21,7	22,3	22,9
12	23,5	24,1	24,7	25,3	25,9	26,5	27,2	27,9	28,6	29,3
13	30,1	30,9	31,6	32,4	33,2	34,0	34,7	35,4	36,1	36,9
14	37,7	38,5	39,3	40,1	40,9	41,7	42,5	43,3	44,1	44,9
15	45,9	46,8	47,7	48,6	49,5	50,4	51,3	52,2	53,1	54,0
16	55,0	56,0	57,0	58,0	59,0	60,0	61,0	62,0	63,0	64,0
17	65,0	66,1	67,2	68,3	69,4	70,5	71,7	72,8	73,9	75,0
18	76,2	77,4	78,6	79,9	81,2	82,4	83,6	84,8	86,0	87,2
19	88,4	89,6	90,8	92,0	93,2	94,4	95,6	96,8	98,0	99,3
20	101	102	103	104	105	107	109	110	112	113
21	115	116	118	119	120	121	122	123	125	127
22	129	131	133	135	137	139	141	143	145	147
23	149	151	153	155	157	159	161	163	165	167
24	169	171	173	175	177	179	182	184	186	188
25	190	192	194	196	199	201	203	205	207	210
26	212	214	216	219	221	223	225	227	229	231
27	233	235	238	240	242	245	247	250	252	255
28	258	260	263	266	269	272	275	278	281	284
29	287	290	293	296	299	302	305	308	311	314

$$30 \leqslant Z\lambda \leqslant 149$$

$Z\lambda$	0,0	1,0	2,0	3,0	4,0	5,0	6,0	7,0	8,0	9,0
30	318	352	387	417	450	488	522	564	605	647
40	694	744	795	848	906	968	1 030	1 110	1 180	1 230
50	1 290	1 370	1 450	1 530	1 610	1 690	1 770	1 850	1 940	2 030
60	2 120	2 220	2 320	2 430	2 530	2 630	2 740	2 840	2 940	3 050
70	3 160	3 270	3 380	3 500	3 650	3 800	3 950	4 100	4 250	4 400
80	4 550	4 700	4 850	5 000	5 150	5 300	5 450	5 600	5 750	5 900
90	6 050	6 200	6 350	6 500	6 650	6 800	6 950	7 100	7 250	7 400
100	7 600	7 800	8 000	8 100	8 300	8 500	8 700	8 900	9 100	9 300
110	9 500	9 700	10 000	10 200	10 400	10 700	11 000	11 200	11 500	11 700
120	12 000	12 300	12 500	12 800	13 100	13 400	13 700	14 000	14 300	14 600
130	14 900	15 200	15 500	15 900	16 300	16 600	16 900	17 300	17 600	17 900
140	18 300	18 700	19 100	19 500	19 800	20 200	20 600	21 000	21 300	21 700

$$150 \leqslant Z\lambda \leqslant 790$$

$Z\lambda$	0,0	10,0	20,0	30,0	40,0	50,0	60,0	70,0	80,0	90,0
100	—	—	—	—	—	22 000	25 000	28 000	31 000	34 500
200	38 500	43 000	48 000	53 500	59 000	64 500	70 000	76 000	82 000	89 000
300	95 000	101 000	107 000	113 000	119 000	125 000	132 000	138 000	145 000	153 000
400	161 000	168 000	176 000	183 000	191 000	199 000	207 000	216 000	224 000	233 000
500	242 000	251 000	260 000	270 000	280 000	290 000	300 000	315 000	330 000	345 000
600	360 000	375 000	390 000	405 000	420 000	435 000	450 000	460 000	480 000	500 000
700	520 000	530 000	550 000	570 000	590 000	600 000	620 000	640 000	660 000	680 000

1-5c. Nomogram for Determining Absorption Coefficients

The nomogram (Fig. 1) gives τ/ρ from Z and λ [249]; to do this, one joins the point for the wavelength on the left-hand scale to the Z for the specimen on the middle scale, the line being produced to meet the τ/ρ scale on the right.

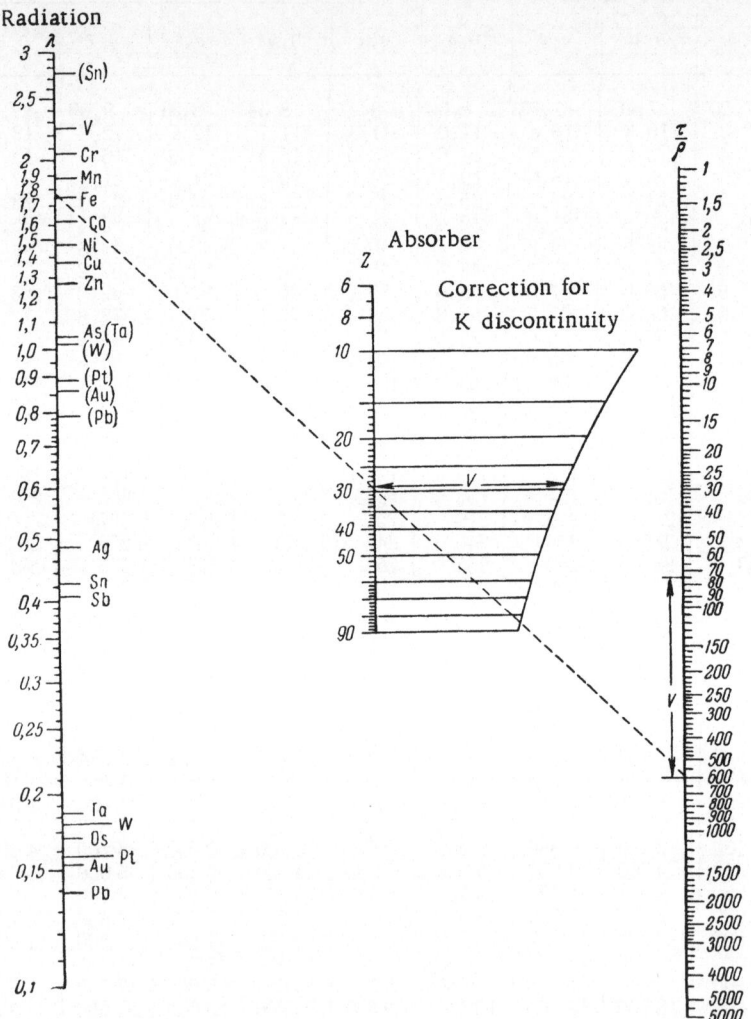

Fig. 1. Nomogram for absorption coefficients.

Correction for the K-edge discontinuity is made upwards on the right-hand scale, the distance being that shown horizontally to the right of the middle scale (see example). The elements in parentheses give their L series at the wavelengths shown.

The nomogram gives the most accurate results for $5 \leq Z\lambda \leq 160$.

1-6. Total Attenuation of X-Rays

1-6a. Atomic Attenuation Coefficients

The table gives the atomic attenuation coefficients [109] for Kα radiations:

$$\mu_0 = \frac{\mu}{\varrho} \, \frac{A}{N} \; ,$$

where μ is the linear coefficient, ρ is the density, A is the atomic weight, and N is Avogadro's number.

Element	Ag $K\alpha$	Rh $K\alpha$	Mo $K\alpha$	Cu $K\alpha$	Ni $K\alpha$	Fe $K\alpha$	Cr $K\alpha$
He	0,11	0,12	0,12				
Li	0,21	0,23	0,34				
Be	0,33	0,37	0,45	2,02	2,70	4,85	7,94
B	0,54	0,63	0,81	5,50	6,72	10,43	16,86
C	0,83	1,01	1,38	10,87	13,36	21,4	35,4
N	1,39	1,62	2,54	19,6	24,9	39,8	63,8
O	2,13	2,56	4,00	33,6	42,5	66,4	105,5
F	3,13	4,13	6,05	54,7	67,3	103,2	160
Ne	4,70	6,00	8,90	81,9	100	153	242
Na	6,65	8,5	12,8	117	143	215	350
Mg	9,1	11,7	17,6	163	192	303	481
Al	12,2	16,1	23,6	217	260	414	655
Si	16,6	21,0	31,2	281	353	542	896
P	21,5	27,4	40,8	373	462	721	1140
S	27,2	35,1	53,0	482	589	926	1441
Cl	34,2	43,8	68,0	604	734	1160	1800
Ar	42,0	52,6	82,5	742	924	1427	2220
K	51,9	69,0	107,2	923	1151	1730	2640
Ca	63,9	84,6	131	1138	1390	2100	3360
Sc	78,0	102,6	157	1380	1650	2510	4050
Ti	93,5	125	188	1616	1960	2980	4780
V	111,8	149	222	1910	2310	3560	650
Cr	134	175	260	2220	2700	4190	769
Mn	157	204	303	2570	3240	576	900
Fe	183	237	352	2980	3650	669	1051
Co	212	272	404	3430	527	782	1220
Ni	242	312	458	475	590	900	1410
Cu	282	362	529	562	693	1051	1640
Zn	303	406	590	635	775	1180	1820
Ga	356	459	663	731	888	1350	2060
Ge	401	513	759	830	1010	1540	2340
As	450	568	858	945	1160	1750	2690
Se	503	640	967	1070	1320	1960	3070
Br	550	695	1067	1202	1460	2190	3440
Kr	610	780	1190	1360	1650	2460	3860
Rb	678	883	1330	1540	1870	2760	4350

Radiation Element	Ag $K\alpha$	Rh $K\alpha$	Mo $K\alpha$	Cu $K\alpha$	Ni $K\alpha$	Fe $K\alpha$	Cr $K\alpha$
Sr	751	985	1460	1710	2090	3090	4810
Y	815	1089	1600	1900	2320	3460	5300
Zr	911	1208	257	2120	2570	3880	5830
Nb	1008	1320	286	2340	2810	4270	6350
Mo	1118	1450	320	2600	3120	4730	6940
Ru { α_1 1340 α_2 203		255	392	3090	3700	5650	8100
Rh	222	282	429	3360	4060	6110	8840
Pd	242	309	469	3640	4460	6600	9580
Ag	263	339	507	3960	4900	7140	10400
Cd	288	373	555	4340	5360	7730	11290
In	312	410	600	4760	5800	8310	12240
Sn	340	446	650	5160	6290	8930	13300
Sb	373	493	706	5690	6850	9650	14550
Te	401	525	758	6060	7290	10270	16500
I	436	570	819	6560	7830	11000	16900
Xe	475	613	879	7100	8420	11800	18300
Cs	516	655	946	7590	8960	12650	18400
Ba	555	705	1022	8130	9600	13560	18500
La	595	755	1098	8650	10180	14500	5000
Ce	656	827	1200	9410	11000	14700	5420
Pr	682	864	1265	9780	11310	14500	5820
Nd	726	926	1360	10400	12150	15500	6260
Sm	820	1020	1540	11580	12900	4540	7160
Eu	876	1114	1650	11560	12500	4840	7660
Gd	930	1186	1760	12200	13200	5160	8200
Tb	985	1260	1880	11400	3680	5540	8740
Dy	1048	1335	2000	12400	3905	5890	9230
Ho	1113	1420	2140	3450	4120	6250	9710
Er	1177	1510	2260	3690	4390	6680	10210
Tu	1245	1600	2400	3860	4670	7140	10750
Yb	1320	1700	2530	4100	4960	7560	11310
Lu	1400	1810	2690	4360	5310	8110	11920
Hf	1490	1910	2850	4610	5610	8550	12500
Ta	1560	2020	3010	4900	5970	9110	13150
W	1656	2140	3200	5180	6340	9700	13800
Os	1840	2400	3540	5850	7100	10900	15100
Ir	1950	2540	3760	6180	7540	11510	15900
Pt	2060	2690	3960	6600	7980	12100	16650
Au	2160	2830	4160	6950	8450	12700	17450
Hg	2290	2970	4370	7360	9000	13320	18200
Tl	2410	3100	4560	7760	9500	14000	19100
Pb	2540	3270	4810	8220	10020	15000	19800
Bi	2680	3440	4970	8680	10630	15400	21000
Rn	3100	4000	5830	10200	12500	17400	24200
Ra	3400	4370	6400	11340	13800	19000	26400
Th	3710	4550	5460	12500	15300	20500	28800
U	4100	5070	6000	13800	16700	22200	31600

1-6b. Mass Attenuation Coefficients μ/ρ for Long Wavelengths [111]

λ, kX	H	Be	C	O	Al	Ni	Cu	Ag	Pt	Au
1,537		6,0	4,87		49,7	48,1	50,4	217	206	216
1,655		8,15	6,18		61,3	60,7	65,1	264	228	259
1,932		10,5	9,38		93,5	94,3	100	394	351	387
2,498		20,4	20,0		194	180	202	779	596	720
3,025		34,7	34,0		346	319	321	1320	939	1252
3,352		45,1	43,8		459	384	404	1450	1120	1470
4,146		73,2			822	627	621	517	1290	1910
4,359		90,8			941	735	730	629		2450
4,718		99,0			1108			764		2550
5,395		152			1630	1250	1300	1012	1640	
6,057		203			2130			1380		1210
6,973						2000	2120		1190	
7,111		257			3170			2145		1730
8,321		381			459	3140	3450	3070	1530	2450
9,868						4540	5030		2440	
11,9					767		7550			
13,3					2180		4340	9920		
14,6					2290		2470	10050		
17,6					3520		3770			
23,7					7330		6870			
44,5	1000			5650						
68	2980			12650						

1-6c. Mass Attenuation Coefficients μ/ρ for Short Wavelengths [111]

λ, kX	C	Na	Al	K	Ni	Cu	Mo	Ag	Sn	Ta	Pb
0,0228	0,084		0,081			0,084			0,093		0,152
0,0284						0,087			0,098		0,191
0,0363	0,100		0,097								
0,0384	0,102		0,099						0,104		0,350
0,040	0,105										0,394
0,0416						0,116					0,398
0,0485						0,132					0,595
0,0500			0,113	0,115	0,126	0,129	0,177	0,229	0,246	0,458	0,600
0,0504			0,119								0,636
0,060	0,120		0,121	0,136	0,163	0,160	0,227	0,283	0,314	0,689	0,900
0,070	0,126	0,126	0,131	0,144	0,187	0,182	0,319	0,400	0,430	1,018	1,455
0,080	0,132	0,134	0,138	0,162	0,218	0,215	0,412	0,544	0,607	1,357	1,829
0,090	0,137	0,140	0,144	0,180	0,250	0,259	0,535	0,715	0,800	1,819	2,490
0,100	0,142	0,147	0,150	0,194	0,297	0,306	0,650	0,900	1,020	2,435	3,248
0,110	0,147	0,154	0,158	0,202	0,349	0,371	0,821	1,081	1,222	3,014	4,079
0,120	0,151	0,161	0,168	0,228	0,420	0,434	1,036	1,362	1,587	3,770	5,150
0,130	0,154	0,168	0,176	0,256	0,490	0,513	1,243	1,656	1,912	4,553	6,412
0,140	0,158	0,174	0,191	0,297	0,594	0,606	1,513	2,038	2,310	5,476	
0,160	0,163	0,186	0,213	0,366	0,771	0,830	2,108				
0,184	0,171	0,208	0,240	0,459	1,10	1,17	3,193				
0,209	0,177	0,231	0,279	0,606	1,56	1,63					5,108

1-6d. Mass Attenuation Coefficients μ/ρ for Elements

The table gives μ/ρ in cm^2/g for Z from 1 to 92 and also for certain compounds

Z	Element	Rad. 0.200 W	0.5608 Ag	0.6147 Rh	0.7107 Mo	1.436 Zn	1.542 Cu	1.659 Ni	1.790 Co	1.937 Fe
1	H	0.326	0.370	0.37	0.38	0.44	0.46	0.47	0.48	0.49
2	He	0.165	0.191	0.199	0.18	0.31	0.37	0.43	0.52	0.64
3	Li	0.143	0.187	0.20	0.22	0.54	0.68	0.87	1.13	1.48
4	Be	0.149	0.22	0.25	0.30	1.02	1.35	1.80	2.42	3.24
5	B	0.157	0.30	0.35	0.45	2.51	3.06	3.79	4.67	5.80
6	C	0.177	0.42	0.51	0.70	4.43	5.50	6.76	8.50	10.7
7	N	0.181	0.60	0.70	1.10	6.85	8.51	10.7	13.6	17.3
8	O	0.189	0.80	1.00	1.50	11.4	12.7	16.2	20.2	25.2
9	F	0.191	1.00	1.32	1.93	14.4	17.5	21.5	36.6	33.0
10	Ne	0.215	1.41	1.80	2.67	20.2	24.6	30.2	37.2	46.0
11	Na	0.225	1.75	2.25	3.36	25.6	30.9	37.9	46.2	56.9
12	Mg	0.254	2.27	2.93	4.38	33.0	40.6	47.9	60.0	75.7
13	Al	0.274	2.74	3.60	5.30	40.0	48.7	58.4	73.4	92.8
14	Si	0.310	3.44	4.52	6.70	49.5	60.3	75.8	94.1	116
15	P	0.337	4.20	5.36	7.98	59.4	73.0	90.5	113	141
16	S	0.390	5.15	6.65	10.03	75.0	91.3	112	139	175
17	Cl	0.422	5.86	7.50	11.62	85.0	103	126	158	199
18	A	0.446	6.40	8.00	12.55	93.0	113	141	174	217
19	K	0.522	8.05	10.7	16.7	119	143	179	218	269
20	Ca	0.606	9.66	12.8	19.8	142	172	210	257	317
21	Sc	0.660	10.5	13.8	21.1	153	185	222	273	338
22	Ti	0.726	11.8	15.8	23.7	167	204	247	304	377
23	V	0.825	13.3	17.7	26.5	186	227	275	339	422
24	Cr	0.910	15.7	20.4	30.4	213	259	316	392	490
25	Mn	1.01	17.4	22.6	33.5	234	284	348	431	63.6
26	Fe	1.134	19.9	25.8	38.3	270	324	397	59.5	72.8
27	Co	1.26	21.8	28.1	41.6	292	354	54.4	65.9	80.6
28	Ni	1.420	25.0	32.3	47.4	325	49.3	61.0	75.1	93.1
29	Cu	1.495	26.4	34.0	49.7	42.0	52.7	65.0	79.8	98.8
30	Zn	1.65	28.2	37.7	54.8	49.3	59.0	72.1	88.5	109
31	Ga	1.75	30.8	39.7	57.3	52.4	63.3	76.9	94.3	116
32	Ge	1.90	33.5	42.8	63.4	57.6	69.4	84.2	104	128
33	As	2.04	36.5	46.0	69.5	63.5	76.5	93.8	115	142
34	Se	2.198	38.5	49.0	74.0	69.4	82.8	101	125	152
35	Br	2.43	42.3	53.5	82.2	77.0	92.6	112	137	169
36	Kr	2.58	45.0	57.5	88.1	83.0	100	122	148	182
37	Rb	2.80	48.2	62.8	94.4	91.5	109	133	161	197
38	Sr	3.03	52.1	68.3	101.1	100	119	145	176	214
39	Y	3.25	55.5	74.0	109.9	107	129	158	192	235
40	Zr	3.50	61.1	80.9	17.2	118	143	173	211	260
41	Nb	3.74	65.8	86.0	18.7	126	153	183	225	279
42	Mo	4.05	70.7	91.6	20.2	136	164	197	242	299
44	Ru	4.60	79.9(α_1) 12.2(α_2)	15.4	23.4	153	185	221	272	337
45	Rh	4.86	13.1	16.6	25.3	165	198	240	293	361
46	Pd	5.13	13.8	17.6	26.7	173	207	254	308	376
47	Ag	5.50	14.8	19.1	28.6	192	223	276	332	402
48	Cd	5.78	15.5	20.1	29.9	202	234	289	352	417
49	In	6.06	16.5	21.7	31.8	214	252	307	366	440
50	Sn	6.32	17.4	22.9	33.3	230	265	322	382	457
51	Sb	6.55	18.6	24.6	35.3	245	284	342	404	482
52	Te	6.76	19.1	25.0	36.1	248	289	347	410	488
53	I	7.29	20.9	27.3	39.2	269	314	375	442	527
54	Xe	7.54	22.1	28.5	41.3	283	330	392	463	552
55	Cs	7.75	23.6	30.0	43.3	298	347	410	486	579
56	Ba	8.12	24.5	31.1	45.2	307	359	423	501	599
57	La	8.25	26.0	33.0	47.9	325	378	444	–	632
58	Ce	8.45	28.4	35.8	52.0	358	407	476	549	636
59	Pr	8.69	29.4	37.2	54.5	–	422	493	–	642
60	Nd	8.88	30.5	38.8	57.0	–	437	510	–	651
62	Sm	9.55	33.1	41.2	62.3	–	467	519	–	183
63	Eu	9.87	35.0	44.5	65.9	–	461	498	–	193
64	Gd	10.2	35.8	45.7	68.0	–	470	509	–	199
65	Tb	10.64	37.5	47.9	71.7	–	435	140	–	211
66	Dy	–	39.1	49.9	75.0	–	462	146	–	220

* Based on values given by J. A. Victoreen, J. Appl. Phys. 20:1141, 1949 and (for the expanded version given here)

(for Kα radiation unless otherwise noted) from 0.200 to 27.4 Å [102].*

2.103	2.291	2.50	3.57	5.17	8.34	9.87	13.37	17.6	21.7	27.4
Mn	Cr	V	Co:K2α	Mo:L1β₁	Al:Kα	Mg:Kα	Cu:L1α	Fe:L1α	Cr:Lα	Ti:Lα
0.50	0.55	0.59	1.7	3.4	7.5	11.5	30	70	130	260
0.74	0.86	1.00	4.5	12	30	44	120	275	500	1000
1.76	2.11	2.49	12	30	78	110	280	640	1200	2300
3.90	4.74	5.79	24	60	152	220	581	1288	2292	4532
7.36	9.37	9.52	45	120	324	440	1233	2711	4784	9200
13.8	17.9	18.8	55.2	160	656	1063	2170	4912	8440	15760
21.8	27.7	30.9	96.	273	1109	1796	3838	8008	13120	22590
32.2	40.1	47.3	150	413	1589	2500	5456	10000	16610	1473
41.1	51.6	64.66	240	680	1913	2780	6340	11600	1015	1949
57.6	72.7	93.29	275	763	2750	4310	8500	1079	1863	3575
72.3	92.5	119	340	880	3129	4450	661	1402	2429	4651
95.2	120	159	430	1200	3797	70	981	2085	3601	6830
117	149	195	500	1370	330	500	1146	2441	4189	7840
146	192	244	540	1600	510	970	1813	3812	6420	11510
177	223	289	625	1800	640	1180	2259	4661	7670	13280
217	273	356	900	221	794	1320	2839	5710	9160	15520
245	308	402	1020	277	962	1570	3364	6530	10210	17330
270	341	437	1210	324	1157	1865	3795	7110	11070	18820
330	425	537	50	550	1429	2300	4504	8310	12960	22030
400	508	619	85	600	1706	2600	5150	9450	14800	24910
428	545	1050	120	680	1819	2750	5280	9750	15210	—
475	603	—	160	750	2002	2820	5680	10480	16300	—
530	27.3	44	220	800	2168	2900	6100	11230	—	—
70.5	89.9	120	280	970	2409	3300	6740	12360	—	—
79.6	99.4	138	325	1080	2556	3500	7160	—	—	—
90.9	115	147	375	1100	2799	3700	7850	—	—	—
102	126	174	420	1130	2956	3980	—	—	—	—
116	145	180	450	1150	3140	4540	—	—	—	—
123	154	197	495	1190	3450	5036	—	—	—	—
135	169	228	575	1460	3685	—	—	—	—	—
144	179	—	—	—	—	—	—	—	—	—
158	196	—	—	—	—	—	—	—	—	—
175	218	—	—	—	—	—	—	—	—	—
188	235	—	—	—	—	—	—	—	—	—
206	264	—	—	—	—	—	—	—	—	—
226	285	—	—	—	—	—	—	—	—	—
246	309	—	—	—	—	—	—	—	—	—
266	334	—	—	—	—	—	—	—	—	—
289	360	—	—	—	—	—	—	—	—	—
317	391	—	—	—	—	—	—	—	—	—
338	415	—	—	—	—	—	—	—	—	—
360	439	—	—	—	—	—	—	—	—	—
404	488	—	—	—	—	—	—	—	—	—
432	522	—	—	—	—	—	—	—	—	—
450	545	—	—	—	—	—	—	—	—	—
486	585	710	—	—	—	—	—	—	—	—
500	608	—	—	—	—	—	—	—	—	—
531	648	—	—	—	—	—	—	—	—	—
555	681	850	—	—	—	—	—	—	—	—
589	727	—	—	—	—	—	—	—	—	—
598	742	—	—	—	—	—	—	—	—	—
650	808	—	—	—	—	—	—	—	—	—
680	852	—	—	—	—	—	—	—	—	—
715	844	—	—	—	—	—	—	—	—	—
777	819	—	—	—	—	—	—	—	—	—
35	218	—	—	—	—	—	—	—	—	—
70	235	—	—	—	—	—	—	—	—	—
—	251	—	—	—	—	—	—	—	—	—
—	263	—	—	—	—	—	—	—	—	—
—	289	—	—	—	—	—	—	—	—	—
—	306	—	—	—	—	—	—	—	—	—
—	316	—	—	—	—	—	—	—	—	—
—	333	—	—	—	—	—	—	—	—	—
—	345	—	—	—	—	—	—	—	—	—

L. H. Burton, R. White, and B. Lundberg, J. Appl. Phys. 28 : 98, 1957.

Z	Element	λ 2,50 / V	2,291 / Cr	2,103 / Mn	1,937 / Fe	1,790 / Co	1,659 / Ni	1,542 / Cu	1,436 / Zn	0,7107 / Mo	0,6147 / Rh	0,5608 / Ag	0,200 / W
67	Ho	—	361	—	232	—	153	128	—	79,3	52,7	41,3	—
68	Er	—	370	—	242	—	159	133	—	82,0	54,6	42,6	—
69	Tu	—	387	—	257	—	168	139	—	86,3	57,6	44,8	—
70	Yb	—	396	—	265	—	174	144	—	88,7	59,4	46,1	—
71	Lu	—	414	—	281	—	184	151	—	93,2	62,6	48,4	—
72	Hf	—	426	—	291	—	191	157	—	96,9	65,0	50,6	3,4
73	Ta	—	440	364	305	246	200	164	136	100,7	67,7	52,2	3,5
74	W	—	456	380	320	258	209	171	143	105,4	70,7	54,6	—
76	Os	—	480	406	346	278	226	186	152	112,9	76,3	58,6	—
77	Ir	—	498	422	362	292	237	194	160	117,9	80,0	61,2	4,25
78	Pt	596	518	436	376	304	248	205	172	123	83,8	64,2	4,4
79	Au	—	537	456	390	317	260	214	179	128	87,1	66,7	—
80	Hg	—	552	471	404	330	272	223	186	132	90,1	69,3	—
81	Tl	—	568	484	416	341	282	231	194	136	92,4	71,7	4,9
82	Pb	—	585	499	429	354	294	241	202	141	95,8	74,4	5,1
83	Bi	—	612	522	448	372	310	253	214	145	100,4	78,1	—
86	Rn	—	657	—	476	—	341	278	—	159	109,1	84,7	—
88	Ra	—	708	598	509	433	371	304	258	172	117	91,1	—
90	Th	—	755	633	536	460	399	327	286	143	119	97,0	—
92	U	560	805	672	566	488	423	352	310	153	129	104,2	5,4
	Air	40	—	24	20	15	—	9,8	8,9	1	0,62	0,6	0,186
	Water	42,1	—	25	21	17	—	10,2	9,2	1,2	0,72	0,7	0,21
	Nylon	22,5	—	14	11	8	—	5,00	4,8	0,7	0,55	0,4	0,19
	Polyethylene	16,2	—	10	8	6	—	4,2	3,4	0,6	0,45	0,38	0,20
	Polystyrene	17,4	—	11	9	7	—	4,5	3,5	0,62	0,46	0,4	0,19

1-6e. Mass Attenuation Coefficients μ/ρ for Some Compounds

The attenuation coefficients for compounds and solid solutions are given by

$$\frac{\mu}{\varrho} = \sum_i \alpha_i \frac{\mu_i}{\varrho_i} \, ,$$

in which α_i is the proportion by weight of component i.

The table gives μ/ρ for various compounds and wavelengths.

Most of the compounds listed occur as nonmetallic inclusions in alloys and are examined by absorption microradiography [5].

$K\alpha$ radiation	Mo	Zn	Cu	Ni	Co	Fe	Mn	Cr
λ, Å Compound	0,707	1,432	1,537	1,655	1,785	1,932	2,098	2,285
$FeCr_2O$	109	7,43	925	1120	918	1150	283	350
$MgAl_2O$	12,95	84,5	110	133	173	207	260	288
$MnFe_2O$	129	856	1250	1360	742	219	343	428
$FeAl_2O$	63,6	412	567	675	217	272	342	380
$NiFe_2O$	153	965	915	—	—	310	—	540
Fe_3C	272	1800	2320	2900	446	508	670	876
WC	1500	2050	2560	3100	3850	4360	5680	6750
FeS	130	940	1100	1365	410	505	635	795
MnS	100	735	835	1070	1445	425	535	670
Al_2O_3	5	100	120	145	180	225	285	360
Fe_2O_3	145	1025	1230	1480	950	310	395	485
Fe_3O_4	145	1050	1240	1530	255	310	390	490
MnO	145	1020	1230	1510	2060	310	375	480
SiO_2	10	65	80	100	125	155	175	255
FeO	180	1290	1530	1880	310	380	470	590
$FeO \cdot SiO_2$	65	470	555	680	190	235	295	375
$MnO \cdot SiO_2$	55	405	490	610	820	220	275	355
Fe_3P	235	1730	2040	2490	495	600	750	945

1-6f. Half-Value Layers for Some Elements

The table gives the thicknesses (in mm) that absorb half of the incident radiation in transmission and 75% in back-scattering.

Element	Radiation					
	Cr	Fe	Co	Ni	Cu	Mo
Be	0,795	1,16		2,09	2,80	12,6
C	.11	0,18		0,29	0,36	2,82
Mg	.033	.053		.083	.098	0,910
Al	.017	.028	0,031	.044	.053	.485
Ti	.00255	.00409		.00624	.00755	.065
Cr	.00107	.00196		.00305	.00372	.032
Mn	.0094	.0147		.00268	.00329	.028
Fe	.0077	.0121	.0160	.00222	.00273	.023
Co	.0062	.0097		.0143	.00220	.019
Ni	.0056	.0087		.0132	.0164	.017
Cu	.0051	.0079		.0120	.0148	.016
Zn	.0056	.0090		.0137	.0167	.0180
Ge	.0065	.0099		.0151	.0183	.0200
Mo	.00155	.0023		.0035	.0041	.034
Ag	.00114	.0059		.0024	.0030	.023
Sn	.00140	.0021		.0030	.0036	.029
W	.00082	.0012		.0018	.0022	.0035
Pt	.00063	.00086		.0013	.0016	.0026
Pb	.00105	.00143		.0021	.0025	.0044

1-6g. Half-Value Layers as Functions of Angle of Incidence

The attenuation by a layer of material depends on the angles of incidence and reflection.

See [6, 173] for geometric relationships for the penetration of rays into materials in the various methods of examination.

The depth of the layer participating in the diffraction pattern is

$$x = \frac{K_x \sin \vartheta}{2\mu}$$

for Bragg focusing (ionization equipment), or

$$x = \frac{K_x \sin \beta}{\mu\,(1+\sin \beta)}$$

for back-scattering; $\beta = 2\vartheta - 90°$ and ϑ is the Bragg angle. Values of K_x are given below; g_x is the proportion of the scattered intensity associated with the layer of thickness x.

g_x	0,50	0,75	0,90	0,95	0,99	0,999
K_x	0,69	1,39	2,30	3,00	4,61	6,91

The values of x given in the following table should be halved for transmission measurements.

The table gives the thicknesses that absorb 50% in transmission and 75% in reflection, α being the angle between the primary beam and the surface, which corresponds to the angle used with sections in the powder camera and with ionization equipment.

These values can be used in studies on the surface layers of electroplated or polished metals, semiconductors, and so on [6, 389].

Kα radiation / α°	Cr	Fe	Co	Ni	Cu	Mo
Fe						
10	0,0013	0,0021	0,0028			0,004
20	.0026	.0041	.0055			.008
30	.0039	.0061	.0080			.012
40	.0049	.0078	.0103			.015
50	.0059	.0093	.0123			.018
60	.0067	.0105	.0139			.020
70	.0072	.0114	.0150			.022
80	.0076	.0119	.0158			.023
90	.0077	.0121	.0160			.023
Cr						
10	0,003				0,0092	0,084
20	.004				.0181	.166
30	.009				.0265	.243
40	.011				.0341	.318
50	.013				.0406	.372
60	.015				.0459	.420
70	.016				.0498	.456
80	.0167				.0522	.478
90	.017				.053	.485
Cu						
10	0,0009	0,0014		0,0021	0,0026	0,003
20	.0017	.0027		.0041	.0051	.005
30	.0026	.0040		.0060	.0074	.008
40	.0033	.0051		.0077	.0095	.010
50	.0039	.0061		.0092	.0113	.012
60	.0044	.0068		.0104	.0128	.014
70	.0048	.0074		.0113	.0139	.015
80	.0050	.0078		.0118	.0146	.016
90	.0051	.0079		.0120	.0148	.016
Be						
10	0,138	0,201	0,363	0,486	2,19	
20	.272	0,397	0,506	0,958	4,31	
30	.398	0,580	1,045	1,400	6,300	
40	.511	0,746	1,343	1,800	8,099	
50	.609	0,889	1,601	2,145	9,652	
60	.688	1,005	1,810	2,425	10,91	
70	.747	1,090	1,964	2,631	11,84	
80	.783	1,142	2,058	2,757	12,41	
90	.795	1,16	2,09	2,80	12,6	

Kα radiation α°	Cr	Fe	Co	Ni	Cu	Mo

Ti

α°	Cr	Fe	Co	Ni	Cu
10		0,00071	0,00108	0,00131	0,011
20		.00140	.00213	.00258	.0222
30		.00205	.00312	.00378	.0325
40		.00263	.00401	.00485	.0418
50		.00313	.00478	.00578	.0498
60		.00354	.00540	.00654	.0563
70		.00384	.00586	.00709	.0611
80		.00403	.00615	.00744	.0640
90		.00409	.00624	.00755	.065

Ni

α°	Cr	Fe	Co	Ni	Cu
10	0,0010	0,0015	0,0023	0,0028	0,003
20	.0019	.0030	.0045	.0056	.006
30	.0028	.0044	.0066	.0082	.009
40	.0036	.0056	.0085	.0105	.011
50	.0043	.0067	.0101	.0126	.013
60	.0048	.0075	.0114	.0142	.015
70	.0053	.0082	.0124	.0154	.016
80	.0055	.0086	.0130	.0162	.017
90	.0056	.0087	.0132	.0164	.017

Ge

α°	Cr	Fe	Co	Ni	Cu
10	0,0011	0,0017	0,0026	0,0032	0,0035
20	.0022	.0034	.0052	.0063	.0068
30	.0033	.00495	.0076	.0092	.0100
40	.0042	.0064	.0097	.0112	.0129
50	.0050	.0076	.0116	.0140	.0153
60	.0056	.0086	.0131	.0158	.0173
70	.0061	.0093	.0142	.0172	.0188
80	.0064	.0097	.0149	.0180	.0197
90	.0065	.0099	.0151	.0183	.0200

1-7. Ionizing Effects of X-Rays

The table gives the relative ionization coefficients for gases [251].

These coefficients relate to equal incident intensities and equal numbers of molecules in the irradiated volume; the value for air is taken as unity.

λ, Å	Air	N₂	O₂	CO₂	N₂O	H₂S	SO₂	H₂Se	C₂H₅Br	CH₃I
2,29	1	—	—	1,40	—	—	—	—	—	—
1,93	1	—	1,37	1,41	1,32	14,7	—	30,3	41,2	—
1,66	1	—	1,35	1,39	1,33	14,9	11,5	—	—	162
1,54	1	0,71	1,38	1,40	1,30	14,7	11,7	29,2	42	—
1,43	1	—	1,42	1,36	1,30	14,3	11,1	—	41,6	—
1,17	1	0,71	1,27	1,38	1,33	14,8	11,2	—	42,2	158
1,11	1	—	1,31	1,35	1,37	15,0	11,7	30,6	41,7	—
0,87	1	—	1,28	1,40	1,31	15,3	11,7	122	153	—

λ, Å	Air	N₂	O₂	CO₂	N₂O	H₂S	SO₂	H₂Se	C₂H₅Br	CH₃I
0,71	1	—	1,28	1,43	1,38	15,2	12,2	190	213	188
0,61	1	—	—	1,41	—	15,3	12,3	—	—	—
0,59	1	—	—	1,39	—	15,4	12,7	—	—	—
0,56	1	0,72	1,32	1,39	1,34	15,4	12,6	231	272	198
0,49	1	—	1,29	1,41	1,31	15,7	—	250	335	205
0,47	1	—	1,28	1,43	1,32	—	—	—	—	—
0,44	1	0,73	—	—	—	—	—	286	—	211
0,39	1	—	—	—	—	—	—	—	—	251

1-8. Refraction of X-Rays

1-8a. Unit Decrements of Refractive Index

The unit is $\delta = \mu - 1$, where μ is the real refractive index for x-rays.

The table gives δ for various wavelengths and materials; it also gives δ/λ^2, which is useful in making refraction corrections in precision lattice-constant measurements[251].

Section 6-12 deals in detail with refraction corrections.

Substance	λ, Å	$\delta \cdot 10^6$	$\frac{\delta}{\lambda^2} \cdot 10^6$	Substance	λ, Å	$\delta \cdot 10^6$	$\frac{\delta}{\lambda^2} \cdot 10^6$
Graphite	0,708	1,23	2,46	Calcite	1,932	13,89	3,72
Sulfur	0,708	1,39	2,78		2,499	22,37	3,59
Aluminum	0,708	1,68	3,36		2,509	23,26	3,70
	1,537	8,4	3,56		2,774	27,05	3,51
Nickel	1,081	11,05	9,40		2,931	28,54	3,32
	1,274	14,8	9,11		3,025	29,34	3,21
	1,389	17,0	8,80		3,040	28,67	3,10
	1,473	17,5	8,07		3,070	30,18	3,20
	1,497	18,7	8,35		3,083	32,02	3,37
	1,537	18,7	7,92		3,218	35,98	3,47
	1,655	25,0	9,13		3,379	39,57	3,47
	1,932	35,5	9,25		3,447	41,87	3,52
Copper	0,708	5,95	11,9		3,734	49,19	3,53
Silver	0,708	5,85	11,7	Quartz	1,537	8,62	3,64
	1,279	21,5	11,32		1,753	11,19	3,64
Celluloid	0,708	0,98	1,96		1,932	13,63	3,65
	1,537	4,78	2,02		2,280	19,06	3,67
Glycerine	1,537	4,41	1,87		2,500	22,89	3,66
Paraffin	0,708	0,70	1,40		3,353	41,66	3,71
	1,537	3,28	1,39		3,735	51,77	3,71
Pyrite	0,631	2,87	7,21		4,719	81,44	3,65
	0,708	3,35	6,70		5,362	104,1	3,62
	1,389	13,2	6,85		7,111	170,1	3,36
	1,537	17,6	7,46		8,323	240,1	3,46
Glass	0,631	1,22	3,1		9,868	346,0	3,55
	0,708	1,64	3,3	Mica	1,537	8,94	3,78
	1,389	6,65	3,44		2,498	24,6	3,95
	1,537	8,12	3,43		3,378	46,6	4,06
	1,753	10,00	3,26		3,447	49,1	4,14
	1,932	12,38	3,31		5,166	103	3,85
Calcite	0,708	2,00	4,00		7,111	182	3,6
	1,537	8,80	3,72		8,320	262	3,79

1-8b. Angle of Total Internal Reflection

Total internal reflection is a consequence of the refraction of x-rays in material media; but the refractive index for x-rays is small, so the angles of total internal reflection are also small.

The effect has been used to produce monochromatic beams [424, 425] and to focus x-rays [426]; the critical angle hardly varies with wavelength, so it is best to use a selectively absorbing filter in conjunction with the reflector [427]. The filter reduces the intensity of the Kβ region, while the reflector tends to suppress the short-wave region.

The table gives angles of total internal reflection α for certain materials [108].

Substance	Density	Wave-length	α
Glass	2,52	1,279 Å	10′
Glass	2,52	0,52 Å	4′
Silver	10,5	1,279 Å	225′
Lacquer	—	1,279 Å	11′

CHAPTER 2

PRODUCTION AND MEASUREMENT OF X-RAY PATTERNS

This chapter gives details of x-ray equipment and of methods for selecting conditions; it also indicates methods of adjusting x-ray cameras and of measuring patterns recorded by photographic and ionization methods.

See [6-13, 298] for detailed information on methods of producing and measuring patterns.

2-1. Equipment for X-Ray Laboratories*

1. X-Ray Equipment

a. The URS-70K1 Universal X-Ray Equipment. This is meant for structural and spectroscopic studies on materials; its maximum output voltage is 70 kV, and the main type of tube is the BSV1 (formerly denoted by BSV-4). This unit uses a half-wave rectifier. Figure 2 shows the unit. Figure 3 shows the electrical system of the URS-70K1 [252]. The main parts are as follows: the high-voltage transformer, the primary-voltage control transformer, the transformers for the filament and rectifier heaters, the KRM-150 rectifier, the x-ray tube, the relay, and the filament-voltage stabilizer. Figure 3 shows that the tube current flows during a half-cycle only, when the filament is negative. The voltage and current vary continuously during this half-cycle.

b. The URS-55 Bench X-Ray Equipment. This small apparatus (Fig. 4) has a maximum working voltage of 55 kV and is intended for use with tubes of type BSV2 (formerly denoted by BSV-L). There is no rectifier and the tube operates with the anode grounded. The high-voltage and filament transformers are contained in oil. The tube is set vertical in a porcelain housing on the top of the oil tank. Figure 5 shows the electrical system [253]. The system allows alteration of the range of variation in the high-voltage.

c. The URS-60 X-Ray Equipment. This apparatus has been in production since 1960; it differs in certain ways from the URS-70 [255]. The control panel is at one side of the working table. The apparatus is designed to operate with a variety of tubes and can be used simultaneously with two tubes working at the same voltage. The voltage ranges up to 60 ±2 kV and the tube current up to 20 mA. The doubler system with smoothing is used, the kenotrons being of type KRM-80.

The URS-60 can be used with photographic or electrical recording. The operation is stabilized by an SN-2 electronic stabilizer together with an anode-current stabilizer. Line-voltage changes of −15 to +7% from nominal produce changes of only ±0.25% at the terminals on the control panel. Interlocks disconnect the high-voltage if the water fails, if the door of the working table is opened, if the cover on the tube is removed, or if the tube current is excessive. The working ranges are up to 40 kV (from 10 kV) and up to 60 kV (from 30 kV).

---

*Only the main Soviet types of equipment are discussed here. See [390] for apparatus and methods of examining x-ray spectra.

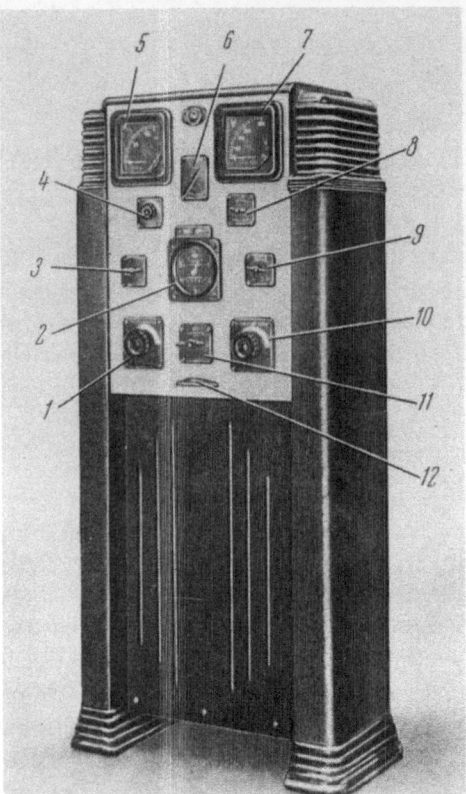

Fig. 2. The URS-70K1 x-ray equipment. 1) Rheostat
for high-voltage transformer; 2) timer; 3) high-voltage
trip; 4) heater rheostat; 5) milliammeter; 6) graph for
reading tube voltage; 7) voltmeter; 8) voltmeter range
switch; 9) trip on current-limiting resistor; 10) line-
voltage corrector; 11) line switch; 12) switch for chang-
ing point of application of high voltage.

Figure 6a shows the working table. The control section has inner and outer panels.
The inner one bears parts that are not adjusted for control purposes, while the outer bears
the control units. The voltmeter 14 is used to set the voltage and adjust the apparatus,
while the milliammeter 8 is used to set the anode current. The meter 11 reads the potential difference across the tube directly; this distinguishes the URS-60 from other types.
Button 13 switches on the high-voltage, and button 9 switches it off, lamps 12 and 10 being
green and red, respectively. Lamp 15, "ready," lights when switch 1 (tube voltage) is in
its extreme-left position; lamp 16, "no flow," lights when the water supply fails. Lamp 6
indicates "up to 40 kV," and lamp 7 "up to 60 kV." Knob 17 switches the filament voltage, 18 being the filament rheostat for tube I and 5 that for tube II; knob 4 controls the
trip limits for the tubes. The clock 2 records the working time and the exposure; it is
fitted with a relay to cut off the high-voltage after a specified time. Switch 20 is used to
correct the input voltage to the autotransformer, while 21 and 22 provide, respectively,
stepwise and smooth control of the anode current. The tube is contained in the jacket 23
or 23a in accordance with its type. The equipment is linked to the line via the stationary

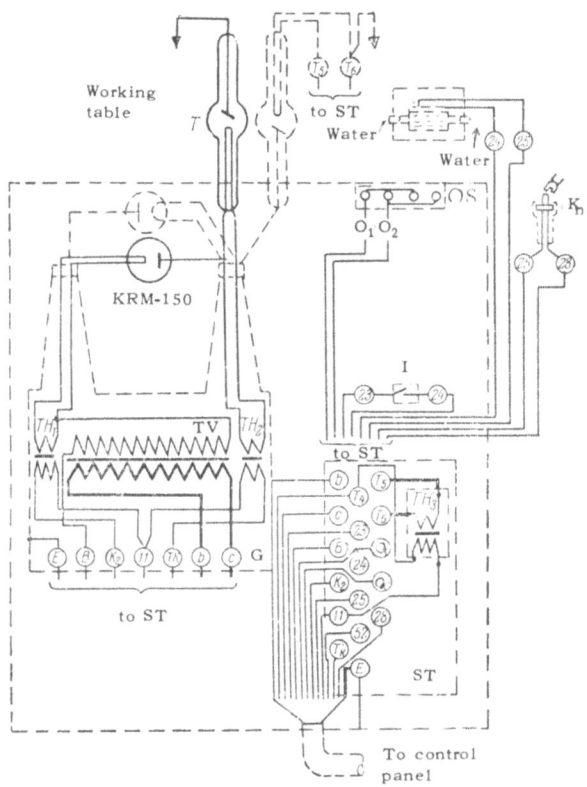

Fig. 3. Electrical system of the URS-70K1.

Lead symbols: L—line; E—ground; b and c—primary of the high-voltage transformer, B—ground lead for end of secondary of high-voltage transformer; M—meter circuit; T—tube filament circuit; K—kenotron heater circuit; O—leads to outlets; 11, 12, etc.— other circuits.

Circuit components. Control panel (diagram on p. 34): CB—control box; CH—constant-voltage adjuster; F_{21} and F_{22}—fuses in relay circuit; F_{11} and F_{12}—fuses in heater and signal-lamp circuits; TT—tapped (control) transformer; TC—transformer for signal lamps; MS—magnetic starter (relay); C—coil on relay; MC—maximum current trip; CO—hold-off coil of trip; RO_0 to RO_4—current-limiting resistors; R—cut-in resistor; RK_1 and RK_2—kenotron heater control resistors; RT_1—tube-filament control resistor; RT_2—tube-filament rheostat; BB—trip in high-voltage relay circuit; BO—trip for current-limiting resistors; SM—switch for meter; PS—switch for point of application; RS—red signal lamp; GS—green signal lamp; Ti—timer, with recording of total time of operation; SD-2—synchronous motor; V—moving-coil voltmeter; mA—milliammeter.

Working table: ST—stage top; G—generator unit; TV—high-voltage transformer; TH_1 to TH_3—heater transformers; T—x-ray tube; KRM-150—kenotron; I—interlock on door; IW—interlock on water flow; K_n—starter button to turn on high voltage; OS—outlet socket to camera drive.

Control Panel

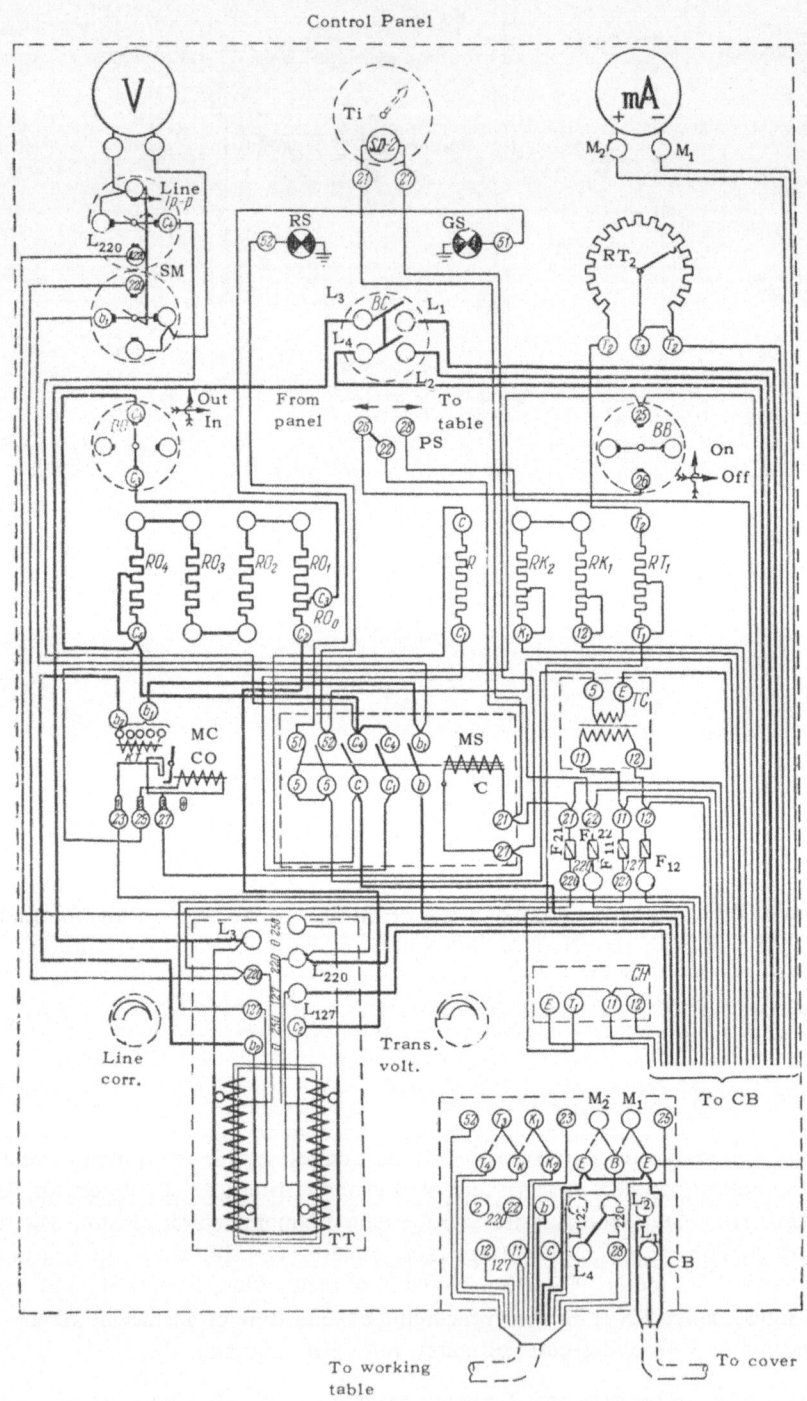

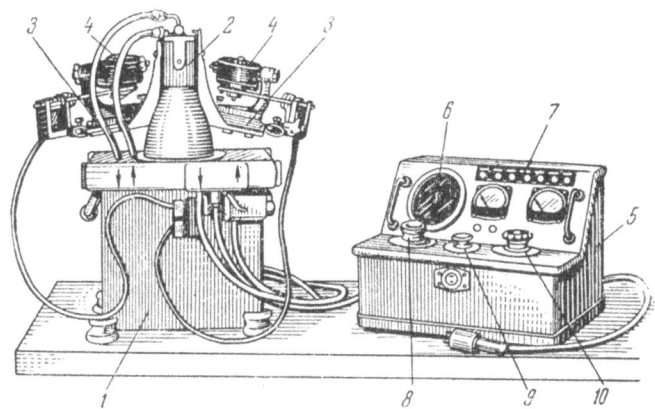

Fig. 4. The URS-55 equipment: 1) protective case; 2) x-ray
tube; 3) supports for cameras; 4) cameras; 5) control panel; 6)
timer; 7) signal lamps; 8) line-voltage switch; 9) filament
control rheostat; 10) high-voltage switch.

rack shown in Fig. 6b. Lamp 7 lights when the stabilizer is operating, and lamp 6 when
the control panel is turned on. The rack also carries the "off" button 4, the signal lamps
19 and 3 (the latter lights when the high-voltage is turned on), the neon lamp 9 ("no flow"),
and the line switch 1. The ammeter 8 monitors the load current; the milliammeter 2
records the tube current, and the voltmeter 5 shows the line voltage. The modes of use are
as described above for the URS-70 and URS-55.

    d. The URS-50I Ionization Equipment. The URS-50I diffractometer differs consider-
ably from the URS-70 and URS-55, for the latter two are equipped only to generate x-rays,
whereas the URS-50I is also fitted to record them [254]. There is also a simplified model,
the URS-25I [268].

    The main feature of the URS-50I is that the diffracted rays are recorded by ionization
methods; the rays may enter a gas-filled counter and cause the gas to become a conductor
of electricity. The current drawn is a function of the radiation intensity.

    The method is particularly advantageous if the rays are recorded by a scintillation
counter (a special scintillator crystal working into a photomultiplier). Figure 7 illustrates
the major gain in sensitivity from the use of such a counter by reference to the (211) and
(444) lines of α-iron recorded with Mo Kα radiation with a scintillation counter (curves a)
and with a Geiger counter (curves b) [256]. The gain is especially notable for weak lines.
Figure 8 gives a general view of the URS-50I; Fig. 9 shows a block diagram of it.

    The URS-50I is used as follows. The x-rays fall on the specimen, which is mounted
on a goniometer of type GUR-3, which rotates the specimen and the counter at a specified
rate. The diffracted rays enter the counter and produce pulses of current, which are ampli-
fied and fed to a scaler. The counts are read either as a total (from an electromechanical
register) or as a rate, which is recorded by a pointer instrument or by a pen recorder, the
last providing a means of obtaining a permanent record of the intensity as a function of angle.

    The basic components are the voltage stabilizer, the high-voltage generator with its
anode-current stabilizer, the goniometer, the scaler or ratemeter, and the electronic
recorder.

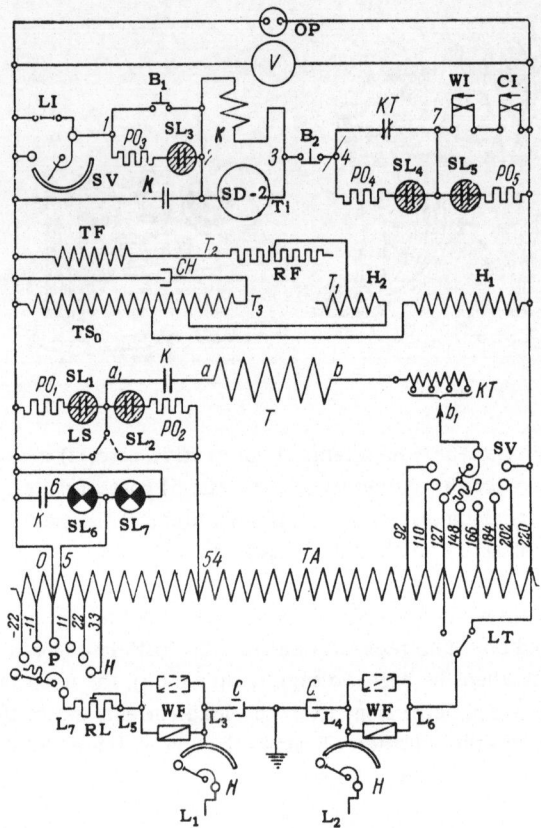

Fig. 5. Electrical system of the URS-55. TA — Autotransformer; T — high-voltage transformer; TF — filament transformer; TS_0— autotransformer of stabilizer;
H_1 and H_2 — windings on stabilizer choke; RF — filament power rheostat; PO_1 to PO_5 — current-limiting
resistances for neon tubes; RL — line-current resistor;
P — power switching resistor; C — capacitor to prevent
radio interference from reaching the line; CH —
capacitor in stabilizer; H — line corrector; SV — volt-
age-control switch; SL_1 to SL_5 — neon signal lamps;
SL_6 and SL_7 — filament signal lamps; B_1 — button to
bring on high-voltage; B_2 — button to cut off high-
voltage; LT — line-voltage tap; LS — switch for limits
of voltage variation; LI — limit interlock; WI —water
interlock; CI — cover interlock; WF — wire fuses; K —
contactor; KT — maximum-current trip; Ti — timer and
total operating time recorder; OP — outlet point for
power to camera motors; V — voltmeter. The high-
voltage part of the system is very simple; it consists
of the secondary windings of the filament transformers,
the tube, the discharge gap, and the milliammeter.

Fig. 6. The URS-60 equipment. a) Working table; b) voltage stabilizer.

The generator is placed under the working table and consists of a high-voltage transformer, a filament transformer, two transformers for the kenotron filaments, two high-voltage capacitors, a current-limiting resistor, two kenotrons, and a relay to discharge the
capacitors. All parts, except for the kenotrons and the relay, are enclosed in a tank filled
with transformer oil. The relay is enclosed in a box under the top of the table. The
armature falls when the current is cut off and closes the discharge circuit of the capacitors.

The SN-1 stabilizer is placed at the bottom of the rack containing the measuring
equipment; it provides a constant input voltage to all units in the URS-50I.

Goniometer. This (Fig. 10) serves to measure the angular positions of specimen
and counter; it is also fitted with means of rotating these at a specified rate. The main
parts of the goniometer are the body, the base, the synchronous motor fitted with reduction
gear and controls, the monochromator, the stage carrying interchangeable holders and the
goniometer head, the sighting tube, and a set of slits. The body 8 is a metal box enclosing
the mechanism for rotating the stage and counter as well as the main optical components.
This body is set up on the base 15 on three vertical legs, which serve to keep the instrument level. Two ball-ended legs that rotate above vertical axes work in guides and serve
to displace the instrument; the third leg is set in a stop. The angle of the goniometer
relative to the base is read from the scale 13 engraved on the base. The upper part of the
body carries the stage 6, which bears the holder and counter; the specimen and counter
rotate about the vertical axis. The instrument allows the specimen and counter to be
rotated separately or together by hand or by power; wheels 9 and 11 serve to rotate the

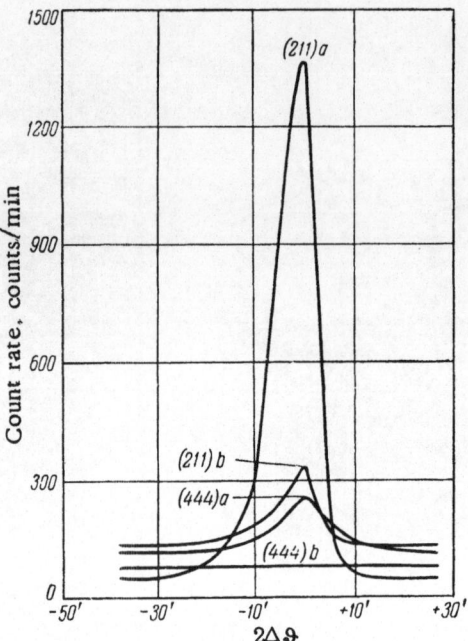

Fig. 7. Curves for the (211) and (444) lines of
α-iron recorded with molybdenum radiation:
a) with a scintillation counter; b) with a Geiger
counter.

specimen and counter (separately if screw 10 is slackened off, together if it is tight-
ened up).

The motor provides speeds of 0.5, 1, 2, and 4 deg/min at the counter, the speed at
the specimen being half this. The motor is switched off and a red signal lamp is lit when
the counter reaches the maximum angle (about 150°). The angular positions are read from
the scales as seen in projection on the matte glass screen 1; handle 14 serves to change
over from one scale to the other. The speed-set control 12 is rotated and clamped with
special knobs. The motor is started, stopped, and reversed with a toggle switch. The
chart drive on the recorder has a separate switch. The specimen holder lies at the center
of the stage. The main types of holder are for polished sections and for cylinders. The
goniometer head provides means of centering and adjustment relative to the axis of rotation. Lens 2 is used in centering the specimen. Monochromatic radiation is provided by
a special holder bearing a plate, which is set to the required angle with the control knob.
The counter is enclosed in the housing 16, which is supported on an arm having its axis of
rotation coincident with that of the stage.

The slits isolate a narrow beam. The slit in front of the counter is adjusted in height
with the screw 7. The x-rays pass through the entrance slit 3 to the beam-catcher 5, the
height of which is regulated by screw 4; the slit 3 and beam-catcher 5 are mounted on the
body of the goniometer. A set of slits of various widths is attached to the apparatus. The
goniometer is supplied with holders for massive specimens and is fitted with a sighting tube
and light source, which provide means of checking the setting of cylindrical specimens.

The electrical system of the goniometer is controlled by a master switch on the
control panel; the motor and lamps are fed via the working table.

Fig. 8. The URS-50I diffractometer: 1) Enclosed
line trip; 2) working table with generator; 3) x-ray
tube in housing; 4) goniometer; 5) radiation counter;
6) rack carrying measuring equipment. 7) EPP-09
electronic potentiometer; 8) integrating circuit and
supply for RE-1 counter; 9) panel bearing electromechanical register and time switch; 10) PS-64
scaler; 11) SN-1 voltage stabilizer.

Mechanical Details of the Goniometer. The specimen can rotate
between limits of −180 and +180°. The counter can rotate between −120 and +150°. The
slits are adjustable from 0 to 8 mm in height and are 0.1, 0.25, 0.5, 1, 2, and 4 mm wide.
The entrance slit is 118 mm from the common axis; the counter slit, 160 mm. The
angular speeds at the counter are 0.5, 1, 2, and 4 deg/min; the specimen rotates at half
these speeds. The angles can be measured to ± 1'.

The RE-1 unit (ratemeter) records the intensity; it consists of a separate head
amplifier (single-stage, resistance-capacity coupled, with negative feedback) mounted on
the goniometer, the ratemeter proper, and a separate high-voltage power supply, the latter
two contained in the rack. The PS-64 scaler is used to reduce the pulse rate by a known
factor; it is placed in the rack with the RE-1. The electromechanical register records the
output from the scaler, and the time switch allows the scaler to operate for a specified
period. The EPP-09 recorder gives a permanent record of the intensity. These units have
complex circuits, which cannot be dealt with here; the manuals on the units must be
consulted.

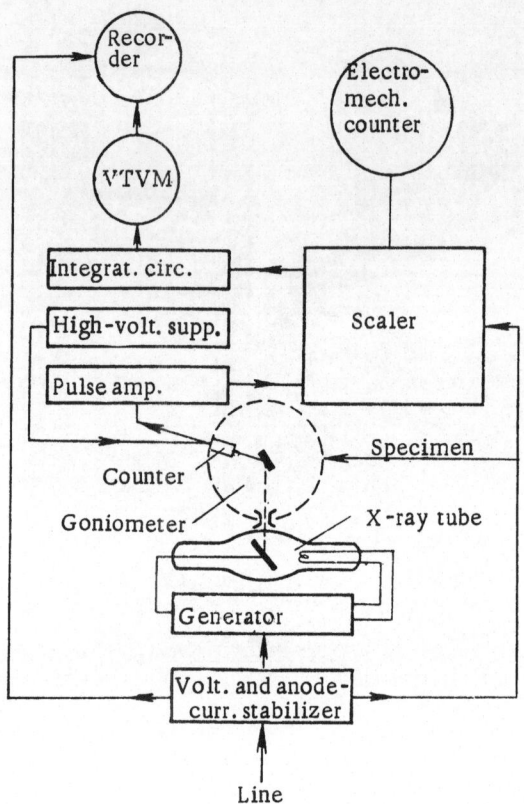

Fig. 9. Block diagram of the URS-50I diffractometer.

Installation and Adjustment of the Goniometer. The manufacturer's instructions of the usual sequence are as follows:

1. The goniometer is set up to bring its entrance slits opposite the window of the x-ray tube (opposite the protection tube). The legs are adjusted to bring the slits to the height of the window, and then the goniometer is moved sideways and rotated to bring the slits opposite the window.

2. The units are adjusted to allow the x-ray beam to extend 2 mm to both sides of the slit; this is done by opening up the slides (setting the entrance slits to "open") with the third slit closed or the counter turned out of the straight-through position (to avoid damage from excessive counting rates); the high voltage is then turned on, and the voltage and current are increased until a clear image is obtained on a fluorescent screen placed behind the slits. A slit 2 mm wide is put into the first slit-holder to provide an image of the whole beam at the bottom and of the slit only at the top. If the slit image is at the edge of the beam, the system must be adjusted to bring it roughly to the middle. To do this, the goniometer is moved sideways and is turned about its axis; the slit is then placed in the second holder and the image is examined again. Any deviation from the central position is removed in the same way. The goniometer is brought within an inch or so of the tube.

3. If the 2-mm slit is properly covered by the beam in both slit holders, this slit is placed in the first holder and the 0.1-mm slit is placed halfway in the second holder. The

Fig. 10. The GUR-3 x-ray goniometer.

base of the goniometer is clamped, and the adjustment screws are used to bring the image of the 0.1-mm slit to the middle of the 2-mm slit. The 0.1-mm slits are then inserted fully into the first and second holders; the goniometer is adjusted as necessary to give a bright image of the 0.1-mm slit. A test is made to see whether this image falls roughly halfway down the third slit; to do this, the counter is placed in the straight-through position and the image is viewed on the fluorescent screen placed in front of the counter slit. If the image falls slightly above the center, the right-hand side of the goniometer is raised slightly with the rear leveling screws or the x-ray tube is rotated slightly clockwise. A further check on the position of the image of the 0.1-mm slit is then made.

4. The first two 0.1-mm slits are then supplemented with a third 0.1-mm slit and the x-ray intensity is minimized; units RE and PS-64 are turned on, and the Geiger counter is adjusted to give the maximum reading on the ratemeter. If the reading goes above full-scale, the stop on the second or third slit is closed down to bring the reading to roughly half-scale. The counter is displaced 5' and the goniometer is adjusted slightly until the maximum reading is obtained; if this maximum is larger than the previous one, the count-er is displaced a further 5' in the same direction, while if it is less it is displaced the other way, and so on until the largest maximum is obtained. This position should correspond to the counter set at zero; if this is not so, the counter is set to zero without altering the other adjustments and the position of the third slit is adjusted (with a special key) to give the maximum reading. The slits are gripped on both sides by screws, so one of these must be

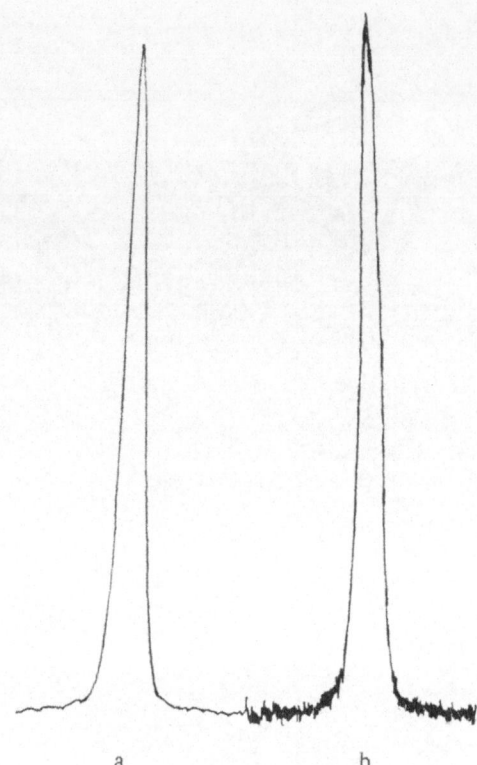

Fig. 11. Curves for a (110) line recorded with
the URS-50I with the following time-constants:
a) IV; b) II.

slackened off and the other turned to drive the slit. The goniometer is then clamped to
the base (with the clamping knob), and the zero is set more precisely by turning the counter 5' to one side and then 5' to the other; the reading in both cases should be the same.
Any difference in the readings is eliminated by setting the counter in the position giving
the smaller reading and then adjusting the front leg of the goniometer to bring the needle
roughly halfway between the two readings; the 5' test is then repeated, and the process is
continued until the zero setting is correct. If adjustment of the leg fails to produce any
appreciable increase in reading, the position of the counter must be adjusted with the two
screws to give the maximum reading.

5. The specimen holder is fitted and clamped; a flat specimen is inserted and the
control knob is turned to give the maximum ratemeter reading. This sets the plane of the
specimen (and the working plane of the holder) parallel to the beam. If the reading is
thereby reduced, the specimen must be blocking out part of the beam. The test is repeated with the specimen (and holder) turned through 180°. If the beam is transmitted completely in one position (if the reading is the same as when there is no specimen) but not at
all in the 180° position, then the holder must be adjusted to allow the beam to be transmitted completely; the beam then passes through the slot in the holder.

The beam is displaced towards the specimen by moving the second slit in the same
sense, which is done by slackening off one screw by half a turn and tightening up the other.

The third slit is closed, the fluorescent screen is fitted, and the intensity is turned up; the goniometer is then adjusted slightly to give a bright image of the 0.1-mm slit. The specimen is then removed and the zero is reset as in Section 4 above. The specimen is then re-fitted and the above sequence is repeated; if the beam is then transmitted fully, the second slit has not been displaced sufficiently; if it is completely cut off, the displacement has been too great. Successive adjustments in this fashion give the situation in which the beam is half cut off with the specimen on either side. The counter is then adjusted to give the maximum reading.

Displacement of the second slit causes the position of maximum reading to move away from zero; when the adjustments to the second slit have reached the point where the maximum lies not more than 5' from zero, the second slit can be clamped and any further adjustments done on the third slit. For this purpose the holder is set to transmit the beam fully and the third slit is displaced towards the specimen to reduce the reading to about half. Then the specimen is removed, the zero is reset (without touching the slits), the specimen is replaced, and so on by successive approximation. It is usually possible to effect this stage in the adjustment with the third slit only; adjustment of the second slit is needed only in exceptional circumstances.

If the beam is completely transmitted or completely cut off in both positions of the specimen, the holder needs adjustment, for the axis of rotation does not lie in the working plane. The holder is adjusted by displacing its upper part (head), which is fixed by three screws to the supporting rod; the adjustment is made to cause the specimen to cut off part of the beam. The screws are not slackened off for this purpose. The large holder is adjusted with two screws which drive the upper plate (the three screws holding this plate are slackened off).

6. The holder is set to bring the working plane parallel to the beam when the specimen scale (green) reads zero. This zero setting is checked by placing a standard specimen in the holder; the counter is turned through 2ϑ, the voltage applied to the x-ray tube is increased, and the specimen control knob is turned to give maximum reading at the rate-meter (the voltage is reduced as required if the needle goes off the scale). The specimen is then set at an angle ϑ to the beam, but the specimen scale may read a different angle; the reading is therefore set to ϑ and the two fixing screws on the holder are slackened off. The holder is turned (with the control knob untouched) to give maximum reading and then is refixed.

The RE-1 unit is turned on by the line switch; the voltage on the counter rises gradually to its normal value (usually 1400-1500 V), any adjustment being made with the "set high voltage" knob. The ratemeter should read zero in the absence of the beam; any deviation is eliminated by operating the "set zero" control. The button "discharge high voltage" should be pressed (to discharge the capacitors) when the unit is switched off. The zero of the RE-1 is adjusted to that of the recorder by means of the mechanical zero set on the milliammeter.

The time constant is set (with the range switch) in accordance with the sensitivity needed; Fig. 11 illustrates the effects of the time constant on the line shape by reference to the (110) line for 18KhGT steel. Line a was recorded with time constant IV and line b with time constant II.

The pen of the EPP-09 sometimes gives a sudden kick; this corresponds to automatic setting of the working current and indicates that the instrument is working correctly. The pen may perform several such kicks when it is first switched on. The response time can be adjusted with the control knob on the electronic unit in the EPP-09.

The units are tested by turning on the "test" switch on the PS-64 and switching off the x-ray tube; all instruments should then show the line frequency (50 cps). The PS-64 is also fitted with switches to vary the scaling factor.

The following example relates to use on steel 20 with Fe Kα radiation with an upper limit to the high voltage of 50 kV; the object is to examine the block sizes and lattice distortion.

The counter voltage is set to 1500 V; time constant III is used, with a chart speed of 1200 mm/hr. The counter rotates at 0.5 deg/min (in step with the specimen), all slits being 0.5 mm wide and 8 mm high. The voltage and current switches are set to position 3-3 for the (110) line (ϑ = 28.5°) and to position 6-6 for the (220) line (ϑ = 72.5°).

Setting Counter Voltage. It is usual to use an MSTR-4 (RM-4) Geiger counter; see [257] for the characteristics of this and of other similar counters for the purpose. The plateau must be plotted before a new counter is used, this being the counting rate as a function of voltage for a constant radiation flux. A counter filled with Kr + Xe is useful for hard radiations.

The apparatus is prepared for plateau plotting by testing the PS-64 as above; then the RE-1 is turned off and the cover on the "set high voltage" control is removed, the control being turned as far as possible to the left. The RE-1 is then turned on and the high-voltage is increased gradually until counting just starts; the counting rate is then recorded for a series of fixed voltages. The high voltage should not be reduced during this series, as this may introduce an error; the charge on the capacitors is not removed when the control is turned down, so the actual voltage remains high although the meter reading may be low. Points at the low-voltage end may be checked by switching off the RE-1, pressing the discharge button, and then repeating the above.

Care should be taken to determine the voltages corresponding to the upper and lower ends of the plateau; the best working voltage is that corresponding to the center of the plateau, for then slight changes in voltage will not affect the readings. The slope of the plateau should not exceed 5%. The plateau should be checked from time to time or when the apparatus has not been used for some while. The position of the plateau is given roughly in the data sheet (recommended working conditions).

e. Faults and Methods of Correction — This section has been omitted from the translation as it would be of value only to those individuals actually using the equipment described.

f. Safety. The hazards are those of exposure to high and low voltages, x-rays, and gases (ozone and oxides of nitrogen produced around high-voltage points); the x-rays can cause burns and general biological damage.

The following rules must be observed in x-ray laboratories:

1. Any apparatus may be switched on only after all parts (especially grounds and interlocks) have been checked.

2. Faults may be corrected and components replaced only with the x-ray equipment turned off.

3. Working x-ray equipment must be enclosed in protective lead screens.

4. The x-ray beam must not be accessible if the window of the tube is open.

5. Care must be taken to check (with the special discharge check) that the high-voltage capacitors in the URS-50I have been completely discharged on switching off.

6. No one should remain unnecessarily in any location in which x-ray equipment is operating.

Special treatises [14] should be consulted for details of safety precautions.

2. X-Ray Tubes and Rectifiers

Two main types of tube are at present produced in the USSR for use in structure analysis: the BSV2 and the BSV3 [238].

The BSV2 [All-Union State Standard (GOST) 8600-57] tube is sealed and has partial shielding, two output beams, and a water-cooled anode. The tube is designed to work in air in a shielded safety housing, with the anode fed by a-c (URS-55 equipment). The heater voltage and current are as follows: (a) anode voltage 25 kV and anode current 1 mA, 2.5 ± 0.8 V and current not less than 1.6 A; (b) anode voltage 25 kV and anode current 30 mA (anodes of iron, cobalt, and nickel, 27 mA; chromium, 24 mA), 4.6 ± 1 V and current not more than 3.2 A.

Table 1 gives the electrical characteristics of the BSV2 series. The limiting anode currents for other voltages can be deduced by dividing the maximum continuous power by the nominal voltage and by a nominal power factor (here 0.7); for low voltages, the current can be increased to the limit set by the above and by the maximum permissible heater current, but it must not exceed 30 mA for tubes with anodes of tungsten, molybdenum, silver, and copper or 25 mA for tubes with iron, cobalt, nickel, and chromium.

TABLE 1

Type	Anode	kV, nom.	Max. cont. power, kW	Current (mA) for rectified voltage (kV)		
				$U_{max} = 30$	40	50
0.8 BSV2 W	Tungsten	55	0.8	38	28	22
0.8 BSV2 Mo	Molybdenum	55	0.8	38	28	22
0.7 BSV2 Ag	Silver	55	0.7	30	25	20
0.7 BSV2 Cu	Copper	50	0.7	30	25	20
0.5 BSV2 Fe	Iron	50	0.5	23	17	14
0.5 BSV2 Co	Cobalt	50	0.5	23	17	14
0.5 BSV2 Ni	Nickel	50	0.5	23	17	14
0.4 BSV2 Cr	Chromium	50	0.4	20	16	12

The vertex angle of the cone of rays is not less than 10°; a line focus is used. The projection of the optical focal spot along the axis of the beam (at 6° to the plane of the anode) is not more than 1.2 mm wide or 1.3 mm long. The differences in output for different windows do not exceed 10%. The guaranteed working life is 300 hr; secondary lines account for not more than 1.5% of the intensity during the first 200 hr.

The BSV3 (GOST 8491-57) is as above, except that it is designed to operate with highly smoothed supplies (as in the URS-50I). The heater voltage and current are as follows: (a) anode voltage 20 kV and anode current 1 mA, 1.2 ± 0.25 V and not less than 1.8 A; (b) 20 kV and 14 mA (12 mA for chromium anodes), 1.9 ± 0.4 V and not more than 3.0 A.

Table 2 gives the nominal anode voltage and maximum powers for use with highly smoothed supplies. The maximum anode currents are deduced by dividing the maximum power by the nominal voltage and the nominal power factor (here 1.0); for low voltages, the current can be increased to the limit set by the power or by the maximum permissible filament current, but it must not be more than 14 mA.

TABLE 2

Type	Anode	kV, nom.	Max. cont. power, kW
0.4 BSV3 Mo	Molybdenum	45	0.45
0.4 BSV3 Cu	Copper	45	0.40
0.3 BSV3 Fe	Iron	40	0.28
0.3 BSV3 Co	Cobalt	40	0.28
0.3 BSV3 Ni	Nickel	40	0.28
0.3 BSV3 Cr	Chromium	40	0.24

The beam has a vertex angle not less than 10°; a line focus is used, whose projection along the beam axis (at 2° to the plane of the anode) is not more than 2.5 mm wide and 0.3 mm long. The guaranteed working life is 300 hr; secondary lines account for not more than 1.5% of the intensity during the first 200 hr.

Many laboratories use BSV1 tubes, which are intended for use with the URS-70K1; these have four output beams and a focus 5 mm in diameter. These tubes work with un-smoothed rectified supplies in air in a screening case; the anode or cathode may be grounded. The voltage is 50-70 kV, the maximum power 0.25-0.8 kW, the heater voltage 3.5-9 V, and the heater current 3.2-4.5 A; the anode material determines the operating parameters. Table 3 gives the limiting values for the anode current.

TABLE 3

Anode	Current (mA) for rectified voltage (kV)				
	U_{max} = 30	40	50	60	70
Tungsten	38	28	23	19	16
Molybdenum	33	25	20	16.5	—
Silver	24	18	14	12	—
Copper	24	18	14	12	—
Nickel	14	10.5	8.7	7	—
Cobalt	14	10.5	8.7	7	—
Iron	14	10.5	8.7	7	—
Chromium	12	9	7	—	—

Microfocus tubes (focal spot about 0.01 mm in diameter) have considerable advantages, for they provide much shorter exposures and enable one to examine small parts of the specimen [10, 258].

The rectifiers (kenotrons) used in structure analysis are the KRM-110 (URS-50I) and the KRM-150 (URS-70K1), whose peak inverse voltages (maximal) are 110 and 150 kV respectively (GOST 6919-54). The maximal parameters for the KRM-110 are as follows: cathode emission at 20 kV of 300 mA, heater current of not more than 14 A at a maximum heater voltage of 10 V, mean rectified current up to 60 mA at heater voltages up to 8 V; the corresponding figures for the KRM-150 are 3.0 kV and 300 mA, 9 A and 13 V, and 30 mA and 12 V.

Before a tube is installed it should be examined for defects, and the body should be cleaned with alcohol. The water-cooling attachment and protection cylinder are removed from the BSV4 before examination. The vacuum is then tested by applying the high voltage briefly with the filament cold. Any reading on the milliammeter indicates a low vacuum in the tube, which is then usually unserviceable.

Undamaged tubes are run in at not more than 1/3 of the nominal voltage and anode currents of 1-5 mA for a short while; the voltage is raised by steps of 3-5 kV every 30 min until the nominal value is reached. Sometimes the milliammeter shows wild fluctuations during this process; this is caused by release of gas. In such cases the voltage should be reduced and kept low for several hours.

The windows should be coated with lacquer not less than once a month; the previous layer is first removed carefully with cotton wool soaked in acetone, and the fresh coating is applied with a soft brush.

The characteristic radiation may be accompanied by emission from impurities in the anode or from material sputtered from other parts of the tube. These interfering radiations may be as much as 10% of the main radiation in the case of the BSV1. The main impurities for Cu anodes are Fe and W; for Fe, Co, Mo, Ni, and Cr they are Cu and W. The radiation is examined for contaminants by the use of a substance giving few lines (Fe, Ni, Cu), which are indexed by methods described in later parts of this book; any excess lines are identified as to origin and radiation.

The anode in the KRM-150 or KRM-110 must not be allowed to overheat, nor must the cathode be underrun; damage may occur or the cathode may be caused to emit if these precautions are not observed. In particular, the anode of the kenotron should not be allowed to become a bright cherry red.

3. X-Ray Cameras

RPK-2. This has a cylindrical cassette 57.3 mm in diameter and is designed to record powder patterns from cylinders and polished sections. Angles of 10 to 80° are accommodated.

RDK. This is an improved model of the above (also 57.3 mm). Provision is made for automatic centering of sections and optical centering of cylinders on a magnetic support, the stop settings being fixed. The range 4-84° is covered.

VRS-3. This is of the above type but has a diameter of 143.25 mm; means are provided for examining two specimens simultaneously with one film. Range 3-87°.

KROS-1. This is meant for back-reflection work with flat specimens; it has several flat cassettes and a cylindrical one. Range 54-85°.

RKE. This is meant for high-speed examination and employs focusing; a flat film is used. Ranges 10-30° and 60-86°.

KMSP Camera and Monochromator. This is designed to produce very low background levels; various methods are used with cassettes 57.3 and 171.89 mm in diameter. The monochromator is a bent quartz crystal; the camera can be evacuated. The effective diameter is determined with a special device. The range is 3-87°. See [15] regarding current nonstandard and foreign instruments.

4. Microphotometers

a. MF-4 Recording Microphotometer. The MF-4 is used to measure line intensities on patterns, which is done either visually or photoelectrically with photographic recording. The light from a filament lamp passes through the film to a photocell, whose output drives a reflecting galvanometer. In addition, a blackening scale is seen in projection, which is used in visual estimation. Photographic recording is used in conjunction with the galvanometer. Figure 12 gives a general view of the instrument.

Fig. 12. The MF-4 recording microphotometer.

b. Installation and Use of the Microphotometer. After assembly and completion of
electrical connections, the bubble gauge 15 is used with the screws 18 to set the instrument level. The lamp setting is checked by unscrewing the lower lens and placing a
piece of thin paper over the top of the tube. The lamp is adjusted to give a bright, sharp,
and symmetrically placed image of the filament.

The galvanometer is adjusted by reference to its own leveling system; the suspension
is freed by turning the clamping knob maximally clockwise. Shutter 20 is closed with the
knob to the left of tube 11, and then the zero (∞ on the logarithmic scale) is brought into
coincidence with the line. The set zero on the galvanometer can be used if the scale adjustment is insufficient.

The appropriate scale is brought to the middle of the screen with control 1; the galvanometer must be set up to give no tilt in the scales and to bring all three scales on the
screen. Then shutter 20 is opened and the slit is adjusted (for width with knob 8, for
height with knob 9) to give full-scale deflection (to 0 on scale 2). Knob 16 is used to
focus the image of the slit on the screen in the tube.

The emulsion on the film is brought into focus on the screen with screw 4; screws
25 and 22, together with knob 26, provide control. Checks are performed to see that the
image remains sharp as the stage is displaced lengthwise and transversely and that the
image of a line parallel to the lengthwise motion does not move transversely as the stage
is moved lengthwise.

The other parts are as follows: 5 — cassette with film; 7 — range-change knob; 10 —
rear cover; 12 — lamphouse; 14 — interchangeable stops for light source; 29 — interchangeable lenses; 30 — objective.

c. Photometric Technique. The film is placed in a special holder or between sheets
of glass to bring the equatorial plane parallel to the lengthwise slides. A scratch or dot is
made 2-3 mm on each side of the line precisely in the middle of the field of view. Trans-
verse displacement brings these dots to the middle of the slit. The emulsion is brought into

focus, and the line between the points is checked for parallelism with the lengthwise slides. Then the film is placed to allow the ray to pass through the hole. The spot is set to ∞ on scale 2 with the shutter closed, and then the shutter is opened; the origin (0 on the logarithmic scale) is set with knob 27. The position of the zero must be checked from time to time, because it shows a tendency to drift. The film is set to cause the beam to pass around one of the initial points; then screw 13 is clamped, and the micrometer screw 28 is used to record the intensity at equal intervals (usually 0.05 or 0.10 mm). These measurements are continued until the second point on the equator is reached; 20 to 70 measurements may be needed.

If the film is symmetrical, the best method is to repeat the measurements on the other side, although this usually involves a preliminary adjustment of the stage.

Before automatic recording can be used, the scale of the recording must be selected and must be set up with an engraved glass scale.

The appropriate motor speed is then set in accordance with the requirement that the time for which any signal acts on the photocell must be greater than the time to reach a steady deflection (here 0.7 sec). The intensity in automatic recording is such that full-scale deflection corresponds to 600-650 divisions on the millimeter scale. This requires the insertion of neutral filters and a circular wedge to reduce the deflection at the hole in the film to 600-650 mm. The starting point on the matte glass is set roughly by means of handle 21 and by sliding the stage with the clamp 23 released. The point is then set precisely with the micrometer screw 28. Finally, the matte glass is replaced by the loaded plate-holder; the motor is switched on with switch 19 to start the recording. Several curves can be recorded on one plate by displacing the zero.

The magnification is chosen in accordance with the line shape; small values are used for broad lines and large ones for narrow lines. Upper lenses of magnification × 6 and × 12 are used with condenser lenses of × 0.2; the × 21 (or any other lens) is used with a × 0.1. The focus must be reset when the lenses are changed; any resulting displacement of the image of the slit is balanced out by turning screw 3. The galvanometer should not be disturbed unless there are serious defects in the working of the microphotometer, because its position is such as to give the highest accuracy. The shunt resistance across the galvanometer must be adjusted if the photocell or the galvanometer is replaced.

Any pattern to be photometered must be clean and free from fogging or changes in color resulting from incorrect processing or prolonged storage. The regions to be examined must be straight edges parallel to the equatorial line in order to provide precise setting for the path to be photometered. The film must fit into the holder.

Several improvements can be made to the MF-4 to facilitate visual and automatic measurements. For example, the gear 17 at the end of the micrometer screw may be fitted with a stop to facilitate the use of the screw; advance by one tooth corresponds to a displacement of 0.05 mm. A mirror system can be used to take readings directly around the site of examination; the system consists of a mirror at the level of tube 6 and of another mirror 24, which images the scale divisions. Also, the galvanometer recording can be replaced by a fast EPP-09 electronic recorder. In the latter case the recorder is connected directly to the photocell (Fig. 13), the circuit including a logarithmic device to give a direct recording of intensity as a function of position [242]. This system requires only a few minutes to record the curve for one line; moreover, the scale can be varied over a very wide range.

The nonrecording microphotometers of type MF-2 used in some laboratories do not differ essentially from the MF-4, apart from the lack of a recording system. Particular

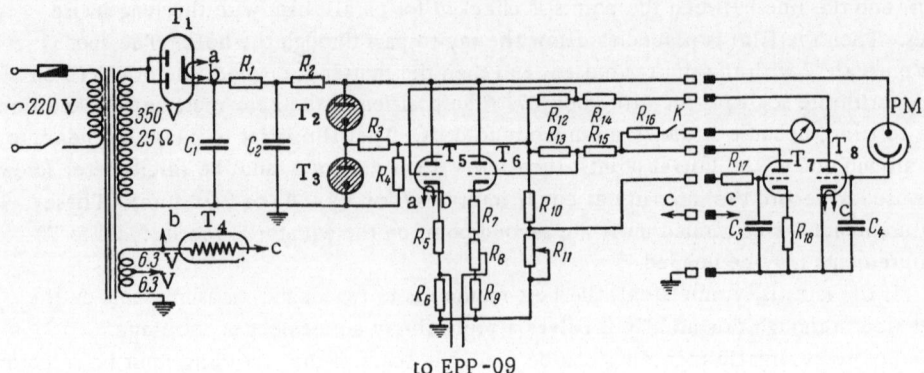

to EPP-09

Fig. 13. Theoretical circuit of a logarithmic electronic unit for use with the MF-4. Symbols: T_1 — 6Ts5S, T_2 — 6G4S, T_3 — 6G3S, T_4 — 0.425B5.5; T_5 to T_8 — 6N8; R_1 = 3.3 kΩ, R_2 = 1 kΩ, R_3 = 15 kΩ, R_4 = 5.6 kΩ, R_5 = R_7 = 5.1 kΩ, R_6 = R_9 = 90 Ω, R_{10} = 75 kΩ, R_8 = R_{11} = 1 kΩ, R_{12} = R_{13} = 27 kΩ, R_{14} = R_{15} = 75 kΩ, R_{16} = 2.2 MΩ, R_{17} = 2.5 MΩ, R_{18} = 910 Ω; C_1 = C_2 = 8 μF, C_3 = C_4 = 1000 pF; PM — STsV-3.

attention must be paid to the zero when the MF-2 is operated with storage batteries, for this tends to drift continually. See [164, 280] for the choice of slit width in photometry.

2-2. Production of Focused Lines

Measurements on lattice distortion, crystal size, etc. demand the measurement of line widths on patterns. Special conditions are required in recording patterns for this purpose; for example, broadening can occur with a cylindrical specimen if it is not properly centered.

The focus condition (minimal line width) is obtained only for a few lines close together when a flat specimen is used with photographic recording. Neglect of the width of the second slit leads to the conclusion that a line for such a specimen having an angle of diffraction ϑ will be in focus if the specimen is rotated with respect to the incident beam through an angle ψ defined by

$$\operatorname{tg} \psi = \frac{\sin 2\vartheta}{\dfrac{R}{b} + \cos 2\vartheta} , \qquad (2)$$

where R is the radius of the cylindrical camera (distance from specimen to film) and b is the distance from the first slit in the collimator to the specimen. In most cameras, R is equal to b, so (2) becomes

$$\operatorname{tg} \psi = \operatorname{tg} \vartheta.$$

That is, the line is in focus for an angle of diffraction ϑ if the section is set at an angle $\varphi = \vartheta$ to the beam in a cylindrical camera (the Bragg-Brentano condition). The condition for flat films in back-reflection cameras is

$$a = A \operatorname{tg}^2 2\vartheta,$$

where A is the distance from specimen to film and a is the distance from the entrance slit of the collimator to the film.

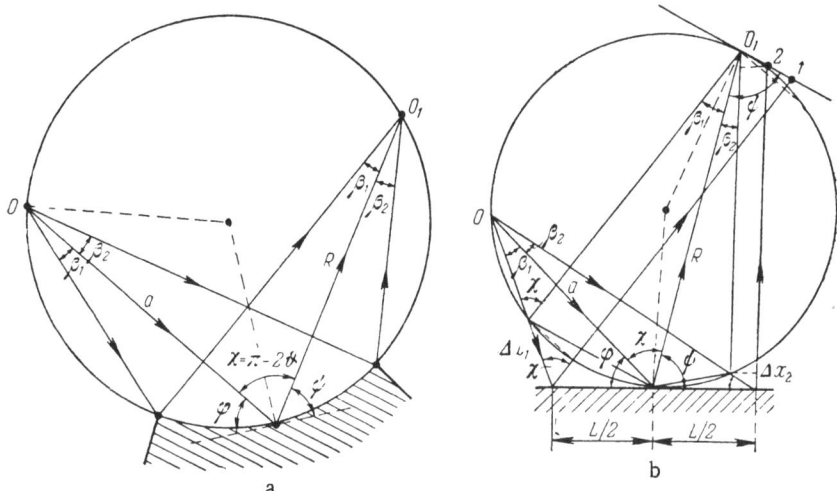

Fig. 14. Focus systems for x-rays. a) Reflection from a curved specimen; b) reflection from a flat specimen.

The use of focus methods with ionization systems gives a considerable increase in intensity. In the URS-50I, the specimen and counter move in step, so the specimen may be set at an angle ϑ and the counter at 2ϑ (which satisfies the Bragg-Brentano condition) and left to rotate in step. This feature of the URS-50I enables one to operate with lines in focus over a large range of angles. Very complicated camera systems and large increases in exposure time are needed to use lines in focus over a large range of angles with photographic recording.

A flat specimen results in a certain loss of sharpness, because only a curved specimen satisfies the focus condition precisely. Figure 14a shows the ray paths for a curved specimen, and Fig. 14b does the same for a flat one [10, 97]. In the diagrams, O is the radiation source (first slit in collimator), O_1 is a line of the pattern, a is the distance from the specimen to the first slit, R is the distance from the specimen to the line, φ is the angle of incidence, ψ is the angle of diffraction, β_1 and β_2 are the angles subtended by the two halves of the specimen at the first slit.

The broadening displaces one of the edges of the line by amounts ΔS_1 and ΔS_2 for the two halves of the specimen (length L); ΔS is given by

$$\Delta S \approx B\beta^2, \tag{3}$$

where

$$B = \frac{2r \sin 2\vartheta}{\sin \varphi \sin \psi}.$$

That is, increase in size causes broadening if β is large.

Convex surfaces (spherical or cylindrical, as on shafts, balls, rings, and so on) are sometimes used, in which case conditions become even worse. The broadening produced by a convex surface can be calculated by analogy with the above.

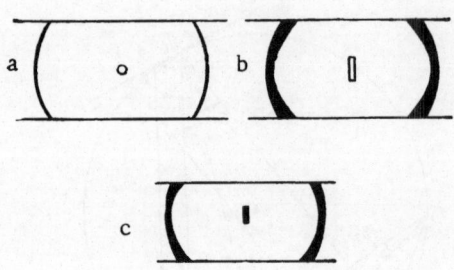

Fig. 15. Line profiles as recorded with the following stops: a) circular; b) rectangular; c) rectangular, beam not normal to axis of specimen.

The precision can be improved in a camera of standard diameter if the first slit is placed closer to the specimen (that is, within the camera) [16]; two lines can then be in focus at once. The displacement S for the slit (as reckoned from the circle bearing the film) is given by

$$S = R\left(1 - \frac{\cos(\varphi_2 + \varphi_1)}{\cos(\varphi_2 - \varphi_1)}\right), \qquad (4)$$

where R is the radius and φ_1 and φ_2 are the angles between the normal to the reflecting plane and the direction of the primary beam for the two lines. For example, this method has been used to determine the lattice constant of Al as a = 4.04142 ± 0.00009 kX.

A circular stop gives lines of equal thickness, but the exposures are usually longer. A rectangular stop along the axis of the specimen reduces the exposure time; the line width at the equator is not substantially altered if the rectangular stop has a width equal to the diameter of the circle, and the exposure is considerably reduced because much more of the surface participates in producing the line. The line width at the edges of the pattern is increased when the stop is circular; this is unimportant if the lines are broad.

Figure 15a, b shows the line shapes produced by circular and rectangular stops. It is important that the beam should be parallel to the axis of the collimator, especially when only one wide slit is used (no stop, focus being for the source). Deviations from the axis of the collimator can occur in any plane; deviations in the horizontal plane produce no marked effect on the line shape and mostly displace all lines through a small angle, but deviations in the vertical plane produce a marked change in line shape (Fig. 15c).

It is very important that the source should not change its position while the pattern is being recorded; this is easily ensured in hard tubes with heated cathodes, but the focal spot of a gas tube can migrate. In the latter case, a narrow slit is placed near the window of the tube; the exposure time is thereby increased, but the lines become much narrower.

Measurement of diffraction angles can be facilitated by fixing two metal strips in the camera or by making two slits in the cassette at a precisely known separation. Then the distance S between a pair of lines gives the exact value of ϑ from

$$\vartheta = \varphi_h \frac{S}{S_k}, \qquad (5)$$

where φ_k is the angle between the projections of the strips on the film and S_k is the distance between the images of the edges on the film. It is assumed for this purpose that film shrinkage is the same at all points; it has been shown [17] that uniformity in this respect is retained over periods of many months.

Particular attention must be given to setting the axis of the objective perpendicular to the film when measurements are made with the comparator; the images of the crosswires and emulsion must lie in the same plane, and the stage or carriage must move exactly along the equatorial line. Line positions should be read at magnifications not exceeding four, the intersection of the crosswires being placed at the center of the line and the measurements being repeated several times from both sides. Measurements on weak lines are facilitated by shutting off part of the field of view (the unexposed part of the film), thus reducing the range in the light intensity. The crosswires should be set diagonally in this case, and the light intensity should be kept low.

2-3.   Methods of Examining Transformations and Lattice States at High and Low Temperatures

Measurements at varying temperatures are needed in studies on phase transitions in metals, thermal-expansion coefficients, temperature factors for intensities, dynamic vibrations, and other related topics. Special cameras are used, which may be classified as high-temperature, low-temperature, and universal; specially adapted ionization systems are also used.

Several methods are used for preparing powder specimens for high temperatures; the material may be enclosed in a thin-walled tube linked to a vacuum pump if the material would evaporate or react with residual gases at the high temperature. Quartz is suitable as a tube material up to 1000°C; refractories such as beryllia, magnesia, and alumina are used at higher temperatures.

It is usually impracticable to prepare the specimen by mixing the powder with Canada balsam or Ramsay's compound and forcing it into a capillary; most such binding agents decompose at high temperatures and so allow the specimen to change its shape. Up to 1000°C one can use specimens made by attaching the powder to a quartz fiber with Canada balsam; the binder decomposes, but the shape is not affected. The fiber is replaced by a wire at higher temperatures, suitable materials being platinum, alloys such as Pt—Rh and Pt—Ir, and other refractory substances. The material must be one that will not react with the specimen; here platinum, silver, and certain other metals also have the advantage that they can act as standards. The best materials for fibers above 1200°C are beryllia or alumina; thin rods are prepared by grinding.

The specimen may be coated with SiO_2 or Al_2O_3 if it has a tendency to react with residual gases or to evaporate.

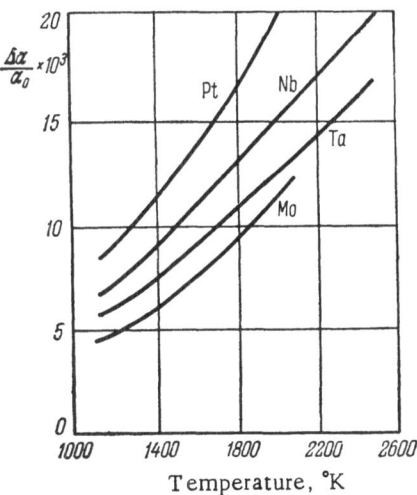

Fig. 16. Lattice constants of some metals as functions of temperature.

The holding plate used at ordinary temperatures is replaced by rigid chucks or screw fixings, or else by a suitable cement; the latter must be allowed to harden before the work can be started. Much work has been done on the design of high-temperature cameras, but it is not possible to consider the design and use of these here. Some very good designs are to be found in [18-45, 282, 288]. Ionization equipment for use at high temperatures is described in [46-52, 284, 286, 287].

The two main ways of providing low temperatures are cooling by conduction through the holder and direct contact between specimen and coolant. The simplest method is to immerse the specimen in a liquid and to pass the stream of cold gas over it. Isaichev's camera [18] has the coolant (usually liquid nitrogen) in a hollow cylinder bearing the holder at its end. The specimen may be placed in a thin-walled glass or plastic tube if it may react with the coolant.

A serious trouble in low-temperature work is the formation of a layer of ice on the specimen; this increases the background and introduces the lines of ice. Icing-up can be prevented if a stream of cold gas is used, provided that the flow is uninterrupted.

The camera, if sealed, can be evacuated or can be rendered rigorously dry with P_2O_5 or other vigorous drying agents.

Some descriptions of cameras for use at fixed or adjustable temperatures are to be found in [51-76, 285].

Thermal-Expansion Measurements. The coefficient of thermal expansion is related to the lattice constant as a function of temperature:

$$\alpha = \frac{1}{a} \frac{da}{dT} \tag{6}$$

Patterns are recorded at various temperatures, and the slope of the curve relating the lattice constant to temperature then gives α [77-83]. Figure 16 shows such curves for the refractory metals platinum, niobium, tantalum, and molybdenum [81].

2-4. Photographic Recording

2-4a. Recording of Patterns from Certain Materials

The table gives exposure times for some materials in the form of wires 0.5-0.8 mm thick in a properly adjusted camera 57 mm in diameter, the source being an evacuated tube having a line focus and with no selective filter between tube and specimen; the film is of XX type [298].

The figures given in the table are very rough; exposure times are commonly 1.5-2 times greater or very much greater if the focal spot is large. Tubes of very small focal-spot size require much shorter exposures; the same applies to the use of a divergent beam in conjunction with focusing by the specimen. The time is inversely proportional to the current and to the film speed, but increases with the radius of the camera.

Material	Anode	Wavelength λ, Å	U_{max}, kV	Current i, mA	Time, min
Aluminum	Mo	0.708	60	10	100—150
Copper*	Mo	0.708	60	10	200—300
Aluminum	Cu	1.539	40	15	15—20
Copper	Cu	1.539	40	15	15—20
Tungsten	Cu	1.539	40	15	30—40
Iron	Fe	1.932	30	12	20—30
Fe_2O_3	Fe	1.932	30	12	60—90

* With an aluminum filter 0.2 mm thick between specimen and film.

2-4b. Nomogram for Setting up Back-Reflection Cameras

The distances a (from the first slit to the film) and A (from specimen to film) are given by

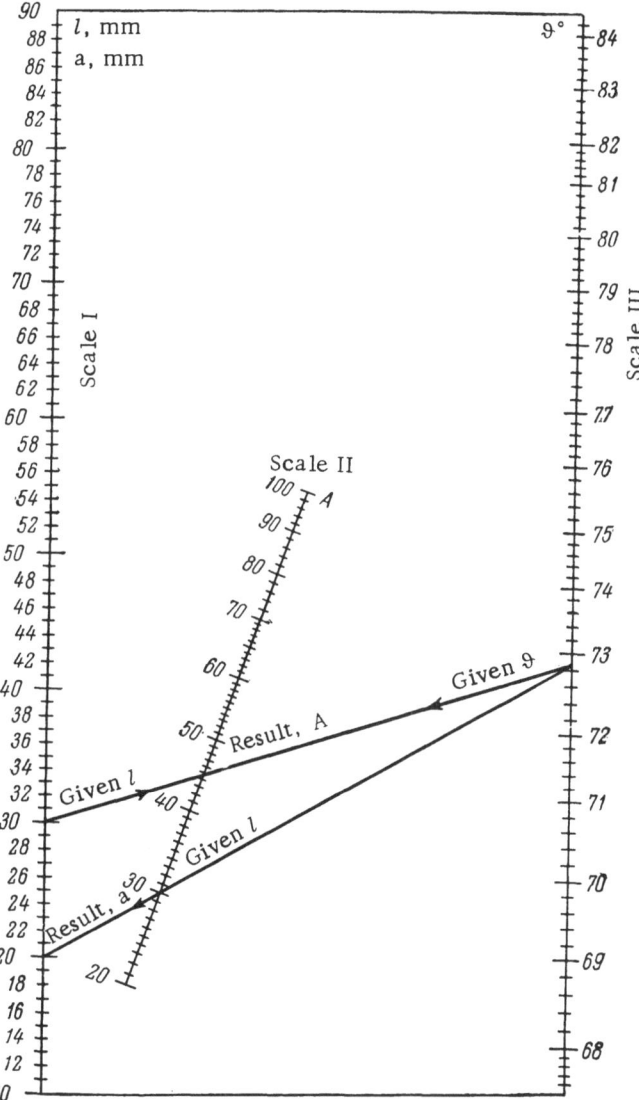

Fig. 17. Nomogram for setting up back-reflection cameras.

$$A = \frac{2l}{\text{tg}\,(180^\circ - 2\vartheta)}, \quad a = A\,\text{tg}^2\,2\vartheta, \tag{7}$$

where $2l$ is the distance between symmetrically placed lines in mm (the diameter of a ring).

The nomogram (Fig. 17) gives a and A as functions of l; scale I is for l and a (mm), II is for A and l (mm), and III is for ϑ [6]. The nomogram is used as follows to determine a and A: (1) l is specified; (2) the point corresponding to l on scale I is joined to the known value of ϑ on scale III, the point where the line meets scale II then giving A; (3) from the point for ϑ on scale III a line is drawn through the point corresponding to l on scale II to meet scale I, which gives a. The nomogram shows an example of the deduction of a and A for the (220) line of Fe in focus with Fe Kα radiation.

2-4c. Nomogram for Setting Cameras for Express Recording

The focus conditions for large ϑ for rapid cameras (RKE) are

$$g = \frac{l}{tg \cdot \beta}, \qquad \gamma = \alpha + \beta,$$

where g is the distance from the specimen (axis of inclination) to the film, γ is the angle of inclination, l is the distance from the axis of the primary beam to the line, β = 180° −2ϑ, α = tg^{-1} (l/f), and f is the distance from the tube focus to film.

Figure 18 gives g and γ for f = 50 mm for ϑ between 59 and 87°; f = 50 mm corresponds to the use of a BSV1 (BSV4) tube.

The parameters are found by specifying g, the known ϑ and g being used to find γ [84]; l must satisfy the design limits for the camera, i.e., must be in the range 15-100 mm (Fig. 18). If the angles are small, f = 230-250 mm, β = 2ϑ, and γ = 180° + β −α.

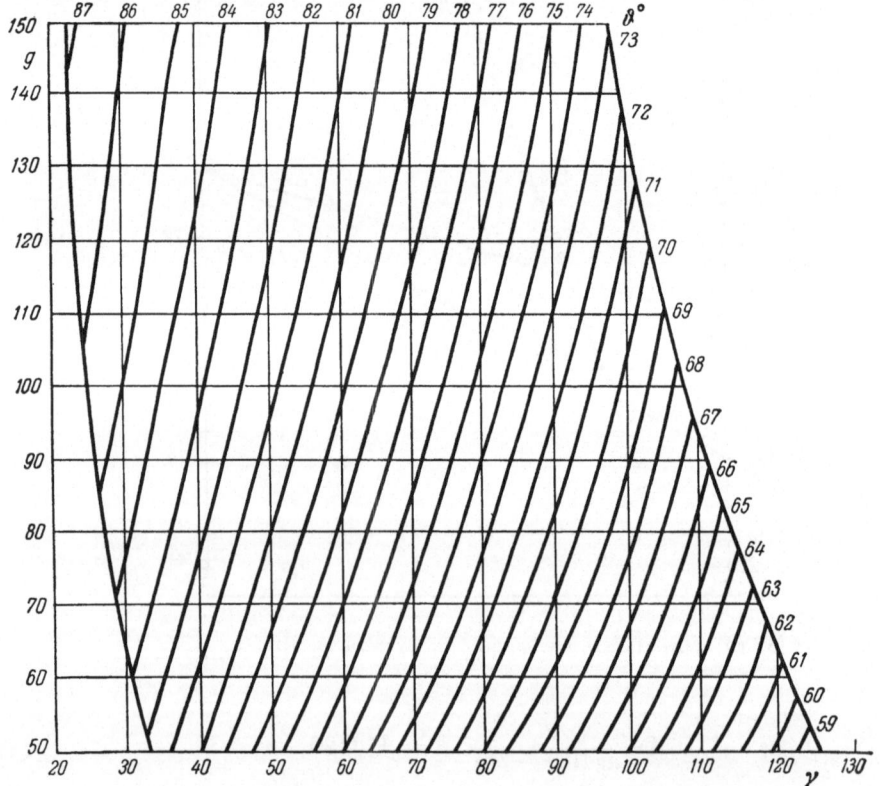

Fig. 18. Nomogram for setting cameras for express recording.

2-5. Ionization Recording

2-5a. Radiation Detectors

The table gives the characteristics of various detectors used in structure analysis [12]. The table is of value in selecting the type of counter for particular uses.

Detector	Lower limit	Upper limit	Efficiency	Stability	Best wavelengths	Additional equipment	Remarks
Film	10^4 quanta per mm^2	10^7 quanta per mm^2	Each absorbed quantum produces some blackening, the efficiency varying from 20% for MoKα to 100% for CrKα	High	Efficiency increases with wavelength, but with discontinuities at the absorption edges of Ag and Br	Microphotometer for intensity measurements	Very low intensities can be recorded if the exposure time is sufficient, and high stability in the primary beam is usually unnecessary
Ionization chamber	200 quanta per min	Very high, not attained in practice	Up to 80% for all radiations	Medium	Ionization current increases as wavelength decreases	Electrometer system, usually complicated	Can almost always be replaced by a Geiger counter
Geiger counter (for exact measurements)	10 quanta per min	50,000 quanta per min	Filled with Ar 65% for soft radiation	Good	Ar-filled counters have a broad sensitivity peak centered on 1.6 Å	Stabilizer, quench unit, ratemeter, recorder	Most widely used in research

Detector	Lower limit	Upper limit	Efficiency	Stability	Best wavelengths	Additional equipment	Remarks
Geiger counter (for rapid measurements)	200 quanta per min	200,000 quanta per min (nonlinear above 10,000 quanta per min)	Up to 70% for Mo radiation if filled with Kr	Good	As above	Simple integrating circuit or galvanometer	Used in routine analyses of low precision
Proportional counter (for high rates)	5 quanta per min	200,000 quanta per min, linear throughout the range		Good	As above	Stabilizer, discriminator, ratemeter, recorder	Has some advantages over the Geiger counter, although the equipment is complex
Proportional counter (for use with monochromator)	Less than 1 quanta per min	100,000 quanta per min	Somewhat less than for Geiger counter	Good	$\lambda/2$ can be entirely suppressed	Complex design of discriminator	Used for very weak reflections
Scintillation counter	5 quanta per min	Linear up to 200,000 quanta per min	70% for Mo $K\alpha$	Medium		As above	Gives the best results with hard radiation; almost always increases precision and sensitivity

2-5b. Absorption of X-Rays in Geiger Counters

Figure 19 gives absorption curves for counters filled with argon and krypton [11].

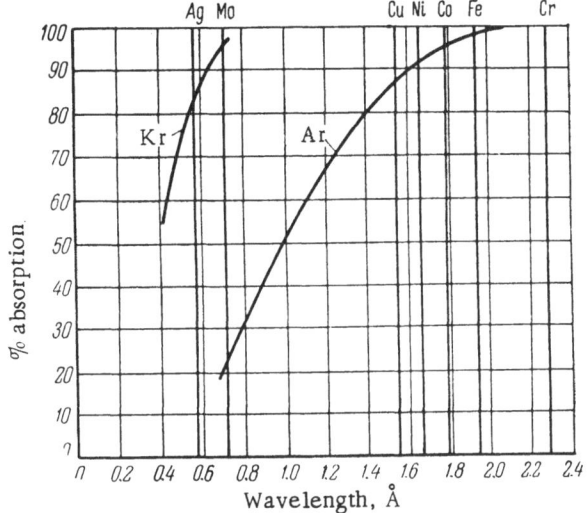

Fig. 19. Absorption of x-rays in Geiger counters with
different fillings.

2-5c. Efficiencies of Counters

Values are given for the various types in the table [114]. The efficiency is defined as the ratio of the quanta recorded to the total entering the counter; the dead time is the time for which the counter is insensitive after a quantum has been recorded. The table also gives the ratio of line to background for graphite for the various types.

Type	Efficiency, %		Dead time, μ sec	Discrimination range	Discrimination loss, %	I_{max}/I_b for graphite (102) line
	CuKα	MoKα				
Geiger: argon-filled	60	13	200			2.3
krypton-filled		30				
Proportional: argon-filled	60	13	1	6-10	20	4.3
krypton-filled		30				
Scintillation	96	99	1	5-10	15	4.5

2-6.   Selective Filters

The table gives the parameters of filters used to absorb the β-components of characteristic spectra and to reduce the background.

The data given are the materials and the mass per unit area (powder) or the thickness (foil) required to reduce the ratio of Kβ to Kα to 1 : 600 [12].

Anode	Filter	Absorption edge, Å	g /cm²	Thickness, mm
Ag	Pd	0.509	0.096	0.079
Pd	Rh	0.534	0.091	0.073
Rh	Ru	0.560	0.077	0.064
Mo	Zr (Nb)	0.689	0.069	0.108
Cu	Ni	1.488	0.019	0.021
Ni	Co	1.608	0.015	0.018
Co	Fe	1.743	0.014	0.018
Fe	Mn	1.896	0.012	0.016
Mn	Cr	2.070	0.011	0.016
Cr	V	2.269	0.009	0.016

A point to be remembered is that a foil is usually better than a powder, because it is more uniform in thickness.

2-7. Characteristics of Monochromator Crystals

These are used whenever a particular section of the characteristic radiation (e.g., $K\alpha_1$) must be isolated or the background reduced. For example, they are used in work on small-angle scattering, on the radial distribution of electron density in amorphous materials and liquids, and in precise measurements of line intensities and widths.

Harmonics (rays with $\lambda' = \lambda/2$, $\lambda/3$, and so on) may persist after reflection from monochromators, but can be suppressed by the use of crystals of low reflectivity for high orders; for example, germanium and silicon suppress the second harmonic.

See [361] for the effects of the energy distribution at the focus of the monochromator on the error of lattice-constant measurements (for Johansson's method).

Methods have also been developed for correcting for polarization effects in work with monochromators [378-381].

2-7a. Reflections and Properties of Monochromator Crystals

The table gives the properties of various monocrystals that are used in flat or bent forms [12]. Additional information is given to assist in the selection of materials for particular uses.

Crystal	Refl.	d, Å	Characteristic		Properties of crystal			Uses
			intensity	width	stability	mech. prop.		
Fluorite	111	3.15	m	medium	good	medium hard		Suppression of harmonics, generally at short wave-lengths
Urea nitrate	002	3.13	s	v. large	v. poor	easily deformed		For large specimens
Calcite	200	3.03	m	small	good	medium soft		For isolation and for small-angle scattering
Sodium chloride	200	2.81	m	large	unstable in moist air	can be bent in warm water		For focusing and general uses
Diamond	111	2.05	w	v. small	good	v. hard		Harmonic suppression
Lithium fluoride	200	2.01	v. s.	medium	good	hard, can be bent		For all uses
β-Alumina	3002 0004	11.22 5.61	w w	— medium	good	hard, brittle		For long wavelengths
Gypsum	020	7.58	m	v. small	poor	soft, can sag		For small-angle scattering and for focusing long wavelengths
Pentaery-thritol	002	4.39	v. s.	medium	poor	soft, readily deformed		For all uses
Quartz	1011	3.34	w	v. small	good	can be bent elastically		For small-angle scattering and focusing
Sodium bromide	200	3.29	m	medium	unstable in moist air	—		—

2-7b. Reflectivity of Monochromator Crystals

The table gives the reflectivity R for CuKα radiation, d being the lattice constant and t the optimal thickness for use with that radiation in transmission [250].

Substance	Refl.	d, Å	$R \cdot 10^5$	t
Calcite	200	3.03	6.7—7.4	—
Aluminum	200	2.02	29.5	0.06
Sodium chloride	200	2.815	31—45	0.06
Quartz	10$\bar{1}$1	3.333	43.5	0.1
Copper	200	1.804	71.5	—
Diamond	111	2.055	86—120	—
Lithium fluoride	200	2.01	93—110	0.3
Pentaerythritol	002	4.365	115	—
Ceylon graphite	002	3.345	500—620	—

2-7c. Optimal Thickness of Monochromator Crystal for Use in Transmission

The table gives the best thickness t for use with CuKα in transmission [250].

Substance	Refl.	Opt. t, mm
Aluminum	111	0.07
Aluminum	400	0.05
Lithium fluoride	200	0.3
Lithium fluoride	400	0.2
Sodium chloride	200	0.06
Quartz	10$\bar{1}$1	0.1
Quartz	1$\bar{3}$40	0.08
Muscovite	060	0.07

2-7d. Properties of Plane Monochromator Crystals

The table gives the parameters of some crystals that are used as plane monochromators; g_{max} is the maximum value of the ratio of intensities of the monochromatic radiation for the asymmetric and symmetric methods:

$$g_{max} = \frac{I_{asym}}{I_{sym}} = \frac{2 \sin \alpha}{\sin \alpha + \sin \beta},$$

where α is the angle of incidence at the monochromator and β is the angle of reflection there; the best value of the latter is

$$\beta_{opt} = \sqrt{\mu t \sin 2\vartheta},$$

where t is the thickness of the surface layer containing defects produced by the cutting and polishing.

Substance	Plane	g_{max}	β_{opt}	μt	t, 10^5 cm
Calcite A	202	1,65	4°	0.01	4
Calcite B	202	1.58	5°	0.01	4
Quartz	101	1,48	6°	0.02	2
Fluorite	111	1.37	9°	0.08	3

2-7e. Reflection Angles for Bent Monochromator Crystals

The table gives the angle ϑ and the range of angles in focus $\Delta\vartheta$ for some crystals used with Mo, Cu, Co, and Fe radiations (Kα doublet) [250].

Material	hkl	d, Å	Mo $K\alpha$ 0.7107 Å		Cu $K\alpha$ 1.5418 Å		Co $K\alpha$ 1.7902 Å		Fe $K\alpha$ 1.9373 Å	
			ϑ	$\Delta\vartheta$,′	ϑ	$\Delta\vartheta$,′	ϑ	$\Delta\vartheta$,′	ϑ	$\Delta\vartheta$,′
Aluminum $a=4.0414$ Å	111	2.3330	8°45′	3.3	19°18′	3.0	22°34′	3.1	24°32′	3.2
	200	2.0207	10°7′	3.8	22°25′	3.5	26°18′	3.7	28°39′	3.9
	222	1.1665	17°44′	6.8	41°21′	7.5	50°5′	8.9	56°24′	10.6
	400	1.0103	20°36′	8.0	49°43′	10.0	62°25′	14.3	73°30′	23.5
Lithium fluoride $a=4.0173$ Å	200	2.0086	10°11′	3.8	22°34′	3.5	26°28′	3.7	28°50′	3.9
	400	1.0043	20°43′	8.0	50°8′	10.2	63°2′	14.7	74°41′	25.5
Sodium chloride $a=5.6287$ Å	200	2.8144	7°15′	2.7	15°54′	·2.4	18°33′	2.5	20°8′	2.6
	400	1.4072	14°37′	5.5	33°13′	5.6	39°30′	6.1	43°30′	6.6
Muscovite $a=5.18$ Å $b=9.02$ Å $c=20.04$Å $\beta=95°30′$	006	3.340	6°6′	2.2	13°21′	2.0	15°33′	2.1	16°52′	2.1
	0010	1.995	10°15′	3.8	22°44′	3.6	26°39′	3.8	29°3′	3.9
	060	1.506	13°39′	5.2	30°48′	5.1	36°28′	5.5	40°2′	5.9
Quartz $a=4.903$ Å $c=5.393$ Å	10$\bar{1}$1	3.336	6°7′	2.3	13°22′	2.0	15°34′	2.1	16°53′	2.1
	11$\bar{2}$2	1.813	11°18′	4.2	25°10′	4.0	29°35′	4.2	32°17′	4.4
	20$\bar{2}$3	1.372	15°1′	5.7	34°11′	5.8	40°43′	6.4	44°54′	7.0
	20$\bar{3}$1	1.368	15°3′	5.7	34°18′	5.8	40°52′	6.4	45°5′	7.0
	13$\bar{4}$0	1.177	17°34′	6.7	40°55′	7.4	49°30′	8.7	55°23′	9.8

2-8. Parameters for Use with Bent Quartz Crystals

The (10$\bar{1}$1) plane of a monocrystal is usually used. Figure 20 shows the type of system used.

Figures 21-25 give data needed in the use of Johann's method (Fig. 20a) and of Johansson's method (Fig. 20b) for the Kα radiations of Mo, Cu, Ni, Co, and Fe [236]. The

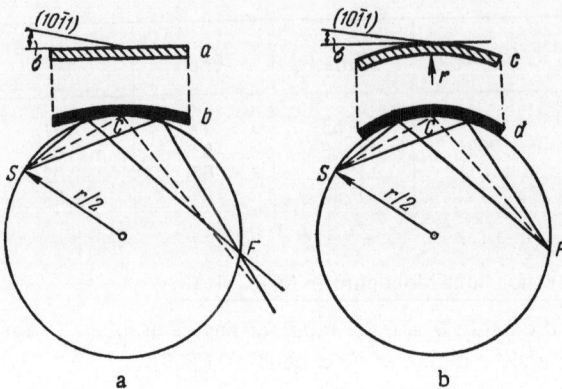

a   b

Fig. 20. Use of bent-crystal monochromators. a) Johann's method; b) Johansson's method. Symbols: a — crystal before bending; b — crystal bent to a radius R equal to the diameter of the focal circle r; c — crystal bent to form part of a cylinder of radius r; d — the same as c but of radius r/2; σ — angle between the surface and the reflecting plane; S — x-ray source; F — focus point; C — center of incident beam.

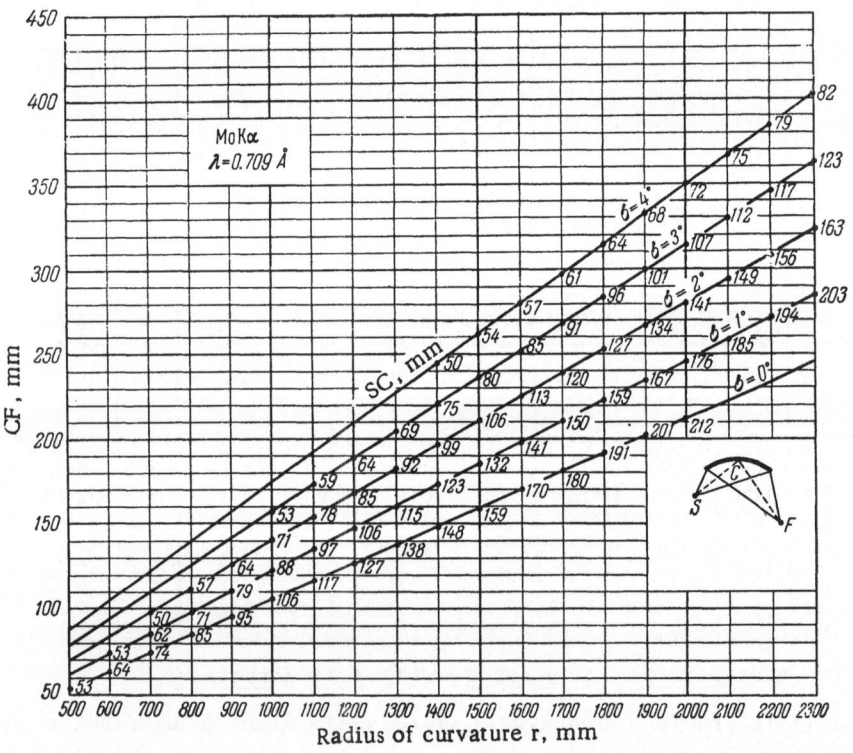

Fig. 21. Graphs for parameters used with bent quartz crystals and molybdenum radiation.

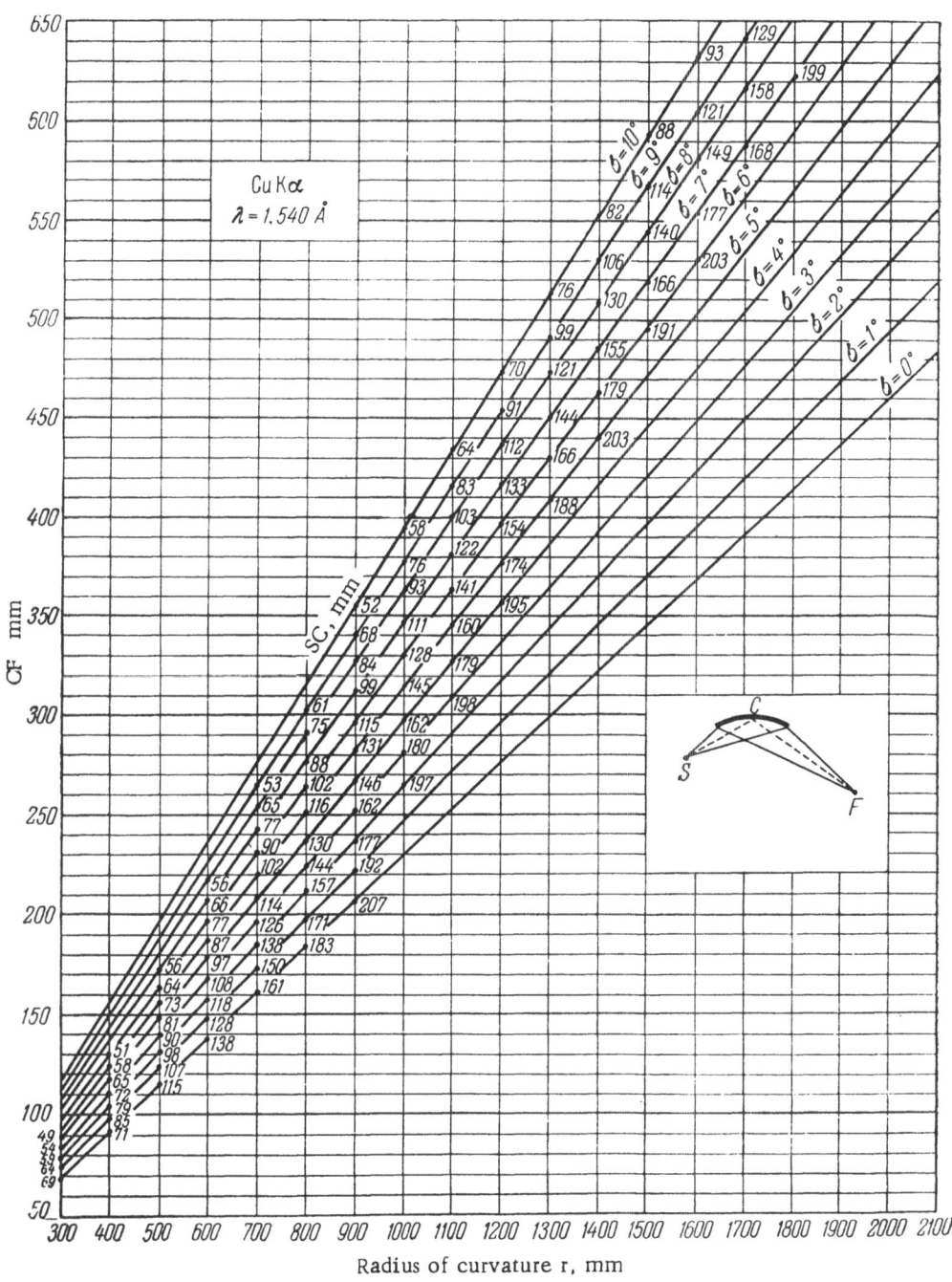

Fig. 22. Graphs for parameters used with bent quartz crystals and copper radiation.

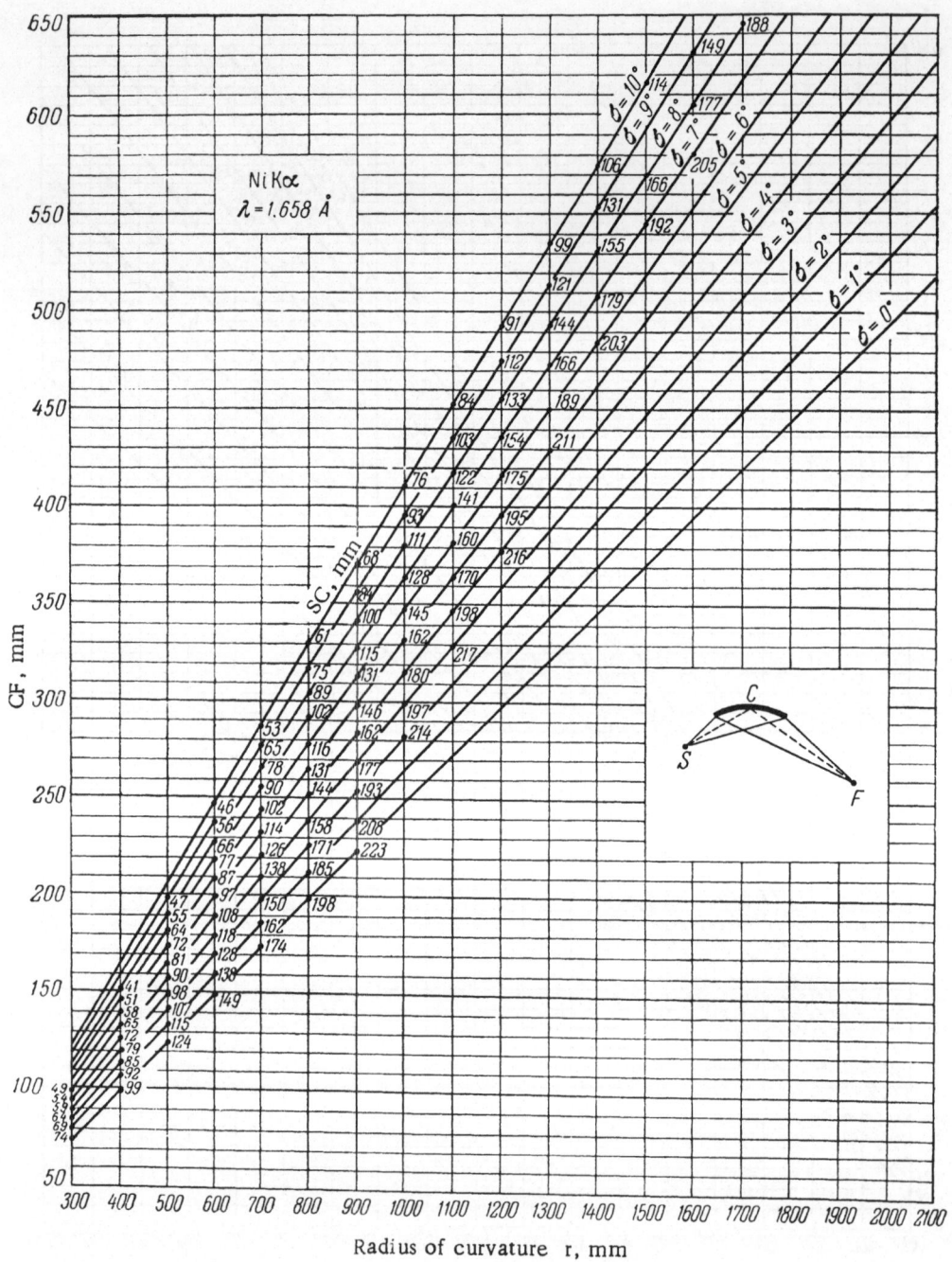

Fig. 23. Graphs for parameters used with bent quartz crystals and nickel radiation.

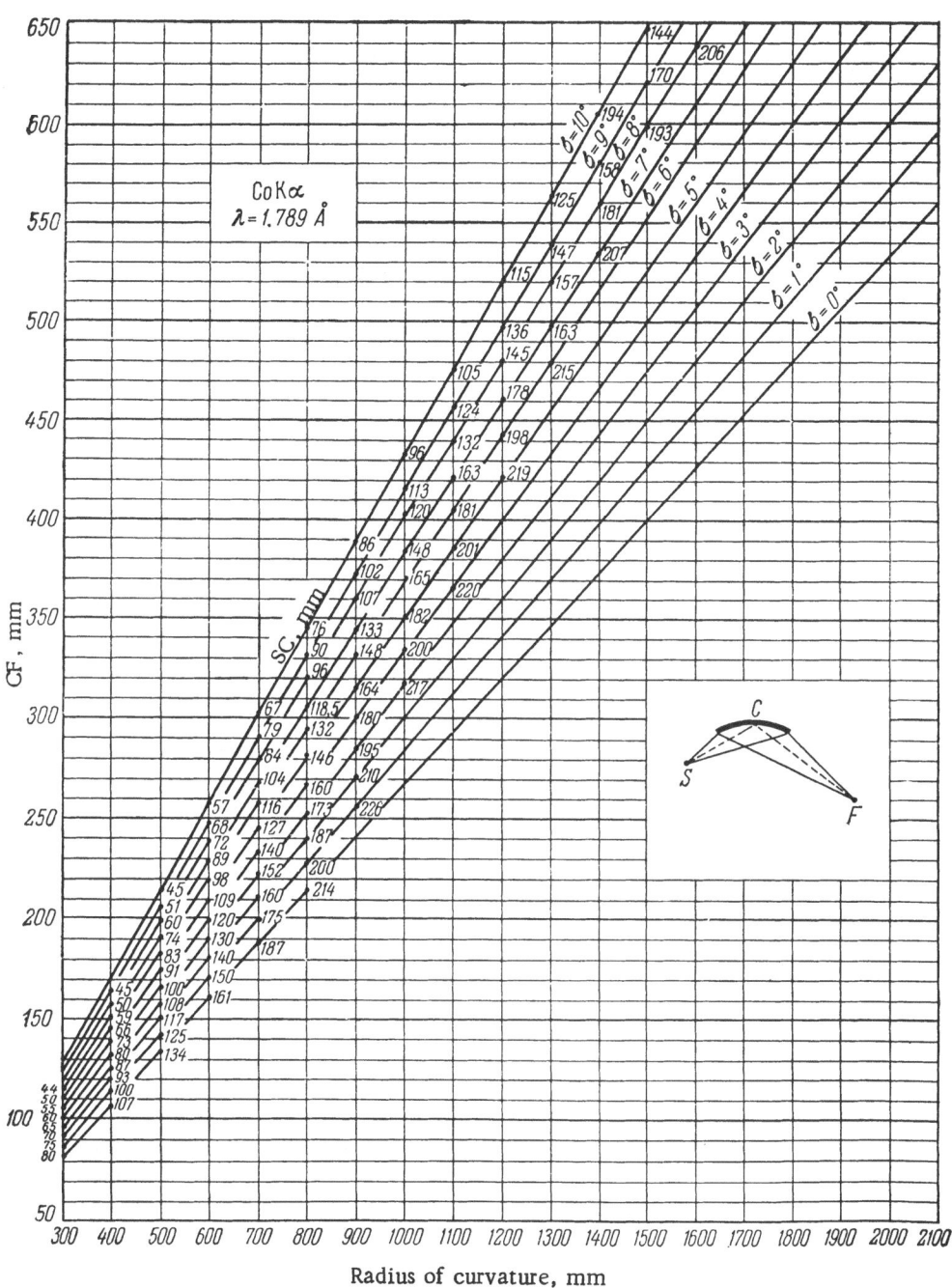

Fig. 24. Graphs for parameters used with bent quartz crystals and cobalt radiation.

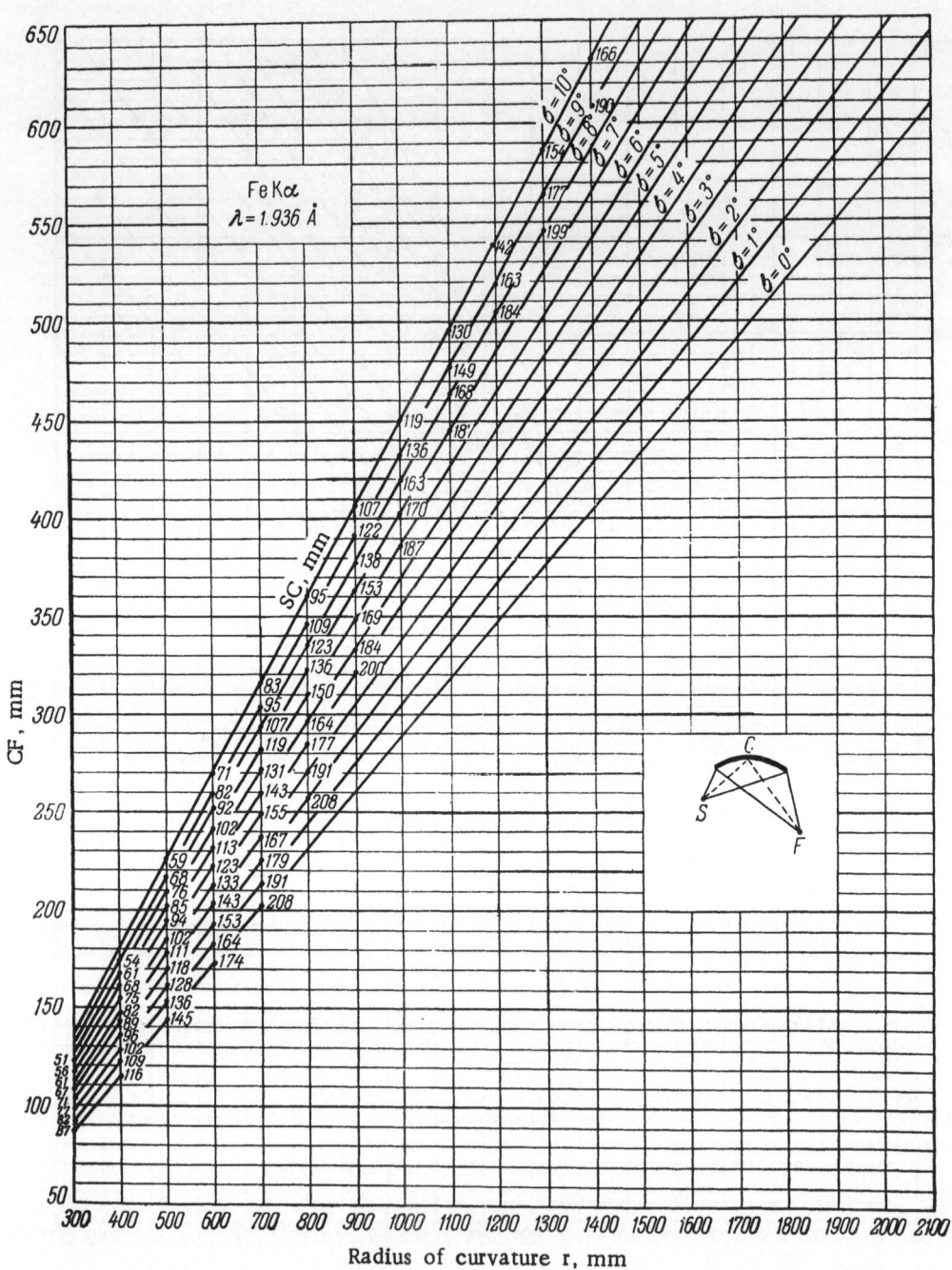

Fig. 25. Graphs for parameters used with bent quartz crystals and iron radiation.

curves give the distance SC from source to center of crystal, the distance CF from this center to the point of focus, the radius of curvature r, and the angle σ between the surface and the (10$\bar{1}$1) plane.

Example 1. A crystal is cut to give a σ of 4° and is bent to R = 1000 mm; it is used with CuKα radiation. Figure 22 gives SC as 160 mm and CF as 297 mm.

Example 2. MoKα radiation is used with CF = 210 mm; Fig. 21 shows that if SC is not less than 100 mm the method of Fig. 20a gives R = r = 1500 mm, whereas Fig. 20b gives R = r/2 = 750 mm. In both cases σ = 2°, SC = 106 mm, and CF = 210 mm.

2-9. Measurement of Diffraction Lines on Patterns

2-9a. Determination of Bragg Angles for a Flat Film

For flat films,

$$\operatorname{tg} 2\vartheta = \frac{S}{R},\tag{8}$$

where R is the distance from object to film and S is the distance to the central spot from the spot in question.

The table gives S for ϑ from 0 to 45° and for R from 10 to 100 mm [109].

The indices of the spots are given by

$$h:k:l = a\,[\cos\omega_1 \operatorname{tg}\vartheta + \sin\omega_1\cos(\varphi_a - \varphi)]:b\,[\cos\omega_2\operatorname{tg}\vartheta +$$
$$+ \sin\omega_2\cos(\varphi_b - \varphi)]:c\,[\cos\omega_3\cdot\operatorname{tg}\vartheta + \sin\omega_3\cos(\varphi_c - \varphi)],$$

where $a:b:c = a:1:c$ are the relative lengths of the axes in the unit cell; ω_1, ω_2, and ω_3 are the angles between the primary beam and the crystallographic axes a, b, and c; ϑ is the Bragg angle; φ is the azimuth angle of the spot; φ_a, φ_b, and φ_c are the inclination angles of planes drawn through the primary and the crystallographic axes with respect to the plane through the beam and the z (vertical) axis.

90−ϑ°	ϑ°	2ϑ°	tg 2ϑ	S							
				R=10 mm	15 mm	20 mm	28.7 mm	30 mm	50.0 mm	57.3 mm	100 mm
90	0	0	0.0000	0.0	0.0	0.0	0.0	0.0	0.0	0.0	0.0
89	1	2	0.0349	0.3	0.5	0.7	1.0	1.0	1.7	2.0	3.5
88	2	4	0.0699	0.7	1.0	1.4	2.0	1.0	1.7	4.0	7.0
87	3	6	0.1051	1.0	1.6	2.1	3.0	3.1	5.3	6.0	10.5
86	4	8	0.1405	1.4	2.1	2.8	4.0	4.2	7.0	8.0	14.0
85	5	10	0.1763	1.8	2.6	3.5	5.1	5.3	8.8	10.1	17.6
84	6	12	0.2126	2.1	3.2	4.3	6.1	6.4	10.1	12.2	21.3
83	7	14	0.2493	2.5	3.7	5.0	7.2	7.5	12.5	14.3	24.9
82	8	16	0.2867	2.9	4.3	5.7	8.2	8.6	14.3	16.4	28.7
81	9	18	0.3249	3.2	4.9	6.5	9.3	9.7	16.2	19.0	32.5

90−ϑ°	ϑ°	2ϑ°	tg 2ϑ	S							
				R=10 mm	15 mm	20 mm	28.7mm	30 mm	50.0mm	57.3mm	100 mm
80	10	20	0.3640	3.6	5.5	7.3	10.4	10.9	18.2	20.8	36.4
79	11	22	0.4040	4.0	6.1	8.1	11.6	12.1	20.2	23.2	40.4
78	12	24	0.4452	4.5	6.7	8.9	12.8	13.3	22.3	25.5	44.5
77	13	26	0.4877	4.9	7.3	9.7	14.0	14.6	24.4	28.0	48.8
76	14	28	0.5317	5.3	8.0	10.6	15.3	16.0	26.6	30.5	53.2
75	15	30	0.5774	5.8	8.7	11.5	16.6	17.3	28.9	33.0	57.7
74	16	32	0.6249	6.2	9.4	12.5	17.9	18.8	31.2	35.8	62.5
73	17	34	0.6745	6.7	10.1	13.5	9.4	20.3	33.7	38.7	67.4
72	18	36	0.7265	7.3	10.9	14.5	20.8	21.8	36.3	41.7	72.6
71	19	38	0.7813	7.8	11.7	15.6	21.6	23.4	39.0	44.8	78.1
70	20	40	0.8391	8.4	12.6	16.8	24.0	25.2	42.0	48.1	83.9
69	21	42	0.9004	9.0	13.5	18.0	25.8	27.0	45.0	51.6	90.0
68	22	44	0.9657	9.7	14.5	19.3	27.7	29.0	48.3	55.4	96.6
67	23	46	1.0355	10.4	15.5	20.7	29.7	31.1	51.8	59.3	103.5
66	24	48	1.1106	11.1	16.6	22.1	31.8	33.3	55.3	63.5	111.1
65	25	50	1.1918	11.9	17.9	23.8	34.2	35.8	59.6	68.4	119.2
64	26	52	1.2799	12.8	19.3	25.8	37.0	38.4	64.0	74.0	128.0
63	27	54	1.3764	13.8	20.6	27.5	39.5	41.3	68.8	78.9	137.6
62	28	56	1.4826	14.8	22.2	29.7	42.5	44.5	74.2	85.0	148.3
61	29	58	1.6003	16.0	24.0	32.0	45.9	48.0	80.0	91.7	160.0
60	30	60	1.7321	17.3	26.0	34.6	49.7	52.0	86.6	99.3	173.2
59	31	62	1.8807	18.8	28.2	37.6	54.0	56.4	94.0	107.9	188.1
58	32	64	2.0503	20.5	30.8	41.0	58.8	61.5	102.5	117.5	205.0
57	33	66	2.2460	22.5	33.7	44.9	64.5	67.4	112.3	131.0	224.6
56	34	68	2.4751	24.8	37.1	49.4	70.9	74.3	123.8	141.8	247.5
55	35	70	2.7475	27.5	41.3	55.0	78.8	82.4	137.4	157.6	274.7
54	36	72	3.0777	30.8	46.1	61.5	88.1	92.3	153.9	176.2	307.8
53	37	74	3.4874	34.9	52.3	69.6	100.0	104.6	174.8	200.0	348.7
52	38	76	4.0108	40.1	60.0	80.0	115.0	120.3	200.5	230.0	401.1
51	39	78	4.7046	47.0	70.5	94.0	135.0	141.2	235.2	268.9	470.5
50	40	80	5.6713	56.7	85.0	113.4	162.8	170.1	283.6	325.6	567.1
49	41	82	7.1154	71.1	—	142.3	—	213.5	355.8	—	711.5
48	42	84	9.5144	95.1	—	190.3	—	285.4	475.7	—	951.4
47	43	86	14.301	143.0	—	286.0	—	429.0	715.0	—	1430.1
46	44	88	28.636	286.4	—	572.7	—	859.1	1431.8	—	2863.6
45	45	90	∞	∞	∞	∞	∞	∞	∞	∞	∞

2-9b. Correction for Deviation of Camera Diameter from Standard

Any deviation from 57.3 mm involves a correction to the measured l :

$$\Delta l = \frac{l_m}{D} (D - D_0),$$ (9)

where l_m is the distance measured on the film, D is the actual diameter, and $D_0 = 57.3$ mm. The table gives Δl for l_m of 6 to 85 mm and for $(D - D_0)$ from 0.05 to 1.3 mm. The correction is subtracted for $D > D_0$ and is added for $D < D_0$.

$l\,m$ \ $D-D_0$	0.05	0.1	0.2	0.3	0.4	0.5	0.6	0.7	0.8	0.9	1.0	1.1	1.2	1.3
6		0.01	0.01	0.03	0.04	0.05	0.06	0.07	0.08	0.09	0.11	0.12	0.13	0.14
7		.01	.03	.04	.05	.06	.07	.09	.10	.11	.12	.14	.15	.16
8		.01	.03	.04	.06	.07	.08	.10	.11	.13	.14	.15	.17	.18
9		.02	.03	.05	.06	.08	.09	.11	.12	.14	.15	.17	.19	.20
10	0.01	.02	.03	.05	.07	.09	.10	.12	.14	.16	.17	.19	.21	.23
11	.01	.02	.04	.06	.08	.10	.11	.13	.15	.17	.19	.21	.23	.25
12	.01	.02	.04	.06	.08	.11	.13	.15	.17	.19	.21	.23	.25	.27
13	.01	.02	.04	.07	.09	.11	.14	.16	.18	.20	.23	.25	.27	.29
14	.01	.02	.05	.07	.10	.12	.15	.17	.20	.22	.25	.27	.29	.32
15	.01	.03	.05	.08	.10	.13	.16	.18	.21	.23	.26	.29	.31	.34
16	.01	.03	.06	.08	.11	.14	.17	.20	.22	.25	.28	.31	.34	.36
17	.01	.03	.06	.09	.12	.15	.18	.21	.24	.27	.30	.33	.36	.38
18	.02	.03	.06	.09	.12	.16	.19	.22	.25	.28	.31	.35	.38	.41
19	.02	.04	.07	.10	.13	.17	.20	.23	.26	.30	.33	.36	.40	.43
20	.02	.04	.07	.11	.14	.17	.21	.24	.28	.31	.35	.38	.42	.45
21	.02	.04	.07	.11	.15	.18	.22	.26	.29	.33	.37	.40	.44	.48
22	.02	.04	.08	.12	.15	.19	.23	.27	.31	.35	.38	.42	.46	.50
23	.02	.04	.08	.12	.16	.20	.24	.29	.32	.36	.40	.44	.48	.52
24	.02	.04	.08	.13	.17	.21	.25	.29	.34	.38	.42	.46	.50	.54
25	.02	.04	.09	.13	.18	.22	.26	.30	.35	.39	.44	.48	.52	.57
26	.02	.04	.09	.14	.18	.23	.27	.32	.36	.41	.45	.50	.54	.59
27	.02	.05	.09	.14	.19	.24	.28	.33	.38	.42	.47	.52	.56	.61
28	.02	.05	.10	.15	.20	.25	.29	.34	.40	.44	.49	.54	.59	.63
29	.02	.05	.10	.15	.20	.25	.30	.36	.41	.46	.51	.56	.61	.66
30	.02	.05	.10	.16	.21	.26	.31	.37	.42	.47	.52	.58	.63	.68
31	.03	.05	.11	.16	.22	.27	.32	.38	.43	.49	.54	.59	.65	.70
32	.03	.06	.11	.17	.22	.28	.33	.39	.45	.50	.56	.61	.67	.73

l_m \ $D-D_0$	0.05	0.1	0.2	0.3	0.4	0.5	0.6	0.7	0.8	0.9	1.0	1.1	1.2	1.3
33	0.03	0.06	0.12	0.17	0.23	0.29	0.34	0.40	0.46	0.52	0.58	0.63	0.69	0.75
34	.03	.06	.12	.18	.24	.30	.36	.42	.48	.53	.59	.65	.71	.77
35	.03	.06	.12	.18	.25	.31	.37	.43	.49	.55	.61	.67	.73	.79
36	.03	.06	.13	.19	.25	.31	.38	.44	.50	.57	.63	.69	.75	.82
37	.03	.07	.13	.19	.26	.32	.39	.45	.52	.58	.65	.71	.77	.84
38	.03	.07	.13	.20	.26	.33	.40	.46	.53	.60	.66	.73	.80	.86
39	.03	.07	.14	.20	.27	.34	.41	.48	.55	.61	.68	.75	.82	.89
40	.03	.07	.14	.21	.28	.35	.42	.49	.56	.63	.70	.77	.84	.91
41	.04	.07	.14	.21	.29	.36	.43	.50	.57	.64	.72	.79	.86	.93
42	.04	.07	.15	.22	.29	.37	.44	.51	.59	.66	.73	.81	.88	.95
43	.04	.08	.15	.22	.30	.38	.45	.53	.62	.68	.75	.82	.90	.98
44	.04	.08	.15	.23	.31	.38	.45	.54	.63	.69	.76	.84	.92	1.00
45	.04	.08	.16	.23	.32	.39	.47	.55	.64	.71	.78	.86	.94	.02
46	.04	.08	.16	.24	.32	.40	.48	.56	.66	.72	.80	.88	.96	.04
47	.04	.08	.16	.24	.33	.41	.49	.57	.67	.74	.82	.90	.98	.06
48	.04	.08	.17	.25	.34	.42	.50	.59	.69	.76	.84	.92	1.00	.09
49	.04	.09	.17	.26	.34	.43	.51	.60	.70	.77	.86	.94	.02	.11
50	.04	.09	.17	.26	.35	.44	.52	.61	.71	.79	.87	.96	.05	.13
51	.04	.09	.18	.27	.36	.45	.53	.62	.73	.80	.89	.98	.07	.16
52	.04	.09	.18	.27	.36	.45	.54	.64	.74	.82	.91	1.00	.09	.18
53	.05	.09	.18	.28	.37	.46	.56	.65	.76	.83	.93	.02	.11	.20
54	.05	.09	.19	.28	.38	.47	.57	.66	.77	.85	.94	.04	.13	.22
55	.05	.10	.19	.29	.39	.48	.58	.67	.78	.86	.96	.06	.15	.25
56	.05	.10	.19	.29	.39	.49	.59	.68	.80	.88	.98	.07	.17	.27
57	.05	.10	.20	.30	.40	.50	.60	.70	.81	.90	1.00	.09	.19	.29
58	.05	.10	.20	.30	.41	.51	.61	.71	.83	.91	.01	.11	.21	.32
59	.05	.10	.21	.31	.41	.52	.62	.72	.84	.93	.03	.13	.24	.34

$\dfrac{D-D_0}{l,m}$	0.05	0.1	0.2	0.3	0.4	0.5	0.6	0.7	0.8	0.9	1.0	1.1	1.2	1.3
60	0.05	0.10	0.21	0.31	0.42	0.52	0.63	0.73	0.85	0.94	1.05	1.15	1.25	1.36
61	.05	.11	.21	.32	.43	.53	.64	.75	.87	.96	.06	.17	.28	.38
62	.05	.11	.22	.33	.43	.54	.65	.76	.87	.97	.08	.19	.30	.41
63	.05	.11	.22	.33	.44	.55	.66	.77	.88	.99	.10	.21	.32	.43
64	.06	.11	.22	.34	.45	.56	.67	.78	.90	1.00	.12	.24	.34	.45
65	.06	.12	.23	.34	.45	.57	.68	.79	.91	.02	.13	.26	.36	.48
66	.06	.12	.23	.35	.45	.58	.69	.81	.92	.04	.15	.28	.38	.51
67	.06	.12	.23	.35	.47	.58	.70	.82	.94	.05	.17	.29	.40	.53
68	.06	.12	.24	.36	.47	.59	.71	.83	.95	.07	.19	.31	.42	.56
69	.06	.12	.24	.36	.48	.60	.72	.84	.96	.08	.20	.32	.44	.58
70	.06	.12	.24	.37	.49	.61	.73	.86	.98	.10	.22	.34	.47	.61
71	.06	.12	.25	.37	.50	.62	.74	.87	.99	.12	.24	.36	.49	.64
72	.06	.13	.25	.38	.50	.63	.75	.88	1.00	.13	.26	.38	.51	.66
73	.06	.13	.25	.38	.51	.64	.76	.89	.02	.15	.27	.40	.53	.68
74	.06	.13	.26	.39	.52	.65	.77	.91	.03	.16	.29	.42	.55	.71
75	.07	.13	.26	.39	.52	.65	.78	.92	.05	.18	.31	.44	.57	.74
76	.07	.13	.26	.40	.53	.66	.79	.93	.06	.19	.33	.46	.59	.76
77	.07	.13	.27	.40	.54	.67	.81	.94	.07	.21	.34	.48	.61	.79
78	.07	.14	.27	.41	.55	.68	.82	.96	.09	.23	.36	.50	.63	.81
79	.07	.14	.27	.41	.55	.69	.83	.97	.10	.24	.38	.52	.66	.83
80	.07	.14	.28	.42	.56	.70	.84	.98	.11	.26	.40	.54	.69	.86
81	.07	.14	.28	.42	.57	.71	.85	.99	.12	.27	.41	.55	.72	.86
82	.07	.14	.29	.43	.57	.71	.86	1.00	.13	.29	.43	.57	.75	.91
83	.07	.14	.29	.43	.58	.72	.87	.01	.14	.30	.45	.59	.77	.93
84	.07	.15	.29	.44	.59	.72	.88	.03	.17	.32	.47	.61	.80	.95
85	.07	.15	.30	.44	.59	.72	.89	.04	.18	.34	.48	.63	.82	.97

2-9c. Correction for Specimen Thickness

Absorption in the specimen tends to displace the lines. The equation most commonly used to determine the correction in measurements between line centers is

$$2l_{corr} = 2l - \varrho\,(1 \pm \cos 2\vartheta), \tag{10}$$

where $2l$ is the distance between centers of symmetrically placed lines as recorded with a cylindrical film from a cylindrical specimen, $2l_{corr}$ is the corrected distance, and ρ is the radius of the specimen. The + sign is used if $2\vartheta < 90°$; the −, if $2\vartheta > 90°$. The tables give $1 \pm \cos 2\vartheta$ for angles from 0 to 90° and also Δl, the correction for specimen thickness, for ρ from 0.15 to 0.60 mm:

$$\Delta l = \frac{\varrho}{2}(1 + \cos 2\vartheta), \quad \text{where} \quad \Delta l = l - l_{corr}.$$

ϑ	$1\pm\cos 2\vartheta$	ϑ	$1\pm\cos 2\vartheta$	ϑ	$1\pm\cos 2\vartheta$	ϑ	$1\pm\cos 2\vartheta$	ϑ	$1\pm\cos 2\vartheta$
0°	2.000	19°	1.788	38°	1.242	57°	0.593	76°	0.117
30'	.000	30'	.777	30'	.225	30'	.577	30'	.109
1°	1.999	20°	.766	39°	.208	58°	.562	77°	.101
30'	.999	30'	.755	30'	.191	30'	.546	30'	.094
2°	.998	21°	.743	40°	.174	59°	.531	78°	.086
30'	.996	30'	.731	30'	.156	30'	.515	30'	.079
3°	.995	22°	.719	41°	.139	60°	.500	79°	.073
30'	.993	30'	.707	30'	.122	30'	.485	30'	.066
4°	.990	23°	.695	42°	.105	61°	.470	80°	.060
30'	.988	30'	.682	30'	.087	30'	.455	30'	.054
5°	.985	24°	.669	43°	.070	62°	.441	81°	.049
30'	.982	30'	.656	30'	.052	30'	.426	30'	.044
6°	.978	25°	.643	44°	.035	63°	.412	82°	.039
30'	.974	30'	.629	30'	.017	30'	.398	30'	.034
7°	.970	26°	.616	45°	1.000	64°	.384	83°	.030
30'	.966	30'	.602	30'	0.983	30'	.371	30'	.026
8°	.961	27°	.588	46°	.965	65°	.357	84°	.022
30'	.956	30'	.574	30'	.948	30'	.344	30'	.018
9°	.951	28°	.559	47°	.930	66°	.331	85°	.015
30'	.946	30'	.545	30'	.913	30'	.318	30'	.012
10°	.940	29°	.530	48°	.895	67°	.305	86°	.010
30'	.934	30'	.515	30'	.878	30'	.293	30'	.007
11°	.927	30°	.500	49°	.871	68°	.281	87°	.005
30'	.921	30'	.485	30'	.844	30'	.269	30'	.004
12°	.914	31°	.469	50°	.826	69°	.257	88°	.002
30'	.906	30'	.454	30'	.809	30'	.245	30'	.001
13°	.899	32°	.438	51°	.792	70°	.234	89°	.001
30'	.891	30'	.423	30'	.775	30'	.223	30'	.000
14°	.883	33°	.407	52°	.758	71°	.212	90°	.000
30'	.875	30'	.391	30'	.741	30'	.201		
15°	.866	34°	.375	53°	.724	72°	.291		
30'	.857	30'	.358	30'	.708	30'	.181		
16°	.848	35°	.342	54°	.601	73°	.171		
30'	.839	30'	.326	30'	.674	30'	.161		
17°	.829	36°	.309	55°	.658	74°	.152		
30'	.819	30'	.292	30'	.642	30'	.143		
18°	.809	37°	.276	56°	.625	75°	.134		
30'	.799	30'	.259	30'	.609	30'	.125		

Δl as a Function of ϑ and ρ

$\vartheta°$ \ ϱ	0.15	0.20	0.25	0.30	0.35	0.40	0.45	0.50	0.55	0.60
10	0.15	0.19	0.24	0.29	0.34	0.39	0.44	0.49	0.53	0.58
12	.14	.19	.24	.29	.34	.38	.43	.48	.53	.58
14	.14	.19	.24	.28	.33	.38	.42	.47	.52	.57
16	.14	.18	.23	.28	.32	.37	.42	.46	.51	.56
18	.14	.18	.23	.27	.32	.36	.41	.45	.50	.54
20	.13	.18	.22	.27	.31	.35	.40	.44	.49	.53
22	.13	.17	.22	.26	.30	.34	.39	.43	·48	.51
24	.13	.17	.21	.25	.29	.33	.38	.42	.47	.50
26	.12	.16	.20	.24	.28	.32	.36	.40	.46	.49
28	.12	.16	.19	.23	.27	.31	.35	.39	.44	.47
30	.11	.15	.18	.22	.26	.30	.34	.38	.42	.45
32	.11	.14	.18	.21	.25	.29	.32	.36	.40	.43
34	.10	.14	.17	.20	.24	.28	.31	.34	.38	.41
36	.10	.13	.16	.19	.23	.26	.29	.33	.36	.38
38	.09	.12	.16	.18	.22	.25	.28	.31	.34	.37
40	.09	.12	.15	.17	.21	.24	.26	.29	.32	.35
42	.08	.11	.14	.17	.20	.22	.25	.28	.30	.33
44	.08	.10	.13	.16	.19	.21	.23	.26	.29	.31
46	.07	.10	.12	.15	.18	.20	.22	.24	.26	.29
48	.07	.09	.11	.14	.17	.18	.20	.22	.24	.27
50	.06	.08	.10	.13	.16	.17	.19	.21	.22	.25
52	.06	.08	.09	.11	.15	.15	.17	.19	.21	.23
54	.05	.07	.09	.10	.13	.14	.16	.17	.18	.20
56	.05	.06	.08	.09	.12	.12	.14	.16	.17	.19
58	.04	.06	.07	.08	.11	.11	.13	.14	.16	.17
60	.04	.05	.06	.07	.10	.10	.11	.12	.14	.15
65	.03	.04	.04	.05	.09	.07	.08	.09	.10	.11
70	.02	.02	.03	.04	.06	.05	.05	.06	.06	.07
75	.01	.01	.02	.02	.02	.03	.03	.03	.04	.04
80	.01	.01	.01	.01	.01	.01	.01	.01	.02	.02

2-9d. Correction for Eccentricity of Specimen in Camera

A displacement of ΔR in the axis of the specimen relative to that of the camera results in the following correction to the measured l:

$$\Delta l = \Delta R \, \sin 2\vartheta. \tag{11}$$

The table gives Δl for ϑ from 4 to 44° and ΔR from 0.05 to 1.0 mm.

$\vartheta°$	$90° - \vartheta°$	ΔR 0.05	0.1	0.2	0.3	0.4	0.5	0.6	0.7	0.8	0.9	1.0
4	86	0.01	0.01	0.03	0.04	0.06	0.07	0.08	0.10	0.12	0.13	0.14
6	84	.01	.02	.04	.06	.08	.10	.12	.15	.17	.19	.21
8	82	.01	.03	.06	.08	.11	.14	.17	.19	.22	.25	.28
10	80	.02	.03	.07	.10	.14	.17	.21	.24	.27	.31	.34
12	78	.02	.04	.08	.12	.16	.20	.24	.28	.33	.37	.41
14	76	.02	.05	.09	.14	.19	.24	.28	.33	.38	.42	.47
16	74	.03	.05	.11	.16	.21	.27	.31	.37	.42	.48	.05
18	72	.03	.06	.12	.18	.24	.29	.35	.41	.47	.53	.59

$\vartheta°$ / $90°-\vartheta°$ (ΔR)	0.50	0.1	0.2	0.3	0.4	0.5	0.6	0.7	0.8	0.9	1.0
20 / 70	0.03	0.06	0.13	0.20	0.26	0.32	0.39	0.45	0.51	0.58	0.64
22 / 68	.04	.07	.14	.21	.28	.35	.42	.49	.56	.63	.69
24 / 66	.04	.07	.15	.22	.30	.37	.45	.52	.58	.68	.74
26 / 64	.04	.08	.16	.24	.32	.39	.47	.55	.63	.71	.79
28 / 62	.04	.08	.17	.25	.33	.41	.50	.58	.66	.75	.83
30 / 60	.04	.09	.17	.26	.35	.43	.52	.61	.69	.78	.87
32 / 58	.05	.09	.18	.27	.36	.45	.54	.63	.72	.81	.90
34 / 56	.05	.09	.18	.28	.37	.47	.56	.65	.74	.83	.93
36 / 54	.05	.10	.19	.29	.38	.47	.57	.67	.76	.86	.95
38 / 52	.05	.10	.19	.29	.39	.48	.58	.68	.78	.87	.97
40 / 50	.05	.10	.20	.30	.39	.48	.59	.69	.79	.87	.98
42 / 48	.05	.10	.20	.30	.40	.50	.60	.70	.80	.90	.99
44 / 46	.05	.10	.20	.30	.40	.50	.60	.70	.80	.90	.99

2-10. Intensity Measurements

2-10a. Number of Counts Needed to Give a Specified Probable Error

The standard deviation of a measured number n from the average n_0 for randomly distributed pulses is given by $\sigma_p = 0.675 \sqrt{n}$; the relative standard deviation is $u_s = 1/\sqrt{n}$, and the relative probable error is $u_p = 0.675/\sqrt{n}$.

The table gives the total numbers of counts needed to give u_p between 0.2 and 5.0% [11].

u_p, %	n	u_p, %	n
0.2	113 900	1.5	2 024
0.4	28 475	2.0	1 139
0.6	12 655	3.0	506
0.8	7 119	4.0	285
1.0	4 556	5.0	182

2-10b. Statistical Error of Count

This error depends on the count n_t taken at the peak for a reflection (including background) and on the background count n_b.

The relative probable error is

$$100 u_p = \frac{67.5}{R-1} \sqrt{\frac{R(R+1)}{n_t}},$$

in which $R = n_t/n_b$, or $100 u = 67.5 (n_t + n_b)^{1/2}/(n_t - n_b)$. Figure 26 shows the mean relative error as a function of total counts recorded; the parameter on the curves is R. The curve for $R = \infty$ can be used if the background is very low.

2-10c. Correction for Particle Size, Fixed Specimen

The intensity variations depend appreciably on crystallite size.
The mean relative deviation in the intensity is

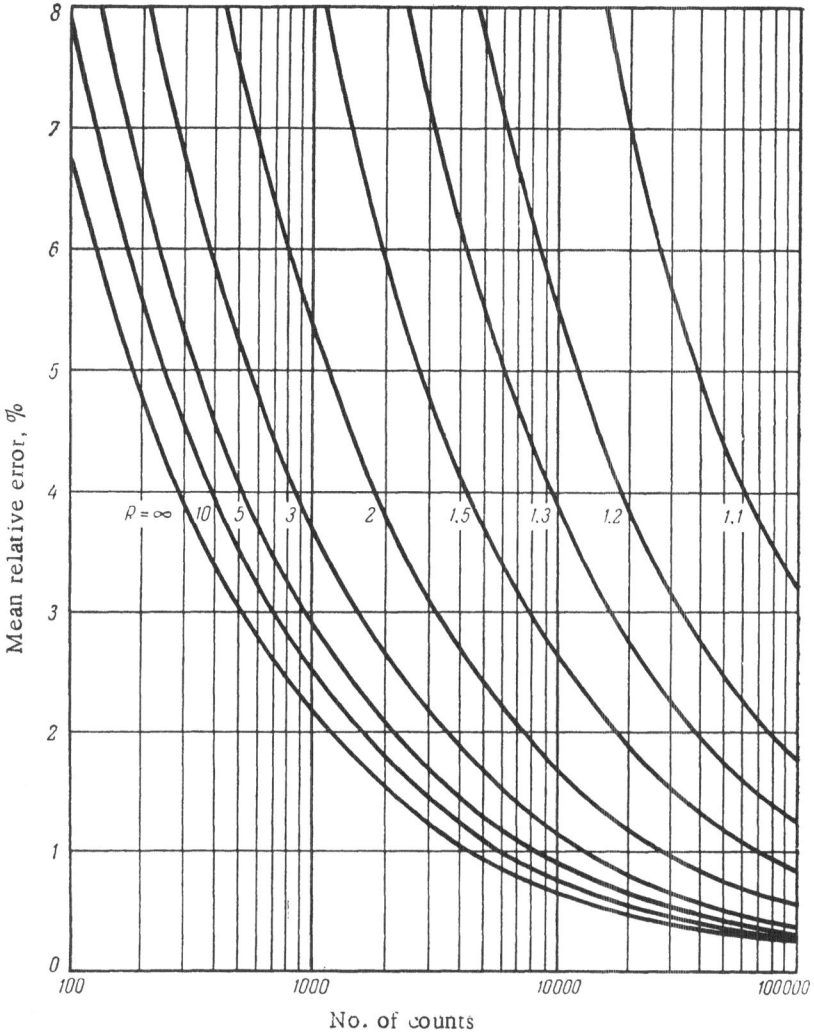

Fig. 26. Evaluation of statistical error of count in ionization recording.

$$u_m = 0.798 \sqrt{\frac{q\mu \sum v_i f_i}{pS}} , \tag{12}$$

where p and q are the proportions of the crystallites in the reflecting and nonreflecting positions, respectively, μ is the linear absorption coefficient for the specimen, S is the area of intersection between the beam and the specimen, and f_i is the volume proportion of crystallites having volume v_i.

Figure 27 shows the mean relative error as a function of crystallite size for several values of μ [11].

The curves are usually applicable to reasonably thick specimens examined in back-reflection.

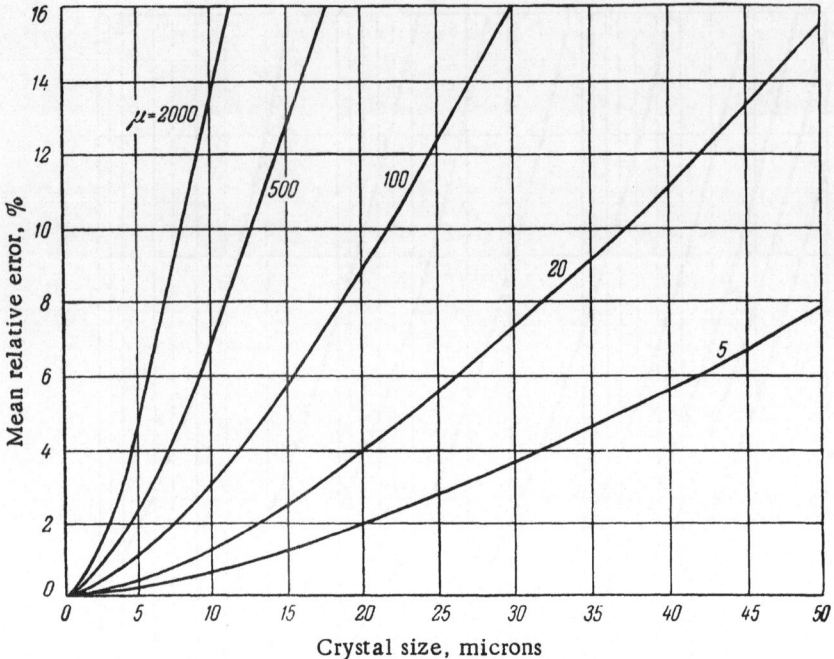

Fig. 27. Correction for particle size for a fixed specimen examined by ionization methods.

2-10d. Correction for Particle Size, Rotating Specimen

A specimen consisting of particles of size a_{av} and having an absorption coefficient μ gives [85] a mean error in the intensity of

$$u = \frac{4R_g \sin \vartheta}{(h_f + h_z)} \sqrt{\frac{\mu a^3_{av}}{Sm}},$$

where R_g is the radius of the goniometer, h_f is the height of the projection of the focus, h_z is the height of the counter slit, m is the multiplicity of the crystal plane, and S is the area of intersection of the primary beam and the specimen.

The table gives values of this error for crystallite sizes between 1 and 100 microns for μ from 5 to 2000 cm^{-1}.

a_{av}, microns \ μ, cm^{-1}	Mean error, %				
	5	20	100	500	2000
1	—	—	0.01	0.02	0.02
2	0.01	0.01	0.03	0.07	0.14
5	0.03	0.05	0.11	0.25	0.50
10	0.07	0.14	0.32	0.72	1.44
20	0.20	0.40	0.90	2.02	4.04
30	0.37	0.73	1.64	3.67	7.34
40	0.56	1.13	2.52	5.64	11.28
50	0.79	1.59	3.54	7.92	15.85
75	1.45	2.90	6.50	14.52	29.0
100	2.24	4.48	10	22.4	44.8

This table has been compiled for the typical conditions R_g = 16 cm, ϑ = 15°, h_f = 0.25 cm, h_z = 0.8 cm, S = 0.42 cm^2, and m = 6.

2-10e. Correction for Lost Counts

Not all the counts are recorded when the counting rate is high. The proportion of counts lost is governed by the dead time, which is the time for which the counter is incapable of responding to a second quantum after a first has been recorded.

The relation of m (the measured rate) to n (the true rate) and τ (the dead time) is

$$n = \frac{m}{1 - m\tau} \, .$$

Figure 28 is intended for use with equipment such as the URS-50I fitted with an RM-4 counter (τ = 200 μ sec).

The true rate is found by drawing a straight line from the origin A to the point on the line DE corresponding to the measured rate m (1000 counts/sec in the example shown by the broken line). This line is produced to meet the top scale, which gives the true rate (here 1250).

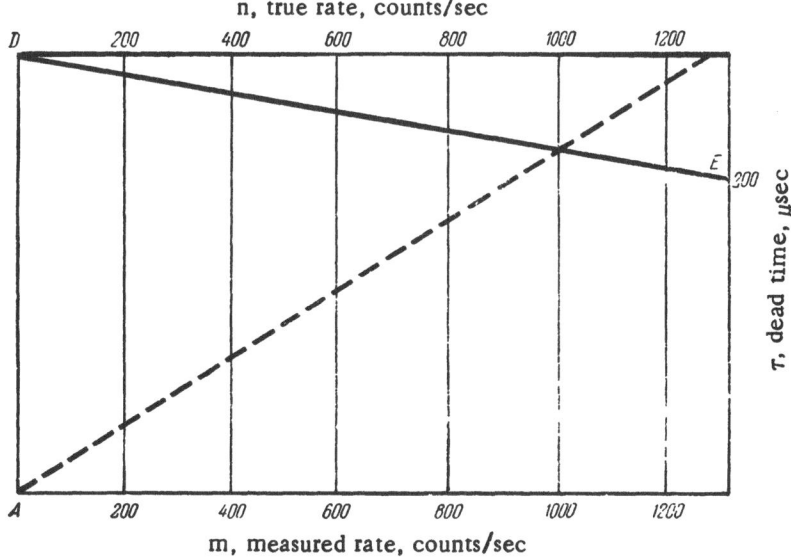

Fig. 28. Correction for lost counts in ionization recording.

2-11.   Doublet Splittings

The table gives the differences $\delta = \vartheta_{\alpha_2} - \vartheta_{\alpha_1}$ for various radiations; δ is given in angular measure for angles between 1 and 86° [2].

$\vartheta°$	$\delta = \vartheta_{\alpha_2} - \vartheta_{\alpha_1}$ in minutes of angle for radiation from:					$\vartheta°$	$\delta = \vartheta_{\alpha_2} - \vartheta_{\alpha_1}$ in minutes of angle for radiation from:				
	Cr	Fe	Ni	Cu	Mo		Cr	Fe	Ni	Cu	Mo
1	0.1	0.1	0.14	0.16	0.4	44	5.1	7.1	7.9	8.5	20.6
2	0.2	0.3	0.3	0.3	0.7	45	5.3	7.3	8.2	8.8	21.4
3	0.3	0.4	0.4	0.5	1.1	46	5.5	7.6	8.5	9.1	22.1
4	0.4	0.5	0.6	0.6	1.5	47	5.7	7.9	8.8	9.4	22.9
5	0.5	0.6	0.7	0.7	1.9	48	5.8	8.2	9.1	9.7	23.7
6	0.6	0.8	0.9	0.9	2.2	49	6.0	8.5	9.4	10.1	24.6

$\vartheta°$	$\delta = \vartheta_{\alpha_2} - \vartheta_{\alpha_1}$ in minutes of angle for radiation from:					$\vartheta°$	$\delta = \vartheta_{\alpha_2} - \vartheta_{\alpha_1}$ in minutes of angle for radiation from:				
	Cr	Fe	Ni	Cu	Mo		Cr	Fe	Ni	Cu	Mo
7	0.6	0.9	1.0	1.1	2.6	50	6.2	8.8	9.7	10.5	25.5
8	0.7	1.0	1.2	1.2	3.0	51	6.4	9.0	10.1	10.9	26.4
9	0.8	1.2	1.3	1.4	3.4	52	6.6	9.4	10.4	11.3	27.3
10	0,9	1.3	1.5	1.6	3.7	53	6.9	9.8	10.8	11.7	28.3
11	1.0	1.4	1.6	1.7	4.1	54	7.2	10.2	11.2	12.1	29.4
12	1.1	1.6	1.7	1.9	4.5	55	7.4	10.5	11.6	12.6	30.5
13	1.2	1.7	1.9	2.0	4.9	56	7.7	10.9	12.1	13.1	31.7
14	1.3	1.8	2.0	2.2	5.3	57	8.0	11.4	16,2	13.6	33.0
15	1.4	2.0	2.2	2.4	5.7	58	8.4	11.9	13.1	14.1	34.3
16	1.5	2.1	2.3	2.5	6.1	59	8.7	12.4	13.7	14.7	35.7
17	1.6	2.3	2.5	2.7	6.5	60	9.0	12.8	14.2	15.3	37.2
18	1.7	2.4	2.6	2.8	6.9	61	9.4	13.3	14.8	16.0	38.8
19	1.8	2.5	2.8	3.0	7.3	62	9.8	13.8	15.4	16.6	40.4
20	1.9	2.7	3.0	3.2	7.7	63	10.2	14.5	16.1	17.3	42.2
21	2.0	2.8	3.1	3.4	8.2	64	10.7	15.2	16.8	18.1	44.2
22	2.1	3.0	3.3	3.6	8.6	65	11.3	15.9	17.6	19.0	46.4
23	2.2	3.1	3.5	3.7	9.0	66	11.8	16.8	18.4	19.9	48.6
24	2.3	3.3	3.6	3.9	9.5	67	12.4	17.6	19.4	22.0	51.2
25	2.4	3.4	3.8	4.1	9.9	68	13.0	18.4	20.4	22.1	53.8
26	2.5	3.6	4.0	4.3	10.4	69	13.7	19.3	21.4	23.1	56.8
27	2.7	3.8	4.2	4.5	10.9	70	14.5	20.4	22.7	24.4	60.0
28	2.8	3.9	4.4	4.7	11.3	71	15.2	21.6	24.0	25.8	63.5
29	2.9	4.1	4.5	4.9	11.8	72	16.2	23.0	25.5	27.5	67.6
30	3.0	4.3	4.7	5.1	12.3	73	17.2	24.5	27.1	29.3	72.2
31	3.1	4.4	4.9	5.3	12.8	74	18.3	26.0	28.8	31.2	77.2
32	3.3	4.6	5.1	5.5	13.3	75	19.5	27.9	30.9	33.3	83.0
33	3.4	4.8	5.3	5.7	13.9	76	21.0	30.0	33.3	36.0	90.0
34	3.5	5.0	5.5	6.0	14.4	77	22.9	32.0	36.0	39.0	98.1
35	3.7	5.2	5.7	6.2	14.9	78	24.8	35.5	39.3	42.4	108.0
36	3.8	5.4	5.9	6.4	15.5	79	27.4	38.2	43.7	46.9	130.5
37	3,9	5.6	6.1	6.6	16.0	80	30.4	43.4	48.5	52.0	136.2
38	4.0	5.8	6.3	6.9	16.6	81	34.0	48.5	54.0	58.8	157.2
39	4.2	6.0	6.6	7.1	17.3	82	38.4	55.3	61.8	67.2	182.0
40	4.3	6.2	6.8	7.3	17.9	83	45.0	65,3	72.8	79,0	244 5
41	4.5	6.4	7,0	7.6	18.5	84	53,0	78.0	88.0	96.0	—
42	4.7	6.6	7.3	7.9	19.2	85	67.0	100.8	112.5	127.5	—
43	4.9	6.9	7.6	8.2	19.9	86	—·	156.0	200.0	—	—

2-12. Data for Calculations on Laue Patterns

2-12a. Net for Calculations on Back-Reflection Patterns

Figure 29 provides direct readings of γ and δ (angular coordinates), normals to reflecting planes, and zone axes. The hyperbolas running from left to right are lines of constant γ, each being the locus of the reflections from planes in the zone whose axis makes an angle γ with the plane of the film in the Oy direction.

The hyperbolas running from the top downwards are lines of constant δ; they correspond to the inclination of the zone axis with respect to the plane of the film in the Ox direction. The lines are spaced at 2°; the scale corresponds to a distance of 3 cm from specimen to film.

The following method is used to deduce γ and δ. The pattern is placed with its center at the center of the net, the edges being parallel to those of the net; then γ and δ are read directly. Figure 30a shows an example of the deduction of γ and δ. The cut left-hand edge of the film lies on the left on sighting along the direction of the incident beam.

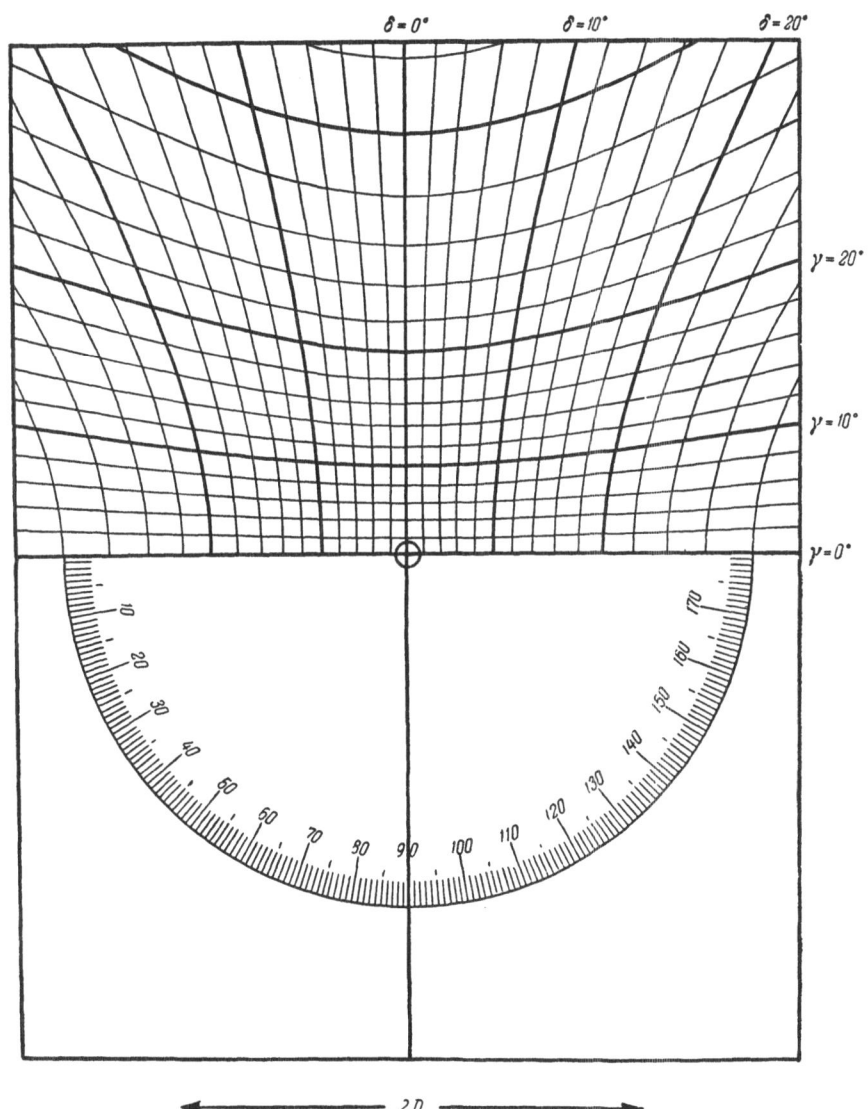

Fig. 29. Net for calculations on Laue back-reflection patterns.

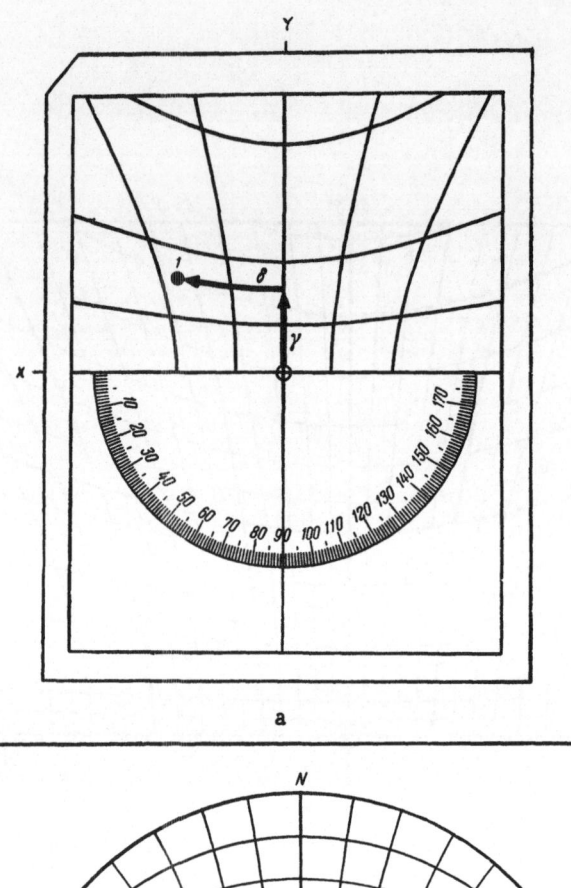

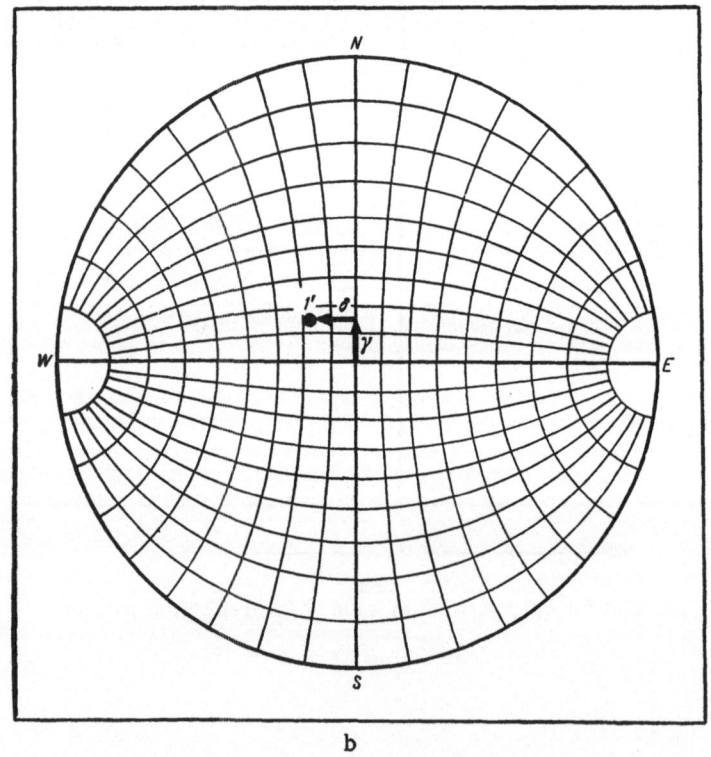

Fig. 30. a) Measurement of a Laue pattern by means of the net;
b) transfer of the results to a stereographic projection.

The γ and δ can then be transferred to a stereographic projection, as shown in Fig. 30b. The latter is set to make its meridians run from left to right, not from top to bottom. The net can be used to determine the orientations of the individual crystals in a polycrystalline specimen and for other such purposes.

2-12b. Net for Calculations on Transmission Patterns

Laue patterns recorded in transmission are used for the purpose mentioned above if the specimen is thin or is of low absorption. The net of Fig. 31 provides a means of deducing φ (the angle formed by the normal to the reflecting plane with the plane of the film) and δ (the inclination of this normal in the xy plane of the film; the ray lies along Oz). This net is analogous to the previous one; it consists of sets of lines of constant δ (full lines) and constant φ (broken lines). The scale corresponds to a distance of 3 cm from specimen to film.

The following method is used to deduce φ and δ. The pattern is placed with its center at the center of the net, the edges being parallel to those of the net; then φ and δ are read directly. Figure 32a shows an example of the deduction of φ and δ. The cut left-hand edge of the film lies on the left on sighting along the direction of the incident beam.

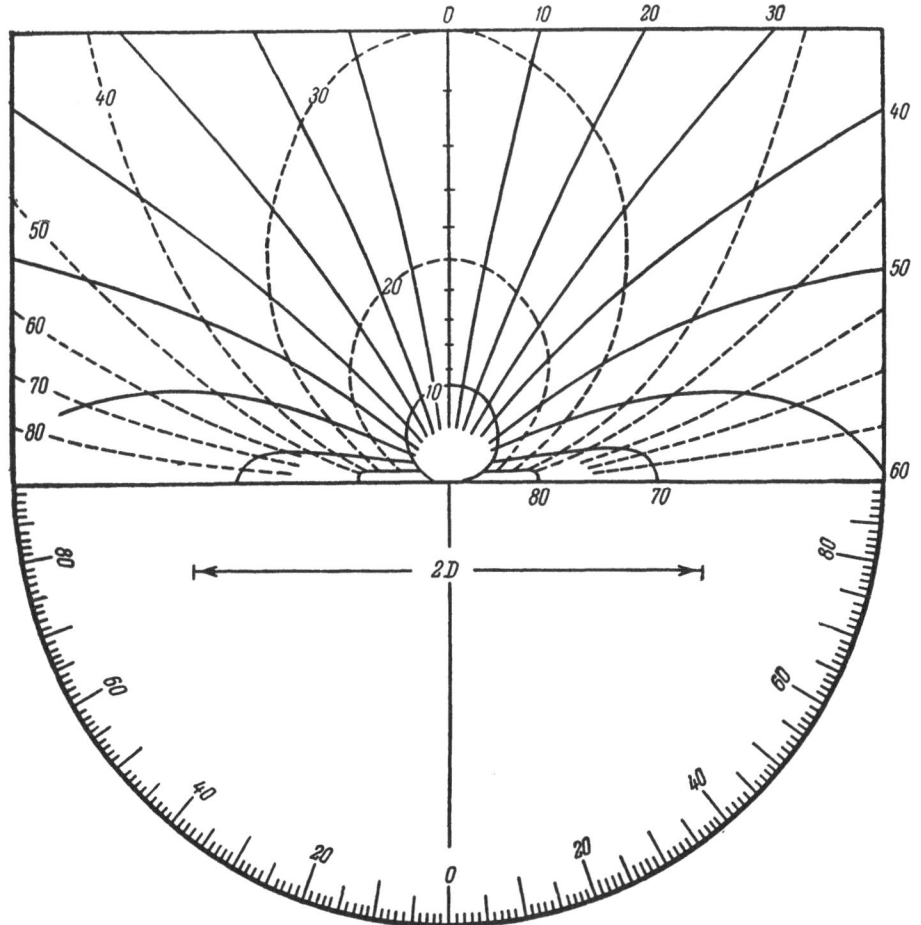

Fig. 31. Net for calculations on Laue transmission patterns.

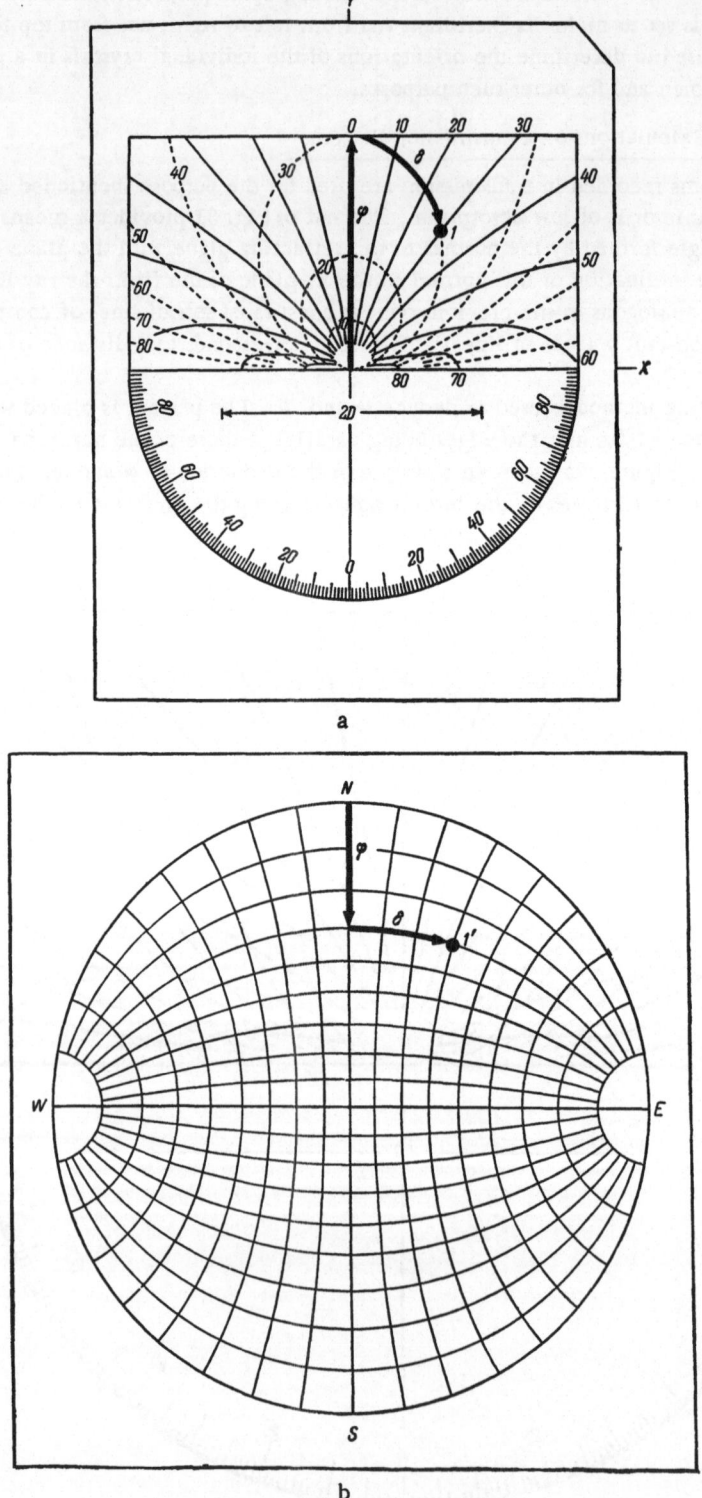

Fig. 32. a) Measurement of a Laue pattern by means of the net;
b) transfer of the results to a stereographic projection.

The φ and δ can then be transferred to a stereographic projection, as shown in Fig. 32b. The latter is set to make its meridians run from left to right, not from top to bottom.

2-12c. Accessory Table for Constructing the Projection of the Crystal from the Laue Pattern

Gnomonic projection is often used to interpret Laue patterns; the planes of the pattern and the projection are brought into coincidence, and the polar coordinates of the reflections are measured (i.e., the length of the radius vector from the reflection to the central spot and the angle φ between this vector and the origin line).

$\vartheta°$	cot ϑ	r		$\vartheta°$	cot ϑ	r	
		$R'=50$ mm	$R'=100$ mm			$R'=50$ mm	$R'=100$ mm
0	—	—	—	25	2.1445	107.2	214.4
1	57.2900	2864.5	5729.0	26	2.0503	102.7	205.0
2	28.6363	1431.8	2863.6	27	1.9626	98.1	196.3
3	19.0811	954.0	1908.1	28	1.8807	94.0	188.1
4	14.3007	715.0	1430.1	29	1.8040	90.2	180.4
5	11.4301	571.5	1143.0	30	1.7320	86.6	173.2
6	9.5144	475.7	951.4	31	1.6643	83.2	166.4
7	8.1443	407.2	814.4	32	1.6003	80.0	160.0
8	7.1154	355.7	711.5	33	1.5399	76.9	154.0
9	6.3137	315.7	631.4	34	1.4826	74.1	148.3
10	5.6713	283.5	567.1	35	1.4281	71.4	142.8
11	5.1445	257.2	514.4	36	1.3764	68.8	137.6
12	4.7046	235.2	470.5	37	1.3270	66.3	132.7
13	4.3315	216.5	433.1	38	1.2799	63.9	128.0
14	4.0108	200.5	401.1	39	1.2349	61.7	123.5
15	3.7320	186.6	373.2	40	1.1917	59.6	119.2
16	3.4874	174.3	348.7	41	1.1504	57.5	115.0
17	3.2708	163.5	327.1	42	1.1106	55.5	111.1
18	3.0777	153.9	307.8	43	1.0724	53.6	107.2
19	2.9042	145.2	290.4	44	1.0355	51.7	103.5
20	2.7475	137.3	274.7	45	1.0000	50.0	100.0
21	2.6051	130.2	260.5				
22	2.4751	123.7	247.5				
23	2.3558	117.8	235.6				
24	2.2460	112.3	224.6				

A spot of coordinates (s, φ) corresponds in projection to a point having coordinates (r, 180° $-\varphi$), where r = R'cot ϑ, R' usually being taken equal to R, the distance from specimen to film. The table gives means of deducing r for various R' [109].

2-13. Determination of Orientation for Large Crystals in Polycrystalline Materials

Laue patterns in back-reflection are used to examine large monocrystals and individual grains in polycrystalline materials.

Such patterns (epigrams) are interpreted as follows [86]: (1) the angular coordinates are transferred to the circle diagram by the method described in Section 2-12; (2) the spots are indexed by means of standard stereographic projections. A tracing bearing the projection is placed over one of the standard projections to bring the two sets of points into the closest possible agreement; the tracing is displaced parallel and is turned within limits of 27.5° (inner circle on the projection) for this purpose.

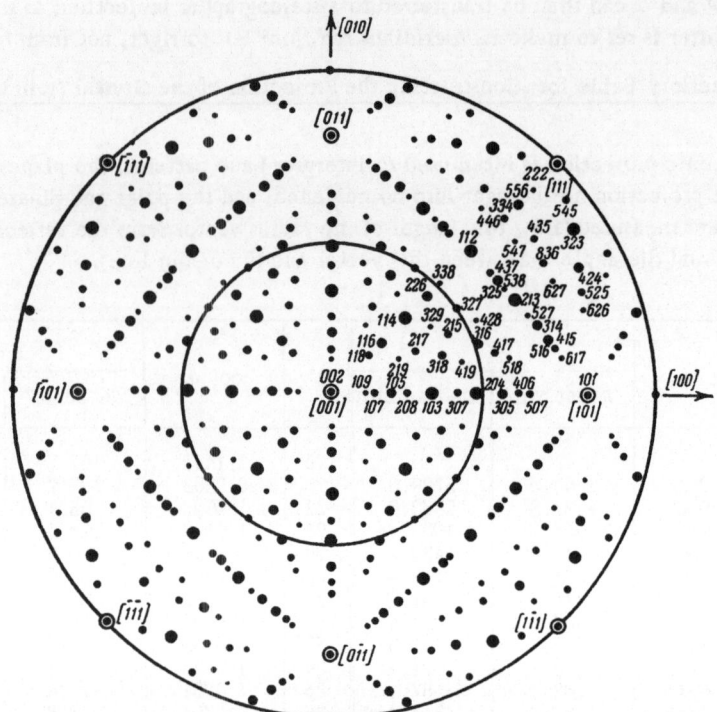

Fig. 33. Standard stereographic projection along [001] for body-
centered cubic crystals.

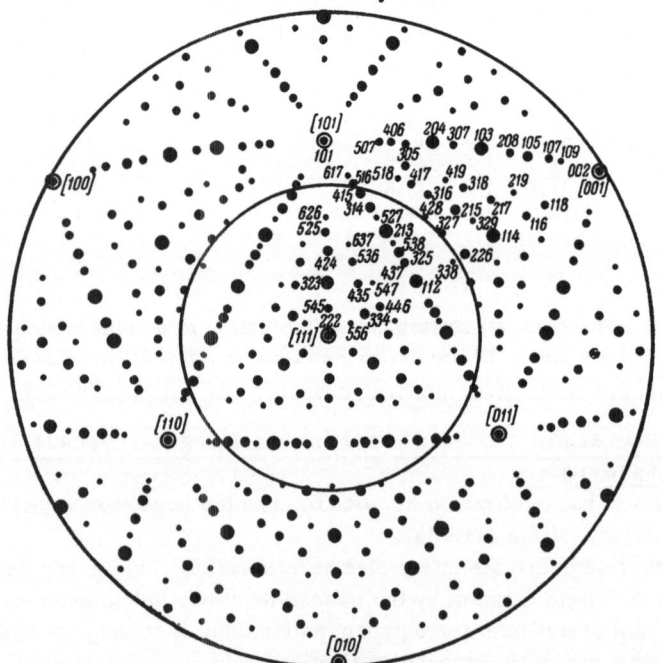

Fig. 34. Standard stereographic projection along [111] for body-
centered cubic crystals.

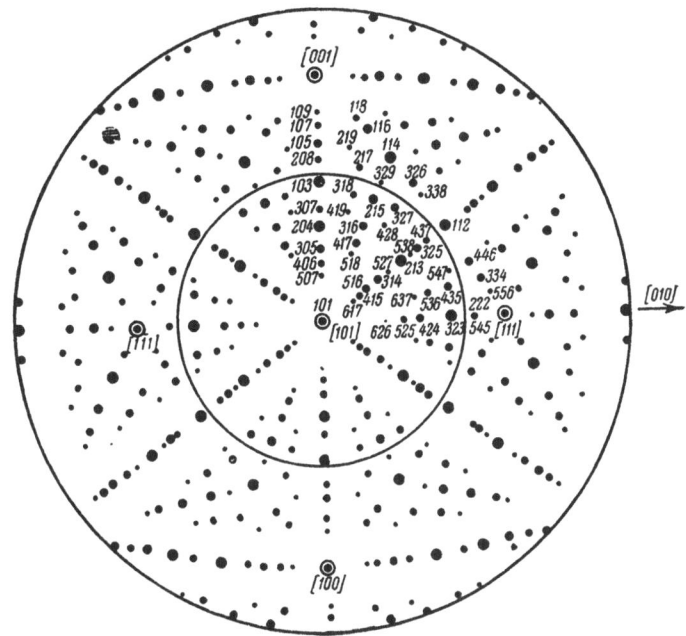

Fig. 35. Standard stereographic projection along [101] for body-centered cubic crystals.

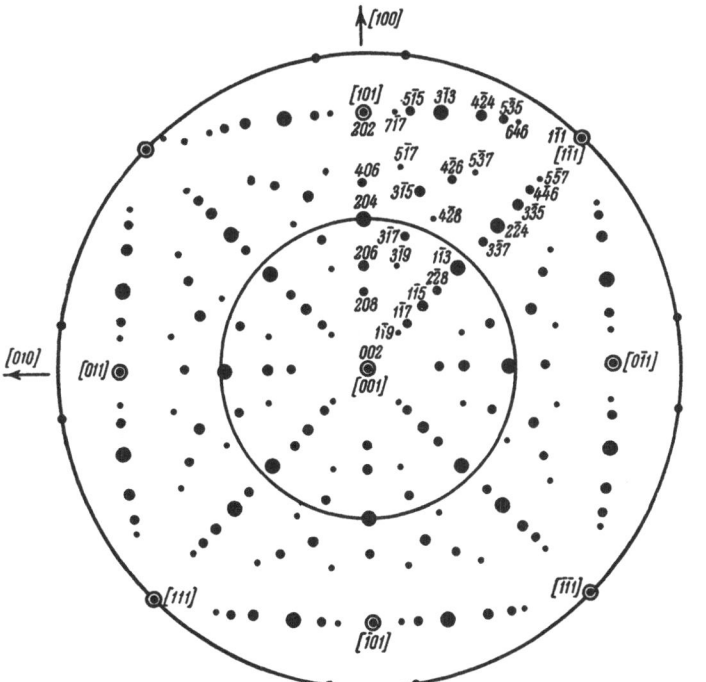

Fig. 36. Standard stereographic projection along [001] for face-centered cubic crystals.

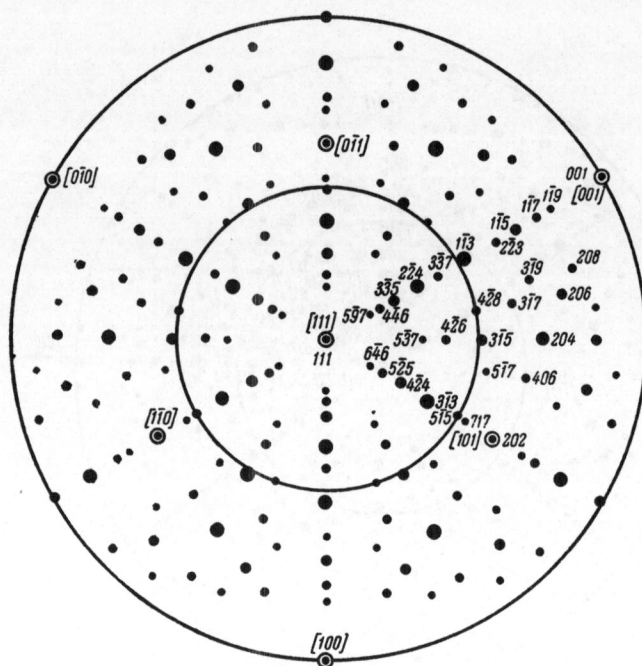

Fig. 37. Standard stereographic projection along [111] for face-centered cubic crystals.

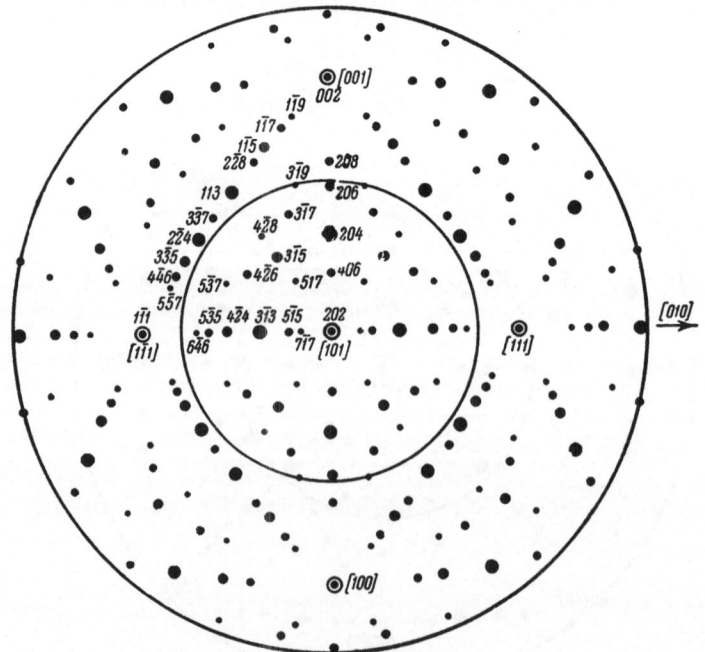

Fig. 38. Standard stereographic projection along [101] for face-centered cubic crystals.

Figures 33-38 give standard projections for the [001], [111], and [101] directions in body-centered and face-centered cubic crystals; these projections have been constructed for $90 - \vartheta < 45°$, as is typical of this mode of use. For use they should be enlarged to half the diameter of the Wulff net to be used.

The points corresponding to the principal crystallographic directions are located after indexing; the corresponding zones (arcs of the large circles on the projections) are produced by means of the Wulff net, and the intersections between two (or more) zones, or the angular relations between directions already known and the principal ones, are used to find the points of emergence of the principal directions. This defines the directions of the principal axes of the crystal with respect to the external XYZ axes; the position of the specimen relative to the latter is usually known, so the position of the principal axes in the crystal can be deduced.

Section 9-3 gives standard projections for ϑ from 0 to 90° for cubic and hexagonal crystals.

CHAPTER 3

INDEXING OF X-RAY PATTERNS

Chapter 3 gives the data needed to index the patterns of cubic, tetragonal, hexagonal, rhombohedral, and orthorhombic crystals.

The relations between the (hk$l$) and the interplanar distances are as follows:

1. Cubic: $\dfrac{1}{d^2} = \dfrac{h^2+k^2+l^2}{a^2}$; $V = a^3$.

2. Tetragonal: $\dfrac{1}{d^2} = \dfrac{h^2+k^2}{a^2} + \dfrac{l^2}{c^2}$; $V = a^2c$.

3. Orthorhombic: $\dfrac{1}{d^2} = \dfrac{h^2}{a^2} + \dfrac{k^2}{b^2} + \dfrac{l^2}{c^2}$; $V = abc$.

4. Trigonal (in trigonal axes):

$$\frac{1}{d^2} = \frac{(h^2+k^2+l^2)\sin^2\alpha + 2(hk+kl+hl)(\cos^2\alpha - \cos\alpha)}{a^2(1-3\cos^2\alpha + 2\cos^3\alpha)};\quad V = a^3\sqrt{1-3\cos^2\alpha + 2\cos^3\alpha}.$$

5. Hexagonal (hexagonal axes, indices hki$l$):

$$\frac{1}{d^2} = \frac{4}{3}\left(\frac{h^2+hk+k^2}{a^2}\right) + \frac{l^2}{c^2}:\quad V = \frac{\sqrt{3}}{2}a^2c = 0.866\ a^2c.$$

6. Monoclinic:

$$\frac{1}{d^2} = \frac{h^2}{a^2\sin^2\beta} + \frac{k^2}{b^2} + \frac{l^2}{c^2\sin^2\beta} - \frac{2hl\cos\beta}{ac\sin^2\beta}:\quad V = abc\sin\beta.$$

Here a, b, and c are the lattice constants, β is the axial angle, V is the volume of the unit cell, and α is the angle of the rhombohedron.

See [6-13] for analytic and graphical methods of indexing.

3-1. Accessory Tables

3-1a. Some Composite Trigonometric Functions

Values are given for the following functions often needed in connection with x-ray patterns:

$$\sin\vartheta,\ \sin^2\vartheta,\ \lg\sin^2\vartheta,\ \frac{1}{\sin\vartheta},\ \frac{1}{\cos\vartheta},\ \operatorname{tg}2\vartheta,$$
$$\lg 4\vartheta_{rad}, 4\vartheta_{rad}\quad \lg\sin\vartheta,\ \lg\operatorname{tg}2\vartheta\ [2,\ 102].$$

The functions are given at intervals of 0.1°; the common differences D are also given at intervals of 1' in ϑ (not 2ϑ or 4ϑ) for $\lg\operatorname{tg}2\vartheta$, $\lg 4\vartheta$, and $\lg\sin\vartheta$, or at intervals of for $\sin\vartheta$, $\sin^2\vartheta$, and $\lg\sin^2\vartheta$.

90

θ°	sin θ	D(0.1°)	sin²θ	D(0.1°)	lg sin²θ	D(0.1°)	1/sin θ	1/cos θ	tg 2θ	lg 4θ_rad	D(1')	4θ_rad	lg sin θ	D(1')	lg tg 2θ	D(1')
0.0	0.00000	175	0.00000	3	—∞		∞	1.00000	0.00000	—		—	—		—	
1	.00175	174	.000003	9	4.47712	60206	571.42857	1.00001	0.00349	7.8439		0.00698	7.2419		7.5429	
2	.00349	175	.000012	13	5.07918	35218	286.53295	1.00001	0.00698	8.1450		0.01395	7.5429		7.8439	
3	.00524	174	.000027	22	5.43136	25884	190.83969	1.00001	0.01047	8.3211		0.02094	7.7190		8.0200	
4	.00698	175	.000049	27	5.69020	19061	143.26648	1.00002	0.01396	8.4460		0.02793	7.8439		8.1450	
5	.00873	174	.000076	33	5.88081	15662	114.54754	1.00004	0.01746	8.5429		0.03491	7.9408		8.2419	
6	.01047	175	.000109	40	6.03743	13576	95.51098	1.00005	0.02095	8.6221		0.04189	8.0200		8.3211	
7	.01222	174	.000149	46	6.17319	11684	81.83306	1.00007	0.02444	8.6890		0.04887	8.0870		8.3881	
8	.01396	175	.000195	52	6.29003	10267	71.63324	1.00010	0.02793	8.7470		0.05585	8.1450		8.4461	
9	.01571	174	.000247	5	6.39270	8442	63.65372	1.00012	0.03143.	8.7982		0.06283	8.1961		8.4973	
1.0	0.01745	175	0.00030	7	6.47712	9108	57.30659	1.00015	0.03492	8.8439		0.06981	8.2419		8.5431	
1	.01920	174	.00037	7	6.56820	7525	52.08333	1.00018	0.03842	8.8853		0.07679	8.2832		8.5845	
2	.02094	175	.00044	7	6.64345	6412	47.75549	1.00022	0.04191	8.9231		0.08378	8.3210		8.6223	
3	.02269	174	.00051	9	6.70757	7058	44.07228	1.00026	0.04541	8.9579		0.09076	8.3558		8.6571	
4	.02443	175	.00060	9	6.77815	6070	40.93328	1.00030	0.04891	8.9901		0.09774	8.3880		8.6894	
5	.02618	174	.00069	9	6.83885	5324	38.19710	1.00034	0.05241	9.0200		0.10472	8.4179		8.7194	
6	.02792	175	.00078	10	6.89209	5239	35.81662	1.00039	0.05591	9.0481		0.11170	8.4459		8.7475	
7	.02967	174	.00088	11	6.94448	5116	33.70408	1.00044	0.05941	9.0744	41	0.11868	8.4723	41	8.7739	

$D(1')$	$\lg \mathrm{tg}\, 2\vartheta$	$D(1')$	$\lg \sin \vartheta$	$4\vartheta_{rad}$	$D(1')$	$\lg 4\vartheta_{rad}$	$\mathrm{tg}\, 2\vartheta$	$\dfrac{1}{\cos \vartheta}$	$\dfrac{1}{\sin \vartheta}$	$D(0.1°)$	$\lg \sin^2 \vartheta$	$D(0.1°)$	$\sin^2 \vartheta$	$D(0.1°)$	$\sin \vartheta$	$\vartheta°$
39	8.7988	39	8.4971	0.12566	39	9.0992	0.06291	1.00049	31.83699	4575	6.99564	11	0.00099	175	0.03141	1.8
37	8.8223	37	8.5206	0.13265	37	9.1227	0.06642	1.00055	30.15682	4497	7.04139	12	.00110	174	.03316	9
35.5	8.8446	35	8.5428	0.13963	35	9.1450	0.06993	1.00061	28.65330	4074	7.08636	12	0.00122	174	0.03490	2.0
33.8	8.8659	33.7	8.5640	0.14661	33.7	9.1662	0.07344	1.00067	27.29258	4022	7.12710	13	.00134	175	.03664	1
32.3	8.8862	32.2	8.5842	0.15359	32.2	9.1864	0.07695	1.00074	26.04845	3951	7.16732	14	.00147	174	.03839	2
30.8	8.9056	30.8	8.6035	0.16057	30.7	9.2057	0.08046	1.00081	24.91901	3621	7.20683	14	.00161	175	.04013	3
29.8	8.9241	29.5	8.6220	0.16755	29.7	9.2241	0.08397	1.00088	23.87775	3571	7.24304	15	.00175	174	.04188	4
28.5	8.9420	28.3	8.6397	0.17453	28.3	9.2419	0.08749	1.00095	22.92526	3512	7.27875	16	.00190	174	.04362	5
27.5	8.9591	27.3	8.6567	0.18151	27.3	9.2589	0.09101	1.00103	22.04586	3248	7.31387	16	.00206	175	.04536	6
26.5	8.9756	26.3	8.6731	0.18850	26.3	9.2753	0.09453	1.00111	21.22692	3205	7.34635	17	.00222	174	.04711	7
25.5	8.9915	25.3	8.6889	0.19548	25.3	9.2911	0.09805	1.00119	20.47083	2984	7.37840	17	.00239	174	.04885	8
24.7	9.0068	24.5	8.7041	0.20246	24.7	9.3063	0.10158	1.00128	19.76675	2951	7.40824	18	.00256	175	.05059	9
24.0	9.0216	23.7	8.7188	0.20944	23.7	9.3211	0.10510	1.00137	19.10585	2763	7.43775	18	0.00274	174	0.05234	3.0
23.2	9.0360	23.0	8.7330	0.21642	23.0	9.3353	0.10863	1.00146	18.49112	2877	7.46538	20	.00292	174	.05408	1
22.3	9.0499	22.3	8.7468	0.22340	22.3	9.3491	0.11217	1.00156	17.91473	2568	7.49415	21	.00312	174	.05582	2
21.8	9.0633	21.5	8.7602	0.23038	21.5	9.3625	0.11570	1.00166	17.37318	2671	7.51983	21	.00331	175	.05756	3
21.2	9.0764	21.0	8.7731	0.23736	21.0	9.3754	0.11924	1.00176	16.86056	2517	7.54654	21	.00352	174	.05931	4
20.7	9.0891	20.3	8.7857	0.24435	20.3	9.3880	0.12278	1.00187	16.38002	2379	7.57171	21	.00373	174	.06105	5

θ°	sin θ	D(0.1°)	sin²θ	D(0.1°)	lg sin²θ	D(0.1°)	1/sin θ	1/cos θ	tg 2θ	lg 4θ_rad	D(1')	4θ_rad	lg sin θ	D(1')	lg tg 2θ	D(1')
3.6	0.06279	174	0.00394	22	7.59550	2359	15.92610	1.00197	0.12633	9.4002	19.8	0.25133	8.7979	19.8	9.1015	20.0
7	.06453	174	.00416	23	7.61909	2337	15.49667	1.00208	0.12988	9.4121	19.3	0.25831	8.8098	19.2	9.1135	19.5
8	.06627	175	.00439	24	7.64246	2312	15.08978	1.00220	0.13343	9.4237	18.8	0.26529	8.8213	18.8	9.1252	19.2
9	.06802	174	.00463	24	7.66558	2195	14.70156	1.00233	0.13698	9.4350	18.3	0.27227	8.8326	18.3	9.1367	18.5
4.0	0.06976	174	0.00487	24	7.68753	2089	14.33486	1.00245	0.14054	9.4460	17.8	0.27925	8.8436	17.8	9.1478	18.2
1	.07150	174	.00511	25	7.70842	2074	13.98601	1.00257	0.14410	9.4567	17.5	0.28623	8.8543	17.3	9.1587	17.7
2	.07324	174	.00536	26	7.72916	2058	13.65374	1.00270	0.14767	9.4672	17.0	0.29322	8.8647	17.0	9.1693	17.3
3	.07498	174	.00562	27	7.74974	2037	13.33689	1.00282	0.15124	9.4774	16.7	0.30020	8.8749	16.7	9.1797	16.8
4	.07672	174	.00589	27	7.77011	1947	13.03441	1.00296	0.15481	9.4874	16.2	0.30718	8.8849	16.2	9.1898	16.5
5	.07846	174	.00616	27	7.78958	1863	12.74535	1.00309	0.15838	9.4971	16.0	0.31416	8.8946	16.0	9.1997	16.2
6	.08020	174	.00643	28	7.80821	1851	12.46883	1.00323	0.16196	9.5067	15.5	0.32114	8.9042	15.5	9.2094	15.8
7	.08194	174	.00671	29	7.82672	1838	12.20405	1.00337	0.16555	9.5160	15.3	0.32812	8.9135	15.2	9.2189	15.5
8	.08368	174	.00700	30	7.84510	1822	11.95029	1.00352	0.16914	9.5252	14.8	0.33510	8.9226	14.8	9.2282	15.3
9	.08542	174	.00730	29	7.86332	1692	11.70686	1.00366	0.17273	9.5341	14.7	0.34208	8.9315	14.7	9.2374	14.8
5.0	0.08716	173	0.00759	31	7.88024	1739	11.47315	1.00382	0.17633	9.5429	14.33	0.34907	8.9403	14.33	9.2463	14.67
1	.08889	174	.00790	31	7.89763	1671	11.24986	1.00398	0.17993	9.5515	14.00	0.35605	8.9489	14.00	9.2551	14.33
2	.09063	174	.00821	32	7.91434	1661	11.03387	1.00414	0.18353	9.5599	13.83	0.36303	8.9573	13.67	9.2637	14.17
3	.09237	174	.00853	33	7.93095	1648	10.82603	1.00430	0.18714	9.5682	13.50	0.37001	8.9655	13.50	9.2722	13.83

$\theta°$	$\sin\theta$	$D(0.1°)$	$\sin^2\theta$	$D(0.1°)$	$\lg\sin^2\theta$	$D(0.1°)$	$\frac{1}{\sin\theta}$	$\frac{1}{\cos\theta}$	$\operatorname{tg}2\theta$	$\lg 4\theta_{rad}$	$D(1')$	$4\theta_{rad}$	$\lg\sin\theta$	$D(1')$	$\lg\operatorname{tg}2\theta$	$D(1')$
5.4	0.09411	174	0.00886	33	7.94743	1589	10.62586	1.00446	0.19076	9.5763	13.33	0.37699	8.9736	13.33	9.2805	13.67
5.5	.09585	173	.00919	33	7.96332	1532	10.43297	1.00462	0.19438	9.5843	13.00	0.38397	8.9816	13.00	9.2887	13.33
5.6	.09758	174	.00952	34	7.97864	1524	10.24800	1.00479	0.19801	9.5921	12.83	0.39095	8.9894	12.67	9.2967	13.17
5.7	.09932	174	.00986	35	7.99388	1515	10.06847	1.00496	0.20164	9.5998	12.67	0.39794	8.9970	12.67	9.3046	12.83
5.8	.10106	173	.01021	36	8.00903	1505	9.89511	1.00515	0.20527	9.6074	12.33	0.40492	9.0046	12.33	9.3123	12.83
5.9	.10279	174	.01057	36	8.02408	1454	9.72857	1.00533	0.20891	9.6148	12.17	0.41190	9.0120	12.00	9.3200	12.50
6.0	0.10453	173	0.01093	36	8.03862	1407	9.56663	1.00551	0.21256	9.6221	12.00	0.41888	9.0192	12.00	9.3275	12.33
6.1	.10626	174	.01129	37	8.05269	1401	9.41088	1.00569	0.21621	9.6293	11.67	0.42586	9.0264	11.67	9.3349	12.17
6.2	.10800	173	.01166	38	8.06670	1393	9.25926	1.00588	0.21986	9.6363	11.67	0.43284	9.0334	11.50	9.3422	11.83
6.3	.10973	174	.01204	39	8.08063	1384	9.11328	1.00608	0.22353	9.6433	11.33	0.43982	9.0403	11.50	9.3493	11.83
6.4	.11147	173	.01243	39	8.09447	1342	8.97102	1.00627	0.22719	9.6501	11.33	0.44680	9.0472	11.17	9.3564	11.67
6.5	.11320	174	.01282	39	8.10789	1304	8.83392	1.00647	0.23087	9.6569	11.00	0.45379	9.0539	11.00	9.3634	11.33
6.6	.11494	173	.01321	40	8.12090	1296	8.70019	1.00667	0.23455	9.6635	10.83	0.46077	9.0605	10.83	9.3702	11.33
6.7	.11667	173	.01361	41	8.13386	1289	8.57118	1.00688	0.23823	9.6700	10.67	0.46775	9.0670	10.67	9.3770	11.17
6.8	.11840	174	.01402	41	8.14675	1252	8.44595	1.00708	0.24193	9.6764	10.67	0.47473	9.0734	10.50	9.3837	11.00
6.9	.12014	173	.01443	42	8.15927	1246	8.32362	1.00729	0.24562	9.6828	10.33	0.48171	9.0797	10.33	9.3903	10.83
7.0	0.12187	173	0.01485	43	8.17173	1239	8.20546	1.00751	0.24933	9.6890	10.33	0.48869	9.0859	10.17	9.3968	10.67
7.1	.12360	173	.01528	43	8.18412	1206	8.09061	1.00773	0.25304	9.6952	10.17	0.49567	9.0920	10.17	9.4032	10.50

θ°	sin θ	D (0.1°)	sin²θ	D (0.1°)	lg sin²θ	D (0.1°)	$\frac{1}{\sin θ}$	$\frac{1}{\cos θ}$	tg 2θ	lg 4θ_rad	D (1')	4θ_rad	lg sin θ	D (1')	lg tg 2θ	D (1')
7.2	0.12533	173	0.01571	44	8.19618	1199	7.97894	1.00795	0.25676	9.7013	10.00	0.50265	9.0981	9.83	9.4095	10.50
3	.12706	174	.01615	44	8.20817	1168	7.87030	1.00818	0.26048	9.7073	9.83	0.50964	9.1040	9.83	9.4158	10.33
4	.12880	173	.01659	45	8.21985	1162	7.76398	1.00840	0.26421	9.7132	9.67	0.51662	9.1099	9.67	9.4220	10.17
5	.13053	173	.01704	45	8.23147	1132	7.66107	1.00863	0.26795	9.7190	9.67	0.52360	9.1157	9.50	9.4281	10.00
6	.13226	173	.01749	46	8.24279	1127	7.56086	1.00886	0.27169	9.7248	9.33	0.53058	9.1214	9.50	9.4341	9.83
7	.13399	173	.01795	47	8.25406	1123	7.46324	1.00910	0.27545	9.7304	9.33	0.53756	9.1271	9.17	9.4400	9.83
8	.13572	172	.01842	47	8.26529	1094	7.36811	1.00934	0.27920	9.7360	9.33	0.54454	9.1326	9.17	9.4459	9.67
9	.13744	173	.01889	48	8.27623	1090	7.27590	1.00958	0.28297	9.7416	9.00	0.55152	9.1381	9.17	9.4517	9.67
8.0	0.13917	173	0.01937	48	8.28713	1063	7.18546	1.00983	0.28675	9.7470	9.00	0.55851	9.1436	8.83	9.4575	9.50
1	.14090	173	.01985	49	8.29776	1056	7.09723	1.01008	0.29053	9.7524	9.00	0.56549	9.1489	8.83	9.4632	9.33
2	.14263	173	.02034	50	8.30835	1055	7.01115	1.01033	0.29432	9.7578	8.67	0.57247	9.1542	8.67	9.4688	9.33
3	.14436	172	.02084	50	8.31890	1029	6.92713	1.01058	0.29811	9.7630	8.67	0.57945	9.1594	8.67	9.4744	9.17
4	.14608	173	.02134	51	8.32919	1026	6.84556	1.01085	0.30192	9.7682	8.67	0.58643	9.1646	8.50	9.4799	9.00
5	.14781	173	.02185	51	8.33945	1002	6.76544	1.01110	0.30573	9.7734	8.33	0.59341	9.1697	8.33	9.4853	9.00
6	.14954	172	.02236	52	8.34947	999	6.68717	1.01137	0.30955	9.7784	8.50	0.60039	9.1747	8.33	9.4907	9.00
7	.15126	173	.02288	53	8.35946	994	6.61113	1.01164	0.31338	9.7835	8.83	0.60737	9.1797	8.33	9.4961	9.00
8	.15299	172	.02341	53	8.36940	972	6.53637	1.01191	0.31722	9.7888	8.50	0.61436	9.1847	8.00	9.5014	8.83
9	.15471	173	.02394	53	8.37912	951	6.46371	1.01219	0.32106	9.7933	8.17	0.62134	9.1895	8.00	9.5066	8.67

D(1')	lg tg 2θ	D(1')	lg sin θ	4θ rad	D(1')	lg 4θ rad	tg 2θ	1/cos θ	1/sin θ	D(0.1°)	lg sin² θ	D(0.1°)	sin² θ	D(0.1°)	sin θ	θ°
8.50	9.5118	8.00	9.1943	0.62832	8.00	9.7982	0.32492	1.01246	6.39264	948	8.38863	54	0.02447	173	0.15643	9.0
8.50	9.5169	7.83	9.1991	0.63530	7.83	9.8030	0.32878	1.01275	6.32271	945	8.39811	55	.02501	172	.15816	1
8.33	9.5220	7.83	9.2038	0.64228	7.83	9.8077	0.33266	1.01303	6.25469	941	8.40756	56	.02556	172	.15988	2
8.33	9.5270	7.67	9.2085	0.64926	7.83	9.8124	0.33654	1.01331	6.18812	922	8.41697	56	.02612	172	.16160	3
8.33	9.5320	7.50	9.2131	0.65624	7.67	9.8171	0.34043	1.01361	6.12257	902	8.42619	56	.02668	173	.16333	4
8.17	9.5370	7.50	9.2176	0.66323	7.50	9.8217	0.34433	1.01390	6.05877	899	8.43521	57	.02724	172	.16505	5
8.00	9.5419	7.50	9.2221	0.67021	7.50	9.8262	0.34824	1.01420	5.99628	897	8.44420	58	.02781	172	.16677	6
8.17	9.5467	7.33	9.2266	0.67719	7.50	9.8307	0.35216	1.01451	5.93507	878	8.45317	58	.02839	172	.16849	7
7.83	9.5516	7.17	9.2310	0.68417	7.33	9.8352	0.35608	1.01481	5.87510	875	8.46195	59	.02897	172	.17021	8
8.00	9.5563	7.33	9.2353	0.69115	7.17	9.8396	0.36002	1.01512	5.81632	859	8.47070	59	.02956	172	.17193	9
7.83	9.5611	7.00	9.2397	0.69813	7.33	9.8439	0.36397	1.01542	5.75871	856	8.47929	60	0.03015	172	0.17365	10.0
7.67	9.5658	7.17	9.2439	0.70511	7.00	9.8483	0.36793	1.01574	5.70223	853	8.48785	61	.03075	171	.17537	1
7.67	9.5704	7.00	9.2482	0.71209	7.17	9.8525	0.37190	1.01605	5.64717	836	8.49638	61	.03136	172	.17708	2
7.67	9.5750	6.83	9.2524	0.71908	7.00	9.8568	0.37588	1.01638	5.59284	834	8.50474	62	.03197	172	.17880	3
7.67	9.5796	6.83	9.2565	0.72606	6.83	9.8610	0.37986	1.01670	5.53955	819	8.51308	62	.03259	172	.18052	4
7.50	9.5842	6.83	9.2606	0.73304	6.83	9.8651	0.38386	1.01704	5.48727	816	8.52127	63	.03321	171	.18224	5
7.50	9.5887	6.67	9.2647	0.74002	6.83	9.8692	0.38787	1.01736	5.43626	801	8.52943	63	.03384	172	.18395	6
7.33	9.5932	6.67	9.2687	0.74700	6.83	9.8733	0.39190	1.01770	5.38590	799	8.53744	64	.03447	171	.18567	7

θ°	sin θ	D(0.1°)	sin² θ	D(0.1°)	lg sin² θ	D(0.1°)	1/sin θ	1/cos θ	tg 2θ	lg 4θ_rad	D(1')	4θ_rad	lg sin θ	D(1')	lg tg 2θ	D(1')
10.8	0.18738	172	0.03511	65	8.54543	797	5.33675	1.01803	0.39593	9.8774	6.67	0.75398	9.2727	6.67	9.5976	7.33
10.9	0.18910	171	0.03576	65	8.55340	782	5.28821	1.01837	0.39997	9.8814	6.50	0.76096	9.2767	6.50	9.6020	7.33
11.0	0.19081	171	0.03641	65	8.56122	769	5.24082	1.01871	0.40403	9.8853	6.67	0.76794	9.2806	6.50	9.6064	7.33
1	.19252	171	.03706	67	8.56891	778	5.19427	1.01907	0.40809	9.8893	6.50	0.77493	9.2845	6.33	9.6108	7.17
2	.19423	172	.03773	66	8.57669	753	5.14854	1.01941	0.41217	9.8932	6.33	0.78191	9.2883	6.33	9.6151	7.17
3	.19595	171	.03839	68	8.58422	762	5.10334	1.01977	0.41626	9.8970	6.33	0.78889	9.2921	6.33	9.6194	7.17
4	.19766	171	.03907	68	8.59184	750	5.05919	1.02013	0.42036	9.9008	6.33	0.79587	9.2959	6.33	9.6236	7.00
5	.19937	171	.03975	68	8.59934	736	5.01580	1.02049	0.42447	9.9046	6.33	0.80285	9.2997	6.17	9.6279	7.17
6	.20108	171	.04043	69	8.60670	735	4.97315	1.02085	0.42860	9.9084	6.17	0.80983	9.3034	6.00	9.6321	7.00
7	.20279	171	.04112	70	8.61405	733	4.93121	1.02122	0.43274	9.9121	6.17	0.81681	9.3070	6.17	9.6362	6.83
8	.20450	170	.04182	70	8.62138	721	4.88998	1.02159	0.43689	9.9158	6.17	0.82380	9.3107	6.00	9.6404	7.00
9	.20620	171	.04252	71	8.62859	720	4.84966	1.02196	0.44105	9.9195	6.00	0.83078	9.3143	6.00	9.6445	6.83
12.0	0.20791	171	0.04323	71	8.63579	707	4.80977	1.02234	0.44523	9.9231	6.00	0.83776	9.3179	5.83	9.6486	6.83
1	.20962	170	.04394	72	8.64286	706	4.77054	1.02272	0.44942	9.9267	6.00	0.84474	9.3214	6.00	9.6527	6.83
2	.21132	171	.04466	72	8.64992	694	4.73216	1.02310	0.45362	9.9303	5.83	0.85172	9.3250	5.67	9.6567	6.67
3	.21303	171	.04538	73	8.65686	694	4.69417	1.02349	0.45784	9.9338	6.00	0.85870	9.3284	5.83	9.6607	6.67
4	.21474	170	.04611	74	8.66380	691	4.65679	1.02389	0.46206	9.9374	5.67	0.86568	9.3319	5.67	9.6647	6.67
5	.21644	170	.04685	74	8.67071	681	4.62022	1.02428	0.46631	9.9408	5.83	0.87266	9.3353	5.67	9.6687	6.50

θ°	sin θ	D (0.1°)	sin² θ	D (0.1°)	lg sin² θ	D (0.1°)	$\frac{1}{\sin θ}$	$\frac{1}{\cos θ}$	tg 2θ	lg 4θ rad	D (1')	4θ rad	lg sin θ	D (1')	lg tg 2θ	D (1')
12.6	0.21814	171	0.04759	74	8.67752	670	4.58421	1.02467	0.47056	9.9443	5.67	0.87965	9.3387	5.67	9.6726	6.50
7	.21985	170	.04833	75	8.68422	668	4.54856	1.02508	0.47483	9.9477	5.67	0.88663	9.3421	5.67	9.6765	6.50
8	.22155	170	.04908	75	8.69090	668	4.51365	1.02548	0.47912	9.9511	5.67	0.89361	9.3455	5.50	9.6804	6.50
9	.22325	170	.04984	76	8.69758	657	4.47928	1.02589	0.48342	9.9545	5.67	0.90059	9.3488	5.50	9.6843	6.50
13.0	0.22495	170	0.05060	77	8.70415	656	4.44543	1.02630	0.48773	9.9579	5.67	0.90757	9.3521	5.50	9.6882	6.33
1	.22665	170	.05137	77	8.71071	646	4.41209	1.02672	0.49206	9.9612	5.50	0.91455	9.3554	5.33	9.6920	6.33
2	.22835	170	.05214	78	8.71717	645	4.37924	1.02714	0.49640	9.9645	5.50	0.92153	9.3586	5.33	9.6958	6.33
3	.23005	170	.05292	79	8.72362	644	4.34688	1.02756	0.50076	9.9678	5.50	0.92852	9.3618	5.33	9.6996	6.33
4	.23175	170	.05371	79	8.73006	634	4.31499	1.02798	0.50514	9.9710	5.33	0.93550	9.3650	5.33	9.7034	6.33
5	.23345	169	.05450	79	8.73640	625	4.28357	1.02842	0.50953	9.9743	5.50	0.94248	9.3682	5.17	9.7072	6.17
6	.23514	170	.05529	80	8.74265	624	4.25279	1.02885	0.51393	9.9775	5.33	0.94946	9.3713	5.33	9.7109	6.17
7	.23684	169	.05609	81	8.74889	622	4.22226	1.02928	0.51835	9.9807	5.33	0.95644	9.3745	5.00	9.7146	6.17
8	.23853	170	.05690	81	8.75511	614	4.19234	1.02973	0.52279	9.9838	5.17	0.96342	9.3775	5.17	9.7183	6.17
9	.24023	169	.05771	82	8.76125	613	4.16268	1.03016	0.52724	9.9870	5.33	0.97040	9.3806	5.17	9.7220	6.17
14.0	0.24192	170	0.05853	82	8.76738	604	4.13360	1.03061	0.53171	9.9901	5.17	0.97738	9.3837	5.00	9.7257	6.00
1	.24362	169	.05935	83	8.77342	603	4.10475	1.03107	0.53620	9.9932	5.17	0.98437	9.3867	5.17	9.7293	6.17
2	.24531	169	.06018	83	8.77945	595	4.07647	1.03151	0.54070	9.9962	5.00	0.99135	9.3897	5.00	9.7330	6.00
3	.24700	169	.06101	84	8.78540	594	4.04858	1.03197	0.54522	9.9993	5.17	0.99833	9.3927	5.00	9.7366	6.00

$\vartheta°$	$\sin\vartheta$	$D\,(0.1°)$	$\sin^2\vartheta$	$D\,(0.1°)$	$\lg\sin^2\vartheta$	$D\,(0.1°)$	$\dfrac{1}{\sin\vartheta}$	$\dfrac{1}{\cos\vartheta}$	$\operatorname{tg}2\vartheta$	$\lg 4\vartheta_{rad}$	$D\,(1')$	$4\vartheta_{rad}$	$\lg\sin\vartheta$	$D\,(1')$	$\lg\operatorname{tg}2\vartheta$	$D\,(1')$
14.4	0.24869	169	0.06185	84	8.79134	586	4.02107	1.03244	0.54975	0.0023	5.00	1.00531	9.3957	4.83	9.7402	6.00
5	.25038	169	.06269	85	8.79720	585	3.99393	1.03290	0.55431	0.0053	5.00	1.01229	9.3986	4.83	9.7438	5.83
6	.25207	169	.06354	85	8.80305	577	3.96715	1.03337	0.55888	0.0083	5.00	1.01927	9.4015	4.83	9.7473	6.00
7	.25376	169	.06430	86	8.80882	576	3.94073	1.03384	0.56347	0.0113	4.83	1.02625	9.4044	4.83	9.7509	5.83
8	.25545	168	.06525	87	8.81458	575	3.91466	1.03432	0.56808	0.0142	4.83	1.03323	9.4073	4.83	9.7544	5.83
9	.25713	169	.06612	87	8.82033	568	3.88908	1.03479	0.57271	0.0171	4.83	1.04022	9.4102	4.67	9.7579	5.83
15.0	0.25882	168	0.06690	87	8.82601	560	3.86368	1.03527	0.57735	0.0200	4.83	1.04720	9.4130	4.67	9.7614	5.83
1	.26050	169	.06786	88	8.83161	560	3.83877	1.03576	0.58201	0.0229	4.83	1.05418	9.4158	4.67	9.7649	5.83
2	.26219	168	.06874	89	8.83721	559	3.81403	1.03625	0.58670	0.0258	4.67	1.06116	9.4186	4.67	9.7684	5.83
3	.26387	169	.06963	89	8.84280	551	3.78974	1.03674	0.59140	0.0286	4.83	1.06814	9.4214	4.67	9.7719	5.67
4	.26556	168	.07052	90	8.84831	551	3.76563	1.03724	0.59612	0.0315	4.67	1.07512	9.4242	4.50	9.7753	5.83
5	.26724	168	.07142	90	8.85382	544	3.74195	1.03774	0.60086	0.0343	4.67	1.08210	9.4269	4.50	9.7788	5.67
6	.26892	168	.07232	90	8.85926	537	3.71858	1.03825	0.60562	0.0371	4.50	1.08909	9.4296	4.50	9.7822	5.67
7	.27060	168	.07322	92	8.86463	542	3.69549	1.03875	0.61040	0.0398	4.67	1.09607	9.4323	4.50	9.7856	5.67
8	.27228	168	.07414	91	8.87005	530	3.67269	1.03926	0.61520	0.0426	4.50	1.10305	9.4350	4.33	9.7890	5.67
9	.27396	168	.07505	93	8.87535	535	3.65017	1.03978	0.62003	0.0453	4.67	1.11003	9.4377	4.50	9.7924	5.67
16.0	0.27564	167	0.07598	92	8.88070	523	3.62792	1.04030	0.62487	0.0481	4.50	1.11701	9.4403	4.33	9.7958	5.50
1	.27731	168	.07690	94	8.88593	527	3.60607	1.04082	0.62973	0.0508	4.50	1.12399	9.4430		9.7992	

ϑ°	sin ϑ	D(0.1°)	sin² ϑ	D(0.1°)	lg sin² ϑ	D(0.1°)	$\frac{1}{\sin ϑ}$	$\frac{1}{\cos ϑ}$	tg 2ϑ	lg 4ϑ_rad	D(1')	4ϑ_rad	lg sin ϑ	D(1')	lg tg 2ϑ	D(1')
16.2	0.27899	168	0.07784	93	8.89120	516	3.58436	1.04135	0.63462	0.0535	4.33	1.13097	9.4456	4.33	9.8025	5.67
3	.28067	167	.07877	95	8.89636	521	3.56290	1.04187	0.63953	0.0561	4.50	1.13795	9.4482	4.33	9.8059	5.50
4	.28234	168	.07972	94	8.90157	509	3.54183	1.04242	0.64446	0.0588	4.33	1.14494	9.4508	4.17	9.8092	5.50
5	.28402	167	.08066	96	8.90666	514	3.52088	1.04295	0.64941	0.0614	4.33	1.15192	9.4533	4.33	9.8125	5.50
6	.28569	167	.08162	96	8.91180	507	3.50030	1.04349	0.65438	0.0640	4.50	1.15890	9.4559	4.17	9.8158	5.50
7	.28736	167	.08258	96	8.91687	502	3.47996	1.04404	0.65938	0.0667	4.17	1.16588	9.4584	4.17	9.8191	5.50
8	.28903	167	.08354	96	8.92189	502	3.45985	1.04458	0.66440	0.0692	4.33·	1.17286	9.4609	4.17	9.8224	5.50
9	.29070	167	.08451	97	8.92691	495	3.43997	1.04514	0.66944	0.0718	4.33	1.17984	9.4634	4.17	9.8257	5.50
17.0	0.29237	167	0.08548	97	8.93186	496	3.42032	1.04570	0.67451	0.0744	4.17	1.18682	9.4659	4.17	9.8290	5.50
1	.29404	167	.08646	98	8.93682	489	3.40090	1.04625	0.67960	0.0769	4.33	1.19381	9.4684	4.17	9.8323	5.33
2	.29571	166	.08744	98	8.94171	489	3.38169	1.04681	0.68471	0.0795	4.17	1.20079	9.4709	4.00	9.8355	5.50
3	.29737	167	.08843	99	8.94660	488	3.36281	1.04738	0.68985	0.0820	4.17	1.20777	9.4733	4.00	9.8388	5.33
4	.29904	167	.08943	100	8.95148	478	3.34403	1.04795	0.69502	0.0845	4.17	1.21475	9.4757	4.00	9.8420	5.33
5	.30071	166	.09042	99	8.95626	483	3.32546	1.04853	0.70021	0.0870	4.17	1.22173	9.4781	4.00	9.8452	5.33
6	.30237	166	.09143	101	8.96109	477	3.30721	1.04911	0.70542	0.0895	4.00	1.22871	9.4805	4.00	9.8484	5.50
7	.30403	167	.09244	101	8.96586	472	3.28915	1.04969	0.71066	0.0919	4.17	1.23569	9.4829	4.00	9.8517	5.33
8	.30570	166	.09345	101	8.97058	471	3.27118	1.05028	0.71593	0.0944	4.00	1.24267	9.4853	3.83	9.8549	5.33
9	.30736	166	.09447	102	8.97529	467	3.25351	1.05087	0.72122	0.0968	4.00	1.24966	9.4876	4.00	9.8581	5.33

$\varphi°$	$\sin\varphi$	$D(0.1°)$	$\sin^2\varphi$	$D(0.1°)$	$\lg\sin^2\varphi$	$D(0.1°)$	$\dfrac{1}{\sin\varphi}$	$\dfrac{1}{\cos\varphi}$	$\operatorname{tg}2\varphi$	$\lg 4\varphi_{rad}$	$D(1')$	$4\varphi_{rad}$	$\lg\sin\varphi$	$D(1')$	$\lg\operatorname{tg}2\varphi$	$D(1')$
18.0	0.30902	166	0.09549	103	8.97996	466	3.23604	1.05146	0.72654	0.0992	4.00	1.25664	9.4900	3.83	9.8613	5.17
1	.31068	165	.09652	103	8.98462	461	3.21875	1.05206	0.73189	0.1016	4.00	1.26362	9.4923	3.83	9.8644	5.33
2	.31233	166	.09755	104	8.98923	460	3.20174	1.05266	0.73726	0.1040	4.00	1.27060	9.4946	3.83	9.8644	5.33
3	.31399	166	.09859	104	8.99383	456	3.18481	1.05326	0.74267	0.1064	4.00	1.27758	9.4969	3.83	9.8708	5.33
4	.31565	165	.09963	105	8.99839	455	3.16807	1.05387	0.74810	0.1088	3.83	1.28456	9.4992	3.83	9.8740	5.17
5	.31730	166	.10068	106	9.00294	455	3.15159	1.05450	0.75355	0.1111	4.00	1.29154	9.5015	3.67	9.8771	5.33
6	.31896	165	.10174	105	9.00749	446	3.13519	1.05511	0.75904	0.1135	3.83	1.29852	9.5037	3.83	9.8803	5.17
7	.32061	166	.10279	107	9.01195	450	3.11905	1.05573	0.76456	0.1158	3.83	1.30551	9.5060	3.67	9.8834	5.17
8	.32227	165	.10386	106	9.01645	441	3.10299	1.05636	0.77010	0.1181	3.83	1.31249	9.5082	3.67	9.8865	5.33
9	.32392	165	.10492	107	9.02086	440	3.08718	1.05698	0.77568	0.1204	3.83	1.31947	9.5104	3.67	9.8897	5.17
19.0	0.32557	165	0.10599	108	9.02526	441	3.07154	1.05762	0.78129	0.1227	3.83	1.32645	9.5126	3.67	9.8928	5.17
1	.32722	165	.10707	108	9.02967	436	3.05605	1.05826	0.78692	0.1250	3.67	1.33343	9.5148	3.67	9.8959	5.17
2	.32887	164	.10815	109	9.03403	435	3.04072	1.05890	0.79259	0.1272	3.83	1.34041	9.5170	3.67	9.8990	5.33
3	.33051	165	.10924	109	9.03838	431	3.02563	1.05955	0.79829	0.1295	3.67	1.34739	9.5192	3.50	9.9022	5.17
4	.33216	165	.11033	110	9.04269	431	3.01060	1.06020	0.80402	0.1317	3.83	1.35438	9.5213	3.67	9.9053	5.17
5	.33381	164	.11143	110	9.04700	427	2.99572	1.06085	0.80978	0.1340	3.67	1.36136	9.5235	3.50	9.9084	5.17
6	.33545	165	.11253	110	9.05127	422	2.98107	1.06150	0.81558	0.1362	3.67	1.36834	9.5256	3.67	9.9115	5.17
7	.33710	164	.11363	111	9.05549	422	2.96648	1.06217	0.82141	0.1384	3.67	1.37532	9.5278	3.50	9.9146	5.00

$\theta°$	$\sin\theta$	$D(0.1°)$	$\sin^2\theta$	$D(0.1°)$	$\lg\sin^2\theta$	$D(0.1°)$	$\dfrac{1}{\sin\theta}$	$\dfrac{1}{\cos\theta}$	$\operatorname{tg}2\theta$	$\lg 4\theta_{rad}$	$D(1')$	$4\theta_{rad}$	$\lg\sin\theta$	$D(1')$	$\lg\operatorname{tg}2\theta$	$D(1')$
19.8	0.33874	164	0.11474	164	9.05971	112	2.95212	1.06283	0.82727	0.1406	3.67	1.38230	9.5299	3.50	9.9176	5.17
9	.34038	164	.11586	112	9.06393	422	2.93789	1.06350	0.83317	0.1428	3.67	1.38928	9.5320	3.50	9.9207	5.17
20.0	0.34202	164	0.11698	112	9.06811	418	2.92381	1.06418	0.83910	0.1450	3.50	1.39626	9.5341	3.33	9.9238	5.17
1	.34366	164	.11810	113	9.07225	414	2.90985	1.06486	0.84507	0.1471	3.67	1.40324	9.5361	3.50	9.9300	5.00
2	.34530	164	.11923	113	9.07639	414	2.89603	1.06554	0.85107	0.1493	3.50	1.41023	9.5382	3.33	9.9300	5.00
3	.34694	163	.12036	114	9.08048	409	2.88234	1.06622	0.85710	0.1514	3.67	1.41721	9.5402	3.50	9.9330	5.17
4	.34857	164	.12150	114	9.08458	410	2.86886	1.06692	0.86318	0.1536	3.50	1.42419	9.5423	3.53	9.9361	5.17
5	.35021	163	.12264	115	9.08863	405	2.85543	1.06761	0.86929	0.1557	3.50	1.43117	9.5443	3.33	9.9392	5.00
6	.35184	163	.12379	115	9.09269	406	2.84220	1.06831	0.87543	0.1578	3.50	1.43815	9.5463	3.50	9.9422	5.17
7	.35347	164	.12494	116	9.09670	401	2.82909	1.06902	0.88162	0.1599	3.50	1.44513	9.5484	3.33	9.9453	5.00
8	.35511	163	.12610	116	9.10072	402	2.81603	1.06971	0.88784	0.1620	3.50	1.45211	9.5504	3.17	9.9483	5.17
9	.35674	163	.12726	117	9.10469	397	2.80316	1.07043	0.89410	0.1641	3.50	1.45910	9.5523	3.33	9.9514	5.00
21.0	0.35837	163	0.12843	117	9.10867	398	2.79041	1.07115	0.90040	0.1662	3.33	1.46608	9.5543	3.33	9.9544	5.17
1	.36000	162	.12960	117	9.11261	394	2.77778	1.07187	0.90674	0.1682	3.50	1.47306	9.5563	3.33	9.9575	5.00
2	.36162	163	.13077	118	9.11651	390	2.76533	1.07259	0.91313	0.1703	3.33	1.48004	9.5583	3.17	9.9636	5.17
3	.36325	163	.13195	119	9.12041	390	2.75292	1.07332	0.91955	0.1723	3.50	1.48702	9.5602	3.17	9.9636	5.00
4	.36488	162	.13314	118	9.12431	390	2.74063	1.07404	0.92601	0.1744	3.33	1.49400	9.5621	3.33	9.9666	5.17
5	.36650	162	.13432	120	9.12814	383	2.72851	1.07478	0.93252	0.1764	3.33	1.50098	9.5641	3.17	9.9697	5.00

θ°	sin θ	D(0.1°)	sin² θ	D(0.1°)	lg sin² θ	D(0.1°)	1/sin θ	1/cos θ	tg 2θ	lg 4θ_rad	D(1')	4θ_rad	lg sin θ	D(1')	lg tg 2θ	D(1')
21.6	0.36812	163	0.13552	119	9.13200	380	2.71651	1.07552	0.93906	0.1784	3.33	1.50796	9.5660	3.17	9.9727	5.00
7	.36975	162	.13671	120	9.13580	380	2.70453	1.07628	0.94565	0.1804	3.33	1.51495	9.5679	3.17	9.9757	5.17
8	.37137	162	.13791	121	9.13960	379	2.69273	1.07702	0.95229	0.1824	3.33	1.52193	9.5698	3.17	9.9788	5.00
9	.37299	162	.13912	121	9.14339	376	2.68104	1.07777	0.95897	0.1844	3.33	1.52891	9.5717	3.17	9.9818	5.00
22.0	0.37461	161	0.14033	121	9.14715	373	2.66944	1.07854	0.96569	0.1864	3.17	1.53589	9.5736	3.00	9.9848	5.17
1	.37622	162	.14154	122	9.15088	373	2.65802	1.07930	0.97246	0.1883	3.33	1.54287	9.5754	3.17	9.9879	5.00
2	.37784	162	.14276	123	9.15461	372	2.64662	1.08007	0.97927	0.1903	3.17	1.54985	9.5773	3.17	9.9909	5.00
3	.37946	161	.14399	122	9.15833	367	2.63532	1.08084	0.98613	0.1922	3.33	1.55683	9.5792	3.00	9.9939	5.17
4	.38107	161	.14521	124	9.16200	369	2.62419	1.08161	0.99304	0.1942	3.17	1.56382	9.5810	3.00	9.9970	5.00
5	.38268	162	.14645	123	9.16569	363	2.61315	1.08239	1.00000	0.1961	3.17	1.57080	9.5828	3.17	0.0000	5.00
6	.38430	161	.14768	124	9.16932	363	2.60213	1.08318	1.00701	0.1980	3.33	1.57778	9.5847	3.00	0.0030	5.17
7	.38591	161	.14892	125	9.17295	363	2.59128	1.08396	1.01406	0.2000	3.17	1.58476	9.5865	3.00	0.0061	5.00
8	.38752	160	.15017	125	9.17658	360	2.58051	1.08476	1.02117	0.2019	3.17	1.59174	9.5883	3.00	0.0091	5.00
9	.38912	161	.15142	125	9.18018	357	2.56990	1.08555	1.02832	0.2038	3.17	1.59872	9.5901	3.00	0.0121	5.17
23.0	0.39073	161	0.15267	126	9.18375	357	2.55931	1.08637	1.03553	0.2057	3.00	1.60570	9.5919	3.00	0.0152	5.00
1	.39234	160	.15393	126	9.18732	354	2.54881	1.08717	1.04279	0.2075	3.17	1.61268	9.5937	2.83	0.0182	5.00
2	.39394	161	.15519	127	9.19086	354	2.53846	1.08797	1.05010	0.2094	3.17	1.61967	9.5954	3.00	0.0212	5.17
3	.39555	160	.15646	127	9.19440	351	2.52813	1.08879	1.05747	0.2113	3.17	1.62665	9.5972	3.00	0.0243	5.00

$\theta°$	$\sin\theta$	$D(0.1°)$	$\sin^2\theta$	$D(0.1°)$	$\lg\sin^2\theta$	$D(0.1°)$	$\dfrac{1}{\sin\theta}$	$\dfrac{1}{\cos\theta}$	$\operatorname{tg}2\theta$	$\lg 4\theta_{rad}$	$D(1')$	$4\theta_{rad}$	$\lg\sin\theta$	$D(1')$	$\lg\operatorname{tg}2\theta$	$D(1')$
23.4	0.39715	160	0.15773	127	9.19791	349	2.51794	1.08962	1.06489	0.2132	3.00	1.63363	9.5990	2.83	0.0273	5.00
5	.39875	160	.15900	128	9.20140	348	2.50784	1.09044	1.07237	0.2150	3.00	1.64061	9.6007	2.83	0.0303	5.17
6	.40035	160	.16028	128	9.20488	345	2.49781	1.09127	1.07990	0.2168	3.17	1.64759	9.6024	3.00	0.0334	5.00
7	.40195	160	.16156	129	9.20833	346	2.48787	1.09211	1.08749	0.2187	3.00	1.65457	5.6042	2.83	0.0364	5.17
8	.40355	159	.16285	129	9.21179	342	2.47801	1.09294	1.09514	0.2205	3.00	1.66155	9.6059	2.83	0.0395	5.00
9	.40514	160	.16414	129	9.21521	340	2.46828	1.09379	1.10285	0.2223	3.00	1.66853	9.6076	2.83	0.0425	5.17
24.0	0.40674	159	0.16543	130	9.21861	340	2.45857	1.09463	1.11061	0.2241	3.17	1.67552	9.6093	2.83	0.0456	5.00
1	.40833	159	.16673	131	9.22201	340	2.44900	1.09549	1.11844	0.2260	3.00	1.68250	9.6110	2.83	0.0486	5.17
2	.40992	159	.16804	130	9.22541	335	2.43950	1.09635	1.12633	0.2278	2.83	1.68948	9.6127	2.83	0.0517	5.00
3	.41151	159	.16934	132	9.22876	337	2.43007	1.09721	1.13428	0.2295	3.00	1.69646	9.6144	2.83	0.0547	5.17
4	.41310	159	.17066	131	9.23213	332	2.42072	1.09808	1.14929	0.2313	3.00	1.70344	9.6161	2.67	0.0578	5.00
5	.41469	159	.17197	132	9.23545	332	2.41144	1.09895	1.15037	0.2331	3.00	1.71042	9.6177	2.83	0.0608	5.17
6	.41628	159	.17329	132	9.23877	330	2.40223	1.09982	1.15851	0.2349	2.83	1.71740	9.6194	2.67	0.0639	5.00
7	.41787	158	.17461	133	9.24207	329	2.39309	1.10070	1.16672	0.2366	3.00	1.72439	9.6210	2.83	0.0670	5.17
8	.41945	159	.17594	133	9.24536	328	2.38407	1.10159	1.17500	0.2384	2.83	1.73137	9.6227	2.67	0.0700	5.00
9	.42104	158	.17727	134	9.24864	327	2.37507	1.10249	1.18334	0.2401	2.83	1.73835	9.6243	2.83	0.0731	5.17
25.0	0.42262	158	0.17861	134	9.25191	324	2.36619	1.10338	1.19175	0.2419	2.83	1.74533	9.6259	2.83	0.0762	5.17
1	.42420	158	.17995	134	9.25515	322	2.35738	1.10428	1.20024	0.2436	2.83	1.75231	9.6276	2.67	0.0793	5.17

D(1')	lg tg 2θ	D(1')	lg sin θ	4θ_rad	D(1')	lg 4θ_rad	lg 2θ	1/cos θ	1/sin θ	D(0.1°)	lg sin² θ	D(0.1°)	sin² θ	D(0.1°)	sin θ	θ°
5.00	0.0824	2.67	9.6292	1.75929	3.00	0.2453	1.20879	1.10518	2.34863	320	9.25837	134	0.18129	158	0.42578	25.2
5.17	0.0854	2.67	9.6308	1.76627	2.83	0.2471	1.21742	1.10610	2.33995	322	9.26157	136	.18263	158	.42736	3
5.17	0.0885	2.67	9.6324	1.77325	2.83	0.2488	1.22612	1.10700	2.33133	318	9.26479	135	.18399	157	.42894	4
5.17	0.0916	2.67	9.6340	1.78024	2.83	0.2505	1.23490	1.10792	2.32283	317	9.26797	136	.18534	158	.43051	5
5.17	0.0947	2.67	9.6356	1.78722	2.83	0.2522	1.24375	1.10886	2.31433	316	9.27114	136	.18670	158	.43209	6
5.17	0.0978	2.50	9.6371	1.79420	2.83	0.2539	1.25268	1.10978	2.30595	315	9.27430	136	.18806	157	.43366	7
5.33	0.1010	2.67	9.6387	1.80118	2.83	0.2556	1.26169	1.11072	2.29764	313	9.27745	137	.18943	157	.43523	8
5.17	0.1041	2.67	9.6403	1.80816	2.67	0.2572	1.27077	1.11165	2.28938	311	9.28058	137	.19080	157	.43680	9
5.17	0.1072	2.50	9.6418	1.81514	2.83	0.2589	1.27994	1.11261	2.28118	310	9.28369	137	0.19217	157	0.43837	26.0
5.17	0.1103	2.67	9.6434	1.82212	2.83	0.2606	1.28919	1.11355	2.27304	309	9.28679	138	.19355	157	.43994	1
5.33	0.1135	2.50	9.6449	1.82911	2.67	0.2622	1.29853	1.11450	2.26495	306	9.28988	138	.19493	157	.44151	2
5.17	0.1166	2.67	9.6465	1.83609	2.83	0.2639	1.30795	1.11546	2.25698	307	9.29294	138	.19631	156	.44307	3
5.33	0.1197	2.50	9.6480	1.84307	2.67	0.2655	1.31745	1.11643	2.24901	304	9.29601	139	.19770	157	.44464	4
5.17	0.1229	2.50	9.6495	1.85005	2.83	0.2672	1.32704	1.11741	2.24115	304	9.29905	139	.19909	156	.44620	5
5.33	0.1260	2.50	9.6510	1.85703	2.67	0.2688	1.33673	1.11838	2.23334	302	9.30209	140	.20049	156	.44776	6
5.17	0.1292	2.67	9.6526	1.86401	2.67	0.2704	1.34650	1.11936	2.22559	301	9.30511	140	.20189	156	.44932	7
5.33	0.1324	2.50	9.6541	1.87099	2.83	0.2721	1.35637	1.12034	2.21789	300	9.30812	140	.20329	156	.45088	8
5.17	0.1356	2.50	9.6556	1.87797	2.67	0.2737	1.36633	1.12133	2.21029	298	9.31112	141	.20470	155	.45243	9
		2.33			2.67							141		156		

φ°	sin φ	D(0.1°)	sin² φ	D(0.1°)	lg sin² φ	D(0.1°)	1/sin φ	1/cos φ	tg 2φ	lg 4φ_rad	D(1')	4φ_rad	lg sin φ	D(1')	lg tg 2φ	D(1')
27.0	0.45399	155	0.20611	141	9.31410	296	2.20269	1.12232	1.37638	0.2753	2.67	1.88496	9.6570	2.50	0.1387	5.33
1	.45554	156	.20752	142	9.31706	296	2.19520	1.12333	1.38653	0.2769	2.67	1.89194	9.6585	2.50	0.1419	5.33
2	.45710	155	.20894	142	9.32002	294	2.18771	1.12433	1.39679	0.2785	2.67	1.89892	9.6600	2.50	0.1451	5.33
3	.45865	155	.21036	142	9.32296	292	2.18031	1.12534	1.40714	0.2801	2.67	1.90590	9.6615	2.33	0.1483	5.50
4	.46020	155	.21178	143	9.32588	293	2.17297	1.12635	1.41759	0.2817	2.67	1.91288	9.6629	2.50	0.1516	5.33
5	.46175	155	.21321	143	9.32881	290	2.16567	1.12738	1.42815	0.2833	2.67	1.91986	9.6644	2.50	0.1548	5.33
6	46330	154	.21464	144	9.33171	290	2.15843	1.12841	1.43881	0.2848	2.50	1.92684	9.6659	2.33	0.1580	5.33
7	.46484	155	.21608	144	9.33461	289	2.15128	1.12945	1.44958	0.2864	2.67	1.93382	9.6673	2.33	0.1612	5.50
8	.46639	154	.21752	144	9.33750	286	2.14413	1.13048	1.46046	0.2880	2.67	1.94081	9.6687	2.50	0.1645	5.33
9	.46793	154	.21896	144	9.34036	285	2.13707	1.13152	1.47146	0.2895	2.50	1.94779	9.6702	2.33	0.1677	5.50
28.0	0.46947	154	0.22040	145	9.34321	285	2.13006	1.13257	1.48256	0.2911	2.67	1.95477	9.6716	2.33	0.1710	5.50
1	.47101	154	.22185	145	9.34606	283	2.12310	1.13362	1.49378	0.2926	2.50	1.96175	9.6730	2.33	0.1743	5.50
2	.47255	154	.22330	146	9.34889	283	2.11618	1.13469	1.50512	0.2942	2.67	1.96873	9.6744	2.50	0.1776	5.50
3	.47409	153	.22476	146	9.35172	281	2.10930	1.13574	1.51658	0.2957	2.50	1.97571	9.6759	2.33	0.1809	5.50
4	.47562	154	.22622	146	9.35453	279	2.10252	1.13682	1.52816	0.2973	2.67	1.98269	9.6773	2.33	0.1842	5.50
5	.47716	153	.22768	146	9.35732	278	2.09573	1.13789	1.53987	0.2988	2.50	1.98968	9.6787	2.33	0.1875	5.50
6	.47869	153	.22914	147	9.36010	278	2.08903	1.13898	1.55170	0.3003	2.50	1.09666	9.6801	2.17	0.1908	5.50
7	.48022	153	.23061	148	9.36288	278	2.08238	1.14006	1.56366	0.3018	2.50	2.00364	9.6814	2.33	0.1941	5.67

φ°	sin φ	D(0.1°)	sin² φ	D(0.1°)	lg sin² φ	D(0.1°)	1/sin φ	1/cos φ	tg 2φ	lg 4φ_rad	D(1')	4φ_rad	lg sin φ	D(1')	lg tg 2φ	D(1)
28.8	0.48175	153	0.23209	146	9.36566	274	2.07577	1.14115	1.57575	0.3033	2.50	2.01062	9.6828	2.33	0.1975	5.50
9	.48328	153	.23356	148	9.36840	274	2.06919	1.14226	1.58797	0.3048	2.50	2.01760	9.6842	2.33	0.2008	5.67
29.0	0.48481	153	0.23504	148	9.37114	273	2.06266	1.14335	1.60033	0.3063	2.50	2.02458	9.6856	2.17	0.2042	5.67
1	.48634	152	.23652	149	9.37387	273	2.05617	1.14447	1.61283	0.3078	2.50	2.03156	9.6869	2.33	0.2076	5.67
2	.48786	152	.23801	149	9.37660	271	2.04977	1.14558	1.62548	0.3093	2.50	2.03854	9.6883	2.17	0.2110	5.67
3	.48938	152	.23950	149	9.37931	269	2.04340	1.14670	1.63826	0.3108	2.50	2.04553	9.6896	2.33	0.2144	5.67
4	.49090	152	.24099	149	9.38200	268	2.03707	1.14783	1.65120	0.3123	2.50	2.05251	9.6910	2.17	0.2178	5.67
5	.49242	152	.24248	150	9.38468	267	2.03079	1.14895	1.66428	0.3138	2.50	2.05949	9.6923	2.33	0.2212	5.67
6	.49394	152	.24398	150	9.38735	267	2.02454	1.15010	1.67752	0.3152	2.33	2.06647	9.6937	2.17	0.2247	5.83
7	.49546	151	.24548	150	9.39002	264	2.01833	1.15124	1.69091	0.3167	2.50	2.07345	9.6950	2.17	0.2281	5.67
8	.49697	152	.24698	151	9.39266	265	2.01219	1.15238	1.70446	0.3182	2.50	2.08043	9.6963	2.33	0.2316	5.83
9	.49849	151	.24849	151	9.39531	263	2.00606	1.15354	1.71817	0.3196	2.33	2.08741	9.6977	2.17	0.2351	5.83
30.0	0.50000	151	0.25000	151	9.39794	262	2.00000	1.15469	1.73205	0.3211	2.50	2.09440	9.6990	2.17	0.2386	5.83
1	.50151	151	.25151	152	9.40056	261	1.99398	1.15587	1.74610	0.3225	2.33	2.10138	9.7003	2.17	0.2421	5.83
2	.50302	151	.25303	152	9.40317	260	1.98799	1.15705	1.76032	0.3239	2.33	2.10836	9.7016	2.17	0.2456	5.83
3	.50453	150	.25455	152	9.40577	259	1.98204	1.15821	1.77471	0.3254	2.50	2.11534	9.7029	2.17	0.2491	5.83
4	.50603	151	.25607	153	9.40836	259	1.97617	1.15941	1.78929	0.3268	2.33	2.12232	9.7042	2.17	0.2527	6.00
5	.50754	150	.25760	152	9.41095	255	1.97029	1.16059	1.80405	0.3282	2.50	2.12930	9.7055	2.17	0.2562	6.00

θ°	sin θ	D (0.1°)	sin² θ	D (0.1°)	lg sin² θ	D (0.1°)	1/sin θ	1/cos θ	tg 2θ	lg 4θ rad	D (1')	4θ rad	lg sin θ	D (1')	lg tg 2θ	D (1')
30.6	0.50904	150	0.25912	153	9.41350	256	1.96448	1.16179	1.81900	0.3297	2.33	2.13628	9.7068	2.00	0.2598	6.00
7	.51054	150	.26065	154	9.41606	256	1.95871	1.16299	1.83413	0.3311	2.33	2.14326	9.7080	2.17	0.2634	6.00
8	.51204	150	.26219	153	9.41862	252	1.95297	1.16420	1.84946	0.3325	2.33	2.15025	9.7093	2.17	0.2670	6.17
9	.51354	150	.26372	154	9.42114	253	1.94727	1.16542	1.86500	0.3339	2.33	2.15723	9.7106	2.00	0.2707	6.00
31.0	0.51504	149	0.26527	155	9.42367	253	1.94160	1.16663	1.88073	0.3353	2.33	2.16421	9.7118	2.17	0.2743	6.17
1	.51653	150	.26681	154	9.42620	250	1.93600	1.16786	1.89667	0.3367	2.33	2.17119	9.7131	2.17	0.2780	6.17
2	.51803	149	.26835	155	9.42870	250	1.93039	1.16910	1.91282	0.3381	2.33	2.17817	9.7144	2.00	0.2817	6.17
3	.51952	149	.26990	155	9.43120	249	1.92475	1.17033	1.92920	0.3395	2.33	2.18515	9.7156	2.00	0.2854	6.17
4	.52101	149	.27145	155	9.43369	247	1.91935	1.17158	1.94579	0.3409	2.17	2.19213	9.7168	2.17	0.2891	6.17
5	.52250	149	.27300	156	9.43616	248	1.91388	1.17283	1.96261	0.3422	2.33	2.19911	9.7181	2.00	0.2928	6.33
6	.52399	148	.27456	156	9.43864	246	1.90843	1.17408	1.97967	0.3436	2.33	2.20610	9.7193	2.00	0.2966	6.33
7	.52547	149	.27612	156	9.44110	244	1.90306	1.17535	1.99695	0.3450	2.33	2.21308	9.7205	2.17	0.3004	6.33
8	.52696	148	.27768	157	9.44354	245	1.89768	1.17662	2.01449	0.3464	2.17	2.22006	9.7218	2.00	0.3042	6.33
9	.52844	148	.27925	156	9.44599	242	1.89236	1.17790	2.03227	0.3477	2.33	2.22704	9.7230	2.00	0.3080	6.33
32.0	0.52992	148	0.28081	157	9.44841	242	1.88708	1.17918	2.05030	0.3491	2.17	2.23402	9.7242	2.00	0.3118	6.50
1	.53140	148	.28238	158	9.45083	243	1.88182	1.18047	2.06860	0.3504	2.33	2.24100	9.7254	2.00	0.3157	6.50
2	.53288	147	.28396	157	9.45326	239	1.87660	1.18177	2.08716	0.3518	2.17	2.24798	9.7266	2.00	0.3196	6.50
3	.53435	148	.28553	158	9.45565	240	1.87143	1.18307	2.10600	0.3531	2.33	2.25497	9.7278	2.00	0.3235	6.50

θ°	sin θ	D (0.1°)	sin² θ	D (0.1°)	lg sin² θ	D (0.1°)	$\frac{1}{\sin θ}$	$\frac{1}{\cos θ}$	tg 2θ	lg 4θ rad	D (1')	4θ rad	lg sin θ	D (1')	lg tg 2θ	D (1')
32.4	0.53583	147	0.28711	158	9.45805	238	1.86626	1.18437	2.12511	0.3545	2.17	2.26195	9.7290	2.00	0.3274	6.50
5	.53730	147	.28869	158	9.46043	237	1.86116	1.18569	2.14451	0.3558	2.33	2.26893	9.7302	2.00	0.3313	6.67
6	.53877	147	.29027	159	9.46280	237	1.85608	1.18701	2.16420	0.3572	2.17	2.27591	9.7314	2.00	0.3353	6.67
7	.54024	147	.29186	159	9.46517	236	1.85103	1.18834	2.18419	0.3585	2.17	2.28289	9.7326	2.00	0.3393	6.67
8	.54171	146	.29345	159	9.46753	235	1.84601	1.18967	2.20449	0.3598	2.17	2.28987	9.7338	1.83	0.3433	6.67
9	.54317	147	.29504	159	9.46988	234	1.84104	1.19101	2.22510	0.3611	2.33	2.29685	9.7349	2.00	0.3473	6.83
33.0	0.54464	146	0.29663	160	9.47222	233	1.83608	1.19236	2.24604	0.3625	2.17	2.30383	9.7361	2.00	0.3514	6.83
1	.54610	146	.29823	160	9.47455	233	1.83117	1.19372	2.26730	0.3638	2.17	2.31082	9.7373	1.83	0.3555	6.83
2	.54756	146	.29983	160	9.47688	231	1.82628	1.19509	2.28891	0.3651	2.17	2.31780	9.7384	2.00	0.3596	7.00
3	.54902	146	.30143	160	9.47919	230	1.82143	1.19644	2.31086	0.3664	2.17	2.32478	9.7396	1.83	0.3638	6.83
4	.55048	146	.30303	160	9.48149	228	1.81660	1.19782	2.33317	0.3677	2.17	2.33176	9.7407	2.00	0.3679	7.00
5	.55194	145	.30463	161	9.48377	229	1.81179	1.19920	2.35585	0.3690	2.17	2.33874	9.7419	1.83	0.3721	7.17
6	.55339	145	.30624	161	9.48606	228	1.80704	1.20050	2.37891	0.3703	2.17	2.34572	9.7430	2.00	0.3764	7.00
7	.55484	146	.30785	161	9.48834	226	1.80232	1.20200	2.40235	0.3716	2.17	2.35270	9.7442	1.83	0.3806	7.17
8	.55630	145	.30946	162	9.49060	227	1.79759	1.20340	2.42618	0.3729	2.00	2.35969	9.7453	1.83	0.3849	7.17
9	.55775	144	.31108	162	9.49287	226	1.79292	1.20480	2.45043	0.3741	2.17	2.36667	9.7464	2.00	0.3892	7.33
34.0	0.55919	145	0.31270	162	9.49513	224	1.78830	1.20621	2.47509	0.3754	2.17	2.37365	9.7476	1.83	0.3936	7.33
1	.56064	144	.31432	162	9.49737	223	1.78368	1.20764	2.50018	0.3767	2.17	2.38063	9.7487	1.83	0.3980	7.33

ϑ°	sin ϑ	D(0.1°)	sin² ϑ	D(0.1°)	lg sin² ϑ	D(0.1°)	$\frac{1}{\sin\vartheta}$	$\frac{1}{\cos\vartheta}$	tg 2ϑ	lg 4ϑ rad	D(1')	4ϑ rad	'g sin ϑ	D(1')	lg tg 2ϑ	D(1')
34.2	0.56208	145	0.31594	162	9.49960	223	1.77911	1.20907	2.52571	0.3780		2.38761	9.7498	1.83	0.4042	7.33
3	.56353	144	.31756	163	9.50183	222	1.77453	1.21051	2.55170	0.3792	2.00	2.39459	9.7509	1.83	0.4068	7.33
4	.56497	144	.31919	163	9.50405	221	1.77001	1.21196	2.57815	0.3805	2.17	2.40157	9.7520	1.83	0.4113	7.50
5	.56641	143	.32082	163	9.50626	220	1.76551	1.21340	2.60509	0.3818	2.17	2.40855	9.7531	1.83	0.4158	7.67
6	.56784	144	.32245	163	9.50846	219	1.76106	1.21486	2.63252	0.3830	2.00	2.41554	9.7542	1.83	0.4204	7.67
7	.56928	143	.32408	163	9.51065	218	1.75660	1.21634	2.66046	0.3843	2.17	2.42252	9.7553	1.83	0.4250	7.67
8	.57071	144	.32571	164	9.51283	218	1.75220	1.21780	2.68892	0.3855	2.00	2.42950	9.7564	1.83	0.4296	7.67
9	.57215	143	.32735	164	9.51501	217	1.74779	1.21929	2.71792	0.3868	2.17	2.43648	9.7575	1.83	0.4342	7.83
35.0	0.57358	143	0.32899	164	9.51718	216	1.74344	1.22078	2.74748	0.3880	2.00	2.44346	9.7586	1.83	0.4389	8.00
1	.57501	142	.33063	164	9.51934	215	1.73910	1.22227	2.77761	0.3892	2.00	2.45044	9.7597	1.83	0.4437	7 83
2	.57643	143	.33227	165	9.52149	215	1.73482	1.22378	2.80833	0.3905	2.17	2.45742	9.7607	1.67	0.4484	8.17
3	.57786	142	.33392	165	9.52364	214	1.73052	1.22528	2.83965	0.3917	2.00	2.46440	9.7618	1.83	0.4533	8.00
4	.57928	142	.33557	165	9.52578	213	1.72628	1.22680	2.87161	0.3929	2.00	2.47139	9.7629	1.83	0.4581	8.17
5	.58070	142	.33722	165	9.52791	212	1.72206	1.22832	2.90421	0.3942	2.17	2.47837	9.7640	1.83	0.4630	8.33
6	.58212	142	.33887	165	9.53003	211	1.71786	1.22986	2.93748	0.3954	2.00	2.48535	9.7650	1.67	0.4680	8.33
7	.58354	142	.34052	166	9.53214	211	1.71368	1.23141	2.97144	0.3966	2.00	2.49233	9.7661	1.83	0.4730	8.33
8	.58496	141	.34218	165	9.53425	209	1.70952	1.23295	3.00611	0.3978	2.00	2.49931	9.7671	1.67	0.4780	8.50
9	.58637	142	.34383	166	9.53634	210	1.70541	1.23451	3.04152	0.3990	2.00	2.50629	9.7682	1.67	0.4831	8.50

θ°	sin θ	D (0.1°)	sin² θ	D (0.1°)	lg sin² θ	D (0.1°)	1/sin θ	1/cos θ	lg 2θ	lg 4θ rad	D (1')	4θ rad	lg sin θ	D (1')	lg tg 2θ	D (1')
36.0	0.58779	141	0.34549	166	9.53844	208	1.70129	1.23606	3.07768	0,4002	2.00	2.51327	9.7692	1.83	0.4882	8.67
1	.58920	141	.34715	167	9.54052	208	1.69722	1.23764	3.11464	0.4014	2.00	2.52026	9.7703	1.67	0.4934	8.67
2	.59061	140	.34882	166	9.54260	206	1.69316	1.23922	3.15240	0.4026	2.00	2.52724	9.7713	1.67	0.4986	8.83
3	.59201	141	.35048	167	9.54466	207	1.68916	1.24080	3.19100	0.4038	2.00	2.53422	9.7713	1.83	0.5039	9.00
4	.59342	140	.35215	166	9.54673	204	1.68515	1.24241	3.23048	0.4050	2.00	2.54120	9.7734	1.67	0.5093	9.00
5	.59482	140	.35381	167	9.54877	205	1.68118	1.24400	3.27085	0.4062	2.00	2.54818	9.7744	1.67	0.5147	9.00
6	.59622	141	.35548	168	9.55082	204	1.67723	1.24561	3.31216	0.4074	2.00	2.55516	9.7754	1.67	0.5201	9.17
7	.59763	139	.35716	167	9.55286	203	1.67328	1.24722	3.35443	0.4086	2.00	2.56214	9.7764	1.67	0.5256	9.33
8	.59902	140	.35883	167	9.55489	202	1.66939	1.24886	3.39771	0.4098	2.00	2.56912	9.7774	1.83	0.5312	9.33
9	.60042	139	.36050	168	9.55691	201	1.66550	1.25050	3.44202	0.4110	1.83	2.57611	9.7785	1.67	0.5368	9.50
37.0	0.60181	140	0.36218	168	9.55892	201	1.66165	1.25213	3.48741	0.4121	2.00	2.58309	9.7795	1.67	0.5425	9.67
1	.60321	139	.36386	168	9.56093	200	1.65780	1.25379	3.53393	0.4133	2.00	2.59007	9.7805	1.67	0.5483	9.67
2	.60460	139	.36554	168	9.56293	200	1.65399	1.25545	3.58560	0.4145	1.83	2.59705	9.7815	1.67	0.5541	9.83
3	.60599	139	.36722	169	9.56493	199	1.65019	1.25712	3.63048	0.4156	2.00	2.60403	9.7825	1.67	0.5600	9.83
4	.60738	138	.36891	168	9.56692	197	1.64642	1.25880	3.68061	0.4168	2.00	2.61101	9.7835	1.50	0.5659	10.00
5	.60876	139	.37059	169	9.56889	198	1.64268	1.26048	3.73205	0.4180	1.83	2.61799	9.7844	1.67	0.5719	10.17
6	.61015	138	.37228	169	9.57087	197	1.63894	1.26216	3.78485	0.4191	2.00	2.62498	9.7854	1.67	0,5780	10.33
7	.61153	138	.37397	169	9.57284	196	1.63524	1.26387	3.83906	0.4203	1.83	2.63196	9.7864	1.67	0.5842	10.50

θ°	sin θ	D (0.1°)	sin² θ	D (0.1°)	lg sin² θ	D (0.1°)	$\frac{1}{\sin\theta}$	$\frac{1}{\cos\theta}$	tg 2θ	lg 4θ rad	D (1')	4θ rad	lg sin θ	D (1')	lg lg 2θ	D (1')
37.8	0.61291	138	0.37566	169	9.57480	194	**1.63156**	1.26558	3.89474	0.4214	2.00	2.63894	9.7874	1.67	0.5905	10.50
9	.61429	137	.37735	169	9.57674	195	1.62790	1.26730	3.95196	0.4226	1.83	2.64592	9.7884	1.50	0.5968	10.67
38.0	0.61566	138	0.37904	169	9.57869	193	1.62427	1.26902	4.01078	0.4237	2.00	2.65290	9.7893	1.67	0.6032	10.83
1	.61704	137	.38073	170	9.58062	193	1.62064	1.27074	4.07127	0.4249	1.83	2.65988	9.7903	1.67	0.6097	11.00
2	.61841	137	.38243	170	9.58255	193	1.61705	1.27249	4.13350	0.4260	1.83	2.66686	9.7913	1.50	0.6163	11.17
3	.61978	137	.38413	169	9.58448	190	1.61348	1.27424	4.19756	0.4271	1.83	2.67384	9.7922	1.67	0.6230	11.33
4	.62115	136	.38582	170	9.58638	191	**1.60992**	1.27601	4.26352	0.4283	2.00	2.68083	9.7932	1.50	0.6298	11.33
5	.62251	137	.38752	170	9.58829	192	1.60640	1.27778	4.33148	0.4294	1.83	2.68781	9.7941	1.67	0.6366	11.67
6	.62388	136	.38923	171	9.59021	189	1.60287	1.27956	4.40152	0.4305	1.83	2.69479	9.7951	1.50	0.6436	11.83
7	.62524	136	.39093	170	9.59210	188	1.59939	1.28131	4.47374	0.4316	2.00	2.70177	9.7960	1.67	0.6507	11.83
8	.62660	136	.39263	171	9.59398	189	1.59591	1.28314	4.54826	0.4328	1.83	2.70875	9.7970	1.50	0.6578	12.17
9	.62796	136	.39434	170	9.59587	187	1.59246	1.28495	4.62518	0.4339	1.83	2.71573	9.7979	1.67	0.6651	12.33
39.0	0.62932	136	0.39604	171	9.59774	187	1.58902	1.28675	4.70463	0.4350	1.83	2.72271	9.7989	1.50	0.6725	12.50
1	.63068	135	.39775	171	9.59961	186	1.58559	1.28858	4.78673	0.4361	1.83	2.72969	9.7998	1.50	0.6800	12.83
2	.63203	135	.39946	171	9.60147	186	1.58220	1.29042	4.87162	0.4372	1.83	2.73668	9.8007	1.67	0.6877	12.83
3	.63338	135	.40117	171	9.60333	185	1.57883	1.29226	4.95945	0.4383	1.83	2.74366	9.8017	1.50	0.6954	13.17
4	.63473	135	.40288	172	9.60518	185	1.57547	1.29411	5.05037	0.4394	1.83	2.75064	9.8026	1.50	0.7033	13.33
5	.63608	134	.40460	171	9.60703	183	1.57213	1.29597	5.14455	0.4405	1.83	2.75762	9.8035	1.50	0.7113	13.67

$\theta°$	$\sin\theta$	$D(0.1°)$	$\sin^2\theta$	$D(0.1°)$	$\lg\sin^2\theta$	$D(0.1°)$	$\dfrac{1}{\sin\theta}$	$\dfrac{1}{\cos\theta}$	$\lg 2\theta$	$\lg 4\theta_{rad}$	$D(1')$	$4\theta_{rad}$	$\lg\sin\theta$	$D(1')$	$\lg\lg 2\theta$	$D(1')$
39.6	0.63742	135	0.40631	171	9.60886	182	1.56882	1.29784	5.24218	0.4416	1.83	2.76460	9.8044	1.50	0.7195	13.83
7	.63877	134	.40802	172	9.61068	183	1.56551	1.29971	5.34345	0.4427	1.83	2.77158	9.8053	1.67	0.7278	14.17
8	.64011	134	.40974	172	9.61251	182	1.56223	1.30161	5.44857	0.4438	1.83	2.77856	9.8063	1.50	0.7363	14.33
9	.64145	134	.41146	172	9.61433	181	1.55897	1.30349	5.55777	0.4449	1.83	2.78555	9.8072	1.50	0.7449	14.67
40.0	0.64279	133	0.41318	172	9.61614	180	1.55572	1.30541	5.67128	0.4460	1.83	2.79253	9.8081	1.50	0.7537	14.8
1	.64412	134	.41490	172	9.61794	180	1.55251	1.30733	5.78938	0.4471	1.83	2.79951	9.8090	1.50	0.7626	15.3
2	.64546	133	.41662	172	9.61974	179	1.54928	1.30924	5.91236	0.4482	1.67	2.80649	9.8099	1.50	0.7718	15.5
3	.64679	133	.41834	172	9.62153	178	1.54610	1.31118	6.04051	0.4492	1.83	2.81347	9.8108	1.50	0.7811	15.8
4	.64812	133	.42006	172	9.62331	178	1.54292	1.31313	6.17419	0.4503	1.83	2.82045	9.8117	1.33	0.7906	16.2
5	.64915	132	.42178	173	9.62509	177	1.53976	1.31508	6.31375	0.4514	1.83	2.82743	9.8125	1.50	0.8003	16.5
6	.65077	133	.42351	172	9.62686	176	1.53664	1.31705	6.45961	0.4525	1.67	2.83441	9.8134	1.50	0.8102	16.8
7	.65210	132	.42523	173	9.62862	177	1.53351	1.31903	6.61220	0.4535	1.83	2.84140	9.8143	1.50	0.8203	17.3
8	.65342	132	.42696	173	9.63039	175	1.53041	1.32100	6.77199	0.4546	1.83	2.84838	9.8152	1.50	0.8307	17.7
9	.65474	132	.42869	173	9.63214	174	1.52732	1.32301	6.93952	0.4557	1.67	2.85536	9.8161	1.33	0.8413	18.2
41.0	0.65606	132	0.43041	173	9.63388	174	1.52425	1.32501	7.11537	0.4567	1.83	2.86234	9.8169	1.50	0.8522	18.5
1	.65738	131	.43214	173	9.63562	174	1.52119	1.32703	7.30018	0.4578	1.67	2.86932	9.8178	1.50	0.8633	19.2
2	.65869	131	.43387	173	9.63736	173	1.51816	1.32906	7.49465	0.4588	1.83	2.87630	9.8187	1.33	0.8748	19.5
3	.66000	131	.43560	173	9.63909	172	1.51515	1.33110	7.69957	0.4599	1.83	2.88328	9.8195	1.50	0.8865	20.0

θ°	D (1')	lg tg 2θ	D (1')	lg sin θ	4θ rad	D (1')	lg 4θ rad	tg 2θ	1/cos θ	1/sin θ	D (0.1°)	lg sin² θ	D (0.1°)	sin² θ	D (0.1°)	sin θ
41.4	20.7	0.8985	1.50	9.8204	2.89027	1.83	0.4609	7,91582	1.33314	1.51215	172	9.64081	174	0.43733	131	0.66131
5	21.2	0.9109	1.33	9.8213	2.89725	1.67	0.4620	8.14435	1.33518	1.50916	171	9.64253	173	.43907	131	.66262
6	21.8	0.9236	1.50	9.8221	2.90423	1.83	0.4630	8.38625	1.33726	1.50618	170	9.64424	173	.44080	130	.66393
7	22.3	0.9367	1.33	9.8230	2.91121	1.67	0.4641	8.64275	1.33933	1.50324	170	9.64594	173	.44253	130	.66523
8	23.2	0.9501	1.50	9.8238	2.91819	1.83	0.4651	8.91520	1.34142	1.50031	169	9.64764	174	.44426	130	.66653
9	24.0	0.9640	1.33	9.8247	2.92517	1.67	0.4662	9.20516	1.34353	1.49739	170	9.64933	174	.44600	130	.66783
42.0	24.7	0.9784	1.50	9.8255	2.93215	1.67	0.4672	9.51436	1.34564	1.49448	167	9.65103	173	0.44774	130	0.66913
1	25.5	0.9932	1.33	9.8264	2.93913	1.67	0.4682	9.8448	1.34775	1.49158	168	9.65270	174	.44917	129	.67043
2	26	1.0085	1.33	9.8272	2.94612	1.83	0.4692	10.199	1.34989	1.48872	167	9.65438	174	.45121	129	.67172
3	27	1.0244	1.50	9.8280	2.95310	1.67	0.4703	10.579	1.35203	1.48586	166	9.65605	173	.45295	129	.67301
4	29	1.0409	1.33	9.8289	2.96008	1.67	0.4713	10.988	1.35417	1.48302	165	9.65771	174	.45468	129	.67430
5	30	1.0580	1.33	9.8297	2.96706	1.67	0.4723	11.430	1.35634	1.48019	166	9.65936	174	.45642	129	.67559
6	31	1.0759	1.33	9.8305	2.97404	1.83	0.4733	11.909	1.35851	1.47737	164	9.66102	174	.45816	128	.67688
7	32	1.0944	1.50	9.8313	2.98102	1.67	0.4744	12.429	1.36071	1.47458	164	9.66266	174	.45990	128	.67816
8	34	1.1138	1.33	9.8322	2.98800	1.67	0.4754	12.996	1.36290	1.47180	164	9.66430	174	.46164	128	.67944
9	35	1.1341	1.33	9.8330	2.99498	1.67	0.4764	13.617	1.36511	1.46903	163	9.66594	174	.46338	128	.68072
43.0	37	1.1554	1.33	9.8338	3.00197	1.67	0.4774	14.301	1.36733	1.46628	162	9.66757	174	0.46512	127	0.68200
1		1.1777	1.33	9.8346	3.00895	1.67	0.4784	15.056	1.36956	1.46355	161	9.65919	184	.46686	128	.68327

φ°	sin φ	D(0.1°)	sin² φ	D(0.1°)	lg sin φ	D(0.1°)	1/sin φ	1/cos φ	tg 2φ	lg 4φ rad	D(1')	4φ rad	lg sin φ	D(1')	lg tg 2φ	D(1')
43.2	0.68455	127	0.46860	175	9.67080	162	1.46081	1.37180	15.895	0.4794	1.67	3.01593	9.8354	1.33	1.2012	—
3	.68582	127	.47035	174	9.67242	160	1.45811	1.37406	16.832	0.4804	1.67	3.02291	9.8362	1.33	1.2261	—
4	.68709	126	.47209	174	9.67402	160	1.45541	1.37633	17.886	0.4814	1.67	3.02989	9.8370	1.33	1.2525	—
5	.68835	127	.47383	175	9.67562	160	1.45275	1.37861	19.081	0.4824	1.67	3.03687	9.8378	1.33	1.2806	—
6	.68962	126	.47558	174	9.67722	159	1.45007	1.38089	20.446	0.4834	1.67	3.04385	9.8386	1.33	1.3106	—
7	.69088	126	.47732	174	9.67881	158	1.44743	1.38318	22.022	0.4844	1.67	3.05084	9.8394	1.33	1.3429	—
8	.69214	126	.47906	175	9.68039	158	1.44479	1.38550	23.859	0.4854	1.67	3.05782	9.8402	1.33	1.3777	—
9	.69340	126	.48081	174	9.68197	157	1.44217	1.38783	26.031	0.4864	1.67	3.06480	9.8410	1.33	1.4155	—
44.0	0.69466	125	0.48255	174	9.68354	157	1.43955	1.39016	28.636	0.4874	1.67	3.07178	9.8418	1.33	1.4569	—
1	.69591	126	.48429	175	9.68511	156	1.43697	1.39251	31.821	0.4884	1.67	3.07876	9.8426	1.33	1.5027	—
2	.69717	125	.48604	174	9.68667	155	1.43437	1.39488	35.801	0.4894	1.67	3.08574	9.8433	1.17	1.5539	—
3	.69842	124	.48778	175	9.68822	156	1.43180	1.39725	40.917	0.4903	1.50	3.09272	9.8441	1.33	1.6119	—
4	.69966	125	.48953	174	9.68978	154	1.42927	1.39964	47.740	0.4913	1.67	3.09970	9.8449	1.33	1.6789	—
5	.70091	124	.49127	175	9.69132	154	1.42672	1.40203	57.290	0.4923	1.67	3.10669	9.8457	1.33	1.7581	—
6	.70215	124	.49302	174	9.69286	153	1.42420	1.40444	71.617	0.4933	1.67	3.11367	9.8464	1.17	1.8550	—
7	.70339	124	.49476	175	9.69439	154	1.42169	1.40687	95.489	0.4942	1.50	3.12065	9.8464	1.33	1.9800	—
8	.70463	124	.49651	174	9.69593	152	1.41918	1.40930	143.24	0.4952	1.67	3.12763	9.8480	1.33	2.1561	—
9	.70587	124	.49825	175	9.69745	152	1.41669	1.41175	286.48	0 4962	1.67	3.13461	9.8487	1.17	2.4571	—

θ°	sin θ	D (C.1°)	sin² θ	D (0.1°)	lg sin² θ	D (0.1°)	1/sin θ	1/cos θ	tg 2θ	lg 4θ rad	D (1')	4θ rad	lg sin θ	D (1')
45.0	0.70711	123	0.50000	174	9.69897	151	1,41421	1.41421	±∞	0.4971	1.67	3.14159	9.8495	1.17
1	.70834	123	.50174	175	9.70048	151	1,41175	1.41669	−286.48	0.4981	1.67	3.14857	9.8502	1.33
2	.70957	123	.50349	175	9.70199	151	1,40930	1.41918	−143.24	0.4991	1.50	3.15556	9.8510	1.17
3	.71080	123	.50524	174	9.70350	149	1,40687	1.42169	−95.489	0.5000	1.67	3.16254	9.8517	1.33
4	.71203	122	.50698	175	9.70499	150	1.40444	1.42420	−71.615	0,5010	1.50	3.16952	9.8525	1.17
5	.71325	122	.50873	174	9.70649	148	1.40203	1.42672	−57.290	0.5019	1.67	3.17650	9.8532	1.33
6	.71447	122	.51047	175	9.70797	149	1.39964	1.42927	−47.740	0.5029	1.67	3.18348	9.8540	1.17
7	.71569	122	.51222	174	9.70946	147	1.39725	1.43180	−40.917	0,5039	1.50	3.19046	9.8547	1.33
8	.71691	122	.51396	175	9.71093	148	1.39488	1.43437	−35.801	0.5048	1.67	3.19744	9.8555	1.17
9	.71813	121	.51571	174	9.71241	146	1.39251	1.43697	−31.821	0.5058	1.50	3.20442	9.8562	1.17
46.0	0.71934	121	0.51745	174	9.71387	146	1.39016	1.43955	−28.636	0.5067	1.50	3.21141	9.8569	1.33
1	.72055	121	.51919	175	9.71533	146	1.38783	1.44217	−26.031	0.5076	1.67	3.21839	9.8577	1.17
2	.72176	121	.52094	174	9.71679	145	1.38550	1.44479	−23.959	0.5086	1.50	3.22537	9.8584	1.17
3	.72297	120	.52268	174	9.71824	144	1.38318	1.44743	−22.022	0.5095	1.67	3.23235	9.8591	1.17
4	.72417	120	.52442	174	9.71968	145	1.38089	1.45007	−20.446	0.5105	1.50	3.23933	9.8598	1.33
5	.72537	120	.52617	175	9.72113	143	1.37861	1.45275	−19.081	0.5114	1.50	3.24631	9.8606	1.17
6	.72657	120	.52791	174	9.72256	143	1.37633	1.45541	−17.886	0.5123	1.67	3.25329	9.8613	1.17
7	.72777	120	.52965	175	9.72399	143	1.37406	1.45811	−16.832	0.5133	1.50	3.26028	9.8620	1.17

θ°	sin θ	D (0.1°)	sin² θ	D (0.1°)	lg sin² θ	D (0.1°)	1/sin θ	1/cos θ	tg 2θ	lg 4θ_rad	D (1')	4θ_rad	lg sin θ	D (1')
46.8	0.72897	119	0.53140	174	9.72542	142	1.37180	1.46081	−15.895	0.5142	1.50	3.26726	9.8627	1.17
9	.73016	119	.53314	174	9.72684	142	1.36956	1.46355	−15.056	0.5151	1.50	3.27424	9.8634	1.17
47 0	0.73135	119	0.53488	174	9.72826	141	1.36733	1.46628	−14.301	0.5160	1.67	3.28122	9.8641	1.17
1	.73254	119	.53662	174	9.72967	140	1.36511	1.46903	−13.617	0.5170	1.50	3.28820	9.8648	1.17
2	.73373	118	.53836	174	9.73107	140	1.36290	1.47180	−12.996	0 5179	1.50	3.29518	9.8655	1.17
3	.73491	119	.54010	174	9.73247	140	1.36071	1.47458	−12.429	0.5188	1.50	3.30216	9.8662	1.17
4	.73610	118	.54184	174	9.73387	139	1.35851	1.47737	−11.909	0.5197	1.50	3.30914	9.8669	1.17
5	.73728	118	.54358	174	9.73526	139	1.35634	1.48019	−11.430	0.5206	1.50	3.31613	9.8676	1.17
6	.73846	117	.54532	173	9.73665	138	1.35417	1.48302	−10.988	0.5215	1.50	3.32311	9.8683	1.17
7	.73963	117	.54705	174	9.73803	138	1.35203	1.48586	−10.579	0.5225	1.67	3.33009	9.8690	1.17
8	.74080	118	.54879	174	9.73941	137	1.34989	1.48872	−10.199	0.5234	1.50	3.33707	9.8697	1.17
9	.74198	116	.55053	173	9.74078	136	1.34775	1.49158	−9.8448	0.5243	1.50	3.34405	9.8704	1.17
48.0	0.74314	117	0.55226	174	9.74214	137	1.34564	1.49448	−9.51436	0.5252	1.50	3.35103	9.8711	1.17
1	.74431	117	.55400	173	9.74351	135	1.34353	1.49739	−9.20516	0.5261	1.50	3.35801	9.8718	1.00
2	.74548	116	.55573	174	9.74486	136	1.34142	1.50031	−8.91520	0.5270	1.50	3.36499	9.8724	1.17
3	.74664	116	.55747	173	9.74622	135	1.33933	1.50324	−8.64275	0.5279	1.50	3.37198	9.8731	1.17
4	.74780	116	.55920	173	9.74757	134	1.33726	1.50618	−8.38625	0.5288	1.50	3.37896	9.8738	1.17
5	.74896	115	.56093	174	9.74891	134	1.33518	1.50916	−8.14435	0.5297	1.50	3.38594	9.8745	1.00

$\theta°$	$\sin\theta$	$\Delta(0.1°)$	$\sin^2\theta$	$\Delta(0.1°)$	lg $\sin^2\theta$	$\Delta(0.1°)$	$\dfrac{1}{\sin\theta}$	$\dfrac{1}{\cos\theta}$	tg 2θ	lg $4\theta_{rad}$	$\Delta(1')$	$4\theta_{rad}$	lg $\sin\theta$	$\Delta(1')$
48.6	0.75011	116	0.56267	173	9.75025	134	1.33314	1.51215	−7.91582	0.5306	1.50	3.39292	9.8751	1.17
7	.75126	115	.56440	173	9.75159	133	1.33110	1.51515	−7.69957	0.5315	1.50	3.39990	9.8758	1.17
8	.75241	115	.56613	173	9.75292	132	1.32906	1.51816	−7.49465	0.5324	1.33	3.40688	9.8765	1.00
9	.75356	115	.56786	173	9.75424	132	1.32703	1.52119	−7.30018	0.5332	1.50	3.41386	9.8771	1.17
49.0	0.75471	114	0.56959	172	9.75556	131	1.32501	1.52425	−7.11537	0.5341	1.50	3.42085	9.8778	1.00
1	.75585	115	.57131	173	9.75687	131	1.32301	1.52732	−6.93952	0.5350	1.50	3.42783	9.8784	1.17
2	.75700	113	.57304	173	9.75818	131	1.32100	1.53041	−6.77199	0.5359	1.50	3.43481	9.8791	1.00
3	.75813	114	.57477	172	9.75949	130	1.31903	1.53351	−6.61220	0.5368	1.50	3.44179	9.8797	1.17
4	.75927	114	.57649	173	9.76079	130	1.31705	1.53664	−6.45961	0.5377	1.33	3.44877	9.8804	1.00
5	.76041	113	.57822	172	9.76209	129	1.31508	1.53976	−6.31375	0.5385	1.50	3.45575	9.8810	1.17
6	.76154	113	.57994	172	9.76338	129	1.31313	1.54292	−6.17419	0.5394	1.50	3.46273	9.8817	1.00
7	.76267	113	.58166	172	9.76467	128	1.31118	1.54610	−6.04051	0.5403	1.50	3.46971	9.8823	1.17
8	.76380	112	.58338	172	9.76595	128	1.30924	1.54928	−5.91236	0.5412	1.33	3.47670	9.8830	1.00
9	.76492	112	.58510	172	9.76723	127	1.30733	1.55251	−5.78938	0.5420	1.50	3.48368	9.8836	1.17
50.0	0.76604	113	0.58682	172	9.76850	128	1.30541	1.55572	−5.67128	0.5429	1.50	3.49066	9.8843	1.00
1	.76717	111	.58854	172	9.76978	126	1.30349	1.55897	−5.55777	0.5438	1.33	3.49764	9.8849	1.00
2	.76828	112	.59026	172	9.77104	126	1.30161	1.56223	−5.44857	0.5446	1.50	3.50462	9.8855	1.17
3	.76940	111	.59198	171	9.77231	125	1.29971	1.56551	−5.34345	0.5455	1.50	3.51160	9.8862	1.00

θ°	sin θ	D (0.1°)	sin² θ	D (0.1°)	lg sin² θ	D (0.1°)	$\frac{1}{\sin θ}$	$\frac{1}{\cos θ}$	tg 2θ	lg 4θ_rad	D (1')	4θ_rad	lg sin θ	D (1')
50.4	0.77051	111	0.59369	171	9.77356	125	1.29784	1.56882	−5.24218	0.5464	1.33	3.51858	9.8868	1.00
5	.77162	111	.59540	172	9.77481	125	1.29597	1.57213	−5.14455	0.5472	1.50	3.52557	9.8874	1.00
6	.77273	111	.59712	171	9.77606	124	1.29411	1.57547	−5.05037	0.5481	1.33	3.53255	9.8880	1.17
7	.77384	110	.59883	171	9.77730	124	1.29226	1.57883	−4.95945	0.5489	1.50	3.53953	9.8887	1.00
8	.77494	111	.60054	171	9.77854	124	1.29042	1.58220	−4.87162	0.5498	1.50	3.54651	9.8893	1.00
9	.77605	110	.60225	171	9.77978	123	1.28858	1.58559	−4.78673	0.5507	1.33	3.55349	9.8899	1.00
51.0	0.77715	109	0.60396	170	9.78101	122	1.28675	1.58902	−4.70463	0.5515	1.50	3.56047	9.8905	1.00
1	.77824	110	.60566	171	9.78223	122	1.28495	1.59246	−4.62518	0.5524	1.33	3.56745	9.8911	1.00
2	.77934	109	.60737	170	9.78345	122	1.28314	1.59591	−4.54826	0.5532	1.50	3.57443	9.8917	1.00
3	.78043	109	.60907	170	9.78467	121	1.28134	1.59939	−4.47374	0.5541	1.33	3.58142	9.8923	1.00
4	.78152	109	.61077	171	9.78588	121	1.27956	1.60287	−4.40152	0.5549	1.33	3.58840	9.8929	1.00
5	.78261	108	.61248	170	9.78709	121	1.27778	1.60640	−4.33148	0.5557	1.50	3.59538	9.8935	1.00
6	.78369	109	.61418	169	9.78830	119	1.27601	1.60992	−4.26352	0.5566	1.33	3.60236	9.8941	1.00
7	.78478	108	.61587	170	9.78949	120	1.27424	1.61348	−4.19756	0.5574	1.50	3.60934	9.8947	1.00
8	.78586	108	.61757	170	9.79069	119	1.27249	1.61705	−4.13350	0.5583	1.33	3.61632	9.8953	1.00
9	.78694	107	.61927	169	9.79188	118	1.27074	1.62064	−4.07127	0.5591	1.33	3.62330	9.8959	1.00
52.0	0.78801	107	0.62096	169	9.79306	118	1.26902	1.62427	−4.01078	0.5599	1.50	3.63028	9.8965	1.00
1	.78908	107	.62265	169	9.79424	118	1.26730	1.62790	−3.95196	0.5608	1.33	3.63727	9.8971	1.00

$\theta°$	$\sin\theta$	$D\,(0.1°)$	$\sin^2\theta$	$D\,(0.1°)$	$\lg\sin^2\theta$	$D\,(0.1°)$	$\dfrac{1}{\sin\theta}$	$\dfrac{1}{\cos\theta}$	$\operatorname{tg}2\theta$	$\lg 4\theta_{rad}$	$D\,(1')$	$4\theta_{rad}$	$\lg\sin\theta$	$D\,(1')$
52.2	0.79015	107	0.62434	169	9.79542	118	1.26558	1.63156	−3.89474	0.5616		3.64425	9.8977	1.00
3	.79122	107	.62603	169	9.79660	117	1.26387	1.63524	−3.83906	0.5624	1.33	3.65123	9.8983	1.00
4	.79229	106	.62772	169	9.79777	116	1.26216	1.63894	−3.78485	0.5633	1.50	3.65821	9.8989	1.00
5	.79335	106	.62941	168	9.79893	116	1.26048	1.64268	−3.73205	0.5641	1.33	3.66519	9.8995	0.83
6	.79441	106	.63109	169	9.80009	116	1.25880	1.64642	−3.68061	0.5649	1.33	3.67217	9.9000	1.00
7	.79547	106	.63278	168	9.80125	115	1.25712	1.65019	−3.63048	0.5657	1.33	3.67915	9.9006	1.00
8	.79653	105	.63446	168	9.80240	115	1.25545	1.65399	−3.58160	0.5666	1.50	3.68614	9.9012	0.83
9	.79758	106	.63614	168	9.80355	115	1.25379	1.65780	−3.53393	0.5674	1.33	3.69312	9.9018	1.00
53.0	0.79864	104	0.63782	168	9.80470	114	1.25213	1.66165	−3.48741	0 5682	1.33	3.70010	9.9023	1.00
1	.79968	105	.63950	167	9.80584	113	1.25050	1.66550	−3.44202	0.5690	1.33	3.70708	9.9029	1.00
2	.80073	105	.64117	167	9.80697	113	1.24886	1.66939	−3.39771	0.5698	1.50	3.71406	9.9035	0.83
3	.80178	104	.64284	168	9.80810	114	1.24722	1.67328	−3.35443	0.5707	1.33	3.72104	9.9041	1.00
4	.80282	104	.64452	167	9.80924	112	1.24561	1.67723	−3.31216	0.5715	1.33	3.72802	9.9046	1.00
5	.80386	103	.64619	166	9.81036	111	1.24400	1.68118	−3.27085	0.5723	1.33	3.73500	9.9052	0.83
6	.80489	104	.64785	167	9.81147	112	1.24241	1.68515	−3.23048	0.5731	1.33	3.74199	9.9057	1.00
7	.80593	103	.64952	166	9.81259	111	1.24080	1.68916	−3.19100	0.5739	1.33	3.74897	9.9063	0.83
8	.80696	103	.65118	167	9.81370	111	1.23922	1.69316	−3.15240	0.5747	1 33	3.75595	9.9069	1.00
9	.80799	103	.65285	166	9.81481	111	1.23764	1.69722	−3.11464	0 5755	1.33	3.76293	9.9074	1.00

$\varphi°$	$\sin\varphi$	$D\,(0.1°)$	$\sin^2\varphi$	$D\,(0.1°)$	$\lg\sin^2\varphi$	$D\,(0.1°)$	$\dfrac{1}{\sin\varphi}$	$\dfrac{1}{\cos\varphi}$	$\operatorname{tg}2\varphi$	$\lg 4\varphi_{rad}$	$D\,(1')$	$4\,\varphi_{rad}$	$\lg\sin\varphi$	$D\,(1')$
54.0	0.80902	102	0.65451	166	9.81592	110	1.23606	1.70129	−3.07768	0.5763	1.33	3.76991	9.9080	0.83
1	.81004	102	.65617	165	9.81702	109	1.23451	1.70541	−3.04152	0.5771	1.33	3.77689	9.9085	1.00
2	.81106	102	.65782	166	9.81811	109	1.23295	1.70952	−3.00611	0.5779	1.33	3.78387	9.9091	0.83
3	.81208	102	.65948	165	9.81920	109	1.23141	1.71368	−2.97144	0.5787	1.33	3.79086	9.9096	0.83
4	.81310	102	.66113	165	9.82029	108	1.22986	1.71786	−2.93748	0.5795	1.33	3.79784	9.9101	1.00
5	.81412	102	.66278	165	9.82137	108	1.22832	1.72206	−2.90421	0.5803	1.33	3.80482	9.9107	0.83
6	.81513	101	.66443	165	9.82245	108	1.22680	1.72628	−2.87161	0.5811	1.33	3.81180	9.9112	1.00
7	.81614	101	.66608	164	9.82353	106	1.22528	1.73052	−2.83965	0.5819	1.33	3.81878	9.9118	0.83
8	.81714	100	.66772	165	9.82459	108	1.22378	1.73482	−2.80833	0.5827	1.33	3.82576	9.9123	1.00
9	.81815	101	.66937	164	9.82567	106	1.22227	1.73910	−2.77761	0.5835	1.33	3.83274	9.9128	0.83
55.0	0.81915	100	0.67101	164	9.82673	106	1.22078	1.74344	−2.74748	0.5843	1.33	3.83972	9.9134	0.83
1	.82015	100	.67265	164	9.82779	106	1.21929	1.74779	−2.71792	0.5851	1.33	3.84671	9.9139	1.00
2	.82115	99	.67429	163	9.82885	105	1.21780	1.75220	−2.68892	0.5859	1.33	3.85369	9.9144	0.83
3	.82214	100	.67592	163	9.82990	104	1.21634	1.75660	−2.66046	0.5867	1.33	3.86067	9.9149	0.83
4	.82314	99	.67755	163	9.83094	104	1.21486	1.76106	−2.63252	0.5874	1.17	3.86765	9.9155	1.00
5	.82413	98	.67918	163	9.83198	105	1.21340	1.76551	−2.60509	0.5882	1.33	3.87463	9.9160	0.83
6	.82511	99	.68081	163	9.83303	103	1.21196	1.77001	−2.57815	0.5890	1.33	3.88161	9.9165	0.83
7	.82610	98	.68244	162	9.83406	103	1.21051	1.77453	−2.55170	0.5898	1.33	3.88859	9.9170	0.83

D (1')	lg sin θ	4θ_rad	D (1')	lg 4θ_rad	tg 2θ	1/cos θ	1/sin θ	D (0.1°)	lg sin² θ	D (0.1°)	sin² θ	D (0.1°)	sin θ	θ°
1.00	9.9175	3.89557	1.17	0.5906	−2.52571	1.77911	1.20907	103	9.83509	162	0.68406	98	0.82708	55.8
0.83	9.9181	3.90256	1.33	0.5913	−2.50018	1.78368	1.20764	103	9.83612	162	.68568	98	.82806	9
0.83	9.9186	3.90954	1.33	0.5921	−2.47509	1.78830	1.20621	102	9.83715	162	0.68730	97	0.82904	56.0
0.83	9.9191	3.91652	1.33	0.5929	−2.45043	1.79292	1.20480	102	9.83817	162	.68892	97	.83001	1
0.83	9.9196	3.92350	1.33	0.5937	−2.42618	1.79759	1.20340	101	9.83919	162	.69054	97	.83098	2
0.83	9.9201	3.93048	1.17	0.5944	−2.40235	1.80232	1.20200	101	9.84020	161	.69215	97	.83195	3
0.83	9.9206	3.93746	1.33	0.5952	−2.37891	1.80704	1.20060	101	9.84121	161	.69376	97	.83292	4
0.83	9.9211	3.94444	1.33	0.5960	−2.35585	1.81179	1.19920	99	9.84222	161	.69537	97	.83389	5
0.83	9.9216	3.95143	1.33	0.5968	−2.33317	1.81660	1.19782	100	9.84321	160	.69697	96	.83485	6
0.83	9.9221	3.95841	1.17	0.5975	−2.31086	1.82143	1.19644	99	9.84421	160	.69857	96	.83581	7
0.83	9.9226	3.96539	1.33	0.5983	−2.28891	1.82628	1.19509	99	9.84520	160	.70017	95	.83676	8
0.83	9.9231	3.97237	1.17	0.5990	−2.26730	1.83117	1.19372	99	9.84619	160	.70177	96	.83772	9
0.83	9.9236	3.97935	1.33	0.5998	−2.24604	1.83608	1.19236	98	9.84718	160	0.70337	95	0.83867	57.0
0.83	9.9241	3.98633	1.33	0.6006	−2.22510	1.84104	1.19101	98	9.84816	159	.70496	95	.83962	1
0.83	9.9246	3.99331	1.17	0.6013	−2.20449	1.84601	1.18967	98	9.84914	159	.70655	95	.84057	2
0.83	9.9251	4.00029	1.33	0.6021	−2.18419	1.85103	1.18834	97	9.85012	159	.70814	94	.84151	3
0.67	9.9255	4.00728	1.17	0.6028	−2.16420	1.85608	1.18701	97	9.85109	159	.70973	94	.84245	4
0.83	9.9260	4.01426	1.33	0.6036	−2.14451	1.86116	1.18569	96	9.85206	158	.71131	94	.84339	5

ϑ°	sin ϑ	D (0.1°)	sin² ϑ	D (0.1°)	lg sin² ϑ	D (0.1°)	$\frac{1}{\sin ϑ}$	$\frac{1}{\cos ϑ}$	tg 2ϑ	lg 4ϑ rad	D (1')	4 ϑ rad	lg sin ϑ	D (1')
57.6	0.84433	93	0.71289	158	9.85302	96	1.18437	1.86626	−2.12511	0.6044	1.17	4.02124	9.9265	0.83
7	.84526	93	.71447	157	9.85398	96	1.18307	1.87143	−2.10600	0.6051	1.33	4.02822	9.9270	0.83
8	.84619	93	.71604	158	9.85494	95	1.18177	1.87660	−2.08716	0.6059	1.17	4.03520	9.9275	0.67
9	.84712	93	.71762	157	9.85589	95	1.18047	1.88182	−2.06860	0.6066	1.33	4.04218	9.9279	0.83
58.0	0.84805	92	0.71919	156	9.85684	94	1.17918	1.88708	−2.05030	0.6074	1.17	4.04916	9.9284	0.83
1	.84897	92	.72075	157	9.85778	95	1.17790	1.89236	−2.03227	0.6081	1.33	4.05615	9.9289	0.83
2	.84989	92	.72232	156	9.85873	94	1.17662	1.89768	−2.01449	0.6089	1.17	4.06313	9.9294	0.67
3	.85081	92	.72388	156	9.85967	93	1.17535	1.90306	−1.99695	0.6096	1.33	4.07011	9.9298	0.83
4	.85173	91	.72544	156	9.86060	93	1.17408	1.90843	−1.97967	0.6104	1.17	4.07709	9.9303	0.83
5	.85264	91	.72700	155	9.86153	93	1.17283	1.91388	−1.96261	0.6111	1.33	4.08407	9.9308	0.67
6	.85355	91	.72855	155	9.86246	92	1.17158	1.91935	−1.94579	0.6118	1.17	4.09105	9.9312	0.83
7	.85446	90	.73010	155	9.86338	92	1.17033	1.92485	−1.92920	0.6126	1.33	4.09803	9.9317	0.83
8	.85536	91	.73165	154	9.86430	92	1.16910	1.93039	−1.91282	0.6133	1.17	4.10501	9.9322	0.67
9	.85627	90	.73319	155	9.86522	92	1.16786	1.93600	−1.89667	0.6141	1.33	4.11200	9.9326	0.83
59.0	0.85717	89	0.73474	154	9.86613	91	1.16663	1.94160	−1.88073	0.6148	1.17	4.11898	9.9331	0.67
1	.85806	90	.73628	153	9.86704	91	1.16542	1.94727	−1.86500	0.6155	1.17	4.12596	9.9335	0.83
2	.85896	89	.73781	153	9.86794	90	1.16420	1.95297	−1.84946	0.6163	1.33	4.13294	9.9340	0.67
3	.85985	89	.73934	154	9.86884	91	1.16299	1.95871	−1.83413	0.6170	1.17	4.13992	9.9344	0.83

θ°	sin θ	D (0.1°)	sin² θ	D (0.1°)	lg sin² θ	D (0.1°)	1/sin θ	1/cos θ	tg 2θ	lg 4θ rad	D (1')	4 θ rad	lg sin θ	D (1')
59.4	0.86074	89	0.74088	152	9.86975	89	1.16179	1.96448	−1.81900	0.6177	1.33	4.14690	9.9349	0.67
5	.86163	88	.74240	153	9.87064	89	1.16059	1.97029	−1.80405	0.6185	1.17	4.15388	9.9353	0.83
6	.86251	89	.74393	152	9.87153	89	1.15941	1.97617	−1.78929	0.6192	1.17	4.16086	9.9358	0.67
7	.86340	87	.74545	152	9.87242	88	1.15821	1.98204	−1.77471	0.6199	1.17	4.16785	9.9362	0.83
8	.86427	88	.74697	152	9.87330	89	1.15705	1.98799	−1.76032	0.6206	1.17	4.17483	9.9367	0.67
9	.86515	88	.74849	151	9.87419	87	1.15587	1.99398	−1.74610	0.6214	1.17	4.18181	9.9371	0.67
60.0	0.86603	87	0.75000	151	9.87506	87	1.15469	2.00000	−1.73205	0.6221	1.17	4.18879	9.9375	0.83
1	.86690	87	.75151	151	9.87593	88	1.15354	2.00606	−1.71817	0.6228	1.17	4.19577	9.9380	0.67
2	.86777	86	.75302	150	9.87681	86	1.15238	2.01219	−1.70446	0.6235	1.33	4.20275	9.9384	0.67
3	.86863	86	.75452	150	9.87767	86	1.15124	2.01833	−1.69091	0.6243	1.17	4.20973	9.9388	0.83
4	.86949	87	.75602	150	9.87853	86	1.15010	2.02454	−1.67752	0.6250	1.17	4.21672	9.9393	0.67
5	.87036	85	.75752	149	9.87939	86	1.14895	2.03079	−1.66428	0.6257	1.17	4.22370	9.9397	0.67
6	.87121	86	.75901	149	9.88025	85	1.14783	2.03707	−1.65120	0.6264	1.17	4.23068	9.9401	0.83
7	.87207	85	.76050	149	9.88110	85	1.14670	2.04340	−1.63826	0.6271	1.17	4.23766	9.9406	0.67
8	.87292	85	.76199	149	9.88195	85	1.14558	2.04977	−1.62548	0.6278	1.33	4.24464	9.9410	0.67
9	.87377	85	.76348	148	9.88280	84	1.14447	2.05617	−1.61283	0.6286	1.17	4.25162	9.9414	0.67
61.0	0.87462	84	0.76496	148	9.88364	84	1.14335	2.06266	−1.60033	0.6293	1.17	4.25860	9.9418	0.67
1	.87546	85	.76644	147	9.88448	83	1.14226	2.06919	−1.58797	0.6300		4.26558	9.9422	0.83

φ°	sin φ	D(0.1°)	sin² φ	D(0.1°)	lg sin² φ	D(0.1°)	$\frac{1}{\sin φ}$	$\frac{1}{\cos φ}$	tg 2φ	lg 4φ_rad	D(1')	4 φ_rad	lg sin φ	D(1')
61.2	0.87631	84	0.76791	148	9.88531	84	1.14115	2.07577	−1.57575	0.6307	1.17	4.27257	9.9427	0.67
3	.87715	83	.76939	146	9.88615	82	1.14006	2.08238	−1.56366	0.6314	1.17	4.27955	9.9431	0.67
4	.87798	84	.77085	147	9.88697	83	1.13898	2.08903	−1.55170	0.6321	1.17	4.28653	9.9435	0.67
5	.87882	83	.77232	146	9.88780	82	1.13789	2.09573	−1.53987	0.6328	1.17	4.29351	9.9439	0.67
6	.87965	83	.77378	146	9.88862	82	1.13682	2.10252	−1.52816	0.6335	1.17	4.30049	9.9443	0.67
7	.88048	82	.77521	146	9.88914	82	1.13574	2.10930	−1.51658	0.6342	1.17	4.30747	9.9447	0.67
8	.88130	83	.77670	145	9.89025	81	1.13469	2.11618	−1.50512	0.6349	1.17	4.31445	9.9451	0.67
9	.88213	82	.77815	145	9.89106	81	1.13362	2.12310	−1.49378	0.6356	1.17	4.32144	9.9455	0.67
62.0	0.88295	82	0.77960	144	9.89187	81	1.13257	2.13006	−1.48256	0.6363	1.17	4.32842	9.9459	0.67
1	.88377	81	.78104	144	9.89267	80	1.13152	2.13707	−1.47146	0.6370	1.17	4.33540	9.9463	0.67
2	.88458	81	.78248	144	9.89347	80	1.13048	2.14413	−1.46046	0.6377	1.17	4.34238	9.9467	0.67
3	.88539	81	.78392	144	9.89427	80	1.12945	2.15128	−1.44958	0.6384	1.17	4.34936	9.9471	0.67
4	.88620	81	.78536	143	9.89507	80	1.12841	2.15843	−1.43881	0.6391	1.17	4.35634	9.9475	0.67
5	.88701	81	.78679	143	9.89586	79	1.12738	2.16567	−1.42815	0.6398	1.17	4.36332	9.9479	0.67
6	.88782	80	.78822	142	9.89665	79	1.12635	2.17297	−1.41759	0.6405	1.17	4.37030	9.9483	0.67
7	.88862	80	.78964	142	9.89743	78	1.12534	2.18031	−1.40714	0.6412	1.17	4.37729	9.9487	0.67
8	.88942	79	.79106	142	9.89821	78	1.12433	2.18771	−1.39679	0.6419	1.17	4.38427	9.9491	0.67
9	.89021	80	.79248	141	9.89899	77	1.12333	2.19520	−1.38653	0.6426	1.17	4.39125	9.9495	0.67

θ	sin θ	D (0.1°)	sin²θ	D (0.1°)	lg sin²θ	D (0.1°)	$\frac{1}{\sin θ}$	$\frac{1}{\cos θ}$	tg 2θ	lg 4θ_rad	D (1')	4 θ_rad	lg sin θ	D (1')
63.0	0.89101	79	0.79389	141	9.89976	77	1.12232	2.20269	−1.37638	0.6433	1.17	4.39823	9.9499	0.67
1	.89180	79	.79530	141	9.90053	77	1.12133	2.21029	−1.36633	0.6440	1.17	4.40521	9.9503	0.50
2	.89259	78	.79671	140	9.90130	76	1.12034	2.21789	−1.35637	0.6447	1.00	4.41219	9.9506	0.67
3	.89337	78	.79811	140	9.90206	76	1.11936	2.22559	−1.34650	0.6453	1.17	4.41917	9.9510	0.67
4	.89415	78	.79951	140	9.90282	76	1.11838	2.23334	−1.33673	0.6460	1.17	4.42615	9.9514	0.67
5	.89493	78	.80091	139	9.90358	76	1.11741	2.24115	−1.32704	0.6467	1.17	4.43314	9.9518	0.67
6	.89571	78	.80230	139	9.90434	76	1.11643	2.24901	−1.31745	0.6474	1.17	4.44012	9.9522	0.50
7	.89649	77	.80369	138	9.90509	75	1.11546	2.25698	−1.30795	0.6481	1.17	4.44710	9.9525	0.67
8	.89726	77	.80507	138	9.90583	74	1.11450	2.26495	−1.29853	0.6488	1.00	4.45408	9.9529	0.67
9	.89803	76	.80615	138	9.90658	75	1.11355	2.27304	−1.28919	0.6494	1.17	4.46106	9.9533	0.67
64.0	0.89879	77	0.80783	137	9.90732	74	1.11261	2.28118	−1.27994	0.6501	1.17	4.46804	9.9537	0.50
1	.89956	76	.80920	137	9.90806	74	1.11165	2.28938	−1.27077	0.6508	1.17	4.47502	9.9540	0.67
2	.90032	76	.81057	137	9.90879	73	1.11072	2.29764	−1.26169	0.6515	1.00	4.48201	9.9544	0.67
3	.90108	75	.81194	136	9.90952	73	1.10978	2.30595	−1.25268	0.6521	1.00	4.48899	9.9548	0.50
4	.90183	76	.81330	136	9.91025	73	1.10886	2.31433	−1.24375	0.6528	1.17	4.49597	9.9551	0.67
5	.90259	75	.81466	135	9.91098	73	1.10792	2.32283	−1.23490	0.6535	1.17	4.50295	9.9555	0.50
6	.90334	74	.81601	136	9.91170	72	1.10700	2.33133	−1.22612	0.6542	1.00	4.50993	9.9558	0.67
7	90408	75	-.81737	131	9.91242	71	1.10610	2.33995	−1.21742	0.6548	1.17	4.51691	9.9562	0.67

$\vartheta°$	$\sin\vartheta$	$D\,(0.1°)$	$\sin^2\vartheta$	$D\,(0.1°)$	$\lg\sin^2\vartheta$	$D\,(0\,10°)$	$\dfrac{1}{\sin\vartheta}$	$\dfrac{1}{\cos\vartheta}$	$\operatorname{tg}2\vartheta$	$\lg 4\vartheta_{rad}$	$D\,(1')$	$4\,\vartheta_{rad}$	$\lg\sin\vartheta$	$D\,(1')$
64.8	0.90483	74	0.81871	134	9.91313	71	1.10518	2.34863	−1.20879	0.6555	1.17	4.52389	9.9566	0.50
9	.90557	74	.82005	134	9.91384	71	1.10428	2.35738	−1.20024	0.6562	1.17	4.53087	9.9569	0.67
65.0	0.90631	73	0.82139	134	9.91455	71	1.10338	2.36619	−1.19175	0.6569	1.00	4.53786	9.9573	0.50
1	.90704	74	.82273	133	9.91526	70	1.10249	2.37507	−1.18334	0.6575	1.17	4.54484	9.9576	0.67
2	.90778	73	.82406	133	9.91596	70	1.10159	2.38407	−1.17500	0.6582	1.17	4.55182	9.9580	0.50
3	.90851	73	.82539	132	9.91666	69	1.10070	2.39309	−1.16672	0.6589	1.00	4.55880	9.9583	0.67
4	.90924	72	.82671	132	9.91735	70	1.09982	2.40223	−1.15851	0.6595	1.17	4.56578	9.9587	0.50
5	.90996	72	.82803	131	9.91805	68	1.09895	2.41144	−1.15037	0.6602	1.00	4.57276	9.9590	0.67
6	.91068	72	.82934	132	9.91873	69	1.09808	2.42072	−1.14229	0.6608	1.17	4.57974	9.9594	0.50
7	.91140	72	.83066	130	9.91942	68	1.09721	2.43007	−1.13428	0.6615	1.17	4.58673	9.9597	0.67
8	.91212	71	.83196	131	9.92010	69	1.09635	2.43950	−1.12633	0.6622	1.00	4.59371	9.9601	0.50
9	.91283	72	.83327	130	9.92079	67	1.09549	2.44900	−1.11844	0.6628	1.17	4.60069	9.9604	0.50
66.0	0.91355	70	0.83457	129	9.92146	67	1.09463	2.45857	−1.11061	0.6635	1.00	4.60767	9.9607	0.67
1	.91425	71	.83586	129	9.92213	67	1.09379	2.46828	−1.10285	0.6641	1.17	4.61465	9.9611	0.50
2	.91496	70	.83715	129	9.92280	67	1.09294	2.47801	−1.09514	0.6648	1.17	4.62163	9.9614	0.50
3	.91566	70	.83844	128	9 92347	66	1.09211	2.48787	−1.08749	0.6655	1.00	4.62861	9.9617	0.67
4	.91636	70	.83972	128	9.92413	67	1.09127	2.49781	−1.07990	0.6661	1.17	4.63559	9.9621	0.50
5	.91706	69	.84100	127	9.92480	65	1.09044	2.50784	−1.07237	0.6668	1.00	4.64258	9.9624	0.50

$\theta°$	$\sin\theta$	$D(0.1°)$	$\sin^2\theta$	$D(0.1°)$	$\lg\sin^2\theta$	$D(0.1°)$	$\dfrac{1}{\sin\theta}$	$\dfrac{1}{\cos\theta}$	$\operatorname{tg}2\theta$	$\lg 4\theta_{rad}$	$D(1')$	$4\theta_{rad}$	$\lg\sin\theta$	$D(1')$
66.6	0.91775	70	0.84227	127	9.92545	66	1.08962	2.51794	−1.06489	0.6674	1.17	4.64956	9.9627	0.67
7	.91845	69	.84354	127	9.92611	65	1.08879	2.52813	−1.05747	0.6681	1.00	4.65654	9.9631	0.50
8	.91914	68	.84481	126	9.92676	65	1.08797	2.53846	−1.05010	0.6687	1.17	4.66352	9.9634	0.50
9	.91982	68	.84607	126	9.92741	64	1.08717	2.54881	−1.04279	0.6694	1.00	4.67050	9.9637	0.50
67.0	0.92050	69	0.84733	125	9.92805	64	1.08637	2.55931	−1.03553	0.6700	1.17	4.67748	9.9640	0.50
1	.92119	67	.84858	125	9.92869	64	1.08555	2.56990	−1.02832	0.6707	1.00	4.68446	9.9643	0.50
2	.92186	68	.84983	125	9.92933	64	1.08476	2.58051	−1.02117	0.6713	1.17	4.69145	9.9647	0.67
3	.92254	67	.85108	124	9.92997	63	1.08396	2.59128	−1.01406	0.6720	1.00	4.69843	9.9650	0.50
4	.92321	67	.85232	123	9.93060	63	1.08318	2.60213	−1.00701	0.6726	1.00	4.70541	9.9653	0.50
5	.92388	67	.85355	124	9.93123	63	1.08239	2.61315	−1.00000	0.6732	1.17	4.71239	9.9656	0.50
6	.92455	66	.85479	122	9.93186	62	1.08161	2.62419	−0.99304	0.6739	1.00	4.71937	9.9659	0.50
7	.92521	66	.85601	123	9.93248	62	1.08084	2.63532	−0.98613	0.6745	1.17	4.72635	9.9662	0.50
8	.92587	66	.85724	122	9.93310	62	1.08007	2.64662	−0.97927	0.6752	1.00	4.73333	9.9666	0.67
9	.92653	65	.85846	121	9.93372	61	1.07930	2.65802	−0.97246	0.6758	1.00	4.74031	9.9669	0.50
68.0	0.92718	66	0.85967	121	9.93433	61	1.07854	2.66944	−0.96569	0.6764	1.17	4.74730	9.9672	0.50
1	.92784	65	.86088	121	9.93494	61	1.07777	2.68104	−0.95897	0.6771	1.00	4.75428	9.9675	0.50
2	.92849	64	.86209	120	9.93555	61	1.07702	2.69273	−0.95229	0.6777	1.17	4.76126	9.9678	0.50
3	.92913	65	.86329	119	9.93616	60	1.07628	2.70453	−0.94565	0.6784	1.00	4.76824	9.9681	0.50

ϑ°	sin ϑ	D(0.1°)	sin² ϑ	D(0.1°)	lg sin² ϑ	D(0.1°)	$\frac{1}{\sin ϑ}$	$\frac{1}{\cos ϑ}$	tg 2ϑ	lg 4ϑ_rad	D(1')	4ϑ_rad	lg sin ϑ	D(1')
68.4	0.92978	64	0.86448	120	9.93676	60	1.07552	2.71651	−0.93906	0.6790	1.00	4.77522	9.9684	0.50
5	.93042	64	.86568	118	9.93736	59	1.07478	2.72851	−0.93252	0.6796	1.17	4.78220	9.9687	0.50
6	.93106	63	.86686	119	9.93795	59	1.07404	2.74063	−0.92601	0.6803	1.00	4.78918	9.9690	0.50
7	.93169	63	.86805	118	9.93854	59	1.07332	2.75292	−0.91955	0.6809	1.00	4.79616	9.9693	0.50
8	.93232	63	.86923	117	9.93913	59	1.07259	2.76533	−0.91313	0.6815	1.17	4.80315	9.9696	0.50
9	.93295	63	.87040	117	9.93972	58	1.07187	2.77778	−0.90674	0.6822	1.00	4.81013	9.9699	0.50
69.0	0.93358	62	0.87157	117	9.94030	58	1.07115	2.79041	−0.90040	0.6828	1.00	4.81711	9.9702	0.50
1	.93420	63	.87274	116	9.94088	58	1.07043	2.80316	−0.89410	0.6834	1.00	4.82409	9.9704	0.33
2	.93483	61	.87390	116	9.94146	58	1.06971	2.81603	−0.88784	0.6840	1.17	4.83107	9.9707	0.50
3	.93544	62	.87506	115	9.94204	57	1.06902	2.82909	−0.88162	0.6847	1.00	4.83805	9.9710	0.50
4	.93606	61	.87621	114	9.94261	56	1.06831	2.84220	−0.87543	0.6853	1.00	4.84503	9.9713	0.50
5	.93667	61	.87735	115	9.94317	57	1.06761	2.85543	−0.86929	0.6859	1.00	4.85202	9.9716	0.50
6	.93728	61	.87850	114	9.94374	57	1.06692	2.86886	−0.86318	0.6865	1.17	4.85900	9.9719	0.50
7	.93789	60	.87964	113	9.94431	55	1.06622	2.88234	−0.85710	0.6872	1.00	4.86598	9.9722	0.50
8	.93849	60	.88077	113	9.94486	56	1.06554	2.89603	−0.85107	0.6878	1.00	4.87296	9.9724	0.33
9	.93909	60	.88190	112	9.94542	55	1.06486	2.90985	−0.84507	0.6884	1.00	4.87994	9.9727	0.50
70.0	0.93969	60	0.88302	112	9.94597	55	1.06418	2.92381	−0.83910	0.6890	1.17	4.88692	9.9730	0.50
1	.94029	59	.88414	112	9.94652	55	1.06350	2.93789	−0.83317	0.6897	1.00	4.89390	9.9733	0.33

D (1')	lg sin θ	4θ rad	D (1')	lg 4θ rad	tg 2θ	$\frac{1}{\cos θ}$	$\frac{1}{\sin θ}$	D (0.1°)	lg sin² θ	D (0.1°)	sin² θ	D (0.1°)	sin θ	θ°
0.50	9.9735	4.90088	1.00	0.6903	−0.82727	2.95212	1.06283	55	9.94707	111	0.88526	59	0.94088	70.2
0.50	9.9738	4.90787	1.00	0.6909	−0.82141	2.96648	1.06217	53	9.94762	110	.88637	59	.94147	3
0.33	9.9741	4.91485	1.00	0.6915	−0.81558	2.98107	1.06150	54	9.94815	110	.88747	58	.94206	4
0.50	9.9743	4.92183	1.00	0.6921	−0.80978	2.99572	1.06085	54	9.94869	110	.88857	58	.94264	5
0.50	9.9746	4.92881	1.17	0.6927	−0.80402	3.01060	1.06020	53	9.94923	109	.88967	58	.94322	6
0.33	9.9749	4.93579	1.00	0.6934	−0.79829	3.02563	1.05955	53	9.94976	109	.89076	58	.94380	7
0.50	9.9751	4.94277	1.00	0.6940	−0.79259	3.04072	1.05890	53	9.95029	108	.89185	58	.94438	8
0.50	9.9754	4.94975	1.00	0.6946	−0.78692	3.05605	1.05826	52	9.95082	108	.89293	57	.94495	9
0.33	9.9757	4.95674	1.00	0.6952	−0.78129	3.07154	1.05762	52	9.95134	107	0.89401	57	0.94552	71.0
0.50	9.9759	4.96372	1.00	0.6958	−0.77568	3.08718	1.05698	52	9.95186	106	.89508	56	.94609	1
0.33	9.9762	4.97070	1.00	0.6964	−0.77010	3.10299	1.05636	51	9.95238	107	.89614	56	.94665	2
0.50	9.9764	4.97768	1.00	0.6970	−0.76456	3.11905	1.05573	51	9.95289	105	.89721	56	.94721	3
0.50	9.9767	4.98466	1.00	0.6976	−0.75904	3.13519	1.05511	51	9.95340	106	.89826	55	.94777	4
0.33	9.9770	4.99164	1.00	0.6982	−0.75355	3.15159	1.05450	51	9.95391	105	.89932	56	.94832	5
0.50	9.9772	4.99862	1.17	0.6989	−0.74810	3.16807	1.05387	50	9.95442	104	.90037	55	.94888	6
0.33	9.9775	5.00560	1.00	0.6995	−0.74267	3.18481	1.05326	50	9.95492	104	.90141	54	.94943	7
0.50	9.9777	5.01259	1.00	0.7001	−0.73726	3.20174	1.05266	50	9.95542	103	.90245	55	.94997	8
0.33	9.9780	5.01957	1.00	0.7007	−0.73189	3.21875	1.05206	49	9.95592	103	.90348	54	.95052	9

ϑ°	sin ϑ	D (0.1°)	sin² ϑ	D (0.1°)	lg sin² ϑ	D (0.1°)	1/sin ϑ	1/cos ϑ	tg 2ϑ	lg 4ϑ_rad	D (1')	4 ϑ_rad	lg sin ϑ	D (1')
72.0	0.95106	53	0.90451	102	9.95641	49	1.05146	3.23604	−0.72654	0.7013	1.00	5.02655	9.9782	0.50
1	.95159	54	.90553	102	9.95690	49	1.05087	3.25351	−0.72122	0.7019	1.00	5.03353	9.9785	0.33
2	.95213	53	.90655	101	9.95739	49	1.05028	3.27118	−0.71593	0.7025	1.00	5.04051	9.9787	0.33
3	.95266	53	.90756	101	9.95788	48	1.04969	3.28915	−0.71066	0.7031	1.00	5.04749	9.9789	0.50
4	.95319	53	.90857	101	9.95836	48	1.04911	3.30721	−0.70542	0.7037	1.00	5.05447	9.9792	0.33
5	.95372	52	.90958	99	9.95884	47	1.04853	3.32546	−0.70021	0.7043	1.00	5.06145	9.9794	0.50
6	.95424	52	.91057	100	9.95931	48	1.04795	3.34403	−0.69502	0.7049	1.00	5.06844	9.9797	0.33
7	.95476	52	.91157	99	9.95979	47	1.04738	3.36281	−0.68985	0.7055	1.00	5.07542	9.9799	0.33
8	.95528	51	.91256	98	9.96026	47	1.04681	3.38169	−0.68471	0.7061	1.00	5.08240	9.9801	0.50
9	.95579	51	.91354	98	9.96073	46	1.04625	3.40090	−0.67960	0.7067	1.00	5.08938	9.9804	0.33
73.0	0.95630	51	0.91452	97	9.96119	46	1.04570	3.42032	−0.67451	0.7073	1.00	5.09636	9.9806	0.33
1	.95681	51	.91549	97	9.96165	46	1.04514	3.43997	−0.66944	0.7079	1.00	5.10334	9.9808	0.50
2	.95732	50	.91646	96	9.96211	46	1.04458	3.45985	−0.66440	0.7084	0.83	5.11032	9.9811	0.33
3	.95782	50	.91742	96	9.96257	45	1.04404	3.47996	−0.65938	0.7090	1.00	5.11731	9.9813	0.33
4	.95832	50	.91838	96	9.96302	46	1.04349	3.50030	−0.65438	0.7096	1.00	5.12429	9.9815	0.33
5	.95882	49	.91934	94	9.96348	44	1.04295	3.52088	−0.64941	0.7102	1.00	5.13127	9.9817	0.50
6	.95931	50	.92028	95	9.96392	45	1.04242	3.54183	−0.64446	0.7108	1.00	5.13825	9.9820	0.50
7	.95981	48	.92123	93	9.96437	44	1.04187	3.56290	−0.63953	0.7114	1.00	5.14523	9.9822	0.33

$\theta°$	$\sin\theta$	$D(0.1°)$	$\sin^2\theta$	$D(0.1°)$	$\lg\sin^2\theta$	$D(0.1°)$	$\dfrac{1}{\sin\theta}$	$\dfrac{1}{\cos\theta}$	$tg\,2\theta$	$\lg\,4\theta_{rad}$	$D(1')$	$4\theta_{rad}$	$\lg\sin\theta$	$D(1')$
73.8	0.96029	49	0.92216	94	9.96481	44	1.04135	3.58436	−0.63462	0.7120	1.00	5.15221	9.9824	0.33
9	.96078	48	.92310	92	9.96525	43	1.04082	3.60607	−0.62973	0.7126	1.00	5.15919	9.9826	0.33
74.0	.96126	48	0.92402	93	9.96568	44	1.04030	3.62792	−0.62487	0.7132	1.00	5.16617	9.9828	0.50
1	.96174	48	.92495	91	9.96612	43	1.03978	3.65017	−0.62003	0.7138	0.83	5.17316	9.9831	0.33
2	.96222	47	.92586	92	9.96655	43	1.03926	3.67269	−0.61520	0.7143	1.00	5.18014	9.9833	0.33
3	.96269	47	.92678	90	9.96698	42	1.03876	3.69549	−0.61040	0.7149	1.00	5.18712	9.9835	0.33
4	.96316	47	.92768	90	9.96740	42	1.03825	3.71858	−0.60562	0.7155	1.00	5.19410	9.9837	0.33
5	.96363	47	.92858	90	9.96782	42	1.03774	3.74195	−0.60086	0.7161	1.00	5.20108	9.9839	0.33
6	.96410	46	.92948	89	9.96824	42	1.03724	3.76563	−0.59612	0.7167	1.00	5.20806	9.9841	0.33
7	.96456	46	.93037	89	9.96866	41	1.03674	3.78974	−0.59140	0.7173	0.83	5.21504	9.9843	0.33
8	.96502	45	.93126	88	9.96907	41	1.03625	3.81403	−0.58670	0.7178	1.00	5.22203	9.9845	0.33
9	.96547	46	.93214	87	9.96948	41	1.03576	3.83877	−0.58201	0.7184	1.00	5.22901	9.9847	0.33
75.0	0.96593	45	0.93301	87	9.96989	40	1.03527	3.86369	−0.57735	0.7190	1.00	5.23599	9.9849	0.33
1	.96638	44	.93388	87	9.97029	41	1.03479	3.88908	−0.57271	0.7196	1.00	5.24297	9.9851	0.33
2	.96682	45	.93475	86	9.97070	39	1.03432	3.91466	−0.56808	0.7202	0.83	5.24995	9.9853	0.33
3	.96727	44	.93561	85	9.97109	40	1.03384	3.94073	−0.56347	0.7207	1.00	5.25693	9.9855	0.33
4	.96771	44	.93646	85	9.97149	39	1.03337	3.96715	−0.55888	0.7213	1.00	5.26391	9.9857	0.33
5	.96815	43	.93731	84	9.97188	39	1.03290	3.99393	−0.55431	0.7219	1.00	5.27089	9.9859	0.33

θ°	sin θ	D (0.1°)	sin² θ	D (0.1°)	lg sin² θ	D (0.1°)	$\frac{1}{\sin\theta}$	$\frac{1}{\cos\theta}$	tg 2θ	lg 4θrad	D (1′)	4θrad	lg sin θ	D (1′)
75.6	0.96858	44	0.93815	84	9.97227	39	1.03244	4.02107	−0.54975	0.7225	0.83	5.27788	9.9861	0.33
7	.96902	43	.93899	83	9.97266	38	1.03197	4.04858	−0.54522	0.7230	1.00	5.28486	9.9863	0.33
8	.96945	42	.93982	83	9.97304	39	1.03151	4.07647	−0.54070	0.7236	1.00	5.29184	9.9865	0.33
9	.96987	43	.94065	82	9.97343	38	1.03107	4.10475	−0.53620	0.7242	1.00	5.29882	9.9867	0.33
76.0	0.97030	42	0.94147	82	9.97381	37	1.03061	4.13360	−0.53171	0.7248	0.83	5.30580	9.9869	0.33
1	.97072	41	.94229	81	9.97418	38	1.03016	4.16268	−0.52724	0.7253	1.00	5.31278	9.9871	0.33
2	.97113	42	.94310	81	9.97456	37	1.02973	4.19234	−0.52279	0.7259	1.00	5.31976	9.9873	0.33
3	.97155	41	.94391	80	9.97493	37	1.02928	4.22226	−0.51835	0.7265	1.00	5.32674	9.9875	0.33
4	.97196	41	.94471	79	9.97530	36	1.02885	4.25279	−0.51393	0.7270	1.00	5.33373	9.9876	0.17
5	.97237	41	.94550	79	9.97566	36	1.02842	4.28357	−0.50953	0.7276	1.00	5.34071	9.9878	0.33
6	.97278	40	.94629	79	9.97602	37	1.02798	4.31499	−0.50514	0.7282	0.83	5.34769	9.9880	0.33
7	.97318	40	.94708	78	9.97639	35	1.02756	4.34688	−0.50076	0.7287	1.00	5.35467	9.9882	0.33
8	.97358	40	.94786	77	9.97674	36	1.02714	4.37924	−0.49640	0.7293	1.00	5.36165	9.9884	0.17
9	.97398	39	.94863	77	9.97710	35	1.02672	4.41209	−0.49206	0.7299	0.83	5.36863	9.9885	0.33
77.0	0.97437	39	0.94940	76	9.97745	35	1.02630	4.44543	−0.48773	0.7304	1.00	5.37561	9.9887	0.33
1	.97476	39	.95016	76	9.97780	34	1.02589	4.47928	−0.48342	0.7310	1.00	5.38260	9.9889	0.33
2	.97515	38	.95092	75	9.97814	35	1.02548	4.51365	−0.47912	0.7316	0.83	5.38958	9.9891	0.17
3	.97553	39	.95167	74	9.97849	33	1.02508	4.54856	−0.47483	0.7321	1.00	5.39656	9.9892	0.33

θ°	sin θ	D (0.1°)	sin² θ	D (0.1°)	lg sin² θ	D (0.1°)	$\frac{1}{\sin θ}$	$\frac{1}{\cos θ}$	tg 2θ	lg 4θ rad	D (1')	4 θ rad	lg sin θ	D (1')
77.4	0.97592	38	0.95241	74	9.97882	34	1.02467	4.58421	−0.47056	0.7327	0.83	5.40354	9.9894	0.33
5	.97630	37	.95315	74	9.97916	34	1.02428	4.62022	−0.46631	0.7332	1.00	5.41052	9.9896	0.17
6	.97667	38	.95389	73	9.97950	33	1.02389	4.65679	−0.46206	0.7338	1.00	5.41750	9.9897	0.33
7	.97705	37	.95462	72	9.97983	33	1.02349	4.69417	−0.45784	0.7344	0.83	5.42448	9.9899	0.33
8	.97742	36	.95534	72	9.98016	33	1.02310	4.73216	−0.45362	0.7349	1.00	5.43146	9.9901	0.17
9	.97778	37	.95606	71	9.98049	32	1.02272	4.77054	−0.44942	0.7355	0.83	5.43845	9.9902	0.33
78.0	0.97815	36	0.95677	71	9.98081	32	1.02234	4.80977	−0.44523	0.7360	1.00	5.44543	9.9904	0.33
1	.97851	36	.95748	70	9.98113	32	1.02196	4.84966	−0.44105	0.7366	0.83	5.45241	9.9906	0.33
2	.97887	35	.95818	70	9.98145	31	1.02159	4.88998	−0.43689	0.7371	1.00	5.45939	9.9907	0.17
3	.97922	36	.95888	69	9.98176	32	1.02122	4.93121	−0.43274	0.7377	1.00	5.46637	9.9909	0.33
4	.97958	34	.95957	68	9.98208	30	1.02085	4.97315	−0.42860	0.7383	0.83	5.47335	9.9910	0.17
5	.97992	35	.96025	68	9.98238	31	1.02049	5.01580	−0.42447	0.7388	1.00	5.48033	9.9912	0.33
6	.98027	34	.96093	68	9.98269	31	1.02013	5.05919	−0.42036	0.7394	0.83	5.48732	9.9913	0.17
7	.98061	35	.96161	66	9.98300	30	1.01977	5.10334	−0.41626	0.7399	1.00	5.49430	9.9915	0.33
8	.98096	33	.96227	66	9.98330	29	1.01941	5.14854	−0.41217	0.7405	0.83	5.50128	9.9916	0.17
9	.98129	34	.96293	66	9.98359	30	1.01907	5.19427	−0.40809	0.7410	1.00	5.50826	9.9918	0.33
79.0	0.98163	33	0.96359	65	9.98389	30	1.01871	5.24082	−0.40403	0.7416	0.83	5.51524	9.9919	0.17
1	.98196	33	.96424	65	9.98419	29	1.01837	5.28821	−0.39997	0.7421	1.00	5.52222	9.9921	0.33

θ°	sin θ	D (0.1°)	sin² θ	D (0.1°)	lg sin² θ	D (0.1°)	1/sin θ	1/cos θ	tg 2θ	lg 4θ_rad	D (1')	4 θ_rad	lg sin θ	D (1')
79.2	0.98229	32	0.96189	64	9.98448	29	1.01803	5.33675	−0.39593	0.7427	0.83	5.52920	9.9922	0.33
3	.98261	33	.96553	63	9.98477	28	1.01770	5.38590	−0.39190	0.7432	1.00	5.53618	9.9924	0.17
4	.98294	31	.96616	63	9.98505	28	1.01736	5.43626	−0.38787	0.7438	0.83	5.54317	9.9925	0.33
5	.98325	32	.96679	62	9.98533	28	1.01704	5.48727	−0.38386	0.7443	1.00	5.55015	9.9927	0.17
6	.98357	31	.96741	62	9.98561	28	1.01670	5.53955	−0.37986	0.7449	0.83	5.55713	9.9928	0.17
7	.98388	32	.96803	61	9.98589	27	1.01638	5.59284	−0.37588	0.7454	0.83	5.56411	9.9929	0.33
8	.98420	30	.96864	61	9.98616	28	1.01605	5.64717	−0.37190	0.7459	1.00	5.57109	9.9931	0.17
9	.98450	31	.96925	60	9.98644	26	1.01574	5.70223	−0.36793	0.7465	0.83	5.57807	9.9932	0.33
80.0	0.98481	30	0.96985	59	9.98670	27	1.01542	5.75871	−0.36397	0.7470	1.00	5.58505	9.9934	0.17
1	.98511	30	.97044	59	9.98697	26	1.01512	5.81632	−0.36002	0.7476	0.83	5.59203	9.9935	0.17
2	.98541	29	.97103	58	9.98723	26	1.01481	5.87510	−0.35608	0.7481	1.00	5.59902	9.9936	0.17
3	.98570	30	.97161	58	9.98749	26	1.01451	5.93507	−0.35216	0.7487	0.83	5.60600	9.9937	0.33
4	.98600	29	.97219	57	9.98775	26	1.01420	5.99628	−0.34824	0.7492	0.83	5.61298	9.9939	0.17
5	.98629	28	.97276	56	9.98801	25	1.01390	6.05877	−0.34433	0.7497	1.00	5.61996	9.9940	0.17
6	.98657	29	.97332	56	9.98826	25	1.01361	6.12257	−0.34043	0.7503	0.83	5.62694	9.9941	0.33
7	.98686	28	.97388	56	9.98851	25	1.01331	6.18812	−0.33654	0.7508	0.83	5.63392	9.9943	0.17
8	.98714	27	.97444	55	9.98876	24	1.01303	6.25469	−0.33266	0.7513	0.83	5.64090	9.9944	0.17
9	.98741	28	.97499	54	9.98900	24	1.01275	6.32271	−0.32878	0.7519	0.83	5.64789	9.9945	0.17

θ°	sin θ	Δ(0.1°)	sin² θ	Δ(0.1°)	lg sin² θ	Δ(0,1°)	$\frac{1}{\sin θ}$	$\frac{1}{\cos θ}$	tg 2θ	lg 4θ_rad	Δ(1')	4 θ_rad	lg sin θ	Δ(1')
81.0	0.98769	27	0.97553	53	9.98924	24	1.01246	6.39264	−0.32492	0.7524		5.65487	9.9946	0.17
1	.98796	27	.97606	54	9.98948	24	1.01219	6.46371	−0.32106	0.7530	1.00	5.66185	9.9947	0.33
2	.98823	26	.97660	52	9.98972	23	1.01191	6.53637	−0.31722	0.7535	0.83	5.66883	9.9949	0.17
3	.98849	27	.97712	52	9.98995	23	1.01164	6.61113	−0.31338	0.7540	0.83	5.67581	9.9950	0.17
4	.98876	26	.97764	51	9.99018	23	1.01137	6.68717	−0.30955	0.7546	1.00	5.68279	9.9951	0.17
5	.98902	25	.97815	51	9.99041	22	1.01110	6.76544	−0.30573	0.7551	0.83	5.68977	9.9952	0.17
6	.98927	26	.97866	50	9.99063	22	1.01085	6.84556	−0.30192	0.7556	0.83	5.69675	9.9953	0.17
7	.98953	25	.97916	50	9.99085	23	1.01058	6.92713	−0.29811	0.7562	1.00	5.70374	9.9954	0.17
8	.98978	24	.97966	49	9.99108	21	1.01033	7.01115	−0.29432	0.7567	0.83	5.71072	9.9955	0.17
9	.99002	25	.98015	48	9.99129	22	1.01008	7.09723	−0.29053	0.7572	0.83	5.71770	9.9956	0.33
82.0	0.99027	24	0.98063	48	9.99151	21	1.00983	7.18546	−0.28675	0.7578	1.00	5.72468	9.9958	0.17
1	.99051	24	.98111	47	9.99172	21	1.00958	7.27590	−0.28297	0.7583	0.83	5.73166	9.9959	0.17
2	.99075	23	.98158	47	9.99193	20	1.00934	7.36811	−0.27920	0.7588	0.83	5.73864	9.9960	0.17
3	.99098	24	.98205	46	9.99213	21	1.00910	7.46324	−0.27545	0.7593	0.83	5.74562	9.9961	0.17
4	.99122	22	.98251	45	9.99234	20	1.00886	7.56086	−0.27169	0.7599	1.00	5.75261	9.9962	0.17
5	.99144	23	.98296	45	9.99254	19	1.00863	7.66107	−0.26795	0.7604	0.83	5.75959	9.9963	0.17
6	.99167	22	.98341	44	9.99273	20	1.00840	7.76398	−0.26421	0.7609	0.83	5.76657	9.9964	0.17
7	.99189	22	.98385	44	9.99293	19	1.00818	7.87030	−0.26048	0.7614	1.00	5.77355	9.9965	0.17

φ°	sin φ	D (0.1°)	sin² φ	D (0.1°)	lg sin² φ	D (0.1°)	1/sin φ	1/cos φ	tg 2φ	lg 4φ_rad	D (1')	4 φ_rad	lg sin φ	D (1')
82.8	0.99211	22	0.98429	43	9.99312	19	1.00795	7.97894	−0.25676	0.7620	0.83	5.78053	9.9966	0.17
9	.99233	22	.98472	43	9.99331	19	1.00773	8.09061	−0.25304	0.7625	0.83	5.78751	9.9967	0.17
83.0	0.99255	21	0.98515	42	9.99350	19	1.00751	8.20546	−0.24933	0.7630	0.83	5.79449	9.9968	0.00
1	.99276	21	.98557	41	9.99369	18	1.00729	8.32362	−0.24562	0.7635	1.00	5.80147	9.9968	0.17
2	.99297	20	.98598	41	9.99387	18	1.00708	8.44595	−0.24193	0.7641	0.83	5.80846	9.9969	0.17
3	.99317	20	.98639	40	9.99405	17	1.00688	8.57118	−0.23823	0.7646	0.83	5.81544	9.9970	0.17
4	.99337	20	.98679	40	9.99422	18	1.00667	8.70019	−0.23455	0.7651	0.83	5.82242	9.9971	0.17
5	.99357	20	.98719	38	9.99440	17	1.00647	8.83392	−0.23087	0.7656	0.83	5.82940	9.9972	0.17
6	.99377	19	.98757	39	9.99457	17	1.00627	8.97102	−0.22719	0.7661	1.00	5.83638	9.9973	0.17
7	.99396	19	.98796	38	9.99474	17	1.00608	9.11328	−0.22353	0.7667	0.83	5.84336	9.9974	0.17
8	.99415	19	.98834	37	9.99491	16	1.00588	9.25926	−0.21986	0.7672	0.83	5.85034	9.9975	0.17
9	.99434	18	.98871	36	9.99507	16	1.00569	9.41088	−0.21621	0.7677	0.83	5.85732	9.9975	0.00
84.0	0.99452	18	0.98907	36	9.99523	16	1.00551	9.56663	−0.21256	0.7682	0.83	5.86431	9.9976	0.17
1	.99470	18	.98943	36	9.99539	15	1.00533	9.72857	−0.20891	0.7687	0.83	5.87129	9.9977	0.17
2	.99488	18	.98979	35	9.99554	16	1.00515	9.89511	−0.20527	0.7692	1.00	5.87827	9.9978	0.17
3	.99506	17	.99014	34	9.99570	15	1.00496	10.06847	−0.20164	0.7698	0.83	5.88525	9.9978	0.00
4	.99523	17	.99048	33	9.99585	14	1.00479	10.24800	−0.19801	0.7703	0.83	5.89223	9.9979	0.17
5	.99540	16	.99081	33	9.99599	15	1.00462	10.43297	−0.19438	0.7708		5.89921	9.9980	0.17

ϑ°	sin ϑ	D (0.1°)	sin² ϑ	D (0.1°)	lg sin² ϑ	D (0.1°)	1/sin ϑ	1/cos ϑ	tg 2ϑ	lg 4ϑ_rad	D (1')	4 ϑ_rad	lg sin ϑ	D (1')
84.6	0.99556	16	0.99114	33	9.99614	14	1.00446	10.62586	−0.19076	0.7713	0.83	5.90619	9.9981	0.00
7	.99572	16	.99147	32	9.99628	14	1.00430	10.82603	−0.18714	0.7718	0.83	5.91318	9.9981	0.00
8	.99588	16	.99179	31	9.99642	14	1.00414	11.00387	−0.18353	0.7723	0.83	5.92016	9.9982	0.17
9	.99604	15	.99210	30	9.99656	13	1.00398	11.24986	−0.17993	0.7728	1.00	5.92714	9.9983	0.00
85.0	0.99619	16	0.99240	30	9.99669	13	1.00382	11.47315	−0.17633	0.7734	0.83	5.93412	9.9983	0.17
1	.99635	14	.99270	30	9.99682	13	1.00366	11.70686	−0.17273	0.7739	0.83	5.94110	9.9984	0.17
2	.99649	15	.99300	29	9.99695	13	1.00352	11.95029	−0.16914	0.7744	0.83	5.94808	9.9985	0.00
3	.99664	14	.99329	28	9.99708	12	1.00337	12.20405	−0.16555	0.7749	0.83	5.95506	9.9986	0.17
4	.99678	14	.99357	27	9.99720	12	1.00323	12.46883	−0.16196	0.7754	0.83	5.96204	9.9986	0.17
5	.99692	13	.99384	27	9.99732	11	1.00309	12.74535	−0.15838	0.7759	0.83	5.96903	9.9987	0.00
6	.99705	14	.99411	27	9.99743	12	1.00296	13.03441	−0.15481	0.7764	0.83	5.97601	9.9987	0.17
7	.99719	12	.99438	26	9.99755	12	1.00282	13.33689	−0.15124	0.7769	0.83	5.98299	9.9988	0.00
8	.99731	13	.99464	25	9.99767	11	1.00270	13.65374	−0.14767	0.7774	0.83	5.98997	9.9988	0.17
9	.99744	12	.99489	24	9.99778	10	1.00257	13.98601	−0.14410	0.7779	0.83	5.99695	9.9989	0.00
86.0	0.99756	12	0.99513	24	9.99788	10	1.00245	14.33486	−0.14054	0.7784	0.83	6.00393	9.9989	0.17
1	.99768	12	.99537	24	9.99798	11	1.00233	14.70156	−0.13698	0.7789	0.83	6.01091	9.9990	0.00
2	.99780	12	.99561	23	9.99809	10	1.00220	15.08978	−0.13343	0.7794	0.83	6.01790	9.9990	0.17
3	.99792	11	.99584	22	9.99819	10	1.00208	15.49667	−0.12988	0.7799	0.83	6.02488	9.9991	0.00

$\vartheta°$	$\sin\vartheta$	$D(0.1°)$	$\sin^2\vartheta$	$D(0.1°)$	$\lg\sin^2\vartheta$	$D(0.1°)$	$\dfrac{1}{\sin\vartheta}$	$\dfrac{1}{\cos\vartheta}$	$\operatorname{tg}2\vartheta$	$\lg 4\vartheta\,\mathrm{rad}$	$D(1')$	$4\vartheta\,\mathrm{rad}$	$\lg\sin\vartheta$	$D(1')$
86.4	0.99803	10	0.99606	21	9.99829	9	1.00197	15.92610	−0.12633	0.7805	0.83	6.03186	9.9991	0.17
5	.99813	11	.99627	21	9.99838	9	1.00187	16.38002	−0.12278	0.7810	0.83	6.03884	9.9992	0.00
6	.99824	10	.99648	21	9.99847	9	1.00176	16.86056	−0.11924	0.7815	0.83	6.04582	9.9992	0.17
7	.99834	10	.99669	19	9.99856	8	1.00166	17.37318	−0.11570	0.7820	0.83	6.05280	9.9993	0.00
8	.99844	10	.99688	20	9.99864	9	1.00156	17.91473	−0.11217	0.7825	0.83	6.05978	9.9993	0.17
9	.99854	9	.99708	18	9.99873	8	1.00146	18.49112	−0.10863	0.7830	0.83	6.06676	9.9994	0.00
87.0	0.99863	9	0.99726	18	9.99881	8	1.00137	19.10585	−0.10510	0.7835	0.83	6.07375	9.9994	0.00
1	.99872	9	.99744	17	9.99889	7	1.00128	19.76675	−0.10158	0.7840	0.83	6.08073	9.9994	0.17
2	.99881	8	.99761	17	9.99896	7	1.00119	20.47083	−0.09805	0.7845	0.83	6.08771	9.9995	0.00
3	.99889	8	.99778	16	9.99903	7	1.00111	21.22632	−0.09453	0.7850	0.83	6.09469	9.9995	0.17
4	.99897	8	.99794	16	9.99910	7	1.00103	22.04586	−0.09101	0.7854	0.83	6.10167	9.9996	0.00
5	.99905	7	.99810	15	9.99917	7	1.00095	22.92526	−0.08749	0.7859	0.83	6.10865	9.9996	0.17
6	.99912	7	.99825	14	9.99924	6	1.00088	23.87775	−0.08397	0.7864	0.83	6.11563	9.9996	0.00
7	.99919	7	.99839	14	9.99930	6	1.00081	24.91901	−0.08046	0.7869	0.83	6.12262	9.9996	0.00
8	.99926	7	.99853	13	9.99936	6	1.00074	26.04845	−0.07695	0.7874	0.83	6.12960	9.9997	0.00
9	.99933	6	.99866	12	9.99942	5	1.00067	27.29258	−0.07344	0.7879	0.83	6.13658	9.9997	0.17
88.0	0.99939	6	0.99878	12	9.99947	5	1.00061	28.65330	−0.06993	0.7884	0.83	6.14356	9.9997	0.00
1	.99945	6	.99890	11	9.99952	5	1.00055	30.15682	−0.06642	0.7889	0.83	6.15054	9.9998	0.00

θ°	sin θ	D (0.1°)	sin² θ	D (0.1°)	lg sin² θ	D (0.1°)	$\frac{1}{\sin θ}$	$\frac{1}{\cos θ}$	tg 2θ	lg 4θ_rad	D (1')	4 θ_rad	lg sin θ	D (1')
88.2	0.99951	5	0.99901	11	9.99957	5	1.00049	31.83699	−0.06291	0.7894	0.83	6.15752	9.9998	0.00
3	.99956	5	.99912	10	9.99962	4	1.00044	33.70408	−0.05941	0.7899	0.83	6.16450	9.9998	0.00
4	.99961	5	.99922	9	9.99966	4	1.00039	35.81662	−0.05591	0.7904	0.83	6.17148	9.9998	0.17
5	.99966	4	.99931	9	9.99970	4	1.00034	38.19710	−0.05241	0.7909	0.83	6.17847	9.9999	0.00
6	.99970	4	.99940	9	9.99974	4	1.00030	40.93328	−0.04891	0.7914	0.83	6.18545	9.9999	0.00
7	.99974	4	.99949	7	9.99978	3	1.00026	44.07228	−0.04541	0.7919	0.83	6.19243	9.9999	0.00
8	.99978	4	.99956	7	9.99981	3	1.00022	47.75549	−0.04191	0.7924	0.83	6.19941	9.9999	0.00
9	.99982	3	.99963	7	9.99984	3	1.00018	52.08333	−0.03842	0.7928	0.67	6,20639	9.9999	0.00
89.0	0.99985	3	0.99970	5	9.99987	2	1.00015	57.30659	−0.03492	0.7933	0.83	6.21337	9.9999	0.17
1	.99988	2	.99975	6	9.99989	3	1.00012	63.65372	−0.03143	0.7938	0.83	6.22035	9.9999	0.00
2	.99990	3	.99981	4	9.99992	1	1.00010	71.63324	−0.02793	0.7943	0.83	6.22733	0.0000	0.00
3	.99993	2	.99985	4	9.99993	2	1.00007	81.83306	−0.02444	0.7948	0.83	6.23432	0.0000	0.00
4	.99995	1	.99989	3	9.99995	2	1.00005	95.51098	−0.02095	0.7953	0.83	6.24130	0.0000	0.00
5	.99996	2	.99992	3	9.99997	1	1.00004	114.54754	−0.01746	0.7958	0.67	6.24828	0.0000	0.00
6	.99998	1	.99995	2	9.99998	1	1.00002	143.26648	−0.01396	0.7962	0.83	6.25526	0.0000	0.00
7	.99999	1	.99997	2	9.99999	1	1.00001	190.83969	−0.01047	0.7967	0.83	6.26224	0.0000	0.00
8	.99999	1	.99999	1	0.00000	0	1.00001	286.53295	−0.00698	0.7972	0.83	6.26922	0.0000	0.00
9	1.00000	0	1.00000	0	0.00000	0	1.00000	571.42857	−0.00349	0.7977	0.83	6.27620	0.0000	0.00
90.0	1.00000		1.00000		0.00000		1.00000	∞	−0.00000	0.7982		—	0.0000	0.00

3-1b. Values of $1/d^2$

The table gives $1/d^2$, where d is the interplanar distance, for d between 1.000 and 10.000 Å [270].

Values of $1/d^2$ for d outside this range can be found by shifting the decimal point in $1/d^2$ by two places in the opposite sense for each place in d, as in the following:

$$d,\ \text{Å} \quad 3.040 \qquad 0.3040 \qquad 30.40$$

$$\frac{1}{d^2},\ \text{Å}^{-2} \quad 0.1082 \qquad 10.82 \qquad 0.001082$$

See [109, 391] for tables of d as a function of ϑ for various radiations.

d	$1/d^2$	d	$1/d^2$	d	$1/d^2$	d	$1/d^2$	d	$1/d^2$
1.000	1.0000	1.040	0.9246	1.080	0.8573	1.120	0.7972		
1	0.9980	1	.9228	1	.8557	1	.7958		
2	.9960	2	.9210	2	.8542	2	.7944		
3	.9940	3	.9192	3	.8526	3	.7930		
4	.9920	4	.9175	4	.8510	4	7915		
5	.9901	5	.9157	5	.8495	5	.7901		
6	.9881	6	.9140	6	.8479	6	.7887		
7	.9861	7	.9122	7	.8463	7	.7873		
8	.9842	8	.9105	8	.8448	8	.7859		
9	.9822	9	.9088	9	.8432	9	.7845		
1.010	.9803	1.050	.9070	1.090	.8417	1.130	.7831		
1	.9784	1	.9053	1	.8401	1	.7818		
2	.9762	2	.9036	2	.8386	2	.7804		
3	.9745	3	.9019	3	.8370	3	.7790		
4	.9726	4	.9002	4	.8356	4	.7776		
5	.9707	5	.8985	5	.8340	5	.7763		
6	.9688	6	.8968	6	.8325	6	.7749		
7	.9668	7	.8951	7	.8310	7	.7735		
8	.9650	8	.8934	8	.8295	8	.7722		
9	.9631	9	.8917	9	.8280	9	.7708		
1.020	.9612	1.060	.8900	1.100	.8264	1.140	.7695		
1	.9593	1	.8883	1	.8249	1	.7681		
2	.9574	2	.8866	2	.8235	2	.7668		
3	.9555	3	.8850	3	.8220	3	.7654		
4	.9537	4	.8833	4	.8205	4	.7641		
5	.9518	5	.8817	5	.8190	5	.7628		
6	.9499	6	.8800	6	.8175	6	.7614		
7	.9481	7	.8784	7	.8160	7	.7601		
8	.9463	8	.8767	8	.8145	8	.7588		
9	.9444	9	.8751	9	.8131	9	.7575		
1.030	.9426	1.070	.8734	1.110	.8116	1.150	.7561		
1	.9408	1	.8718	1	.8102	1	.7548		
2	.9390	2	.8702	2	.8087	2	.7535		
3	.9371	3	.8686	3	.8073	3	.7522		
4	.9353	4	.8669	4	.8058	4	.7509		
5	.9335	5	.8653	5	.8044	5	.7496		
6	.9317	6	.8637	6	.8029	6	.7483		
7	.9299	7	.8621	7	.8015	7	.7470		
8	.9281	8	.8605	8	.8000	8	.7457		
9	.9263	9	.8589	9	.7986	9	.7444		

d	$1/d^2$	d	$1/d^2$	d	$1/d^2$	d	$1/d^2$
1.160	0.7432	1.220	0.6719	1.280	0.6104	1.340	0.5569
1	.7419	1	.6708	1	.6094	1	.5561
2	.7406	2	.6697	2	.6084	2	.5553
3	.7393	3	.6686	3	.6075	3	.5544
4	.7381	4	.6675	4	.6066	4	.5536
5	.7368	5	.6664	5	.6056	5	.5528
6	.7355	6	.6653	6	.6047	6	.5520
7	.7343	7	.6642	7	.6037	7	.5511
8	.7330	8	.6631	8	.6028	8	.5503
9	.7318	9	.6621	9	.6019	9	.5495
1.170	.7305	1.230	.6610	1.290	.6009	1.350	.5487
1	.7293	1	.6599	1	.6000	1	.5479
2	.7280	2	.6588	2	.5991	2	.5471
3	.7268	3	.6578	3	.5981	3	.5463
4	.7255	4	.6567	4	.5972	4	.5455
5	.7243	5	.6556	5	.5963	5	.5447
6	.7231	6	.6546	6	.5954	6	.5439
7	.7219	7	.6535	7	.5945	7	.5431
8	.7206	8	.6525	8	.5935	8	.5423
9	.7194	9	.6514	9	.5926	9	.5415
1.180	.7182	1.240	.6504	1.300	.5917	1.360	.5407
1	.7169	1	.6493	1	.5908	1	.5399
2	.7158	2	.6483	2	.5899	2	.5391
3	.7145	3	.6472	3	.5890	3	.5383
4	.7133	4	.6462	4	.5881	4	.5375
5	.7121	5	.6452	5	.5872	5	.5367
6	.7109	6	.6441	6	.5863	6	.5359
7	.7097	7	.6431	7	.5854	7	.5351
8	.7085	8	.6421	8	.5845	8	.5344
9	.7074	9	.6410	9	.5836	9	.5336
1.190	.7062	1.250	.6400	1.310	.5827	1.370	.5328
1	.7050	1	.6390	1	.5818	1	.5320
2	.7038	2	.6380	2	.5809	2	.5312
3	.7026	3	.6369	3	.5800	3	.5305
4	.7014	4	.6359	4	.5792	4	.5297
5	.7003	5	.6349	5	.5783	5	.5289
6	.6991	6	.6339	6	.5774	6	.5282
7	.6979	7	.6329	7	.5765	7	.5274
8	.6968	8	.6319	8	.5757	8	.5266
9	.6956	9	.6309	9	.5748	9	.5259
1.200	.6944	1.260	.6299	1.320	.5739	1.380	.5251
1	.6933	1	.6289	1	.5731	1	.5243
2	.6921	2	.6279	2	.5722	2	.5236
3	.6910	3	.6269	3	.5713	3	.5228
4	.6898	4	.6259	4	.5705	4	.5221
5	.6887	5	.6249	5	.5696	5	.5213
6	.6876	6	.6239	6	.5687	6	.5206
7	.6864	7	.6229	7	.5679	7	.5198
8	.6853	8	.6220	8	.5670	8	.5191
9	.6841	9	.6210	9	.5662	9	.5183
1.210	.6830	1.270	.6200	1.330	.5653	1.390	.5176
1	.6819	1	.6190	1	.5645	1	.5168
2	.6808	2	.6181	2	.5636	2	.5161
3	.6796	3	.6171	3	.5628	3	.5153
4	.6785	4	.6161	4	.5619	4	.5146
5	.6774	5	.6151	5	.5611	5	.5139
6	.6763	6	.6142	6	.5603	6	.5131
7	.6752	7	.6132	7	.5594	7	.5124
8	.6741	8	.6123	8	.5586	8	.5117
9	.6730	9	.6113	9	.5577	9	.5109

d	$1/d^2$	d	$1/d^2$	d	$1/d^2$	d	$1/d^2$
1.400	0.5102	1.460	0.4691	1.520	0.4328	1.580	0.4006
1	.5095	1	.4685	1	.4323	1	.4001
2	.5087	2	.4678	2	.4317	2	.3996
3	.5080	3	.4672	3	.4311	3	.3991
4	.5073	4	.4666	4	.4306	4	.3986
5	.5066	5	.4659	5	.4300	5	.3981
6	.5058	6	.4653	6	.4294	6	.3976
7	.5051	7	.4647	7	.4289	7	.3971
8	.5044	8	.4640	8	.4283	8	.3966
9	.5037	9	.4634	9	.4277	9	.3961
1.410	.5030	1.470	.4628	1.530	.4272	1.590	.3956
1	.5023	1	.4621	1	.4266	1	.3951
2	.5016	2	.4615	2	.4261	2	.3946
3	.5009	3	.4609	3	.4255	3	.3941
4	.5002	4	.4603	4	.4250	4	.3936
5	.4994	5	.4596	5	.4244	5	.3931
6	.4987	6	.4590	6	.4239	6	.3926
7	.4980	7	.4584	7	.4233	7	.3921
8	.4973	8	.4578	8	.4228	8	.3916
9	.4966	9	.4572	9	.4222	9	.3911
1.420	.4959	1.480	.4565	1.540	.4217	1.600	.3906
1	.4952	1	.4559	1	.4211	1	.3901
2	.4945	2	.4553	2	.4206	2	.3897
3	.4938	3	.4547	3	.4200	3	.3892
4	.4932	4	.4541	4	.4195	4	.3887
5	.4925	5	.4535	5	.4189	5	.3882
6	.4918	6	.4529	6	.4184	6	.3877
7	.4911	7	.4522	7	.4178	7	.3872
8	.4904	8	.4516	8	.4173	8	.3867
9	.4897	9	.4510	9	.4168	9	.3863
1.430	.4890	1.490	.4504	1.550	.4162	1.610	.3858
1	.4883	1	.4498	1	.4157	1	.3853
2	.4877	2	.4492	2	.4152	2	.3848
3	.4870	3	.4486	3	.4146	3	.3844
4	.4863	4	.4480	4	.4141	4	.3839
5	.4856	5	.4474	5	.4136	5	.3834
6	.4849	6	.4468	6	.4130	6	.3829
7	.4843	7	.4462	7	.4125	7	.3825
8	.4836	8	.4456	8	.4120	8	.3820
9	.4829	9	.4450	9	.4114	9	.3815
1.440	.4823	1.500	.4444	1.560	.4109	1.620	.3810
1	.4816	1	.4439	1	.4104	1	.3806
2	.4809	2	.4433	2	.4099	2	.3801
3	.4802	3	.4427	3	.4093	3	.3796
4	.4796	4	.4421	4	.4088	4	.3792
5	.4789	5	.4415	5	.4083	5	.3787
6	.4783	6	.4409	6	.4078	6	.3782
7	.4776	7	.4403	7	.4073	7	.3778
8	.4769	8	.4397	8	.4067	8	.3773
9	.4763	9	.4392	9	.4062	9	.3768
1.450	.4756	1.510	.4386	1.570	.4057	1.630	.3764
1	.4750	1	.4380	1	.4052	1	.3759
2	.4743	2	.4374	2	.4047	2	.3755
3	.4737	3	.4368	3	.4041	3	.3750
4	.4730	4	.4363	4	.4036	4	.3745
5	.4724	5	.4357	5	.4031	5	.3741
6	.4717	6	.4351	6	.4026	6	.3736
7	.4711	7	.4345	7	.4021	7	.3732
8	.4704	8	.4340	8	.4016	8	.3727
9	.4698	9	.4334	9	.4011	9	.3723

d	$1/d^2$	d	$1/d^2$	d	$1/d^2$	d	$1/d^2$
1.640	0.3718	1.700	0.3460	1.760	0.3228	1.820	0.3019
1	.3713	1	.3456	1	.3225	1	.3016
2	.3709	2	.3452	2	.3221	2	.3012
3	.3704	3	.3448	3	.3217	3	.3009
4	.3700	4	.3444	4	.3214	4	.3006
5	.3695	5	.3440	5	.3210	5	.3002
6	.3691	6	.3436	6	.3206	6	.2999
7	.3686	7	.3432	7	.3203	7	.2996
8	.3682	8	.3428	8	.3199	8	.2993
9	.3678	9	.3424	9	.3196	9	.2989
1.650	.3673	1.710	.3420	1.770	.3192	1.830	.2986
1	.3669	1	.3416	1	.3188	1	.2983
2	.3664	2	.3412	2	.3185	2	.2980
3	.3660	3	.3408	3	.3181	3	.2976
4	.3655	4	.3404	4	.3178	4	.2973
5	.3651	5	.3400	5	.3174	5	.2970
6	.3647	6	.3396	6	.3170	6	.2967
7	.3642	7	.3392	7	.3167	7	.2963
8	.3638	8	.3388	8	.3163	8	.2960
9	.3633	9	.3384	9	.3160	9	.2957
1.660	.3629	1.720	.3380	1.780	.3156	1.840	.2954
1	.3625	1	.3376	1	.3153	1	.2950
2	.3620	2	.3372	2	.3149	2	.2947
3	.3616	3	.3368	3	.3146	3	.2944
4	.3612	4	.3365	4	.3142	4	.2941
5	.3607	5	.3361	5	.3139	5	.2938
6	.3603	6	.3357	6	.3135	6	.2935
7	.3599	7	.3353	7	.3131	7	.2931
8	.3594	8	.3349	8	.3128	8	.2928
9	.3590	9	.3345	9	.3124	9	.2925
1.670	.3586	1.730	.3341	1.790	.3121	1.850	.2922
1	.3581	1	.3337	1	.3118	1	.2919
2	.3577	2	.3334	2	.3114	2	.2916
3	.3573	3	.3330	3	.3111	3	.2912
4	.3569	4	.3326	4	.3107	4	.2909
5	.3564	5	.3322	5	.3104	5	.2906
6	.3560	6	.3318	6	.3100	6	.2903
7	.3556	7	.3314	7	.3097	7	.2900
8	.3552	8	.3311	8	.3093	8	.2897
9	.3547	9	.3307	9	.3090	9	.2894
1.680	.3543	1.740	.3303	1.800	.3086	1.860	.2891
1	.3539	1	.3299	1	.3083	1	.2887
2	.3535	2	.3295	2	.3080	2	.2884
3	.3530	3	.3292	3	.3076	3	.2881
4	.3526	4	.3288	4	.3073	4	.2878
5	.3522	5	.3284	5	.3069	5	.2875
6	.3518	6	.3280	6	.3066	6	.2872
7	.3514	7	.3277	7	.3063	7	.2869
8	.3510	8	.3273	8	.3059	8	.2866
9	.3505	9	.3269	9	.3056	9	.2863
1.690	.3501	1.750	.3265	1.810	.3052	1.870	.2860
1	.3497	1	.3262	1	.3049	1	.2857
2	.3493	2	.3258	2	.3046	2	.2854
3	.3489	3	.3254	3	.3042	3	.2851
4	.3485	4	.3250	4	.3039	4	.2847
5	.3481	5	.3247	5	.3036	5	.2844
6	.3477	6	.3243	6	.3032	6	.2841
7	.3472	7	.3239	7	.3029	7	.2838
8	.3468	8	.3236	8	.3026	8	.2835
9	.3464	9	.3232	9	.3022	9	.2832

d	$1/d^2$	d	$1/d^2$	d	$1/d^2$	d	$1/d^2$
1.880	0.2829	1.940	0.2657	2.000	0.2500	2.060	0.2356
1	.2826	1	.2654	1	.2498	1	.2354
2	.2823	2	.2652	2	.2495	2	.2352
3	.2820	3	.2649	3	.2493	3	.2350
4	.2817	4	.2646	4	.2490	4	.2347
5	.2814	5	.2643	5	.2488	5	.2345
6	.2811	6	.2641	6	.2485	6	.2343
7	.2808	7	.2638	7	.2483	7	.2341
8	.2805	8	.2635	8	.2480	8	.2338
9	.2802	9	.2633	9	.2478	9	.2336
1.890	.2799	1.950	.2630	2.010	.2475	2.070	.2334
1	.2797	1	.2627	1	.2473	1	.2332
2	.2794	2	.2624	2	.2470	2	.2329
3	.2791	3	.2622	3	.2468	3	.2327
4	.2788	4	.2619	4	.2465	4	.2325
5	.2785	5	.2616	5	.2463	5	.2323
6	.2782	6	.2614	6	.2460	6	.2320
7	.2779	7	.2611	7	.2458	7	.2318
8	.2776	8	.2608	8	.2456	8	.2316
9	.2773	9	.2606	9	.2453	9	.2314
1.900	.2770	1.960	.2603	2.020	.2451	2.080	.2311
1	.2767	1	.2600	1	.2448	1	.2309
2	.2764	2	.2598	2	.2446	2	.2307
3	.2761	3	.2595	3	.2443	3	.2305
4	.2758	4	.2592	4	.2441	4	.2303
5	.2756	5	.2590	5	.2439	5	.2300
6	.2753	6	.2587	6	.2436	6	.2298
7	.2750	7	.2585	7	.2434	7	.2296
8	.2747	8	.2582	8	.2431	8	.2294
9	.2744	9	.2579	9	.2429	9	.2292
1.910	.2741	1.970	.2577	2.030	.2427	2.090	.2289
1	.2738	1	.2574	1	.2424	1	.2287
2	.2735	2	.2571	2	.2422	2	.2285
3	.2733	3	.2569	3	.2419	3	.2283
4	.2730	4	.2566	4	.2417	4	.2281
5	.2727	5	.2564	5	.2415	5	.2278
6	.2724	6	.2561	6	.2412	6	.2276
7	.2721	7	.2559	7	.2410	7	.2274
8	.2718	8	.2556	8	.2408	8	.2272
9	.2716	9	.2553	9	.2405	9	.2270
1.920	.2713	1.980	.2551	2.040	.2403	2.100	.2268
1	.2710	1	.2548	1	.2401	1	.2265
2	.2707	2	.2546	2	.2398	2	.2263
3	.2704	3	.2543	3	.2396	3	.2261
4	.2701	4	.2540	4	.2394	4	.2259
5	.2699	5	.2538	5	.2391	5	.2257
6	.2696	6	.2535	6	.2389	6	.2255
7	.2693	7	.2533	7	.2387	7	.2253
8	.2690	8	.2530	8	.2384	8	.2250
9	.2687	9	.2528	9	.2382	9	.2248
1.930	.2685	1.990	.2525	2.050	.2380	2.110	.2246
1	.2682	1	.2523	1	.2377	1	.2244
2	.2679	2	.2520	2	.2375	2	.2242
3	.2676	3	.2518	3	.2373	3	.2240
4	.2674	4	.2515	4	.2370	4	.2238
5	.2671	5	.2513	5	.2368	5	.2236
6	.2668	6	.2510	6	.2366	6	.2233
7	.2665	7	.2508	7	.2363	7	.2231
8	.2663	8	.2505	8	.2361	8	.2229
9	.2660	9	.2503	9	.2359	9	.2227

d	$1/d^2$	d	$1/d^2$	d	$1/d^2$	d	$1/d^2$
2.120	0.2225	2.180	0.2104	2.240	0.1993	2.300	0.1890
1	.2223	1	.2102	1	.1991	1	.1889
2	.2221	2	.2100	2	.1989	2	.1887
3	.2219	3	.2098	3	.1988	3	.1885
4	.2217	4	.2096	4	.1986	4	.1884
5	.2215	5	.2095	5	.1984	5	.1882
6	.2212	6	.2093	6	.1982	6	.1881
7	.2210	7	.2091	7	.1981	7	.1879
8	.2208	8	.2089	8	.1979	8	.1877
9	.2206	9	.2087	9	.1977	9	.1876
2.130	.2204	2.190	.2085	2.250	.1975	2.310	.1874
1	.2202	1	.2083	1	.1974	1	.1872
2	.2200	2	.2081	2	.1972	2	.1871
3	.2198	3	.2079	3	.1970	3	.1869
4	.2196	4	.2077	4	.1968	4	.1868
5	.2194	5	.2076	5	.1967	5	.1866
6	.2192	6	.2074	6	.1965	6	.1864
7	.2190	7	.2072	7	.1963	7	.1863
8	.2188	8	.2070	8	.1961	8	.1861
9	.2186	9	.2068	9	.1960	9	.1860
2.140	.2184	2.200	.2066	2.260	.1958	2.320	.1858
1	.2182	1	.2064	1	.1956	1	.1850
2	.2180	2	.2062	2	.1954	2	.1855
3	.2177	3	.2060	3	.1953	3	.1853
4	.2175	4	.2059	4	.1951	4	.1852
5	.2173	5	.2057	5	.1949	5	.1850
6	.2171	6	.2055	6	.1948	6	.1848
7	.2169	7	.2053	7	.1946	7	.1847
8	.2167	8	.2051	8	.1944	8	.1845
9	.2165	9	.2049	9	.1942	9	.1844
2.150	.2163	2.210	.2047	2.270	.1941	2.330	.1842
1	.2161	1	.2046	1	.1939	1	.1840
2	.2159	2	.2044	2	.1937	2	.1839
3	.2157	3	.2042	3	.1936	3	.1837
4	.2155	4	.2040	4	.1934	4	.1836
5	.2153	5	.2038	5	.1932	5	.1834
6	.2151	6	.2036	6	.1930	6	.1833
7	.2149	7	.2035	7	.1929	7	.1831
8	.2147	8	.2033	8	.1927	8	.1829
9	.2145	9	.2031	9	.1925	9	.1828
2.160	.2143	2.220	.2029	2.280	.1924	2.340	.1826
1	.2141	1	.2027	1	.1922	1	.1825
2	.2139	2	.2025	2	.1920	2	.1823
3	.2137	3	.2024	3	.1919	3	.1822
4	.2135	4	.2022	4	.1917	4	.1820
5	.2133	5	.2020	5	.1915	5	.1819
6	.2131	6	.2018	6	.1914	6	.1817
7	.2130	7	.2016	7	.1912	7	.1815
8	.2128	8	.2015	8	.1910	8	.1814
9	.2126	9	.2013	9	.1909	9	.1812
2.170	.2124	2.230	.2011	2.290	.1907	2.350	.1811
1	.2122	1	.2009	1	.1905	1	.1809
2	.2120	2	.2007	2	.1904	2	.1808
3	.2118	3	.2005	3	.1902	3	.1806
4	.2116	4	.2004	4	.1900	4	.1805
5	.2114	5	.2002	5	.1899	5	.1803
6	.2112	6	.2000	6	.1897	6	.1802
7	.2110	7	.1998	7	.1895	7	.1800
8	.2108	8	.1997	8	.1894	8	.1799
9	.2106	9	.1995	9	.1892	9	.1797

d	$1/d^2$	d	$1/d^2$	d	$1/d^2$	d	$1/d^2$
2.360	0.1795	2.420	0.1708	2.480	0.1626	2.540	0.1550
1	.1794	1	.1706	1	.1625	1	.1549
2	.1792	2	.1705	2	.1623	2	.1548
3	.1791	3	.1703	3	.1622	3	.1546
4	.1789	4	.1702	4	.1621	4	.1545
5	.1788	5	.1700	5	.1619	5	.1544
6	.1786	6	.1699	6	.1618	6	.1543
7	.1785	7	.1698	7	.1617	7	.1541
8	.1783	8	.1696	8	.1615	8	.1540
9	.1782	9	.1695	9	.1614	9	.1539
2.370	.1780	2.430	.1694	2.490	.1613	2.550	.1538
1	.1779	1	.1692	1	.1612	1	.1537
2	.1777	2	.1691	2	.1610	2	.1535
3	.1776	3	.1689	3	.1609	3	.1534
4	.1774	4	.1688	4	.1608	4	.1533
5	.1773	5	.1687	5	.1606	5	.1532
6	.1771	6	.1685	6	.1605	6	.1531
7	.1770	7	.1684	7	.1604	7	.1529
8	.1768	8	.1682	8	.1603	8	.1528
9	.1767	9	.1681	9	.1601	9	.1527
2.380	.1765	2.440	.1680	2.500	.1600	2.560	.1526
1	.1764	1	.1678	1	.1599	1	.1525
2	.1762	2	.1677	2	.1597	2	.1523
3	.1761	3	.1676	3	.1596	3	.1522
4	.1759	4	.1674	4	.1595	4	.1521
5	.1758	5	.1673	5	.1594	5	.1520
6	.1757	6	.1671	6	.1592	6	.1519
7	.1755	7	.1670	7	.1591	7	.1518
8	.1754	8	.1669	8	.1590	8	.1516
9	.1752	9	.1667	9	.1589	9	.1515
2.390	.1751	2.450	.1666	2.510	.1587	2.570	.1514
1	.1749	1	.1665	1	.1586	1	.1513
2	.1748	2	.1663	2	.1585	2	.1512
3	.1746	3	.1662	3	.1583	3	.1510
4	.1745	4	.1661	4	.1582	4	.1509
5	.1743	5	.1659	5	.1581	5	.1508
6	.1742	6	.1658	6	.1580	6	.1507
7	.1740	7	.1656	7	.1578	7	.1506
8	.1739	8	.1655	8	.1577	8	.1505
9	.1738	9	.1654	9	.1576	9	.1503
2.400	.1736	2.460	.1652	2.520	.1575	2.580	.1502
1	.1735	1	.1651	1	.1573	1	.1501
2	.1733	2	.1650	2	·1572	2	.1500
3	.1732	3	.1648	3	.1571	3	.1499
4	.1730	4	.1647	4	.1570	4	.1498
5	.1729	5	.1646	5	.1568	5	.1497
6	.1727	6	.1644	6	.1567	6	.1495
7	.1726	7	.1643	7	.1566	7	.1494
8	.1725	8	.1642	8	.1565	8	.1493
9	.1723	9	.1640	9	.1564	9	.1492
2.410	.1722	2.470	.1639	2.530	.1562	2.590	.1491
1	.1720	1	.1638	1	.1561	1	.1490
2	.1719	2	.1636	2	.1560	2	.1488
3	.1717	3	.1635	3	.1559	3	.1487
4	.1716	4	.1634	4	.1557	4	.1486
5	.1715	5	.1632	5	.1556	5	.1485
6	.1713	6	.1631	6	.1555	6	.1484
7	.1712	7	.1630	7	.1554	7	.1483
8	.1710	8	.1629	8	.1552	8	.1482
9	.1709	9	.1627	9	.1551	9	.1480

d	$1/d^2$	d	$1/d^2$	d	$1/d^2$	d	$1/d^2$
2.600	0.1479	2.660	0.1413	2.720	0.1352	2.780	0.1294
1	.1478	1	.1412	1	.1351	1	.1293
2	.1477	2	.1411	2	.1350	2	.1292
3	.1476	3	.1410	3	.1349	3	.1291
4	.1475	4	.1409	4	.1348	4	.1290
5	.1474	5	.1408	5	.1347	5	.1289
6	.1472	6	.1407	6	.1346	6	.1288
7	.1471	7	.1406	7	.1345	7	.1287
8	.1470	8	.1405	8	.1344	8	.1287
9	.1469	9	.1404	9	.1343	9	.1286
2.610	.1468	2.670	.1403	2.730	.1342	2.790	.1285
1	.1467	1	.1402	1	.1341	1	.1284
2	.1466	2	.1401	2	.1340	2	.1283
3	.1465	3	.1400	3	.1339	3	.1282
4	.1463	4	.1399	4	.1338	4	.1281
5	.1462	5	.1398	5	.1337	5	.1280
6	.1461	6	.1396	6	.1336	6	.1279
7	.1460	7	.1395	7	.1335	7	.1278
8	.1459	8	.1394	8	.1334	8	.1277
9	.1458	9	.1393	9	.1333	9	.1276
2.620	.1457	2.680	.1392	2.740	.1332	2.800	.1276
1	.1456	1	.1391	1	.1331	1	.1275
2	.1455	2	.1390	2	.1330	2	.1274
3	.1453	3	.1389	3	.1329	3	.1273
4	.1452	4	.1388	4	.1328	4	.1272
5	.1451	5	.1387	5	.1327	5	.1271
6	.1450	6	.1386	6	.1326	6	.1270
7	.1449	7	.1385	7	.1325	7	.1269
8	.1448	8	.1384	8	.1324	8	.1268
9	.1447	9	.1383	9	.1323	9	.1267
2.630	.1446	2.690	.1382	2.750	.1322	2.810	.1266
1	.1445	1	.1381	1	.1321	1	.1266
2	.1444	2	.1380	2	.1320	2	.1265
3	.1442	3	.1379	3	.1319	3	.1264
4	.1441	4	.1378	4	.1318	4	.1263
5	.1440	5	.1377	5	.1318	5	.1262
6	.1439	6	.1376	6	.1317	6	.1261
7	.1438	7	.1375	7	.1316	7	.1260
8	.1437	8	.1374	8	.1315	8	.1259
9	.1436	9	.1373	9	.1314	9	.1258
2.640	.1435	2.700	.1372	2.760	.1313	2.820	.1257
1	.1434	1	.1371	1	.1312	1	.1257
2	.1433	2	.1370	2	.1311	2	.1256
3	.1432	3	.1369	3	.1310	3	.1255
4	.1430	4	.1368	4	.1309	4	.1254
5	.1429	5	.1367	5	.1308	5	.1253
6	.1428	6	.1366	6	.1307	6	.1252
7	.1427	7	.1365	7	.1306	7	.1251
8	.1426	8	.1364	8	.1305	8	.1250
9	.1425	9	.1363	9	.1304	9	.1249
2.650	.1424	2.710	.1362	2.770	.1303	2.830	.1249
1	.1423	1	.1361	1	.1302	1	.1248
2	.1422	2	.1360	2	.1301	2	.1247
3	.1421	3	.1359	3	.1300	3	.1246
4	.1420	4	.1358	4	.1300	4	.1245
5	.1419	5	.1357	5	.1299	5	.1244
6	.1418	6	.1356	6	.1298	6	.1243
7	.1417	7	.1355	7	.1297	7	.1242
8	.1415	8	.1354	8	.1296	8	.1242
9	.1414	9	.1353	9	.1295	9	.1241

d	$1/d^2$	d	$1/d^2$	d	$1/d^2$	d	$1/d^2$
2.840	0.1240	2.900	0.1189	2.960	0.1141	3.020	0.1096
1	.1239	1	.1188	1	.1141	1	.1096
2	.1238	2	.1187	2	.1140	2	.1095
3	.1237	3	.1187	3	.1139	3	.1094
4	.1236	4	.1186	4	.1138	4	.1094
5	.1235	5	.1185	5	.1137	5	.1093
6	.1235	6	.1184	6	.1137	6	.1092
7	.1234	7	.1183	7	.1136	7	.1091
8	.1233	8	.1183	8	.1135	8	.1091
9	.1232	9	.1182	9	.1134	9	.1090
2.850	.1231	2.910	.1181	2.970	.1134	3.030	.1089
1	.1230	1	.1180	1	.1133	1	.1088
2	.1229	2	.1179	2	.1132	2	.1088
3	.1229	3	.1178	3	.1131	3	.1087
4	.1228	4	.1178	4	.1131	4	.1086
5	.1227	5	.1177	5	.1130	5	.1086
6	.1226	6	.1176	6	.1129	6	.1085
7	.1225	7	.1175	7	.1128	7	.1084
8	.1224	8	1174	8	.1128	8	.1083
9	.1223	9	.1174	9	.1127	9	.1083
2.860	.1223	2.920	.1173	2.980	.1126	3.040	.1082
1	.1222	1	.1172	1	.1125	1	.1081
2	.1221	2	.1171	2	.1125	2	.1081
3	.1220	3	.1170	3	.1124	3	.1080
4	.1219	4	.1170	4	.1123	4	.1079
5	.1218	5	.1169	5	.1122	5	.1079
6	.1217	6	.1168	6	.1122	6	.1078
7	.1217	7	.1167	7	.1121	7	.1077
8	.1216	8	.1166	8	.1120	8	.1076
9	.1215	9	.1166	9	.1119	9	.1076
2.870	.1214	2.930	.1165	2.990	.1119	3.050	.1075
1	.1213	1	.1164	1	.1118	1	.1074
2	.1212	2	.1163	2	.1117	2	.1074
3	.1212	3	.1162	3	.1116	3	.1073
4	.1211	4	.1162	4	.1116	4	.1072
5	.1210	5	.1161	5	.1115	5	.1071
6	.1209	6	.1160	6	.1114	6	.1071
7	.1208	7	.1159	7	.1113	7	.1070
8	.1207	8	.1159	8	.1113	8	.1069
9	.1206	9	.1158	9	.1112	9	.1069
2.880	.1206	2.940	.1157	3.000	.1111	3.060	.1068
1	.1205	1	.1156	1	.1110	1	.1067
2	.1204	2	.1155	2	.1110	2	.1067
3	.1203	3	.1155	3	.1109	3	.1066
4	.1202	4	.1154	4	.1108	4	.1065
5	.1201	5	.1153	5	.1107	5	.1064
6	.1201	6	.1152	6	.1107	6	.1064
7	.1200	7	.1151	7	.1106	7	.1063
8	.1199	8	.1151	8	.1105	8	.1062
9	.1198	9	.1150	9	.1104	9	.1062
2.890	.1197	2.950	.1149	3.010	.1104	3.070	.1061
1	.1196	1	.1148	1	.1103	1	.1060
2	.1196	2	.1148	2	.1102	2	.1060
3	.1195	3	.1147	3	.1102	3	.1059
4	.1194	4	.1146	4	.1101	4	.1058
5	.1193	5	.1145	5	.1100	5	.1058
6	.1192	6	.1144	6	.1099	6	.1057
7	.1192	7	.1144	7	.1099	7	.1056
8	.1191	8	.1143	8	.1098	8	.1056
9	.1190	9	.1142	9	.1097	9	.1055

d	$1/d^2$	d	$1/d^2$	d	$1/d^2$	d	$1/d^2$
3.080	0.1054	3.140	0.1014	3.200	0.0977	3.260	0,0941
1	.1053	1	.1014	1	.0976	1	.0940
2	.1053	2	.1013	2	.0975	2	.0940
3	.1052	3	.1012	3	.0975	3	.0939
4	.1051	4	.1012	4	.0974	4	.0939
5	.1051	5	.1011	5	.0974	5	.0938
6	.1050	6	.1010	6	.0973	6	.0937
7	.1049	7	.1010	7	.0972	7	.0937
8	.1049	8	.1009	8	.0972	8	.0936
9	.1048	9	.1008	9	.0971	9	.0936
3.090	.1047	3.150	.1008	3.210	.0970	3.270	.0935
1	.1047	1	.1007	1	.0970	1	.0935
2	.1046	2	.1007	2	.0969	2	.0934
3	.1045	3	.1006	3	.0969	3	.0933
4	.1045	4	.1C05	4	.0968	4	.0933
5	.1044	5	.1005	5	.0967	5	.0932
6	.1043	6	.1004	6	.0967	6	.0932
7	.1043	7	.1003	7	.0966	7	0931
8	.1042	8	.1003	8	.0966	8	.0931
9	.1041	9	.1002	9	.0965	9	.0930
3.100	.1041	3.160	.1001	3.220	.0964	3.280	.0930
1	.1040	1	.1001	1	.0964	1	.0929
2	.1039	2	.1000	2	.0963	2	.0928
3	.1039	3	.1000	3	.0963	3	.0928
4	.1038	4	.0999	4	.0962	4	.0927
5	.1037	5	.0998	5	.0961	5	.0927
6	.1037	6	.0998	6	.0961	6	.0926
7	.1036	7	.0997	7	.0960	7	.0926
8	.1035	8	.0996	8	.0960	8	.0925
9	.1035	9	.0996	9	.0959	9	.0924
3.110	.1034	3.170	.0995	3.230	.0959	3.290	.0924
1	.1033	1	.0995	1	.0958	1	.0923
2	.1033	2	.0994	2	.0957	2	.0923
3	.1032	3	.0993	3	.0957	3	.0922
4	.1031	4	.0993	4	.0956	4	.0922
5	.1031	5	.0992	5	.0956	5	.0921
6	.1030	6	.0991	6	.0955	6	.0921
7	.1029	7	.0991	7	.0954	7	.0920
8	.1029	8	.0990	8	.0954	8	.0919
9	.1028	9	.0990	9	.0953	9	.0919
3.120	.1027	3.180	.0989	3.240	.0953	3.300	.0918
1	.1027	1	.0988	1	.0952	1	.0918
2	.1026	2	.0988	2	.0951	2	.0917
3	.1025	3	.0987	3	.0951	3	.0917
4	.1025	4	.0986	4	.0950	4	.0916
5	.1024	5	.0986	5	.0950	5	.0915
6	.1023	6	.C985	6	.0949	6	.0915
7	.1023	7	.0985	7	.0948	7	.0914
8	.1022	8	.0984	8	.0948	8	.0914
9	.1021	9	.0983	9	.0947	9	.0913
3.130	.1021	3.190	.0983	3.250	.0947	3.310	.0913
1	.1020	1	.0982	1	.0946	1	.0912
2	.1019	2	.0981	2	.0946	2	.0912
3	.1019	3	.0981	3	.0945	3	.0911
4	.1018	4	.0980	4	.0944	4	.0911
5	.1017	5	.0980	5	.0944	5	.0910
6	.1017	6	.0979	6	.0943	6	.0909
7	.1016	7	.0978	7	.0943	7	.0909
8	.1016	8	.0978	8	.0942	8	.0908
9	.1015	9	.0977	9	.0942	9	.0908

d	$1/d^2$	d	$1/d^2$	d	$1/d^2$	d	$1/d^2$
3.320	0.0907	3.380	0.0875	3.440	0.0845	3.500	0.0816
1	.0907	1	.0875	1	.0845	1	.0816
2	.0906	2	.0874	2	.0844	2	.0315
3	.0906	3	.0874	3	.0844	3	.0815
4	.0905	4	.0873	4	.0843	4	.0814
5	.0905	5	.0873	5	.0843	5	.0814
6	.0904	6	.0872	6	.0842	6	.0814
7	.0903	7	.0872	7	.0842	7	.0813
8	.0903	8	.0871	8	.0841	8	.0813
9	.0902	9	.0871	9	.0841	9	.0812
3.330	.0902	3.390	.0870	3.450	.0840	3.510	.0812
1	.0901	1	.0870	1	.0840	1	.0811
2	.0901	2	.0869	2	.0839	2	.0811
3	.0900	3	.0869	3	.0839	3	.0810
4	.0900	4	.0868	4	.0838	4	.0810
5	.0899	5	.0868	5	.0838	5	.0809
6	.0899	6	.0867	6	.0837	6	.0809
7	.0898	7	.0867	7	.0837	7	.0808
8	.0897	8	.0866	8	.0836	8	.0808
9	.0897	9	.0866	9	.0836	9	.0808
3.340	.0896	3.400	.0865	3.460	.0835	3.520	.0807
1	.0896	1	.0865	1	.0835	1	.0807
2	.0895	2	.0864	2	.0834	2	.0806
3	.0895	3	.0864	3	.0834	3	.0806
4	.0894	4	.0863	4	.0833	4	.0805
5	.0894	5	.0863	5	.0833	5	.0805
6	.0893	6	.0862	6	.0832	6	.0804
7	.0893	7	.0862	7	.0832	7	.0804
8	.0892	8	.0861	8	.0831	8	.0803
9	.0892	9	.0860	9	.0831	9	.0803
3.350	.0891	3.410	.0860	3.470	.0831	3.530	.0803
1	.0891	1	.0859	1	.0830	1	.0802
2	.0890	2	.0859	2	.0830	2	.0802
3	.0889	3	.0858	3	.0829	3	.0801
4	.0889	4	.0858	4	.0829	4	.0801
5	.0888	5	.0857	5	.0828	5	.0800
6	.0888	6	.0857	6	.0828	6	.0800
7	.0887	7	.0856	7	.0827	7	.0799
8	.0887	8	.0856	8	.0827	8	.0799
9	.0886	9	.0855	9	.0826	9	.0798
3.360	.0886	3.420	.0855	3.480	.0826	3.540	.0798
1	.0885	1	.0854	1	.0825	1	.0798
2	.0885	2	.0854	2	.0825	2	.0797
3	.0884	3	.0853	3	.0824	3	.0797
4	.0884	4	.0853	4	.0824	4	.0796
5	.0883	5	.0852	5	.0823	5	.0796
6	.0883	6	.0852	6	.0823	6	.0795
7	.0882	7	.0851	7	.0822	7	.0795
8	.0882	8	.0851	8	.0822	8	.0794
9	.0881	9	.0850	9	.0821	9	.0794
3.370	.0881	3.430	.0850	3.490	.0821	3.550	.0793
1	.0880	1	.0849	1	.0821	1	.0793
2	.0879	2	.0849	2	.0820	2	.0793
3	.0879	3	.0849	3	.0820	3	.0792
4	.0878	4	.0848	4	.0819	4	.0792
5	.0878	5	.0848	5	.0819	5	.0791
6	.0877	6	.0847	6	.0818	6	.0791
7	.0877	7	.0847	7	.0818	7	.0790
8	.0876	8	.0846	8	.0817	8	.0790
9	.0876	9	.0846	9	.0817	9	.0789

d	$1/d^2$	d	$1/d^2$	d	$1/d^2$	d	$1/d^2$
3.560	0.0789	3.620	0.0763	3.680	0.0738	3.740	0.0715
1	.0789	1	.0763	1	.0738	1	.0715
2	.0788	2	.0762	2	.0738	2	.0714
3	.0788	3	.0762	3	.0737	3	.0714
4	.0787	4	.0761	4	.0737	4	.0713
5	.0787	5	.0761	5	.0736	5	.0713
6	.0786	6	.0761	6	.0736	6	.0713
7	.0786	7	.0760	7	.0736	7	.0712
8	.0786	8	.0760	8	.0735	8	.0712
9	.0785	9	.0759	9	.0735	9	.0711
3.570	.0785	3.630	.0759	3.690	.0734	3.750	.0711
1	.0784	1	.0758	1	.0734	1	.0711
2	.0784	2	.0758	2	.0734	2	.0710
3	.0783	3	.0758	3	.0733	3	.0710
4	.0783	4	.0757	4	.0733	4	.0710
5	.0782	5	.0757	5	.0732	5	.0709
6	.0782	6	.0756	6	.0732	6	.0709
7	.0782	7	.0756	7	.0732	7	.0708
8	.0781	8	.0756	8	.0731	8	.0708
9	.0781	9	.0755	9	.0731	9	.0708
3.580	.0780	3.640	.0755	3.700	.0730	3.760	.0707
1	.0780	1	.0754	1	.0730	1	.0707
2	.0779	2	.0754	2	.0730	2	.0707
3	.0779	3	.0753	3	.0729	3	.0706
4	.0779	4	.0753	4	.0729	4	.0706
5	.0778	5	.0753	5	.0728	5	.0705
6	.0778	6	.0752	6	.0728	6	.0705
7	.0777	7	.0752	7	.0728	7	.0705
8	.0777	8	.0751	8	.0727	8	.0704
9	.0776	9	.0751	9	.0727	9	.0704
3.590	.0776	3.650	.0751	3.710	.0727	3.770	.0704
1	.0775	1	.0750	1	.0726	1	.0703
2	.0775	2	.0750	2	.0726	2	.0703
3	.0775	3	.0749	3	.0725	3	.0702
4	.0774	4	.0749	4	.0725	4	.0702
5	.0774	5	.0749	5	.0725	5	.0702
6	.0773	6	.0748	6	.0724	6	.0701
7	.0773	7	.0748	7	.0724	7	.0701
8	.0772	8	.0747	8	.0723	8	.0701
9	.0772	9	.0747	9	.0723	9	.0700
3.600	.0772	3.660	.0747	3.720	.0723	3.780	.0700
1	.0771	1	.0746	1	.0722	1	.0699
2	.0771	2	.0746	2	.0722	2	.0699
3	.0770	3	.0745	3	.0721	3	.0699
4	.0770	4	.0745	4	.0721	4	.0698
5	.0769	5	.0744	5	.0721	5	.0698
6	.0769	6	.0744	6	.0720	6	.0698
7	.0769	7	.0744	7	.0720	7	.0697
8	.0768	8	.0743	8	.0720	8	.0697
9	.0768	9	.0743	9	.0719	9	.0697
3.610	.0767	3.670	.0742	3.730	.0719	3.790	.0696
1	.0767	1	.0742	1	.0718	1	.0696
2	.0766	2	.0742	2	.0718	2	.0695
3	.0766	3	.0741	3	.0718	3	.0695
4	.0766	4	.0741	4	.0717	4	.0695
5	.0765	5	.0740	5	.0717	5	.0694
6	.0765	6	.0740	6	.0716	6	.0694
7	.0764	7	.0740	7	.0716	7	.0694
8	.0764	8	.0739	8	.0716	8	.0693
9	.0764	9	.0739	9	.0715	9	.0693

d	$1/d^2$	d	$1/d^2$	d	$1/d^2$	d	$1/d^2$
3.800	0.0693	3.860	0.0671	3.920	0.0651	3.980	0.0631
1	.0692	1	.0671	1	.0650	1	.0631
2	.0692	2	.0670	2	.0650	2	.0631
3	.0691	3	.0670	3	.0650	3	.0630
4	.0691	4	.0670	4	.0649	4	.0630
5	.0691	5	.0669	5	.0649	5	.0630
6	.0690	6	.0669	6	.0649	6	.0629
7	.0690	7	.0669	7	.0648	7	.0629
8	.0690	8	.0668	8	.0648	8	.0629
9	.0689	9	.0668	9	.0648	9	.0628
3.810	.0689	3.870	.0668	3.930	.0647	3.990	.0628
1	.0689	1	.0667	1	.0647	1	.0628
2	.0688	2	.0667	2	.0647	2	.0628
3	.0688	3	.0667	3	.0646	3	.0627
4	.0687	4	.0666	4	.0646	4	.0627
5	.0687	5	.0666	5	.0646	5	.0627
6	.0687	6	.0666	6	.0645	6	.0626
7	.0686	7	.0665	7	.0645	7	.0626
8	.0686	8	.0665	8	.0645	8	.0626
9	.0686	9	.0665	9	.0645	9	.0625
3.820	.0685	3.880	.0664	3.940	.0644	4.000	.062500
1	.0685	1	.0664	1	.0644	1	.062469
2	.0685	2	.0664	2	.0644	2	.062438
3	.0684	3	.0663	3	.0643	3	.062406
4	.0684	4	.0663	4	.0643	4	.062375
5	.0683	5	.0663	5	.0643	5	.062344
6	.0683	6	.0662	6	.0642	6	.062313
7	.0683	7	.0662	7	.0642	7	.062282
8	.0682	8	.0662	8	.0642	8	.062251
9	.0682	9	.0661	9	.0641	9	.062220
3.830	.0682	3.890	.0661	3.950	.0641	4.010	.062189
1	.0681	1	.0661	1	.0641	1	.062158
2	.0681	2	.0660	2	.0640	2	.062127
3	.0681	3	.0660	3	.0640	3	.062096
4	.0680	4	.0659	4	.0640	4	.062065
5	.0680	5	.0659	5	.0639	5	.062034
6	.0680	6	.0659	6	.0639	6	.062003
7	.0679	7	.0658	7	.0639	7	.061972
8	.0679	8	.0658	8	.0638	8	.061941
9	.0679	9	.0658	9	.0638	9	.061910
3.840	.0678	3.900	.0657	3.960	.0638	4.020	.061880
1	.0678	1	.0657	1	.0637	1	.061849
2	.0677	2	.0657	2	.0637	2	.061818
3	.0677	3	.0656	3	.0637	3	.061787
4	.0677	4	.0656	4	.0636	4	.061757
5	.0676	5	.0656	5	.0636	5	.061726
6	.0676	6	.0655	6	.0636	6	.061695
7	.0676	7	.0655	7	.0635	7	.061665
8	0675	8	.0655	8	.0635	8	.061634
9	0675	9	.0654	9	.0635	9	.061604
3.850	.0675	3.910	.0654	3.970	.0634	4.030	.061573
1	.0674	1	.0654	1	.0634	1	.061542
2	.0674	2	.0653	2	.0634	2	.061512
3	.0674	3	.0653	3	.0634	3	.061481
4	.0673	4	.0653	4	.0633	4	.061451
5	.0673	5	.0652	5	.0633	5	.061420
6	.0673	6	.0652	6	.0633	6	.061390
7	.0672	7	.0652	7	.0632	7	.061360
8	.0672	8	.0651	8	.0632	8	.061329
9	.0672	9	.0651	9	.0632	9	.061299

d	$1/d^2$	d	$1/d^2$	d	$1/d^2$	d	$1/d^2$
4.040	0.061269	4.100	0.059488	4.160	0.057785	4.220	0.056153
1	.061238	1	.059459	1	.057757	1	.056127
2	.061208	2	.059430	2	.057729	2	.056100
3	.061178	3	.059401	3	.057702	3	.056074
4	.061147	4	.059372	4	.057674	4	.056047
5	.061117	5	.059344	5	.057646	5	.056020
6	.061087	6	.059315	6	.057618	6	.055994
7	.061057	7	.059286	7	.057591	7	.055967
8	.061027	8	.059257	8	.057563	8	.055941
9	.060996	9	.059228	9	.057536	9	.055915
4.050	.060966	4.110	.059199	4.170	.057508	4.230	.055888
1	.060936	1	.059170	1	.057480	1	.055862
2	.060906	2	.059142	2	.057453	2	.055835
3	.060876	3	.059113	3	.057425	3	.055809
4	.060846	4	.059084	4	.057398	4	.055783
5	.060816	5	.059056	5	.057370	5	.055756
6	.060786	6	.059027	6	.057343	6	.055730
7	.060756	7	.058998	7	.057315	7	.055704
8	.060726	8	.058969	8	.057288	8	.055677
9	.060696	9	.058941	9	.057261	9	.055651
4.060	.060666	4.120	.058912	4.180	.057233	4.240	.055625
1	.060636	1	.058884	1	.057206	1	.055599
2	.060607	2	.058855	2	.057178	2	.055572
3	.060577	3	.058827	3	.057151	3	.055546
4	.060547	4	.058798	4	.057124	4	.055520
5	.060517	5	.058770	5	.057096	5	.055494
6	.060487	6	.058741	6	.057069	6	.055468
7	.060458	7	.058713	7	.057042	7	.055442
8	.060428	8	.058684	8	.057015	8	.055415
9	.060398	9	.058656	9	.056987	9	.055389
4.070	.060369	4.130	.058627	4.190	.056960	4.250	.055363
1	.060339	1	.058599	1	.056933	1	.055337
2	.060309	2	.058571	2	.056906	2	.055311
3	.060280	3	.058542	3	.056879	3	.055285
4	.060250	4	.058514	4	.056852	4	.055259
5	.060221	5	.058486	5	.056825	5	.055233
6	.060191	6	.058457	6	.056797	6	.055207
7	.060161	7	.058429	7	.056770	7	.055181
8	.060132	8	.058401	8	.056743	8	.055155
9	.060103	9	.058373	9	.056716	9	.055130
4.080	.060073	4.140	.058344	4.200	.056689	4.260	.055104
1	.060044	1	.058316	1	.056662	1	.055078
2	.060014	2	.058288	2	.056635	2	.055052
3	.059985	3	.058260	3	.056608	3	.055026
4	.059955	4	.058232	4	.056582	4	.055000
5	.059926	5	.058204	5	.056555	5	.054975
6	.059897	6	.058176	6	.056528	6	.054949
7	.059867	7	.058148	7	.056501	7	.054923
8	.059838	8	.058120	8	.056474	8	.054897
9	.059809	9	.058092	9	.056447	9	.054872
4.090	.059780	4.150	.058064	4.210	.056420	4.270	.054846
1	.059750	1	.058036	1	.056394	1	.054820
2	.059721	2	.058008	2	.056367	2	.054795
3	.059692	3	.057980	3	.056340	3	.054769
4	.058663	4	.057952	4	.056313	4	.054743
5	.059634	5	.057924	5	.056287	5	.054718
6	.059605	6	.057896	6	.056260	6	.054692
7	.059576	7	.057868	7	.056233	7	.054667
8	.059546	8	.057840	8	.056207	8	.054641
9	.059517	9	.057813	9	.056180	9	.054615

d	$1/d^2$	d	$1/d^2$	d	$1/d^2$	d	$1/d^2$
4.280	0.054590	4.340	0.053091	4.400	0.051653	4.460	0.050272
1	.054564	1	.053067	1	.051629	1	.050250
2	.054539	2	.053042	2	.051606	2	.050227
3	.054513	3	.053018	3	.051583	3	.050205
4	.054488	4	.052993	4	.051559	4	.050182
5	.054463	5	.052969	5	.051536	5	.050160
6	.054437	6	.052944	6	.051512	6	.050137
7	.054412	7	.052920	7	.051489	7	.050115
8	.054386	8	.052896	8	.051466	8	.050093
9	.054361	9	.052871	9	.051442	9	.050070
4.290	.054336	4.350	.052847	4.410	.051419	4.470	.050048
1	.054310	1	.052823	1	.051396	1	.050025
2	.054285	2	.052799	2	.051372	2	.050003
3	.054260	3	.052774	3	.051349	3	.049981
4	.054235	4	.052750	4	.051326	4	.049958
5	.054209	5	.052726	5	.051303	5	.049936
6	.054184	6	.052702	6	.051279	6	.049914
7	.054159	7	.052677	7	.051256	7	.049891
8	.054134	8	.052653	8	.051233	8	.049869
9	.054108	9	.052629	9	.051210	9	.049847
4.300	.054083	4.360	.052605	4.420	.051187	4.480	.049825
1	.054058	1	.052581	1	.051163	1	.049802
2	.054033	2	.052557	2	.051140	2	.049780
3	.054008	3	.052533	3	.051117	3	.049758
4	.053983	4	.052509	4	.051094	4	.049736
5	.053958	5	.052485	5	.051071	5	.049714
6	.053933	6	.052461	6	.051048	6	.049691
7	.053908	7	.052436	7	.051025	7	.049669
8	.053883	8	.052412	8	.051002	8	.049647
9	.053858	9	.052388	9	.050979	9	.049625
4.310	.053833	4.370	.052365	4.430	.050956	4.490	.049603
1	.053808	1	.052341	1	.050933	1	.049581
2	.053783	2	.052317	2	.050910	2	.049559
3	.053758	3	.052293	3	.050887	3	.049537
4	.053733	4	.052269	4	.050864	4	.049515
5	.053708	5	.052245	5	.050841	5	.049493
6	.053683	6	.052221	6	.050818	6	.049471
7	.053658	7	.052197	7	.050795	7	.049449
8	.053633	8	.052173	8	.050772	8	.049427
9	.053608	9	.052150	9	.050749	9	.049405
4.320	.053584	4.380	.052126	4.440	.050726	4.500	.049383
1	.053559	1	.052102	1	.050704	1	.049361
2	.053534	2	.052078	2	.050681	2	.049339
3	.053509	3	.052054	3	.050658	3	.049317
4	.053485	4	.052031	4	.050635	4	.049295
5	.053460	5	.052007	5	.050612	5	.049273
6	.053435	6	.051983	6	.050590	6	.049251
7	.053410	7	.051959	7	.050567	7	.049229
8	.053386	8	.051936	8	.050544	8	.049208
9	.053361	9	.051912	9	.050521	9	.049186
4.330	.053336	4.390	.051888	4.450	.050499	4.510	.049164
1	.053312	1	.051865	1	.050476	1	.049142
2	.053287	2	.051841	2	.050453	2	.049120
3	.053263	3	.051818	3	.050431	3	.049099
4	.053238	4	.051794	4	.050408	4	.049077
5	.053214	5	.051770	5	.050385	5	.049055
6	.053189	6	.051747	6	.050363	6	.049033
7	.053164	7	.051723	7	.050340	7	.049012
8	.053140	8	.051700	8	.050318	8	.048990
9	.053115	9	.051676	9	.050295	9	.048968

d	$1/d^2$	d	$1/d^2$	d	$1/d^2$	d	$1/d^2$
4.520	0.048947	**4.580**	0.047673	**4.640**	0.046448	**4.700**	0.045269
1	.048925	1	.047652	1	.046428	1	.045250
2	.048903	2	.047631	2	.046408	2	.045231
3	.048882	3	.047610	3	.046388	3	.045212
4	.048860	4	.047589	4	.046368	4	.045192
5	.048839	5	.047569	5	.046348	5	.045173
6	.048817	6	.047548	6	.046328	6	.045154
7	.048795	7	.047527	7	.046308	7	.045135
8	.048774	8	.047507	8	.046288	8	.045116
9	.048752	9	.047486	9	.046268	9	.045096
4.530	.048731	**4.590**	.047465	**4.650**	.046248	**4.710**	.045077
1	.048709	1	.047444	1	.046228	1	.045058
2	.048688	2	.047424	2	.046208	2	.045039
3	.048666	3	.047403	3	.046189	3	.045020
4	.048645	4	.047383	4	.046169	4	.045001
5	.048623	5	.047362	5	.046149	5	.044982
6	.048602	6	.047341	6	.046129	6	.044963
7	.048581	7	.047321	7	.046109	7	.044944
8	.048559	8	.047300	8	.046089	8	.044925
9	.048538	9	.047280	9	.046070	9	.044906
4.540	.048516	**4.600**	.047259	**4.660**	.046050	**4.720**	.044887
1	.048495	1	.047238	1	.046030	1	.044868
2	.048474	2	.047218	2	.046010	2	.044849
3	.048452	3	.047197	3	.045991	3	.044830
4	.048431	4	.047177	4	.045971	4	.044811
5	.048410	5	.047156	5	.045951	5	.044792
6	.048388	6	.047136	6	.045931	6	.044773
7	.048367	7	.047115	7	.045912	7	.044754
8	.048346	8	.047095	8	.045892	8	.044735
9	.048325	9	.047075	9	.045872	9	.044716
4.550	.048303	**4.610**	.047054	**4.670**	.045853	**4.730**	.044697
1	.048282	1	.047034	1	.045833	1	.044678
2	.048261	2	.047013	2	.045814	2	.044659
3	.048240	3	.046993	3	.045794	3	.044640
4	.048219	4	.046973	4	.045774	4	.044621
5	.048197	5	.046952	5	.045755	5	.044603
6	.048176	6	.046932	6	.045735	6	.044584
7	.048155	7	.046912	7	.045716	7	.044565
8	.048134	8	.046891	8	.045696	8	.044546
9	.048113	9	.046871	9	.045677	9	.044527
4.560	.048092	**4.620**	.046851	**4.680**	.045657	**4.740**	.044509
1	.048071	1	.046830	1	.045638	1	.044490
2	.048050	2	.046810	2	.045618	2	.044471
3	.048029	3	.046790	3	.045599	3	.044452
4	.048007	4	.046770	4	.045579	4	.044434
5	.047986	5	.046749	5	.045560	5	.044415
6	.047965	6	.046729	6	.045540	6	.044396
7	.047944	7	.046709	7	.045521	7	.044377
8	.047923	8	.046669	8	.045501	8	.044359
9	.047902	9	.046669	9	.045482	9	.044340
4.570	.047881	**4.630**	.046649	**4.690**	.045463	**4.750**	.044321
1	.047861	1	.046628	1	.045443	1	.044303
2	.047840	2	.046608	2	.045424	2	.044284
3	.047819	3	.046588	3	.045405	3	.044265
4	.047798	4	.046568	4	.045385	4	.044247
5	.047777	5	.046548	5	.045366	5	.044228
6	.047756	6	.046528	6	.045347	6	.044210
7	.047735	7	.046508	7	.045327	7	.044191
8	.047714	8	.046488	8	.045308	8	.044172
9	.047693	9	.046468	9	.045289	9	.044154

d	$1/d^2$	d	$1/d^2$	d	$1/d^2$	d	$1/d^2$
4.760	0.044135	4.820	0.043043	4.880	0.041991	4.940	0.040978
1	.044117	1	.043025	1	.041974	1	.040961
2	.044098	2	.043008	2	.041957	2	.040944
3	.044080	3	.042990	3	.041940	3	.040928
4	.044061	4	.042972	4	.041923	4	.040911
5	.044043	5	.042954	5	.041905	5	.040895
6	.044024	6	.042936	6	.041888	6	.040878
7	.044006	7	.042919	7	.041871	7	.040862
8	.043987	8	.042901	8	.041854	8	.040845
9	.043969	9	.042883	9	.041837	9	.040829
4.770	.043950	4.830	.042865	4.890	.041820	4.950	.040812
1	.043932	1	.042848	1	.041803	1	.040796
2	.043914	2	.042830	2	.041786	2	.040779
3	.043895	3	.042812	3	.041769	3	.040763
4	.043877	4	.042794	4	.041752	4	.040746
5	.043858	5	.042777	5	.041734	5	.040730
6	.043840	6	.042759	6	.041717	6	.040713
7	.043822	7	.042741	7	.041700	7	.040697
8	.043803	8	.042724	8	.041683	8	.040681
9	.043785	9	.042706	9	.041666	9	.040664
4.780	.043767	4.840	.042688	4.900	.041649	4.960	.040648
1	.043748	1	.042671	1	.041632	1	.040631
2	.043730	2	.042653	2	.041615	2	.040615
3	.043712	3	.042635	3	.041598	3	.040599
4	.043694	4	.042618	4	.041581	4	.040582
5	.043675	5	.042600	5	.041564	5	.040566
6	.043657	6	.042583	6	.041548	6	.040550
7	.043639	7	.042565	7	.041531	7	.040533
8	.043621	8	.042548	8	.041514	8	.040517
9	.043602	9	.042530	9	.041497	9	.040501
4.790	.043584	4.850	.042512	4.910	.041480	4.970	.040484
1	.043566	1	.042495	1	.041463	1	.040468
2	.043548	2	.042477	2	.041446	2	.040452
3	.043530	3	.042460	3	.041429	3	.040436
4	.043511	4	.042442	4	.041412	4	.040419
5	.043493	5	.042425	5	.041395	5	.040403
6	.043475	6	.042408	6	.041379	6	.040387
7	.043457	7	.042390	7	.041362	7	.040371
8	.043439	8	.042373	8	.041345	8	.040354
9	.043421	9	.042355	9	.041328	9	.040338
4.800	.043403	4.860	.042338	4.920	.041311	4.980	.040322
1	.043385	1	.042320	1	.041295	1	.040306
2	.043367	2	.042303	2	.041278	2	.040290
3	.043349	3	.042286	3	.041261	3	.040273
4	.043331	4	.042268	4	.041244	4	.040257
5	.043313	5	.042251	5	.041228	5	.040241
6	.043294	6	.042233	6	.041211	6	.040225
7	.043276	7	.042216	7	.041194	7	.040209
8	.043258	8	.042199	8	.041177	8	.040193
9	.043240	9	.042181	9	.041161	9	.040177
4.810	.043223	4.870	.042164	4.930	.041144	4.990	.040160
1	.043205	1	.042147	1	.041127	1	.040144
2	.043187	2	.042129	2	.041111	2	.040128
3	.043169	3	.042112	3	.041094	3	.040112
4	.043151	4	.042095	4	.041077	4	.040096
5	.043133	5	.042078	5	.041061	5	.040080
6	.043115	6	.042060	6	.041044	6	.040064
7	.043097	7	.042043	7	.041027	7	.040048
8	.043079	8	.042026	8	.041011	8	.040032
9	.043061	9	.042009	9	.040994	9	.040016

d	$1/d^2$	d	$1/d^2$	d	$1/d^2$	d	$1/d^2$
5.000	0.040000	5.060	0.039057	5.120	0.038147	5.180	0.037268
1	.039984	1	.039042	1	.038132	1	.037254
2	.039968	2	.039026	2	.038117	2	.037240
3	.039952	3	.039011	3	.038102	3	.037225
4	.039936	4	.038995	4	.038087	4	.037211
5	.039920	5	.038980	5	.038073	5	.037197
6	.039904	6	.038965	6	.038058	6	.037182
7	.039888	7	.038949	7	.038043	7	.037168
8	.039872	8	.038934	8	.038028	8	.037154
9	.039856	9	.038918	9	.038013	9	.037139
5.010	.039840	5.070	.038903	5.130	.037998	5.190	.037125
1	.039825	1	.038888	1	.037984	1	.037111
2	.039809	2	.038872	2	.037969	2	.037096
3	.039793	3	.038857	3	.037954	3	.037082
4	.039777	4	.038842	4	.037939	4	.037068
5	.039761	5	.038826	5	.037924	5	.037053
6	.039745	6	.038811	6	.037910	6	.037039
7	.039729	7	.038796	7	.037895	7	.037025
8	.039714	8	.038781	8	.037880	8	.037011
9	.039698	9	.038765	9	.037865	9	.036996
5.020	.039682	5.080	.038750	5.140	.037851	5.200	.036982
1	.039666	1	.038735	1	.037836	1	.036968
2	.039650	2	.038720	2	.037821	2	.036954
3	.039635	3	.038704	3	.037807	3	.036940
4	.039619	4	.038689	4	.037792	4	.036925
5	.039603	5	.038674	5	.037777	5	.036911
6	.039587	6	.038659	6	.037762	6	.036897
7	.039571	7	.038644	7	.037748	7	.036883
8	.039556	8	.038628	8	.037733	8	.036869
9	.039540	9	.038613	9	.037718	9	.036855
5.030	.039524	5.090	.038598	5.150	.037704	5.210	.036840
1	.039509	1	.038583	1	.037689	1	.036826
2	.039493	2	.038568	2	.037675	2	.035812
3	.039477	3	.038553	3	.037660	3	.036798
4	.039462	4	.038537	4	.037645	4	.036784
5	.039446	5	.038522	5	.037631	5	.036770
6	.039430	6	.038507	6	.037616	6	.036756
7	.039415	7	.038492	7	.037602	7	.036742
8	.039399	8	.038477	8	.037587	8	.036728
9	.039383	9	.038462	9	.037572	9	.036713
5.040	.039368	5.100	.038447	5.160	.037558	5.220	.036699
1	.039352	1	.038432	1	.037543	1	.036685
2	.039336	2	.038417	2	.037529	2	.036671
3	.039321	3	.038402	3	.037514	3	.036657
4	.039305	4	.038387	4	.037500	4	.036643
5	.039290	5	.038371	5	.037485	5	.036629
6	.039274	6	.038356	6	.037471	6	.036615
7	.039258	7	.038341	7	.037456	7	.036601
8	.039243	8	.038326	8	.037442	8	.036587
9	.039227	9	.038311	9	.037427	9	.036573
5.050	.039212	5.110	.038296	5.170	.037413	5.230	.036559
1	.039196	1	.038281	1	.037398	1	.036545
2	.039181	2	.038266	2	.037384	2	.036531
3	.039165	3	.038252	3	.037369	3	.036517
4	.039150	4	.038237	4	.037355	4	.036503
5	.039134	5	.038222	5	.037340	5	.036489
6	.039119	6	.038207	6	.037326	6	.036475
7	.039103	7	.038192	7	.037312	7	.036462
8	.039088	8	.038177	8	.037297	8	.036448
9	.039072	9	.038162	9	.037283	9	.036434

d	$1/d^2$	d	$1/d^2$	d	$1/d^2$	d	$1/d^2$
5.240	0.036420	5.300	0.035600	5.360	0.034807	5.420	0.034041
1	.036406	1	.035586	1	.034794	1	.034028
2	.036392	2	.035573	2	.034781	2	.034016
3	.036378	3	.035560	3	.034768	3	.034003
4	.036364	4	.035546	4	.034755	4	.033991
5	.036350	5	.035533	5	.034742	5	.033978
6	.036337	6	.035519	6	.034730	6	.033966
7	.036323	7	.035506	7	.034717	7	.033953
8	.036309	8	.035493	8	.034704	8	.033941
9	.036295	9	.035479	9	.034691	9	.033928
5.250	.036281	5.310	.035466	5.370	.034678	5.430	.033916
1	.036267	1	.035453	1	.034665	1	.033903
2	.036254	2	.035439	2	.034652	2	.033891
3	.036240	3	.035426	3	.034639	3	.033878
4	.036226	4	.035413	4	.034626	4	.033866
5	.036212	5	.035399	5	.034613	5	.033853
6	.036198	6	.035386	6	.034600	6	.033841
7	.036185	7	.035373	7	.034588	7	.033828
8	.036171	8	.035359	8	.034575	8	.033816
9	.036157	9	.035346	9	.034562	9	.033804
5.260	.036143	5.320	.035333	5.380	.034549	5.440	.033791
1	.036130	1	.035319	1	.034536	1	.033779
2	.036116	2	.035306	2	.034523	2	.033766
3	.036102	3	.035293	3	.034511	3	.033754
4	.036088	4	.035280	4	.034498	4	.033741
5	.036075	5	.035266	5	.034485	5	.033729
6	.036061	6	.035253	6	.034472	6	.033717
7	.036047	7	.035240	7	.034459	7	.033704
8	.036034	8	.035227	8	.034446	8	.033692
9	.036020	9	.035213	9	.034434	9	.033680
5.270	.036006	5.330	.035200	5.390	.034421	5.450	.033667
1	.035993	1	.035187	1	.034408	1	.033655
2	.035979	2	.035174	2	.034395	2	.033643
3	.035965	3	.035161	3	.034383	3	.033630
4	.035952	4	.035147	4	.034370	4	.033618
5	.035938	5	.035134	5	.034357	5	.033606
6	.035924	6	.035121	6	.034344	6	.033593
7	.035911	7	.035108	7	.034332	7	.033581
8	.035897	8	.035095	8	.034319	8	.033569
9	.035884	9	.035082	9	.034306	9	.033556
5.280	.035870	5.340	.035069	5.400	.034292	5.460	.033544
1	.035856	1	.035055	1	.034281	1	.033532
2	.035843	2	.035042	2	.034268	2	.033519
3	.035829	3	.035029	3	.034255	3	.033507
4	.035816	4	.035016	4	.034243	4	.033495
5	.035802	5	.035003	5	.034230	5	.033483
6	.035789	6	.034990	6	.034217	6	.033470
7	.035775	7	.034977	7	.034205	7	.033458
8	.035762	8	.034964	8	.034192	8	.033446
9	.035748	9	.034951	9	.034180	9	.033434
5.290	.035735	5.350	.034938	5.410	.034167	5.470	.033421
1	.035721	1	.034924	1	.034154	1	.033409
2	.035708	2	.034911	2	.034142	2	.033397
3	.035694	3	.034898	3	.034129	3	.033385
4	.035681	4	.034885	4	.034116	4	.033373
5	.035667	5	.034872	5	.034104	5	.033360
6	.035654	6	.034859	6	.034091	6	.033348
7	.035640	7	.034846	7	.034079	7	.033336
8	.035627	8	.034833	8	.034066	8	.033324
9	.035613	9	.034820	9	.034054	9	.033312

d	$1/d^2$	d	$1/d^2$	d	$1/d^2$	d	$1/d^2$
5.480	0.033300	5.540	0.032582	5.600	0.031888	5.660	0.031215
1	.033287	1	.032570	1	.031876	1	.031204
2	.033275	2	.032559	2	.031865	2	.031193
3	.033263	3	.032547	3	.031854	3	.031182
4	.033251	4	.032535	4	.031842	4	.031171
5	.033239	5	.032523	5	.031831	5	.031160
6	.033227	6	.032512	6	.031820	6	.031149
7	.033215	7	.032500	7	.031808	7	.031138
8	.033203	8	.032488	8	.031797	8	.031127
9	.033190	9	.032477	9	.031786	9	.031116
5.490	.033178	5.550	.032465	5.610	.031774	5.670	.031105
1	.033166	1	.032453	1	.031763	1	.031094
2	.033154	2	.032442	2	.031752	2	.031083
3	.033142	3	.032430	3	.031740	3	.031072
4	.033130	4	.032418	4	.031729	4	.031061
5	.033118	5	.032406	5	.031718	5	.031050
6	.033106	6	.032395	6	.031706	6	.031040
7	.033094	7	.032383	7	.031695	7	.031029
8	.033082	8	.032372	8	.031684	8	.031018
9	.033070	9	.032360	9	.031672	9	.031007
5.500	.033058	5.560	.032348	5.620	.031661	5.680	.030996
1	.033046	1	.032337	1	.031650	1	.030985
2	.033034	2	.032325	2	.031639	2	.030974
3	.033022	3	.032313	3	.031627	3	.030963
4	.033010	4	.032302	4	.031616	4	.030952
5	.032998	5	.032290	5	.031605	5	.030941
6	.032986	6	.032279	6	.031594	6	.030930
7	.032974	7	.032267	7	.031582	7	.030920
8	.032962	8	.032255	8	.031571	8	.030909
9	.032950	9	.032244	9	.031560	9	.030898
5.510	.032938	5.570	.032232	5.630	.031549	5.690	.030887
1	.032926	1	.032221	1	.031538	1	.030876
2	.032914	2	.032209	2	.031526	2	.030865
3	.032902	3	.032197	3	.031515	3	.030854
4	.032890	4	.032186	4	.031504	4	.030844
5	.032878	5	.032174	5	.031493	5	.030833
6	.032866	6	.032163	6	.031482	6	.030822
7	.032854	7	.032151	7	.031471	7	.030811
8	.032843	8	.032140	8	.031459	8	.030800
9	.032831	9	.032128	9	.031448	9	.030790
5.520	.032819	5.580	.032117	5.640	.031437	5.700	.030779
1	.032807	1	.032105	1	.031426	1	.030768
2	.032795	2	.032094	2	.031415	2	.030757
3	.032783	3	.032082	3	.031404	3	.030746
4	.032771	4	032071	4	.031393	4	.030736
5	.032759	5	.032059	5	.031381	5	.030725
6	.032748	6	.032048	6	.031370	6	.030714
7	.032736	7	.032036	7	.031359	7	.030703
8	.032724	8	.032025	8	.031348	8	.030692
9	.032712	9	.032013	9	.031337	9	.030682
5.530	.032700	5.590	.032002	5.650	.031326	5.710	.030671
1	.032688	1	.031991	1	.031315	1	.030660
2	.032677	2	.031979	2	.031304	2	.030650
3	.032665	3	.031968	3	.031293	3	.030639
4	.032653	4	.031956	4	.031282	4	.030628
5	.032641	5	.031945	5	.031270	5	.030617
6	.032629	6	.031933	6	.031259	6	.030607
7	.032618	7	.031922	7	.031248	7	.030596
8	.032606	8	.031911	8	.031237	8	.030585
9	.032594	9	.031899	9	.031226	9	.030575

d	$1/d^2$	d	$1/d^2$	d	$1/d^2$	d	$1/d^2$
5.720	0.030564	5.780	0.029933	5,840	0.029321	5.900	0.028727
1	.030553	1	.029922	1	.029311	1	.028718
2	.030542	2	.029912	2	.029301	2	.028708
3	.030532	3	.029902	3	.029291	3	.028698
4	.030521	4	.029891	4	.029281	4	.028688
5	.030510	5	.029881	5	.029271	5	.028679
6	.030500	6	.029871	6	.029261	6	.028669
7	.030489	7	.029860	7	.029251	7	.028659
8	.030479	8	.029850	8	.029241	8	.028650
9	.030468	9	.029840	9	.029231	9	.028640
5.730	.030457	5.790	.029829	5.850	.029221	5.910	.028630
1	.030447	1	.029819	1	.029211	1	.028621
2	.030436	2	.029809	2	.029201	2	.028611
3	.030425	3	.029798	3	.029191	3	.028601
4	.030415	4	.029788	4	.029181	4	.028592
5	.030404	5	.029778	5	.029171	5	.028582
6	.030394	6	.029768	6	.029161	6	.028572
7	.030383	7	.029757	7	.029151	7	.028563
8	.030372	8	.029747	8	.029141	8	.028553
9	.030362	9	.029737	9	.029131	9	.028543
5.740	.030351	5.800	.029727	5.860	.029121	5.920	.028534
1	.030341	1	.029716	1	.029111	1	.028524
2	.030330	2	.029706	2	.029101	2	.028514
3	.030320	3	.029696	3	.029091	3	.028505
4	.030309	4	.029686	4	.029081	4	.028495
5	.030298	5	.029675	5	.029071	5	.028485
6	.030288	6	.029665	6	.029061	6	.028476
7	.030277	7	.029655	7	.029051	7	.028466
8	.030267	8	.029645	8	.029042	8	.028457
9	.030256	9	.029634	9	.029032	9	.028447
5.750	.030246	5.810	.029624	5.870	.029022	5.930	.028437
1	.030235	1	.029614	1	.029012	1	.028428
2	.030225	2	.029604	2	.029002	2	.028418
3	.030214	3	.029594	3	.028992	3	.028409
4	.030204	4	.029584	4	.028982	4	.028399
5	.030193	5	.029573	5	.028972	5	.028390
6	.030183	6	.029563	6	.028963	6	.028380
7	.030172	7	.029553	7	.028953	7	.028370
8	.030162	8	.029543	8	.028943	8	.028361
9	.030151	9	.029533	9	.028933	9	.028351
5.760	.030141	5.820	.029523	5.880	.028923	5.940	.028342
1	.030130	1	.029512	1	.028913	1	.028332
2	.030120	2	.029502	2	.028903	2	.028323
3	.030109	3	.029492	3	.028894	3	.028313
4	.030099	4	.029482	4	.028884	4	.028304
5	.030089	5	.029472	5	.028874	5	.028294
6	.030078	6	.020462	6	.028864	6	.028285
7	.030068	7	.029452	7	.028854	7	.028275
8	.030057	8	.029442	8	.028845	8	.028266
9	.030047	9	.029431	9	.028835	9	.028256
5.770	.030036	5.830	.029421	5.890	.028825	5.950	.028247
1	.030026	1	.029411	1	.028815	1	.028237
2	.030016	2	.029101	2	.028805	2	.028228
3	.030005	3	.029391	3	.028796	3	.028218
4	.029995	4	.029381	4	.028786	4	.028209
5	.029984	5	.029371	5	.028776	5	.028199
6	.029974	6	.029361	6	.028766	6	.028190
7	.029964	7	.029351	7	.028757	7	.028180
8	.029953	8	.029311	8	.028747	8	.028171
9	.029943	9	.029331	9	.028737	9	.028161

d	$1/d^2$	d	$1/d^2$	d	$1/d^2$	d	$1/d^2$
5.960	0.028152	6.020	0.027594	6.080	0.027052	6.140	0.026525
1	.028142	1	.027584	1	.027043	1	.026517
2	.028133	2	.027575	2	.027034	2	.026508
3	.028124	3	.027566	3	.027025	3	.026500
4	.028114	4	.027557	4	.027016	4	.026491
5	.028105	5	.027548	5	.027007	5	.026482
6	.028095	6	.027539	6	.026998	6	.026474
7	.028086	7	.027529	7	.026989	7	.026465
8	.028076	8	.027520	8	.026981	8	.026456
9	.028067	9	.027511	9	.026972	9	.026448
5.970	.028058	6.030	.027502	6.090	.026963	6.150	.026439
1	.028048	1	.027493	1	.026954	1	.026431
2	.028039	2	.027484	2	.026945	2	.026422
3	.028029	3	.027475	3	.026936	3	.026414
4	.028020	4	.027466	4	.026927	4	.026405
5	.028011	5	.027457	5	.026919	5	.026396
6	.028001	6	.027447	6	.026910	6	.026388
7	.027992	7	.027438	7	.026901	7	.026379
8	.027983	8	.027429	8	.026892	8	.026371
9	.027973	9	.027420	9	.026883	9	.026362
5.980	.027964	6.040	.027411	6.100	.026874	6.160	.026354
1	.027955	1	.027402	1	.026866	1	.026345
2	.027945	2	.027393	2	.026857	2	.026336
3	.027936	3	.027384	3	.026848	3	.026328
4	.027927	4	.027375	4	.026839	4	.026319
5	.027917	5	.027366	5	.026830	5	.026311
6	.027908	6	.027357	6	.026822	6	.026302
7	.027899	7	.027348	7	.026813	7	.026294
8	.027889	8	.027339	8	.026804	8	.026285
9	.027880	9	.027330	9	.026795	9	.026277
5.990	.027871	6.050	.027321	6.110	.026787	6.170	.026268
1	.027861	1	.027312	1	.026778	1	.026260
2	.027852	2	.027302	2	.026769	2	.026251
3	.027843	3	.027293	3	.026760	3	.026243
4	.027833	4	.027284	4	.026752	4	.026234
5	.027824	5	.027275	5	.026743	5	.026226
6	.027815	6	.027266	6	.026734	6	.026217
7	.027806	7	.027257	7	.026725	7	.026209
8	.027796	8	.027248	8	.026717	8	.026200
9	.027787	9	.027239	9	.026717	9	.026192
6.000	.027778	6.060	.027230	6.120	.026699	6.180	.026183
1	.027769	1	.027221	1	.026690	1	.026175
2	.027759	2	.027212	2	.026682	2	.026166
3	.027750	3	.027204	3	.026673	3	.026158
4	.027741	4	.027195	4	.026664	4	.026149
5	.027732	5	.027186	5	.026656	5	.026141
6	.027722	6	.027177	6	.026647	6	.026132
7	.027713	7	.027168	7	.026638	7	.026125
8	.027704	8	.027159	8	.026629	8	.026116
9	.027695	9	.027150	9	.026621	9	.026107
6.010	.027685	6.070	.027141	6.130	.026612	6.190	.026099
1	.027676	1	.027132	1	.026603	1	.026090
2	.027667	2	.027123	2	.026595	2	.026082
3	.027659	3	.027114	3	.026586	3	.026073
4	.027649	4	.027105	4	.026577	4	.026065
5	.027639	5	.027096	5	.026569	5	.026057
6	.027630	6	.027087	6	.026560	6	.026048
7	.027621	7	.027078	7	.026551	7	.026040
8	.027612	8	.027069	8	.026543	8	.026031
9	.027603	9	.027060	9	.026534	9	.026023

d	$1/d^2$	d	$1/d^2$	d	$1/d^2$	d	$1/d^2$
6.200	0.026015	6.260	0.025518	6.320	0.025036	6.380	0.024567
1	.026006	1	.025510	1	.025028	1	.024560
2	.025998	2	.025502	2	.025020	2	.024552
3	.025989	3	.025494	3	.025012	3	.024544
4	.025981	4	.025486	4	.025004	4	.024537
5	.025973	5	.025478	5	.024996	5	.024529
6	.025964	6	.025469	6	.024989	6	.024521
7	.025956	7	.025461	7	.024981	7	.024514
8	.025948	8	.025453	8	.024973	8	.024506
9	.025939	9	.025445	9	.024965	9	.024498
6.210	.025931	6.270	.025437	6.330	.024957	6.390	.024491
1	.025923	1	.025429	1	.024949	1	.024483
2	.025914	2	.025421	2	.024941	2	.024475
3	.025906	3	.025413	3	.024933	3	.024468
4	.025897	4	.025405	4	.024926	4	.024460
5	.025889	5	.025396	5	.024918	5	.024452
6	.025881	6	.025388	6	.024910	6	.024445
7	.025872	7	.025380	7	.024902	7	.024437
8	.025864	8	.025372	8	.024894	8	.024429
9	.025856	9	.025364	9	.024886	9	.024422
6.220	.025848	6.280	.025356	6.340	.024878	6.400	.024414
1	.025839	1	.025348	1	.024870	1	.024406
2	.025831	2	.025340	2	.024863	2	.024399
3	.025823	3	.025332	3	.024855	3	.024391
4	.025814	4	.025324	4	.024847	4	.024384
5	.025806	5	.025316	5	.024839	5	.024376
6	.025798	6	.025308	6	.024831	6	.024368
7	.025789	7	.025300	7	.024823	7	.024361
8	.025781	8	.025292	8	.024816	8	.024353
9	.025773	9	.025283	9	.024808	9	.024346
6.230	.025765	6.290	.025275	6.350	.024800	6.410	.024338
1	.025756	1	.025267	1	.024792	1	.024330
2	.025748	2	.025259	2	.024784	2	.024323
3	.025740	3	.025251	3	.024777	3	.024315
4	.025732	4	.025243	4	.024769	4	.024308
5	.025723	5	.025235	5	.024761	5	.024300
6	.025715	6	.025227	6	.024753	6	.024292
7	.025707	7	.025219	7	.024745	7	.024285
8	.025699	8	.025211	8	.024738	8	.024277
9	.025690	9	.025203	9	.024730	9	.024270
6.240	.025682	6.300	.025195	6.360	.024722	6.420	.024262
1	.025674	1	.025187	1	.024714	1	.024255
2	.025666	2	.025179	2	.024707	2	.024247
3	.025657	3	.025171	3	.024699	3	.024240
4	.025649	4	.025163	4	.024691	4	.024232
5	.025641	5	.025155	5	.024683	5	.024224
6	.025633	6	.025147	6	.024676	6	.024217
7	.025625	7	.025139	7	.024668	7	.024209
8	.025616	8	.025131	8	.024660	8	.024202
9	.025608	9	.025123	9	.024652	9	.024194
6.250	.025600	6.310	0.25115	6.370	.024645	6.430	.024187
1	.025592	1	0.25108	1	.024637	1	.024179
2	.025584	2	0.25100	2	.024629	2	.024172
3	.025575	3	0.25092	3	.024621	3	.024164
4	.025567	4	0.25084	4	.024614	4	.024157
5	.025559	5	0.25076	5	.024606	5	.024149
6	.025551	6	0.25068	6	.024598	6	.024142
7	.025543	7	0.25060	7	.024590	7	.024134
8	.025535	8	0.25052	8	.024583	8	.024127
9	.025527	9	0.25044	9	.024575	9	.024119

d	$1/d^2$	d	$1/d^2$	d	$1/d^2$	d	$1/d^2$
6.440	0.024112	**6.500**	0.023669	**6.560**	0.023238	**6.620**	0.022818
1	.024104	1	.023661	1	.023231	1	.022811
2	.024097	2	.023654	2	.023223	2	.022805
3	.024089	3	.023647	3	.023216	3	.022798
4	.024082	4	.023640	4	.023209	4	.022791
5	.024074	5	.023632	5	.023202	5	.022784
6	.024067	6	.023625	6	.023195	6	.022777
7	.024059	7	.023618	7	.023188	7	.022770
8	.024052	8	.023610	8	.023181	8	.022763
9	.024044	9	.023603	9	.023174	9	.022756
6.450	.024037	**6.510**	.023596	**6.570**	.023167	**6.630**	.022750
1	.024030	1	.023589	1	.023160	1	.022743
2	.024022	2	.023581	2	.023153	2	.022736
3	.024015	3	.023574	3	.023146	3	.022729
4	.024007	4	.023567	4	.023139	4	.022722
5	.024000	5	.023560	5	.023132	5	.022715
6	.023992	6	.023553	6	.023125	6	.022708
7	.023985	7	.023545	7	.023118	7	.022702
8	.023978	8	.023538	8	.023111	8	.022695
9	.023970	9	.023531	9	.023104	9	.022688
6.460	.023963	**6.520**	.023524	**6.580**	.023097	**6.640**	.022681
1	.023955	1	.023516	1	.023090	1	.022674
2	.023948	2	.023509	2	.023083	2	.022667
3	.023940	3	.023502	3	.023076	3	.022661
4	.023933	4	.023495	4	.023069	4	.022654
5	.023926	5	.023488	5	.023062	5	.022647
6	.023918	6	.023480	6	.023055	6	.022640
7	.023911	7	.023473	7	.023048	7	.022633
8	.023903	8	.023466	8	.023041	8	.022627
9	.023896	9	.023459	9	.023034	9	.022620
6.470	.023889	**6.530**	.023452	**6.590**	.023027	**6.650**	.022613
1	.023881	1	.023444	1	.023020	1	.022606
2	.023874	2	.023437	2	.023013	2	.022599
3	.023867	3	.023430	3	.023006	3	.022593
4	.023859	4	.023423	4	.022999	4	.022586
5	.023852	5	.023416	5	.022992	5	.022579
6	.023844	6	.023409	6	.022985	6	.022572
7	.023837	7	.023401	7	.022978	7	.022565
8	.023830	8	.023394	8	.022971	8	.022559
9	.023822	9	.023387	9	.022964	9	.022552
6.480	.023815	**6.540**	.023380	**6.600**	.022957	**6.660**	.022545
1	.023808	1	.023373	1	.022950	1	.022538
2	.023800	2	.023366	2	.022943	2	.022532
3	.023793	3	.023359	3	.022936	3	.022525
4	.023786	4	.023351	4	.022929	4	.022511
5	.023778	5	.023344	5	.022922	5	.022505
6	.023771	6	.023337	6	.022915	6	.022500
7	.023764	7	.023330	7	.022908	7	.022498
8	.023756	8	.023323	8	.022901	8	.022491
9	.023749	9	.023316	9	.022894	9	.022484
6.490	.023742	**6.550**	.023309	**6.610**	.022887	**6.670**	.022478
1	.023734	1	.023302	1	.022881	1	.022471
2	.023727	2	.023294	2	.022874	2	.022464
3	.023720	3	.023287	3	.022867	3	.022457
4	.023712	4	.023280	4	.022860	4	.022451
5	.023705	5	.023273	5	.022853	5	.022444
6	.023698	6	.023266	6	.022846	6	.022437
7	.023691	7	.023259	7	.022839	7	.022430
8	.023683	8	.023252	8	.022832	8	.022424
9	.023676	9	.023245	9	.022825	9	.022417

d	$1/d^2$	d	$1/d^2$	d	$1/d^2$	d	$1/d^2$
6.680	0.022410	**6.740**	0.022013	**6.800**	0.021626	**6.860**	0.021250
1	.022404	1	.022007	1	.021620	1	.021243
2	.022397	2	.022000	2	.021614	2	.021237
3	.022390	3	.021993	3	.021607	3	.021231
4	.022383	4	.021987	4	.021601	4	.021225
5	.022377	5	.021980	5	.021595	5	.021219
6	.022370	6	.021974	6	.021588	6	.021213
7	.022363	7	.021967	7	.021582	7	.021206
8	.022357	8	.021961	8	.021576	8	.021200
9	.022350	9	.021954	9	.021569	9	.021194
6.690	.022343	**6.750**	.021948	**6.810**	.021563	**6.870**	.021188
1	.022337	1	.021941	1	.021556	1	.021182
2	.022330	2	.021935	2	.021550	2	.021176
3	.022323	3	.021928	3	.021544	3	.021169
4	.022317	4	.021922	4	.021538	4	.021163
5	.022310	5	.021915	5	.021531	5	.021157
6	.022303	6	.021909	6	.021525	6	.021151
7	.022297	7	.021902	7	.021519	7	.021145
8	.022290	8	.021896	8	.021512	8	.021139
9	.022283	9	.021889	9	.021506	9	.021126
6.700	.022277	**6.760**	.021883	**6.820**	.021500	**6.880**	.021126
1	.022270	1	.021877	1	.021493	1	.021120
2	.022263	2	.021870	2	.021487	2	.021114
3	.022257	3	.021864	3	.021481	3	.021108
4	.022250	4	.021857	4	.021474	4	.021102
5	.022243	5	.021851	5	.021468	5	.021096
6	.022237	6	.021844	6	.021462	6	.021089
7	.022230	7	.021838	7	.021456	7	.021083
8	.022224	8	.021831	8	.021449	8	.021077
9	.022217	9	.021825	9	.021443	9	.021071
6.710	.022210	**6.770**	.021818	**6.830**	.021437	**6.890**	.021065
1	.022204	1	.021812	1	.021430	1	.021059
2	.022197	2	.021806	2	.021424	2	.021053
3	.022190	3	.021799	3	.021418	3	.021047
4	.022184	4	.021793	4	.021412	4	.021041
5	.022177	5	.021786	5	.021405	5	.021034
6	.022171	6	.021780	6	.021399	6	.021028
7	.022164	7	.021773	7	.021393	7	.021022
8	.022157	8	.021767	8	.021387	8	.021016
9	.022151	9	.021760	9	.021380	9	.021010
6.720	.022144	**6.780**	.021754	**6.840**	.021374	**6.900**	.021004
1	.022138	1	.021748	1	.021368	1	.020998
2	.022131	2	.021741	2	.021362	2	.020992
3	.022125	3	.021735	3	.021355	3	.020986
4	.022118	4	.021728	4	.021349	4	.020980
5	.022111	5	.021722	5	.021343	5	.020974
6	.022105	6	.021716	6	.021337	6	.020968
7	.022098	7	.021709	7	.021330	7	.020961
8	.022092	8	.021703	8	.021324	8	.020955
9	.022085	9	.021696	9	.021318	9	.020949
6.730	.022079	**6.790**	.021690	**6.850**	.021312	**6.910**	.020943
1	.022072	1	.021684	1	.021306	1	.020937
2	.022065	2	.021677	2	.021299	2	.020931
3	.022059	3	.021671	3	.021293	3	.020925
4	.022052	4	.021665	4	.021287	4	.020915
5	.022046	5	.021658	5	.021281	5	.020913
6	.022039	6	.021652	6	.021274	6	.020907
7	.022033	7	.021645	7	.021268	7	.020901
8	.022026	8	.021639	8	.021262	8	.020895
9	.022020	9	.021633	9	.021256	9	.020889

d	$1/d^2$	d	$1/d^2$	d	$1/d^2$	d	$1/d^2$
6.920	0.020883	6.980	0.020525	7.040	0.020177	7.100	0.019837
1	.020877	1	.020519	1	.020171	1	.019832
2	.020871	2	.020514	2	.020165	2	.019826
3	.020865	3	.020508	3	.020160	3	.019821
4	.020859	4	.020502	4	.020154	4	.019815
5	.020853	5	.020496	5	.020148	5	.019809
6	.020847	6	.020490	6	.020143	6	.019804
7	.020841	7	.020484	7	.020137	7	.019798
8	.020835	8	.020478	8	.020131	8	.019793
9	.020829	9	.020472	9	.020125	9	.019787
6.930	.020823	6.990	.020467	7.050	.020120	7.110	.019782
1	.020817	1	.020461	1	.020114	1	.019776
2	.020811	2	.020455	2	.020108	2	.019770
3	.020805	3	.020449	3	.020103	3	.019765
4	.020799	4	.020443	4	.020097	4	.019759
5	.020793	5	.020437	5	.020091	5	.019754
6	.020787	6	.020432	6	.020086	6	.019748
7	.020781	7	.020426	7	.020080	7	.019743
8	.020775	8	.020420	8	.020074	8	.019737
9	.020769	9	.020414	9	.020068	9	.019732
6.940	.020763	7.000	.020408	7.060	.020063	7.120	.019726
1	.020757	1	.020402	1	.020057	1	.019721
2	.020751	2	.020397	2	.020051	2	.019715
3	.020745	3	.020391	3	.020046	3	.019709
4	.020739	4	.020385	4	.020040	4	.019704
5	.020733	5	.020379	5	.020034	5	.019698
6	.020727	6	.020373	6	.020029	6	.019693
7	.020721	7	.020367	7	.020023	7	.019687
8	.020715	8	.020362	8	.020017	8	.019682
9	.020709	9	.020356	9	.020012	9	.019676
6.950	.020703	7.010	.020350	7.070	.020006	7.130	.019671
1	.020697	1	.020344	1	.020000	1	.019665
2	.020691	2	.020338	2	.019995	2	.019660
3	.020685	3	.020333	3	.019989	3	.019654
4	.020679	4	.020327	4	.019983	4	.019649
5	.020673	5	.020321	5	.019978	5	.019643
6	.020667	6	.020315	6	.019972	6	.019638
7	.020661	7	.020309	7	.019966	7	.019632
8	.020655	8	.020304	8	.019961	8	.019627
9	.020649	9	.020298	9	.019955	9	.019621
6.960	.020643	7.020	.020292	7.080	.019950	7.140	.019616
1	.020637	1	.020286	1	.019944	1	.019610
2	.020632	2	.020280	2	.019938	2	.019605
3	.020626	3	.020275	3	.019933	3	.019599
4	.020620	4	.020269	4	.019927	4	.019594
5	.020614	5	.020263	5	.019921	5	.019588
6	.020608	6	.020257	6	.019916	6	.019583
7	.020602	7	.020252	7	.019910	7	.019577
8	.020596	8	.020246	8	.019905	8	.019572
9	.020590	9	.020240	9	.019899	9	.019566
6.970	.020584	7.030	.020234	7.090	.019893	7.150	$.01956_1$
1	.020578	1	.020229	1	.019888	1	$.01955_5$
2	.020572	2	.020223	2	.019882	2	$.01955_0$
3	.020567	3	.020217	3	.019877	3	$.01954_4$
4	.020561	4	.020211	4	.019871	4	$.01953_9$
5	.020555	5	.020206	5	.019865	5	$.01953_4$
6	.020549	6	.020200	6	.019860	6	$.01952_4$
7	.020543	7	.020194	7	.019854	7	$.01952_8$
8	.020537	8	.020188	8	.019849	8	$.01951_3$
9	.020531	9	.020183	9	.019843	9	$.01951_7$

d	$1/d^2$	d	$1/d^2$	d	$1/d^2$	d	$1/d^2$
7.160	0.019506	7.220	0.019183	7.280	0.018868	7.340	0.018561
1	.019501	1	.019178	1	.018863	1	.018556
2	.019495	2	.019173	2	.018858	2	.018551
3	.019490	3	.019167	3	.018853	3	.018546
4	.019484	4	.019162	4	.018848	4	.018541
5	.019479	5	.019157	5	.018843	5	.018536
6	.019474	6	.019152	6	.018837	6	.018531
7	.019468	7	.019146	7	.018832	7	.018526
8	.019463	8	.019141	8	.018827	8	.018521
9	.019457	9	.019136	9	.018822	9	.018516
7.170	.019452	7.230	.019130	7.290	.018817	7.350	.018511
1	.019446	1	.019125	1	.018812	1	.018506
2	.019441	2	.019120	2	.018806	2	.018501
3	.019436	3	.019115	3	.018801	3	.018496
4	.019430	4	.019109	4	.018796	4	.018491
5	.019425	5	.019104	5	.018791	5	.018486
6	.019419	6	.019099	6	.018786	6	.018481
7	.019414	7	.019093	7	.018781	7	.018476
8	.019409	8	.019088	8	.018776	8	.018471
9	.019403	9	.019083	9	.018770	9	.018466
7.180	.019398	7.240	.019078	7.300	.018765	7.360	.018461
1	.019392	1	.019072	1	.018760	1	.018456
2	.019387	2	.019067	2	.018755	2	.018451
3	.019382	3	.019062	3	.018750	3	.018445
4	.019376	4	.019056	4	.018745	4	.018440
5	.019371	5	.019051	5	.018740	5	.018435
6	.019365	6	.019046	6	.018734	6	.018430
7	.019360	7	.019041	7	.018729	7	.018425
8	.019355	8	.019035	8	.018724	8	.018420
9	.019349	9	.019030	9	.018719	9	.018415
7.190	.019344	7.250	.019025	7.310	.018714	7.370	.018410
1	.019338	1	.019020	1	.018709	1	.018405
2	.019333	2	.019014	2	.018704	2	.018400
3	.019328	3	.019009	3	.018699	3	.018395
4	.019322	4	.019004	4	.018693	4	.018391
5	.019317	5	.018999	5	.018688	5	.018386
6	.019312	6	.018994	6	.018683	6	.018381
7	.019306	7	.018988	7	.018678	7	.018376
8	.019301	8	.018983	8	.018673	8	.018371
9	.019295	9	.018978	9	.018668	9	.018366
7.200	.019290	7.260	.018973	7.320	.018663	7.380	.018361
1	.019285	1	.018967	1	.018658	1	.018356
2	.019279	2	.018962	2	.018653	2	.018351
3	.019274	3	.018957	3	.018648	3	.018346
4	.019269	4	.018952	4	.018642	4	.018341
5	.019263	5	.018946	5	.018637	5	.018336
6	.019258	6	.018941	6	.018632	6	.018331
7	.019253	7	.018936	7	.018627	7	.018326
8	.019247	8	.018931	8	.018622	8	.018321
9	.019242	9	.018926	9	.018617	9	.018316
7.210	.019237	7.270	.018920	7.330	.018612	7.390	.018311
1	.019231	1	.018915	1	.018607	1	.018306
2	.019226	2	.018910	2	.018602	2	.018301
3	.019221	3	.018905	3	.018597	3	.018296
4	.019215	4	.018900	4	.018592	4	.018291
5	.019210	5	.018894	5	.018587	5	.018286
6	.019205	6	.018889	6	.018582	6	.018281
7	.019199	7	.018884	7	.018576	7	.018276
8	.019194	8	.018879	8	.018571	8	.018271
9	.019189	9	.018874	9	.018566	9	.018266

d	$1/d^2$	d	$1/d^2$	d	$1/d^2$	d	$1/d^2$
7.400	0.018262	7.460	0.017969	7.520	0.017683	7.580	0.017405
1	.018257	1	.017964	1	.017679	1	.017400
2	.018252	2	.017959	2	.017674	2	.017395
3	.018247	3	.017954	3	.017669	3	.017391
4	.018242	4	.017950	4	.017665	4	.017386
5	.018237	5	.017945	5	.017660	5	.017382
6	.018232	6	.017940	6	.017655	6	.017377
7	.018227	7	.017935	7	.017650	7	.017372
8	.018222	8	.017930	8	.017646	8	.017368
9	.018217	9	.017926	9	.017641	9	.017363
7.410	.018212	7.470	.017921	7.530	.017636	7.590	.017359
1	.018207	1	.017916	1	.017632	1	.017354
2	.018202	2	.017911	2	.017627	2	.017350
3	.018198	3	017906	3	.017622	3	.017345
4	.018193	4	.017902	4	.017618	4	.017340
5	.018188	5	.017897	5	.017613	5	.017335
6	.018183	6	.017892	6	.017608	6	.017331
7	.018178	7	.017887	7	.017604	7	.017327
8	.018173	8	.017883	8	.017599	8	.017322
9	.018168	9	.017878	9	.017594	9	.017318
7.420	.018163	7.480	.017873	7.540	.017590	7.600	.017313
1	.018158	1	.017868	1	.017585	1	.017308
2	.018153	2	.017863	2	.017580	2	.017304
3	.018149	3	.017859	3	.017576	3	.017299
4	.018144	4	.017854	4	.017571	4	.017295
5	.018139	5	.017849	5	.017566	5	.017290
6	.018134	6	.017844	6	.017562	6	.017286
7	.018129	7	.017840	7	.017557	7	.017281
8	.018124	8	.017835	8	.017552	8	.017277
9	.018119	9	.017830	9	.017548	9	.017272
7.430	.018114	7.490	.017825	7.550	.017543	7.610	.017268
1	.018109	1	.017821	1	.017538	1	.017263
2	.018105	2	.017816	2	.017534	2	.017258
3	.018100	3	.017811	3	.017529	3	.017254
4	.018095	4	.017806	4	.017525	4	.017249
5	.018090	5	.017802	5	.017520	5	.017245
6	.018085	6	.017797	6	.017515	6	.017240
7	.018080	7	.017792	7	.017511	7	.017236
8	.018075	8	.017787	8	.017506	8	.017231
9	.018071	9	.017783	9	.017501	9	.017227
7.440	.018066	7.500	.017778	7.560	.017497	7.620	.017222
1	.018061	1	.017773	1	.017492	1	.017218
2	.018056	2	.017768	2	.017487	2	.017213
3	.018051	3	.017764	3	.017483	3	.017209
4	.018046	4	.017759	4	.017478	4	.017204
5	.018041	5	.017754	5	.017474	5	.017200
6	.018037	6	.017749	6	.017469	6	.017195
7	.018032	7	.017745	7	.017464	7	.017191
8	.018027	8	.017740	8	.017460	8	.017186
9	.018022	9	.017735	9	.017455	9	.017182
7.450	.018017	7.510	.017730	7.570	.017451	7.630	.017177
1	.018012	1	.017726	1	.017446	1	.017173
2	.018008	2	.017721	2	.017441	2	.017168
3	.018003	3	.017716	3	.017437	3	.017164
4	.017998	4	.017712	4	.017432	4	.017159
5	.017993	5	.017707	5	.017427	5	.017155
6	.017988	6	.017702	6	.017423	6	.017150
7	.017983	7	.017697	7	.017418	7	.017146
8	.017979	8	.017693	8	.017414	8	.017141
9	.017974	9	.017688	9	.017409	9	.017137

d	$1/d^2$	d	$1/d^2$	d	$1/d^2$	d	$1/d^2$
7.640	0.017132	7.700	0.016866	7.760	0.016606	7.820	0.016353
1	.017128	1	.016862	1	.016602	1	.016348
2	.017123	2	.016857	2	.016598	2	.016344
3	.017119	3	.016853	3	.016594	3	.016340
4	.017114	4	.016849	4	.016589	4	.016336
5	.017110	5	.016844	5	.016585	5	.016332
6	.017105	6	.016840	6	.016581	6	.016328
7	.017101	7	.016836	7	.016577	7	.016323
8	.017096	8	.016831	8	.016572	8	.016319
9	.017092	9	.016827	9	.016568	9	.016315
7.650	.017087	7.710	.016823	7.770	.016564	7.830	.016311
1	.017083	1	.016818	1	.016559	1	.016307
2	.017079	2	.016814	2	.016555	2	.016303
3	.017074	3	.016809	3	.016551	3	.016298
4	.017070	4	.016805	4	.016547	4	.016294
5	.017065	5	.016801	5	.016542	5	.016290
6	.017061	6	.016796	6	.016538	6	.016286
7	.017056	7	.016792	7	.016534	7	.016282
8	.017052	8	.016788	8	.016530	8	.016278
9	.017047	9	.016783	9	.016525	9	.016273
7.660	.017043	7.720	.016779	7.780	.016521	7.840	.016269
1	.017038	1	.016775	1	.016517	1	.016265
2	.017034	2	.016770	2	.016513	2	.016261
3	.017031	3	.016766	3	.016508	3	.016257
4	.017025	4	.016762	4	.016504	4	.016253
5	.017021	5	.016757	5	.016500	5	.016249
6	.017016	6	.016753	6	.016496	6	.016244
7	.017012	7	.016749	7	.016491	7	.016240
8	.017007	8	.016744	8	.016487	8	.016236
9	.017003	9	.016740	9	.016483	9	.016232
7.670	.016998	7.730	.016736	7.790	.016479	7.850	.016228
1	.016994	1	.016731	1	.016475	1	.016224
2	.016990	2	.016727	2	.016470	2	.016220
3	.016985	3	.016723	3	.016466	3	.016215
4	.016981	4	.016718	4	.016462	4	.016211
5	.016976	5	.016714	5	.016458	5	.016207
6	.016972	6	.016710	6	.016453	6	.016203
7	.016967	7	.016705	7	.016449	7	.016199
8	.016963	8	.016701	8	.016445	8	.016195
9	.016959	9	.016697	9	.016441	9	.016191
7.680	.016954	7.740	.016692	7.800	.016437	7.860	.016187
1	.016950	1	.016688	1	.016432	1	.016182
2	.016945	2	.016684	2	.016428	2	.016178
3	.016941	3	.016679	3	.016424	3	.016174
4	.016937	4	.016675	4	.016420	4	.016170
5	.016932	5	.016671	5	.016416	5	.016166
6	.016928	6	.016667	6	.016411	6	.016162
7	.016923	7	.016662	7	.016407	7	.016158
8	.016919	8	.016658	8	.016403	8	.016154
9	.016915	9	.016654	9	.016399	9	.016150
7.690	.016910	7.750	.016649	7.810	.016394	7.870	.016145
1	.016906	1	.016645	1	.016390	1	.016141
2	.016901	2	.016641	2	.016386	2	.016137
3	.016897	3	.016636	3	.016382	3	.016133
4	.016893	4	.016632	4	.016378	4	.016129
5	.016888	5	.016624	5	.016374	5	.016125
6	.016884	6	.016624	6	.016369	6	.016121
7	.016879	7	.016619	7	.016365	7	.016117
8	.016875	8	.016615	8	.016361	8	.016113
9	.016871	9	.016611	9	.016357	9	.016109

d	$1/d^2$	d	$1/d^2$	d	$1/d^2$	d	$1/d^2$
7.880	0.016105	7.940	0.015862	8.000	0.015625	8.060	0.015393
.1	.016100	1	.015858	1	.015621	1	.015389
2	.016096	2	.015854	2	.015617	2	.015386
3	.016092	3	.015850	3	.015613	3	.015382
4	.016088	4	.015846	4	.015609	4	.015378
5	.016084	5	.015842	5	.015605	5	.015374
6	.016080	6	.015838	6	.015602	6	.015370
7	.016076	7	.015834	7	.015598	7	.015367
8	.016072	8	.015830	8	.015594	8	.015363
9	.016068	9	.015826	9	.015590	9	.015359
7.890	.016064	7.950	.015822	8.010	.015586	8.070	.015355
1	.016060	1	.015818	1	.015582	1	.015351
2	.016056	2	.015814	2	.015578	2	.015348
3	.016052	3	.015810	3	.015574	3	.015344
4	.016047	4	.015806	4	.015570	4	.015340
5	.016043	5	.015802	5	.015567	5	.015336
6	.016039	6	.015798	6	.015563	6	.015332
7	.016035	7	.015794	7	.015559	7	.015329
8	.016031	8	.015790	8	.015555	8	.015325
9	.016027	9	.015786	9	.015551	9	.015321
7.900	.016023	7.960	.015782	8.020	.015547	8.080	.015317
1	.016019	1	.015778	1	.015543	1	.015313
2	.016015	2	.015775	2	.015539	2	.015310
3	.016011	3	.015771	3	.015536	3	.015306
4	.016007	4	.015767	4	.015532	4	.015302
5	.016003	5	.015763	5	.015528	5	.015298
6	.015999	6	.015759	6	.015524	6	.015294
7	.015995	7	.015755	7	.015520	7	.015291
8	.015991	8	.015751	8	.015516	8	.015287
9	.015987	9	.015747	9	.015512	9	.015283
7.910	.015983	7.970	.015743	8.030	.015508	8.090	.015279
1	.015979	1	.015739	1	.015505	1	.015276
2	.015975	2	.015735	2	.015501	2	.015272
3	.015970	3	.015731	3	.015497	3	.015268
4	.015966	4	.015727	4	.015493	4	.015264
5	.015962	5	.015723	5	.015489	5	.015260
6	.015958	6	.015719	6	.015485	6	.015257
7	.015954	7	.015715	7	.015481	7	.015253
8	.015950	8	.015711	8	.015478	8	.015249
9	.015946	9	.015707	9	.015474	9	.015245
7.920	.015942	7.980	.015703	8.040	.015470	8.100	.015242
1	.015938	1	.015699	1	.015466	1	.015238
2	.015934	2	.015696	2	.015462	2	.015234
3	.015930	3	.015692	3	.015458	3	.015230
4	.015926	4	.015688	4	.015455	4	.015227
5	.015922	5	.015684	5	.015451	5	.015223
6	.015918	6	.015680	6	.015447	6	.015219
7	.015914	7	.015676	7	.015443	7	.015215
8	.015910	8	.015672	8	.015439	8	.015212
9	.015906	9	.015668	9	.015435	9	.015208
7.930	.015902	7.990	.015664	8.050	.015432	8.110	.015204
1	.015898	1	.015660	1	.015428	1	.015200
2	.015894	2	.015656	2	.015424	2	.015197
3	.015890	3	.015652	3	.015420	3	.015193
4	.015886	4	.015648	4	.015416	4	.015189
5	.015882	5	.015645	5	.015412	5	.015185
6	.015878	6	.015641	6	.015409	6	.015182
7	.015874	7	.015637	7	.015405	7	.015178
8	.015870	8	.015633	8	.015401	8	.015174
9	.015866	9	.015629	9	.015397	9	.015170

d	$1/d^2$	d	$1/d^2$	d	$1/d^2$	d	$1/d^2$
8.120	0.015167	8.180	0.014945	8.240	0.014728	8.300	0.014516
1	.015163	1	.014941	1	.014724	1	.014512
2	.015159	2	.014938	2	.014721	2	.014509
3	.015155	3	.014934	3	.014717	3	.014505
4	.015152	4	.014930	4	.014714	4	.014502
5	.015148	5	.014927	5	.014710	5	.014498
6	.015144	6	.014923	6	.014707	6	.014495
7	.015140	7	.014919	7	.014703	7	.014491
8	.015137	8	.014916	8	.014700	8	.014488
9	.015133	9	.014912	9	.014696	9	.014484
8.130	.015129	8.190	.014908	8.250	.014692	8.310	.014481
1	.015126	1	.014905	1	.014689	1	.014477
2	.015122	2	.014901	2	.014685	2	.014474
3	.015118	3	.014898	3	.014682	3	.014471
4	.015114	4	.014894	4	.014678	4	.014467
5	.015111	5	.014890	5	.014675	5	.014464
6	.015107	6	.014887	6	.014671	6	.014460
7	.015103	7	.014883	7	.014667	7	.014457
8	.015100	8	.014879	8	.014664	8	.014453
9	.015096	9	.014876	9	.014660	9	.014450
8.140	.015092	8.200	.014872	8.260	.014657	8.320	.014446
1	.015088	1	.014868	1	.014653	1	.014443
2	.015085	2	.014865	2	.014650	2	.014439
3	.015081	3	.014861	3	.014646	3	.014436
4	.015077	4	.014858	4	.014643	4	.014432
5	.015074	5	.014854	5	.014639	5	.014429
6	.015070	6	.014850	6	.014636	6	.014425
7	.015066	7	.014847	7	.014632	7	.014422
8	.015063	8	.014843	8	.014628	8	.014418
9	.015059	9	.014840	9	.014625	9	.014415
8.150	.015055	8.210	.014836	8.270	.014621	8.330	.014412
1	.015051	1	.014832	1	.014618	1	.014408
2	.015048	2	.014829	2	.014614	2	.014405
3	.015044	3	.014825	3	.014611	3	.014401
4	.015040	4	.014821	4	.014607	4	.014398
5	.015037	5	.014818	5	.014604	5	.014394
6	.015033	6	.014814	6	.014600	6	.014391
7	.015029	7	.014811	7	.014597	7	.014387
8	.015026	8	.014807	8	.014593	8	.014384
9	.015022	9	.014803	9	.014590	9	.014380
8.160	.015018	8.220	.014800	8.280	.014586	8.340	.014377
1	.015015	1	.014796	1	.014583	1	.014374
2	.015011	2	.014793	2	.014579	2	.014370
3	.015007	3	.014789	3	.014576	3	.014367
4	.015004	4	.014785	4	.014572	4	.014363
5	.015000	5	.014782	5	.014569	5	.014360
6	.014996	6	.014778	6	.014565	6	.014356
7	.014993	7	.014775	7	.014561	7	.014353
8	.014989	8	.014771	8	.014558	8	.014349
9	.014985	9	.014767	9	.014554	9	.014346
8.170	.014982	8.230	.014764	8.290	.014551	8.350	.014343
1	.014978	1	.014760	1	.014547	1	.014339
2	.014974	2	.014757	2	.014544	2	.014336
3	.014971	3	.014753	3	.014540	3	.014332
4	.014967	4	.014750	4	.014537	4	.014329
5	.014963	5	.014746	5	.014533	5	.014325
6	.014960	6	.014742	6	.014530	6	.014322
7	.014956	7	.014739	7	.014526	7	.014319
8	.014952	8	.014735	8	.014523	8	.014315
9	.014949	9	.014732	9	.014519	9	.014312

d	$1/d^2$	d	$1/d^2$	d	$1/d^2$	d	$1/d^2$
8.360	0.014308	8.420	0.014102	8.480	0.013906	8.540	0.013711
1	.014305	1	.014102	1	.013903	1	.013708
2	.014301	2	.014098	2	.013900	2	.013705
3	.014298	3	.014095	3	.013896	3	.013702
4	.014295	4	.014092	4	.013893	4	.013699
5	.014291	5	.014088	5	.013890	5	.013695
6	.014288	6	.014085	6	.013887	6	.013692
7	.014284	7	.014082	7	.013883	7	.013689
8	.014281	8	.014078	8	.013880	8	.013686
9	.014278	9	.014075	9	.013877	9	.013683
8.370	.014274	8,430	.014072	8.490	.013873	8.550	.013679
1	.014271	1	.014068	1	.013870	1	.013676
2	.014267	2	.014065	2	.013867	2	.013673
3	.014264	3	.014062	3	.013864	3	.013670
4	.014260	4	.014058	4	.013860	4	.013667
5	.014257	5	.014055	5	.013857	5	.013663
6	.014254	6	.014052	6	.013854	6	.013660
7	.014250	7	.014048	7	.013851	7	.013657
8	.014247	8	.014045	8	.013847	8	.013654
9	.014243	9	.014042	9	.013844	9	.013651
8.380	.014240	8.440	.014038	8.500	.013841	8.560	.013647
1	.014237	1	.014035	1	.013838	1	.013644
2	.014233	2	.014032	2	.013834	2	.013641
3	.014230	3	.014028	3	.013831	3	.013638
4	.014226	4	.014025	4	.013828	4	.013635
5	.014223	5	.014022	5	.013825	5	.013632
6	.014220	6	.014018	6	.013821	6	.013628
7	.014216	7	.014015	7	.013818	7	.013625
8	.014213	8	.014012	8	.013815	8	.013622
9	.014210	9	.014008	9	.013812	9	.013619
8.390	.014206	8.450	.014005	8,510	.013808	8,570	.013616
1	.014203	1	.014002	1	.013805	1	.013612
2	.014199	2	.013998	2	.013802	2	.013609
3	.014196	3	.013995	3	.013799	3	.013606
4	.014193	4	.013992	4	.013795	4	.013603
5	.014189	5	.013989	5	.013792	5	.013600
6	.014186	6	.013985	6	.013789	6	.013597
7	.014182	7	.013982	7	.013786	7	.013593
• 8	.014179	8	.013979	8	.013782	8	.013590
9	.014176	9	.013975	9	.013779	9	.013587
8.400	.014172	8.460	.013972	8,520	.013776	8.580	.013584
1	.014169	1	.013969	1	.013773	1	.013581
2	.014166	2	.013965	2	.013769	2	.013578
3	.014162	3	.013962	3	.013766	3	.013574
4	.014159	4	.013959	4	.013763	4	.013571
5	.014155	5	.013956	5	.013760	5	.013568
6	.014152	6	.013952	6	.013757	6	.013565
7	.014149	7	.013949	7	.013753	7	.013562
8	.014145	8	.013946	8	.013750	8	.013559
9	.014142	9	.013942	9	.013747	9	.013555
8.410	.014139	8.470	.013939	8.530	.013744	8.590	.013552
1	.014135	1	.013936	1	.013740	1	.013549
2	.014132	2	.013932	2	.013737	2	.013546
3	.014129	3	.013929	3	.013734	3	.013543
4	.014125	4	.013926	4	.013731	4	.013540
5	.014122	5	.013923	5	.013728	5	.013537
6	.014118	6	.013919	6	.013724	6	.013533
7	.014115	7	.013916	7	.013721	7	.013530
8	.014112	8	.013913	8	.013718	8	.013527
9	.014108	9	.013909	9	.013715	9	.013524

d	$1/d^2$	d	$1/d^2$	d	$1/d^2$	d	$1/d^2$
8.600	0.013521	8.660	0.013334	8.720	0.013151	8.780	0.012972
1	.013518	1	.013331	1	.013148	1	.012969
2	.013515	2	.013328	2	.013145	2	.012966
3	.013511	3	.013325	3	.013142	3	.012963
4	.013508	4	.013322	4	.013139	4	.012960
5	.013505	5	.013319	5	.013136	5	.012957
6	.013502	6	.013316	6	.013133	6	.012954
7	.013499	7	.013313	7	.013130	7	.012951
8	.013496	8	.013310	8	.013127	8	.012949
9	.013493	9	.013306	9	.013124	9	.012946
8.610	.013489	8.670	.013303	8.730	.013121	8.790	.012943
1	.013486	1	.013300	1	.013118	1	.012940
2	.013483	2	.013297	2	.013115	2	.012937
3	.013480	3	.013294	3	.013112	3	.012934
4	.013477	4	.013291	4	.013109	4	.012931
5	.013474	5	.013288	5	.013106	5	.012928
6	.013471	6	.013285	6	.013103	6	.012925
7	.013468	7	.013282	7	.013100	7	.012922
8	.013464	8	.013279	8	.013097	8	.012919
9	.013461	9	.013276	9	.013094	9	.012916
8.620	.013458	8.680	.013273	8.740	.013091	8.800	.012913
1	.013455	1	.013270	1	.013088	1	.012910
2	.013452	2	.013267	2	.013085	2	.012907
3	.013449	3	.013264	3	.013082	3	.012904
4	.013446	4	.013261	4	.013079	4	.012901
5	.013443	5	.013257	5	.013076	5	.012899
6	.013439	6	.013254	6	.013073	6	.012896
7	.013436	7	.013251	7	.013070	7	.012893
8	013433	8	.013248	8	.013067	8	.012890
9	.013430	9	.013245	9	.013064	9	.012887
8.630	.013427	8.690	.013242	8.750	.013061	8.810	.012884
1	.013424	1	.013239	1	.013058	1	.012881
2	.013421	2	.013236	2	.013055	2	.012878
3	.013418	3	.013233	3	.013052	3	.012875
4	.013415	4	.013230	4	.013049	4	.012872
5	.013411	5	.013227	5	.013046	5	.012869
6	.013408	6	.013224	6	.013043	6	.012866
7	.013405	7	.013221	7	.013040	7	.012863
8	.013402	8	.013218	8	.013037	8	.012861
9	.013399	9	.013215	9	.013034	9	.012858
8.640	.013396	8.700	.013212	8.760	.013031	8.820	.012855
1	.013393	1	.013209	1	.013028	1	.012852
2	.013390	2	.013206	2	.013025	2	.012849
3	.013387	3	.013203	3	.013023	3	.012846
4	.013384	4	.013200	4	.013020	4	.012843
5	.013380	5	.013197	5	.013017	5	.012840
6	.013377	6	.013194	6	.013014	6	.012837
7	.013374	7	.013191	7	.013011	7	.012834
8	.013371	8	.013188	8	.013008	8	.012831
9	.013368	9	.013184	9	.013005	9	.012829
8.650	.013365	8.710	.013181	8.770	.013002	8.830	.012826
1	.013362	1	.013178	1	.012999	1	012823
2	.013359	2	.013175	2	.012996	2	.012820
3	.013356	3	.013172	3	.012993	3	.012817
4	.013353	4	.013169	4	.012990	4	.012814
5	.013350	5	.013166	5	.012987	5	.012811
6	.013346	6	.013163	6	.012984	6	.012808
7	.013343	7	.013160	7	.012981	7	.012805
8	.013340	8	.013157	8	.012978	8	.012802
9	.013337	9	.013154	9	.012975	9	.012800

d	$1/d^2$	d	$1/d^2$	d	$1/d^2$	d	$1/d^2$
8.840	0.012797	8.900	0.012625	8.960	0.012456	9.020	0.012291
1	.012794	1	.012622	1	.012453	1	.012288
2	.012791	2	.012619	2	.012451	2	.012286
3	.012788	3	.012616	3	.012448	3	.012283
4	.012785	4	.012613	4	.012445	4	.012280
5	.012782	5	.012610	5	.012442	5	.012277
6	.012779	6	.012608	6	.012439	6	.012275
7	.012776	7	.012605	7	.012437	7	.012272
8	.012773	8	.012602	8	.012434	8	.012269
9	.012771	9	.012599	9	.012431	9	.012267
8.850	.012768	8.910	.012596	8.970	.012428	9.030	.012264
1	.012765	1	.012594	1	.012426	1	.012261
2	.012762	2	.012591	2	.012423	2	.012258
3	.012759	3	.012588	3	.012420	3	.012256
4	.012756	4	.012585	4	.012417	4	.012253
5	.012753	5	.012582	5	.012415	5	.012250
6	.012750	6	.012579	6	.012412	6	.012248
7	.012748	7	.012577	7	.012409	7	.012245
8	.012745	8	.012574	8	.012406	8	.012242
9	.012742	9	.012571	9	.012403	9	.012239
8.860	.012739	8.920	.012568	8.980	.012401	9.040	.012237
1	.012736	1	.012565	1	.012398	1	.012234
2	.012733	2	.012562	2	.012395	2	.012231
3	.012730	3	.012560	3	.012392	3	.012229
4	.012727	4	.012557	4	.012390	4	.012226
5	.012725	5	.012554	5	.012387	5	.012223
6	.012722	6	.012551	6	.012384	6	.012220
7	.012719	7	.012548	7	.012381	7	.012218
8	.012716	8	.012546	8	.012379	8	.012215
9	.012713	9	.012543	9	.012376	9	.012212
8.870	.012710	8.930	.012540	8.990	.012373	9.050	.012210
1	.012707	1	.012537	1	.012371	1	.012207
2	.012704	2	.012534	2	.012368	2	.012204
3	.012702	3	.012532	3	.012365	3	.012202
4	.012699	4	.012529	4	.012362	4	.012199
5	.012696	5	.012526	5	.012359	5	.012196
6	.012693	6	.012523	6	.012357	6	.012193
7	.012690	7	.012520	7	.012354	7	.012191
8	.012687	8	.012518	8	.012351	8	.012188
9	.012684	9	.012515	9	.012348	9	.012185
8.880	.012682	8.940	.012512	9.000	.012346	9.060	.012183
1	.012679	1	.012509	1	.012343	1	.012180
2	.012676	2	.012506	2	.012340	2	.012177
3	.012673	3	.012504	3	.012337	3	.012175
4	.012670	4	.012501	4	.012335	4	.012172
5	.012667	5	.012498	5	.012332	5	.012169
6	.012664	6	.012495	6	.012329	6	.012167
7	.012662	7	.012492	7	.012326	7	.012164
8	.012659	8	.012490	8	.012324	8	.012161
9	.012656	9	.012487	9	.012321	9	.012159
8.890	.012653	8.950	.012484	9.010	.012318	9.070	.012156
1	.012650	1	.012481	1	.012316	1	.012153
2	.012647	2	.012478	2	.012313	2	.012150
3	.012645	3	.012476	3	.012310	3	.012148
4	.012642	4	.012473	4	.012307	4	.012145
5	.012639	5	.012470	5	.012305	5	.012142
6	.012636	6	.012467	6	.012302	6	.012140
7	.012633	7	.012464	7	.012299	7	.012137
8	.012630	8	.012462	8	.012296	8	.012134
9	.012628	9	.012459	9	.012294	9	.012132

d	$1/d^2$	d	$1/d^2$	d	$1/d^2$	d	$1/d^2$
9.080	0.012129	9.140	0.011970	9.200	0.011815	9.260	0.011662
1	.012126	1	.011968	1	.011812	1	.011660
2	.012124	2	.011965	2	.011810	2	.011657
3	.012121	3	.011963	3	.011807	3	.011655
4	.012118	4	.011960	4	.011804	4	.011652
5	.012116	5	.011957	5	.011802	5	.011650
6	.012113	6	.011955	6	.011799	6	.011647
7	.012110	7	.011952	7	.011797	7	.011645
8	.012108	8	.011949	8	.011794	8	.011642
9	.012105	9	.011947	9	.011792	9	.011640
9.090	.012102	9.150	.011944	9.210	.011789	9.270	.011637
1	.012100	1	.011942	1	.011787	1	.011634
2	.012097	2	.011939	2	.011784	2	.011632
3	.012094	3	.011936	3	.011781	3	.011629
4	.012092	4	.011934	4	.011779	4	.011627
5	.012089	5	.011931	5	.011776	5	.011624
6	.012086	6	.011929	6	.011774	6	.011622
7	.012084	7	.011926	7	.011771	7	.011619
8	.012081	8	.011923	8	.011769	8	.011617
9	.012078	9	.011921	9	.011766	9	.011614
9.100	.012076	9.160	.011918	9.220	.011764	9.280	.011612
1	.012073	1	.011916	1	.011761	1	.011609
2	.012071	2	.011913	2	.011758	2	.011607
3	.012068	3	.011910	3	.011756	3	.011604
4	.012065	4	.011908	4	.011753	4	.011602
5	.012063	5	.011905	5	.011751	5	.011599
6	.012060	6	.011903	6	.011748	6	.011597
7	.012057	7	.011900	7	.011746	7	.011594
8	.012055	8	.011897	8	.011743	8	.011592
9	.012052	9	.011895	9	.011741	9	.011589
9.110	.012049	9.170	.011892	9.230	.011738	9.290	.011587
1	.012047	1	.011890	1	.011736	1	.011584
2	.012044	2	.011887	2	.011733	2	.011582
3	.012041	3	.011884	3	.011730	3	.011579
4	.012039	4	.011882	4	.011728	4	.011577
5	.012036	5	.011879	5	.011725	5	.011574
6	.012033	6	.011877	6	.011723	6	.011572
7	.012031	7	.011874	7	.011720	7	.011569
8	.012028	8	.011871	8	011718	8	.011567
9	.012026	9	.011869	9	.011715	9	.011565
9.120	.012023	9.180	.011866	9.240	.011713	9.300	.011562
1	.012020	1	.011864	1	.011710	1	.011560
2	.012018	2	.011861	2	.011708	2	.011557
3	.012015	3	.011859	3	.011705	3	.011555
4	.012012	4	.011856	4	.011703	4	.011552
5	.012010	5	.011853	5	.011700	5	.011550
6	.012007	6	.011851	6	.011697	6	.011547
7	.012004	7	.011848	7	.011695	7	.011545
8	.012002	8	.011846	8	.011692	8	.011542
9	.011999	9	.011843	9	.011690	9	.011540
9.130	.011997	9.190	.011840	9.250	.011687	9.310	.011537
1	.011994	1	.011838	1	.011685	1	.011535
2	.011991	2	.011835	2	.011682	2	.011532
3	.011989	3	.011833	3	.011680	3	.011530
4	.011986	4	.011830	4	.011677	4	.011527
5	.011983	5	.011828	5	.011675	5	.011525
6	.011981	6	.011825	6	.011672	6	.011522
7	.011978	7	.011822	7	.011670	7	.011520
8	.011976	8	.011820	8	.011667	8	.011517
9	.011973	9	.011817	9	.011665	9	.011515

d	$1/d^2$	d	$1/d^2$	d	$1/d^2$	d	$1/d^2$
9.320	0.011512	9.380	0.011366	9.440	0.011222	9.500	0.011080
1	.011510	1	.011363	1	.011219	1	.011078
2	.011508	2	.011361	2	.011217	2	.011076
3	.011505	3	.011358	3	.011215	3	.011073
4	.011503	4	.011356	4	.011212	4	.011071
5	.011500	5	.011354	5	.011210	5	.011069
6	.011498	6	.011351	6	.011207	6	.011066
7	.011495	7	.011349	7	.011205	7	.011064
8	.011493	8	.011346	8	.011203	8	.011062
9	.011490	9	.011344	9	.011200	9	.011059
9.330	.011488	9.390	.011341	9.450	.011198	9.510	.011057
1	.011485	1	.011339	1	.011196	1	.011055
2	.011483	2	.011337	2	.011193	2	.011052
3	.011480	3	.011334	3	.011191	3	.011050
4	.011478	4	.011332	4	.011188	4	.011048
5	·011475	5	.011329	5	.011186	5	.011045
6	.011473	6	.011327	6	.011184	6	.011043
7	.011471	7	.011325	7	.011181	7	.011041
8	.011468	8	.011322	8	.011179	8	.011038
9	.011466	9	.011320	9	.011177	9	.011036
9.340	.011463	9.400	.011317	9.460	.011174	9.520	.011034
1	.011461	1	.011315	1	.011172	1	.011C32
2	.011458	2	.011313	2	.011170	2	.011029
3	.011456	3	.011310	3	.011167	3	.011027
4	.011453	4	.011308	4	.011165	4	.011025
5	.011451	5	.011305	5	.011162	5	.011022
6	.011448	6	.011303	6	.011160	6	.011020
7	.011446	7	.011301	7	.011158	7	.011018
8	.011444	8	.011298	8	.011155	8	.011015
9	.011441	9	.011296	9	.011153	9	.011013
9.350	.011439	9.410	.011293	9.470	.011151	9.530	.011011
1	.011436	1	.011291	1	.011148	1	·.011008
2	.011434	2	.011288	2	.011146	2	.011006
3	.011431	3	.011286	3	.011144	3	.011004
4	.011429	4	.011284	4	.011141	4	.011001
5	.011426	5	.011281	5	.011139	5	.010999
6	.011424	6	.011278	6	.011137	6	.010997
7	.011422	7	.011276	7	.011134	7	.010995
8	.011419	8	.011274	8	.011132	8	.010992
9	.011417	9	.011272	9	.011129	9	.010990
9.360	.011414	9.420	.011269	9.480	.011127	9.540	.010988
1	.011412	1	.011267	1	.011125	1	.010985
2	.011409	2	.011265	2	.011122	2	.010983
3	.011407	3	.011262	3	.011120	3	.010981
4	.011405	4	.011260	4	.011118	4	.010978
5	.011402	5	.011257	5	.011115	5	.010976
6	.011400	6	.011255	6	.011113	6	.010974
7	.011397	7	.011253	7	.011111	7	.010972
8	.011395	8	.011250	8	.011108	8	.010969
9	.011392	9	.011248	9	.011106	9	.010967
9.370	.011390	9.430	.011245	9.490	.011104	9.550	.010965
1	.011387	1	.011243	1	.011101	1	.010962
2	.011385	2	.011241	2	.011099	2	.010960
3	.011383	3	.011238	3	.011097	3	.010958
4	.011380	4	.011236	4	.011094	4	.010955
5	.011378	5	.011234	5	.011092	5	.010953
6	.011375	6	.011231	6	.011090	6	.010951
7	.011373	7	.011230	7	.011087	7	.010949
8	.011370	8	.011226	8	.011085	8	.010946
9	.011368	9	.011224	9	.011083	9	.010944

d	$1/d^2$	d	$1/d^2$	d	$1/d^2$	d	$1/d^2$
9,560	0.010942	9,620	0.010806	9,680	0.010672	9,740	0.010541
1	.010939	1	.010803	1	.010670	1	.010539
2	.010937	2	.010801	2	.010668	2	.010537
3	.010935	3	.010799	3	.010665	3	.010534
4	.010933	4	.010797	4	.010663	4	.010532
5	.010930	5	.010794	5	.010661	5	.010530
6	.010928	6	.010792	6	.010659	6	.010528
7	.010926	7	.010790	7	.010657	7	.010526
8	.010923	8	.010788	8	.010654	8	.010524
9	.010921	9	.010785	9	.010652	9	.010522
9,570	.010919	9,630	.010783	9,690	.010650	9,750	.010519
1	.010916	1	.010781	1	.010648	1	.010517
2	.010914	2	.019779	2	.010646	2	.010515
3	.010912	3	.010776	3	.010643	3	.010513
4	.010910	4	.010774	4	.010641	4	.010511
5	.010909	5	.010772	5	.010639	5	.010509
6	.010905	6	.010770	6	.010637	6	.010506
7	.010903	7	.010768	7	.010635	7	.010504
8	.010901	8	.010765	8	.010632	8	.010502
9	.010898	9	.010763	9	.010630	9	.010500
9,580	.010896	9,640	.010762	9,700	.010628	9,760	.010498
1	.010894	1	.010759	1	.010626	1	.010496
2	.010891	2	.010756	2	.010624	2	.010494
3	.010889	3	.010754	3	.010622	3	.010491
4	.010887	4	.010752	4	.010619	4	.010489
5	.010885	5	.010750	5	.010617	5	.010487
6	.010882	6	.010747	6	.010615	6	.010485
7	.010880	7	.010745	7	.010613	7	.010483
8	.010878	8	.010743	8	.010611	8	.010481
9	.010876	9	.010741	9	.010608	9	.010479
9,590	.010873	9,650	.010738	9,710	.010606	9,770	.010476
1	.010871	1	.010736	1	.010604	1	.010474
2	.010869	2	.010734	2	.010602	2	.010472
3	.010866	3	.010732	3	.010600	3	.010470
4	.010864	4	.010730	4	.010598	4	.010468
5	.010862	5	.010727	5	.010595	5	.010466
6	.010860	6	.010725	6	.010593	6	.010464
7	.010857	7	.010723	7	.010591	7	.010461
8	.010855	8	.010721	8	.010589	8	.010459
9	.010853	9	.010718	9	.010587	9	.010457
9,600	.010851	9,660	.010716	9,720	.010584	9,780	.010455
1	.010848	1	.010714	1	.010582	1	.010453
2	.010846	2	.010712	2	.010580	2	.010451
3	.010844	3	.010710	3	.010578	3	.010448
4	.010842	4	.010707	4	.010576	4	.010446
5	.010839	5	.010705	5	.010574	5	.010444
6	.010837	6	.010703	6	.010571	6	.010442
7	.010835	7	.010701	7	.010569	7	.010440
8	.010833	8	.010698	8	.010567	8	.010438
9	.010830	9	.010696	9	.010565	9	.010436
9,610	.010828	9,670	.010694	9,730	.010563	9,790	.010434
1	.010827	1	.010692	1	.010560	1	.010431
2	.010824	2	.010690	2	.010558	2	.010429
3	.010821	3	.010688	3	.010556	3	.010427
4	.010819	4	.010685	4	.010554	4	.010425
5	.010817	5	.010683	5	.010552	5	.010423
6	.010815	6	.010681	6	.010550	6	.010421
7	.010812	7	.010679	7	.010548	7	.010419
8	.010810	8	.010676	8	.010545	8	.010416
9	.010808	9	.010674	9	.010543	9	.010414

d	$1/d^2$	d	$1/d^2$	d	$1/d^2$	d	$1/d^2$
9.800	0.010412	9.850	0.010307	9.900	0.010203	9.950	0.010101
1	.010410	1	.010305	1	.010201	1	.010099
2	.010408	2	.010303	2	.010199	2	.010097
3	.010406	3	.010301	3	.010197	3	.010095
4	.010404	4	.010298	4	.010195	4	.010093
5	.010402	5	.010296	5	.010193	5	.010091
6	.010400	6	.010294	6	.010191	6	.010088
7	.010397	7	.010292	7	.010189	7	.010086
8	.010395	8	.010290	8	.010186	8	.010084
9	.010393	9	.010288	9	.010184	9	.010082
9.810	.010391	9.860	.010286	9.910	.010182	9.960	.010080
1	.010389	1	.010284	1	.010180	1	.010078
2	.010387	2	.010282	2	.010178	2	.010076
3	.010385	3	.010280	3	.010176	3	.010074
4	.010383	4	.010278	4	.010174	4	.010072
5	.010381	5	.010276	5	.010172	5	.010070
6	.010378	6	.010273	6	.010170	6	.010068
7	.010376	7	.010271	7	.010168	7	.010066
8	.010374	8	.010269	8	.010166	8	.010064
9	.010372	9	.010267	9	.010164	9	.010062
9.820	.010370	9.870	.010265	9.920	.010162	9.970	.010060
1	.010368	1	.010263	1	.010160	1	.010058
2	.010366	2	.010261	2	.010158	2	.010056
3	.010364	3	.010259	3	.010156	3	.010054
4	.010362	4	.010257	4	.010154	4	.010052
5	.010359	5	.010255	5	.010152	5	.010050
6	.010357	6	.010253	6	.010150	6	.010048
7	.010355	7	.010251	7	.010148	7	.010046
8	.010353	8	.010248	8	.010146	8	.010044
9	.010351	9	.010246	9	.010144	9	.010042
9.830	.010349	9.880	.010244	9.930	.010141	9.980	.010040
1	.010347	1	.010242	1	.010139	1	.010038
2	.010345	2	.010240	2	.010137	2	.010036
3	.010342	3	.010238	3	.010135	3	.010034
4	.010340	4	.010236	4	.010133	4	.010032
5	.010338	5	.010234	5	.010131	5	.010030
6	.010336	6	.010232	6	.010129	6	.010028
7	.010334	7	.010230	7	.010127	7	.010026
8	.010332	8	.010228	8	.010125	8	.010024
9	.010330	9	.010226	9	.010123	9	.010022
9.840	.010328	9.890	.010224	9.940	.010121	9.990	.010020
1	.010326	1	.010222	1	.010119	1	.010018
2	.010324	2	.010220	2	.010117	2	.010016
3	.010322	3	.010217	3	.010115	3	.010014
4	.010320	4	.010215	4	.010113	4	.010012
5	.010317	5	.010213	5	.010111	5	.010010
6	.010315	6	.010211	6	.010109	6	.010008
7	.010313	7	.010209	7	.010107	7	.010006
8	.010311	8	.010207	8	.010105	8	.010004
9	.010309	9	.010205	9	.010103	9	.010002
						10.000	.010000

3-1c. Values of nλ and lg (nλ/2)

The tables give the values of nλ and lg (nλ/2) for some radiations used with poly-crystalline substances [110] for n from 1 to 30.

The choice of anode material must be made with allowance for the strong absorption from the following elements:

Material of anode	Z	Absorbed strongly by	
		Kα	Kβ
Ti	22	Sc	Sc
V	23	Sc	Ti
Cr	24	Ti	V
Mn	25	V	Cr
Fe	26	Cr	Mn
Co	27	Mn	Fe
Ni	28	Fe	Co
Cu	29	Co	Ni
Zn	30	Ni	Cu
Zr	40	Ru	Y
Nb	41	Sr	Zr
Mo	41	Y	Nb

and also from elements of lower Z (the effect is less marked the greater the difference between the Z of the anode and absorber).

$n\lambda$

n

Radiation	1	2	3	4	5	6	7	8	9	10	11	12	13	14	15
Tiα_2	2.747	5.49	8.24	10.99	13.73	16.48	19.23	21.97	24.72	27.47	30.21	32.96	35.71	38.46	41.20
Tiα_1	2.743	5.49	8.23	10.97	13.72	16.46	19.20	21.95	24.69	27.43	30.18	32.92	35.66	38.40	41.15
Tiβ_1	2.509	5.02	7.53	10.04	12.55	15.05	17.56	20.07	22.58	25.09	27.60	30.11	32.62	35.13	37.64
Vα_2	2.502	5.00	7.51	10.01	12.51	15.01	17.51	20.02	22.52	25.02	27.52	30.03	32.53	35.03	37.53
Vα_1	2.498	4.997	7.50	9.99	12.49	14.99	17.49	19.99	22.49	24.98	27.48	29.98	32.48	34.98	37.48
Crα_2	2.289	4.578	6.87	9.16	11.44	13.73	16.02	18.31	20.60	22.89	25.18	27.47	29.76	32.04	34.33
Crα_1	2.285	4.570	6.86	9.14	11.43	13.71	16.00	18.28	20.57	22.85	25.14	27.42	29.71	31.99	34.28
Vβ_1	2.280	4.559	6.84	9.12	11.40	13.68	15.96	18.24	20.52	22.80	25.08	27.36	29.64	31.92	34.20
Mnα_2	2.102	4.203	6.30	8.41	10.51	12.61	14.71	16.81	18.91	21.02	23.12	25.22	27.32	29.42	31.52
Mnα_1	2.098	4.195	6.29	8.39	10.49	12.59	14.68	16.78	18.88	20.98	23.07	25.17	27.27	29.37	31.46
Crβ_1	2.081	4.161	6.24	8.32	10.40	12.48	14.56	16.64	18.73	20.81	22.89	24.97	27.05	29.13	31.21
Feα_2	1.936	3.872	5.81	7.74	9.68	11.62	13.55	15.49	17.42	19.36	21.30	23.23	25.17	27.10	29.04
Feα_1	1.932	3.864	5.80	7.73	9.66	11.59	13.52	15.46	17.39	19.32	21.25	23.19	25.12	27.05	28.98
Mnβ_1	1.906	3.812	5.72	7.62	9.53	11.44	13.34	15.25	17.16	19.06	20.97	22.87	24.78	26.69	28.59
Coα_2	1.789	3.578	5.37	7.16	8.95	10.74	12.52	14.31	16.10	17.85	19.68	21.47	23.26	25.05	26.84
Coα_1	1.785	3.571	5.36	7.14	8.93	10.71	12.50	14.28	16.07	17.85	19.64	21.42	23.21	24.99	26.78
Feβ_1	1.753	3.506	5.26	7.01	8.77	10.52	12.27	14.02	15.78	17.53	19.28	21.04	22.79	24.54	26.30
Niα_2	1.658	3.317	4.975	6.63	8.29	9.95	11.61	13.27	14.93	16.58	18.24	19.90	21.56	23.22	24.88
Niα_1	1.655	3.309	4.964	6.62	8.27	9.93	11.58	13.24	14.89	16.55	18.20	19.85	21.51	23.16	24.82
Coβ_1	1.617	3.235	4.852	6.47	8.09	9.70	11.32	12.94	14.56	16.17	17.79	19.41	21.03	22.64	24.26
Cuα_2	1.541	3.082	4.624	6.16	7.71	9.25	10.79	12.33	13.87	15.41	16.95	18.49	20.04	21.58	23.12
Cuα_1	1.537	3.075	4.612	6.15	7.69	9.22	10.76	12.30	13.84	15.37	16.91	18.45	19.99	21.52	23.06
Niβ_1	1.497	2.994	4.491	5.99	7.49	8.98	10.48	11.98	13.47	14.97	16.47	17.97	19.46	20.96	22.46
Znα_2	1.436	2.872	4.308	5.74	7.18	8.62	10.05	11.49	12.92	14.36	15.80	17.23	18.67	20.10	21.54
Znα_1	1.432	2.864	4.297	5.73	7.16	8.59	10.03	11.46	12.89	14.32	15.75	17.19	18.62	20.05	21.48
Cuβ_1	1.389	2.779	4.168	5.56	6.95	8.34	9.73	11.12	12.50	13.89	15.28	16.67	18.06	19.45	20.84
Znβ_1	1.293	2.585	3.878	5.17	6.46	7.76	9.05	10.34	11.63	12.93	14.22	15.51	16.80	18.10	19.39
Zrα_2	0.789	1.577	2.366	3.154	3.943	4.731	5.52	6.31	7.10	7.89	8.67	9.46	10.25	11.04	11.83
Zrα_1	0.784	1.569	2.353	3.137	3.922	4.706	5.49	6.27	7.06	7.84	8.63	9.41	10.20	10.98	11.76
Nbα_2	0.749	1.498	2.247	2.996	3.745	4.493	5.24	5.99	6.74	7.49	8.24	8.99	9.74	10.48	11.23
Nbα_1	0.745	1.489	2.234	2.979	3.724	4.468	5.21	5.96	6.70	7.45	8.19	8.94	9.68	10.43	11.17
Moα_2	0.713	1.426	2.138	2.851	3.564	4.277	4.990	5.70	6.42	7.13	7.84	8.55	9.27	9.98	10.69
Moα_1	0.708	1.416	2.123	2.831	3.539	4.247	4.955	5.66	6.37	7.08	7.79	8.49	9.20	9.91	10.62
Zrβ_1	0.700	1.401	2.101	2.801	3.502	4.202	4.902	5.60	6.30	7.00	7.70	8.40	9.10	9.80	10.50
Nbβ_1	0.664	1.329	1.993	2.658	3.322	3.986	4.651	5.32	5.98	6.64	7.31	7.97	8.64	9.30	9.97
Moβ_1	0.631	1.262	1.893	2.524	3.155	3.786	4.417	5.05	5.68	6.31	6.94	7.57	8.20	8.83	9.47

$n\lambda$

Radiation	16	17	18	19	20	21	22	23	24	25	26	27	28	29	30
$Ti\alpha_2$	43.95	46.70	49.44	52.2	54.9	57.7	60.4	63.2	65.9	68.7	71.4	74.2	76.9	79.7	82.4
$Ti\alpha_1$	43.89	46.63	49.38	52.1	54.9	57.6	60.4	63.1	65.8	68.6	71.3	74.1	76.8	79.6	82.3
$Ti\beta_1$	40.14	42.65	45.16	47.67	50.2	52.7	55.2	57.7	60.2	62.7	65.2	67.6	70.3	72.8	75.3
$V\alpha_2$	40.03	42.54	45.04	47.54	50.0	52.5	55.0	57.5	60.1	62.6	65.1	67.5	70.1	72.6	75.1
$V\alpha_1$	39.97	42.47	44.97	47.47	49.97	52.5	55.0	57.5	60.0	62.5	65.0	67.5	70.0	72.5	75.0
$Cr\alpha_2$	36.62	38.91	41.20	43.49	45.78	48.07	50.4	52.6	54.9	57.2	59.5	61.8	64.1	66.4	68.7
$Cr\alpha_1$	36.56	38.85	41.13	43.42	45.70	47.99	50.3	52.6	54.8	57.1	59.4	61.7	64.0	66.3	68.6
$V\beta_1$	36.48	38.75	41.03	43.31	45.59	47.87	50.2	52.4	54.7	57.0	59.3	61.6	63.8	66.1	68.4
$Mn\alpha_2$	33.62	35.73	37.83	39.93	42.03	44.13	46.23	48.33	50.4	52.5	54.6	56.7	58.8	60.9	63.0
$Mn\alpha_1$	33.56	35.66	37.76	39.85	41.95	44.05	46.15	48.24	50.3	52.4	54.5	56.6	58.7	60.8	62.9
$Cr\beta_1$	33.29	35.37	37.45	39.53	41.61	43.69	45.77	47.85	49.93	52.0	54.1	56.2	58.3	60.3	62.4
$Fe\alpha_2$	30.98	32.91	34.85	36.78	38.72	40.66	42.59	44.53	46.46	48.40	50.3	52.3	54.2	56.1	58.1
$Fe\alpha_1$	30.91	32.85	34.78	36.71	38.64	40.57	42.51	44.44	46.37	48.30	50.2	52.2	54.1	56.0	58.0
$Mn\beta_1$	30.50	32.41	34.31	36.22	38.12	40.03	41.94	43.84	45.75	47.66	49.56	51.5	53.4	55.3	57.2
$Co\alpha_2$	28.63	30.42	32.21	33.99	35.78	37.57	39.36	41.15	42.94	44.73	46.52	48.31	50.1	51.9	53.7
$Co\alpha_1$	28.56	30.35	32.14	33.92	35.71	37.49	39.28	41.06	42.85	44.63	46.42	48.20	49.99	51.8	53.6
$Fe\beta_1$	28.05	29.80	31.55	33.31	35.06	36.81	38.57	40.32	42.07	43.83	45.58	47.33	49.08	50.8	52.6
$Ni\alpha_2$	26.53	28.19	29.85	31.51	33.17	34.83	36.48	38.14	39.80	41.46	43.12	44.78	46.44	48.09	49.75
$Ni\alpha_1$	26.47	28.13	29.78	31.44	33.09	34.74	36.40	38.05	39.71	41.36	43.02	44.67	46.33	47.98	49.64
$Co\beta_1$	25.88	27.50	29.11	30.73	32.35	33.97	35.58	37.20	38.82	40.44	42.05	43.67	45.29	46.90	48.52
$Cu\alpha_2$	24.66	26.20	27.74	29.28	30.82	32.37	33.91	35.45	36.99	38.53	40.07	41.61	43.15	44.69	46.24
$Cu\alpha_1$	24.60	26.14	27.67	29.21	30.75	32.29	33.82	35.36	36.90	38.44	39.97	41.51	43.05	44.58	46.12
$Ni\beta_1$	23.95	25.45	26.95	28.44	29.94	31.44	32.94	34.43	35.93	37.43	38.92	40.42	41.92	43.42	44.91
$Zn\alpha_2$	22.98	24.41	25.85	27.28	28.72	30.16	31.59	33.03	34.46	35.90	37.34	38.77	40.21	41.64	43.08
$Zn\alpha_1$	22.92	24.35	25.78	27.21	28.64	30.08	31.51	32.94	34.37	35.81	37.24	38.67	40.10	41.53	42.97
$Cu\beta_1$	22.23	23.62	25.01	26.40	27.79	29.18	30.57	31.96	33.35	34.74	36.12	37.51	38.90	40.29	41.68
$Zn\beta_1$	20.68	21.97	23.27	24.56	25.85	27.14	28.44	29.73	31.02	32.32	33.61	34.90	36.19	37.49	38.78
$Zr\alpha_2$	12.62	13.40	14.19	14.98	15.77	16.56	17.35	18.14	18.92	19.71	20.50	21.29	22.08	22.87	23.66
$Zr\alpha_1$	12.55	13.33	14.12	14.90	15.69	16.47	17.25	18.04	18.82	19.61	20.39	21.18	21.96	22.74	23.53
$Nb\alpha_2$	11.98	12.73	13.48	14.23	14.98	15.73	16.48	17.22	17.97	18.72	19.47	20.22	20.97	21.72	22.47
$Nb\alpha_1$	11.92	12.66	13.40	14.15	14.89	15.64	16.38	17.13	17.87	18.62	19.36	20.11	20.85	21.60	22.34
$Mo\alpha_2$	11.40	12.12	12.83	13.54	14.26	14.97	15.68	16.39	17.11	17.82	18.53	19.25	19.96	20.67	21.38
$Mo\alpha_1$	11.32	12.03	12.74	13.45	14.16	14.86	15.57	16.28	16.99	17.70	18.40	19.11	19.82	20.53	21.23
$Zr\beta_1$	11.20	11.91	12.61	13.31	14.01	14.71	15.41	16.11	16.81	17.51	18.21	18.91	19.61	20.31	21.01
$Nb\beta_1$	10.63	11.29	11.96	12.62	13.29	13.95	14.62	15.28	15.95	16.61	17.27	17.94	18.60	19.27	19.93
$Mo\beta_1$	10.10	10.73	11.36	11.99	12.62	13.25	13.88	14.51	15.14	15.78	16.41	17.04	17.67	18.30	18.93

$$\lg \frac{n}{2}\lambda$$

Radiation	n						
	1	2	3	4	5	6	7
$Ti\alpha_2$	0.1377986	0.4388286	0.6149199	0.7398586	0.8367686	0.9159499	0.9828966
$Ti\alpha_1$	0.1371227	0.4382527	0.6143440	0.7392827	0.8361927	0.9153740	0.9823207
$Ti\beta_1$	0.0984707	0.3995007	0.5755920	0.7005307	0.7974407	0.8766220	0.9435687
$V\alpha_2$	0.0972799	0.3983099	0.5744012	0.6993399	0.7962499	0.8754312	0.9423779
$V\alpha_1$	0.0966233	0.3976533	0.5737446	0.6986833	0.7955933	0.8747746	0.9417213
$V\beta_1$	0.0568477	0.3578777	0.5339690	0.6589077	0.7558177	0.8349990	0.9019457
$Cr\alpha_2$	0.0585987	0.3596287	0.5357200	0.6606587	0.7575687	0.8367500	0.9036967
$Cr\alpha_1$	0.0578619	0.3588919	0.5349832	0.6599219	0.7568319	0.8360132	0.9029599
$Cr\beta_1$	0.0171586	0.3181886	0.4942799	0.6192186	0.7161286	0.7953099	0.8622566
$Mn\alpha_2$	0.0214973	0.3225273	0.4986186	0.6235573	0.7204673	0.7996486	0.8665953
$Mn\alpha_1$	0.0206741	0.3217041	0.4977954	0.6227341	0.7196441	0.7988254	0.8657721
$Mn\beta_1$	9.9791385	0.2801685	0.4562598	0.5811985	0.6781085	0.7572898	0.8242365
$Fe\alpha_2$	9.9858781	0.2869081	0.4629994	0.5879381	0.6848481	0.7640294	0.8309761
$Fe\alpha_1$	9.9849942	0.2860242	0.4621155	0.5870542	0.6839642	0.7631455	0.8300922
$Fe\beta_1$	9.9427551	0.2437851	0.4198764	0.5448151	0.6417251	0.7209064	0.7878531
$Co\alpha_2$	9.9516265	0.2526565	0.4287478	0.5536865	0.6505965	0.7297778	0.7967245
$Co\alpha_1$	9.9506788	0.2517088	0.4278001	0.5527388	0.6496488	0.7288301	0.7957768
$Co\beta_1$	9.9077982	0.2088282	0.3849195	0.5098582	0.6067682	0.6859495	0.7528962
$Ni\alpha_2$	9.9186462	0.2196762	0.3957675	0.5207062	0.6176162	0.6967975	0.7637442
$Ni\alpha_1$	9.9176368	0.2186668	0.3947581	0.5196968	0.6166068	0.6957881	0.7627348
$Ni\beta_1$	9.8742063	0.1752363	0.3513276	0.4762663	0.5731763	0.6523576	0.7193043
$Cu\alpha_2$	9.8868380	0.1878680	0.3639593	0.4888980	0.5858080	0.6649893	0.7319360
$Cu\alpha_1$	9.8857555	0.1867855	0.3628768	0.4878155	0.5847255	0.6639068	0.7308535
$Cu\beta_1$	9.8417817	0.1428117	0.3189030	0.4438417	0.5407517	0.6199330	0.6868797
$Zn\alpha_2$	9.8561335	0.1571635	0.3332548	0.4581935	0.5551035	0.6342848	0.7012315
$Zn\alpha_1$	9.8549646	0.1559946	0.3320859	0.4570246	0.5539346	0.6331159	0.7000626
$Zn\beta_1$	9.8104174	0.1114474	0.2875387	0.4124774	0.5093874	0.5885687	0.6555154
$Zr\alpha_2$	9.5957772	9.8968072	0.0728985	0.1978372	0.2947472	0.3739285	0.4408752
$Zr\alpha_1$	9.5934522	9.8944822	0.0705735	0.1955122	0.2924222	0.3716035	0.4385502
$Zr\beta_1$	9.5442417	9.8452717	0.0213630	0.1463017	0.2432117	0.3223930	0.3893397
$Nb\alpha_2$	9.5733880	9.8744180	0.0505093	0.1754480	0.2723580	0.3515393	0.4184860
$Nb\alpha_1$	9.5709222	9.8719522	0.0480435	0.1729822	0.2698922	0.3490735	0.4160202
$Nb\beta_1$	9.5213866	9.8224166	9.9985079	0.1234466	0.2203566	0.2995379	0.3664846
$Mo\alpha_2$	9.5519408	9.8529708	0.0290621	0.1540008	0.2509108	0.3300921	0.3970388
$Mo\alpha_1$	9.5488996	9.8499296	0.0260209	0.1509596	0.2478696	0.3270509	0.3939976
$Mo\beta_1$	9.4989842	9.8000142	9.9761055	0.1010442	0.1979542	0.2771355	0.3440822

$$\lg \frac{n}{2}\lambda$$

			n					Radiation
8	9	10	11	12	13	14	15	
1.0408886	1.0920411	1.1377986	1.1791913	1.2169799	1.2517420	1.2839266	1.3138449	Tiα_2
1.0403127	1.0914652	1.1372227	1.1786154	1.2164040	1.2511661	1.2833507	1.3132690	Tiα_1
1.0015607	1.0527132	1.0984707	1.1398634	1.1776520	1.2124141	1.2445987	1.2745170	Tiβ_1
1.0003699	1.0515224	1.0972799	1.1386726	1.1764612	1.2112233	1.2434079	1.2733262	Vα_2
0.9997133	1.0508658	1.0966233	1.1380160	1.1758046	1.2105667	1.2427513	1.2726696	Vα_1
0.9599377	1.0110902	1.0568477	1.0982404	1.1360290	1.1707911	1.2029757	1.2328940	Vβ_1
0.9616887	1.0128412	1.0585987	1.0999914	1.1377800	1.1725421	1.2047267	1.2346450	Crα_2
0.9609519	1.0121044	1.0578619	1.0992546	1.1370432	1.1718053	1.2039899	1.2339082	Crα_1
0.9202486	0.9714011	1.0171586	1.0585513	1.0963399	1.1311020	1.1632866	1.1932049	Crβ_1
0.9245873	0.9757398	1.0214973	1.0628900	1.1006786	1.1354407	1.1676253	1.1975436	Mnα_2
0.9237641	0.9749166	1.0206741	1.0620668	1.0998554	1.1346175	1.1668021	1.1967204	Mnα_1
0.8822285	0.9333810	0.9791385	1.0205312	1.0583198	1.0930819	1.1252665	1.1551848	Mnβ_1
0.8889681	0.9401206	0.9858781	1.0272708	1.0650594	1.0998215	1.1320061	1.1619244	Feα_2
0.8880842	0.9392367	0.9849942	1.0263869	1.0641755	1.0989376	1.1311222	1.1610405	Feα_1
0.8458451	0.8969976	0.9427551	0.9841478	1.0219364	1.0566985	1.0888831	1.1188014	Feβ_1
0.8547165	0.9058690	0.9516265	0.9930192	1.0308078	1.0655699	1.0977545	1.1276728	Coα_2
0.8537688	0.9049213	0.9506788	0.9920715	1.0298601	1.0646222	1.0968068	1.1267251	Coα_1
0.8108882	0.8620407	0.9077982	0.9491909	0.9869795	1.0217416	1.0539262	1.0838445	Coβ_1
0.8217362	0.8728887	0.9186462	0.9600389	0.9978275	1.0325896	1.0647742	1.0946925	Niα_2
0.8207268	0.8718793	0.9176368	0.9590295	0.9968181	1.0315802	1.0637648	1.0936831	Niα_1
0.7772963	0.8284488	0.8742063	0.9155990	0.9533876	0.9881497	1.0203343	1.0502526	Niβ_1
0.7899280	0.8410805	0.8868380	0.9282307	0.9660193	1.0007814	1.0329660	1.0628843	Cuα_2
0.7888455	0.8399980	0.8857555	0.9271482	0.9649368	0.9996989	1.0318835	1.0618018	Cuα_1
0.7448717	0.7960242	0.8417817	0.8831744	0.9209630	0.9557251	0.9879097	1.0178280	Cuβ_1
0.7592235	0.8103760	0.8561335	0.8975262	0.9353148	0.9700769	1.0022615	1.0321798	Znα_2
0.7580546	0.8092071	0.8549646	0.8963573	0.9341459	0.9689080	1.0010926	1.0310109	Znα_1
0.7135074	0.7646599	0.8104174	0.8518101	0.8895987	0.9243608	0.9565454	0.9864637	Znβ_1
0.4988672	0.5500197	0.5957772	0.6371699	0.6749585	0.7097206	0.7419052	0.7718235	Zrα_2
0.4965422	0.5476947	0.5934522	0.6348449	0.6726335	0.7073956	0.7395802	0.7694985	Znα_1
0.4473317	0.4984842	0.5442417	0.5856344	0.6234230	0.6581851	0.6903697	0.7202880	Zrβ_1
0.4764780	0.5276305	0.5733880	0.6147807	0.6525693	0.6873314	0.7195160	0.7494343	Nbα_2
0.4740122	0.5251647	0.5709222	0.6123149	0.6501035	0.6848656	0.7170502	0.7469685	Nbα_1
0.4244766	0.4756291	0.5213866	0.5627793	0.6005679	0.6353300	0.6675146	0.6974329	Nbβ_1
0.4550308	0.5061833	0.5519408	0.5933335	0.6311221	0.6658842	0.6980688	0.7279871	Moα_2
0.4519896	0.5031421	0.5488996	0.5902923	0.6280809	0.6628430	0.6950276	0.7249459	Moα_1
0.4020742	0.4532267	0.4989842	0.5403769	0.5781655	0.6129276	0.6451122	0.6750305	Moβ_1

$$\lg \frac{n}{2} \lambda$$

Radiation	n						
	16	17	18	19	20	21	22
$Ti\alpha_2$	1.3419186	1.3682475	1.3930711	1.4165522	1.4388286	1.4600179	1.4802213
$Ti\alpha_1$	1.3413427	1.3676716	1.3924952	1.4159763	1.4382527	1.4594420	1.4796454
$Ti\beta_1$	1.3025907	1.3289196	1.3537432	1.3772243	1.3995007	1.4206900	1.4408934
$V\alpha_2$	1.3013999	1.3277288	1.3525524	1.3760335	1.3983099	1.4194992	1.4397026
$V\alpha_1$	1.3007433	1.3270722	1.3518958	1.3753769	1.3976533	1.4188426	1.4390460
$V\beta_1$	1.2609677	1.2872966	1.3121202	1.3356013	1.3578777	1.3790670	1.3992704
$Cr\alpha_2$	1.2627187	1.2890476	1.3138712	1.3373523	1.3596287	1.3808180	1.4010214
$Cr\alpha_1$	1.2619819	1.2883108	1.3131344	1.3366155	1.3588919	1.3800812	1.4002846
$Cr\beta_1$	1.2212786	1.2476075	1.2724311	1.2959122	1.3181886	1.3393779	1.3595813
$Mn\alpha_2$	1.2256173	1.2519462	1.2767698	1.3002509	1.3225273	1.3437166	1.3639200
$Mn\alpha_1$	1.2247941	1.2511230	1.2759466	1.2994277	1.3217041	1.3428934	1.3630968
$Mn\beta_1$	1.1832585	1.2095874	1.2344110	1.2578921	1.2801685	1.3013578	1.3215612
$Fe\alpha_2$	1.1899981	1.2163270	1.2411506	1.2646317	1.2869081	1.3080974	1.3283008
$Fe\alpha_1$	1.1891142	1.2154431	1.2402667	1.2637478	1.2860242	1.3072135	4.3274169
$Fe\beta_1$	1.1468751	1.1732040	1.1980276	1.2215087	1.2437851	1.2649744	1.2851778
$Co\alpha_2$	1.1557465	1.1820754	1.2068990	1.2303801	1.2526565	1.2738458	1.2940492
$Co\alpha_1$	1.1547988	1.1811277	1.2059513	1.2294324	1.2517088	1.2728981	1.2931015
$Co\beta_1$	1.1119182	1.1382471	1.1630707	1.1865518	1.2088282	1.2300175	1.2502209
$Ni\alpha_2$	1.1227662	1.1490951	1.1739187	1.1973998	1.2196762	1.2408655	1.2610689
$Ni\alpha_1$	1.1217568	1.1480857	1.1729093	1.1963904	1.2186668	1.2398561	1.2600595
$Ni\beta_1$	1.0783263	1.1046552	1.1294788	1.1529599	1.1752363	1.1964256	1.2166290
$Cu\alpha_2$	1.0909580	1.1172869	1.1421105	1.1655916	1.1878680	1.2090573	1.2292607
$Cu\alpha_1$	1.0898755	1.1162044	1.1410280	1.1645091	1.1867855	1.2079748	1.2281782
$Cu\beta_1$	1.0459017	1.0722306	1.0970542	1.1205353	1.1428117	1.1640010	1.1842044
$Zn\alpha_2$	1.0602535	1.0865824	1.1114060	1.1348871	1.1571635	1.1783528	1.1985562
$Zn\alpha_1$	1.0590846	1.0854135	1.1102371	1.1337182	1.1559946	1.1771839	1.1973873
$Zn\beta_1$	1.0145374	1.0408663	1.0656899	1.0891710	1.1114474	1.1326367	1.1528401
$Zr\alpha_2$	0.7998972	0.8262261	0.8510497	0.8745308	0.8968072	0.9179965	0.9381999
$Zr\alpha_1$	0.7975722	0.8239011	0.8487247	0.8722058	0.8944822	0.9156715	0.9358749
$Zr\beta_1$	0.7483617	0.7746906	0.7995142	0.8229953	0.8452717	0.8664610	0.8866644
$Nb\alpha_2$	0.7775080	0.8038369	0.8286605	0.8521416	0.8744180	0.8956073	0.9158107
$Nb\alpha_1$	0.7750422	0.8013711	0.8261947	0.8496758	0.8719522	0.8931415	0.9133449
$Nb\beta_1$	0.7255066	0.7518355	0.7766591	0.8001402	0.8224166	0.8436059	0.8638093
$Mo\alpha_2$	0.7560608	0.7823897	0.8072133	0.8306944	0.8529708	0.8741601	0.8943635
$Mo\alpha_1$	0.7530196	0.7793485	0.8041721	0.8276532	0.8499296	0.8711189	0.8913223
$Mo\beta_1$	0.7031042	0.7294331	0.7542567	0.7777378	0.8000142	0.8212035	0.8414069

$$\lg \frac{n}{2} \lambda$$

23	24	25	26	27	28	29	30	Radiation
1.4995264	1.5180098	1.5357386	1.5527720	1.5691624	1.5849566	1.6001966	1.6149199	$\text{Ti}\alpha_2$
1.4989505	1.5174339	1.5351627	1.5521961	1.5685865	1.5843807	1.5996207	1.6143440	$\text{Ti}\alpha_1$
1.4601985	1.4786819	1.4964107	1.5134441	1.5298345	1.5456287	1.5608687	1.5755920	$\text{Ti}\beta_1$
1.4590077	1.4774911	1.4952199	1.5122533	1.5286437	1.5444379	1.5596779	1.5744012	$\text{V}\alpha_2$
1.4583511	1.4768345	1.4945633	1.5115967	1.5279871	1.5437813	1.5590213	1.5737446	$\text{V}\alpha_1$
1.4185755	1.4370589	1.4547877	1.4718211	1.4882115	1.5040057	1.5192457	1.5339690	$\text{V}\beta_1$
1.4203265	1.4388099	1.4565387	1.4735721	1.4899625	1.5057567	1.5209967	1.5357200	$\text{Cr}\alpha_2$
1.4195897	1.4380731	1.4558019	1.4728353	1.4892257	1.5050199	1.5202599	1.5349832	$\text{Cr}\alpha_1$
1.3788864	1.3973698	1.4150986	1.4321320	1.4485224	1.4643166	1.4795566	1.4942799	$\text{Cr}\beta_1$
1.3832251	1.4017085	1.4194373	1.4364707	1.4528611	1.4686553	1.4838953	1.4986187	$\text{Mn}\alpha_2$
1.3824019	1.4008853	1.4186141	1.4356475	1.4520379	1.4678321	1.4830721	1.4977954	$\text{Mn}\alpha_1$
1.3408663	1.3593497	1.3770785	1.3941119	1.4105023	1.4262965	1.4415365	1.4562598	$\text{Mn}\beta_1$
1.3476059	1.3660893	1.3838181	1.4008515	1.4172419	1.4330361	1.4482761	1.4629994	$\text{Fe}\alpha_2$
1.3467220	1.3652054	1.3829342	1.3999676	1.4163580	1.4321522	1.4473922	1.4621155	$\text{Fe}\alpha_1$
1.3044829	1.3229663	1.3406951	1.3577285	1.3741189	1.3899131	1.4051531	1.4198764	$\text{Fe}\beta_1$
1.3133543	1.3318377	1.3495665	1.3665999	1.3829903	1.3987845	1.4140245	1.4287478	$\text{Co}\alpha_2$
1.3124066	1.3308900	1.3486188	1.3656522	1.3820426	1.3978368	1.4130768	1.4278001	$\text{Co}\alpha_1$
1.2695260	1.2880094	1.3057382	1.3227716	1.3391620	1.3549562	1.3701962	1.3849195	$\text{Co}\beta_1$
1.2803740	1.2988574	1.3165862	1.3336196	1.3500100	1.3658042	1.3810442	1.3957675	$\text{Ni}\alpha_2$
1.2793646	1.2978480	1.3155768	1.3326102	1.3490006	1.3647948	1.3800348	1.3947581	$\text{Ni}\alpha_1$
1.2359341	1.2544175	1.2721463	1.2891797	1.3055701	1.3213643	1.3366043	1.3513276	$\text{Ni}\beta_1$
1.2485658	1.2670492	1.2847780	1.3018114	1.3182018	1.3339960	1.3492360	1.3639593	$\text{Cu}\alpha_2$
1.2474833	1.2659667	1.2836955	1.3007289	1.3171193	1.3329135	1.3481535	1.3628768	$\text{Cu}\alpha_1$
1.2035095	1.2219929	1.2397217	1.2567551	1.2731455	1.2889397	1.3041797	1.3189030	$\text{Cu}\beta_1$
1.2178613	1.2363447	1.2540735	1.2711069	1.2874973	1.3032915	1.3185315	1.3332548	$\text{Zn}\alpha_2$
1.2166924	1.2351758	1.2529046	1.2699380	1.2863284	1.3021226	1.3173626	1.3320859	$\text{Zn}\alpha_1$
1.1721452	1.1906286	1.2083574	1.2253908	1.2417812	1.2575754	1.2728154	1.2875387	$\text{Zn}\beta_1$
0.9575050	0.9759884	0.9937172	1.0107506	1.0271410	1.0429352	1.0581752	1.0728985	$\text{Zr}\alpha_2$
0.9551800	0.9736634	0.9913922	1.0084256	1.0248160	1.0406102	1.0558502	1.0705735	$\text{Zr}\alpha_1$
0.9059695	0.9244529	0.9421817	0.9592151	0.9756055	0.9913997	1.0066397	1.0213630	$\text{Zr}\beta_1$
0.9351158	0.9535992	0.9713280	0.9883614	1.0047518	1.0205460	1.0357860	1.0505093	$\text{Nb}\alpha_2$
0.9326500	0.9511334	0.9688622	0.9858956	1.0022860	1.0180802	1.0333202	1.0480435	$\text{Nb}\alpha_l$
0.8831144	0.9015978	0.9193266	0.9363600	0.9527504	0.9685446	0.9837846	0.9985079	$\text{Nb}\beta_1$
0.9136686	0.9321520	0.9498808	0.9669142	0.9833046	0.9990988	1.0143388	1.0290621	$\text{Mo}\alpha_2$
0.9106274	0.9291108	0.9468396	0.9638730	0.9802634	0.9960576	1.0112976	1.0260209	$\text{Mo}\alpha_1$
0.8607120	0.8791954	0.8969242	0.9139576	0.9303480	0.9461422	0.9613822	0.9761055	$\text{Mo}\beta_1$

3-1d. Values of λ, $\lambda/2$, $\lambda^2/4$, lg $(\lambda^2/4)$, and $(\lambda/2)\sqrt{n}$

These are given (in Å) for the α_1, α_2, and β_1 lines of the K series and for the α line (fo

Z	Element	K line	λ	$\lambda/2$	$\lambda^2/4$	lg $\lambda^2/4$	n: 2	3	4
23	V	α_1	2.50340	1.25170	1.56675	0.19500	1.77017	2.16801	2.50340
		α_2	2.50718	1.25359	1.57149	0.19631	1.77284	2.17128	2.50718
		α	2.50466	1.25233	1.56833	0.19544	1.77106	2.16910	2.50466
		β_1	2.28431	1.14216	1.30453	0.11545	1.61525	1.97828	2.28432
24	Cr	α_1	2.28962	1.14481	1.31059	0.11747	1.61900	1.98287	2.28962
		α_2	2.29351	1.14676	1.31506	0.11895	1.62176	1.98625	2.29352
		α	2.29092	1.14546	1.31209	0.11796	1.61992	1.98399	2.29092
		β_1	2.08480	1.04240	1.08660	0.03607	1.47417	1.80549	2.08480
25	Mn	α_1	2.10175	1.05088	1.10435	0.04311	1.48617	1.82018	2.10176
		α_2	2.10569	1.05285	1.10848	0.04473	1.48895	1.82359	2.10570
		α	2.10306	1.05153	1.10572	0.04365	1.48708	1.82130	2.10306
		β_1	1.91015	0.95508	0.91218	9.96008†	1.35068	1.65425	1.91016
26	Fe	α_1	1.93597	0.96799	0.93700	9.97174	·1.36894	1.67661	1.93598
		α_2	1.93991	0.96996	0.94082	9.97351	1.37173	1.68002	1.93992
		α	1.93728	0.96864	0.93826	9.97232	1.36986	1.67773	1.93728
		β_1	1.75653	0.87827	0.77136	9.88726	1.24206	1,52121	1.75654
27	Co	α_1	1.78892	0.89446	0.80006	9.90312	1.26495	1.54925	1.78892
		α_2	1.79278	0.89639	0.80352	9.90500	1.26768	1.55259	1.79278
		α	1.79021	0.89511	0.80122	9.90375	1.26587	1.55038	1.79022
		β_1	1.62075	0.81038	0.65672	9.81738	1.14605	1.40362	1.62076
28	Ni	α_1	1.65784	0.82892	0.68711	9.83703	1.17227	1.43573	1.65784
		α_2	1.66169	0.83085	0.69031	9.83904	1.17500	1.43907	1.66170
		α	1.65912	0.82956	0.68817	9.83770	1.17317	1.43684	1.65912
		β_1	1.50010	0.75005	0.56258	9.75018	1.06073	1.29912	1.50010
29	Cu	α_1	1.54051	0.77026	0.59330	9.77327	1.08931	1.33413	1.54052
		α_2	1.54433	0.77217	0.59625	9.77543	1.09201	1.33744	1.54434
		α	1.54178	0.77089	0.59427	9.77398	1.09020	1.33522	1.54178
		β_1	1.39217	0.69609	0.48454	9.68533	0.98442	1.20566	1.39218
30	Zn	α_1	1.43511	0.71756	0.51489	9.71171	1.01478	1.24285	1.43512
		α_2	1.43894	0.71947	0.51764	9.71403	1.01748	1.24616	1.43894
		α	1.43639	0.71820	0.51581	9.71249	1.01569	1.24396	1.43640
		β_1	1.29522	0.64761	0.41940	9.62263	0.91586	1.12169	1.29522
42	Mo	α_1	0.70926	0.35463	0.12576	9.09954	0.50152	0.61424	0.70926
		α_2	0.71354	0.35677	0.12728	9.10476	0.50455	0.61794	0.71354
		α	0.71069	0.35535	0.12627	9.10130	0.50254	0.61548	0.71070
		β_1	0.63225	0.31613	0.09994	8.99974	0.44707	0.54755	0.63226
45	Rh	α_1	0.61325	0.30663	0.09402	8.97322	0.43364	0.53110	0.61326
		α_2	0.61761	0.30881	0.09536	8.97937	0.43672	0.53487	0.61762
		α	0.61470	0.30735	0.09446	8.97525	0.43466	0.53235	0.61470
		β_1	0.54559	0.27280	0.07442	8.87169	0.38580	0.47250	0.54560
46	Pd	α_1	0.58542	0.29271	0.08568	8.93288	0.41395	0.50699	0.58542
		α_2	0.58980	0.29490	0.08697	8.93937	0.41705	0.51078	0.58980
		α	0.58688	0.29344	0.08611	8.93505	0.41499	0.50825	0.58688
		β_1	0.52052	0.26026	0.06774	8.83085	0.36806	0.45078	0.52052
47	Ag	α_1	0.55936	0.27968	0.07822	8.89332	0.39553	0.48442	0.55936
		α_2	0.56378	0.28189	0.07946	8.90015	0.39865	0.48825	0.56378
		α	0.56083	0.28042	0.07864	8.89564	0.39657	0.48570	0.56084
		β_1	0.49701	0.24851	0.06176	8.79071	0.35145	0.43043	0.49702
74	W	α_1	0.20904	0.10452	0.01092	8.03822	0.14781	0.18103	0.20904
		α_2	0.21388	0.10694	0.01144	8.05843	0.15124	0.18523	0.21388
		α	0.21065	0.10533	0.01109	8.04493	0.14896	0.18244	0.21066
		β_1	0.18459	0.09230	0.00852	7.93044	0.13053	0.15987	0.18460

* Obtained from values given by Y. Cauchois and H. Hulubei, Longueurs d'onde d'emissions X
† From here on, value given is 10 + log $\lambda^2/4$.

small angles or when the diffraction lines are too broad for the doublet to be resolved)[102].*

$(\lambda / 2)\sqrt{n}$							
5	6	7	8	9	10	11	12
2.79889	3.06603	3.31169	3.54035	3.75510	3.95823	4.15141	4.33601
2.80535	3.07066	3.31669	3.54569	3.76077	3.96420	4.15768	4.34256
2.80030	3.06757	3.31335	3.54213	3.75699	3.96022	4.15350	4.33820
2.55395	2.79771	3.02187	3.23052	3.42648	3.61183	3.78811	3.95656
2.55988	2.80420	3.02888	3.23801	3.43443	3.62021	3.79690	3.96574
2.56424	2.80898	2.80898	3.24353	3.44028	3.62638	3.80337	3.97249
2.56133	2.80579	2.80579	3.23985	3.43638	3.62227	3.79906	3.96799
2.33088	2.55335	2.55335	2.94836	3.12720	3.29636	3.45724	3.61098
2.34984	2.57412	2.78037	2.97234	3.15264	3.32318	3.48537	3.64035
2.35425	2.57895	2.78558	2.97791	3.15855	3.32941	3.49190	3.64718
2.35129	2.57571	2.78209	2.97418	3.15459	3.32523	3.48753	3.64261
2.13563	2.33946	2.52690	2.70138	2.86524	3.02023	3.16764	3.30849
2.16449	2.37108	2.56106	2.73789	2.90397	3.06106	3.21045	3.35321
2.16890	2.37591	2.56627	2.74346	2.90988	3.06729	3.21699	3.36004
2.16595	2.37267	2.56278	2.73973	2.90592	3.06311	3.21261	3.35547
1.96387	2.15131	2.32368	2.48413	2.63481	2.77734	2.91289	3.04242
2.00008	2.19097	2.36652	2.52992	2.68338	2.82853	2.96658	3.09850
2.00439	2.19570	2.37162	2.53538	2.68917	2.83464	2.97299	3.10518
2.00153	2.19256	2.36824	2.53176	2.68533	2.83059	2.96874	3.10075
1.81207	1.98502	2.14406	2.29210	2.43114	2.56265	2.68772	2.80724
1.85352	2.03043	2.19312	2.34454	2.48676	2.62128	2.74921	2.87146
1.85784	2.03516	2.19822	2.35000	2.49255	2.62738	2.75561	2.87815
1.85495	2.03200	2.19481	2.34635	2.48868	2.62331	2.75134	2.87368
1.67716	1.83724	1.98444	2.12146	2.25015	2.37187	2.48763	2.59825
1.72236	1.88674	2.03792	2.17863	2.31078	2.43578	2.55466	2.66826
1.72663	1.89142	2.04297	2.18403	2.31651	2.44182	2.56099	2.67482
1.72376	1.88829	2.03958	2.18041	2.31267	2.43777	2.55675	2.67044
1.55651	1.70507	1.84168	1.96884	2.08827	2.20123	2.30867	2.41133
1.60451	1.75766	1.89848	2.02957	2.15268	2.26913	2.37987	2.48570
1.60879	1.76233	1.90354	2.03497	2.15841	2.27517	2.38621	2.49232
1.60595	1.75922	1.90018	2.03138	2.15460	2.27115	2.38200	2.48792
1.44810	1.58631	1.71341	1.83172	1.94283	2.04792	2.14788	2.24339
0.79298	0.86866	0.93826	1.00305	1.06389	1.12144	1.17617	1.22847
0.79776	0.87390	0.94392	1.00910	1.07031	1.12821	1.18327	1.23589
0.79459	0.87043	0.94017	1.00508	1.06605	1.12372	1.17856	1.23097
0.70689	0.77436	0.83640	0.89415	0.94839	0.99969	1.04848	1.09511
0.68565	0.75109	0.81127	0.86728	0.91989	0.96965	1.01698	1.06220
0.69052	0.75643	0.81703	0.87345	0.92643	0.97654	1.02421	1.06975
0.68726	0.75285	0.81317	0.86932	0.92205	0.97193	1.01936	1.06469
0.61000	0.66822	0.72176	0.77160	0.81840	0.86267	0.90477	0.94501
0.65452	0.71699	0.77444	0.82791	0.87813	0.92563	0.97081	1.01398
0.65942	0.72235	0.78023	0.83410	0.88470	0.93256	0.97807	1 02156
0.65615	0.71878	0.77637	0.82997	0.88032	0.92794	0.97323	1.01651
0.58196	0.63750	0.68858	0.73613	0.78078	0.82301	0.86318	0.90157
0.62538	0.68507	0.73996	0.79106	0.83904	0.88443	0.92759	0.96884
0.63033	0.69049	0.74581	0.79731	0.84567	0.89142	0.93492	0.97650
0.62704	0.68689	0.74192	0.79315	0.84126	0.88677	0.93005	0.97140
0.55569	0.60872	0.65750	0.70289	0.74553	0.78586	0.82421	0.86086
0.23371	0.25602	0.27653	0 29563	0.31356	0.33052	0.34665	0.36207
0.23913	0.26195	0.28294	0.30247	0.32082	0.33817	0.35468	0.37045
0.23553	0.25800	0.27868	0.29792	0.31599	0.33308	0.34934	0.36487
0.20639	0.22609	0.24420	0.25106	0.27690	0.29188	0.30612	0.31974

et des discontinutés d'absorption, Paris, 1941, and by K. Lonsdale, Acta Cryst. 3:400, 1950.

3-2. Space-Group Symbols

Crystals are classified as to symmetry by means of the following symmetry elements in the international nomenclature: rotation axes (1-, 2-, 3-, 4-, and 6-fold), screw axes (the number denoting the axis has a subscript denoting the fraction of the translation, as in 2_1 and 4_2), inversion axes (joint action of rotation and inversion at an inversion center; symbols: $\bar{1}, \bar{2}, \bar{3}, \bar{4}, \bar{6}$), and symmetry planes, of which there are several types: mirror planes m, c glide planes (direction of motion parallel to the vertical z axis of the crystal), a and b glide planes (parallel to the horizontal x and y axes), n glide planes (diagonal glide representing the vector sums a + b, a + c, and b + c), and d glide planes (glide on diagonals). Several additional symbols are needed for the cubic system.

Examples

6m, a 6-fold axis lying in an m plane;

6/m, a 6-fold axis lying normal to an m plane;

$P2_1/c$, a primitive cell in which a 2-fold screw axis is normal to a c glide plane.

See [87, 88] for detailed discussion of space groups. The table gives the relations between the group symbols [111].

Schönflies symbols	Internatl. symbols		Symbols for other orientations		
	1935 (abb.)	1952 (std.)	*abc*	*acb*	*cba*
			Triclinic system		
C_1^1		$P1$		$P1$	
C_1^i		$P\bar{1}$		$P\bar{1}$	
			Monoclinic system		
C_2^1	$P2$	$P2$	$P112$	$P121$	$P211$
C_2^2	$P2_1$	$P2_1$	$P112_1$	$P12_11$	$P2_111$
C_2^3	$C2$	$B2$	$\begin{cases} B112; & A112 \\ 2_1 & 2_1 \end{cases}$	$\begin{cases} C121 & A121 \\ 2_1 & 2_1 \end{cases}$	$\begin{cases} B211; & C211 \\ 2_1 & 2_1 \end{cases}$
C_s^1	Pm	Pm	$P11m$	$P1m1$	$Pm11$
C_s^2	Pc	Pb	$P11b; \; P11a$	$P1c1; \; P1a1$	$Pb11; \; Pc11$
C_s^3	Cm	Bm	$\begin{cases} B11m; & A11m \\ a & b \end{cases}$	$\begin{cases} C1m1; & A1m1 \\ a & c \end{cases}$	$\begin{cases} Bm11; & Cm11 \\ c & b \end{cases}$
C_s^4	Cc	Bb	$\begin{cases} B11b; & A11a \\ n & n \end{cases}$	$\begin{cases} C1c1; & A1a1 \\ n & n \end{cases}$	$\begin{cases} Bb11; & Cc11 \\ n & n \end{cases}$
C_{2h}^1	$P2/m$	$P2/m$	$P11\frac{2}{m}$	$P1\frac{2}{m}1$	$P\frac{2}{m}11$
C_{2h}^2	$P2_1/m$	$P2_1/m$	$P11\frac{2_1}{m}$	$P1\frac{2_1}{m}1$	$P\frac{2_1}{m}11$
C_{2h}^3	$C2/m$	$B2/m$	$\begin{cases} B11\frac{2}{m}; \; A11\frac{2}{m} \\ \frac{2_1}{a} \quad\; \frac{2_1}{b} \end{cases}$	$\begin{cases} C1\frac{2}{m}1; \; A1\frac{2}{m}1 \\ \frac{2_1}{a} \quad\; \frac{2_1}{c} \end{cases}$	$\begin{cases} B\frac{2}{m}11; \; C\frac{2}{m}11 \\ \frac{2_1}{c} \quad\; \frac{2_1}{b} \end{cases}$

Schönflies symbols	Internatl. symbols		Symbols for other orientations		
	1935 (abb.)	1952 (std.)	abc	acb	cba
C_{2h}^4	$P2/c$	$P2/b$	$P11\frac{2}{b}$; $P11\frac{2}{a}$	$P1\frac{2}{c}1$; $P1\frac{2}{a}1$	$P\frac{2}{b}11$; $P\frac{2}{c}11$
C_{2h}^5	$P2_1/c$	$P2_1/b$	$P11\frac{2_1}{b}$; $P11\frac{2_1}{a}$	$P1\frac{2_1}{c}1$; $P1\frac{2_1}{a}1$	$P\frac{2_1}{b}11$; $P\frac{2_1}{c}11$
C_{2h}^6	$C2/c$	$B2/b$	$\left\{\begin{array}{l}B11\frac{2}{b};\ A11\frac{2}{a}\\ \frac{2_1}{n}\qquad\frac{2_1}{n}\end{array}\right.$	$\left\{\begin{array}{l}C1\frac{2}{c}1;\ A1\frac{2}{a}1\\ \frac{2_1}{n}\qquad\frac{2_1}{n}\end{array}\right.$	$\left\{\begin{array}{l}B\frac{2}{b}11;\ C\frac{2}{c}11\\ \frac{2_1}{n}\qquad\frac{2_1}{n}\end{array}\right.$

Orthorhombic system

Schönflies symbols	Internatl. symbols		Symbols for other orientations				
	1935 (abb.)	1952 (std.)	cab	bca	$a\bar{c}b$	$ba\bar{c}$	$\bar{c}ba$
C_{2v}^1	Pmm	$Pmm2$	$P2mm$	$Pm2m$	$Pm2m$	$Pmm2$	$P2mm$
C_{2v}^2	Pmc	$Pmc2_1$	$P2_1ma$	$Pb2_1m$	$Pm2_1b$	$Pcm2_1$	$P2_1am$
C_{2v}^3	Pcc	$Pcc2$	$P2aa$	$Pb2b$	$Pb2b$	$Pcc2$	$P2aa$
C_{2v}^4	Pma	$Pma2$	$P2mb$	$Pc2m$	$Pm2a$	$Pbm2$	$P2cm$
C_{2v}^5	Pca	$Pca2_1$	$P2_1ab$	$Pc2_1b$	$Pb2_1a$	$Pbc2_1$	$P2_1ca$
C_{2v}^6	Pnc	$Pnc2$	$P2na$	$Pb2n$	$Pn2b$	$Pcn2$	$P2an$
C_{2v}^7	Pmn	$Pmn2_1$	$P2_1mn$	$Pn2_1m$	$Pm2_1n$	$Pnm2_1$	$P2_1nm$
C_{2v}^8	Pba	$Pba2$	$P2cb$	$Pc2a$	$Pc2a$	$Pba2$	$P2cb$
C_{2v}^9	Pna	$Pna2_1$	$P2_1nb$	$Pc2_1n$	$Pn2_1a$	$Pbn2_1$	$P2_1cn$
C_{2v}^{10}	Pnn	$Pnn2$	$P2nn$	$Pn2n$	$Pn2n$	$Pnn2$	$P2nn$
C_{2v}^{11}	Cmm	$Cmm2$ $ba2$	$A2mm$ $2cb$	$Bm2m$ $c2a$	$Bm2m$ $c2a$	$Cmm2$ $ba2$	$A2mm$ $2cb$
C_{2v}^{12}	Cmc	$Cmc2_1$ $bn2_1$	$A2_1ma$ 2_1cn	$Bb2_1m$ $n2_1a$	$Bm2_1b$ $c2_1n$	$Ccm2_1$ $na2_1$	$A2_1am$ 2_1nb
C_{2v}^{13}	Ccc	$Ccc2$ $nn2$	$A2aa$ $2nn$	$Bb2b$ $n2n$	$Bb2b$ $n2n$	$Ccc2$ $nn2$	$A2aa$ $2nn$
C_{2v}^{14}	Amm	$Amm2$ $nc2_1$	$B2mm$ 2_1na	$Cm2m$ $b2_1n$	$Am2m$ $n2_1b$	$Bmm2$ $cn2_1$	$C2mm$ 2_1an
C_{2v}^{15}	Abm	$Abm2$ $cc2_1$	$B2cm$ 2_1aa	$Cm2a$ $b2_1b$	$Ac2m$ $b2_1b$	$Bma2$ $cc2_1$	$C2mb$ 2_1aa
C_{2v}^{16}	Ama	$Ama2$ $nn2_1$	$B2mb$ 2_1nn	$Cc2m$ $n2_1n$	$Am2a$ $n2_1n$	$Bbm2$ $nn2_1$	$C2cm$ 2_1nn
C_{2v}^{17}	Aba	$Aba2$ $cn2_1$	$B2cb$ 2_1an	$Cc2a$ $n2_1b$	$Ac2a$ $b2_1n$	$Bba2$ $nc2_1$	$C2cb$ 2_1na
C_{2v}^{18}	Fmm	$Fmm2$ $bc2$	$F2mm$ 2_1ca	$Fm2m$ $b2_1a$	$Fm2m$ $c2_1b$	$Fmm2$ $ca2_1$	$F2mm$ 2_1ab
C_{2v}^{19}	Fdd	$Fdd2$ $dd2_1$	$F2dd$ 2_1dd	$Fd2d$ $d2_1d$	$Fd2d$ $d2_1d$	$Fdd2$ $dd2_1$	$F2dd$ 2_1dd
C_{2v}^{20}	Imm	$Imm2$ $nn2_1$	$I2mm$ 2_1nn	$Im2m$ $n2_1n$	$Im2m$ $n2_1n$	$Imm2$ $nn2_1$	$I2mm$ 2_1nn

Schönflies symbols	Internatl. symbols		Symbols for other orientations				
	1935 (abb.)	1952 (std.)	cab	bca	$a\bar{c}b$	$b\bar{a}c$	$\bar{c}ba$
C_{2v}^{21}	Iba	$Iba2$ $cc2_1$	$I2cb$ 2_1aa	$Ic2a$ $b2_1b$	$Ic2a$ $b2_1b$	$Iba2$ $cc2_1$	$I2cb$ 2_1aa
C_{2v}^{22}	Ima	$Ima2$ $nc2_1$	$I2mb$ 2_1na	$Ic2m$ $b2_1n$	$Im2a$ $n2_1b$	$Ibm2$ $cn2_1$	$I2cm$ 2_1an
$D_2^1 = V^1$	$P222$	$P222$	$P222$	$P222$	$P222$	$P222$	$P222$
$D_2^2 = V^2$	$P222_1$	$P222_1$	$P2_122$	$P22_12$	$P22_12$	$P222_1$	$P2_122$
$D_2^3 = V^3$	$P2_12_12$	$P2_12_12$	$P22_12_1$	$P2_122_1$	$P2_122_1$	$P2_12_12$	$P22_12_1$
$D_2^4 = V^4$	$P2_12_12_1$	$P2_12_12_1$	$P2_12_12_1$	$P2_12_12_1$	$P2_12_12_1$	$P2_12_12_1$	$P2_12_12_1$
$D_2^5 = V^5$	$C222_1$	$C222_1$ $2_12_12_1$	$A2_122$ $2_12_12_1$	$B22_12$ $2_12_12_1$	$B22_12$ $2_12_12_1$	$C222_1$ $2_12_12_1$	$A2_122$ 2_12_12
$D_2^6 = V^6$	$C222$	$C222$ 2_12_12	$A222$ 22_12_1	$B222$ 2_122_1	$B222$ 2_122_1	$C222$ 2_12_12	$A222$ 22_12_1
$D_2^7 = V^7$	$F222$	$F222$ $2_12_12_1$	$F222$ $2_12_12_1$	$F222$ $2_12_12_1$	$F222$ $2_12_12_1$	$F222$ $2_12_12_1$	$F222$ $2_12_12_1$
$D_2^8 = V^8$	$I222$	$I222$ $2_12_12_1$	$I222$ $2_12_12_1$	$I222$ $2_12_12_1$	$I222$ $2_12_12_1$	$I222$ $2_12_12_1$	$I222$ $2_12_12_1$
$D_2^9 = V^9$	$I2_12_12_1$	$I2_12_12_1$ 222	$I2_12_12_1$ 222	$I2_12_12_1$ 222	$I2_12_12_1$ 222	$I2_12_12_1$ 222	$I2_12_12_1$ 222
$D_{2h}^1 = V_h^1$	$Pmmm$	$P\dfrac{2}{m}\dfrac{2}{m}\dfrac{2}{m}$	$Pmmm$	$Pmmm$	$Pmmm$	$Pmmm$	$Pmmm$
$D_{2h}^2 = V_h^2$	$Pnnn$	$P\dfrac{2}{n}\dfrac{2}{n}\dfrac{2}{n}$	$Pnnn$	$Pnnn$	$Pnnn$	$Pnnn$	$Pnnn$
$D_{2h}^3 = V_h^3$	$Pccm$	$P\dfrac{2}{c}\dfrac{2}{c}\dfrac{2}{m}$	$Pmaa$	$Pbmb$	$Pbmb$	$Pccm$	$Pmaa$
$D_{2h}^4 = V_h^4$	$Pban$	$P\dfrac{2}{b}\dfrac{2}{a}\dfrac{2}{n}$	$Pncb$	$Pcna$	$Pcna$	$Pban$	$Pncb$
$D_{2h}^5 = V_h^5$	$Pmma$	$P\dfrac{2_1}{m}\dfrac{2}{m}\dfrac{2}{a}$	$Pbmm$	$Pmcm$	$Pmam$	$Pmmb$	$Pcmm$
$D_{2h}^6 = V_h^6$	$Pnna$	$P\dfrac{2}{n}\dfrac{2_1}{n}\dfrac{2}{a}$	$Pbnn$	$Pncn$	$Pnan$	$Pnnb$	$Pcnn$
$D_{2h}^7 = V_h^7$	$Pmna$	$P\dfrac{2}{m}\dfrac{2}{n}\dfrac{2_1}{a}$	$Pbmn$	$Pncm$	$Pman$	$Pnmb$	$Pcnm$
$D_{2h}^8 = V_h^8$	$Pcca$	$P\dfrac{2_1}{c}\dfrac{2}{c}\dfrac{2}{a}$	$Pbaa$	$Pbcb$	$Pbab$	$Pccb$	$Pcaa$
$D_{2h}^9 = V_h^9$	$Pbam$	$P\dfrac{2_1}{b}\dfrac{2_1}{a}\dfrac{2}{m}$	$Pmcb$	$Pcma$	$Pcma$	$Pbam$	$Pmcb$
$D_{2h}^{10} = V_h^{10}$	$Pccn$	$P\dfrac{2_1}{c}\dfrac{2_1}{c}\dfrac{2}{n}$	$Pnaa$	$Pbnb$	$Pbnb$	$Pccn$	$Pnaa$
$D_{2h}^{11} = V_h^{11}$	$Pbcm$	$P\dfrac{2}{b}\dfrac{2_1}{c}\dfrac{2_1}{m}$	$Pmca$	$Pbma$	$Pcmb$	$Pcam$	$Pmab$
$D_{2h}^{12} = V_h^{12}$	$Pnnm$	$P\dfrac{2_1}{n}\dfrac{2_1}{n}\dfrac{2}{m}$	$Pmnn$	$Pnmn$	$Pnmn$	$Pnnm$	$Pmnn$
$D_{2h}^{13} = V_h^{13}$	$Pmmn$	$P\dfrac{2_1}{m}\dfrac{2_1}{m}\dfrac{2}{n}$	$Pnmm$	$Pmnm$	$Pmnm$	$Pmmn$	$Pnmm$
$D_{2h}^{14} = V_h^{14}$	$Pbcn$	$P\dfrac{2_1}{b}\dfrac{2}{c}\dfrac{2_1}{n}$	$Pnca$	$Pbna$	$Pcnb$	$Pcan$	$Pnab$

Schönflies symbols	Internatl. symbols		Symbols for other orientations				
	1935 (abb.)	1952 (std.)	cab	bca	$a\bar{c}b$	$ba\bar{c}$	$\bar{c}ba$
$D_{2h}^{15}=V_h^{15}$	Pbca	$P\frac{2_1}{b}\frac{2_1}{c}\frac{2_1}{a}$	Pbca	Pbca	Pcab	Pcab	Pcab
$D_{2h}^{16}=V_h^{16}$	Pnma	$P\frac{2_1}{n}\frac{2_1}{m}\frac{2_1}{a}$	Pbnm	Pmcn	Pnam	Pmnb	Pcmn
$D_{2h}^{17}=V_h^{17}$	Cmcm Cbnn	$C\frac{2}{m}\frac{2}{c}\frac{2_1}{m}$	Amma Ancn	Bbmm Bnna	Bmmb Bcnn	Ccmm Cnan	Amam Annb
$D_{2h}^{18}=V_h^{18}$	Cmca Cbnb	$C\frac{2}{m}\frac{2}{c}\frac{2_1}{a}$	Abma Accn	Bbcm Bnaa	Bmab Bccn	Ccmb Cnaa	Acam Abnb
$D_{2h}^{19}=V_h^{19}$	Cmmm Cban	$C\frac{2}{m}\frac{2}{m}\frac{2}{m}$	Ammm Ancb	Bmmm Bcna	Bmmm Bcna	Cmmm Cban	Ammm Ancb
$D_{2h}^{20}=V_h^{20}$	Cccm Cnnn	$C\frac{2}{c}\frac{2}{c}\frac{2}{m}$	Amaa Annn	Bbmb Bnnn	Bbmb Bnnn	Cccm Cnnn	Amaa Annn
$D_{2h}^{21}=V_h^{21}$	Cmma Cbab	$C\frac{2}{m}\frac{2}{m}\frac{2}{a}$	Abmm Accb	Bmcm Bcaa	Bmam Bcca	Cmmb Cbaa	Acmm Abcb
$D_{2h}^{22}=V_h^{22}$	Ccca Cnnb	$C\frac{2}{c}\frac{2}{c}\frac{2}{a}$	Abaa Acnn	Bbcb Bnan	Bbab Bncn	Cccb Cnna	Acaa Abnn
$D_{2h}^{23}=V_h^{23}$	Fmmm	$F\frac{2}{m}\frac{2}{m}\frac{2}{m}$	Fmmm	Fmmm	Fmmm	Fmmm	Fmmm
	Fbca Fcab Fnnn		Fbca Fcab Fnnn	Fbca Fcab Fnnn	Fcab Fbca Fnnn	Fcab Fbca Fnnn	Fcab Fbca Fnnn
$D_{2h}^{24}=V_h^{24}$	Fddd	$F\frac{2}{d}\frac{2}{d}\frac{2}{d}$	Fddd	Fddd	Fddd	Fddd	Fddd
$D_{2h}^{25}=V_h^{25}$	Immm Innn	$I\frac{2}{m}\frac{2}{m}\frac{2}{m}$	Immm Innn	Immm Innn	Immm Innn	Immm Innn	Immm Innn
$D_{2h}^{26}=V_h^{26}$	Ibam Iccn	$I\frac{2}{b}\frac{2}{a}\frac{2}{m}$	Imcb Inaa	Icma Ibnb	Icma Ibnb	Ibam Iccn	Imcb Inaa
$D_{2h}^{27}=V_h^{27}$	Ibca Icab	$I\frac{2}{b}\frac{2}{c}\frac{2}{a}$	Ibca Icab	Ibca Icab	Icab Ibca	Icab Ibca	Ica Ibca
$D_{2h}^{28}=V_h^{28}$	Imma Innb	$I\frac{2}{m}\frac{2}{m}\frac{2}{a}$	Ibmm Icnn	Imcm Inan	Imam Incn	Immb Inna	Icmm Ibnn

Tetragonal system

Schönflies symbols	Internatl. symbols		Other orientations $(a\pm b)$ $(b\mp a)$ c	Schönflies symbols	Internatl. symbols		Other orientations $(a\pm b)$ $(b\mp a)$ c
	1935 (abb.)	1952 (std.)			1935 (abb.)	1952 (std.)	
C_4^1	P4		C4	D_4^1	P42	P422	C422
C_4^2	$P4_1$		$C4_1$	D_4^2	$P42_1$	$P42_12$	$C422_1$
C_4^3	$P4_2$		$C4_2$	D_4^3	$P4_12$	$P4_122$	$C4_122$
C_4^4	$P4_3$		$C4_3$	D_4^4	$P4_12_1$	$P4_12_12$	$C4_122_1$
C_4^5	I4		F4	D_4^5	$P4_22$	$P4_222$	$C4_222$
C_4^6	$I4_1$		$F4_1$	D_4^6	$P4_22_1$	$P4_22_12$	$C4_222_1$
				D_4^7	$P4_32$	$P4_322$	$C4_322$

Schönflies symbols	Internatl. symbols		Other orientations (a ± b) (b ∓ a)c	Schönflies symbols	Internatl. symbols		Other orientations (a ± b) (b ∓ a)c
	1935 (abb.)	1952 (std.)			1935 (abb.)	1952 (std.)	
S_4^1	$P\bar{4}$		$C\bar{4}$	D_4^8	$P4_32_1$	$P4_32_12$	$C4_322_1$
S_4^2	$I\bar{4}$		$F\bar{4}$	D_4^9	$I42$	$I422$	$F422$
				D_4^{10}	$I4_12$	$I4_122$	$F4_122$
C_{4h}^1	$P4/m$		$C4/m$				
C_{4h}^2	$P4_3/m$		$C4_3/m$	D_{4h}^4	$P4/nnc$	$P\dfrac{4}{n}\dfrac{2}{n}\dfrac{2}{c}$	$C\dfrac{4}{n}\dfrac{2}{c}\dfrac{2}{n}$
C_{4h}^3	$P4/n$		$C4/n$				
C_{4h}^4	$P4_2/n$		$C4_2/n$	D_{4h}^5	$P4/mbm$	$P\dfrac{4}{m}\dfrac{2_1}{b}\dfrac{2}{m}$	$C\dfrac{4}{m}\dfrac{2}{m}\dfrac{2_1}{b}$
C_{4h}^5	$I4/m$		$F4/m$				
C_{4h}^6	$I4_1/a$		$F4_1/a$	D_{4h}^6	$P4/mnc$	$P\dfrac{4}{m}\dfrac{2_1}{n}\dfrac{2}{c}$	$C\dfrac{4}{m}\dfrac{2}{c}\dfrac{2_1}{n}$
C_{4v}^1	$P4mm$	$P4mm$	$C4mm$	D_{4h}^7	$P4/nmm$	$P\dfrac{4}{n}\dfrac{2_1}{m}\dfrac{2}{m}$	$C\dfrac{4}{n}\dfrac{2}{m}\dfrac{2_1}{m}$
C_{4v}^2	$P4bm$	$P4bm$	$C4mb$				
C_{4v}^3	$P4cm$	$P4_2cm$	$C4_2mc$	D_{4h}^8	$P4/ncc$	$P\dfrac{4}{n}\dfrac{2_1}{c}\dfrac{2}{c}$	$C\dfrac{4}{n}\dfrac{2}{c}\dfrac{2_1}{c}$
C_{4v}^4	$P4nm$	$P4_2nm$	$C4_2mn$				
C_{4v}^5	$P4cc$	$P4cc$	$C4cc$	D_{4h}^9	$P4/mmc$	$P\dfrac{4_2}{m}\dfrac{2}{m}\dfrac{2}{c}$	$C\dfrac{4_2}{m}\dfrac{2}{c}\dfrac{2}{m}$
C_{4v}^6	$P4nc$	$P4nc$	$C4cn$				
C_{4v}^7	$P4mc$	$P4_2mc$	$C4_2cm$	D_{4h}^{10}	$P4/mcm$	$P\dfrac{4_2}{m}\dfrac{2}{c}\dfrac{2}{m}$	$C\dfrac{4_2}{m}\dfrac{2}{m}\dfrac{2}{c}$
C_{4v}	$P4bc$	$P4_2bc$	$C4_2cb$				
C_{4v}^9	$I4mm$	$I4mm$	$F4mm$	D_{4h}^{11}	$P4/nbc$	$P\dfrac{4_2}{n}\dfrac{2}{b}\dfrac{2}{c}$	$C\dfrac{4_2}{n}\dfrac{2}{c}\dfrac{2}{b}$
C_{4v}^{10}	$I4cm$	$I4cm$	$F4mc$				
C_{4v}^{11}	$I4md$	$I4_1md$	$F4_1dm$	D_{4h}^{12}	$P4/nnm$	$P\dfrac{4_2}{n}\dfrac{2}{n}\dfrac{2}{m}$	$C\dfrac{4_2}{n}\dfrac{2}{m}\dfrac{2}{n}$
C_{4v}^{12}	$I4cd$	$I4_1cd$	$F4_1dc$				
$D_{2d}^1 = V_d^1$	$P\bar{4}2m$	$P\bar{4}2m$	$C\bar{4}m2$	D_{4h}^{13}	$P4/mbc$	$P\dfrac{4_2}{m}\dfrac{2_1}{b}\dfrac{2}{c}$	$C\dfrac{4_2}{m}\dfrac{2}{c}\dfrac{2_1}{b}$
$D_{2d}^2 = V_d^2$	$P\bar{4}2c$	$P\bar{4}2c$	$C\bar{4}c2$				
$D_{2d}^3 = V_d^3$	$P\bar{4}2_1m$	$P\bar{4}2_1m$	$C\bar{4}m2_1$	D_{4h}^{14}	$P4/mnm$	$P\dfrac{4_2}{m}\dfrac{2_1}{n}\dfrac{2}{m}$	$C\dfrac{4_2}{m}\dfrac{2}{m}\dfrac{2_1}{n}$
$D_{2d}^4 = V_d^4$	$P\bar{4}2_1c$	$P\bar{4}2_1c$	$C\bar{4}c2_1$				
$D_{2d}^5 = V_d^5$	$C\bar{4}2m$	$P\bar{4}m2$	$C\bar{4}2m$	D_{4h}^{15}	$P4/nmc$	$P\dfrac{4_2}{n}\dfrac{2_1}{m}\dfrac{2}{c}$	$C\dfrac{4_2}{n}\dfrac{2}{c}\dfrac{2_1}{m}$
$D_{2d}^6 = V_d^6$	$C\bar{4}2c$	$P\bar{4}c2$	$C\bar{4}2c$				
$D_{2d}^7 = V_d^7$	$C\bar{4}2b$	$P\bar{4}b2$	$C\bar{4}2b$	D_{4h}^{16}	$P4/ncm$	$P\dfrac{4_2}{n}\dfrac{2_1}{c}\dfrac{2}{m}$	$C\dfrac{4_2}{n}\dfrac{2}{m}\dfrac{2_1}{c}$
$D_{2d}^8 = V_d^8$	$C\bar{4}2n$	$P\bar{4}n2$	$C\bar{4}2n$				
$D_{2d}^9 = V_d^9$	$F\bar{4}2m$	$I\bar{4}m2$	$F\bar{4}2m$	D_{4h}^{17}	$I4/mmm$	$I\dfrac{4}{m}\dfrac{2}{m}\dfrac{2}{m}$	$F\dfrac{4}{m}\dfrac{2}{m}\dfrac{2}{m}$
$D_{2d}^{10} = V_d^{10}$	$F\bar{4}2c$	$I\bar{4}c2$	$F\bar{4}2c$				
$D_{2d}^{11} = V_d^{11}$	$I\bar{4}2m$	$I\bar{4}2m$	$F\bar{4}m2$				
$D_{2d}^{12} = V_d^{12}$	$I\bar{4}2d$	$I\bar{4}2$	$F\bar{4}d2$				

Schönflies symbols	Internatl. symbols		Other orientations (a ± b) (b ∓a)c	Schönflies symbols	Internatl. symbols		Other orientations (a ± b) (b ∓a)c
	1935 (abb.)	1952 (std.)			1935 (abb.)	1952 (std.)	
D_{4h}^1	$P4/mmm$	$P\dfrac{4}{m}\dfrac{2}{m}\dfrac{2}{m}$	$C\dfrac{4}{m}\dfrac{2}{m}\dfrac{2}{m}$	D_{4h}^{18}	$I4/mcm$	$I\dfrac{4}{m}\dfrac{2}{c}\dfrac{2}{m}$	$F\dfrac{5}{m}\dfrac{2}{m}\dfrac{2}{c}$
D_{4h}^2	$P4/mcc$	$P\dfrac{4}{m}\dfrac{2}{c}\dfrac{2}{c}$	$C\dfrac{4}{m}\dfrac{2}{c}\dfrac{2}{c}$	D_{4h}^{19}	$I4/amd$	$I\dfrac{4_1}{a}\dfrac{2}{m}\dfrac{2}{d}$	$F\dfrac{4_1}{a}\dfrac{2}{d}\dfrac{2}{m}$
D_{4h}^3	$P4/nbm$	$P\dfrac{4}{n}\dfrac{2}{b}\dfrac{2}{m}$	$C\dfrac{4}{n}\dfrac{2}{m}\dfrac{2}{b}$	D_{4h}^{20}	$I4/acd$	$I\dfrac{4_1}{a}\dfrac{2}{c}\dfrac{2}{d}$	$F\dfrac{4_1}{a}\dfrac{2}{d}\dfrac{2}{c}$

Trigonal (rhombohedral) and hexagonal systems

Schönflies symbols	Internatl. symbols			Schönflies symbols	Internatl. symbols		
	1935 (abb.)	1952 (abb.)	1952 (full)		1935 (abb.)	1952 (abb.)	1952 (full)
C_3^1	$C3$	$P3$		C_{3v}^1	$C3m$	$P3m1$	
C_3^2	$C3_1$	$P3_1$		C_{3v}^2	$H3m$	$P31m$	
C_3^3	$C3_2$	$P3_2$		C_{3v}^3	$C3c$	$P3c1$	
C_3^4	$R3$	$R3$		C_{3v}^4	$H3c$	$P31c$	
				C_{3v}^5	$R3m$	$R3m$	
C_{3i}^1	$C\bar3$	$P\bar3$		C_{3v}^6	$R3c$	$R3c$	
C_{3i}^2	$R\bar3$	$R\bar3$		D_{3d}^1	$H\bar3m$	$P\bar31m$	$P\bar31\dfrac{2}{m}$
D_3^1	$H32$	$P312$		D_{3d}^2	$H\bar3c$	$P\bar31c$	$P\bar31\dfrac{2}{c}$
D_3^2	$C32$	$P321$		D_{3d}^3	$C\bar3m$	$P\bar3m1$	$P\bar3\dfrac{2}{m}1$
D_3^3	$H3_12$	$P3_112$		D_{3d}^4	$C\bar3c$	$P\bar3c1$	$P\bar3\dfrac{2}{c}1$
D_3^4	$C3_12$	$P3_121$					
D_3^5	$H3_22$	$P3_212$		D_6^5	$C6_42$	$P6_422$	
D_3^6	$C3_22$	$P3_221$		D_6^6	$C6_32$	$P6_322$	
D_3^7	$R32$	$R32$		C_{6v}^1	$C6mm$	$P6mm$	
D_{3d}^5	$R\bar3m$	$R\bar3m$	$R\bar3\dfrac{2}{m}$	C_{6v}^2	$C6cc$	$P6cc$	
D_{3d}^6	$R\bar3c$	$R\bar3c$	$R\bar3\dfrac{2}{c}$	C_{6v}^3	$C6cm$	$P6_3cm$	
				C_{6v}^4	$C6mc$	$P6_3mc$	
C_6^1	$C6$	$P6$					
C_6^2	$C6_1$	$P6_1$		D_{3h}^1	$C\bar6m2$	$P\bar6m2$	
C_6^3	$C6_5$	$P6_5$		D_{3h}^2	$C\bar6c2$	$P\bar6c2$	
C_6^4	$C6_2$	$P6_2$		D_{3h}^3	$H\bar6m2$	$P\bar62m$	
C_6^5	$C6_4$	$P6_4$		D_{3h}^4	$H\bar6c2$	$P\bar62c$	
C_6^6	$C6_3$	$P6_3$					

Schönflies symbols	Internatl. symbols			Schönflies symbols	Internatl. symbols		
	1935 (abb.)	1952 (abb.)	1952 (full)		1935 (abb.)	1952 (abb.)	1952 (full)
C_{3h}^1	$C\bar6$	$P\bar6$		D_{6h}^1	$C6/mmm$	$P6/mmm$	$P\dfrac{6}{m}\dfrac{2}{m}\dfrac{2}{m}$
C_{6h}^1	$C6/m$	$P6/m$		D_{6h}^2	$C6/mcc$	$P6/mcc$	$P\dfrac{6}{m}\dfrac{2}{c}\dfrac{2}{c}$
C_{6h}^2	$C6_3/m$	$P6_3/m$		D_{6h}^3	$C6/mcm$	$P6_3/mcm$	$P\dfrac{6_3}{m}\dfrac{2}{c}\dfrac{2}{m}$
D_6^1	$C62$	$P622$		D_{6h}^4	$C6/mmc$	$P6_3/mmc$	$P\dfrac{6_3}{m}\dfrac{2}{m}\dfrac{2}{c}$
D_6^2	$C6_12$	$P6_122$					
D_6^3	$C6_52$	$P6_522$					
D_6^4	$C6_22$	$P6_222$					

Cubic system

Schönflies symbols	Internatl. symbols			Schönflies symbols	Internatl. symbols	
	1935 (abb.)	1952 (abb.)	1952 (full)		1952 (abb.)	1952 (full)
T^1		$P\overline{23}$		T_d^1	$P\bar43m$	
T^2		$F23$		T_d^2	$F\bar43m$	
T^3		$I23$		T_d^3	$I\bar43m$	
T^4		$P2_13$		T_d^4	$P\bar43n$	
T^5		$I2_13$		T_d^5	$F\bar43c$	
T_h^1	$Pm3$		$P\dfrac{2}{m}\bar3$	T_d^6	$I\bar43d$	
T_h^2	$Pn3$		$P\dfrac{2}{n}\bar3$			
T_h^3	$Fm3$		$F\dfrac{2}{m}\bar3$	O_h^1	$Pm3m$	$P\dfrac{4}{m}\bar3\dfrac{2}{m}$
T_h^4	$Fd3$		$F\dfrac{2}{d}\bar3$	O_h^2	$Pn3n$	$P\dfrac{4}{n}\bar3\dfrac{2}{n}$
T_h^5	$Im3$		$I\dfrac{2}{m}\bar3$	O_h^3	$Pm3n$	$P\dfrac{4_2}{m}\bar3\dfrac{2}{n}$
T_h^6	$Pa3$		$P\dfrac{2_1}{a}\bar3$	O_h^4	$Pn3m$	$P\dfrac{4_2}{n}\bar3\dfrac{2}{m}$
T_h^7	$Ia3$		$I\dfrac{2_1}{a}\bar3$	O_h^5	$Fm3m$	$F\dfrac{4}{m}\bar3\dfrac{2}{m}$
O^1	$P43$		$P432$	O_h^6	$Fm3c$	$F\dfrac{4}{m}\bar3\dfrac{2}{c}$
O^2	$P4_23$		$P4_232$	O_h^7	$Fd3m$	$F\dfrac{4_1}{d}\bar3\dfrac{2}{m}$
O^3	$F43$		$F432$	O_h^8	$Fd3c$	$F\dfrac{4_1}{d}\bar3\dfrac{2}{c}$
O^4	$F4_13$		$F4_132$	O_h^9	$Im3m$	$I\dfrac{4}{m}\bar3\dfrac{2}{m}$
O^5	$I43$		$I432$	O_h^{10}	$Ia3d$	$I\dfrac{4_1}{a}\bar3\dfrac{2}{d}$
O^6	$P4_33$		$P4_332$			
O^7	$P4_13$		$P4_132$			
O^8	$I4_13$		$I4_132$			

3-3. Tables of Absences for Determining X-Ray Groups

The tables are for deducing x-ray groups (that is, space groups having systematic absences in common) [89]. These groups are divided by crystallographic system and Laue class. The tables give the numbers of the x-ray groups, the diffraction symbols, the systematic absences, and the space groups; a dash indicates that there are no absences, while h, k, and l, or sums of these, denote the systematic absences. If no special indication is given, a symbol in the absence column denotes that reflections are absent for which h, k, l, or a sum of these, is odd. Absences of special types are mentioned; Σ denotes the systematic absences for a face-centered lattice:

$$h+k=2n+1; \quad k+l=2n+1; \quad l+h=2n+1.$$

The boxes enclose the independent absences necessary and sufficient to define the x-ray groups; the others are derived from the independent ones.

I. Triclinic system   Laue class $C_i - \bar{1}$

№	Diffraction symbol	Absences	Space groups
1	$\bar{1}P\bar{1}$		$C_1 = P_1$ $C_i^1 = P\bar{1}$

II. Monoclinic system   Laue class $C_{2h} - 2/m$

№	Diffraction symbol	Absences			Space groups
		hkl	$h0l$	$0k0$	
2	$2/mP—/—$	—	—	—	$C_{2h}^1 = P2/m$; $C_2^1 = P2$; $C_s^1 = Pm$
3	$2/mP2_1/—$	—	—	$\boxed{k}$	$C_{2h}^2 = P2_1/m$; $C_2^2 = P2_1$
4	$2/mP—/m$	—	$\boxed{l}$	—	$C_{2h}^4 = P2/c$; $C_s^2 = Pc$
5	$2/mP2_1/c$	—	$\boxed{l}$	$\boxed{k}$	$C_{2h}^5 = P2_1/c$
6	$2/mC—/—$	$\boxed{h+k}$	h	k	$C_{2h}^3 = C2/m$; $C_2^3 = C2$; $C_s^3 = Cm$
7	$2/mC—/c$	$\boxed{h+k}$	h $\boxed{l}$	k	$C_{2h}^6 = C2/c$; $C_s^4 = Cc$

III. Orthorhombic system

Laue class $V_h - D_{2h} - mmm$

№	Diffraction symbol	\	\	Abs	ences	\	\	\	Space groups
		hkl	$0kl$	$h0l$	$hk0$	$h00$	$0k0$	$00l$	
8	$mmmP$ - - -	—	—	—	—	—	—	—	$D_{2h}^1 = Pmmm$; $D_2^1 = P222$; $C_{2v}^1 = Pmm2$
9	$mmmP2_1$ - -	—	—	—	—	$\boxed{h}$	—	—	$D_2^2 = P2_122$
10	$mmmP2_12_1$ -	—	—	—	—	$\boxed{h}$	$\boxed{k}$	—	$D_2^3 = P2_12_12$
11	$mmmP2_12_12_1$	—	—	—	—	$\boxed{h}$	$\boxed{k}$	$\boxed{l}$	$D_2^4 = P2_12_12_1$
12	$mmmPc$ - -	—	$\boxed{l}$	—	—	—	—	l	$D_{2h}^5 = Pcmm$; $C_{2v}^2 = Pcm2$; $C_{2v}^4 = Pc2m$
13	$mmmPn$ - -	—	$\boxed{k+l}$	—	—	—	k	l	$D_{2h}^{13} = Pnmm$; $C_{2v}^7 = Pnm2$
14	$mmmPba$ -	—	$\boxed{k}$	$\boxed{h}$	—	h	k	—	$D_{2h}^9 = Pbam$; $C_{2v}^8 = Pba2$
15	$mmmPca$ -	—	$\boxed{l}$	$\boxed{h}$	—	h	—	l	$D_{2h}^{11} = Pcam$; $C_{2v}^5 = Pca2$
16	$mmmPcc$ -	—	$\boxed{l}$	$\boxed{l}$	—	—	—	l	$D_{2h}^3 = Pccm$; $C_{2v}^3 = Pcc2$
17	$mmmPna$ -	—	$\boxed{k+l}$	$\boxed{h}$	—	h	k	l	$D_{2h}^{16} = Pnam$; $C_{2v}^9 = Pna2$
18	$mmmPnc$ -	—	$\boxed{k+l}$	$\boxed{l}$	—	—	k	l	$D_{2h}^7 = Pncm$; $C_{2v}^6 = Pnc2$
19	$mmmPnn$ -	—	$\boxed{k+l}$	$\boxed{h+l}$	—	h	k	l	$D_{2h}^{12} = Pnnm$; $C_{2v}^{10} = Pnn2$
20	$mmmPbca$	—	$\boxed{k}$	$\boxed{l}$	$\boxed{h}$	h	k	l	$D_{2h}^{15} = Pbca$
21	$mmmPcca$	—	$\boxed{l}$	$\boxed{l}$	$\boxed{h}$	h	—	l	$D_{2h}^8 = Pcca$
22	$mmmPban$	—	$\boxed{k}$	$\boxed{h}$	$\boxed{h+k}$	h	k	—	$D_{2h}^4 = Pban$
23	$mmmPbcn$	—	$\boxed{k}$	$\boxed{l}$	$\boxed{h+k}$	h	k	l	$D_{2h}^{14} = Pbcn$

№	Diffraction symbol	hkl	0kl	h0l	hk0	h00	0k0	100	Space groups
24	$mmmPccn$	—	l	l	$h+k$	h	k	l	$D_{2h}^{10}=Pccn$
25	$mmmPnna$	—	$k+l$	$h+l$	h	h	k	l	$D_{2h}^{6}=Pnna$
26	$mmmPnnn$	—	$k+l$	$h+l$	$h+k$	h	k	l	$D_{2h}^{2}=Pnnn$
27	$mmmA---$	$k+l$	$k+l$	l	k	—	k	l	$D_{2h}^{19}=Ammm$; $D_{2}^{6}=A222$; $C_{2v}^{11}=A2mm$
28	$mmmA--2_1$	$k+l$	$k+l$	l	k	h	k	l	$D_{2}^{5}=A2_122$; $C_{2v}^{14}=Amm3$
29	$mmmA-a-$	$k+l$	$k+l$	h	k	—	k	l	$D_{2h}^{17}=Amam$; $C_{2v}^{12}=Ama2$; $C_{2v}^{16}=A2am$
30	$mmmAb--$	$k+l$	k	l	k	h	k	l	$D_{2h}^{21}=Abmm$; $C_{2v}^{15}=Abm2$
31	$mmmAba-$	$k+l$	k	h	k	h	k	l	$D_{2h}^{18}=Abam$; $C_{2v}^{17}=Aba2$
32	$mmmA-aa$	$k+l$	k	h	k	h	k	l	$D_{2h}^{20}=Amaa$; $C_{2v}^{13}=A2aa$
33	$mmmAbaa$	$k+l$	k	h	k	h	k	l	$D_{2h}^{22}=Abaa$
34	$mmmI---$	$h+k+l$	$k+l$	$h+l$	$h+k$	h	k	l	$D_{2h}^{25}=Immm$; $D_{2}^{8}=I222$; $D_{2}^{9}=I2_12_12_1$; $C_{2v}^{20}=Imm2$
35	$mmmIb--$	$h+k+l$	k	$h+l$	$h+k$	h	k	l	$D_{2h}^{28}=Ibmm$; $C_{2v}^{22}=Ibm2$
36	$mmmIba-$	$h+k+l$	k	h	$h+k$	h	k	l	$D_{2h}^{26}=Ibam$; $C_{2v}^{21}=Iba2$
37	$mmmIbca$	$h+k+l$	k	l	h	h	k	l	$D_{2h}^{27}=Ibca$
38	$mmmF---$	Σ	$k+l$	$h+l$	$h+k$	h	k	l	$D_{2h}^{23}=Fmmm$; $D_{2}^{7}=F222$; $C_{2v}^{18}=Fmm2$

№	Diffraction symbol	Absences							Space groups
		hkl	$0kl$	$h0l$	$hk0$	$h00$	$0k0$	$00l$	
39	$mmmFdd-$	Σ	$k+l=4n$	$h+l=4n$	$h+k$	$h=4n$	$k=4n$	$l=4n$	$C_{2v}^{19}=Fdd2$
40	$mmmFddd$	Σ	$k+l=4n$	$h+l=4n$	$h+k=4n$	$h=4n$	$k=4n$	$l=4n$	$D_{2h}^{24}=Fddd$

Laue class $C_{4h}-4/m$

IV. Tetragonal system

№	Diffraction symbol	Absences			Space groups
		hkl	$hk0$	$00l$	
41	$4/mP\ -/-$	—	—	—	$C_{4h}^1=P4/m;\quad C_4^1=P4;\quad S_4^1=P\bar{4}$
42	$4/mP4_2/-$	—	—	l	$C_{4h}^2=P4_2/m;\quad C_4^3=P4_2$
43	$4/mP4_1/\cdot$	—	—	$l=4n$	$C_4^2=P4_1\ [C_4^4=P4_3]$
44	$4/mP\ -/n$	—	$h+k$	—	$C_{4h}^3=P4/n$
45	$4/mP4_2/n$	—	$h+k$	l	$C_{4h}^4=P4_2/n$
46	$4/mI\ -/-$	$h+k+l$	$h+k$	l	$C_{4h}^5=I4/m;\quad C_4^5=I4;\quad S_4^2=I\bar{4}$
47	$4/mI4_1/-$	$h+k+l$	$h+k$	$l=4n$	$C_4^6=I4_1$
48	$4/mI4_1/a$	$h+k+l$	h	$l=4n$	$C_{4h}^6=I4_1/a$

V. Tetragonal system

Laue class D_{4h}–$4/mmm$

№	Diffraction symbol	Absences						Space groups
		hkl	$hk0$	$0kl$	hhl	$h00$	$00l$	
49	$4/mmmP\!-\!/\!-\!-\!-$	—	—	—	—	—	—	$D_{4h}^1 = P4/mmm$; $D_{2d}^1 = P\bar{4}2m$; $C_{4v}^1 = P4mm$; $D_4^1 = P42$; $D_{2d}^5 = P\bar{4}m2$;
50	$4/mmmP\!-\!/\!-\!2_1\!-$	—	—	—	—	h	—	$D_4^2 = P42_1$
51	$4/mmmP4_2/\!-\!-\!-$	—	—	—	—	—	l	$D_{2d}^3 = P\bar{4}2_1m$; $D_4^5 = P4_22$
52	$4/mmmP4_1/\!-\!-\!-$	—	—	—	—	—	$l = 4n$	$D_4^3 = P4_12$; $[D_4^7 = P4_32]$
53	$4/mmmP4_2/\!-\!2_1\!-$	—	—	—	—	h	l	$D_4^6 = P4_22_1$
54	$4/mmmP4_1/\!-\!2_1\!-$	—	—	—	—	h	$l = 4n$	$D_4^4 = P4_12_1\ [D_4^8 = P4_32_1]$
55	$4/mmmP\!-\!/\!-\!b\!-$	—	—	k	—	h	—	$D_{4h}^5 = P4/mbm$; $D_{2d}^7 = P\bar{4}b2$; $C_{4v}^2 = P4bm$
56	$4/mmmP\!-\!/\!-\!c\!-$	—	—	l	—	h	—	$D_{4h}^{10} = P4/mcm$; $D_{2d}^6 = P\bar{4}c2$; $C_{4v}^3 = P4cm$
57	$4/mmmP\!-\!/\!-\!n\!-$	—	—	$k+l$	—	h	—	$D_{4h}^{14} = P4/mnm$; $D_{2d}^8 = P\bar{4}n2$; $C_{4v}^4 = P4nm$
58	$4/mmmP\!-\!/n\!-\!-$	—	$h+k$	—	—	—	—	$D_{4h}^7 = P4/nmm$
59	$4/mmmP\!-\!/\!-\!-\!c$	—	—	—	l	—	l	$D_{4h}^9 = P4/mmc$; $D_{2d}^2 = P\bar{4}2c$; $C_{4v}^7 = P4mc$
60	$4/mmmP\!-\!/\!-\!2_1c$	—	—	—	l	h	l	$D_{2d}^4 = P\bar{4}2_1c$
61	$4/mmmP\!-\!/\!-\!bc$	—	—	k	l	h	l	$D_{4h}^{13} = P4/mbc$; $C_{4v}^8 = P4bc$
62	$4;mmmP\!-\!/\!-\!cc$	—	—	l	l	—	l	$D_{4h}^2 = P4/mcc$; $C_{4v}^5 = P4cc$

№	Diffraction symbol	Absences						Space groups
		hkl	$hk0$	$0kl$	hhl	$h00$	$l00$	
63	$4/mmmP-/-nc$	—	—	$k+l$	l	h	l	$D_{4h}^{6} = P4/mnc$; $C_{4v}^{6} = P4nc$
64	$4/mmmP-/nb-$	—	$h+k$	k	—	h	—	$D_{4h}^{3} = P4/nbm$
65	$4/mmmP-/nc-$	—	$h+k$	l	—	h	l	$D_{4h}^{16} = P4/ncm$
66	$4/mmmP-/nn-$	—	$h+k$	$k+l$	—	h	l	$D_{4h}^{12} = P4/nnm$
67	$4/mmmP-/n-c$	—	$h+k$	—	l	h	l	$D_{4h}^{15} = P4/nmc$
68	$4/mmmP-/nbc$	—	$h+k$	k	l	h	l	$D_{4h}^{11} = P4/nbc$
69	$4/mmmP-/ncc$	—	$h+k$	l	l	h	l	$D_{4h}^{8} = P4/ncc$
70	$4/mmmP-/nnc$	—	$h+k$	$k+l$	l	h	l	$D_{4h}^{4} = P4/nnc$
71	$4/mmmI-/---$	$h+k+l$	$h+k$	$k+l$	$2h+l$	h	$l=4n$	$D_{2d}^{9} = I4m2$; $D_{2d}^{11} = \bar{I}42m$; $D_{4h}^{17} = I4/mmm$; $C_{4v}^{9} = I4mm$; $D_{4}^{9} = I42$
72	$4/mmmI4_1/---$	$h+k+l$	$h+k$	$k+l$	$2h+l$	h	l	$D_{4}^{10} = I4_{1}2$
73	$4/mmmI-/-c--$	$h+k+l$	$h+k$	k	$2h+l$	h		$D_{4h}^{18} = I4/mcm$
74	$4/mmmI-/--d$	$h+k+l$	$h+k$	$k+l$	$2h+l=4n$	h	$l=4n$	$D_{2d}^{12} = \bar{I}42c$; $D_{2d}^{10} = \bar{I}4c2$; $C_{4v}^{10} = I4cm$
75	$4/mmmI-/-cd$	$h+k+l$	$h+k$	l	$2h+l=4n$	h	$l=4n$	$C_{4v}^{12} = I4cd$; $C_{4v}^{14} = I4md$
76	$4/mmmI-/a-d$	$h+k+l$	k	$k+l$	$2h+l=4n$	h	$l=4n$	$D_{4h}^{19} = I4/amd$
77	$4/mmmI-/acd$	$h+k+l$	k	l	$2h-l=4n$	h	$l=4n$	$D_{4h}^{20} = I4/acd$

VI. Trigonal system

Laue class C_{3i} – $\bar{3}$

№	Diffraction symbol	Absences			Space groups
		hkil	$h\bar{h}0l$	$000l$	
78	$\bar{3}C$——	—	—	—	$C_{3i}^1 = C\bar{3}$; $C_3^1 = C3$
79	$\bar{3}C3_1$—	—	—	$l=3n$	$C_3^2 = C3_1$; $[C_3^3 = C3_2]$
80	$\bar{3}R$—	$h-k+l=3n$	$2h+l=3n$	$l=3n$	$C_{3i}^2 = R\bar{3}$; $C_3^4 = R3$

VII. Trigonal system

Laue class D_{3d} – $\bar{3}m$

№	Diffraction symbol	Absences			Space groups
		hkil	$h\bar{h}0l$	$000l$	
81	$\bar{3}mC$———	—	—	—	$D_{3d}^1 = C\bar{3}m1$; $D_3^1 = C312$; $C_{3v}^1 = C3m1$ $D_{3d}^3 = C\bar{3}m1$; $D_3^2 = C321$; $C_{3v}^2 = C31m$
82	$\bar{3}mC3_1$——	—	—	$l=3n$	$D_3^3 = C3_112$; $[D_3^5 = C3_212]$ $D_3^4 = C3_121$; $[D_3^6 = C3_221]$
83	$\bar{3}mC$—c—	—	—	l	$D_{3d}^4 = C\bar{3}c$; $C_{3v}^3 = C3c$
84	$\bar{3}mH$——c	$h-k=3n$	$l, h=3n$	l	$D_{3d}^2 = H\bar{3}m$; $C_{3v}^4 = H3c$
85	$\bar{3}mR$——	$h-k+l=3n$	$2h+l=3n$	$l=3n$	$D_{3d}^5 = R\bar{3}m$; $D_3^7 = R32$; $C_{3v}^5 = R3m$
86	$\bar{3}mR$—c	$h-k+l=3n$	$2h+l=3n$	$l=6n$	$D_{3d}^6 = R\bar{3}c$; $C_{3v}^6 = R3c$

VIII. Hexagonal system

Laue class C_{6h} – 6/m

№	Diffraction symbol	Absences 000l	Space groups
87	6/mC—/—	—	$C_{6h}^1 = C6/m$; $C_6^1 = C6$; $C_{3h}^1 = C\bar6$
88	6/mC6₃/—	l	$C_{6h}^2 = C6_3/m$; $C_6^6 = C6_3$
89	6/mC6₂/—	l=3n	$C_6^4 = C6_2$; $[C_6^5 = C6_4]$
90	6/mC6₁/—	l=6n	$C_6^2 = C6_1$; $[C_6^3 = C6_5]$

IX. Hexagonal system

Laue class D_{6h} – 6/mmm

№	Diffraction symbol	Absences 000l	Absences $\bar{h}h0l$	Absences $hh2\bar{h}l$	Space groups
91	6/mmmC—/———	—	—	—	$D_{6h}^1 = C6/mmm$; $D_6^1 = C62$; $C_{6v}^1 = C6mm$; $D_{3h}^1 = C\bar6m2$; $D_{3h}^3 = C\bar62m$
92	6/mmmC6₃/———	l	—	—	$D_6^6 = C6_32$
93	6/mmmC6₂/———	l=3n	—	—	$D_6^4 = C6_22$; $[D_6^5 = C6_42]$
94	6/mmmC6₁/———	l=6n	—	—	$D_6^2 = C6_12$; $[D_6^3 = C6_52]$
95	6/mmmC—/——c	l	—	l	$D_{6h}^4 = C6/mmc$; $C_{6v}^4 = C6mc$; $D_{3h}^4 = C\bar62c$
96	6/mmmC—/—c—	l	l	—	$D_{6h}^3 = C6/mcm$; $C_{6v}^3 = C6cm$; $D_{3h}^2 = C\bar6c2$
97	6/mmmC—/—cc	l	l	l	$D_{6h}^2 = C6/mcc$; $C_{6v}^2 = C6cc$

X. Cubic system

Laue class T_h — m3

№	Diffraction symbol	Absences			Space groups
		hkl	$hk0$	$h00$	
98	$m3P$——	—	—	—	$T_h^1 = Pm3$; $T^1 = P23$
99	$m3P2_1$—	—	—	$\boxed{h}$	$T^4 = P2_13$
100	$m3P$—a	—	$\boxed{h}$	h	$T_h^6 = Pa3$
101	$m3P$—n	—	$\boxed{h+k}$	h	$T_h^2 = Pn3$
102	$m3$: – –	$\boxed{h+k+l}$	$h+k$	h	$T^3 = I23$ $T_h^5 = Im3$; $T^5 = I2_13$
103	$m3Ia$—	$\boxed{h+k+l}$	$\boxed{h}$ k	h	$T_h^7 = Ia3$
104	$m3F$——	$\boxed{\Sigma}$	h k	h	$T_h^3 = Fm3$; $T^2 = F23$
105	$m3Fd$—	$\boxed{\Sigma}$	$\boxed{h+k=4n}$	$h=4n$	$T_h^4 = Fd3$

XI. Cubic system

Laue class O_h — m3m

№	Diffraction symbol	Absences				Space groups
		hkl	$hk0$	hhl	$h00$	
106	$m3mP$———	—	—	—	—	$O_h^1 = Pm3m$; $O^1 = P43$; $T_d^1 = P\bar{4}3m$
107	$m3mP4_2$—	—	—	—	$\boxed{h}$	$O^2 = P4_23$
108	$m3mP4_1$—	—	—	—	$\boxed{h=4n}$	$O^7 = P4_13\,[O^6 = P4_33]$

№	Diffraction symbol	Absences				Space groups
		hkl	$hk0$	hhl	$h00$	
109	$m3mPn$— —	—	$\boxed{h+k}$	—	h	$O_h^4 = Pn3m$
110	$m3mP$— —n	—	—	$\boxed{l}$	—	$O_h^3 = Pm3n$; $T_d^4 = P\bar{4}3n$
111	$m3mPn$—n	—	$\boxed{h+k}$	$\boxed{l}$	h	$O_h^2 = Pn3n$
112	$m3mI$— — —	$\boxed{h+k+l}$	$h+k$	$2h+l$	h	$O_h^9 = Im3m$; $O^5 = I43$; $T_d^3 = I\bar{4}3m$
113	$m3mI4_1$— —	$\boxed{h+k+l}$	$h+k$	$2h+l$	$\boxed{h=4n}$	$O^8 = I4_13$
114	$m3mI$— —d	$\boxed{h+k+l}$	$h+k$	$\boxed{2h+l=4n}$	h	$T_d^6 = I\bar{4}3d$
115	$m3mIa$—d	$\boxed{h+k+l}$	$\boxed{h}$	$\boxed{2h+l=4n}$	h	$O_h^{10} = Ia3d$
116	$m3mF$— — —	$\boxed{\Sigma}$	h	$h+l$	h	$O_h^5 = Fm3m$; $O^3 = F43$; $T_d^2 = F\bar{4}3m$
117	$m3mF4_1$— —	$\boxed{\Sigma}$	h	$h+l$	h	$O^4 = F4_13$
118	$m3mF$— —c	$\boxed{\Sigma}$	h	$\boxed{l}$	h	$O_h^6 = Fm3c$; $T_d^5 = F\bar{4}3c$
119	$m3mFd$— —	$\boxed{\Sigma}$	$\boxed{h+k=4n}$	$h+l$	$h=4n$	$O_h^7 = Fd3m$
120	$m3mFd$—c	$\boxed{\Sigma}$	$\boxed{h+k=4n}$	$\boxed{l}$	$h=4n$	$O_h^8 = Fd3c$

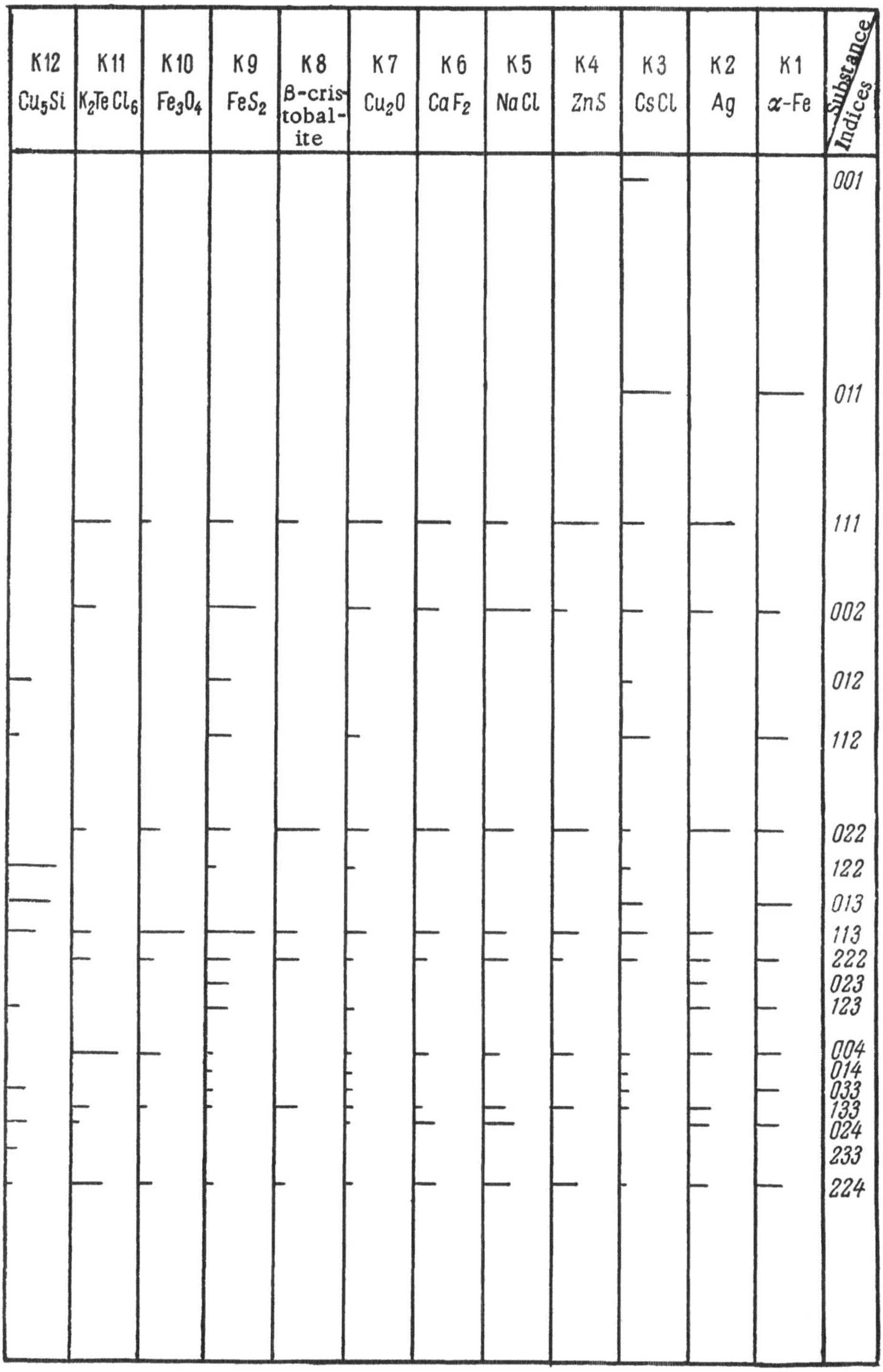

Fig. 39. Schemes for x-ray patterns of cubic crystals (structures K1 to K12).

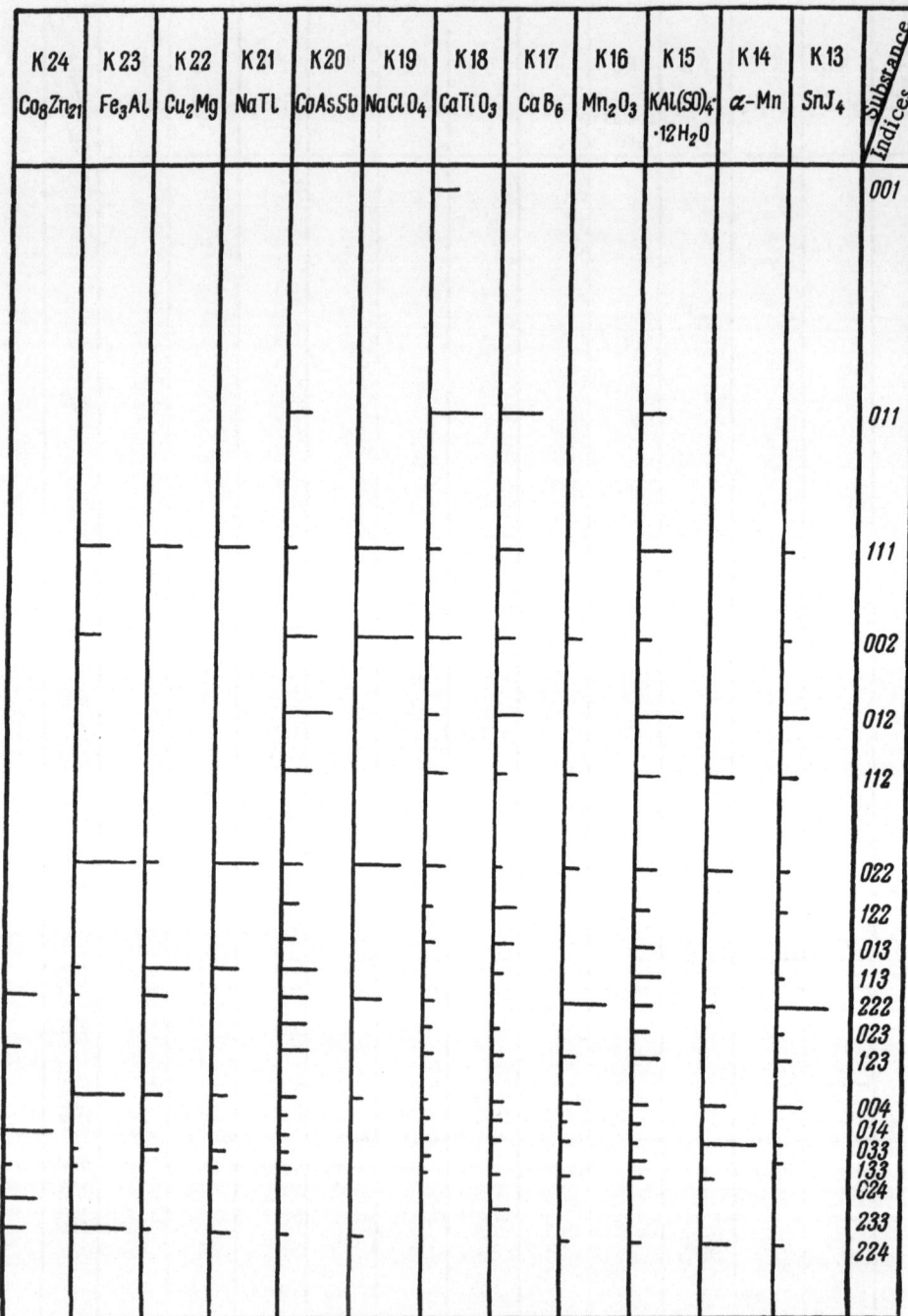

Fig. 40. Schemes for x-ray patterns of cubic crystals (structures K13 to K24).

CUBIC SYSTEM

3-4. Schemes for X-Ray Patterns

Figures 39 and 40 give theoretical patterns for some cubic materials; the lines denote lines on the patterns, the lengths being proportional to the line strengths [102].

3-5. Quadratic Forms

The quadratic form for the cubic system is

$$Q = \frac{1}{a^2}(h^2 + k^2 + l^2). \tag{13}$$

The table gives the (hk$l$) of planes that can be determined from values of ($h^2 + k^2 + l^2$) or $\sqrt{h^2 + k^2 + l^2}$; values of $\lg(h^2 + k^2 + l^2)$ are also given, these being desirable in exact calculations [2]. The indices are arranged in the sequence h $\geq$ k $\geq$ l.

A single asterisk to ($h^2 + k^2 + l^2$) indicates that the preceding number cannot be the sum of the squares of three integers; sometimes there are several (hk$l$) corresponding to one ($h^2 + k^2 + l^2$).

Examples: ($h^2 + k^2 + l^2$) = 34; (hk$l$) can be (530) or (433). The 16* in the ($h^2 + k^2 + l^2$) column indicates that the preceding number, 15, cannot be the sum of the squares of three integers.

$h^2+k^2+l^2$	$\sqrt{h^2+k^2+l^2}$	$\lg(h^2+k^2+l^2)$	h, k, l		
1	1.000	0.0000	1,0,0		
2	.414	.3010	1,1,0		
3	.732	.4771	1,1,1		
4	2.000	.6021	2,0,0		
5	.236	.6990	2,1,0		
6	0.449	0.7782	2,1,1		
8*	.828	.9031	2,2,0		
9	3.000	.9542	3,0,0	2,2,1	
10	.162	1.0000	3,1,0		
11	.317	.0414	3,1,1		
12	.464	.0792	2,2,2		
13	.606	.1139	3,2,0		
14	.742	.1461	3,2,1		
16*	4.000	.2041	4,0,0		
17	.123	.2304	4,1,0	3,2,2	
18	.243	.2553	4,1,1	3,3,0	
19	.359	.2788	3,3,1		
20	.472	.3010	4,2,0		
21	.583	.3222	4,2,1		
22	.690	.3424	3,3,2		
24*	.899	.3802	4,2,2		
25	5.000	.3979	5,0,0	4,3,0	
26	.099	.4150	5,1,0	4,3,1	

$h^2+k^2+l^2$	$\sqrt{h^2+k^2+l^2}$	$\lg(h^2+k^2+l^2)$	h, k, l			
27	5.196	1.4314	5,1,1	3,3,3		
29*	.385	.4624	5,2,0	4,3,2		
30	.477	.4771	5,2,1			
32*	.657	.5051	4,4,0			
33	.745	.5185	5,2,2	4,4,1		
34	.831	.5315	5,3,0	4,3,3		
35	.916	.5441	5,3,1			
36	6.000	.5563	6,0,0	4,4,2		
37	.083	.5682	6,1,0			
38	.164	.5798	6,1,1	5,3,2		
40*	.325	.6021	6,2,0			
41	.403	.6128	6,2,1	5,4,0	4,4,3	
42	.481	.6232	5,4,1			
43	.557	.6335	5,3,3			
44	.633	.6435	6,2,2			
45	.708	.6532	6,3,0	5,4,2		
46	.782	.6628	6,3,1			
48*	.928	.6812	4,4,4			
49	7.000	.6902	7,0,0	6,3,2		
50	.071	.6990	7,1,0	5,5,0	5,4,3	
51	.141	.7076	7,1,1	5,5,1		
52	.211	.7160	6,4,0			
53	.280	.7243	7,2,0	6,4,1		
54	.348	.7324	7,2,1	6,3,3	5,5,2	
56*	.483	.7482	6,4,2			
57	.550	.7559	7,2,2	5,4,4		
58	.616	.7634	7,3,0			
59	.681	.7709	7,3,1	5,5,3		
61*	.810	.7853	6,5,0	6,4,3		
62	.874	.7924	7,3,2	6,5,1		
64*	8.000	.8062	8,0,0			
65	.062	.8129	8,1,0	7,4,0	6,5,2	
66	.124	.8195	8,1,1	7,4,1	5,5,4	
67	.185	.8261	7,3,3			
68	.246	.8325	8,2,0	6,4,4		
69	.307	.8388	8,2,1	7,4,2		
70	.367	.8451	6,5,3			
72*	.485	.8573	8,2,2	6,6,0		
73	.544	.8633	8,3,0	6,6,1		
74	.602	.8692	8,3,1	7,5,0	7,4,3	
75	.660	.8751	7,5,1	5,5,5		
76	.718	.8808	6,6,2			
77	.775	.8865	8,3,2	6,5,4		
78	.832	.8921	7,5,2			
80*	.944	.9031	8,4,0			
81	9.000	.9085	9,0,0	8,4,1	7,4,4	6,6,3
82	.055	.9138	9,1,0	8,3,3		
83	.110	.9191	9,1,1	7,5,3		
84	.165	.9243	8,4,2			
85	.220	.9294	9,2,0	7,6,0		
86	.274	.9345	9,2,1	7,6,1	6,5,5	
88*	.381	.9445	6,6,4			
89	.434	.9494	9,2,2	8,5,0	8,4,3	7,6,2
90	.487	.9542	9,3,0	8,5,1	7,5,4	
91	.539	.9590	9,3,1			
93*	.644	.9685	8,5,2			

$h^2+k^2+l^2$	$\sqrt{h^2+k^2+l^2}$	$\lg(h^2+k^2+l^2)$	$h,\ k,\ l$			
94	9.695	1.9731	9,3,2	7,6,3		
96*	.798	.9823	8,4,4			
97	.849	.9868	9,4,0	6,6,5		
98	.899	.9912	9,4,1	8,5,3	7,7,0	
99	.950	.9956	9,3,3	7,7,1	7,5,5	
100	10.00	2.0000	10,0,0	8,6,0		
101	.05	.0043	10,1,0	9,4,2	8,6,1	7,6,4
102	.10	.0086	10,1,1	7,7,2		
104*	.20	.0170	10,2,0	8,6,2		
105	.25	.0212	10,2,1	8,5,4		
106	.30	.0253	9,5,0	9,4,3		
107	.34	.0294	9,5,1	7,7,3		
108	.39	.0334	10,2,2	6,6,6		
109	.44	.0374	10,3,0	8,6,3		
110	.49	.0414	10,3,1	9,5,2	7,6,5	
113*	.63	.0531	10,3,2	9,4,4	8,7,0	
114	.68	.0569	8,7,1	8,5,5	7,7,4	
115	.72	.0607	9,5,3			
116	.77	.0645	10,4,0	8,6,4		
117	.82	.0682	10,4,1	9,6,0	8,7,2	
118	.86	.0719	10,3,3	9,6,1		
120*	.95	.0792	10,4,2			
121	11.00	.0828	11,0,0	9,6,2	7,6,6	
122	.05	.0864	11,1,0	9,5,4	8,7,3	
123	.09	.0899	11,1,1	7,7,5		
125*	.18	.0969	11,2,0	10,5,0	10,4,3	8,6,5
126	.22	.1004	11,2,1	10,5,1	9,6,3	
128*	.31	.1072	8,8,0			
129	.36	.1106	11,2,2	10,5,2	8,8,1	8,7,4
130	.40	.1139	11,3,0	9,7,0		
131	.44	.1173	11,3,1	9,7,1	9,5,5	
132	.49	.1206	10,4,4	8,8,2		
133	.53	.1239	9,6,4			
134	.58	.1271	11,3,2	10,5,3	9,7,2	7,7,6
136*	.66	.1335	10,6,0	8,6,6		
137	.70	.1367	11,4,0	10,6,1	8,8,3	
138	.75	.1399	11,4,1	8,7,5		
139	.79	.1430	11,3,3	9,7,3		
140	.83	.1461	10,6,2			
141	.87	.1492	11,4,2	10,5,4		
142	.92	.1523	9,6,5			
144*	12.00	.1584	12,0,0	8,8,4		
145	.04	.1614	12,1,0	10,6,3	9,8,0	
146	.08	.1644	12,1,1	11,5,0	11,4,3	9,8,1 9,7,4
147	.12	.1673	11,5,1	7,7,7		
148	.17	.1703	12,2,0			
149	.21	.1732	12,2,1	10,7,0	9,8,2	8,7,6
150	.25	.1761	11,5,2	10,7,1	10,5,5	
152*	.33	.1818	12,2,2	10,6,4		
153	.37	.1847	12,3,0	11,4,4	10,7,2	9,6,6 8,8,5
154	.41	.1875	12,3,1	9,8,3		
155	.45	.1903	11,5,3	9,7,5		
157*	.53	.1959	12,3,2	11,6,0		
158	.57	.1987	11,6,1	10,7,3		
160*	.65	.2041	12,4,0			
161	.69	.2068	12,4,1	11,6,2	10,6,5	9,8,4

$h^2+k^2+l^2$	$\sqrt{h^2+k^2+l^2}$	$\lg(h^2+k^2+l^2)$			h, k, l		
162	12.73	2.2095	12,3,3	11,5,4	9,9,0	8,7,7	
163	.77	.2122	9,9,1				
164	.81	.2148	12,4,2	10,8,0	8,8,6		
165	.85	.2175	10,8,1	10,7,4			
166	.88	.2201	11,6,3	9,9,2	9,7,6		
168*	.96	.2253	10,8,2				
169*	13.00	.2279	13,0,0	12,5,0	12,4,3		
170	.04	.2304	13,1,0	12,5,1	11,7,0	9,8,5	
171	.08	.2330	13,1,1	11,7,1	11,5,5	9,9,3	
172	.11	.2355	10,6,6				
173	.15	.2380	13,2,0	12,5,2	11,6,4	10,8,3	
174	.19	.2405	13,2,1	11,7,2	10,7,5		
176*	.27	.2455	12,4,4				
177	.30	.2480	13,2,2	8,8,7			
178	.34	.2504	13,3,0	12,5,3	9,9,4		
179	.38	.2529	13,3,1	11,7,3	9,7,7		
180	.42	.2553	12,6,0	10,8,4			
181	.45	.2577	12,6,1	10,9,0	9,8,6		
182	.49	.2601	13,3,2	11,6,5	10,9,1		
184*	.56	.2648	12,6,2				
185	.60	.2672	13,4,0	12,5,4	11,8,0	10,9,2	10,7,6
186	.64	.2695	13,4,1	11,8,1	11,7,4		
187	.67	.2718	13,3,3	9,9,5			
189*	.75	.2765	13,4,2	12,6,3	11,8,2	10,8,5	
190	.78	.2788	10,9,3				
192*	.86	.2833	8,8,8				
193	.89	.2856	12,7,0	11,6,6			
194	.93	.2878	13,5,0	13,4,3	12,7,1	12,5,5	11,8,3 9,8,7
195	.96	.2900	13,5,1	11,7,5			
196	14.00	.2923	14,0,0	12,6,4			
197	.04	.2945	14,1,0	12,7,2	10,9,4		
198	.07	.2967	14,1,1	13,5,2	10,7,7	9,9,6	
200*	.14	.3010	14,2,0	10,10,0	10,8,6		
201	.18	.3032	14,2,1	13,4,4	11,8,4	10,10,1	
202	.21	.3054	12,7,3	11,9,0			
203	.25	.3075	13,5,3	11,9,1			
204	.28	.3096	14,2,2	10,10,2			
205	.32	.3118	14,3,0	13,6,0	12,6,5		
206	.35	.3139	14,3,1	13,6,1	11,9,2	11,7,6	10,9,5
208*	.42	.3181	12,8,0				
209	.46	.3201	14,3,2	13,6,2	12,8,1	12,7,4	10,10,3 9,8,8
210	.49	.3222	13,5,4	11,8,5			
211	.53	.3243	11,9,3	9,9,7			
212	.56	.3263	14,4,0	12,8,2			
213	.59	.3284	14,4,1	10,8,7			
214	.63	.3304	14,3,3	13,6,3			
216*	.70	.3345	14,4,2	12,6,6	10,10,4		
217	.73	.3365	12,8,3	10,9,6			
218	.76	.3385	13,7,0	12,7,5	11,9,4		
219	.80	.3404	13,7,1	13,5,5	11,7,7		
221*	.87	.3444	14,5,0	14,4,3	13,6,4	11,10,0	11,8,6
222	.90	.3464	14,5,1	13,7,2	11,10,1		
224*	.97	.3502	12,8,4				
225	15.00	.3522	15,0,0	14,5,2	12,9,0	11,10,2	10,10,5
226	.03	.3541	15,1,0	12,9,1	9,9,8		
227	.07	.3560	15,1,1	13,7,3	11,9,5		

$h^2+k^2+l^2$	$\sqrt{h^2+k^2+l^2}$	$\lg(h^2+k^2+l^2)$	h, k, l
228	15.10	2.3579	14,4,4 10,8,8
229	.13	.3598	15,2,0 12,9,2 12,7,6
230	.17	.3617	15,2,1 14,5,3 13,6,5 11,10,3 10,9,7
232*	.23	.3655	14,6,0
233	.26	.3674	15,2,2 14,6,1 13,8,0 12,8,5
234	.30	.3692	15,3,0 13,8,1 13,7,4 12,9,3 11,8,7
235	.33	.3711	15,3,1
236	.36	.3729	14,6,2 10,10,6
237	.39	.3747	14,5,4 13,8,2 11,10,4
238	.43	.3766	15,3,2 11,9,6
241**	.52	.3820	15,4,0 14,6,3 13,6,6 12,9,4
242	.56	.3838	15,4,1 13,8,3 12,7,7 11,11,0
243	.59	.3856	15,3,3 13,7,5 11,11,1 9,9,9
244	.62	.3874	12,10,0 12,8,6
245	.65	.3892	15,4,2 14,7,0 12,10,1 10,9,8
246	.68	.3909	14,7,1 14,5,5 11,11,2 11,10,5
248*	.75	.3945	14,6,4 12,10,2
249	.78	.3962	14,7,2 13,8,4 11,8,8 10,10,7
250	.81	.3979	15,5,0 15,4,3 13,9,0 12,9,5
251	.84	.3997	15,5,1 13,9,1 11,11,3 11,9,7
253*	.91	4031	12,10,3
254	.94	.4048	15,5,2 14,7,3 13,9,2 13,7,6
256*	16.00	.4082	16,0,0
257	.03	.4099	16,1,0 15,4,4 14,6,5 12,8,7 11,10,6
258	.06	.4116	16,1,1 13,8,5 11,11,4
259	.09	.4133	15,5,3 13,9,3
260	.12	.4150	16,2,0 14,8,0 12,10,4
261	.16	.4166	16,2,1 15,6,0 14,8,1 14,7,4 12,9,6
262	.19	.4183	15,6,1 10,9,9
264*	.25	.4216	16,2,2 14,8,2 10,10,8
265	.28	.4232	16,3,0 15,6,2 12,11,0
266	.31	.4249	16,3,1 15,5,4 13,9,4 12,11,1 11,9,8
267	.34	.4265	13,7,7 11,11,5
268	.37	.4281	14,6,6
269	.40	.4298	16,3,2 14,8,3 13,10,0 13,8,6 12,11,2 12,10,5
270	.43	.4314	15,6,3 14,7,5 13,10,1 11,10,7
272*	.49	.4346	16,4,0 12,8,8
273	.52	.4362	16,4,1 13,10,2
274	.55	.4378	16,3,3 15,7,0 12,11,3 12,9,7
275	.58	.4393	15,7,1 15,5,5 13,9,5
276	.61	.4409	16,4,2 14,8,4
277	.64	.4425	15,6,4 14,9,0
278	.67	.4440	15,7,2 14,9,1 13,10,3 11,11,6
280*	.73	.4472	12,10,6
281	.76	.4787	16,5,0 16,4,3 14,9,2 14,7,6 12,11,4 10,10,9
282	.79	.4502	16,5,1 13,8,7
283	.82	.4518	15,7,3 11,9,9
285*	.88	.4548	16,5,2 14,8,5 13,10,4 11,10,8
286	.91	.4564	15,6,5 14,9,3 13,9,6
288*	.97	.4594	16,4,4 12,12,0
289	17.00	.4609	17,0,0 15,8,0 12,12,1 12,9,8
290	.03	.4624	17,1,0 16,5,3 15,8,1 15,7,4 13,11,0 12,11,5
291	.06	.4639	17,1,1 13,11,1 11,11,7
292	.09	.4654	16,6,0 12,12,2
293	.12	.4669	17,2,0 16,6,1 15,8,2 14,9,4 12,10,7

$h^2+k^2+l^2$	$\sqrt{h^2+k^2+l^2}$	$\lg(h^2+k^2+l^2)$	h, k, l
294	17.15	2.4683	17,2,1 14,7,7 13,11,2 13,10,5
296*	.20	.4713	16,6,2 14,10,0 14,8,6
297	.23	.4728	17,2,2 16,5,4 15,6,6 14,10,1 13,8,8 12,12,3
298	.26	.4742	17,3,0 15,8,3
299	.29	.4757	17,3,1 15,7,5 13,11,3 13,9,7
300	.32	.4771	14,10,2 10,10,10
301	.35	.4786	16,6,3 12,11,6
302	.38	.4800	17,3,2 14,9,5 11,10,9
304*	.44	.4829	12,12,4
305	.46	.4843	17,4,0 16,7,0 15,8,4 14,10,3 13,10,6
306	.49	.4857	17,4,1 16,7,1 16,5,5 15,9,0 13,11,4 12,9,9 11,11,8
307	.52	.7871	17,3,3 15,9,1
308	.55	.4886	16,6,4 12,10,8
309	.58	.4900	17,4,2 16,7,2 14,8,7
310	.61	.4914	15,9,2 15,7 6
312*	.66	.4942	14,10,4
313	.69	.4955	14,9,6 13,12,0 12,12,5
314	.72	.4969	17,5,0 17,4,3 16,7,3 15,8,5 13,12,1 13,9,8 12,11,7
315	.75	.4983	17,5,1 15,9,3 13,11,5
317*	.80	.5011	16,6,5 14,11,0 13,12,2
318	.83	.5024	17,5,2 14,11,1 13,10,7
320*	.89	.5051	16,8,0
321	.92	.5065	17,4,4 16,8,1 16,7,4 14,11,2 14,10,5 11,10,10
322	.94	.5079	15,9,4 13,12,3
323	.97	.5092	17,5,3 15,7,7 11,11,9
234	18.00	.5105	18,0,0 16,8,2 14,8,8 12,12,6
325	.03	.5119	18,1,0 17,6,0 15,10,0 15,8,6 12,10,9
326	.06	.5132	18,1,1 17,6,1 15,10,1 14,11,3 14,9,7 13,11,6
328*	.11	.5159	18,2,0 16,6,6
329	.14	.5172	18,2,1 17,6,2 16,8,3 15,10,2 13,12,4 12,11,8
330	.17	.5185	17,5,4 16,7,5
331	.19	.5198	15,9,5 13,9,9
332	.22	.5211	18,2,2 14,10,6
333	.25	.5224	18,3,0 14,11,4 13,10,8
334	.28	.5237	18,3,1 17,6,3 15,10,3
336*	.33	.5263	16,8,4
337	.36	.5276	18,3,2 16,9,0 12,12,7
338	.38	.5289	17,7,0 16,9,1 15,8,7 13,13,0 13,12,5
339	.41	.5302	17,7,1 17,5,5 13,13,1 13,11,7
340	.44	.5315	18,4,0 14,12,0
341	.47	.5328	18,4,1 17,6,4 16,9,2 16,7,6 15,10,4 14,12,1 14,9,8
342	.49	.5340	18,3,3 17,7,2 15,9,6 14,11,5 13,13,2 11,11,10
344*	.55	.5366	18,4,2 14,12,2 12,10,10
345	.57	.5378	16,8,5 14,10,7
346	.60	.5391	16,9,3 15,11,0 12,11,9
347	.63	.5403	17,7,3 15,11,1 13,13,3
349*	.68	.5428	18,5,0 18,4,3 14,12,3 13,12,6
350	.71	.5441	18,5,1 17,6,5 15,11,2 15,10,5 13,10,9
352*	.76	.5465	12,12,8
353	.79	.5478	18,5,2 17,8,0 16,9,4 15,8,8 14,11,6
354	.81	.5490	17,8,1 17,7,4 16,7,7 13,13,4 13,11,8
355	.84	.5502	15,11,3 15,9,7
356	.87	.5514	18,4,4 16,10,0 16,8,6 14,12,4
357	.89	.5527	17,8,2 16,10,1
358	.92	.5539	18,5,3 14,9,9

$h^2+k^2+l^2$	$\sqrt{h^2+k^2+l^2}$	$\lg(h^2+k^2+l^2)$		h, k, l			
360*	18.97	2.5563	18,6,0	16,10,2	14,10,8		
361	19.00	.5575	19,0,0	18,6,1	17,6,6	15,10,6	
362	.03	.5587	19,1,0	17,8,3	16,9,5	15,11,4	13,12,7
363	.05	.5599	19,1,1	17,7,5	13,13,5	11,11,11	
364	.08	.5611	18,6,2				
365	.10	.5623	19,2,0	18,5,4	16,10,3	14,13,0	14,12,5
			12,11,10				
366	.13	.5635	19,2,1	14,13,1	14.11.7		
369**	.21	.5670	19,2,2	18,6,3	17,8,4	16,8,7	15,12,0
			14,13,2	13,10,10	12,12,9		
370	.24	.5682	19,3,0	17,9,0	15,12,1	15,9,8	
371	.26	.5694	19,3,1	17,9,1	15,11,5	13,11,9	
372	.29	.5705	16,10,4				
373	.31	.5717	18,7,0	16,9,6	15,12,2		
374	.34	.5729	19,3,2	18,7,1	18,5,5	17,9,2	17,7,6
			15,10,7	14.13,3	13,13,6		
376*	.39	.5752	18,6,4	14.12,6			
377	.42	.5763	19,4,0	18,7,2	16,11,0	14,10,9	13,12,8
378	.44	.5775	19,4,1	17,8,5	16,11,1	15,12,3	
379	.47	.5786	19,3,3	17,9 3			
381*	.52	.5809	19,4,2	16,11,2	16.10,5	14,13,4	14,11,8
382	.54	.5821	18,7,3	15,11,6			
384*	.60	.5843	16,8,8				
385	62	.5855	18,6,5	15,12,4			
386	.65	.5866	19,5,0	19,4,3	17,9,4	16,11.3	16,9.7
			12,11,11				
387	.67	.5877	19,5,1	17,7.7	15,9,9	13,13.7	
388	.70	.5888	18,8,0	12,12,10			
389	.72	.5899	18,8,1	18,7,4	17,10.0	17,8,6	15,10,8
			14,12.7				
390	.75	.5911	19,5,2	17,10,1	14,13,5	13,11,10	
392*	.80	.5933	18,8,2	16,10,6	14,14,0		
393	.82	.5944	19,4,4	17,10.2	16,11,4	14,14,1	
394	.85	.5955	15,13,0	15,12.5	13,12,9		
395	.87	.5966	19,5,3	17.9,5	15,13,1	15,11,7	
396	.90	.5977	18,6,6	14,14.2	14,10,10		
397	.92	.5988	19,6,0	18,8,3			
398	.95	.5999	19,6,1	18,7,5	17.10.3	15,13.2	14,11,9
400*	20.00	.6021	20,0,0	16,12,0			
401	.02	.6031	20,1,0	19,6,2	16,12.1	16,9,8	14,14,3
			14,13,6				
402	.05	.6042	20,1.1	19,5,4	17,8,7	16,11,5	13,13,8
403	.07	.6053	15,13,3				
404	.10	.6064	20,2,0	18,8,4	16,12,2	14,12,8	
405	.12	.6075	20,2,1	18,9,0	17,10,4	16,10,7	15,12,6
406	.15	.6085	19,6,3	18,9.1	17,9,6	15,10,9	
408*	.20	.6107	20,2,2	14,14,4			
409	.22	.6117	20,3,0	18,9,2	18,7,6	16,12,3	12,12,11
410	.25	.6128	20,3,1	19,7,0	17,11,0	15,13,4	15,11,8
411	.27	.6138	19,7,1	19,5,5	17,11,1	13,11,11	
413*	.32	.6160	20,3,2	19,6,4	18.8.5	16,11,6	13,12,10
414	.35	.6170	19,7,2	18,9,3	17.11.2	17,10.5	14.13.7
416*	.40	.6191	20,4,0	16,12,4			
417	.42	.6201	20,4.1	17.8,8	14,14,5	14,11,10	
418	.45	.6212	20,3,3	16,9,9	15,12,7		
419	.47	.6222	19,7,3	17.11.3	17,9,7	15,13,5	13,13.9
420	.49	.6232	20,4,2	16.10,8			
421	.52	.6243	18,9,4	15,14.10	14,12.9		
422	.54	.6253	19,6.5	18,7,7	15,14,1		
424* .	.59	.6274	18,10,0	18,8,6			
425	.62	.6284	20,5,0	20,4,3	19,8,0	18,10,1	17,10,6
			16,13,0	16,12,5	15,14,2	15,10,10	

$h^2+k^2+l^2$	$\sqrt{h^2+k^2+l^2}$	$\lg(h^2+k^2+l^2)$	h, k, l
426	20.64	2.6294	20,5,1 19,8,1 19,7,4 17,11,4 16,13,1 16,11,7
427	.66	.6304	15,11,9
428	.69	.6314	18,10,2 14,14,6
429	.71	.6325	20,5,2 19,8,2 16,13,2 14,13,8
430	.74	.6335	18,9,5 15,14,3 15,13,6
432*	.78	.6355	20,4,4 12,12,12
433	.81	.6365	19.6.6 18,10,3 17,12.0 15,12.8
434	.83	.6375	20,5,3 19.8.3 17,12,1 17,9,8 16,13,3 13.12,11
435	.86	.6385	19,7,5 17.11,5
436	.88	.6395	20,6.0 16,12,6
437	.90	.6405	20,6,1 18,8,7 17,12.2 16,10,9 15,14,4
438	.93	.6415	17,10,7 14,11,11 13,13,10
440*	.98	.6435	20,6.2 18,10,4 14,12,10
441	21.00	.6444	21,0,0 20,5,4 19,8,4 18,9.6 16,13,4 16,11,8 14,14,7
442	.02	.6454	21,1,0 19,9,0 17,12,3
443	.05	.6464	21,1,1 19,9,1 15,13,7
445*	.10	.6484	21,2,0 20,6,3 18,11.0
446	.12	.6493	21,2,1 19,9,2 19.7.6 18,11.1 17,11,6 15,14,5 15,11,10 14.13,9
449**	.19	.6522	21,2,2 20,7,0 18.11,2 18,10.5 17,12.4 16,12.7
450	.21	.6532	21,3,0 20,7,1 20,5,5 19,8,5 16,13,5 15,15.0 15.12,9
451	.24	.6542	21,3.1 19,9,3 17,9,9 15.15,1
452	.26	.6551	20,6.4 18,8,8 16,14,0
453	.28	.6561	20,7.2 17,10,8 16,14,1
454	.31	.6571	21,3,2 18,11,3 18,9.7 15,15.2
456*	.35	.6590	16,14,2 16,10.10 14,14,8
457	.38	.6599	21,4,0 15,14,6 13.12.12
458	.40	.6609	21,4,1 20,7,3 19.9,4 17,13,0 17.12,5 16,11,9 15,13,8
459	.42	.6618	21,3.3 19,7,7 17,13,1 17,11,7 15,15,3 13,13,11
460	.45	.6628	18,10,6
461	.47	.6637	21,4,2 20,6,5 19,10,0 19,8,6 18,11.4 16,14,3 16,13,6 14.12,11
462	.49	.6646	19,10,1 17,13,2
464*	.54	.6665	20,8,0 16,12,8
465	.56	.6675	20,8,1 20,7,4 19,10.2 14,13,10
466	.59	.6684	21,5,0 21,4,3 15.15,4
467	.61	.6693	21,5,1 19,9,5 17.13.3 15,11,11
468	.63	.6702	20,8,2 18,12,0 16,14,4
469	.66	.6712	18,12,1 18,9,8 17,12,6 15.12.10
470	.68	.6721	21,5,2 19,10,3 18,11,5 17,10,9 15,14,7
472*	.73	.6739	20,6,6 18,12,2
473	.75	.6749	21,4,4 20,8,3 18,10,7 14,14.9
474	.77	.6758	20,7,5 19,8,7 17,13,4 17,11.8 16,13,7
475	.79	.6767	21,5,3 15.15,5 15,13,9
477*	.84	.6785	21,6,0 19.10,4 18.12.3 16,14.5 16,11.10
478	.86	.6794	21,6,1 19,9,6
480*	.91	.6812	20,8,4
481	.93	.6821	21,6,2 20,9,0 18,11,6 16,15,0 16.12,9
482	.95	.6830	21,5,4 20,9,1 19,11,0 17,12,7 16.15.1 13,13,12
483	.98	.6839	19,11.1 17.13.5
484	22.00	.6848	22,0,0 18,12,4 14.12,12
485	.02	.6857	22,1.0 20,9.2 20,7,6 17,14,0 16,15,2 15,14,8
486	.05	.6866	22,1,1 21,6,3 19,11,2 19,10,5 18,9,9 17,14.1 15,15,6 14.13,11
488*	.09	.6884	22.2.0 18,10,8 16,14,6
489	.11	.6893	22,2,1 20,8,5 19,8.8 17,14,2 17,10,10 16,13,8

$h^2+k^2+l^2$	$\sqrt{h^2+k^2+l^2}$	$\lg(h^2+k^2+l^2)$	h, k, l
490	22.14	2.6902	21,7,0 20,9,3 16,15,3 15,12,11
491	.16	.6911	21,7,1 21,5,5 19,11,3 19,9,7 17,11,9
492	.18	.6920	22,2,2 14,14,10
493	.20	.6928	22,3,0 21,6,4 18,13,0 18,12,5
494	.23	.6937	22,3,1 21,7,2 18,13,1 18,11,7 17.14.3 17,13,6 15,13,10
497**	.29	.6964	22,3,2 20,9.4 19.10.6 18,13,2 17,12,8 16,15,4
498	.32	.6972	20,7,7 19,11,4 16,11,11
499	.34	.6981	21,7,3 15,15,7
500	.36	.6990	22,4,0 20,10,0 20,8,6 16,12,10
501	.38	.6998	22,4,1 20,10,1 17,14,4 16,14,7
502	.41	.7007	22,3,3 21.6.5 18.13,3 15,14,9
504*	.45	.7024	22,4,2 20,10,2 18,12,6
505	.47	.7033	21,8,0 19,12,0 18,10,9
506	.49	.7042	21,8,1 21,7,4 20,9,5 19,12,1 19.9.8 16,15,5 16,13,9
507	.52	.7050	19,11,5 17,13.7 13,13,13
509*	.56	.7067	22,5,0 22,4,3 21,8,2 20,10,3 19,12.2 18.13,4 18,11,8 14,13,12
510	.58	.7076	22.5,1 19,10,7 17,14,5 17,11,10
512*	.63	.7093	16,16.0
513	.65	.7101	22,5,2 21,6,6 20,8,7 16,16,1 15.12,12 14,14,11
514	.67	.7110	21.8,3 19,12,3 17,15,0 17,12.9 15,15.8
515	.69	.7118	21,7,5 17,15,1 15,13,11
516	.72	.7126	22,4,4 20.10,4 16.16,2 16.14.8
517	.74	.7135	20.9,6 18.12,7 16,15,6
518	.76	.7143	22.5.3 19,11,6 18,13,5 17,15,2
520*	.80	.7160	22,6,0 18.14,0
521	.83	.7168	22,6,1 21,8,4 20,11,0 19,12,4 18,14,1 17,14,6 16,16,3 16,12,11 15,14,10
522	.85	.7177	21,9,0 20,11,1 17,13,8
523	.87	.7185	21,9,1 19,9,9 17,15,3
524	.89	.7193	22,6,2 18,14,2 18,10,10
525	.91	.7202	22,5,4 20,11,2 20,10.5 19,10,8 16,13,10
526	.93	.7210	21,9,2 21,7,6 18.11.9
528*	.98	.7226	20,8,8 16,16,4
529	23,00	.7235	23,0,0 22,6,3 18,14,3 18,13,6
530	.02	.7243	23,1,0 21,8,5 20,11,3 20,9,7 19,13.0 19,12,5 17,15,4 16,15,7
531	.04	.7251	23,1,1 21,9,3 19,13,1 19,11.7 17.11.11 15,15.9
532	.07	.7259	18,12.8
533	.09	.7267	23,2,0 22,7,0 17,12,10 16,14,9
534	.11	.7275	23,2,1 22,7,1 22,5.5 19,13,2 17,14,7 14,13,13
536*	.15	.7292	22.6.4 20,10.6 18.14.4 14.14,12
537	.17	.7300	23,2,2 22,7,2 20,11,4 16,16,5
538	.19	.7308	23,3,0 21,9,4 15,13,12
539	.22	.7316	23,3,1 21,7,7 19,13,3 17,15.5 17.13.9
541*	.26	.7332	21,10,0 21,8,6 19,12,6
542	.28	.7340	23,3,2 22,7,3 21,10,1 19,10,9 18,13,7 15,14,11
544*	.32	.7356	20,12,0 16,12,12
545	.35	.7364	23,4,0 22,6,5 21,10,2 20,12,1 20.9.8 18,14,5 18,11.10 17,16,0 16,15.8
546	.37	.7372	23,4,1 20,11,5 19,13,4 19,11,8 17.16.1 16,13,11
547	.39	.7380	23,3,3 21,9,5
548	.41	.7388	22,8,0 20,12,2 16,16.6
549	.43	.7396	23,4,2 22,8,1 22,7,4 20,10,7 18,15,0 18,12,9 17,16,2 17,14,8

$h^2+k^2+l^2$	$\sqrt{h^2+k^2+l^2}$	$\lg(h^2+k^2+l^2)$	h, k, l
550	23.45	2.7404	21,10,3 18,15,1 17,15,6 15,15,10
552*	.49	.7419	22,8,2 16,14,10
553	.52	.7427	20,12,3 18,15,2
554	.54	.7435	23,5,0 23.4.3 21,8,7 19,12.7 17,16,3 17,12,11
555	.56	.7443	23.5.1 19.13.5
556	.58	.7451	22,6,6 18.14,6
557	.60	.7459	22,8,3 21,10.4 20,11,6 19,14,0 18,13,8
558	.62	.7466	23,5,2 22.7,5 21,9.6 19,14,1 18,15,3 17,13,10
560*	.66	.7482	20,12,4
561	.69	.7490	23,4,4 19,14,2 19,10,10 17,16,4 16,16,7 14,14,13
562	.71	.7497	21,11,0 20,9,9 16,15,9
563	.73	.7505	23,5,3 21,11,1 19,11,9 17,15,7 15,13.13
564	.75	.7513	22,8,4 20,10.8
565	.77	.7520	23,6,0 22,9,0 18,15,4 15,14,12
566	.79	.7528	23,6,1 22,9,1 21,11,2 21,10,5 19,14.3 19.13,6 18,11,11 17,14,9
568*	.83	.7543	18,12,10
569	.85	.7551	23,6,2 22,9,2 22.7.6 21.8.8 20.13.0 20,12,5 19,12,8 18.14.7 16.13.12
570	.87	.7559	23,5,4 20,13,1 20,11,7 17,16.5
571	.90	.7566	21,11,3 21,9,7 15,15,11
573*	.94	.7582	22,8,5 20,13,2 19,14,4 16,14,11
574	.96	.7589	23,6,3 22,9,3 18,15,5 18.13.9
576*	24.00	.7604	24,0,0 16,16,8
577	.02	.7612	24,1,0 21,10,6 17,12,12
578	.04	.7619	24,1,1 23,7,0 21,11,4 20,13,3 17,17,0 17,15,8
579	.06	.7627	23,7,1 23.5.5 19,13.7 17.17,1 17,13,11
580	.08	.7634	24,2,0 20,12,6 18,16.0
581	.10	.7642	24,2,1 23.6.4 22,9,4 20,10,9 18,16,1 17,16,6 16,15,10
582	.12	.7649	23.7,2 22,7,7 19,14,5 19,11,10 17,17,2
584*	.17	.7664	24,2.2 22,10,0 22,8,6 18,16,2 18,14,8
585	.19	.7672	24,3,0 22,10,1 21,12,0 20,13,4 20,11,8 18,15.6 17,14,10
586	.21	.7679	24,3,1 21.12.1 21,9.8 19,15,0 19,12,9
587	.23	.7686	23,7,3 21,11,5 19,15,1 17,17,3
588	.25	.7694	22,10,2 14,14,14
589	.27	.7701	24.3,2 21,12,2 18,16,3 18,12.11
590	.29	.7709	23,6,5 22,9,5 21,10,7 19,15,2 15,14,13
592*	.33	.7723	24,4,0
593	.35	.7731	24,4.1 23,8,0 22,10,3 20,12,7 19,14,6 18,13,10 16,16,9
594	.37	.7738	24.3.3 23,8,1 23,7,4 21,12,3 20,13,5 19,13,8 17,17,4 17,16,7 16,13,13 15,15,12
595	.39	.7745	19,15,3 17,15,9
596	.41	.7752	24,4,2 20,14,0 18,16,4 16,14,12
597	.43	.7760	23,8,2 22,8,7 20,14,1
598	.45	.7767	21,11,6 18,15,7
600*	.49	.7782	22,10,4 20,14,2 20,10,10
601	.52	.7789	24,5,0 24,4,3 23,6,6 22,9,6 21,12,4 18,14,9
602	.54	.7796	24,5,1 23,8,3 20,11,9 19,15,4 17,13,12 16,15,11
603	.56	.7803	23,7,5 21,9,9 19,11,11 17,17,5
605*	.60	.7818	24,5,2 22,11,0 21,10,8 20,14,3 20,13,6 19,12,10 18,16,5
606	.62	.7825	22,11,1 19,14,7 17,14,11
608*	.66	.7839	24,4,4 20,12,8
609	.68	.7846	23,8,4 22,11,2 22,10,5 17,16,8

$h^2+k^2+l^2$	$\sqrt{h^2+k^2+l^2}$	$\lg(h^2+k^2+l^2)$			$h,\ k,\ l$		
610	24.70	2.7853	24,5,3	23,9,0	21,13,0	21,12,5	
611	.72	.7860	23,9,1	21,13,1	21,11,7	19,15,5	19,13,9
612	.74	.7868	24,6,0	22,8,8	20,14,4	18,12,12	16,16,10
613	.76	.7875	24,6,1	18,17,0	18,15,8		
614	.78	.7882	23,9,2	23,7,6	22,11,3	22,9,7	21,13,2
			18,17,1	18,13,11	17,17,6	17,15,10	
616*	.82	.7896	24,6,2	18,16,6			
617	.84	.7903	24,5,4	19,16,0	18,17,2	15,14,14	
618	.86	.7910	23,8,5	20,13,7	19,16,1		
619	.88	.7917	23,9,3	21,13,3	15,15,13		
620	.90	.7924	22,10,6	18,14,10			
621	.92	.7931	24,6,3	22,11,4	21,12,6	20,14,5	20,11,10
			19,16,2	19,14,8	16,14,13		
622	.94	.7938	21,10,9	19,15,6	18,17,3		
625**	25.00	.7959	25,0,0	24,7,0	20,15,0	20,12,9	16,15,12
626	.02	.7966	25,1,0	24,7,1	24,5,5	23,9,4	21,13,4
			21,11,8	20 15,1	19,16,3	19,12,11	17,16,9
627	.04	.7973	25,1,1	23,7,7	17,17,7	17,13,13	
628	.06	.7980	24,6,4	22,12,0			
629	.08	.7987	25,2,0	24,7,2	23,10,0	23,8,6	22,12,1
			22,9,8	20,15,2	18,17,4	18,16,7	17,14,12
630	.10	.7993	25,2,1	23,10,1	22,11,5	19,13,10	18,15,9
632*	.14	.8007	22,12,2	20,14,6			
633	.16	.8014	25,2,2	23,10,2	22,10,7	20,13,8	19,16,4
			16,16,11				
634	.18	.8021	25,3,0	24,7,3	21,12,7	20,15,3	
635	.20	.8028	25,3,1	23,9,5	21,13,5	19,15,7	17,15,11
637*	.24	.8041	24,6,5	22,12,3	21,14,0	18,13,12	
638	.26	.8048	25,3,2	23,10,3	21,14,1	19,14,9	18,17,5
640*	.30	.8062	24,8,0				
641	.32	.8069	25,4,0	24,8,1	24,7,4	22,11,6	21,14,2
			21,10,10	20,15,4	18,14,11		
642	.34	.8075	25,4,1	23,8,7	20,11,11	19,16,5	17,17,8
643	.36	.8082	25,3,3	21,11,9			
644	.38	.8089	24,8,2	22,12,4	20,12,10	18,16,8	
645	.40	.8096	25,4,2	23,10,4	20,14,7	17,16,10	
646	.42	.8102	23,9,6	22,9,9	21,14,3	21,13,6	15,15,14
648*	.46	.8116	24,6,6	22,10,8	18,18,0	16,14,14	
649	.48	.8122	24,8,3	21,12,8	19,12,12	18,18,1	18,17,6
			18,15,10				
650	.50	.8129	25,5,0	25,4,3	24,7,5	23,11,0	20,15,5
			20,13,9	19,17,0	19,15,8	16,15,13	
651	.51	.8136	25,5,1	23,11,1	19,17,1	19,13,11	
652	.53	.8142	18,18,2				
653	.55	.8149	22,13,0	22,12,5	21,14,4	19,16,6	
654	.57	.8156	25,5,2	23,11,2	23,10,5	22,13,1	22,11,7
			19,17,2	17,14,13			
656*	.61	.8169	24,8,4	20,16,0	16,16,12		
657	.63	.8176	25,4,4	24,9,0	23,8,8	22,13,2	20,16,1
			19,14,10	18,18,3			
658	.65	.8182	24,9,1	17,15,12			
659	.67	.8189	25,5,3	23,11,3	23,9,7	21,13,7	19,17,3 17,17,9
660	.69	.8195	20,16,2	20,14,8			
661	.71	.8202	25,6,0	24,9,2	24,7,6	20,15,6	18,16,9
662	.73	.8209	25,6,1	22,13,3	21,14,5	21,11,10	18,17,7
			18,13,13				
664*	.77	.8222	22,12,6	18,18,4	18,14,12		
665	.79	.8228	25,6,2	24,8,5	23,10,6	22,10,9	20,16,3
			20,12,11				
666	.81	.8235	25,5,4	24,9,3	23,11,4	21,15,0	21,12,9
			19,17,4	19,16,7	17,16,11		
667	.83	.8241	21,15,1	19,15,9			

$h^2+k^2+l^2$	$\sqrt{h^2+k^2+l^2}$	$\lg(h^2+k^2+l^2)$	h, k, l				
669*	25.87	2.8254	22,13,4	22,11,8	20,13,10		
670	.88	.8261	25,6,3	21,15,2	18,15,11		
672*	.92	.8274	20,16,4				
673	.94	.8280	24,9,4	23,12,0	21,14,6	18,18,5	
674	.96	.8287	25,7,0	24,7,7	23,12,1	23,9,8	21,13,8
			20,15,7	19,13,12			
675	.98	.8293	25,7,1	25,5,5	23,11,5 21,15,3 19,17,5 15,15,15		
676	26.00	.8299	26,0,0	24,10,0	24,8,6		
677	.02	.8306	26,1,0	25,6,4	24,10,1	23,12,2	22,12,7
			20,14,9	18,17,8	16,15,14		
678	.04	.8312	26,1,1	25,7,2	23,10,7	22,13,5	19,14,11
			17,17,10				
680*	.08	.8325	26,2,0	24,10,2	22,14,0	18,16,10	
681	.10	.8331	26,2,1	22,14,1	20,16,5	19,16,8	17,14,14
			16,16,13				
682	.12	.8338	24,9,5	23,12,3	21,15,4		
683	.13	.8344	25,7,3	21,11,11	17,15,13		
684	.15	.8351	26,2,2	22,14,2	22,10,10	18,18,6	
685	.17	.8357	26,3,0	24,10,3	21,12,10	19,18,0	
686	.19	.8363	26,3,1	25,6,5	23,11,6	22,11,9	21,14,7
			19,18,1	19,17,6	19,15,10		
688*	.23	.8376	20,12,12				
689	.25	.8382	26,3,2	25,8,0	24,8,7	23,12,4	22,14,3
			22,13,6	20,17,0	20,15,8	19,18,2	18,14,13
			17,16,12				
690	.27	.8388	25,8,1	25,7,4	20,17,1	20,13,11	
691	.29	.8395	23,9,9	21,15,5	21,13,9		
692	.31	.8401	26,4,0	24,10,4	22,12,8	20,16,6	
693	.32	.8407	26,4,1	25,8,2	24,9,6	23,10,8	20,17,2
			18,15,12				
694	.34	.8414	26,3,3	19,18,3	18,17,9		
696*	.38	.8426	26,4,2	22,14,4	20,14,10		
697	.40	.8432	25,6,6	24,11,0	21,16,0	18,18,7	
698	.42	.8439	25,8,3	24,11,1	23,13,0	23,12,5	21,16,1
			20,17,3	19,16,9			
699	.44	.8445	25,7,5	23,13,1	23,11,7	19,17,7	19,13,13
			17,17,11				
701*	.48	.8457	26,5,0	26,4,3	24,11,2	24,10,5	21,16,2
			21,14,8	19,18,4	19,14,12 18,16,11		
702	.50	.8463	26,5,1	23,13,2	22,13,7	21,15,6	
704*	.53	.8476	24,8,8				
705	.55	.8482	26,5,2	25,8,4	22,14,5	22,11,10 20,17,4	
			20,16,7				
706	.57	.8488	25,9,0	24,11,3	24,9,7	21,16,3	21,12,11
			20,15,9	16,15,15			
707	.59	.8494	25,9,1	23,13,3	19,15,11		
708	.61	.8500	26,4,4	16,16,14			
709	.63	.8506	23,12,6	22,15,0	22,12,9		
710	.65	.8513	26,5,3	25,9,2	25,7,6	23,10,9	22,15,1
			21,13,10	19,18,5	17,15,14		
712*	.68	.8525	26,6,0	24,10,6	18,18,8		
713	.70	.8531	26,6,1	24,11,4	22,15,2	21,16,4	20,13,12
			18,17,10				
714	.72	.8537	25,8,5	23,13,4	23,11,8	20,17,5	19,17,8
			17,16,13				
715	.74	.8543	25,9,3	21,15,7			
716	.76	.8549	26,6,2	22,14,6	18,14,14		
717	.78	.8555	26,5,4	22,13,8	20,14,11	19,16,10	
718	.80	.8561	22,15,3	21,14,9	18,15,13		
720*	.83	.8573	24,12,0	20,16,8			
721	.85	.8579	26,6,3	24,12,1	24,9,8	19,18,6	
722	.87	.8585	25,9,4	24,11,5	23,12,7	21,16,5	19,19,0
			17,17,12				

$h^2+k^2+l^2$	$\sqrt{h^2+k^2+l^2}$	$\lg(h^2+k^2+l^2)$	h, k, l
723	26 89	2.8591	25,7,7 23,13,5 19,19,1
724	.91	.8597	24,12,2 20,18,0 18,16,12
725	.93	.8603	26,7,0 25,10,0 25,8,6 24,10,7 23,14,0 22,15,4 20,18,1 20,17,6 20,15,10
726	.94	.8609	26,7,1 26,5,5 25,10,1 23,14,1 22,11,11 19,19,2 19,14,13
728*	.98	.8621	26,6,4 22,12,10 20,18,2
729	27.00	.8627	27,0,0 26,7,2 25,10,2 24,12,3 23,14,2 23,10,10 22,14,7 21,12,12 18,18,9
730	.02	.8633	27,1,0 21,17,0 21,15,8 19,15,12
731	.04	.8639	27,1,1 25,9,5 23,11,9 21,17,1 21,13,11 19,19,3 19,17,9
733*	.07	.8651	27,2,0 24,11,6 21,16,6 20,18,3
734	.09	.8657	27,2,1 26,7,3 25,10,3 23,14,3 23,13,6 22,15,5 22,13,9 21,17,2 19,18,7 18,17,11
736*	.13	.8669	24,12,4
737	.15	.8675	27,2,2 26,6,5 23,12,8 21,14,10 20,16,9 16,16,15
738	.17	.8681	27,3,0 25,8,7 24,9,9 20,17,7 20,13,13 19,19,4 19,16,11
739	.18	.8686	27,3,1 21,17,3 17,15,15
740	.20	.8692	26,8,0 24,10,8 22,16,0 20,18,4 20,14,12
741	.22	.8698	26,8,1 26,7,4 25,10,4 23,14,4 22,16,1 17,16,14
742	.24	.8704	27,3,2 25,9,6
744*	.28	.8716	26,8,2 22,16,2 22,14,8
745	.29	.8722	27,4,0 24,13,0 24,12,5 22,15,6 18,15,14
746	.31	.8727	27,4,1 25,11,0 24,13,1 24,11,7 21,17,4 21,16,7 20,15,11
747	.33	.8733	27,3,3 25,11,1 23,13,7 21,15,9 19,19,5 17,17,13
748	.35	.8739	26,6,6 18,18,10
749	.37	.8745	27,4,2 26,8,3 24,13,2 22,16,3 22,12,11 20,18,5 19,18,8 18,16,13
750	.39	.8751	26,7,5 25,11,2 25,10,5 23,14,5 23,11,10 19,17,10
753**	.44	.8768	25,8,8 22,13,10 20,17,8 19,14,14
754	.46	.8774	27,5,0 27,4,3 24,13,3 23,15,0 23,12,9 21,13,12
755	.48	.8779	27,5,1 25,11,3 25,9,7 23,15,1 21,17,5 19,15,13
756	.50	.8785	26,8,4 24,12,6 22,16,4 20,16,10
757	.51	.8791	26,9,0 24,10,9 18,17,12
758	.53	.8797	27,5,2 26,9,1 23,15,2 22,15,7 21,14,11 19,19,6
760*	.57	.8808	20,18,6
761	.59	.8814	27,4,4 26,9,2 26,7,6 25,10,6 24,13,4 24,11,8 23,14,6 22,14,9 21,16,8 20,19,0 19,16,12
762	.60	.8820	25,11,4 23,13,8 20,19,1
763	.62	.8825	27,5,3 23,15,3
765*	.66	.8837	27,6,0 26,8,5 22,16,5 21,18,0 20,19,2 20,14,13
766	.68	.8842	27,6,1 26,9,3 21,18,1 21,17,6 21,15,10 19,18,9
768*	.71	.8854	16,16,16
769	.73	.8859	27,6,2 25,12,0 24,12,7 21,18,2 20,15,12, 18,18,11
770	.75	.8865	27,5,4 25,12,1 25,9,8 24,13,5 23,15,4 20,19,3 20,17,9 17,16,15
771	.77	.8871	25,11,5 23,11,11 19,19,7 19,17,11
772	.78	.8876	24,14,0 22,12,12
773	.80	.8882	26,9,4 25,12,2 24,14,1 23,12,10 22,17,0 22,15,8 20,18,7

$h^2+k^2+l^2$	$\sqrt{h^2+k^2+l^2}$	$\lg(h^2+k^2+l^2)$	h, k, l
774	27.82	2.8887	27,6,3 26,7,7 25,10,7 23,14,7 22,17,1 22,13,11 21,18,3 18,15,15 17,17,14
776*	.86	.8899	26,10,0 26,8,6 24,14,2 24,10,10 22,16,6 18,16,14
777	.87	.8904	26,10,1 22,17,2 20,19,4 20,16,11
778	.89	.8910	27,7,0 25,12,3 24,11,9 21,16,9
779	.91	.8915	27,7,1 27,5,5 23,15,5 23,13,9 21,17,7 21,13,13
780	.93	.8921	26,10,2 22,14,10
781	.95	.8927	27,6,4 24,14,3 24,13,6 21,18,4 21,14,12
782	.96	.8932	27,7,2 26,9,5 25,11,6 22,17,3 19,15,14 18,17,13
784*	28.00	.8943	28,0,0 24,12,8
785	.02	.8949	28,1,0 26,10,3 25,12,4 23,16,0 19,18,10
786	.04	.8954	28,1,1 23,16,1 20,19,5 19,19,8 19,16,13
787	.05	.8960	27,7,3 25,9,9 21,15,11
788	.07	.8965	28,2,0 24,14,4 20,18,8
789	.09	.8971	28,2,1 26,8,7 25,10,8 23,16,2 23,14,8 22,17,4 22,16,7 20,17,10
790	.11	.8976	27,6,5 23,15,6 22,15,9 21,18,5
792*	.14	.8987	28,2,2 26,10,4 20,14,14 18,18,12
793	.16	.8993	28,3,0 27,8,0 26,9,6
794	.18	.8998	28,3,1 27,8,1 27,7,4 25,13,0 25,12,5 24,13,7 23,16,3 23,12,11 21,17,8 20,15,13 19,17,12
795	.20	.9004	25,13,1 25,11,7
797*	.23	.9015	28,3,2 27,8,2 26,11,0 24,14,5 24,11,10 22,13,12 21,16,10 20,19,6
798	.25	.9020	26,11,1 25,13,2 23,13,10 22,17,5
800*	.28	.9031	28,4,0 20,20,0 20,16,12
801	.30	.9036	28,4,1 27,6,6 26,11,2 26,10,5 24,15,0 24,12,9 23,16,4 22,14,11 21,18,6 20,20,1 17,16,16
802	.32	.9042	28,3,3 27,8,3 24,15,1 21,19,0
803	.34	.9047	27,7,5 25,13,3 23,15,7 21,19,1 19,19,9 17,17,15
804	.35	.9053	28,4,2 26,8,8 22,16,8 20,20,2
805	.37	.9058	25,12,6 24,15,2 20,18,9 18,16,15
806	.39	.9063	26,11,3 26,9,7 25,10,9 23,14,9 21,19,2 21,14,13 19,18,11
808*	.43	.9074	24,14,6 22,18,0
809	.44	.9079	28,5,0 28,4,3 27,8,4 24,13,8 22,18,1 22,17,6 22,15,10 20,20,3 18,17,14
810	.46	.9085	28,5,1 27,9,0 25,13,4 25,11,8 24,15,3 23,16,5 21,15,12 20,19,7 20,17,11
811	.48	.9090	27,9,1 21,19,3 21,17,9 19,15,15
812	.50	.9096	26,10,6 22,18,2
813	.51	.9101	28,5,2 26,11,4 19,16,14
814	.53	.9106	27,9,2 27,7,6 21,18,7
816*	.57	.9117	28,4,4 20,20,4
817	.58	.9122	24,15,4 23,12,12 22,18,3 18,18,13
818	.60	.9128	28,5,3 27,8,5 25,12,7 24,11,11 23,17,0 23,15,8 21,19,4 21,16,11
819	.62	.9133	27,9,3 25,13,5 23,17,1 23,13,11 19,17,13
820	.64	.9138	28,6,0 26,12,0 24,12,10
821	.65	.9143	28,6,1 26,12,1 26,9,8 25,14,0 24,14,7 23,16,6 22,16,9 20,15,14
822	.67	.9149	26,11,5 25,14,1 23,17,2 22,17,7 22,13,13 19,19,10
824*	.71	.9159	28,6,2 26,12,2 22,18,4 22,14,12 20,18,10
825	.72	.9165	28,5,4 26,10,7 25,14,2 25,10,10 23,14,10 20,20,5 20,19,8 20,16,13
826	.74	.9170	27,9,4 24,15,5 24,13,9

$h^2+k^2+l^2$	$\sqrt{h^2+k^2+l^2}$	$\lg(h^2+k^2+l^2)$	h, k, l
827	28.76	2.9175	27,7,7 25,11,9 23,17,3 21,19,5
829*	.79	.9186	28,6,3 27,10,0 27,8,6 26,12,3 21,18,8 19,18,12
830	.81	.9191	27,10,1 25,14,3 25,13,6 22,15,11 21,17,10
832*	.84	.9201	24,16,0
833	.86	.9206	28,7,0 27,10,2 26,11,6 25,12,8 24,16,1 22,18,5 21,14,14 20,17,12
834	.88	.9212	28,7,1 28,5,5 23,17,4 23,16,7 17,17,16
835	.90	.9217	27,9,5 23,15,9 21,15,13
836	.91	.9222	28,6,4 26,12,4 24,16,2 24,14,8 20,20,6 18,16,16
837	.93	.9227	28,7,2 25,14,4 24,15,6 22,17,8
838	.95	.9232	27,10,3 26,9,9 21,19,6 18,17,15
840*	.98	.9243	26,10,8 22,16,10
841	29.00	.9248	29,0,0 24,16,3 24,12,11 21,20,0 21,16,12
842	.02	.9253	29,1,0 28,7,3 27,8,7 23,13,12 21,20,1 20,19,9 19,16,15
843	.03	.9258	29,1,1 25,13,7 23,17,5 19,19,11
844	.05	.9263	22,18,6 18,18,14
845	.07	.9269	29,2,0 28,6,5 27,10,4 26,13,0 26,12,5 24,13,10 22,19,0 21,20,2 20,18,11
846	.09	.9274	29,2,1 27,9,6 26,13,1 26,11,7 25,14,5 25,11,10 23,14,11 22,19,1 21,18,9 19,17,14
848*	.12	.9284	28,8,0 24,16,4
849	.14	.9289	29,2,2 28,8,1 28,7,4 26,13,2 23,16,8 22,19,2 22,14,13 20,20,7
850	.15	.9294	29,3,0 27,11,0 25,15,0 25,12,9 24,15,7 21,20,3 20,15,15
851	.17	.9299	29,3,1 27,11,1 25,15,1 21,19,7 21,17,11
852	.19	.9304	28,8,2 20,16,14
853	.21	.9309	24,14,9 23,18,0 22,15,12
854	.22	.9315	29,3,2 27,11,2 27,10,5 26,13,3 25,15,2 23,18,1 23,17,6 23,15,10 22,19,3 22,17,9 19,18,13
856*	.26	.9325	28,6,6 26,12,6
857	.27	.9330	29,4,0 28,8,3 27,8,8 26,10,9 25,14,6 24,16,5 23,18,2 22,18,7 21,20,4
858	.29	.9335	29,4,1 28,7,5 25,13,8 20,17,13
859	.31	.9340	29,3,3 27,11,3 27,9,7 25,15,3
861*	.34	.9350	29,4,2 26,13,4 26,11,8 22,19,4 22,16,11 20,19,10
862	.36	.9355	23,18,3 21,15,14
864*	.39	.9365	28,8,4 24,12,12 20,20,8
865	.41	.9370	28,9,0 27,10,6 24,17,0 24,15,8 21,18,10
866	.43	.9375	29,5,0 29,4,3 28,9,1 27,11,4 25,15,4 24,17,1 24,13,11 23,16,9 21,20,5 21,19,8 21,16,13 19,19,12
867	.44	.9380	29,5,1 25,11,11 23,17,7 23,13,13 17,17,17
868	.46	.9385	24,16,6 20,18,12
869	.48	.9390	28,9,2 28,7,6 26,12,7 25,12,10 24,17,2 23,18,4 23,14,12 18,17,16
870	.50	.9395	29,5,2 26,13,5 25,14,7 22,19,5
872*	.53	.9405	26,14,0 24,14,10 22,18,8
873	.55	.9410	29,4,4 28,8,5 27,12,0 26,14,1 22,17,10 19,16,16 18,18,15
874	.56	.9415	28,9,3 27,12,1 27,9,8 24,17,3 21,17,12
875	.58	.9420	29,5,3 27,11,5 25,15,5 25,13,9 23,15,11 19,17,15
876	.60	.9425	26,14,2 26,10,10 22,14,14
877	.61	.9430	29,6,0 27,12,2 21,20,6
878	.63	.9435	29,6,1 27,10,7 26,11,9 23,18,5 22,15,13

$h^2+k^2+l^2$	$\sqrt{h^2+k^2+l^2}$	$\lg(h^2+k^2+l^2)$	h, k, l				
881**	29.68	2.9450	29,6,2	28,9,4	26,14,3	26,13,6	25,16,0
			24,17,4	24,16,7	22,19,6	20,20,9	20,16,15
			19,18,14				
882	.70	.9455	29,5,4	28,7,7	27,12,3	25,16,1	24,15,9
			23,17,8	21,21,0	20,19,11		
883	.72	.9460	21,21,1	21,19,9			
884	.73	.9465	28,10,0	28,8,6	26,12,8	22,20,0	22,16,12
885	.75	.9469	28,10,1	25,16,2	25,14,8	23,16,10	22,20,1
			20,17,14				
886	.77	.9474	29,6,3	27,11,6	25,15,6	21,21,2	21,18,11
888*	.80	.9484	28,10,2	26,14,4	22,20,2		
889	.82	.9489	27,12,4	24,13,12	23,18,6	22,18,9	
890	.83	.9494	29,7,0	28,9,5	25,16,3	25,12,11	24,17,5
			23,19,0	21,20,7			
891	.85	.9499	29,7,1	29,5,5	27,9,9	23,19,1	21,21,3
			21,15,15	19,19,13			
893*	.88	.9509	29,6,4	28,10,3	27,10,8	24,14,11	22,20,3
			21,16,14	20,18,13			
894	.90	.9513	29,7,2	26,13,7	25,13,10	23,19,2	23,14,13
			22,19,7	22,17,11			
896*	.93	.9523	24,16,8				
897	.95	.9528	28,8,7	26,14,5	26,11,10	25,16,4	
898	.97	.9533	27,13,0	27,12,5	23,15,12	21,21,4	
899	.98	.9538	29,7,3	27,13,1	27,11,7	25,15,7	23,19,3
			23,17,9	21,17,13			
900	30.00	.9542	30,0,0	28,10,4	24,18,0	22,20,4	20,20,10
901	.02	.9547	30,1,0	28,9,6	26,15,0	26,12,9	24,18,1
			24,17,6	24,15,10			
902	.03	.9552	30,1,1	29,6,5	27,13,2	26,15,1	25,14,9
			23,18,7	21,19,10	18,17,17		
904*	.07	.9562	30,2,0	24,18,2	18,18,16		
905	.08	.9566	30,2,1	29,8,0	28,11,0	26,15,2	22,15,14
			21,20,8	20,19,12			
906	.10	.9571	29,8,1	29,7,4	28,11,1	25,16,5	23,19,4
			23,16,11	19,17,16			
907	.12	.9576	27,13,3	21,21,5			
908	.13	.9581	30,2,2	26,14,6	22,18,10		
909	.15	.9586	30,3,0	29,8,2	28,11,2	28,10,5	27,12,6
			26,13,8	24,18,3	22,20,5	22,19,8	22,16,13
			21,18,12				
910	.17	.9590	30,3,1	27,10,9	26,15,3	19,18,15	
912*	.20	.9600	28,8,8	20,16,16			
913	.22	.9605	30,3,2	29,6,6	25,12,12	24,16,9	
914	.23	.9609	29,8,3	28,11,3	28,9,7	27,13,4	27,11,8
			25,17,0	25,15,8	24,17,7	24,13,13	20,17,15
915	.25	.9614	29,7,5	25,17,1	25,13,11	23,19,5	
916	.27	.9619	30,4,0	24,18,4	24,14,12		
917	.28	.9624	30,4,1	26,15,4	25,16,6	23,18,8	22,17,12
918	.30	.9628	30,3,3	26,11,11	25,17,2	23,17,10	21,21,6
			19,19,14				
920*	.33	.9638	30,4,2	28,10,6	26,12,10	22,20,6	20,18,14
921	.35	.9643	29,8,4	28,11,4	26,14,7	25,14,10	23,14,14
			20,20,11				
922	.36	.9647	29,9,0	27,12,7	24,15,11	21,20,9	21,16,15
923	.38	.9652	29,9,1	27,13,5	25,17,3	23,15,13	21,19,11
925*	.41	.9661	30,5,0	30,4,3	27,14,0	24,18,5	22,21,0
926	.43	.9666	30,5,1	29,9,2	29,7,6	27,14,1	26,15,5
			26,13,9	23,19,6	22,21,1	22,19,9	21,17,14
928*	.46	.9675	28,12,0				
929	.48	.9680	30,5,2	28,12,1	28,9,8	27,14,2	27,10,10
			24,17,8	23,20,0	23,16,12	22,21,2	22,18,11
930	.50	.9685	29,8,5	28,11,5	25,17,4	25,16,7	23,20,1
			20,19,13				

$h^2+k^2+l^2$	$\sqrt{h^2+k^2+l^2}$	$\lg(h^2+k^2+l^2)$	h, k, l
931	30.51	2.9689	29,9,3 27,11,9 25,15,9 21,21,7
932	.53	.9694	30,4,4 28,12,2 26,16,0 24,16,10
933	.55	.9699	28,10,7 26,16,1 23,20,2 22,20,7
934	.56	.9703	30,5,3 27,14,3 27,13,6 23,18,9 22,21,3 22,15,15 21,18,13
936*	.59	.9713	30,6,0 26,16,2 26,14,8 24,18,6 22,16,14
937	.61	.9717	30,6,1 28,12,3 27,12,8 26,15,6 24,19,0 18,18,17
938	.63	.9722	29,9,4 25,13,12 24,19,1 23,20,3
939	.64	.9727	29,7,7 25,17,5 23,19,7 23,17,11 19,17,17
940	.66	.9731	30,6,2
941	.68	.9736	30,5,4 29,10,0 29,8,6 28,11,6 27,14,4 26,16,3 26,12,11 24,19,2 24,14,13 22,21,4 21,20,10 19,18,16
942	.69	.9741	29,10,1 25,14,11 22,17,13
944*	.72	.9750	28,12,4 20,20,12
945	.74	.9754	30,6,3 29,10,2 26,13,10 25,16,8 24,15,12 23,20,4 22,19,10 20,17,16
946	.76	.9759	28,9,9 24,19,3 24,17,9 21,21,8 21,19,12
947	.77	.9763	29,9,5 27,13,7 19,19,15
948	.79	.9768	28,10,8 26,16,4 22,20,8
949	.81	.9773	30,7,0 25,18,0 24,18,7 20,18,15
950	.82	.9777	30,7,1 30,5,5 29,10,3 27,14,5 27,11,10 26,15,7 25,18,1 25,17,6 25,15,10 23,15,14 22,21,5
952*	.85	.9786	30,6,4 22,18,12
953	.87	.9791	30,7,2 28,13,0 28,12,5 26,14,9 25,18,2 24,19,4 24,16,11 23,18,10 21,16,16
954	.89	.9795	29,8,7 28,13,1 28,11,7 27,15,0 27,12,9 23,20,5 23,19,8 23,16,13
955	.90	.9800	27,15,1 21,17,15
957*	.94	.9809	29,10,4 28,13,2 26,16,5 20,19,14
958	.95	.9814	30,7,3 29,9,6 27,15,2 25,18,3
961**	31.00	.9827	31,0,0 30,6,5 27,14,6 22,21,6 21,18,14
962	.02	.9832	31,1,0 29,11,0 28,13,3 27,13,8 25,16,9 24,19,5 23,17,12 21,20,11
963	.03	.9836	31,1,1 29,11,1 27,15,3 25,17,7 25,13,13 21,21,9
964	.05	.9841	30,8,0 28,12,6 26,12,12 24,18,8
965	.06	.9845	31,2,0 30,8,1 30,7,4 28,10,9 26,17,0 26,15,8 25,18,4 25,14,12 24,17,10 23,20,6 22,20,9 22,16,15
966	.08	.9850	31,2,1 29,11,2 29,10,5 26,17,1 26,13,11 22,19,11
968*	.11	.9859	30,8,2 26,16,6 24,14,14 22,22,0
969	.13	.9863	31,2,2 29,8,8 28,13,4 28,11,8 26,17,2 22,22,1 22,17,14 20,20,13
970	.14	.9868	31,3,0 27,15,4 24,15,13 23,21,0
971	.16	.9872	31,3,1 29,11,3 29,9,7 27,11,11 25,15,11 23,21,1 23,19,9 21,19,13
972	.18	.9877	30,6,6 26,14,10 22,22,2 18,18,18
973	.19	.9881	30,8,3 27,12,10 24,19,6
974	.21	.9886	31,3,2 30,7,5 27,14,7 26,17,3 25,18,5 23,21,2 23,18,11 22,21,7 19,18,17
976*	.24	.9894	24,20,0 24,16,12
977	.26	.9899	31,4,0 29,10,6 28,12,7 24,20,1 22,22,3 22,18,13
978	.27	.9903	31,4,1 29,11,4 28,13,5 25,17,8 23,20,7 20,17,17 19,19,16
979	.29	.9908	31,3,3 27,15,5 27,13,9 23,21,3 23,15,15

$h^2+k^2+l^2$	$\sqrt{h^2+k^2+l^2}$	$\lg(h^2+k^2+l^2)$	h, k, l				
980	31.30	2.9912	30,8,4	28,14,0	24,20,2	20,18,16	
981	.32	.9917	31,4,2	30,9,0	28,14,1	26,17,4	26,16,7
			25,16,10	24,18,9	23,16,14		
982	.34	.9921	30,9,1	26,15,9	21,21,10		
984*	.37	.9930	28,14,2	28,10,10	22,22,4	22,20,10	
985	.38	.9934	30,9,2	30,7,6	29,12,0	27,16,0	25,18,6
			24,20,3	21,20,12			
986	.40	.9939	31,5,0	31,4,3	29,12,8	28,11,9	27,16,1
			25,19,0	24,19,7	24,17,11	23,21,4	21,17,16
			20,19,15				
987	.42	.9943	31,5,1	29,11,5	25,19,1	23,17,13	
989*	.45	.9952	30,8,5	29,12,2	28,14,3	28,13,6	27,16,2
			27,14,8	26,13,12	22,21,8	22,19,12	
990	.46	.9956	31,5,2	30,9,3	29,10,7	27,15,6	26,17,5
			25,19,2	25,14,13	23,19,10	21,18,15	
992*	.50	.9965	28,12,8	24,20,4			
993	.51	.9969	31,4,4	26,14,11	23,20,8	22,22,5	
994	.53	.9974	29,12,3	27,16,3	27,12,11	25,15,12	
995	.54	.9978	31,5,3	25,19,3	25,17,9	23,21,5	
996	.56	.9983	28,14,4	26,16,8	22,16,16	20,20,14	
997	.58	.9987	31,6,0	30,9,4	24,15,14	23,18,12	
998	.59	.9991	31,6,1	30,7,7	29,11,6	27,13,10	25,18,7
			22,17,15	21,19,14			

3-6. Graphs for Indexing X-Ray Patterns

The relation for line indices for cubic materials is

$$\frac{\sin^2\vartheta_i}{\sin^2\vartheta_1} = \frac{h_i^2+k_i^2+l_i^2}{h_1^2+k_1^2+l_1^2}.$$

The $\sin^2\vartheta_i$ are deduced, and lines associated with the Kβ radiation are distinguished; the $\sin^2\vartheta_i$ are divided by $\sin^2\vartheta_1$, and then the schemes (Figs. 39 and 40) are used to choose the indices for the first line. The indices for the other lines are then deduced from the above relation.

The graphs of Fig. 41 [8] are used to index with respect to $\sin\vartheta$; the following stages are involved:

1. The $\sin\vartheta$ for the observed lines are entered on a strip of paper on a scale corresponding to that of the ordinates in Fig. 41.

2. The strip is moved along while parallel to the ordinate axis to bring all $\sin\vartheta$ into coincidence with the lines in Fig. 41.

3. The indices are determined for each line, λ/a being read from the position of the strip; then a (the lattice constant) is deduced.

Figure 42 is used if the interplanar spacings are known; this applies to d between 0 and 20.00 Å [270]. The d are entered in the scale used there on a strip of paper, which is moved while remaining parallel to the abscissa to bring all marks into coincidence with lines. The numbers given at the top define the indices, while the position of the strip defines a.

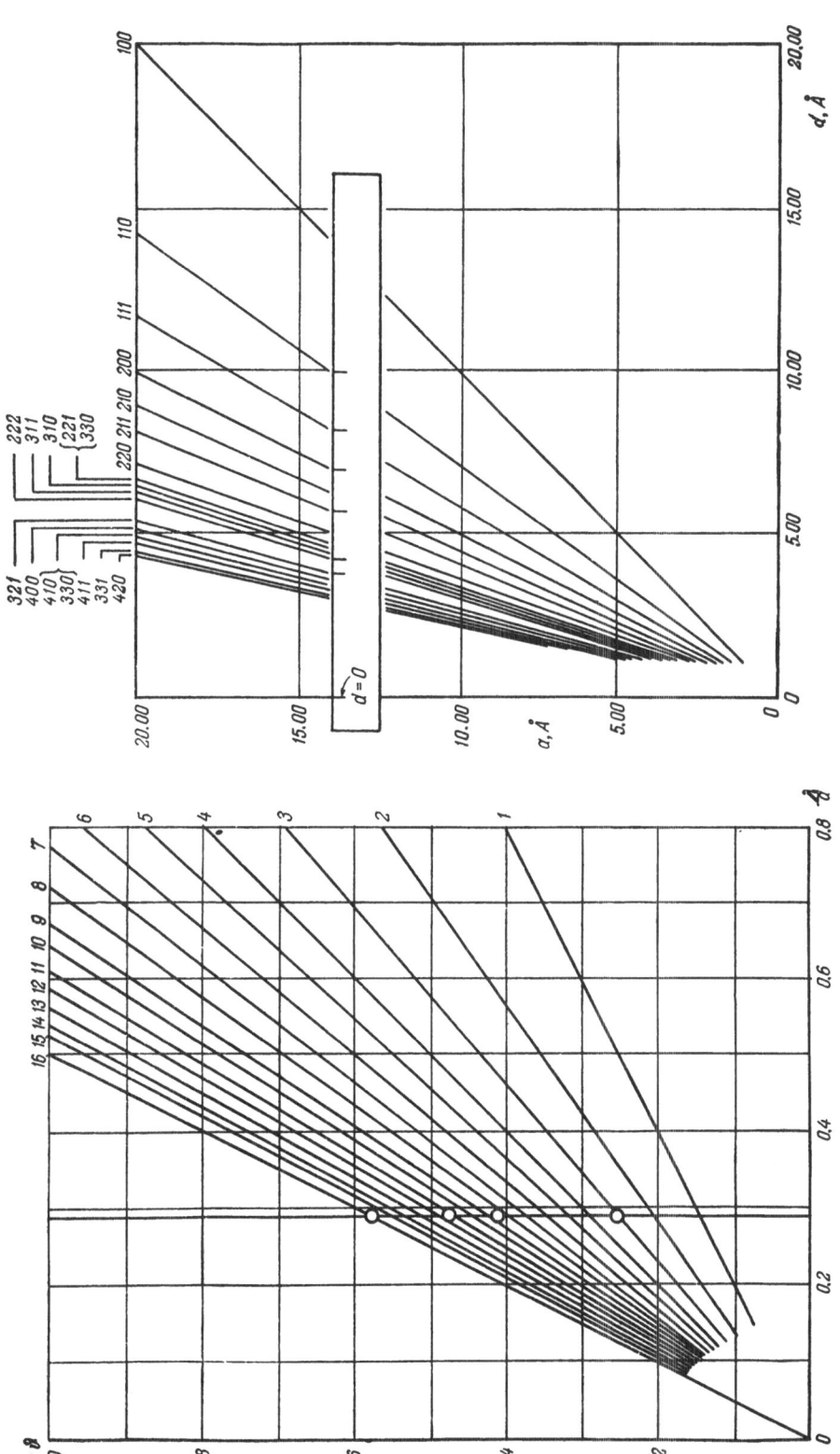

Fig. 42. Graphs for indexing patterns for cubic crystals from known values of d.

Fig. 41. Graphs for indexing patterns for cubic crystals from known values of sin θ.

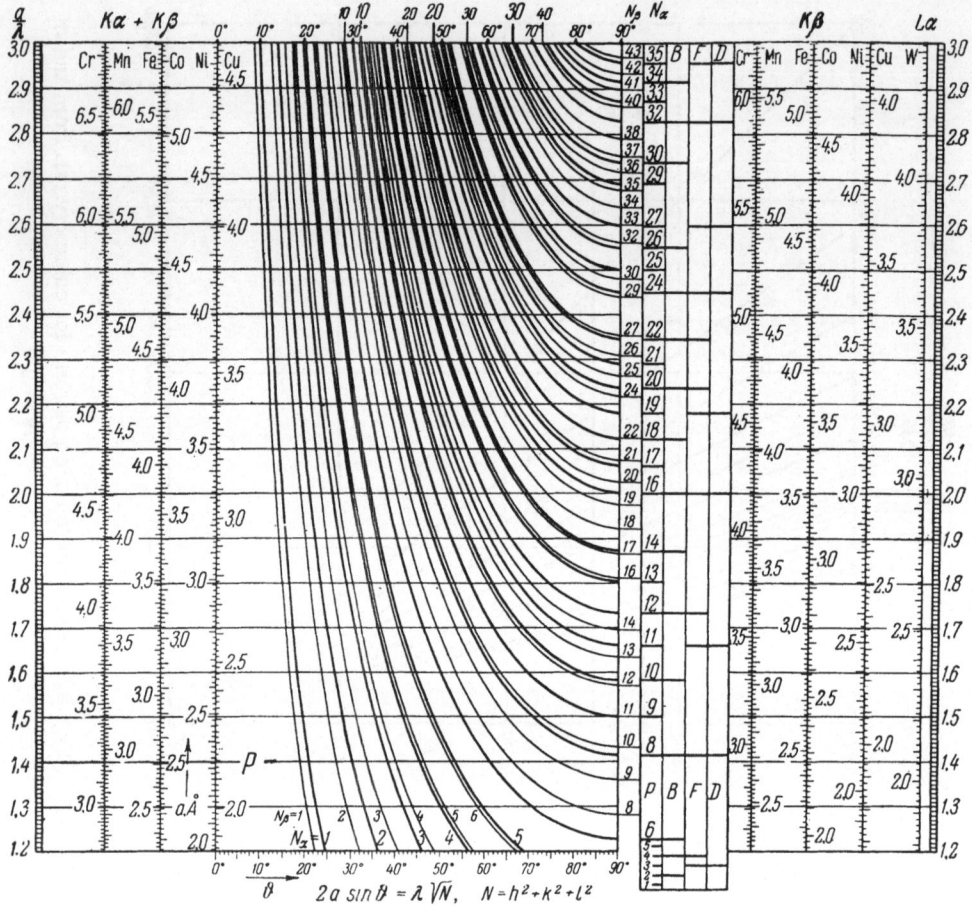

Fig. 43. Nomogram (I-P) for indexing patterns for crystals with simple cubic lattices.

Figure 41 gives values for $h^2 + k^2 + l^2 \leq 16$. Figures 43-50, or analytic methods, are used if the unit cell is very large.

Figures 43-50 [90] are graphical representations of the Wulff-Bragg relations for the cubic system in the form

$$\frac{a}{\lambda} = \frac{1}{2}\sqrt{N} \operatorname{cosec} \vartheta,$$

where $N = h^2 + k^2 + l^2$; the nomograms are applicable to a large range of lattice constants.

The nomograms are all of the same type and are adapted to the various divisions of the cubic system; nomogram I-P (Fig. 43) applies to simple cubic lattices, the field of the nomogram being bounded on both sides by scales of a/λ, which are followed by scales of lattice constants. Values for $K\alpha$ are given on the left, and approximate values for the $K\beta$ and $L\alpha$ components of W radiation to the right. The scales for $K\beta$ are next to schemes for the patterns from primitive (P), body-centered (B), face-centered (F), and diamond (D) lattices. The thick lines are for $K\alpha$; the thin ones, for $K\beta$. The numbers on the curves are the sums of the squares of the indices. A horizontal line drawn through any point on the

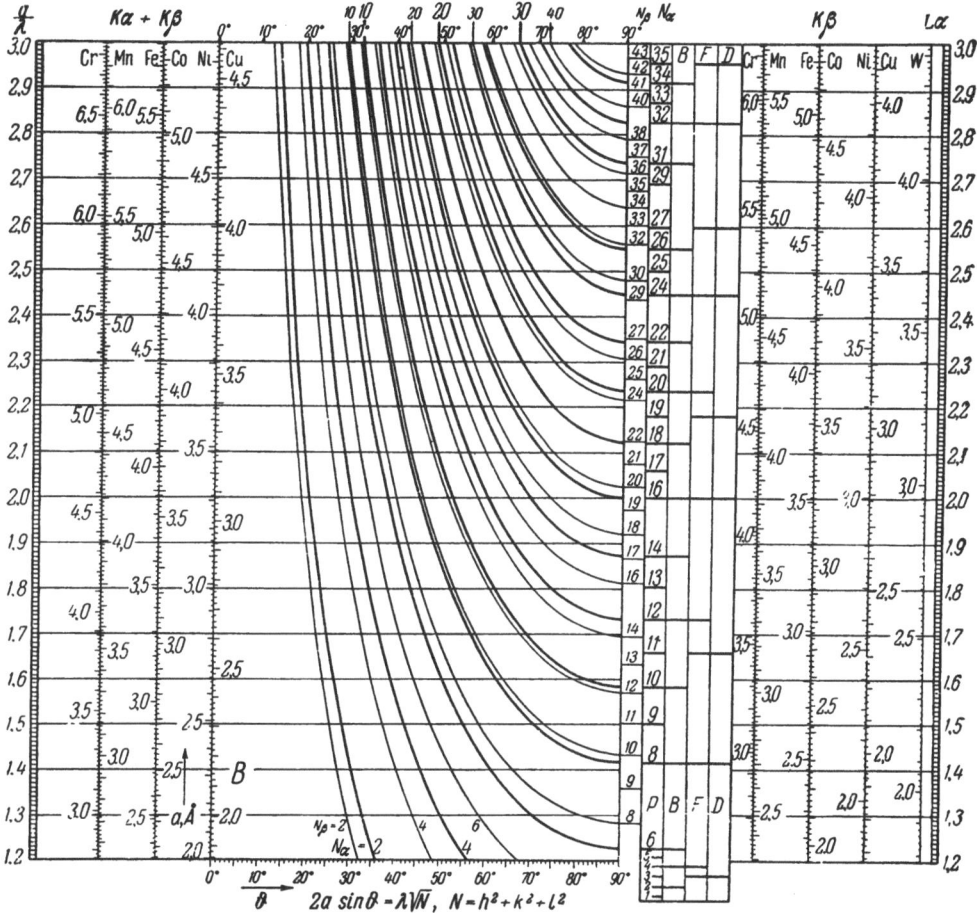

Fig. 44. Nomogram (I-B) for indexing patterns for crystals with body-centered cubic lattices.

lattice-constant scale meets the nomogram lines at points corresponding to the Bragg angles for the pattern lines.

Nomogram I-B (Fig. 44) applies to body-centered cubic lattices; I-F (Fig. 45), to face-centered ones; and I-D (Fig. 46), to diamond ones.

Nomogram II-PB (Fig. 47) is for indexing patterns from mixtures of crystals with simple cubic and body-centered cubic lattices for reflections whose sums of squares of indices range up to 80; the thick lines represent reflections given only by substances with body-centered structures.

Nomogram II-FD (Fig. 48) is for indexing patterns from mixtures of crystals with face-centered cubic and diamond lattices for reflections whose sums of squares of indices range up to 80; here the thin lines represent reflections given only by substances with the diamond structure.

Nomogram III-PB (Fig. 49) is for indexing patterns from mixtures of crystals with simple cubic and body-centered cubic lattices for reflections whose sums of squares of indices range up to 142; the thick lines represent reflections given only by substances with body-centered structures.

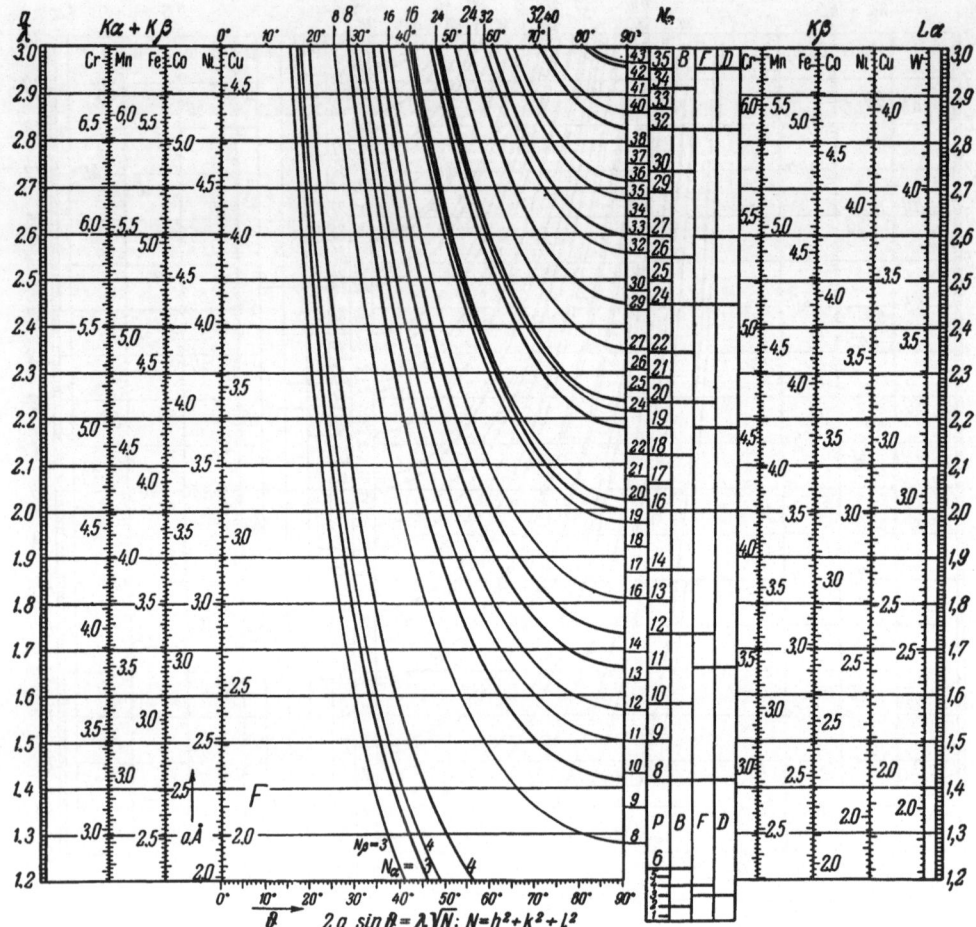

Fig. 45. Nomogram (I-F) for indexing patterns for crystals with face-centered cubic lattices.

Nomogram III-FD (Fig. 50) is for indexing patterns from mixtures of crystals with face-centered cubic and diamond lattices for reflections whose sums of squares of indices range up to 142.

These nomograms can be used to index patterns, to determine lattice constants, to distinguish lines caused by $K\beta$, to select suitable combinations of radiations for determining block sizes, and so on.

The error in the deduction of angles from the nomograms will be clear from the following example for the determination of line positions for diamond (a = 3.56 Å) examined with unfiltered copper radiation:

A strip of paper is placed parallel to the abscissa on nomogram I-D at the proper level (3.56 Å) on the scale for Cu radiation in the left part of the nomogram.

The points where the strip meets the graphs correspond to the positions of the lines on the pattern.

The Bragg angles are read from the top or bottom scale of angles; calculation and measurement give the following:

$N=h^2+k^2+l^2$	Line	Bragg angles	
		from nomogram	calculated
3	111 (β)	19°54′	19°48′
3	111 (α)	22°06′	22°02′
8	220 (β)	33°48′	33°35′
8	220 (α)	38°00′	37°46′
11	311 (β)	40°36′	40°26′
11	311 (α)	46°18′	45°54′
16	400 (β)	51°36′	51°27′
19	331 (β)	58°24′	58°28′
16	400 (α)	60°00′	60°01′
19	331 (α)	70°54′	70°43′
24	422 (β)	73°36′	73°19′

The accuracy is clearly sufficient for most practical purposes.

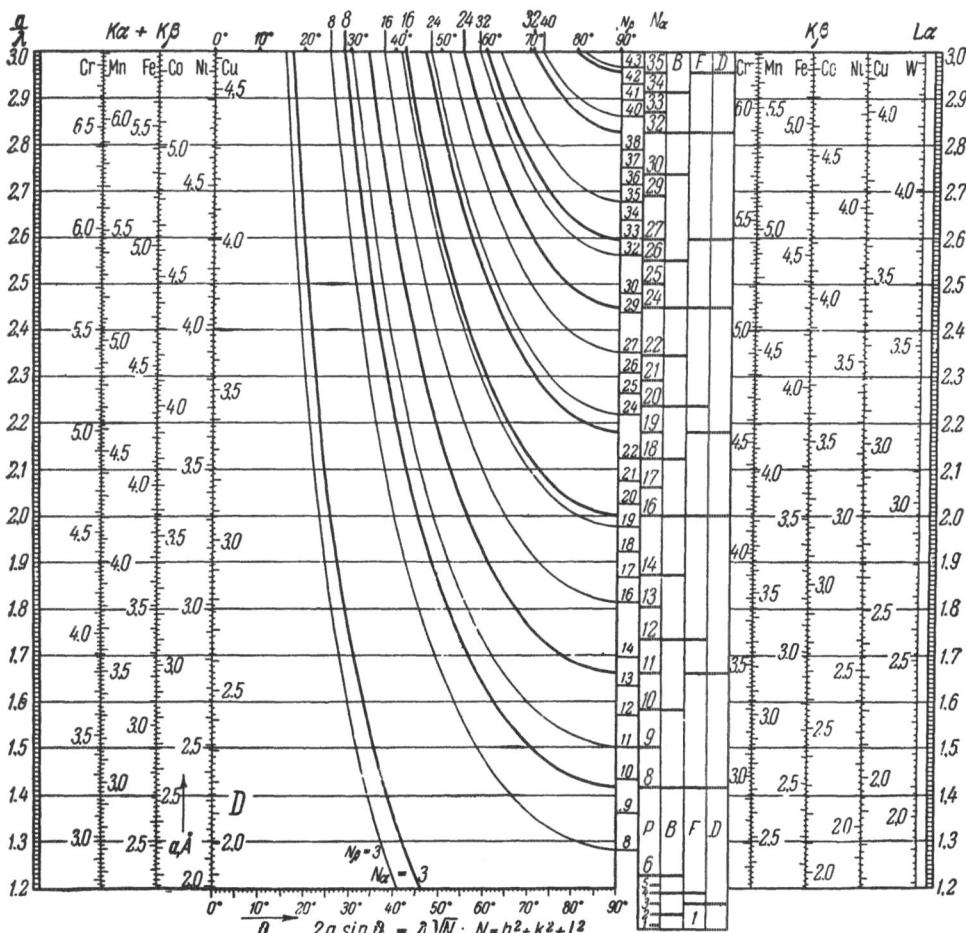

Fig. 46. Nomogram (I-D) for indexing patterns for crystals with the diamond structure.

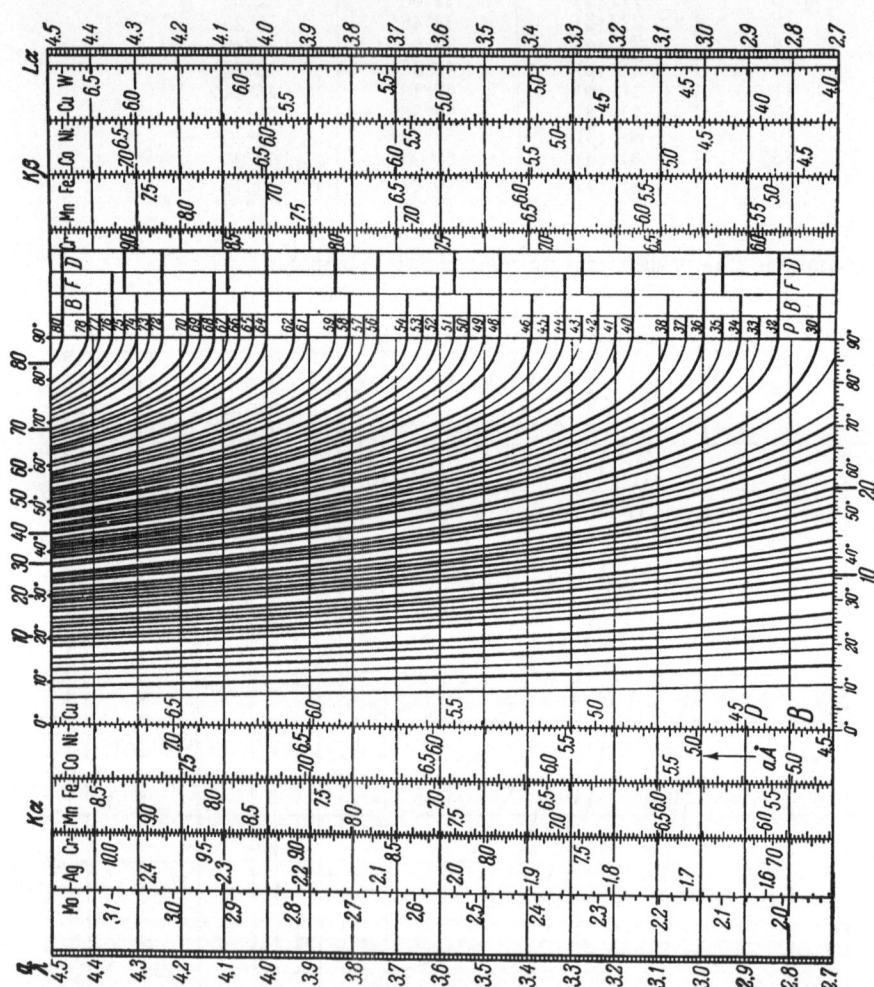

Fig. 47. Nomogram (II–PB) for indexing patterns from mixtures of crystals with simple cubic and body-centered cubic lattices for reflections whose sums of squares of indices range up to 80.

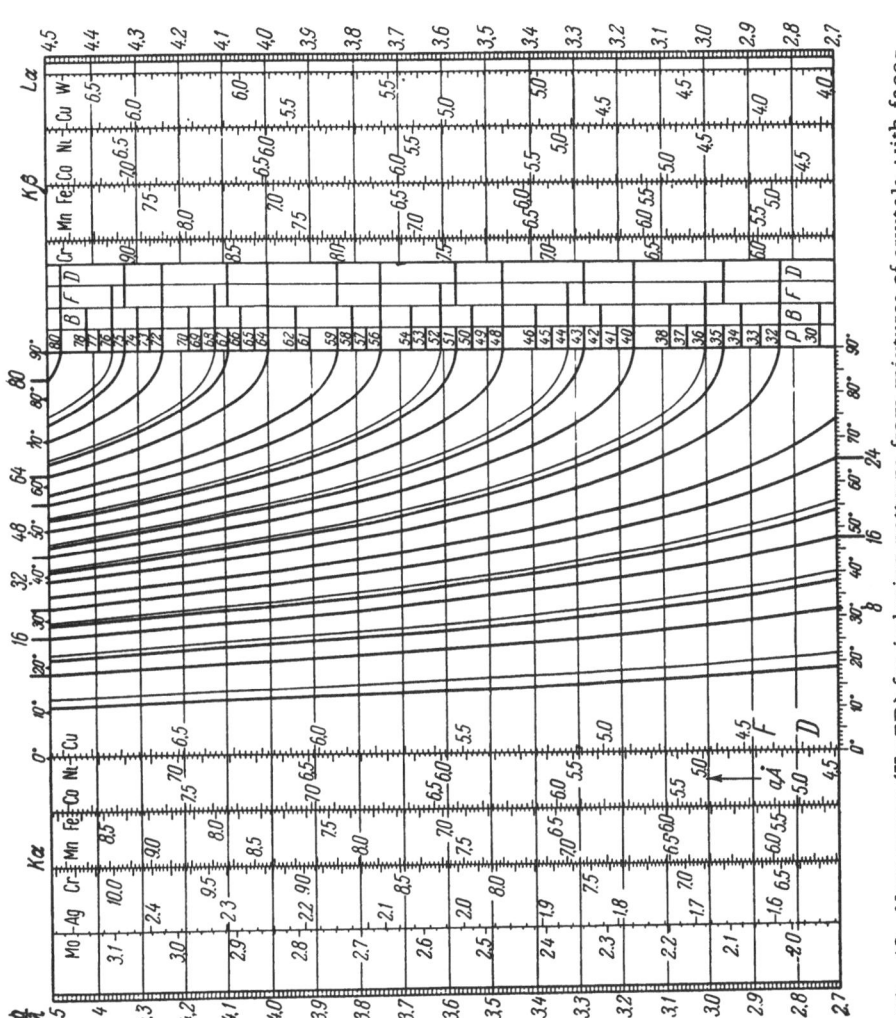

Fig. 48. Nomogram (II–FD) for indexing patterns from mixtures of crystals with face-centered cubic and diamond lattices for reflections whose sums of squares of indices range up to 80.

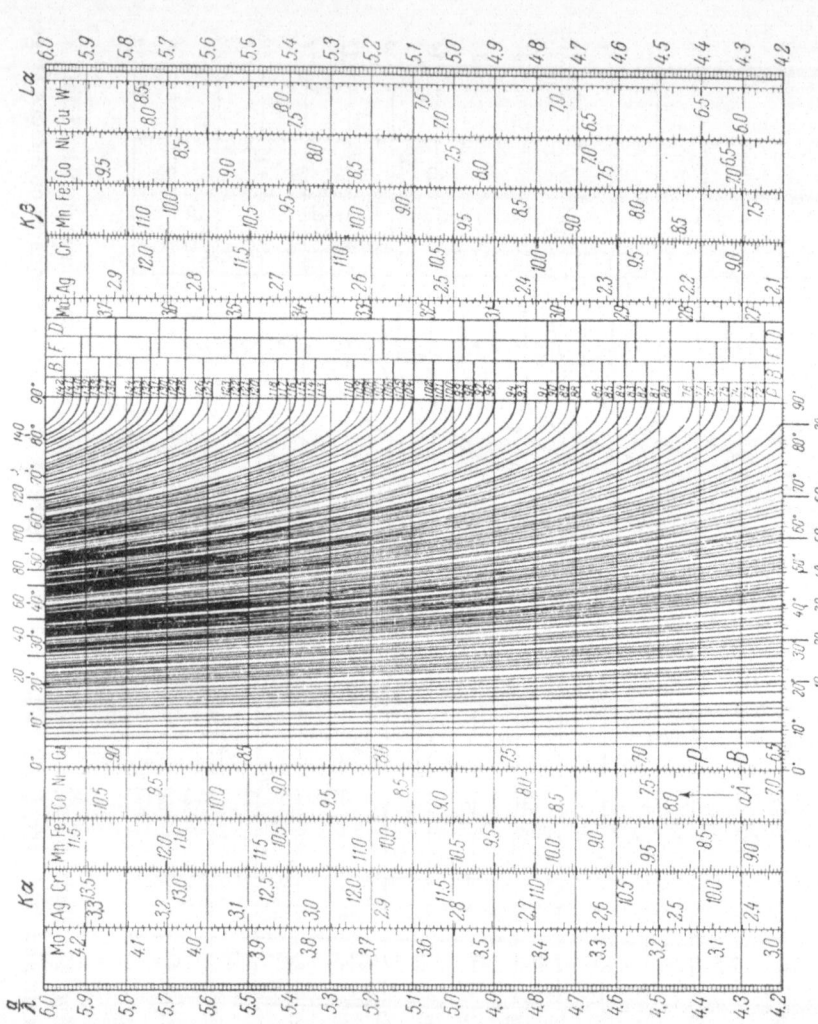

Fig. 49. Nomogram (III—PB) for indexing patterns from mixtures of crystals with simple cubic and body-centered cubic lattices for reflections whose sums of squares of indices range up to 142.

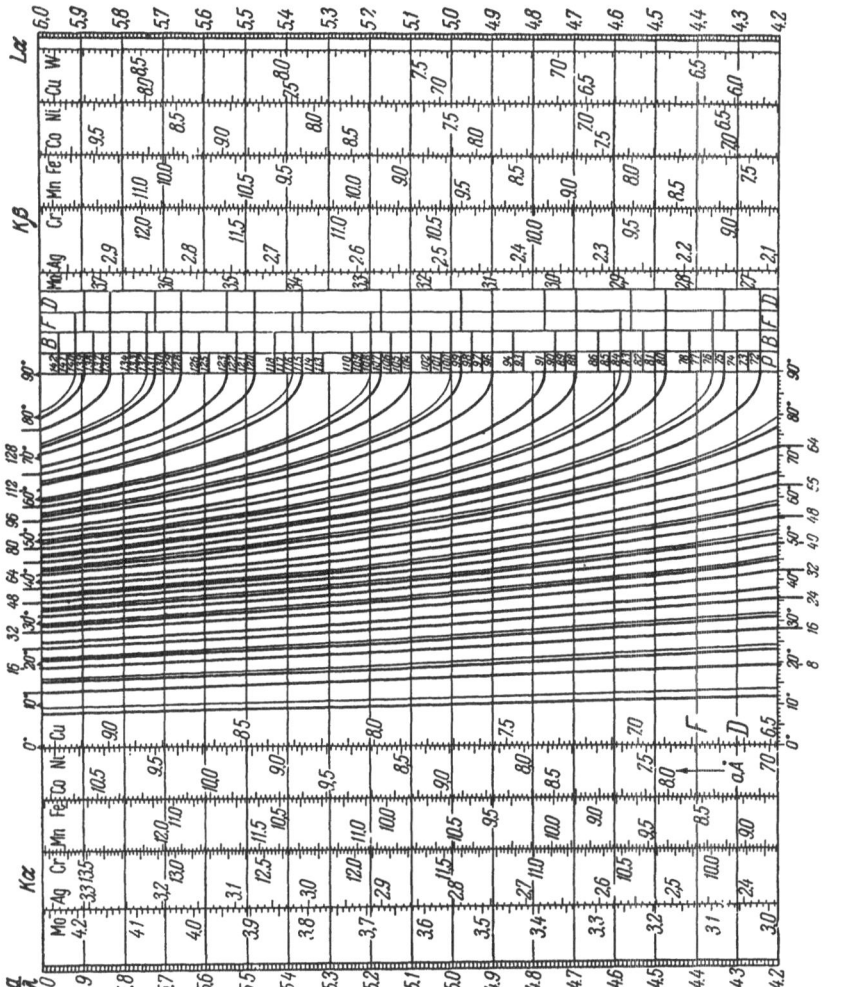

Fig. 50. Nomogram (III–FD) for indexing patterns from mixtures of crystals with face-centered cubic and diamond lattices for reflections whose sums of squares of indices range up to 142.

3-7. Graphs for Determining Whether a Material is Cubic

For a cubic lattice we have

$$Q_{hkl} = \frac{1}{d^2_{hkl}} = (h^2 + k^2 + l^2)\, a^{*2} = N a^{*2},\tag{14}$$

where a* is the repeat distance in the reciprocal lattice and N = h² + k² + l².

Figure 51 gives graphs [270] for determining whether a material is cubic. The abscissa is Q and the ordinate (a*)².

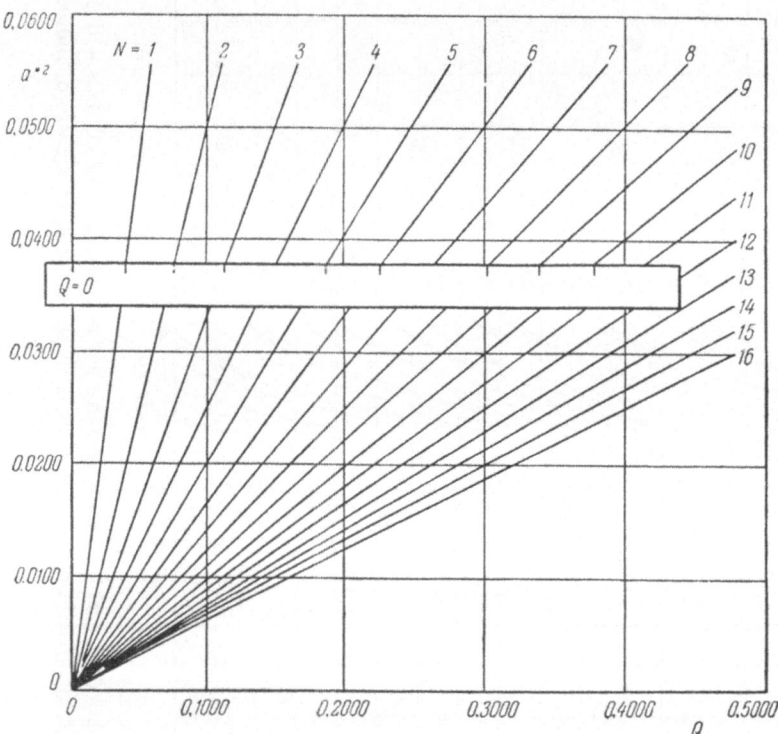

Fig. 51. Graphs for determining whether a material is cubic.

The steps are as follows:

1. Q is determined for each line on the pattern.
2. The β lines are eliminated.
3. The Q are entered on a strip of paper in the scale of the graphs.
4. The strip is moved to bring all marks into coincidence with the graphs.

Complete coincidence shows that the material is cubic; it also gives a* and (from the table of quadratic forms in Section 3-5) the indices of the lines.

3-8. Limiting Values for Sums of the Squares of Indices for Various Unit-Cell Volumes and Radiations

The table gives limiting values for $(h^2 + k^2 + l^2)$ for unit-cell volumes corresponding to lattice constants from 2.15 to 7.65 Å [271].

The $(h^2 + k^2 + l^2)$ relate to Cr, Co, and Cu radiations (K series).

E x a m p l e . Iron, a = 2.86 Å, has a cell volume V of about 23.5 Å^3; the table indicates that Cr Kα gives a limiting sum of about 6, so the possible lines are (110), (200), and (211).

V, Å³	Radiation		
	Cr (λ=2.22850₃ Å)	Co (λ=1.78529 Å)	Cu (λ=1.53739₅ Å)
10	3.551	5.817	7.844
20	5.643	9 244	12.466
30	7.395	12.114	16.336
40	8.960	14.678	19.794
50	10.397	17.033	22.969
60	11.700	19.166	25.845
70	13.010	21.314	28.741
80	14.224	23.302	31.422
90	15.382	25.199	33.981
100	16.508	27.043	36.467
110	17.585	28.807	38.846
120	18.635	30.528	41.166
130	19.661	32.208	43.432
140	20.659	33.843	45.637
150	21.625	35.426	47.772
160	22.580	36.990	49.880
170	23.513	38.518	51.941
180	24.420	40.006	53.947
190	25.320	41.479	55.934
200	26.199	42.920	57.877
250	30.406	49.811	67.169
300	34.328	56.236	75.840
350	38.044	62.323	84.042
400	41.588	69.130	91.872
450	44.986	73.696	99.378

TETRAGONAL SYSTEM

3-9. Schemes for X-Ray Patterns

Figure 52 gives theoretical patterns for tetragonal substances [102].

The dashes correspond to pattern lines, and the lengths are proportional to the intensity. The numbers are the indices.

The scheme applies to the basic types of tetragonal structure.

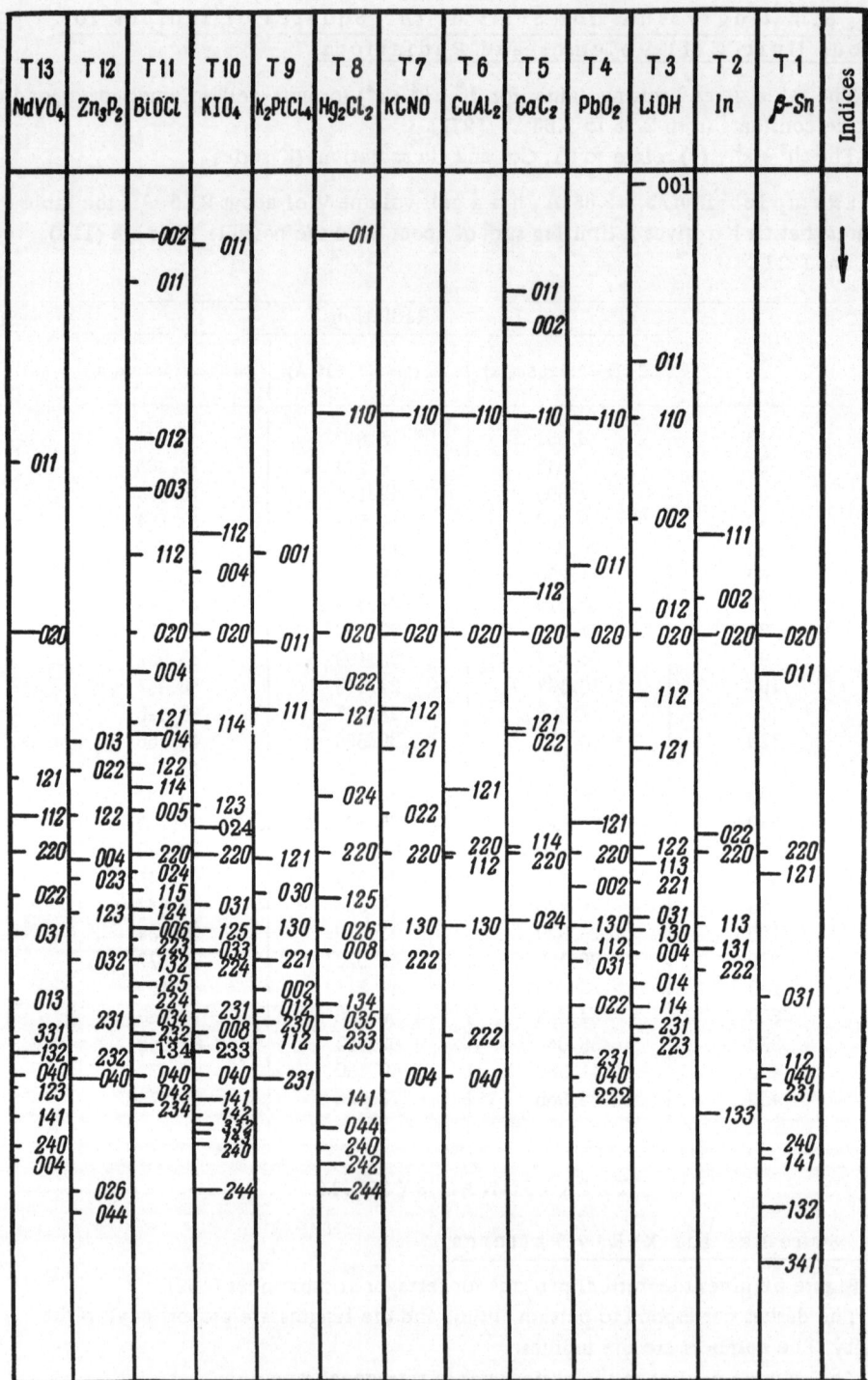

Fig. 52. Schemes for the x-ray patterns of tetragonal crystals.

3-10. Quadratic Forms

The quadratic form for the tetragonal system is

$$Q = \frac{1}{a^2}\left[(h^2 + k^2) + \left(\frac{a}{c}\right)^2 l^2 \right],\tag{15}$$

where a and c are the lattice constants in the Ox and Oz directions. The table gives h and k for $h^2 + k^2$ from 1 to 997 [2].

h^2+k^2	$h,\ k$	h^2+k^2	$h,\ k$	h^2+k^2	$h,\ k$	h^2+k^2	$h,\ k$
1	1, 0	100	10, 0	221	14, 5	338	13, 13
2	1, 1	100	8, 6	221	11, 10	340	18, 4
4	2, 0	101	10, 1	225	15, 0	340	14, 12
5	2, 1	104	10, 2	225	12, 9	346	15, 11
8	2, 2	106	9, 5	226	15, 1	349	18, 5
9	3, 0	109	10, 3	229	15, 2	353	17, 8
10	3, 1	113	8, 7	232	14, 6	356	16, 10
13	3, 2	116	10, 4	233	13, 8	360	18, 6
16	4, 0	117	9, 6	234	15, 3	361	19, 0
17	4, 1	121	11, 0	241	15, 4	362	19, 1
18	3, 3	122	11, 1	242	11, 11	365	19, 2
20	4, 2	125	11, 2	244	12, 10	365	14, 13
25	5, 0	125	10, 5	245	14, 7	369	15, 12
25	4, 3	128	8, 8	250	15, 5	370	19, 3
26	5, 1	130	11, 3	250	13, 9	370	17, 9
29	5, 2	130	9, 7	256	16, 0	373	18, 7
32	4, 4	136	10, 6	257	16, 1	377	19, 4
34	5, 3	137	11, 4	260	16, 2	377	16, 11
36	6, 0	144	12, 0	260	14, 8	386	19, 5
37	6, 1	145	12, 1	261	15, 6	388	18, 8
40	6, 2	145	9, 8	265	16, 3	389	17, 10
41	5, 4	146	11, 5	269	13, 10	392	14, 14
45	6, 3	148	12, 2	272	16, 4	394	15, 13
49	7, 0	149	10, 7	274	15, 7	397	19, 6
50	7, 1	153	12, 3	277	14, 9	400	20, 0
50	5, 5	157	11, 6	281	16, 5	400	16, 12
52	6, 4	160	12, 4	288	12, 12	401	20, 1
53	7, 2	162	9, 9	289	17, 0	404	20, 2
58	7, 3	164	10, 8	289	15, 8	405	18, 9
61	6, 5	169	13, 0	290	17, 1	409	20, 3
64	8, 0	169	12, 5	290	13, 11	410	19, 7
65	8, 1	170	13, 1	292	16, 6	410	17, 11
65	7, 4	170	11, 7	293	17, 2	416	20, 4
68	8, 2	173	13, 2	296	14, 10	421	15, 14
72	6, 6	178	13, 3	298	17, 3	424	18, 10
73	8, 3	180	12, 6	305	17, 4	425	20, 5
74	7, 5	181	10, 9	305	16, 7	425	19, 8
80	8, 4	185	13, 4	306	15, 9	425	16, 13
81	9, 0	185	11, 8	313	13, 12	433	17, 12
82	9, 1	193	12, 7	314	17, 5	436	20, 6
85	9, 2	194	13, 5	317	14, 11	441	21, 0
85	7, 6	196	14, 0	320	16, 8	442	21, 1
89	8, 5	197	14, 1	324	18, 0	442	19, 9
90	9, 3	200	14, 2	325	18, 1	445	21, 2
97	9, 4	200	10, 10	325	17, 6	445	18, 11
98	7, 7	202	11, 9	325	15, 10	449	20, 7
		205	14, 3	328	18, 2	450	21, 3
		205	13, 6	333	18, 3	450	15, 15
		208	12, 8	337	16, 9	452	16, 14
		212	14, 4	338	17, 7	457	21, 4
		218	13, 7			458	17, 13

h^2+k^2	h, k	h^2+k^2	h, k	h^2+k^2	h, k	h^2+k^2	h, k
461	19, 10	601	24, 5	740	26, 8	872	26, 14
464	20, 8	605	22, 11	740	22, 16	873	27, 12
466	21, 5	610	23, 9	745	27, 4	877	29, 6
468	18, 12	610	21, 13	745	24, 13	881	25, 16
477	21, 6	612	24, 6	746	25, 11	882	21, 21
481	20, 9	613	18, 17	754	27, 5	884	28, 10
481	16, 15	617	19, 16	754	23, 15	884	22, 20
482	19, 11	625	25, 0	757	26, 9	890	29, 7
484	22, 0	625	24, 7	761	20, 19	890	23, 19
485	22, 1	625	20, 15	765	27, 6	898	27, 13
485	17, 14	626	25, 1	765	21, 18	900	30, 0
488	22, 2	628	22, 12	769	25, 12	900	24, 18
490	21, 7	629	25, 2	772	24, 14	901	30, 1
493	22, 3	629	23, 10	773	22, 17	901	26, 15
493	18, 13	634	25, 3	776	26, 10	904	30, 2
500	22, 4	637	21, 14	778	27, 7	905	29, 8
500	20, 10	640	24, 8	784	28, 0	905	28, 11
505	21, 8	641	25, 4	785	28, 1	909	30, 3
505	19, 12	648	18, 18	785	23, 16	914	25, 17
509	22, 5	650	25, 5	788	28, 2	916	30, 4
512	16, 16	650	23, 11	793	28, 3	922	29, 9
514	17, 15	650	19, 17	793	27, 8	925	30, 5
520	22, 6	653	22, 13	794	25, 13	925	27, 14
520	18, 14	656	20, 16	797	26, 11	925	22, 21
521	20, 11	657	24, 9	800	28, 4	928	28, 12
522	21, 9	661	25, 6	800	20, 20	929	23, 20
529	23, 0	666	21, 15	801	24, 15	932	26, 16
530	23, 1	673	23, 12	802	21, 19	936	30, 6
530	19, 13	674	25, 7	808	22, 18	937	24, 19
533	23, 2	676	26, 0	809	28, 5	941	29, 10
533	22, 7	676	24, 10	810	27, 9	949	30, 7
538	23, 3	677	26, 1	818	23, 17	949	25, 18
541	21, 10	680	26, 2	820	28, 6	953	28, 13
544	20, 12	680	22, 14	820	26, 12	954	27, 15
545	23, 4	685	26, 3	821	25, 14	961	31, 0
545	17, 16	685	19, 18	829	27, 10	962	31, 1
548	22, 8	689	25, 8	832	24, 16	962	29, 11
549	18, 15	689	20, 17	833	28, 7	964	30, 8
554	23, 5	692	26, 4	841	29, 0	965	31, 2
557	19, 14	697	24, 11	841	21, 20	965	26, 17
562	21, 11	697	21, 16	842	29, 1	968	22, 22
565	23, 6	698	23, 13	845	29, 2	970	31, 3
565	22, 9	701	26, 5	845	26, 13	970	23, 21
569	20, 13	706	25, 9	845	22, 19	976	24, 20
576	24, 0	709	22, 15	848	28, 8	977	31, 4
577	24, 1	712	26, 6	850	29, 3	980	28, 14
578	23, 7	720	24, 12	850	27, 11	981	30, 9
578	17, 17	722	19, 19	850	25, 15	985	29, 12
580	24, 2	724	20, 18	853	23, 18	985	27, 16
580	18, 16	725	26, 7	857	29, 4	986	31, 5
584	22, 10	725	25, 10	865	28, 9	986	25, 19
585	24, 3	725	23, 14	865	24, 17	997	31, 6
585	21, 12	729	27, 0	866	29, 5		
586	19, 15	730	27, 1				
592	24, 4	730	21, 17				
593	23, 8	733	27, 2				
596	20, 14	738	27, 3				

3-11. Values of $(a/c)^2 l^2$

The quadratic forms for the tetragonal and hexagonal systems contain $(a/c)^2 l^2$; the table is for c/a from 0.30 to 2.90 and for l from 1 to 9 [102]. The Q of (15) is found by first determining c/a and l; the intersections of the corresponding verticals and horizontals give the $(a/c)^2 l^2$, which are added to the function of h and k as in (15).

c/a \ l	1	2	3	4	5	6	7	8	9
0.30	11.11	44.44	100.0	177.8	277.8	400.0	544.4	711.1	900.0
1	0.41	41.64	93.69	166.6	260.2	374.8	510.1	666.2	843.2
2	9.766	39.06	87.89	156.3	244.2	351.6	478.5	625.0	791.0
3	9.183	36.73	82.64	146.9	229.6	330.6	450.0	587.7	743.8
4	8.651	34.60	77.86	138.4	216.3	311.4	423.9	553.7	700.7
5	8.163	32.65	73.47	130.6	204.1	293.9	400.0	522.4	661.2
6	7.716	30.86	69.44	123.5	192.9	277.8	378.1	493.8	625.0
7	7.305	29.22	65.74	116.9	182.6	263.0	357.9	467.5	591.7
8	6.925	27.70	62.32	110.8	173.1	249.3	339.3	443.2	560.9
9	6.575	26.30	59.17	105.2	164.4	236.7	322.2	420.8	532.6
0.40	6.250	25.00	56.25	100.0	156.2	225.0	306.2	400.0	506.2
1	5.949	23.80	53.54	95.18	148.7	214.2	291.5	380.7	481.9
2	5.669	22.68	51.02	90.70	141.7	204.1	277.8	362.8	459.2
3	5.408	21.63	48.67	86.53	135.2	194.7	265.0	346.1	438.0
4	5.165	20.66	46.48	82.64	129.1	185.9	253.1	330.6	418.4
5	4.938	19.75	44.44	79.01	123.5	177.8	242.0	316.0	400.0
6	4.726	18.90	42.53	75.62	118.1	170.1	231.6	302.5	382.8
7	4.527	18.11	40.74	72.43	113.2	163.0	221.8	289.7	366.7
8	4.340	17.36	39.06	69.44	108.5	156.2	212.7	277.8	351.5
9	4.165	16.66	37.48	66.64	104.1	149.9	204.1	266.6	337.4
0.50	4.000	16.00	36.00	64.00	100.0	144.0	196.0	256.0	324.0
1	3.845	15.38	34.60	61.52	96.12	138.4	188.4	246.1	311.4
2	3.698	14.79	33.28	59.17	92.45	133.1	181.2	236.7	299.5
3	3.560	14.24	32.04	56.96	89.00	128.2	174.4	227.8	288.4
4	3.429	13.72	30.86	54.86	85.72	123.4	168.0	219.5	277.7
5	3.306	13.22	29.75	52.90	82.65	119.0	162.0	211.6	267.8
6	3.189	12.76	28.70	51.02	79.72	114.8	156.3	204.1	258.3
7	3.078	12.31	27.70	49.25	76.95	110.8	150.8	197.0	249.3
8	2.973	11.89	26.76	47.57	74.32	107.0	145.7	190.3	240.8
9	2.873	11.49	25.86	45.97	71.82	103.4	140.8	183.9	232.7
0.60	2.778	11.11	25.00	44.45	69.45	100.0	136.1	177.8	225.0
2	2.601	10.40	23.41	41.62	65.03	93.64	127.4	166.5	210.7
4	2.441	9.764	21.97	39.06	61.02	87.88	119.6	156.2	197.7
6	2.296	9.184	20.66	36.74	57.40	82.66	112.5	146.9	186.0
8	2.163	8.652	19.47	34.61	54.07	77.87	106.0	138.4	175.2
0.70	2.041	8.164	18.37	32.66	51.02	73.48	100.0	130.6	165.3
2	1.929	7.716	17.36	30.86	48.23	69.44	94.52	123.5	156.2
4	1.826	7.304	16.43	29.22	45.65	65.74	89.47	116.9	147.9
6	1.731	6.924	15.58	27.70	43.28	62.32	84.82	110.8	140.2
8	1.644	6.576	14.80	26.30	41.10	59.18	80.56	105.2	133.2
0.80	1.562	6.250	14.06	25.00	39.06	56.25	76.56	100.0	126.6
2	1.487	5.949	13.38	23.80	37.18	53.54	72.87	95.18	120.5
4	1.417	5.669	12.75	22.68	35.43	51.02	69.44	90.70	114.8
6	1.352	5.408	12.17	21.63	33.80	48.68	66.25	86.53	109.5
8	1.291	5.165	11.62	20.66	32.28	46.49	63.27	82.64	104.6

c/a \ l	1	2	3	4	5	6	7	8	9
0.90	1.235	4.938	11.11	19.75	30.86	44.44	60.49	79.01	100.0
2	1.181	4.726	10.63	18.90	29.54	42.53	57.89	75.62	95.70
4	1.132	4.527	10.19	18.11	28.29	40.74	55.45	72.43	91.67
6	1.085	4.340	9.766	17.36	27.13	39.06	53.17	69.45	87.89
8	1.041	4.165	9.371	16.66	26 03	37.48	51.02	66.64	84.34
1.00	1.000	4.000	9.000	16.00	25.00	36.00	49.00	64.00	81.00
05	0.9070	3.628	8.163	14.51	22.67	32.65	44.44	58.05	73.47
10	0.8264	3.306	7.438	13.22	20.66	29.75	40.49	52.89	66.94
15	0.7561	3.024	6.805	12.10	18.90	27.22	37.05	48.39	61.24
20	0.6944	2.778	6.250	11.11	17.36	25.00	34.03	44.44	56.25
25	0.6400	2.560	5.760	10.24	16.00	23.04	31.36	40.96	51.84
30	0.5917	2.367	5.325	9.467	14.79	21.30	28.99	37.87	47.93
35	0.5487	2.195	4.938	8.779	13.72	19.75	26.89	35.12	44.44
40	0.5102	2.041	4.592	8.163	12.75	18.37	25.00	32.65	41.33
45	0.4756	1.902	4.280	7.610	11.89	17.12	23.30	30.44	38.52
50	0.4444	1.778	4.000	7.110	11.11	16.00	21.78	28.44	36.00
55	0.4162	1.665	3.746	6.659	10.40	14.98	20.39	26.64	33.71
60	0.3906	1.562	3.515	6.250	9.765	14.06	19.14	25.00	31.64
65	0.3673	1.469	3.306	5.877	9.182	13.22	18.00	23.51	29.75
70	0.3460	1.384	3.114	5.536	8.650	12.46	16.95	22.14	28.03
75	0.3265	1.306	2.939	5.224	8.163	11.76	16.00	20.90	26.45
80	0.3086	1.234	2.777	4.938	7.715	11.11	15.12	19.75	25.00
85	0.2922	1.169	2.630	4.675	7.305	10.52	14.32	18.70	23.67
90	0.2770	1.108	2.493	4.432	6.925	9.972	13.57	17.73	22.44
95	0.2630	1.052	2.367	4.208	6.575	9.468	12.89	16.83	21.30
2.00	0.2500	1.000	2.250	4.000	6.250	9.000	12.25	16.00	20.25
10	0.2268	0.9072	2.041	3.629	5.670	8.165	11.11	14.52	18.37
20	0.2066	0.8264	1.859	3.306	5.165	7.438	10.12	13.22	16.73
30	0.1890	0.7560	1.701	3.024	4.725	6.804	9.261	12.10	15.31
40	0.1736	0.6944	1.562	2.778	4.340	6.250	8.506	11.11	14.06
50	0.1600	0.6400	1.440	2.560	4.000	5.760	7.840	10.24	12.96
60	0.1479	0.5916	1.331	2.366	3.697	5.324	7.247	9.466	11.98
70	0.1372	0.5488	1.235	2.195	3.430	4.939	6.723	8.781	11.11
80	0.1276	0.5104	1.148	2.042	3.190	4.594	6.252	8.166	10.34
90	0.1189	0.4756	1.070	1.902	2.972	4.280	5.826	7.610	9.631

3-12. Analytic Method of Indexing X-Ray Patterns

This method is based on a relation between $\sin^2 \vartheta_{hkl}$ and the indices:

$$\sin^2 \vartheta_{hkl} = A(h^2 + k^2) + Cl^2. \tag{16}$$

The value for $\sin^2 \vartheta_{100}$ (for $l = 0$) must be A; $\sin^2 \vartheta_{110} = 2A$, $\sin^2 \vartheta_{210} = 5A$, $\sin^2 \vartheta_{220} = 8A$, and so on.

If the phase is not cubic, and if the ratio of the $\sin^2 \vartheta$ for the first two lines at small angles is 2, it can be assumed that the phase is tetragonal and that the lines are (100) and (110) or (110) and (200). This is tested to find A and hence to deduce the indices for all lines of (hk0) type [92].

The method of deducing C is illustrated by an example. The first 9 lines in the pattern for $CuAl_2$ (tetragonal) gave the $\sin^2 \vartheta$ listed in Table I.

TABLE I

Line	$\sin^2 \vartheta$	$\sin^2 \vartheta - A$	$\sin^2 \vartheta - 2A$	$\sin^2 \vartheta - 4A$	$\sin^2 \vartheta - 5A$
1	0.0445	0.0001	—	—	—
2	0.0888	0.0444	0.0000	—	—
3	0.1449	0.1005	0.0561	—	—
4	0.1767	0.1323	0.0879	—	—
5	0.1811	0.1367	0.0823	0.0035	—
6	0.2204	0.1760	0.1316	0.0418	—
7	0.2245	0.1801	0.1357	0.0469	—
8	0.3117	0.2673	0.2229	0.1341	0.0897
9	0.3554	0.3110	0.2668	0.1777	0.1334

The ratio of the $\sin^2 \vartheta$ for the first two lines is close to 2, so it is assumed that the substance is tetragonal, line 1 being (100) and line 2 being (110). Then A = 0.0444, 2A = 0.0888, 4A = 0.1776, 5A = 0.2220, and 8A = 0.3552. Comparison with the $\sin^2 \vartheta$ shows that line 4 is (200), line 6 is (210), and line 9 is (220); then A,..., 5A are subtracted from the $\sin^2 \vartheta$ (Table I).

If the lines actually are (100) and (110), each line in Table I should contain the corresponding Cl^2.

For example, the differences for lines 4, 6, and 8 contain the value 0.1329 ±0.012, but lines 3, 5, and 7 do not give this. We assume that 0.1329 corresponds to l = 2, so C = 0.0322.

Substitution into (16) for all lines does not give satisfactory agreement with the observed $\sin^2 \vartheta$; a similar calculation with the 0.1354 ± 0.0013 given by lines 5, 7, and 8 also gives poor results, so the indices for lines 1 and 2 must have been chosen wrongly.

We therefore assume that line 1 is (110) and line 2 is (200), which gives A = 0.0222, 2A = 0.0444, 4A = 0.0888, 5A = 0.1110, 8A = 0.1776, and 9A = 0.1998. The differences in this case have the values given in Table II.

TABLE II

Line	$\sin^2 \vartheta - A$	$\sin^2 \vartheta - 2A$	$\sin^2 \vartheta - 4A$	$\sin^2 \vartheta - 5A$	$\sin^2 \vartheta - 8A$	$\sin^2 \vartheta - 9A$
1	0.0223	0.0001	—	—	—	—
2	0.0666	0.0444	0.0000	—	—	—
3	0.1227	0.1005	0.0561	0.0339	—	—
4	0.1545	0.1323	0.0879	0.0657	0.0035	—
5	0.1589	0.1367	0.0923	0.0701	0.0418	0.0206
etc.						

Table II shows that one of the differences for line 3 is 0.0339, which is close to the C found in the first trial; 4C (for l = 2) in this case is 0.1356, which is close to the mean value of 0.1354 given by Table I.

Then we have A = 0.0222 and C = 0.0339; the indexing is completed by finding the $\sin^2 \vartheta$ for all (hkl) from (16) and comparing these with the measured values.

3-13. Graphs for Indexing X-Ray Patterns

The Wulff-Bragg relation for the tetragonal system is

$$2 \lg d_{hkl} = -\lg \left[(h^2 + k^2) + l^2 \frac{1}{\left(\frac{c}{a}\right)^2} \right].$$

(17)

Graphs representing this enable us to index patterns for tetragonal crystals if the d are known.

The logarithms of the interplanar distances are plotted on a strip of paper in the scale of the graph. This strip is moved parallel to the axis of abscissas to bring all marks into coincidence with lines on the graph. The numbers at the ends of the lines are the indices, while the position on the vertical scale gives c/a.

Figure 53 relates to simple tetragonal lattices; Fig. 54, to body-centered ones [273]; and Fig. 55, also to body-centered ones, but in terms of $\sin^2 \vartheta$ and c/a [173].

Figure 56 applies to face-centered lattices [273].

Figure 57 (Schwarz and Summa [464]) provides the indices and lattice constants from ϑ and $\sin^2 \vartheta$.

Various types of graphs on logarithmic scales are used to index reflections with high indices and in certain other cases.

Figure 58 gives Bunn curves for c/a from 0.4 to 3.0 [274]; the coordinates are lg (c/a) on the abscissa, lg $(h^2 + k^2)$ on the left, and lg l^2 on the right. The mode of use is as for Figs. 53-56, but it is not necessary to reverse the strips carrying the lg d. The graph is used by first constructing a scale for the vertical axis by inserting the data for some substance with known indices.

Figure 59 [91] is usable up to (413) reflections; here lg (c/a) is the vertical scale , lg $(h^2 + k^2)$ and lg l^2 being horizontal. The scale of d in logarithmic form lies in the lower part.

Bond graphs [270] (Fig. 60) are used for $(h^2 + k^2)$ up to 9 and l^2 up to 16; here the abscissa is c/a, the ordinates are the same as in Fig. 58, and the scale of lg d is on the left.

Reflections up to (603) are accommodated by Harrington curves (Fig. 61) [270], in which c/a is used in logarithmic form (a scale of d/n must first be compiled).

Reflections above (603) can appear if molybdenum (or harder) radiation is used. In this case one has to construct graphs as in Fig. 62 [12].

The possible $h^2 + k^2$ are laid out along the left vertical axis, and the l^2 along the right one; the points are joined by straight lines.

The relation between the indices and d can be written as

$$\frac{1}{d_{hkl}^2} = \frac{h^2 + k^2}{a^2} + \frac{l^2}{c^2}.$$

The scale for c/a is chosen as follows:

$$\frac{1}{a^2} = 1 \text{ (left scale)} \qquad \frac{c}{a} = \infty,$$

$$\frac{1}{a^2} = 0 \text{ (right scale)} \qquad \frac{c}{a} = 0,$$

$$\frac{1}{a^2} = \frac{1}{2} \qquad \frac{c}{a} = 1, \text{ etc.}$$

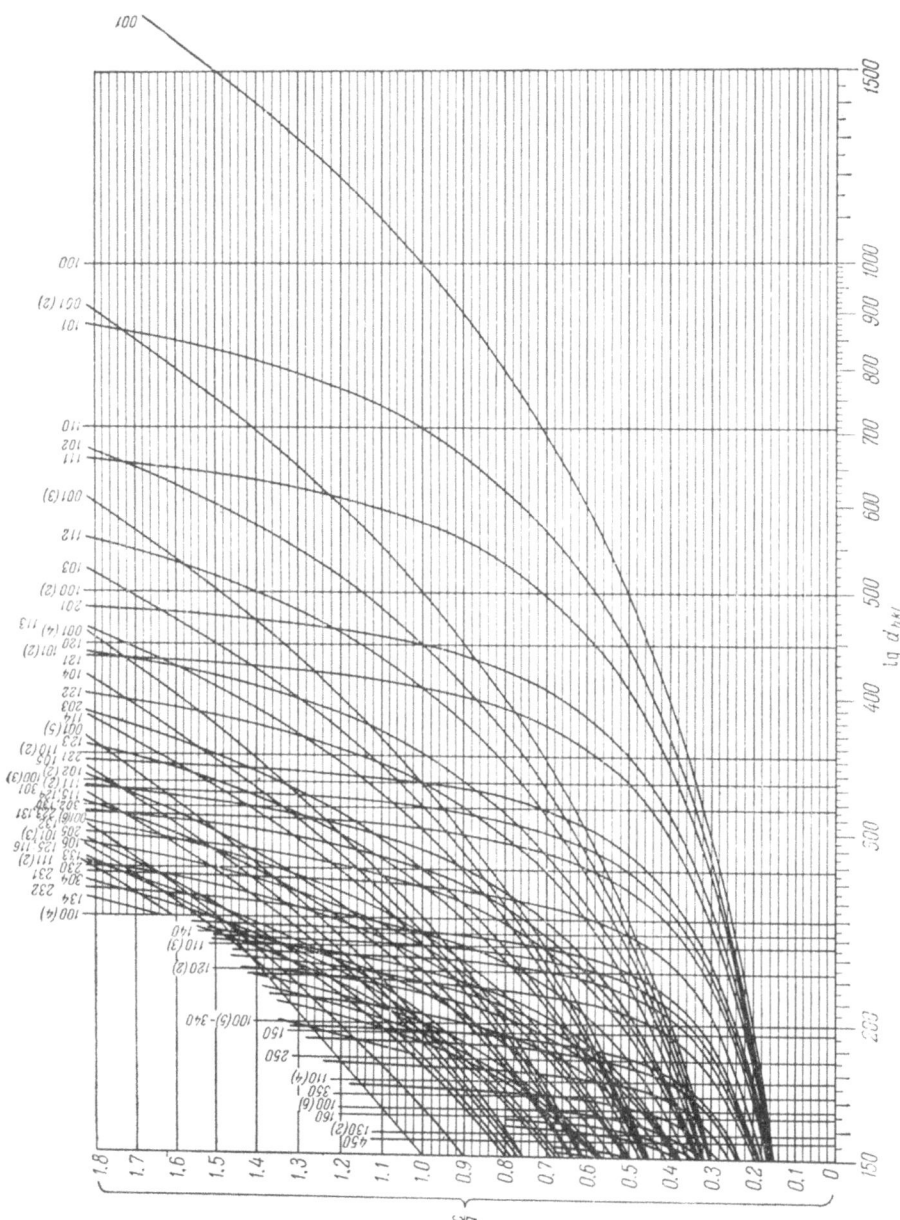

Fig. 53. Hull-Davey chart for indexing patterns from crystals of simple tetragonal structure.

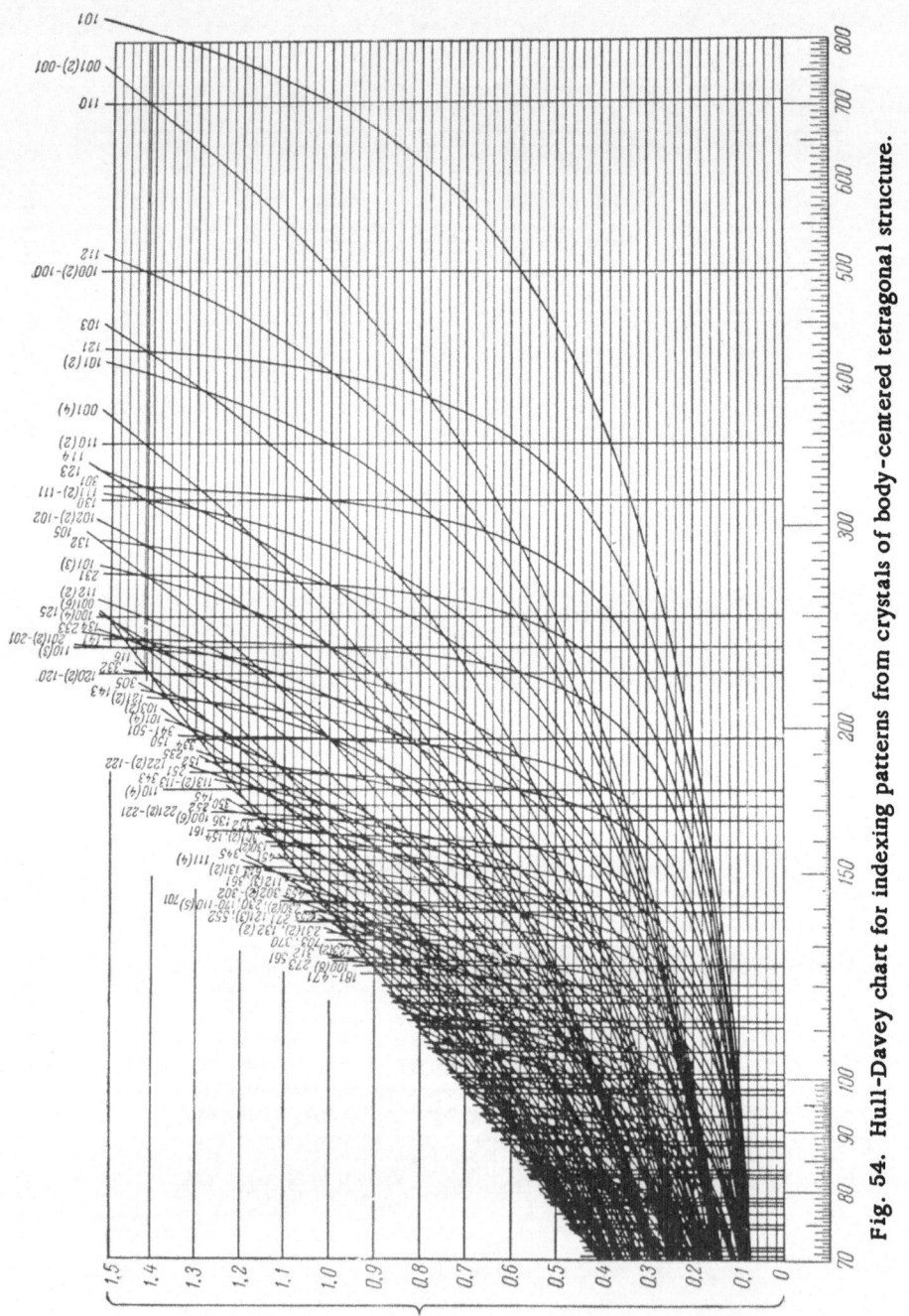

Fig. 54. Hull-Davey chart for indexing patterns from crystals of body-centered tetragonal structure.

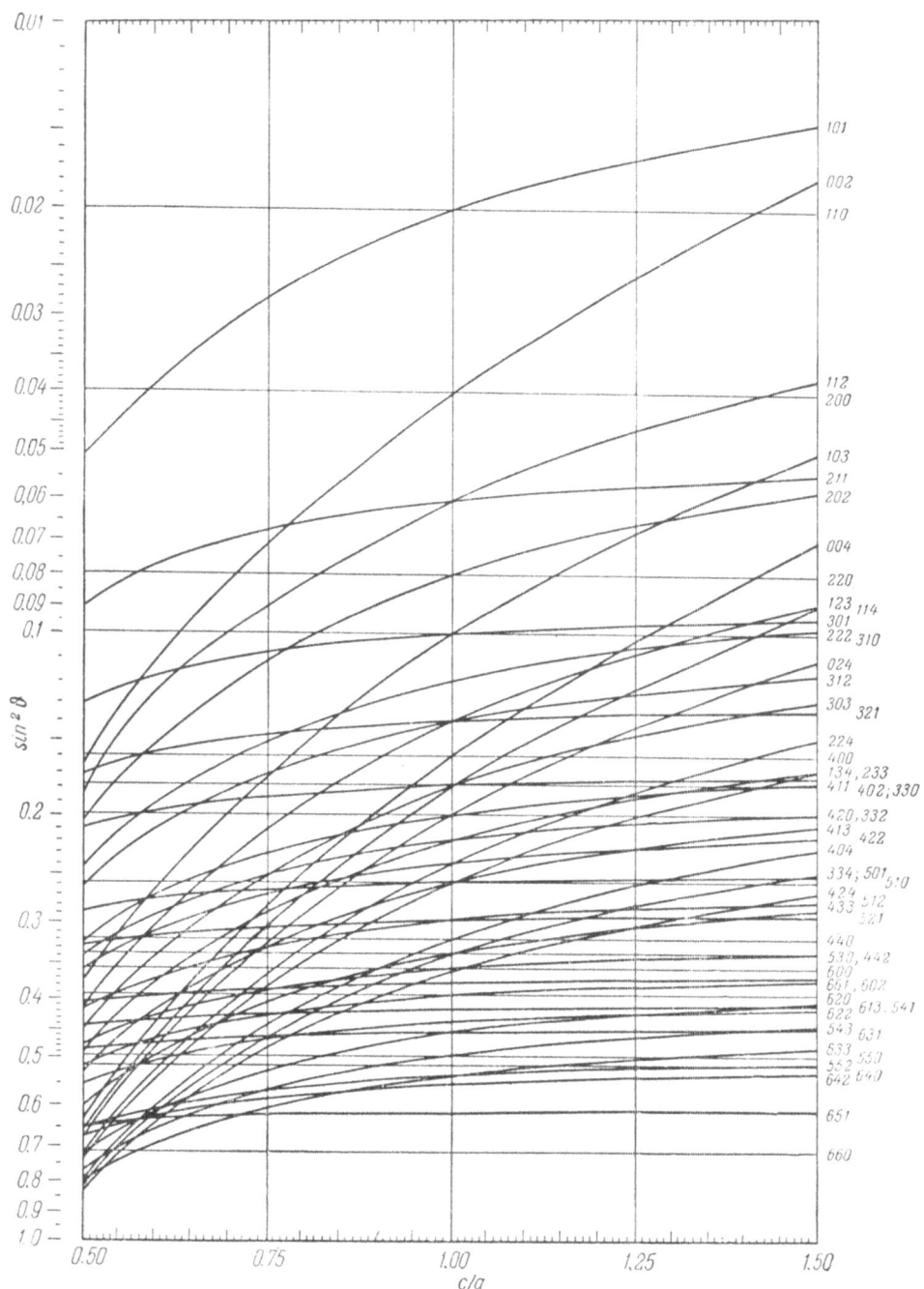

Fig. 55. Hull-Davey chart for indexing patterns from crystals of body-centered tetragonal structure by reference to $\sin^2 \vartheta$.

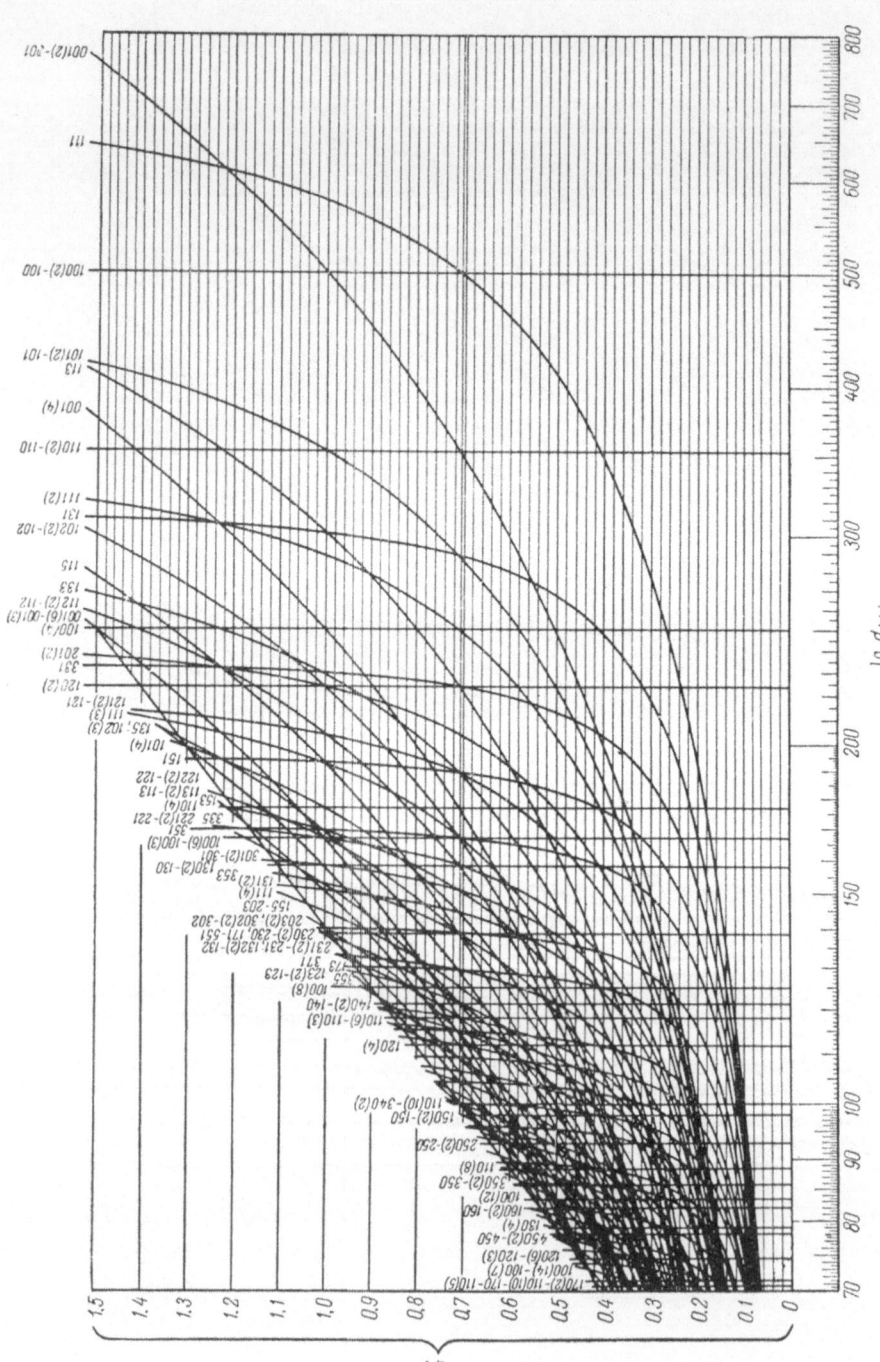

Fig. 56. Hull-Davey chart for indexing patterns from crystals of face-centered tetragonal structure.

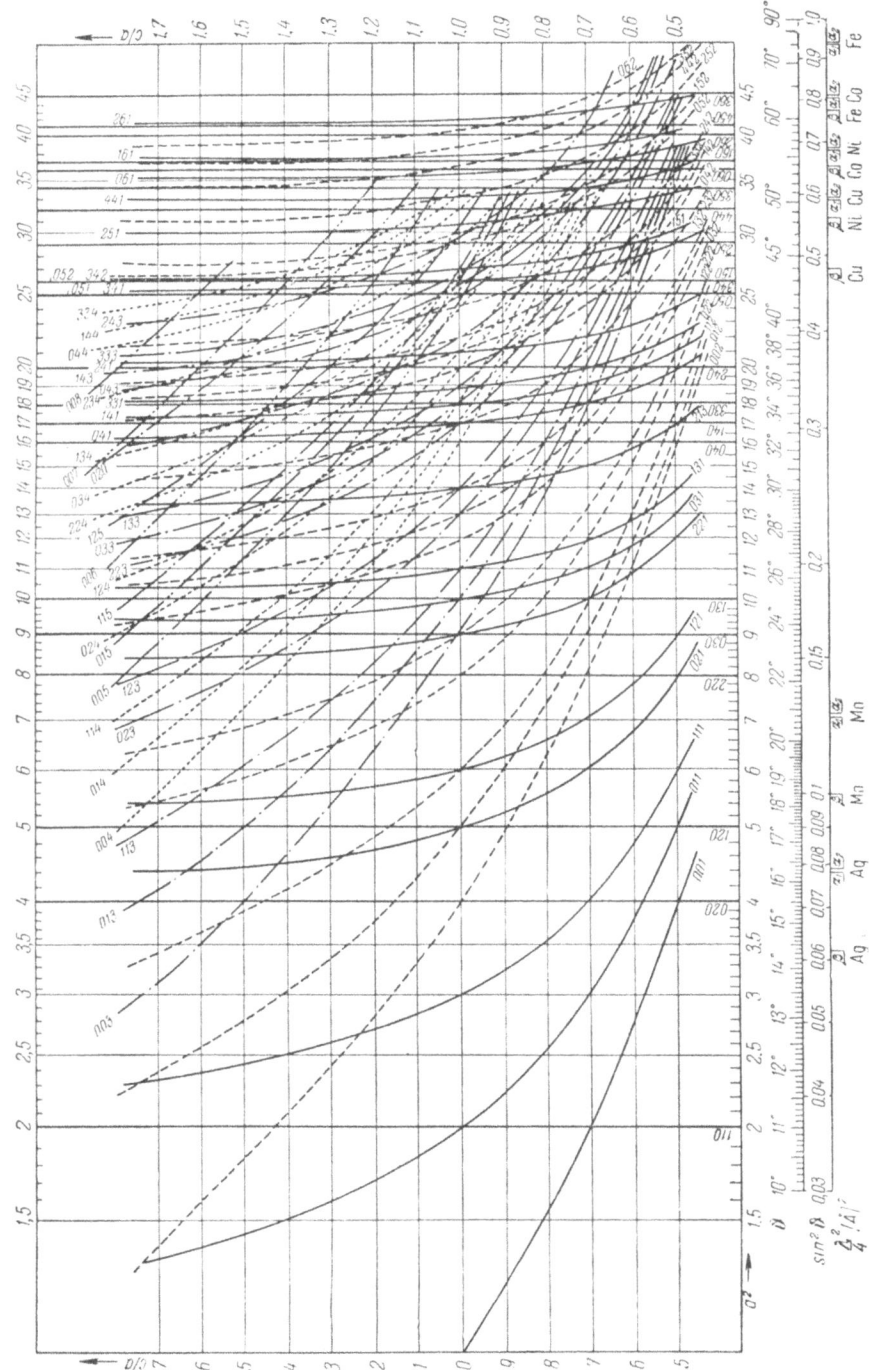

Fig. 57. Graphs for indexing patterns from crystals of tetragonal structure (Schwarz and Summa [464]).

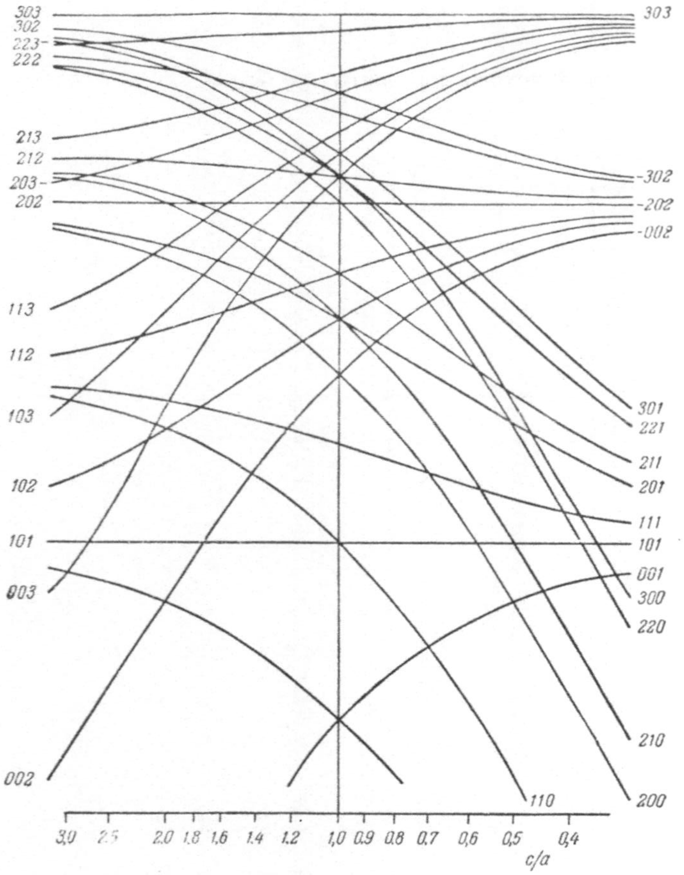

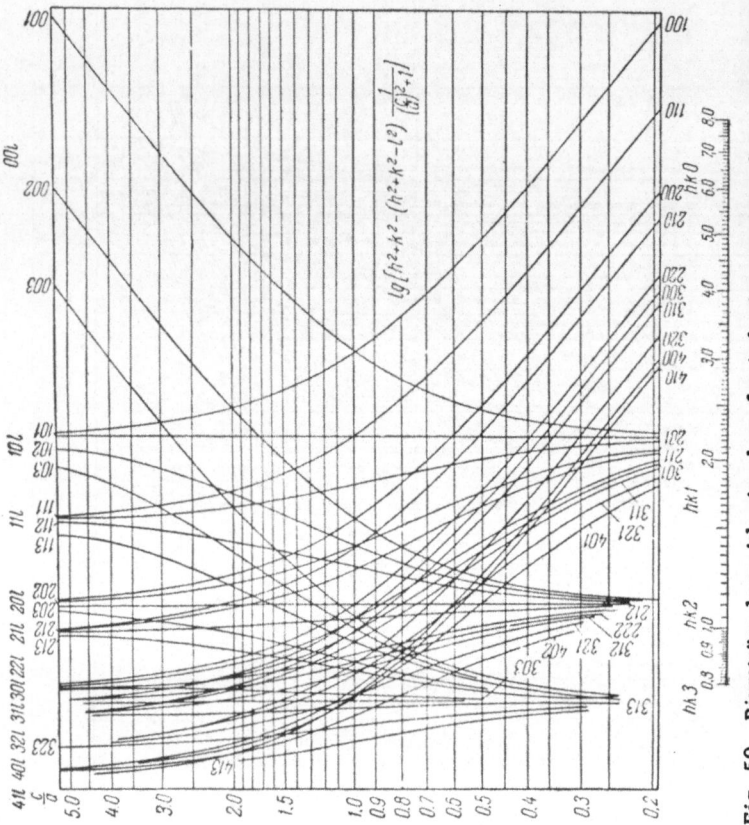

Fig. 59. Bjurström logarithmic chart for indexing patterns from crystals of tetragonal structure.

Fig. 58. Bunn chart for indexing patterns from crystals of tetragonal structure.

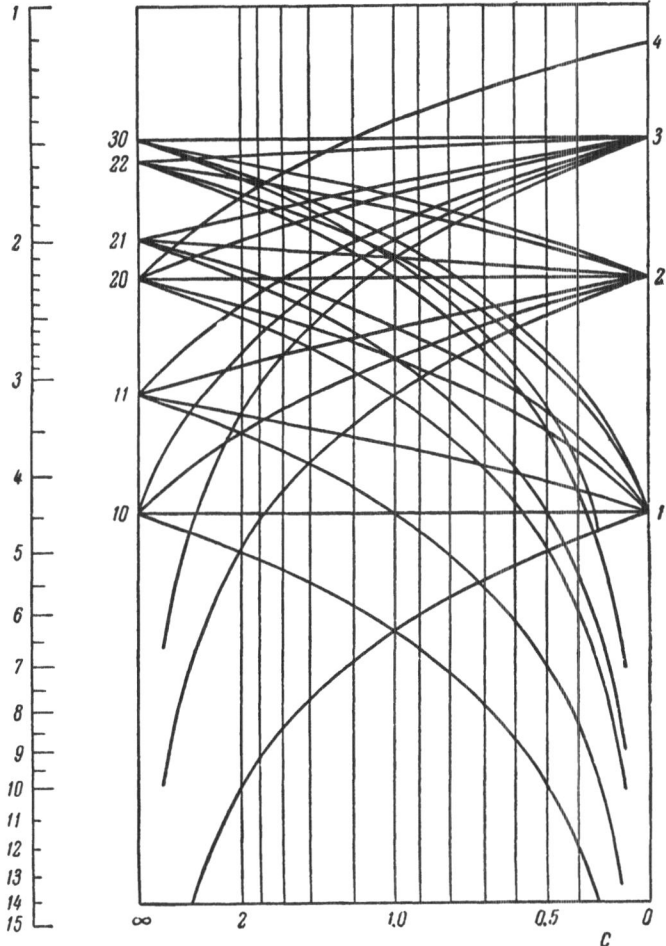

Fig. 60. Bond graphs for indexing patterns from crystals of
tetragonal structure.

Figure 62 illustrates this for $h^2 + k^2$ up to 9 and l^2 up to 9.

The graphs are constructed in arbitrary units, so the additional scale curves of Fig. 63 must be constructed. A vertical line is drawn on a sheet of tracing paper, and this is marked with the $1/d^2$ as deduced from measurements on the pattern; these points are joined to an arbitrary point on the abscissa. This pattern of lines is moved over Fig. 62 to find the position at which all lines of both sets meet on the same vertical line; the indexing is then done as for Figs. 53-56.

The graphs are easier to use if c/a is known, because the mark representing the first line on the pattern is brought onto the curve having those indices.

HEXAGONAL SYSTEM

3-14. Schemes for X-Ray Patterns

Figures 64 and 65 give theoretical patterns for hexagonal substances [102].

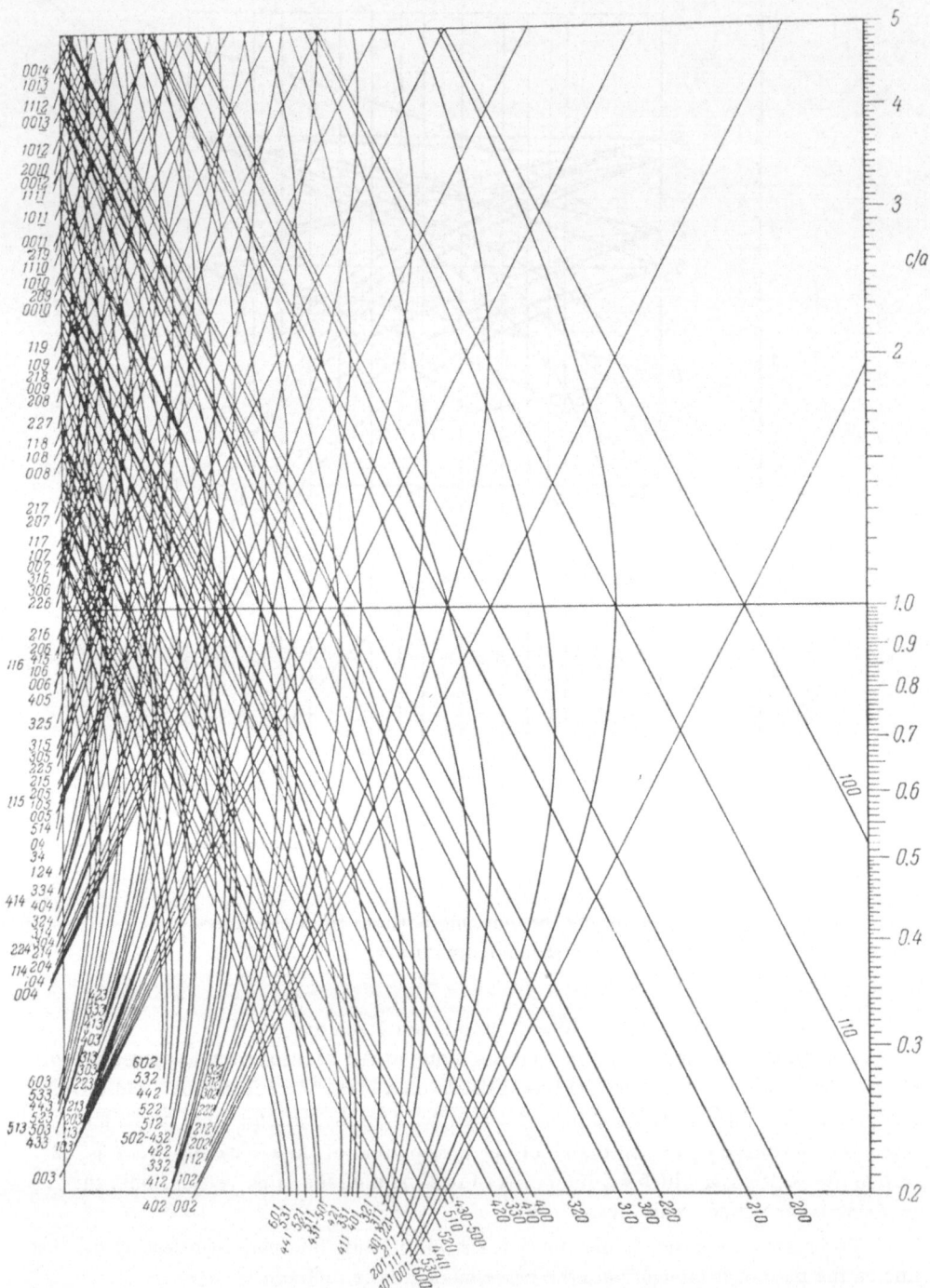

Fig. 61. Harrington graphs for indexing patterns from crystals of tetragonal structure.

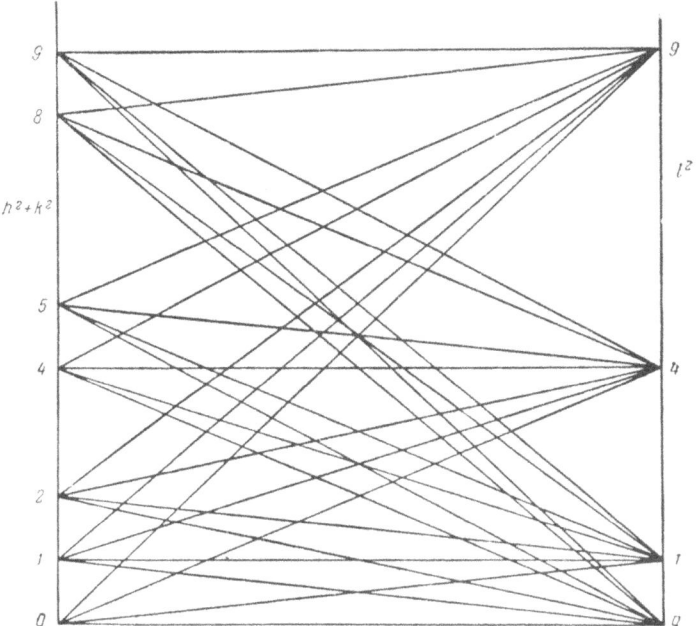

Fig. 62. Bjurström graphs for indexing patterns from crystals of
tetragonal structure.

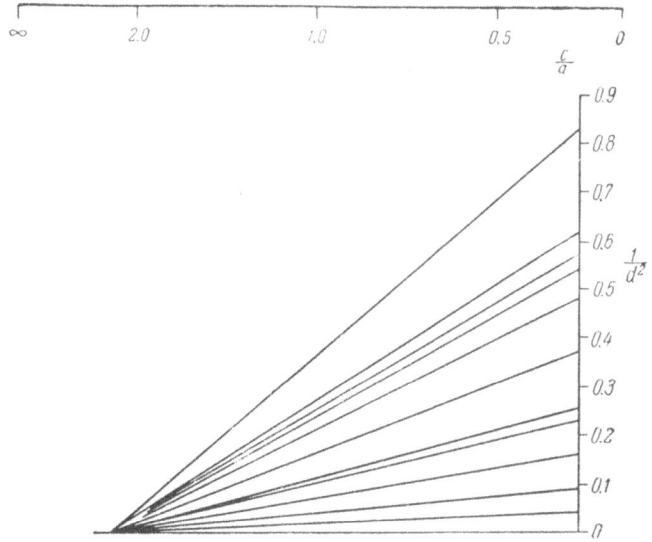

Fig. 63. Diagram for use with the previous figure.

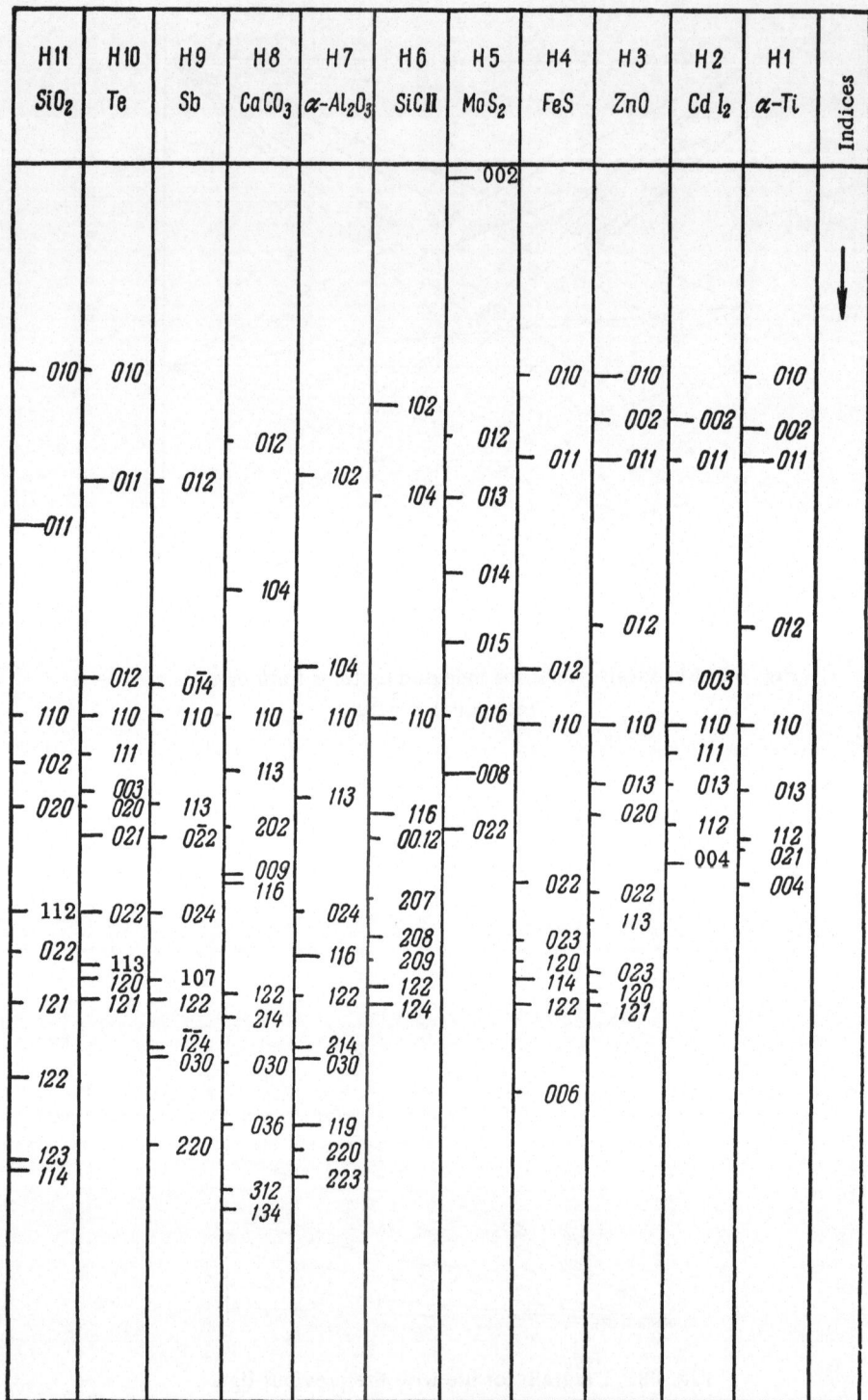

Fig. 64. Schemes for the x-ray patterns of hexagonal crystals (structures H1 to H11).

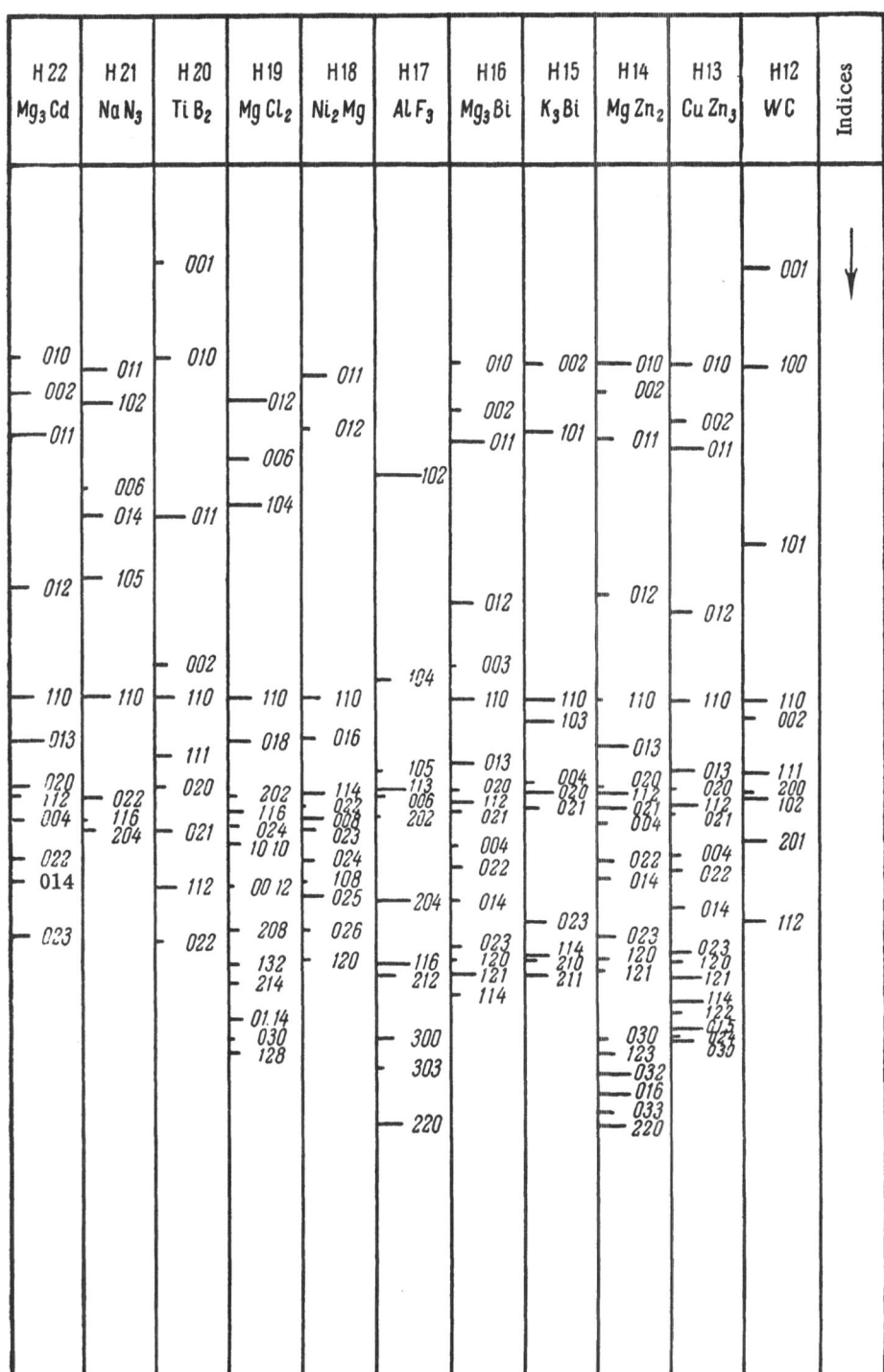

Fig. 65. Schemes for the x-ray patterns of hexagonal crystals (structures H12 to H22).

The dashes correspond to pattern lines, and the lengths are proportional to the intensity. The numbers are the indices.

3-15. Quadratic Forms

The quadratic form for the hexagonal system is

$$Q=\frac{1}{a^2}\left[\frac{4}{3}(h^2+k^2+hk)+\frac{l^2}{\left(\frac{c}{a}\right)^2}\right]=\frac{1}{a^2}\left[\frac{4}{3}s+\frac{l^2}{\left(\frac{c}{a}\right)^2}\right],\qquad(18)$$

where

$$s=h^2+k^2+hk=k^2+i^2+ki=i^2+h^2+ih,\quad i=-(h+k).$$

The table gives s, 4s/3, and the indices corresponding to the quadratic forms for each s [2].

The table covers the range in s from 1 to 999.

s	4s/3	h, k, i	l	s	4s/3	h, k, i	l
1	1.3	1, 0, $\bar{1}$	1	73	97.3	8, 1, $\bar{9}$	1
3	4.0	1, 1, $\bar{2}$	0	75	100.0	5, 5, $\bar{10}$	0
4	5.3	2, 0, $\bar{2}$	2	76	101.3	6, 4, $\bar{10}$	2
7	9.3	2, 1, $\bar{3}$	1	79	105.3	7, 3, $\bar{10}$	1
9	12.0	3, 0, $\bar{3}$	0	81	108.0	9, 0, $\bar{9}$	0
12	16.0	2, 2, $\bar{4}$	0	84	112.0	8, 2, $\bar{10}$	0
13	17.3	·3, 1, $\bar{4}$	2	91	121.3	6, 5, $\bar{11}$	1
16	21.3	4, 0, $\bar{4}$	1	91	121.3	9, 1, $\bar{10}$	2
19	25.3	3, 2, $\bar{5}$	1	93	124.0	7, 4, $\bar{11}$	0
21	28.0	4, 1, $\bar{5}$	0	97	129.3	8, 3, $\bar{11}$	2
25	33.3	5, 0, $\bar{5}$	2	100	133.3	10, 0, $\bar{10}$	1
27	36.0	3, 3, $\bar{6}$	0	103	137.3	9, 2, $\bar{11}$	1
28	37.3	4, 2, $\bar{6}$	2	108	144.0	6, 6, $\bar{12}$	0
31	41.3	5, 1, $\bar{6}$	1	109	145.3	7, 5, $\bar{12}$	2
36	48.0	6, 0, $\bar{6}$	0	111	148.0	10, 1, $\bar{11}$	0
37	49.3	4, 3, $\bar{7}$	1	112	149.3	8, 4, $\bar{12}$	1
39	52.0	5, 2, $\bar{7}$	0	117	156.0	9, 3, $\bar{12}$	0
43	57.3	6, 1, $\bar{7}$	2	121	161.3	11, 0, $\bar{11}$	2
48	64.0	4, 4, $\bar{8}$	0	124	165.3	10, 2, $\bar{12}$	2
49	65.3	7, 0, $\bar{7}$	1	127	169.3	7, 6, $\bar{13}$	1
49	65.3	5, 3, $\bar{8}$	2	129	172.0	8, 5, $\bar{13}$	0
52	69.3	6, 2, $\bar{8}$	1	133	177.3	11, 1, $\bar{12}$	1
57	76.0	7, 1, $\bar{8}$	0	133	177.3	9, 4, $\bar{13}$	2
61	81.3	5, 4, $\bar{9}$	1	139	185.3	10, 3, $\bar{13}$	1
63	84.0	6, 3, $\bar{9}$	0	144	192.0	12, 0, $\bar{12}$	0
64	85.3	8, 0, $\bar{8}$	2	147	196.0	11, 2, $\bar{13}$	0
67	89.3	7, 2, $\bar{9}$	2	147	196.0	7, 7, $\bar{14}$	0
				148	197.3	8, 6, $\bar{14}$	2

s	$4s/3$	$h,\quad k,\quad i$	l	s	$4s/3$	$h,\quad k,\quad i$	l
151	201.3	9, 5, $\overline{14}$	1	291	388.0	14, 5, $\overline{19}$	0
156	208.0	10, 4, $\overline{14}$	0	292	389.3	16, 2, $\overline{18}$	2
157	209.3	12, 1, $\overline{13}$	2	300	400.0	10, 10, $\overline{20}$	0
163	217.3	11, 3, $\overline{14}$	2	301	401.3	15, 4, $\overline{19}$	2
169	225.3	13, 0, $\overline{13}$	1	301	401.3	11, 9, $\overline{20}$	2
169	225.3	8, 7, $\overline{15}$	1	304	405.3	12, 8, $\overline{20}$	1
171	228.0	9, 6, $\overline{15}$	0	307	409.3	17, 1, $\overline{18}$	1
172	229.3	12, 2, $\overline{14}$	1	309	412.0	13, 7, 20	0
175	233.3	10, 5, $\overline{15}$	2	313	417.3	16, 3, $\overline{19}$	1
181	241.3	11, 4, $\overline{15}$	1	316	421.3	14, 6, $\overline{20}$	2
183	244.0	13, 1, $\overline{14}$	0	324	432.0	18, 0, $\overline{18}$	0
189	252.0	12, 3, $\overline{15}$	0	325	433.3	15, 5, $\overline{20}$	1
192	256.0	8, 8, $\overline{16}$	0	327	436.0	17, 2, $\overline{19}$	0
193	257.3	9, 7, $\overline{16}$	2	331	441.3	11, 10, $\overline{21}$	1
196	261.3	14, 0, $\overline{14}$	2	333	444.0	12, 9, $\overline{21}$	0
196	261.3	10, 6, $\overline{16}$	1	336	448.0	16, 4, $\overline{20}$	0
199	265.3	13, 2, $\overline{15}$	2	337	449.3	13, 8, $\overline{21}$	2
201	268.0	11, 5, $\overline{16}$	0	343	457.3	18, 1, $\overline{19}$	2
208	277.3	12, 4, $\overline{16}$	2	343	457.3	14, 7, $\overline{21}$	1
211	281.3	14, 1, $\overline{15}$	1	349	465.3	17, 3, 20	2
217	289.3	13, 3, $\overline{16}$	1	351	468.0	15, 6, $\overline{21}$	0
217	289.3	9, 8, $\overline{17}$	1	361	481.3	19, 0, $\overline{19}$	1
219	292.0	10, 7, $\overline{17}$	0	361	481.3	16, 5, $\overline{21}$	2
223	297.3	11, 6, $\overline{17}$	2	363	484.0	11, 11, $\overline{22}$	0
225	300.0	15, 0, $\overline{15}$	0	364	485.3	18, 2, $\overline{20}$	1
228	304.0	14, 2, $\overline{16}$	0	364	485.3	12, 10, $\overline{22}$	2
229	305.3	12, 5, $\overline{17}$	1	367	489.3	13, 9, $\overline{22}$	1
237	316.0	13, 4, $\overline{17}$	0	372	496.0	14, 8, $\overline{22}$	0
241	321.3	15, 1, $\overline{16}$	2	373	497.3	17, 4, $\overline{21}$	1
243	324.0	9, 9, $\overline{18}$	0	379	505.3	15, 7, $\overline{22}$	2
244	325.3	10, 8, $\overline{18}$	2	381	508.0	19, 1, $\overline{20}$	0
247	329.3	14, 3, $\overline{17}$	2	387	516.0	18, 3, $\overline{21}$	0
247	329.3	11, 7, $\overline{18}$	1	388	517.3	16, 6, $\overline{22}$	1
252	336.0	12, 6, $\overline{18}$	0	397	529.3	12, 11, $\overline{23}$	1
256	341.3	16, 0, $\overline{16}$	1	399	532.0	17, 5, $\overline{22}$	0
259	345.3	15, 2, $\overline{17}$	1	399	532.0	13, 10, $\overline{23}$	0
259	345.3	13, 5, $\overline{18}$	2	400	533.3	20, 0, $\overline{20}$	2
268	357.3	14, 4, $\overline{18}$	1	403	537.3	19, 2, $\overline{21}$	2
271	361.3	10, 9, $\overline{19}$	1	403	537.3	14, 9, $\overline{23}$	2
273	364.0	16, 1, $\overline{17}$	0	409	545.3	15, 8, $\overline{23}$	1
273	364.0	11, 8, $\overline{19}$	0	412	549.3	18, 4, $\overline{22}$	2
277	369.3	12, 7, $\overline{19}$	2	417	556.0	16, 7, $\overline{23}$	0
279	372.0	15, 3, $\overline{18}$	0	421	561.3	20, 1, $\overline{21}$	1
283	377.3	13, 6, $\overline{19}$	1	427	569.3	19, 3, $\overline{22}$	1
289	385.3	17, 0, $\overline{17}$	2	427	569.3	17, 6, $\overline{23}$	2

s	4s/3	h, k, i	l	s	4s/3	h, k, i	l
432	576.0	12, 12, $\overline{24}$	0	571	761.3	21, 5, $\overline{26}$	1
433	577.3	13, 11, $\overline{24}$	2	576	768.0	24, 0, $\overline{24}$	0
436	581.3	14, 10, $\overline{24}$	1	577	769.3	19, 8, $\overline{27}$	2
439	585.3	18, 5, $\overline{23}$	1	579	772.0	23, 2, $\overline{25}$	0
441	588.0	21, 0, $\overline{21}$	0	588	784.0	22, 4, $\overline{26}$	0
441	588.0	15, 9, $\overline{24}$	0	588	784.0	14, 14, $\overline{28}$	0
444	592.0	20, 2, $\overline{22}$	0	589	785.3	20, 7, $\overline{27}$	1
448	597.3	16, 8, $\overline{24}$	2	589	785.3	15, 13, $\overline{28}$	2
453	604.0	19, 4, $\overline{23}$	0	592	789.3	16, 12, $\overline{28}$	1
457	609.3	17, 7, $\overline{24}$	1	597	796.0	17, 11, $\overline{28}$	0
463	617.3	21, 1, $\overline{22}$	2	601	801.3	24, 1, $\overline{25}$	2
468	624.0	18, 6, $\overline{24}$	0	603	804.0	21, 6, $\overline{27}$	0
469	625.3	20, 3, $\overline{23}$	2	604	805.3	18, 10, $\overline{28}$	2
469	625.3	13, 12, $\overline{25}$	1	607	809.3	23, 3, $\overline{26}$	2
471	628.0	14, 11, $\overline{25}$	0	613	817.3	19, 9, $\overline{28}$	1
475	633.3	15, 10, $\overline{25}$	2	619	825.3	22, 5, $\overline{27}$	2
481	641.3	19, 5, $\overline{24}$	2	624	832.0	20, 8, $\overline{28}$	0
481	641.3	16, 9, $\overline{25}$	1	625	833.3	25, 0, $\overline{25}$	1
484	645.3	22, 0, $\overline{22}$	1	628	837.3	24, 2, $\overline{26}$	1
487	649.3	21, 2, $\overline{23}$	1	631	841.3	15, 14, $\overline{29}$	1
489	652.0	17, 8, $\overline{25}$	0	633	844.0	16, 13, $\overline{29}$	0
496	661.3	20, 4, $\overline{24}$	1	637	849.3	23, 4, $\overline{27}$	1
499	665.3	18, 7, $\overline{25}$	2	637	849.3	21, 7, $\overline{28}$	2
507	676.0	22, 1, $\overline{23}$	0	637	849.3	17, 12, $\overline{29}$	2
507	676.0	13, 13, $\overline{26}$	0	643	857.3	18, 11, $\overline{29}$	1
508	677.3	14, 12, $\overline{26}$	2	651	868.0	25, 1, $\overline{26}$	0
511	681.3	19, 6, $\overline{25}$	1	651	868.0	19, 10, $\overline{29}$	0
511	681.3	15, 11, $\overline{26}$	1	652	869.3	22, 6, $\overline{28}$	1
513	684.0	21, 3, $\overline{24}$	0	657	876.0	24, 3, $\overline{27}$	0
516	688.0	16, 10, $\overline{26}$	0	661	881.3	20, 9, $\overline{29}$	2
523	697.3	17, 9, $\overline{26}$	2	669	892.0	23, 5, $\overline{28}$	0
525	700.0	20, 5, $\overline{25}$	0	673	897.3	21, 8, $\overline{29}$	1
529	705.3	23, 0, $\overline{23}$	2	675	900.0	15, 15, $\overline{30}$	0
532	709.3	22, 2, $\overline{24}$	2	676	901.3	26, 0, $\overline{26}$	2
532	709.3	18, 8, $\overline{26}$	1	676	901.3	16, 14, $\overline{30}$	2
541	721.3	21, 4, $\overline{25}$	2	679	905.3	25, 2, $\overline{27}$	2
543	724.0	19, 7, $\overline{26}$	0	679	905.3	17, 13, $\overline{30}$	1
547	729.3	14, 13, $\overline{27}$	1	684	912.0	18, 12, $\overline{30}$	0
549	732.0	15, 12, $\overline{27}$	0	687	916.0	22, 7, $\overline{29}$	0
553	737.3	23, 1, $\overline{24}$	1	688	917.3	24, 4, $\overline{28}$	2
553	737.3	16, 11, $\overline{27}$	2	691	921.3	19, 11, $\overline{30}$	2
556	741.3	20, 6, $\overline{26}$	2	700	933.3	20, 10, $\overline{30}$	1
559	745.3	22, 3, $\overline{25}$	1	703	937.3	26, 1, $\overline{27}$	1
559	745.3	17, 10, $\overline{27}$	1	703	937.3	23, 6, $\overline{29}$	2
567	756.0	18, 9, $\overline{27}$	0	709	945.3	25, 3, $\overline{28}$	1
				711	948.0	21, 9, $\overline{30}$	0

s	$4s/3$	$h,\ k,\ i$	l	s	$4s/3$	$h,\ k,\ i$	l
721	961.3	24, 5, $\overline{29}$	1	867	1156.0	17, 17, $\overline{34}$	0
721	961.3	16, 15, $\overline{31}$	1	868	1157.3	26, 6, $\overline{32}$	2
723	964.0	17, 14, $\overline{31}$	0	868	1157.3	18, 16, $\overline{34}$	2
724	965.3	22, 8, $\overline{30}$	2	871	1161.3	29, 1, $\overline{30}$	1
727	969.3	18, 13, $\overline{31}$	2	871	1161.3	19, 15, $\overline{34}$	1
729	972.0	27, 0, $\overline{27}$	0	873	1164.0	24, 9, $\overline{33}$	0
732	976.0	26, 2, $\overline{28}$	0	876	1168.0	20, 14, $\overline{34}$	0
733	977.3	19, 12, $\overline{31}$	1	877	1169.3	28, 3, $\overline{31}$	1
739	985.3	23, 7, $\overline{30}$	1	883	1177.3	21, 13, $\overline{34}$	2
741	988.0	25, 4, $\overline{29}$	0	889	1185.3	27, 5, $\overline{32}$	1
741	988.0	20, 11, $\overline{31}$	0	889	1185.3	25, 8, $\overline{33}$	2
751	1001.3	21, 10, $\overline{31}$	2	892	1189.3	22, 12, $\overline{34}$	1
756	1008.0	24, 6, $\overline{30}$	0	900	1200.0	30, 0, $\overline{30}$	0
757	1009.3	27, 1, $\overline{28}$	2	903	1204.0	29, 2, $\overline{31}$	0
763	1017.3	26, 3, $\overline{29}$	2	903	1204.0	23, 11, $\overline{34}$	0
763	1017.3	22, 9, $\overline{31}$	1	907	1209.3	26, 7, $\overline{33}$	1
768	1024.0	16, 16, $\overline{32}$	0	912	1216.0	28, 4, $\overline{32}$	0
769	1025.3	17, 15, $\overline{32}$	2	916	1221.3	24, 10, $\overline{34}$	2
772	1029.3	18, 14, $\overline{32}$	1	919	1225.3	18, 17, $\overline{35}$	1
775	1033.3	25, 5, $\overline{30}$	2	921	1228.0	19, 16, $\overline{35}$	0
777	1036.0	23, 8, $\overline{31}$	0	925	1233.3	20, 15, $\overline{35}$	2
777	1036.0	19, 13, $\overline{32}$	0	927	1236.0	27, 6, $\overline{33}$	0
784	1045.3	28, 0, $\overline{28}$	1	931	1241.3	30, 1, $\overline{31}$	2
784	1045.3	20, 12, $\overline{32}$	2	931	1241.3	25, 9, $\overline{34}$	1
787	1049.3	27, 2, $\overline{29}$	1	931	1241.3	21, 14, $\overline{35}$	1
793	1057.3	24, 7, $\overline{31}$	2	937	1249.3	29, 3, $\overline{32}$	2
793	1057.3	21, 11, $\overline{32}$	1	939	1252.0	22, 13, $\overline{35}$	0
796	1061.3	26, 4, $\overline{30}$	1	948	1264.0	26, 8, $\overline{34}$	0
804	1072.0	22, 10, $\overline{32}$	0	949	1265.3	28, 5, $\overline{33}$	2
811	1081.3	25, 6, $\overline{31}$	1	949	1265.3	23, 12, $\overline{35}$	2
813	1084.0	28, 1, $\overline{29}$	0	961	1281.3	31, 0, $\overline{31}$	1
817	1089.3	23, 9, $\overline{32}$	2	961	1281.3	24, 11, $\overline{35}$	1
817	1089.3	17, 16, $\overline{33}$	1	964	1285.3	30, 2, $\overline{32}$	1
819	1092.0	27, 3, $\overline{30}$	0	967	1289.3	27, 7, $\overline{34}$	2
819	1092.0	18, 15, $\overline{33}$	0	972	1296.0	18, 18, $\overline{36}$	0
823	1097.3	19, 14, $\overline{33}$	2	973	1297.3	29, 4, $\overline{33}$	1
829	1105.3	20, 13, $\overline{33}$	1	973	1297.3	19, 17, $\overline{36}$	2
831	1108.0	26, 5, $\overline{31}$	0	975	1300.0	25, 10, $\overline{35}$	0
832	1109.3	24, 8, $\overline{32}$	1	976	1301.3	20, 16, $\overline{36}$	1
837	1116.0	21, 12, $\overline{33}$	0	981	1308.0	21, 15, $\overline{36}$	0
841	1121.3	29, 0, $\overline{29}$	2	988	1317.3	28, 6, $\overline{34}$	1
844	1125.3	28, 2, $\overline{30}$	2	988	1317.3	22, 14, $\overline{36}$	2
847	1129.3	22, 11, $\overline{33}$	2	991	1321.3	26, 9, $\overline{35}$	2
849	1132.0	25, 7, $\overline{32}$	0	993	1324.0	31, 1, $\overline{32}$	0
853	1137.3	27, 4, $\overline{31}$	2	997	1329.3	23, 13, $\overline{36}$	1
859	1145.3	23, 10, $\overline{33}$	1	999	1332.0	30, 3, $\overline{33}$	0

3-16. Analytic Method of Indexing for the Hexagonal and Trigonal Systems

The method of differences can be used for the hexagonal system; the relation of $\sin^2 \vartheta$ to the indices is

$$\sin^2 \vartheta_{hkl} = A(h^2 + hk + k^2) + Cl^2, \tag{19}$$

where $A = \lambda^2/3a^2$ and $C = \lambda^2/4c^2$.

For reflections of (hk0) type we have

$$\sin^2 \vartheta_{100} = A; \quad \sin^2 \vartheta_{110} = 3A; \quad \sin^2 \vartheta_{200} = 4A;$$

$$\sin^2 \vartheta_{210} = 7A; \quad \sin^2 \vartheta_{3)0} = 9A; \quad \sin^2 \vartheta_{220} = 12A, \text{etc.}$$

The ratio of the $\sin^2 \vartheta$ is 3 for several line pairs, which does not occur in the tetragonal system; this enables one to deduce that the substance is hexagonal.

Equation (19) applies also to trigonal crystals; relatively few substances are trigonal, so compliance with (19) primarily shows that the substance is hexagonal.

A point to note here is that certain carbides (including $Cr_{23}C_6$) are trigonal; the trigonal system can be distinguished from the hexagonal by the absence of certain lines that should be present for the hexagonal system.

The systematic absences for this case are to be found in [87].

The methods of handling the differences in the $\sin^2 \vartheta$ and of indexing are as for the tetragonal system.

3-17. Graphs for Indexing X-Ray Patterns

The relation of d_{hkl} to hkl may be written as

$$d_{hkl} = \frac{1}{\sqrt{\dfrac{4}{3}\left(\dfrac{h^2 + hk + k^2}{a^2} + \dfrac{l^2}{c^2}\right)}} \tag{20}$$

or

$$\frac{1}{d_{hkl}^2} = \frac{4}{3}(h^2 + hk + k^2)\frac{1}{a^2} + l^2 \frac{1}{c^2}.$$

Graphs representing these relations have been constructed as follows: Figures 66 and 67, for the simple hexagonal lattice for c/a (ordinate) from 0 to 3.6 against lg d; Figs. 68-70, for the hexagonal close-packed lattice for c/a from 0 to 5.4 (same coordinates); Fig. 71, for this same lattice for a/c from 0 to 0.9 (c/a from 1.1 to 20) against ϑ, $\sin^2 \vartheta$, and $\lambda^2/4$ [273].

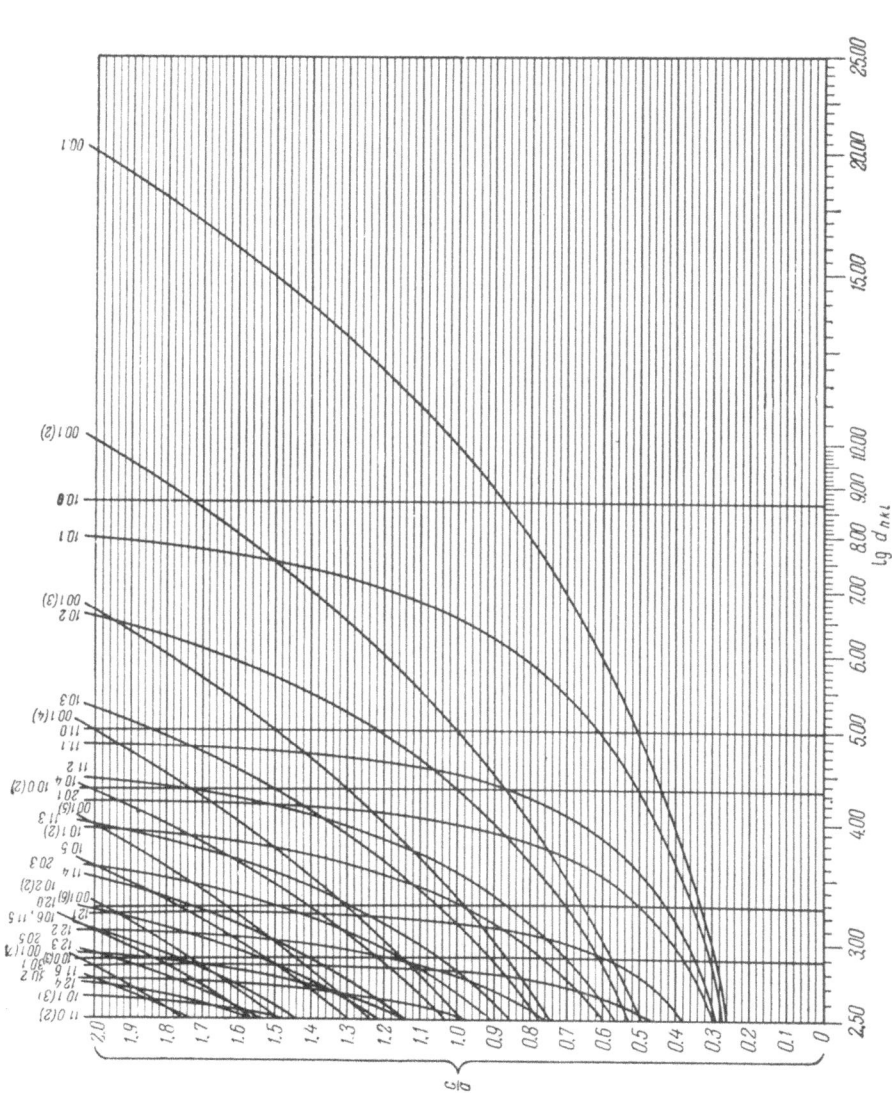

Fig. 66. Graphs for indexing patterns from crystals of simple hexagonal structure (c/a from 0 to 2.0).

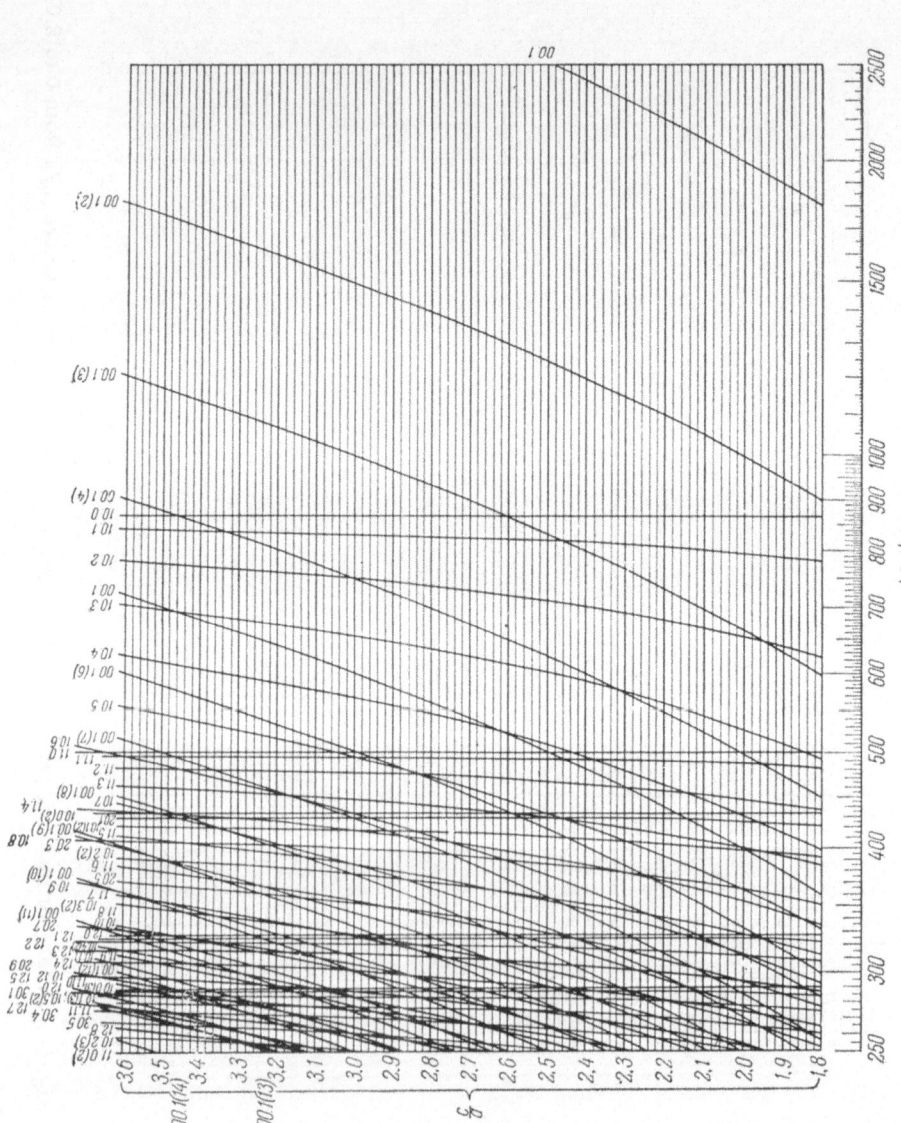

Fig. 67. Graphs for indexing patterns from crystals of simple hexagonal structure (c/a from 1.8 to 3.6).

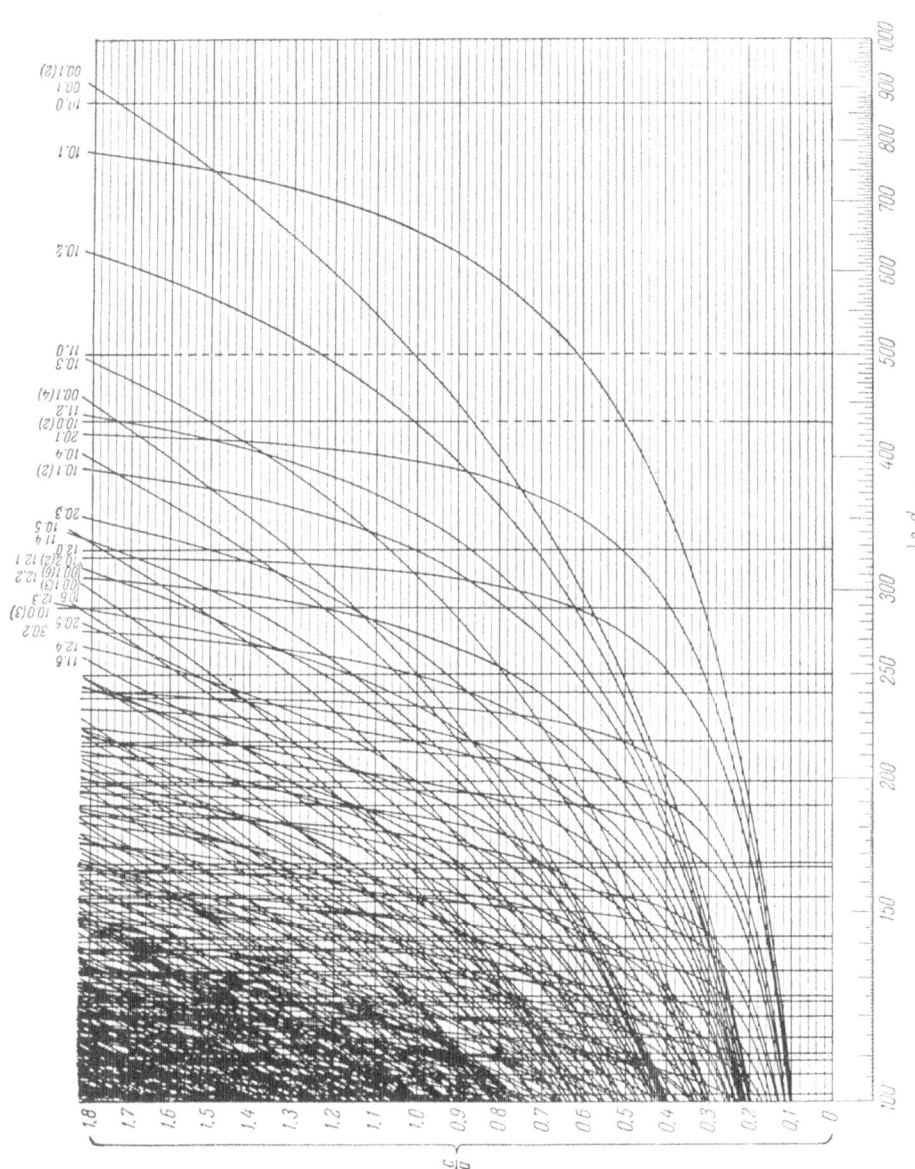

Fig. 68. Graphs for indexing patterns from crystals of close-packed hexagonal structure (c/a from 0 to 1.8).

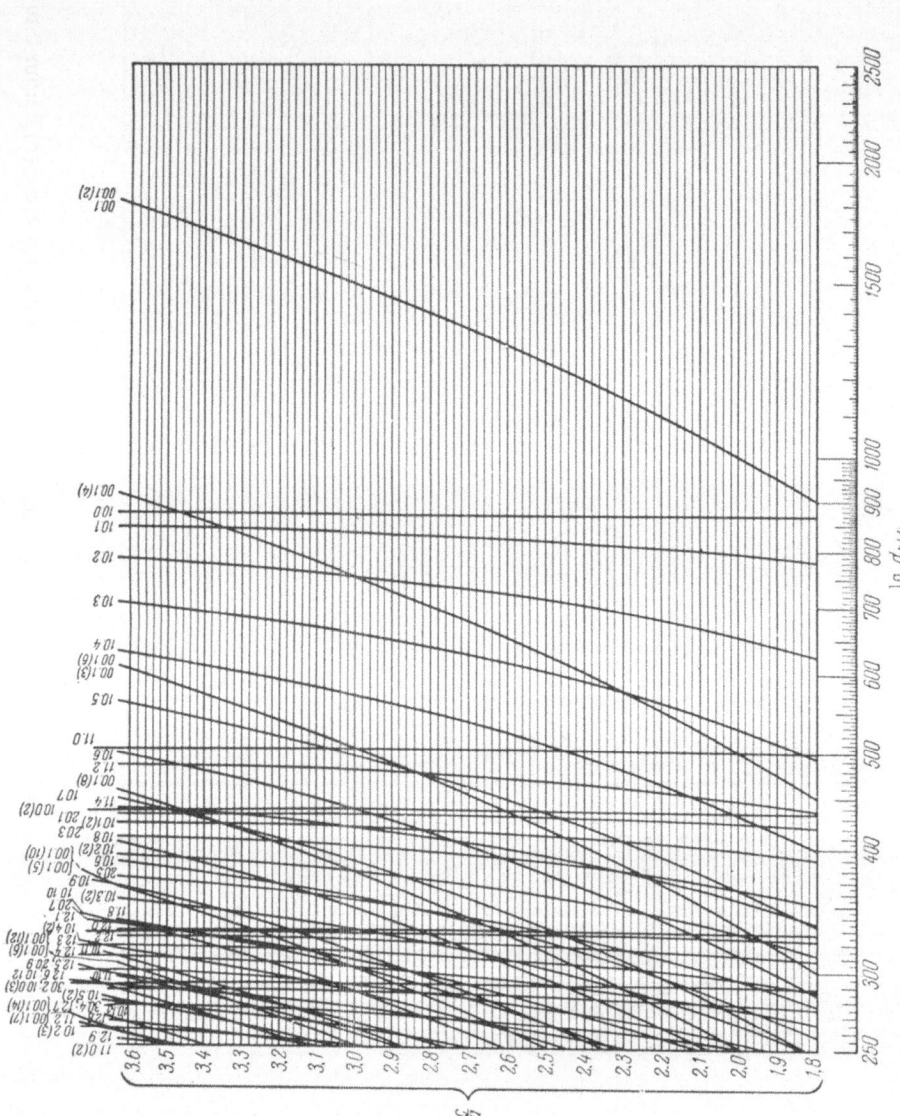

Fig. 69. Graphs for indexing patterns from crystals of close-packed hexagonal structure (c/a from 1.8 to 3.6).

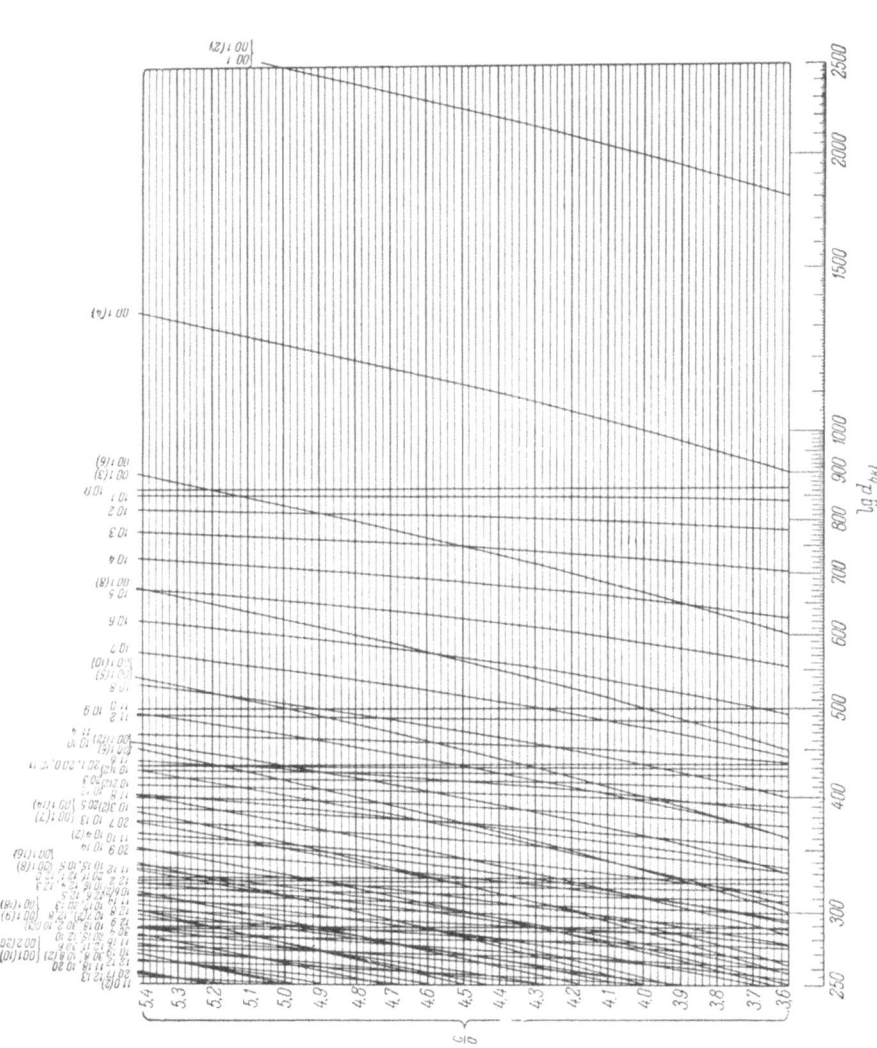

Fig. 70. Graphs for indexing patterns from crystals of close-packed hexagonal structure (c/a from 3.6 to 5.4).

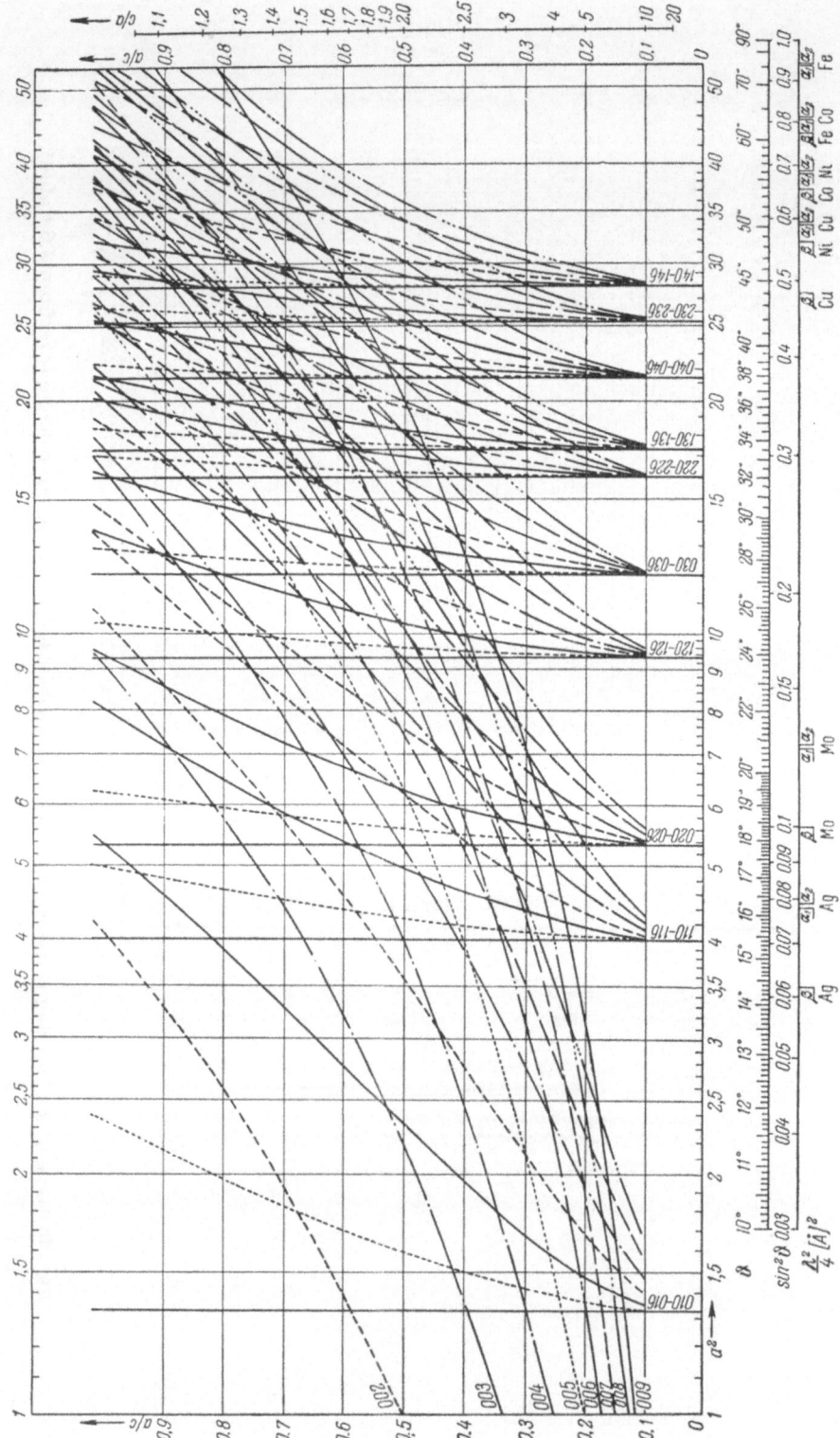

Fig. 71. Schwarz and Summa graphs for indexing patterns from hexagonal crystals.

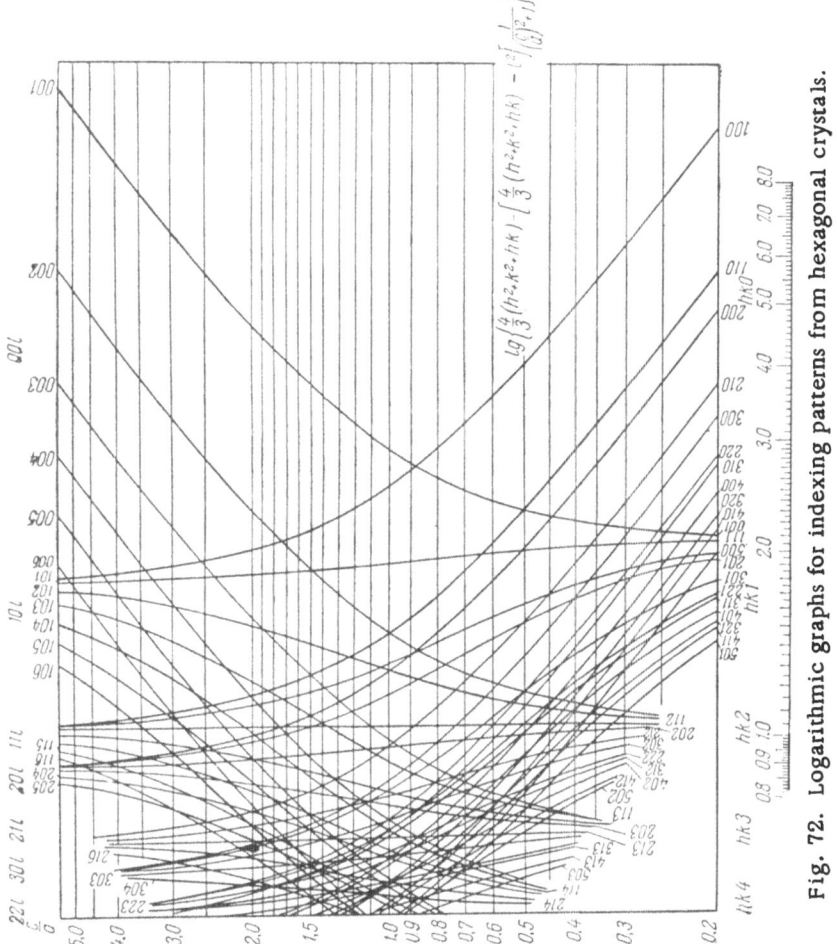

Fig. 72. Logarithmic graphs for indexing patterns from hexagonal crystals.

Figures 66-71 are analogous to those for the tetragonal system and are used in the same way.

They are also applicable to indexing trigonal substances in the hexagonal system of indices; the pattern from a trigonal substance lacks lines satisfying

$$-h+k+l=3n,$$
$$n=0,\ 1,\ 2,\ 3.$$

Various types of graphs on logarithmic scales are used to index reflections with high indices and in certain other cases.

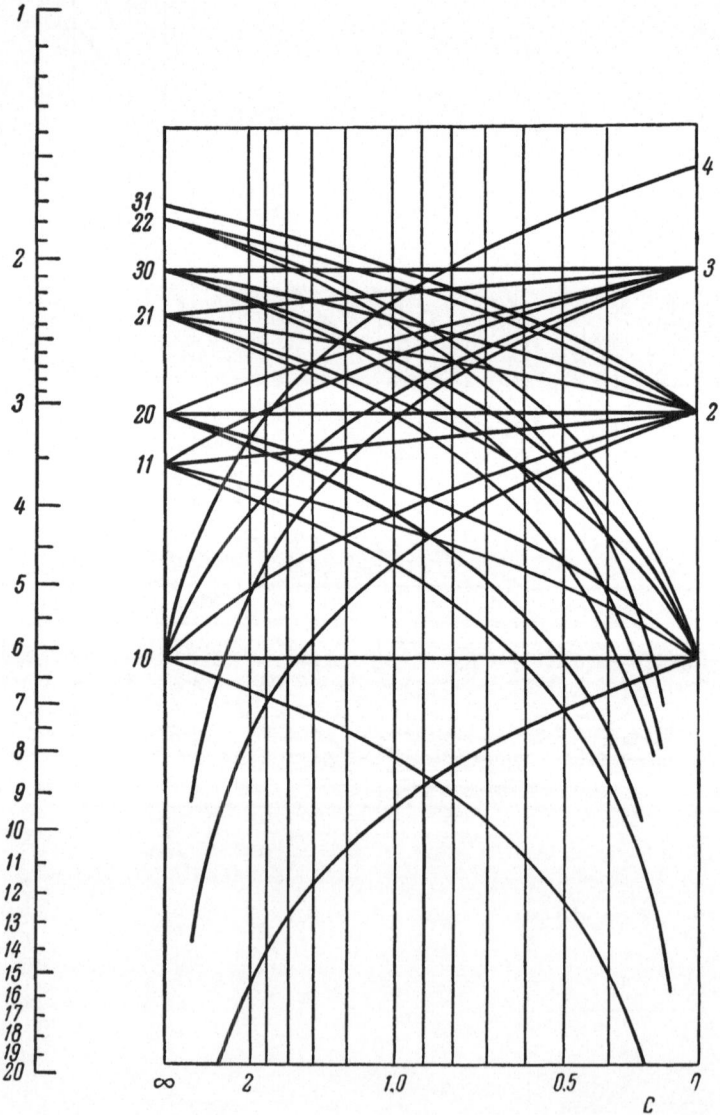

Fig. 73. Bond graphs for indexing patterns for hexagonal crystals.

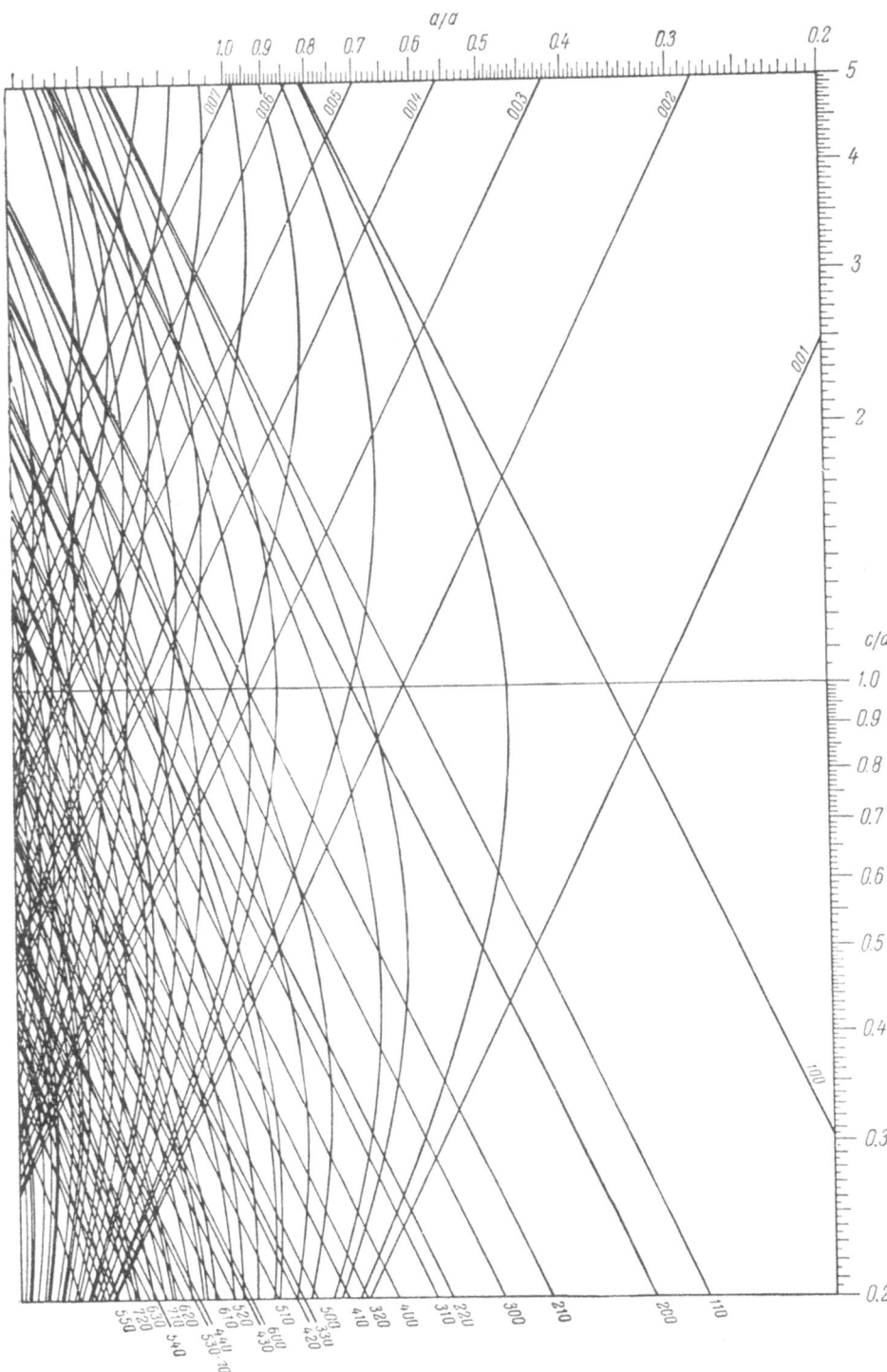

Fig. 74. Harrington graphs for indexing patterns from hexagonal crystals.

Figures 72-74 [91, 270] are as for the tetragonal case; they are arranged in order of increasing highest indices. Figure 75 is from [12].

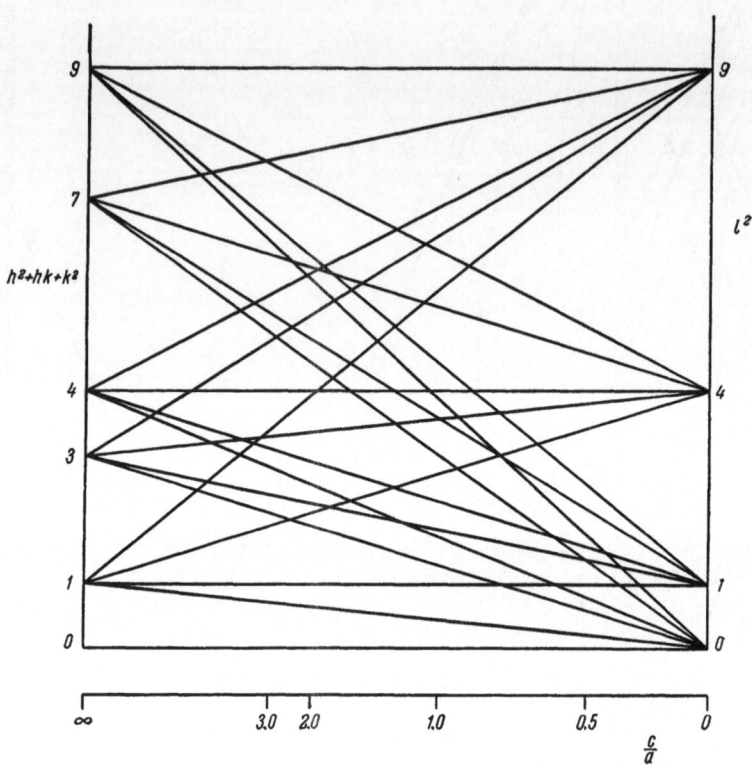

Fig. 75. Bjurström graphs for indexing patterns from crystals of hexagonal structure.

The methods of using these are as described for the tetragonal system in Section 3-9.

3-18. Relations Between Indices in Indexing for the Hexagonal, Trigonal, and Orthohexagonal Systems

Crystals belonging to the hexagonal system can be indexed by means of a set of four indices; the unit cell then has the shape shown in Fig. 76.

Three-place indexing is also possible for the hexagonal (Fig. 76a), orthohexagonal (Fig. 76c),.and trigonal (Fig. 76b) systems. Figure 76d shows the hexagonal and trigonal (rhombohedral) cells together.

The following relationships apply between the lattice constants in the various methods of indexing (the subscripts are H, hexagonal; O, orthohexagonal; R, rhombohedral or trigonal):

$$a_O = a_H \sqrt{3}; \quad a_H = 2a_R \frac{\sin \alpha_R}{2}; \quad \frac{c_H}{a_H} = \sqrt{\frac{9}{4 \sin^2 \frac{\alpha_R}{2}} - 3};$$

$$b_O = b_H; \quad b_H = 2a_R \frac{\sin \alpha_R}{2};$$

$$c_O = c_H; \quad c_H = a_R \sqrt{\frac{9 - 12 \sin^2 \alpha_R}{2}}; \quad \sin \frac{\alpha_R}{2} = \frac{3}{2\sqrt{3 + \left(\frac{c}{a}\right)^2}}.$$

The table provides means of passing from the four-place system to the others [109]. Values are given for various mutual positions of the axes.

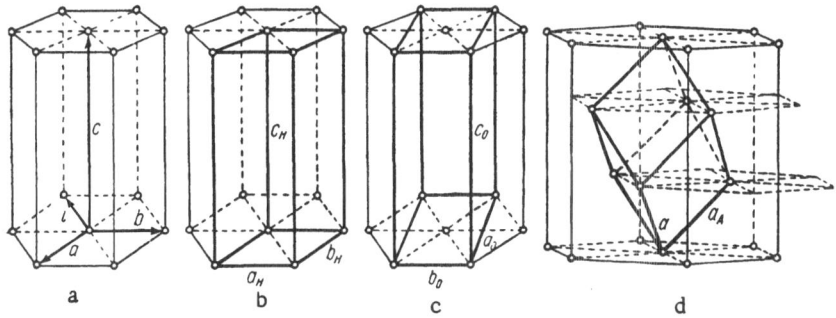

Fig. 76. Unit cells in the hexagonal system. (a) Four-place indexing; (b) three-place indexing; (c) orthohexagonal three-place indexing; (d) hexagonal and trigonal cells.

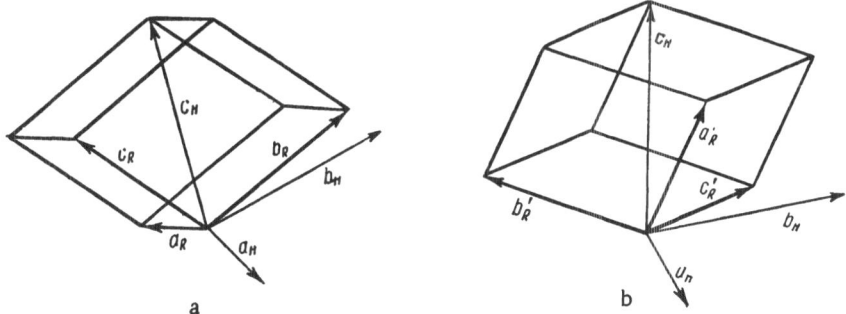

Fig. 77. Conversion from hexagonal axes to trigonal ones.

Two dispositions of the axes are possible in the transition from hexagonal to trigonal (Fig. 77).

Case I (Fig. 77a):

$$3h_R = h_H - k_H + l_H; \quad 3k_R = k_H - i_H + l_H; \quad 3l_R = i_H - h_H + l_H,$$

where $i_H = -(h_H + k_H)$.

Case II (Fig. 77b):

$$3h'_R = -h_H + k_H + l_H; \quad 3k'_R = -k_H + i_H + l_H; \quad 3l'_R = -i_H + h_H + l_H.$$

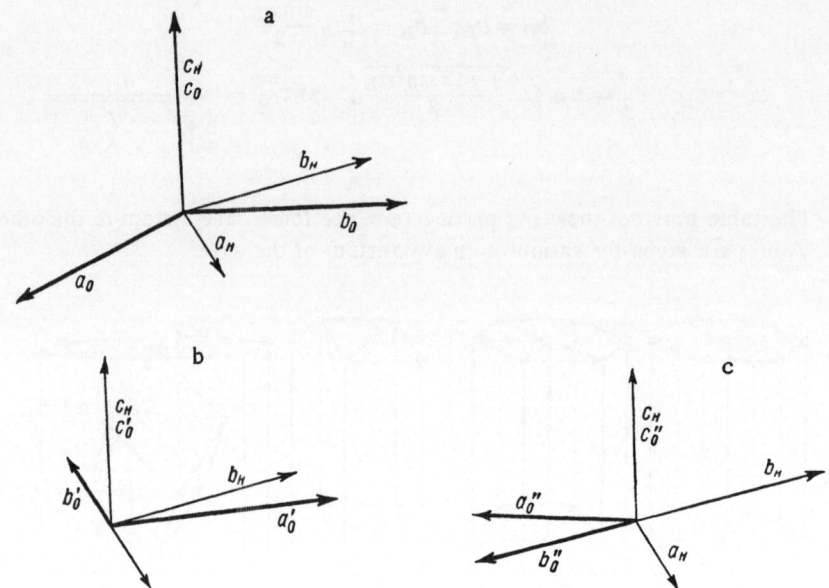

Fig. 78. Conversion from hexagonal axes to orthohexagonal ones.

Three dispositions of the axes are possible in the transition from hexagonal to ortho-hexagonal (Fig. 78).

Case A (Fig. 78a):

$$h_O = h_H - k_H; \quad k_O = h_H + k_H; \quad l_O = l_H.$$

Case B (Fig. 78b):

$$h'_O = k_H - i_H; \quad k'_O = k_H + i_H; \quad l'_O = l_H.$$

Case C (Fig. 78c):

$$h''_O = i_H - h_H; \quad k''_O = i_H + h_H; \quad l''_O = l_H.$$

Hexagonal $h_H k_H i_H l_H$	Trigonal I $3h_R \; 3k_R \; 3l_R$			Trigonal II $3h'_R \; 3k'_R \; 3l'_R$			Orthohexagonal A $h_O \; k_O \; l_O$		Orthohexagonal B $h'_O \; k'_O \; l'_O$		Orthohexagonal C $h''_O \; k''_O \; l''_O$	
$1, 0, \bar{1}, 0$	$1,$	$1,$	$\bar{2}$	$\bar{1},$	$\bar{1},$	2	$1, 1, 0$		$1, \bar{1}, 0$		$\bar{2}, 0, 0$	
$1, 0, \bar{1}, l$	$1+l,$	$1+l,$	$\bar{2}+l$	$l-1,$	$l-1,$	$2+l$	$1, 1, l$		$1, \bar{1}, l$		$\bar{2}, 0, l$	
$2, 0, \bar{2}, 0$	$2,$	$2,$	$\bar{4}$	$\bar{2},$	$\bar{2},$	4	$2, 2, 0$		$2, \bar{2}, 0$		$\bar{4}, 0, 0$	
$2, 0, \bar{2}, l$	$2+l,$	$2+l,$	$l-4$	$l-2,$	$l-2,$	$4+l$	$2, 2, l$		$2, \bar{2}, l$		$\bar{4}, 0, l$	
$3, 0, \bar{3}, 0$	$3,$	$3,$	$\bar{6}$	$\bar{3},$	$\bar{3},$	6	$3, 3, 0$		$3, \bar{3}, 0$		$\bar{6}, 0, 0$	
$3, 0, \bar{3}, l$	$3+l,$	$3+l,$	$l-6$	$l-3,$	$l-3,$	$6+l$	$3, 3, l$		$3, \bar{3}, l$		$\bar{6}, 0, l$	
$4, 0, \bar{4}, 0$	$\bar{4},$	$\bar{4},$	8	$\bar{4},$	$\bar{4},$	8	$4, 4, 0$		$4, \bar{4}, 0$		$\bar{8}, 0, 0$	
$4, 0, \bar{4}, l$	$4+l,$	$4+l,$	$l-8$	$l-4,$	$l-4,$	$8+l$	$4, 4, l$		$4, \bar{4}, l$		$\bar{8}, 0, l$	
$5, 0, \bar{5}, 0$	$5,$	$5,$	$\overline{10}$	$\bar{5},$	$\bar{5},$	10	$5, 5, 0$		$5, \bar{5}, 0$		$\overline{10}, 0, 0$	
$5, 0, \bar{5}, l$	$5+l,$	$5+l,$	$l-10$	$l-5,$	$l-5,$	$10+l$	$5, 5, l$		$5, \bar{5}, l$		$\overline{10}, 0, l$	
$6, 0, \bar{6}, 0$	$6,$	$6,$	$\overline{12}$	$\bar{6},$	$\bar{6},$	12	$6, 6, 0$		$6, \bar{6}, 0$		$\overline{12}, 0, 0$	
$6, 0, \bar{6}, l$	$6+l,$	$6+l,$	$l-12$	$l-6,$	$l-6,$	$12+l$	$6, 6, l$		$6, \bar{6}, l$		$\overline{12}, 0, l$	

Hexagonal	Trigonal		Orthohexagonal		
$h_H k_H i_H l_H$	I — $3h_R$ $3k_R$ $3l_R$	II — $3h'_R$ $3k'_R$ $3l'_R$	A — $h_O k_O l_O$	B — $h'_O k'_O l'_O$	C — $h''_O k''_O l''_O$
1, 1, $\bar 2$, 0	0, 3, $\bar 3$	0, $\bar 3$, 3	0, 2, 0	3, $\bar 1$, 0	$\bar 3$, $\bar 1$, 0
1, 1, $\bar 2$, l	l, 3+l, $l$$-$3	l, $l$$-$3, l+3	0, 2, l	3, $\bar 1$, l	$\bar 3$, $\bar 1$, l
1, 2, $\bar 3$, 0	$\bar 1$, 5, $\bar 4$	1, $\bar 5$, 4	$\bar 1$, 3, 0	5, $\bar 1$, 0	$\bar 4$, $\bar 2$, 0
1, 2, $\bar 3$, l	$l$$-$1, l+5, $l$$-$4	l+1, $l$$-$5, l+4	$\bar 1$, 3, l	5, $\bar 1$, l	$\bar 4$, $\bar 2$, l
1, 3, $\bar 4$, 0	$\bar 2$, 7, $\bar 5$	2, $\bar 7$, 5	$\bar 2$, 4, 0	7, $\bar 1$, 0	$\bar 5$, $\bar 3$, 0
1, 3, $\bar 4$, l	$l$$-$2, l+7, $l$$-$5	l+2, $l$$-$7, l+5	2, 4, l	7, $\bar 1$, l	$\bar 5$, $\bar 3$, l
1, 4, $\bar 5$, 0	3, $\bar 9$, 6	3, $\bar 9$, 6	$\bar 3$, 5, 0	9, $\bar 1$, 0	$\bar 6$, $\bar 4$, 0
1, 4, $\bar 5$, l	$l$$-$3, l+9, $l$$-$6	l+3, $l$$-$9, l+6	$\bar 3$, 5, l	9, $\bar 1$, l	$\bar 6$, $\bar 4$, l
1, 5, $\bar 6$, 0	$\bar 4$, 11, $\bar 7$	4, $\overline{11}$, 7	$\bar 4$, 6, 0	11, $\bar 1$, 0	$\bar 7$, $\bar 5$, 0
1, 5, $\bar 6$, l	$l$$-$4, l+11, $l$$-$7	l+4, $l$$-$11, l+7	$\bar 4$, 6, l	11, $\bar 1$, l	$\bar 7$, $\bar 5$, l
2, 1, $\bar 3$, 0	1, 4, $\bar 5$	$\bar 1$, $\bar 4$, 5	1, 3, 0	4, $\bar 2$, 0	$\bar 5$, $\bar 1$, 0
2, 1, $\bar 3$, l	l+1, l+4, $l$$-$5	$l$$-$1, $l$$-$4, l+5	1, 3, l	4, $\bar 2$, l	$\bar 5$, $\bar 1$, l
2, 2, $\bar 4$, 0	0, 6, $\bar 6$	0, $\bar 6$, 6	0, 4, 0	6, $\bar 2$, 0	$\bar 6$, $\bar 2$, 0
2, 2, $\bar 4$, l	l, l+6, $l$$-$6	l, $l$$-$6, l+6	0, 4, l	6, $\bar 2$, l	$\bar 6$, $\bar 2$, l
2, 3, $\bar 5$, 0	$\bar 1$, 8, $\bar 7$	1, $\bar 8$, 7	$\bar 1$, 5, 0	8, $\bar 2$, 0	$\bar 7$, $\bar 3$, 0
2, 3, $\bar 5$, l	$l$$-$1, l+8, $l$$-$7	l+1, $l$$-$8, l+7	$\bar 1$, 5, l	8, $\bar 2$, l	$\bar 7$, $\bar 3$, l
2, 4, $\bar 6$, 0	$\bar 2$, 10, $\bar 8$	2, $\overline{10}$, 8	$\bar 2$, 6, 0	10, $\bar 2$, 0	$\bar 8$, $\bar 4$, 0
2, 4, $\bar 6$, l	$l$$-$2, l+10, $l$$-$8	l+2, $l$$-$10, l+8	$\bar 2$, 6, l	10, $\bar 2$, l	$\bar 8$, $\bar 4$, l
3, 1, $\bar 4$, 0	2, 5, $\bar 7$	$\bar 2$, $\bar 5$, 7	2, 4, 0	5, $\bar 3$, 0	$\bar 7$, $\bar 1$, 0
3, 1, $\bar 4$, l	l+2, l+5, $l$$-$7	$l$$-$2, $l$$-$5, l+7	2, 4, l	5, $\bar 3$, l	$\bar 7$, $\bar 1$, l
3, 2, $\bar 5$, 0	1, 7, $\bar 8$	$\bar 1$, $\bar 7$, 8	1, 5, 0	7, $\bar 3$, 0	$\bar 8$, $\bar 2$, 0
3, 2, $\bar 5$, l	l+1, l+7, $l$$-$8	$l$$-$1, $l$$-$7, l+8	1, 5, l	7, $\bar 3$, l	$\bar 8$, $\bar 2$, l
3, 3, $\bar 6$, 0	0, 9, $\bar 9$	0, $\bar 9$, 9	0, 6, 0	9, $\bar 3$, 0	$\bar 9$, $\bar 3$, 0
3, 3, $\bar 6$, l	l, l+9, $l$$-$9	l, $l$$-$9, l+9	0, 6, l	9, $\bar 3$, l	$\bar 9$, $\bar 3$, l
3, 4, $\bar 7$, 0	$\bar 1$, 11, $\overline{10}$	1, $\overline{11}$, 10	$\bar 1$, 7, 0	11, $\bar 3$, 0	$\overline{10}$, $\bar 4$, 0
3, 4, $\bar 7$, l	$l$$-$1, l+11, $l$$-$10	l+1, $l$$-$11, l+10	$\bar 1$, 7, l	11, $\bar 3$, l	$\overline{10}$, $\bar 4$, l
4, 1, $\bar 5$, 0	3, 6, $\bar 9$	$\bar 3$, $\bar 6$, 9	3, 5, 0	6, $\bar 4$, 0	$\bar 9$, $\bar 1$, 0
4, 1, $\bar 5$, l	l+3, l+6, $l$$-$9	$l$$-$3, $l$$-$6, l+9	3, 5, l	6, $\bar 4$, l	$\bar 9$, $\bar 1$, l
4, 2, $\bar 6$, 0	2, 8, $\overline{10}$	$\bar 2$, $\bar 8$, 10	2, 6, 0	8, $\bar 4$, 0	$\overline{10}$, $\bar 2$, 0
4, 2, $\bar 6$, l	l+2, l+8, $l$$-$10	$l$$-$2, $l$$-$8, l+10	2, 6, l	8, $\bar 4$, l	$\overline{10}$, $\bar 2$, l
4, 3, $\bar 7$, 0	1, 10, $\overline{11}$	$\bar 1$, $\overline{10}$, 11	1, 7, 0	10, $\bar 4$, 0	$\overline{11}$, $\bar 3$, 0
4, 3, $\bar 7$, l	l+1, l+10, $l$$-$11	$l$$-$1, $l$$-$10, l+11	1, 7, l	10, $\bar 4$, l	$\overline{11}$, $\bar 3$, l
4, 4, $\bar 8$, 0	0, 12, $\overline{12}$	0, $\overline{12}$, 12	0, 8, 0	12, $\bar 4$, 0	$\overline{12}$, $\bar 4$, 0
4, 4, $\bar 8$, l	l, l+12, $l$$-$12	l, $l$$-$12, l+12	0, 8, l	12, $\bar 4$, l	$\overline{12}$, $\bar 4$, l
5, 1, $\bar 6$, 0	4, 7, $\overline{11}$	$\bar 4$, $\bar 7$, 11	4, 6, 0	7, $\bar 5$, 0	$\overline{11}$, $\bar 1$, 0
5, 1, $\bar 6$, l	l+4, l+7, $l$$-$11	$l$$-$4, $l$$-$7, l+11	4, 6, l	7, $\bar 5$, l	$\overline{11}$, $\bar 1$, l
5, 2, $\bar 7$, 0	3, 9, $\overline{12}$	$\bar 3$, $\bar 9$, 12	3, 7, 0	9, $\bar 5$, 0	$\overline{12}$, $\bar 2$, 0
5, 2, $\bar 7$, l	l+3, l+9, $l$$-$12	$l$$-$3, $l$$-$9, l+12	3, 7, l	9, $\bar 5$, l	$\overline{12}$, $\bar 2$, l
5, 3, $\bar 8$, 0	2, 11, $\overline{13}$	$\bar 2$, $\overline{11}$, 13	2, 8, 0	11, $\bar 5$, 0	$\overline{13}$, $\bar 3$, 0
5, 3, $\bar 8$, l	l+2, l+11, $l$$-$13	$l$$-$2, $l$$-$11, l+13	2, 8, l	11, $\bar 5$, l	$\overline{13}$, $\bar 3$, l
6, 1, $\bar 7$, 0	5, 8, $\overline{13}$	$\bar 5$, $\bar 8$, 13	5, 7, 0	8, $\bar 6$, 0	$\overline{13}$, $\bar 1$, 0
6, 1, $\bar 7$, l	l+5, l+8, $l$$-$13	$l$$-$5, $l$$-$8, l+13	5, 7, l	8, $\bar 6$, l	$\overline{13}$, $\bar 1$, l

TRIGONAL SYSTEM

3-19. Quadratic Forms

The quadratic form for the trigonal system is

$$Q = \frac{\cos^2\frac{\alpha}{2}}{a^2 \sin\frac{\alpha}{2}\sin\frac{3\alpha}{2}}\left[(h^2+k^2+l^2)-\left(1-\operatorname{tg}^2\frac{\alpha}{2}\right)(kl+lh+hk)\right],$$

where α is the rhombohedral angle.

The table gives values of $kl + lh + hk$ for all possible combinations of signs in the three-place system.

The $(h^2 + k^2 + l^2)$ range from 1 to 99 [2].

$h^2+k^2+l^2$	hkl	$kl+lh+hk$			
		$\begin{matrix}+\,+\,+\\-\,-\,-\end{matrix}$	$\begin{matrix}-\,+\,+\\+\,-\,-\end{matrix}$	$\begin{matrix}+\,-\,+\\-\,+\,-\end{matrix}$	$\begin{matrix}+\,+\,-\\-\,-\,+\end{matrix}$
1	100	0	0	0	0
2	110	1	−1	−1	1
3	111	3	−1	−1	−1
4	200	0	0	0	0
5	210	2	−2	−2	2
6	211	5	−3	−1	−1
8	220	4	−4	−4	4
9	300	0	0	0	0
9	221	8	−4	−4	0
10	310	3	−3	−3	3
11	311	7	−5	−1	−1
12	222	12	−4	−4	−4
13	320	6	−6	−6	6
14	321	11	−7	−5	1
16	400	0	0	0	0
17	410	4	−4	−4	4
17	322	16	−8	−4	−4
18	411	9	−7	−1	−1
18	330	9	−9	−9	9
19	331	15	−9	−9	3
20	420	8	−8	−8	8
21	421	14	−10	−6	2
22	332	21	−9	−9	−3
24	422	20	−12	−4	−4
25	500	0	0	0	0
25	430	12	−12	−12	12
26	510	5	−5	−5	5
26	431	19	−13	−11	5
27	511	11	−9	−1	−1
27	333	27	−9	−9	−9
29	520	10	−10	−10	10
29	432	26	−14	−10	−2
30	521	17	−13	−7	3
32	440	16	−16	−16	16
33	522	24	−16	−4	−4
33	441	24	−16	−16	8
34	530	15	−15	−15	15
34	433	33	−15	−9	−9
35	531	23	−17	−13	7
36	600	0	0	0	0
36	442	32	−16	−16	0
37	610	6	−6	−6	6
38	611	13	−11	−1	−1
38	532	31	−19	−11	−1

$h^2+k^2+l^2$	hkl	$kl+lh+hk$			
		$\begin{matrix}+\ +\ +\\-\ -\ -\end{matrix}$	$\begin{matrix}-\ +\ +\\+\ -\ -\end{matrix}$	$\begin{matrix}+\ -\ +\\-\ +\ -\end{matrix}$	$\begin{matrix}+\ +\ -\\-\ -\ +\end{matrix}$
40	620	12	−12	−12	12
41	621	20	−16	−8	4
41	540	20	−20	−20	20
41	443	40	−16	−16	−8
42	541	29	−21	−19	11
43	533	39	−21	−9	−9
44	622	28	−20	−4	−4
45	630	18	−18	−18	18
45	542	38	−22	−18	2
46	631	27	−21	−15	9
48	444	48	−16	−16	−16
49	700	0	0	0	0
49	632	36	−24	−12	0
50	710	7	−7	−7	7
50	550	25	−25	−25	25
50	543	47	−23	−17	−7
51	711	15	−13	−1	−1
51	551	35	−25	−25	15
52	640	24	−24	−24	24
53	720	14	−14	−14	14
53	641	34	−26	−22	14
54	721	23	−19	−9	5
54	633	45	−27	−9	−9
54	552	45	−25	−25	5
56	642	44	−28	−20	4
57	722	32	−24	−4	−4
57	544	56	−24	−16	−16
58	730	21	−21	−21	21
59	731	31	−25	−17	11
59	553	55	−25	−25	−5
61	650	30	−30	−30	30
61	643	54	−30	−18	−6
62	732	41	−29	−13	1
62	651	41	−31	−29	19
64	800	0	0	0	0
65	810	8	−8	−8	8
65	740	28	−28	−28	28
65	652	52	−32	−28	8
66	811	17	−15	−1	−1
66	741	39	−31	−25	17
66	554	65	−25	−25	−15
67	733	51	−33	−9	−9
68	820	16	−16	−16	16
68	644	64	−32	−16	−16
69	821	26	−22	−10	6
69	742	50	−34	−22	6
70	653	63	−33	−27	−3
72	822	36	−28	−4	−4
72	660	36	−36	−36	36
73	830	24	−24	−24	24
73	661	48	−36	−36	24
74	831	35	−29	−19	13
74	750	35	−35	−35	35
74	743	61	−37	−19	−5
75	751	47	−37	−33	23
75	555	75	−25	−25	−25
76	662	60	−36	−36	12
77	832	46	−34	−14	2
77	654	74	−34	−26	−14
78	752	59	−39	−31	11

$h^2+k^2+l^2$	hkl	$kl+lh+hk$			
		$\begin{matrix}+\,+\,+\\-\,-\,-\end{matrix}$	$\begin{matrix}-\,+\,+\\+\,-\,-\end{matrix}$	$\begin{matrix}+\,-\,+\\-\,+\,-\end{matrix}$	$\begin{matrix}+\,+\,-\\-\,-\,+\end{matrix}$
80	840	32	−32	−32	32
81	900	0	0	0	0
81	841	44	−36	−28	20
81	744	72	−40	−16	−16
81	663	72	−36	−36	0
82	910	9	−9	−9	9
82	833	57	−39	−9	−9
83	911	19	−17	−1	−1
83	753	71	−41	−29	−1
84	842	56	−40	−24	8
85	920	18	−18	−18	18
85	760	42	−42	−42	42
86	921	29	−25	−11	7
86	761	55	−43	−41	29
86	655	85	−35	−25	−25
88	664	84	−36	−36	−12
89	922	40	−32	−4	−4
89	850	40	−40	−40	40
89	843	68	−44	−20	−4
89	762	68	−44	−40	16
90	930	27	−27	−27	27
90	851	53	−43	−37	27
90	754	83	−43	−27	−13
91	931	39	−33	−21	15
93	852	66	−46	−34	14
94	932	51	−39	−15	3
94	763	81	−45	−39	3
96	844	80	−48	−16	−16
97	910	36	−36	−36	36
97	665	96	−36	−36	−24
98	941	49	−41	−31	23
98	853	79	−49	−31	1
98	770	49	−49	−49	49
99	933	63	−45	−9	−9
99	771	63	−49	−49	35
99	755	95	−45	−25	−25

3-20. Values of $(2\cos\alpha)/(1+\cos\alpha) = 1-tg^2(\alpha/2)$

The table gives values of the function appearing in the quadratic form [2]. The range of angles is from 10 to 110°.

$\alpha°$	0′	10′	20′	30′	40′	50′
10	0.99234	0.99209	C.99182	0.99155	0,99129	0.99101
1	.99073	.99045	.99015	.98986	.98956	.98926
2	.98895	.98864	.98833	.98801	.98761	.98735
3	.98702	.98668	.98634	.98599	.98564	.98528
4	.98493	.98456	.98419	.98382	.98343	.98306
5	.98267	.98227	.98188	.98148	.98108	.98066
6	.98025	.97983	.97910	.97898	.97854	.97810
7	.97766	.97722	.97676	.97631	.97585	.97538
8	.97492	.97444	.97396	.97348	.97299	.97250
9	.97200	.97149	.97099	.97047	.96995	.96994

α°	0'	10'	20'	30'	40'	50'
20	0.96891	0.96838	0.96785	0.96730	0.96675	0.96621
1	.96564	.96509	.96452	.96396	.96337	.96280
2	.96221	.96163	.96104	.96043	.95982	.95922
3	.95860	.95799	.95736	.95673	.95611	.95545
4	.95481	.95416	.95353	.95286	.95220	.95152
5	.95084	.95018	.94949	.94881	.94810	.94740
6	.94669	.94600	.94527	.94456	.94382	.94309
7	.94235	.94163	.94087	.94013	.93935	.93860
8	.93783	.93705	.93629	.93549	.93472	.93393
9	.93310	.93230	.93151	.93069	.92985	.92904
30	.92819	.92736	.92651	.92566	.92480	.92397
1	.92309	.92223	.92133	.92046	.91957	.91868
2	.91778	.91686	.91598	.91504	.91414	.91318
3	.91226	.91133	.91038	.90943	.90846	.90749
4	.90655	.90555	.90459	.90357	.90259	.90749
5	.90060	.8996	.8986	.8975	.8965	.8955
6	.8944	.8934	.8923	.8913	.8902	.8891
7	.8881	.8870	.8859	.8848	.8836	.8826
8	.8814	.8803	.8792	.8780	.8769	.8758
9	.8746	.8734	.8723	.8711	.8699	.8687
40	.8675	.8663	.8651	.8639	.8627	.8615
1	.8602	.8589	.8577	.8565	.8552	.8539
2	.8526	.8513	.8501	.8488	.8475	.8462
3	.8448	.8435	.8422	.8409	.8395	.8381
4	.8386	.8354	.8340	.8327	.8313	.8299
5	.8284	.8270	.8256	.8241	.8227	.8213
6	.8197	.8184	.8169	.8154	.8140	.8124
7	.8109	.8095	.8080	.8064	.8048	.8033
8	.8018	.8002	.7986	.7970	.7955	.7939
9	.7923	.7907	.7891	.7875	.7858	.7842
50	.7825	.7809	.7792	.7776	.7759	.7742
1	.7725	.7708	.7691	.7673	.7656	.7638
2	.7621	.7603	.7586	.7568	.7550	.7532
3	.7515	.7496	.7478	.7459	.7441	.7422
4	.7403	.7385	.7366	.7348	.7322	.7310
5	.7290	.7271	.7251	.7232	.7213	.7193
6	.7173	.7153	.7133	.7113	.7093	.7072
7	.7052	.7031	.7010	.6990	.6969	.6962
8	.6927	.6907	.6885	.6864	.6842	.6820
9	.6800	.6777	.6755	.6733	.6711	.6689
60	.6667	.6644	.6621	.6599	.6575	.6553
1	.6531	.6507	.6484	.6460	.6437	.6414
2	.6389	.6366	.6342	.6317	.6293	.6269
3	.6245	.6221	.6195	.6170	.6145	.6120
4	.6095	.6070	.6045	.6019	.5993	.5967
5	.5941	.5915	.5889	.5862	.5837	.5810
6	.5783	.5856	.5728	.5701	.5675	.5647
7	.5619	.5590	.5564	.5535	.5506	.5479
8	.5450	.5423	.5393	.5363	.5336	.5305
9	.5227	.5247	.5218	.5187	.5158	.5127
70	.5098	.5066	.5036	.5004	.4974	.4944
1	.4911	.4881	.4850	.4817	.4786	.4754
2	.4720	.4689	.4657	.4625	.4590	.4557
3	.4525	.4492	.4459	.4423	.4390	.4356
4	.4322	.4288	.4254	.4216	.4182	.4147
5	.4112	.4076	.4041	.4005	.3969	.3933
6	.3896	.3860	.3823	.3786	.3748	.3711
7	.3673	.3635	.3597	.3558	.3520	.3481
8	.3442	.3402	.3363	.3326	.3286	.3245
9	3205	.3164	.3123	.3082	.3043	.3002

α°	0′	10′	20′	30′	40′	50′
80	0.2960	0.2917	0.2875	0.2832	0.2792	0.2749
1	.2705	.2662	.2618	.2577	.2532	.2487
2	.2442	.2400	.2355	.2309	.2263	.2220
3	.2173	.2126	.2079	.2035	.1987	.1939
4	.1894	.1845	.1796	.1751	.1701	.1652
5	.1602	.1555	.1504	.1453	.1406	.1354
6	.1302	.1254	.1202	.1149	.1100	.1046
7	.0997	.0943	.0888	.0838	.0783	.0727
8	.0676	.0620	.0564	.0511	.0454	.0397
9	.0344	.0286	.0232	.0173	.0114	.0060
90	.000	—.006	—.012	—.018	—.024	—.029
1	—.036	—.041	—.048	—.054	—.060	—.066
2	—.073	—.078	—.085	—.091	—.097	—.104
3	—.111	—.117	—.124	—.130	—.137	—.143
4	—.150	—.157	—.164	—.170	—.177	—.184
5	—.191	—.198	—.205	—.212	—.219	—.226
6	—.234	—.241	—.248	—.255	—.262	—.270
7	—.278	—.285	—.292	—.300	—.308	—.316
8	—.323	—.331	—.339	—.347	—.355	—.363
9	—.371	—.379	—.387	—.395	—.403	—.412
100	—.420	—.429	—.437	—.445	—.454	—.463
1	—.472	—.480	—.489	—.498	—.507	—.516
2	—.525	—.534	—.543	—.552	—.562	—.571
3	—.581	—.590	—.600	—.609	—.619	—.629
4	—.638	—.648	—.658	—.666	—.678	—.688
5	—.698	—.708	—.719	—.729	—.740	—.751
6	—.761	—.772	—.782	—.793	—.805	—.816
7	—.826	—.837	—.848	—.860	—.872	—.883
8	—.894	—.906	—.918	—.929	—.942	—.953
9	—.965	—.978	—.990	—1.002	—1.015	—1.027
110	—1.040	—1.052	—1.065	—1.078	—1.091	—1.104

3-21. Relation of α to c/a for the Trigonal System

The relation of α to c/a is

$$\left(\frac{c}{a}\right)_{hex} = \sqrt{\frac{9}{4\sin^2\frac{\alpha}{2}} - 3}$$

or

$$\sin\frac{\alpha}{2} = \frac{1}{2\sqrt{\frac{1}{3} + \frac{1}{9}\left(\frac{c}{a}\right)^2}}.$$

The relations between lattice constants for indexing in the trigonal and hexagonal systems are as follows:

$$a_{hex} = a_{trig}\sin\frac{\alpha}{2},$$

$$a_{trig} = a_{hex}\sqrt{\frac{1}{3} + \frac{1}{9}\left(\frac{c}{a}\right)^2}.$$

The table gives values of c/a and $\sqrt{\frac{1}{3} + \frac{1}{9}\left(\frac{c}{a}\right)^2}$ for α between 0 and 119° [2].

Let $F = \sqrt{\dfrac{1}{3}+\dfrac{1}{9}\left(\dfrac{c}{a}\right)^2}$

α	c/a	F	α	c/a	F	α	c/a	F	α	c/a	F	α	c/a	F
0	∞	∞	24	7.00	2.40	48	3.26	1.23	72	1.87	0.850	96	1.04	0.674
1	172	57.3	25	6.71	2.31	49	3.17	1.204	73	1.83	0.840	97	1.01	0.668
2	85.9	28.6	26	6.44	2.22	50	3.10	1.18	74	1.79	0.830	98	0.975	0.662
3	57.3	19.1	27	6.19	2.14	51	3.02	1.16	75	1.75	0.821	99	0.944	0.658
4	42.9	14.3	28	5.95	2.07	52	2.95	1.14	76	1.71	0.811	100	0.913	0.653
5	33.3	10.7	29	5.73	2.00	53	2.88	1.12	77	1.67	0.802	101	0.883	0.648
6	28.6	9.53	30	5.53	1.93	54	2.81	1.10	78	1.64	0.795	102	0.852	0.6435
7	24.5	8.19	31	5.34	1.87	55	2.75	1.08	79	1.60	0.786	103	0.821	0.639
8	21.4	7.16	32	5.16	1.81	56	2.69	1.065	80	1.56	0.777	104	0.791	0.635
9	18.0	6.03	33	4.99	1.76	57	2.61	1.04	81	1.53	0.770	105	0.758	0.630
10	17.1	5.72	34	4.83	1.71	58	2.56	1.03	82	1.49	0.762	106	0.727	0.626
11	15.5	5.20	35	4.68	1.66	59	2.51	1.016	83	1.46	0.755	107	0.694	0.622
12	14.2	4.73	36	4.53	1.61	60	2.45	1.00	84	1.42	0.747	108	0.662	0.618
13	13.1	4.41	37	4.40	1.57	61	2.39	0.985	85	1.39	0.740	109	0.628	0.614
14	12.2	4.10	38	4.27	1.53	62	2.34	0.970	86	1.35	0.732	110	0.593	0.610
15	11.4	3.84	39	4.15	1.50	63	2.29	0.957	87	1.32	0.726	111	0.560	0.606
16	10.6	3.58	40	4.03	1.45	64	2.24	0.944	88	1.29	0.720	112	0.523	0.603
17	10.0	3.38	41	3.82	1.40	65	2.19	0.931	89	1.26	0.714	113	0.485	0.5995
18	9.43	3.20	42	3.81	1.395	66	2.14	0.918	90	1.22	0.706	114	0.446	0.596
19	8.92	3.03	43	3.71	1.365	67	2.09	0.905	91	1.19	0.700	115	0.404	0.593
20	8.46	2.88	44	3.61	1.335	68	2.05	0.895	92	1.16	0.695	116	0.361	0.590
21	8.05	2.74	45	3.52	1.31	69	2.00	0.882	93	1.13	0.689	117	0.308	0.586
22	7.67	2.62	46	3.43	1.28	70	1.96	0.872	94	1.10	0.684	118	0.250	0.583
23	7.32	2.51	47	3.34	1.255	71	1.92	0.862	95	1.07	0.6785	119	0.175	0.580

3-22. Relation of Sum of the Squares of the Indices to α for
Fixed Cell Volume

The sum for any angle, $(h^2 + k^2 + l^2)_\alpha$, for fixed cell volume is related to the corresponding quantity for $\alpha = 90°$ by

$$(h^2 + k^2 + l^2)_\alpha = f\,(h^2 + k^2 + l^2)_{90°},$$

where

$$f = \frac{(1 - 3\cos^2\alpha + 2\cos^3\alpha)^{2/3}}{\sin^2\alpha}\ .$$

The table gives f for α between 20 and 114° [271].

α°	f	α°	f	α°	f	α°	f
20	0.4093	44	0.6899	68	0.9032	92	0.9987
22	0.4359	46	0.7102	70	0.9174	94	0.9946
24	0.4619	48	0.7300	72	0.9308	96	0.9873
26	0.4873	50	0.7494	74	0.9434	98	0.9761
28	0.5119	52	0.7684	76	0.9551	100	0.9605
30	0.5359	54	0.7870	78	0.9658	102	0.9396
32	0.5593	56	0.8051	80	0.9754	104	0.9124
34	0.5823	58	0.8228	82	0.9836	106	0.8775
36	0.6047	60	0.8400	84	0.9904	108	0.8334
38	0.6267	62	0.8566	86	0.9956	110	0.7776
40	0.6482	64	0.8727	88	0.9988	112	0.7070
42	0.6693	66	0.8883	90	1.0000	114	0.6165

3-23. Graphs for Indexing X-Ray Patterns

The relation of d_{hkl} to the indices is

$$d_{hkl} = \frac{1}{\dfrac{1}{a}\sqrt{\dfrac{(h^2 + k^2 + l^2)\sin^2\alpha + 2\,(hk + kl + lh)\,(\cos^2\alpha - \cos\alpha)}{(1 - 3\cos^2\alpha + 2\cos^3\alpha)}}} \tag{22}$$

This relationship has been used to compile the corresponding full curves [273] (Figs. 79 and 80) and also simplified curves (Fig. 81).

The graphs are used as described above (Section 3-13) for the tetragonal system.

Ebert curves are a modified form of Bjurström curves and are used in the same way.

The lower horizontal axis in Fig. 82 [271] gives $M^2 = l^2$ for indexing in the hexagonal system; the next one gives the sum of the squares for body-centered cubic crystals in trigonal coordinates; the third does the same for simple cubic crystals; the fourth, the same for face-centered cubic; and finally, the top one gives $M_1 = h^2 + hk + k^2$ in the hexagonal system of indexing. The vertical axis has α and s, the latter being given by $(4/3)\,(c^2/a^2) = s/(1-s)$.

Indexing is much facilitated if the structure and number n of atoms in the unit cell are known. The cell volume V is given by $V = nM/N\rho$, where M is the molecular weight, ρ is the density, and N is Avogadro's number ($6.03 \cdot 10^{23}$).

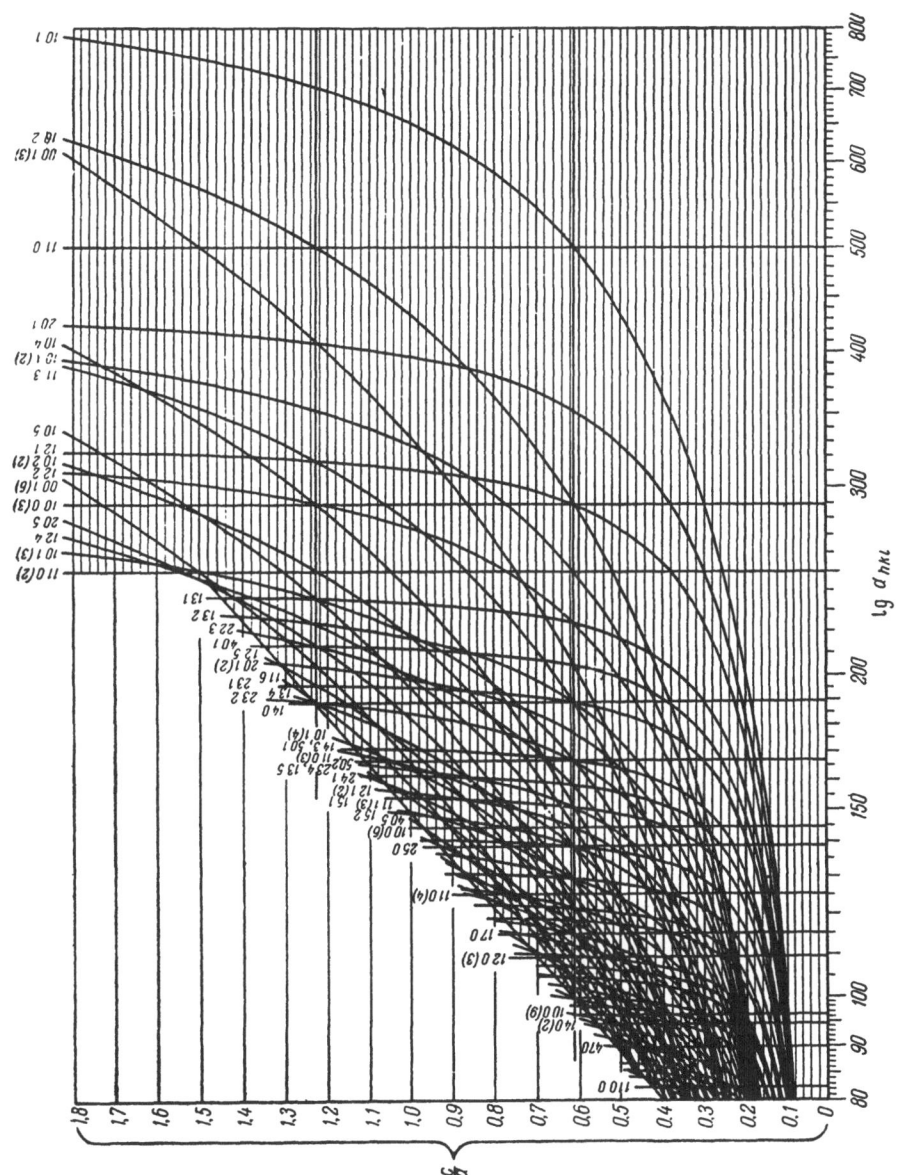

Fig. 79. Graphs for indexing patterns from crystals of trigonal structure (c/a from 0 to 1.8).

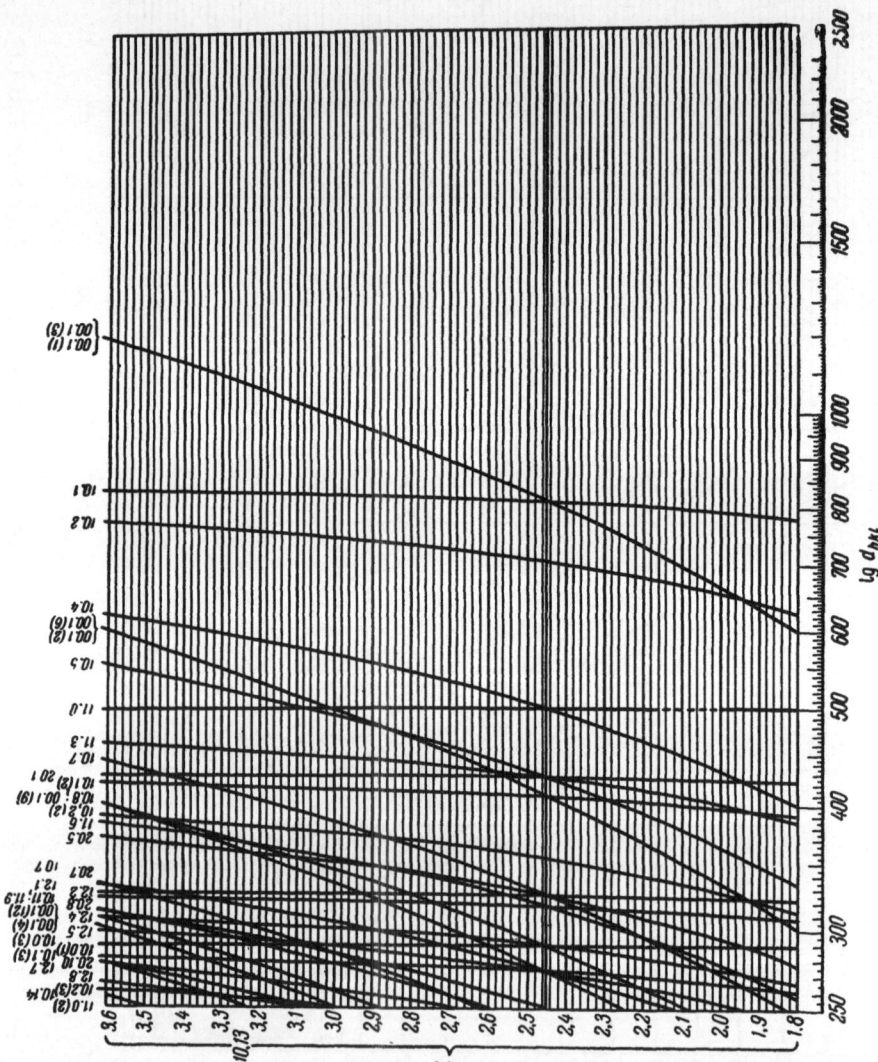

Fig. 80. Graphs for indexing patterns from crystals of trigonal structure (c/a from 1.8 to 3.6).

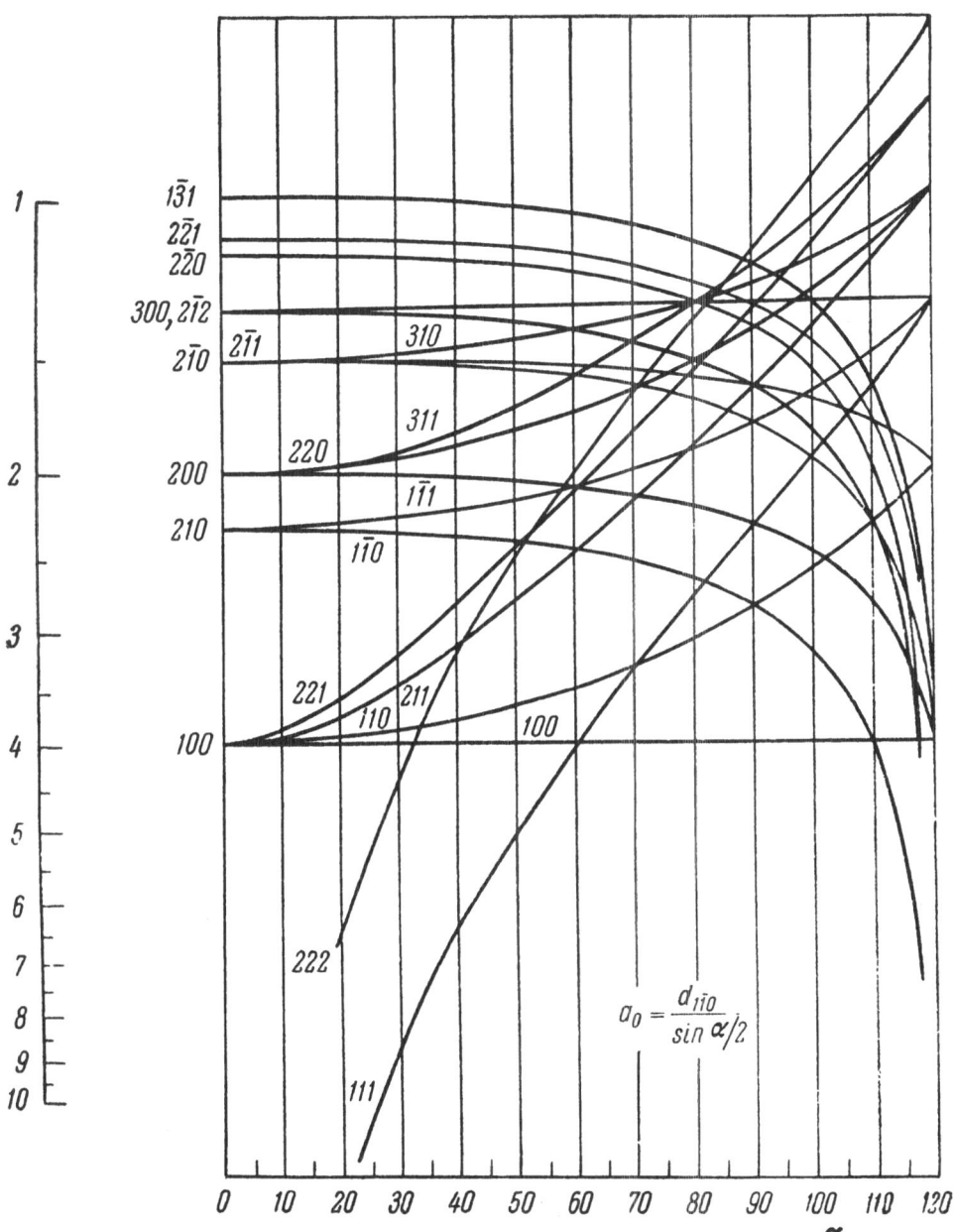

Fig. 81. Bond graphs for indexing patterns from crystals of trigonal structure.

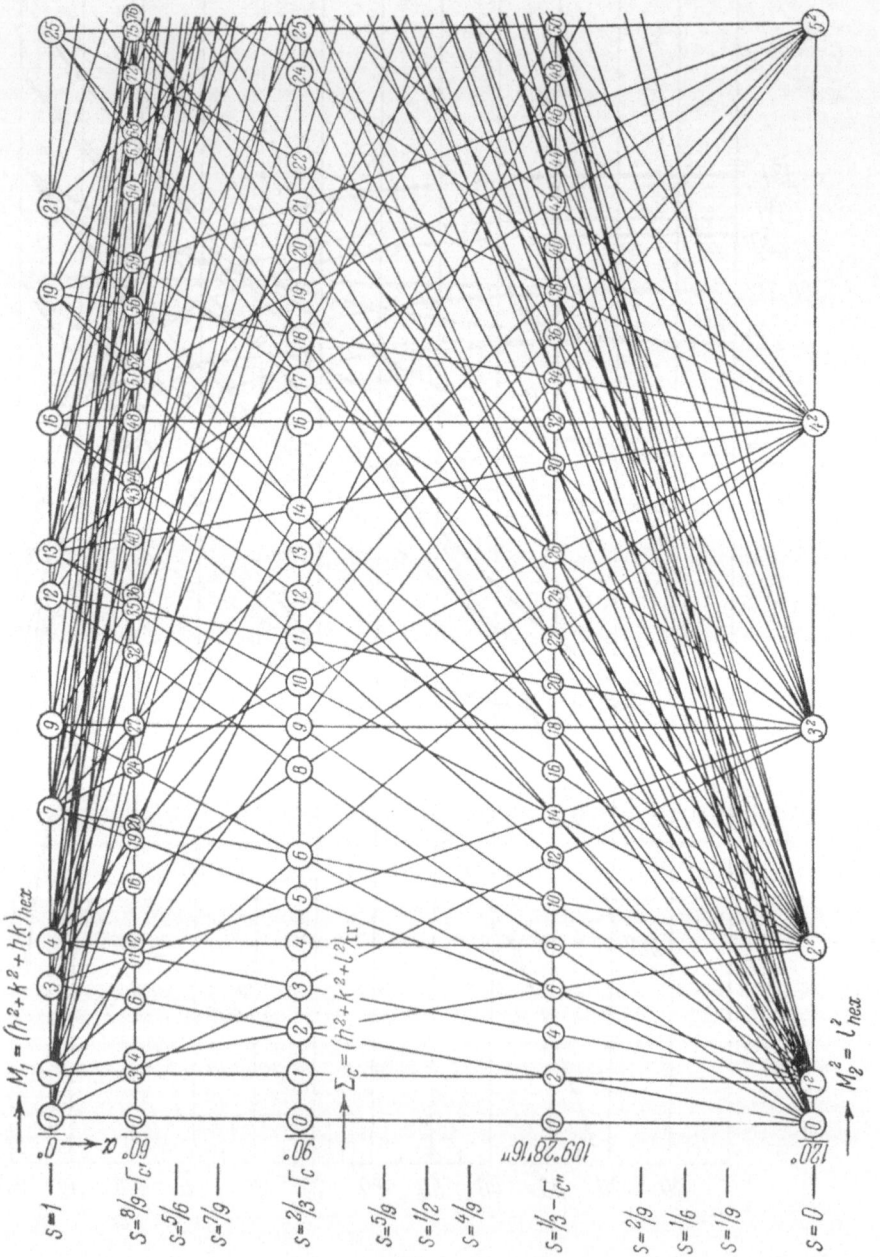

Fig. 82. Ebert graphs for indexing patterns from crystals of trigonal structure.

Ebert graphs are used as for the graphs given for higher systems (see above) if the cell volume is not known. The steps are as follows if the volume is known:

1. Table 3-7 is used to find $(h^2 + k^2 + l^2)$ for the given cell volume for $\alpha = 90°$.

2. Table 3-19 is used to find $(h^2 + k^2 + l^2)$ for the angles.

3. The $(h^2 + k^2 + l^2)$ for the various α are transferred to Fig. 82 and the corresponding curve is drawn in.

4. The line diagram (Fig. 57) is compiled and applied to find the point where the extreme line meets the above curve.

5. A horizontal line is drawn through this point.

The ordinate of this line gives the angle; the points where this line meets the sloping straight lines on the graphs and the line diagram enable one to deduce the indices.

Figure 83 shows the application to Fe_2O_3 (trigonal), for which $V = 100.5$ Å^6.

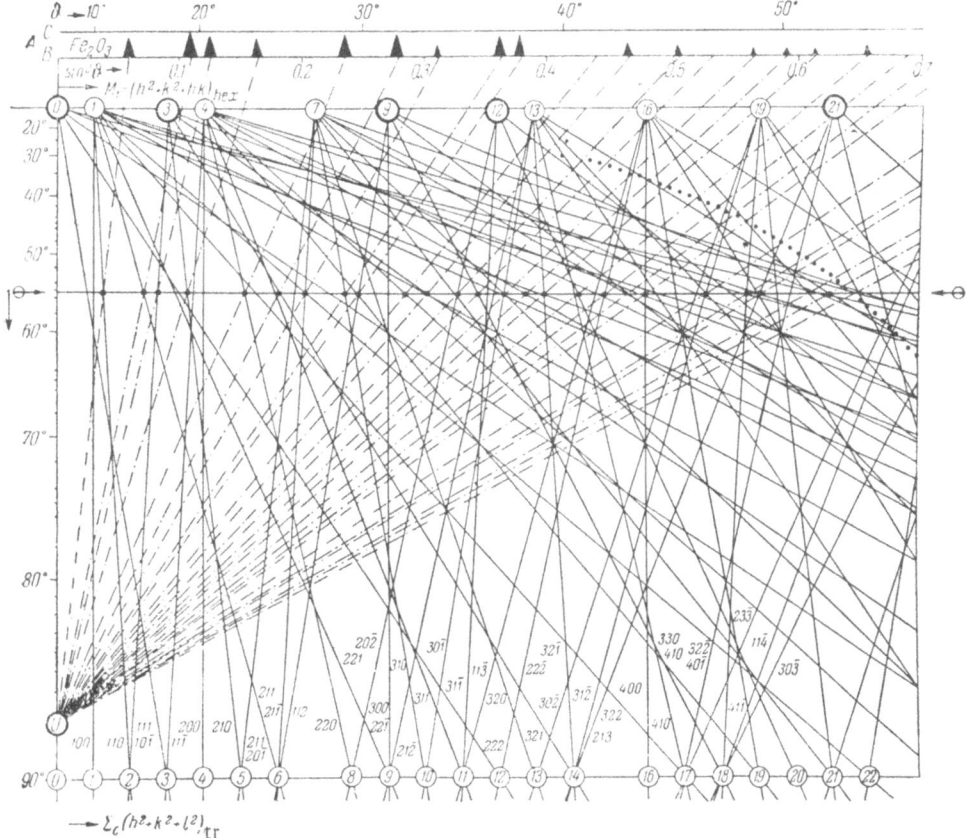

Fig. 83. Example of the use of Ebert graphs for indexing patterns from crystals of trigonal structure.

ORTHORHOMBIC SYSTEM

3-24. Analytic Method of Indexing X-Ray Patterns

The method of differences becomes troublesome to use here, because (16) is replaced by

$$\sin^2 \vartheta_{hkl} = Ah^2 + Bk^2 + Cl^2. \tag{23}$$

The approximate values of A, B, and C can be deduced from the theoretical relation

$$A \approx B \approx C \approx 0.4 \frac{\sin^2 \vartheta_{max}}{M^{2/3}}, \tag{24}$$

where M is the number of lines in the given part of the pattern and ϑ_{max} is the largest Bragg angle for these.

From (23) we have

$$\left.\begin{aligned}
\sin^2 \vartheta_{100} &= A, \quad \sin^2 \vartheta_{010} = B, \quad \sin^2 \vartheta_{001} = C, \\
\sin^2 \vartheta_{011} &= B + C = \sin^2 \vartheta_{010} + \sin^2 \vartheta_{001}, \\
\sin^2 \vartheta_{101} &= A + C = \sin^2 \vartheta_{100} + \sin^2 \vartheta_{001}, \\
\sin^2 \vartheta_{110} &= A + B = \sin^2 \vartheta_{100} + \sin^2 \vartheta_{010}, \\
\sin^2 \vartheta_{111} &= A + B + C = \sin^2 \vartheta_{100} + \sin^2 \vartheta_{010} + \sin^2 \vartheta_{001}.
\end{aligned}\right\} \tag{25}$$

Sometimes (25) will enable one to select the indices by reference to the $\sin^2 \vartheta$; but this becomes impossible if (100), (010), or (001) should happen to be lacking.

We can put (25) in the form

$$C = \sin^2 \vartheta_{001} = \sin^2 \vartheta_{101} - \sin^2 \vartheta_{100} = \sin^2 \vartheta_{011} - \sin^2 \vartheta_{010} = \sin^2 \vartheta_{hk1} - \sin^2 \vartheta_{hk0} \tag{26}$$

Similar expressions apply for A and B, so the coefficients may be found by reference to a table of differences for all possible pairs.

The following example illustrates the process for $NiAl_3$ [92], which often occurs in alloys with unusual physical properties.

The pattern for $NiAl_3$ had 20 lines whose sin ϑ did not exceed 0.5; (24) gives

$$A \approx B \approx C \approx 0.4 \frac{0.25}{20^{2/3}} = 0.014.$$

Table I gives the differences of the $\sin^2 \vartheta$ for the line pairs; for example, the value found at the intersection of row b and column a is $\sin^2 \vartheta_b - \sin^2 \vartheta_a = 0.0530 - 0.0496 = 0.0034$, and so on. The differences are taken over the range from 0 to 0.10; they are used in constructing the diagram shown in Fig. 84. The abscissa represents the differences in the $\sin^2 \vartheta$; the ordinate, the line number or symbol. The lengths of the dashes represent the errors in the $\sin^2 \vartheta$ (here ±0.0005).

A ruler is then placed parallel to the vertical axis and is moved along, vertical lines being drawn in where the ruler meets larger numbers of dashes. Figure 84 shows that 0.0591 corresponds to 7 intersections; 0.0437, 0.0205, 0.0153, and 0.0146 each give 5 intersections, and so on. This means that one of the coefficients (A, say) may be

TABLE I

Line	sin²θ	a	b	c	d	c	l	q	l	m	n	o	p	q	r
a	0.0496	—													
b	0.0530	0.0034	—												
c	0.0591	0.0095	0.0061	—											
d	0.0678	0.0182	0.0148	0.0087	—										
e	0.0882	0.0386	0.0352	0.0291	0.0204	—									
f	0.1082	0.0586	0.0552	0.0491	0.0404	0.0200	—								
g	0.1230	0.0734	0.0700	0.0639	0.0552	0.0348	0.0148								
l	0.1325	0.0829	0.0795	0.0734	0.0647	0.0443	0.0243								
m	0.1386	0.0890	0.0856	0.0795	0.0708	0.0504	0.0304	0.0156	0.0061						
n	0.1565	—	—	0.0974	0.0887	0.0683	0.0483	0.0335	0.0240	0.0179					
o	0.1672			—	0.0994	0.0790	0.0590	0.0442	0.0347	0.0286	0.0107				
p	0.1716				—	0.0834	0.0634	0.0486	0.0391	0.0330	0.0151	0.0044			
b	0.1858					0.0976	0.0776	0.0628	0.0533	0.0472	0.0293	0.0186	0.0142		
r	0.1980					—	0.0898	0.0750	0.0655	0.0594	0.0415	0.0308	0.0264	0.0122	
s	0.1999						0.0917	0.0769	0.0674	0.0613	0.0434	0.0327	0.0283	0.0141	0,0019
t	0.2063						0.0981	0.0833	0.0738	0.0677	0.0498	0.0391	0.0347	0.0205	0.0083
u	0.2156							0.0926	0.0831	0.0770	0.0591	0.0484	0.0440	0.0298	0.0176
v	0.2266								0.0941	0.0880	0.0701	0.0594	0.0550	0.0408	0.0286
w	0.2366									0.0980	0.0801	0.0694	0.0650	0.0508	0.0386
x	0.2587											0.0915	0.0871	0.0729	0.6607

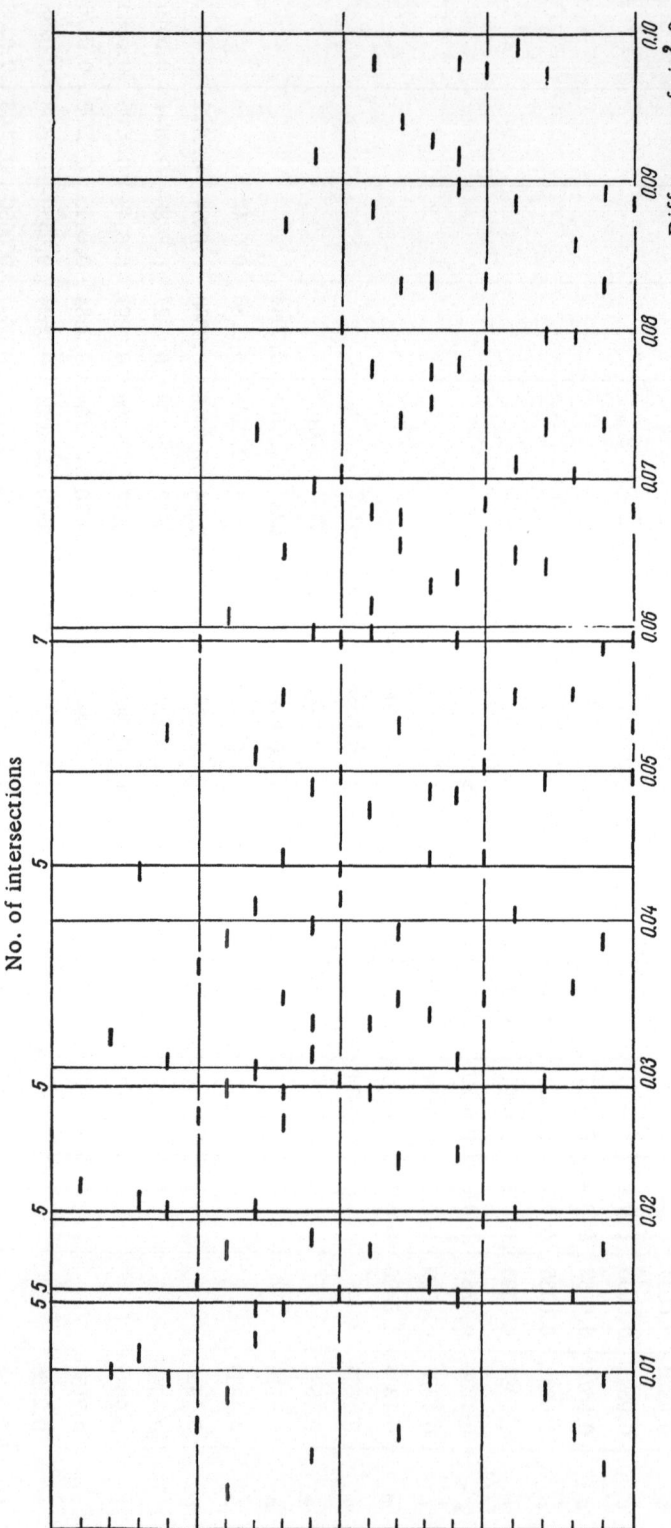

Fig. 84. Diagram for the analytic method of indexing the pattern for an orthorhombic crystal (NiAl₃).

0.0591/2^2 = 0.0148, so line c, for which $\sin^2 \vartheta$ = 0.0591, should be taken as (200); $\sin^2 \vartheta_b - \sin^2 \vartheta_a$ = 0.0148, so these lines have the same k and l, h being 0 and 1 respectively. Also, $\sin^2 \vartheta_f - \sin^2 \vartheta_g$ = 0.0148 and $\sin^2 \vartheta_f - \sin^2 \vartheta_n$ = 0.0590 = 4A, so lines f, g, and n have the same k and l, h being 0, 1, 2, and so on. Continuing in this way, the indexing is completed.

The following is a detailed treatment for the first seven lines.

We assume that h for line a is not zero, for almost none of the differences of the $\sin^2 \vartheta$ involving line a have values close to 0.0148 or 0.0591; $\sin^2 \vartheta_f - \sin^2 \vartheta_a$ = 0.0586, which is close to 0.0591, but line f has h = 0; so line a cannot have h as zero. We assume that the line is (110); then B = $\sin^2 \vartheta_a - A$ = 0.0496 − 0.0148 = 0.0348, which occurs 4 times in the table. Further, 4 × 0.0348 = 0.1392, which corresponds to the $\sin^2 \vartheta$ for line m, which therefore is (020). Again, $\sin^2 \vartheta_e - \sin^2 \vartheta_b$ = 0.0352, so these lines have the same h and k, l being 0 and 1. But we have already found that line b has h = 0, so the indices can be (001) or (002); then C may be $\sin^2 \vartheta_b$ = 0.0530 or 0.0530/4 = 0.0132. Both values occur once in the table. We assume that line b is (011); then C = $\sin^2 \vartheta_b - B$ = 0.0530 − 0.0348 = 0.0182, which occurs 4 times. Line b is therefore (011). We now have A = 0.0148, B = 0.0348, and C = 0.0182, so the indices for the other lines are easily established. A test for correct choice is that the measured $\sin^2 \vartheta$ agree with the calculated ones. Table II gives the results for the first eight lines; it shows that agreement on the $\sin^2 \vartheta$ is not exact, although the indexing is correct, so it is best to revise the values for the coefficients by solving equations of the type of (25) for A, B, and C for any three lines. If this is not done, the higher reflections may show large discrepancies between the experimental and theoretical $\sin^2 \vartheta$. The revised values are A = 0.0148, B = 0.0346, and C = 0.0183.

TABLE II

Line	hkl	$\sin^2 \vartheta_c$	$\sin^2 \vartheta_e$
a	110	0.0496	0.0496
b	011	0.0530	0.0530
c	200	0.0590	0.0591
d	111	0.0678	0.0678
e	102	0.0876	0.0882
f	012	0.1076	0.1082
g	112	0.1224	0.1230
	(300)	(0.1328)	
j			0.1325
	(202)	(0.1318)	

3-25. Graphs for Indexing X-Ray Patterns

Figures 85-92 [273] enable one to index the patterns given by orthorhombic substances.

The use of these graphs is discussed in detail above.

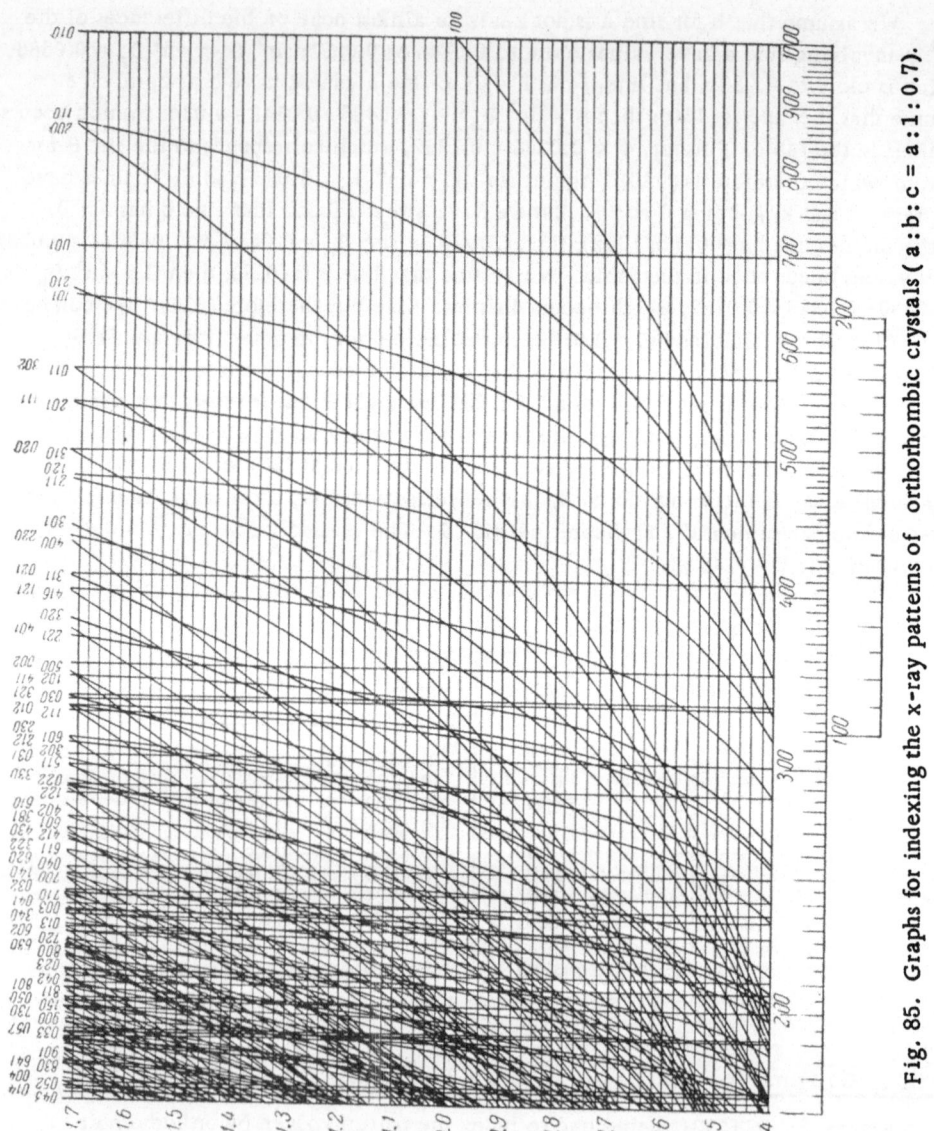

Fig. 85. Graphs for indexing the x-ray patterns of orthorhombic crystals (a : b : c = a : 1 : 0.7).

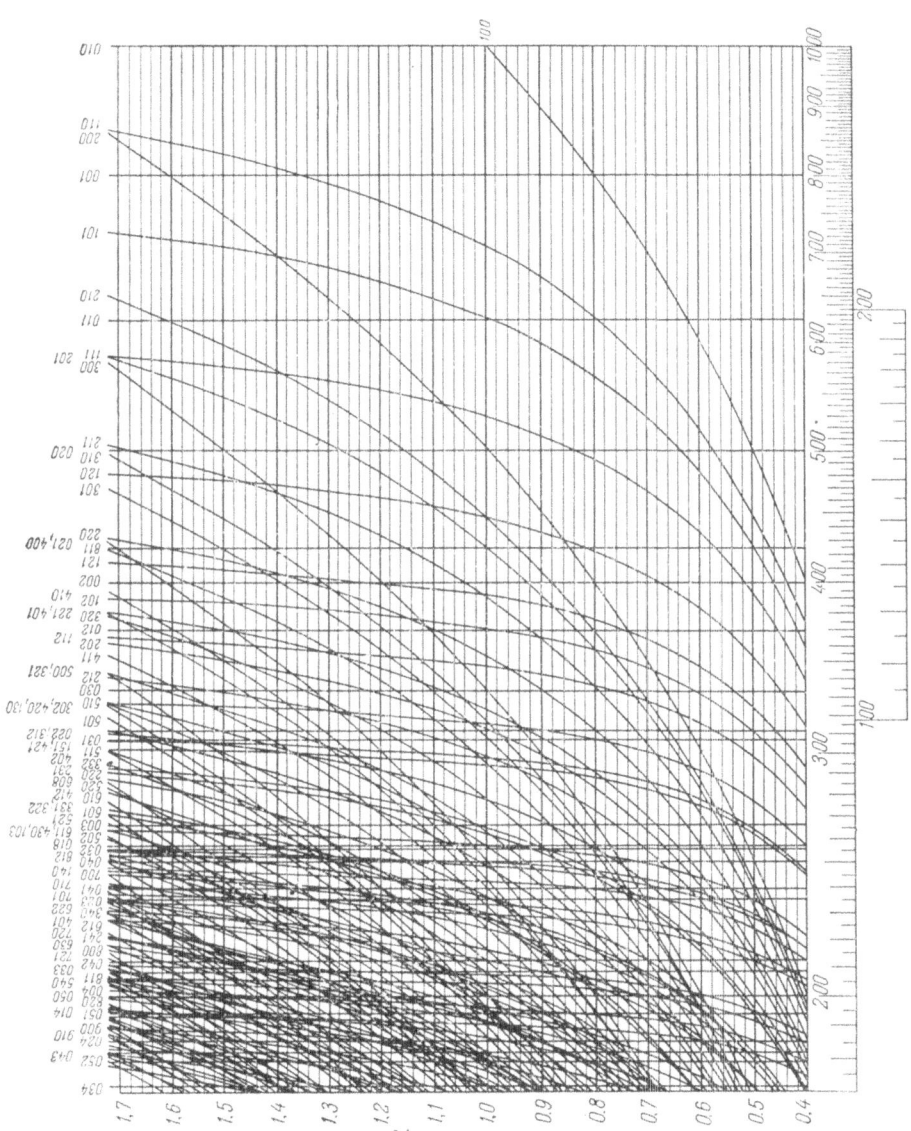

Fig. 86. Graphs for indexing the x-ray patterns of orthorhombic crystals (a : b : c = a : 1 : 0.8).

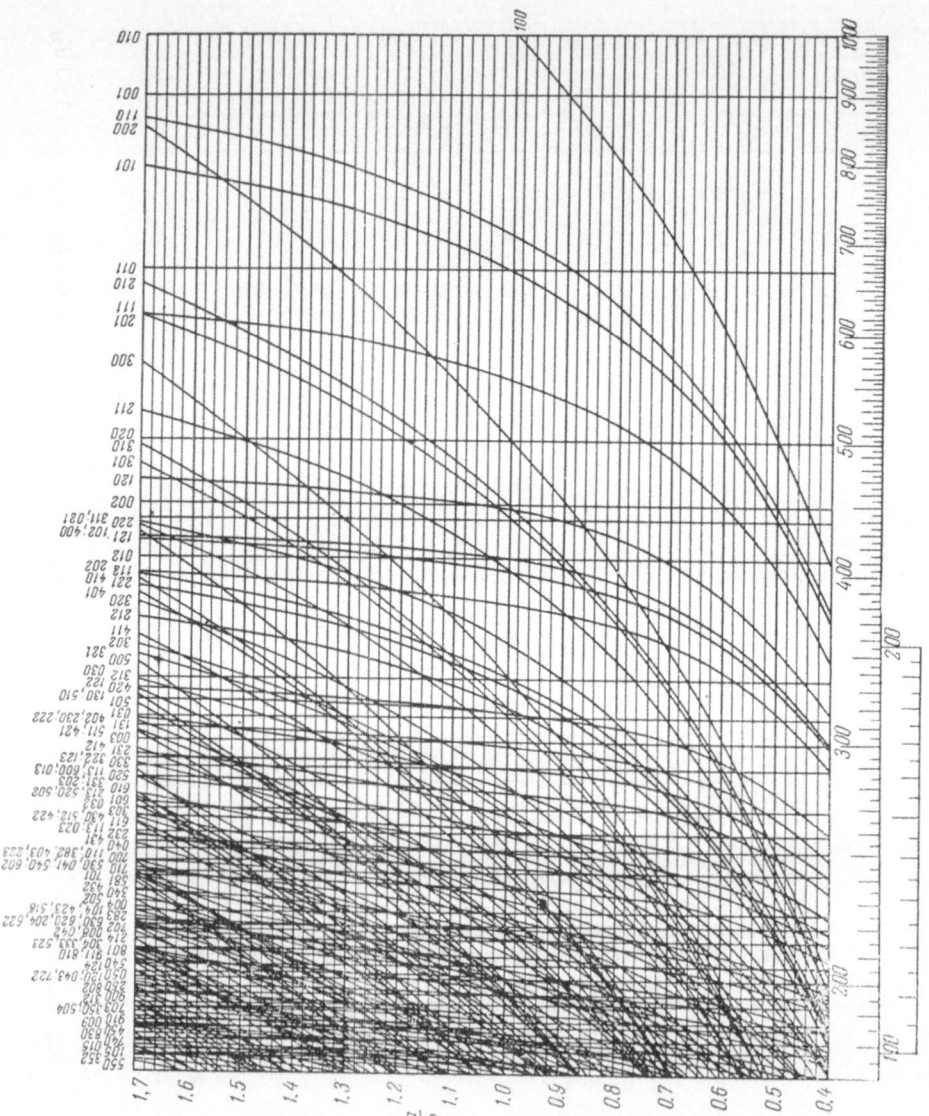

Fig. 87. Graphs for indexing the x-ray patterns of orthorhombic crystals (a : b : c = a : 1 : 0.9).

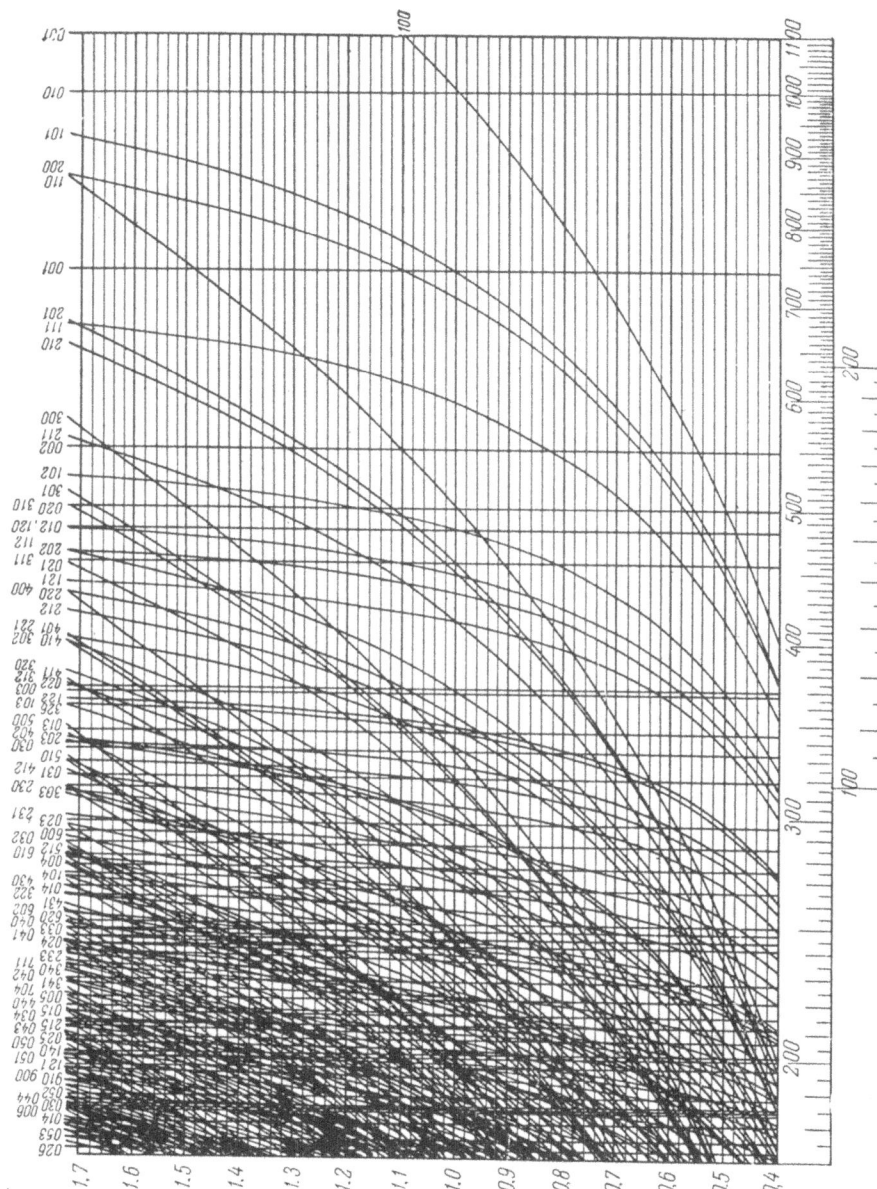

Fig. 88. Graphs for indexing the x-ray patterns of orthorhombic crystals (a:b:c = a:1:1.1).

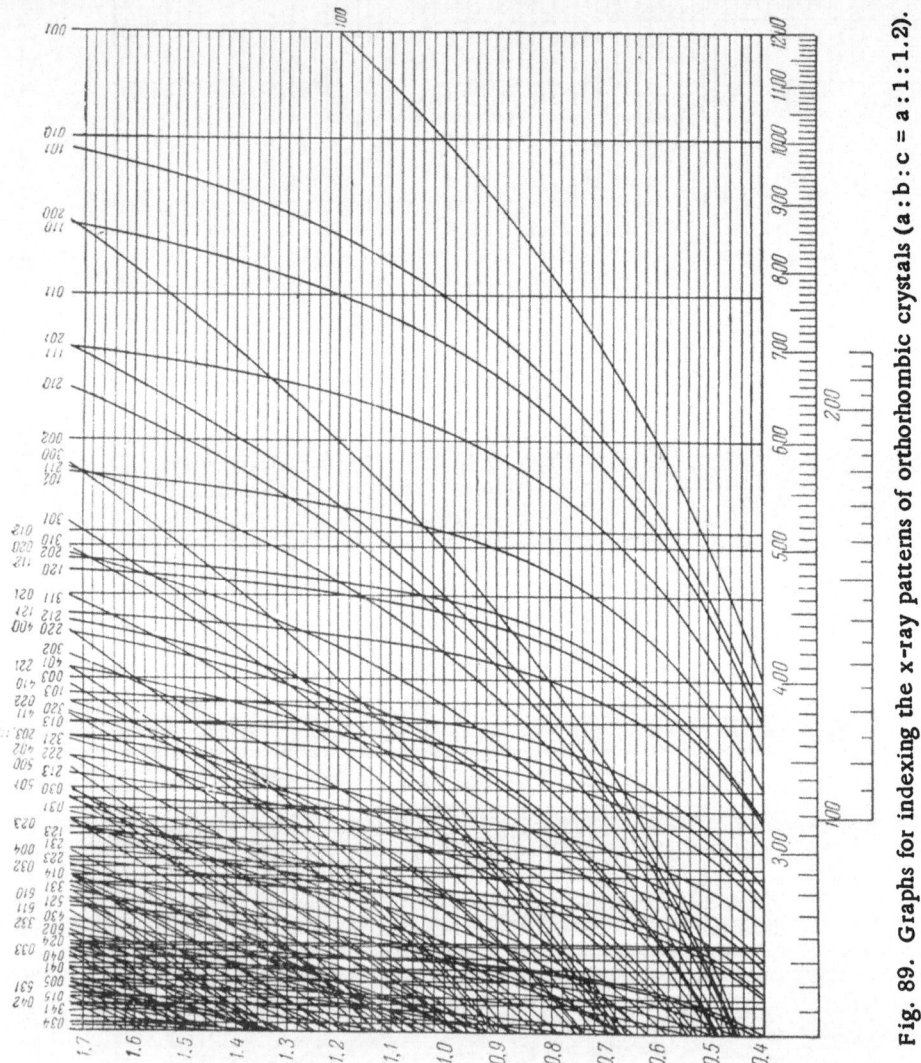

Fig. 89. Graphs for indexing the x-ray patterns of orthorhombic crystals (a : b : c = a : 1 : 1.2).

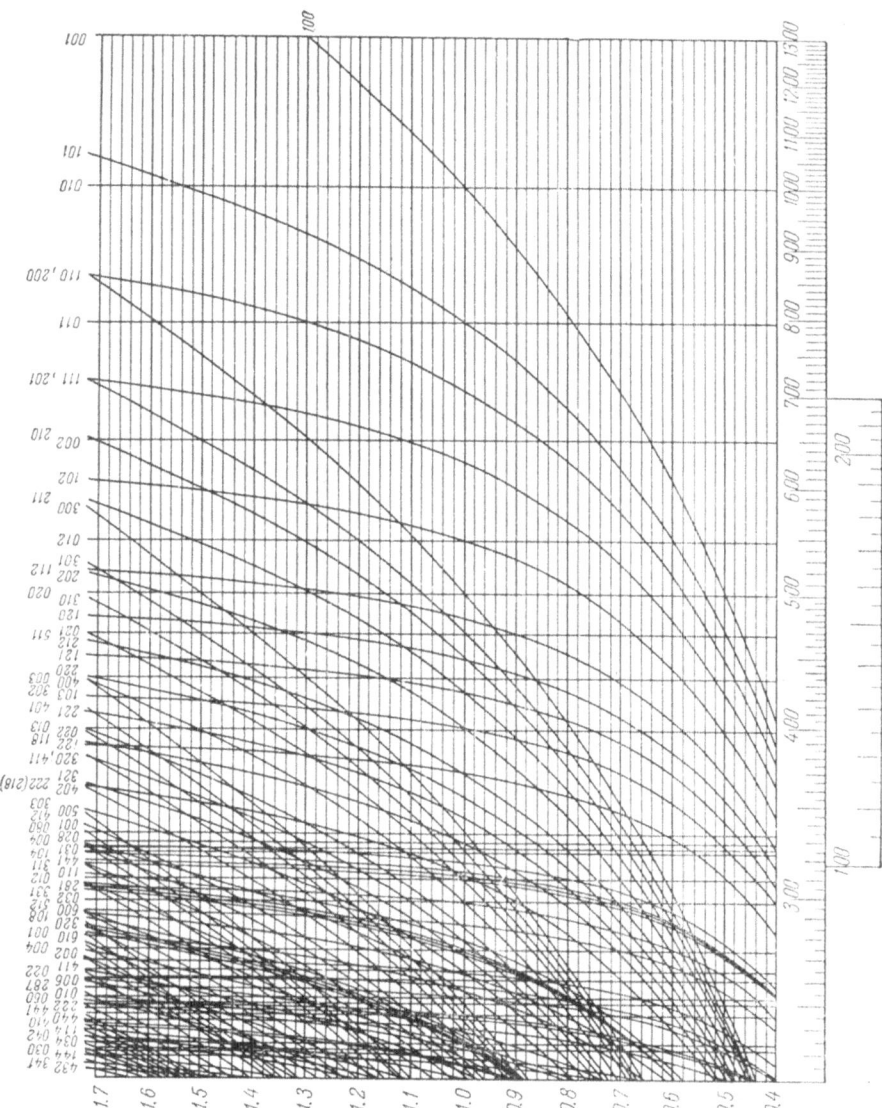

Fig. 90. Graphs for indexing the x-ray patterns of orthorhombic crystals (a : b : c = a : 1 : 1.3).

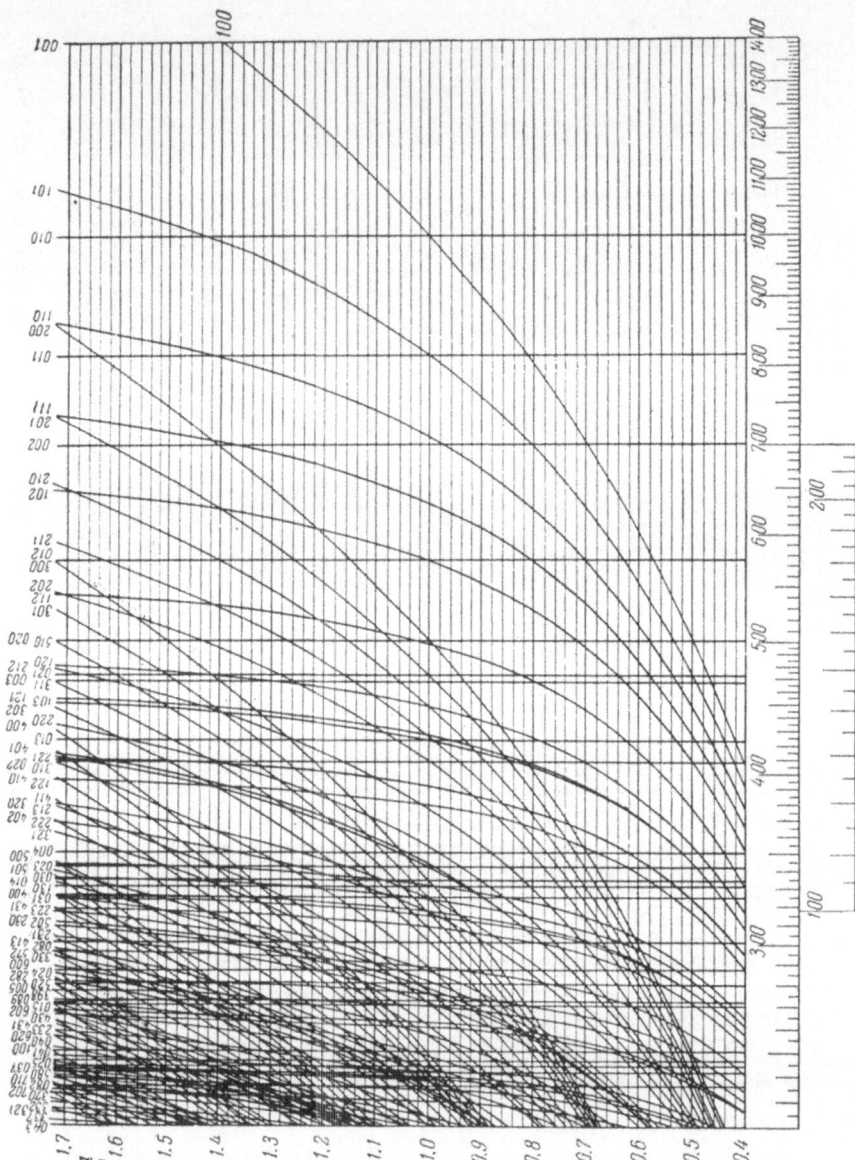

Fig. 91. Graphs for indexing the x-ray patterns of orthorhombic crystals (a : b : c = a : 1 : 1.4).

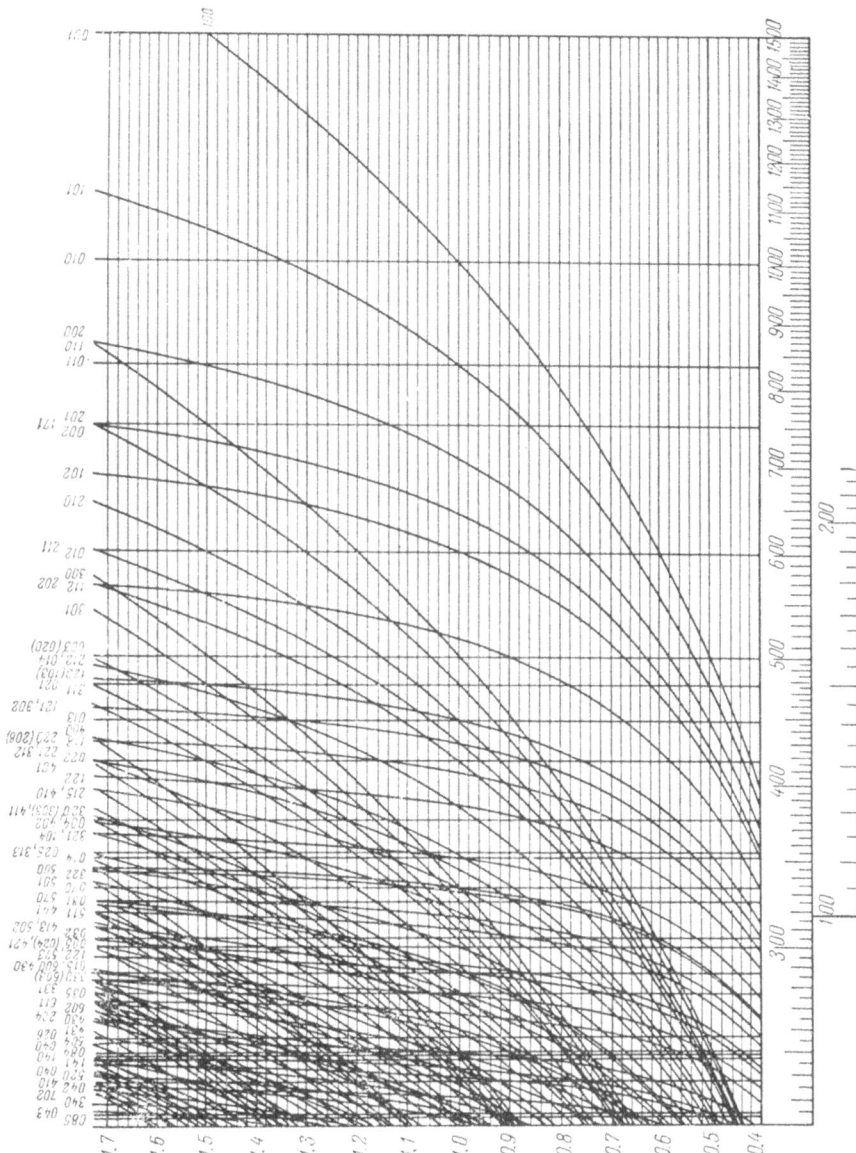

Fig. 92. Graphs for indexing the x-ray patterns of orthorhombic crystals (a : b : c = a : 1 : 1.5).

LOWER SYSTEMS

3-26. Graphical Method of Indexing X-Ray Patterns

Several methods have been proposed recently for the patterns from crystals of low symmetry; one of the simplest has the following steps: (1) the $\sin^2 \vartheta$ are calculated, these being proportional to $d*^2 = 1/d^2$; (2) $d*^2_{001}$ is chosen by trial and error; (3) a graph is constructed with $[(d*_i/c*)^2 - l^2]$ as ordinate and l as abscissa; (4) the resulting points are joined by straight lines; (5) h and k are chosen by trial and error.

Figure 93 illustrates the method for $Na_2S_2O_3$ (anhydrous); the pattern was recorded with Cu Kα radiation; The second and third columns of Table I give the measured and calculated $\sin^2 \vartheta$.

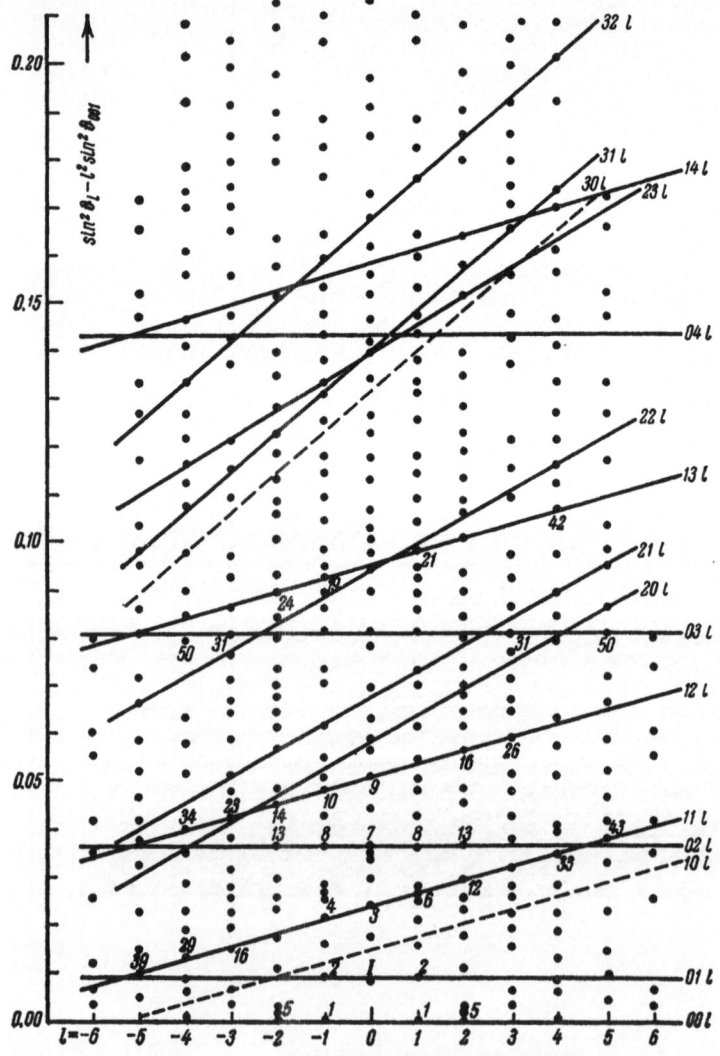

Fig. 93. Example of analytic indexing for a monoclinic substance
($Na_2S_2O_3$).

TABLE I

Line	$\sin^2\theta_e$	$\sin^2\theta_C$	hkl
1	0.00830	0.00830	001
2	.01716	.01728	011
3	.02354	.02356	110
4	.02910	.02903	$11\bar{1}$
5	.03316	.03320	002
6	.03476	.03473	111
7	.03595	.03592	020
8	.04394	.04422	021
9	.05060	.05052	120
10	.05602	.05597	$12\bar{1}$
11	.05840	.05840	200
12	.06244	.06280	112
13	.06910	.06912	022
14	.07805	.07802	$12\bar{2}$
15	.08126	.08138	211
16	.08962	.08942 / .08973	122 / $11\bar{3}$
17	.09406	.09432	220
18	.09729	.09692	$22\bar{1}$
19	.10058	.10087	$13\bar{1}$
20	.1029	.1030	202
21	.1065	.1066	131
22	.1128	.1120	212
23	.1171	.1167 / .1179	$12\bar{3}$ / $22\bar{2}$
24	.1226	.1229	$13\bar{2}$
25	.1260	.1250	213

Line	$\sin^2\theta_e$	$\sin^2\theta_C$	hkl
26	0.1336	0.1338	123
27	.1390	.1389 / .1392	222 / 230
28	.1417	.1418 / .1418	014 / $23\bar{1}$
29	.1460	.1450	114
30	.1516	.1520	041
31	.1555	.1555	033
32	.1612	.1616	$13\bar{3}$
33	.1674	.1673 / .1671 / .1678	320 / $32\bar{1}$ / 114
34	.1724	.1710	$12\bar{4}$
35	.1842	.1838 / .1842 / .1845	232 / 321 / 303
36	.1902	.1907	312
37	.1961	.1968	$23\bar{3}$
38	.2120	.2122 / .2120	330 / $33\bar{1}$
39	.2173	.2176 / .2168 / .2168	322 / $13\bar{4}$ / $11\bar{5}$
40	.2232	.2230	214
41	.2304	.2310	233
42	.2398	.2391 / .2396	150 / 134

Line	$\sin^2\theta_e$	$\sin^2\theta_C$	hkl
43	0.2448	0.2445 / .2453	$15\bar{1}$ / $11\bar{5}$
44	.2490	.2492	$23\bar{4}$
45	.2539	.2533	401
46	.2503	.2597	$24\bar{3}$
47	.2659	.2650	$32\bar{4}$
48	.2738	.2741 / .2733	$40\bar{3}$ / $22\bar{5}$
49	.2780	.2780	152
50	.2883	.2883	035
51	.2932	.2939	243
52	.3022	.3025	144
53	.3053	.3052 / .3053 / .3057	$15\bar{3}$ / $11\bar{6}$ / 315
54	.3106	.3100 / .3108	$42\bar{3}$ / $33\bar{4}$
55	.3243	.3248	$43\bar{2}$
56	.3337	.3341 / .3343	431 / 324
57	.3403	.3405	$25\bar{3}$
58	.3540	.3549	$34\bar{4}$
59	.3589	.358 / .3590	$22\bar{6}$ / $50\bar{1}$
60	.3725	.3720	351
61	.3788	.3784 / .3787 / .3787 / .3792	423 / $50\bar{2}$ / $44\bar{1}$ / 334

The first line has $d_1 = 8.56$ Å ; we assume this to be (001), and so

$$\sin^2 \vartheta_{001} = \sin^2 \vartheta_1 = 0.00830.$$

We plot a graph of $\sin^2 \vartheta_i - l^2 \sin^2 \vartheta_{001}$ (ordinate) against l (abscissa), each line on the pattern giving 13 points ($l = -6$ to $l = 6$). Trial shows that points 13, 8, 7, 8, and 13 lie on a horizontal line.

The general expression for $d*^2$ for a triclinic lattice is

$$d_{hkl}^{*2} = h^2 a^{*2} + k^2 b^{*2} + l^2 c^{*2} + 2hka^*b^* \cos \gamma^* + 2klb^*c^* \cos \alpha^* + 2hla^*c^* \cos \beta^*.$$

We put

$$p = \frac{1}{c^*} (2ha^* \cos \beta^* + 2kb^* \cos \alpha^*),$$

$$q = \frac{1}{c^{*2}} (h^2 a^{*2} + k^2 b^{*2} + 2hka^*b^* \cos \gamma^*),$$

and obtain

$$\frac{d_{hkl}^{*2}}{c^{*2}} - l^2 = pl + q = \sin^2 \vartheta_i - l^2 \sin^2 \vartheta_{001}.$$

We have found that point 7 lies on a horizontal line and represents $p_7 = 0$; this is possible only if h = 0 and $\alpha* = 90°$ or if k = 0 and $\beta* = 90°$; therefore the substance must be monoclinic.

Further, $d_7 = 4.06$, so $k \neq 1$ ($d_7 \neq d_{010}$); we assume that k = 2, which means that points corresponding to (02l) reflections lie on this horizontal line. The corresponding line for (01l) passes through points 2 and 2, its distance from the abscissa being 1/4 of that for the (02l) reflections. The horizontal lines for (03l) and (04l) are found similarly; they have ordinates, respectively, 9 and 16 times that for (01l).

Then the indices can be entered in Table I for points that lie on horizontal lines.

A similar argument for the inclined lines enables one to identify reflections such as (1kl), (2kl), and (3kl), from which the lattice constants are found as

$$a = 6.43 \pm 0.02 \text{ Å}, \quad b = 8.13 \pm 0.01 \text{ Å}, \quad c = 8.54 \pm 0.02 \text{ Å}, \quad \beta = 97.4 \pm 0.4°.$$

See [93, 108] for the detailed theory of the method.

3-27. Use of the Theory of Homology in Indexing Patterns for Crystals of Low Symmetry

The theory of homology is an extension of the theory of symmetry [94]. The method is based on the consideration of low-symmetry crystals as ones of high symmetry deformed by stretching along one of the symmetry axes. This enables one to correlate the lines from crystals in different systems; some of the lines split as the symmetry is reduced.

Figure 94 gives schemes for patterns from face-centered structures; Fig. 95, from body-centered ones; Fig. 96, from simple (hexahedral) ones; and Fig. 97, from hexagonal close-packed ones. These schemes show the splitting that occurs during transitions from the cubic and hexagonal systems to lower ones.

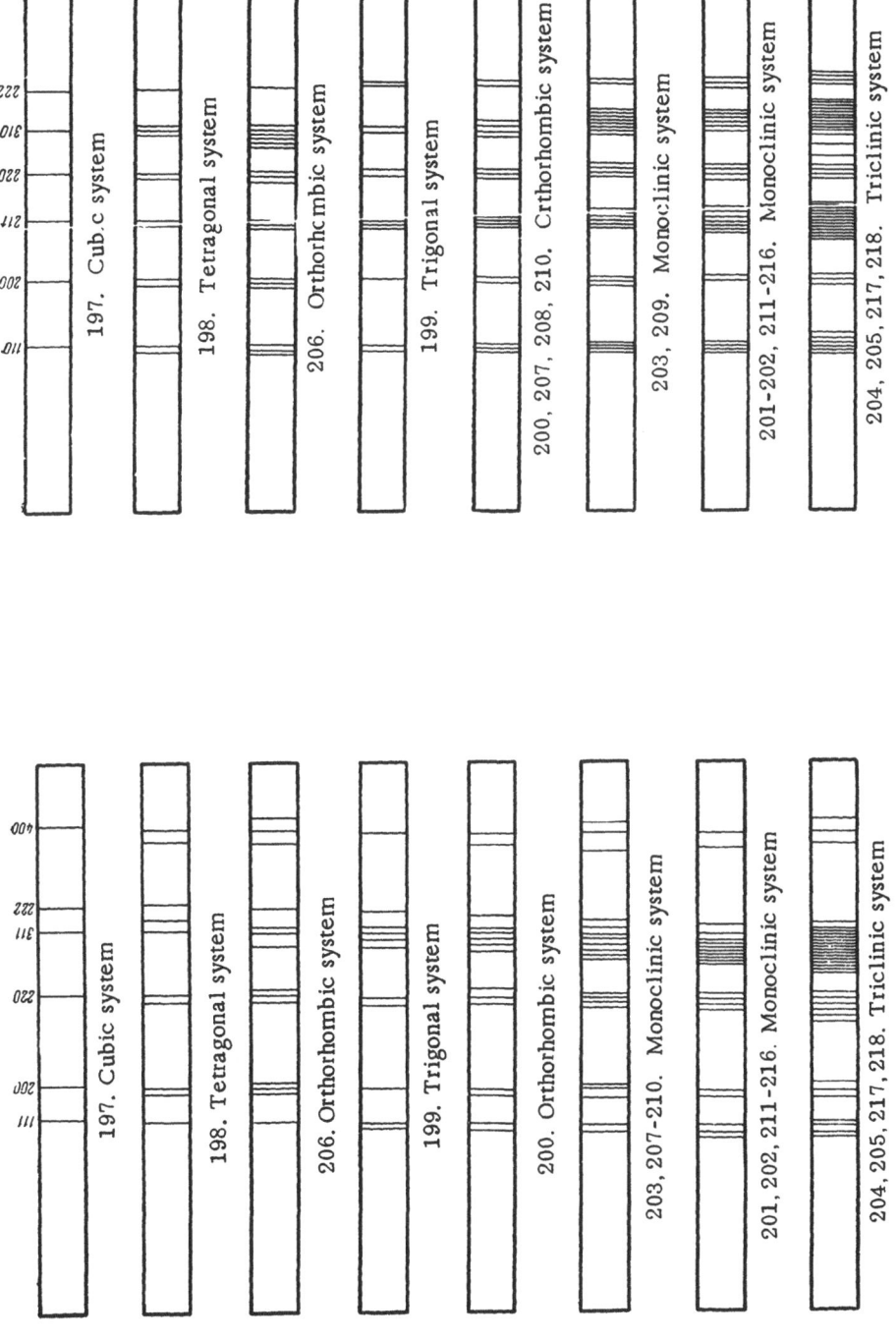

Fig. 95. Line splitting for x-ray patterns from crystals of initial body-centered structure.

Fig. 94. Line splitting for x-ray patterns from crystals of initial face-centered structure.

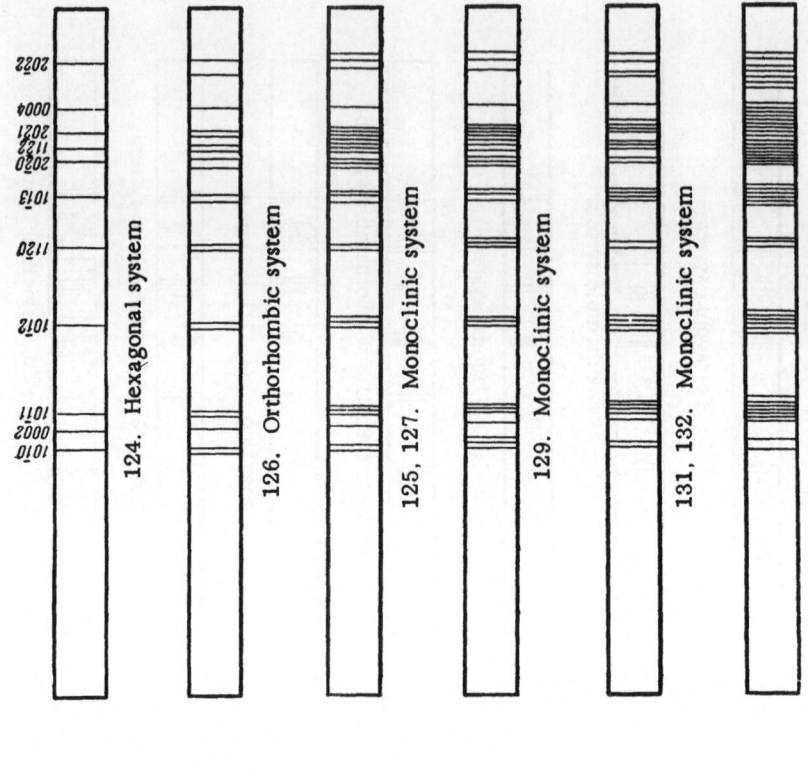

Fig. 97. Line splitting for x-ray patterns from crystals of initial hexagonal close-packed structure.

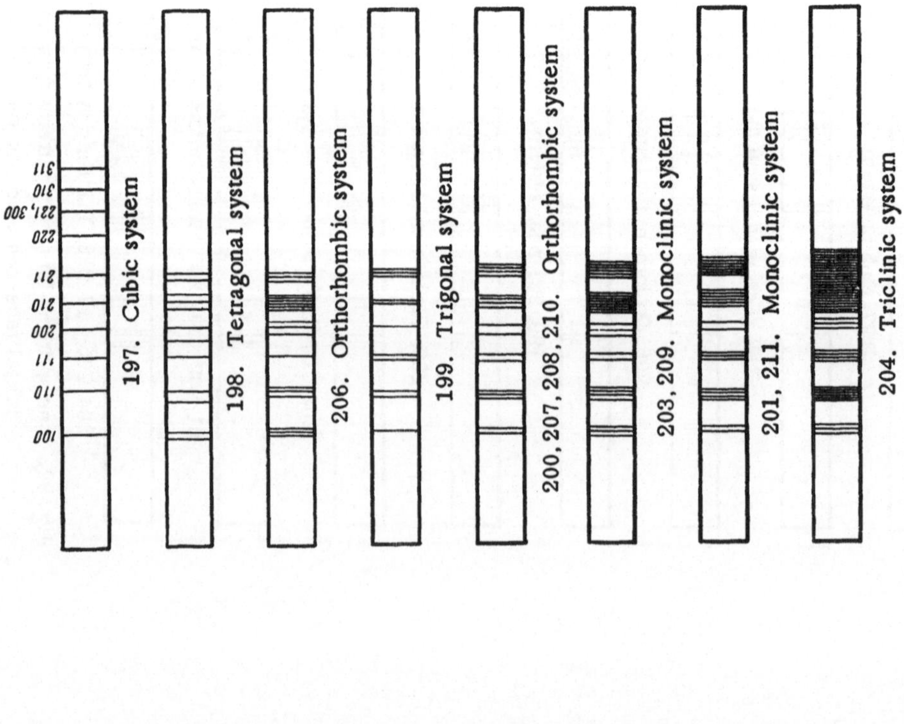

Fig. 96. Line splitting for x-ray patterns from crystals of initial simple (hexahedral) structure.

Table 3-27a gives the number of lines resulting from splitting of the first lines for face-centered structures; Table 3-27b gives the indices of these lines.

Tables 3-27c and 3-27d do the same for body-centered structures; 3-27e and 3-27f, for simple (hexahedral) ones; and 3-27g and 3-27h, for hexagonal ones.

For example, the first column in Table 3-27b gives Mikheev's number for the species of homology, as well as the system and the species of symmetry to which the species of homology relates. The second column gives the determinant for converting from the Fedorov setting to the one usually used, which takes account only of the symmetry. The other columns give the indices of the lines resulting from splitting; the numbers underneath the indices are those of the pairs of plane nets forming the given line.

The pattern is indexed as follows:

1. The structure type is established by comparing the pattern with Figs. 94-97 or by analogy with the structures of chemically homologous structures (for example, the structure of the solvent may be used as first approximation for a solid solution).

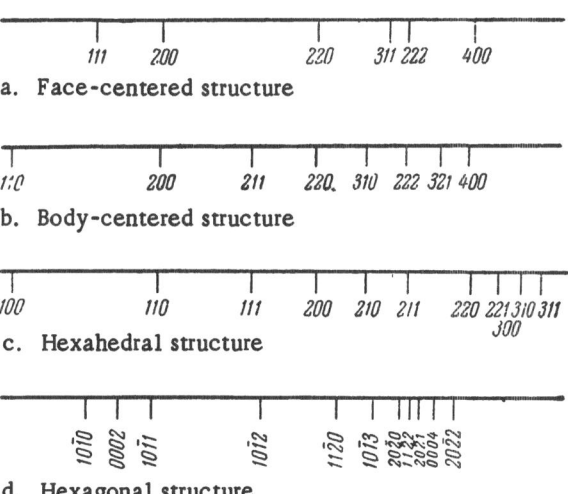

Fig. 98. Logarithmic scales for indexing.

2. The type of line splitting is determined by means of Fig. 98, which shows the array of lines for each type of structure on a logarithmic scale. The interplanar distances given by the pattern are plotted on a logarithmic scale on a strip of paper, which is placed over the scheme for the corresponding structure to bring the lines as nearly as possible into coincidence. The type of splitting is determined from the part of Table 3-27 appropriate to the structure.

3. The indices are deduced for the main lines in the pattern; parts b, d, f, and h of Table 3-27 are used in conjunction with the number for the species of homology. No particular difficulty arises if several such species are possible, because the indices are usually the same. Two permutations of the indices are possible for monoclinic and tri- clinic structures, but these merely represent different settings of the structures.

4. The dimensions of the unit cell are deduced by reference to the indices of the first lines; they are found directly from reflections of (h00), (0k0), and (00l) types for orthogonal crystals, while in other cases they are found by solving sets of equations based as far as possible on reflections of (hk0), (h0l), and (0kl) types.

TABLE I

№	I	$\frac{d}{n}$	Symbols in cubic setting	Symbols in tetragonal setting
1	8	2.992	202	112
2	6	2.974	220	200
3	4	2.539	113; 311	100; 211
4	6	2.436	222	202
5	7	2.119	004	004
6	9	2.103	400	220
7	4	1.937	331; 313	301; 213
8	6	1.748	224	204
9	7	1 719	422	312
10	4	1.620	333; 511	303; 321
11	7	1.493	404	224
12	5	1.487	440	400
13	2	1.404	442; 600	402; 330
14	4	1.339	620	420
15	5	1.337	602	332
16	3	1.269	622	422
17	3	1.217	444	404

Example. Potassium cryolite, K_3AlF_6. Table I, columns 1-3, give the x-ray results; it is assumed that the structure type is as for the related (pseudocubic) compounds cryolite Na_3AlF_6 (monoclinic) and elpasolite K_2NaAlF_6 (cubic). Both compounds have face-centered structures. The d/n are marked off on a strip of paper, which is compared with the pattern in Fig. 98 to bring the lines as closely as possible into coincidence (Fig. 99). The (220) clearly splits into two, (400) into two, and (222) and (311) not at all.

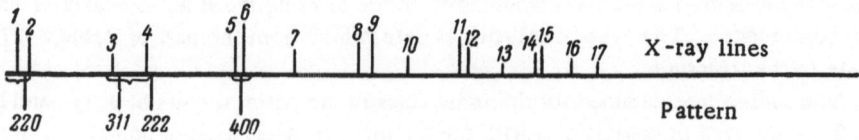

Fig. 99. Deduction of splitting type for potassium cryolite.

Table 3-27a then gives the splitting type; splitting of (220) into two can occur in the tetragonal and trigonal systems, and of (400) into two in the tetragonal, orthorhombic, and monoclinic systems; (222) does not split in the tetragonal and orthorhombic systems.

The sole possibility is then the tetragonal system; (311) does not split, as it should for this system, because the intensity is low and the second (weaker) line is not detected.

The indices are now deduced.

Table 3-27b shows that species 198 of homology and the tetragonal system together require that the first and second lines in the x-ray pattern should be (202) and (220), respectively; this disposition of the lines, and not the reverse one, is implied by the intensity, which is higher for the first, as the table shows: line (202) is formed by four plane nets, and (220) by two only.

The indices of lines 3-6 (fourth column of Table I) are found similarly; the lattice constants must be determined in order to find the indices of the other lines.

First, a is found from the (220) and (400) lines, c is found from (004), the values (kX) being 5.948 and 8.476, respectively.

The indices in the cubic system are converted to those in a simple tetragonal lattice for the first lines by means of

$$\begin{vmatrix} 1/2 & 1/2 & 0 \\ 1/2 & 1/2 & 0 \\ 0 & 0 & 1 \end{vmatrix}.$$

The fifth column in Table I gives the final values of the indices. See [97, 109, 110] for the principles of transition to other settings.

The method of homology is difficult to apply if the deviation from orthogonality is large, because the splittings are then large; there are also difficulties if many superlattice lines are present.

TABLE 3-27a

Split-ting type	Main lines on pattern						Species number and system
	111	200	220	311	222	400	
1	1	1	1	1	1	1	197. Cubic
2	1	2	2	2	1	2	198. Tetragonal
3	1	3	3	3	1	3	206. Orthorhombic
4	1	1	2	3	2	1	199. Trigonal
5	2	2	3	4	2	2	200, 207, 208, 210. Orthorhombic
6	2	3	4	6	2	3	203, 209. Monoclinic
7	3	2	4	7	3	2	201, 202, 211, 212, 213, 214, 215, 216. Monoclinic
8	4	3	6	12	4	3	204, 205, 217, 218. Triclinic

TABLE 3-27b

No. of homology species, system, and species of symmetry	Determinant	Main lines on pattern					
		111	200	220	311	222	400
197. Cubic, planaxial	100 010 001	111 4	200 3	220 6	311 12	222 4	400 3
198. Tetragonal, planaxial	100 010 001	111 4	002 200 1　2	202 220 4　2	113 311 4　8	222 4	004 400 1　2
199. Trigonal, planaxial	110 011 111	0003 1 02̄21 3	202̄2 3	022̄4 22̄4̄0 3　3	202̄5 27̄4̄3 3　6 04̄4̄1 3	0006 1 04̄4̄2 3	404̄4 3
200. Orthorhombic, planaxial	101̄ 010 101	012 210 2　2	202 020 2　1	004 400 1　1 222 4	214 412 4　4 032 230 2　2	024 2 420 2	404 2 040 1
201, 202. Monoclinic, planaxial	11̄1̄ 011̄ 111	1̄03 1 121 2	202 1 2̄22 2	021 420 2　2 040 404̄ 1　1	3̄25 323 2　2 5̄23 105 2　1 501 341 1　2 543 2	2̄06 1 242 2 202 1	404 1 4̄44 2
203. Monoclinic, planaxial	100 010 011	111 2 1̄11 2	200 020 1　1 002 1	202 2 022 220 2　1 2̄20 1	311 131 2　2 113 2 3̄11 1̄31 2　2 1̄13 2	222 2 2̄22 2	400 040 1　1 004 1
204, 218. Triclinic, central	101 011 110 111	0003 1 202̄1 1 02̄2̄1 1 0221 1	2̄022 1 02̄2̄2 1 22̄02 1	022̄4 1 022̄4 1 2̄204 1 42̄2̄0 1 22̄4̄0 1 2̄4̄20 1	2̄025 02̄2̄5 1　1 22̄05 4̄223 1　1 2̄2̄43 22̄4̄3 1　1 2̄4̄23 4̄2̄23 1　1 24̄23 4̄041 1　1 04̄4̄1 44̄01 1　1	0006 1 404̄2 1 04̄4̄2 1 04̄4̄2 1	404̄4 1 04̄4̄4 1 44̄04 1

TABLE 3-27b (cont'd)

No. of homology species, system, and species of symmetry	Determinant	Main lines on pattern					
		111	200	220	311	222	400
205, 217. Triclinic, central	100 010 001	111 $\bar{1}$11 1 1 $\bar{1}\bar{1}$1 1$\bar{1}$1 1 1	200 1 020 1 002 1	220 2$\bar{2}$0 1 1 202 202 1 1 022 022 1 1	311 $\bar{3}$11 1 1 131 1 $\bar{1}$31 113 1 1 $\bar{1}$13 1 3$\bar{1}$1 $\bar{3}$11 1 1 131 $\bar{1}$31 1 1 1$\bar{1}$3 $\bar{1}$13 1 1	222 $\bar{2}$22 1 1 $\bar{2}$22 222 1 1	400 1 040 1 004 1
206. Orthorhombic, planaxial	100 010 001	111 4	002 020 1 1 200 1	202 2 022 220 2 2	113 131 4 4 311 4	222 4	004 1 040 1 004 1
207, 208, 210. Orthorhombic, planaxial	110 $\bar{1}$10 001	201 2 021 2	002 1 022 2	040 222 1 4 400 1	023 203 2 2 241 421 4 4	402 042 2 2	004 1 044 2
209. Monoclinic, planaxial	110 010 001	111 2 11$\bar{1}$ 2	002 020 1 1 200 1	$\bar{2}$02 022 1 2 220 202 2 1	$\bar{1}$13 113 2 2 $\bar{1}$31 131 2 2 $\bar{3}$11 311 2 2	222 $\bar{2}$22 2 2	004 1 004 1 040 1
211, 216. Monoclinic, planaxial	110 $\bar{1}$10 001	$\bar{2}$01 021 1 2 201 1	002 220 1 2	$\bar{2}$22 222 2 2 040 400 1 1	$\bar{2}$03 023 1 2 203 $\bar{2}$41 1 2 $\bar{4}$21 241 2 2 421 1	402 $\bar{4}$02 1 1 042 1	004 1 440 2
212, 213, 214, 215 Monoclinic, planaxial	10$\bar{1}$ $\bar{1}$10 01$\bar{1}$	0003 1 $\bar{2}$201 2 20$\bar{2}$1 1	$\bar{2}$022 1 20$\bar{2}$2 2	$\bar{2}$204 2 $\bar{2}$4$\bar{2}$0 1 22$\bar{4}$0 2 2$\bar{2}$04 1	20$\bar{2}$5 $\bar{4}$223 2 2 40$\bar{4}$1 02$\bar{2}$5 1 2 $\bar{2}$4$\bar{2}$3 22$\bar{4}$3 2 2 04$\bar{4}$1 2	0006 1 $\bar{4}$402 1 40$\bar{4}$2 1	$\bar{4}$044 1 04$\bar{4}$4 2

TABLE 3-27c

Splitting type	Main lines on pattern						Species number and system
	110	200	211	220	310	222	
1	1	1	1	1	1	1	197. Cubic
2	2	2	2	2	3	2	198. Tetragonal
3	3	3	3	3	6	1	206. Orthorhombic
4	2	1	3	2	2	2	199. Trigonal
5	3	4	2	3	4	2	200, 207, 208, 210. Orthorhombic
6	4	3	6	4	8	2	203, 209. Monoclinic
7	4	2	7	4	6	3	201, 202, 211, 212, 213, 214, 215, 216. Monoclinic
8	6	3	12	6	12	4	204, 205, 217, 218. Triclinic

TABLE 3-27d

No. of homology species, system, and species of symmetry	Determinant	Main lines on pattern					
		110	200	211	220	310	222
197. Cubic, planaxial	100 010 001	110 6	200 3	211 12	220 6	310 12	222 6
198. Tetragonal, planaxial	100 010 001	101 110 4 2	002 200 1 2	112 211 4 8	202 220 4 2	103 301 4 4 130 4	222 4
206. Orthorhombic, planaxial	100 010 001	101 011 2 2 110 2	002 020 1 1 200 1	112 121 4 4 211 4	202 022 2 2 220 2	103 301 2 2 130 310 2 2 013 031 2 2	222 4
199. Trigonal, planaxial	$1\bar{1}0$ 011 101 111	$01\bar{1}2$ 3 $11\bar{2}0$ 3	$20\bar{2}1$ 3	$10\bar{1}4$ $12\bar{3}2$ 3 6 $03\bar{3}0$ 3	$02\bar{2}4$ $22\bar{4}0$ 3 3	$21\bar{3}1$ 6 $41\bar{3}2$ 6	$04\bar{4}2$ 3 0006 1
207, 208, 210. Orthorhombic, planaxial	110 $\bar{1}10$ 001	020 111 1 4 200 1	002 022 2 2	202 022 2 2 311 131 4 4	040 222 1 4 400 1	113 331 4 4 240 420 2 2	402 2 042 2

TABLE 3-27d (cont'd)

No. of homology species, system, and species of symmetry	Determinant	Main lines on pattern					
		110	200	211	220	310	222
203, 209. Monoclinic, planaxial	$\bar{1}$10 010 011	101 011 1 2 110 101 2 1	002 020 1 1 200 1	$\bar{1}$12 112 2 2 $\bar{1}$21 121 2 2 $\bar{2}$11 211 2 2	$\bar{2}$02 022 1 2 220 202 2 1	$\bar{1}$03 103 1 1 031 013 2 2 310 130 2 2 $\bar{3}$01 301 1 1	222 2 $\bar{2}\bar{2}$2 2
211. 216. Monoclinic, planaxial	110 $\bar{1}$10 001	$\bar{1}$11 111 2 2 020 200 1 1	220 002 2 1	$\bar{2}$02 222 1 2 202 311 1 2 $\bar{3}$11 131 2 2 $\bar{1}$31 2	$\bar{2}$22 222 2 2 040 400 1 1	$\bar{1}$13 113 2 2 331 331 2 2 420 240 2 2	402 1 042 2 402 1
201, 202. Monoclinic, planaxial	1$\bar{1}\bar{1}$ 01$\bar{1}$ 011	012 210 2 2 020 20$\bar{2}$ 1 1	202 1 $\bar{2}$22 2	$\bar{2}$14 004 2 1 222 $\bar{2}$32 2 2 400 $\bar{2}$30 1 2 412 2	024 420 2 2 040 $\bar{4}$04 1 1	214 $\bar{2}$34 2 2 $\bar{2}$42 412 2 2 $\bar{4}$24 $\bar{4}$32	$\bar{2}$06 2 242 2 202 2
212, 213, 214, 215. Monoclinic, planaxial	10$\bar{1}$ $\bar{1}$10 01$\bar{1}$ 111	$\bar{1}$102 2 10$\bar{1}$2 1 $\bar{1}$2$\bar{1}$0 1 11$\bar{2}$0 2	02$\bar{2}$2 2 $\bar{2}$022 1	01$\bar{1}$4 $\bar{1}$014 2 1 30$\bar{3}$0 $\bar{2}$3$\bar{1}$1 1 2 21$\bar{3}$2 03$\bar{3}$0 2 2 $\bar{3}$212 2	$\bar{2}$204 2 20$\bar{2}$4 $\bar{2}$4$\bar{2}$0 1 1 22$\bar{4}$0 2	12$\bar{3}$4 2 $\bar{1}$3$\bar{2}$4 2 $\bar{3}$124 2 $\bar{1}$4$\bar{3}$2 2 13$\bar{4}$2 $\bar{4}$132 2	0006 1 $\bar{4}$403 1 40$\bar{4}$2 2
205, 217. Triclinic, central	100 010 001	$\bar{1}$01 101 1 1 011 0$\bar{1}$1 1 1 $\bar{1}$10 110 1 1	102 020 1 1 200 1	$\bar{1}$12 1$\bar{1}$2 1 1 $\bar{1}\bar{1}$2 112 1 1 $\bar{1}$21 1$\bar{2}$1 1 1 $\bar{1}\bar{2}$1 121 1 1 $\bar{2}$11 2$\bar{1}$1 1 1 $\bar{2}\bar{1}$1 211 1 1	$\bar{2}$02 202 1 1 022 0$\bar{2}$2 1 1 $\bar{2}\bar{2}$0 220 1 1	$\bar{1}$03 103 1 1 013 0$\bar{1}$3 1 1 031 0$\bar{3}$1 1 1 $\bar{1}$30 130 1 1 $\bar{3}$01 301 1 1 $\bar{3}\bar{1}$0 310 1 1	$\bar{2}\bar{2}$2 1 2$\bar{2}$2 1 222 1 $\bar{2}$2$\bar{2}$ 1

TABLE 3-27d (cont'd)

No. of homology species, system, and species of symmetry	Determinant	Main lines on pattern					
		110	200	211	220	310	222
204, 218. Triclinic, central	101	01$\bar{1}$2 1	$\bar{2}$022 1	$\bar{3}$122 1 21$\bar{3}$2 1	0$\bar{2}$24 1 20$\bar{2}$4 1	$\bar{4}$132 1	0006 1
	01$\bar{1}$	10$\bar{1}$2 1	02$\bar{2}$2 1	30$\bar{3}$0 1 $\bar{3}$212 1	$\bar{2}$204 1 4$\bar{2}\bar{2}$0 1	31$\bar{4}$2 1	40$\bar{4}$2 1
	110	$\bar{1}$102 1	2$\bar{2}$02 1	$\bar{1}$232 1 03$\bar{3}$0 1	22$\bar{4}$0 1 $\bar{2}$4$\bar{2}$0 1	13$\bar{4}$2 1	0$\bar{4}$42 1
	$\bar{1}$11	21$\bar{1}$0 1		01$\bar{1}$4 1 1$\bar{1}$04 1		12$\bar{3}$4 1	04$\bar{4}$2 1
		11$\bar{2}$0 1		3$\bar{3}$00 1 $\bar{1}$014 1		3$\bar{2}\bar{1}$4 1	
		$\bar{1}$2$\bar{1}$0 1		$\bar{2}$31$\bar{2}$ 1 1$\bar{3}$22 1		4$\bar{3}\bar{1}$2 1	
						$\bar{3}$124 1	
						2$\bar{1}$34 1	
						$\bar{1}$4$\bar{3}$2 1	
						$\bar{1}$3$\bar{2}$4 1	
						23$\bar{1}$4 1	
						3$\bar{4}$12 1	

TABLE 3-27e

Splitting type	Main lines on pattern						Species number and system
	100	110	111	200	210	211	
1	1	1	1	1	1	1	197. Cubic
2	2	2	1	2	3	2	198. Tetragonal
3	3	3	1	3	6	3	206. Orthorhombic
4	1	2	2	1	2	3	199. Trigonal
5	2	3	2	2	4	4	200, 207, 208, 210. Orthorhombic
6	3	4	2	3	8	6	203, 209. Monoclinic
7	2	4	3	2	6	7	201, 202, 211, 212, 213, 214, 215, 216. Monoclinic
8	3	6	4	3	12	12	204, 205, 217, 218. Triclinic

TABLE 3-27f

No. of homology species, system, and species of symmetry	Determinant	Main lines on pattern					
		100	110	111	200	210	211
197. Cubic, planaxial	100 010 001	100 3	100 6	111 4	200 3	210 12	211 12
198. Tetragonal, planaxial	100 010 001	001 1 100 2	101 110 4 2	111 4	002 1 200 2	102 201 120 4 4 4	112 211 4 8

TABLE 3-27f (cont'd)

No. of homology species, system, and species of symmetry	Determinant	Main lines on pattern					
		100	110	111	200	210	211
206. Orthorhombic, planaxial	100 010 001	011 1 010 1	101 011 2 2 110 2 100 1	111 4	002 1 020 1 200 1	102 201 120 2 2 2 210 012 021 2 2 2	112 121 4 4 211 4
199. Trigonal, planaxial	1̄10 011̄ 101̄ 111	101̄1 3	011̄2 3 112̄0 3	022̄1 3 0003 1	202̄2 3	32̄1̄2 336̄0 6 6	101̄4 123̄2 3 6 033̄0 3
200. Orthorhombic, planaxial	101̄ 010 101	101 2 010 1	002 200 1 1 111 4	012 2 210 2	202 2 020 1	103 212 301 2 4 2 121 4	113 311 4 4 024 220 2 2
207, 208, 210. Orthorhombic, planaxial	110 1̄10 001	001 1 011 2	020 111 1 1 200 1	201 2 021 2	002 1 022 2	112 221 310 4 4 2 130 2	202 022 2 2 311 131 4 4
203, 209. Monoclinic, planaxial	100 010 001	001 1 010 1 100 1	1̄01 011 1 2 110 101 2 1	2̄2̄2 2 222 2	002 1 020 1 200 1	1̄02 012 102 1 2 1 2̄01 021 201 1 2 1 120 210 2 2	1̄12 112 2 2 1̄21 121 2 2 2̄11 211 2 2
211, 216. Monoclinic, planaxial	110 1̄10 001	001 1 110 2	1̄11 111 2 2 020 200 1 1	2̄01 1 021 2 201 1	002 1 220 2	1̄12 112 2̄21 2 2 2 221 210 130 2 2 2	2̄02 022 202 1 2 1 311 1̄31 131 2 2 2 3̄11 2
212, 213, 214, 215. Monoclinic, planaxial	101̄ 1̄10 011̄ 111	011̄1 2 1̄011 1	1̄102 101̄2 1 1 1̄21̄0 112̄0 1 2	0003 1 2̄201 2 202̄1 1	022̄2 2 2̄022 1	2̄113 112̄3 2 2 1̄21̄3 132̄1̄ 2 2 123̄1 312̄1 2 2	1̄014 011̄4 1 2 2̄312 3̄212 2 2 303̄0 033̄0 1 2 213̄2 2

TABLE 3-27f (cont'd)

No. of homology species, system, and species of symmetry	Determinant	Main lines on pattern					
		100	110	111	200	210	211
205, 217. Triclinic, central	100 010 001	001 1 010 1 100 1	$\bar{1}01$ 010 1 1 011 $0\bar{1}1$ 1 1 $\bar{1}10$ 110 1 1	$\bar{1}11$ 1 $1\bar{1}1$ 1 111 1 $\bar{1}\bar{1}1$ 1	002 1 020 1 200 1	$\bar{1}02$ $0\bar{1}2$ 102 1 1 1 012 $\bar{2}01$ $0\bar{2}1$ 1 1 1 201 021 120 1 1 1 $\bar{1}20$ 210 $21\bar{0}$ 1 1 1	$\bar{1}12$ $1\bar{1}2$ $\bar{1}\bar{1}2$ 1 1 1 112 $\bar{1}21$ $1\bar{2}1$ 1 1 1 $1\bar{2}1$ 121 $\bar{2}11$ 1 1 1 $2\bar{1}1$ $\bar{2}\bar{1}1$ 211 1 1 1
204, 218. Triclinic, central	101 $01\bar{1}$ 110 111	$\bar{1}011$ 1 $01\bar{1}1$ 1 1101 1	$01\bar{1}2$ $10\bar{1}2$ 1 1 $\bar{1}102$ $2\bar{1}\bar{1}0$ 1 1 $\bar{1}1\bar{2}0$ $1\bar{2}10$ 1 1	0003 1 $20\bar{2}1$ 1 $0\bar{2}21$ 1 $02\bar{2}1$ 1	$\bar{2}022$ 1 $02\bar{2}2$ 1 $2\bar{2}02$ 1	$\bar{3}121$ $\bar{1}\bar{1}23$ 1 1 $\bar{2}\bar{1}31$ $\bar{2}113$ 1 1 $1\bar{2}13$ $3\bar{2}\bar{1}1$ 1 1 $2\bar{3}11$ $2\bar{1}\bar{1}3$ 1 1 $1\bar{1}23$ $\bar{1}3\bar{2}1$ 1 1 $\bar{1}2\bar{1}3$ $12\bar{3}1$ 1 1	$\bar{3}122$ $21\bar{3}2$ 1 1 $\bar{3}212$ $\bar{1}2\bar{3}2$ 1 1 $1\bar{2}32$ $03\bar{3}0$ 1 1 $01\bar{1}4$ $1\bar{1}04$ 1 1 $3\bar{3}00$ $\bar{1}014$ 1 1 $\bar{2}3\bar{1}2$ $23\bar{1}2$ 1 1

TABLE 3-27g

Splitting type	Main lines on pattern								Species number and system
	$10\bar{1}0$	0002	$10\bar{1}1$	$10\bar{1}2$	$11\bar{2}0$	$10\bar{1}3$	$20\bar{2}0$	$11\bar{2}2$	
1	1	1	1	1	1	1	1	1	124. Hexagonal
2	2	1	2	2	2	2	2	2	126. Orthorhombic
3	2	1	3	3	2	2	3	4	125, 127. Monoclinic
4	3	1	3	3	3	3	3	3	129. Monoclinic
5	2	1	4	4	2	4	2	3	131, 132. Monoclinic
6	3	1	6	6	3	3	3	6	128, 130. Triclinic

TABLE 3-27h

No. of homology species, system, and species of symmetry	Determinant	Main lines on pattern							
		$10\bar{1}0$	0002	$10\bar{1}1$	$10\bar{1}2$	$11\bar{2}0$	$10\bar{1}3$	$20\bar{2}0$	$11\bar{2}2$
124. Hexagonal, heroidal-planal	100 010 011	$10\bar{1}0$ 3	0002 1	$10\bar{1}1$ 6	$10\bar{1}2$ 6	$11\bar{2}0$ 3	$10\bar{1}3$ 6	$20\bar{2}0$ 3	$11\bar{2}2$ 6
126. Orthorhombic, planal	$1\bar{1}0$ 001 $\bar{1}\bar{1}0$	101 2 200 1	020 1	111 210 4 2	121 220 4 2	002 1 301 2	131 4 230 2	202 2 400 1	022 2 321 4

TABLE 3-27h (cont'd)

No. of homology species, system, and species of symmetry	Determinant	Main lines on pattern							
		$10\bar{1}0$	0002	$10\bar{1}1$	$10\bar{1}2$	$11\bar{2}0$	$10\bar{1}3$	$20\bar{2}0$	$11\bar{2}2$
125, 127. Monoclinic, planal	110 $\bar{1}$10 001	110 2 020 1	002 1	111 $\bar{1}$11 2 2 021 2	112 $\bar{1}\bar{1}$2 2 2 022 2	130 2 200 1	113 2 $\bar{1}$13 2 023 2	220 1 $\bar{2}$20 1 040 1	131 2 $\bar{1}$32 2 202 1 $\bar{2}$02 1
129. Monoclinic, planal	100 001 $\bar{1}$10	$\bar{1}$01 1 001 1 100 1	020 1	$\bar{1}$11 011 2 2 110 2	$\bar{1}$21 021 2 2 120 2	$\bar{2}$01 1 $\bar{1}$02 1 101 1	$\bar{1}$31 2 031 2 130 2	$\bar{2}$02 1 002 1 200 1	$\bar{2}$21 2 $\bar{1}$22 2 121 2
131, 132. Monoclinic, axial	210 010 001	200 1 110 2	002 1	$\bar{2}$01 201 1 1 $\bar{1}$11 111 2 2	$\bar{2}$02 202 1 1 $\bar{1}$12 112 2 2	310 2 020 1	$\bar{2}$03 1 203 1 $\bar{1}$13 2 113 2	400 1 220 2	$\bar{3}$12 2 312 2 022 2
128, 130. Triclinic, primitive	100 010 001	$10\bar{1}0$ 1 $01\bar{1}0$ 1 $\bar{1}100$ 1	0002 1	$10\bar{1}1$ $\bar{1}011$ 1 1 $01\bar{1}1$ $0\bar{1}11$ 1 1 $\bar{1}101$ $1\bar{1}01$ 1 1	$10\bar{1}2$ $\bar{1}012$ 1 1 $01\bar{1}2$ $0\bar{1}12$ 1 1 $\bar{1}102$ $1\bar{1}02$ 1 1	$11\bar{2}0$ 1 $\bar{1}2\bar{1}0$ 1 $\bar{2}110$ 1	$10\bar{1}3$ 1 $01\bar{1}3$ 1 $\bar{1}103$ 1 $\bar{1}013$ 1 $0\bar{1}13$ 1 $1\bar{1}03$ 1	$20\bar{2}0$ 1 $0\bar{2}20$ 1 $\bar{2}200$ 1	$11\bar{2}2$ 1 $\bar{1}2\bar{1}2$ 1 $2\bar{1}\bar{1}2$ 1 $\bar{1}\bar{1}22$ 1 $1\bar{2}12$ 1 $\bar{2}112$ 1

CHAPTER 4

LINE INTENSITIES ON X-RAY PATTERNS

Chapter 4 gives data needed to calculate integral intensities for lines as recorded in various ways.

See [95-100, 275] for a detailed description of methods of measuring and calculating line intensities.

4-1. Some Formulas for Line Intensities

1. The absolute intensity of the diffracted radiation from an ideal mosaic crystal or from a crystalline powder is, in the absence of extinction,

$$P' = I_0 \frac{N^2 e^4 \lambda^3}{m^2 c^4} V_s PLGHF^2 f_T, \tag{27}$$

where I_0 is the intensity of the incident beam of unpolarized radiation, N is the number of unit cells in unit volume, V_s is the volume exposed to the beam, P is the polarization factor, L is the Lorentz factor, G is the geometric factor for the conditions used, H is the multiplicity factor, F is the structure factor, and f_T is the temperature factor.

2. The absolute intensity of the Bragg reflection from the face of a mosaic crystal is

$$E\omega = I_0 \frac{N^2}{2\mu} \frac{e^4 \lambda^3}{2m^2 c^4} F^2 \frac{(1+\cos^2 2\vartheta)}{\sin 2\vartheta}, \tag{28}$$

where E is the energy scattered in the direction ϑ, ϑ is the inclination of the crystal to the primary beam, μ is the linear attenuation coefficient, and ω is the angular velocity of the crystal.

3. The absolute intensity scattered by a mosaic crystal examined in transmission for an angle of inclination $\psi = \vartheta$ is

$$E\omega = I_0 N^2 \frac{e^4 \lambda^3}{2m^2 c^4} F^2 \frac{1+\cos^2 2\vartheta}{\sin 2\vartheta} \frac{d}{\cos \vartheta} e^{-\frac{\mu d}{\cos \vartheta}}, \tag{29}$$

where d is the thickness in centimeters.

4. The absolute intensity scattered by a powder specimen with a flat surface for an inclination $\psi = 90° - \vartheta$ as recorded with ionization equipment is

$$P'' = I_0 \frac{N^2}{18\mu r} \frac{e^4 \lambda^3}{2m^2 c^4} F^2 H l \frac{1+\cos^2 2\vartheta}{\sin^2 \vartheta \cos \vartheta}, \tag{30}$$

where P" is the proportion of the radiation scattered in the given direction that is recorded, r is the radius of the goniometer, and l is the distance traveled by the reflected ray in the counter.

5. The absolute intensity scattered by a plane powder specimen in transmission for $\psi = \vartheta$ is

$$P_t = I_0 \frac{N^2}{4\pi^2} \frac{e^4\lambda^3}{2m^2c^4} F^2 H l \frac{1+\cos^2 2\vartheta}{\sin 2\vartheta} \frac{\varrho'}{\varrho} , \qquad (31)$$

where ρ' and ρ are the densities of powder and monocrystal, respectively.

6. The relative intensity recorded in a Debye camera (powder and cylindrical film) is

$$I = a \frac{1+\cos^2 2\vartheta}{\sin^2 \vartheta \cos \vartheta} F^2 A(\vartheta) f_T, \qquad (32)$$

where a is a coefficient related to $I_0 N^2 e^4 V_s/m^2 c^4$.

7. The relative intensity as recorded by a flat film normal to the primary beam is

$$I = a \frac{(1+\cos^2 2\vartheta)\cos 2\vartheta}{\sin^2 \vartheta \cos \vartheta} F^2 A(\vartheta) f_T. \qquad (33)$$

8. The relative intensity as recorded with a monochromator (Guinier's system) is [101]

$$I = \frac{1+\cos^2 2\alpha \cos^2 \vartheta}{(1+\cos^2 2\alpha) \sin 2\vartheta \cos 2\vartheta} F^2 A(\vartheta) f_T, \qquad (34)$$

where α is the angle of reflection for the monochromator.

4-2. Values of $e^4\lambda^3/2m^2c^4$

The table gives values of $C = e^4\lambda^3/2m^2c^4$ for various $K\alpha_1$ and $K\alpha_2$ [102]; this factor appears in the expression for the absolute intensities of the interference spots in all methods. The value of $e^4/2m^2c^4$ is $7.83 \cdot 10^{-26}$ cm^2.

Anode	$C(\alpha_1)$	$C(\alpha_2)$	Anode	$C(\alpha_1)$	$C(\alpha_2)$	Anode	$C(\alpha_1)$	$C(\alpha_2)$
V	61,42	61,70	Ni	17,83	17,96	Pd	0,785	0,803
Cr	46,99	47,23	Cu	14,31	14,42	Ag	0,685	0,702
Mn	36,35	36,55	Zn	11,57	11,66	W	0,357	0,383
Fe	28,40	28,58	Mo	1,400	1,422			
Co	22,41	22,56	Rh	0,903	0,922			

ANGULAR INTENSITY FACTORS

4-3. Product of the Polarization, Lorentz, and Geometric Factors for Patterns Recorded without Monochromators

The polarization factor is

$$P = \frac{1}{2}(1 + \cos^2 2\vartheta). \tag{35}$$

The Lorentz factor is

$$L = \frac{1}{\sin 2\vartheta}. \tag{36}$$

The geometric factor G is dependent on the method. Sections 4-3a to 4-3c give PLG for ϑ between 2 and 87° [102].

4-3a. Debye Patterns in Cylindrical Cameras

$$PLG = \frac{1 + \cos^2 2\vartheta}{\sin^2 \vartheta \cos \vartheta}. \tag{37}$$

$\vartheta°$	0	.1	.2	.3	.4	.5	.6	.7	.8	.9
2	1639	1486	1354	1239	1138	1048	968,9	898,3	835,1	778,4
3	727,2	680,9	638,8	600,5	565,6	533,6	504,3	477,3	452,3	429,3
4	408,0	388,2	369,9	352,7	336,8	321,9	308,0	294,9	282,6	271,1
5	260,3	250,1	240,5	231,4	222,9	214,7	207,1	199,8	192,9	186,3
6	180,1	174,2	168,5	163,1	158,0	153,1	148,4	144,0	139,7	135,6
7	131,7	128,0	124,4	120,9	117,6	114,4	111,4	108,5	105,6	102,9
8	100,3	97,80	95,37	93,03	90,78	88,60	86,51	84,48	82,52	80,63
9	78,79	77,02	75,31	73,66	72,05	70,49	68,99	67,53	66,12	64,74
10	63,41	62,12	60,87	59,65	58,46	57,32	56,20	55,11	54,06	53,03
11	52,04	51,06	50,12	49,19	48,30	47,43	46,58	45,75	44,94	22,16
12	43,39	42,64	41,91	41,20	40,50	39,82	39,16	38,51	37,88	37,27
13	36,67	36,08	35,50	34,94	34,39	33,85	33,33	32,81	32,31	31,82
14	31,34	30,87	30,41	29,96	29,51	29,08	28,66	28,24	27,83	27,44
15	27,05	26,66	26,29	25,92	25,56	25,21	24,86	24,52	24,19	23,86
16	23,54	23,23	22,92	22,61	22,32	22,02	21,74	21,46	21,18	20,91
17	20,64	20,38	20,12	19,87	19,62	19,38	19,14	18,90	18,67	18,44
18	18,22	18,00	17,78	17,57	17,36	17,15	16,95	16,75	16,56	16,36
19	16,17	15,99	15,80	15,62	15,45	15,27	15,10	14,93	14,76	14,60
20	14,44	14,28	14,12	13,97	13,81	13,66	13,52	13,37	13,23	13 09
21	12,95	12,81	12,68	12,54	12,41	12,28	12,15	12,03	11,91	11,78
22	11,66	11,54	11,43	11,31	11,20	11,09	10,98	10,87	10,76	10,65
23	10,55	10,45	10,35	10,24	10,15	10,05	9,951	9,857	9,763	9,671
24	9,579	9,489	9,400	9,313	9,226	9,141	9,057	8,973	8,891	8,810
25	8,730	8,651	8,573	8,496	8,420	8,345	8,271	8,198	8,126	8,054
26	7,984	7,915	7,846	7,778	7,711	7,645	7,580	7,515	7,452	7,389
27	7,327	7,266	7,205	7,145	7,086	7,027	6,969	6,912	6,856	6,800
28	6,745	6,692	6,637	6,584	6,532	6,480	6,429	6,379	6,329	6,279
29	6,230	6,183	6,135	6,088	6,042	5,995	5,950	5,905	5,861	5,817
30	5,774	5,731	5,688	5,647	5,605	5,564	5,524	5,484	5,445	5,406
31	5,367	5,329	5,292	5,254	5,218	5,181	5,145	5,110	5,075	5,040
32	5,006	4,972	4,939	4,906	4,873	4,841	4,809	4,777	4,746	4,715
33	4,685	4,655	4,625	4,595	4,566	4,538	4,509	4,481	4,453	4,426
34	4,399	4,372	4,346	4,320	4,294	4,268	4,243	4,218	4,193	4,169
35	4,145	4,121	4,097	4,074	4,052	4,029	4,006	3,984	3,962	3,941
36	3,919	3,898	3,877	3,857	3,836	3,816	3,797	3,777	3,758	3,739
37	3,720	3,701	3,683	3,665	3,647	3,629	3,612	3,594	3,577	3,561
38	3,544	3,527	3,513	3,497	3,481	3,465	3,449	3,434	3,419	3,404
39	3,389	3,375	3,361	3,347	3,333	3.320	3,306	3,293	3,280	3,268

$\vartheta°$	0	.1	.2	.3	.4	.5	.6	.7	.8	.9
40	3,255	3,242	3,230	3,218	3,206	3,194	3,183	3,171	3,160	3,149
41	3,138	3,127	3,117	3,106	3,096	3,086	3,076	3,067	3,057	3,048
42	3,038	3,029	3,020	3,012	3,003	2,994	2,986	2,978	2,970	2,962
43	2,954	2,945	2,939	2,932	2,925	2,918	2,911	2,904	2,897	2,891
44	2,884	2,878	2,872	2,866	2,860	2,855	2,849	2,844	2,838	2,833
45	2,828	2,824	2,819	2,814	2,810	2,805	2,801	2,797	2,793	2,789
46	2,785	2,782	2,778	2,775	2,772	2,769	2,766	2,763	2,760	2,757
47	2,755	2,752	2,750	2,748	2,746	2,744	2,742	2,740	2,738	2,737
48	2,736	2,735	2,733	2,732	2,731	2,730	2,730	2,729	2,729	2,728
49	2,728	2,728	2,728	2,728	2,728	2,728	2,729	2,729	2,730	2,730
50	2,731	2,732	2,733	2,734	2,735	2,737	2,738	2,740	2,741	2,743
51	2,745	2,747	2,749	2,751	2,753	2,755	2,758	2,760	2,763	2,766
52	2,769	2,772	2,775	2,778	2,782	2,785	2,788	2,792	2,795	2,799
53	2,803	2,807	2,811	2,815	2,820	2,824	2,828	2,833	2,838	2,843
54	2,848	2,853	2,858	2,863	2,868	2,874	2,879	2,885	2,890	2,896
55	2,902	2,908	2,914	2,921	2,927	2,933	2,940	2,946	2,953	2,960
56	2,967	2,974	2,981	2,988	2,996	3,004	3,011	3,019	3,026	3,034
57	3,042	3,050	3,059	3,067	3,075	3,084	3,092	3,101	3,110	3,119
58	3,128	3,137	3,147	3,156	3,166	3,175	3,185	3,195	3,205	3,215
59	3,225	3,235	3,246	3,256	3,267	3,278	3,289	3,300	3,311	3,322
60	3,333	3,345	3,356	3,368	3,380	3,392	3,404	3,416	3,429	3,441
61	3,454	3,466	3,479	3,492	3,505	3,518	3,532	3,545	3,559	3,573
62	3,587	3,601	3,615	3,629	3,643	3,658	3,673	3,688	3,703	3,718
63	3,733	3,749	3,764	3,780	3,796	3,812	3,828	3,844	3,861	3,878
64	3,894	3,911	3,928	3,946	3,963	3,980	3,998	4,016	4,034	4,052
65	4,071	4,090	4,108	4,127	4,147	4,166	4,185	4,205	4,225	4,245
66	4,265	4,285	4,306	4,327	4,348	4,369	4,390	4,412	4,434	4,456
67	4,478	4,500	4,523	4,546	4,569	4,592	4,616	4,640	4,664	4,688
68	4,712	4,737	4,762	4,787	4,812	4,838	4,864	4,890	4,916	4,943
69	4,970	4,997	5,024	5,052	5,080	5,109	5,137	5,166	5,195	5,224
70	5,254	5,284	5,315	5,345	5,376	5,408	5,440	5,471	5,504	5,536
71	5,569	5,602	5,636	5,670	5,705	5,740	5,775	5,810	5,846	5,883
72	5,919	5,956	5,994	6,032	6,071	6,109	6,149	6,189	6,229	6,270
73	6,311	6,352	6,394	6,437	6,480	6,524	6,568	6,613	6,658	6,703
74	6,750	6,797	6,844	6,892	6,941	6,991	7,041	7,091	7,142	7,194
75	7,247	7,300	7,354	7,409	7,465	7,521	7,578	7,636	7,694	7,753
76	7,813	7,874	7,936	7,999	8,063	8,128	8,193	8,259	8,327	8,395
77	8,465	8,536	8,607	8,680	8,754	8,829	8,905	8,982	9,061	9,142
78	9,223	9,305	9,389	9,474	9,561	9,649	9,739	9,831	9,924	10,02
79	10,12	10,21	10,31	10,41	10,52	10,62	10,73	10,84	10,95	11,06
80	11,18	11,30	11,42	11,54	11,67	11,80	11,93	12,06	12,20	12,34
81	12,48	12,63	12,78	12,93	13,08	13,24	13,40	13,57	13,74	13,92
82	14,10	14,28	14,47	14,66	14,86	15,07	15,28	15,49	15,71	15,94
83	16,17	16,41	16,66	16,91	17,17	17,44	17,72	18,01	18,31	18,61
84	18,93	19,25	19,59	19,94	20,30	20,68	21,07	21,47	21,89	22,32
85	22,77	23,24	23,73	24,24	24,78	25,34	25,92	26,52	27,16	27,83
86	28,53	29,27	30,04	30,86	31,73	32,64	33,60	34,63	35,72	36,88
87	38,11	39,43	40,84	42,36	44,00	45,76	47,68	49,76	52,02	54,50

4-3b. Recording on Flat Film

$$PLG = \frac{(1+\cos^2 2\vartheta)\cos 2\vartheta}{\sin^2 \vartheta \cos \vartheta}.$$

(38)

$\vartheta°$	.0	.2	.4	.6	.8
2	1635	1350	1134	964,9	831,1
3	723,2	634,8	561,6	500,6	448,3
4	404,0	365,9	332,8	304,0	278,6
5	256,3	236,5	219,0	203,2	189,0

$\vartheta°$	.0	.2	.4	.6	.8
6	176,2	164,6	154,1	144,5	135,8
7	127,8	120,5	113,7	107,5	101,7
8	96,41	91,49	86,91	82,64	78,66
9	74,93	71,46	68,20	65,15	62,28
10	59,58	56,96	54,65	52,39	50,26
11	48,25	46,33	44,52	42,81	41,18
12	39,63	38,16	36,76	35,43	34,16
13	32,95	31,79	30,69	29,64	28,63
14	27,67	26,75	25,86	25,01	24,19
15	23,42	22,67	21,95	21,26	20,60
16	19,96	19,35	18,76	18,19	17,64
17	17,11	16,60	16,11	15,64	15,18
18	14,74	14,31	13,90	13,50	13,12
19	12,74	12,38	12,04	11,70	11,37
20	11,06	10,75	10,45	10,17	9,893
21	9,624	9,364	9,106	8,857	8,625
22	8,388	8,166	7,947	7,737	7,528
23	7,329	7,138	6,948	6,761	6,583
24	6,410	6,241	6,077	5,918	5,762
25	5,612	5,465	5,322	5,183	5,047
26	4,915	4,787	4,662	4,541	4,422
27	4,307	4,194	4,085	3,977	3,873
28	3,772	3,673	3,577	3,483	3,391
29	3,301	3,215	3,130	3,047	2,966
30	2,887	2,810	2,734	2,661	2,590
31	2,520	2,452	2,385	2,320	2,257
32	2,194	2,134	2,075	2,017	1,961
33	1,906	1,852	1,799	1,747	1,697
34	1,648	1,600	1,553	1,507	1,462
35	1,418	1,374	1,333	1,291	1,251
36	1,211	1,172	1,134	1,098	1,061
37	1,025	0,990	0,956	0,923	0,890
38	0,857	0,826	0,795	0,764	0,734
39	0,705	0,676	0,647	0,619	0,592
40	0,565	0,539	0,513	0,487	0,462
41	0,437	0,412	0,388	0,364	0,341
42	0,318	0,295	0,272	0,250	0,228
43	0,206	0,185	0,163	0,142	0,121
44	0,101	0,080	0,060	0,040	0,020
45	0,000	0,020	0,039	0,059	0,078
46	0,097	0,116	0,135	0,154	0,173
47	0,192	0,211	0,230	0,249	0,267
48	0,286	0,305	0,323	0,342	0,361
49	0,380	0,399	0,417	0,436	0,455
50	0,474	0,493	0,512	0,532	0,551
51	0,571	0,590	0,610	0,630	0,650
52	0,670	0,690	0,711	0,731	0,752
53	0,773	0,794	0,815	0,836	0,858
54	0,880	0,902	0,924	0,947	0,969
55	0,993	1,016	1,039	1,063	1,087
56	1,111	1,136	1,161	1,186	1,211
57	1,237	1,264	1,290	1,317	1,344
58	1,371	1,399	1,427	1,456	1,485
59	1,514	1,544	1,574	1,605	1,635
60	1,667	1,698	1,731	1,763	1,797
61	1,830	1,864	1,899	1,934	1,970
62	2,006	2,042	2,079	2,117	2,156
63	2,194	2,234	2,274	2,314	2,356
64	2,397	2,440	2,483	2,527	2,571
65	2,617	2,662	2,710	2,757	2,805

$\vartheta°$	.0	.2	.4	.6	.8
66	2,854	2,904	2,954	3,005	3,058
67	3,111	3,165	3,219	3,275	3,332
68	3,390	3,448	3,508	3,569	3,630
69	3,693	3,757	3,822	3,889	3,956
70	4,025	4,095	4,166	4,240	4,313
71	4,388	4,465	4,544	4,624	4,705
72	4,789	4,874	4,961	5,049	5,140
73	5,232	5,326	5,422	5,521	5,622
74	5,724	5,829	5,937	6,048	6,160
75	6,276	6,394	6,516	6,641	6,768
76	6,898	7,033	7,171	7,313	7,459
77	7,608	7,762	7,921	8,084	8,252
78	8,426	8,604	8,788	8,978	9,175
79	9,383	9,586	9,808	10,031	10,263
80	10,51	10,76	11,02	11,29	11,58
81	11,87	12,18	12,50	12,83	13,18
82	13,55	13,94	14,34	14,77	15,22
83	15,69	16,19	16,72	17,28	17,88
84	18,52	19,19	19,91	20,70	21,53
85	22,42	23,40	24,46	25,61	26,87
86	28,25	29,78	31,48	33,36	35,50
87	37,90	40,65	43,82	47,51	51,87
88	57,10				

4-3c. Reflection from a Monocrystal

$$PLG = \frac{1+\cos^2 2\vartheta}{\sin 2\vartheta} \ . \tag{39}$$

$\vartheta°$	.0	.2	.4	.6	.8
2	28,60	25,99	23,83	21,97	20,40
3	19,03	17,83	16,77	15,84	14,99
4	14,23	13,55	12,92	12,35	11,82
5	11,34	10,90	10,49	10,10	9,747
6	9,413	9,099	8,806	8,529	8,270
7	8,025	7,796	7,573	7,367	7,166
8	6,980	6,801	6,631	6,468	6,312
9	6,163	6,020	5,884	5,753	5,627
10	5,506	5,381	5,277	5,169	5,065
11	4,965	4,867	4,773	4,683	4,595
12	4,511	4,428	4,348	4,271	4,196
13	4,124	4,053	3,985	3,919	3,853
14	3,791	3,730	3,669	3,612	3,555
15	3,501	3,446	3,394	3,343	3,293
16	3,244	3,197	3,151	3,105	3,061
17	3,017	2,975	2,934	2,894	2,854
18	2,815	2,777	2,740	2,703	2,668
19	2,633	2,598	2,566	2,533	2,500
20	2,469	2,438	2,407	2,378	2,349
21	2,320	2,293	2,264	2,236	2,212
22	2,184	2,159	2,134	2,110	2,085
23	2,061	2,039	2,016	1,992	1,970
24	1,948	1,927	1,906	1,885	1,865
25	1,845	1,825	1,806	1,787	1,768
26	1,750	1,732	1,714	1,697	1,680
27	1,663	1,647	1,630	1,614	1,599

ϑ^{c}	.0	.2	.4	.6	.8
28	1,583	1,568	1,553	1,539	1,524
29	1,510	1,497	1,483	1,469	1,456
30	1,443	1,431	1,418	1,406	1,394
31	1,382	1,371	1,359	1,348	1,337
32	1,326	1,316	1,306	1,295	1,285
33	1,276	1,266	1,257	1,248	1,239
34	1,230	1,221	1,213	1,205	1,196
35	1,189	1,181	1,174	1,166	1,159
36	1,152	1,145	1,138	1,132	1,126
37	1,119	1,113	1,108	1,102	1,096
38	1,091	1,086	1,081	1,076	1,071
39	1,066	1,062	1,058	1,054	1,050
40	1,046	1,042	1,039	1,036	1,032
41	1,029	1,027	1,024	1,021	1,019
42	1,016	1,014	1,012	1,011	1,009
43	1,007	1,006	1,005	1,004	1,003
44	1,002	1,001	1,001	1,000	1,000
45	1,000	1,000	1,000	1,001	1,001
46	1,002	1,003	1,004	1,005	1,006
47	1,007	1,009	1,011	1,012	1,014
48	1,017	1,019	1,021	1,024	1,027
49	1,029	1,033	1,036	1,039	1,043
50	1,046	1,050	1,054	1,058	1,062
51	1,067	1,071	1,076	1,081	1,086
52	1,091	1,096	1,102	1,107	1,113
53	1,119	1,125	1,132	1,138	1,145
54	1,152	1,159	1,166	1,173	1,181
55	1,189	1,196	1,205	1,213	1,221
56	1,230	1,239	1,248	1,257	1,266
57	1,276	1,286	1,295	1,305	1,316
58	1,326	1,337	1,348	1,359	1,371
59	1,382	1,394	1,406	1,418	1,431
60	1,443	1,456	1,469	1,483	1,497
61	1,510	1,524	1,539	1,553	1,568
62	1,584	1,599	1,614	1,630	1,647
63	1,663	1,680	1,698	1,714	1,732
64	1,750	1,768	1,787	1,806	1,825
65	1,845	1,865	1,885	1,906	1,927
66	1,948	1,970	1,992	2,014	2,038
67	2,061	2,085	2,109	2,134	2,159
68	2,184	2,211	2,237	2,264	2,292
69	2,320	2,348	2,378	2,407	2,438
70	2,469	2,500	2,532	2,566	2,599
71	2,633	2,668	2,704	2,740	2,777
72	2,815	2,854	2,893	2,934	2,975
73	3,018	3,061	3,105	3,150	3,197
74	3,244	3,293	3,343	3,394	3,446
75	3,500	3,555	3,612	3,670	3,729
76	3,791	3,853	3,918	3,985	4,054
77	4,124	4,197	4,272	4,349	4,428
78	4,511	4,595	4,683	4,773	4,868
79	4,967	5,064	5,170	5,277	5,388
80	5,505	5,627	5,753	5,885	6,022
81	6,163	6,315	6,466	6,628	6,800
82	6,981	7,168	7,365	7,576	7,793
83	8,025	8,271	8,528	8,805	9,101
84	9,413	9,745	10,10	10,48	10,90
85	11,34	11,82	12,35	12,92	13,54
86	14,23	14,99	15,83	16,77	17,83
87	19,03	20,40	21,98	23,82	25,99
88	28,60				

4-4. Product of the Polarization, Lorentz, and Geometric Factors for Symmetrical Recording with a Monochromator

The polarization factor is calculated as follows: (a) for kinematic scattering

$$P_k = \frac{1 + \cos^2 2\alpha \cos^2 2\vartheta}{1 + \cos^2 2\alpha} \tag{40}$$

and (b) for dynamic scattering

$$P_d = \frac{1 + \cos 2\alpha \cos^2 2\vartheta}{1 + \cos 2\alpha} , \tag{41}$$

where α is the angle of reflection for the monochromator. The difference between P_k and P_d does not exceed 6%.

The Lorentz factor L takes the form

$$L = \frac{1}{\sin 2\vartheta} . \tag{42}$$

Then PLG for this case is

$$PLG = \frac{1 + \cos^2 2\alpha \cos^2 \vartheta}{(1 + \cos^2 2\alpha) \sin 2\vartheta \cos 2\vartheta} \tag{43}$$

The table gives PLG for Guinier's symmetrical method with a quartz crystal ($\beta = 0$) [102].

The table applies to Cu Kα ($\alpha = 13°24'$), Co Kα ($\alpha = 15°37'$), Cr Kα ($\alpha = 20°09'$), and Mo Kα ($\alpha = 6°43'$).

The PLG for unsymmetrical recording should be multiplied by the following factors, taking into account β which is the inclination of the camera with respect to the primary beam:

$$\frac{\sin 4\vartheta}{\sin 3\vartheta} \quad \text{for} \quad \beta = 30°,$$

$$\frac{\sin 4\vartheta}{\sin \frac{8}{3} \vartheta} \quad \text{for} \quad \beta = 45°,$$

$$\frac{\sin 4\vartheta}{\sin \frac{12}{5} \vartheta} \quad \text{for} \quad \beta = 60°.$$

Cu Kα

$\vartheta°$	.0	.2	.4	.6	.8
2	14.334	13.041	11.963	11.034	10.252
3	9.573	8.974	8.451	7.985	7.567
4	7.193	6.853	6.546	6.263	6.006
5	5.769	5.551	5.349	5.162	4.986
6	4.822	4.671	4.527	4.394	4.268
7	4.150	4.037	3.932	3.832	3.737

$\vartheta°$	.0	.2	.4	.6	.8
8	3.647	3.562	3.480	3.403	3.329
9	3.259	3.191	3.127	3.066	3.007
10	2.950	2.896	2.845	2.794	2.746
11	2.699	2.656	2.613	2.572	2.532
12	2.494	2.457	2.421	2.387	2.354
13	2.322	2.291	2.261	2.232	2.204
14	2.177	2.150	2.125	2.100	2.076
15	2.053	2.031	2.009	1.988	2.968
16	1.948	1.929	1.910	1.892	1.875
17	1.858	1.842	1.826	1.810	1.795
18	1.781	1.767	1.753	1.739	1.727
19	1.715	1.703	1.691	1.680	1.669
20	1.659	1.648	1.639	1.630	1.621
21	1.612	1.603	1.595	1.587	1.580
22	1.573	1.567	1.559	1.553	1.548
23	1.542	1.536	1.531	1.527	1.523
24	1.519	1.515	1.511	1.508	1.505
25	1.502	1.500	1.498	1.496	1.495
26	1.494	1.493	1.492	1.492	1.492
27	1.493	1.493	1.495	1.496	1.498
28	1.499	1.502	1.504	1.508	1.512
29	1.516	1.520	1.525	1.530	1.536
30	1.541	1.548	1.555	1.562	1.570
31	1.578	1.587	1.597	1.607	1.618
32	1.629	1.641	1.654	1.667	1.680
33	1.695	1.711	1.727	1.744	1.762
34	1.782	1.802	1.823	1.845	1.869
35	1.893	1.920	1.946	1.976	2.005
36	2.038	2.072	2.108	2.145	2.185
37	2.228	2.272	2.320	2.371	2.425
38	2.481	2.543	2.608	2.676	2.751
39	2.831	2.916	3.009	3.108	3.216
40	3.332	3.460	3.598	3.750	3.917
41	4.100	4.305	4.533	4.787	5.074
42	5.340	5.774	6.206	6.712	7.309
43	8.029	8.910	10.011	11.430	13.320
44	15.972	19.951	26.593	39.878	79.712

Co Kα

$\vartheta°$	.0	.2	.4	.6	.8
2	14.338	13.049	11.967	11.038	10.253
3	9.574	8.980	8.452	7.987	7.570
4	7.196	6.857	6.549	6.267	6.010
5	5.772	5.555	5.353	5.165	4.990
6	4.827	4.675	4.532	4.399	4.273
7	4.155	4.043	3.937	3.837	3.743
8	3.653	3.568	3.487	3.409	3.335
9	3.265	3.198	3.134	3.073	3.014
10	2.958	2.904	2.853	2.802	2.755
11	2.708	2.665	2.622	2.581	2.541
12	2.503	2.467	2.431	2.397	2.364
13	2.332	2.301	2.272	2.243	2.215
14	2.188	2.162	2.137	2.112	2.089
15	2.065	2.044	2.022	2.001	1.980
16	1.961	1.942	1.923	1.906	1.889
17	1.872	1.856	1.840	1.825	1.810
18	1.796	1.782	1.769	1.756	1.743
19	1.731	1.719	1.708	1.697	1.686
20	1.677	1.666	1.657	1.648	1.639
21	1.631	1.623	1.615	1.607	1.600
22	1.594	1.587	1.581	1.575	1.569
23	1.564	1.559	1.554	1.550	1.546
24	1.542	1.539	1.535	1.533	1.530
25	1.527	1.525	1.524	1.523	1.522
26	1.520	1.521	1.521	1.521	1.522
27	1.522	1.523	1.524	1.526	1.529
28	1.530	1.534	1.537	1.540	1.545
29	1.549	1.554	1.559	1.565	1.571
30	1.579	1.585	1.592	1.600	1.609
31	1.618	1.627	1.638	1.647	1.660
32	1.672	1.685	1.699	1.712	1.727
33	1.743	1.759	1.776	1.794	1.814
34	1.834	1.855	1.877	1.900	1.926
35	1.951	1.978	2.007	2.037	2.070
36	2.103	2.138	2.175	2.216	2.257
37	2.301	2.348	2.398	2.450	2.507
38	2.566	2.629	2.697	2.770	2.847
39	2.930	3.020	3.115	3.219	3.331

$\vartheta°$	.0	.2	.4	.6	.8
40	3.452	3.585	3.728	3.886	4.059
41	4.249	4.463	4.699	4.964	5.262
42	5.601	5.989	6.438	6.962	7.584
43	8.331	9.244	10.387	11.860	13.823
44	16.576	20.706	27.598	41.395	82.755

Cr Kα

$\vartheta°$	.0	.2	.4	.6	.8
2	14.330	13.057	11.968	11.048	10.259
3	9.581	8.982	8.460	7.992	7.577
4	7.202	6.866	6.558	6.275	6.019
5	5.782	5.566	5.363	5.176	5.001
6	4.839	4.687	4.545	4.411	4.287
7	4.168	4.057	3.951	3.852	3.758
8	3.669	3.584	3.503	3.426	3.353
9	3.283	3.216	3.153	3.092	3.033
10	2.977	2.924	2.874	2.823	2.776
11	2.730	2.687	2.645	2.604	2.565
12	2.527	2.491	2.456	2.423	2.390
13	2.359	2.328	2.299	2.271	2.243
14	2.217	2.191	2.167	2.143	2.120
15	2.097	2.075	2.054	2.034	2.015
16	1.996	1.977	1.959	1.942	1.925
17	1.909	1.893	1.878	1.864	1.849
18	1.836	1.823	1.809	1.798	1.785
19	1.774	1.763	1.752	1.742	1.732
20	1.722	1.713	1.705	1.696	1.687
21	1.680	1.672	1.665	1.659	1.652
22	1.646	1.641	1.635	1.629	1.625
23	1.621	1.616	1.612	1.609	1.605
24	1.602	1.600	1.598	1.596	1.594
25	1.592	1.592	1.591	1.590	1.590
26	1.590	1.591	1.592	1.593	1.595
27	1.597	1.599	1.601	1.605	1.608
28	1.612	1.616	1.621	1.626	1.631
29	1.637	1.643	1.649	1.657	1.664
30	1.672	1.681	1.690	1.700	1.710
31	1.721	1.732	1.744	1.756	1.770
32	1.784	1.799	1.814	1.831	1.847

$\vartheta°$	.0	.2	.4	.6	.8
33	1.865	1.884	1.904	1.925	1.946
34	1.968	1.993	2.018	2.044	2.072
35	2.100	2.132	2.164	2.198	2.233
36	2.271	2.310	2.352	2.396	2.443
37	2.492	2.544	2.598	2.657	2.719
38	2.785	2.856	2.931	3.010	3.095
39	3.187	3.285	3.391	3.505	3.628
40	3.762	3.908	4.065	4.239	4.429
41	4.640	4.872	5.132	5.422	5.748
42	6.121	6.547	7.038	7.613	8.293
43	9.111	10.113	11.364	12.977	15.126
44	18.139	22.661	30.204	45.306	90.572

Mo Kα

$\vartheta°$	.0	.2	.4	.6	.8
2	14.334	13.035	11.951	11.035	10.249
3	9.568	9.225	8.448	7.981	7.563
4	7.186	6.848	6.539	6.257	5.999
5	5.761	5.579	5.385	5.175	4.987
6	4.812	4.646	4.491	4.358	4.239
7	4.139	4.018	3.912	3.810	3.719
8	3.635	3.551	3.470	3.392	3.315
9	3.244	3.178	3.112	3.051	2.990
10	2.935	2.887	2.823	2.775	2.724
11	2.683	2.632	2.589	2.549	2.510
12	2.474	2.436	2.398	2.362	2.329
13	2.299	2.271	2.236	2.208	2.179
14	2.153	2.125	2.100	2.075	2.049
15	2.029	2.008	1.987	1.966	1.945
16	1.924	1.905	1.885	1.867	1.848
17	1.830	1.813	1.795	1.780	1.763
18	1.749	1.734	1.718	1.704	1.690
19	1.679	1.665	1.652	1.641	1.630
20	1.621	1.609	1.598	1.589	1.579
21	1.569	1.562	1.554	1.546	1.537
22	1.528	1.523	1.516	1.510	1.504
23	1.500	1.493	1.487	1.481	1.476
24	1.472	1.467	1.463	1.460	1.457

$\vartheta°$	.0	.2	.4	.6	.8
25	1.454	1.449	1.445	1.443	1.440
26	1.438	1.437	1.436	1.435	1.435
27	1.435	1.435	1.435	1.435	1.436
28	1.437	1.438	1.439	1.440	1.442
29	1.445	1.449	1.453	1.458	1.462
30	1.470	1.473	1.476	1.480	1.486
31	1.494	1.500	1.508	1.517	1.526
32	1.537	1.548	1.559	1.571	1.584
33	1.601	1.612	1.627	1.643	1.659
34	1.678	1.696	1.715	1.735	1.756
35	1.778	1.800	1.824	1.849	1.875
36	1.904	1.932	1.964	1.998	2.036
37	2.082	2.118	2.159	2.204	2.253
38	2.315	2.370	2.429	2.491	2.560
39	2.631	2.702	2.780	2.864	2.960
40	3.085	3.178	3.295	3.431	3.594
41	3.792	4.005	4.243	4.507	4.774
42	4.992	5.334	5.733	6.199	6.751
43	7.406	8.226	9.240	10.548	12.294
44	14.741	18.412	24.536	36.787	73.564

4-5. Some Trigonometric Functions

The table gives values of functions needed in corrections for angle in the intensity formulas [109]:

$$\frac{1+\cos^2 2\vartheta}{\sin 2\vartheta} \cdots (a), \qquad \frac{1+\cos^2 2\vartheta}{\cos^2 \vartheta \sin \vartheta} \cdots (b), \qquad \frac{1+\cos^2 2\vartheta}{\sin^2 \vartheta \cos \vartheta} \cdots (c).$$

This table differs from those of 4-3a to 4-3c in that the variable is $\sin \vartheta$, the cases being (a) reflection from a monocrystal, (b) powder specimen with cylindrical film, and (c) powder with flat film.

$\sin\vartheta$	$\dfrac{1+\cos^2 2\vartheta}{\sin 2\vartheta}$	$\dfrac{1+\cos^2 2\vartheta}{\cos^2\vartheta \sin\vartheta}$	$\dfrac{1+\cos^2 2\vartheta}{\sin^2\vartheta \cos\vartheta}$
0	∞	∞	∞
0,025	39,962	79,95	3197
0,050	19,925	39,90	797,0
0,075	13,221	26,52	352,6
0,100	9,851	19,80	197,0
0,125	7,815	15,75	125,0
0,150	6,446	13,04	85,95
0,20	4,711	9,616	47,11
0,25	3,647	7,533	29,17
0,30	2,922	6,126	19,48
0,35	2,394	5,111	13,68
0,40	1,995	4,353	9,973

$\sin\vartheta$	$\dfrac{1+\cos^2 2\vartheta}{\sin 2\vartheta}$	$\dfrac{1+\cos^2 2\vartheta}{\cos^2\vartheta\,\sin\vartheta}$	$\dfrac{1+\cos^2 2\vartheta}{\sin^2\vartheta\,\cos\vartheta}$
0,45	1,685	3,774	7,487
0,50	1,443	3,333	5,774
0,55	1,258	3,012	4,576
0,60	1,123	2,808	3,744
0,65	1,037	2,730	3,189
0,70	1,004	2,812	2,869
0,75	1,024	3,095	2,730
0,80	1,123	3,744	2,808
0,85	1,338	5,079	3,148
0,90	1,764	8,096	3,921
0,95	2,778	17,792	5,848
1,00	∞	∞	∞

ATOMIC INTENSITY FACTORS

4-6. Accessory Table for Calculating Atomic Factors

The Thomas-Fermi method gives, for Z > 17,

$$F = Z\Phi(U), \tag{44}$$

where $U = \dfrac{4\pi \cdot 0,468}{\frac{1}{3}\lambda Z}\sin\vartheta$, Z being the atomic number.

The table gives $\Phi(U)$ and $\Phi^2(U)$ for U from 0 to 3.11 [111].

U	Φ	Φ^2	U	Φ	Φ^2
0,00	1,000	1,000	1,71	0,284	0,081
0,16	0,922	0,850	1,86	0,264	0,067
0,31	0,796	0,634	2,02	0,240	0,058
0,47	0,684	0,468	2,17	0,224	0,050
0,62	0,589	0,347	2,33	0,205	0,042
0,78	0,522	0,272	2,48	0,189	0,036
0,93	0,469	0,220	2,64	0,175	0,031
1,09	0,422	0,178	2,80	0,167	0,028
1,24	0,378	0,143	2,95	0,156	0,024
1,40	0,342	0,117	3,11	0,147	0,022
1,55	0,309	0,096			

4-7. Atomic Scattering Factors for Atoms and Ions

The table gives the atomic factors f_0 for neutral atoms and also for positive and negative ions for Z between 1 and 100 and for $(\sin\vartheta)/\lambda$ from 0 to 1.30 (up to 1.50 for $25 \leq Z \leq 80$) [102].*

The table includes results given elsewhere [103-107].

See [453, 454] for factors calculated by the Hartree-Fock method for numerous atoms and ions of the transition elements.

---

*Dr. K. Sagel has provided additions to this table for the English edition — Ed.

sin θ/λ — Element	0,00	0,05	0,10	0,15	0,20	0,25	0,30	0,35	0,40	0,50	0,60	0,70	0,80	0,90	1,00	1,10	1,20	1,30
H	1,00	0,945	0,802	0,638	0,481	0,350	0,250	0,180	0,132	0,078	0,040	0,023	0,018	0,01	0	—	—	—
He	2,00	1,956	1,850	1,700	1,510	1,306	1,104	0,920	0,766	0,512	0,345	0,236	0,167	0,120	0,085	0,063	0,048	0,036
Li+	2,0	1,98	1,96	1,88	1,8	1,7	1,5	1,4	1,3	1,0	0,8	0,6	0,5	0,4	0,3	0,3	—	—
Li	3,000	2,710	2,215	1,904	1,741	1,627	1,512	1,394	1,269	1,032	0,823	0,650	0,513	0,404	0,320	0,255	0,205	0,104
Be+	2,0	2,0	2,0	1,9	1,9	1,8	1,7	1,7	1,6	1,4	1,2	1,0	0,9	0,7	0,6	0,5		
Be	4,000	3,706	3,065	2,462	2,059	1,827	1,693	1,600	1,520	1,362	1,195	1,030	0,877	0,739	0,621	0,521	0,438	0,349
B+3	2,0	2,0	1,99	1,95	1,9	1,9	1,8	1,8	1,7	1,6	1,4	1,3	1,2	1,0	0,9	0,7		
B	5,00	4,752	4,063	3,348	2,601	2,162	1,864	1,772	1,620	1,501	1,399	1,283	1,168	1,036	0,901	0,795	0,696	0,600
C	6,000	5,764	5,141	4,362	3,612	3,003	2,538	2,212	1,983	1,707	1,548	1,423	1,313	1,202	1,096	0,992	0,896	0,802
*C	5,89	5,65	5,41	4,32	4,23	3,72	3,20	2,88	2,55	2,20	1,99	1,83	1,71	1,63	1,55			
N+5	2,0	2,0	2,0	2,0	2,0	2,0	1,9	1,9	1,9	1,8	1,7	1,6	1,5	1,4	1,3	1,16	—	—
N+3	4,0	3,9	3,7	3,4	3,0	2,7	2,4	2,2	2,0	1,8	1,66	1,56	1,49	1,39	1,28	1,17		
N	7,000	6,781	6,203	5,420	4,600	3,856	3,241	2,760	2,397	1,944	1,698	1,550	1,444	1,350	1,263	1,175	1,083	1,005
*O	8,26	8,03	7,80	7,30	6,80	6,20	5,60	5,09	4,57	3,75	3,10	2,58	2,20	1,90	1,71			
O	8,000	7,796	7,250	6,482	5,634	4,814	4,094	3,492	3,010	2,338	1,944	1,714	1,566	1,462	1,374	1,296	1,220	1,144
O-2	10,0	9,0	8,0	6,8	5,5	4,7	3,8	3,3	2,7	2,1	1,8	1,5	1,4	1,4	1,35	1,26		
F	9,000	8,790	8,208	7,396	6,501	5,625	4,837	4,100	3,598	2,769	2,252	1,926	1,725	1,587	1,484	1,404	1,333	1,263
F-	10,000	9,630	8,733	7,656	6,597	5,643	4,820	4,129	3,566	2,751	2,237	1,921	1,723	1,583	1,485	1,406	1,334	1,264
Ne	10,000	9,812	9,295	8,546	7,665	6,768	5,905	5,128	4,454	3,403	2,692	2,234	1,934	1,737	1,601	1,496	1,418	1,345
Na+	10,0	9,8	9,5	8,9	8,2	7,5	6,7	5,98	5,25	4,05	3,2	2,65	2,25	1,95	1,75	1,6		
Na	11,000	10,56	9,76	9,02	8,34	7,62	6,89	6,16	5,47	4,29	3,40	2,76	2,31	2,00	1,78	1,63	1,52	1,44
Mg+3	10,00	9,91	9,66	9,26	8,75	8,15	7,51	6,85	6,20	4,99	4,03	3,28	2,71	2,30	2,01	1,81	1,65	1,54
Mg	12,0	11,3	10,5	9,6	8,6	7,6	7,25	6,60	5,95	4,8	3,85	3,15	2,55	2,2	2,0	1,8		
Al+3	10,00	9,93	9,72	9,38	8,94	8,42	7,85	7,26	6,65	5,51	4,53	3,72	3,10	2,62	2,27	2,01	1,82	1,68

Element \ $\frac{\sin\theta}{\lambda}$	0,00	0,05	0,10	0,15	0,20	0,25	0,30	0,35	0,40	0,50	0,60	0,70	0,80	0,90	1,00	1,10	1,20	1,30	1,40	1,50
Al	13,0	12,0	11,0	9,98	8,95	8,35	7,75	7,18	6,6	5,5	4,5	3,7	3,1	2,65	2,3	2,0	—	—	—	—
Si+4	10,00	9,95	9,79	9,54	9,20	8,79	8,33	7,83	7,31	6,26	5,28	4,42	3,71	3,13	2,68	2,33	2,06	1,86	—	—
Si	14,0	12,68	11,35	10,38	9,4	8,8	8,2	7,68	7,15	6,1	5,1	4,2	3,4	2,95	2,6	2,3	—	—	—	—
P+5	10,0	9,9	9,8	9,53	9,25	8,85	8,45	7,98	7,5	6,55	5,65	4,8	4,05	3,4	3,0	2,6	—	—	—	—
P	15,0	13,7	12,4	11,2	10,0	9,23	8,45	7,95	7,45	6,5	5,65	4,8	4,05	3,4	3,0	2,6	—	—	—	—
P-3	18,0	15,4	12,7	11,3	9,8	9,1	8,4	7,9	7,45	6,5	5,65	4,85	4,05	3,4	3,0	2,6	—	—	—	—
S+6	10,0	9,93	9,85	9,63	9,4	9,1	8,7	8,2	7,85	6,85	6,05	5,25	4,5	3,9	3,35	2,9	—	—	—	—
S	16,00	15,8	15,0	14,1	12,3	10,7	8,95	8,40	7,85	6,85	6,05	5,25	4,5	3,9	3,35	2,9	—	—	—	—
S-	18,0	16,2	14,3	12,5	10,7	9,8	8,9	8,38	7,85	6,85	6,0	5,25	4,5	3,9	3,35	2,9	—	—	—	—
Cl	17,0	15,8	14,6	13,0	11,3	10,3	9,25	8,65	8,05	7,25	6,5	5,75	5,05	4,4	3,85	3,35	—	—	—	—
Cl-	18,00	17,33	15,68	13,74	11,97	10,57	9,51	8,74	8,15	7,30	6,60	5,91	5,24	4,60	4,01	3,49	3,06	2,69	—	—
A	18,00	17,54	16,30	14,65	12,93	11,42	10,20	9,25	8,54	7,56	6,86	6,23	5,61	5,01	4,43	3,90	3,43	3,03	—	—
K-	18,00	17,65	16,68	15,30	13,76	12,27	10,96	9,89	9,04	7,86	7,11	6,51	5,94	5,39	4,84	4,43	3,83	3,40	—	—
K	19,0	17,8	16,5	14,9	13,3	12,1	10,8	9,0	9,2	7,9	6,7	5,9	5,2	4,6	4,2	3,7	3,3	—	—	—
Ca++	18,0	17,4	16,8	15,4	14,0	12,8	11,5	10,4	9,3	8,1	7,35	6,7	6,2	5,7	5,1	4,6	—	—	—	—
Ca	20,00	19,09	17,33	15,73	14,32	12,98	11,71	10,59	9,64	8,26	7,38	6,75	6,21	5,70	5,19	4,69	4,21	3,77	—	—
Sc+	18,0	17,4	16,7	15,4	14,0	12,7	11,4	10,4	9,4	8,3	7,6	6,9	6,4	5,8	5,35	4,85	—	—	—	—
Sc	21,00	19,08	17,21	15,80	14,29	13,02	11,79	10,71	9,80	8,41	7,52	6,85	6,30	5,80	5,37	4,91	4,45	4,02	—	—
Ti+4	18,0	17,5	17,0	15,7	14,4	13,2	11,9	10,9	9,9	8,5	7,85	7,3	6,7	6,15	5,65	5,05	4,77	4,38	—	—
Ti	22,00	20,05	18,05	16,42	14,97	13,52	12,23	11,22	10,23	8,77	7,81	7,12	6,52	6,05	5,70	5,19	5,03	4,64	—	—
V	23,00	21,30	19,19	17,62	15,81	14,22	12,90	11,76	10,79	9,19	8,09	7,41	6,77	6,28	5,82	5,42	5,31	4,91	—	—
Cr2+	22,00	21,65	20,67	19,27	17,67	16,04	14,50	13,10	11,87	9,93	8,60	7,69	7,06	6,56	6,13	5,72	4,6	—	—	—
Cr	24,0	22,6	21,1	19,3	17,4	15,8	14,2	13,2	12,1	10,6	9,2	8,0	7,1	6,3	5,7	5,1	—	—	—	—
Mn	25,00	24,38	22,77	20,78	18,88	17,25	15,84	14,56	13,41	11,54	10,04	8,84	7,85	7,03	6,34	5,75	5,25	4,82	4,44	4,11
Mn+	24,00	23,59	22,44	20,82	19,02	17,30	15,79	14,51	13,42	11,55	10,04	8,84	7,85	7,03	6,34	5,75	5,25	4,82	4,44	4,11
Mn2+	23,00	22,70	21,84	20,55	19,01	17,42	15,90	14,55	13,38	11,53	10,06	8,84	7,85	7,03	6,34	5,75	5,25	4,82	4,44	4,11
Mn3+	22,06	21,77	21,10	20,08	18,80	17,40	15,99	14,65	13,45	11,50	10,03	8,85	7,85	7,03	6,34	5,75	5,25	4,82	4,44	4,11
Mn4+	21,00	20,82	20,30	19,47	18,42	17,23	15,97	14,72	13,54	11,53	10,00	8,82	7,85	7,03	6,34	5,75	5,25	4,82	4,44	4,11

Element $\frac{\sin\theta}{\lambda}$	0,00	0,05	0,10	0,15	0,20	0,25	0,30	0,35	0,40	0,50	0,60	0,70	0,80	0,90	1,00	1,10	1,20	1,30	1,40	1,50
Fe	26,00	25,36	23,71	21,66	19,71	18,03	16,56	15,24	14,05	12,11	10,54	9,29	8,25	7,39	6,67	6,06	5,53	5,08	4,68	4,33
Fe$^+$	25,00	24,57	23,39	21,71	19,85	18,08	16,52	15,20	14,05	12,12	10,54	9,29	8,25	7,39	6,67	6,06	5,53	5,08	4,68	4,33
Fe^{2+}	24,00	23,68	22,79	21,44	19,85	18,19	16,62	15,22	14,02	12,09	10,56	9,29	8,25	7,39	6,67	6,06	5,53	5,08	4,68	4,33
Fe^{3+}	23,00	22,76	22,06	20,98	19,65	18,19	16,71	15,33	14,08	12,06	10,54	9,30	8,25	7,39	6,67	6,06	5,53	5,08	4,68	4,33
Fe^{4+}	22,00	21,81	21,26	20,39	19,28	18,03	16,71	15,40	14,18	12,09	10,50	9,28	8,25	7,39	6,67	6,06	5,53	5,08	4,68	4,33
Co	27,00	26,34	24,65	22,55	20,54	18,81	17,29	15,92	14,69	12,67	11,05	9,74	8,66	7,77	7,01	6,37	5,82	5,34	4,93	4,56
Co$^+$	26,00	25,56	24,33	22,60	20,68	18,85	17,25	15,88	14,70	12,68	11,04	9,74	8,66	7,77	7,01	6,37	5,82	5,34	4,93	4,56
Co^{2+}	25,00	24,67	23,74	22,34	20,69	18,97	17,35	15,90	14,66	12,66	11,07	9,74	8,66	7,77	7,01	6,37	5,82	5,34	4,93	4,56
Co^{3+}	24,00	23,75	23,01	21,87	20,50	18,97	17,44	16,01	14,72	12,63	11,04	9,76	8,66	7,77	7,01	6,37	5,82	5,34	4,93	4,56
Co^{4+}	23,00	22,80	22,22	21,30	20,14	18,82	17,45	16,09	14,81	12,65	11,00	9,73	8,66	7,77	7,01	6,37	5,82	5,34	4,93	4,56
Ni	28,00	27,33	25,60	23,44	21,37	19,59	18,03	16,61	15,34	13,25	11,56	10,20	9,08	8,14	7,35	6,68	6,11	5,61	5,17	4,79
Ni$^+$	27,00	26,54	25,28	23,49	21,52	19,63	17,98	16,57	15,34	13,25	11,55	10,20	9,08	8,14	7,35	6,68	6,11	5,61	5,17	4,79
Ni^{2+}	26,00	25,66	24,69	23,24	21,53	19,75	18,08	16,59	15,30	13,24	11,58	10,19	9,08	8,14	7,35	6,68	6,11	5,61	5,17	4,79
Ni^{3+}	25,00	24,73	23,97	22,80	21,35	19,76	18,18	16,69	15,36	13,20	11,56	10,21	9,08	8,14	7,35	6,68	6,11	5,61	5,17	4,79
Ni^{4+}	24,00	23,79	23,18	22,22	21,00	19,63	18,19	16,78	15,45	13,22	11,52	10,19	9,08	8,14	7,35	6,68	6,11	5,61	5,17	4,79
Cu	29,00	28,31	26,54	24,33	22,10	20,38	18,76	17,30	15,98	13,82	12,07	10,66	9,49	8,52	7,70	7,00	6,40	5,88	5,43	5,02
Cu$^+$	28,00	27,53	26,22	24,38	22,35	20,42	18,71	17,26	15,99	13,83	12,07	10,66	9,49	8,52	7,70	7,00	6,40	5,88	5,43	5,02
Cu^{2+}	27,00	26,64	25,64	24,14	22,37	20,54	18,81	17,27	15,95	13,81	12,09	10,65	9,49	8,52	7,70	7,00	6,40	5,88	5,43	5,02
Cu^{3+}	26,00	25,72	24,93	23,71	22,20	20,56	18,91	17,38	16,00	13,77	12,07	10,68	9,49	8,52	7,70	7,00	6,40	5,88	5,43	5,02
Cu^{4+}	25,00	24,78	24,14	23,13	21,86	20,43	18,93	17,47	16,10	13,79	12,03	10,66	9,49	8,52	7,70	7,00	6,40	5,88	5,43	5,02
Zn	30,00	29,30	27,48	25,22	23,05	21,17	19,50	17,99	16,64	14,40	12,59	11,12	9,91	8,90	8,05	7,32	6,70	6,15	5,68	5,26
Zn$^+$	29,00	28,51	27,17	25,28	23,19	21,20	19,45	17,95	16,65	14,41	12,58	11,13	9,91	8,90	8,05	7,32	6,70	6,15	5,68	5,26
Zn^{2+}	28,00	27,63	26,59	25,04	23,22	21,33	19,55	17,97	16,60	14,40	12,61	11,12	9,91	8,90	8,05	7,32	6,70	6,15	5,68	5,26
Zn^{3+}	27,00	26,71	25,88	24,61	23,05	21,35	19,65	18,07	16,65	14,35	12,59	11,14	9,91	8,90	8,05	7,32	6,70	6,15	5,68	5,26
Zn^{4+}	26,00	25,77	25,10	24,05	22,73	21,23	19,68	18,16	16,75	14,36	12,55	11,12	9,91	8,90	8,05	7,32	6,70	6,15	5,68	5,26

$\frac{\sin\theta}{\lambda}$ Element	0,05	0,00	0,10	0,15	0,20	0,25	0,30	0,35	0,40	0,50	0,60	0,70	0,80	0,90	1,00	1,10	1,20	1,30	1,40	1,50
Ga	31,00	30,28	28,43	26,11	23,89	21,96	20,25	18,69	17,29	14,98	13,11	11,59	10,33	9,29	8,40	7,64	6,99	6,43	5,94	5,50
Ga+	30,00	29,50	28,12	26,17	24,03	21,99	20,19	18,65	17,30	14,99	13,10	11,60	10,33	9,29	8,40	7,64	6,99	6,43	5,94	5,50
Ga2+	29,00	28,62	27,54	25,95	24,06	22,12	20,29	18,66	17,26	14,98	13,13	11,59	10,33	9,29	8,40	7,64	6,99	6,43	5,94	5,50
Ga3+	28,00	27,70	26,84	25,52	23,90	22,14	20,39	18,76	17,30	14,93	13,11	11,61	10,33	9,29	8,40	7,64	6,99	6,43	5,94	5,50
Ga4+	27,00	26,76	26,06	24,97	23,59	22,04	20,42	18,86	17,40	14,94	13,07	11,60	10,33	9,29	8,40	7,64	6,99	6,43	5,94	5,50
Ge	32,00	31,26	29,37	27,00	24,73	22,75	20,99	19,39	17,95	15,57	13,63	12,06	10,76	9,68	8,76	7,97	7,29	6,71	6,19	5,74
Ge+	31,00	30,49	29,07	27,07	24,87	22,78	20,94	19,35	17,96	15,57	13,63	12,07	10,76	9,68	8,76	7,97	7,29	6,71	6,19	5,74
Ge2+	30,00	29,61	28,50	26,85	24,91	22,91	21,03	19,36	17,92	15,57	13,65	12,06	10,76	9,68	8,76	7,97	7,29	6,71	6,19	5,74
Ge3+	29,00	28,69	27,80	26,43	24,76	22,94	21,14	19,46	17,96	15,52	13,64	12,08	10,76	9,68	8,76	7,97	7,29	6,71	6,19	5,74
Ge4+	28,00	27,75	27,02	25,89	24,45	22,84	21,18	19,55	18,05	15,53	13,60	12,07	10,76	9,68	8,76	7,97	7,29	6,71	6,19	5,74
As	33,00	32,25	30,32	27,90	25,58	23,54	21,74	20,09	18,61	16,16	14,16	12,54	11,19	10,07	9,11	8,30	7,60	6,99	6,46	5,98
As+	32,00	31,47	30,02	27,97	25,72	23,57	21,68	20,06	18,63	16,16	14,16	12,54	11,19	10,07	9,11	8,30	7,60	6,99	6,46	5,98
As2+	31,00	30,60	29,45	27,76	25,76	23,70	21,77	20,06	18,58	16,16	14,18	12,53	11,19	10,07	9,11	8,30	7,60	6,99	6,46	5,98
As3+	30,00	29,68	28,75	27,35	25,62	23,74	21,88	20,16	18,61	16,11	14,17	12,55	11,19	10,07	9,11	8,30	7,60	6,99	6,46	5,98
As4+	29,00	28,74	27,98	26,80	25,32	23,65	21,93	20,26	18,71	16,11	14,13	12,55	11,19	10,07	9,11	8,30	7,60	6,99	6,46	5,98
Se	34,00	33,23	31,26	28,80	26,42	24,34	22,49	20,80	19,28	16,75	14,69	13,02	11,62	10,46	9,47	8,63	7,91	7,27	6,72	6,23
Se+	33,00	32,46	30,97	28,87	26,57	24,37	22,43	20,76	19,29	16,76	14,69	13,02	11,62	10,46	9,47	8,63	7,91	7,27	6,72	6,23
Se2+	32,00	31,58	30,41	28,66	26,61	24,50	22,52	20,76	19,24	16,75	14,71	13,01	11,62	10,46	9,47	8,63	7,91	7,27	6,72	6,23
Se3+	31,00	30,67	29,71	28,26	26,47	24,54	22,63	20,86	19,28	16,71	14,70	13,03	11,62	10,46	9,47	8,63	7,91	7,27	6,72	6,23
Se4+	30,00	29,73	28,94	27,72	26,18	24,46	22,68	20,96	19,37	16,70	14,66	13,03	11,62	10,46	9,47	8,63	7,91	7,27	6,72	6,23
Br	35,00	34,22	32,21	29,70	27,27	25,14	23,24	21,51	19,95	17,35	15,22	13,50	12,06	10,86	9,84	8,97	8,21	7,56	6,99	6,48
Br+	34,00	33,45	31,92	29,77	27,41	25,17	23,18	21,47	19,96	17,35	15,22	13,50	12,06	10,86	9,84	8,97	8,21	7,56	6,99	6,48
Br2+	33,00	32,57	31,36	29,57	27,46	25,30	23,27	21,47	19,91	17,35	15,24	13,49	12,06	10,86	9,84	8,97	8,21	7,56	6,99	6,48
Br3+	32,00	31,66	30,67	29,17	27,33	25,35	23,38	21,56	19,94	17,30	15,24	13,51	12,06	10,86	9,84	8,97	8,21	7,56	6,99	6,48
Br4+	31,00	30,72	29,90	28,64	27,05	25,27	23,44	21,67	20,03	17,29	15,20	13,51	12,06	10,86	9,84	8,97	8,21	7,56	6,99	6,48

$\frac{\sin\theta}{\lambda}$ Element	0,00	0,05	0,10	0,15	0,20	0,25	0,30	0,35	0,40	0,50	0,60	0,70	0,80	0,90	1,00	1,10	1,20	1,30	1,40	1,50
Kr	36,00	35,21	33,16	30,60	28,12	25,94	24,00	22,22	20,62	17,95	15,76	13,98	12,50	11,26	10,21	9,31	8,53	7,85	7,26	6,73
Kr+	35,00	34,43	32,87	30,67	28,26	25,97	23,94	22,18	20,63	17,95	15,76	13,99	12,50	11,26	10,21	9,31	8,53	7,85	7,26	6,73
Kr++	34,00	33,56	32,31	30,47	28,31	26,10	24,02	22,18	20,58	17,95	15,78	13,97	12,50	11,26	10,21	9,31	8,53	7,85	7,26	6,73
Kr3+	33,00	32,65	31,63	30,08	28,19	26,15	24,14	22,27	20,61	17,90	15,78	14,00	12,50	11,26	10,21	9,31	8,53	7,85	7,26	6,73
Kr4+	32,00	31,71	30,87	29,56	27,91	26,08	24,19	22,37	20,70	17,89	15,73	14,00	12,50	11,26	10,21	9,31	8,53	7,85	7,26	6,73
Rb	37,00	36,19	34,11	31,50	28,97	26,75	24,75	22,93	21,29	18,55	16,30	14,47	12,94	11,66	10,58	9,65	8,84	8,14	7,53	6,99
Rb+	36,00	35,42	33,82	31,58	29,11	26,77	24,70	22,90	21,31	18,55	16,30	14,48	12,94	11,66	10,58	9,65	8,84	8,14	7,53	6,99
Rb3+	35,00	34,55	33,27	31,38	29,17	26,90	24,78	22,89	21,26	18,55	16,32	14,46	12,94	11,66	10,58	9,65	8,84	8,14	7,53	6,99
Rb3+	34,00	33,63	32,59	31,00	29,05	26,96	24,89	22,98	21,28	18,50	16,32	14,48	12,94	11,66	10,58	9,65	8,84	8,14	7,53	6,99
Rb4+	33,00	32,70	31,83	30,48	28,78	26,89	24,95	23,09	21,37	18,49	16,28	14,49	12,94	11,66	10,58	9,65	8,84	8,14	7,53	6,99
Sr	38,00	37,18	35,06	32,40	29,83	27,55	25,51	23,65	21,96	19,15	16,84	14,96	13,39	12,07	10,95	9,99	9,16	8,44	7,80	7,24
Sr+	37,00	36,41	34,77	32,48	29,97	27,57	25,45	23,61	21,98	19,15	16,84	14,97	13,39	12,07	10,95	9,99	9,16	8,44	7,80	7,24
Sr++	36,00	35,54	34,23	32,29	30,03	27,71	25,53	23,61	21,93	19,16	16,86	14,95	13,39	12,07	10,95	9,99	9,16	8,44	7,80	7,24
Sr3+	35,00	34,62	33,55	31,91	29,91	27,77	25,65	23,69	21,95	19,11	16,86	14,97	13,39	12,07	10,95	9,99	9,16	8,44	7,80	7,24
Sr4+	34,00	33,69	32,79	31,40	29,65	27,71	25,72	23,80	22,04	19,09	16,82	14,98	13,39	12,07	10,95	9,99	9,16	8,44	7,80	7,24
Y	39,00	38,16	36,01	33,30	30,68	28,36	26,28	24,37	22,64	19,76	17,39	15,46	13,84	12,48	11,32	10,34	9,48	8,73	8,08	7,50
Y+	38,00	37,39	35,73	33,39	30,82	28,38	26,22	24,33	22,66	19,76	17,39	15,46	13,84	12,48	11,32	10,34	9,48	8,73	8,08	7,50
Y++	37,00	36,52	35,18	33,20	30,88	28,51	26,29	24,32	22,61	19,77	17,41	15,45	13,84	12,48	11,32	10,34	9,48	8,73	8,08	7,50
Y3+	36,00	35,61	34,51	32,83	30,78	28,58	26,41	24,41	22,63	19,72	17,41	15,47	13,84	12,48	11,32	10,34	9,48	8,73	8,08	7,50
Y4+	35,00	34,68	33,75	32,32	30,52	28,52	26,48	24,51	22,71	19,70	17,37	15,47	13,84	12,48	11,32	10,34	9,48	8,73	8,08	7,50
Zr	40,00	39,15	36,96	34,21	31,54	29,17	27,04	25,09	23,32	20,37	17,94	15,95	14,29	12,89	11,70	10,68	9,80	9,03	8,36	7,76
Zr+	39,00	38,38	36,68	34,29	31,68	29,19	26,98	25,06	23,35	20,37	17,93	15,96	14,29	12,89	11,70	10,68	9,80	9,03	8,36	7,76
Zr++	38,00	37,51	36,14	34,11	31,74	29,32	27,05	25,04	23,29	20,38	17,95	15,94	14,29	12,89	11,70	10,68	9,80	9,03	8,36	7,76
Zr3+	37,00	36,60	35,47	33,74	31,64	29,39	27,17	25,13	23,81	20,33	17,96	15,96	14,29	12,89	11,70	10,68	9,80	9,03	8,36	7,76
Zr4+	36,00	35,67	34,72	33,24	31,39	29,34	27,25	25,23	23,39	20,31	17,92	15,97	14,29	12,89	11,70	10,68	9,80	9,03	8,36	7,76

Element \ $\frac{\sin\theta}{\lambda}$	0,00	0,05	0,10	0,15	0,20	0,25	0,30	0,35	0,40	0,50	0,60	0,70	0,80	0,90	1,00	1,10	1,20	1,30	1,40	1,50
Nb	41,00	40,14	37,91	35,11	32,40	29,98	27,81	25,81	24,01	20,98	18,49	16,45	14,74	13,31	12,08	11,04	10,13	9,33	8,64	8,02
Nb+	40,00	39,37	37,63	35,20	32,53	30,00	27,74	25,78	24,03	20,98	18,49	16,46	14,74	13,31	12,08	11,04	10,13	9,33	8,64	8,02
Nb²+	39,00	38,50	37,10	35,02	32,60	30,13	27,82	25,77	23,98	20,99	18,50	16,44	14,74	13,31	12,08	11,04	10,13	9,33	8,64	8,02
Nb³+	38,00	37,59	36,43	34,66	32,51	30,20	27,94	25,85	23,99	20,94	18,51	16,46	14,74	13,31	12,08	11,04	10,13	9,33	8,64	8,02
Nb⁴+	37,00	36,66	35,68	34,16	32,26	30,16	28,01	25,95	24,07	20,92	18,47	16,47	14,74	13,31	12,08	11,04	10,13	9,33	8,64	8,02
Mo	42,00	41,12	38,86	36,02	33,25	30,79	28,57	26,53	24,69	21,60	19,04	16,95	15,20	13,73	12,46	11,39	10,45	9,64	8,92	8,29
Mo+	41,00	40,36	38,59	36,11	33,39	30,81	28,51	26,51	24,72	21,59	19,04	16,96	15,20	13,73	12,46	11,39	10,45	9,64	8,92	8,29
Mo²+	40,00	39,49	38,05	35,94	33,46	30,94	28,58	26,49	24,66	21,61	19,06	16,94	15,20	13,73	12,46	11,39	10,45	9,64	8,92	8,29
Mo³+	39,00	38,58	37,39	35,58	33,37	31,02	28,70	26,57	24,67	21,56	19,07	16,96	15,20	13,73	12,46	11,39	10,45	9,64	8,92	8,29
Mo⁴+	38,00	37,65	36,64	35,08	33,14	30,98	28,78	26,68	24,75	21,53	19,03	16,97	15,20	13,73	12,46	11,39	10,45	9,64	8,92	8,29
Tc	43,00	42,11	39,81	36,92	34,12	31,61	29,34	27,26	25,38	22,21	19,60	17,46	15,65	14,15	12,85	11,74	10,78	9,94	9,21	8,55
Tc+	42,00	41,34	39,54	37,02	34,25	31,62	29,28	27,24	25,41	22,21	19,60	17,46	15,65	14,15	12,85	11,74	10,78	9,94	9,21	8,55
Tc²+	41,00	40,48	39,01	36,85	34,33	31,75	29,35	27,22	25,35	22,22	19,61	17,45	15,65	14,15	12,85	11,74	10,78	9,94	9,21	8,55
Tc³+	40,00	39,57	38,35	36,49	34,24	31,83	29,47	27,29	25,36	22,18	19,62	17,46	15,65	14,15	12,85	11,74	10,78	9,94	9,21	8,55
Tc⁴+	39,00	38,64	37,60	36,01	34,01	31,80	29,55	27,40	25,43	22,15	19,59	17,48	15,65	14,15	12,85	11,74	10,78	9,94	9,21	8,55
Ru	44,00	43,10	40,76	37,83	34,98	32,43	30,12	27,99	26,07	22,83	20,16	17,96	16,12	14,57	13,24	12,10	11,11	10,25	9,49	8,82
Ru+	43,00	42,33	40,50	37,93	35,12	32,44	30,05	27,97	26,10	22,83	20,16	17,97	16,12	14,57	13,24	12,10	11,11	10,25	9,49	8,82
Ru²+	42,00	41,47	39,97	37,76	35,19	32,57	30,12	27,94	26,04	22,84	20,17	17,95	16,12	14,57	13,24	12,10	11,11	10,25	9,49	8,82
Ru³+	41,00	40,56	39,31	37,41	35,11	32,65	30,24	28,02	26,05	22,80	20,18	17,97	16,12	14,57	13,24	12,10	11,11	10,25	9,49	8,82
Ru⁴+	40,00	39,63	38,57	36,93	34,88	32,62	30,32	28,13	26,12	22,77	20,15	17,98	16,12	14,57	13,24	12,10	11,11	10,25	9,49	8,82
Rh	45,00	44,08	41,72	38,74	35,84	33,24	30,89	28,72	26,76	23,46	20,72	18,47	16,58	14,99	13,63	12,46	11,45	10,56	9,78	9,09
Rh+	44,00	43,32	41,45	38,84	35,98	33,25	30,83	28,70	26,79	23,45	20,72	18,48	16,58	14,99	13,63	12,46	11,45	10,56	9,78	9,09
Rh²+	43,00	42,46	40,93	38,68	36,06	33,38	30,80	28,67	26,74	23,47	20,73	18,46	16,58	14,99	13,63	12,46	11,45	10,56	9,78	9,09
Rh³+	42,00	41,55	40,27	38,33	35,98	33,47	31,01	28,75	26,74	23,42	20,75	18,48	16,58	14,99	13,63	12,46	11,45	10,56	9,78	9,09
Rh⁴+	41,00	40,62	39,53	37,85	35,76	33,45	31,10	28,86	26,81	23,39	20,71	18,49	16,58	14,99	13,63	12,46	11,45	10,56	9,78	9,09

Element	0,00	0,05	0,10	0,15	0,20	0,25	0,30	0,35	0,40	0,50	0,60	0,70	0,80	0,90	1,00	1,10	1,20	1,30	1,40	1,50
Pd	46,00	45,07	42,87	39,65	36,70	34,06	31,67	29,46	27,46	24,08	21,28	18,98	17,05	15,42	14,02	12,82	11,78	10,87	10,07	9,37
Pd^{+}	45,00	44,31	42,41	39,75	36,84	34,07	31,60	29,43	27,49	24,08	21,28	18,99	17,05	15,42	14,02	12,82	11,78	10,87	10,07	9,37
Pd^{2+}	44,00	43,45	41,89	39,59	36,92	34,20	31,66	29,41	27,43	24,09	21,30	18,97	17,05	15,42	14,02	12,82	11,78	10,87	10,07	9,37
Pd^{3+}	43,00	42,54	41,23	39,25	36,85	34,29	31,78	29,48	27,43	24,05	21,31	18,99	17,05	15,42	14,02	12,82	11,78	10,87	10,07	9,37
Pd^{4+}	42,00	41,61	40,50	38,78	36,63	34,27	31,87	29,59	27,50	24,01	21,28	19,01	17,05	15,42	14,02	12,82	11,78	10,87	10,07	9,37
Ag	47,00	46,06	43,63	40,56	37,57	34,88	32,44	30,19	28,16	24,71	21,85	19,50	17,52	15,85	14,42	13,19	12,12	11,19	10,37	9,64
Ag^{+}	46,00	45,30	43,37	40,66	37,71	34,89	32,38	30,17	28,18	24,70	21,85	19,50	17,52	15,85	14,42	13,19	12,12	11,19	10,37	9,64
Ag^{2+}	45,00	44,44	42,85	40,51	37,79	35,02	32,44	30,14	28,13	24,72	21,86	19,49	17,52	15,85	14,42	13,19	12,12	11,19	10,37	9,64
Ag^{3+}	44,00	43,53	42,19	40,17	37,72	35,11	32,56	30,21	28,13	24,68	21,88	19,50	17,52	15,85	14,42	13,19	12,12	11,19	10,37	9,64
Ag^{4+}	43,00	42,60	41,46	39,70	37,51	35,10	32,65	30,32	28,19	24,64	21,85	19,52	17,52	15,85	14,42	13,19	12,12	11,19	10,37	9,64
Cd	48,00	47,04	44,58	41,47	38,44	35,71	33,22	30,93	28,85	25,34	22,42	20,02	17,99	16,28	14,81	13,56	12,46	11,51	10,66	9,92
Cd^{+}	47,00	46,28	44,32	41,58	38,57	35,71	33,16	30,91	28,88	25,33	22,42	20,02	17,99	16,28	14,81	13,56	12,46	11,51	10,66	9,92
Cd^{2+}	46,00	45,42	43,80	41,43	38,66	35,84	33,22	30,88	28,83	25,35	22,43	20,00	17,99	16,28	14,81	13,56	12,46	11,51	10,66	9,92
Cd^{3+}	45,00	44,52	43,16	41,09	38,59	35,93	33,34	30,95	28,82	25,31	22,45	20,02	17,99	16,28	14,81	13,56	12,46	11,51	10,66	9,92
Cd^{4+}	44,00	43,59	42,43	40,63	38,39	35,92	33,43	31,05	28,89	25,27	22,42	20,04	17,99	16,28	14,81	13,56	12,46	11,51	10,66	9,92
In	49,00	48,03	45,53	42,39	39,31	36,53	34,00	31,67	29,56	25,97	22,99	20,53	18,46	16,71	15,21	13,93	12,80	11,82	10,96	10,19
In^{+}	48,00	47,27	45,28	42,49	39,44	36,53	33,94	31,65	29,58	25,96	22,99	20,54	18,46	16,71	15,21	13,93	12,80	11,82	10,96	10,19
In^{2+}	47,00	46,41	44,76	42,34	39,53	36,66	33,99	31,62	29,53	25,98	23,00	20,52	18,46	16,71	15,21	13,93	12,80	11,82	10,96	10,19
In^{3+}	46,00	45,51	44,12	42,01	39,47	36,76	34,11	31,68	29,52	25,94	23,02	20,54	18,46	16,71	15,21	13,93	12,80	11,82	10,96	10,19
In^{4+}	45,00	44,59	43,39	41,55	39,27	36,75	34,21	31,79	29,59	25,90	22,99	20,55	18,46	16,71	15,21	13,93	12,80	11,82	10,96	10,19
Sn	50,00	49,02	46,49	43,30	40,17	37,36	34,78	32,41	30,26	26,60	23,56	21,05	18,93	17,15	15,61	14,30	13,15	12,14	11,26	10,47
Sn^{+}	49,00	48,26	46,24	43,41	40,31	37,36	34,72	32,39	30,29	26,60	23,56	21,06	18,93	17,15	15,61	14,30	13,15	12,14	11,26	10,47
Sn^{2+}	48,00	47,40	45,72	43,26	40,40	37,49	34,77	32,36	30,26	26,61	23,57	21,04	18,93	17,15	15,61	14,30	13,15	12,14	11,26	10,47
Sn^{3+}	47,00	46,50	45,08	42,94	40,34	37,58	34,89	32,42	30,22	26,57	23,59	21,06	18,93	17,15	15,61	14,30	13,15	12,14	11,26	10,47
Sn^{4+}	46,00	45,58	44,35	42,48	40,15	37,58	34,99	32,53	30,28	26,53	23,56	21,08	18,93	17,15	15,61	14,30	13,15	12,14	11,26	10,47

Column headers: $\frac{\sin \theta}{\lambda}$

$\frac{\sin\theta}{\lambda}$ Element	0,00	0,05	0,10	0,15	0,20	0,25	0,30	0,35	0,40	0,50	0,60	0,70	0,80	0,90	1,00	1,10	1,20	1,30	1,40	1,50
Sb	51,00	50,01	47,45	44,21	41,05	38,18	35,57	33,15	30,96	27,24	24,14	21,58	19,41	17,59	16,02	14,67	13,49	12,47	11,56	10,76
Sb$^+$	50,00	49,25	47,20	44,32	41,18	38,18	35,50	33,13	30,99	27,23	24,14	21,58	19,41	17,59	16,02	14,67	13,49	12,47	11,56	10,76
Sb^{2+}	49,00	48,39	46,69	44,18	41,27	38,31	35,56	33,10	30,94	27,25	24,15	21,57	19,41	17,59	16,02	14,67	13,49	12,47	11,56	10,76
Sb^{3+}	48,00	47,49	46,04	43,86	41,22	38,41	35,67	33,16	30,93	27,21	24,17	21,58	19,41	17,59	16,02	14,67	13,49	12,47	11,56	10,76
Sb^{4+}	47,00	46,57	45,32	43,40	41,03	38,41	35,77	33,27	30,99	27,17	24,14	21,60	19,41	17,59	16,02	14,67	13,49	12,47	11,56	10,76
Te	52,00	51,00	48,40	45,13	41,92	39,01	36,35	33,90	31,67	27,87	24,71	22,10	19,89	18,03	16,42	15,05	13,84	12,79	11,86	11,04
Te$^+$	51,00	50,24	48,15	45,24	42,05	39,01	36,29	33,88	31,70	27,87	24,71	22,11	19,89	18,03	16,42	15,05	13,84	12,79	11,86	11,04
Te^{2+}	50,00	49,38	47,65	45,10	42,14	39,14	36,34	33,84	31,64	27,89	24,72	22,09	19,89	18,03	16,42	15,05	13,84	12,79	11,86	11,04
Te^{3+}	49,00	48,48	47,01	44,78	42,09	39,24	36,46	33,90	31,63	27,85	24,74	22,10	19,89	18,03	16,42	15,05	13,84	12,79	11,86	11,04
Te^{4+}	48,00	47,56	46,28	44,33	41,91	39,24	36,56	34,01	31,69	27,81	24,72	22,12	19,89	18,03	16,42	15,05	13,84	12,79	11,86	11,04
J	53,00	51,98	49,36	46,05	42,79	39,84	37,14	34,64	32,38	28,51	25,29	22,63	20,37	18,47	16,83	15,42	14,19	13,12	12,16	11,32
J$^+$	52,00	51,23	49,11	46,15	42,92	39,84	37,08	34,62	32,41	28,51	25,29	22,63	20,37	18,47	16,83	15,42	14,19	13,12	12,16	11,32
J^{2+}	51,00	50,37	48,61	46,02	43,02	39,97	37,12	34,59	32,35	28,53	25,30	22,62	20,37	18,47	16,83	15,42	14,19	13,12	12,16	11,32
J^{3+}	50,00	49,47	47,97	45,70	42,97	40,07	37,24	34,65	32,34	28,49	25,32	22,63	20,37	18,47	16,83	15,42	14,19	13,12	12,16	11,32
J^{4+}	49,00	48,55	47,25	45,26	42,79	40,08	37,35	34,75	32,39	28,45	25,30	22,65	20,37	18,47	16,83	15,42	14,19	13,12	12,16	11,32
Xe	54,00	52,97	50,32	46,96	43,66	40,67	37,93	35,39	33,09	29,16	25,87	23,16	20,86	18,92	17,24	15,80	14,54	13,44	12,47	11,61
Xe$^+$	53,00	52,22	50,07	47,07	43,80	40,67	37,86	35,37	33,12	29,15	25,88	23,16	20,86	18,92	17,24	15,80	14,54	13,44	12,47	11,61
Xe^{2+}	52,00	51,36	49,57	46,94	43,89	40,80	37,91	35,34	33,06	29,17	25,88	23,15	20,86	18,92	17,24	15,80	14,54	13,44	12,47	11,61
Xe^{3+}	51,00	50,46	48,93	46,63	43,85	40,90	38,03	35,39	33,05	29,13	25,90	23,16	20,86	18,92	17,24	15,80	14,54	13,44	12,47	11,61
Xe^{4+}	50,00	49,54	48,22	46,18	43,67	40,91	38,13	35,50	33,10	29,09	25,88	23,18	20,86	18,92	17,24	15,80	14,54	13,44	12,47	11,61
Cs	55,00	53,96	51,27	47,88	44,54	41,50	38,72	36,14	33,80	29,80	26,46	23,69	21,34	19,36	17,65	16,18	14,90	13,77	12,78	11,90
Cs$^+$	54,00	53,21	51,03	47,99	44,67	41,50	38,65	36,12	33,83	29,79	26,46	23,69	21,34	19,36	17,65	16,18	14,90	13,77	12,78	11,90
Cs^{2+}	53,00	52,35	50,53	47,86	44,77	41,63	38,70	36,08	33,78	29,81	26,47	23,68	21,34	19,36	17,65	16,18	14,90	13,77	12,78	11,90
Cs^{3+}	52,00	51,45	49,90	47,55	44,72	41,73	38,81	36,14	33,76	29,77	26,49	23,69	21,34	19,36	17,65	16,18	14,90	13,77	12,78	11,90
Cs^{4+}	51,00	50,53	49,18	47,11	44,55	41,75	38,92	36,24	33,81	29,73	26,47	23,71	21,34	19,36	17,65	16,18	14,90	13,77	12,78	11,90

$\frac{\sin\theta}{\lambda}$ / Element	0,00	0,05	0,10	0,15	0,20	0,25	0,30	0,35	0,40	0,50	0,60	0,70	0,80	0,90	1,00	1,10	1,20	1,30	1,40	1,50
Ba	56,00	54,95	52,23	48,80	45,41	42,33	39,51	36,89	34,51	30,44	27,04	24,22	21,83	19,81	18,07	16,57	15,25	14,11	13,09	12,19
Ba+	55,00	54,20	51,99	48,91	45,54	42,33	39,44	36,87	34,54	30,43	27,04	24,23	21,83	19,81	18,07	16,57	15,25	14,11	13,09	12,19
Ba2+	54,00	53,34	51,49	48,78	45,64	42,46	39,49	36,83	34,49	30,46	27,05	24,21	21,83	19,81	18,07	16,57	15,25	14,11	13,09	12,19
Ba3+	53,00	52,44	50,86	48,48	45,60	42,56	39,60	36,89	34,47	30,42	27,07	24,22	21,83	19,81	18,07	16,57	15,25	14,11	13,09	12,19
Ba4+	52,00	51,52	50,15	48,04	45,43	42,58	39,71	36,99	34,52	30,38	27,05	24,24	21,83	19,81	18,07	16,57	15,25	14,11	13,09	12,19
La	57,00	55,94	53,19	49,71	46,29	43,17	40,30	37,64	35,23	31,09	27,63	24,76	22,32	20,26	18,48	16,95	15,61	14,44	13,40	12,48
La+	56,00	55,18	52,95	49,83	46,42	43,16	40,23	37,63	35,26	31,08	27,63	24,76	22,32	20,26	18,48	16,95	15,61	14,44	13,40	12,48
La2+	55,00	54,33	52,45	49,71	46,52	43,29	40,28	37,59	35,20	31,10	27,64	24,75	22,32	20,26	18,48	16,95	15,61	14,44	13,40	12,48
La3+	54,00	53,43	51,82	49,40	46,48	43,39	40,39	37,64	35,18	31,07	27,66	24,76	22,32	20,26	18,48	16,95	15,61	14,44	13,40	12,48
La4+	53,00	52,51	51,11	48,97	46,32	43,42	40,50	37,74	35,23	31,02	27,64	24,78	22,32	20,26	18,48	16,95	15,61	14,44	13,40	12,48
Ce	58,00	56,93	54,15	50,63	47,16	44,00	41,09	38,40	35,94	31,74	28,22	25,30	22,81	20,71	18,90	17,34	15,97	14,77	13,71	12,77
Ce+	57,00	56,17	53,91	50,75	47,29	43,99	41,03	38,38	35,97	31,73	28,22	25,30	22,81	20,71	18,90	17,34	15,97	14,77	13,71	12,77
Ce2+	56,00	55,32	53,42	50,63	47,40	44,12	41,07	38,34	35,92	31,75	28,23	25,29	22,81	20,71	18,90	17,34	15,97	14,77	13,71	12,77
Ce3+	55,00	54,42	52,79	50,33	47,37	44,23	41,19	38,39	35,90	31,72	28,25	25,29	22,81	20,71	18,90	17,34	15,97	14,77	13,71	12,77
Ce4+	54,00	53,50	52,08	49,90	47,20	44,26	41,29	38,49	35,95	31,67	28,23	25,32	22,81	20,71	18,90	17,34	15,97	14,77	13,71	12,77
Pr	59,00	57,91	55,11	51,55	48,04	44,84	41,89	39,15	36,66	32,39	28,81	25,84	23,31	21,17	19,32	17,72	16,33	15,11	14,03	13,07
Pr+	58,00	57,16	54,87	51,67	48,17	44,83	41,82	39,14	36,69	32,38	28,81	25,84	23,31	21,17	19,32	17,72	16,33	15,11	14,03	13,07
Pr2+	57,00	56,31	54,38	51,55	48,28	44,96	41,86	39,10	36,64	32,40	28,82	25,83	23,31	21,17	19,32	17,72	16,33	15,11	14,03	13,07
Pr3+	56,00	55,42	53,75	51,25	48,25	45,07	41,98	39,14	36,62	32,37	28,84	25,83	23,31	21,17	19,32	17,72	16,33	15,11	14,03	13,07
Pr4+	55,00	54,50	53,05	50,83	48,09	45,09	42,09	39,25	36,66	32,32	28,83	25,85	23,31	21,17	19,32	17,72	16,33	15,11	14,03	13,07
Nd	60,00	58,90	56,07	52,47	48,92	45,68	42,69	39,91	37,38	33,04	29,40	26,38	23,80	21,62	19,74	18,11	16,69	15,45	14,34	13,37
Nd+	59,00	58,15	55,83	52,59	49,05	45,66	42,62	39,89	37,41	33,03	29,40	26,38	23,80	21,62	19,74	18,11	16,69	15,45	14,34	13,37
Nd2+	58,00	57,30	55,34	52,48	49,16	45,79	42,66	39,85	37,36	33,06	29,41	26,37	23,80	21,62	19,74	18,11	16,69	15,45	14,34	13,37
Nd3+	57,00	56,41	54,72	52,18	49,13	45,90	42,77	39,90	37,34	33,02	29,43	26,37	23,80	21,62	19,74	18,11	16,69	15,45	14,34	13,37
Nd4+	56,00	55,49	54,01	51,76	48,97	45,94	42,88	40,00	37,38	32,98	29,42	26,39	23,80	21,62	19,74	18,11	16,69	15,45	14,34	13,37

Element \ $\frac{\sin\theta}{\lambda}$	0,00	0,05	0,10	0,15	0,20	0,25	0,30	0,35	0,40	0,50	0,60	0,70	0,80	0,90	1,00	1,10	1,20	1,30	1,40	1,50
Pm	61,00	59,89	57,02	53,39	49,80	46,51	43,48	40,67	38,10	33,69	29,99	26,92	24,30	22,08	20,16	18,51	17,05	15,79	14,66	13,66
Pm+	60,00	59,14	56,79	53,51	49,93	46,50	43,42	40,65	38,13	33,68	30,00	26,92	24,30	22,08	20,16	18,51	17,05	15,79	14,66	13,66
Pm2+	59,00	58,29	56,31	53,40	50,04	46,63	43,45	40,61	38,08	33,71	30,00	26,91	24,30	22,08	20,16	18,51	17,05	15,79	14,66	13,66
Pm3+	58,00	57,40	55,68	53,11	50,01	46,74	43,57	40,66	38,06	33,68	30,03	26,91	24,30	22,08	20,16	18,51	17,05	15,79	14,66	13,66
Pm4+	57,00	56,48	54,98	52,69	49,86	46,78	43,68	40,76	38,10	33,63	30,01	26,94	24,30	22,08	20,16	18,51	17,05	15,79	14,66	13,66
Sm	62,00	60,88	57,98	54,32	50,68	47,35	44,28	41,43	38,82	34,35	30,59	27,46	24,80	22,54	20,58	18,90	17,42	16,13	14,98	13,96
Sm+	61,00	60,13	57,75	54,43	50,81	47,34	44,21	41,41	38,86	34,34	30,59	27,46	24,80	22,54	20,58	18,90	17,42	16,13	14,98	13,96
Sm2+	60,00	59,28	57,27	54,33	50,92	47,47	44,25	41,37	38,81	34,37	30,60	27,45	24,80	22,54	20,58	18,90	17,42	16,13	14,98	13,96
Sm3+	59,00	58,39	56,65	54,04	50,90	47,58	44,37	41,41	38,78	34,33	30,62	27,46	24,80	22,54	20,58	18,90	17,42	16,13	14,98	13,96
Sm4+	58,00	57,47	55,95	53,62	50,75	47,62	44,48	41,51	38,82	34,29	30,61	27,48	24,80	22,54	20,58	18,90	17,42	16,13	14,98	13,96
Eu	63,00	61,87	58,94	55,24	51,56	48,19	45,08	42,19	39,55	35,01	31,19	28,01	25,30	23,00	21,01	19,29	17,79	16,47	15,30	14,26
Eu+	62,00	61,12	58,72	55,36	51,69	48,18	45,01	42,17	39,58	34,99	31,19	28,01	25,30	23,00	21,01	19,29	17,79	16,47	15,30	14,26
Eu2+	61,00	60,27	58,23	55,25	51,80	48,31	45,05	42,13	39,53	35,02	31,19	28,00	25,30	23,00	21,01	19,29	17,79	16,47	15,30	14,26
Eu3+	60,00	59,38	57,61	54,96	51,78	48,42	45,16	42,17	39,50	34,99	31,22	28,00	25,30	23,00	21,01	19,29	17,79	16,47	15,30	14,26
Eu4+	59,00	58,46	56,92	54,55	51,64	48,46	45,28	42,27	39,54	34,94	31,21	28,03	25,30	23,00	21,01	19,29	17,79	16,47	15,30	14,26
Gd	64,00	62,86	59,91	56,16	52,45	49,03	45,88	42,95	40,27	35,66	31,79	28,56	25,80	23,46	21,44	19,69	18,16	16,81	15,62	14,67
Gd+	63,00	62,11	59,68	56,28	52,57	49,02	45,81	42,94	40,31	35,65	31,79	28,56	25,80	23,46	21,44	19,69	18,16	16,81	15,62	14,67
Gd2+	62,00	61,26	59,20	56,18	52,68	49,15	45,85	42,89	40,26	35,68	31,79	28,55	25,80	23,46	21,44	19,69	18,16	16,81	15,62	14,67
Gd3+	61,00	60,37	58,58	55,89	52,67	49,26	45,96	42,93	40,23	35,65	31,82	28,55	25,80	23,46	21,44	19,69	18,16	16,81	15,62	14,67
Gd4+	60,00	59,45	57,88	55,48	52,52	49,30	46,08	43,03	40,27	35,60	31,81	28,57	25,80	23,46	21,44	19,69	18,16	16,81	15,62	14,67
Tb	65,00	63,85	60,87	57,08	53,33	49,88	46,68	43,71	41,00	36,33	32,39	29,11	26,31	23,93	21,87	20,09	18,53	17,16	15,95	14,87
Tb+	64,00	63,10	60,64	57,20	53,45	49,86	46,62	43,70	41,03	36,31	32,39	29,11	26,31	23,93	21,87	20,09	18,53	17,16	15,95	14,87
Tb2+	63,00	62,25	60,16	57,10	53,57	49,99	46,65	43,66	40,98	36,34	32,39	29,10	26,31	23,93	21,87	20,09	18,53	17,16	15,95	14,87
Tb3+	62,00	61,36	59,55	56,82	53,55	50,10	46,76	43,69	40,95	36,31	32,42	29,10	26,31	23,93	21,87	20,09	18,53	17,16	15,95	14,87
Tb4+	61,00	60,45	58,85	56,42	53,41	50,15	46,88	43,79	40,99	36,26	32,41	29,12	26,31	23,93	21,87	20,09	18,53	17,16	15,95	14,87

$\frac{\sin\theta}{\lambda}$ Element	0,00	0,05	0,10	0,15	0,20	0,25	0,30	0,35	0,40	0,50	0,60	0,70	0,80	0,90	1,00	1,10	1,20	1,30	1,40	1,50
Dy	66,00	64,84	61,83	58,01	54,21	50,72	47,49	44,48	41,73	36,99	32,99	29,66	26,81	24,39	22,30	20,49	18,90	17,51	16,27	15,18
Dy+	65,00	64,09	61,60	58,13	54,34	50,70	47,42	44,47	41,76	36,97	33,00	29,66	26,81	24,39	22,30	20,49	18,90	17,51	16,27	15,18
Dy++	64,00	63,24	61,12	58,03	54,45	50,83	47,45	44,42	41,71	37,00	33,00	29,65	26,81	24,39	22,30	20,49	18,90	17,51	16,27	15,18
Dy3+	63,00	62,35	60,51	57,75	54,44	50,95	47,56	44,46	41,68	36,97	33,02	29,65	26,81	24,39	22,30	20,49	18,90	17,51	16,27	15,18
Dy4+	62,00	61,44	59,82	57,35	54,30	50,99	47,68	44,55	41,72	36,93	33,02	29,67	26,81	24,39	22,30	20,49	18,90	17,51	16,27	15,18
Ho	67,00	65,83	62,79	58,93	55,10	51,56	48,29	45,24	42,46	37,65	33,59	30,21	27,32	24,86	22,73	20,89	19,27	17,86	16,60	15,48
Ho+	66,00	65,08	62,57	59,06	55,22	51,55	48,22	45,23	42,49	37,63	33,60	30,21	27,32	24,86	22,73	20,89	19,27	17,86	16,60	15,48
Ho++	65,00	64,23	62,09	58,96	55,34	51,67	48,26	45,19	42,44	37,67	33,60	30,20	27,32	24,86	22,73	20,89	19,27	17,86	16,60	15,48
Ho3+	64,00	63,34	61,48	58,68	55,33	51,79	48,37	45,22	42,41	37,64	33,63	30,20	27,32	24,86	22,73	20,89	19,27	17,86	16,60	15,48
Ho4+	63,00	62,43	60,79	58,28	55,19	51,84	48,48	45,32	42,45	37,59	33,62	30,23	27,32	24,86	22,73	20,89	19,27	17,86	16,60	15,48
Er	68,00	66,82	63,75	59,86	55,98	52,41	49,10	46,01	43,19	38,31	34,20	30,76	27,83	25,33	23,17	21,29	19,65	18,21	16,93	15,79
Er+	67,00	66,07	63,53	59,98	56,11	52,39	49,03	46,00	43,22	38,30	34,21	30,77	27,83	25,33	23,17	21,29	19,65	18,21	16,93	15,79
Er++	66,00	65,23	63,05	59,88	56,22	52,52	49,06	45,95	43,18	38,33	34,21	30,76	27,83	25,33	23,17	21,29	19,65	18,21	16,93	15,79
Er2+	65,00	64,34	62,45	59,61	56,22	52,64	49,17	45,99	43,14	38,30	34,23	30,76	27,83	25,33	23,17	21,29	19,65	18,21	16,93	15,79
Er4+	64,00	63,42	61,75	59,21	56,08	52,69	49,29	46,08	43,17	38,26	34,23	30,78	27,83	25,33	23,17	21,29	19,65	18,21	16,93	15,79
Tm	69,00	67,80	64,71	60,78	56,87	53,26	49,90	46,78	43,92	38,98	34,81	31,32	28,34	25,80	23,60	21,70	20,03	18,56	17,26	16,10
Tm+	68,00	67,06	64,49	60,91	56,99	53,24	49,84	46,77	43,96	38,96	34,82	31,32	28,34	25,80	23,60	21,70	20,03	18,56	17,26	16,10
Tm++	67,00	66,22	64,02	60,81	57,11	53,36	49,87	46,72	43,91	39,00	34,81	31,31	28,34	25,80	23,60	21,70	20,03	18,56	17,26	16,10
Tm3+	66,00	65,33	63,41	60,54	57,10	53,48	49,97	46,76	43,88	38,97	34,84	31,31	28,34	25,80	23,60	21,70	20,03	18,56	17,26	16,10
Tm4+	65,00	64,41	62,72	60,15	56,98	53,53	50,09	46,85	43,90	38,92	34,84	31,34	28,34	25,80	23,60	21,70	20,03	18,56	17,26	16,10
Yb	70,00	68,79	65,67	61,71	57,75	54,10	50,71	47,55	44,66	39,65	35,42	31,88	28,85	26,28	24,04	22,11	20,40	18,91	17,59	16,41
Yb+	69,00	68,05	65,46	61,83	57,88	54,08	50,64	47,54	44,69	39,35	35,43	31,88	28,85	26,28	24,04	22,11	20,40	18,91	17,59	16,41
Yb3+	68,00	67,21	64,98	61,74	58,00	54,21	50,67	47,49	44,64	39,66	35,42	31,87	28,85	26,28	24,04	22,11	20,40	18,91	17,59	16,41
Yb3+	67,00	66,32	64,38	61,47	57,99	54,33	50,78	47,52	44,61	39,64	35,45	31,87	28,85	26,28	24,04	22,11	20,40	18,91	17,59	16,41
Yb4+	66,00	65,40	63,69	61,08	57,87	54,38	50,90	47,62	44,64	39,59	35,45	31,89	28,85	26,28	24,04	22,11	20,40	18,91	17,59	16,41

$\frac{\sin\theta}{\lambda}$ Element	0,00	0,05	0,10	0,15	0,20	0,25	0,30	0,35	0,40	0,50	0,60	0,70	0,80	0,90	1,00	1,10	1,20	1,30	1,40	1,50
Cp	71,00	69,78	66,64	62,63	58,64	54,95	51,52	48,32	45,39	40,32	36,03	32,44	29,37	26,75	24,48	22,51	20,78	19,27	17,92	16,72
Cp+	70,00	69,04	66,42	62,76	58,76	54,93	51,45	48,31	45,43	40,30	36,04	32,44	29,37	26,75	24,48	22,51	20,78	19,27	17,92	16,72
Cp²⁺	69,00	68,20	65,95	62,67	58,88	55,06	51,48	48,26	45,38	40,33	36,03	32,43	29,37	26,75	24,48	22,51	20,78	19,27	17,92	16,72
Cp³⁺	68,00	67,31	65,35	62,41	58,88	55,18	51,59	48,29	45,34	40,31	36,06	32,43	29,37	26,75	24,48	22,51	20,78	19,27	17,92	16,72
Cp⁴⁺	67,00	66,40	64,66	62,01	58,76	55,23	51,71	48,39	45,37	40,26	36,06	32,45	29,37	26,75	24,48	22,51	20,78	19,27	17,92	16,72
Hf	72,00	70,77	67,60	63,56	59,53	55,80	52,33	49,09	46,13	40,99	36,64	33,00	29,88	27,23	24,92	22,92	21,17	19,62	18,25	17,04
Hf+	71,00	70,03	67,38	63,69	59,65	55,78	52,26	49,08	46,16	40,97	36,65	33,00	29,88	27,23	24,92	22,92	21,17	19,62	18,25	17,04
Hf²⁺	70,00	69,19	66,91	63,60	59,77	55,90	52,29	49,03	46,12	41,00	36,64	32,99	29,88	27,23	24,92	22,92	21,17	19,62	18,25	17,04
Hf³⁺	69,00	68,30	66,31	63,34	59,77	56,02	52,40	49,06	46,08	40,98	36,67	32,99	29,88	27,23	24,92	22,92	21,17	19,62	18,25	17,04
Hf⁴⁺	68,00	67,39	65,63	62,95	59,65	56,08	52,51	49,16	46,10	40,93	36,67	33,01	29,88	27,23	24,92	22,92	21,17	19,62	18,25	17,04
Ta	73,00	71,76	68,56	64,49	60,42	56,65	53,14	49,86	46,86	41,66	37,25	33,56	30,40	27,70	25,36	23,33	21,55	19,98	18,59	17,35
Ta+	72,00	71,02	68,35	64,62	60,54	56,63	53,07	49,86	46,90	41,64	37,26	33,56	30,40	27,70	25,36	23,33	21,55	19,98	18,59	17,35
Ta²⁺	71,00	70,18	67,88	64,53	60,66	56,75	53,10	49,81	46,85	41,68	37,26	33,56	30,40	27,70	25,36	23,33	21,55	19,98	18,59	17,35
Ta³⁺	70,00	69,29	67,28	64,27	60,67	56,87	53,20	49,84	46,82	41,65	37,29	33,55	30,40	27,70	25,36	23,33	21,55	19,98	18,59	17,35
Ta⁴⁺	69,00	68,38	66,60	63,88	60,55	56,93	53,32	49,93	46,84	41,61	37,29	33,58	30,40	27,70	25,36	23,33	21,55	19,98	18,59	17,35
W	74,00	72,75	69,52	65,42	61,31	57,50	53,95	50,64	47,60	42,33	37,87	34,12	30,92	28,18	25,80	23,74	21,93	20,34	18,92	17,67
W+	73,00	72,01	69,31	65,54	61,43	57,48	53,88	50,63	47,64	42,32	37,88	34,13	30,92	28,18	25,80	23,74	21,93	20,34	18,92	17,67
W²⁺	72,00	71,17	68,85	65,46	61,55	57,60	53,91	50,58	47,59	42,35	37,87	34,12	30,92	28,18	25,80	23,74	21,93	20,34	18,92	17,67
W³⁺	71,00	70,28	68,25	65,20	61,56	57,72	54,01	50,61	47,56	42,33	37,90	34,11	30,92	28,18	25,80	23,74	21,93	20,34	18,92	17,67
W⁴⁺	70,00	69,37	67,57	64,82	61,44	57,78	54,13	50,70	47,58	42,28	37,90	34,14	30,92	28,18	25,80	23,74	21,93	20,34	18,92	17,67
Re	75,00	73,74	70,49	66,34	62,20	58,35	54,76	51,41	48,34	43,01	38,48	34,69	31,44	28,66	26,25	24,16	22,32	20,70	19,26	17,98
Re+	74,00	73,00	70,28	66,47	62,32	58,33	54,70	51,41	48,38	42,99	38,49	34,69	31,44	28,66	26,25	24,16	22,32	20,70	19,26	17,98
Re²⁺	73,00	72,16	69,81	66,39	62,44	58,45	54,72	51,36	48,33	43,02	38,49	34,68	31,44	28,66	26,25	24,16	22,32	20,70	19,26	17,98
Re³⁺	72,00	71,27	69,22	66,13	62,45	58,57	54,83	51,38	48,29	43,00	38,52	34,68	31,44	28,66	26,25	24,16	22,32	20,70	19,26	17,98
Re⁴⁺	71,00	70,36	68,54	65,75	62,34	58,64	54,95	51,47	48,32	42,95	38,52	34,70	31,44	28,66	26,25	24,16	22,32	20,70	19,26	17,98

$\dfrac{\sin\theta}{\lambda}$ / Element	0,00	0,05	0,10	0,15	0,20	0,25	0,30	0,35	0,40	0,50	0,60	0,70	0,80	0,90	1,00	1,10	1,20	1,30	1,40	1,50
Os	76,00	74,73	71,45	67,27	63,09	59,20	55,58	52,19	49,08	43,68	39,10	35,26	31,96	29,14	26,70	24,57	22,70	21,06	19,60	18,30
Os+	75,00	73,99	71,24	67,40	63,21	59,18	55,51	52,19	49,12	43,67	39,11	35,26	31,96	29,14	26,70	24,57	22,70	21,06	19,60	18,30
Os2+	74,00	73,15	70,78	67,32	63,33	59,30	55,53	52,13	49,07	43,70	39,11	35,25	31,96	29,14	26,70	24,57	22,70	21,06	19,60	18,30
Os3+	73,00	72,27	70,18	67,07	63,34	59,43	55,64	52,16	49,04	43,08	39,13	35,25	31,96	29,14	26,70	24,57	22,70	21,06	19,60	18,30
Os4+	72,00	71,35	69,51	66,69	63,23	59,49	55,76	52,25	49,06	43,63	39,14	35,27	31,96	29,14	26,70	24,57	22,70	21,06	19,60	18,30
Ir	77,00	75,72	72,42	68,20	63,98	60,06	56,39	52,96	49,83	44,36	39,72	35,82	32,48	29,63	27,14	24,99	23,09	21,43	19,94	18,62
Ir+	76,00	74,98	72,21	68,33	64,10	60,03	56,33	52,96	49,86	44,34	39,73	35,83	32,48	29,63	27,14	24,99	23,09	21,43	19,94	18,62
Ir2+	75,00	74,14	71,75	68,25	64,22	60,15	56,35	52,91	49,82	44,38	39,72	35,82	32,48	29,63	27,14	24,99	23,09	21,43	19,94	18,62
Ir3+	74,00	73,26	71,15	68,00	64,24	60,28	56,45	52,94	49,78	44,36	39,75	35,81	32,48	29,63	27,14	24,99	23,09	21,43	19,94	18,62
Ir4+	73,00	72,35	70,48	67,63	64,13	60,34	56,57	53,02	49,80	44,31	39,76	35,84	32,48	29,63	27,14	24,99	23,09	21,43	19,94	18,62
Pt	78,00	76,71	73,38	69,13	64,87	60,91	57,21	53,74	50,57	45,04	40,34	36,39	33,01	30,11	27,59	25,41	23,48	21,79	20,28	18,94
Pt+	77,00	75,97	73,17	69,26	64,99	60,88	57,14	53,74	50,61	45,02	40,35	36,40	33,01	30,11	27,59	25,41	23,48	21,79	20,28	18,94
Pt2+	76,00	75,13	72,71	69,18	65,12	61,01	57,16	53,69	50,56	45,05	40,35	36,39	33,01	30,11	27,59	25,41	23,48	21,79	20,28	18,94
Pt3+	75,00	74,25	72,12	68,94	65,13	61,13	57,26	53,71	50,52	45,04	40,37	36,38	33,01	30,11	27,59	25,41	23,48	21,79	20,28	18,94
Pt4+	74,00	73,34	71,45	68,56	65,03	61,20	57,39	53,80	50,54	44,99	40,38	36,41	33,01	30,11	27,59	25,41	23,48	21,79	20,28	18,94
Au	79,00	77,70	74,35	70,06	65,77	61,76	58,02	54,52	51,31	45,72	40,96	36,96	33,53	30,60	28,04	25,83	23,87	22,16	20,63	19,27
Au+	78,00	76,97	74,14	70,19	65,88	61,74	57,96	54,52	51,35	45,70	40,97	36,97	33,53	30,60	28,04	25,83	23,87	22,16	20,63	19,27
Au2+	77,00	76,13	73,68	70,12	66,01	61,86	57,98	54,47	51,31	45,73	40,97	36,96	33,53	30,60	28,04	25,83	23,87	22,16	20,63	19,27
Au3+	76,00	75,24	73,09	69,87	66,03	61,99	58,08	54,49	51,27	45,72	41,00	36,95	33,53	30,60	28,04	25,83	23,87	22,16	20,63	19,27
Au4+	75,00	74,33	72,42	69,50	65,92	62,05	58,20	54,54	51,28	45,67	41,00	36,98	33,53	30,60	28,04	25,83	23,87	22,16	20,63	19,27
Hg	80,00	78,69	75,31	70,99	66,66	62,62	58,84	55,30	52,06	46,40	41,59	37,54	34,06	31,08	28,50	26,25	24,27	22,52	20,97	19,59
Hg+	79,00	77,96	75,10	71,12	66,78	62,59	58,78	55,30	52,10	46,38	41,60	37,54	34,06	31,08	28,50	26,25	24,27	22,52	20,97	19,59
Hg2+	78,00	77,12	74,65	71,05	66,90	62,71	58,79	55,25	52,05	46,41	41,59	37,53	34,06	31,08	28,50	26,25	24,27	22,52	20,97	19,59
Hg3+	77,00	76,23	74,06	70,81	66,92	62,84	58,90	55,27	52,01	46,40	41,62	37,53	34,06	31,08	28,50	26,25	24,27	22,52	20,97	19,59
Hg4+	76,00	75,32	73,38	70,43	66,82	62,91	59,02	55,36	52,03	46,35	41,63	37,55	34,06	31,08	28,50	26,25	24,27	22,52	20,97	19,59

Element	$\frac{\sin\theta}{\lambda}$ 0,00	0,05	0,10	0,15	0,20	0,25	0,30	0,35	0,40	0,50	0,60	0,70	0,80	0,90	1,00	1,10	1,20	1,30
Tl	81,0	78,3	75,5	71,1	66,7	62,7	58,7	55,0	51,2	45,0	41,1	37,4	34,1	31,1	28,3	26,0	24,1	—
Pb	82,0	79,3	76,5	72,0	67,5	63,5	59,5	55,7	51,9	45,7	41,6	37,9	34,6	31,5	28,8	26,4	24,5	—
Bi	83,0	80,3	77,5	73,0	68,4	64,4	60,4	56,6	52,7	46,4	42,2	38,5	35,1	32,0	29,2	26,8	24,8	—
Po	84,0	81,2	78,4	73,9	69,4	65,4	61,3	57,4	53,5	47,1	42,8	39,1	35,6	32,6	29,7	27,2	25,2	—
At	85,0	82,2	79,4	74,9	70,3	66,2	62,1	58,2	54,2	47,7	43,4	39,6	36,2	33,1	30,1	27,6	25,6	—
Rn	86,0	83,2	80,3	75,8	71,3	67,2	63,0	59,1	55,1	48,4	44,0	40,2	36,8	33,5	30,5	28,0	26,0	—
Fr	87,0	84,2	81,3	76,8	72,2	68,0	63,8	59,8	55,8	49,1	44,5	40,7	37,3	34,0	31,0	28,4	26,4	—
Ra	88,0	85,1	82,2	77,7	73,2	68,9	64,6	60,6	56,5	49,8	45,1	41,3	37,8	34,6	31,5	28,8	26,7	—
Ac	89,0	86,1	83,2	78,7	74,1	69,8	65,5	61,4	57,3	50,4	45,8	41,8	38,3	35,1	32,0	29,2	27,1	—
Th	90,0	87,1	84,1	79,6	75,1	70,7	66,3	62,2	58,1	51,1	46,5	42,4	38,8	35,5	32,4	29,6	27,5	—
Pa	91,0	88,1	85,1	80,6	76,0	71,6	67,1	63,0	58,8	51,7	47,1	43,0	39,3	36,0	32,8	30,1	27,9	—
U	92,0	89,0	86,0	81,5	76,9	72,4	67,9	63,8	59,6	52,4	47,7	43,5	39,8	36,5	33,3	30,6	28,3	—
Np	93	90	87	83	78	74	69	65	60	53	48	44	40	37	34	31	29	—
Pu	94	91	88	84	79	74	69	65	61	54	49	44	41	38	34	31	29	—
Am	95	92	89	84	79	75	70	66	62	55	50	45	42	38	35	32	30	—
Cm	96	93	90	85	80	76	71	67	62	55	50	46	42	39	35	32	30	—
Bk	97	94	91	86	81	77	72	68	63	56	51	46	43	39	36	33	30	—
Cf	98	95	92	87	82	78	73	69	64	57	52	47	43	40	36	33	31	—
Ei	99	96	93	88	83	79	74	70	65	57	52	48	44	40	37	34	31	—
Fm	100	97	94	89	84	80	75	71	66	58	53	48	44	41	37	34	31	—

4-8. Correction for Anomalous Dispersion

It is usual to take the frequency of the incident radiation as large relative to the natural frequency in approximate calculations on scattering functions; precise calculations require correction for the anomalous dispersion introduced by the K shell, the scattering function being

$$|f| = f_0 + \Delta f_K' + \frac{1}{2} \frac{(\Delta f_K'')^2}{(f_0 + \Delta f_K')} ,$$

where f_0 is the scattering function for wavelengths much shorter than the absorption edge. Table 4-7 gives f_0.

The $\Delta f_K'$ and $\Delta f_K''$ are given by

$$\Delta f_K' = \frac{2^7 e^{-4}}{9} \left[\frac{4}{(1-\delta_K)^2} \frac{1}{x^2} \ln|x^2 - 1| - \frac{1}{(1-\delta_K)^3} \left(\frac{2}{x^2} + \frac{1}{x^3} \ln\left|\frac{x-1}{x+1}\right| \right) \right] ,$$

$$\Delta f_K'' = \frac{2^7 e^{-4}}{9} \pi \left[\frac{4}{x^2(1-\delta_K)^2} - \frac{1}{x^3(1-\delta_K)^3} \right] ,$$

where $x = \omega/\omega_K$, ω is the frequency of the incident radiation, ω_K is the frequency of the K edge, and δ_K is the atomic scattering parameter for the element.

Table 4-8a gives $\Delta f_K'$ as a function of λ/λ_K and δ_K, where λ is the incident wavelength and λ_K is the wavelength for the K edge [98]. Table 4-8b gives $\Delta f_K''$ [98]. The value of $\Delta f_K'$ can be positive or negative; account must be taken of this in deducing f. Table 4-8c gives δ_K [102].

4-8a. Values of $\Delta f_K'$

$\dfrac{\lambda}{\lambda_K}$ \ δ_K	0,12	0,14	0,16	0,18	0,20	0,22	0,24	0,26	0,28	0,30
0,05	0,02	0,02	0,02	0,02	0,02	0,02	0,03	0,03	0,03	0,03
0,10	0,05	0,06	0,06	0,06	0,07	0,07	0,07	0,08	0,08	0,08
0,15	0,10	0,10	0,11	0,11	0,12	0,12	0,13	0,13	0,14	0,15
0,20	0,14	0,15	0,15	0,16	0,17	0,18	0,18	0,19	0,20	0,21
0,25	0,18	0,19	0,20	0,21	0,22	0,23	0,24	0,25	0,26	0,27
0,30	0,22	0,23	0,24	0,25	0,26	0,27	0,28	0,29	0,30	0,32
0,35	0,25	0,25	0,26	0,28	0,29	0,30	0,31	0,32	0,34	0,35
0,40	0,26	0,27	0,28	0,29	0,30	0,31	0,32	0,33	0,35	0,36
0,45	0,25	0,26	0,27	0,28	0,29	0,30	0,32	0,33	0,34	0,35
0,50	0,23	0,24	0,24	0,25	0,26	0,27	0,28	0,29	0,29	0,30
0,55	0,19	0,19	0,20	0,20	0,21	0,22	0,22	0,23	0,23	0,24
0,60	0,12	0,12	0,12	0,12	0,12	0,12	0,12	0,12	0,12	0,12
0,65	0,02	0,02	0,01	0,01	0,00	−0,00	−0,01	−0,02	−0,03	−0,04
0,70	−0,12	−0,13	−0,14	−0,15	−0,16	−0,18	−0,19	−0,21	−0,23	−0,25
0,75	−0,30	−0,32	−0,34	−0,35	−0,38	−0,40	−0,43	−0,46	−0,49	−0,53
0,80	−0,52	−0,54	−0,57	−0,60	−0,63	−0,66	−0,70	−0,74	−0,78	−0,83
0,85	−0,89	−0,93	−0,97	−1,02	−1,07	−1,12	−1,18	−1,24	−1,31	−1,39
0,90	−0,38	−1,44	−1,50	−1,57	−1,65	−1,72	−1,81	−1,90	−2,00	−2,10
0,95	−2,22	−2,31	−2,41	−2,51	−2,62	−2,74	−2,87	−3,00	−3,14	−3,30
0,975	−2,93	−3,04	−3,17	−3,30	−3,44	−3,59	−3,75	−3,92	−4,10	−4,29
0,980	−3,17	−3,30	−3,43	−3,57	−3,72	−3,88	−4,06	−4,23	−4,43	−4,54
0,985	−3,46	−3,60	−3,75	−3,90	−4,06	−4,23	−4,42	−4,61	−4,82	−5,04
0,990	−3,91	−4,06	−4,23	−4,40	−4,57	−4,76	−4,99	−5,21	−5,44	−5,69
1,005	−4,72	−4,89	−5,20	−5,30	−5,52	−5,75	−5,99	−6,26	−6,53	−6,83
1,010	−4,09	−4,25	−4,42	−4,59	−4,79	−4,99	−5,20	−5,43	−5,67	−5,93

$\frac{\lambda}{\lambda_K}$ \ δ_K	0,12	0,14	0,16	0,18	0,20	0,22	0,24	0,26	0,28	0,30
1,015	−3,73	−3,88	−4,03	−4,19	−4,37	−4,55	−4,75	−4,96	−5,18	−5,45
1,020	−3,49	−3,62	−3,77	−3,92	−4,08	−4,26	−4,54	−4,64	−4,85	−5,07
1,025	−3,28	−3,41	−3,55	−3,69	−3,85	−4,01	−4,19	−4,37	−4,57	−4,78
1,030	−3,13	−2,25	−3,39	−3,52	−3,67	−3,83	−3,99	−4,17	−4,36	−4,56
1,035	−3,00	−3,12	−3,25	−3,38	−3,52	−3,67	−3,83	−4,00	−4,18	−4,38
1,040	−2,90	−3,02	−3,14	−3,27	−3,40	−3,55	−3,71	−3,87	−4,04	−4,23
1,045	−2,81	−2,92	−3,03	−3,16	−3,29	−3,43	−3,58	−3,74	−3,91	−4,10
1,050	−2,72	−2,83	−2,95	−3,07	−3,20	−3,33	−3,48	−3,63	−3,80	−3,98
1,055	−2,65	−2,76	−2,87	−2,99	−3,12	−3,25	−3,39	−3,53	−3,70	−3,88
1,060	−2,59	−2,69	−2,80	−2,92	−3,04	−3,17	−3,31	−3,46	−3,62	−3,79
1,065	−2,53	−2,63	−2,74	−2,85	−2,97	−3,10	−3,23	−3,38	−3,53	−3,70
1,070	−2,47	−2,57	−2,68	−2,79	−2,91	−3,03	−3,17	−3,31	−3,46	−3,62
1,08	−2,38	−2,47	−2,57	−2,68	−2,80	−2,91	−3,04	−3,18	−3,33	−3,48
1,09	−2,30	−2,39	−2,50	−2,59	−2,70	−2,82	−2,94	−3,08	−3,22	−3,37
1,10	−2,23	−2,32	−2,41	−2,51	−2,62	−2,73	−2,86	−2,98	−3,12	−3,27
1,15	−1,98	−2,05	−2,14	−2,23	−2,32	−2,42	−2,53	−2,65	−2,77	−2,92
1,20	−1,82	−1,89	−1,97	−2,05	−2,14	−2,23	−2,33	−2,44	−2,55	−2,69
1,25	−1,70	−1,77	−1,85	−1,92	−2,01	−2,09	−2,19	−2,29	−2,40	−2,51
1,30	−1,64	−1,71	−1,78	−1,86	−1,94	−2,02	−2,11	−2,21	−2,32	−2,43
1,40	−1,50	−1,56	−1,63	−1,69	−1,77	−1,85	−1,93	−2,02	−2,11	−2,22
1,50	−1,42	−1,48	−1,54	−1,61	−1,68	−1,75	−1,83	−1,92	−2,01	−2,11
1,60	−1,37	−1,43	−1,49	−1,55	−1,62	−1,69	−1,76	−1,85	−1,93	−2,03
1,80	−1,29	−1,35	−1,40	−1,46	−1,53	−1,59	−1,67	−1,75	−1,83	−1,92
2,00	−1,25	−1,30	−1,35	−1,41	−1,47	−1,54	−1,61	−1,68	−1,76	−1,85
3,0	−1,15	−1,20	−1,25	−1,30	−1,36	−1,42	−1,49	−1,56	−1,63	−1,71
∞	−1,09	−1,14	−1,18	−1,23	−1,29	−1,35	−1,41	−1,47	−1,55	−1,62

4-8b. Values of $\Delta f''_K$

$\frac{\lambda}{\lambda_K}$ \ δ_K	0,12	0,14	0,16	0,18	0,20	0,22	0,24	0,26	0,28	0,30
0,0	0,00	0,00	0,00	0,00	0,00	0,00	0,00	0,00	0,00	0,00
0,1	0,04	0,04	0,05	0,05	0,05	0,05	0,05	0,06	0,06	0,06
0,2	0,16	0,17	0,17	0,18	0,19	0,20	0,21	0,22	0,23	0,25
0,3	0,35	0,36	0,38	0,40	0,42	0,44	0,46	0,48	0,51	0,54
0,4	0,60	0,63	0,65	0,68	0,72	0,75	0,79	0,83	0,87	0,92
0,5	0,91	0,95	0,99	1,03	1,08	1,13	1,18	1,24	1,31	1,37
0,6	1,26	1,32	1,37	1,43	1,50	1,56	1,64	1,72	1,80	1,89
0,7	1,66	1,73	1,80	1,88	1,96	2,04	2,14	2,24	2,34	2,46
0,8	2,09	2,17	2,26	2,35	2,45	2,56	2,67	2,79	2,92	3,05
0,9	2,55	2,65	2,75	2,86	2,98	3,10	3,23	3,37	3,52	3,67
1,0	3,03	3,14	3,26	3,38	3,52	3,65	3,80	3,96	4,12	4,30

4-8c. Values of δ_K for Some Elements

The relation for δ_K is

$$\delta_K = \left(A - \frac{911}{\lambda_K} \right) A,$$

where λ_K is for the K edge and $A = (Z - 0.3)^2 + 1.33 \cdot 10^{-5} \cdot (Z - 0.3)^4$, Z being the atomic number of the scattering element.

Values of δ_K are given for Z from 20 to 92.

z	Ele-ment	δ_K	z	Ele-ment	δ_K	z	Ele-ment	δ_K	z	Ele-ment	δ_K
20	Ca	0,240	39	Y	0,188	58	Ce	0,156	77	Ir	0,141
21	Sc	0,233	40	Zr	0,186	59	Pr	0,155	78	Pt	0,140
22	Ti	0,227	41	Nb	0,184	60	Nd	0,154	79	Au	0,140
23	V	0,223	42	Mo	0,182	61	Pm	0,153	80	Hg	0,139
24	Cr	0,218	43	Tc	0,180	62	Sm	0,152	81	Tl	0,138
25	Mn	0,216	44	Ru	0,179	63	Eu	0,151	82	Pb	0,138
26	Fe	0,215	45	Rh	0,177	64	Gd	0,150	83	Bi	0,137
27	Co	0,212	46	Pd	0,176	65	Tb	0,150	84	Po	0,136
28	Ni	0,209	47	Ag	0,174	66	Dy	0,149	85	At	0,136
29	Cu	0,207	48	Cd	0,172	67	Ho	0,148	86	Rn	0,135
30	Zn	0,205	49	In	0,170	68	Er	0,147	87	Fr	0,134
31	Ga	0,203	50	Sn	0,169	69	Tu	0,147	88	Ra	0,134
32	Ge	0,201	51	Sb	0,167	70	Yb	0,146	89	Ac	0,133
33	As	0,200	52	Te	0,166	71	Lu	0,145	90	Th	0,132
34	Se	0,198	53	I	0,163	72	Hf	0,144	91	Pa	0,132
35	Br	0,196	54	Xe	0,162	73	Ta	0,144	92	U	0,131
36	Kr	0,194	55	Cs	0,160	74	W	0,143			
37	Rb	0,192	56	Ba	0,159	75	Re	0,142			
38	Sr	0,190	57	La	0,158	76	Os	0,142			

STRUCTURE FACTORS

4-9. Accessory Table for Calculating Structure Factors

This table is designed to give $\cos 2\pi x$ and $\sin 2\pi x$ [2], whose absolute values are given in the middle column for the x given in the left and right columns; A, B, C, D on the left relate to the sine, while E, F, G, H on the right relate to the cosine. Columns A and E give values of the argument from 0 to 0.25; B and F, from 0.25 to 0.50; C and G, from 0.50 to 0.75; and D and H, from 0.75 to 1.00. See [102] for a more detailed table.

				sin 2πx	cos 2 πx		x		
A	B	C	D	A B C D + + — —	E F G H + — — +	E	F	G	H
0,000	0,500	0,500	1,000	0,0000		0,250	0,250	0,750	0,750
0,001	0,499	0,501	0,999	0,0063		0,249	0,251	0,749	0,751
0,002	0,498	0,502	0,998	0,0126		0,248	0,252	0,748	0,752
0,003	0,497	0,503	0,997	0,0188		0,247	0,253	0,747	0;753
0,004	0,496	0,504	0,996	0,0251		0,246	0,254	0,746	0,754
0,005	0,495	0,505	0,995	0,0314		0,245	0,255	0,745	0,755
0,006	0,494	0,506	0,994	0,0377		0,244	0,256	0,744	0,756
0,007	0,493	0,507	0,993	0,0440		0,243	0,257	0,743	0,757
0,008	0,492	0,508	0,992	0,0502		0,242	0,258	0,742	0,758
0,009	0,491	0,509	0,991	0,0565		0,241	0,259	0,741	0,759
0,010	0,490	0,510	0,990	0,0628		0,240	0,260	0,740	0,760
0,011	0,489	0,511	0,989	0,0691		0,239	0,261	0,739	0,761
0,012	0,488	0,512	0,988	0,0753		0,238	0,262	0,738	0,762
0,013	0,487	0,513	0,987	0,0816		0,237	0,263	0,737	0,763
0,014	0,486	0,514	0,986	0,0879		0,236	0,264	0,736	0,764
0,015	0,485	0,515	0,985	0,0941		0,235	0,265	0,735	0,765
0,016	0,484	0,516	0,984	0,1004		0,234	0,266	0,734	0,766
0,017	0,483	0,517	0,983	0,1066		0,233	0,267	0,733	0,767
0,018	0,482	0,518	0,982	0,1129		0,232	0,268	0,732	0,768
0,019	0,481	0,519	0,981	0,1191		0,231	0,269	0,731	0,769

A	B	C	D	sin 2πx ABCD ++--	cos 2πx EFGH +--+	E	F	G	H
0,020	0,480	0,520	0,980	0,1253		0,230	0,270	0,730	0,770
0,021	0,479	0,521	0,979	0,1316		0,229	0,271	0,729	0,771
0,022	0,478	0,522	0,978	0,1378		0,228	0,272	0,728	0,772
0,023	0,477	0,523	0,977	0,1440		0,227	0,273	0,727	0,773
0,024	0,476	0,524	0,976	0,1502		0,226	0,274	0,726	0,774
0,025	0,475	0,525	0,975	0,1564		0,225	0,275	0,725	0,775
0,026	0,474	0,526	0,974	0,1626		0,224	0,276	0,724	0,776
0,027	0,473	0,527	0,973	0,1688		0,223	0,277	0,723	0,777
0,028	0,472	0,528	0,972	0,1750		0,222	0,278	0,722	0,778
0,029	0,471	0,529	0,971	0,1812		0,221	0,279	0,721	0,779
0,030	0,470	0,530	0,970	0,1874		0,220	0,280	0,720	0,780
0,031	0,469	0,531	0,969	0,1935		0,219	0,281	0,719	0,781
0,032	0,468	0,532	0,968	0,1997		0,218	0,282	0,718	0,782
0,033	0,467	0,533	0,967	0,2059		0,217	0,283	0,717	0,783
0,034	0,466	0,534	0,966	0,2120		0,216	0,284	0,716	0,784
0,035	0,465	0,535	0,965	0,2181		0,215	0,285	0,715	0,785
0,036	0,464	0,536	0,964	0,2243		0,214	0,286	0,714	0,786
0,037	0,463	0,537	0,963	0,2304		0,213	0,287	0,713	0,787
0,038	0,462	0,538	0,962	0,2365		0,212	0,288	0,712	0,788
0,039	0,461	0,539	0,961	0,2426		0,211	0,289	0,711	0,789
0,040	0,460	0,540	0,960	0,2487		0,210	0,290	0,710	0,790
0,041	0,459	0,541	0,959	0,2548		0,209	0,291	0,709	0,791
0,042	0,458	0,542	0,958	0,2608		0,208	0,292	0,708	0,792
0,043	0,457	0,543	0,957	0,2669		0,207	0,293	0,707	0,793
0,044	0,456	0,544	0,956	0,2730		0,206	0,294	0,706	0,794
0,045	0,455	0,545	0,955	0,2790		0,205	0,295	0,705	0,795
0,046	0,454	0,546	0,954	0,2850		0,204	0,296	0,704	0,796
0,047	0,453	0,547	0,953	0,2910		0,203	0,297	0,703	0,797
0,048	0,452	0,548	0,952	0,2970		0,202	0,298	0,702	0,798
0,049	0,451	0,549	0,951	0,3030		0,201	0,299	0,701	0,799
0,050	0,450	0,550	0,950	0,3090		0,200	0,300	0,700	0,800
0,051	0,449	0,551	0,949	0,3150		0,199	0,301	0,699	0,801
0,052	0,448	0,552	0,948	0,3209		0,198	0,302	0,698	0,802
0,053	0,447	0,553	0,947	0,3269		0,197	0,303	0,697	0,803
0,054	0,446	0,554	0,946	0,3328		0,196	0,304	0,696	0,804
0,055	0,445	0,555	0,945	0,3387		0,195	0,305	0,695	0,805
0,056	0,444	0,556	0,944	0,3446		0,194	0,306	0,694	0,806
0,057	0,443	0,557	0,943	0,3505		0,193	0,307	0,693	0,807
0,058	0,442	0,558	0,942	0,3564		0,192	0,308	0,692	0,808
0,059	0,441	0,559	0,941	0,3623		0,191	0,309	0,691	0,809
0,060	0,440	0,560	0,940	0,3681		0,190	0,310	0,690	0,810
0,061	0,439	0,561	0,939	0,3740		0,189	0,311	0,689	0,811
0,062	0,438	0,562	0,938	0,3798		0,188	0,312	0,688	0,812
0,063	0,437	0,563	0,937	0,3856		0,187	0,313	0,687	0,813
0,064	0,436	0,564	0,936	0,3914		0,186	0,314	0,686	0,814
0,065	0,435	0,565	0,935	0,3971		0,185	0,315	0,685	0,815
0,066	0,434	0,566	0,934	0,4029		0,184	0,316	0,684	0,816
0,067	0,433	0,567	0,933	0,4086		0,183	0,317	0,683	0,817
0,068	0,432	0,568	0,932	0,4144		0,182	0,318	0,682	0,818
0,069	0,431	0,569	0,931	0,4201		0,181	0,319	0,681	0,819
0,070	0,430	0,570	0,930	0,4258		0,180	0,320	0,680	0,820
0,071	0,429	0,571	0,929	0,4315		0,179	0,321	0,679	0,821
0,072	0,428	0,572	0,928	0,4371		0,178	0,322	0,678	0,822
0,073	0,427	0,573	0,927	0,4428		0,177	0,323	0,677	0,823
0,074	0,426	0,574	0,926	0,4484		0,176	0,324	0,676	0,824
0,075	0,425	0,575	0,925	0,4540		0,175	0,325	0,675	0,825
0,076	0,424	0,576	0,924	0,4596		0,174	0,326	0,674	0,826

x				sin 2πx	cos 2πx	x			
				A B C D + + − −	E F G H + − − +				
A	**B**	**C**	**D**			**E**	**F**	**G**	**H**
0,077	0,423	0,577	0,923	0,4652		0,173	0,327	0,673	0,827
0,078	0,422	0,578	0,922	0,4707		0,172	0,328	0,672	0,828
0,079	0,421	0,579	0,921	0,4762		0,171	0,329	0,671	0,829
0,080	0,420	0,580	0,920	0,4818		0,170	0,330	0,670	0,830
0,081	0,419	0,581	0,919	0,4873		0,169	0,331	0,669	0,831
0,082	0,418	0,582	0,918	0,4927		0,168	0,332	0,668	0,832
0,083	0,417	0,583	0,917	0,4982		0,167	0,333	0,667	0,833
0,084	0,416	0,584	0,916	0,5036		0,166	0,334	0,666	0,834
0,085	0,415	0,585	0,915	0,5090		0,165	0,335	0,665	0,835
0,086	0,414	0,586	0,914	0,5144		0,164	0,336	0,664	0,836
0,087	0,413	0,587	0,913	0,5198		0,163	0,337	0,663	0,837
0,088	0,412	0,588	0,912	0,5252		0,162	0,338	0,662	0,838
0,089	0,411	0,589	0,911	0,5305		0,161	0,339	0,661	0,839
0,090	0,410	0,590	0,910	0,5358		0,160	0,340	0,660	0,840
0,091	0,409	0,591	0,909	0,5411		0,159	0,341	0,659	0,841
0,092	0,408	0,592	0,908	0,5464		0,158	0,342	0,658	0,842
0,093	0,407	0,593	0,907	0,5516		0,157	0,343	8,657	0,843
0,094	0,406	0,594	0,906	0,5569		0,156	0,344	0,656	0,844
0,095	0,405	0,595	0,905	0,5621		0,155	0,345	0,655	0,845
0,096	0,404	0,596	0,904	0,5673		0,154	0,346	0,654	0,846
0,097	0,403	0,597	0,903	0,5724		0,153	0,347	0,653	0,847
0,098	0,402	0,598	0,902	0,5776		0,152	0,348	0,652	0,848
0,099	0,401	0,599	0,901	0,5827		0,151	0,349	0,651	0,849
0,100	0,400	0,600	0,900	0,5878		0,150	0,350	0,650	0,850
0,101	0,399	0,601	0,899	0,5929		0,149	0,351	0,649	0,851
0,102	0,398	0,602	0,898	0,5979		0,148	0,352	0,648	0,852
0,103	0,397	0,603	0,897	0,6029		0,147	0,353	0,647	0,853
0,104	0,396	0,604	0,896	0,6079		0,146	0,354	0,646	0,854
0,105	0,395	0,605	0,895	0,6129		0,145	0,355	0,645	0,855
0,106	0,394	0,606	0,894	0,6179		0,144	0,356	0,644	0,856
0,107	0,393	0,607	0,893	0,6228		0,143	0,357	0,643	0,857
0,108	0,392	0,608	0,892	0,6277		0,142	0,358	0,642	0,858
0,109	0,391	0,609	0,891	0,6326		0,141	0,359	0,641	0,859
0,110	0,390	0,610	0,890	0,6374		0,140	0,360	0,640	0,860
0,111	0,389	0,611	0,889	0,6423		0,139	0,361	0,639	0,861
0,112	0,388	0,612	0,888	0,6471		0,138	0,362	0,638	0,862
0,113	0,387	0,613	0,887	0,6518		0,137	0,363	0,637	0,863
0,114	0,386	0,614	0,886	0,6566		0,136	0,364	0,636	0,864
0,115	0,385	0,615	0,885	0,6613		0,135	0,365	0,635	0,865
0,116	0,384	0,616	0,884	0,6660		0,134	0,366	0,634	0,866
0,117	0,383	0,617	0,883	0,6707		0,133	0,367	0,633	0,867
0,118	0,382	0,618	0,882	0,6753		0,132	0,368	0,632	0,868
0,119	0,381	0,619	0,881	0,6800		0,131	0,369	0,631	0,869
0,120	0,380	0,620	0,880	0,6845		0,130	0,370	0,630	0,870
0,121	0,379	0,621	0,879	0,6891		0,129	0,371	0,629	0,871
0,122	0,378	0,622	0,878	0,6937		0,128	0,372	0,628	0,872
0,123	0,377	0,623	0,877	0,6982		0,127	0,373	0,627	0,873
0,124	0,376	0,624	0,876	0,7026		0,126	0,374	0,626	0,874
0,125	0,375	0,625	0,875	0,7071		0,125	0,375	0,625	0,875
0,126	0,374	0,626	0,874	0,7115		0,124	0,376	0,624	0,876
0,127	0,373	0,627	0,873	0,7159		0,123	0,377	0,623	0,877
0,128	0,372	0,628	0,872	0,7203		0,122	0,378	0,622	0,878
0,129	0,371	0,629	0,871	0,7247		0,121	0,379	0,621	0,879
0,130	0,370	0,630	0,870	0,7290		0,120	0,380	0,620	0,880
0,131	0,369	0,631	0,869	0,7333		0,119	0,381	0,619	0,881
0,132	0,368	0,632	0,868	0,7375		0,118	0,382	0,618	0,882
0,133	0,367	0,633	0,867	0,7417		0,117	0,383	0,617	0,883

A	B	C	D	sin 2πx A B C D + + − −	cos 2πx E F G H + − − +	E	F	G	H
0,134	0,366	0,634	0,866	0,7459		0,116	0,384	0,616	0,884
0,135	0,365	0,635	0,865	0,7501		0,115	0,385	0,615	0,885
0,136	0,364	0,636	0,864	0,7543		0,114	0,386	0,614	0,886
0,137	0,363	0,637	0,863	0,7584		0,113	0,387	0,613	0,887
0,138	0,362	0,638	0,862	0,7624		0,112	0,388	0,612	0,888
0,139	0,361	0,639	0,861	0,7665		0,111	0,389	0,611	0,889
0,140	0,360	0,640	0,860	0,7705		0,110	0,390	0,610	0,890
0,141	0,359	0,641	0,859	0,7745		0,109	0,391	0,609	0,891
0,142	0,358	0,642	0,858	0,7785		0,108	0,392	0,608	0,892
0,143	0,357	0,643	0,857	0,7824		0,107	0,393	0,607	0,893
0,144	0,356	0,644	0,856	0,7863		0,106	0,394	0,606	0,894
0,145	0,355	0,645	0,855	0,7902		0,105	0,395	0,605	0,895
0,146	0,354	0,646	0,854	0,7940		0,104	0,396	0,604	0,896
0,147	0,353	0,647	0,853	0,7978		0,103	0,397	0,603	0,897
0,148	0,352	0,648	0,852	0,8016		0,102	0,398	0,602	0,898
0,149	0,351	0,649	0,851	0,8053		0,101	0,399	0,601	0,899
0,150	0,350	0,650	0,850	0,8090		0,100	0,400	0,600	0,900
0,151	0,349	0,651	0,849	0,8127		0,099	0,401	0,599	0,901
0,152	0,348	0,652	0,848	0,8163		0,098	0,402	0,598	0,902
0,153	0,347	0,653	0,847	0,8199		0,097	0,403	0,597	0,903
0,154	0,346	0,654	0,846	0,8235		0,096	0,404	0,596	0,904
0,155	0,345	0,655	0,845	0,8271		0,095	0,405	0,595	0,905
0,156	0,344	0,656	0,844	0,8306		0,094	0,406	0,594	0,906
0,157	0,343	0,657	0,843	0,8341		0,093	0,407	0,593	0,907
0,158	0,342	0,658	0,842	0,8375		0,092	0,408	0,592	0,908
0,159	0,341	0,659	0,841	0,8409		0,091	0,409	0,591	0,909
0,160	0,340	0,660	0,840	0,8443		0,090	0,410	0,590	0,910
0,161	0,339	0,661	0,839	0,8477		0,089	0,411	0,589	0,911
0,162	0,338	0,662	0,838	0,8510		0,088	0,412	0,588	0,912
0,163	0,337	0,663	0,837	0,8543		0,087	0,413	0,587	0,913
0,164	0,336	0,664	0,836	0,8575		0,086	0,414	0,586	0,914
0,165	0,335	0,665	0,835	0,8607		0,085	0,415	0,585	0,915
0,166	0,334	0,666	0,834	0,8639		0,084	0,416	0,584	0,916
0,167	0,333	0,667	0,833	0,8671		0,083	0,417	0,583	0,917
0,168	0,332	0,668	0,832	0,8702		0,082	0,418	0,582	0,918
0,169	0,331	0,669	0,831	0,8733		0,081	0,419	0,581	0,919
0,170	0,330	0,670	0,830	0,8763		0,080	0,420	0,580	0,920
0,171	0,329	0,671	0,829	0,8793		0,079	0,421	0,579	0,921
0,172	0,328	0,672	0,828	0,8823		0,078	0,422	0,578	0,922
0,173	0,327	0,673	0,827	0,8852		0,077	0,423	0,577	0,923
0,174	0,326	0,674	0,826	0,8881		0,076	0,424	0,576	0,924
0,175	0,325	0,675	0,825	0,8910		0,075	0,425	0,575	0,925
0,176	0,324	0,676	0,824	0,8938		0,074	0,426	0,574	0,926
0,177	0,323	0,677	0,823	0,8966		0,073	0,427	0,573	0,927
0,178	0,322	0,678	0,822	0,8994		0,072	0,428	0,572	0,928
0,179	0,321	0,679	0,821	0,9021		0,071	0,429	0,571	0,929
0,180	0,320	0,680	0,820	0,9048		0,070	0,430	0,570	0,930
0,181	0,319	0,681	0,819	0,9075		0,069	0,431	0,569	0,931
0,182	0,318	0,682	0,818	0,9101		0,068	0,432	0,568	0,932
0,183	0,317	0,683	0,817	0,9127		0,067	0,433	0,567	0,933
0,184	0,316	0,684	0,816	0,9152		0,066	0,434	0,566	0,934
0,185	0,315	0,685	0,815	0,9178		0,065	0,435	0,565	0,935
0,186	0,314	0,686	0,814	0,9202		0,064	0,436	0,564	0,936
0,187	0,313	0,687	0,813	0,9227		0,063	0,437	0,563	0,937
0,188	0,312	0,688	0,812	0,9251		0,062	0,438	0,562	0,938
0,189	0,311	0,689	0,811	0,9274		0,061	0,439	0,561	0,939
0,190	0,310	0,690	0,810	0,9298		0,060	0,440	0,560	0,940
0,191	0,309	0,691	0,809	0,9321		0,059	0,441	0,559	0,941
0,192	0,308	0,692	0,808	0,9343		0,058	0,442	0,558	0,942

	x			sin 2πx	cos 2πx		x		
A	B	C	D	A B C D + + − −	E F G H + − − +	E	F	G	H
0,193	0,307	0,693	0,807	0,9365		0,057	0,443	0,557	0,943
0,194	0,306	0,694	0,806	0,9387		0,056	0,444	0,556	0,944
0,195	0,305	0,695	0,805	0,9409		0,055	0,445	0,555	0,945
0,196	0,304	0,696	0,804	0,9430		0,054	0,446	0,554	0,946
0,197	0,303	0,697	0,803	0,9451		0,053	0,447	0,553	0,947
0,198	0,302	0,698	0,802	0,9471		0,052	0,448	0,552	0,948
0,199	0,301	0,699	0,801	0,9491		0,051	0,449	0,551	0,949
0,200	0,300	0,700	0,800	0,9511		0,050	0,450	0,550	0,950
0,201	0,299	0,701	0,799	0,9530		0,049	0,451	0,549	0,951
0,202	0,298	0,702	0,798	0,9549		0,048	0,452	0,548	0,952
0,203	0,297	0,703	0,797	0,9567		0,047	0,453	0,547	0,953
0,204	0,296	0,704	0,796	0,9585		0,046	0,454	0,546	0,954
0,205	0,295	0,705	0,795	0,9603		0,045	0,455	0,545	0,955
0,206	0,294	0,706	0,794	0,9620		0,044	0,456	0,544	0,956
0,207	0,293	0,707	0,793	0,9637		0,043	0,457	0,543	0,957
0,208	0,292	0,708	0,792	0,9654		0,042	0,458	0,542	0,958
0,209	0,291	0,709	0,791	0,9670		0,041	0,459	0,541	0,959
0,210	0,290	0,710	0,790	0,9686		0,040	0,460	0,540	0,960
0,211	0,289	0,711	0,789	0,9701		0,039	0,461	0,539	0,961
0,212	0,288	0,712	0,788	0,9716		0,038	0,462	0,538	0,962
0,213	0,287	0,713	0,787	0,9731		0,037	0,463	0,537	0,963
0,214	0,286	0,714	0,786	0,9745		0,036	0,464	0,536	0,964
0,215	0,285	0,715	0,785	0,9759		0,035	0,465	0,535	0,965
0,216	0,284	0,716	0,784	0,9773		0,034	0,466	0,534	0,966
0,217	0,283	0,717	0,783	0,9786		0,033	0,467	0,533	0,967
0,218	0,282	0,718	0,782	0,9799		0,032	0,468	0,532	0,968
0,219	0,281	0,719	0,781	0,9811		0,031	0,469	0,531	0,969
0,220	0,280	0,720	0,780	0,9823		0,030	0,470	0,530	0,970
0,221	0,279	0,721	0,779	0,9834		0,029	0,471	0,529	0,971
0,222	0,278	0,722	0,778	0,9846		0,028	0,472	0,528	0,972
0,223	0,277	0,723	0,777	0,9856		0,027	0,473	0,527	0,973
0,224	0,276	0,724	0,776	0,9867		0,026	0,474	0,526	0,974
0,225	0,275	0,725	0,775	0,9877		0,025	0,475	0,525	0,975
0,226	0,274	0,726	0,774	0,9887		0,024	0,476	0,524	0,976
0,227	0,273	0,727	0,773	0,9896		0,023	0,477	0,523	0,977
0,228	0,272	0,728	0,772	0,9905		0,022	0,478	0,522	0,978
0,229	0,271	0,729	0,771	0,9913		0,021	0,479	0,521	0,979
0,230	0,270	0,730	0,770	0,9921		0,020	0,480	0,520	0,980
0,231	0,269	0,731	0,769	0,9929		0,019	0,481	0,519	0,981
0,232	0,268	0,732	0,768	0,9936		0,018	0,482	0,518	0,982
0,233	0,267	0,733	0,767	0,9943		0,017	0,483	0,517	0,983
0,234	0,266	0,734	0,766	0,9943		0,016	0,484	0,516	0,984
0,235	0,265	0,735	0,765	0,9956		0,015	0,485	0,515	0,985
0,236	0,264	0,736	0,764	0,9961		0,014	0,486	0,514	0,986
0,237	0,263	0,737	0,763	0,9967		0,013	0,487	0,513	0,987
0,238	0,262	0,738	0,762	0,9972		0,012	0,488	0,512	0,988
0,239	0,261	0,739	0,761	0,9976		0,011	0,489	0,511	0,989
0,240	0,260	0,740	0,760	0,9980		0,010	0,490	0,510	0,990
0,241	0,259	0,741	0,759	0,9984		0,009	0,491	0,509	0,991
0,242	0,258	0,742	0,758	0,9987		0,008	0,492	0,508	0,992
0,243	0,257	0,743	0,757	0,9990		0,007	0,493	0,507	0,993
0,244	0,256	0,744	0,756	0,9993		0,006	0,494	0,506	0,994
0,245	0,255	0,745	0,755	0,9995		0,005	0,495	0,505	0,995
0,246	0,254	0,746	0,754	0,9997		0,004	0,496	0,504	0,996
0,247	0,253	0,747	0,753	0,9998		0,003	0,497	0,503	0,997
0,248	0,252	0,748	0,752	0,9999		0,002	0,498	0,502	0,998
0,249	0,251	0,749	0,751	1,0000		0,001	0,499	0,501	0,999
0,250	0,250	0,750	0,750	1,0000		0,000	0,500	0,500	1,000

4-10. Nomogram for Calculating Structure Factors

Approximate calculations are conveniently performed with the nomogram in Fig. 100 [237].

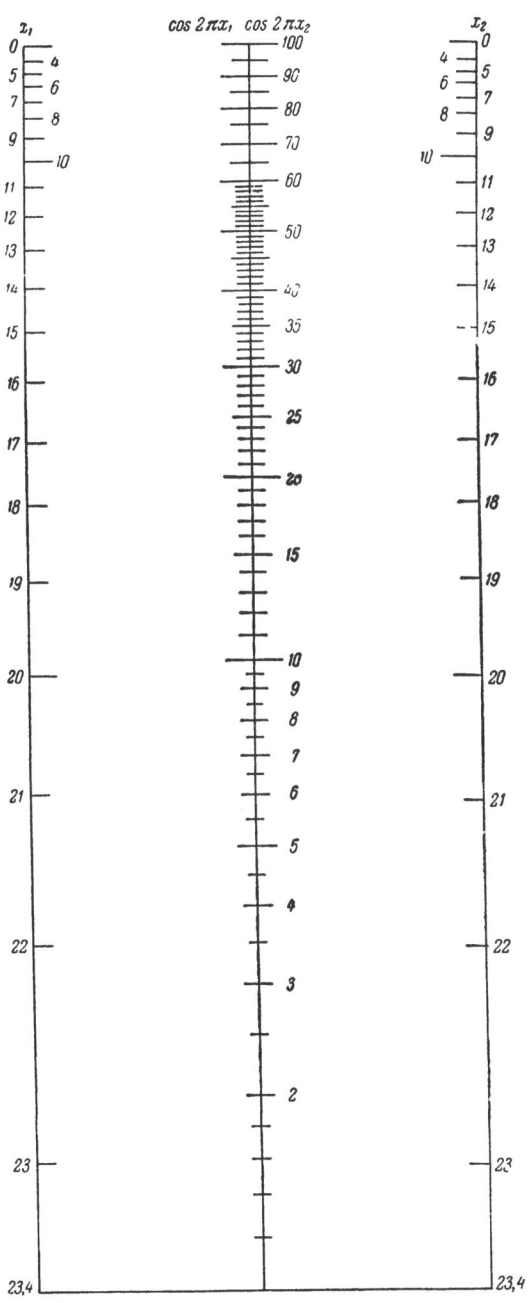

Fig. 100. Nomogram for structure factors.

This nomogram reads from x_1 and x_2 (left and right scales) to $\cos 2\pi x_1 \cdot \cos 2\pi x_2$ (center scale); the x_1 and x_2 are joined by a straight line to take the reading. Here x_1 and x_2 are given as hundredths of a full circle; the product is normalized to 100. The product $\sin 2\pi x_1 \cdot \cos 2\pi x_2$ is found from $\sin \alpha = \cos (90° - \alpha)$.

Convenient nomograms have been given for $\cos 2\pi(hx + lz)$ and $\sin 2\pi(hx + lz)$ [276], for $\cos 2\pi (hx + ky + lz)$ and $\sin 2\pi(hx + ky + lz)$ [275], and for $\cos 2\pi hx \cdot \cos 2\pi ky \cdot \cos 2\pi lz$ [277]; other rapid methods of calculation are considered in [275 , 383]. A simple device of slide-rule type has also been described [446].

4-11. Array of Atoms in Some Types of Crystal Structure

The table gives the atomic coordinates for some structure types in the cubic, tetragonal, and hexagonal systems. Figures 101-109 give schemes for the unit cells [102, 112, 113].

The numbers for the structures in each system are those used in the schemes for the x-ray patterns (Chapter 3) and in Table 5-2 for the crystal structures (Chapter 5).

See [111-113] for the data on other structure types.

Symbol	Structure type	Space group	No. of atoms in cell	Coordinates
				Cubic System
K 1	bcc	$O_h^9 - Im3m$	2	$(0, 0, 0)$, $\left(\frac{1}{2}, \frac{1}{2}, \frac{1}{2}\right)$.
K 2	fcc	$O_h^5 - Fm3m$	4	$(0, 0, 0)$, $\left(\frac{1}{2}, \frac{1}{2}, 0\right)$. (d)
K 3	CsCl	$O_h^1 - Pm3m$	1	Cs: $(0, 0, 0)$; Cl: $\left(\frac{1}{2}, \frac{1}{2}, \frac{1}{2}\right)$.
K 4	ZnS (sphalerite)	$T_d^2 - F\bar{4}3m$	4	$(0, 0, 0)$, $\left(\frac{1}{2}, \frac{1}{2}, 0\right)$, (d) + 4Zn: $(0, 0, 0)$, + 4S; $\left(\frac{1}{4}, \frac{1}{4}, \frac{1}{4}\right)$.
K 5	NaCl	$O_h^5 - Fm3m$	4	4Na: $(0, 0, 0)$, $\left(\frac{1}{2}, \frac{1}{2}, 0\right)$, (d) 4Cl: $\left(\frac{1}{2}, \frac{1}{2}, \frac{1}{2}\right)$, $\left(11\frac{1}{2}\right)$. (d)
K 6	CaF_2	$O_h^5 - Fm3m$	4	$(0, 0, 0)$, $\left(\frac{1}{2}, \frac{1}{2}, 0\right)$, (d) + 4Ca: $(0, 0, 0)$, + 8F: $\pm\left(\frac{1}{4}, \frac{1}{4}, \frac{1}{4}\right)$.

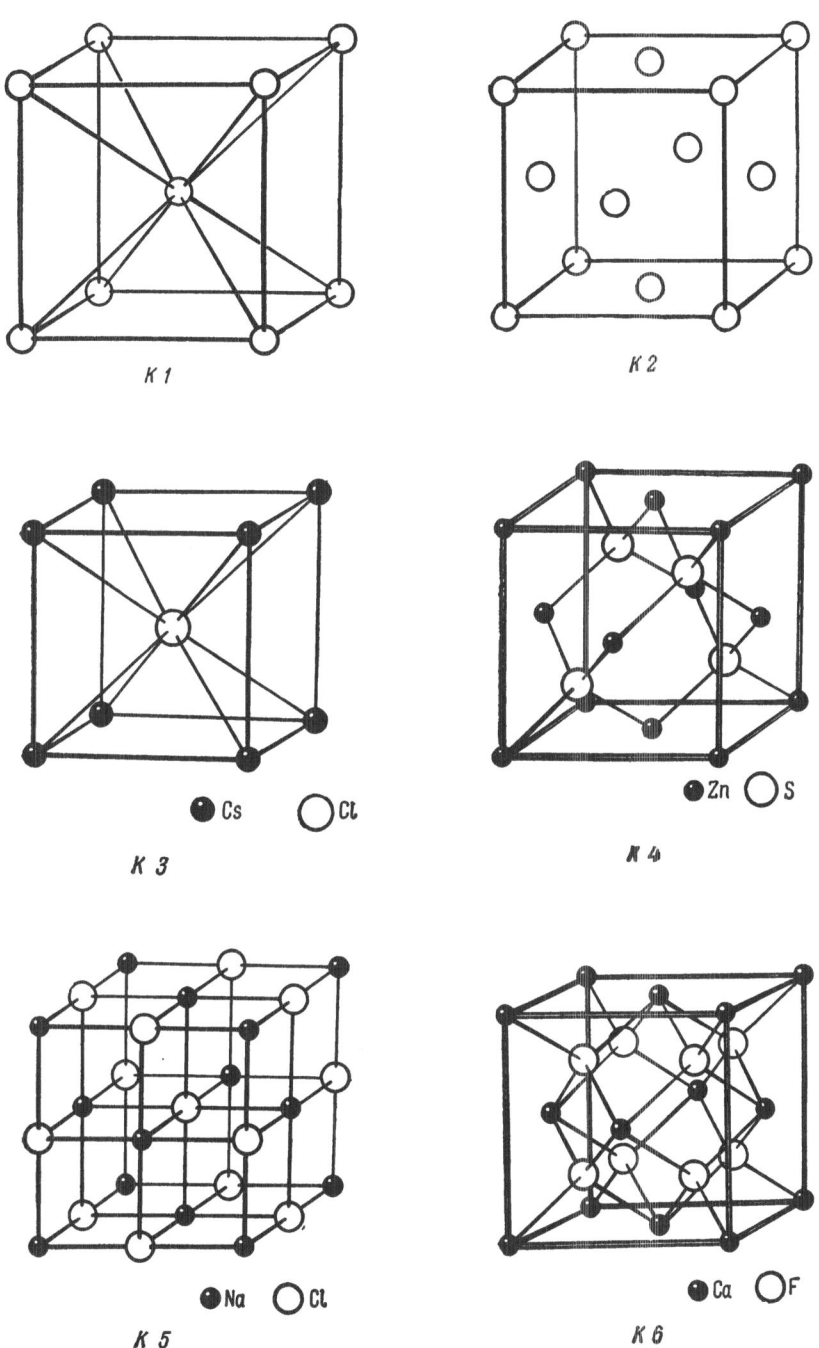

Fig. 101. Arrays of atoms in structures of types K1 to K6.

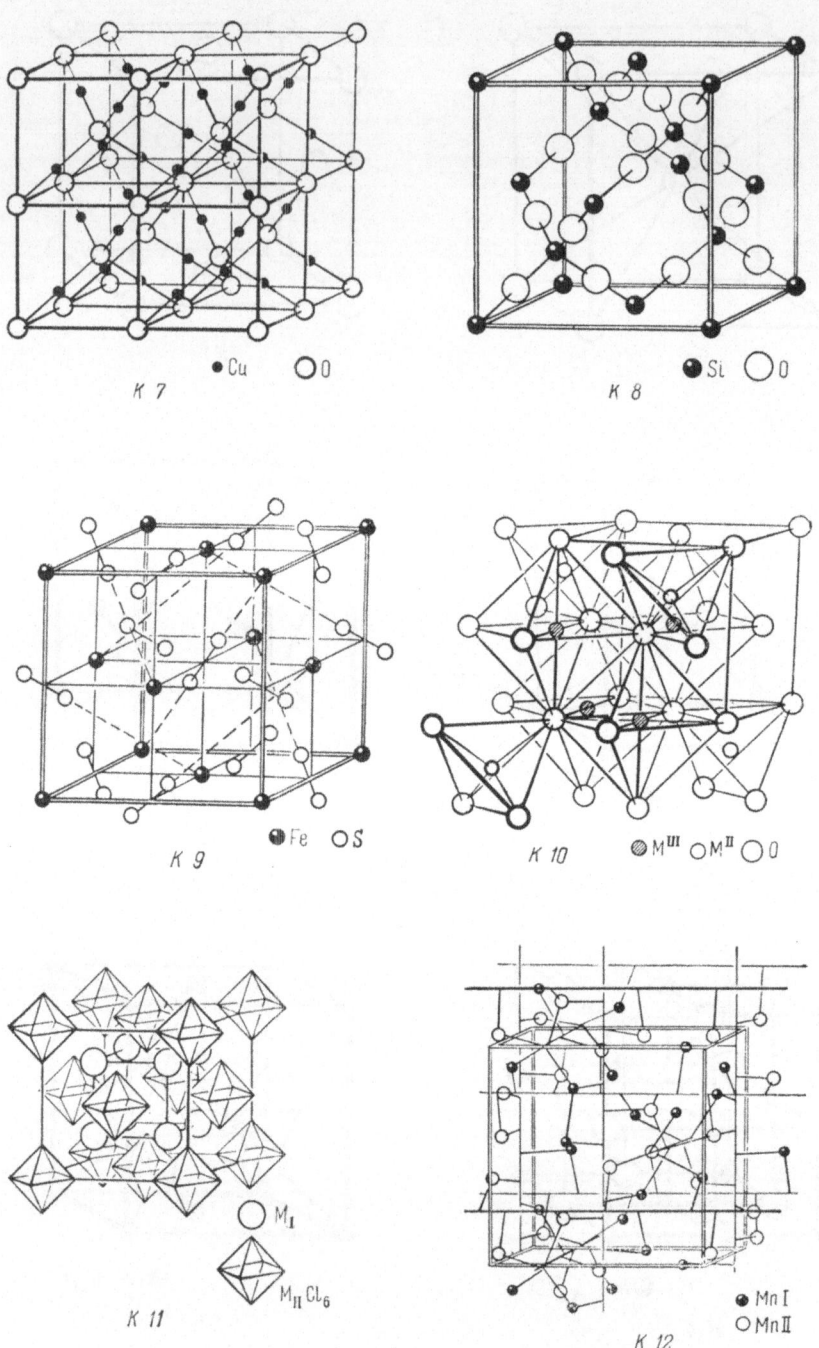

Fig. 102. Arrays of atoms in structures of types K7 to K12.

Symbol	Structure type	Space group	No. of atoms in cell	Coordinates
K 7	Cu_2O (cuprite)	$O_h^4 - Pn3m$	2	2O: $(0, 0, 0)$, $\left(\frac{1}{2}, \frac{1}{2}, \frac{1}{2}\right)$; 4Cu: $\left(\frac{1}{4}, \frac{1}{4}, \frac{1}{4}\right)$, $\left(\frac{3}{4}, \frac{3}{4}, \frac{1}{4}\right)$. (d)
K 8	SiO_2 (β-cristobalite)	$O_h^7 - Fd3m$	8	$(0, 0, 0)$, $\left(\frac{1}{2}, \frac{1}{2}, 0\right)$ $+$ 8Si: $(0, 0, 0)$, $\left(\frac{1}{4}, \frac{1}{4}, \frac{1}{4}\right)$; $+$ 16O: $\left(\frac{1}{8}, \frac{1}{8}, \frac{1}{8}\right)$, $\left(\frac{1}{8}, \frac{3}{8}, \frac{3}{8}\right)$. (d)
K 9	FeS_2 (pyrite)	$T_h^6 - Pa3$	4	4Fe: $(0, 0, 0)$, $\left(\frac{1}{2}, \frac{1}{2}, 0\right)$. (d) 8Si: $\pm(x, x, x)$, $\pm\left(\frac{1}{2}+x, \frac{1}{2}-x, \bar{x}\right)$, (d) $x = 0,386$.
K 10	Al_2MgO_4 (spinel)	$O_h^7 - Fd3m$	8	$(0, 0, 0)$, $\left(\frac{1}{2}, \frac{1}{2}, 0\right)$, (d) $+$ 8Mg: $(0, 0, 0)$, $\left(\frac{1}{4}, \frac{1}{4}, \frac{1}{4}\right)$, $+$ 16Al: $\left(\frac{5}{8}, \frac{5}{8}, \frac{5}{8}\right)$, $\left(\frac{5}{8}, \frac{7}{8}, \frac{7}{8}\right)$, (d) $+$ 32O: (x, x, x), $(x, \bar{x}, \bar{x})$, (d) $\left(\frac{1}{4}-x, \frac{1}{4}-x, \frac{1}{4}-x\right)$, $\left(\frac{1}{4}-x, \frac{1}{4}+x, \frac{1}{4}+x\right)$, (d) $x = -\frac{1}{8}$.
K 11	K_2PtCl_6	$O_h^5 - Fm3m$	4	$(0, 0, 0)$, $\left(\frac{1}{2}, \frac{1}{2}, 0\right)$, (d) $+$ 4Pt: $(0, 0, 0)$, $+$ 8K: $\left(\frac{1}{4}, \frac{1}{4}, \frac{1}{4}\right)$, $\left(\frac{3}{4}, \frac{3}{4}, \frac{3}{4}\right)$, $+$ 24Cl: $\pm(x, 0, 0)$. (d)

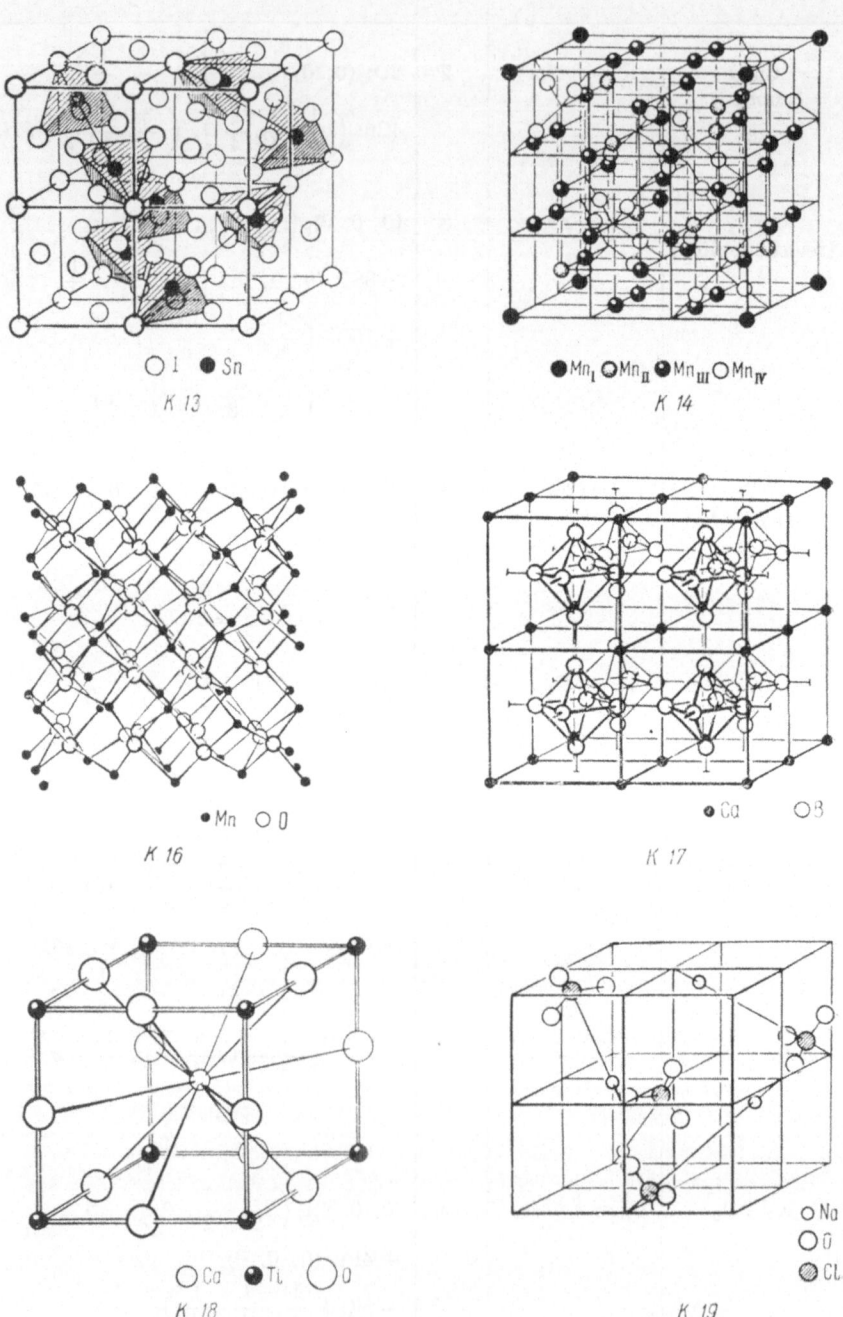

Fig. 103. Arrays of atoms in structures of types K13 to K19.

Symbol	Structure type	Space group	No. of atoms in cell	Coordinates
K 12	β-Mn	$\begin{cases} O^6 - P4_33 \\ O^7 - P4_13 \end{cases}$	20	8Mn: $(x,\ x,\ x)$, $\left(\frac{1}{2}+x,\ \frac{1}{2}-x,\ x\right)$, (d) $\left(\frac{3}{4}-x,\ \frac{3}{4}-x,\ \frac{3}{4}-x\right)$, $\left(\frac{1}{4}-x,\ \frac{3}{4}+x,\ \frac{1}{4}+x\right)$, (d) 12Mn: $\left(\frac{3}{8},\ \bar{y},\ \frac{3}{4}+y\right)$, (d) $\left(\frac{7}{8},\ \frac{1}{2}+y,\ \frac{1}{4}-y\right)$, (d) $\left(\frac{1}{8},\ y,\ \frac{1}{y}+y\right)$, (d) $\left(\frac{5}{8},\ \frac{1}{2}-y,\ \frac{3}{4}-y\right)$, (d) $x=0{,}061,\ y=0{,}206.$
K 13	SnI$_4$	$T_h^6 - Pa3$	8	8Sn: $\pm(x,\ x,\ x)$, $\left(x+\frac{1}{2},\ \frac{1}{2}-x,\ \bar{x}\right)$, (d) 8I: $\pm(y,\ y,\ y)$, $\left(y+\frac{1}{2},\ \frac{1}{2}-y,\ \bar{y}\right)$, (d) 24I: $\pm(u,\ v,\ w)$, $\left(u+\frac{1}{2},\ \frac{1}{2}-u,\ \bar{w}\right)$, $\left(\bar{u},\ v+\frac{1}{2},\ \frac{1}{2}-w\right)$, $\left(\frac{1}{2}-u,\ \bar{v},\ w+\frac{1}{2}\right)$, $(w,\ u,\ v)$, $\left(\bar{w},\ u+\frac{1}{2},\ \frac{1}{2}-v\right)$, $\left(\frac{1}{2}-w,\ \bar{u},\ w+\frac{1}{2}\right)$, $\left(w+\frac{1}{2},\ \frac{1}{2}-u,\ \bar{v}\right)$, (v,w,u), $\left(\frac{1}{2}-v,\ \bar{w},\ u+\frac{1}{2}\right)$, $\left(v+\frac{1}{2},\ \frac{1}{2}-w,\ \bar{u}\right)$, $\left(\bar{v},\ w+\frac{1}{2},\ \frac{1}{2}-u\right)$, $x=0{,}129,\ y=0{,}253,\ u=0{,}009,$ $v=0{,}001,\ w=0{,}253.$

Symbol	Structure type	Space group	No. of atoms in cell	Coordinates

K 14	α-Mn	$T_d^3 - I\bar{4}3m$	58	$(0, 0, 0)$, $\left(\frac{1}{2}, \frac{1}{2}, \frac{1}{2}\right)$
				$+$ 2Mn: $(0, 0, 0)$,
				$+$ 8Mn: (x, x, x), $(\bar{x}, \bar{x}, x)$, (d)
				$+$ 24Mn: (x, x, y), (d)
				          $(\bar{x}, \bar{x}, y)$
				          $(\bar{x}, x, \bar{y})$, (d)
				          $(x, \bar{x}, \bar{y})$, (d)
				$+$ 24Mn: the same coordinates, but another parameter (y).
				$x = 0,317$, $y = 0,356$.

K 15	FeSi	$T^4 - P2_1 3$	4	4Fe: (x, x, x), $\left(\frac{1}{2}+x, \frac{1}{2}-x, \bar{x}\right)$, (d) 4Si: the same coordinates, but another parameter, $x_{Fe} = 0,137$, $x_{Si} = 0,842$.

K 16	Mn_2O_3	$T_h^7 - Ia3$	16	$(0, 0, 0)$, $\left(\frac{1}{2}, \frac{1}{2}, \frac{1}{2}\right)$,
				$+$ 8Mn: $\left(\frac{1}{4}, \frac{1}{4}, \frac{1}{4}\right), \left(\frac{1}{4}, \frac{3}{4}, \frac{3}{4}\right) \cdot$ (d)
				$+$ 24Mn: $\pm\left(x, 0, \frac{1}{4}\right)$, (d)
				        $\pm\left(x, \frac{1}{2}, \frac{3}{4}\right)$, (d)
				$+$ 48O: $\pm(x, y, z)$, (d)
				       $\pm\left(x, \bar{y}, \frac{1}{2}-z\right)$, (d)
				       $\pm\left(x, \frac{1}{2}+y, \bar{z}\right)$, (d)
				$x = 0,385$, $y = 0,145$, $z = 0,380$.

K 17	CaB_6	$O_h^1 - Pm3m$	1	1Ca: $(0, 0, 0)$, 6B: $\pm\left(\frac{1}{2}, \frac{1}{2}, x\right)$, (d) $x = 0,207$.
K 18	$CaTiO_3$	$O_h^1 - Pm3m$		1Ca: $(0, 0, 0)$, 1Ti: $\left(\frac{1}{2}, \frac{1}{2}, \frac{1}{2}\right)$, 3O: $\left(\frac{1}{2}, 0, 0\right)$. (d)

Symbol	Structure type	Space group	No. of atoms in cell	Coordinates
K 19	NaClO$_3$	T^4	4	4Na: (x,x,x), $\left(x+\dfrac{1}{2},\dfrac{1}{2}-x,\ \bar{x}\right)$, (d) 4Cl: $(y,\ y,\ y)$, $\left(y+\dfrac{1}{2},\dfrac{1}{2}-y,\ \bar{y}\right)$, (d) 12O: $(u,\ v,\ w)$, $\left(u+\dfrac{1}{2},\ \dfrac{1}{2}-v,\ \bar{w}\right)$, $\left(\bar{u},\ v+\dfrac{1}{2},\ \dfrac{1}{2}-w\right)$, $\left(\dfrac{1}{2}-u,\ \bar{v},\ w+\dfrac{1}{2}\right)$, $(w,\ u,\ v)$, $\left(\bar{w},\ u+\dfrac{1}{2},\ \dfrac{1}{2}-v\right)$, $\left(\dfrac{1}{2}-w,\ \bar{u},\ v+\dfrac{1}{2}\right)$, $\left(w+\dfrac{1}{2},\ \dfrac{1}{2}-u,\ \bar{v}\right)$, $(v,\ w,\ u)$, $\left(\dfrac{1}{2}-v,\ \bar{w},\ u+\dfrac{1}{2}\right)$, $\left(v+\dfrac{1}{2},\ \dfrac{1}{2}-w,\ \bar{u}\right)$, $\left(\bar{v},\ w+\dfrac{1}{2},\ \dfrac{1}{2}-u\right)$.
K 20	NiSbS (CoAsS)	$T^4 - P2_13$	4	4Ni: $(x,\ x,\ x)$, $\left(\dfrac{1}{2}+x,\ \dfrac{1}{2}-x,\ x\right)$, (d) 4Sb and 4S: the same coordinates, but another parameter.
K 21	NaTl	$O_h^7 - Fd3m$	8	$(0,\ 0,\ 0)$, $\left(\dfrac{1}{2},\ \dfrac{1}{2},\ 0\right)$, (d) $+$ 8Na: $(0,\ 0,\ 0)$, $\left(\dfrac{1}{4},\ \dfrac{1}{4},\ \dfrac{1}{4}\right)$, $+$ 8Tl: $\left(\dfrac{1}{2},\dfrac{1}{2},\dfrac{1}{2}\right)$, $\left(\dfrac{3}{4},\dfrac{3}{4},\dfrac{3}{4}\right)$.
K 22	β-Cu$_2$Mg	$O_h^7 - Fd3m$	8	$(0,\ 0,\ 0)$, $\left(\dfrac{1}{2},\ \dfrac{1}{2},\ 0\right)$, (d) $+$ 8Mg: $(0,\ 0,\ 0)$, $\left(\dfrac{1}{4},\ \dfrac{1}{4},\ \dfrac{1}{4}\right)$, $+$ 16Cu: $\left(\dfrac{5}{8},\dfrac{5}{8},\dfrac{5}{8}\right)$, $\left(\dfrac{7}{8},\dfrac{7}{8},\dfrac{7}{8}\right)$. (d)
K 23	Cu$_2$AlMn	$O_h^5 - Fm3m$		$(0,\ 0,\ 0)$, $\left(\dfrac{1}{2},\ \dfrac{1}{2},\ 0\right)$, (d) $+$ 4Al: $(0,\ 0,\ 0)$, $+$ 8Cu: $\left(\dfrac{1}{4},\ \dfrac{1}{4},\ \dfrac{1}{4}\right)$, $\left(\dfrac{3}{4},\dfrac{3}{4},\dfrac{3}{4}\right)$. $+$ 4Mn: $\left(\dfrac{1}{2},\ \dfrac{1}{2},\ \dfrac{1}{2}\right)$.

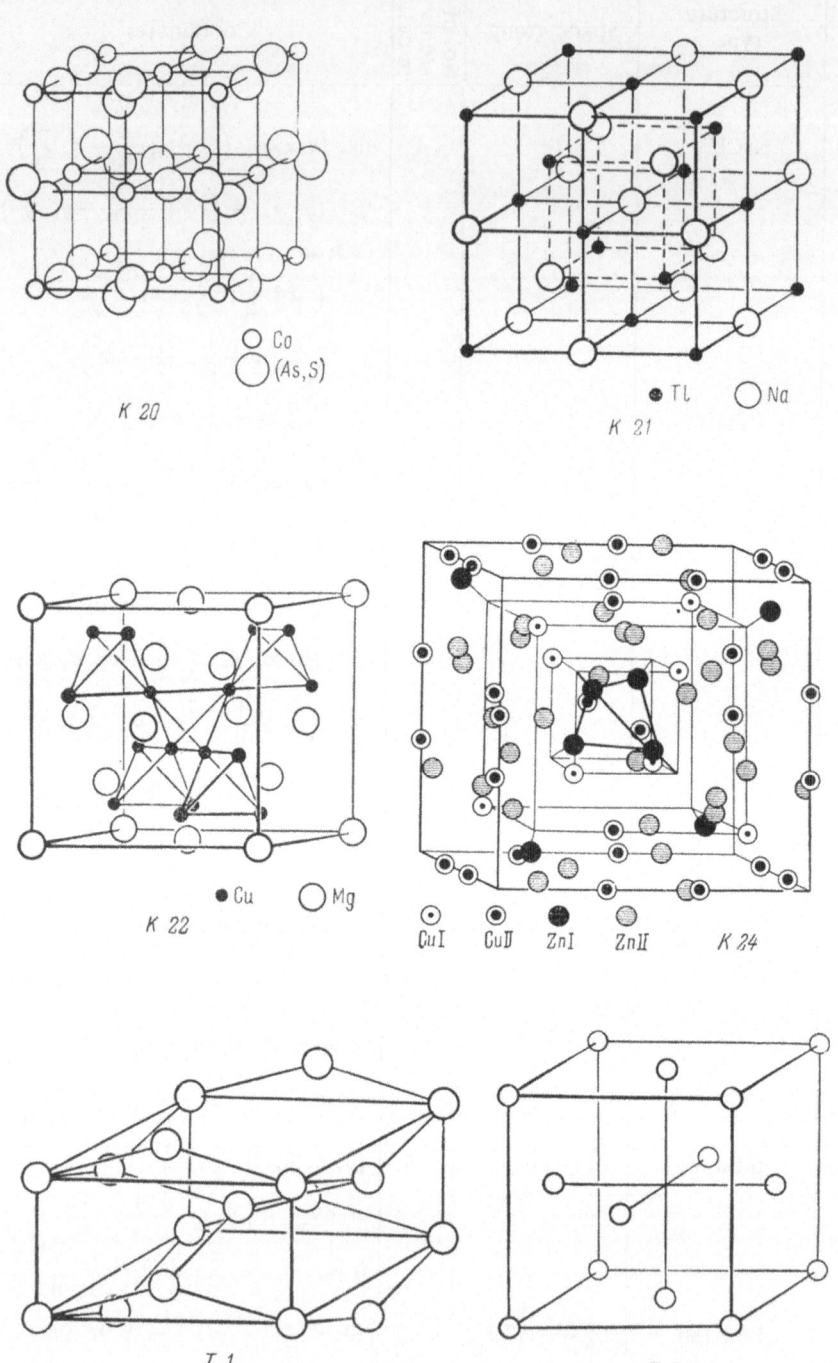

Fig. 104. Arrays of atoms in structures of types K20 to T2.

Symbol	Structure type	Space group	No. of atoms in cell	Coordinates
K 24	Fe_3Zn_{10} (γ-phase)	$T_d^3 - I\bar{4}3m$	4	$(0, 0, 0)$, $\left(\frac{1}{2}, \frac{1}{2}, \frac{1}{2}\right)$, $+ 12Fe$: $\pm (x, 0, 0)$, (d) $+ 16Zn$: $\pm (y, y, y)$, $(y, \bar{y}, \bar{y})$, (d) $+ 24Zn$: $\pm (z, z, 0)$, $(z, \bar{z}, 0)$, (d) $x = 0,355$, $y = 0,172$, $y_z = 0,110$, $z_1 = 0,313$, $z_2 = 0$.

Tetragonal System

T 1	β-Sn	$D_{4h}^{19} - \dfrac{I4}{amd}$	4	$4Sn$: $(0, 0, 0)$, $\left(\frac{1}{2}, \frac{1}{2}, \frac{1}{2}\right)$, $\left(\frac{1}{2}, 0, \frac{1}{4}\right)$, $\left(0, \frac{1}{2}, \frac{3}{4}\right)$.
T 2	Zn (γ-Mn)	$D_{4h}^{17} - \dfrac{I4}{mmm}$	2	$4Zn$: $(0, 0, 0)$, $\left(\frac{1}{2}, \frac{1}{2}, 0\right)$. (d)
T 3	PbO (SnO)	$D_{4h}^{7} - \dfrac{P4}{nmm}$	2	$2Pb$: $\left(0, \frac{1}{2}, z\right)$, $\left(\frac{1}{2}, 0, z\right)$, $z = 0,238$, $2O$: $(0, 0, 0)$, $\left(\frac{1}{2}, \frac{1}{2}, 0\right)$.
T 4	TiO_2 (rutile)	$D_{4h}^{14} - \dfrac{P4}{mnm}$	6	$2Ti$: $(0, 0, 0)$ $\left(\frac{1}{2}, \frac{1}{2}, \frac{1}{2}\right)$ $4O$: $(p, p, 0)$ $(\bar{p}, \bar{p}, 0)$ $\left(\frac{1}{2} - p, \frac{1}{2} + p, \frac{1}{2}\right)$ $\left(\frac{1}{2} + p, \frac{1}{2} - p, \frac{1}{2}\right)$
T 5	CaC_2	D_{4h}^{18}		$(0, 0, 0)$, $\left(1.2, 1.2, \frac{1}{2}\right)$ $+ 2Ca$: $(0, 0, 0)$ $+ 4C$: $(0, 0, z)$
T 6	$CuAl_2$	$D_{4h}^{18} - \dfrac{I4}{mcm}$		$(0, 0, 0)$ $\left(\frac{1}{2}, \frac{1}{2}, \frac{1}{2}\right)$ $+ 4Cu$: $\pm \left(0, 0, \frac{1}{4}\right)$ $+ 8Al$: $\pm \left(x, \frac{1}{2} + x, 0\right)$ $\pm \left(\frac{1}{2} + x, \bar{x}, 0\right)$.

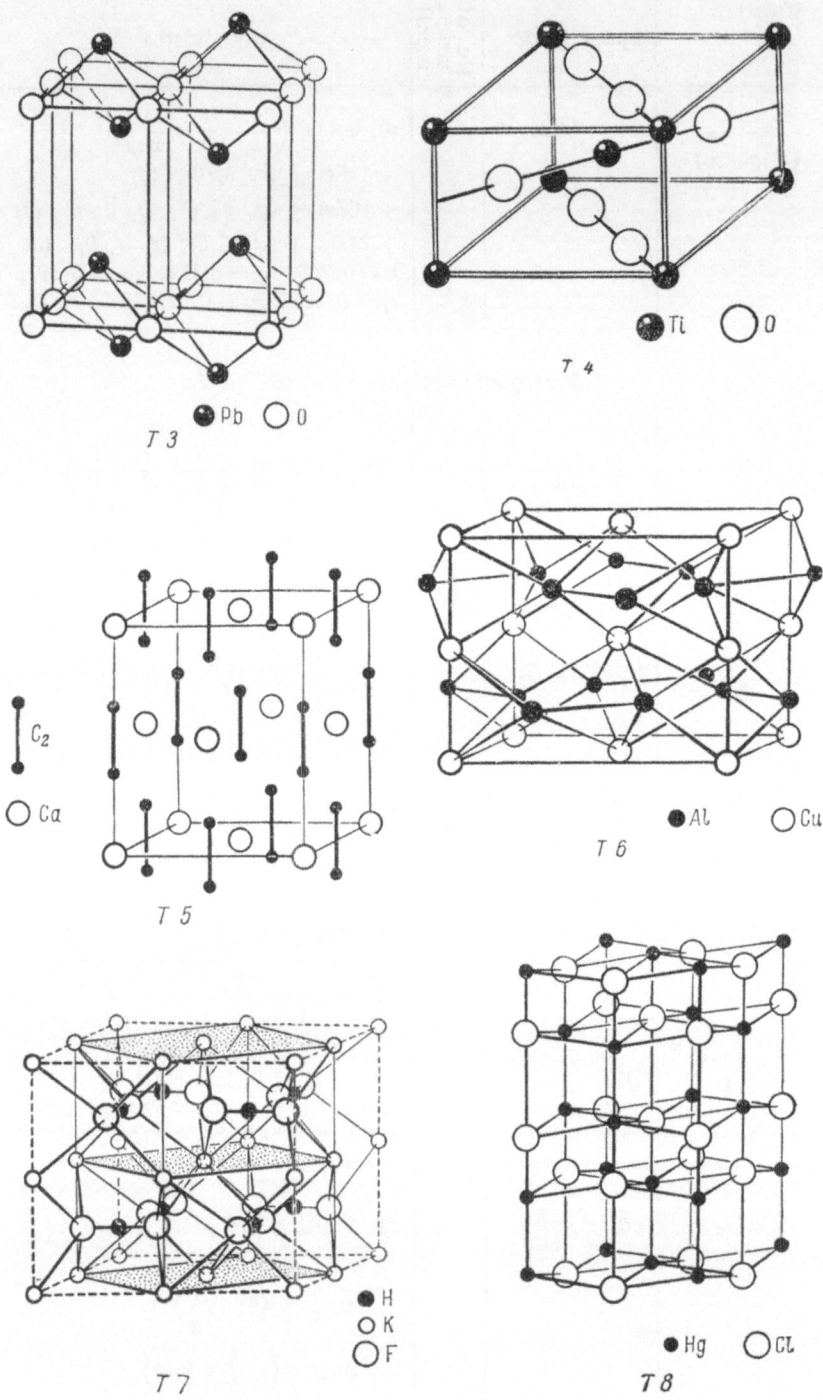

Fig. 105. Arrays of atoms in structures of types T3 to T8.

Symbol	Structure type	Space group	No. of atoms in cell	Coordinates
T 7	KHF_2	$D_{4h}^{18} - \dfrac{I4}{mcm}$	16	$0, 0, 0, \left(\dfrac{1}{2}, \dfrac{1}{2}, \dfrac{1}{2}\right),$ $+\left(0, 0, \dfrac{1}{4}\right),\left(\dfrac{1}{2}, \dfrac{1}{2}, \dfrac{1}{4}\right)$ $+ 4H: \left(0, \dfrac{1}{2}, 0\right),\left(\dfrac{1}{2}, 0, 0\right)$ $+ 8F: \left(p, p + \dfrac{1}{2}, 0\right),$ $\left(\dfrac{1}{2} -p, p, 0\right),\left(\bar{p}, \dfrac{1}{2} -p, 0\right),$ $\left(p + \dfrac{1}{2}, \bar{p}, 0\right).$
T 8	Hg_2Cl_2 (calomel)	$D_{4h}^4 - \dfrac{P4}{nnc}$	16	$(0, 0, 0),\left(\dfrac{1}{2}, \dfrac{1}{2}, 0\right) \quad (d)$ $+ 8Hg: (0, 0, p), (0, 0, \bar{p})$ $+ 8Cl: \left(\dfrac{1}{2}, 0, q\right), \left(\dfrac{1}{2}, 0, \bar{q}\right).$
T 9	K_2PtCl_4	$D_{4h}^1 - \dfrac{P4}{mmm}$	2	Pt: $(0, 0, 0),$ 2K: $\left(0, \dfrac{1}{2}, \dfrac{1}{2}\right),$ $\left(\dfrac{1}{2}, 0, \dfrac{1}{2}\right);$ 4Cl: $(x, x, 0), (x, \bar{x}, 0),$ $(\bar{x}, x, 0), (\bar{x}, \bar{x}, 0).$
T 10	KIO_4 (NH_4IO_4)	$C_{4h}^6 - \dfrac{T4_1}{a}$	4	4K: $\left(0, 0, \dfrac{1}{2}\right), \left(\dfrac{1}{2}, 0, \dfrac{1}{4}\right),$ $\pm\left(\dfrac{1}{2}, \dfrac{1}{2}, \dfrac{1}{2}\right),$ 4I: $(0, 0, 0), \left(0, \dfrac{1}{2}, \dfrac{1}{4}\right),$ $\pm\left(\dfrac{1}{2}, \dfrac{1}{2}, \dfrac{1}{2}\right),$ 16O: $(x, y, z), (\bar{x}, \bar{y}, z), (\bar{y}, x, \bar{z}),$ $(y, \bar{x}, \bar{z}),$ $\left(x, y + \dfrac{1}{2}, \dfrac{1}{4} - z\right),$ $\left(\bar{x}, \dfrac{1}{2} - y, \dfrac{1}{4} - z\right),$ $\left(\bar{y}, x + \dfrac{1}{2}, z + \dfrac{1}{4}\right),$ $\left(y, \dfrac{1}{2} - x, z + \dfrac{1}{4}\right), \left(\dfrac{1}{2}, \dfrac{1}{2}, \dfrac{1}{2}\right).$

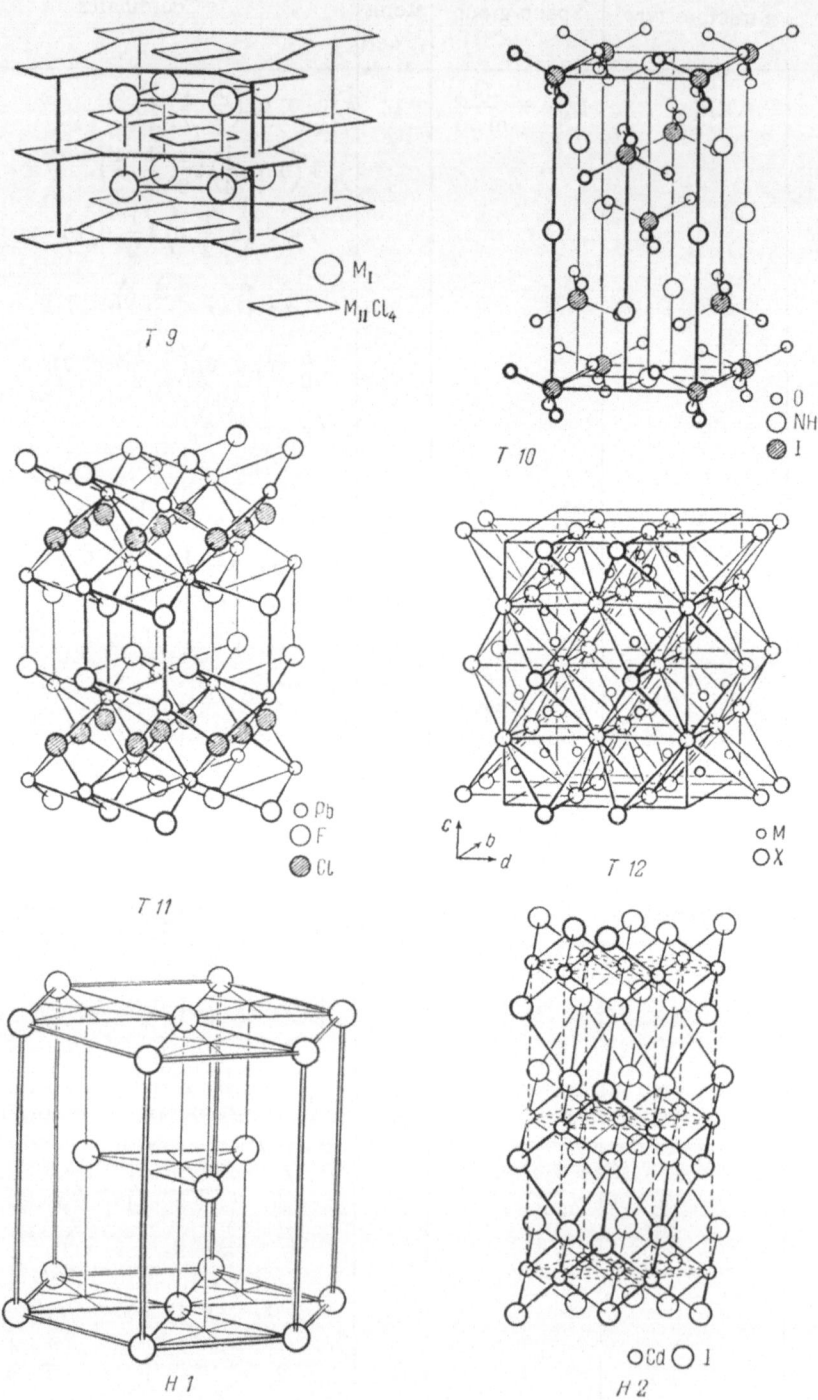

Fig. 106. Arrays of atoms in structures of types T9 to H2.

Symbol	Structure type	Space group	No. of atoms in cell	Coordinates
T 11	PbFCl	$D_{4h}^{7} - \dfrac{P4}{nmm}$	2	2F: $(0,\ 0,\ 0)$, $\left(\dfrac{1}{2},\ \dfrac{1}{2},\ 0\right)$, 2Cl: $\left(0,\ \dfrac{1}{2},\ x\right)$, $\left(\dfrac{1}{2},\ 0,\ \bar{x}\right)$, 2Pb: $\left(0,\ \dfrac{1}{2},\ y\right)$, $\left(\dfrac{1}{2},\ 0,\ \bar{y}\right)$.
T 12	Zn_3P_2	$D_{4h}^{15} - \dfrac{P4}{nmc}$	8	4P: $\pm\,(0,\ 0,\ z_1)$, $\pm\left(\dfrac{1}{2},\ \dfrac{1}{2},\ \dfrac{1}{2}+z_1\right)$, 4P: $\left(0,\ \dfrac{1}{2},\ z_2\right)$, $\left(\dfrac{1}{2},\ 0,\ z_2\right)$, $\left(0,\ \dfrac{1}{2},\ \dfrac{1}{2}+z_2\right)$, $\left(\dfrac{1}{2},\ 0,\ \dfrac{1}{2}-z_2\right)$, 8P: $\pm\,(x,\ x,\ 0)$, $(\bar{x},\ x,\ 0)$, $\left(\dfrac{1}{2}+x,\ \dfrac{1}{2}+x,\dfrac{1}{2}\right)$, $\left(\dfrac{1}{2}-x,\ \dfrac{1}{2}+x,\dfrac{1}{2}\right)$, 8Zn: $(0,\ x_2\ z_3)$, $(0,\ \bar{x}_2,\ z_3)$, $(x_2,\ 0,\ \bar{z}_3)$, $(\bar{x}_2,\ 0,\ \bar{z}_3)$, $\left(\dfrac{1}{2},\ \dfrac{1}{2}+x_2,\ \dfrac{1}{2}-z_3\right)$, $\left(\dfrac{1}{2},\ \dfrac{1}{2}-x_2,\ x_2-z_3\right)$, 16Zn: doubled coordinates with another parameter, $z_1 = 0,25$, $z_2 = 0,239$, $x = 0,261$, $x_2 = 0,283$, $z_3 = 0,386$.

Hexagonal System

Symbol	Structure type	Space group	No. of atoms in cell	Coordinates
H 1	Hexagonal close-packed (Mg)	$D_{6h}^{4} - \dfrac{C6}{mmc}$	2	2Mg: $\left(\dfrac{2}{3},\ \dfrac{1}{3},\ 0\right)$, $\left(\dfrac{1}{3},\ \dfrac{2}{3},\ \dfrac{1}{3}\right)$.
H 2 (R)	CdI_2	$D_{3d}^{3} - C3m$	1	1Cd: $(0,\ 0,\ 0)$, 2 I: $\left(\dfrac{1}{3},\ \dfrac{2}{3},\ z\right)$, $\left(\dfrac{2}{3},\ \dfrac{1}{3},\ \bar{z}\right)$, $z \cong 0,25$.
H 3	ZnS 66 (wurtzite)	$C_{6v}^{4} - C6mc$	2	2Zn: $\left(\dfrac{1}{3},\ \dfrac{2}{3},\ 0\right)$, $\left(\dfrac{2}{3},\ \dfrac{1}{3},\ \dfrac{1}{2}\right)$; 2S: $\left(\dfrac{1}{2},\ \dfrac{2}{3},\ z\right)$, $\left(\dfrac{2}{3},\ \dfrac{1}{3},\ \dfrac{1}{2}+z\right)$, $z = 0,375$.

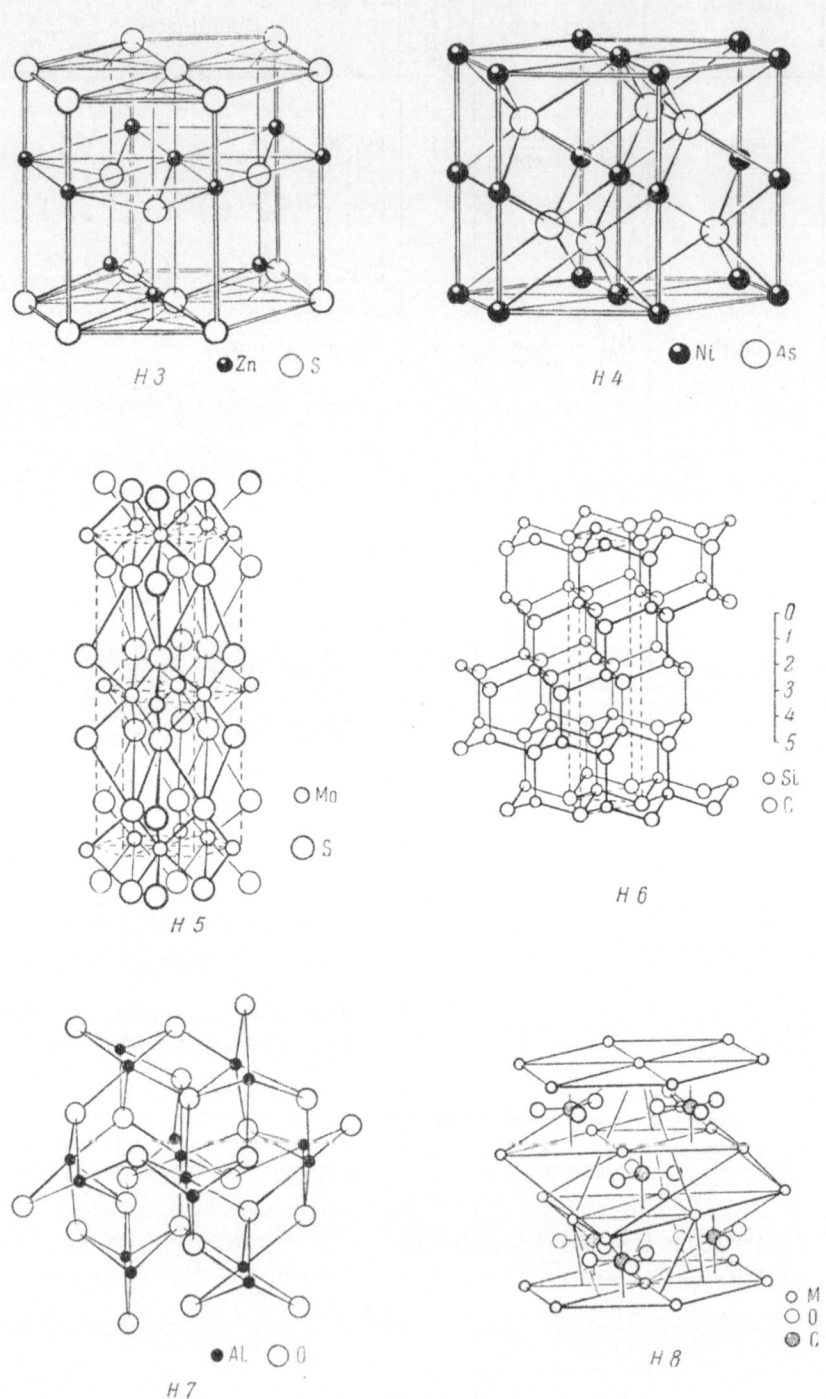

Fig. 107. Arrays of atoms in structures of types H3 to H8.

Symbol	Structure type	Space group	No. of atoms in cell	Coordinates
H 4	NiAs	$D_{6h}^4 - \dfrac{C6}{mmc}$	2	2Ni: $(0,\ 0,\ 0)$, $\left(0,\ 0,\ \dfrac{1}{2}\right)$, 2As: $\left(\dfrac{1}{3},\ \dfrac{2}{3},\ \dfrac{1}{4}\right)$, $\left(\dfrac{2}{3},\ \dfrac{1}{3},\ \dfrac{3}{4}\right)$.
H 5	MoSr	$D_{6h}^4 - \dfrac{C6}{mmc}$	2	2Mo: $\left(\dfrac{1}{2},\ \dfrac{2}{3},\ \dfrac{1}{3}\right)$, $\left(\dfrac{2}{3},\ \dfrac{1}{3},\ \dfrac{1}{4}\right)$, 4Sr: $\left(\dfrac{1}{3},\ \dfrac{2}{3},\ z\right)$, $\left(\dfrac{2}{3},\ \dfrac{1}{3},\ \bar{z}\right)$, $\left(\dfrac{1}{3},\ \dfrac{2}{3},\ \dfrac{1}{3} - z\right)$, $\left(\dfrac{2}{3},\ \dfrac{1}{3},\ \dfrac{1}{2} + z\right)$, $z = 0,379.$
H 6	SiC (carborundum)			
	SiC—I	$C_{3v}^5 - R3m$		
	SiC—II	$C_{6v}^4 - C6mc$		
	SiC—III	$C_{6v}^4 - C6mc$		
H7	Al_2O_3 (corundum)	$D_{3d}^6 - R\bar{3}c$	2	4Al: $(x,\ x,\ x)$, $(\bar{x},\ \bar{x},\ x)$, $\pm\left(\dfrac{1}{2},\ \dfrac{1}{2},\ \dfrac{1}{2}\right)$, 6O: $(y,\ \bar{y},\ 0)$, $(\bar{y},\ 0,\ y)$, $(0,\ y, \bar{y})$, $\left(\dfrac{1}{2} - y,\ y + \dfrac{1}{2},\ \dfrac{1}{2}\right)$, $\left(y + \dfrac{1}{2},\ \dfrac{1}{2},\ y - \dfrac{1}{2}\right)$, $\left(\dfrac{1}{2},\ \dfrac{1}{2} - y,\ y + \dfrac{1}{2}\right)$, $x = 0,105,\quad y = 0,303.$
H 8	$CaCO_3$	$D_{3d}^6 - R\bar{3}c$	2	2Ca: $\left(\dfrac{1}{4},\ \dfrac{1}{4},\ \dfrac{1}{4}\right)$, $\left(\dfrac{3}{4},\ \dfrac{3}{4},\ \dfrac{3}{4}\right)$, 2C: $(0,\ 0,\ 0)$, $\left(\dfrac{1}{2},\ \dfrac{1}{2},\ \dfrac{1}{2}\right)$, 6O: $(x,\ \bar{x},\ 0)$, (d) $\left(\dfrac{1}{2} - x,\ x + \dfrac{1}{2},\ \dfrac{1}{2}\right)$. (d).
H 9 (R)	As	$D_{3d}^5 - R\bar{3}m$	2	2As: $\pm (x,\ x,\ x)$, $x = 0,226.$
H 10	Se (gray)	$\begin{cases} D_3^4 - C3_1 2 \\ D_3^6 - C3_2 2 \end{cases}$	3	3Se: $\left(x,\ 0,\ \dfrac{1}{3}\right)$, $\left(0,\ x, \dfrac{2}{3}\right)$, $(\bar{x},\ \bar{x},\ 0)$, $x = 0,217.$

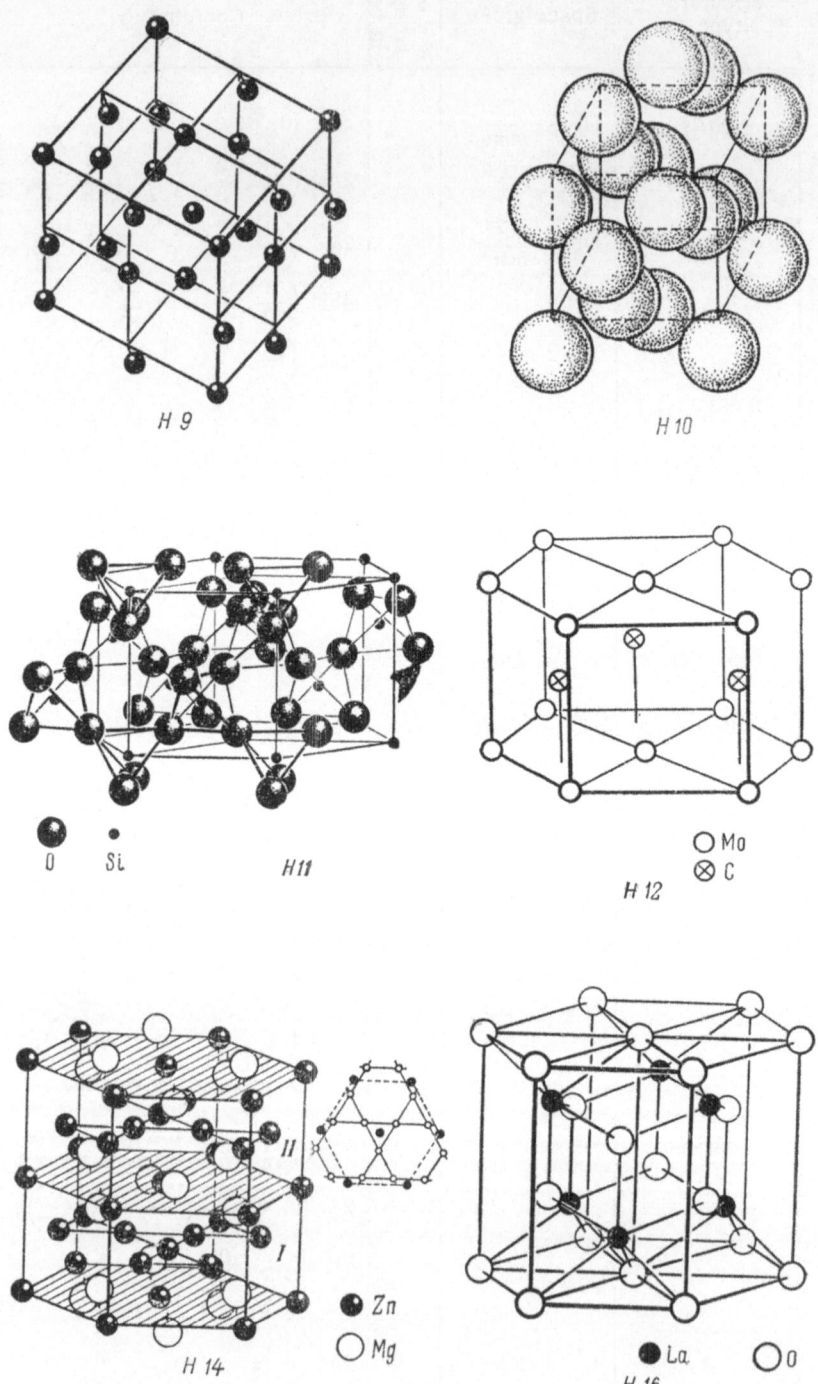

Fig. 108. Arrays of atoms in structures of types H9 to H16.

Symbol	Structure type	Space group	No. of atoms in cell	Coordinates
H 11	SiO_2 (β-quartz)	$\begin{cases} D_6^4 - C6_2 2 \\ D_6^5 - C6_4 2 \end{cases}$	3	3Si: $\left(\frac{1}{2}, \frac{1}{2}, \frac{1}{3}\right)$, $\left(\frac{1}{2}, 0, 0\right)$, $\left(0, \frac{1}{2}, \frac{2}{3}\right)$, 6O: $\left(x, \bar{x}, \frac{5}{6}\right)$, $\left(\bar{x}, x, \frac{5}{6}\right)$, $\left(x, 2x, \frac{1}{2}\right)$, $\left(\bar{x}, 2\bar{x}, \frac{1}{2}\right)$, $\left(2x, x, \frac{1}{6}\right)$, $\left(2\bar{x}, \bar{x}, \frac{1}{6}\right)$, $x = 0,197$.
H 12	WC (MoC)	$D_{6h}^1 - \dfrac{C6}{mmn}$	1	1W: $(0, 0, 0)$, 1C: $\left(\frac{1}{3}, \frac{2}{3}, \frac{1}{2}\right)$.
H 13	$AgZn_3$ (ε-phase)	$D_{6h}^4 - \dfrac{C6}{mmc}$		Hexagonal close-packed
H 14	$MgZn_2$	$D_{6h}^4 - \dfrac{C6}{mmc}$	4	4Mg: $\pm\left(\frac{1}{3}, \frac{2}{3}, z\right)$, $\pm\left(\frac{1}{3}, \frac{2}{3}, \frac{1}{2}-z\right)$, 2Zn: $(0, 0, 0)$, $\left(0, 0, \frac{1}{2}\right)$, 6Zn: $\pm\left(x, 2x, \frac{1}{4}\right)$, $\pm\left(2\bar{x}, \bar{x}, \frac{1}{4}\right)$, $\pm\left(x, \bar{x}, \frac{1}{4}\right)$, $z = 0,062$, $x = -0,170$.
H 15	Na_3As	$D_{6h}^4 - \dfrac{C6}{mmc}$	2	2As: $\pm\left(\frac{1}{3}, \frac{2}{3}, \frac{1}{4}\right)$, 2Na: $\pm\left(0, 0, \frac{1}{4}\right)$, 4Na: $\pm\left(\frac{1}{2}, \frac{2}{3}, z\right)$, $\pm\left(\frac{2}{3}, \frac{1}{3}, \frac{1}{2}+z\right)$, $z = 0,583$.
H 16	La_2O_3	$D_{3d}^9 - C\bar{3}m$	1	2La: $\left(\frac{1}{3}, \frac{2}{3}, z\right)$, $\left(\frac{2}{3}, \frac{1}{3}, \bar{z}\right)$, 1O: $(0, 0, 0)$, 2O: $\left(\frac{1}{3}, \frac{2}{3}, y\right)$, $\left(\frac{2}{3}, \frac{1}{3}, \bar{y}\right)$, $z = 0,23$, $y = 0,63$.

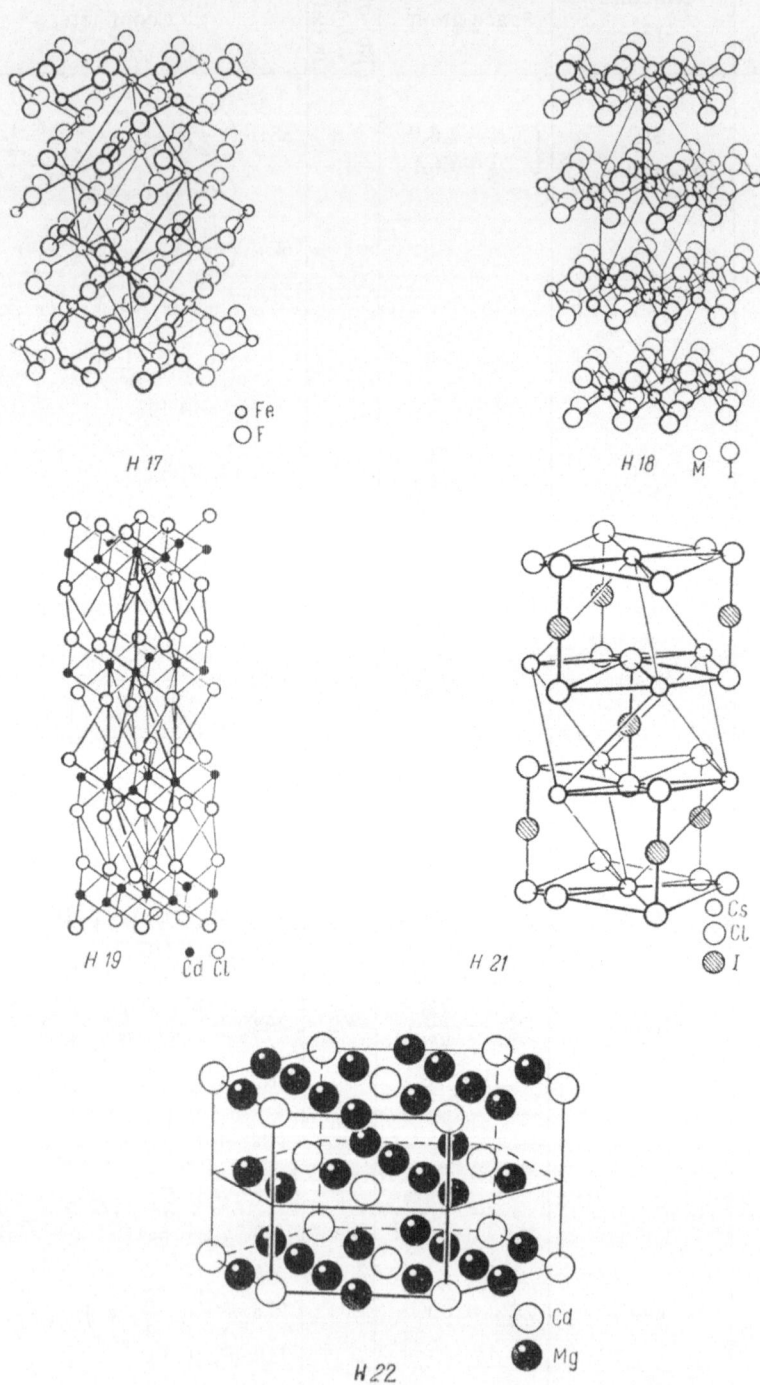

Fig. 109. Arrays of atoms in structures of types H17 to H22.

Symbol	Structure type	Space group	No. of atoms in cell	Coordinates
H 17	FeF_3	$\begin{cases} D_3^7 - R32 \\ D_{3d}^6 - R\bar{3}C \end{cases}$	2	2Fe: (x, x, x), $(\bar{x}, \bar{x}, \bar{x})$, 3F: $(0, y, \bar{y})$, (d) 3F: $\left(\frac{1}{2}, z, \bar{z} \right)$, (d) $x = 0,250$, $y = 0,333$, $z = 0,167$.
H 18 (R)	BiI_3	$C_3^2 - R\bar{3}$	2	2Bi: $\pm (u, u, u)$, 6I: $\pm (x, y, z)$, (d), $u = \frac{1}{6}$, $x = \frac{5}{12}$, $y = \frac{1}{12}$, $z = \frac{1}{4}$.
H 19 (R)	$CdCl_2$	$D_{3d}^5 - R\bar{3}m$	1	1Cd: $(0, 0, 0)$, 2Cl: (x, x, x), $(\bar{x}, \bar{x}, \bar{x})$, $x = 0,25$.
H 20	AlB_2	$D_{6h}^1 - \dfrac{C6}{mmn}$		1Al: $(0, 0, 0)$, 2B: $\left(\frac{1}{3}, \frac{2}{3}, \frac{1}{2} \right)$, $\left(\frac{2}{3}, \frac{1}{3}, \frac{1}{2} \right)$.
H 21	$Cs ICl_2$	$D_{3d}^5 - R\bar{3}m$	1	1Cs: $(0, 0, 0)$, 1I: $\left(\frac{1}{2}, \frac{1}{2}, \frac{1}{2} \right)$, 2Cl: $\pm (x, x, x)$.
H 22	Mg_3Cd	$D_{6h}^4 - \dfrac{C6}{mmc}$	2	2Cd: $\pm \left(\frac{1}{3}, \frac{2}{3}, \frac{1}{4} \right)$, 6Mg: $\pm \left(2x, x, \frac{1}{3} \right)$, $\pm \left(\bar{x}, x, \frac{1}{3} \right)$, $\pm \left(\bar{x}, 2x, \frac{1}{4} \right)$, $x = \frac{5}{6}$,

4-12. Structure Factors for Some Cubic Space Groups

Structure factors are given for the space groups to which belong the lattices of certain structure types in the cubic system; A is the real part of the factor and B is the imaginary one.

The symbols are as in Table 4-11; the structure factors for groups not given in that table are to be found in [88]; See [433] on allowance for anisotropy in the thermal motions of the atoms.

K 1. $Im3m - O_h^9$

$$A = 16 \cos^2 2\pi \frac{h+k+l}{4} \{\cos 2\pi hx \, [\cos 2\pi ky \cos 2\pi lz + \cos 2\pi ly \cos 2\pi kz] +$$
$$+ \cos 2\pi hy \, [\cos 2\pi kz \cos 2\pi lx + \cos 2\pi lz \cos 2\pi kx] +$$
$$+ \cos 2\pi hz \, [\cos 2\pi kx \cos 2\pi ly + \cos 2\pi lx \cos 2\pi ky]\},$$

$B = 0.$

K 2, K 5, K 6, K 11, K 23. $Fm3m - O_h^5$

$$A = 32 \cos^2 2\pi \frac{h+k}{4} \{\cos 2\pi hx \, [\cos 2\pi ky \cos 2\pi lz + \cos 2\pi ly \cos 2\pi kz] +$$
$$+ \cos 2\pi hy \, [\cos 2\pi kz \cos 2\pi lx + \cos 2\pi lz \cos 2\pi kx] +$$
$$+ \cos 2\pi hz \, [\cos 2\pi kx \cos 2\pi ly + \cos 2\pi lx \cos 2\pi ky]\},$$

$B = 0.$

K 3. $Pm3m - O_h^1$

$$A = 8 \{\cos 2\pi hx \, [\cos 2\pi ky \cos 2\pi lz + \cos 2\pi ly \cos 2\pi kz] +$$
$$+ \cos 2\pi hy \, [\cos 2\pi kz \cos 2\pi lx + \cos 2\pi lz \cos 2\pi kx] +$$
$$+ \cos 2\pi hz \, [\cos 2\pi kx \cos 2\pi ly + \cos 2\pi lx \cos 2\pi ky]\},$$

$B = 0.$

K 4. $F\bar{4}3m - T_d^2$

$$A = 16 \cos^2 2\pi \frac{h+k}{4} \cos^2 2\pi \frac{k+l}{4} \{\cos 2\pi lz \, [\cos 2\pi hx \cos 2\pi ky + \cos 2\pi kx \cos 2\pi hy] +$$
$$+ \cos 2\pi hz \, [\cos 2\pi kx \cos 2\pi ly + \cos 2\pi lx \cos 2\pi ky] +$$
$$+ \cos 2\pi kz \, [\cos 2\pi lx \cos 2\pi hy + \cos 2\pi hx \cos 2\pi ly]\},$$

$$B = - 16 \cos^2 2\pi \frac{h+k}{4} \cos^2 2\pi \frac{k+l}{4} \{\sin 2\pi lz \, [\sin 2\pi hx \sin 2\pi ky + \sin 2\pi kx \sin 2\pi hy] +$$
$$+ \sin 2\pi hz \, [\sin 2\pi kx \sin 2\pi ly + \sin 2\pi lx \sin 2\pi ky] +$$
$$+ \sin 2\pi kz \, [\sin 2\pi lx \sin 2\pi hy + \sin 2\pi hx \sin 2\pi ly]\},$$

$B = 0$ for $h = 0$ (or $k = 0$, or $l = 0$).

K 7. $Pn3m - O_h^4$

$$A = 8 \cos^2 2\pi \frac{h+k+l}{4} \{\cos 2\pi hx \, [\cos 2\pi ky \cos 2\pi lz + \cos 2\pi ly \cos 2\pi kz] +$$
$$+ \cos 2\pi hy \, [\cos 2\pi kz \cos 2\pi lx + \cos 2\pi lz \cos 2\pi kx] +$$
$$+ \cos 2\pi hz \, [\cos 2\pi kx \cos 2\pi ly + \cos 2\pi lx \cos 2\pi ky]\},$$

$$B = - 8 \sin^2 2\pi \frac{h+k+l}{4} \{\sin 2\pi hx \, [\sin 2\pi ky \sin 2\pi lz + \sin 2\pi ly \sin 2\pi kz] +$$
$$+ \sin 2\pi hy \, [\sin 2\pi kz \sin 2\pi lx + \sin 2\pi lz \sin 2\pi kx] +$$
$$+ \sin 2\pi hz \, [\sin 2\pi kz \sin 2\pi ly + \sin 2\pi lx \sin 2\pi ky]\}.$$

K 8, K 10, K 21, K 22. $Fd3m - O_h^7$

$$A = G \cos 2\pi \frac{h+k+l}{8} ; \quad B = G \sin 2\pi \frac{h+k+l}{8} , \text{ where}$$

$$G = 32 \cos^2 2\pi \frac{h+k}{4} \cos^2 2\pi \frac{k+l}{4} \Big\{ \cos 2\pi \frac{h+k+l}{8} \, [\cos 2\pi hx \, (\cos 2\pi ky \cos 2\pi lz +$$
$$+ \cos 2\pi ly \cos 2\pi kz) + \cos 2\pi hy \, (\cos 2\pi kz \cos 2\pi lx + \cos 2\pi lz \cos 2\pi kx) +$$
$$+ \cos 2\pi hz \, (\cos 2\pi kx \cos 2\pi ly + \cos 2\pi lx \cos 2\pi ky)] -$$
$$- \sin 2\pi \frac{h+k+l}{8} \, [\sin 2\pi hx \, (\sin 2\pi ky \sin 2\pi lz + \sin 2\pi ly \sin 2\pi kz) +$$
$$+ \sin 2\pi hy \, (\sin 2\pi kz \sin 2\pi lx + \sin 2\pi lz \sin 2\pi kx) +$$
$$+ \sin 2\pi hz \, (\sin 2\pi kx \sin 2\pi ly + \sin 2\pi lx \sin 2\pi ky)] \Big\} .$$

K 9, K 13. $Pa3 - T_h^6$

$$A = 8 \left\{ \cos 2\pi \left(hx + \tfrac{k-h}{4} \right) \cos 2\pi \left(ky + \tfrac{l-k}{4} \right) \cos 2\pi \left(lz + \tfrac{h-l}{4} \right) + \right.$$
$$+ \cos 2\pi \left(hy + \tfrac{k-h}{4} \right) \cos 2\pi \left(kz + \tfrac{l-k}{4} \right) \cos 2\pi \left(lx + \tfrac{h-l}{4} \right) +$$
$$\left. + \cos 2\pi \left(hz + \tfrac{k-h}{4} \right) \cos 2\pi \left(kx + \tfrac{l-k}{4} \right) \cos 2\pi \left(ly + \tfrac{h-l}{4} \right) \right\},$$

$B = 0.$

K 12. $P4_332 - O^6$

$$A = 4 \left\{ \cos 2\pi \left(hx + \tfrac{k-h}{4} \right) \cos 2\pi \left(ky + \tfrac{l-k}{4} \right) \cos 2\pi \left(lz + \tfrac{h-l}{4} \right) + \right.$$
$$+ \cos 2\pi \left(kx - \tfrac{l}{4} \right) \cos 2\pi \left(hy - \tfrac{k}{4} \right) \cos 2\pi \left(lz - \tfrac{h}{4} \right) +$$
$$+ \cos 2\pi \left(hy + \tfrac{k-h}{4} \right) \cos 2\pi \left(kz + \tfrac{l-k}{4} \right) \cos 2\pi \left(lx + \tfrac{h-l}{4} \right) +$$
$$+ \cos 2\pi \left(ky - \tfrac{l}{4} \right) \cos 2\pi \left(hz - \tfrac{k}{4} \right) \cos 2\pi \left(lx - \tfrac{h}{4} \right) +$$
$$+ \cos 2\pi \left(hz + \tfrac{k-h}{4} \right) \cos 2\pi \left(kx + \tfrac{l-k}{4} \right) \cos 2\pi \left(ly + \tfrac{h-l}{4} \right) +$$
$$\left. + \cos 2\pi \left(kz - \tfrac{l}{4} \right) \cos 2\pi \left(hx - \tfrac{k}{4} \right) \cos 2\pi \left(ly - \tfrac{h}{y} \right) \right\},$$

$$B = -4 \left\{ \sin 2\pi \left(hx + \tfrac{k-h}{4} \right) \sin 2\pi \left(ky + \tfrac{l-k}{4} \right) \sin 2\pi \left(lz + \tfrac{h-l}{4} \right) - \right.$$
$$- \sin 2\pi \left(kx - \tfrac{l}{4} \right) \sin 2\pi \left(hy - \tfrac{k}{4} \right) \sin 2\pi \left(lz - \tfrac{h}{4} \right) + \sin 2\pi \left(hy + \tfrac{k-h}{4} \right) \times$$
$$\times \sin 2\pi \left(kz + \tfrac{l-k}{4} \right) \sin 2\pi \left(lx + \tfrac{h-l}{4} \right) - \sin 2\pi \left(ky - \tfrac{l}{4} \right) \sin 2\pi \left(hz - \tfrac{k}{4} \right) \times$$
$$\times \sin 2\pi \left(lx - \tfrac{h}{4} \right) + \sin 2\pi \left(hz + \tfrac{k-h}{4} \right) \sin 2\pi \left(kx + \tfrac{l-k}{4} \right) \sin 2\pi \left(ly + \tfrac{h-l}{4} \right) -$$
$$\left. - \sin 2\pi \left(kz - \tfrac{l}{4} \right) \sin 2\pi \left(hx - \tfrac{k}{4} \right) \sin 2\pi \left(ly - \tfrac{h}{4} \right) \right\}.$$

K 12. $P4_332 - O^7$

$$A = 4 \left\{ \cos 2\pi \left(hx + \tfrac{k-h}{4} \right) \cos 2\pi \left(ky + \tfrac{l-k}{4} \right) \cos 2\pi \left(lz + \tfrac{h-l}{4} \right) + \right.$$
$$+ \cos 2\pi \left(kx + \tfrac{l}{4} \right) \cos 2\pi \left(hy + \tfrac{k}{4} \right) \cos 2\pi \left(lz + \tfrac{h}{4} \right) + \cos 2\pi \left(hy + \tfrac{k-h}{4} \right) \times$$
$$\times \cos 2\pi \left(kz + \tfrac{l-k}{4} \right) \cos 2\pi \left(lx + \tfrac{h-l}{4} \right) + \cos 2\pi \left(ky + \tfrac{l}{4} \right) \cos 2\pi \left(hz + \tfrac{k}{4} \right) \times$$
$$\times \cos 2\pi \left(lx + \tfrac{h}{4} \right) + \cos 2\pi \left(hz + \tfrac{k-h}{4} \right) \cos 2\pi \left(kx + \tfrac{l-k}{4} \right) \cos 2\pi \left(ly + \tfrac{h-l}{4} \right) +$$
$$\left. + \cos 2\pi \left(kz + \tfrac{l}{4} \right) \cos 2\pi \left(hx + \tfrac{k}{4} \right) \cos 2\pi \left(ly + \tfrac{h}{4} \right) \right\},$$

$$B = -4 \left\{ \sin 2\pi \left(hx + \tfrac{k-h}{4} \right) \sin 2\pi \left(ky + \tfrac{l-k}{4} \right) \sin 2\pi \left(lz + \tfrac{h-l}{4} \right) - \right.$$
$$- \sin 2\pi \left(kx + \tfrac{l}{4} \right) \sin 2\pi \left(hy + \tfrac{k}{4} \right) \sin 2\pi \left(lz + \tfrac{h}{4} \right) + \sin 2\pi \left(hy + \tfrac{k-h}{4} \right) \times$$
$$\times \sin 2\pi \left(kz + \tfrac{l-k}{4} \right) \sin 2\pi \left(lx + \tfrac{h-l}{4} \right) - \sin 2\pi \left(ky + \tfrac{l}{4} \right) \sin 2\pi \left(hz + \tfrac{k}{4} \right) \times$$
$$\times \sin 2\pi \left(lx + \tfrac{h}{4} \right) + \sin 2\pi \left(hz + \tfrac{k-h}{4} \right) \sin 2\pi \left(kx + \tfrac{l-k}{4} \right) \sin 2\pi \left(ly + \tfrac{h-l}{4} \right) -$$
$$\left. - \sin 2\pi \left(kz + \tfrac{l}{4} \right) \sin 2\pi \left(hx + \tfrac{k}{4} \right) \sin 2\pi \left(ly + \tfrac{h}{4} \right) \right\}.$$

K 14, K 24. $I\bar{4}3m - T_d^3$

$$A = 8 \cos^2 2\pi \frac{h+k+l}{4} \{\cos 2\pi lz \, [\cos 2\pi hx \cos 2\pi ky + \cos 2\pi kx \cos 2\pi hy] +$$
$$+ \cos 2\pi hz \, [\cos 2\pi kx \cos 2\pi ly + \cos 2\pi lx \cos 2\pi ky] +$$
$$+ \cos 2\pi kz \, [\cos 2\pi lx \cos 2\pi hy + \cos 2\pi hx \cos 2\pi ly]\},$$

$$B = -8 \cos^2 2\pi \frac{h+k+l}{4} \{\sin 2\pi lz \, [\sin 2\pi hx \sin 2\pi ky + \sin 2\pi kx \sin 2\pi hy] +$$
$$+ \sin 2\pi hz \, [\sin 8\pi kx \sin 2\pi ly + \sin 2\pi lx \sin 2\pi ky] +$$
$$+ \sin 2\pi kz \, [\sin 2\pi lx \sin 2\pi hy + \sin 2\pi hx \sin 2\pi ly]\},$$

$B = 0$, if $h = 0$ (or $k = 0$, or $l = 0$).

K 15, K 19, K 20. $P2_13 - T^4$

$$A = 4 \left\{ \cos 2\pi \left(hx + \frac{k-h}{4} \right) \cos 2\pi \left(ky + \frac{l-k}{4} \right) \cos 2\pi \left(lz + \frac{h-l}{4} \right) + \right.$$
$$+ \cos 2\pi \left(kx + \frac{l-k}{4} \right) \cos 2\pi \left(ly + \frac{h-l}{4} \right) \cos 2\pi \left(hz + \frac{k-h}{4} \right) +$$
$$\left. + \cos 2\pi \left(lx + \frac{h-l}{4} \right) \cos 2\pi \left(hy + \frac{k-h}{4} \right) \cos 2\pi \left(kz + \frac{l-k}{4} \right) \right\},$$

$$B = -4 \left\{ \sin 2\pi \left(hx + \frac{k-h}{4} \right) \sin 2\pi \left(ky + \frac{l-k}{4} \right) \sin 2\pi \left(lz + \frac{h-l}{4} \right) + \right.$$
$$+ \sin 2\pi \left(kx + \frac{l-k}{4} \right) \sin 2\pi \left(ly + \frac{h-l}{4} \right) \sin 2\pi \left(hz + \frac{k-h}{4} \right) +$$
$$\left. + \sin 2\pi \left(lx + \frac{h-l}{4} \right) \sin 2\pi \left(hy + \frac{k-h}{4} \right) \sin 2\pi \left(kz + \frac{l-k}{4} \right) \right\}.$$

K 16. $Ia3 - T_h^7$

$$A = 16 \cos 2\pi \frac{h+k+l}{4} \left\{ \cos 2\pi \left(hx + \frac{l}{4} \right) \cos 2\pi \left(ky + \frac{h}{4} \right) \cos 2\pi \left(lz + \frac{k}{4} \right) + \right.$$
$$+ \cos 2\pi \left(hy + \frac{l}{4} \right) \cos 2\pi \left(kz + \frac{h}{4} \right) \cos 2\pi \left(lx + \frac{k}{4} \right) +$$
$$\left. + \cos 2\pi \left(hz + \frac{l}{4} \right) \cos 2\pi \left(kx + \frac{h}{4} \right) \cos 2\pi \left(ly + \frac{k}{4} \right) \right\},$$

$B = 0$.

K 17, K 18. $Pm3m - O_h^1$

$$A = 8 \{\cos 2\pi hx \, [\cos 2\pi ky \cos 2\pi lz + \cos 2\pi ly \cos 2\pi kz] +$$
$$+ \cos 2\pi hy \, [\cos 2\pi kz \cos 2\pi lx + \cos 2\pi lz \cos 2\pi kx] +$$
$$+ \cos 2\pi hz \, [\cos 2\pi kx \cos 2\pi ly + \cos 2\pi lx \cos 2\pi ky]\},$$

$B = 0$.

$P\bar{4}3m - T_d^1$

$$A = 4 \{\cos 2\pi lz \, [\cos 2\pi hx \cos 2\pi ky + \cos 2\pi kx \cos 2\pi hy] +$$
$$+ \cos 2\pi hz \, [\cos 2\pi kx \cos 2\pi ly + \cos 2\pi lx \cos 2\pi ky] +$$
$$+ \cos 2\pi kz \, [\cos 2\pi lx \cos 2\pi hy + \cos 2\pi hx \cos 2\pi ly]\},$$

$$B = -4 \{\sin 2\pi lz \, [\sin 2\pi hx \sin 2\pi ky + \sin 2\pi kx \sin 2\pi hy] +$$
$$+ \sin 2\pi hz \, [\sin 2\pi kx \sin 2\pi ly + \sin 2\pi lx \sin 2\pi ky] +$$
$$+ \sin 2\pi kz \, [\sin 2\pi lx \sin 2\pi hy + \sin 2\pi hx \sin 2\pi ly]\},$$

$B = 0$, if $h = 0$ (or $k = 0$, or $l = 0$).

4-13. Structure Factors for Some Cubic Structure Types

The table gives F and p for Debye photographs for some types of structure occurring in the cubic system [109]. The data are for structure types A1 (copper type), A2 (tungsten type), A4 (diamond type), B1 (NaCl type), B2 (CsCl type), and C1 (CaF$_2$ type) for Σh^2 from 1 to 49. The symbols are f, atomic factor for the element; f_1 and f_2, factors for the elements in a compound (e.g., f_1 is for Ca and f_2 is for F in CaF$_2$).

hkl	A1		A2		A4		B1		B2		C1 *	
	F^2	p	F^2	p	F^2	p	F^2	p	F^2	p	F^2	p
100	0	—	0	—	0	—	0	—	$(f_1-f_2)^2$	6	0	—
110	0	—	$4f^2$	12	0	—	0	—	$(f_1+f_2)^2$	12	0	—
111	$16f^2$	8	0	—	$32f^2$	8	$16(f_1-f_2)^2$	8	$(f_1-f_2)^2$	8	$16f_1^2$	8
200	$16f^2$	6	$4f^2$	6	0	—	$16(f_1+f_2)^2$	6	$(f_1+f_2)^2$	6	$16(f_1-2f_2)^2$	6
210	0	—	0	—	0	—	0	—	$(f_1-f_2)^2$	24	0	—
211	0	—	$4f^2$	24	0	—	0	—	$(f_1+f_2)^2$	24	0	—
220	$16f^2$	12	$4f^2$	12	$64f^2$	12	$16(f_1+f_2)^2$	12	$(f_1+f_2)^2$	12	$16(f_1+2f_2)^2$	12
300; 221	0	—	0	—	0	—	0	—	$2(f_1-f_2)^2$	15	0	—
310	0	—	$4f^2$	24	0	—	0	—	$(f_1-f_2)^2$	24	$16(f_1+2f_2)^2$	24
311	$16f^2$	24	0	—	$32f^2$	24	$16(f_1-f_2)^2$	24	$(f_1-f_2)^2$	24	$16f_1^2$	24
222	$16f^2$	8	$4f^2$	8	0	—	$16(f_1+f_2)^2$	8	$(f_1+f_2)^2$	8	$16(f_1-2f_2)^2$	8
320	0	—	0	—	0	—	0	—	$(f_1-f_2)^2$	24	0	—
321	0	—	$4f^2$	48	0	—	0	—	$(f_1+f_2)^2$	48	$16(f_1-2f_2)^2$	48
400	$16f^2$	6	$4f^2$	6	$64f^2$	6	$16(f_1+f_2)^2$	6	$(f_1+f_2)^2$	6	$16(f_1+2f_2)^2$	6
410; 322	0	—	0	—	0	—	0	—	$2(f_1-f_2)^2$	24	0	—
411; 330	0	—	$4f^2$	18	0	—	0	—	$2(f_1+f_2)^2$	18	0	—
331	$16f^2$	24	0	—	$32f^2$	24	$16(f_1-f_2)^2$	24	$(f_1-f_2)^2$	24	$16f_1^2$	24
420	$16f^2$	24	$4f^2$	24	0	—	$16(f_1+f_2)^2$	24	$(f_1+f_2)^2$	24	$16(f_1-2f_2)^2$	24
421	0	—	0	—	0	—	0	—	$(f_1-f_2)^2$	48	0	—
332	0	—	$4f^2$	24	0	—	0	—	$(f_1+f_2)^2$	24	0	—
422	$16f^2$	24	$4f^2$	24	$64f^2$	24	$16(f_1+f_2)^2$	24	$(f_1+f_2)^2$	24	$16(f_1+2f_2)^2$	24
500; 430	0	—	0	—	0	—	0	—	$2(f_1-f_2)^2$	15	0	—
510; 431	0	—	$8f^2$	36	0	—	0	—	$2(f_1+f_2)^2$	36	0	—
511; 333	$32f^2$	16	0	—	$64f^2$	16	$32(f_1-f_2)^2$	16	$2(f_1-f_2)^2$	16	$32f_1^2$	16
520; 432	0	—	0	—	0	—	0	—	$2(f_1-f_2)^2$	36	0	—
521	0	—	$4f^2$	48	0	—	0	—	$(f_1+f_2)^2$	48	0	—
440	$16f^2$	12	$4f^2$	12	$64f^2$	12	$16(f_1+f_2)^2$	12	$(f_1+f_2)^2$	12	$16(f_1+2f_2)^2$	12
522; 441	0	—	0	—	0	—	0	—	$2(f_1-f_2)^2$	24	$32f_1^2$	24
530; 433	0	—	$8f^2$	24	0	—	0	—	$2(f_1+f_2)^2$	24	0	—
531	$16f^2$	48	0	—	$32f^2$	48	$16(f_1-f_2)^2$	48	$(f_1-f_2)^2$	48	$16f_1^2$	48
600; 442	$32f^2$	15	$8f^2$	15	0	—	$32(f_1+f_2)^2$	15	$2(f_1+f_2)^2$	15	$32(f_1-2f_2)^2$	15
610	0	—	0	—	0	—	0	—	$(f_1-f_2)^2$	24	0	—
611; 532	0	—	$8f^2$	36	0	—	0	—	$2(f_1+f_2)^2$	36	0	—
620	$16f^2$	24	$4f^2$	24	$64f^2$	24	$16(f_1+f_2)^2$	24	$(f_1+f_2)^2$	24	$16(f_1+2f_2)^2$	24
621; 540; 443	0	—	0	—	0	—	0	—	$2(f_1-f_2)^2$	32	0	—
541	0	—	$4f^2$	48	0	—	0	—	$(f_1+f_2)^2$	48	0	—
533	$16f^2$	24	0	—	$32f^2$	24	$16(f_1-f_2)^2$	24	$(f_1-f_2)^2$	24	$16f_1^2$	24
622	$16f^2$	24	$4f^2$	24	0	—	$16(f_1+f_2)^2$	24	$(f_1+f_2)^2$	24	$16(f_1-2f_2)^2$	24
630; 542	0	—	0	—	0	—	0	—	$2(f_1-f_2)^2$	36	0	—
631	0	—	$4f^2$	48	0	—	0	—	$(f_1+f_2)^2$	48	0	—
444	$16f^2$	8	$4f^2$	8	$64f^2$	8	$16(f_1+f_2)^2$	8	$(f_1+f_2)^2$	8	$16(f_1-2f_2)^2$	8
700; 632	0	—	0	—	0	—	0	—	$2(f_1-f_2)^2$	27	0	—

*f_1 for Ca, f_2 for F.

4-14.   Structure Factors for Some Tetragonal Space Groups

Structure factors are given for the space groups to which belong the lattices of certain structure types in the tetragonal system [88].

T 1. $\dfrac{I4_1}{amd} - D_{4h}^{19}$

$$A = G \cos 2\pi \frac{2h+l}{8}, \quad B = -G \sin 2\pi \frac{2h+l}{8}, \quad \text{where}$$

$$G = 16 \cos^2 2\pi \frac{h+k+l}{4} \left[\cos 2\pi hx \cos 2\pi ky \cos 2\pi \left(lz + \frac{2h+l}{8} \right) + \right.$$
$$\left. + \cos 2\pi kx \cos 2\pi hy \cos 2\pi \left(lz - \frac{2h+l}{8} \right) \right].$$

T 2,   T 5. $\dfrac{I4}{mmm} - D_{4h}^{17}$

$$A = 16 \cos^2 2\pi \frac{h+k+l}{4} \cos 2\pi lz [\cos 2\pi hx \cos 2\pi ky + \cos 2\pi kx \cos 2\pi hy],$$
$$B = 0.$$

T 3,   T 11. $\dfrac{P4}{nmm} - D_{4h}^{7}$

$$A = 8 \cos^2 2\pi \frac{h+k}{4} \cos 2\pi lz [\cos 2\pi hx \cos 2\pi ky + \cos 2\pi kx \cos 2\pi hy],$$
$$B = 8 \sin^2 2\pi \frac{h+k}{4} \sin 2\pi lz [\cos 2\pi hx \cos 2\pi ky - \cos 2\pi kx \cos 2\pi hy].$$

T 4. $\dfrac{P4}{mnm} - D_{4h}^{14}$

$$A = 8 \cos 2\pi lz \left[\cos 2\pi \left(hx - \frac{h+k+l}{4} \right) \cos 2\pi \left(ky + \frac{h+k+l}{4} \right) + \right.$$
$$\left. + \cos 2\pi \left(kx + \frac{h+k+l}{4} \right) \cos 2\pi \left(hy - \frac{h+k+l}{4} \right) \right],$$
$$B = 0.$$

T 6,   T 7. $\dfrac{I4}{mcm} - D_{4h}^{18}$

$$A = 16 \cos^2 2\pi \frac{h+k+l}{4} \cos 2\pi lz \left[\cos 2\pi \left(hx + \frac{l}{4} \right) \cos 2\pi \left(ky - \frac{l}{4} \right) + \right.$$
$$\left. + \cos 2\pi \left(kx + \frac{l}{4} \right) \cos 2\pi \left(hy + \frac{l}{4} \right) \right],$$
$$B = 0.$$

T 8. $\dfrac{P4}{nnc} - D_{4h}^{4}$

$$A = 8 \cos^2 2\pi \frac{h+k+l}{4} \cos 2\pi lz [\cos 2\pi hx \cos 2\pi ky + \cos 2\pi kx \cos 2\pi hy],$$
$$B = -8 \sin^2 2\pi \frac{h+k+l}{4} \sin 2\pi lz [\sin 2\pi hx \sin 2\pi ky - \sin 2\pi kx \sin 2\pi hy].$$

T 9. $\dfrac{P4}{mmm} - D_{4h}^{1}$

$$A = 8 \cos 2\pi lz [\cos 2\pi hx \cos 2\pi ky + \cos 2\pi kx \cos 2\pi hy],$$
$$B = 0.$$

T 10. $\dfrac{I4_1}{a} - C_{4h}^6$

$$A = G \cos \frac{2k+l}{8}, \quad B = G \sin \frac{2k+l}{8}, \quad \text{where}$$

$$G = 8 \cos^2 2\pi \frac{h+k+l}{4} \left[\cos 2\pi (hx + ky) \cos 2\pi \left(lz - \frac{2k+l}{8} \right) + \right.$$

$$\left. + \cos 2\pi (hy - kx) \cos 2\pi \left(lz + \frac{2k+l}{8} \right) \right] =$$

$$= 16 \cos^2 2\pi \frac{h+k+l}{4} \left\{ \cos 2\pi \frac{2k+l}{8} \cos \pi \left[(h-k) x + (h+k) y \right] \times \right.$$

$$\times \cos \pi \left[(h+k) x - (h-k) y \right] \cos 2\pi lz - \sin 2\pi \frac{2k+l}{8} \sin \pi \times$$

$$\left. \times \left[(h-k) x + (h+k) y \right] \sin \pi \left[(h+k) x - (h-k) y \right] \sin 2\pi lz \right\}.$$

T 12. $\dfrac{P4_2}{nmc} - D_{4h}^{15}$

$$A = 8 \cos^2 2\pi \frac{h+k+l}{4} \cos 2\pi lz \left[\cos 2\pi hx \cos 2\pi ky + \cos 2\pi kx \cos 2\pi hy \right],$$

$$B = 8 \sin^2 2\pi \frac{h+k+l}{4} \sin 2\pi lz \left[\cos 2\pi hx \cos 2\pi ky - \cos 2\pi kx \cos 2\pi hy \right].$$

4-15.   Structure Factors for Some Hexagonal Space Groups

Structure factors are given for the space groups to which belong the lattices of certain structure types in the hexagonal system [88] (from Table 3-3).

H 1, H 4, H 5, H 22. $\dfrac{P6_3}{mmc} - D_{6h}^4$

$$A = 8 \cos 2\pi \left(lz + \frac{l}{4} \right) \left\{ \cos \pi i (x+y) \cos \pi \left[(h-k)(x-y) - \frac{l}{2} \right] + \right.$$

$$+ \cos \pi h (x+y) \cos \pi \left[(k-i)(x-y) - \frac{l}{2} \right] +$$

$$\left. + \cos \pi k (x+y) \cos \pi \left[(i-h)(x-y) - \frac{l}{2} \right] \right\},$$

$B = 0.$

H 2, H 16. $P\bar{3}m1 - D_{3d}^3$

$$A = 4 \left\{ \cos \pi i (x+y) \cos \pi \left[(h-k)(x-y) + 2lz \right] + \right.$$

$$+ \cos \pi h (x+y) \cos \pi \left[(k-i)(x-y) + 2lz \right] +$$

$$\left. + \cos \pi k (x+y) \cos \pi \left[(i-h)(x-y) + 2lz \right] \right\},$$

$B = 0.$

H 3, H 6. $P6_3mc - C_{6v}^4$

$$A = 4 \cos 2\pi \left(lz - \frac{l}{4} \right) \left\{ \cos \pi i (x+y) \cos \pi \left[(h-k)(x-y) + \frac{l}{2} \right] + \right.$$

$$+ \cos \pi h (x+y) \cos \pi \left[(k-i)(x-y) + \frac{l}{2} \right] +$$

$$\left. + \cos \pi k (x+y) \cos \pi \left[(i-h)(x-y) + \frac{l}{2} \right] \right\},$$

$$B = 4 \sin 2\pi \left(lz - \frac{l}{4} \right) \left\{ \cos \pi i\, (x+y) \cos \pi \left[(h-k)(x-y) + \frac{l}{2} \right] + \right.$$

$$+ \cos \pi h\, (x+y) \cos \pi \left[(k-i)(x-y) + \frac{l}{2} \right] +$$

$$\left. + \cos \pi k\, (x+y) \cos \pi \left[(i-h)(x-y) + \frac{l}{2} \right] \right\}.$$

H 7, H 8. $R\bar{3}c - D_{3d}^6$

$$= 4 \left\{ \cos \pi [(h+k)(x+y) + 2lz] \cos \pi \left[(h-k)(x-y) - \frac{h+k+l}{2} \right] + \right.$$

$$+ \cos \pi \left[(k+l)(x+y) + 2hz + \frac{h+k+l}{2} \right] \cos \pi \left[(k-l)(x-y) - \frac{h+k+l}{2} \right] +$$

$$\left. + \cos \pi \left[(l+h)(x+y) + 2kz + \frac{h+k+l}{2} \right] \cos \pi \left[(l-h)(x-y) - \frac{h+k+l}{2} \right] \right\},$$

$B = 0.$

H 9, H 19, H 21. $R\bar{3}m - D_{3d}^5$

$$A = 4 \{ \cos \pi [(h+k)(x+y) + 2lz] \cos \pi\, (h-k)(x-y) +$$

$$+ \cos \pi [(k+l)(x+y) + 2hz] \cos \pi\, (k-l)(x-y) +$$

$$+ \cos \pi [(l+h)(x+y) + 2kz] \cos \pi\, (l-h)(x-y) \},$$

$B = 0.$

H 10. $P3_121 - D_3^4$

$$A = 2 \left\{ \cos \pi [(h-k)(x-y) + 2lz] \cos \pi [(k-i)(x-y) + 2lz] \times \right.$$

$$\times \cos \pi \left[h\, (x+y) - 2\frac{l}{3} \right] + \cos \pi [(i-h)(x-y) + 2lz] \cos \pi \left[k\, (x+y) + 2\frac{l}{3} \right] \right\},$$

$$B = -2 \left\{ \cos \pi [(h-k)(x-y) + 2lz] \sin \pi\, (x+y) + \right.$$

$$+ \cos \pi [(k-i)(x-y) + 2lz] \sin \pi \left[h\, (x+y) - 2\frac{l}{3} \right] +$$

$$\left. + \cos \pi [(i-h)(x-y) + 2lz] \sin \pi \left[k\, (x+y) + 2\frac{l}{3} \right] \right\}.$$

H 10. $P3_221 - D_3^6$

$$A = 2 \left\{ \cos \pi [(h-k)(x-y) + 2lz] \cos \pi [(k-i)(x-y) + 2lz] \times \right.$$

$$\times \cos \pi \left[h\, (x+y) + 2\frac{l}{3} \right] + \cos \pi [(i-h)(x-y) + 2lz] \cos \pi \left[k(x+y) - 2\frac{l}{3} \right] \right\},$$

$$B = -2 \left\{ \cos \pi [(h-k)(x-y) + 2lz] \sin \pi i\, (x+y) + \right.$$

$$+ \cos \pi [(k-i)(x-y) + 2lz] \sin \pi \left[h\, (x+y) + 2\frac{l}{3} \right] +$$

$$\left. + \cos \pi [(i-h)(x-y) + 2lz] \sin \pi \left[k\, (x+y) - 2\frac{l}{3} \right] \right\}.$$

H 11, H 13, H 14, H 15. $P6_222 - D_6^{14}$

$$A = 2 \left\{ \cos 2\pi lz \left[\cos 2\pi (hx + ky) + \cos 2\pi (hx + iy) \right] + \right.$$
$$+ \cos 2\pi \left(lz - \frac{l}{3} \right) \left[\cos 2\pi (kx + iy) + \cos 2\pi (ix + ky) \right] +$$
$$\left. + \cos 2\pi \left(lz + \frac{l}{3} \right) \left[\cos 2\pi (ix + hy) + \cos 2\pi (kx + hy) \right] \right\},$$

$$B = 2 \left\{ \sin 2\pi lz \left[\cos 2\pi (hx + ky) - \cos 2\pi (hx + iy) \right] + \right.$$
$$+ \sin 2\pi \left(lz - \frac{l}{3} \right) \left[\cos 2\pi (kx + iy) - \cos 2\pi (ix + ky) \right] +$$
$$\left. + \sin 2\pi \left(lz + \frac{l}{3} \right) \left[\cos 2\pi (ix + hy) - \cos 2\pi (kx + hy) \right] \right\}.$$

H 11. $P6_422 - D_6^5$

$$A = 2 \left\{ \cos 2\pi \, lz \left[\cos 2\pi (hx + ky) + \cos 2\pi (hx + iy) \right] + \right.$$
$$+ \cos 2\pi \left(lz + \frac{l}{3} \right) \left[\cos 2\pi (kx + iy) + \cos 2\pi (ix + ky) \right] +$$
$$\left. + \cos 2\pi \left(lz - \frac{l}{3} \right) \left[\cos 2\pi (ix + hy) + \cos 2\pi (kx + hy) \right] \right\},$$

$$B = 2 \left\{ \sin 2\pi lz \left[\cos 2\pi (hx + ky) - \cos 2\pi (hx + iy) \right] + \right.$$
$$+ \sin 2\pi \left(lz + \frac{l}{3} \right) \left[\cos 2\pi (kx + iy) - \cos 2\pi (ix + ky) \right] +$$
$$\left. + \sin 2\pi \left(lz - \frac{l}{3} \right) \left[\cos 2\pi (ix + hy) - \cos 2\pi (kx + hy) \right] \right\}.$$

H 12, H 20. $\dfrac{P6}{mmm} - D_{6h}^1$

$$A = 8 \cos 2\pi lz \left\{ \cos \pi i \, (x + y) \cos \pi (h - k) \, (x - y) + \right.$$
$$\left. + \cos \pi h \, (x + y) \cos \pi (k - i) \, (x - y) + \cos \pi k \, (x + y) \cos \pi (i - h) \, (x - y) \right\},$$
$$B = 0.$$

H 17. $P32 - D_3^7$

$$A = 2 \left\{ \cos \pi (h - k) \, (x - y) \cos \pi \left[(h + k) \, (x + y) + 2lz \right] + \right.$$
$$+ \cos \pi (k - l) \, (x - y) \cos \pi \left[(k + l) \, (x + y) + 2hz \right] +$$
$$\left. + \cos \pi (l - h) \, (x - y) \cos \pi \left[(l + h) \, (x + y) + 2kz \right] \right\},$$
$$B = 2 \left\{ \sin \pi (h - k) \, (x - y) \cos \pi \left[(h + k) \, (x + y) + 2lz \right] + \right.$$
$$+ \sin \pi (k - l) \, (x - y) \cos \pi \left[(k + l) \, (x + y) + 2hz \right] +$$
$$\left. + \sin \pi (l - h) \, (x - y) \cos \pi \left[(l + h) \, (x + y) + 2kz \right] \right\}.$$

4-16. Structure Factors for Some Hexagonal Structure Types

The table gives F and p for Debye photographs for the hexagonal close-packed lattice of A3 type (magnesium type), the values being for hk$l$ from 011 to 333, f being the atomic factor for the element [109].

hkl	F^2	p	hkl	F^2	p	hkl	F^2	p	hkl	F^2	p
00.1	0	—	00.6	$4f^2$	2	02.7	$3f^2$	8	05.1	$3f^2$	8
01.0	f^2	4	03.2	$4f^2$	8	00.8	$4f^2$	8	14.4	$4f^2$	16
00.2	$4f^2$	2	02.5	$3f^2$	8	04.3	$3f^2$	8	22.7	0	—
01.1	$3f^2$	8	01.6	f^2	8	01.8	f^2	8	11.9	0	—
01.2	f^2	8	12.4	f^2	16	22.5	0	—	23.5	$3f^2$	16
00.3	0	—	03.3	0	—	23.0	f^2	4	04.6	f^2	8
11.0	$4f^2$	4	22.0	f^2	4	03.6	$4f^2$	8	05.2	f^2	8
11.1	0	—	22.1	$3f^2$	8	23.1	$3f^2$	16	02.9	$3f^2$	8
01.3	$3f^2$	8	13.0	f^2	4	13.5	$3f^2$	16	13.7	$3f^2$	16
02.0	f^2	4	11.6	$4f^2$	8	23.2	f^2	16	03.8	$4f^2$	8
11.2	$4f^2$	8	13.1	$3f^2$	16	04.4	f^2	8	33.0	$4f^2$	4
02.1	$3f^2$	8	22.2	$4f^2$	8	12.7	$3f^2$	16	05.3	$3f^2$	8
00.4	$4f^2$	2	03.4	$4f^2$	8	11.8	$4f^2$	8	33.1	0	—
20.2	f^2	8	02.5	f^2	8	14.0	f^2	8	14.5	0	—
10.4	f^2	8	00.7	0	—	14.1	0	—	33.2	$4f^2$	8
11.3	0	—	12.5	$3f^2$	16	02.8	f^2	8	24.0	f^2	8
20.3	$3f^2$	8	13.2	f^2	16	23.3	$3f^2$	16	00.10	$4f^2$	2
12.0	f^2	4	01.7	$3f^2$	8	22.6	$4f^2$	8	24.1	$3f^2$	16
00.5	0	—	22.3	0	—	14.2	$4f^2$	16	23.6	f^2	16
12.1	$3f^2$	16	13.3	$3f^2$	16	00.9	0	—	24.2	f^2	16
11.4	$4f^2$	8	03.5	0	—	03.7	0	—	01.10	f^2	8
01.5	$3f^2$	8	04.0	f^2	4	13.6	f^2	16	05.4	f^2	8
12.2	f^2	16	04.1	$3f^2$	8	01.9	$3f^2$	8	22.8	$4f^2$	8
02.4	f^2	8	22.4	$4f^2$	8	23.4	f^2	16	04.7	$3f^2$	8
05.0	$4f^2$	4	11.7	0	—	04.5	$3f^2$	8	12.9	$3f^2$	16
03.1	0	—	12.6	f^2	16	12.8	f^2	16	33.3	0	—
12.3	$3f^2$	16	04.2	f^2	8	14.3	0	—			
11.5	0	—	13.4	f^2	16	.05.0	f^2	4			

4-17. The Temperature Factor

This factor takes the form

$$e^{-2M} = e^{-2B \left(\frac{\sin \vartheta}{\lambda} \right)^2},$$

where $B = 8\pi^2 \overline{u^2}$ and $\overline{u^2}$ is the mean square displacement of the atoms from their ideal positions as a result of thermal motion.

The relation for B is

$$B = \frac{6h^2}{m_a k\Theta} \left[\frac{\Phi(x)}{x} + \frac{1}{4} \right] = B' \frac{1}{m_a \Theta}, \qquad (45)$$

where m_a is the atomic mass of the substance, h is Planck's constant, k is Boltzmann's constant, Θ is the characteristic temperature for the crystal (this can be put as $\Theta = h\nu_{max}/k$), $x = \Theta/T$, and $\Phi(x)$ is Debye's function, which is

$$\Phi(x) = \frac{1}{x} \int_0^x \frac{\xi \, d\xi}{(e^\xi - 1)}.$$

Methods of calculating the factor from the dynamic theory of interference in an ideal crystal [431] give the factor as e^{-M}.

4-17a. Debye's Function

This is used in calculations on dynamic displacements from the lattice nodes and on temperature factors.

The table gives $\Phi(x)$ for x from 0 to 20 [98].

$\frac{\Theta}{T}$	$\Phi\left(\frac{\Theta}{T}\right)$	$\frac{\Theta}{T}$	$\Phi\left(\frac{\Theta}{T}\right)$	$\frac{\Theta}{T}$	$\Phi\left(\frac{\Theta}{T}\right)$	$\frac{\Theta}{T}$	$\Phi\left(\frac{\Theta}{T}\right)$
0	1,000	1,2	0,740	3	0,483	9	0,183
0,2	0,951	1,4	0,704	4	0,388	10	0,164
0,4	0,904	1,6	0,669	5	0,321	12	0,137
0,6	0,860	1,8	0,637	6	0,271	14	0,114
0,8	0,818	2,0	0,607	7	0,234	16	0,103
1,0	0,778	2,5	0,540	8	0,205	20	0,0022

4-17b. Values of the Temperature Factor for Various B and ϑ

The table gives $e^{-M} = e^{-B\,[(\sin\theta)/\lambda]^2}$ for $B\cdot10^{16}$ from 0.0 to 10.0 and for $[(\sin\vartheta)/\lambda]\cdot10^{-8}$ from 0.0 to 1.2 [2].

$B\cdot10^{16}$ \ $\frac{\sin\vartheta}{\lambda}\cdot10^{-8}$	0,0	0,1	0,2	0,3	0,4	0,5	0,6	0,7	0,8	0,9	1,0	1,1	1,2
0,0	1,000	1,000	1,000	1,000	1,000	1,000	1,000	1,000	1,000	1,000	1,000	1,000	1,000
0,1	.000	0,999	0,996	0,991	0,984	0,975	0,964	0,952	0,938	0,923	0,905	0,886	0,866
0,2	.000	.998	.992	.982	.968	.951	.931	.906	.880	.850	.819	.785	.750
0,3	.000	.997	.988	.973	.953	.928	.898	.863	.826	.784	.741	.695	.649
0,4	.000	.996	.984	.964	.938	.905	.866	.821	.774	.724	.670	.616	.562
0,5	.000	.995	.980	.955	.924	.882	.834	.782	.726	.667	.607	.548	.487
0,6	.000	.994	.976	.947	.909	.860	.804	.745	.681	.615	.549	.484	.421
0,7	.000	.993	.972	.939	.894	.839	.776	.710	.639	.567	.497	.429	.365
0,8	.000	.992	.968	.931	.880	.818	.750	.676	.599	.523	.449	.380	.314
0,9	.000	.991	.964	.923	.866	.798	.724	.644	.561	.482	.406	.336	.273
1,0	.000	.990	.960	.915	.852	.779	.698	.613	.527	.445	.368	.298	.236
1,1	.000	.989	.957	.907	.839	.759	.672	.584	.494	.410	.333	.264	.205
1,2	.000	.988	.953	.898	.826	.740	.649	.556	.464	.378	.301	.234	.178
1,3	.000	.987	.950	.890	.813	.722	.626	.529	.435	.349	.273	.207	.154
1,4	.000	.986	.946	.882	.800	.704	.604	.503	.408	.322	.247	.184	.133
1,5	.000	.985	.942	.874	.787	.687	.582	.479	.383	.297	.223	.167	.116
1,6	.000	.984	.938	.866	.774	.670	.562	.458	.359	.274	.202	.144	.100
1,7	.000	.983	.935	.858	.762	.654	.543	.436	.337	.252	.183	.128	.086
1,8	.000	.982	.931	.850	.750	.638	.523	.414	.316	.233	.165	.113	.075
1,9	.000	.981	.927	.842	.739	.622	.505	.394	.296	.215	.149	.100	.065
2,0	.000	.980	.924	.834	.727	.607	.487	.375	.278	.198	.135	.089	.056
2,2	.000	.978	.916	.820	.719	.577	.452	.340	.245	.169	.110		
2,4	.000	.976	.908	.806	.698	.549	.421	.327	.215	.144	.090		
2,6	.000	.974	.901	.791	.677	.522	.391	.283	.190	.122	.074		
2,8	.000	.972	.894	.777	.657	.497	.361	.254	.167	.108	.060		
3,0	.000	.970	.887	.763	.638	.472	.348	.230	.147	.089	.049		
3,5	.000	.966	.869	.730	.592	.419	.284	.180	.106	.059	.036		
4,0	.000	.961	.852	.698	.549	.368	.237	.141	.078	.039	.018		
4,5	.000	.956	.835	.667	.487	.325	.198	.111	.056				
5,0	.000	.951	.819	.638	.449	.287	.165	.106	.041				
5,5	.000	.946	.793	.610	.415	.253	.138	.068	.027				
6,0	.000	.942	.786	.583	.383	.223	.115	.054	.011				
7,0	.000	.932	.763	.533	.326	.174	.080						
8,0	.000	.923	.726	.487	.278	.135	.056						
9,0	.000	.914	.698	.445	.237	.105	.039						
10,0	.000	.905	.670	.407	.202	.082	.027						

4-17c. The Functions e^{-x} and $\lg e^{-x}$

The table gives e^{-x} and $\lg e^{-x}$ for x from 0.00 to 7.70 [2].

The function e^{-x} is used in expressions for temperature factors, statistical displacements, and certain other purposes in structure analysis.

x	e^{-x}	$10+\lg e^{-x}$	x	e^{-x}	$10+\lg e^{-x}$	x	e^{-x}	$10+\lg e^{-x}$
0,00	1,0000	10,0000	**0,50**	0,6065	9,7829	**1,00**	0,3679	9,5657
0,01	0,9900	9,9957	0,51	.6005	.7785	1,01	.3642	.5614
0,02	.9802	.9913	0,52	.5945	.7742	1,02	.3606	.5570
0,03	.9704	.9870	0,53	.5886	.7698	1,C3	.3570	.5527
0,04	.9608	.9826	0,54	.5827	.7655	1,04	.3535	.5483
0,05	.9512	.9783	0,55	.5769	.7611	1,05	.3499	.5440
0,06	.9418	.9739	0,56	.5712	.7568	1,06	.3465	.5396
0,07	.9324	.9696	0,57	.5655	.7525	1,07	.3430	.5353
0,08	.9231	.9653	0,58	.5599	.7481	1,08	.3396	.5310
0,09	.9139	.9609	0,59	.5543	.7438	1,09	.3362	.5266
0,10	.9048	.9566	**0,60**	.5488	.7394	**1,10**	.3329	.5223
0,11	.8958	.9522	0,61	.5434	.7351	1,11	.3296	.5179
0,12	.8869	.9479	0,62	.5379	.7307	1,12	.3263	.5136
0,13	.8781	.9435	0,63	.5326	.7264	1,13	.3230	.5092
0,14	.8694	.9392	0,64	.5273	.7221	1,14	.3198	.5049
0,15	.8607	.9349	0,65	.5220	.7177	1,15	.3166	.5006
0,16	.8521	.9305	0,66	.5169	.7134	1,16	.3135	.4962
0,17	.8437	.9262	0,67	.5117	.7090	1,17	.3104	.4919
0,18	.8353	.9218	0,68	.5066	.7047	1,18	.3073	.4875
0,19	.8270	.9175	0.69	.5016	.7003	1,19	.3042	.4832
0,20	.8187	.9131	**0,70**	.4966	.6960	**1,20**	.3012	.4788
0,21	.8106	.9088	0,71	.4916	.6917	1,21	.2982	.4745
0,22	.8025	.9045	0,72	.4868	.6873	1,22	.2952	.4702
0,23	.7945	.9001	0,73	.4819	.6830	1,23	.2923	.4658
0,24	.7866	.8958	0,74	.4771	.6786	1,24	.2894	.4615
0,25	.7788	.8914	0,75	.4724	.6743	1,25	.2865	.4571
0,26	.7711	.8871	0,76	.4677	.6699	1,26	.2837	.4528
0,27	.7634	.8827	0,77	.4630	.6656	1,27	.2808	.4484
0,28	.7558	.8784	0,78	.4584	.6613	1,28	.2780	.4441
0,29	.7483	.8741	0,79	.4538	.6569	1,29	.2753	.4398
0,30	.7408	.8697	**0,80**	.4493	.6526	**1,30**	.2725	.4354
0,31	.7334	.8654	0,81	.4449	.6482	1,31	.2698	.4311
0,32	.7261	.8610	0,82	.4404	.6439	1,32	.2671	.4267
0,33	.7189	.8567	0,83	.4360	.6395	1,33	.2645	.4224
0,34	.7118	.8523	0,84	.4317	.6352	1,34	.2618	.4180
0,35	.7047	.8480	0,85	.4274	.6308	1,35	.2592	.4137
0,36	.6977	.8437	0,86	.4232	.6265	1,36	.2567	.4094
0,37	.6907	.8393	0,87	.4190	.6222	1,37	.2541	.4050
0,38	.6839	.8350	0,88	.4148	.6178	1,38	.2516	.4007
0,39	.6771	.8306	0,89	.4107	.6135	1,39	.2491	.3963
0,40	.6703	.8263	**0,90**	.4066	.6091	**1,40**	.2466	.3920
0,41	.6637	.8219	0,91	.4025	.6048	1,41	.2441	.3876
0,42	.6570	.8176	0,92	.3985	.6004	1,42	.2417	.3833
0,43	.6505	.8133	0,93	.3946	.5961	1,43	.2393	.3790
0,44	.6440	.8089	0,94	.3906	.5918	1,44	.2369	.3746
0,45	.6376	.8046	0,95	.3867	.5874	1,45	.2346	.3703
0,46	.6313	.8002	0,96	.3829	.5831	1,46	.2322	.3659
0,47	.6250	.7959	0,97	.3791	.5787	1,47	.2299	.3616
0,48	.6188	.7915	0,98	.3753	.5744	1,48	.2276	.3572
0,49	.6126	.7872	0,99	.3716	.5700	1,49	.2254	.3529

x	e^{-x}	$10+\lg e^{-x}$
1,50	0,2231	9,3486
1,51	.2209	.3442
1,52	.2187	.3399
1,53	.2165	.3355
1,54	.2144	.3312
1,55	.2122	.3268
1,56	.2101	.3225
1,57	.2080	.3182
1,58	.2060	.3138
1,59	.2039	.3095
1,60	.2019	.3051
1,61	.1999	.3008
1,62	.1979	.2964
1,63	.1959	.2921
1,64	.1940	.2878
1,65	.1920	.2834
1,66	.1901	.2791
1,67	.1882	.2747
1,68	.1864	.2704
1,69	.1845	.2660
1,70	.1827	.2617
1,71	.1809	.2574
1,72	.1791	.2530
1,73	.1773	.2487
1,74	.1755	.2443
1,75	.1738	.2400
1,76	.1720	.2356
1,77	.1703	.2313
1,78	.1686	.2270
1,79	.1670	.2226
1,80	.1653	.2183
1,81	.1637	.2139
1,82	.1620	.2096
1,83	.1604	.2052
1,84	.1588	.2009
1,85	.1572	.1966
1,86	.1557	.1922
1,87	.1541	.1879
1,88	.1526	.1835
1,89	.1511	.1792
1,90	.1496	.1748
1,91	.1481	.1705
1,92	.1466	.1662
1,93	.1451	.1618
1,94	.1437	.1575
1,95	.1423	.1531
1,96	.1409	.1488
1,97	.1395	.1444
1,98	.1381	.1401
1,99	.1367	.1358

x	e^{-x}	$10+\lg e^{-x}$
2,00	0,1353	9,1314
2,02	.1327	.1227
2,04	.1300	.1140
2,06	.1275	.1054
2,08	.1249	.0967
2,10	.1225	.0880
2,12	.1200	.0793
2,14	.1177	.0706
2,16	.1153	.0619
2,18	.1130	.0532
2,20	.1108	.0446
2,22	.1086	.0359
2,24	.1065	.0272
2,26	.1044	.0185
2,28	.1023	.0098
2,30	.1003	.0011
2,32	.0983	8,9924
2,34	.0963	.9838
2,36	.0944	.9751
2,38	.0926	.9664
2,40	.0907	.9577
2,42	.0889	.9490
2,44	.0872	.9403
2,46	.0854	.9316
2,48	.0837	.9229
2,50	.0821	.9143
2,52	.0805	.9056
2,54	.0789	.8969
2,56	.0773	.8882
2,58	.0758	.8795
2,60	.0743	.8708
2,62	.0728	.8621
2,64	.0714	.8535
2,66	.0699	.8448
2,68	.0686	.8361
2,70	.0672	.8274
2,72	.0659	.8187
2,74	.0646	.8100
2,76	.0633	.8013
2,78	.0620	.7927
2,80	.0608	.7840
2,82	.0596	.7753
2,84	.0584	.7666
2,86	.0573	.7579
2,88	.0561	.7492
2,90	.0550	.7405
2,92	.0539	.7319
2,94	.0529	.7232
2,96	.0518	.7145
2,98	.0508	.7058

x	e^{-x}	$10+\lg e^{-x}$
3,00	0,0498	8,6971
3,05	.0474	.6754
3,10	.0450	.6537
3,15	.0429	.6320
3,20	.0408	.6103
3,25	.0388	.5885
3,30	.0369	.5668
3,35	.0351	.5451
3,40	.0334	.5234
3,45	.0317	.5017
3,50	.0302	.4800
3,55	.0287	.4583
3,60	.0273	.4365
3,65	.0260	.4148
3,70	.0247	.3931
3,75	.0235	.3714
3,80	.0224	.3497
3,85	.0213	.3280
3,90	.0202	.3063
3,95	.0193	.2845
4,00	.0183	.2628
4,05	.0174	.2411
4,10	.0166	.2194
4,15	.0158	.1977
4,20	.0150	.1760
4,25	.0143	.1542
4,30	.0136	.1325
4,35	.0129	.1108
4,40	.0123	.0891
4,45	.0117	.0674
4,50	.0111	.0457
4,60	.0101	.0022
4,70	.0091	7,9588
4,80	.0082	.9154
4,90	.0074	.8720
5,00	.0067	.8285
5,25	.0052	.7200
5,50	.0041	.6114
6,00	.0025	.3942
7,00	.0009	6,9599
7,70	.0005	.6559

n	$n \lg e$
1	0,4342945
2	0,8685890
3	1,3028834
4	1,7371779
5	2,1714724
6	2,6057669
7	3,0400614
8	3,4743559
9	3,9086503

4-17d. Values of B' in the Expression for the Temperature Factor

The table gives $B' = \dfrac{6h^2}{k}\left[\dfrac{\Phi(x)}{x} + \dfrac{1}{4}\right]$, with $x = \Theta/T$, for x from 0 to 20 [2].

x	$B' \cdot 10^{36}$	x	$B' \cdot 10^{36}$	x	$B' \cdot 10^{36}$	x	$B' \cdot 10^{36}$
0	∞	0,7	2,72	1,8	1,13	7,0	0,533
0,1	18,79	0,8	·2,39	2,0	1,04	8,0	0,518
0,2	8,98	0,9	2,13	2,5	0,876	9,0	0,508
0,3	6,29	1,0	1,93	3,0	0,772	10,0	0,501
0,4	4,72	1,2	1,62	4,0	0,652	12,0	0,491
0,5	3,78	1,4	1,41	5,0	0,590	14,0	0,485
0,6	3,16	1,6	1,26	6,0	0,555	16,0	0,482
						20,0	0,477

4-18. Multiplicity Factors for the Various Crystal Systems

The p factor for any reflection on a powder pattern is numerically equal to the number of sets of planes belonging to the given form.

The magnitude depends on the Laue class and on the combinations of the indices for the reflecting planes.

The table gives p for various forms and systems.

E x a m p l e . The material is cubic and has space group O_h^5 (face-centered cubic) [102]; the table shows that the factor for the (111) line, which corresponds to a combination of (hhh) type, is 8; for (200), which is of ($00l$) type, $p = 6$.

System	Symbols for sets of planes						
Cubic O_h, O, T_d, T_h, T	(hkl) 48 24	(hhl) 24 24	$(0kl)$ 24 12	$(0kl)$ 12 12	(hhh) 8 8	$(00l)$ 6 6	
Hexagonal D_{6h}, D_6, C_{6v}, D_{3h}, C_{6h}, C_6, C_{3h}	$(hkil)$ 24 12	$(hh\overline{2h}l)$ 12 12	$(0k\overline{k}l)$ 12 12	$(hki0)$ 12 6	$(hki0)$ 6 6	$(0k\overline{k}0)$ 6 6	$(000l)$ 6 6
Trigonal D_{3d}, D_3, C_{3v}, C_{3i}, C_3	$(hkil)$ 12 6	$(hh\overline{2h}l)$ 12 6	$(0\overline{k}kl)$ 6 6	$(hki0)$ 12 6	$(hh\overline{2h}0)$ 6 6	$(0k\overline{k}0)$ 6 6	$(000l)$ 2 2
Tetragonal D_{4h}, D_4, C_{4v}, V_d, C_{4h}, C_4, S_4	(hkl) 16 8	(hhl) 8 8	$(0kl)$ 8 8	$(hk0)$ 8 4	$(hh0)$ 4 4	$(0k0)$ 4 4	$(00l)$ 2 2 .
Orthorhombic V_h, V, C_{2v}	(hkl) 8	$(0kl)$ 4	$(h0l)$ 4	$(hk0)$ 4	$(h00)$· 2	$(0k0)$ 2	$(00l)$ 2

ABSORPTION FACTORS

4-19. Absorption Factors for Cylindrical Specimens

The absorption factor for strong absorption ($\mu r > 5.0$) can be found from

$$A(\vartheta) = \sum_{n=1}^{n=\infty} \frac{a_n}{(\mu r)^n} \tag{46}$$

or from the approximation

$$A(\vartheta) = \frac{a_1}{\mu r} + \frac{a_2}{(\mu r)^2} + \frac{a_3}{(\mu r)^3} + \frac{a_5}{(\mu r)^5} + \frac{a_7}{(\mu r)^7}.$$

The table gives the a_n in (46) as a function of ϑ [109].

$\vartheta°$	a_1	a_2	a_3	a_5	a_7
0	—	—	0,318	0,48	2,67
22,5	0,031	0,080	0,240	—	—
45	0,114	0,180	—	—	—
67,5	0,232	0,100	—	—	—
90	0,311	0,014	—	—	—

See [385-388] for graphical methods.

Programs for computer calculation of the absorption factor have been compiled for certain methods [432].

4-19a. Absolute Values for Homogeneous Specimens

The table gives the absolute $A(\vartheta)$ for homogeneous cylindrical specimens for μr from 0 to 5.0, r being the radius and μ the absorption coefficient [102].

$\sin^2\vartheta$	0	0,0302	0,1170	0,2500	0,4132	0,5868	0,7500	0,8830	0,9699	1,00
ϑ / μr	0°	10°	20°	30°	40°	50°	60°	70°	80°	90°
0,0	1,0000	1,0000	1,0000	1,0000	1,0000	1,0000	1,0000	1,0000	1,0000	1,0000
0,1	0,847	0,8475	0,9481	0,8486	0,8493	0,8499	0,850	0,8502	0,8505	0,851
0,2	.712	.7135	.7150	.7165	.7181	.7200	.7222	.7245	.7270	.729
0,3	.600	.6022	.6050	.6082	.6120	.6170	.6221	.6252	.6310	.635
0,4	.510	.5135	.5162	.5200	.5245	.5308	.5390	.5460	.5510	.556
0,5	.435	.4362	.4401	.4465	.4540	.4626	.4720	.4800	.4875	.490
0,6	.639	.3709	.3759	.3832	.3910	.4020	.4145	.4255	.4330	.436
0,7	.314	.3160	.3220	.3312	.3420	.3555	.3690	.3801	.3899	.393
0,8	.268	.2701	.2762	.2862	.2985	.3130	.3278	.3410	.3520	.356
0,9	.230	.2320	.2385	.2500	.2640	.2792	.2945	.3088	.3198	324

sin²θ	0	0,0302	0,1170	0,2500	0,4132	0,5868	0,7500	0,8830	0,9699	1,00
θ μr	0°	10°	20°	30°	40°	50°	60°	70°	80°	90°
1,0	0,1977	0,2002	0,2075	0,2190	0,2338	0,2507	0,2672	0,2810	0,2910	0,295
1,1	.1698	.1722	.1800	.1920	.2070	.2250	.2434	.2582	.2685	.2715
1,2	.1459	.1487	.1571	.1702	.1865	.2052	.2232	.2381	.2473	.2510
1,3	.1256	.1285	.1375	.1512	.1680	.1870	.2050	.2202	.2303	.2335
1,4	.1084	.1115	.1203	.1342	.1518	.1710	.1892	.2044	.2148	.2180
1,5	.0938	.0967	.1060	.1200	.1374	.1569	.1749	.1900	.2012	.2050
1,6	.0811	.0841	.0940	.1085	.1260	.1452	.1632	.1808	.1900	.1932
1,7	.0710	.0744	.0839	.0980	.1153	.1345	.1525	.1679	.1783	.1824
1,8	.0615	.0695	.0747	.0888	.1063	.1250	.1426	.1580	.1692	.1730
1,9	.0537	.0571	.0670	.0812	.0983	.1171	.1346	.1496	.1605	.1644
2,0	.0471	.0502	.0600	.0741	.0914	.1099	.1271	.1420	.1528	.1567
2,1	.0416	.0450	.0545	.0683	.0856	.1039	.1205	.1348	.1455	.1493
2,2	.0367	.0402	.0500	.0636	.0800	.0961	.1146	.1277	.1388	.1426
2,3	.0324	.0356	.0453	.0588	.0748	.0901	.1083	.1225	.1330	.1365
2,4	.0287	.0317	.0412	.0548	.0706	.0859	.1037	.1169	.1271	.1309
2,5	.0255	.0288	.0380	.0510	.0665	.0812	.0987	.1120	.1220	.1256
2,6	.0227	.0258	.0349	.0478	.0631	.0777	.0947	.1073	.1173	.1211
2,7	.0202	.0233	.0322	.0447	.0594	.0737	.0903	.1032	.1131	.1167
2,8	.01803	.0212	.0300	.0420	.0563	.0702	.0870	.0998	.1095	.1127
2,9	.01607	.0190	.0277	.0395	.0534	.0671	.0833	.0960	.1056	.1089
3,0	.01436	.0173	.0262	.0375	.0510	.0640	.0797	.0914	.0993	.1054
3,1	.01288	.0158	.0244	.0356	.0490	.0627	.0766	.0890	.0984	.1021
3,2	.01159	.0142	.0228	.0338	.0468	.0604	.0740	.0862	.0956	.0990
3,3	.01049	.0130	.0215	.0321	.0447	.0582	.0715	.0836	.0928	.0961
3,4	.00955	.0121	.0205	.0306	.0430	.0561	.0691	.0810	.0900	.0983
3,5	.00871	.0111	.0192	.0293	.0413	.0541	.0670	.0786	.0874	.0906
3,6	.00796	.0106	.0179	.0281	.0399	.0521	.0649	.0762	.0850	.0881
3,7	.00729	.00988	.0171	.0270	.0384	.0504	.0628	.0742	.0828	.0858
3,8	.00670	.00928	.0162	.0260	.0370	.0489	.0611	.0722	.0806	.0836
3,9	.00617	.00867	.0155	.0250	.0358	.0473	.0595	.0701	.0786	.0815
4,0	.00568	.00810	.0147	.0239	.0347	.0458	.0576	.0682	.0764	.0794
4,1	.00525	.00755	.0140	.0230	.0335	.0445	.0559	.0663	.0745	.0774
4,2	.00488	.00715	.0134	.0222	.0324	.0432	.0544	.0645	.0726	.0755
4,3	.00453	.00678	.0128	.0215	.0315	.0420	.0528	.0630	.0710	.0738
4,4	.00420	.00641	.0124	.0207	.0305	.0408	.0517	.0615	.0692	.0721
4,5	.00391	.00604	.0119	.0201	.0297	.0398	.0502	.0600	.0677	.0705
4,6	.00364	.00569	.0114	.0195	.0289	.0388	.0492	.0587	.0662	.0689
4,7	.00340	.00539	.0110	.0188	.0281	.0378	.0479	.0574	.0650	.0675
4,8	.00316	.00518	.0106	.0183	.0274	.0370	.0467	.0560	.0636	.0661
4,9	.00294	.00492	.0103	.0178	.0267	.0361	.0457	.0550	.0622	.0647
5,0	.00275	.00468	.0100	.0173	.0260	.0352	.0448	.0540	.0610	.0635

4-19b. Relative Values for Homogeneous Specimens

The table gives the relative $A(\vartheta)$ for homogeneous cylindrical specimens for μr from 5.0 to ∞, r being the radius. The $A(\vartheta)$ for $\vartheta = 90°$ is taken as 100 [102].

$\sin^2 \vartheta$	0	0,0302	0,1170	0,2500	0,4132	0,5868	0,7500	0,8830	0,9699
μr ϑ	0°	10°	20°	30°	40°	50°	60°	70°	80°
5,0	4,34	7,28	15,83	27,33	40,85	55,54	70,8	84,8	94,7
5,5	3,50	6,44	14,90	26,35	39,90	54,76	70,0	84,3	94,5
6,0	2,91	5,81	14,19	25,57	39,06	53,89	69,3	83,9	94,4
6,5	2,44	5,32	13,59	24,92	38,40	53,27	68,9	83,6	94,3
7,0	2,12	4,96	13,12	24,35	37,82	52,74	68,4	83,4	94,2
7,5	1,83	4,67	12,70	23,88	37,47	52,25	68,0	83,1	94,1
8,0	1,61	4,39	12,36	23,47	36,85	51,83	67,7	82,9	94,0
8,5	1,44	4,17	12,05	23,11	36,45	51,48	67,4	82,7	93,95
9,0	1,26	3,98	11,78	22,76	36,10	51,14	67,1	82,5	93,9
9,5	1,14	3,82	11,57	22,50	35,80	50,88	66,9	82,3	93,85
10	1,02	3,69	11,37	22,24	35,54	50,60	66,7	82,2	93,8
11	0,84	3,50	11,06	21,84	35,09	50,16	66,3	82,0	93,7
12	0,69	3,32	10,78	21,50	34,71	49,75	66,0	81,8	93,6
13	0,59	3,15.	10,53	21,18	34,35	49,40	65,7	81,6	93,5
14	0,50	3,03	10,30	20,92	34,09	49,17	65,5	81,4	93,45
15	0,44	2,93	10,08	20,72	33,85	48,94	65,3	81,3	93,4
16	0,39	2,86	9,90	20,52	33,66	48,73	65,1	81,2	93,4
17	0,35	2,80	9,73	20,37	33,47	48,57	65,0	81,1	93,35
18	0,31	2,75	9,57	20,22	33,30	48,42	64,9	81,0	93,35
19	0,28	2,72	9,45	20,06	33,16	48,28	64,8	80,95	93,3
20	0,25	2,70	9,35	19,93	33,01	48,16	64,7	80,9	93,3
25	0,16	2,60	9,17	19,55	32,58	47,86	64,3	80,7	93,3
30	0,11	2,51	9,03	19,26	32,20	47,55	64,0	80,5	93,3
35	0,08	2,42	8,90	19,04	31,94	47,29	63,8	80,3	93,2
40	0,06	2,35	8,80	18,78	31,75	47,07	63,6	80,2	93,2
45	0,05	2,30	8,72	18,65	31,61	46,90	63,5	80,1	93,2
50	0,04	2,26	8,64	18,54	31,50	46,75	63,4	79,95	93,1
60	0,03	2,22	8,56	18,42	31,37	46,60	63,3	79,9	93,1
70	0,02	2,20	8,49	18,33	31,27	46,46	63,2	79,9	93,1
80	0,015	2,18	8,43	18,27	31,20	46,35	63,1	79,8	93,0
90	0,013	2,17	8,37	18,23	31,15	46,27	63,05	79,8	93,3
100	0,01	2,16	8,33	18,19	31,10	46,20	63,0	79,8	92,9
∞	0,00	2,07	8,03	17,77	30,55	45,80	62,6	79,5	92,8

4-19c. Absolute Values for Powder Specimens Attached to Fibers

The specimen is commonly a crystalline powder glued to a wire or glass fiber.

The table gives the absolute values for various thicknesses of specimen and fiber [102].

Here μ_g is the linear attenuation coefficient for the fiber, μ is the same for the specimen, r_g is the radius of the fiber, and r is the overall radius.

sin² θ	0	0,0302	0,1170	0,2500	0,4132	0,5868	0,7500	0,8830	0,9699	1
μ_g \ θ \ μ_r	0°	10°	20°	30°	40°	50°	60°	70°	80°	90°

$r_g/r = 0,4$

μ_g		0°	10°	20°	30°	40°	50°	60°	70°	80°	90°
200	0,1	0,559	0,560	0,566	0,570	0,573	0,601	0,650	0,688	0,712	0,721
	0,25	.434	.440	.450	.458	.467	.493	.538	.566	.590	.597
	0,50	.317	.320	.330	.342	.359	.386	.420	.447	.465	.473
	0,75	.224	.228	.239	.256	.276	.302	.329	.352	.367	.372
	1,00	.160	.167	.178	.196	.216	.245	.273	.296	.311	.316
	1,25	.119	.124	.135	.153	.174	.202	.226	.249	.262	.269
	1,50	.089	.092	.104	.121	.144	.170	.193	.216	.226	.232
	1,75	.063	.068	.082	.097	.125	.143	.169	.187	.199	.204
	2,00	.045	.0484	.063	.080	.108	.124	.146	.165	.176	.181
	2,25	.0360	.0399	.053	.068	.093	.108	.127	.147	.156	.162
	2,50	.0268	.0310	.0430	.058	.080	.096	.115	.132	.142	.146
	2,75	.0207	.0241	.0359	.051	.069	.087	.105	.121	.129	.132
	3,00	.0155	.0188	.0305	.044	.060	.078	.095	.110	.120	.123
	3,50	.0102	.0132	.0225	.0342	.048	.065	.083	.095	.107	.108
	4,00	.0067	.0095	.0175	.0281	.041	.056	.071	.084	.094	.094
	4,50	.0045	.0071	.0136	.0244	.0352	.049	.063	.074	.083	.085
	5,00	.0034	.0055	.0115	.0208	.0310	.044	.056	.066	.074	.077
	6,00	.0020	.0038	.0089	.0162	.0247	.0351	.046	.055	.061	.063
	7,00	.0012	.0030	.0069	.0126	.0201	.0292	.0388	.047	.052	.054
	8,00	.0008	.0021	.0058	.0104	.0172	.0255	.0341	.041	.046	.047
	10,00	.0004	.0012	.0043	.0087	.0133	.0200	.0270	.0329	.037	.038

$r_g/r = 0,4$

μ_g		0°	10°	20°	30°	40°	50°	60°	70°	80°	90°
30	0,1	.759	.760	.763	.765	.766	.771	.779	.786	.791	.793
	0,25	.596	.603	.606	.612	.615	.621	.632	.642	.644	.646
	0,50	.435	.439	.450	.458	.464	.475	.486	.496	.503	.505
	0,75	.308	.311	.323	.335	.346	.361	.376	.389	.397	.400
	1,00	.222	.225	.238	.250	.264	.281	.300	.315	.325	.329
	1,25	.156	.165	.173	.184	.205	.226	.245	.263	.273	.280
	1,50	.112	.118	.126	.141	.162	.184	.208	.218	.232	.236
	1,75	.082	.090	.098	.114	.133	.154	.174	.192	.203	.211
	2,00	.062	.066	.077	.092	.111	.132	.152	.169	.180	.184
	2,25	.049	.053	.064	.078	.094	.117	.134	.150	.159	.166
	2,50	.0356	.0397	.051	.065	.082	.101	.120	.136	.145	.149
	2,75	.0273	.0318	.042	.055	.074	.091	.108	.122	.132	.138
	3,00	.0213	.0243	.0348	.048	.064	.082	.098	.113	.122	.125
	3,50	.0133	.0151	.0248	.0369	.050	.066	.084	.105	.106	.108
	4,00	.0080	.0098	.0184	.0291	.042	.056	.071	.084	.094	.094
	4,50	.0054	.0073	.0144	.0250	.0364	.048	.063	.074	.083	.085
	5,00	.0036	.0054	.0119	.0213	.0317	.043	.056	.066	.074	.077
	6,00	.0019	.0035	.0089	.0162	.0247	.0351	.046	.055	.061	.063
	7,00	.0013	.0026	.0069	.0125	.0203	.0290	.0388	.047	.052	.054
	8,00	.0008	.0021	.0058	.0104	.0172	.0255	.0341	.041	.046	.047
	10,00	.0004	.0012	.0043	.0087	.0133	.0200	.0270	.0329	.037	.038

$r_g/r = 0,4$

μ_g		0°	10°	20°	30°	40°	50°	60°	70°	80°	90°
5	0,1	.850	.850	.849	.850	.850	.851	.853	.855	.856	.858
	0,25	.668	.669	.670	.676	.676	.678	.685	.692	.693	.693
	0,50	.490	.495	.500	.504	.505	.514	.522	.529	.534	.535
	0,75	.356	.359	.361	.367	.375	.386	.400	.412	.419	.423
	1,00	.251	.253	.257	.265	.282	.301	.316	.329	.338	.341
	1,25	.175	.181	.196	.205	.222	.243	.262	.271	.281	.286
	1,50	.128	.133	.148	.162	.175	.194	.214	.229	.239	.243
	1,75	.097	.102	.116	.129	.143	.163	.181	.196	.207	.211
	2,00	.070	.075	.089	.103	.117	.135	.155	.172	.182	.186
	2,25	.055	.059	.074	.087	.101	.113	.136	.151	.162	.166
	2,50	.0398	.044	.058	.071	.085	.103	.122	.137	.147	.150
	2,75	.0301	.0353	.047	.060	.073	.092	.111	.123	.131	.136

sin² ϑ	0	0,0302	0,1170	0,2500	0,4132	0,5868	0,7500	0,8830	0,9699	1
μ_g \ μ_r \ ϑ	0°	10°	20°	30°	40°	50°	60°	70°	80°	90°
5 — 3,00	0,0231	0,0269	0,0378	0,051	0,065	0,082	0,100	0,114	0,123	0,126
3,50	.0135	.0170	.0262	.0396	.050	.067	.084	.105	.106	.109
4,00	.0086	.0112	.0194	.0320	.044	.056	.071	.084	.094	.094
4,50	.0055	.0075	.0150	.0258	.0359	.049	.063	.074	.083	.085
5,00	.0036	.0055	.0124	.0225	.0315	.043	.056	.066	.074	.077
6,00	.0019	.0033	.0089	.0162	.0247	.0351	.046	.055	.061	.063
7,00	.0011	.0026	.0069	.0125	.0203	.0294	.0388	.047	.052	.054
8,00	.0008	.0021	.0058	.0104	.0172	.0255	.0341	.041	.046	.047
10,00	.0004	.0012	.0043	.0087	.0133	.0200	.0270	.0329	.037	.038

$r_g/r = 0,5$

μ_r	0°	10°	20°	30°	40°	50°	60°	70°	80°	90°
0,1	.486	.489	.496	.502	.509	.527	.556	.625	.667	.686
0,25	.384	.386	.396	.410	.422	.434	.475	.531	.564	.581
0,50	.287	.289	.302	.318	.336	.364	.397	.437	.462	.473
0,75	.210	.214	.227	.246	.269	.298	.330	.364	.385	.395
1,00	.152	.157	.171	.192	.216	.248	.275	.305	.323	.330
1,25	.113	.119	.133	.153	.183	.212	.237	.266	.285	.293
1,50	.086	.088	.101	.124	.151	.180	.204	.233	.252	.260
1,75	.065	.069	.081	.102	.128	.153	.176	.203	.219	.226
2,00	.048	.052	.066	.085	.108	.133	.156	.178	.192	.198
2,25	.0378	.043	.056	.073	.095	.118	.137	.159	.174	.180
200 — 2,50	.0295	.0328	.046	.063	.082	.105	.126	.144	.156	.161
2,75	.0230	.0267	.0394	.055	.072	.094	.113	.131	.144	.149
3,00	.0180	.0216	.0332	.048	.066	.086	.105	.122	.134	.138
3,50	.0114	.0152	.0250	.0376	.057	.070	.089	.105	.117	.121
4,00	.0076	.0101	.0194	.0311	.050	.060	.076	.091	.102	.105
4,50	.0052	.0084	.0159	.0264	.042	.053	.067	.081	.091	.096
5,00	.0037	.0068	.0135	.0229	.0361	.047	.060	.072	.081	.086
6,00	.0024	.0046	.0106	.0184	.0276	.0385	.049	.060	.068	.071
7,00	.0014	.0032	.0084	.0146	.0247	.0333	.041	.051	.058	.061
8,00	.0009	.0026	.0068	.0125	.0195	.0297	.0357	.045	.051	.053
10,00	.0004	.0018	.0046	.0095	.0150	.0217	.0281	.0357	.041	.042

$r_g/r = 0,5$

μ_r	0°	10°	20°	30°	40°	50°	60°	70°	80°	90°
0,1	.726	.724	.725	.726	.727	.731	.737	.755	.766	.771
0,25	.586	.586	.586	.587	594	.600	.612	.627	.637	.642
0,50	.438	.439	.442	.450	.460	.472	.487	.501	.510	.515
0,75	.320	.320	.329	.343	.355	.368	.380	.400	.412	.420
1,00	.238	.239	.249	.263	.279	.295	.312	.330	.342	.348
1,25	.175	.180	.192	.208	.226	.246	.263	.285	:297	.302
1,50	.133	.136	.146	.163	.184	.205	.223	.243	.255	.260
1,75	.100	.106	.117	.132	.150	.171	.189	.210	.223	.231
2,00	.076	.080	.090	.106	.126	.145	.163	.183	.196	.201
2,25	.057	.062	.073	.088	.106	.138	.144	.159	.174	.180
30 — 2,50	.044	.048	.060	.076	.094	.113	.129	.141	.157	.162
2,75	.0346	.0378	.049	.064	.083	.099	.115	.130	.143	.148
3,00	.0268	.0292	.041	.055	.071	.089	.106	.123	.135	.139
3,50	.0164	.0204	.0314	.041	.057	.074	.089	.106	.116	.121
4,00	.0108	.0135	.0240	.0330	.046	.062	.076	.094	.102	.105
4,50	.0069	.0094	.0189	.0281	.0397	.053	.067	.082	.092	.095
5,00	.0044	.0070	.0150	.0239	.0347	.047	.060	.072	.082	.086
6,00	.0023	.0045	.0104	.0183	.0275	.0384	.049	.060	.068	.071
7,00	.0014	.0033	.0083	.0146	.0231	.0328	.041	.051	.058	.061
8,00	.0009	.0026	.0068	.0125	.0195	.0297	.0357	.045	.051	.053
10,00	.0004	.C018	.0046	.0095	.0150	.0217	.0281	.0357	.041	.042

$\sin^2\vartheta$	0	0,0302	0,1170	0,2500	0,4132	0,5868	0,7500	0,8830	0,9699	1
μ_g / μr ϑ	0°	10°	20°	30°	40°	50°	60°	70°	80°	90°

$$r_g/r = 0,5$$

μ_g	μr	0°	10°	20°	30°	40°	50°	60°	70°	80°	90°
	0,1	0,845	0,845	0,843	0,844	0,845	0,847	0,850	0,853	0,856	0,858
	0,25	.682	.678	.682	.683	.684	.690	.695	.705	.707	.710
	0,50	.517	.514	.517	.521	.529	.537	.546	.552	.556	.560
	0,75	.385	.385	.386	.395	.406	.418	.430	.440	.448	.450
	1,00	.284	.283	.294	.304	.314	.327	.339	.352	.362	.369
	1,25	.210	.210	.222	.236	.247	.266	.283	.300	.309	.313
	1,50	.160	.161	.169	.183	.198	.217	.236	.251	.262	.264
	1,75	.122	.123	.133	.147	.162	.183	.197	.218	.229	.233
5	2,00	.091	.093	.103	.117	.135	.153	.169	.189	.201	.206
	2,25	.068	.072	.083	.098	.114	.148	.149	.166	.177	.183
	2,50	.052	.056	.067	.080	.097	.116	.133	.149	.159	.163
	2,75	.040	.045	.056	.068	.085	.103	.119	.136	.144	.150
	3,00	.0318	.0350	.045	.058	.073	.091	.109	.125	.136	.140
	3,50	.0196	.0237	.0325	.044	.058	.074	.092	.105	.117	.123
	4,00	.0126	.0151	.0236	.0343	.047	.062	.077	.092	.103	106
	4,50	.0075	.0101	.0185	.0285	.040	.053	.067	.081	.092	.096
	5,00	.0049	.0072	.0148	.0244	.0352	.047	.060	.073	.082	.086
	6,00	.0024	.0042	.0102	.0182	.0274	.0384	.049	.060	.068	.071
	7,00	.0014	.0031	.0078	.0145	.0224	.0328	.041	.051	.058	.061
	8,00	.0085	.0026	.0068	.0125	.0195	.0297	.057	.045	.051	.053
	10,00	.0043	.0018	.0046	.0095	.0150	.0217	.0281	.0357	.041	.042

4-20. Absorption Factors for Plane Specimens

The general expression for the absorption factor is

$$A = \frac{1}{V} \int_V e^{-\mu(l'+l'')}\, dV,$$

where V is the volume of the crystal, and l and l'' are the paths of the incident and reflected rays in the crystal.

For an infinite plate

$$A = \frac{b}{V\mu} \frac{\cos\delta}{\cos\psi + \cos\delta}\left[1 - e^{-\mu d\,\frac{\cos\psi + \cos\delta}{\cos\psi\,\cos\delta}}\right],$$

where b is the cross section of the incident beam, ψ is the angle between that beam and the normal to the specimen, δ is the same for the reflected beam, and d is the thickness.

For strongly absorbing materials,

$$A \approx \frac{b}{V\mu} \frac{\cos\delta}{\cos\psi + \cos\delta}.$$

The deviation from unity for the intensity factor VA/S depends on ψ and δ; the strength of the diffracted beam increases as δ increases and as ψ decreases. An increase in ψ causes less of the primary energy to reach unit area; an increase in δ produces an increased proportion of the diffracted radiation per unit area of the beam and so an increase in the blackening.

The focus condition is usually obeyed at all angles when polished sections and flat powder specimens are used with ionization equipment; the specimen is rotated at half the angular velocity of the counter, so

$$\psi = \delta$$

and

$$\frac{VA}{S_2} = \frac{1}{2\mu} \, ,$$

which means that the absorption factor is independent of angle. This factor can therefore be neglected in relative measurements with ionization equipment.

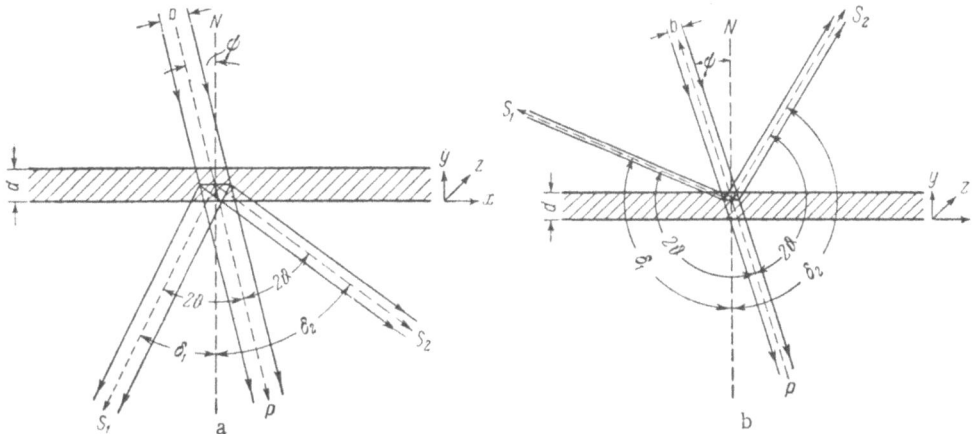

Fig. 110. Ray paths for flat specimens: (a) in transmission; (b) in reflection.

Figure 110 shows ray paths for flat specimens examined in transmission and reflection; δ_1 and δ_2 are the angles between the diffracted beams and the normal to the plane of the specimen.

The absorption factors for beams S_1 and S_2 in transmission (see Fig. 110a) are, respectively,

$$\mu A\,(\vartheta)_t^{S_1} = \frac{\cos(2\vartheta - \psi)}{\cos\psi - \cos(2\vartheta - \psi)} \, [e^{-\frac{\mu d}{\cos\psi}} - e^{-\frac{\mu d}{\cos(2\vartheta - \psi)}}],$$

$$\mu A\,(\vartheta)_t^{S_2} = \frac{\cos(2\vartheta + \psi)}{\cos\psi - \cos(2\vartheta - \psi)} \, [e^{-\frac{\mu d}{\cos\psi}} + e^{-\frac{\mu d}{\cos(2\vartheta - \psi)}}].$$

For reflection (Fig. 110b),

$$\mu A\,(\vartheta)_r^{S_1} = \frac{\cos(2\vartheta - \psi)}{\cos\psi - \cos(2\vartheta + \psi)} \, [e^{-\frac{\mu d}{\cos\psi}} e^{-\frac{\mu d}{\cos(2\vartheta - \psi)}} + 1],$$

$$\mu A\,(\vartheta)_r^{S_2} = \frac{\cos(2\vartheta + \psi)}{\cos\psi - \cos(2\vartheta + \psi)} \, [e^{-\frac{\mu d}{\cos\psi}} e^{-\frac{\mu d}{\cos(2\vartheta - \psi)}} + 1].$$

The table gives $\mu A(\vartheta)$ for ψ from 0 to 80° as a function of μd [102].

The first line (for ϑ) in this table corresponds to S_1 and the second (for ϑ') to S_2.

The list below gives the angles for which the $\mu A(\vartheta)$ are equal: $\mu A(\vartheta_1) = \mu A(\vartheta_2)$ and $\mu A(\vartheta_1') = \mu A(\vartheta_2')$.

$\psi°$	$\vartheta_1°$	$\vartheta_2°$	$\vartheta_1'°$	$\vartheta_2'°$
30	20	10	90	60
			85	65
40	30	10	80	60
	40	0	90	50
50	30	20	70	60
	40	10	80	50
	50	0	90	40
60	40	20	70	50
	50	10	80	40
	60	0	90	30
70	50	20	70	40
	60	10	80	30
	70	0	90	20
80	50	30	60	40
	60	20	70	30
	70	10	80	20
	80	0	90	10

It is assumed that $\mu d = \infty$ for the usual methods of examining a polished section with ionization equipment, with a cylindrical camera, or with a back-reflection camera.

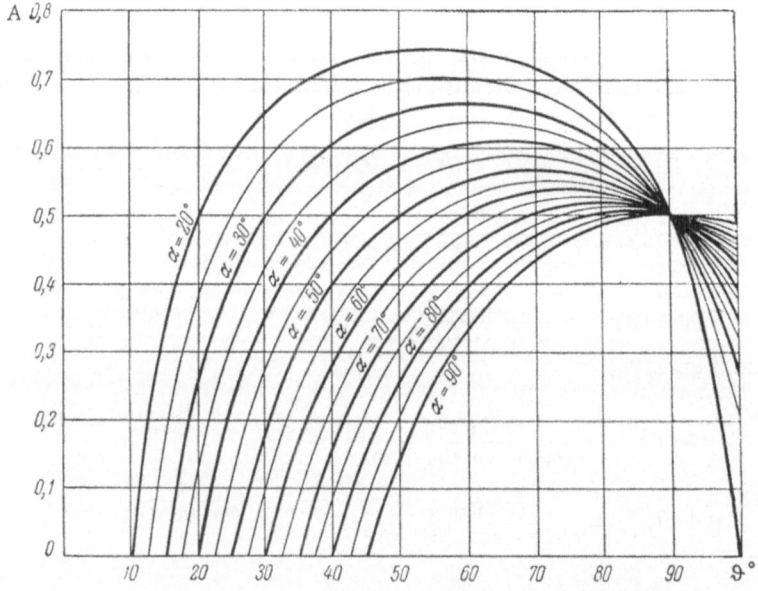

Fig. 111. Absorption factors for plane specimens.

Figure 111 gives approximate values for the absorption factor, α being the angle between the incident beam and the plane of the specimen [12].

$$\psi = 0°$$

ϑ	0°	10°	20°	25°	30°		40°	45°	50°	60°	70°	80°	90°
ϑ'	0°	10°	20°	25°	30°	35°	40°	45°	50°	60°	70°	80°	90°
μd													
0,1	0,090	0,082	0,076	0,074	0,073	0,061	0,057	0	0,068	0,090	0,092	0,093	0,094
0,2	.164	.161	.159	.155	.150	.132	.105	0	.096	.150	.158	.162	.165
0,3	.218	.217	.213	.196	.183	.153	.115	0	.126	.194	.219	.224	.226
0,4	.268	.265	.253	.238	.210	.175	.119	0	.138	.233	.270	.274	.275
0,5	.300	.296	.282	.256	.234	.185	.116	0	.142	.256	.300	.312	.315
0,6	.330	.324	.301	.275	.246	.194	.110	0	.145	.277	.325	.345	.350
0,7	.347	.339	.312	.279	.248	.190	.102	0	.146	.289	.345	.371	.378
0,8	.359	.351	.319	.283	.245	.185	.098	0	.147	.303	.365	.392	.400
0,9	.362	.355	.318	.279	.238	.173	.084	0	.147	.309	.376	.409	.415
1,0	.364	.356	.315	.276	.231	.162	.076	0	.148	.316	.390	.423	.431
1,1	.362	.355	.309	.263	.219	.145	.069	0	.148	.322	.397	.435	.443
1,2	.360	.344	.292	.250	.210	.128	.064	0	.149	.325	.405	.445	.455
1,3	.354	.342	.293	.244	.198	.123	.058	0	.149	.327	.410	.455	.462
1,4	.345	.331	.282	.238	.186	.118	.052	0	.149	.328	.415	.460	.470
1,5	.335	.318	.270	.225	.173	.111	.047	0	.149	.329	.419	.463	.475
1,6	.324	.308	.256	.213	.160	.104	.042	0	.149	.330	.422	.465	.480
1,7	.312	.293	.243	.199	.149	.095	.039	0	.149	.331	.425	.468	.483
1,8	.298	.281	.230	.185	.139	.086	.035	0	.149	.332	.426	.470	.485
1,9	.285	.266	.216	.173	.127	.080	.030	0	.149	.333	.428	.473	.489
2,0	.273	.255	.202	.160	.117	.075	.028	0	.150	.334	.430	.476	.492
2,1	.259	.240	.191	.151	.107	.068	.024	0	.150	.334	.430	.477	.492
2,2	.245	.226	.176	.142	.100	.061	.022	0	.150	.334	.431	.479	.494
2,3	.228	.216	.166	.131	.091	.058	.019	0	.150	.334	.431	.480	.495
2,4	.217	.202	.155	.120	.083	.053	.018	0	.150	.334	.431	.481	.496
2,5	.203	.190	.142	.111	.076	.047	.016	0	.150	.334	.431	.481	.496
2,6	.192	.178	.134	.102	.069	.041	.015	0	.150	.334	.432	.481	.496
2,7	.180	.167	.122	.094	.062	.038	.013	0	.150	.334	.432	.481	.496
2,8	.172	.157	.115	.087	.056	.035	.012	0	.150	.334	.432	.481	.497
2,9	.160	.146	.105	.081	.052	.032	.011	0	.150	.334	.432	.481	.497
3,0	.150	.136	.099	.075	.048	.030	.010	0	.150	.334	.433	.482	.497
3,2	.132	.118	.083	.063	.039	.025	.008	0	.150	.334	.433	.482	.498
3,4	.116	.103	.069	.050	.032	.020	.006	0	.150	.334	.433	.482	.498
3,6	.099	.086	.059	.041	.025	.015	.004	0	.150	.334	.433	.482	.498
3,8	.086	.074	.050	.036	.021	.012	.003	0	.150	.334	.433	.482	.498
4,0	.073	.065	.042	.028	.018	.008	.002	0	.150	.334	.433	.482	.498
4,5	.052	.048	.027	.020	.010	—	.001	0	.150	.334	.433	.482	.498
5,0	.040	.033	.018	.015	.006	—	.001	0	.150	.334	.433	.482	.498
∞	0	0	0	0	0	0	0	0	.151	.334	.433	.483	.500

$$\psi = 10°$$

ϑ	0°	5°	20°	30°	35°	40°	45°	50°	60°	70°	80°	90°	90°
ϑ'	0°	0°	10°	20°	25°	30°	35°	40°	50°	60°	70°	80°	90°
μd													
0,1	0,090	0,091	0,090	0,089	0,083	0,075	0,103	0	0,066	0,090	0,091	0,091	0,092
0,2	.164	.165	.163	.157	.150	.136	.108	0	.134	.160	.165	.165	.166
0,3	.225	.226	.224	.220	.205	.176	.114	0	.169	.211	.222	.230	.231
0,4	.270	.271	.264	.243	.225	.189	.120	0	.193	.255	.270	.278	.279
0,5	.302	.305	.293	.276	.236	.192	.115	0	.213	.286	.307	.320	.325
0,6	.328	.330	.320	.286	.250	.191	.110	0	.225	.312	.341	.353	.355
0,7	.346	.348	.333	.290	.251	.185	.101	0	.233	.327	.365	.376	.380
0,8	.358	.360	.340	.291	.250	.176	.092	0	.239	.340	.385	.400	.405
0,9	.364	.367	.344	.289	.241	.168	.085	0	.243	.354	.400	.416	.424
1,0	.365	.369	.346	.284	.234	.160	.078	0	.247	.364	.414	.433	.439
1,1	.363	.367	.342	.274	.222	.149	.071	0	.248	.370	.422	.445	.450
1,2	.359	.362	.334	.264	.210	.139	.064	0	.250	.375	.430	.454	.459
1,3	.350	.352	.322	.251	.198	.129	.058	0	.250	.378	.435	.460	.464

$\psi = 10°$

μd \ θ	0°	5°	20°	30°	35°	40°	45°	50°	60°	70°	80°	90°	90°
θ'	0°	0°	10°	20°	25°	30°	35°	40°	50°	60°	70°	80°	90°
1,4	0,338	0,340	0,311	0,238	0,185	0,116	0,052	0	0,251	0,381	0,440	0,465	0,470
1,5	.326	.329	.298	.225	.172	.107	.047	0	.251	.384	.445	.471	.476
1,6	.314	.320	.287	.213	.160	.097	.042	0	.252	.387	.450	.475	.480
1,7	.303	.307	.272	.199	.149	.090	.038	0	.252	.390	.454	.478	.485
1,8	.291	.300	.258	.185	.138	.083	.035	0	.252	.393	.457	.483	.489
1,9	.278	.284	.244	.171	.126	.074	.031	0	.252	.395	.458	.486	.493
2,0	.266	.273	.233	.159	.115	.066	.028	0	.253	.397	.460	.489	.496
2,1	.252	.258	.219	.147	.108	.061	.026	0	.253	.398	.462	.492	.499
2,2	.237	.244	.206	.138	.100	.058	.025	0	.253	.398	.463	.495	.503
2,3	.223	.230	.192	.127	.091	.050	.022	0	.253	.398	.463	.496	.503
2,4	.208	.216	.180	.117	.081	.046	.020	0	.253	.398	.464	.497	.504
2,5	.197	.204	.169	.108	.074	.042	.017	0	.253	.398	.464	.497	.504
2,6	.186	.193	.159	.100	.067	.039	.015	0	.253	.398	.465	.498	.505
2,7	.175	.183	.147	.091	.062	.034	.013	0	.253	.398	.465	.498	.505
2,8	.163	.170	.137	.083	.056	.029	.012	0	.253	.398	.465	.498	.505
2,9	.158	.160	.126	.076	.052	.026	.011	0	.253	.398	.465	.498	.505
3,0	.145	.150	.117	.069	.048	.024	.010	0	.253	.399	.466	.499	.506
3,2	.125	.130	.101	.059	.040	.019	.009	0	.253	.399	.466	.499	.506
3,4	.109	.113	.087	.050	.036	.015	.008	0	.253	.399	.466	.499	.506
3,6	.094	.098	.074	.041	.030	.012	.006	0	.253	.399	.466	.499	.506
3,8	.081	.085	.063	.035	.017	.009	.004	0	.253	.399	.466	.499	.506
4,0	.068	.072	.053	.029	.013	.006	—	0	.253	.399	.466	.499	.506
4,5	.048	.050	.037	.019	.008	.003	—	0	.253	.399	.466	.499	.506
5,0	.031	.032	.022	.012	—	.002	—	0	.253	.399	.466	.499	.506
∞	0	0	0	0	—	0	—	0	.254	.400	.467	.500	.507

$\psi = 20°$

μd \ θ	0°	10°	30°	40°	45°	50°	65°	55°	70°	80°	90°	90°	90°
θ'	0°	0°	10°	20°	25°	30°	45°	35°	50°	60°	70°	80°	90°
0,1	0,086	0,089	0,085	0,070	0,063	0,055	0,048	0	0,076	0,082	0,090	0,091	0,090
0,2	.171	.173	.169	.155	.139	.110	.140	0	.159	.169	.174	.175	.174
0,3	.226	.230	.219	.197	.159	.124	.180	0	.204	.224	.238	.240	.238
0,4	.279	.281	.265	.230	.180	.126	.220	0	.240	.275	.288	.289	.288
0,5	.309	.318	.294	.248	.188	.123	.225	0	.269	.315	.330	.332	.330
0,6	.335	.344	.315	.256	.197	.114	.230	0	.292	.341	.360	.366	.360
0,7	.349	.360	.326	.258	.193	.103	.237	0	.307	.365	.386	.392	.386
0,8	.360	.371	.332	.255	.190	.094	.245	0	.318	.382	.409	.416	.409
0,9	.366	.376	.331	.247	.178	.085	.251	0	.325	.395	.423	.433	.423
1,0	.367	.378	.327	.239	.165	.077	.258	0	.331	.406	.441	.450	.441
1,1	.362	.376	.319	.225	.155	.071	.262	0	.334	.414	.450	.461	.450
1,2	.355	.372	.309	.214	.145	.065	.264	0	.338	.423	.460	.471	.460
1,3	.347	.362	.296	.199	.135	.061	.265	0	.339	.428	.466	.479	.466
1,4	.335	.352	.285	.186	.125	.055	.266	0	.340	.432	.472	.487	.472
1,5	.321	.338	.267	.174	.115	.049	—	0	.342	.436	.476	.490	.476
1,6	.308	.326	.252	.161	.105	.044	—	0	.343	.438	.480	.494	.480
1,7	.294	.312	.239	.148	.095	.039	—	0	.344	.440	.484	.498	.484
1,8	.284	.299	.229	.136	.085	.035	—	0	.345	.442	.488	.501	.488
1,9	.268	.284	.213	.126	.075	.030	0,267	0	.345	.443	.491	.503	.491
2,0	.254	.271	.201	.115	.065	.025	—	0	.346	.445	.494	.506	.494
2,1	.238	.254	.189	.105	.060	.023	—	0	.346	.446	.495	.508	.495
2,2	.225	.241	.175	.095	.055	.022	—	0	.346	.447	.496	.510	.496
2,3	.212	.226	.163	.086	.047	.021	—	0	.346	.447	.496	.512	.496
2,4	.200	.215	.151	.079	.043	.019	—	0	.346	.447	.497	.513	.497
2,5	.184	.200	.140	.072	.038	.017	—	0	.346	.447	.497	.514	.497
2,6	.174	.189	.129	.065	.034	.015	—	0	.346	.448	.498	.514	.498
2,7	.161	.176	.119	.057	.031	.013	—	0	.346	.448	.498	.515	.498
2,8	.150	.165	.110	.054	.028	.012	—	0	.347	.448	.499	.515	.499

$$\psi = 20°$$

ϑ	0°	10°	30°	40°	45°	50°	55°	65°	70°	80°	90°	90°	90°
ϑ' / μd	0°	0°	10°	20°	25°	30°	35°	45°	50°	60°	70°	80°	90°
2,9	0,139	0,154	0,102	0,049	0,025	0,010	0	—	0,347	0,448	0,499	0,515	0,499
3,0	.129	.145	.094	.044	.022	.008	0	—	.347	.449	.500	.516	.500
3,2	.103	.126	.080	.035	.019	.007	0	—	.347	.449	.500	.516	.500
3,4	.097	.107	.067	.029	.016	.007	0	—	.347	.449	.500	.516	.500
3,6	.084	.094	.055	.024	.011	.007	0	—	.347	.449	.500	.516	.500
3,8	.072	.082	.047	.019	.009	.006	0	—	.347	.449	.500	.516	.500
4,0	.060	.069	.039	.015	.007	.006	0	—	.347	.449	.500	.516	.500
4,5	.040	.047	.024	.010	—	.006	0	—	.347	.449	.500	.516	.500
5,0	.025	.031	.015	.005	—	.006	0	—	.347	.449	.500	.516	.500
∞	0	0	0	0	—	0	0	—	.348	.450	.500	.517	.500

$$\psi = 30°$$

ϑ	0°	10°	15°	30°	40°	45°	50°	55°	60°	70°	80°	90°	90°	90°
ϑ' / μd	0°	0°	0°	0°	10°	15°	20°	25°	30°	40°	50°	60°	65°	75°
0,1	0,092	0,093	0,093	0,092	0,090	0,085	0,076	0,060	0	0,077	0,090	0,092	0,092	0,093
0,2	.184	.185	.186	.183	.179	.170	.153	.120	0	.155	.177	.184	.186	.188
0,3	.245	.248	.250	.241	.230	.206	.190	.127	0	.194	.235	.251	.252	.252
0,4	.288	.299	.301	.290	.268	.246	.210	.134	0	.220	.283	.306	.307	.310
0,5	.314	.332	.335	.321	.293	.260	.212	.125	0	.238	.316	.343	.346	.350
0,6	.343	.362	.364	.346	.309	.273	.209	.116	0	.250	.339	.375	.384	.389
0,7	.349	.381	.382	.362	.314	.270	.199	.107	0	.258	.356	.397	.410	.414
0,8	.360	.389	.390	.368	.312	.267	.189	.097	0	.263	.371	.415	.431	.441
0,9	.361	.390	.393	.369	.306	.254	.177	.088	0	.269	.383	.433	.450	.458
1,0	.358	.388	.393	.364	.299	.244	.167	.078	0	.272	.393	.450	.465	.474
1,1	.352	.385	.389	.355	.286	.235	.152	.070	0	.272	.398	.459	.475	.486
1,2	.345	.376	.381	.345	.276	.226	.140	.062	0	.273	.404	.468	.485	.496
1,3	.331	.364	.371	.341	.262	.208	.128	.056	0	.273	.406	.473	.491	.502
1,4	.317	.352	.358	.318	.248	.190	.118	.050	0	.274	.408	.478	.495	.508
1,5	.302	.338	.344	.303	.257	.175	.110	.045	0	.274	.411	.483	.499	.512
1,6	.290	.325	.330	.290	.214	.160	.100	.040	0	.274	.412	.486	.502	.515
1,7	.274	.311	.317	.273	.200	.147	.090	.035	0	.275	.414	.489	.504	.519
1,8	.259	.297	.300	.259	.186	.135	.081	.030	0	.275	.415	.491	.507	.522
1,9	.242	.280	.284	.244	.169	.122	.074	.028	0	.275	.417	.493	.509	.525
2,0	.228	.265	.269	.228	.155	.110	.065	.025	0	.276	.419	.495	.514	.529
2,1	.214	.250	.253	.214	.140	.101	.059	.023	0	.276	.419	.495	.515	.530
2,2	.199	.234	.239	.199	.125	.092	.054	.020	0	.276	.420	.496	.516	.531
2,3	.185	.221	.224	.186	.114	.084	.049	.018	0	.277	.420	.496	.516	.532
2,4	.172	.206	.210	.173	.105	.075	.042	.016	0	.277	.421	.497	.517	.532
2,5	.162	.194	.197	.160	.097	.067	.037	.014	0	.277	.421	.498	.517	.532
2,6	.150	.180	.184	.148	.091	.060	.033	.012	0	.278	.421	.499	.518	.533
2,7	.141	.169	.172	.137	.082	.055	.030	.010	0	.278	.422	.499	.518	.533
2,8	.133	.157	.160	.126	.075	.050	.026	.008	0	.278	.422	.500	.519	.534
2,9	.122	.146	.149	.117	.069	.045	.024	—	0	.278	.422	.500	.519	.534
3,0	.112	.135	.138	.107	.062	.040	.021	—	0	.279	.423	.500	.520	.535
3,2	.095	.117	.127	.091	.051	.036	.018	—	0	.279	.423	.500	.520	.535
3,4	.078	.100	.100	.077	.041	.028	.015	—	0	.279	.423	.500	.522	.535
3,6	.063	.085	.086	.065	.034	.022	.012	—	0	.279	.423	.500	.522	.536
3,8	.052	.073	.073	.054	.028	.017	.008	—	0	.279	.423	.500	.522	.536
4,0	.044	.060	.064	.045	.022	.012	.005	—	0	.279	.423	.500	.524	.538
4,5	.034	.039	.090	.038	.014	.006	.003	—	0	.280	.424	.500	.524	.538
5,0	.028	.025	.026	.017	.006	.003	.001	—	0	.280	.424	.500	.524	.538
∞	0	0	0	0	0	—	0	—	0	.282	.425	.500	.526	.540

$$\psi = 40°$$

ϑ	0°	10°	20°	50°	60°	65°	70°	80°	90°	0°	0°
ϑ' μd	0°	0°	0°	10°	20°	25°	30°	40°	50°	60°	70°
0,1	0,102	0,103	0,104	0,096	0,078	0	0,085	0,094	0,105	0,107	0,108
0,2	.201	.206	.207	.188	.133	0	.140	.189	.204	.208	.209
0,3	.259	.271	.275	.235	.144	0	.160	.240	.273	.279	.283
0,4	.309	.325	.329	.271	.145	0	.172	.287	.324	.338	.341
0,5	.340	.357	.365	.286	.136	0	.179	.313	.362	.383	.386
0,6	.357	.385	.393	.292	.125	0	.182	.330	.396	.418	.424
0,7	.365	.401	.409	.290	.111	0	.184	.349	.419	.444	.455
0,8	.366	.406	.416	.282	.100	0	.185	.362	.438	.468	.477
0,9	.363	.405	.417	.268	.089	0	.185	.372	.452	.484	.494
1,0	.354	.400	.414	.255	.079	0	.185	.378	.463	.500	.509
1,1	.342	.389	.405	.238	.069	0	.185	.383	.471	.510	.520
1,2	.328	.379	.395	.222	.061	0	.185	.385	.478	.519	.530
1,3	.310	.362	.381	.205	.053	0	.185	.389	.483	.525	.537
1,4	.294	.350	.367	.188	.047	0	.185	.391	.487	.531	.543
1,5	.274	.333	.350	.173	.042	0	.185	.392	.491	.535	.547
1,6	.256	.316	.334	.156	.036	0	.185	.393	.493	.539	.552
1,7	.237	.297	.315	.142	.032	0	.185	.394	.495	.541	.555
1,8	.222	.281	.299	.128	.028	0	.185	.395	.496	.543	.557
1,9	.205	.264	.281	.116	.024	0	.185	.395	.497	.545	.559
2,0	.190	.247	.265	.104	.022	0	.185	.396	.498	.546	.560
2,1	.176	.229	.248	.093	.020	0	.185	.396	.498	.547	.562
2,2	.162	.214	.232	.084	.017	0	.185	.396	.499	.548	.563
2,3	.150	.198	.216	.075	.015	0	.185	.396	.499	.548	.563
2,4	.136	.185	.202	.066	.013	0	.185	.397	.499	.548	.564
2,5	.124	.171	.186	.061	.011	0	.185	.397	.499	.549	.564
2,6	.112	.159	.174	.053	.010	0	.185	.397	.500	.549	.564
2,7	.101	.146	.159	.046	.009	0	.185	.397	.500	.549	.565
2,8	.092	.135	.149	.041	.008	0	.185	.397	.500	.549	.565
2,9	.084	.124	.138	.037	.007	0	.185	.397	.500	.549	.565
3,0	.076	.115	.128	.033	.006	0	.185	.398	.500	.550	.566
3,2	.063	.096	.109	.027	.005	0	.185	.398	.500	.550	.566
3,4	.054	.081	.093	.021	.004	0	.185	.398	.500	.550	.566
3,6	.045	.067	.078	.017	.003	0	.185	.398	.500	.550	.566
3,8	.039	.057	.065	.014	.002	0	.185	.398	.500	.550	.566
4,0	.031	.047	.055	.010	.002	0	.185	.398	.500	.550	.566
4,5	.018	.030	.037	.006	.001	0	.185	.398	.500	.550	.566
5,0	.011	.017	.026	.004	.001	0	.185	.398	.500	.550	.566
∞	0	0	0	0	0	0	.186	.399	.500	.551	.567

$$\psi = 50°$$

ϑ	0°	10°	20°	60°	70°	80°	90°	90°	90°
ϑ' μd	0°	0°	0°	10°	20°	30°	40°	50°	60°
0,1	0,106	0,116	0,118	0,098	0	0,104	0,115	0,117	0,118
0,2	.215	.238	.240	.178	0	.202	.229	.239	.244
0,3	.284	.313	.315	.230	0	.253	.300	.320	.324
0,4	.335	.364	.374	.253	0	.296	.355	.380	.389
0,5	.360	.393	.411	.257	0	.322	.387	.423	.435
0,6	.367	.415	.434	.252	0	.339	.420	.462	.476
0,7	.364	.423	.447	.235	0	.349	.440	.487	.500
0,8	.357	.424	.450	.215	0	.355	.456	.509	.520
0,9	.343	.415	.448	.194	0	.356	.466	.523	.539
1,0	.329	.400	.436	.176	0	.357	.474	.535	.553
1,1	.309	.384	.420	.155	0	.357	.481	.544	.566
1,2	.289	.368	.402	.138	0	.357	.485	.552	.576
1,3	.267	.348	.386	.123	0	.357	.488	.556	.582

$\psi = 50°$

ϑ	0°	10°	20°	60°	70°	80°	90°	90°	90°
ϑ' μd	0°	0°	0°	10°	20°	30°	40°	50°	60°
1,4	0,246	0,331	0,370	0,110	0	0,357	0,490	0,559	0,587
1,5	.224	.310	.347	.099	0	.357	.492	.562	.591
1,6	.204	.289	.328	.090	0	.357	.493	.564	.593
1,7	.185	.267	.307	.078	0	.357	.494	.566	.595
1,8	.168	.249	.288	.067	0	.357	.495	.568	.597
1,9	.153	.230	.268	.058	0	.357	.496	.570	.599
2,0	.139	.211	.249	.049	0	.358	.497	.571	.600
2,1	.124	.196	.231	.043	0	.358	.497	.571	.602
2,2	.108	.180	.214	.038	0	.358	.498	.572	.603
2,3	.097	.165	.198	.032	0	.358	.496	.572	.603
2,4	.088	.149	.182	.028	0	.358	.499	.572	.604
2,5	.078	.136	.168	.024	0	.358	.499	.573	.604
2,6	.069	.125	.155	.019	0	.358	.500	.573	.604
2,7	.062	.113	.142	.017	0	.358	.500	.574	.605
2,8	.056	.102	.130	.015	0	.358	.500	.574	.605
2,9	.050	.093	.117	.013	0	.358	.500	.574	.605
3,0	.044	.085	.108	.011	0	.358	.500	.575	.606
3,2	.034	.070	.090	.011	0	.358	.500	.575	.606
3,4	.026	.057	.075	.011	0	.358	.500	.575	.606
3,6	.021	.048	.064	.011	0	.358	.500	.575	.606
3,8	.017	.039	.052	.011	0	.358	.500	.575	.606
4,0	.014	.031	.044	.010	0	.358	.500	.575	.606
4,5	.007	.019	.027	.010	0	.358	.500	.575	.606
5,0	.003	.010	.017	.008	0	.359	.500	.575	.606
∞	0	0	0	0	0	.359	.500	.577	.608

$\psi = 60°$

ϑ	0°	10°	20°	30°	70°	75°	90°	90°	90°	90°
ϑ' μd	0°	0°	0°	0°	10°	15°	30°	40°	50°	60°
0,1	0,137	0,149	0,156	0,157	0,115	0	0,139	0,154	0,159	0,159
0,2	.268	.288	.295	.296	.189	0	.275	.293	.300	.301
0,3	.334	.357	.372	.380	.191	0	.340	.375	.385	.393
0,4	.359	.415	.436	.442	.186	0	.399	.444	.461	.466
0,5	.363	.440	.470	.479	.165	0	.429	.486	.506	.517
0,6	.360	.448	.485	.495	.144	0	.455	.522	.548	.557
0,7	.339	.445	.485	.498	.120	0	.470	.544	.573	.582
0,8	.311	.432	.481	.495	.102	0	.480	.562	.595	.606
0,9	.285	.413	.467	.482	.085	0	.486	.573	.610	.620
1,0	.264	.391	.448	.465	.070	0	.491	.583	.622	.634
1,1	.237	.364	.425	.443	.058	0	.493	.590	.630	.642
1,2	.215	.340	.402	.420	.048	0	.496	.594	.638	.649
1,3	.194	.312	.376	.395	.042	0	.496	.595	.642	.653
1,4	.173	.289	.353	.372	.035	0	.497	.595	.646	.655
1,5	.153	.261	.327	.346	.029	0	.497	.595	.648	.658
1,6	.132	.239	.304	.324	.024	0	.497	.596	.650	.660
1,7	.116	.216	.279	.299	.020	0	.497	.596	.651	.662
1,8	.100	.195	.255	.275	.016	0	.498	.596	.652	.663
1,9	.086	.175	.230	.252	.013	0	.498	.596	.652	.664
2,0	.074	.158	.215	.234	.010	0	.498	.596	.653	.665
2,1	.065	.143	.195	.215	.009	0	.498	.598	.653	.665
2,2	.058	.127	.179	.196	.007	0	.499	.598	.653	.665
2,3	.050	.114	.164	.182	.006	0	.499	.598	.658	.665
2,4	.044	.103	.150	.170	.005	0	.499	.599	.658	.665
2,5	.039	.091	.136	.154	.004	0	.500	.599	.654	.665
2,6	.035	.082	.124	.143	.003	0	.500	.599	.654	.665
2,7	.030	.072	.111	.127	.003	0	.500	.600	.654	.665

$$\psi = 60°$$

ϑ	0°	10°	20°	30°	70°	75°	90°	90°	90°	90°
ϑ' μd	0°	0°	0°	0°	10°	15°	30°	40°	50°	60°
2,8	0,026	0,065	0,101	0,116	0,002	0	0,500	0,600	0,654	0,665
2,9	.024	.056	.091	.105	.002	0	.500	.600	.654	.665
3,0	.022	.050	.083	.094	.001	0	.500	.601	.655	.667
3,2	.018	.040	.067	.078	.001	0	.500	.601	.655	.667
3,4	.013	.032	.055	.065	.001	0	.500	.601	.655	.667
3,6	.009	.025	.044	.054	.001	0	.500	.601	.655	.667
3,8	.006	.020	.035	.044	.001	0	.500	.601	.655	.667
4,0	.003	.015	.029	.035	.001	0	.500	.601	.655	.667
4,5	.001	.009	.018	.024	.001	0	.500	.601	.655	.667
5,0	.001	.004	.011	.013	.001	0	.500	.601	.655	.667
∞	0	0	0	0	0	0	.500	.602	.656	.668

$$\psi = 70°$$

ϑ	0°	10°	20°	35°	80°	90°	90°	90°	90°
ϑ' μd	0°	0°	0°	0°	10°	20°	30°	40°	50°
0,1	0,176	0,196	0,204	0,208	0	0,180	0,199	0,208	0,211
0,2	.319	.375	.392	.398	0	.345	.386	.400	.405
0,3	.352	.466	.476	.490	0	.405	.479	.492	.500
0,4	.355	.482	.520	.546	0	.452	.544	.576	.589
0,5	.334	.482	.539	.570	0	.471	.580	.622	.639
0,6	.290	.470	.539	.572	0	.482	.607	.655	.673
0,7	.255	.443	.523	.562	0	.491	.623	.674	.695
0,8	.222	.411	.496	.537	0	.496	.634	.690	.712
0,9	.185	.369	.462	.506	0	.498	.641	.698	.721
1,0	.152	.335	.430	.477	0	.499	.644	.705	.729
1,1	.127	.299	.397	.446	0	.499	.646	.708	.733
1,2	.104	.268	.365	.412	0	.499	.647	.710	.737
1,3	.086	.237	.334	.380	0	.499	.648	.711	.740
1,4	.074	.209	.299	.350	0	.500	.648	.712	.741
1,5	.057	.186	.271	.322	0	.500	.649	.713	.742
1,6	.045	.156	.242	.293	0	.500	.649	.714	.742
1,7	.040	.139	.220	.270	0	.500	.649	.714	.742
1,8	.034	.122	.200	.245	0	.500	.649	.714	.743
1,9	.027	.104	.178	.223	0	.500	.649	.714	.743
2,0	.021	.089	.157	.202	0	.500	.650	.715	.743
2,1	.017	.076	.142	.184	0	.500	.650	.715	.743
2,2	.014	.066	.127	.166	0	.500	.650	.715	.743
2,3	.011	.058	.114	.152	0	.500	.650	.715	.743
2,4	.008	.051	.102	.136	0	.500	.650	.716	.743
2,5	.005	.044	.091	.124	0	.500	.651	.716	.744
2,6	.003	.038	.081	.112	0	.500	.651	.716	.744
2,7	.002	.032	.072	.102	0	.500	.651	.716	.744
2,8	.001	.029	.065	.092	0	.500	.651	.716	.744
2,9	.001	.024	.056	.084	0	.500	.651	.716	.744
3,0	.001	.020	.050	.075	0	.500	.652	.717	.744
3,2	.001	.016	.038	.061	0	.500	.652	.717	.744
3,4	.001	.013	.030	.049	0	.500	.652	.717	.744
3,6	.001	.010	.024	.041	0	.500	.652	.717	.744
3,8	.001	.007	.020	.034	0	.500	.652	.717	.744
4,0	.001	.005	.016	.028	0	.500	.652	.717	.744
4,5	.001	.003	.010	.019	0	.500	.652	.717	.744
5,0	.001	.001	.005	.010	0	.500	.652	.717	.744
∞	0	0	0	0	0	.500	.653	.718	.745

$$\psi = 80°$$

ϑ	0°	10°	20°	30°	40°	85°	90°	90°	90°	90°	90°
ϑ' μd	0°	0°	0°	0°	0°	5°	10°	20°	30°	40°	50°
0,1	0,392	0,320	0,296	0,310	0,312	0	0,236	0,301	0,319	0,324	0,330
0,2	.384	.543	.587	.604	.608	0	.450	.585	.613	.629	.631
0,3	.320	.554	.640	.666	.675	0	.480	.675	.699	.730	.742
0,4	.246	.535	.638	.679	.690	0	.495	.709	.762	.789	.795
0,5	.182	.468	.601	.658	.670	0	.497	.725	.787	.816	.822
0,6	.127	.413	.550	.609	.625	0	.499	.735	.798	.830	.837
0,7	.090	.354	.492	.563	.580	0	.499	.738	.802	.836	.843
0,8	.060	.294	.442	.512	.533	0	.499	.741	.807	.841	.848
0,9	.040	.247	.393	.465	.489	0	.499	.741	.808	.842	.850
1,0	.028	.202	.346	.419	.440	0	.500	.742	.809	.843	.851
1,1	.018	.169	.310	.382	.400	0	.500	.742	.809	.843	.851
1,2	.013	.138	.269	.341	.363	0	.500	.742	.809	.843	.851
1,3	.009	.114	.240	.306	.325	0	.500	.742	.809	.843	.851
1,4	.006	.093	.208	.276	.297	0	.500	.742	.809	.843	.851
1,5	.005	.076	.185	.249	.269	0	.500	.742	.810	.843	.852
1,6	.003	.062	.160	.223	.245	0	.500	.742	.810	.843	.852
1,7	.002	.052	.141	.201	.220	0	.500	.742	.810	.843	.852
1,8	.001	.042	.123	.181	.195	0	.500	.742	.809	.843	.851
1,9	.001	.034	.108	.162	.180	0	.500	.742	.809	.843	.851
2,0	.001	.028	.095	.146	.164	0	.500	.742	.809	.843	.851
2,1	.001	.023	.082	.133	.150	0	.500	.742	.809	.843	.851
2,2	.001	.019	.073	.118	.136	0	.500	.742	.809	.843	.851
2,3	.001	.016	.064	.108	.124	0	.500	.742	.809	.843	.851
2,4	.001	.013	.056	.095	.110	0	.500	.742	.809	.843	.851
2,5	.001	.011	.049	.087	.100	0	.500	.742	.809	.843	.851
2,6	.001	.009	.043	.077	.090	0	.500	.742	.809	.843	.851
2,7	.001	.007	.038	.068	.081	0	.500	.742	.809	.843	.851
2,8	.001	.006	.033	.062	.074	0	.500	.742	.809	.843	.851
2,9	.001	.005	.029	.055	.065	0	.500	.742	.809	.843	.851
3,0	.001	.004	.026	.050	.060	0	.500	.743	.809	.843	.851
3,2	.001	.003	.021	.042	.050	0	.500	.743	.809	.843	.851
3,4	.001	.002	.017	.033	.041	0	.500	.743	.809	.843	.851
3,6	.001	.002	.012	.027	.033	0	.500	.743	.809	.843	.851
3,8	.001	.001	.009	.022	.027	0	.500	.743	.809	.843	.851
4,0	.001	.001	.007	.018	.022	0	.500	.743	.809	.843	.851
4,5	.001	.001	.004	.012	.014	0	.500	.743	.809	.843	.851
5,0	.001	.001	.002	.006	.008	0	.500	.743	.809	.843	.851
∞	0	0	0	0	0	0	.500	.743	.811	.845	.853

4-21.   Absorption Factors for Spherical Specimens

4-21a. Absolute Values

Spherical specimens are used in studies on preferred orientation and in certain other cases.

The table gives absolute values of A(ϑ) for ϑ from 0 to 90° and for μ r from 0.1 to 5.0, r being the radius of the sphere [278].

sin² θ	0	0,302	0,1170	0,2500	0,4132	0,5868	0,7500	0,8830	0,9699	1
μr θ	0°	10°	20°	30°	40°	50°	60°	70°	80°	90°
0,1	0,862	0,862	0,862	0,863	0,863	0,865	0,868	0,871	0,872	0,872
0,2	.742	.742	.742	.742	.743	.744	.747	.749	.752	.753
0,3	.646	.646	.646	.647	.647	.649	.653	.657	.660	.661
0,4	.560	.560	.560	.562	.566	.570	.576	.582	.597	.589
0,5	.489	.489	.490	.493	.499	.506	.515	.522	.529	.531
0,6	.422	.422	.423	.428	.436	.446	.456	.465	.473	.476
0,7	.368	.369	.371	.377	.386	.397	.408	.419	.429	.432
0,8	.321	.322	.325	.332	.342	.354	.366	.378	.389	.393
0,9	.281	.282	.286	.294	.307	.320	.332	.344	.356	.359
1,0	.245	.246	.250	.260	.274	.288	.301	.315	.324	.330
1,1	.215	.217	.222	.233	.248	.263	.277	.289	.302	.306
1,2	.189	.191	.198	.210	.225	.241	.254	.268	.282	.286
1,3	.167	.169	.177	.189	.206	.222	.236	.250	.262	.267
1,4	.147	.149	.156	.170	.187	.204	.219	.233	.245	.250
1,5	.131	.133	.140	.154	.171	.189	.205	.219	.232	.236
1,6	.115	.118	.126	.139	.156	.175	.191	.206	.221	.222
1,7	.102	.104	.113	.127	.144	.162	.179	.194	.206	.210
1,8	.0910	.0939	.102	.116	.134	.152	.169	.183	.195	.199
1,9	.0814	.0841	.0925	.109	.125	.143	.159	.173	.185	.189
2,0	.0731	.0760	.0846	.0982	.116	.134	.150	.165	.177	.181
2,1	.0653	.0683	.0772	.0905	.108	.126	.142	.156	.169	.173
2,2	.0585	.0610	.0700	.0834	.100	.118	.134	.149	.161	.165
2,3	.0528	.0557	.0646	.0776	.0941	.112	.128	.142	.155	.159
2,4	.0476	.0505	.0593	.0726	.0891	.106	.122	.136	.148	.152
2,5	.0430	.0458	.0547	.0681	.0838	.101	.116	.131	.143	.147
2,6	.0388	.0415	.0503	.0632	.0794	.0962	.111	.125	.137	.141
2,7	.0352	.0382	.0466	.0594	.0754	.0922	.106	.120	.132	.136
2,8	.0321	.0350	.0435	.0561	.0715	.0876	.103	.116	.127	.131
2,9	.0290	.0321	.0405	.0528	.0679	.0838	.0986	.112	.124	.126
3,0	.0267	.0294	.0377	.0496	.0646	.0802	.0948	.108	.118	.122
3,1	.0244	.0272	.0356	.0474	.0619	.0771	.0912	.104	.113	.118
3,2	.0224	.0251	.0331	.0446	.0590	.0742	.0881	.101	.111	.114
3,3	.0205	.0230	.0311	.0422	.0562	.0711	.0850	.0974	.108	.111
3,4	.0189	.0214	.0290	.0401	.0540	.0688	.0825	.0948	.105	.108
3,5	.0174	.0199	.0277	.0386	.0518	.0660	.0798	.0920	.102	.105
3,6	.0161	.0187	.0265	.0370	.0500	.0634	.0775	.0892	.0987	.102
3,7	.0149	.0174	.0250	.0353	.0482	.0616	.0752	.0868	.0962	.0991
3,8	.0138	.0164	.0236	.0338	.0463	.0593	.0728	.0843	.0935	.0967
3,9	.0128	.0153	.0225	.0324	.0448	.0570	.0685	.0818	.0911	.0941
4,0	.0119	.0142	.0210	.0310	.0431	.0556	.0674	.0798	.0888	.0918
4,1	.0111	.0134	.0202	.0299	.0418	.0544	.0664	.0777	.0866	.0895
4,2	.0103	.0126	.0192	.0287	.0406	.0531	.0648	.0752	.0843	.0873
4,3	.00960	.0118	.0181	.0278	.0394	.0518	.0633	.0736	.0821	.0852
4,4	.00899	.0113	.0176	.0266	.0377	.0502	.0616	.0720	.0805	.0833
4,5	.00841	.0107	.0169	.0257	.0368	.0488	.0600	.0702	.0786	.0815
4,6	.00787	.0100	.0162	.0248	.0357	.0474	.0585	.0684	.0767	.0797
4,7	.00738	.0095	.0156	.0240	.0347	.0462	.0572	.0670	.0752	.0780
4,8	.00693	.0091	.0150	.0232	.0336	.0450	.0559	.0657	.0736	.0765
4,9	.00650	.0087	.0144	.0226	.0328	.0439	.0545	.0643	0722	.0749
5,0	.00613	.0083	.0139	.0219	.0320	.0428	.0533	.0629	0708	.0734

4-21b. Relative Values

The table gives relative values of A(ϑ) for μr from 5.0 to ∞, the A(ϑ) for 90° being taken as 100.

sin² ϑ	0	0,0302	0,1170	0,2500	0,4132	0,5868	0,7500	0,8830	0,9699
ϑ	0°	10°	20°	30°	40°	50°	60°	70°	80°
μr									
5,0	8,35	11,0	18,8	30,0	43,6	58,4	72,6	85,4	94,9
5,5	6,87	9,56	17,6	28,6	42,3	47,3	71,9	84,9	94,7
6,0	5,75	8,46	16,3	27,6	41,2	56,3	71,2	84,5	94,6
6,5	4,89	7,72	15,5	26,7	40,4	55,6	70,7	84,2	94,5
7,0	4,23	6,92	14,7	25,9	39,6	54,9	70,2	83,9	94,4
7,5	3,67	6,44	14,2	25,3	39,0	54,3	69,7	83,7	94,3
8,0	3,26	6,00	13,7	24,7	38,4	53,8	69,4	83,5	94,2
8,5	2,88	5,58	13,3	24,2	37,8	53,3	69,0	83,3	94,2
9,0	2,56	5,17	12,9	23,8	37,4	52,9	68,7	83,1	94,1
9,5	2,32	4,91	12,5	23,4	37,0	52,5	68,4	82,9	94,0
10,0	2,07	4,69	12,2	23,1	36,6	52,2	68,2	82,7	94,0
11,0	1,72	4,28	11,7	22,6	36,0	51,7	67,7	82,4	93,9
12,0	1,44	3,97	11,2	22,2	35,5	51,2	67,4	82,1	93,8
13,0	1,22	3,76	10,9	21,8	35,1	50,9	67,1	81,9	93,7
14,0	1,06	3,56	10,6	21,5	34,8	50,6	66,9	81,6	93,7
15,0	0,923	3,37	10,4	21,2	34,5	50,3	66,7	81,7	93,6
16,0	0,800	3,25	10,2	21,0	34,2	50,0	66,6	81,7	93,6
17,0	0,692	3,16	10,0	20,8	33,9	49,7	66,5	81,6	93,6
18,0	0,620	3,05	9,85	20,6	33,7	49,5	66,4	81,6	93,6
19,0	0,565	2,95	9,70	20,3	33,5	49,3	66,3	81,5	93,6
20,0	0,513	2,86	9,59	20,2	33,3	49,2	66,2	81,5	93,6
25,0	0,317	2,60	9,25	19,9	32,8	48,8	65,9	81,3	93,5
30,0	0,227	2,49	8,99	19,4	32,4	48,4	65,6	81,2	93,4
35,0	0,172	2,38	8,74	19,1	32,1	48,1	65,4	81,1	93,4
40,0	0,124	2,31	8,57	18,9	31,8	47,9	65,3	81,0	93,4
45,0	0,094	2,27	8,48	18,7	31,6	47,7	65,2	80,9	93,3
50,0	0,078	2,24	8,40	18,6	31,4	47,5	65,1	80,9	93,3
60,0	0,061	2,20	8,31	18,5	31,2	47,2	65,0	80,8	93,3
70,0	0,047	2,18	8,25	18,4	31,0	47,1	64,9	80,8	93,3
80,0	0,036	2,16	8,20	18,3	30,9	47,0	64,8	80,7	93,3
90,0	0,027	2,14	8,17	18,2	30,8	46,9	64,7	80,7	93,3
100,0	0,019	2,12	8,14	18,2	30,8	46,9	64,6	80,7	93,2
∞	0	2,06	7,16	17,8	30,3	46,4	63,9	80,4	93,2

4-22. Absorption Factors for Powder Mixed with Binder

The powder is usually mixed with a binder such as Canada balsam or Ramsay's compound; the absorption coefficient is affected.

In this case the A(ϑ) given by the table in Section 4-20 is multiplied by a factor τ, which takes account of the difference between the absorption coefficient μ for the powder and the absorption coefficient for the binder μ_b, and also of the particle size.

The table gives $100 \cdot \tau$ (microabsorption factor) for ϑ from 0 to 90° as a function of $(\mu - \mu_b)\alpha$, where α is the radius of a particle [247].

sin² ϑ	0	0,1464	0,5000	0,8536	1,0000
ϑ $(\mu-\mu_b)a$	0°	22,5°	45°	67,5°	90°
0,00	100,0	100,0	100,0	100,0	100,0
0,01	98,6	98,6	98,6	98,6	98,6
0,02	97,2	97,2	97,3	97,3	97,3
0,03	95,9	95,9	96,0	96,0	96,0
0,04	94,5	94,5	94,6	94,6	94,7
0,05	93,2	93,2	93,3	93,3	93,4
0,06	91,8	91,8	91,9	92,0	92,1
0,07	90,5	90,5	90,6	90,6	90,8
0,08	89,2	89,2	89,3	89,4	89,5
0,09	87,8	87,8	87,9	88,0	88,2
0,10	86,5	86,5	86,6	86,8	87,0
0,20	74,2	75,0	75,3	76,0	76,0
0,30	64,0	65,1	65,3	65,8	67,1
0,40	54,5	56,6	56,9	58,0	58,7
0,50	46,8	48,9	49,6	51,6	52,9
0,60	41,0	42,2	43,6	46,0	47,7
0,70	36,0	37,2	38,7	41,8	43,4
0,80	31,5	32,6	34,7	37,9	39,9
0,90	27,6	28,7	31,1	34,5	36,4
1,00	23,9	25,1	28,2	31,8	33,4
1,10	20,9	22,2	25,5	29,1	30,9
1,20	18,4	19,7	23,2	26,8	28,7
1,30	16,2	16,6	21,2	24,8	26,8
1,40	14,5	16,0	19,4	23,2	25,0
1,50	12,9	14,3	17,9	21,8	23,5
1,60	11,4	12,9	16,5	20,7	22,1
1,70	10,1	11,8	15,2	19,6	20,9
1,80	9,0	10,6	14,1	18,5	19,8
1,90	8,0	9,6	13,1	17,6	18,8
2,00	7,1	8,7	12,3	16,8	18,0
2,10	6,2	7,8	11,4	16,0	17,2
2,20	5,6	7,1	10,8	15,3	16,4
2,30	5,0	6,5	10,2	14,5	15,7
2,40	4,5	6,0	9,6	13,8	15,2
2,50	4,0	5,5	9,1	13,2	14,6
2,60	3,6	5,1	8,7	12,6	14,1
2,70	3,3	4,8	8,3	12,0	13,6
2,80	3,0	4,6	7,9	11,4	13,0
2,90	2,8	4,3	7,6	11,0	12,6
3,00	2,6	4,1	7,3	10,6	12,2
3,50	1,7	2,8	5,9	8,9	10,5
4,00	1,2	2,1	5,0	7,81	9,3
4,50	0,9	1,9	4,2	7,10	8,3
5,00	0,6	1,7	3,7	6,0	7,4

Part 2

Some Special Problems and Methods in X-Ray Structure Analysis

CHAPTER 5

PHASE ANALYSIS

This chapter gives tables of the structures of elements and compounds together with accessory tables for use in certain methods of phase analysis.

See [6-9, 11-13] for methods of phase analysis and [114-116] for the use of ionization equipment for this purpose. Many applications in various branches of industry are dealt with in [317-319, 393].

5-1. Methods

5-1a. Qualitative Methods

Qualitative methods can be divided into two groups:

A. The substance is unknown, in which case the steps are as follows: (1) the structure is deduced from the line intensities and positions; (2) the lines are indexed; (3) the lattice constants are deduced; (4) the tables are used to identify a substance whose constants agree with the measured values; (5) the conclusions are checked by spectral or chemical analysis.

B. If it may be assumed that certain substances can be present, the actual patterns and intensities are compared with the standard ones given in Section 5-2. A point to note here is that these results apply only to material of stoichiometric composition; interplanar distances vary in a regular fashion when solid solutions are formed.

Further, the relative line intensities are much dependent on the radiation used, on account of the angular intensity factors.

Small time constants should be used in ionization recording in order to resolve close lines.

Background fluctuations affect the precision even in qualitative analysis.

It is reasonably reliable to assume that a peak whose height exceeds three times the mean fluctuation in the background is a diffraction peak, because the probability of such a peak arising by accident does not exceed 1/300.

5-1b. Quantitative Methods Involving Intensity Measurement

1. **Mixture Consisting of n Components with Roughly Equal Absorption Coefficients.**
The intensity of a line is then proportional to the amount of the corresponding component.
The method is applicable to mixtures of austenite with ferrite, of allotropic forms, and so
on. Curve 1 in Fig. 112 relates to a mixture of quartz and cristobalite, two forms of SiO_2.
The points (from experiments) fall very close to the line. There are three steps in the deduction of the proportion of quartz: (1) determination of the intensity for the stronger
line of pure quartz (d = 3.34 Å); (2) the same for this line in the mixture under the same
conditions; (3) deduction from the calibration curve.

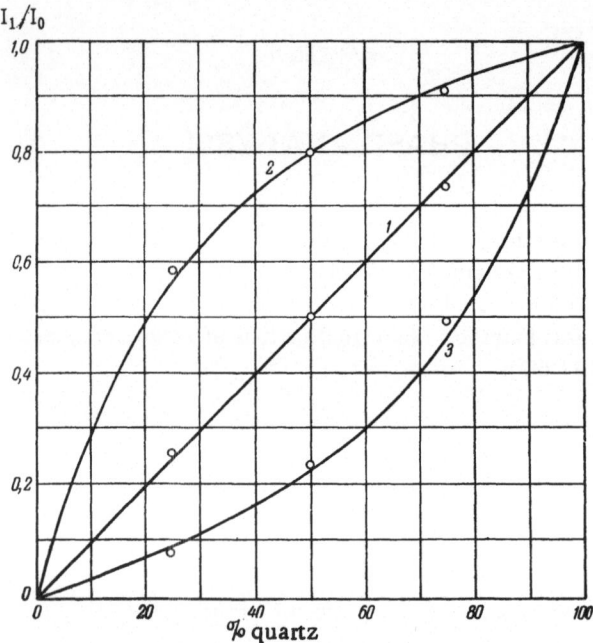

Fig. 112. Calibration curves for quantitative analysis of
two-phase mixtures: (1) quartz−cristobalite; (2) quartz−
BeO; (3) quartz−KCl.

2. **Mixture of Two Components Differing in Absorption Coefficient.** The calibration curve is drawn up from

$$\frac{I_1}{I_0} = \frac{x_1\mu_1}{x_1(\mu_1 - \mu_2) + \mu_2},$$

where I_0 is the intensity of the strongest line for the component to be determined (in pure
form), I_1 is the same but for the mixture, μ_1 and μ_2 are the absorption coefficients, and x_1
is the concentration.

Curves 2 and 3 in Fig. 112 are for mixtures of quartz (μ_1 = 34.9) with BeO (μ_2 = 8.6) or KCl (μ_2 = 124), respectively. The experimental points lie close to the curves. The graphs are used as for the previous case.

3. Mixture of n Components Differing in Absorption Coefficient. Here the concentration of any one component is deduced by the use of mixtures with a standard substance, from which a calibration curve is drawn up. The formula is

$$x_1 = k \frac{I_1}{I_s},$$

where I_1 is the intensity of the strongest line for that component and I_s is the line intensity of the standard.

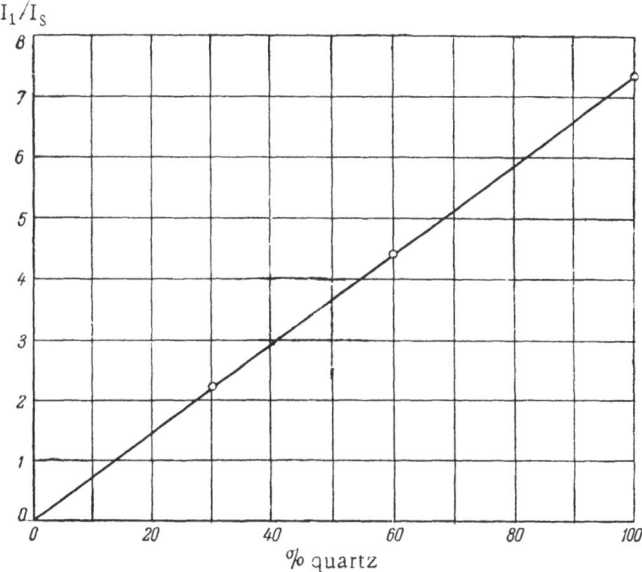

Fig. 113. Calibration curve for analysis of mixtures of quartz with $CaCO_3$, with CaF_2 as standard.

An example is the deduction of the proportion of quartz in mixtures with $CaCO_3$, CaF_2 being used as standard. Mixtures are made up containing 30, 60, and 100% quartz, the CaF_2 being constant at 20%. The relative line intensities are measured for quartz (d = 3.34 Å) and fluorite (d = 3.16 Å). Figure 113 shows the resulting calibration curve.

The standard must satisfy the following requirements: (1) its lines must be sharp and strong; (2) there must be a strong line near the strongest line of the component to be determined.

The intensity corresponding to a given proportion of a component varies from one line to another.

Figure 114 gives the intensities of lines of cristobalite in mixtures with quartz; their relation to proportion is always linear, but the lines differ in slope. This means that a calibration curve must be constructed for each line.

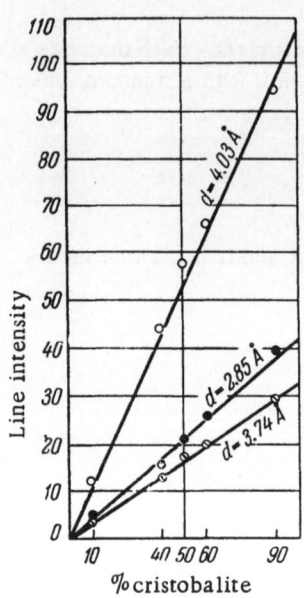

Fig. 114. Intensities of lines of cristobalite in mixtures with quartz.

Any preferred orientation in the specimen has a pronounced effect if ionization methods are used, because only a small part of the diffraction ring is viewed. Special methods of preparation designed to eliminate preferred orientations [117] are used or, alternatively, the material is mixed with an amorphous substance and the intensity is extrapolated to zero concentration [118].

Powders prepared in a ball mill, with a pestle and mortar, or by filing may have preferred orientations, so it is usually best to record a preliminary pattern with an ordinary x-ray camera (in order to check for any texture) before ionization equipment is used.

Ionization methods in conjunction with a standard have been used to determine the proportion of quartz in dust with an error of only 2% for contents up to 50% [119]. Ionization methods also enable one to determine residual austenite in steel and to perform phase analyses on alloys generally.

Quantitative analyses for known phases can be performed via intensity measurements at the peaks of suitable lines; for example, the counter can be set at the corresponding angles in the URS-50I and the counting rates recorded. More accurate results are obtained if the counts are recorded for a specified period, as is done with the scaler and timer in the URS-50I. The proportion of any given phase is deduced from previously constructed calibration curves.

The method of constructing such curves is analogous to that for photographic recording. The accuracy can be made very high; for example, the discrepancies between x-ray and chemical analyses for bauxites may be made only 1-2% [120].

Particular attention should be given to producing a smooth surface with coarse-grained specimens and to setting the surface of the specimen perpendicular to the axis of rotation in the vertical plane.

The following procedure is used to obtain a smooth surface on a powder specimen. The powder (mixed with binder) is poured onto a holder in the form of a plate having a recess; shellac in alcohol or Ramsay's compound may be used. The material outside the groove is removed with a scalpel and the surface is made smooth by pressing on glass. Care should be taken to make the surface of the specimen exactly coincident with that of the plate.

Specimens may be coated with lacquer for prolonged storage; this is removed with acetone before use.

Materials that react with the binder or that do not allow of pressing into the holder can be handled by pouring the powder into the recess and holding it in place with a thin sheet of cellophane. Correction must be made for the absorption and scattering in the cellophane if high precision is needed. The accuracy is improved by rotating the specimen.

Phase analysis of a mixture of two phases can be performed by reference to the absorption coefficient [121]. The line shape is unimportant in phase analysis, so broad slits can be used to employ much of the surface of the specimen.

A difficulty often encountered is overlap between the lines of the various phases. See [279] for methods for such mixtures; the latest methods are described in [243].

After the substance has been identified from the structure type and lattice constants, the results may be checked by comparing the actual density with the x-ray density ρ, which is given by

$$\varrho = \frac{1.6604 \, Mn}{V} \, ,$$

where 1.6604 ($\times \ 10^{-24}$ g) is the weight of a hypothetical atom of atomic weight 1.000, M is the molecular weight of the compound, and n is the number of molecules in the unit cell; V is the volume (in $\mathring{A}^3$) of the unit cell; see Chapter 3 for unit-cell volumes.

It is usual to use monocrystals if the positions of the atoms in the structure (or structures) are needed in addition to the phase composition.

It is sometimes possible to deduce structures and atomic positions from the patterns given by polycrystals.

5-2. Crystal Structures of Elements and Compounds

The table gives the structure for nearly 1200 substances [102].

A theoretical pattern is constructed and compared with the actual pattern; the steps are as follows:

(1) The structure type (second column) is used with the schemes for x-ray patterns to determine the indices and relative intensities of the lines.

(2) The quadratic forms (Chapter 3) are used to compute the interplanar distances for all reflections.

(3) The results are used to compile tables of the type given in Section 5-3.

The symbols used for the crystal systems are: K — cubic, T — tetragonal, H —hexagonal, R — orthorhombic, and M — monoclinic. The numbers correspond to those in the scheme of Chapter 3.

Compound	Structure type	Lattice constants			α or β
		a	b	c	
Ag	K 2	4,08624	—	—	—
Ag₃Al	K 12	6,934	—	—	—
Ag₅Al₃	H 13	2,876	—	4,439	—
Ag₃As	H 13	2,896	—	4,709	—
AgAsMg	K 6	6,253	—	—	—
Ag₃AsO₄	K 5	5,772	—	—	—
AgAuTe₄	M	8,958	4,489	145,62	14°20′
AgBe₂	K 22	6,2997	—	—	—
AgBiS₂	R	8,096	7,836	5,661	—
AgBr	K 5	5,78	—	—	—
AgBrO₃	T	8,61	—	8,096	—
AgCd	K 19	6,934	—	—	—
AgCd₃	M	4,84	9,54	3,237	92°42′
AgCl	K 5	4,93	—	—	—
AgClO₂	H 2	2,995	—	5,712	—
AgFeO₂	H 21	4,62	—	—	85°28′
Ag₅Hg₈	K 24	10,03	—	—	—
Ag I	H 3	4,589	—	7,509	—
Ag IO₄	T 10	5,381	—	12,044	—
AgLa	K 3	3,768	—	—	—
AgLi	K 3	3,174	—	—	—
AgMg	K 3	3,287	—	—	—
AgMnO₄	M	5,671	8,287	7,134	92°30′
Ag₂MoO₄	K 10	9,279	—	—	—
AgN₃	R	5,591	5,942	6,052	—
AgNO₃	R	3,517	6,152	5,170	—
Ag₂O	K 7	4,73	—	—	—
Ag₂O₃	K	9,84	—	—	—
Ag₃PO₄	K	5,044	—	—	—
AgReO₄	T 10	5,36	—	11,944	—
AgReO₄	K	4,89	—	—	—
Ag₂S	R	4,78	6,934	7,004	—
Ag₂SO₄	R	5,831	12,676	10,271	—
Ag₃Sb	H 13	2,93—2,97	—	4,76—4,79	—
AgSbS₂	M	13,97	4,399	12,846	98°37′
AgSeO₄	R	6,082	12,836	10,23	—
Ag₅Sn	H 13	2,94—2,96	—	4,77—4,78	—
AgZn	K 3	3,162	—	—	—
AgZn₃	H 13	2,826	—	4,469	—
Ag₅Zn₈	K 24	9,349	—	—	—
Al	K 2	4,0496	—	—	—
AlAs	K 4	5,631	—	—	—
AlAsO₄	H	5,040	—	11,24	—
Al₄Ba	T	4,539	—	11,17	—
AlBr	H 20	3,006	—	3,247	—
Al₂BeO₄	R	4,429	9,409	5,481	—
Al₄C₃	H	8,547	—	—	20°28′
Al₅C₃N	H	3,287	—	21,634	—
AlCl₃	H 17	5,922	—	17,56	—
Al₂CoO₄	K 10	8,117	—	—	—
Al₂Cu	T 6	6,052	—	4,87	—
Al₂CuO₄	K 10	8,080	—	—	—
AlF₃	H 17	5,039	—	—	58°31′
Al₂F₂SiO₄	R	4,65	8,79	8,897	—
Al₂FeO₄	K 10	8,135	—	—	—
AlLi	K	6,373	—	—	—
Al₂Mg₃	K 14	10,561	—	—	—
Al₂MgO₄	K 10	8,075	—	—	—
Al₂MnO₄	K 10	8,288	—	—	—
Al₂Mn₃(SiO₄)₃	K 15	11,57	—	—	—
AlN	H 3	3,110	—	4,975	—
AlNd	K 3	3,74	—	—	—

Compound	Structure type	Lattice constants			α or β
		a	b	c	
AlNi	K 3	2,887	—	—	—
Al_2NiO_4	K 10	8,066	—	—	—
α-Al_2O_3	H 7	5,140	—	—	55°6'
β-Al_2O_3	H	5,571	—	22,645	—
γ-Al_2O_3	K 10	7,926	—	—	—
AlO_2H	R	4,409	9,409	2,846	—
$Al(OH)_3$	M	8,641	5,070	9,719	95°26 '
AlP	K 4	5,431	—	—	—
$AlPO_4$	H	4,94	—	10,96	—
AlSb	K 4	6,142	—	—	—
$AlSbO_4$	T 4	4,519	—	2,967	—
Al_2SiO_5	R	7,415	7,615	5,712	—
Al_3Ti	T	5,435	—	8,577	—
Al_2ZnO_4	K 10	8,078	—	—	—
As	H 9	4,131	—	—	54°10'
AsI	H 17	8,267	—	—	51°20'
As_2NiO_4	T	8,237	—	5,631	—
As_2O_3	K 4	11,08	—	—	—
As_2PbS_4	M	58,498	7,81	83,47	90°
AsS	M	9,29	13,53	6,57	73°27'
Au	K 2	4,07856	—	—	—
Au_3Al	K 12	6,924	—	—	—
$AuAl_2$	K 6	6,012	—	—	—
Au_2Bi	K 22	7,958	—	—	—
AuCdS	K 3	3,347	—	—	—
AuCdS	R	3,146	4,86	4,76	—
AuCu	T 2	3,99	—	3,73	—
$AuCu_3$	K 2	3,76	—	—	—
$AuGa_2$	K 6	6,095	—	—	—
Au_3Hg	H 13	2,916	—	4,78	—
$AuIn_2$	K 6	6,391	—	—	—
AuMg	K 3	3,266	—	—	—
AuMn	T	3,287	—	3,126	—
Au_2Pb	K 22	7,926	—	—	—
$AuSb_2$	K 9	6,64	—	—	—
AuSn	H 4	4,328	—	5,523	—
Au_5Sn	H 13	2,906—2,936	—	4,79—4,77	—
AuZn	K 3	3,196	—	—	—
$AuZn_3$	H 13	2,816	—	4,389	—
Au_5Zn_8	K 24	9,29—9,24	—	—	—
$BAsO_4$	T	4,467	—	6,79	—
BN	H 4	2,504	—	6,66	—
B_2O_3	K	10,055	—	—	—
BPO_4	T	4,341	—	6,653	—
Ba	K 1	5,019	—	—	—
$BaAl_2O_4$	H	5,22	—	8,808	—
BaB_6	K 17	4,289	—	—	—
$BaBr_2$	R	9,858	8,264	4,958	—
BaC_2	T 5	6,233	—	7,064	—
$BaCO_3$	R	8,852	6,562	5,266	—
$BaCeO_3$	K 18	4,386	—	—	—
$BaCl_2$	R	9,352	7,839	4,715	—
BaF_2	K 6	6,199	—	—	—
$BaHPO_4$	R	4,619	14,11	17,13	—
BaI_2	R	10,587	8,880	5,279	—
$BaMoO_4$	T 10	5,57	—	12,786	—
BaNH	K 5	5,852	—	—	—
$Ba(NO_3)_2$	K 9	8,126	—	—	—
BaO	K 5	5,534	—	—	—
BaO_2	T 5	5,351	—	6,784	—
$BaO_2 \cdot TiO_2$	M	9,429	3,938	16,926	103°2'
BaS	K 5	6,363	—	—	—

Compound	Structure type	Lattice constants			α or β
		a	b	c	
BaS$_3$	R	8,337	9,659	4,83	—
BaSO$_4$	R	8,881	5,452	7,154	—
BaSe	K 5	6,633	—	—	—
Ba$_2$SiO$_4$	R	5,772	10,19	7,575	—
BaSnO$_3$	K 18	4,108	—	—	—
BaTe	K 5	7,000	—	—	—
BaThO$_3$	K 18	4,489	—	—	—
BaTiO$_3$	K 18	3,978	—	—	—
BaWO$_4$	T 10	5,65	—	12,725	—
BaZrO$_3$	K 18	4,185	—	—	—
Be	H 1	2,28606	—	3,58429	—
Be$_2$C	K 6	4,339	—	—	—
BeCo	K 3	2,611	—	—	--
BeCu	K 3	2,703	—	—	—
Be$_2$GeO$_4$	H	7,905	—	—	108°6'
Be$_3$N$_2$	K 16	8,146	—	—'	—
BeO	H 3	2,700	—	4,384	—
Be$_3$P$_2$	K 16	10,17	—	—	--
BePd	K 3	2,819	—	—	...
BeS	K 4	4,86	—	—	—
BeSe	K 4	5,08	—	—	—
Be$_2$SiO$_4$	H	7,696	—	—	108°1'
BeTe	K 4	5,551	—	—	—
Bi	H 9	4,7459	—	—	57°14,5
BiAsO$_4$	T 10	5,09	—	11,724	—
BiF$_3$	K	5,86	—	—	—
BiI$_3$	H 17	8,146	—	—	54°50'
BiOBr	T 11	3,928	—	8,116	—
BiOCl	T 11	3,898	—	7,395	—
BiOI	T 11	4,018	—	9,168	—
α-Bi$_2$O$_3$	M	5,842	8,156	7,495	67°4'
β-Bi$_2$O$_3$	T	10,952	—	5,631	—
BiTi	K 3	3,988	—	—	—
BiVO$_4$	R	3,391	5,050	12,004	—
Br	R	4,489	6,683	8,737	—
C	K 4	3,5668	—	—	—
C	H 9	3,642	—	—	39°30'
CO	K	5,641	—	—	—
CO$_2$	K 9	5,641	—	—	—
CSi	K 4	4,357	—	—	—
Ca	K 2	5,576	—	—	—
Ca	H 1	3,988	—	6,533	—
CaB$_6$	K 17	4,153	—	—	—
CaC$_2$	T 5	3,878	—	6,383	—
CaCN$_3$	H 21	5,41	—	—	39°55'
CaCO$_3$	R 7	7,956	5,732	4,95	—
CaCO$_3$	H 8	6,374	—	—	46°6'
CaCl$_2$	R	6,253	6,443	4,208	—
CaCrO$_4$	T 13	7,265	—	6,363	—
CaF$_2$	K 6	5,462	—	—	—
CaGa$_2$	H 20	4,323	—	4,323	—
CaI$_2$	H 2	4,489	—	6,964	—
CaMg(CO$_3$)$_2$	H 8	6,062	—	—	46°54'
CaMoO$_4$	T 10	5,24	—	11,46	—
Ca$_3$N$_2$	K 16	10,421	—	—	—
CaNH	K 5	5,016	—	—	—
CaO	K 5	4,807	—	—	—
CaO$_2$	T 5	5,02	—	5,932	—
Ca(OH)$_2$	H 2	8,5916	—	4,9060	—
CaO·SiO$_2$	M	15,36	7,295	7,084	95°24'
CaPb$_2$	K 2	4,90	—	—	—
CaS	K 5	5,69	—	—	—

Compound	Structure type	Lattice constants			α or β
		a	b	c	
CaSe	K 5	5,992	—	—	—
CaSn$_3$	K 2	4,74	—	—	—
CaSn(BO$_3$)$_2$	H 8	6,013	—	—	47°42′
CaSnO$_3$	K 18	3,928	—	—	—
CaTe	K 5	6,358	—	—	—
CaTiO$_3$	K 18	—	—	—	—
CaTl	K 3	3,494	—	—	—
CaTl$_3$	K 2	4,804	—	—	—
CaWO$_3$	T 10	5,251	—	11,403	—
CaZrO$_3$	K 18	3,998	—	—	—
Cd	H 1	2,9851	—	5,6206	—
Cd$_3$As$_2$	T 12	8,963	—	12,676	—
CdBr$_2$	H 19	6,643	—	—	34°42′
CdCO$_3$	H 8	6,124	—	—	47°24′
CdCl	K 3	3,868	—	—	—
CdCl$_2$	H 19	6,243	—	—	36°2′
CdCrO$_4$	R	5,685	8,692	6,907	—
CdF$_2$	K 6	5,411	—	—	—
CdHg	T	3,938	—	2,916	—
CdI$_2$	H 2	4,249	—	6,854	—
CdLa	K 3	3,908	—	—	—
CdMg$_3$	H 22	3,136	—	5,080	—
Cd$_3$Mg	H 22	2,936	—	5,521	—
CdMoO$_4$	T 10	5,150	—	11,192	—
Cd$_3$N$_2$	K 16	10,811	—	—	—
CdO	K 5	4,698	—	—	—
Cd(OH)$_2$	H 2	3,487	—	4,679	—
CdP$_2$	T	5,291	—	19,74	—
Cd$_3$P$_2$	T 12	8,764	—	12,405	—
CdPr	K 3	3,828	—	—	—
CdS	K 4	5,832	—	—	—
CdS	H 3	4,139	—	6,7045	—
CdSb$_2$	M	7,2145	13,537	6,172	100°14′
CdSe	K 4	6,052	—	—	—
CdSe	H 3	4,309	—	7,034	—
CdTe	K 4	6,423	—	—	—
CdTiO$_2$	H 7	5,882	—	—	53°36′
Ce	K 2	5,150	—	—	—
Ce	H 1	3,657	—	5,972	—
CeB$_6$	K 17	4,138	—	—	—
CeC$_2$	T 5	3,878	—	6,473	—
CeF$_3$	H	7,128	—	7,288	—
CeGa$_2$	H 20	4,312	—	4,316	—
CeMg$_3$	K 21	7,385	—	—	—
CeO$_2$	K 6	5,420	—	—	—
Ce$_2$O$_3$	H 16	3,888	—	6,062	—
CePO$_4$	M	6,774	7,014	6,453	—
CePb$_3$	K 2	4,870	—	—	—
CeSn$_3$	K 2	4,720	—	—	—
Co	K 2	3,561	—	—	—
Co	H 1	2,519	—	4,113	—
CoAl	K 3	2,854	—	—	—
Co$_2$Al$_5$	H	7,671	—	7,620	—
CoAs	R	5,972	5,160	3,517	—
CoAsS	K 20	5,611	—	—	—
Co$_2$B	T 6	5,016	—	4,220	—
CoBr$_2$	H 2	3,687	—	6,132	—
CoCO$_3$	H 8	5,685	—	—	48°14′
CoCl$_2$	H 19	6,172	—	—	33°26′
CoCrO$_4$	R	5,516	8,298	6,219	—
Co$_2$CuO$_4$	K 10	8,055	—	—	—
Co$_2$CuS$_4$	K 10	9,477	—	—	—

Compound	Structure type	Lattice constants			α or β
		a	b	c	
CoF_2	T 4	4,669	—	3,196	—
CoF_3	H 17	5,311	—	—	57°0'
CoI_2	H 2	3,968	—	6,663	—
Co_2MgO_4	K 10	8,123	—	—	—
CoO	K 5	4,259	—	—	—
Co_3O_4	K 10	8,126	—	—	—
$Co(OH)_2$	H 2	3,179	—	4,609	—
CoP	R 9	5,599	5,076	3,280	—
CoS	H 4	3,374	—	5,170	—
CoS_2	K 9	5,535	—	—	—
Co_3S_4	K	9,399	—	—	—
Co_9S_8	K	9,930	—	—	—
$CoSO_4$	R	4,660	6,723	8,467	—
CoSb	H 4	3,874	—	5,198	—
CoSe	H 4	3,621	—	5,289	—
$CoSe_2$	K 9	5,866	—	—	—
CoSi	K 15	4,447	—	—	—
Co_2Si	R	7,114	4,920	3,738	—
CoSn	H	5,279	—	4,259	—
$CoSn_2$	T	6,361	—	5,450	—
$CoSnO_4$	K 16	8,617	—	—	—
CoTe	H 4	3,888	—	6,371	—
$CoTe_2$	R	5,312	6,311	3,890	—
$CoTiO_3$	H 7	5,481	—	—	54°42'
$CoTiO_4$	K 10	8,437	—	—	—
$CoWO_4$	M	4,669	5,710	4,990	90°
$CoZn_3$	K 12	4,553	—	—	—
Co_5Zn_2	K 24	8,898—8,988	—	—	—
Co_2ZnO_4	K 10	8,124	—	—	—
Cr	K 1	2,885	—	—	—
Cr	H 1	2,722	—	4,427	—
γ-Cr	K 14	8,738	—	—	—
Cr_2Al	T	3,004	—	8,637	—
Cr_5Al_8	H	7,804	—	—	109°8'
CrAs	R	3,486	6,223	5,742	—
Cr_2As	T	3,620	—	6,353	—
Cr_2Be_3	H 14	4,249	—	6,934	—
$CrBr_3$	H 17	7,064	—	—	32°36'
Cr_3C_2	R	2,826	5,531	11,483	—
Cr_2CdO_4	K 10	8,584	—	—	—
Cr_2CdS_4	K 10	10,210	—	—	—
Cr_2Cl_3	H 17	6,012	—	17,33	—
Cr_2CoO_4	K 10	8,336	—	—	—
Cr_2FeO_4	K 10	8,361	—	—	—
Cr_2MgO_4	K 10	8,322	—	—	—
Cr_2MnO_4	K 10	8,453	—	—	—
Cr_2MnS_4	K 10	10,07	—	—	—
Cr_2N	H 13	2,756—2,776	—	4,459—4,449	—
$CrNbO_4$	T 4	4,644	—	3,011	—
Cr_2NiO_4	K 10	8,316	—	—	—
CrO_2	T 4	4,78	—	3,968	—
CrO_3	R	8,477	4,78	5,711	—
Cr_2O_3	H 7	5,391	—	—	54°50'
CrP	R	5,942	5,366	3,126	—
Cr_3P	T	9,144	—	4,569	—
CrS	H 4	3,455	—	5,766	—
CrSb	H 4	4,118	—	5,471	—
$CrSbO_4$	T 4	4,586	—	3,048	—
CrSe	H 4	3,691	—	6,031	—
CrSi	K 15	4,629	—	—	—
$CrSi_2$	H	4,431	—	6,371	—
$CrTaO_4$	T 4	4,635	—	3,015	—

Compound	Structure type	Lattice constants			α or β
		a	b	c	
CrTe	H 4	3,989	—	6,223	—
Cr_2ZnO_4	K 10	8,313	—	—	—
$CrZnSn_4$	K 10	9,940	—	—	—
Cs	K 1	6,062	—	—	—
CsBr	K 3	4,296	—	—	—
CsCl	K 3	4,118	—	—	—
CsCl	K 5	7,034	—	—	—
$CsClO_4$	K 19	7,976	—	—	—
Cs_2CrO_4	R	11,157	8,3799	6,239	—
$CsCrO_4F$	T 10	5,7265	—	1,453	—
CsF	K 5	6,020	—	—	—
CsH	K 5	6,389	—	—	—
Cs I	K 3	4,571	—	—	—
$Cs I_3$	R	6,834	9,970	11,042	—
$Cs ICl_2$	H 21	5,471	—	—	70°42'
$Cs IO_3$	K 18	4,669	—	—	—
$Cs IO_4$	R	5,850	6,026	14,393	—
CsO_2	T 5	6,293	—	7,215	—
Cs_2O	H 19	6,74	—	—	36°59'
$CsReO_4$	R	5,634	5,980	14,269	—
Cs_2SO_4	R 3	10,905	8,215	6,231	—
$CsSO_3F$	T 10	5,622	—	11,155	—
Cu	K 2	3,6149	—	—	—
$CuAl_2$	T 6	6,062	—	4,890	—
Cu_3Al	K 23	6,954	—	—	—
Cu_4Al_3	K 24	8,717—8,697	—	—	—
Cu_2AlMn	K 23	5,912	—	—	—
Cu_3As	H	7,102	—	7,235	—
CuAsS	R	3,788	5,481	11,493	—
CuBe	K 3	2,695	—	—	—
$CuBe_2$	K 22	5,952	—	—	—
CuBiMg	K 6	6,269	—	—	—
CuBr	K 4	5,691	—	—	—
Cu_5Cd_8	K 24	9,659	—	—	—
CuCdSb	K 6	6,275	—	—	—
CuCl	K 4	5,417	—	—	—
$CuCrO_4$	R	5,437	8,943	5,890	—
CuF	K 4	4,264	—	—	—
CuF_2	K 6	5,417	—	—	—
$CuFeO_2$	H 21	5,972	—	—	29°26'
$CuFeS_2$	T	5,251	—	10,311	—
CuH	H 3	2,899	—	4,623	—
Cu_4Hg_3	K 24	9,419	—	—	—
Cu I	K 4	6,054	—	—	—
$CuMg_2$	R	5,280	9,068	18,247	—
Cu_2Mg	K 22	7,044	—	—	—
CuMgSb	K 6	6,164	—	—	—
Cu_2MnSn	K 23	6,178	—	—	—
CuO	M	4,662	3,417	5,118	99°29'
Cu_2O	K 7	4,263	—	—	—
Cu_3P	H	7,084	—	7,144	—
CuPd	K 3	2,994	—	—	—
CuS	H	3,808	—	16,463	—
Cu_2S	K 6	5,601	—	—	—
Cu_2S	R	11,823	2,695	13,427	—
Cu_2Sb	T	4,000	—	6,092	—
Cu_3Sb	H 13	2,725—2,755	—	4,339—4,350	—
Cu_2Se	K 6	5,761	—	—	—
Cu_5Si	K 12	6,223	—	—	—
Cu_5Si	H 13	2,595	—	4,238	—
$Cu_{15}Si_4$	K	9,709	—	—	—
CuSn	H 4	4,198	—	5,096	—

Compound	Structure type	Lattice constants			α or β
		a	b	c	
Cu_3Sn	H 13	2,756	—	4,319	—
$Cu_{31}Sn_8$	K 24	17,956	—	—	—
$CuZn$	K 3	2,951	—	—	—
$CuZn_3$	H 13	2,756	—	4,299	—
Cu_5Zn_8	K 24	8,87—8,91	—	—	—
α-Fe	K 1	2,86647	—	—	—
β-Fe	K 1	2,906	—	—	—
γ-Fe	K 2	3,637	—	—	—
δ-Fe	K 1	2,936	—	—	—
Fe_3Al	K 23	5,812	—	—	—
$FeAs$	R	5,794	5,187	3,095	—
$FeAs_2$	R	5,261	5,932	2,856	—
Fe_2As	T	3,634	—	5,985	—
$FeAsS$	M	9,529	5,661	6,433	90°
FeB	R	4,061	5,506	2,952	—
Fe_2B	T 6	5,109	—	4,249	—
FeB_2	H 2	3,747	—	6,182	—
Fe_3C	R	4,526	5,089	6,743	—
$FeCO_3$	H 8	5,766	—	—	47°25′
Fe_2CdO_4	K 10	8,748	—	—	—
$FeCl_2$	H 19	6,212	—	—	33°33′
$FeCl_3$	H 17	6,703	—	—	52°30′
Fe_2CoO_4	K 10	8,367	—	—	—
Fe_2CuO_4	K 10	8,457	—	—	—
Fe_2CuO_4	T	8,297	—	8,697	—
Fe_2CuS_3	R	6,443	11,062	6,202	—
FeF_2	T 4	4,679	—	3,304	—
FeF_3	H 17	5,401	—	—	58°0′
FeI_2	H 2	4,048	—	6,764	—
Fe_2MgO_4	K 10	8,383	—	—	—
Fe_2MnO_4	K 10	8,474	—	—	—
Fe_7Mo_6	H	8,988	—	—	36°39′
Fe_2N	H 13	2,675—2,776	—	4,369—4,439	—
Fe_4N	K 2	3,803	—	—	—
$FeNbO_4$	T 4	4,689	—	3,056	—
Fe_2NiO_4	K 10	8,357	—	—	—
FeO	K 5	8,357	—	—	—
α-Fe_2O_3	H 7	5,4243	—	—	55°17′
γ-Fe_2O_3	K 10	8,337	—	—	—
δ-Fe_2O_3	H	5,100	—	4,419	—
Fe_3O_4	K 10	8,391	—	—	—
$FeOCl$	R	3,758	7,665	3,307	—
$Fe(OH)_2$	H 2	3,246	—	4,479	—
α-FeO_2H	R	4,609	10,03	3,046	—
β-FeO_2H	R	10,581	10,261	3,036	—
γ-FeO_2H	R	3,878	12,535	3,066	—
FeP	R	5,794	5,187	3,092	—
FeP_2	R	2,730	4,985	5,668	—
Fe_2P	H	5,864	—	3,460	—
Fe_3P	T	9,108	—	4,505	—
$FePO_4$	H	5,045	—	11,200	—
Fe_2PbO_3	K 10	7,826	—	—	—
FeS	H 4	3,460	—	5,681	—
FeS_2	K 9	5,416	—	—	—
FeS_2	R	4,445	5,425	3,388	—
$FeSO_4$	M	15,37	13,00	20,06	104°15′
$FeSb$	H 4	4,068	—	5,140	—
$FeSb_2$	R	5,831	6,533	3,195	—
Fe_3Sb_2	H 4	4,118	—	5,180	—
$FeSbO_4$	T 4	4,632	—	3,017	—
$FeSe$	H 4	3,644	—	5,970	—
$FeSe_2$	R	4,800	5,726	3,582	—

Compound	Structure type	Lattice constants			α or β
		a	b	c	
FeSi	K 15	4,447	—	—	—
FeSi$_2$	T	2,692	—	5,140	—
FeSn	H	5,303	—	4,449	—
FeTaO$_4$	T 4	4,681	—	3,048	—
FeTe	H 4	3,808	—	5,662	—
FeTe$_2$	R	5,351	6,273	3,857	—
FeTiO$_3$	H 7	5,531	—	—	54°49′
Fe$_2$W	H	4,739	—	7,716	—
Fe$_7$W$_6$	H	9,038	—	—	30°30′
Fe$_3$W$_3$C	K	11,06	—	—	—
FeWO$_4$	M	4,709	5,701	4,940	90°
FeZn$_7$	H 13	2,796	—	4,459	—
Fe$_3$Zn$_{10}$	K 24	8,948	—	—	—
Fe$_5$Zn$_{21}$	K 24	8,978—9,008	—	—	—
Fe$_2$ZnO$_4$	K 10	8,440	—	—	—
Ga	R	4,526	4,520	7,660	—
GaAs	K 4	5,646	—	—	—
Ga$_2$MgO$_4$	K 10	8,295	—	—	—
GaN	H 3	3,185	—	5,180	—
Ga$_2$O$_3$	H 7	5,290	—	—	55°35′
GaP	K 4	5,447	—	—	—
GaSb	K 4	6,130	—	—	—
GaSbO$_4$	T 4	4,599	—	3,036	—
Ga$_2$ZnO$_4$	K 10	8,340	—	—	—
Ge	K 4	5,631	—	—	—
Ge I$_2$	H 2	4,138	—	6,803	—
Ge I$_4$	K 13	11,91	—	—	—
GeMg$_2$	K 6	6,391	—	—	—
Ge$_3$N$_4$	H	8,582	—	—	107°46′
GeO$_2$	T 4	4,403	—	2,866	—
GeS	R	4,299	10,44	3,647	—
Hf	H 1	3,327	—	5,471	—
HfC	K 5	4,46671	—	—	—
HfF$_4$	M	9,469	9,860	7,635	94°29′
HfO$_2$	K 6	5,125	—	—	—
HfSi	H	6,874	—	12,62	—
HfSi$_2$	R	3,677	14,589	3,647	—
Hg	H	3,016	—	—	70°32′
HgBr$_2$	R	4,629	6,814	12,475	—
Hg$_2$Br$_2$	T 8	4,659	—	11,132	—
HgCl$_2$	R	5,972	12,765	4,339	—
Hg$_2$Cl$_2$	T 8	4,459	—	11,00	—
HgF$_2$	K 6	5,551	—	—	—
Hg$_2$I$_2$	T 8	4,930	—	11,16	—
HgO	R	3,303	3,520	5,515	—
HgS	K 4	5,815	—	—	—
HgSe	K 4	6,082	—	—	—
HgTe	K 4	6,373	—	—	—
In	T 2	4,592	—	4,940	—
InAs	K 4	6,048	—	—	—
InBO$_3$	H 8	5,853	—	—	48°10′
In$_2$BaO$_4$	T	8,247	—	8,17	—
In$_2$CaO$_4$	T	6,213	—	9,842	—
In$_2$CdO$_4$	T	6,129	—	9,895	—
InCl$_2$	R	6,864	9,659	10,561	—
In$_2$MgO$_4$	K 10	8,828	—	—	—
InN	H 3	3,540	—	5,704	—
In$_2$O$_3$	K 16	10,14	—	—	—
InP	K 4	5,873	—	—	—
InSb	K 4	6,474	—	—	—
In$_2$SrO$_4$	T	7,996	—	7,996	—
Ir	K 2	3,83886	—	—	—

Compound	Structure type	Lattice constants			α or β
		a	b	c	
IrO$_2$	T 4	4,499	—	3,146	—
Ir$_2$P	K 6	5,546	—	—	—
I	R	4,805	7,269	9,800	—
K	K 1	5,211	—	—	—
KAl(SO$_4$)$_2$	K	12,225	—	—	—
KAl(SO$_4$)$_2$	H	4,719	—	7,976	—
K$_3$As	H 15	5,794	—	10,242	—
KBF$_4$	R 1	7,856	5,691	7,395	—
KBF$_4$	K 19	7,485	—	—	—
KBO$_2$	H	7,776	—	—	110°36'
KBaPO$_4$	T	9,860	—	8,357	—
K$_3$Bi	H 15	6,190	—	10,955	—
KBi$_2$	K 22	9,519	—	—	—
KBr	K 5	6,599	—	—	—
KBrO$_3$	H 8	4,409	—	—	86°
KCN	K 5	6,523	—	—	—
KCNO	T 7	6,082	—	7,044	—
KCNS	R	6,673	7,595	6,653	—
KCaPO$_4$	H	10,62	—	5,862	—
KCaPO$_4$	R	9,76	7,625	5,491	—
K$_2$Cd(CN)$_4$	K 10	12,866	—	—	—
KCl	K 5	6,283	—	—	—
KClO$_3$	M	4,659	5,601	7,104	109°38'
KClO$_4$	K 19	7,515	—	—	—
KClO$_4$	R	9,840	6,012	7,806	—
K$_2$CrO$_4$	R	10,421	7,625	5,932	—
KF	K 5	5,351	—	—	—
KH	K 5	5,711	—	—	—
KHC$_2$	T 5	6,062	—	8,447	—
KHF$_2$	T 7	5,681	—	6,824	—
K$_2$Hg(CN)$_4$	K 10	12,786	—	—	—
KI	K 5	7,066	—	—	—
KIO$_3$	K 18	4,469	—	—	—
KIO$_4$	T 10	5,734	—	12,655	—
KMgF$_3$	K 18	4,008	—	—	—
KMnO$_4$	R	9,108	5,731	7,425	—
KN$_3$	T 7	6,106	—	7,070	—
KNO$_2$	M	4,459	5,000	7,325	114°50'
KNO$_3$	R	5,431	9,188	6,463	—
KNbO$_3$	K 18	4,018	—	—	—
KNiF$_3$	K 18	4,018	—	—	—
K$_2$O	K 6	6,453	—	—	—
KO$_2$	T 5	5,711	—	6,764	—
KOH	K 5	5,792	—	—	—
K$_2$PdCl$_4$	T 9	7,054	—	4,108	—
K$_2$PtCl$_4$	T 9	7,004	—	4,068	—
K$_2$PtCl$_6$	K 11	9,750	—	—	—
KReO$_4$	T 10	5,626	—	12,525	—
K$_2$S	K 6	7,406	—	—	—
KSH	H	4,383	—	—	68°51'
K$_2$SO$_4$	R	10,028	7,435	4,743	—
K$_2$S$_2$O$_5$	M	6,964	6,202	7,565	102°41'
K$_3$Sb	H 15	6,037	—	10,716	—
K$_2$Se	K 6	7,691	—	—	—
K$_2$SeO$_4$	R	6,032	10,421	7,615	—
K$_2$SnCl$_6$	K 11	10,000	—	—	—
KTaO$_3$	K 18	3,980	—	—	—
K$_2$Te	K 6	8,168	—	—	—
K$_2$TeO$_4$	R	10,521	7,916	6,263	—
K$_2$Zn(CN)$_4$	K 10	12,565	—	—	—
KZnF$_3$	K 18	4,058	—	—	—
La	K 2	5,305	—	—	—

Compound	Structure type	Lattice constants			α or β
		a	b	c	
La	H 1	3,758	—	6,072	—
LaAs	K 5	6,137	—	—	—
LaB$_6$	K 17	4,153	—	—	—
LaBO$_3$	R	8,237	5,842	5,110	—
LaBi	K 5	6,578	—	—	—
LaC$_2$	T 5	5,551	—	6,543	—
LaCrO$_3$	K 18	3,687	—	—	—
LaF$_3$	H	7,134	—	7,265	—
LaFeO$_3$	K 18	3,898	—	—	—
LaGa$_2$	H 20	4,329	—	—	—
LaMg$_3$	K 21	7,495	—	—	—
LaMnO$_3$	K 18	3,896	—	—	—
LaN	K 5	5,286	—	—	—
La$_2$O$_3$	H 16	3,938	—	6,142	—
LaP	K 5	6,025	—	—	—
LaPO$_4$	M	6,904	7,064	6,493	103°34′
LaSb	K 5	6,488	.—	—	—
Li	K 1	3,5087	—	—	—
LiAg	K 3	3,174	—	—	—
LiAl	K 21	6,313	—	—	—
Li$_3$As	H 15	4,396	—	7,826	—
LiBH$_4$	R	6,834	4,449	7,736	—
Li$_2$BeF$_4$	H	8,166	—	—	107°40′
LiBr	K 5	5,501	—	—	—
LiCd	K 21	6,703	—	—	—
LiCl	K 5	5,1398	—	—	—
LiD	K 5	4,073	—	—	—
LiF	K 5	4,025	—	—	—
LiGa	K 21	6,202	—	—	—
LiH	K 5	4,093	—	—	—
LiHg	K 3	3,294	—	—	—
Li I	K 5	6,012	—	—	—
Li$_2$MoO$_4$	H	8,788	—	—	108°10′
Li$_3$N	H	3,665	—	3,890	—
LiNO$_3$	H 8	5,752	—	—	48°3′
Li$_2$O	K 6	4,628	—	—	—
LiOH	K 3	3,557	—	4,349	—
Li$_3$P	H 15	4,273	—	7,594	—
Li$_2$S	K 6	5,719	—	—	—
Li$_2$SO$_4$	M	8,267	4,960	8,457	107°54′
α-Li$_3$Sb	H 15	4,710	—	8,326	—
Li$_2$Se	K 6	6,017	—	—	—
Li$_2$Te	K 6	6,517	—	—	—
Li$_2$WO$_4$	H	8,788	—	—	108°10′
LiZn	K 21	6,221	—	—	—
Mg	H 1	3,2092	—	5,2102	—
Mg$_3$As$_2$	K 16	12,355	—	—	—
MgAu	K 3	3,266	—	—	—
Mg$_3$Bi$_2$	H 16	4,675	—	7,390	—
MgBr$_2$	H 2	3,818	—	6,202	—
MgCO$_4$	H 8	5,621	—	—	48°12′
MgCaSiO$_4$	R	4,830	11,102	6,383	—
MgCe	K 3	3,906	—	—	—
MgCd$_3$	H 1	2,936	—	5,521	—
Mg$_3$Cd	H 1	6,273	—	5,080	—
MgCl$_2$	H 19	6,233	—	—	33°3 6′
MgCrO$_4$	R	11,914	12,034	6,904	—
MgF$_2$	T 4	4,670	—	3,086	—
MgHg	K 3	3,447	—	—	—
Mg I$_2$	H 2	4,148	—	6,894	—
MgLa	K 3	3,973	—	—	—
Mg$_3$N$_2$	K 16	9,970	—	—	—

Compound	Structure type	Lattice constants			α or β
		a	b	c	
$MgNi_2$	H	4,815	—	15,802	—
MgO	K 5	4,211	—	—	—
$Mg(OH)_2$	H 2	3,116	—	4,780	—
Mg_3P_2	K 16	12,034	—	—	—
Mg_2Pb	K 6	6,850	—	—	—
MgPr	K 3	3,888	—	—	—
MgS	K 5	5,200	—	—	—
$MgSO_4$	R	11,924	12,034	6,874	—
Mg_3Sb_2	H 16	4,582	—	7,244	—
MgSe	K 5	5,462	—	—	—
MgSi	K 6	6,403	—	—	—
Mg_2SiO_4	R	4,765	10,231	5,997	—
Mg_2Sn	K 6	6,784	—	—	—
Mg_2SnO_4	K 10	8,597	—	—	—
MgSr	K 3	3,908	—	—	—
MgTe	H 3	4,529	—	7,345	—
$MgTiO_3$	H 7	5,551	—	—	54°39′
Mg_2TiO_4	K 10	8,457	—	—	—
MgTl	K 3	3,635	—	—	—
Mg_2VO_4	K 10	8,403	—	—	—
$MgWO_4$	M	4,689	5,671	4,930	—
MgZn	H	10,681	—	17,165	—
$MgZn_2$	H 14	5,160	—	8,497	—
$MgZn_5$	H	9,940	—	16,513	—
α-Mn	K 14	8,908	—	—	—
β-Mn	K 12	6,303	—	—	—
γ-Mn	T 2	3,788	—	3,527	—
MnAs	H 4	3,723	—	5,716	—
MnAs	R	6,373	5,641	3,627	—
MnB	R	2,956	11,523	4,108	—
$MnBe_2$	H 14	4,239	—	6,924	—
MnBi	H 4	4,309	—	6,132	—
$MnCO_3$	H 8	5,852	—	—	47°45′
$MnCl_2$	H 19	6,213	—	—	34°35′
MnF_2	T 4	4,875	—	—	—
$MnFe_5$	H 13	2,545	—	4,088	—
MnI_2	H 2	4,168	—	6,834	—
MnO	K 5	4,444	—	—	—
MnO_2	T 4	4,449	—	2,896	—
Mn_2O_3	K 16	9,429	—	—	—
Mn_3O_4	T	5,762	—	9,439	—
$Mn(OH)_2$	H 2	3,116	—	4,749	—
MnP	R	5,917	5,260	3,173	—
Mn_2P	H	6,092	—	3,457	—
MnS	K 5	5,223	—	—	—
MnS	K 4	5,611	—	—	—
MnS	H 3	3,984	—	6,445	—
MnS_2	K 9	6,109	—	—	—
MnSb	H 4	4,128	—	5,796	—
Mn_2Sb	T	4,088	—	6,623	—
Mn_3Sb_2	H 4	4,138	—	5,752	—
MnSe	K 5	5,459	—	—	—
MnSe	K 4	5,832	—	—	—
MnSe	H 3	4,128	—	6,734	—
MnSi	K 15	4,557	—	—	—
Mn_5Si_3	H	6,912	—	4,810	—
MnTe	H 4	4,132	—	6,712	—
$MnTe_2$	K 9	6,957	—	—	—
$MnTiO_3$	H 7	5,631	—	—	54°16′
Mn_2TiO_4	K 10	8,687	—	—	—
$MnWO_4$	M	4,850	5,772	4,800	89°7′
$MnZn_7$	H 13	2,756	—	4,409	—

Compound	Structure type	Lattice constants			α or β
		a	b	c	
Mo	K 1	3,147	—	—	—
$MoBe_2$	H 14	4,439	—	7,295	—
MoC	H 12	2,907	—	2,792	—
Mo_2C	H 13	3,000	—	4,739	—
MoN	H 12	2,866	---	2,806	—
MoO_2	T 4	4,870	—	2,796	—
MoO_3	R	3,928	13,968	3,667	---
MoS_2	H	3,156	—	12,304	---
$MoSi_2$	T 5	3,206	—	7,877	---
NH_4BF_4	K 19	7,565	—	—	—
NH_4Br	K 5	6,914	---	—	—
NH_4Br	K 3	4,055	—	—	—
NH_4Br	T	6,019	—	4,264	---
NH_4Cl	K 3	3,874	---	—	---
NH_4Cl	K 5	6,543	—	—	—
NH_4ClO_2	T	6,343	—	3,758	—
NH_4ClO_4	R 1	9,221	5,828	7,464	—
NH_4ClO_4	K 19	7,645	—	—	—
NH_4F	H 3	4,399	—	7,034	---
NH_4I	K 3	4,379	—	—	—
NH_4I	K 5	7,259	—	—	—
NH_4I_3	R	6,653	9,679	10,842	---
NH_4IO_4	T 10	5,950	—	12,816	—
NH_4OsO_3N	R	5,551	8,572	13,567	—
$(NH_4)_2PbCl_6$	K 11	10.16	—	—	—
$(NH_4)_2PdCl_4$	T 9	7,225	—	4,269	—
NH_4SH	T 3	6,023	—	4,018	—
$(NH_4)_2SO_4$	R	5,982	10,621	7,776	—
$(NH_4)_2S_2O_8$	M	7,846	8,056	6,142	95°9′
$(NH_4)_2SiF_6$	K 11	8,397	—	—	—
$(NH_4)_2SnCl_6$	K 11	10,06	—	—	—
Na	K 1	4,2906	---	—	—
Na_3As	H 15	5,098	—	9,000	—
Na_2Au	T 6	7,417	—	5,522	—
$NaBO_2$	H	7,235	—	—	111°29′
$NaBaPO_4$	T	7,976	—	8,287	—
NaBi	T	3,467	—	4,820	—
Na_3Bi	H 15	5,499	—	9,674	---
NaBr	K 5	5,973	—	—	—
$NaBrO_3$	K 5	6,733	—	—	—
NaCH	K 5	5,842	—	—	—
$NaCdPO_4$	H	10,572	—	5,772	—
NaCl	K 5	5,63995	—	—	—
$NaClO_3$	K 5	6,583	—	—	---
$NaClO_4$	K 19	7,094	—	—	—
Na_2CrO_4	R	5,922	9,249	7,215	—
$NaCrS_2$	H 21	6,884	—	—	29°48′
NaF	K 5	4,629	---	—	—
$NaFeO_2$	H 21	5,601	—	—	31°20′
NaH	K 5	4,890	—	—	—
$NaHC_2$	T 5	5,411	—	—	—
$NaHCO_3$	M	7,525	9,720	3,537	93°19′
$NaHF_2$	H 21	5,060	—	—	46°2′
NaIn	K 21	7,315	—	—	—
NaI	K 5	6,433	—	—	—
$NaIO_4$	T 10	5,331	—	11,954	—
NaN_3	H 21	5,499	—	—	38°43′
$NaNO_2$	R	3,557	3,571	5,391	—
$NaNO_3$	H 8	6,333	—	—	47°13′
$NaNbO_3$	K 18	3,898	—	—	—
Na_2O	K 6	5,561	—	—	—
Na_3P	H 15	4,990	---	8,815	—

Compound	Structure type	Lattice constants			α or β
		a	b	c	
$Na_{15}Pb_4$	K	13,317	—	—	—
$NaPb_3$	K 2	4,880	—	—	—
Na_2S	K 6	6,243	—	—	—
$NaSH$	H	3,994	—	—	68°5′
Na_2SO_3	H	5,451	—	6,152	—
Na_2SO_4	R	5,862	12,315	9,770	—
Na_3Sb	H 15	5,366	—	9,515	—
Na_2Se	K 6	6,823	—	—	—
$NaTaO_3$	K 18	3,888	—	—	—
Na_2Te	K 6	7,329	—	—	—
$NaTl$	K 21	7,485	—	—	—
$NaWO_3$	K 18	3,838	—	—	—
$NaZn_{13}$	K	12,295	—	—	—
Nb	K 1	3,301	—	—	—
NbC	K 5	4,409	—	—	—
NbN	K 5	4,419	—	—	—
NbO_2	T 4	4,780	—	2,926	—
Nd	H 1	3,664	—	5,882	—
NdC_2	T 5	5,421	—	6,243	—
NdF_3	H	7,035	—	7,211	—
Nd_2O_3	H 16	3,848	—	6,002	—
$NdPO_4$	M	6,724	6,934	6,373	103°28′
Ni	K 2	3,5238	—	—	—
Ni	H 1	2,655	—	4,329	—
$NiAl$	K 3	2,826	—	—	—
$NiAl_3$	R	6,611	7,367	4,812	—
Ni_2Al_3	H	2,846	—	—	90°20′
$NiAs$	H 4	3,619	—	5,049	—
$NiAs_2$	R	5,752	5,822	11,42	—
$NiAsS$	K 20	5,691	—	—	—
Ni_2B	T 6	4,990	—	4,245	—
$NiBi$	H 4	4,078	—	5,371	—
$NiBr_2$	H 19	6,478	—	—	33°20″
Ni_3C	H 13	2,651	—	4,349	—
Ni_5Cd_{21}	K 24	9,781	—	—	—
$NiCl_2$	H 19	6,142	—	—	33°36″
$NiCrO_4$	R	5,503	8,236	6,125	—
NiF_2	T 4	4,719	—	3,124	—
Ni_2FeS_4	K 10	9,464	—	—	—
Ni_2GeO_4	K 10	8,216	—	—	—
NiI_2	H 19	6,934	—	—	32°40″
Ni_2Mg	H 18	4,815	—	15,802	—
$NiMgZn$	K 22	6,974	—	—	—
NiO	K 5	4,1767	—	—	—
$Ni(OH)_2$	H 2	3,123	—	4,604	—
Ni_2P	H	5,862	—	3,367	—
NiS	H 4	3,427	—	5,311	—
NiS	H	5,666	—	—	110°36″
NiS_2	K 9	5,752	—	—	—
Ni_3S_4	K 10	9,476	—	—	—
$NiSO_4$	R	11,884	12,104	6,824	—
$NiSb$	H 4	3,948	—	5,150	—
$NiSbS$	K 20	5,922	—	—	—
$NiSe$	H 4	3,667	—	5,341	—
$NiSe_2$	K 9	4,446	—	—	—
$NiSn$	H 4	4,089	—	5,184	—
Ni_3Sn	H 22	5,286	—	4,249	—
$NiTe$	H 4	3,965	—	5,365	—
$NiTe$	H 2	3,869	—	5,303	—
$NiTiO_3$	H 7	5,461	—	—	55°8′
$NiWO_4$	M	4,689	5,671	4,940	89°40″
$NiZn$	T	2,751	—	3,211	—

Compound	Structure type	Lattice constants			α or β
		a	b	c	
Ni_5Zn_{21}	K 24	8,922	—	—	—
Os	H 1	2,7353	—	4,3190	—
OsO_2	T 4	4,519	—	3,196	—
OsS_2	K 9	5,6187	—	—	—
$OsSe_2$	K 9	5,945	—	—	—
$OsTe_2$	K 9	6,382	—	—	—
P (white)	K	7,184	—	—	—
P (black)	R	3,317	4,389	10,52	—
PH_4I	T 3	6,353	—	4,629	—
Pb	K 2	4,9497	—	—	—
$PbBr_2$	R	9,537	8,054	4,726	—
$PbCO_3$	R	8,485	6,158	5,176	—
$PbCl_2$	R	9,048	7,623	4,535	—
$PbCrO_4$	M	7,114	7,415	6,814	—
PbF_2	K 6	5,954	—	—	—
PbFBr	T 11	4,188	—	7,605	—
$PbHPO_4$	M	4,659	6,643	5,742	—
PbI_2	H 2	4,549	—	6,874	—
$PbMg_2$	K 6	6,850	—	—	—
$PbMoO_4$	T 10	5,421	—	12,104	—
$Pb(NO_3)_2$	K 9	7,856	—	—	—
PbO	T 3	3,955	—	4,999	—
PbO (yellow)	R	5,470	4,733	5,871	—
PbO_2	T 4	4,941	—	3,374	—
Pb_2O	M	7,044	5,631	3,938	82°
Pb_2O	K 7	5,391	—	—	—
Pb_3O_4	T	8,806	—	6,564	—
Pb_2O_5	R	16,333	8,136	5,261	—
Pb_2O_5	H	7,445	—	—	87°
$Pb(OH)_2$	H	5,271	—	14,73	—
PbS	K 5	5,9351	—	—	—
$PbSO_4$	R	8,467	5,391	6,944	—
PbSe	K 5	6,152	—	—	—
Pb_2SnO_4	T	8,738	—	6,613	—
PbTe	K 5	6,353	—	—	—
$PbTiO_3$	K 18	3,898	—	—	—
$PbWO_4$	T 10	5,451	—	12,034	—
$PbZrO_3$	K 18	9,299	—	—	—
Pd	K 2	3,8902	—	—	—
PdH	K 9	5,982	—	—	—
PdBe	K 3	2,819	—	—	—
$PdCl_2$	R	3,818	3,347	11,022	—
PdCu	K 3	3,004	—	—	—
$PdCu_3$	K 2	3,697	—	—	—
PdF_2	T 4	4,940	—	3,387	—
PdF_3	H 17	5,571	—	—	—
Pd_2H	K 7	4,008	—	—	54°0″
PdO	T 3	3,035	—	5,325	—
PdS	T	6,443	—	6,623	—
PdSb	H 4	4,078	—	5,591	—
$PdSb_2$	K 9	6,452	—	—	—
PdTe	H 4	4,135	—	5,674	—
$PdTe_2$	H 2	4,036	—	5,128	—
Pd_5Zn_{21}	K 24	8,105	—	—	—
Pr	H 1	3,664	—	5,892	—
PrC_2	T 5	5,451	—	6,363	—
PrF_3	H	7,075	—	7,233	—
PrO_2	K 6	5,401	—	—	—
PrO_3	H 16	3,858	—	6,012	—
$PrPO_4$	M	6,764	6,954	6,413	103°21″
$PrVO_4$	T 4	7,305	—	6,423	—
Pt	K 2	3,9236	—	—	—

Compound	Structure type	Lattice constants			α or β
		a	b	c	
$PtAl_2$	K 6	5,922	—	—	—
PtAs	K 9	5,694	—	—	—
$PtBi_2$	K 9	6,696	—	—	—
PtCu	K 2	7,716	—	—	—
PtCu	H	7,575	—	—	90°54′
$PtCu_3$	K 2	3,717	—	—	—
$PtGa_2$	K 6	5,923	—	—	—
$PtIn_2$	K 6	6,366	—	—	—
PtO	T 3	3,046	—	5,351	—
PtP_2	K 9	5,694	—	—	—
PtS	T	3,477	—	4,359	—
PtS_2	H 2	3,544	—	5,029	—
PtSb	H 4	4,138	—	5,471	—
$PtSb_2$	K 9	6,440	—	—	—
$PtSe_2$	H 2	3,732	—	5,072	—
PtSn	H 4	4,111	—	5,439	—
$PtSn_2$	K 6	6,426	—	—	—
PtTe	H 4	4,138	—	5,461	—
$PtTe_2$	H 2	4,018	—	5,212	—
PtTl	H	5,616	—	4,659	—
Pt_5Zn_{21}	K 24	18,116	—	—	—
Rb	K 1	5,631	—	—	—
$RbBF_4$	R	9,088	5,611	7,245	—
RbBr	K 5	6,868	—	—	—
RbCN	K 5	5,561	—	—	—
RbCl	K 5	6,553	—	—	—
$RbClO_4$	K 19	7,716	—	—	—
$RbClO_4$	R	9,289	5,822	7,545	—
RbF	K 5	5,651	—	—	—
RbH	K 5	6,052	—	—	—
RbI	K 5	7,341	—	—	—
$RbIO_4$	T 10	5,886	—	12,964	—
RbN_3	T 7	6,372	—	7,435	—
Rb_2O	K 6	6,754	—	—	—
RbO_2	T 5	6,012	—	7,044	—
Rb_2PdCl_6	K 11	10,215	—	—	—
$RbReO_4$	T 10	5,815	—	13,193	—
Rb_2S	K 6	7,665	—	—	—
RbSH	H	4,534	—	—	69°20′
Rb_2SO_4	R	5,982	10,451	7,826	—
Re	H 1	2,7608	—	—	—
Rh	K 2	2,8012	—	—	—
RhF_3	H 17	5,351	—	—	54°20′
Rh_2MgO_4	K 10	8,527	—	—	—
$RhNbO_4$	T 4	4,695	—	3,0201	—
Rh_2O_3	H 7	5,481	—	—	55°40′
Rh_2P	K 6	5,516	—	—	—
RhS_2	K 9	5,585	—	—	—
$RhSbO_4$	T 4	4,610	—	3,106	—
$RhTaO_4$	T 4	4,693	—	3,026	—
$RhVO_4$	T 4	4,616	—	2,929	—
Rh_2ZnO_4	K 10	8,537	—	—	—
S	M	10,922	10,982	11,042	83°16′
S	R	10,501	12,946	24,048	—
SH_2	K 9	5,801	—	—	—
Sb	H 9	4,5066	—	—	57°6,6′
Sb_2CO_3	T	8,507	—	5,922	—
Sb_2FeO_4	T	8,609	—	5,917	—
SbI_3	H 17	8,197	—	—	54°14′
Sb_2MgO_4	T	8,462	—	5,919	—
Sb_2MnO_4	T	8,703	—	5,992	—
Sb_2NiO_4	T	8,367	—	5,922	—

Compound	Structure type	Lattice constants			α or β
		a	b	c	
Sb_2O_3	K 4	11,162	—	—	—
Sb_2O_4	K	10,261	—	—	—
Sb_2S_3	R	11,223	11,303	3,838	—
$SbTaO_4$	R	4,926	5,553	11,804	—
Sb_2Tl_7	K	11,613	—	—	—
$SbZnO_4$	T	8,508	—	5,932	—
SbC_2O_3	K 16	9,810	—	—	—
α-Se	M	9,010	8,991	11,543	91°34'
β-Se	M	12,766	8,056	9,269	93°4'
β-Se	H	4,364	—	4,965	—
SeH_2	K 9	6,062	—	—	—
SeO_2	T	8,370	—	5,062	—
Si	K 4	5,4306	—	—	—
SiC	K 4	4,357	—	—	—
SiF_4	K	5,421	—	—	—
SiI_4	K 13	12,010	—	—	—
$SiMg_2$	K 6	6,403	—	—	—
SiO	K	6,413	—	—	—
SiO_2 (α- quartz)	H	4,910	—	5,401	—
SiO_2 (β- quartz)	H	5,020	—	5,511	—
SiO_2 (α- cristobalite)	T	4,970	—	6,934	—
SiO_2 (β- cristobalite)	K 8	7,134	—	—	—
SiO_2 (α- tridymite)	R	9,900	17,134	16,333	—
SiO_2 (β- tridymite)	H	5,040	—	8,227	—
SiP_2O_7	K	7,475	—	—	—
SiS_2	R	5,611	5,541	9,569	—
Sn (white)	T 1	5,831	—	3,176	—
Sn (gray)	K 4	6,473	—	—	—
SnAs	K 5	5,692	—	—	—
$SnCl_2$	R	6,623	9,359	10,000	—
SnI_4	K 13	12,225	—	—	—
$SnMg_2$	K 6	6,779	—	—	—
SnO	T 3	3,804	—	4,826	—
SnO_2	T 4	4,747	—	3,191	—
SnS	R	4,339	11,202	3,988	—
SnS_2	H 2	3,646	—	5,880	—
SnSb	K 5	6,142	—	—	—
SnTe	K 5	6,298	—	—	—
Sr	K 2	6,087	—	—	—
SrB_6	K 17	4,198	—	—	—
SrC_2	T 5	5,822	—	6,693	—
$SrCO_3$	R	5,130	8,417	6,092	—
$SrCeO_3$	K 18	4,279	—	—	—
$SrCl_2$	K 6	6,994	—	—	—
SrF_2	K 6	5,792	—	—	—
SrH_2	R	6,377	7,358	3,883	—
$SrHfO_3$	K 18	4,279	—	—	—
$SrMoO_4$	T 10	5,371	—	11,964	—
SrNH	K 5	5,461	—	—	—
$Sr(NO_3)_2$	K 9	7,826	—	—	—
SrO	K 5	5,156	—	—	—
SrO_2	T 5	5,030	—	6,563	—
SrS	K 5	5,882	—	—	—
$SrSO_4$	R 1	8,377	5,371	6,854	—
SrSe	K 5	6,022	—	—	—
$SrSnO_3$	K 18	4,033	—	—	—
SrTl	K 3	4,032	—	—	—
SrTe	K 5	6,483	—	—	—
$SrTiO_3$	K 18	3,907	—	—	—
$SrWO_4$	T 10	5,411	—	11,924	—
$SrZrO_3$	K 18	4,088	—	—	—
Ta	K 1	3,303	—	—	—

Compound	Structure type	Lattice constants			α or β
		a	b	c	
TaC	K 5	4,454	—	—	—
Ta$_2$C	H	3,097	—	4,940	—
TaN	H 3	3,056	—	4,950	—
TaS$_2$	H	3,397	—	5,411	—
Te	H	4,456	—	5,922	—
TeO$_2$	T 4	4,800	—	3,778	—
Th	K 2	5,090	—	—	—
ThB$_6$	K 17	4,158	—	—	—
ThC$_2$	T	5,862	—	5,291	—
ThO$_2$	K 6	5,601	—	—	—
α-Ti	H 1	2,959	—	4,689	—
β-Ti	K 1	3,307	—	—	—
TiAg	T 2	4,104	—	4,077	—
TiAl	T 2	3,986	—	4,085	—
TiAl$_3$	T	5,435	—	8,577	—
TiAu$_2$	H 1	2,786	—	4,760	—
TiAu$_6$	T 2	4,068	—	4,188	—
TiB	K 4	4,210	—	—	—
TiB$_2$	H 20	3,032	—	3,222	—
TiBe$_2$	K 22	6,448	—	—	—
TiBr$_4$	K 13	11,272	—	—	—
TiC	K 5	4,329	—	—	—
TiCo	K 1	2,994	—	—	—
TiCo$_2$	K 22	6,704	—	—	—
TiCo$_2$	H	4,725	—	15,401	—
TiCr$_2$	K 22	6,943	—	—	—
TiCr$_2$	H 14	4,932	—	9,469	—
TiCu	T 1	3,108—3,118	—	5,887—5,921	—
Ti$_3$Cu	T 2	4,127	—	3,588	—
TiFe	K 3	2,975	—	—	—
TiFe$_2$	H 14	4,780	—	7,806	—
TiH$_2$	T 1	3,126	—	4,188	—
Ti$_3$Hg	K	5,192	—	—	—
TiI$_2$	H 2	4,118	—	6,834	—
TiI$_4$	K 13	12,026	—	—	—
TiN	K 5	4,244	—	—	—
TiO	K 5	4,244	—	—	—
TiO$_2$	T 4	4,603	—	2,965	—
TiO$_2$	R	9,189	5,451	5,150	—
Ti$_2$O$_3$	H 7	5,381	—	—	56°48'
TiS$_2$	H 2	3,407	—	5,701	—
TiSb	H 4	4,070	—	6,306	—
TiSb$_2$	T 6	6,666	—	5,817	—
TiSe	H 4	3,566	—	6,233	—
TiSe$_2$	H 2	3,540	—	6,007	—
Ti$_5$Si$_3$	H	7,465	—	5,162	—
Ti$_5$Sn$_3$	H	8,049	—	5,454	—
TiTe	H 2	3,842	—	6,403	—
TiTe$_2$	H 2	3,782	—	6,552	—
TiU$_2$	H 20	4,838	—	2,853	—
TiZn$_2$	H 14	5,074	—	8,227	—
TiZn$_3$	K 2	3,940	—	—	—
Tl	H 1	3,4565	—	5,5249	—
Tl	K 2	4,851	—	—	—
TlAsS$_2$	M	15,050	11,333	6,112	127°45'
TlBF$_4$	R	9,489	5,822	7,415	—
TlBi	K 3	3,988	—	—	—
TlBr	K 3	3,978	—	—	—
TlCl	K 3	3,842	—	—	—
TlClO$_4$	K 19	7,716	—	—	—
TlClO$_4$	R 1	9,439	5,892	7,515	—
TlF	R	5,190	5,506	6,092	—

Compound	Structure type	Lattice constants			α or β
		a	b	c	
TlI	K 3	4,206	—	—	—
TlI	R	5,251	4,579	12,946	—
TlNO₃	R	6,182	12,295	3,988	—
Tl₂O₃	K 16	10,591	—	—	—
TlReO₄	T 10	5,773	—	13,357	—
TlReO₄	R	5,634	5,803	13,322	—
TlSe	T	8,036	—	7,014	—
Tu	H 1	3,530	—	5,575	—
Tu₂O₃	K 16	10,541	—	—	—
TuVO₄	T 4	7,014	—	6,223	—
U	R	2,858	5,877	4,955	—
UCl₄	K	14,609	—	—	—
UO₂	K 6	5,481	—	—	—
V	K 1	3,0399	—	—	—
VBe₂	H 14	4,399	—	7,145	—
VBr	H 2	3,776	—	6,192	—
VC	K 5	4,158	—	—	—
V₂C	H 13	2,861	—	4,529	—
VCrO₄	R	5,579	8,225	5,989	—
V₂FeO₄	K 10	8,485	—	—	—
VI₂	H 2	4,008	—	6,689	—
V₂MgO₄	K 10	8,411	—	—	—
VN	K 5	4,137	—	—	—
VO	K 5	4,108	—	—	—
VO₂	T 4	4,549	—	2,886	—
V₂O₃	H 7	5,441	—	—	53°53'
V₂O₅	R	11,503	4,369	3,557	—
VS	H 4	3,367	—	5,825	—
VSe	H 4	3,587	—	5,989	—
W	K 1	3,1647	—	—	—
β-W	K	5,048	—	—	—
WBe₂	H 14	4,449	—	7,285	—
WC	H 12	2,916	—	2,844	—
W₂C	H	5,862	—	5,291	—
WO₂	T 4	4,870	—	2,776	—
WO₃	R	7,295	7,495	3,828	—
W₂P	H	6,192	—	6,794	—
WS₂	H	3,187	—	12,525	—
WSi₂	T 5	3,218	—	7,900	—
W₂Zr	K 22	7,625	—	—	—
Y	H 1	3,670	—	5,827	—
YAlO₃	K 18	3,677	—	—	—
YAsO₄	T 13	6,904	—	6,282	—
YC₂	H	3,798	—	6,593	—
YNbO₄	T	7,776	—	11,343	—
Y₂O₃	K 16	10,621	—	—	—
YPO₄	T 13	6,894	—	6,052	—
YTaO₄	T	7,766	—	11,433	—
YVO₄	T 13	7,144	—	6,363	—
YbCl₂	R	6,543	6,693	6,924	—
Yb₂O₃	K 16	10,411	—	—	—
YbVO₄	T 13	7,034	—	6,243	—
Zn	H 1	2,6648	—	4,9456	—
ZnAs₂	R	7,736	8,006	36,353	—
Zn₃As₂	T 12	8,333	—	11,784	—
Zn(CN)₂	K 7	5,902	—	—	—
ZnCO₄	H 8	5,680	—	—	48°26'
ZnCe	K 3	3,707	—	—	—
ZnCl₂	H 19	6,323	—	—	34°48'
ZnCrO₄	R	5,516	8,400	6,232	—
ZnF₂	T 4	4,730	—	3,146	—
Zn₃Hg	H	2,705	—	5,451	—

Compound	Structure type	Lattice constants			α or β
		a	b	c	
ZnLa	K 3	3,758	—	—	—
Zn_3N_2	K 16	9,763	—	—	—
ZnO	H 3	3,2491	—	5,2052	—
$Zn(OH)_2$	R	5,170	8,547	4,930	—
ZnP_2	T	5,080	—	18,668	—
Zn_3P_2	T 12	8,113	—	11,473	—
ZnPr	K 3	3,677	—	—	—
ZnS	K 4	5,423	—	—	—
ZnS	H 3	3,819	—	6,247	—
$ZnSO_4$	R	11,874	12,114	6,844	—
ZnSe	K 4	5,661	—	—	—
Zn_2SiO_4	H	8,637	—	—	107°44′
Zn_2SnO_4	K 10	8,627	—	—	—
ZnTe	K 4	6,082	—	—	—
Zn_2TiO_4	K 10	8,462	—	—	—
$ZnWO_4$	M	4,689	5,742	4,960	89°30′
α-Zr	H 1	3,237	—	5,150	—
β-Zr	K 1	3,617	—	—	—
$ZrAl_2$	R	10,421	7,225	4,980	—
$ZrAl_3$	T	16,934	—	4,315	—
ZrB	K 2	4,659	—	—	—
ZrB_2	H 20	3,156	—	3,537	—
ZrB_{12}	K 2	7,423	—	—	—
ZrC	K 5	4,696	—	—	—
$ZrCO_2$	K 22	6,901	—	—	—
$ZrCl_4$	K 13	10,341	—	—	—
$ZrCr_2$	H 14	5,089	—	8,279	—
Zr_2Cu	T 2	4,545	—	3,724	—
$ZrGe_2$	R	3,808	15,040	3,768	—
Zr_3Ge	H 22	6,533	—	5,391	—
ZrH_2	T 2	4,974	—	4,449	—
$ZrMn_2$	H 14	5,039	—	8,240	—
ZrN	K 5	4,619	—	—	—
ZrO_2	K 6	5,080	—	—	—
$ZrOs_2$	H 14	5,189	—	8,526	—
ZrP_2O_7	K	8,206	—	—	—
$ZrPt_3$	H 22	5,644	—	9,229	—
$ZrRu_2$	H 14	5,141	—	8,507	—
ZrS_2	H 2	3,687	—	5,862	—
Zr_2Sb	H	8,417	—	5,611	—
$ZrSe_2$	H 2	3,798	—	6,192	—
$ZrSi_2$	R	3,728	14,640	3,677	—
$ZrSiO_4$	T 13	6,593	—	5,942	—
ZrSn	R	7,448	5,834	5,167	—
Zr_2U	K 1	10,699	—	—	—
ZrV_2	H 14	5,288	—	8,664	—
ZrW_2	K 2	7,625	—	—	—
$ZrZn_2$	K 12	7,410	—	—	—

5-3.   Interplanar Distances and Line Intensities for Elements and Compounds

Publisher's Note

There follows at this point in the Russian original text a 126-page tabulation of interplanar distances and line intensities, which inspection revealed to be taken in most part from the ASTM Powder Data File, circa 1957.

These tables have been omitted from the translation, in recognition that more and more exact data are conveniently available in the current ASTM Powder Data File, and also in respect to the copyright to this material held by the Joint Committee on Chemical Analysis by Powder Diffraction Methods, the continuance of whose nonprofit operations depends on the income from sale of the card file.

Full information on the X-Ray Powder Data File can be obtained from X-Ray Department, American Society for Testing Materials, 1916 Race Street, Philadelphia, Pennsylvania, 19103.

5-4. Tables for Phase Analysis of Isomorphous Compounds

5-4a. Cubic System

Figures 115 and 116 show schemes for the x-ray patterns of 33 types of cubic structure; Table 5-4a gives the lattice constants for some common compounds belonging to these types.

Table 5-4a is used as follows in phase analysis: (1) the lg d and the relative intensities are plotted on a strip of paper in the scale of the figures; (2) the upper scale in the figure is used to check whether the substance is cubic; (3) the schemes are used to identify the structure type; (4) the indices and interplanar distances are used to calculate the lattice constants; and (5) the composition is deduced from Table 5-4a.

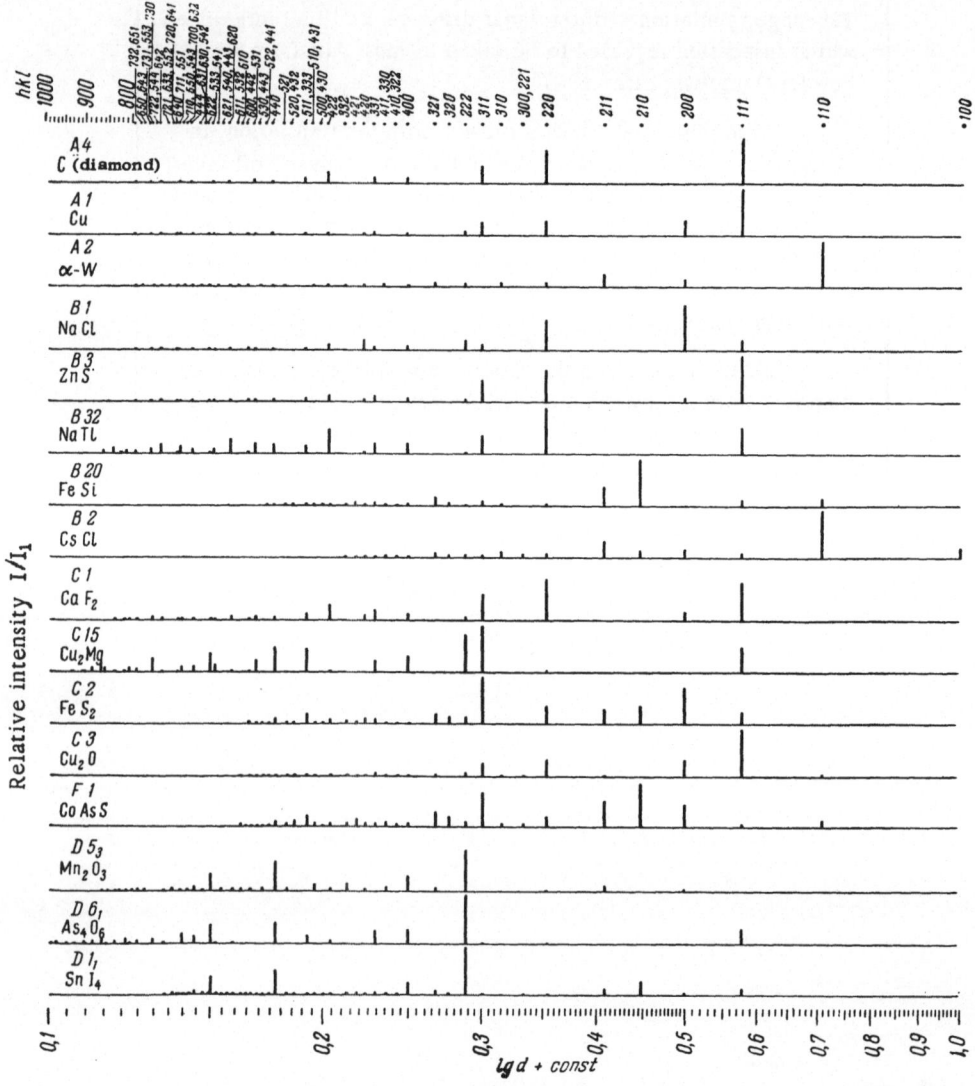

Fig. 115. Schemes for the x-ray patterns of crystals in the cubic system (structures from A4 to D1₁).

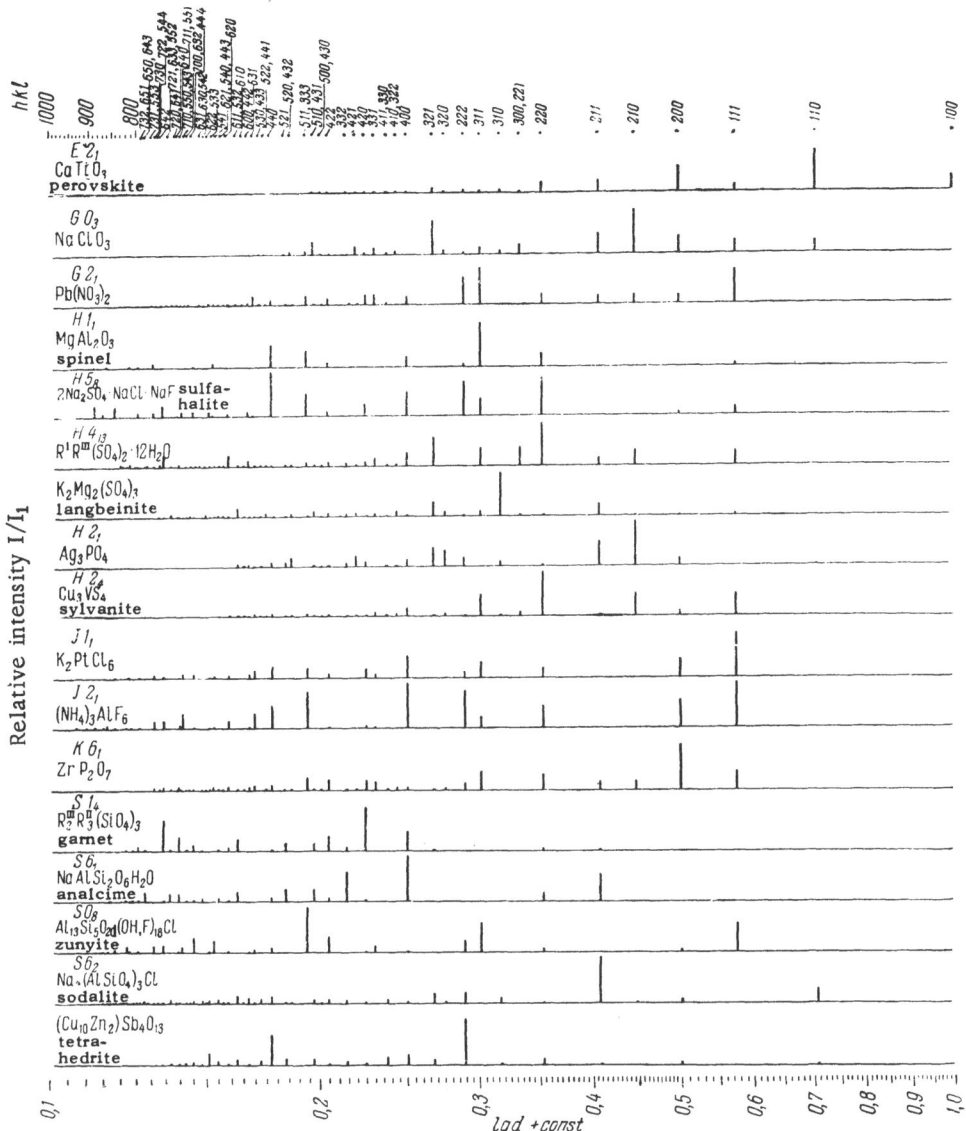

Fig. 116. Schemes for the x-ray patterns of crystals in the cubic system (structures from E2₁ to tetrahedrite).

If the lattice constants are not given in the table (as for solid solutions), use is made of Vegard's law (lattice constant proportional to concentration).

Consider the determination of the phase composition of an unknown substance [125]. Table 1 (first two columns) gives the results from the x-ray pattern of the given unknown substance, and comparison with standards shows that the material consists of two cubic phases isomorphous with NaCl (structure B1). The third and fourth columns give the indices for the two phases A and B, whose lattice constants are found to be 6.96 and 6.64 Å, respectively; Table 5-4a shows these lattice constants to be close to those of BaTe(6.99Å)

TABLE 1. Calculation of the X-Ray Pattern for the Substance

d, Å	I	hkl		I/I₁	
		Phase A	Phase B	Phase A	Phase B
4,02	10	111	—	0,50	—
3,83	6	—	111	—	0,30
3,48	20	200	—	1,00	—
3,315	20	—	200	—	1,00
2,460	17,5	220	—	0,88	—
2,350	17,5	—	220	—	0,88
2,100	6	311	—	0,30	—
2,010	8	222	311	0,40	—
1,920	4	—	222	—	0,20
1,740	4	400	—	0,10	—
1,660	2	—	400	—	0,10
1,599	1,5	331	—	0,08	—
1,559	6	420	—	0,80	—
1,487	6	—	420	—	0,30
1,421	4	422	—	0,20	—

TABLE 2. X-Ray Analysis of the Precipitate

d, Å	I	hkl		I/I₁	
		Ag (Br, I)	Ag I	Ag (Br, I)	Ag I
3,77	3	—	111	—	1,0
2,96	75	200	—	1,00	—
2,29	2	—	220	—	0,7
2,087	40	220	—	0,53	—
1,961	1	—	311	—	0,3
1,710	10	222	—	0,13	—
1,479	3	400	—	0,04	—
1,324	7	420	—	0,09	—
1,207	4	422	—	0,05	—

and SrTe (6.65 Å). However, qualitative spectral analysis shows that potassium is the only metal present. Table 5-4a then indicates that KCl, KCN, KBr, and KI have B1 structures and lattice constants close to the values found. Tests for Cl⁻ and CN⁻ are negative, but a test with $AgNO_3$ gives a typical halide precipitate. Table 2 gives the results from the patterns for the precipitate, which indicate that it consists of a very little AgI together with AgBr·AgI (lattice constant 5.916 ± 0.009Å), whose lattice constant lies between the 5.76 Å of AgBr and the 5.92 Å of AgI. Vegard's law indicates that the original material consisted of a phase consisting of 80% KI and 20% KBr together with one consisting of 90% KBr and 10% KI. An artificial mixture of these phases gives the pattern of the original material.

TABLE 5-4a

Type	**A4**	4,14	CrN
		4,14	VC
3,56	C (diamond)	4,142	$63Li_2Fe_2O_4 \cdot 37Li_2TiO_3$
5,42	Si	4,173	NiO
5,62	Ge	4,207	MgO
6,46	α-Sn	4,282	MgO (1570° K)
		4,225	TiN
Type	**A1**	4,235	TiO
		4,24	80TiN—20TiC
3,517	Ni	4,27	CoO
3,554	α-Co	4,28	V—N
3,60	(Taenite)(57,7% Fe, 40,8% Ni, 0,5% P)	4,283	FeO (160° K)
3,608	Cu	4,290	FeO (299° K)
3,63	γ-Fe (1370° K)	4,30	VC (ε- phase)
3,797	Rh	4,315	TiC
3,831	Ir	4,40	NbC
3,880	Pd	4,41	NbN
3,88—4,04	Pd—H	4,426	MnO (117°K)
3,912	Pt	4,436	MnO (299° K)
4,041	Al	4,44	ScN
4,070	Au	4,446	TaC
4,077	Ag	4,458	HfC
4,30	Co—N	4,615	NaF
4,40—4,46	Ti—H	4,62	ZrN
4,52	Ne (4° K)	4,69	CdO
4,66	Zr—H	4,69	ZrC
4,84	β-Tl	4,80	CaO
4,939	Pb	4,82	Na_2CeO_3
5,08	Th	4,84	Na_2PrO_3
5,14	α-Ce	4,88	NaH
5,296	β-La	4,92	AgF
5,43	Ar (4° K)	5,006	CaNH
5,56	Ca	5,13	SrO
5,59	Kr (20° K)	5,14	LiCl
5,70	Kr (92° K)	5,14	NdN
6,05	Sr	5,19	MgS
6,20	Xe (88° K)	5,192	MnS (130° K)
		5,210	MnS (299° K)
Type	**A2**	5,33	KF
		5,45	MgSe
2,861	α-Fe	5,45	MnSe
2,875	α-Cr	5,45	SrNH
2,90	β-Fe (1070° K)	5,49	LiBr
2,93	δ-Fe (1700° K)	5,52	BaO
3,03	V	5,545	AgCl
3,03—3,41	V—C	5,55—5,76	AgCl—AgBr
3,140	Mo	5,627	NaCl
3,157	W	5,63	RbF
3,295	Nb	5,68	CaS
3,30	Ta	5,69	SnAs
3,32	β-Ti (1200° K)	5,70	KH
3,46	Li (~80° K)	5,755	AgBr
3,50	Li	5,70—5,92	AgBr—AgI
3,61	β-Zr(1120° K)	5,83	NdP
4,24	Na (~80° K)	5,83	NaCN
4,29	Na	5,84	BaNH
5,02	Ba	5,87	SrS
5,20	K (120° K)	5,91	CaSe
5,33	K	5,94	PbS
5,62	Rb (~80° K)	5,95	NaBr
6,05	Cs (~80° K)	5,957	EuS
		5,96	NdAs
Type	**B1**	6,00	PrAs
		6,00	LiI
4,018	LiF	6,01	CsF
4,065	LiD	6,04	RbH
4,08	VO	6,05	β-NaSH (>360° K)
4,09	LiH		
4,12	Li_2TiO_3		
4,12—4,20	$Li_2TiO_3 \cdot MgO$		
4,13	VN		

6,06	CeAs
6,13	LaAs
6,14	PbSe
6,23	SrSe
6,278	KCl
6,285	SnTe
6,31	NdSb
6,345	CaTe
6,35	PrSb
6,36	BaS
6,38	CsH
6,40	CeSb
6,44	PbTe
6,462	NaI
6,45	PrBi
6,48	LaSb
6,49	CeBi
6,53	KCN
6,53	NH_4Cl ($>457°$ K)
6,56	RbCl
6,57	LaBi
6,58	KBr
6,59	BaSe
6,60	β-KSH ($>440°$ K)
6,65	SrTe
6,82	RbCN
6,86	RbBr
6,90	NH_4Br ($>411°$ K)
6,93	β-RbSH ($470°$ K)
6,99	BaTe
7,052	KI
7,10	β-CsCl ($>730°$ K)
7,24	NH_4I ($>255°$ K)
7,325	RbI

Type HO_5

$6,96 \pm 0,04$	$AgClO_4$ ($453 \pm 20°$ K)
$7,16 \pm 0,10$	$NaClO_4$ ($618 \pm 35°$ K)
$7,49 \pm 0,02$	$KClO_4$ ($598 \pm 15°$ K)
$7,65 \pm 0,05$	$TiClO_4$ ($553°$ K)
$7,65 \pm 0,02$	NH_4ClO_4 ($528 \pm 15°$ K)
$7,68 \pm 0,03$	$RbClO_4$ ($513 \pm 10°$ K)
$7,97 \pm 0,01$	$CsClO_4$ ($513 \pm 10°$ K)

Type B3

4,255	CuF
4,36	SiC-IV
4,855	BeS
5,10	BeSe
5,304	Colusite$(Cu, Fe, Mo, Sn)_4$ $(S, As, Te)_{3-4}$
5,41	CuCl
5,425	β-ZnS
5,43	AlP
5,44	GaP
5,58	BeTe
5,60	MnS (red)
5,63	AlAs
5,635	GaAs
5,655	ZnSe
5,68	CuBr
5,82	β-CdS
5,84	HgS
5,86	InP
6,04	CdSe
6,04	InAs
6,05	CuI
6,07	HgSe
6,08	ZnTe

6,103	α-Cu_2HgI_4
6,12	AlSb
6,12	GaSb
6,13	SnSb
6,383	α-Ag_2HgI_4
6,40	HgTe
6,43	CdTe
6,45	InSb
6,48	AgI

Type B32

6,195	LiGa
6,209	LiZn
6,36	LiAl
6,687	LiCd
6,786	LiIn
7,297	NaIn
7,373	$CeMg_3$
7,373	$PrMg_3$
7,473	NaTl

Type B20

4,437	NiSi
4,438	FeSi
4,438	CoSi
4,548	MnSi
4,620	CrSi

Type B2

2,603	NiBe
2,606	CoBe
2,69	CuBe
2,813	PdBe
2,82	AlNi
2,945	CuZn
2,989	CuPd
3,146	AuZn
3,156	AgZn
3,168	AgLi
3,259	AuMg
3,275	AgMg
3,287	HgLi
3,325	AgCd
3,34	AuCd ($670°$K)
3,424	LiTl
3,442	HgMg
3,628	MgTl
3,67	PrZn
3,70	CeZn
3,73	AlNd
3,74	α-RbCl ($83°$ K)
3,75	LaZn
3,82	TlCN
3,82	PrCd
3,84	TlSb
3,835	TiCl
3,847	CaTl
3,86	NH_4Cl ($<457°$ K)
3,86	CeCd
3,88	MgPr
3,90	LaCd
3,97	TlBr
3,98	TlBi
4,024	SrTl
4,05	NH_4Br ($<411°$ K)
4,112	CsCl
4,20	TiI
4,20	CsCl ($<720°$ K)
4,25	CsCN

4,287	CsBr
4,29	CsSH
4,37	NH_4I (290° K)
4,56	CsI

Type $D2_1$

4,07	YB_6
4,07	ErB_6
4,10	NdB_6
4,12	GdB_6
4,12	PrB_6
4,13	CeB_6
4,13	YbB_6
4,14	CaB_6
4,15	LaB_6
4,15	ThB_6
4,19	SrB_6
4,33	BaB_6

Type C1

4,33	Be_2C
4,619	Li_2O
5,06	$(3ZrO_2 \cdot MgO)$
5,07	ZrO_2
5,08	$(95ZrO_2 \cdot 5CeO_2)$
5,13	$(95HfO_2 \cdot 5CeO_2)$
5,38	PrO_2
5,40	CeO_2
5,40	CdF_2
5,406	CuF_2
5,45	CaF_2
5,47	UO_2
5,526	$(66CaF_2 \cdot 33YF_3)$
5,53	$(91CaFe_2 \cdot 9ThF_4)$
5,54	HgF_2
5,55	Na_2O
5,58	ThO_2
5,59	Cu_2S
5,704	Li_2S
5,749	Cu_2Se
5,782	SrF_2
5,796	EuF_2
5,838	$(66SrF_2 \cdot 33LaF_3)$
5,91	$PtAl_2$
5,91	$PtGa_2$
5,935	$\beta\text{-}PbF_2$ (520° K)
5,99	Al_2Au
6,005	Li_2Se
6,06	$AuGa_2$
6,19	BaF_2
6,34	Mg_2Si
6,35	$PtIn_2$
6,368	RaF_2
6,379	Mg_2Ge
6,436	K_2O
6,50	Li_2Te
6,50	$AuIn_2$
6,526	Na_2S
6,763	Mg_2Sn
6,809	Na_2Se
6,81	Mg_2Pb
6,98	$SrCl_2$
7,314	Na_2Te
7,38	K_2S
7,65	RbS_2
7,676	K_2Se
8,152	K_2Te

Type C15

5,94	Be_2Cu
6,287	Be_2Ag
6,435	Be_2Ti
6,96	$MgNiZn$
7,03	Cu_2Mg
7,61	W_2Zr
7,79	Au_2Na
7,91	Au_2Pb
7,94	Au_2Bi
8,02	Al_2Ca
8,04	Al_2Ce
8,16	Al_2La
9,50	Bi_2K

Type C2

5,41	FeS_2
4,42	$(Fe, Ni)S_2$ (6,5% Ni)
5,57	RhS_2
5,57	RuS_2
5,57	Bravoite (53,8% NiS_2, 39,1% FeS_2, 7,1% CoS_2)
5,62	OsS_2
5,64	CoS_2
5,65	$(Cu, Ni, Co, Fe) (S, Se)_2$
5,68	PtP_2
5,74	NiS_2
5,85	$CoSe_2$
5,92	$RuSe_2$
5,93	$OsSe_2$
5,94	$PtAs_2$
5,97	$PdAs_2$
6,02	$NiSe_2$
6,096	MnS_2
6,36	$RuTe_2$
6,37	$OsTe_2$
6,43	$PtSb_2$
6,44	$PdSb_2$
6,64	$AuSb_2$
6,94	$MnTe_2$

Type C3

4,25	Cu_2O
4,73	Ag_2O

Type F1

5,55	$CoAsS$
5,68	$NiAsS$
5,90	$NiSbS$
	$(Ni, Fe)AsS$ (plessite)
	$Ni(As, Sb)S$ (corynite)
	$Ni(Sb, Bi)S$ (kallilite)
	$(Co, Ni)SbS$ (willyamite)

Type $D5_3$

8,13	Be_3N_2
9,37	$(Mn, Fe)_2O_3$
9,42	Mn_2O_3
9,74	Zn_3N_2
9,79	Sc_2O_3
9,94	Mg_3N_2
10,12	In_2O_3
10,15	Be_3P_2
10,37	Lu_2O_3
10,39	Yb_2O_3
10,52	Tu_2O_3
10,54	Er_2O_3
10,57	Tl_2O_3

10,58	Ho_2O_3
10,60	Y_2O_3
10,63	Dy_2O_3
10,70	Tb_2O_3
10,79	Gd_2O_3
10,79	Cd_3N_2
10,84	Eu_2O_3
10,85	Sm_2O_3
11,05	Nd_2O_3
11,40	$\alpha\text{-}Ca_3N_2$
12,02	Mg_3P_2
12,33	Mg_3As_2

Type D6₁

11,05	As_4O_6
11,14	Sb_4O_6

Type D1₁

10,32	$ZrCl_4$
11,25	$TiBr_4$
(11,34)	CBr_4 (>320° K)
(11,62)	CI_4
11,89	GeI_4
11,99	SiI_4
12,00	TiI_4
12,23	SnI_4

Type E2₁

3,67	$YAlO_3$
3,75	$CdTiO_3$
3,78	$LaAlO_3$
3,80	$CaTiO_3$
3,83	$NaWO_3$
3,85	$(Na, Ce, Ca)(Ti, Nb)O_3$
3,88	$NaTaO_3$ (loparite)
3,89	$LaGaO_3$
3,89	$NaNbO_3$
3,91	$SrTiO_3$
3,92	$CaSnO_3$
3,97	$BaTiO_3$
3,98	$KTaO_3$
3,99	$CaZrO_3$
4,00	$KMgF_3$
4,005	$KNiF_3$
4,01	$KNbO_3$
4,03	$SrSnO_3$
4,05	$KZnF_3$
4,07	$KCoF_3$
4,07	$SrHfO_3$
4,09	$SrZrO_3$
4,18	$BaZrO_3$
4,35	$BaPrO_3$
4,38	$BaCeO_3$
4,46	KIO_3
4,48	$BaThO_3$
4,5	NH_4IO_3
4,52	$RbIO_3$
4,66	$CsIO_3$
5,12	$MgZrO_3$
5,20	$CsCdCl_3$
5,33	$CsCdBr_3$
5,44	$CsHgCl_3$
5,77	$CsHgBr_3$

Type G0₃

6,57	$NaClO_3$
6,71	$NaBrO_3$

Type G2₁

7,60	$Ca(NO_3)_2$
7,81	$Sr(NO_3)_2$
7,84	$Pb(NO_3)_2$
8,11	$Ba(NO_3)_2$

Type H1₁

8,045	$NiAl_2O_4$
8,07	$CuAl_2O_4$
8,07	$CoCo_2O_4$
8,07	$MgAl_2O_4$
8,08	$CoAl_2O_4$
8,08	$ZnAl_2O_4$
8,10	$FeAl_2O_4$
8,11	$(Ni, Co)(Co, Ni)_2O_4$
8,11	$(Zn, Co)Co_2O_4$
8,11	$MgCo_2O_4$
8,27	$MnAl_2O_4$
8,27	$(Mn, Co)(Co, Mn)_2O_4$
8,28	$MgGa_2O_4$
8,30	$NiCr_2O_4$
8,30	$MgCr_2O_4$
8,31	$ZnCr_2O_4$
8,32	$CoCr_2O_4$
8,32	$ZnGa_2O_4$
8,35	$NiFe_2O_4$
8,35	$Cu_2Cr_2O_4$
8,35	$FeCr_2O_4$
8,36	$MgFe_2O_4$
8,38	$CoFe_2O_4$
8,38	$NiMn_2O_4$
8,40	$ZnFe_2O_4$
8,40	$FeFe_2O_4$
8,42	$(Mn, Mg)Fe_2O_4$
8,42	$TiCo_2O_4$
8,43	$MnCr_2O_4$
8,43	$TiMg_2O_4$
8,43	$TiZn_2O_4$
8,44	$CuFe_2O_4$
8,47	FeV_2O_4
8,49	$MnCr_2O_4$
8,50	$TiFe_2O_4$
8,54	$MnFe_2O_4$
8,58	$CdCr_2O_4$
8,58	$SnMg_2O_4$
8,61	$SnCo_2O_4$
8,63	$SnZn_2O_4$
8,67	$CdFe_2O_4$
8,67	$TiMn_2O_4$
8,81	$MgIn_2O_4$
9,26	Ag_2MoO_4
9,4	$CoCo_2S_4$
9,45	$(Co, Ni)_3S_4$
9,46	$CuCo_2S_4$
9,5	NiN_2S_4
9,92	$ZnCr_2S_4$
10,05	$MnCr_2S_4$
10,19	$CdCr_2S_4$
12,54	$K_2Zn(CN)_4$
12,76	$K_2Hg(CN)_4$
12,84	$K_2Cd(CN)_4$

Type H5₈

10,08	$2Na_2SO_4NaClNaF$

Type H4$_{13}$

12,11	$KCr(SO_4)_2 \cdot 12H_2O$
12,12	$KAl(SO_4)_2 \cdot 12H_2O$
12,15	$NH_4Al(SO_4)_2 \cdot 12H_2O$
12,15	$NH_3Fe(SO_4)_2 \cdot 12H_2O$
12,20	$RbAl(SO_4)_2 \cdot 12H_2O$
12,21	$TlAl(SO_4)_2 \cdot 12H_2O$
12,31	$CsAl(SO_4)_2 \cdot 12H_2O$

Langbeinite type

9,93	$K_2Mg(SO_4)_3$
10,2	$K_2(Ca, Mg)(SO_3)_3$

Type H2$_1$

6,00	Ag_3PO_4
6,120	Ag_3AsO_4 (90° K)
6,130	Ag_3AsO_4 (380° K)

Type H2$_4$

5,37	Cu_3VS_4

Type J1$_1$

8,17	K_2SiF_6
8,35	$(NH_4)_2SiF_6$
8,38	$Rb_2CrF_5 \cdot H_2O$
8,41	$Tl_2CrF_5 \cdot H_2O$
8,42	$(NH_4)_2VF_5 \cdot H_2O$
8,42	$Rb_2VF_5 \cdot H_2O$
8,45	$Tl_2VF_5 \cdot H_2O$
8,45	Rb_2SiF_6
8,58	Tl_2SiF_6
8,87	Cs_2SiF_6
8,99	Cs_2GeF_6
9,73	K_2PtCl_6
9,73	K_2OsCl_6
9,76	Tl_2PtCl_6
9,84	$(NH_4)_2PtCl_6$
9,86	K_2ReCl_6
9,88	Rb_2PtCl_6
9,92	Rb_2TiCl_6
9,94	$(NH_4)_2SeCl_6$
9,97	K_2SnCl_6
9,97	Tl_2SnCl_6
9,98	Rb_2SeCl_6
10,02	Rb_2PdBr_6
10,04	$(NH_4)_2SnCl_6$
10,08	$Ni(NH_3)_6Cl_6$
10,10	Rb_2SnCl_2
10,10	$Co(NH_3)_6Cl_6$
10,11	Tl_2TeCl_6
10,14	$(NH_4)_2PbCl_6$
10,14	K_2TeCl_6
10,15	$Fe(NH_3)_6Cl_2$
10,16	$Mg(NH_3)_6Cl_2$
10,17	Cs_2PtCl_6
10,18	Rb_2ZrCl_6
10,18	$(NH_4)_2TeCl_3$
10,20	$Mn(NH_3)_6Cl_2$
10,20	Rb_2PbCl_6
10,22	Cs_2TiCl_6
10,23	Rb_2TeCl_6
10,25	$Zn(NH_3)_6(ClO_4)_2$
10,26	Cs_2SeCl_6
10,30	K_2OsBr_6
10,35	Cs_2SnCl_6
10,36	K_2SeBr_6

10,36	K_2PtBr_6
10,39	$Co(NH_3)_6Br_2$
10,4	$Ni(NH_3)_6Br_2$
10,41	Cs_2ZrCl_6
10,42	Cs_2PbCl_6
10,45	Cs_2TeCl_6
10,45	$Co(NH_3)_5H_2OSO_4Br$
10,46	$(NH_4)_2SeBr_6$
10,46	$Zn(NH_3)_6Br_2$
10,47	$Fe(NH_3)_6Br_2$
10,47	$Mg(NH_3)_6Br_2$
10,48	K_2SnBr_6
10,51	$Co(NH_3)_6SO_4Br$
10,52	$Mn(NH_3)_6Br_2$
10,54	$Sr_2Ni(NO_2)_6$
10,55	$Pb_2Ni(NO_2)_6$
10,57	$(NH_4)_2SnBr_6$
10,58	Rb_2SnBr_6
10,62	$Co(NH_3)_5H_2OSO_4I$
10,63	$Co(NH_3)_6SeO_4Br$
10,67	$Ba_2Ni(NO_2)_6$
10,71	$Ca(NH_3)_6Br_2$
10,71	$Co(NH_3)_6SO_4I$
10,77	Cs_2SnBr_6
10,79	$Co(NH_3)_6SeO_4I$
10,9	$Ni(NH_3)_6I_2$
10,91	$Co(NH_3)_6I_2$
10,96	$Zn(NH_3)_6I_2$
10,97	$Fe(NH_3)_6I_2$
10,98	$Mg(NH_3)_6I_2$
11,04	$Mn(NH_3)_6I_2$
11,04	$Cd(NH_3)_6I_2$
11,24	$Ca(NH_3)_6I_2$
11,27	$Ni(NH_3)(BF_4)_2$
11,3	$Co(NH_3)_6(BF_4)_2$
(11,3)	$Zn(NH_3)_6(ClO_4)_2$
11,34	$Mg(NH_3)_6(BF_4)_2$
11,34	$Fe(NH_3)_6(BF_4)_2$
11,37	$Mn(NH_3)_6(BF_4)_2$
11,38	$Cd(NH_3)_6(BF_4)_2$
11,41	$Ni(NH_3)_6(ClO_4)_2$
11,43	$Co(NH_3)_6(ClO_4)_2$
11,46	$Ni(NH_3)_6(SO_3F)_2$
11,49	$Co(NH_3)_6(SO_3F)_2$
11,52	$Fe(NH_3)_6(ClO_4)_2$
11,53	$Mg(NH_3)_6(ClO_4)_2$
11,54	$Cd(NH_3)_6Br_2$
11,54	$Fe(NH_3)_6(SO_3F)_2$
11,58	$Mn(NH_3)_6(ClO_4)_2$
11,59	$Cd(NH_3)_6(ClO_4)_2$
11,59	$Mn(NH_3)_6(SO_3F)_2$
11,62	$Cd(NH_3)_6(SO_3F)_2$
11,91	$Ni(NH_3)_6(PF_6)_2$
11,94	$Co(NH_3)_6(PF_6)_2$

Type J2$_1$ and similar structures

8,88	Li_3FeF_6
8,90	$(NH_4)_3AlF_6$
9,01	$(NH_4)_3CrF_6$
9,04	$(NH_4)_3VF_6$
9,10	$(NH_4)_3FeF_6$
9,10	$(NH_4)_3MoO_3F_3$
9,26	Na_3FeF_6
9,93	K_3FeF_6
9,96	$CuLi_2Fe(CN)_6$
10,0	$CuR_2Fe(CN)_6$
	R-Na, K, Rb, NH_4, Tl

10,15	$K_2CdFe(NO_2)_6$
10,17	$K_2CaCo(NO_2)_6$
10,19	$K_2CaFe(NO_2)_6$
10,2	$Fe^{+++}RFe^{++}(CN)_6$
	R-Na, K, Rb, NH_4
10,22	$K_2HgFe(NO_2)_6$
10,23	$K_2SrCo(NO_2)_6$
10,25	$(NH_4)_2CdFe(NO_2)_6$
10,25	$NaTl_2Co(NO_2)_6$
10,28	$K_2CdNi(NO_2)_6$
10,28	$(NH_4)_2CdFe(NO_2)_6$
10,29	$K_2HgNi(NO_2)_6$
10,30	$K_2SrFe(NO_2)_6$
10,30	$Tl_2CaFe(NO_2)_6$
10,31	$K_2PbFe(NO_2)_6$
10,32	$K_2CaNi(NO_2)_6$
10,34	$(NH_4)_2SrFe(NO_2)_6$
10,37	$(NH_4)_2\ PbFe(NO_2)_6$
10,37	$Tl_2CdNi(NO_2)_6$
10,39	$Tl_2PbFe(NO_2)_6$
10,39	$NaRb_2Co(NO_2)_6$
10,40	$Tl_2SrFe(NO_2)_6$
10,4	$K_2PbCo(NO_2)_6$
10,41	$(NH_4)_2CdNi(NO_2)_6$
10,42	$Tl_2HgNi(NO_2)_6$
10,43	$K_2BaFe(NO_2)_6$
10,45	$K_2BaCo(NO_2)_6$
10,45	$K_3Co(NO_2)_6$
10,46	$(NH_4)_2HgNi(NO_2)_6$
10,47	$Rb_2HgNi(NO_2)_6$
10,49	$K_2SrNi(NO_2)_6$
10,49	$K_4Ni(NO_2)_6$
10,50	$(NH_4)_2BaFe(NO_2)_6$
10,54	$K_2LiBi(NO_2)_6$
10,55	$K_2PbNi(NO_2)_6$
10,55	$Tl_2BaFe(NO_2)_6$
10,58	$Rb_2CdNi, Cd(NO_2)_6$
10,58	$K_3Ir(NO_2)_6$
10,59	$Rb_2LiBi(NO_2)_6$
10,6	$K_2PbCu(NO_2)_6$
10,63	$K_3Rh(NO_2)_6$
10,63	$(NH_4)_2LiBi(NO_2)_6$
10,64	$Tl_2LiBi(NO_2)_6$
10,67	$K_2BaNi(NO_2)_6$
10,70	$NaCs_2Co(NO_2)_6$
10,70	$Ba_3[Rh(NO_2)_6]_2$
10,72	$Tl_3Co(NO_2)_6$
10,73	$Rb_3Co(NO_2)_6$
10,73	$(NH_4)_3Ir(NO_2)_6$
10,73	$Tl_3Ir(NO_2)_6$
10,77	$Rb_3Ir(NO_2)_6$
10,8	$(NH_4)_2Co(NO_2)_6$
10,81	$Cs_2Cd[Ni, Cd(NO_2)_6]$
10,82	$Co(NH_3)_5H_2O\,I_3$
10,83	$Rb_3Rh(NO_2)_6$

10,88	$K_2NaBi(NO_2)_6$
10,89	$Co(NH_3)_6\,I_3$
10,91	$Tl_3Rh(NO_2)_6$
10,91	$(NH_4)_3Rh(NO_2)_6$
10,94	$Cs_2LiBi(NO_2)_6$
10,95	$K_2AgBi(NO_2)_6$
10,98	$Rb_2NaBi(NO_2)_6$
10,99	$(NH_4)_2NaBi(NO_2)_6$
11,01	$Tl_2NaBi(NO_2)_6$
11,05	$Rb_2AgBi(NO_2)_6$
11,06	$Tl_2AgBi(NO_2)_6$
11,10	$(NH_4)_2AgBi(NO_2)_6$
11,15	$Cs_2NaBi(NO_2)_6$
11,15	$Cs_3Co(NO_2)_6$
11,17	$Cs_3Ir(NO_2)_6$
11,19	$Cs_3Bi(NO_2)_6$
11,19	$Cs_2AgBi(NO_2)_6$
11,21	$Co(NH_3)_6(BF_4)_3$
11,30	$Cs_3Rh(NO_2)_6$
11,32	$[Co(NH_3)_5 \cdot H_2O](ClO_4)_3$
11,39	$Co(NH_3)_6(ClO_4)_3$
11,67	$Co(NH_3)_6(PF_6)$

Type K6₁

7,46	SiP_2O_7
7,80	TiP_2O_7
7,98	SnP_2O_7
8,18	HfP_2O_7
8,20	ZrP_2O_7
8,61	UP_2O_7

Type S1₄

11,51	$Al_2(Mg,\ Fe)_3(SiO_4)_3$ (pyrope)
11,51	$Al_2Fe_3(SiO_4)_3$ (almandine)
11,60	$Al_2Mn_3(SiO_4)_3$ (spessartine)
11,87	$Al_2Ca_3(SiO_4)_3$ (grossular)
11,89	$(Al,\ Fe)_2Ca_3(SiO_4)_3$ (hessonite)
11,95	$Cr_2Ca_3(SiO_4)_3$ (uvarovite)
12,03	$Fe_2Ca_3(SiO_4)_3$ (andradite)
12,10	$(Na,\ Li)_3AlF_6$ (cryolithionite)
12,35 12,46	$(Mg,\ Mn)_2(Ca,\ Na)_3\ (AlSO_4)_3$ (berzeliite)

Type S6₁

13,68	$NaAlSi_2O_6H_2O$

Type S0₈

13,82	$Al_{13}Si_5O_{20}(OH,\ F)_{18}Cl$ (zunyite)

Type S6₂

8,87	$Na_4(AlSiO_4)_3Cl$ (sodalite)

Tetrahydrite
Type

10,19	$(Cu,\ Fe)_{12}As_4S_{13}$ (binnite)
10,2—10,6	$(Cu,\ Ag)_{10}(Zn,\ Fe)_2(Sb,\ As)_4S_{13}$

5-4b. Tetragonal System

Figures 117-120 give schemes for 40 structure types and the appropriate parts of the graphs for indexing patterns; the latter are given for the range of c/a typical of phases of the given structure type.

Table 5-4b-1 lists about 300 tetragonal compounds by structure types; Table 5-4b-2 gives the lattice constants a, c, and c/a for 400 compounds arranged in order of increasing c/a.

A mixture of tetragonal compounds is indexed as follows: (1) the d and the relative intensities are plotted on a strip of paper in the scale used here; (2) comparison with the patterns for the cubic system is used to show that the substance is not cubic; (3) comparison with Figs. 117-120 is used to find the structure type; (4) the x-ray pattern is indexed (from the indices given under the scheme); (5) the lattice constants are calculated; (6) Tables 5-4b-1 and 5-4b-2 are used to find the phase composition from these constants; (7) the results are checked by qualitative spectral analysis or spot tests.

E x a m p l e . Table 3 gives values of d and relative intensities recorded with Mo Kα radiation [126]. Two methods can be used. In the first, the lg d are plotted on a strip of paper in the appropriate scale, which is then compared with the scale for the cubic system (top scale in Figs. 117-120) to show that the substance is not cubic. The strip is then tested against the patterns for the tetragonal system.

TABLE 3. Substance with a Tetragonal Structure

d, Å	I/I_1	hkl
7,25	0,06	001
3,75	1,00 (100)	101
3,60	0,15	002
3,10	1,00 (100)	110
(2,98)	(0,01)	
2,78	0,63	102
2,35	0,75 (75)	112
2,19	0,63	200
2,11	0,25	201, 103
1,90	0,40	211, 113
1,72	0,25	212
1,68	0,20	104
1,62	0,02	203
(1,59)	(0,01)	
1,55	0,15	220, 114
1,52	0,15	221, 213
1,43	0,08	222, 301
1,39	0,10	204, 310
1,355	0,02	311, 302

This test shows that structure C11, for which c/a = 1.64 (a = 4.4 Å, c = 7.2 Å), gives a similar pattern, but some lines (d = 7.25, 2.78, 1.72, 1.68, and 1.62 Å) cannot be indexed. Structure C38 with c/a = 1.65 (a = 4.38 Å, c = 7.23 Å) enables us to index all lines, but the intensities are in marked disagreement. Further, the table shows that C38 does not have a compound with these lattice constants. The same occurs for D3₁ (c/a

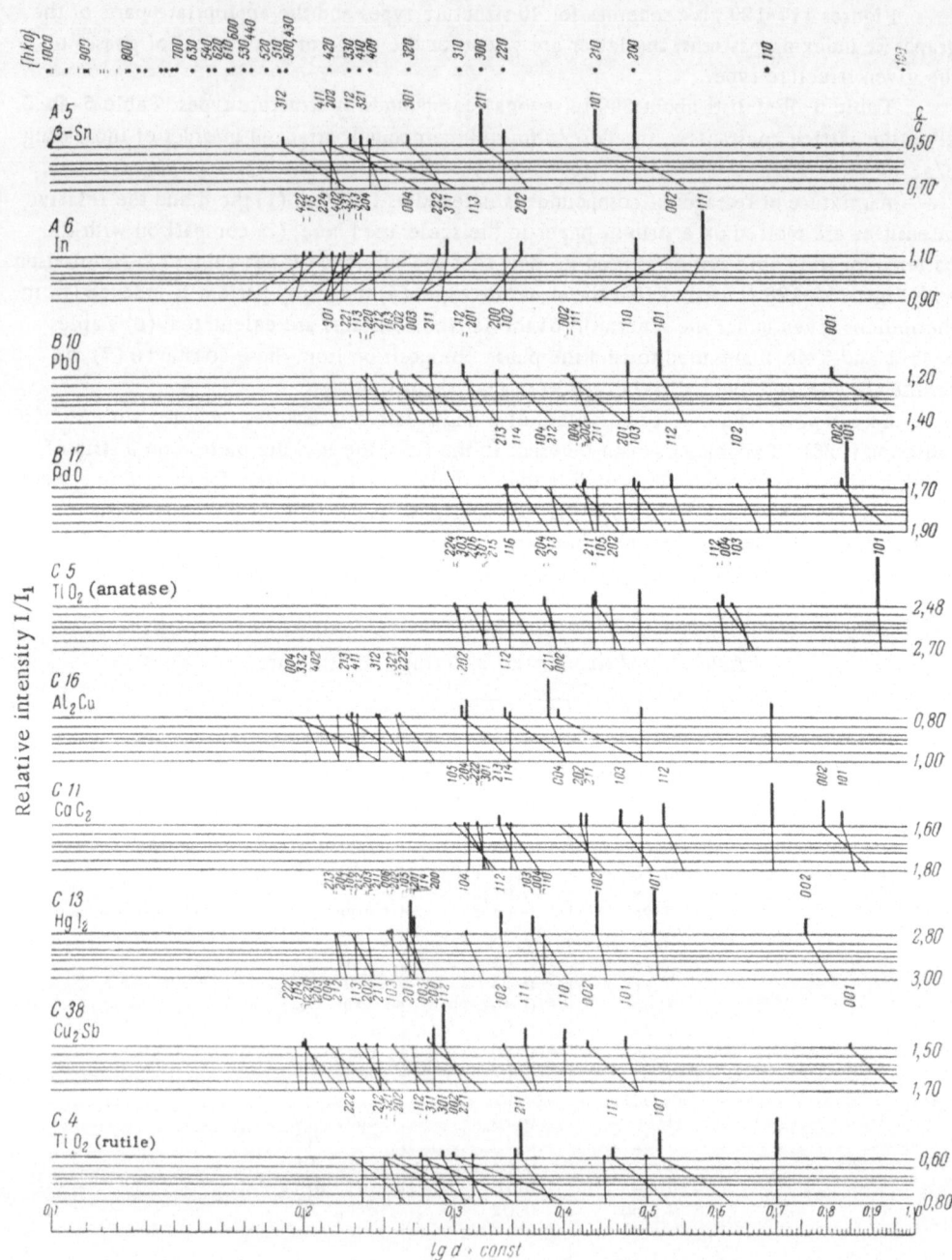

Fig. 117. Graphs for phase analysis of crystals in the tetragonal system (structures A5 to C4).

Fig. 118. Graphs for phase analysis of crystals in the tetragonal system (structures $D3_1$ to phosgenite).

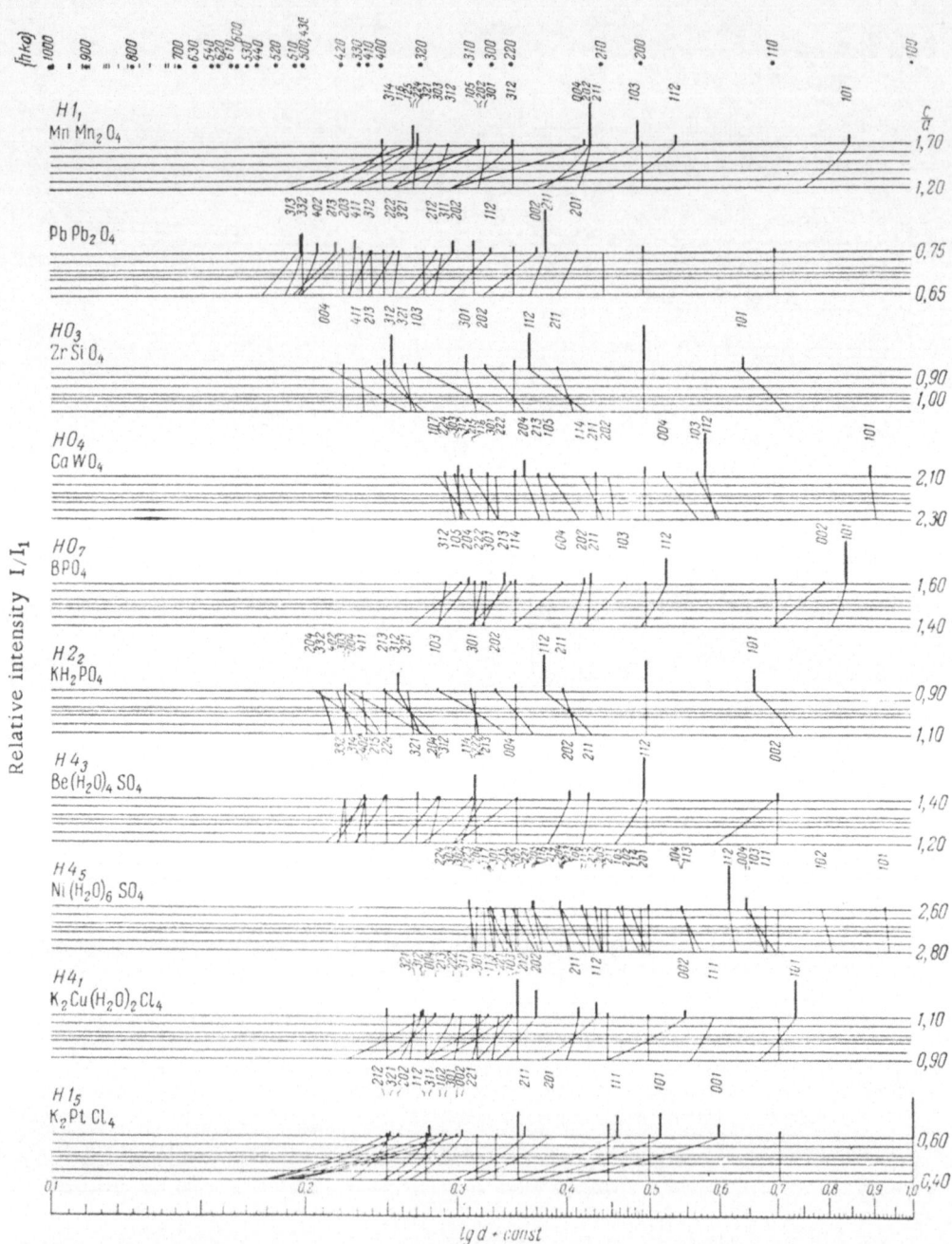

Fig. 119. Graphs for phase analysis of crystals in the tetragonal system (structures $H1_1$ to $H1_5$).

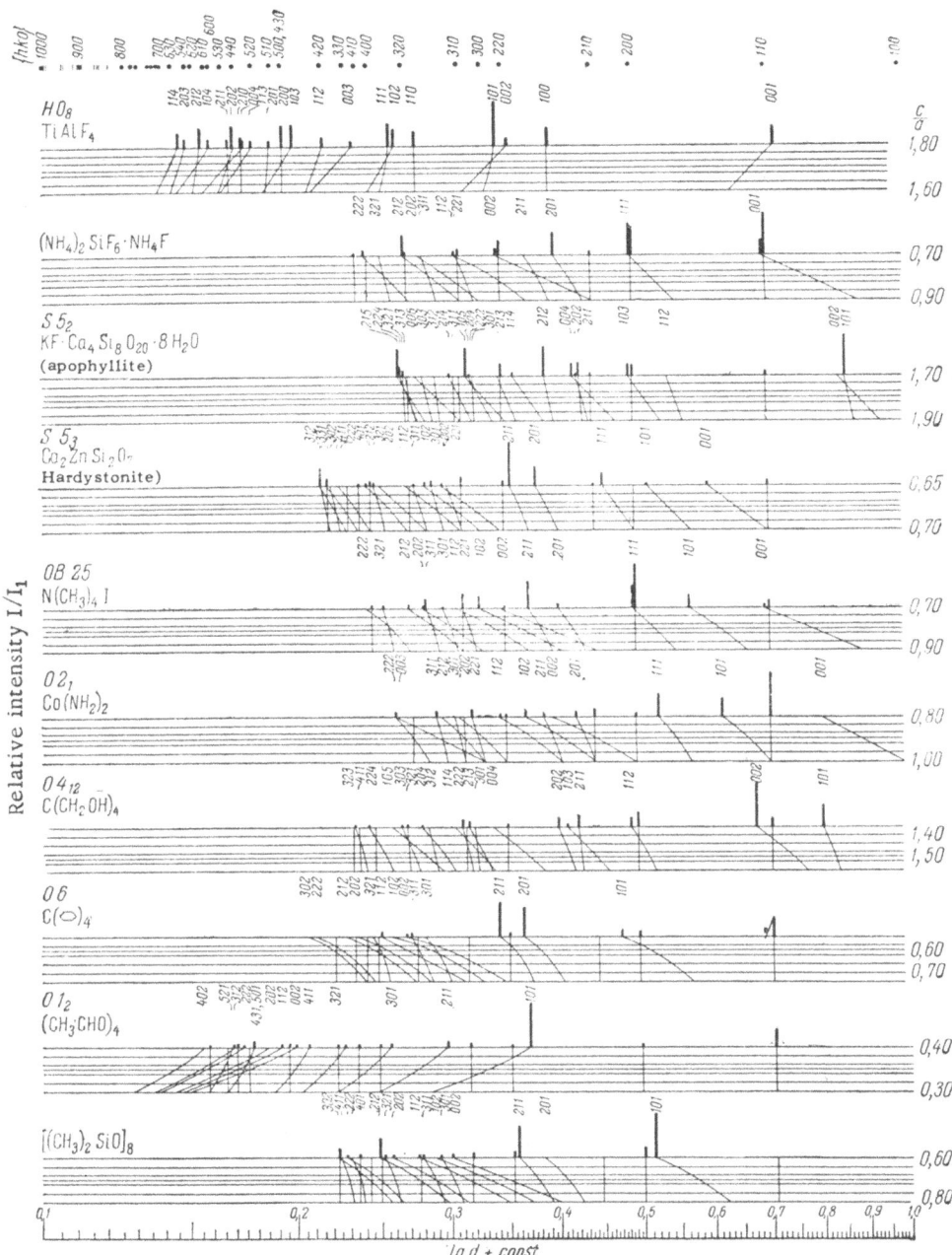

Fig. 120. Graphs for phase analysis of crystals in the tetragonal system (structures HO_8 to $[(CH_3)_2SiO]_8$).

= 2.32, a = 6.20 Å , c = 14.5 Å) and DO_{22} (c/a = 1.65, a = 8.75 Å , c = 14.5 Å). Finally, EO_1 gives good agreement for position and intensity for c/a = 1.65, a = 4.38Å , and c = 7.23 Å . Table 5-4b-1 shows that BaFCl has these lattice constants; spectral analysis shows that Ba is a major component. Further tests on the other structures give only partial agreement. The weak lines with d = 2.98 and 1.59 Å are not indexed; they probably relate to an impurity.

The second method is based on the use of Table 5-4b-2, which gives the compounds in order of increasing c/a. The initial steps here are as before; the difference is that a substance with known c/a and lattice constants is found from Table 5-4b-2. Checks must be made for c/a multiplied by 2, $\sqrt{2}$, $1/\sqrt{2}$, and 1/2.

TABLE 5-4b-1

c/a	a, kX	c, kX	Substance
		Type A5	
0,5456	5,819	3,175	β-Sn
		Type A6	
0,936	3,774	3,533	γ-Mn
0,952	3,76	3,58	95Mn·5Cu
0,962	3,767	3,624	89Mn·11Cu
0,981	3,864	3,693	79Mn·21Cu
0,998	3,752	3,744	66Mn·34Cu
1,077	4,585	4,937	In
		Type B10	
1,22	3,55	4,33	LiOH
1,26	3,98	5,01	PbO (red)
1,27	3,80	4,81	SnO
		Type B17	
1,74	3,03	5,26	PdO
1,76	3,03	5,32	PtO
1,76	3,47	6,10	PtS
		Types B25 and 0B25	
0,667	6,011	4,009	NH_4SH
0,707	6,18	4,37	γ-NH_4I
0,709	5,70	4,04	γ-NH_4Br (~173° K)
0,729	6,34	4,62	PH_4I
		Type B34	
1,03	6,37	6,58	(Pd, Pt, Ni) S
1,04	6,35	6,60	PdS
		Type B37	
0,873	8,02	7,00	TiSe

c/a	a, kX	c, kX	Substance
		Type NaBi	
1,35	3,24	4,38	$MgIn$
1,39	3,46	4,80	$NaBi$
		Type C4	
0,570	(4,86)	(2,77)	WO_2
0,574	(4,86)	(2,79)	MoO_2
0,621	4,77	2,96	NbO_2
0,633	4,61	2,92	$RhVO_4$
0,634	4,54	2,88	VO_2
0,642	4,69	3,01	$RhNbO_4$
0,644	4,58	2,95	TiO_2
0,645	4,68	3,02	$RhTaO_4$
0,649	4,41	2,86	CrO_2
0,649	4,64	3,01	$CrCbO_4$
0,650	4,63	3,01	$CrTaO_4$
0,651	4,39	2,86	GeO_2
0,652	4,40	2,87	MnO_2
0,652	4,62	3,01	$FeSbO_4$
0,652	4,67	3,04	$FeTaO_4$
0,652	4,68	3,05	$FeNbO_4$
0,656	4,51	2,96	$AlSbO_4$
0,659	4,64	3,06	MgF_2
0,660	4,59	3,03	$GaSbO_4$
0,660	4,71	3,11	NiF_2
0,664	4,58	3,04	$CrSbO_4$
0,665	4,72	3,14	ZnF_2
0,672	4,72	3,17	SnO_2
0,674	4,60	3,10	$RhSbO_4$
0,678	4,87	3,30	MnF_2
0,679	4,70	3,19	CoF_2
0,683	4,95	3,38	PbO_2
0,686	4,93	3,38	PdF_2
0,690	4,51	3,11	RuO_2
0,696	4,83	3,36	FeF_2
0,699	4,49	3,14	IrO_2
0,707	4,51	3,19	OsO_2
0,787	(4,79)	(3,77)	TeO_2
		Type C5	
2,51	3,75	9,43	TiO_2
		Type C11	
1,15	6,27	7,22	Cs_2O
1,17	5,99	7,02	Rb_2O
1,18	5,70	6,73	K_2O
1,20	—	—	UC_2
1,28	(4,14)	(5,28)	ThC_2
1,61	4,39	7,05	BaC_2
1,63	4,11	6,68	SrC_2
1,65	3,82	6,30	NdC_2
1,65	3,87	6,37	CaC_2
1,66	3,85	6,38	PrC_2
1,67	3,54	5,91	CaO_2
1,67	3,75	6,28	SmC_2
1,67	3,87	6,48	CeC_2
1,67	3,92	6,55	LaC_2
1,79	3,78	6,77	BaO_2

c/a	a, kX	c, kX	Substance
1,85	3,55	6,55	SrO_2
1,97	(4,28)	(8,42)	KHC_2
2,10	(3,89)	(8,17)	$NaHC_2$

Type C13

2,83	4,36	12,36	HgI_2

Type C16

0,806	6,052	4,878	Al_2Cu
0,814	6,52	5,31	Sn_2Fe
0,818	6,647	5,434	Sn_2Mn
0,832	5,099	4,240	Fe_2B
0,841	5,006	4,212	Co_2B
0,846	5,899	4,991	Ge_2Fe
0,851	4,980	4,236	Ni_2B
0,852	6,835	5,821	Pb_2Pd
0,857	6,348	5,441	Sn_2Co
0,880	6,651	5,853	Pb_2Rh

Type C20

2,45	3,21	7,88	WSi_2
2,46	3,20	7,86	$MoSi_2$

Type C30

1,40	4,96	6,92	SiO_2
1,41	5,00	7,06	$AlPO_4$

Type C38

1,53	3,992	6,091	Cu_2Sb
1,61	4,08	6,56	Mn_2Sb
1,65	3,627	5,973	Fe_2As
1,67	3,76	6,27	Mn_2As
1,75	3,613	6,333	Cr_2As

Type C47

0,605	8,35	5,05	SeO_2

Type C48

2,88	2,998	8,630	Cr_2Al

Type ZnP_2

3,68	5,07	18,65	ZnP_2
3,73	5,28	19,70	CdP_2

Type $D0_{22}$

1,46	5,548	8,093	$TiGa_3$
1,55	5,605	8,712	$ZrGa_3$
1,56	5,334	8,305	VAl_3
1,57	5,422	8,536	$TaAl_3$
1,58	5,425	8,579	$TiAl_3$
1,58	5,427	8,584	$NbAl_3$

Type DO_{23}

4,32	4,003	17,29	$ZrAl_3$

c/a	a, kX	c, kX	Substance
		Type D1$_3$	
2,46	4,53	11,14	Al_4Ba
2,48	4,45	11,04	Al_4Sr
2,54	4,35	11,07	Al_4Ca
		Type D3$_1$	
2,36	4,92	11,62	Hg_2I_2
2,39	4,65	11,10	Hg_2Br_2
2,44	4,46	10,89	Hg_2Cl_2
2,98	3,66	10,9	Hg_2F_2
		Type D5$_9$	
1,40	8,75	12,28	Cd_3P_2
1,41	8,10	11,45	Zn_3P_2
1,41	8,32	11,76	Zn_3As_2
1,41	8,95	12,65	Cd_3As_2
		Type Fe$_3$P	
0,490	9,09	4,45	Fe_3P
0,491	9,0i	4,42	$(Fe, Ni, Co)_3P$
0,492	8,92	4,39	Ni_3P
0,499	9,13	4,56	Cr_3P
0,502	9,16	4,60	Mn_3P
		Type E0$_1$	
1,65	4,38	7,22	$BaFCl$
1,68	4,10	6,88	$SrFCl$
1,71	4,65	7,93	$BaFI$
1,76	3,89	6,83	$CaFCl$
1,76	4,09	7,21	$PbFCl$
1,82	4,18	7,59	$PbFBr$
1,89	3,89	7,37	$BiOCl$
2,07	3,92	8,11	$BiOBr$
2,28	4,01	9,14	$BiOI$
		Type E1$_1$	
1,82	5,66	10,30	$AgFeS_2$
1,97	5,26	10,37	$CuFeS_2$
0,550	10,81	5,94	$(ZnCl_2)_4$
		Type E2$_5$	
1,89	4,19	7,94	NH_4HgCl_3
		Type E2$_6$	
0,502	13,30—13,52	6,69—6,78	$KMg(H_2O)_6(Cl, Br)_3$
		Type E3$_1$	
1,00	6,34	6,34	Ag_2HgI_4
1,01	6,08	6,14	Cu_2HgI_4
		Type E6$_1$	
0,906	(6,41)	(5,81)	$Sr(OH)_2 \cdot 8H_2O$

c/a	a, kX	c, kX	Substance

Type $E6_2$

0,880	(6,32)	(5,56)	$SrO_2 \cdot 8H_2O$

Type $FeTa_2O_6$

1,94	4,70	9,10	$NiTa_2O_6$
1,94	4,71	9,12	$Fe(Nb, Ta)_2O_6$
1,94	4,73	9,16	$CoTa_2O_6$
1,95	4,70	9,18	$MgTa_2O_6$
1,95	4,71	9,18	$FeTa_2O_6$

Type $F1_1$

0,921	9,69	8,92	$Hg(CN)_2$

Type $F5_2$

1,16	6,07	7,03	$KNCO$
1,16	6,09	7,06	KN_3
1,17	6,36	7,41	RbN_3
1,20	5,67	6,81	$KFHF$

Type $F5_4$

0,593	6,31	3,74	NH_4ClO_2

Type $G0_9$

0,871	5,74	5,00	NH_4NO_3 (357—398° K)

Phosgenite Type

1,086	8,13	8,83	$Pb_2Cl_2CO_3$
1,086	8,34	9,06	$Pb_2Br_2CO_3$

Type $H0_3$

0,867	7,13	6,18	YVO_4
0,874	7,25	6,34	$CaCrO_4$
0,888	6,87	6,10	YPO_4
0,901	6,58	5,93	$ZrSiO_4$
0,910	6,89	6,27	$YAsO_4$
1,46	(7,74)	(11,31)	$Y(Nb, Ta)O_4$
1,46	(7,76)	(11,32)	$YNbO_4$
1,47	(7,75)	(11,41)	$YTaO_4$

Type $H0_4$

2,15	5,94	12,80	NH_4IO_4
2,17	5,15	11,17	$CdMoO_4$
2,17	5,24	11,38	$CaWO_4$
2,17	5,35	11,63	$NaLa(WO_4)_2$
2,18	5,32	11,59	$NaCe(WO_4)_2$
2,18	5,34	11,63	$LiLa(WO_4)_2$
2,19	5,23	11,44	$CoMoO_4$
2,19	5,27	11,55	$NaBi(MoO_4)_2$
2,19	5,36	11,72	$NaReO_4$
2,20	5,23	11,50	$LiBi(MoO_4)_2$
2,20	5,31	11,67	$LiLa(MoO_4)_2$
2,20	5,33	11,70	$NaLa(MoO_4)_2$
2,20	5,40	11,90	$SrWO_4$
2,20	5,75	12,63	KIO_4
2,20	5,87	12,94	$RbIO_4$

c/a	a, kX	c, kX	Substance
2,20	5,87	12,94	NH_4ReO_4
2,21	5,44	12,01	$RbWO_4$
2,21	5,44	12,03	$KLa(WO_4)_2$
2,22	5,38	11,92	$KBi(MoO_4)_2$
2,22	5,39	11,94	$KCe(WO_4)_2$
2,22	5,62	12,50	$KReO_4$
2,23	5,35	11,92	$AgReO_4$
2,23	5,37	11,96	$SrMoO_4$
2,23	5,41	12,08	$RbMoO_4$
2,23	5,42	12,11	$KLa(MoO_4)_2$
2,24	5,32	11,93	$NaIO_4$
2,24	5,37	12,01	$AgIO_4$
2,26	5,62	12,70	$BaWO_4$
2,27	5,80	13,17	$RbReO_4$
2,29	5,56	12,76	$BaMoO_4$
2,30	5,08	11,69	$BiAsO_4$
2,31	5,76	13,33	$\beta\text{-}TlReO_4$ (400° K)
2,32	5,65	13,08	$KOsO_3N$
2,36	5,46	12,89	$KCrO_3F$
2,52	5,61	14,13	$CsSO_3F$
2,53	5,72	14,50	$CsCrO_3F$

Type H0₇

c/a	a, kX	c, kX	Substance
1,524	4,459	6,796	$BAsO_4$
1,533	4,332	6,640	BPO_4

Type H0₈

c/a	a, kX	c, kX	Substance
1,73	3,55	6,14	$KAlF_4$
1,73	3,62	6,26	$RbAlF_4$
1,76	3,61	6,37	$TlAlF_4$
1,77	3,59	6,35	NH_4AlF_4

Type H1₁

c/a	a, kX	c, kX	Substance
1,48	5,85	8,68	$CuFe_2O_4$
1,58	6,20	9,82	$CaIn_2O_4$
1,59	5,74	9,15	$ZnMn_2O_4$
1,61	6,12	9,87	$CdIn_2O_4$
1,64	5,75	9,42	$MnMn_2O_4$

Type PbPb₂O₄

c/a	a, kX	c, kX	Substance
0,685	8,22	5,62	$NiAsO_4$
0,687	8,592	5,905	$FeSb_2O_4$
0,689	8,685	5,980	$MoSb_2O_4$
0,696	8,49	5,91	$CoSb_2O_4$
0,697	8,491	5,920	$ZnSb_2O_4$
0,699	8,445	5,907	$MgSb_2O_4$
0,708	8,35	5,91	$NiSb_2O_4$
0,722	8,72	6,30	$SnPb_2O_4$
0,742	8,85	6,57	$PbPb_2O_4$

Type H1₅

c/a	a, kX	c, kX	Substance
0,582	7,04	4,10	K_2PdCl_4
0,591	6,99	4,13	K_2PtCl_4
0,591	7,21	4,26	$(NH_4)_2PdCl_4$

c/a	a, kX	c, kX	Substance
		Type H2$_2$	
0,936	7,61	7,12	KH_2AsO_4
0,938	7,43	6,97	KH_2PO_4
1,003	7,52	7,54	$NH_4H_2PO_4$
1,005	7,70	7,74	$NH_4H_2AsO_4$
		Type H2$_6$	
1,97	5,46	10,73	Cu_2FeSnS_4
		Type H4$_1$	
1,024	7,81	8,00	$Rb_2CuCl_2 \cdot 2H_2O$
1,05	7,58	7,96	$(NH_4)_2CuCl_4 \cdot 2H_2O$
1,05	7,9	8,3	$(NH_4)_2CuBr_4 \cdot 2H_2O$
1,06	7,45	7,88	$K_2CuCl_4 \cdot 2H_2O$
1,09	7,50	8,16	$(NH_4)_2 FeCl_4 \cdot 2H_2O$
		Type H4$_9$	
0,403	10,40	4,19	$Pt(NH_3)_4Cl_2 \cdot H_2O$
0,423	10,21	4,31	$Pd(NH_3)_4Cl_2 \cdot H_2O$
		Type H4$_{17}$	
0,753	8,43	6,35	$Ag_2SO_4 \cdot 4NH_3$
		Type H5$_9$	
2,95	6,99	20,63	$Ca(UO_2)_2 (PO_4)_2 \cdot 10,5H_2O$
		Type H5$_{10}$	
1,21	6,98	8,42	$Ca(UO_2)_2 (PO_4)_2 \cdot 6,5H_2O$
		Type La$_2$(MoO$_4$)$_3$	
2,18	5,32	11,60	$Pr_2(MoO_4)_3$
2,19	5,29	11,58	$Nd_2(MoO_4)_3$
2,21	5,33	11,78	$Ce_2(MoO_4)_3$
2,21	5,35	11,84	$La_2(MoO_4)_3$
2,23	5,22	11,62	$Sm_2(MoO_4)_3$
		Type NaBaPO$_4$	
0,847	9,76	8,27	$NaBaPO_4$
0,848	9,84	8,34	$KBaPO_4$
0,857	9,25	7,93	$NaSrPO_4$
0,859	9,50	8,16	$KSrPO_4$
		Type S1$_5$	
1,25	6,99	8,75	$K_2OsO_2Cl_4$
		Type S1$_9$	
1,50	6,97	10,43	$AgCo(NH_3)_2 (NO_2)_4$
		Type S1$_{11}$	
0,974	8,12	7,91	$AgSb(OH)_6$
0,984	8,01	7,88	$NaSb(OH)_6$

c/a	a, kX	c, kX	Substance
		Type S3$_1$	
1,14	15,84	18,01	$K_3TlCl_6 \cdot 2H_2O$
1,15	16,95	19,45	$7Rb_3TlBr_6 \cdot 8H_2O$
		Type K3$_1$	
1,58	9,18	14,47	Cs_3CoCl_5
1,61	8,7	14,0	Rb_3CoCl_5
1,71	8,39	14,34	$NH_4Pb_2Br_5$
1,72	8,41	14,5	$RbPb_2Br_5$
1,73	8,14	14,1	KPb_2Br_5
		Type K7$_5$	
1,48	7,01	10,36	$Na_5Al_3F_{14}$
		Type K7$_6$	
1,45	7,49	10,87	$Cs_2AuAuCl_6$
1,49	7,38	11,01	$Cs_2AgAuCl_6$
		Type K_3CrO_8	
1,13	6,70	7,60	K_3CrO_8
1,13	7,37	8,34	Cs_2TaO_8
1,14	7,05	8,05	Rb_3TaO_8
1,16	6,78	7,86	K_3NbO_8
1,16	6,78	7,88	K_3TaO_8
		Type $LiBi_3O_4Cl_2$	
3,13	3,840	12,03	$LiBi_3O_4Cl_2$
3,13	3,877	12,13	$NaBi_3O_4Cl_2$
3,20	3,925	12,55	$NaBi_3O_4Br_2$
3,20	3,943	12,62	$Cd_2Bi_2O_4Br_2$
3,22	3,876	12,47	$LiBi_3O_4Br_2$
3,34	3,970	13,24	$Cd_2Bi_2O_4I_2$
3,34	3,990	13,31	$NaBi_3O_4I_2$
3,35	3,941	13,19	$LiBi_3O_4I_2$
		Type L10	
0,825	3,89	3,21	NiZn
0,935	3,98	3,72	AuCu
0,966	3,85	3,72	FePd
0,967	3,66	3,54	NiMn
		Type S2$_3$	
0,757	15,63	11,83	$Ca_{10}Mg_2Al_4Si_9O_{34}(OH)_4$
		Type S5$_2$	
1,76	9,00	15,84	$KCa_4Si_8O_{20}F \cdot 8H_2O$
		Type S5$_3$	
0,637	7,83	4,99	$Ca_2ZnSi_2O_7$
0,651	7,76	5,05	$(Ca, Na)_2(Mg, Al)(Al, Si)_2O_7$
0,655	7,75	5,08	$Ca_2Al_2SiO_7$
0,659	(7,47)	(4,92)	$(Ca, Na)_2Be(Al, Si)_2(O, F)_7$
0,675	(7,38)	(4,98)	$(Ca, Na)_2BeSi_2(O, OH, F)_7$
		Type S6$_4$	
0,624	12,27	7,66	$Ca_4Al_6Si_6O_{24}(SO_4, CO_3)$
0,627	12,09	7,58	$Na_4Al_3SiO_{24}Cl$

TABLE 5-4b-2

c/a	a, kX	c, kX	Structure type	Substance
0,311	12,2	3,79		$Na_{0,2-0,4}WO_3$
0,367	11,44	4,20	$S_4^2 - 1\bar{4}$	$Cd[Hg(CNS)_4]$
0,394	11,09	4,37	$S_4^2 - 1\bar{4}$	$Co[Hg(CNS)_4]$
0,401	11,06	4,43	$S_4^2 - 1\bar{4}$	$Zn[Hg(CNS)_4]$
0,403	10,40	4,19	$H4_9$	$Pt(NH_3)_4Cl_2 \cdot H_2O$
0,417	10,12	4,22		$Be(W, Mo)$
0,422	10,21	4,31	$H4_9$	$Pd(NH_3)_4Cl_2 \cdot H_2O$
0,429	14,6	6,26		$MgPt(CN)_4 \cdot 7H_2O$
0,467	8,12	3,77		$CS_2 (\sim 100° K)$
0,490	9,09	4,45	Fe_3P	Fe_3P
0,491	9,01	4,42	Fe_3P	$(Fe, Ni, Co)_3P$
0,492	8,92	4,39	Fe_3P	Ni_3P
0,495	7,56	3,74		W_4O_{11}
0,499	9,13	4,56	Fe_3P	Cr_3P
0,502	9,16	4,60	Fe_3P	Mn_3P
0,502	13,30—13,52	6,69—6,78	$E2_6$	$KMg(H_2O)_6(Cl, Br)_3$
0,514	34,04	17,49	$D_{4h}^{14}-P4/mnm$	$NaK(Ca, Mg, Mn) \cdot Al_4Si_5O_{18} \cdot 8H_2O$
0,518	5,56	2,88		$CdHg$
0,546	5,819	3,175	$A5$	β-Sn
0,550	10,81	5,94	$E1_4$	$[PNCl_2]_4$
0,550	12,17	6,69		$AgClO_2$
0,552	7,82	4,32		$ZnHg(CNS)_4$
0,570	(4,86)	(2,77)	$C4$	WO_2
0,574	(4,86)	(2,79)	$C4$	MoO_2
0,582	7,04	4,10	$H1_5$	K_2PdCl_4
0,591	6,99	4,13	$H1_5$	K_2PtCl_4
0,591	7,21	4,26	$H1_5$	$(NH_4)_2PdCl_4$
0,593	6,31	3,74	$F5_4$	NH_4ClO_2
0,603	9,22	5,56		$Na_2Co(CNS)_4 \cdot 8H_2O$
0,605	8,35	5,05	$C47$	SeO_2
0,617	4,81	2,97	$C4$	NbO_2
0,622	(5,72)	3,56	C_{4h}^5-14/m	Ni_4Mo
0,624	12,27	7,66	$S6_4$	$Ca_4Al_6Si_6O_{24}(SO_4, CO_3)$
0,627	12,09	7,58	$S6_4$	$Na_4Al_3Si_9O_{24}Cl$
0,633	4,61	2,92	$C4$	$RhVO_4$
0,634	4,54	2,88	$C4$	VO_2
0,637	7,83	4,99	$S5_3$	$Ca_2ZnSi_2O_7$
0,642	4,69	3,01	$C4$	$RhNbO_4$
0,644	4,58	2,95	$C4$	TiO_2
0,645	4,68	3,02	$C4$	$RhTaO_4$
0,649	4,41	2,86	$C4$	CrO_2
0,649	4,64	3,01	$C4$	$CrNbO_4$
0,650	4,63	3,01	$C4$	$CrTaO_4$
0,651	4,39	2,86	$C4$	GeO_2
0,651	4,67	3,04	$C4$	$FeTaO_4$
0,651	7,76	5,05	$S5_3$	$(Ca, Na)_2(Mg, Al)(Al_2Si)_2O_7$

c/a	a, kX	c, kX	Structure type	Substance
0,652	4,40	2,87	$C4$	MnO_2
0,652	4,62	3,01	$C4$	$FeSbO_4$
0,652	4,68	3,05	$C4$	$FeNbO_4$
0,655	7,75	5,08	$S5_3$	$Ca_2Al_2SiO_7$
0,656	4,51	2,96	$C4$	$AlSbO_4$
0,659	4,64	3,06	$C4$	MgF_2
0,659	(7,47)	(4,92)	$S5_3$	$(Ca, Na)_2 Be(Al, Si)_2 (O, F)_7$
0,660	4,59	3,03	$C4$	$GaSbO_4$
0,660	4,71	3,11	$C4$	NiF_2
0,664	4,58	3,04	$C4$	$CrSbO_4$
0,665	4,72	3,14	$C4$	ZnF_2
0,667	6,011	4,009	$B25$	NH_4SH
0,672	4,72	3,17	$C4$	SnO_2
0,674	4,60	3,10	$C4$	$RhSbO_4$
0,675	(7,38)	(4,98)	$S5_3$	$(Ca, Na)_2 BeSi_2 (O, OH, F)_7$
0,678	4,87	3,30	$C4$	MnF_2
0,679	4,70	3,19	$C4$	CoF_2
0,683	4,95	3,38	$C4$	PbO_2
0,684	8,22	5,62	$PbPb_2O_4$	$NiAs_2O_4$
0,686	4,93	3,38	$C4$	PdF_2
0,687	8,592	5,905	$PbPb_2O_4$	$FeSb_2O_4$
0,689	8,685	5,980	$PbPb_2O_4$	$MnSb_2O_4$
0,690	4,51	3,11	$C4$	RuO_2
0,696	4,83	3,36	$C4$	FeF_2
0,696	8,49	5,91	$PbPb_2O_4$	$CoSb_2O_4$
0,697	8,491	5,920	$PbPb_2O_4$	$ZnSb_2O_4$
0,699	4,49	3,14	$C4$	IrO_2
0,699	8,445	5,907	$PbPb_2O_4$	$MgSb_2O_4$
0,707	4,51	3,19	$C4$	OsO_2
0,707	6,18	4,37	$B25$	γ-NH_4I
0,708	8,35	5,91	$PbPb_2O_4$	$NiSb_2O_4$
0,709	5,70	4,04	$B25$	γ-$NH_4Br(\sim173°$ K$)$
0,718	10,15	7,29	D_4^6—$P4_22_1$	$OsO_5C_4(CH_3)_8$
0,718	12,04	8,65		$Ca(OCl)_2 \cdot 3H_2O$
0,722	8,28	5,98	$OB25$	$N(CH_3)_4ClO_4$
0,722	8,72	6,30	$PbPb_2O_4$	$SnPb_2O_4$
0,724	7,94	5,75	$OB25$	$N(CH_3)_4I$
0,729	6,34	4,62	$B25$	PH_4I
0,731	16,53	12,09		Cd_3Hg
0,739	10,76	7,95		$Na_2(TiFe)Si_4O_{11}$
0,742	8,85	6,57	$PbPb_2O_4$	$PbPb_2O_4$
0,752	9,74	7,32		Cu_3Pd
0,753	8,43	6,35	$H4_{17}$	$Ag_2SO_4 \cdot 4NH_3$
0,757	15,63	11,83	$S2_3$	$Ca_{10}Mg_2Al_4Si_9O_{34}(OH)_4$
0,787	(4,79)	(3,77)	$C4$	TeO_2
0,803	9,29	7,46	C_{4h}^3—$P4/n$	PCl_5
0,806	6,052	4,878	$C16$	Al_2Cu
0,814	6,52	5,31	$C16$	Sn_2Fe
0,818	6,647	5,434	$C16$	Sn_2Mn
0,825	3,89	3,21	$L10$	$NiZn$
0,832	5,099	4,240	$C16$	Fe_2B
0,841	5,006	4,212	$C16$	Co_2B
0,846	5,899	4,991	$C16$	Ge_2Fe
0,847	9,76	8,27	$NaBaPO_4$	$NaBaPO_4$
0,848	9,84	8,34	$NaBaPO_4$	$KBaPO_4$
0,851	4,980	4,236	$C16$	Ni_2B
0,852	6,835	5,821	$C16$	Pb_2Pd
0,857	6,348	5,441	$C16$	Sn_2Co
0,857	9,25	7,93	$NaBaPO_4$	$NaSrPO_4$
0,859	9,50	8,16	$NaBaPO_4$	$KSrPO_4$
0,867	7,13	6,18	$H0_3$	YVO_4

c/a	a, kX	c, kX	Structure type	Substance
0,871	5,74	5,00	$G0_9$	NH_4NO_3 (357—398° K)
0,873	8,02	7,00	B_{37}	TlSe
0,874	7,25	6,34	$H0_3$	$CaCrO_4$
0,880	(6,32)	(5,56)	$(E6_2)$	$SrO_2 \cdot 8H_2O$
0,880	6,651	5,853	$C16$	Pb_2Rh
0,880	(11,36)	(9,96)		Ag_3Ca
0,888	6,87	6,10	$H0_3$	YPO_4
0,895	4,96	4,44		$\sim ZrH_2$
0,901	6,58	5,93	$H0_3$	$ZrSiO_4$
0,904	6,13	5,54	$D_{4h}^7 - P4/nmm$	$CuB_2O_4 \cdot CuCl_2 \cdot 4H_2O$
0,906	(6,41)	(5,81)	$(E6_1)$	$Sr(OH)_2 \cdot 8H_2O$
0,910	6,89	6,27	$H0_3$	$YAsO_4$
0,917	5,83	5,35		$\sim MnBi_2$
0,921	9,69	8,92	Fl_1	$Hg(CN)_2$
0,930	4,85	4,51		$PbIn_{2-3}$
0,933	8,48	7,91	$C_{4h}^5 - 14/m$	$AgClO_3$
0,934	10,58	9,88		$(Ca, Na)_2Be(Si, Al)_2(O, F)_7$.
0,935	3,98	3,72	$L10$	AuCu
0,936	3,774	3,533	$A6$	γ-Mn
0,936	7,61	7,12	$H2_2$	KH_2AsO_4
0,937	5,83	5,46	$C_{4v}^1 - P4/mm$	$2Pb(OH)_2 \cdot CuCl_2$.
0,938	7,43	6,97	$H2_2$	KH_2PO_4
0,941	8,59	8,08		$AgBrO_3$
0,948	(3,85)	(3,65)		$W_{12}O_{32}(OH)_x$.
0,952	3,76	3,58	$A6$	95Mn·5Cu
0,960	3,77	3,62		96Mn·4Pd
0,962	3,767	3,624	$A6$	89Mn·11Cu
0,962	4,18	4,02		$\sim$70Mo—30N
0,966	3,85	3,72	$L10$	FePd
0,967	3,66	3,54	$L10$	NiMn
0,971	4,20	4,08		62Mn·38N
0,974	8,12	7,91	Sl_{11}	$AgSb(OH)_6$
0,975	8,00	7,80		trans -$Pd(NH_3)_2Cl_2$.
0,976	3,77	3,68		92Mn·8N
0,977	8,7	8,5		$Pd(NH_3)_2I_2$
0,981	3,764	3,693	$A6$	79Mn·21Cu
0,984	8,01	7,88	Sl_{11}	$NaSb(OH)_6$
0,986	(3,61)	3,56		Ni_4Mo
0,998	3,752	3,744	$A6$	66Mn·34Cu
1,00	6,34	6,34	$(E3_1)$	Ag_2HgI_4
1,003	7,52	7,54	$H2_2$	$NH_4H_2PO_4$
1,005	7,70	7,74	$H2_2$	$NH_4H_2AsO_4$
1,01	3,99	4,02	$E2_1$	$BaTiO_3$
1,01	4,96	5,03	$(L1_2)$	$SrPb_3$
1,01	(6,08)	(6,14)	$(E3_1)$	Cu_2HgI_4
1,02	5,07	5,16		ZrO_2 (<1273° K)
1,02	6,29	6,42	$C_4^1 - P_4$	$Pt(NH_3)_4PtCl_4$
1,02	7,81	8,00	$H4_1$	$Rb_2CuCl_4 \cdot 2H_2O$
1,03	6,37	6,58	$B34$	(Pd, Pt, Ni)S
1,04	5,79	6,00		$\sim Ni_2Sb$
1,04	6,35	6,60	$B34$	PdS
1,04	10,29	10,55		$Mg(ClO_2)_2 \cdot 6H_2O$
1,04	10,64	11,07	$D_4^{10} - \bar{1}4_12$	Cr_2Ni
1,05	7,58	7,96	$H4_1$	$(NH_4)_2CuCl_4 \cdot 2H_2O$
1,05	7,9	8,3	$H4_1$	$(NH_4)_2CuBr_4 \cdot 2H_2O$
1,05	12,95	13,65	$C_{4h}^6 - 14_1/a$	$KAlSi_2O_6$
1,06	2,84	3,01		(α- martensite)
1,06	3,89	4,13	$(E2_1)$	$PbTiO_3$

c/a	a, kX	c, kX	Structure type	Substance
1,06	7,45	7,88	$H4_1$	$K_2CuCl_4 \cdot 2H_2O$
1,06	22,0	23,3		$Al_2C_{12}O_{12} \cdot 18H_2O$
1,08	4,585	4,937	$A6$	In
1,09	7,50	8,16	$H4_1$	$(NH_4)_2FeCl_4 \cdot 2H_2O$
1,09	8,13	8,83	Phosgenite	$Pb_2Cl_2CO_3$
1,09	8,34	9,06	Phosgenite	$Pb_2Br_2CO_3$
1,13	6,70	7,60	K_3CrO_8	K_3CrO_8
1,13	7,37	8,34	K_3CrO_8	Cs_3TaO_8
1,14	(5,33)	(6,08)		$AgFO_3$
1,14	7,05	8,05	K_3CrO_8	Rb_3TaO_8
1,14	15,84	18,01	$S3_1$	$K_3TlCl_6 \cdot 2H_2O$
1,15	6,27	7,22	$C11$	CsO_2
1,15	16,95	19,45	$S3_1$	$7Rb_3TlBr_6 \cdot 8H_2O$
1,16	5,99	7,02	$C11$	RbO_2
1,16	6,07	7,03	$F5_2$	$KNCO$
1,16	6,09	7,06	$F5_2$	KN_3
1,16	6,78	7,86	$K_3Cr_2O_8$	K_3NbO_8
1,16	6,78	7,88	K_3CrO_8	K_3TaO_8
1,17	2,75	3,21	$L10$	NiZn
1,17	6,36	7,41	$F5_2$	RbN_3
1,18	5,70	6,73	$C11$	KO_2
1,20	5,67	6,81	$F5_2$	KFHF
1,21—1,31	4,05—3,98	4,90—5,10	$D_{4h}^7 - P4/nmm$	$PbO - Bi_2O_3$
1,21	6,98	8,42	$H5_{10}$	$Ca(UO_2)_2(PO_4)_2 \cdot 6.5H_2O$
1,22	3,55	4,33	$B10$	LiOH
1,25	6,99	8,75	$S1_2$	$K_2OsO_2Cl_4$
1,26	3,36	4,25		α-LiBi
1,26	3,98	5,01	$B10$	PbO
1,26	12,43	15,6		$Ca_4NaAl_3Si_5O_{19}$
1,27	3,80	4,81	$B10$	SnO
1,28	(4,14)	(5,28)	$C11$	ThC_2
1,28	8,81	11,27		$C_2(CH_3)_2Br_4$
1,28	11,70	14,95		$Fe_2(TeO_3)_3 \cdot xH_2O$
1,29	8,97	11,55	$D_{4h}^2 - P4/mcc$	$Sr(OH)_2 \cdot 8H_2O$
1,32	2,669	3,533	$A6$	γ-Mn
1,32	2,78	3,66		Ni-N
1,32	2,81	3,72	$L10$	AuCu
1,33	4,73	6,29		$5PbCrO_4 \cdot 3PbMoO_4 \cdot 10PbSO_4$
1,35	2,66	3,58	$A6$	$95Mn \cdot 5Cu$
1,35	3,24	4,38	NaBi	MgIn
1,36	2,66	3,62	$A6$	$89Mn \cdot 11Cu$
1,37	2,59	3,54	$L10$	NiMn
1,37	2,72	3,72	$L10$	FePd
1,39	2,66	3,69	$A6$	$79Mn \cdot 21Cu$
1,39	3,46	4,80	NaBi	NaBi
1,40	4,96	6,92	$C30$	SiO_2
1,40	8,75	12,28	$D5_9$	Cd_3P_2
1,41	2,65	3,74	$A6$	$66Mn \cdot 34Cu$
1,41	5,00	7,06	$C30$	$AlPO_4$
1,41	5,48	7,74		Li_2O_2
1,41	8,10	11,39		Ni_9Sb_4
1,41	8,10	11,45	$D5_9$	Zn_3P_2
1,41	8,32	11,76	$D5_9$	Zn_3As_2
1,41	8,95	12,65	$D5_9$	Cd_3As_2
1,43	12,83	18,38	$D_{4h}^6 - P4/mnc$	$B_2O_3 \cdot 24WO_3 \cdot 66H_2O$
1,43	12,98	18,52		$H_4SiW_{12}O_{40} \cdot 31H_2O$
1,44	12,80	18,40		$(NH_4)_5BW_{12}O_{40} \cdot 26H_2O$
1,45	7,49	10,87	$K7_6$	$Cs_2AuAuCl_6$
1,45	9,50	13,81	$D_{4h}^1 - P4/mmm$	$CuCl \cdot 3SC(NH_2)_2$
1,46	5,548	8,093	$D0_{22}$	$TiGa_3$

c/a	a, kX	c, kX	Structure type	Substance
1,46	(7,74)	(11,31)	$H0_3$	$Y(Nb, Ta)O_4$
1,46	(7,76)	(11,32)	$H0_3$	$YNbO_4$
1,47	3,77	5,52	$B10$	$FeSe$
1,47	(7,75)	(11,41)	HC_3	$YTaO_4$
1,48	5,85	8,68	$(H1_1)$	$CuFe_2O_4$
1,48	7,01	10,36	$K7_5$	$Na_5Al_3F_{14}$
1,49	6,65	9,91		Na_2O_2
1,49	7,38	11,01	$K7_6$	$Cs_2AgAuCl_6$
1,50	6,97	10,43	$S1_9$	$AgCo(NH_3)_2(NO_2)_4$
1,51	4,14	6,25		$Pb(ClO_2)_2$
1,51	4,85	7,33	$OB21$	$C_3H_7NH_3I$
1,52	3,24	4,94	$A6$	In
1,52	4,459	6,796	HO_7	$BAsO_4$
1,53	3,992	6,091	$C38$	Cu_2Sb
1,53	4,332	6,640	HO_7	BPO_4
1,55	5,605	8,712	$D0_{22}$	$ZrGa_3$
1,56	5,334	8,305	$D0_{22}$	VAl_3
1,57	5,17	8,12		$\sim Fe_3Ti$
1,57	5,422	8,536	$D0_{22}$	$TaAl_3$
1,58	5,425	8,579	$D0_{22}$	$TiAl_3$
1,58	5,427	8,584	$D0_{22}$	$NbAl_3$
1,58	6,20	9,82	$(H1_1)$	$CaIn_2O_4$
1,58	9,18	14,47	$K3_1$	Cs_3CoCl_5
1,59	5,74	9,15	$(H1_1)$	$ZnMn_2O_4$
1,61	4,08	6,56	$C38$	Mn_2Sb
1,61	4,39	7,05	$C11$	BaC_2
1,61	4,57	7,36	$OB21$	$C_3H_7NH_3Br$
1,63	6,12	9,87	$(H1_1)$	$CdIn_2O_4$
1,61	8,7	14,0	$K3_1$	Rb_3CoCl_5
1,61	4,11	6,68	$C11$	SrC_2
1,63	15,0	24,4	$D_{4h}^{17}-14/mmm$	$\sim PbCl_2 \cdot Cu(OH)_2$
1,64	5,75	9,42	$(H1_1)$	$MnMn_2O_4$
1,65	3,627	5,973	$C38$	Fe_2As
1,65	3,82	6,30	$C11$	NdC_2
1,65	3,87	6,37	$C11$	CaC_2
1,65	4,38	7,22	$E0_1$	$BaFCl$
1,66	3,85	6,38	$C11$	PrC_2
1,67	3,54	5,91	$C11$	CaO_2
1,67	3,75	6,28	$C11$	SmC_2
1,67	3,76	6,27	$C38$	Mn_2As
1,67	3,87	6,48	$C11$	CeC_2
1,67	3,92	6,55	$C11$	LaC_2
1,68	4,10	6,88	$E0_1$	$SrFCl$
1,69	5,83	9,88	$D_{4h}^{19}-14/amd$	$6CuO \cdot Cu_2O$
1,71	4,65	7,93	$E0_1$	$BaFI$
1,71	8,39	14,34	$K3_4$	$NH_4Pb_2Br_5$
1,72	5,09	8,76	$OB20$	CH_3NH_3Br
1,72	8,41	14,5	$K3_4$	$RbPb_2Br_5$
1,73	3,55	6,14	HO_8	$KAlF_4$
1,73	3,62	6,26	HO_8	$RbAlF_4$
1,73	8,14	14,1	$K3_4$	KPb_2Br_5
1,74	3,03	5,26	$B17$	PdO
1,75	3,613	6,333	$C38$	Cr_2As
1,75	5,11	8,96	$OB20$	CH_3NH_3I
1,76	3,03	5,32	$B17$	PtO
1,76	3,47	6,10	$B17$	PtS
1,76	3,61	6,37	HO_8	$TlAlF_4$
1,76	3,89	6,83	$E0_1$	$CaFCl$
1,76	4,09	7,21	$E0_1$	$PbFCl$
1,76	9,00	15,84	$S5_2$	$KCa_4Si_8O_{20}F \cdot 8H_2O$
1,77	3,59	6,35	HO_8	NH_4AlF_4

c/a	a, kX	c, kX	Structure type	Substance
1,79	3,78	6,77	$C11$	BaO_2
1,81	5,72	10,37		α-$Pt(NH_3)_2Cl_4$
1,82	4,18	7,59	$E0_1$	$PbFBr$
1,82	5,66	10,30	$E1_1$	$AgFeS_2$
1,83	4,16	7,61	D_{4h}^{10}—$P4/mcm$	NH_4CN
1,85	3,55	6,55	$C11$	SrO_2
1,89	3,89	7,37	$E0_1$	$BiOCl$
1,89	4,19	7,94	$E2_5$	NH_4HgCl_3
1,90	2,69	5,10		$FeSi_2$
1,94	4,70	9,10	$FeTa_2O_6$	$NiTa_2O_6$
1,94	4,71	9,12	$FeTa_2O_6$	$Fe(Nb, Ta)_2O_6$
1,94	4,73	9,16	$FeTa_2O_6$	$CoTa_2O_6$
1,95	4,70	9,18	$FeTa_2O_6$	$MgTa_2O_6$
1,95	4,71	9,18	$FeTa_2O_6$	$FeTa_2O_6$
1,95	7,80	15,23		$Pb(Cl, OH)_2 \cdot 4PbO \cdot 2Fe_2O_3$
1,97	(4,28)	(8,42)	$C11$	KHC_2
1,97	5,26	10,37	$E1_1$	$CuFeS_2$
1,97	5,46	10,73	$H2_6$	Cu_2FeSnS_4
1,98	13,99	27,70	D_4^{10}—14_12	$KUO_2(CH_3COO)_3$
1,99	4,02	8,02		H_2O_2
1,99	9,50	18,93	D_{4h}^{20}—$14/acd$	$3Mn_2O_3MnSiO_3$
2,00	13,79	27,60	D_4^{10}—14_12	$NH_4UO_2(CH_3COO)_3$
2,03	15,4	31,2		$Pb_5Cu_4Cl_{10}O_4 \cdot 6H_2O$
2,06	2,83	5,82		$\sim CuGa_2$
2,06	3,92	8,09	$D0_{22}$	$TiGa_3$
2,07	3,92	8,11	$E0_1$	$BiOBr$
2,08	5,42	11,3	D_{2d}^{12}—$1\bar{4}2d$	$(Bi, W)_8 \cdot nO_{12}$
2,10	(3,89)	(8,17)	$C11$	$NaHC_2$
2,14	7,50	16,05	D_{4h}^8—$P4/ncc$	$BaFeSi_4O_{10}$
2,15	5,94	12,80	$H0_4$	NH_4IO_4
2,17	5,15	11,17	$H0_4$	$CdMoO_4$
2,17	5,24	11,38	$H0_4$	$CaWO_4$
2,17	5,35	11,63	$H0_4$	$NaLa(WO_4)_2$
2,18	5,32	11,59	$H0_4$	$NaCe(WO_4)_2$
2,18	5,32	11,60	$La_2(MoO_4)_3$	$Pr_2(MoO_4)_3$
2,18	5,34	11,63	$H0_4$	$LiLa(WO_4)_2$
2,19	5,23	11,44	$H0_4$	$CaMoO_4$
2,19	5,27	11,55	$H0_4$	$NaBi(MoO_4)_2$
2,19	5,29	11,58	$La_2(MoO_4)_3$	$Nd_2(MoO_4)_3$
2,19	5,36	11,72	$H0_4$	$NaReO_4$
2,20	3,77	8,31	$D0_{22}$	VAl_3
2,20	3,96	8,71	$D0_{22}$	$ZrGa_3$
2,20	5,23	11,50	$H0_4$	$LiBi(MoO_4)_2$
2,20	5,31	11,67	$H0_4$	$LiLa(MoO_4)_2$
2,20	5,33	11,70	$H0_4$	$NaLa(MoO_4)_2$
2,20	5,40	11,90	$H0_4$	$SrWO_4$
2,20	5,75	12,63	$H0_4$	KIO_4
2,20	5,87	12,94	$H0_4$	$RbIO_4$
2,20	5,87	12,94	$H0_4$	NH_4ReO_4
2,21	5,33	11,78	$La_2(MoO_4)_3$	$Ce_2(MoO_4)_3$
2,21	5,35	11,84	$La_2(MoO_4)_3$	$La_2(MoO_4)_3$
2,21	5,44	12,01	$H0_4$	$PbWO_4$
2,21	5,44	12,03	$H0_4$	$KLa(WO_4)_2$
2,22	5,38	11,92	$H0_4$	$KBi(MoO_4)_2$
2,22	5,39	11,94	$H0_4$	$KCe(WO_4)_2$
2,22	5,62	12,50	$H0_4$	$KReO_4$
2,23	3,83	8,54	$D0_{22}$	$TaAl_3$
2,23	5,22	11,62	$La_2(MoO_4)_3$	$Sn_2(MoO_4)_3$
2,23	5,35	11,92	$H0_4$	$AgReO_4$

c/a	a, kX	c, kX	Structure type	Substance
2,23	5,37	11,96	HO_4	$SrMoO_4$
2,23	5,41	12,08	HO_4	$PbMoO_4$
2,23	5,42	12,11	HO_4	$KLa(MoO_4)_2$
2,24	3,836	8,579	DO_{22}	$TiAl_3$
2,24	3,837	8,584	DO_{22}	$NbAl_3$
2,24	5,32	11,93	HO_4	$Na\,IO_4$
2,24	5,37	12,01	HO_4	$Ag\,IO_4$
2,26	5,62	12,70	HO_4	$BaWO_4$
2,27	5,80	13,17	HO_4	$RbReO_4$
2,28	4,01	9,14	EO_1	$BiOI$
2,29	5,56	12,76	HO_4	$BaMoO_4$
2,30	5,08	11,69	HO_4	$BiAsO_4$
2,31	5,76	13,33	HO_4	$\beta\text{-}TlReO_4$ (400° K)
2,32	5,65	13,08	HO_4	$KOsO_3N$
2,36	4,92	11,62	$D3_1$	HgI_2
2,36	5,46	12,89	HO_4	$KCrO_3F$
2,39	4,65	11,10	$D3_1$	Hg_2Br_2
2,42	12,5	30,25		$6Pb(S,\ Tl)_2AuTl_2$
2,44	4,46	10,89	$D3_1$	Hg_2Cl_2
2,45	3,21	7,88	$C20$	WSi_2
2,46	3,20	7,86	$C20$	$MoSi_2$
2,46	4,53	11,14	Dl_3	Al_4Ba
2,47	6,29	15,55		$(CH_2CO)_2NI$
2,48	4,45	11,04	Dl_3	Al_4Sr
2,51	3,75	9,43	$C5$	TiO_2
2,52	5,61	14,13	HO_4	$CsSO_3F$
2,53	5,72	14,50	HO_4	$CsCrO_3F$
2,54	4,35	11,07	Dl_3	Al_4Ca
2,68	7,04	18,88		$CaNa_4Al_{12}(PO_4)_8(OH)_{18} \cdot 6H_2O$
2,80	6,95	19,45		$[(NH_2)_2CNH]_2H_2CO_3$
2,83	4,36	12,36	$C13$	HgI_2
2,88	2,998	8,630	$C48$	Cr_2Al
2,91	7,05	20,5	$D_{4h}^{17}-14/mmm$	$Cu(UO_2)_2(PO_4)_2 \cdot 8H_2O$
2,95	6,99	20,63	$H5_9$	$Ca(UO_2)_2(PO_4)_2 \cdot 10^{1}/_2H_2O$
2,98	3,66	10,9	$D3_1$	Hg_2F_2
3,13	3,840	12,03	$D_{4h}^{17}-14/mmm$	$LiBi_3O_4Cl_2$
3,13	3,877	12,13	$LiBi_3O_4Cl_2$	$NaBi_3O_4Cl_2$
3,16	5,513	17,422		$MnSi_2$
3,20	3,925	12,55	$LiBi_3O_4Cl_2$	$NaBi_3O_4Br_2$
3,20	3,943	12,62	$LiBi_3O_4Cl_2$	$Cd_2Bi_2O_4Br_2$
3,22	3,876	12,47	$LiBi_3O_4Cl_2$	$LiBi_3O_4Br_2$
3,34	3,97	13,24	$LiBi_3O_4Cl_2$	$Cd_2Bi_2O_4I_2$
3,34	3,990	13,31	$LiBi_3O_4Cl_2$	$NaBi_3O_4I_2$
3,35	3,941	13,19	$LiBi_3O_4Cl_2$	$LiBi_3O_4I_2$
3,47	4,13	14,35	$D_{4h}^{19}-14/amd$	$TiSi_2$
3,68	5,07	18,65	ZnP_2	ZnP_2
3,73	5,28	19,70	ZnP_2	CdP_2
3,91	4,09	15,99		La_2MoO_6
4,03	15,4	62,0		$Pb_9Cu_8Ag_3Cl_{21}O_8 \cdot 9H_2O$
4,32	4,003	17,29	DO_{23}	$ZrAl_3$
5,76	3,78	21,77		Bayerite

5-4c. Hexagonal System

Figures 121-126 give theoretical x-ray patterns and parts of the indexing curves; Tables 5-4c-1 and 5-4c-2 give data for phase analysis (distribution with respect to structures and c/a) [127].

Phase analysis is performed as for the tetragonal system.

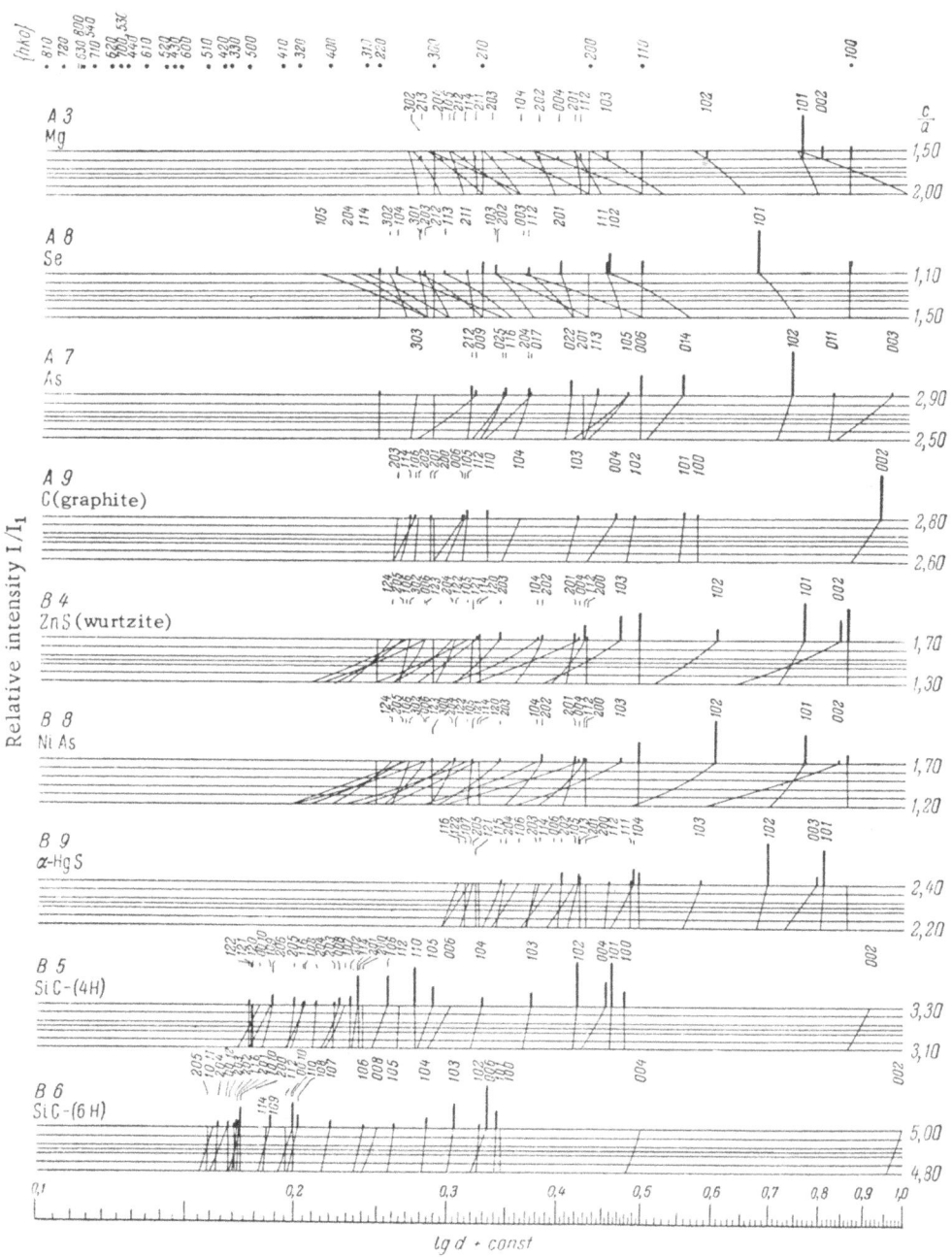

Fig. 121. Graphs for phase analysis of crystals in the hexagonal system (structures A3 to B6).

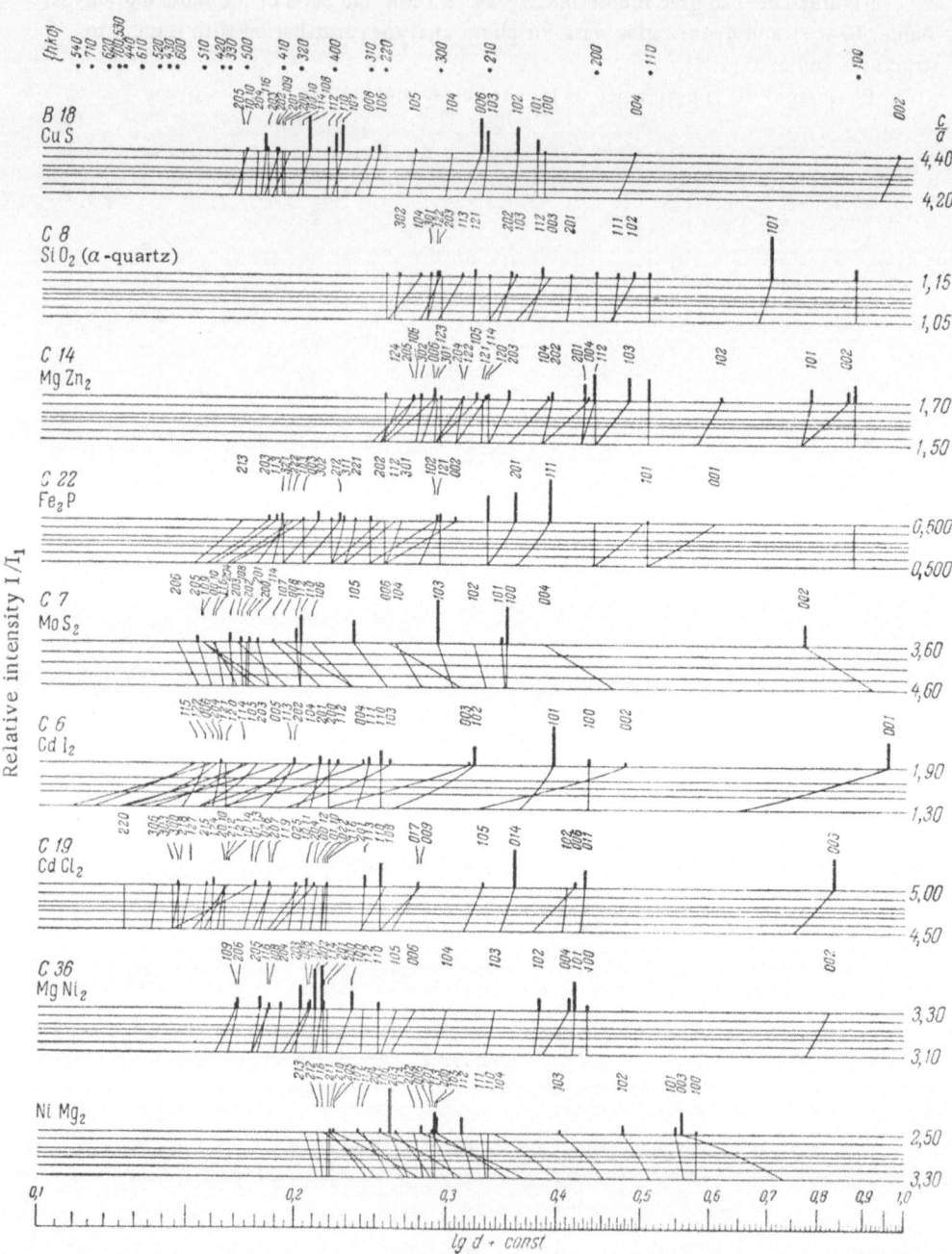

Fig. 122. Graphs for phase analysis of crystals in the hexagonal system (structures B18 to NiMg₂).

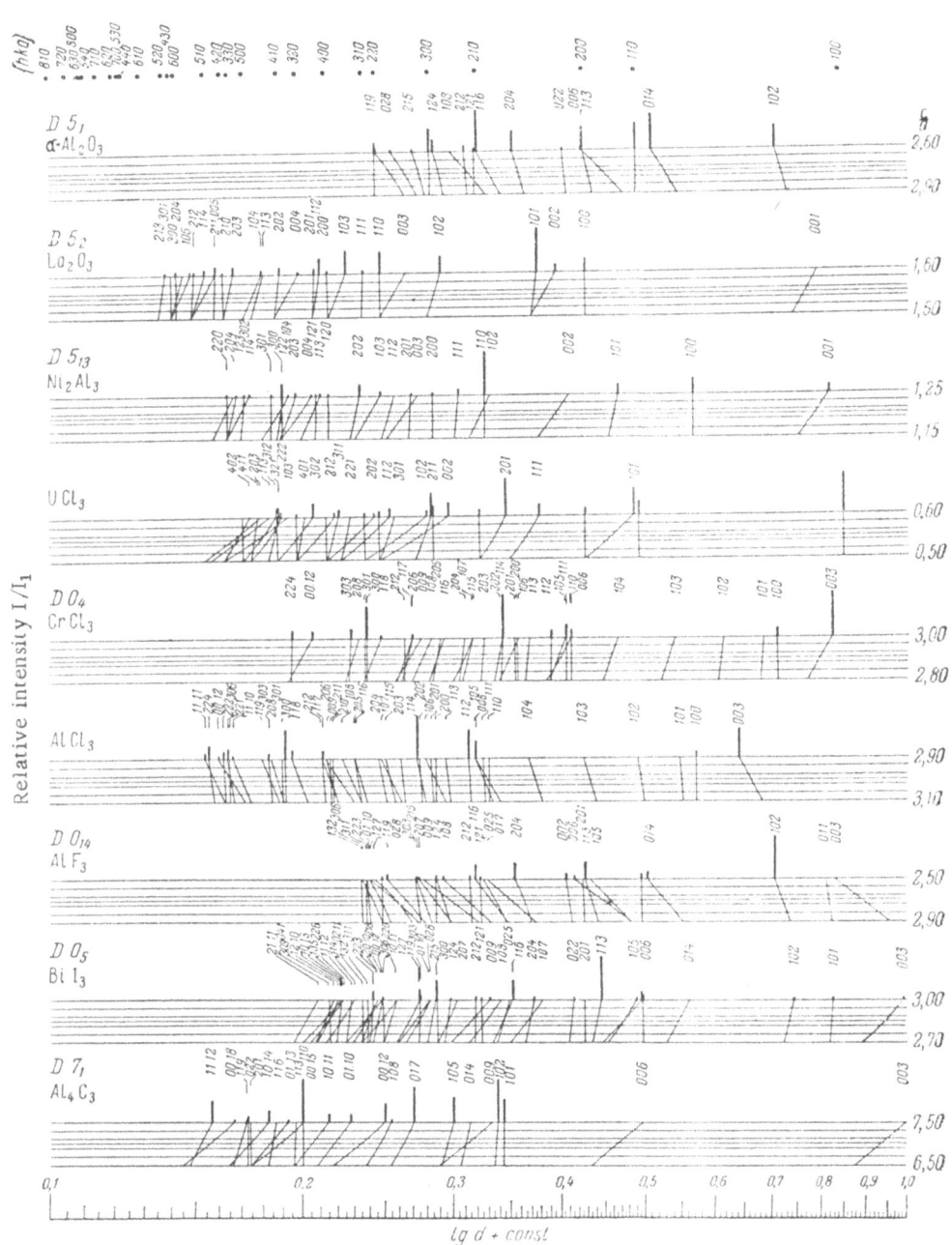

Fig. 123. Graphs for phase analysis of crystals in the hexagonal system (structures D5₁ to D7₁).

Fig. 123. Graphs for phase analysis of crystals in the hexagonal system (structures $D5_1$ to $D7_1$).

Fig. 124. Graphs for phase analysis of crystals in the hexagonal system [structures E2₂ to Mg(H₂O)₆SO₃].

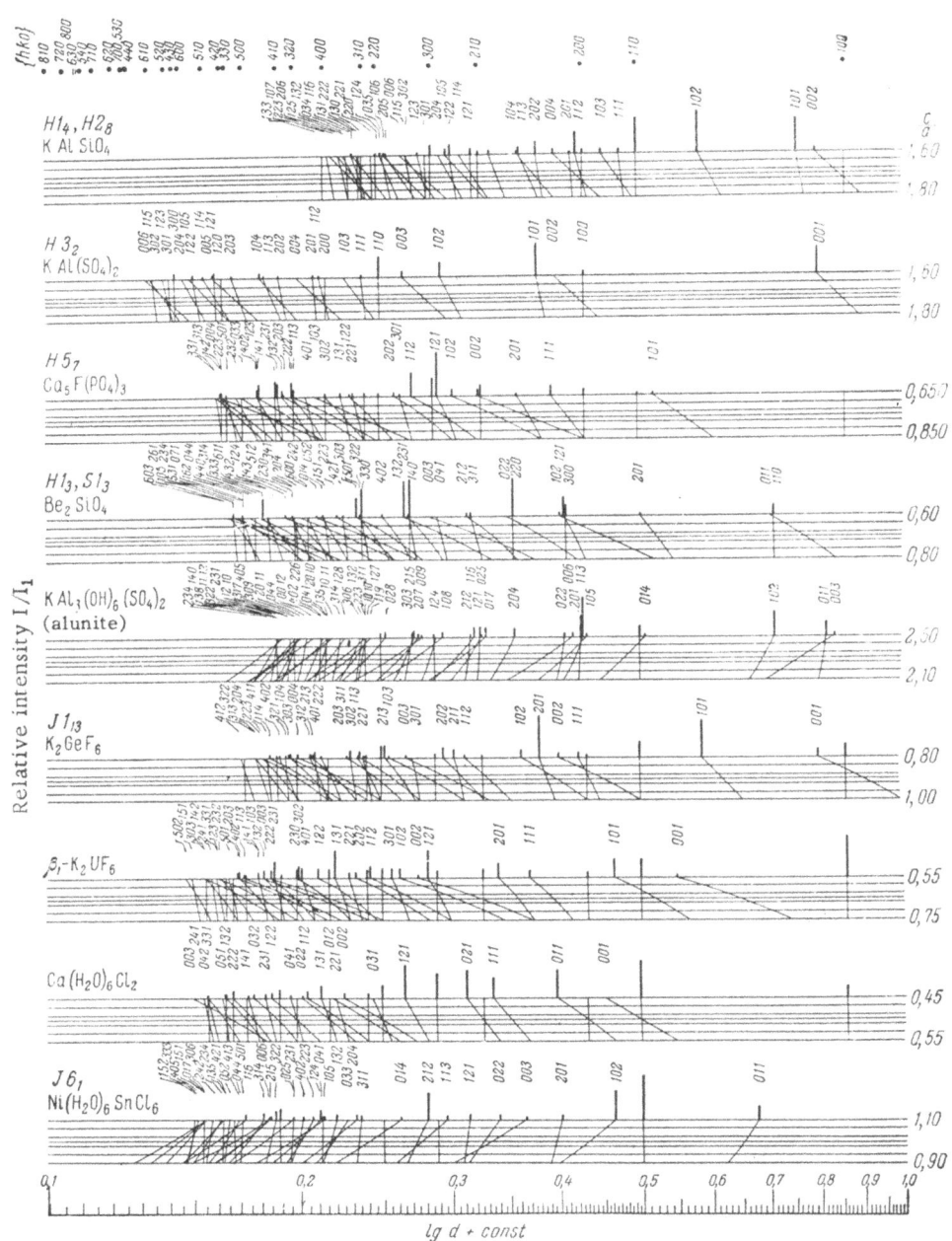

Fig. 125. Graphs for phase analysis of crystals in the hexagonal system (structures H1$_4$ and H2$_8$ to J6$_1$).

Fig. 126. Graphs for phase analysis of crystals in the hexagonal system (structures J2₂ to O1₃).

TABLE 5-4c-1

c/a	a, Å	c, Å	Substance
		Type A3	
1,55—1,56	2,76—2,77	4,30	$Cu_{1-x}Zn_{3+x}$
1,58	2,29	3,61	α-Be
1,58	3,52	5,57	Lu
1,58	3,54	5,60	Er
1,58	3,56	5,63	Ho
1,58	3,59	5,66	Dy
1,58	3,60	5,67	Tb
1,59	2,70	4,28	Ru
1,59	2,72	4,32	Os
1,59	3,21	5,09	Hf
1,59	3,23	5,15	α-Zr
1,59	3,31	5,25	Sc
1,59	3,53	5,57	Tu
1,59	3,63	5,76	Gd
1,59	3,65	5,81	Y
1,60	2,74	4,39	Tc
1,60	2,95	4,70	Ti
1,60	3,45	5,52	Tl
1,51	2,76	4,46	Re
1,61	2,80	4,49	$Fe_{1-x}Zn_{7+x}$
1,61	3,67	5,89	Nd
1,62	2,51	4,07	α-Co
1,62—1,63	3,19—3,10	5,16—5,04	Mg—Cd(80—42 at. % Mg)
1,62	3,20	5,18	δ-(Mg, Ag)(3 at. % Ag)
1,62—1,63	3,20—3,21	5,19—5,21	Mg—Zn(3—0 at. % Zn)
1,62	3,66	5,91	Pr
1,62	3,66	5,94	β-Ce
1,62	3,75	6,07	α-La
1,63	2,65	4,33	(Ni)
1,63	2,73	4,43	β-Cr
1,63	3,07—3,10	4,99—5,04	$MgCd_{2-x}$
1,63	3,13	5,09	$Cd(Mg, Al)_2$
1,63	3,21—3,17	5,21—5,15	Mg—Al(2—11 at. % Al)
1,63	3,21	5,20	Mg—Mn (1,5 at. % Mn)
1,63	3,21	5,21	Mg
1,63	3,58	5,84	He (1,45° K, 37 atm)
1,63	3,76	6,13	H_2 (4° K)
1,63	4,32	7,06	β-Sr (521° K)
1,64	3,97	6,50	β-Ca
1,64	4,04	6,62	β-N_2 ($\sim$40° K)
1,86	2,67	4,95	Zn
1,86	2,67	4,96	Zn—Cd (1 at. % Cd)
1,86	2,67	4,97—4,98	Zn—Al (0—5,0 at. % Al)
1,86	2,67	4,97—4,99	Hg—Zn (1,51—5,87 at.% Hg)
1,89	2,98	5,62	Cd
		Type A7	
2,61	4,55	11,86	Bi
2,62	4,30	11,28	Sb
2,80	3,77	10,56	As
2,89	3,77	10,89	As—Sn (70,9 at. % As)
		Type A8	
1,14	4,35	4,96	Se
1,33	4,46	5,92	Te
		Type A9	
2,74	2,46	6,75	C (graphite)

c/a	a, Å	c, Å	Substance

Type A10

1,94	3,48	6,74	Hg (177° K)

Type B4

c/a	a, Å	c, Å	Substance
1,60	2,90	4,62	CuH
1,60	3,12	4,99	AlN
1,60	3,26	5,21	ZnO
1,60	3,26	5,23	(Zn, Cd)O (5 mol.% CdO)
1,60	3,28	5,26	(Zn, Mn)O (22 mol. % MnO)
1,60	4,40	7,03	NH_4F
1,61	3,54	5,70	ZnN
1,62	3,06	4,95	TaN
1,62	3,98	6,44	MnS
1,62	4,15	6,72	CdS
1,62	4,53	7,36	MgTe
1,63	2,70	4,38	BeO
1,63	3,19	5,18	GaN
1,63	4,13	6,73	MnSe
1,63	4,20	6,86	Cd (S, Se)
1,63	4,31	7,03	CdSe
1,64	3,82	6,25	ZnS
1,64	4,06	6,66	CuBr (664—743° K)
1,64	4,59	7,52	Ag I
1,65	4,31	7,09	Cu I (675—713° K)

Type B5

3,27	3,08	10,08	SiC—(4H)
3,27	3,81	12,45	ZnS—(4H)

Type B6

4,90	3,08	15,12	SiC—(6H)
4,90	3,82	18,71	ZnS—(6H)

Type B7

12,23	3,83	46,84	ZnS—(15R)
12,28	3,08	37,82	SiC—(15R)

Type B8

c/a	a, Å	c, Å	Substance
1,21	4,20	5,10	CuSn
1,23	4,18—4,19	5,13—5,16	$NiIn_{1-x}$
1,23	4,23	5,21	γ-FeSn (~44 at. % Sn)
1,23	4,28	5,25	$CuIn_{1-x}$
1,24—1,25	4,40—4,38	5,47—5,49	$MnSn_{1-x}$
1,25	4,04	5,03	$FeGe_{1-x}$
1,26	4,05—4,15	5,12—5,21	$CoSn_{1-x}$
1,26	4,03—4,11	5,14—5,18	$FeSb_{1-x}$
1,26—1,27	4,15—4,05	5,21—5,12	$NiSn_{1-x}$
1,28	3,93	5,01	$CoGe_{1-x}$
1,28—1,30	3,96—3,85	5,05—4,99	$NiGe_{1-x}$
1,28	4,25	5,46	PtPb
1,28	4,34	5,56	RhSn
1,28	4,35	5,55	AuSn
1,28	4,46—4,49	5,69—5,74	$PdPb_{1-x}$
1,29	4,38—4,40	5,63—5,66	$PdSn_{1-x}$
1,31	3,92	5,14	Ni(Sb, As)
1,31	3,94	5,15	NiSb

c/a	a, Å	c, Å	Substance
1,32	4,08	5,36	NiBi
1,32	4,11	5,44	PtSn
1,32	4,14	5,46	PtTe
1,33	4,12	5,48	CrSb
1,33	4,14	5,48	PtSb
1,34	3,88	5,19	CoSb
1,36	3,76	5,10	NiAs·NiSb
1,36	3,96	5,37	NiTe
1,37	4,08	5,59	PdSb
1,37	4,14	5,67	PdTe
1,38	3,90	5,37	CoTe
1,39	3,61—3,63	5,02—5,05	Ni(As, Sb)
1,40	3,62	5,05	NiAs
1,40	3,99	5,57	IrSn
1,40	4,13	5,79	MnSb
1,42	4,31	6,13	MnBi
1,46	3,62	5,29	CoSe
1,46	3,67	5,34	NiSe
1,49	3,81	5,66	FeTe
1,53	3,39	5,19	CoS
1,53	3,72	5,71	MnAs
1,55—1,56	3,42—3,43	5,32—5,34	β-NiS
1,56	3,98	6,21	CrTe
1,57	3,37	5,30	(Fe, Co)S
1,58	3,51	5,56	$Fe_{1-x}S$c
1,60	3,42	5,47	(Ni, Fe)S
1,61	3,80	6,12	VTe
1,62—1,63	3,60—3,69	5,82—6,03	$Cr_{1-x}Se$
1,62	4,13	6,71	MnTe
1,63	3,43	5,56	$Cr_{1-x}S$
1,63	3,63	5,92	FeSe
1,66—1,68	3,44—3,46	5,69—5,80	$Fe_{0,83-0,96}S$
1,66	3,45	5,71	CrS
1,67	3,59	5,99	VSe
1,67	3,83	6,39	TiTe
1,68	3,45	5,81	FeS
1,73	3,36	5,81	VS
1,75	3,56	6,22	TiSe

Type B9

2,29	4,16	9,52	HgS

TypeB12

2,66	2,52	6,69	BN

Type B13

0,32	9,86	3,19	γ-NiSe
0,328—0,329	9,59—9,63	3,15—3,17	NiS

Type B18

1,32	3,79	16,38	CuS
1,38	3,94	17,25	CuSe

Type B22

1,94	5,39	10,45	RbSeH
1,99	5,15	10,24	KSeH
1,99	5,17	10,30	RbSH

c/a	a, Å	c, Å	Substance
2,01	4,96	9,95	KSH
2,05	4,47	9,16	NaSH
2,05	4,66	9,54	NaSeH

Type B30

c/a	a, Å	c, Å	Substance
1,61	10,68	17,19	MgZn

Type B35

c/a	a, Å	c, Å	Substance
0,807	5,28	4,26	CoSn
0,828	5,61	4,65	PtTl
0,839	5,30	4,45	β-FeSn
0,956	4,55	4,35	ϵ-InNi

Type C6

c/a	a, Å	c, Å	Substance
1,27	4,04	5,13	$PdTe_2$
1,30	4,02	5,21	$PtTe_2$
1,34	3,48	4,67	$Cd(OH)_2$
1,36	3,73	5,07	$PtSe_2$
1,37	3,41	4,67	$(Cd, Mn)(OH)_2$
1,37	3,86	5,29	$NiTe_2$
1,38	3,25	4,48	$Fe(OH)_2$
1,38	3,50—3,52	4,83—4,85	$(Ca, Cd)(OH)_2$
1,38	3,59	4,93	$Ca(OH)_2$
1,38—1,43	3,89—3,79	5,38—5,41	$CoTe_{2-x}$
1,40	3,59	5,01	$Cd(OH)_{1,6}Cl_{0,4}$
1,41	3,33	4,69	$Mn(OH)_2$
1,42	3,55	5,03	PtS_2
1,43	3,79	5,41	$CoTe_2$
1,46	3,19	4,65	$(Co, Zn)(OH)_2$
1,46	3,19	4,66	$Zn(OH)_2$
1,46	3,20	4,66	$Co(OH)_2$
1,47	3,14	4,62	$(Ni, Zn)(OH)_2$
1,47—1,49	3,17—3,12	4,66—4,64	$(Co, Ni)(OH)_2$
1,48	3,11	4,61	$Ni(OH)_2$
1,48	3,16	4,67	$(Co, Mg)(OH)_2$
1,48	3,16	4,67	$(Co, Zn)(OH)_2$
1,51	3,10	4,68	$(Ni, Mg)(OH)_2$
1,51	4,55	6,89	PbI_2
1,52	3,05	4,67	$(Ni, Zn)(OH)_2$
1,52	3,13	4,74	$Mg(OH)_2$
1,54	4,25	6,54	ZnI_2
1,55	3,59	5,55	$Cd(OH, Cl)_2$
1,55	4,49	6,97	CaI_2
1,55	4,49	6,97	YbI_2
1,59	3,69	5,86	ZrS_2
1,60	3,10	4,94	Ta_2C
1,61	3,65	5,88	SnS_2
1,61	4,26	6,86	CdI_2
1,62	3,83	6,20	$MnBr_2$
1,63	3,80	6,19	$ZrSe_2$
1,64	3,77	6,18	VBr_2
1,64	3,83	6,27	$MgBr_2$
1,64	4,14	6,80	GeI_2
1,64	4,17	6,83	MnI_2
1,65	3,56	5,88	$TiCl_2$
1,65	3,75	6,18	$FeBr_2$
1,66	3,70	6,13	$CoBr_2$
1,66	4,11	6,82	TiI_2
1,66	4,15	6,89	MgI_2

c/a	a, Å	c, Å	Substance
1,67	3,41	5,70	TiS_2
1,67	4,00	6,67	VI_2
1,67	4,05	6,76	FeI_2
1,68	3,97	6,66	CoI_2
1,69	3,54	5,99	$TiSe_2$
1,70	3,16	5,37	$Ni(OH)_{1,5}Cl_{0,5}$
1,71	3,23	5,51	$Co(OH)_{1,5}Cl_{0,5}$
1,71	3,28	5,60	$Co(OH)_{1,5}Br_{0,33}Cl_{0,17}$
1,73	3,41	5,91	TaS_2
1,73	3,77	6,51	$TiTe_2$
1,81	3,65	6,61	$Cd(OH)_{1,5}I_{0,5}$
1,83	3,24	5,92	$Co(OH)_{1,5}Br_{0,5}$
1,83	3,36	6,13	VSe_2
1,84	3,18	5,83	$Ni(OH)_{1,5}Br_{0,5}$
1,91	3,01	5,72	Ag_2F
2,19	3,18	6,97	$Co(OH)_{1,5}(NO_3)_{0,5}$

Type C7

c/a	a, Å	c, Å	Substance
3,90	3,16	12,32	MoS_2
3,90—3,92	3,16	12,29—12,32	$WS_{1,9-2,0}$
3,94	3,29	12,97	WSe_2

Type C8

c/a	a, Å	c, Å	Substance
1,10	(4,73)	(5,19)	BeF_2
1,10	4,91	5,40	α-SiO_2
1,11	5,04	(5,60)	$FePO_4$
1,12	4,90	(5,47)	$AlPO_4$
1,12	5,04	(5,62)	$AlAsO_4$
1,13	4,99	5,65	α-GeO_2

Type C10

c/a	a, Å	c, Å	Substance
1,63	4,51	7,35	H_2O (273° K)
1,63	4,54	7,36	D_2O (272° K)
1,63	5,04	8,24	SiO_2 (β-tridymite)

Type C12

c/a	a, Å	c, Å	Substance
7,85	3,89	30,53	$CaSi_2$

Type C14

c/a	a, Å	c, Å	Substance
1,61	6,64	10,66	$BaMg_2$
1,62	4,22	6,84	$FeBe_2$
1,62	6,23	10,12	$CaMg_2$
1,63	4,24	6,92	$MnBe_2$
1,63	4,26	6,96	$CrBe_2$
1,63	4,36	7,10	$ReBe_2$
1,63	4,40	7,14	VBe_2
1,63	4,78	7,81	$TeFe_2$
1,63	4,81	7,86	$TaFe_2$
1,63	4,84	7,88	$NbFe_2$
1,63	4,87	7,95	$TaMn_2$
(1,63)	4,96	$8,08\times2$	$TiMn_2$
1,63	5,14	8,39	$ZrCr_2$
1,63	5,26	8,60	$ZrRe_2$
1,63	5,48	8,95	$ThMn_2$
1,63	6,43	10,47	$SrMg_9$
1,64	4,44	7,29	$MoBe_2$
1,64	4,45	7,28	WBe_2

c/a	a, Å	c, Å	Substance
1,64—1,65	4,87—4,82	7,81—7,97	$Mg(Ni, Cu)_2$
1,64	4,87	7,97	$NbMn_2$
1,64	5,01	7,89	$MgCu_{1,5}Si_{0,5}$
1,64	5,04	8,24	$ZrMn_2$
1,64	5,10	8,36	$MgCuAl—MgCu_2$
1,64—1,65	5,14—5,22	8,44—8,62	$Mg(Zn, Cu)_2$
1,64	5,17	8,51	$MgZn_2$
1,64	5,19	8,53	$ZrOs_2$
1,64	5,28	8,66	ZrV_2
1,64	6,26	10,25	$CaLi_2$
1,64	7,50	12,29	KNa_2
1,65	5,20—5,22	8,60—8,62	$Mg(Zn, Ni)_2$
1,66	5,14	8,51	$ZrRu_2$

Type C19

c/a	a, Å	c, Å	Substance
4,41	4,28	18,23	Cs_2O
4,54	3,86	17,50	$CdCl_2$
4,59	3,60	16,53	$Cd(OH)_{1,25}Cl_{0,75}$
4,71	3,78	17,82	$ZnCl_2$
4,75	3,69	17,51	$MnCl_2$
4,76	3,96	18,80	$CdBr_2$
4,78	3,93	18,77	$ZnBr_2$
4,90	3,55	17,36	$NiCl_2$
4,90	3,56	17,44	$CoCl_2$
4,90	3,58	17,55	$FeCl_2$
4,90	3,59	17,58	$Cd(OH)_{1,4}Br_{0,6}$
4,90	3,60	17,63	$MgCl_2$
4,93	3,73	18,34	$NiBr_2$
5,04	3,91	19,67	$Ni I_2$

Type C22

c/a	a, Å	c, Å	Substance
0,511	6,77	3,46	Pd_2Ge
0,526	6,75	3,55	Pt_2Ge
0,532	6,49	3,45	Pd_2Si
0,567	6,09	3,46	Mn_2P
0,575	5,86	3,37	Ni_2P
0,583	5,85	3,41	$(Ni,Cu)_2P$
0,589	5,87	3,46	Fe_2P

Type C27

c/a	a, Å	c, Å	Substance
3,22	4,25	13,70	CdI_2
3,22—4,71	4,25—3,99	13,70—18,78	$Cd(Br, I)_2$

Type C32

c/a	a, Å	c, Å	Substance
0,647	4,99	3,23	UHg_2
1,00	4,28	4,28	$PrGa_2$
1,00	4,31	4,31	$CeGa_2$
1,00	4,32	4,32	$CaGa_2$
1,02	3,00	3,06	VB_2
1,02	3,05	3,11	MoB_2
1,02	4,33	4,40	$LaGa_2$
1,05	3,09	3,24	TaB_2
1,05	3,86	4,07	$\beta\text{-}VSi_2$
1,07	3,03	3,23	TiB_2
1,07	3,09	3,31	NbB_2
1,08	3,01	3,25	AlB_2
1,11	3,17	3,53	ZrB_2

c/a	a, Å	c, Å	Substance
		Type C33	
6,95	4,32	30,07	Bi_2Te_2S
6,96	4,37	30,42	Bi_2Te_3
		Type C36	
3,26	4,73	15,40	$CoTi_2$
3,26	4,73	15,42	$TaCo_2$
3,26	4,74	15,46	$NbCo_2$
3,26	4,96	16,15	$ZrFe_2$
3,29	4,82	15,86	$MgNi_2$
3,30	4,82—4,87	15,94—16,10	$Mg(Ni, Zn)_2$
3,30	5,10	16,80	$MgCuAl$
		Type C40	
1,44	4,43	6,36	$CrSi_2$
		Type C41	
1,63	4,73	7,73	Fe_2W
		Type DO_4	
2,90	5,99	17,38	$CrCl_3$
		Type DO_5	
2,75	7,50	20,65	$Bi I_3$
2,79	6,39	17,82	$ScCl_3$
2,79	7,50	20,93	$Sb I_3$
2,86	6,13	17,54	$TiCl_3$
2,87	6,05	17,38	$FeCl_3$
2,87	6,42	18,40	FeB_3
2,88	6,46	18,64	$TiBr_3$
2,89	6,02	17,38	VCl_3
2,91	6,29	18,30	$CrBr_3$
2,97	7,19	21,38	$As I_3$
		Type DO_6	
1,02	6,99	7,16	SmF_3
1,02	7,07	7,23	PrF_3
1,02	7,12	7,28	CeF_3
1,02	7,13	7,29	$(Ce, La)F_3$
1,02	7,14	7,29	$(La, Ce, Pr)F_3$
1,02	7,17	7,34	LaF_3
1,03	7,03	7,21	NdF_3
$1,75 \cdot \sqrt{3}$	$4,20 \cdot \sqrt{3}$	7,41	$Pb_{0,5}Th_{0,5}F_3$
1,75	4,28	7,48	$Ba_{0,5}U_{0,5}F_3$
1,76	4,19	7,35	$Pb_{0,5}U_{0,5}F_3$
1,76	4,28	7,54	AcF_3
1,76	4,20	7,54	$Ba_{0,5}Th_{0,5}F_3$
1,77	4,09	7,25	PuF_3
1,77	4,12	7,29	NpF_3
1,77	4,14	7,34	UF_3
1,78	4,04	7,19	$Ca_{0,5}Th_{0,5}F_3$
1,78	4,08	7,24	AmF_3
1,78	4,11	7,30	$Sr_{0,5}U_{0,5}F_3$
1,78	4,14	7,34	$Sr_{0,5}Th_{0,5}F_3$
1,80	4,05	7,30	$ThOF_2$

c/a	a, Å	c, Å	Substance
		Type DO_{12}	
2,56	5,23	13,38	FeF_3
2,59	5,17	13,40	VF_3
2,62	5,07	13,30	CoF_3
2,79	4,89	13,65	RhF_3
2,81	5,06	14,22	PdF_3
		Type DO_{14}	
2,54	4,92	12,48	AlF_3
		Type DO_{15}	
(2,97)	(5,92)	(17,56)	$AlCl_3$
		Type DO_{18}	
1,77	4,27	7,60	Li_3P
1,77	4,71	8,33	Li_3Sb
1,77	4,99	8,82	Na_3P
1,77	5,10	9,00	Na_3As
1,77	5,37	9,52	Na_3Sb
1,77	5,46	9,68	Na_3Bi
1,77	5,79	10,24	K_3As
1,77	6,19	10,95	K_3Bi
1,78	4,40	7,83	Li_3As
1,78	4,87	8,65	Mg_3Hg
1,78	6,04	10,71	K_3Sb
1,78	6,04	10,72	$K_3(Sb, Te)$
1,82	4,64	8,46	Mg_3Au
		Type DO_{19}	
0,796	5,33	4,25	Ni_3In
0,799	5,46	4,36	Fe_3Sn
0,800—0,801	5,66—5,67	4,53—4,54	$Mn_{3+x}Sn_{1-x}$
0,802	5,29—5,31	4,24—4,26	$Ni_{3+x}Sn_{1-x}$
0,805	5,13	4,13	Co_3W
0,805	5,96	4,80	$Ti_3(Sb_{0,8}Ti_{0,2})$
0,809	6,28	5,08	Mg_3Cd
0,810	6,23	5,05	Cd_3Mg
0,817	5,35	4,38	$Mn_{3+x}Ge_{1-x}$
		Type DO_{21}	
1,02	6,99	7,13	Cu_3P
		Type DO_{24}	
1,63	5,11	8,32	Ni_3Ti
		Type UCl_3	
0,553	7,94	4,39	$PrBr_3$
0,554	7,93	4,39	$NpBr_3$
0,558	7,95	4,44	$CeBr_3$
0,559	7,94	4,44	UBr_3
0,563	6,27	3,53	$Dy(OH)_3$
0,565	6,25	3,53	$Er(OH)_3$
0,565	6,27	3,54	$Gd(OH)_3$
0,565	6,27	3,54	$Sm(OH)_3$
0,566	(6,24)	(3,53)	$Y(OH)_3$

c/a	a, Å	c, Å	Substance
0,566	7,97	4,51	$LaBr_3$
0,573	7,40	4,24	$NdCl_3$
0,574	7,40	4,25	$PuCl_3$
0,574	7,42	4,26	$PrCl_3$
0,576	7,38	4,25	$AmCl_3$
0,577	7,42	4,28	$NpCl_3$
0,579	7,45	4,31	$CeCl_3$
0,580	8,08	4,69	$AcBr_3$
0,581	6,48	3,77	$Pr(OH)_3$
0,581	7,44	4,32	UCl_3
0,583	6,43	3,75	$Nd(OH)_3$
0,585	7,48	4,38	$LaCl_3$
0,590	6,53	3,86	$La(OH)_3$
0,597	7,64	4,56	$AcCl_3$

Type D2₂

1,66	9,94	16,51	$MgZn_5$

Type D4₁

1,91	4,55	8,71	B_2H_6

Type D5₁

2,64	5,14	13,58	Ti_2O_3
3,71	4,96	13,42	α-Ga_2O_3
2,71	5,11	13,83	Rh_2O_3
2,73	4,79—4,99	13,09—13,64	$(Al, Fe)_2O_3$
2,73	4,76	13,00	α-Al_2O_3
2,74	4,98—5,01	13,65—13,71	$(Fe, Cr)_2O_3$
2,74	5,03	13,77	Fe_2O_3
2,75	4,78—4,93	13,08—13,55	$(Al, Cr)_2O_3$
2,75	4,95	13,61	Cr_2O_3
2,82	4,94	13,95	V_2O_3

(see $E2_2$)

Type D5₂

1,55	4,08	6,30	Ac_2O_3
1,56	3,85	6,02	Nd_2O_3
1,56	3,86	6,01	Pr_2O_3
1,56	3,89	6,07	Ce_2O_3
1,56	3,94	6,14	La_2O_3
1,58	4,58	7,24	Sb_2Mg_3
1,59	3,88	6,19	Th_2N_3
1,59	4,68	7,42	Bi_2Mg_3

(see Ce_2O_2S)

Type D5₆

4,00	5,93	23,73	$K_2O \cdot 11Fe_2O_3$
4,02	5,59	22,50	$Na_2O \cdot 11Al_2O_3$
4,03	5,93	22,88	$Rb_2O \cdot 11Fe_2O_3$
4,06	5,59	22,67	$(K, Na)_2O \cdot 11Al_2C_3$
4,06	5,59	22,72	$K_2O \cdot 11Al_2O_3$

Type D5₁₃

1,13	4,53	5,51	In_3Pt_2
1,18	4,40	5,25	In_3Ni_2

c/a	a, Å	c, Å	Substance
1,21	4,04	4,90	Al_3Ni_2
1,21	4,06	4,90	Ga_3Ni_2
1,21	4,53	5,50	In_3Pd_2
1;23	4,23	5,18	Ga_3Pt_2

Type D7₁

7,51	3,33	24,98	Al_4C_3

Type D8₅

5,41	4,74	25,63	Co_7W_6
5,41	4,75	25,68	Fe_7Mo_6
5,43—5,61	4,76—4,75	25,83—26,68	$Fe_{7+x}W_{6-x}$

Type D8₈

0,696	6,91	4,81	Mh_5Si_3
0,698	6,75	4,72	Fe_5Si_3
0,718	8,26	5,93	Mg_5Hg_3

Type D8₁₀

0,623	12,72	7,93	Cr_5Al_8

Type D8₁₁

0,992	7,68	7,61	Co_2Al_5

Type Zn₅Ca

0,779	5,43	4,23	Zn_5La
0,781	5,40	4,22	Zn_5Ca
0,796	4,96	3,95	Ni_5Ca
0,796	5,17	4,12	Cu_5La
0,800	5,10	4,08	Cu_5Ca
0,803	4,95	3,97	Ni_5Pr
0,804	5,14	4,13	$Cu_{5-x}Ce_{1+x}$
0;808	4,96	4,01	Ni_5La
0,811	4,91	3,98	Ni_5Gd
0,811	4,92	3,99	Ni_5Th
0,816	4,95	4,04	Co_5Th
0,818	4,96	4,06	Co_5Ce
0,822	4,87	4,00	Ni_5Ce
0,846	5,24	4,45	$Zn_{5+x}Th_{1-x}$

Type E0₃

2,81	3,67	10,29	$Cd(OH)Cl$

Type E2₂

2,70	5,12	13,85	$LiCbO_3$
2,74	5,05	13,85	$NiTiO_3$
2,76	5,05	13,93	$MgTiO_3$
2,76	5,09	14,06	$FeTiO_3$
2,77	5,05	13,72	$CoTiO_3$
2,80	5,14	14,36	$MnTiO_3$
2,82	4,96	13,98	$Fe_{1-x}Ti_{1+x}O$
2,84	5,26	14,94	$CdTiO_3$
3,00	5,33	15,98	$NaSbO_3$
3,40	5,30	18,25	$KSbO_3$

(see D5₁)

c/a	a, Å	c, Å	Substance
		Type E2$_3$	
0,943	5,48	5,17	$LiIO_3$
		Type E9$_2$	
1,63	5,43	8,85	$4BeO \cdot NaSbO_3$
		Type E9$_4$	
6,59	3,29	21,64	Al_5C_3N
		Type F0$_2$	
0,945	6,21	5,87	OCS ($<135°$ K)
		Type F4$_1$	
2,46	6,46	15,92	$Fe_2(CO)_9$
		Type F5$_1$	
1,37	6,26	8,61	$AgFeO_2$
1,93	6,33	12,21	$CsICl_2$
4,05	3,68	14,88	$CaCN_2$
4,04	3,50	14,12	$NaHF_2$
4,18	3,64	15,20	NaN_3
4,22	3,59	15,13	NaOCN
4,78	3,39	16,20	$RbCrS_2$
5,27	3,03	15,96	$NaFeO_2$
5,47	3,71	20,29	$NaCrSe_2$
5,55	3,53	19,57	$NaCrS_2$
5,64	3,04	17,12	$CuFeO_2$
5,74	2,98	17,01	$CuCrO_2$
5,86	3,62	21,20	$KCrS_2$
7,03	3,44	24,20	$KCrSe_2$
7,84	3,43	26,90	$RbCrSe_2$
		Type F5$_{10}$	
2,38	7,39	17,58	$KAg(CN)_2$
		Type F5$_{13}$	
0,544	11,94	6,49	$Na_3(B_3O_6)$
0,573	12,79	7,33	$K_3(B_3O_6)$
		Type Ce$_2$O$_2$S	
1,71	4,01	6,83	Ce_2O_2S
1,71	4,04	6,89	La_2O_2S
1,73	3,93	6,77	Pu_2O_2S
		(see D5$_2$)	
		Type PbSb$_2$O$_6$	
0,913	5,27	4,81	$HgSb_2O_6$
0,915	5,24	4,80	$CdSb_2O_6$
0,943	4,78	4,50	$CoAs_2O_6$
0,960	5,23	5,02	$CaSb_2O_6$
0,985	5,26	5,34	$SrSb_2O_6$
1,01	4,83	4,87	$CdAs_2O_6$

c/a	a, Å	c, Å	Substance
1,01	5,30	5,37	$PbSb_2O_6$
1,03	4,84	4,98	$HgAs_2O_6$
1,05	4,83	5,08	$CaAs_2O_6$
1,08	5,30	5,75	$BaSb_2O_6$
1,11	4,85	5,41	$SrAs_2O_6$
1,13	4,87	5,49	$PbAs_2O_6$

Type KUF_5

0,664	14,71	9,77	$NaUF_5$
0,679	14,43	9,80	$NaPuF_5$
0,685	15,33	10,50	$KThF_5$
0,686	15,15	10,39	KUF_5
0,693	14,94	10,36	$KPuF_5$
0,696	15,22	10,61	$RbPuF_5$

Type $G0_1$

3,22	4,76	15,30	$ScBO_3$
3,23	4,64	14,98	$ZnCO_3$
3,24	4,60	14,91	$MgCO_3$
3,24	4,66	15,08	$CoCO_3$
3,24	4,78	15,49	$InBO_2$
3,25	4,68	15,23	$LiNO_3$
3,27	4,82	15,74	$MnCO_3$
3,27—3,29	4,91—4,95	16,07—16,28	$(Cd, Mn)CO_3$
3,28	4,68	15,36	$FeCO_3$
3,29	4,93	16,22	$(Mn, Ca)CO_3$
3,29	4,96	16,33	$CdCO_3$
3,32	5,06	16,81	$NaNO_3$
3,39	4,97	16,83	$(Ca, Cd)CO_3$
3,40	4,96	16,85	$(Ca, Mn)CO_3$
3,40	5,07	17,25	YBO_3
3,42	4,99	17,06	$CaCO_3$

Type $G0_7$

1,26	6,47	8,18	$CsBrO_3$
1,30	6,20	8,09	NH_4BrO_3
1,31	6,16	8,09	$TiBrO_3$
1,31	6,20	8,12	$RbBrO_3$
1,34	6,06	8,15	$TiClO_3$
1,34	6,09	8,13	$RbClO_3$
1,36	6,02	8,16	$KBrO_3$

Type $G1_1$

3,28	4,86	15,95	$CaSn(BO_3)_2$
3,31	4,84	16,02	$CaMg(CO_3)_2$
3,34	4,83	16,15	$Ca(Mg, Fe)(CO_3)_2$

Type $G2_2$

0,576	11,75	6,77	$Nd(BrO_3)_3 \cdot 9H_2O$

Type $G3_2$

1,13	5,45	6,14	Na_2SO_3

c/a	a, Å	c, Å	Substance

Type G7₁

1,37	7,10	9,74	$(Ce, La,...)FCO_3$
5,24	4,36	22,80	$BaCO_3 \cdot 2(Ce, La,...)FCO_3$
6,82	4,10	27,99	$CaCO_3 \cdot 2(Ce, La,...)FCO_3$

Type RbNO₃

0,706	10,47	7,39	$RbNO_3$
0,715	10,76	7,70	$CsNO_3$

Type H1₃

0,660	12,63	8,33	Be_2GeO_4
0,661	14,29	9,45	$(Zn, Mn)_2SiO_4$
0,662	12,46	8,24	Be_2SiO_4
0,665	14,23	9,47	Li_2WO_4
0,665	14,24	9,47	Li_2MoO_4
0,667	14,22	9,48	Zn_2GeO_4
0,668	13,53	9,04	$LiAlSiO_4$
0,670	13,96	9,35	Zn_2SiO_4
0,673	13,19	8,87	Li_2BeF_4

(see Si_3)

Type H1₄

1,68	5,14	8,62	$KLiSO_4$

(see $H2_8$)

Type H2₈

1,68	5,17	8,67	$KAlSiO_4$
1,68	5,22	8,78	$BaAl_2O_4$

(see $H1_4$)

Type H3₂

1,69	4,72	7,98	$KAl(SO_4)_2$
1,70	4,75	8,05	$KCr(SO_4)_2$
1,72	4,84	8,33	$(NH_4)Fe(SO_4)_2$
1,74	4,73	8,25	$(NH_4)Al(SO_4)_2$
1,75	4,79	8,39	$(NH_4)Cr(SO_4)_2$

Type H4₁₁

0,332	15,97	5,30	$Cd(MnO_4)_2 \cdot 6H_2O$
0,334	15,49	5,18	$Ni(ClO_4)_2 \cdot 6H_2O$
0,334	15,62	5,21	$Ni(MnO_4)_2 \cdot 6H_2O$
0,335	15,55	5,21	$Co(ClO_4)_2 \cdot 6H_2O$
0,336	15,55	5,21	$Zn(ClO_4)_2 \cdot 6H_2O$
0,336	15,61	5,25	$Fe(ClO_4)_2 \cdot 6H_2O$
0,336	15,61	5,24	$Zn(MnO_4) \cdot 6H_2O$
0,336	15,69	5,27	$Mg(MnO_4)_2 \cdot 6H_2O$
0,338	15,73	5,31	$Mn(ClO_4)_2 \cdot 6H_2O$
0,339	15,55	5,27	$Mg(ClO_4)_2 \cdot 6H_2O$
0,340	15,62	5,31	$Cd(ClO_4)_2 \cdot 6H_2O$

c/a	a, Å	c, Å	Substance
		Type H4$_{18}$	
0,693	7,80	5,40	$LiMnO_4 \cdot 3H_2O$
0,704	7,73	5,43	$LiClO_4 \cdot 3H_2O$
		Type H5$_7$	
0,705	9,56	6,74	$Ca_{4,2}Mn_{0,6}Fe_{0,2}(OH)(PO_4)_3$
0,711	10,41	7,41	$Pb_5Cl(VO_4)_3$
0,712	9,57	6,81	$Ca_5Cl(PO_4)_3$
0,713	9,66	6,88	$Ca_9BaCl_2(PO_4)_6$
0,713	9,66	6,88	$Ca_9PbCl_2(PO_4)_6$
0,716	9,62	6,89	$Ca_9MgCl_2(PO_4)_6$
0,717	9,61	6,89	$Ca_9NiCl_2(PO_4)_6$
0,725	9,55	6,92	$Ca_{10}(F, OH)_2(SO_4, SiO_4)_3$
0,726	10,22	7,43	$Pb_2Cl(AsO_4)_3$
0,729	9,35	6,81	$(Ca, Mn)_5F(PO_4)_3$
0,729	9,43	6,87	$(Ca, Mg)_5OH(PO_4)_3$
0,729	9,45	6,89	$Ca_9PbO(PO_4)_6$
0,729	9,50	6,92	Wilkeite
0,729	9,62	7,01	$(Ca, Sr)_5(F, OH)[(P, As)O_4]_3$
0,730	9,43	6,89	$Ca_9NiO(PO_4)_6$
0,730	9,45	6,89	$Ca_9BaO(PO_4)_6$
0,730	9,45	6,89	$Ca_9SrO(PO_4)_6$
0,730	9,47	6,89	$Ca_9ZnCO_3(PO_4)_6$
0,731	9,42	6,89	$Ca_{10}CO_3(PO_4)_6H_2O$
0,731	9,43	6,89	$Ca_9PbCO_3(PO_4)_6$
0,731	9,45	6,90	$Ca_9SrCO_3(PO_4)_6$
0,732	9,39	6,87	$Ca_5(Cl, F)(PO_4)_3$
0,732	9,45	6,91	$Ca_9BaCO_3(PO_4)_6$
0,733	9,44	6,92	$Ca_{10}O(PO_4)_6$
0,733	9,41	6,91	$Ca_5OH(PO_4)_3$
0,733	9,56	7,00	$Ca_{10}(OH)_2(SO_4, SiO_4)_3$
0,734	9,56	7,00	$Ca_{10}F_2(SiO_4, SO_4)_3$
0,735	9,37	6,89	Staffelite I
0,735	9,38	6,89	$Ca_9CdF_2(PO_4)_6$
0,735	9,38	6,89	$Ca_5F(PO_4)_3$
0,735	9,63	7,03	Britholite
0,736	9,14	6,72	$Ca_5Cd_5F_2(PO_4)_6$
0,736	9,35	6,88	Osteolite
0,736	9,62	7,08	$Ca_5Pb_5(OH)_2(PO_4)_6$
0,736	9,90	7,29	$Pb_5OH(PO_4)_3$
0,737	9,34	6,88	Podolite
0,737	9,35	6,89	Francolite
0,737	9,37	6,90	Lewistonite
0,737	9,53	7,02	$Ca_4Na_6F_2(SO_4)_6$
0,738	9,33	6,88	Stafelite II
0,738	9,33	6,88	Derpite
0,739	9,76	7,21	$Sr_5OH(PO_4)_3$
0,743	9,87	7,33	$Pb_5(Cl, F)(PO_4)_3$
0,744	9,27	6,89	$Ca_9(H_2O)_2(PO_4)_6$
0,750	9,29	6,96	$Ca_{10}CO_3H_2O(PO_4)_6$
0,751	10,22	7,68	$Ba_5OH(PO_4)_3$
		Type CePO$_4$	
0,908	6,99	6,35	$NdPO_4$
0,913	7,07	6,45	$CePO_4$
0,913	7,10	6,48	$LaPO_4$
0,928	6,97	6,47	$BiPO_4$

$c/$	a, Å	c, Å	Substance

Alunite type

2,27	7,19	16,33	$NaFe_3(SO_4)_2(OH)_6$
2,27	7,23	16,43	$AgFe_3(SO_4)_2(OH)_6$
2,34	7,21	16,83	$Pb_{0,5}Fe_3(SO_4)_2(OH)_6$
2,36	7,17	16,93	$(H_2O)Fe_3(SO_4)_2(OH)_5(H_2O)$
2,36	7,21	17,03	$KFe_3(SO_4)_2(OH)_6$
2,36	7,21	17,03	$NH_4Fe_3(SO_4)_2(OH)_6$
2,50	6,96	17,41	$KAl_3(SO_4)_3(OH)_6$

Type J1$_2$

1,96	6,53	12,78	$K_2Sn(OH)_6$
2,00	6,41	12,85	$K_2Pt(OH)_6$
2,38	5,96	14,20	$Na_2Sn(OH)_6$

Type J1$_3$

0,497	7,88	3,92	$CaCl_2 \cdot 6H_2O$
0,498	7,99	3,98	$CaBr_2 \cdot 6H_2O$
0,504	8,53	4,30	$Sr I_2 \cdot 6H_2O$
0,505	8,23	4,16	$SrBr_2 \cdot 6H_2O$
0,506	8,40	4,26	$Ca I_2 \cdot 6H_2O$
0,515	7,93	4,08	$SrCl_2 \cdot 6H_2O$
0,517	8,92	4,61	$Ba I_2 \cdot 6H_2O$
1,53	6,74	10,28	$K_2Pt(SCN)_6$
1,54	6,78	10,47	$(NH_4)_2Pt(SCN)_6$
1,55	6,76	10,49	$Rb_2Pt(SCN)_6$

Type J1$_6$

0,828	5,77	4,78	$(NH_4)_2SiF_6$

(see J1$_{13}$)

Type J1$_{12}$

1,91	5,24	10,00	$NaSbF_4(OH)_2$

Type J1$_{13}$

0,806	5,76	4,64	K_2PtF_6
0,812	7,44	6,04	Cs_2PuCl_6
0,814	5,71	4,65	K_2MnF_6
0,815	5,72	4,66	K_2TiF_6
0,816	5,86	4,79	$(NH_4)_2GeF_6$
0,823	5,83	4,80	Rb_2GeF_6
0,827	5,63	4,66	K_2GeF_6

(see J1$_6$)

Type J2$_2$

0,999	11,80	11,79	$Al(H_2O)_6Cl_3$
0,999	11,92	11,91	$Cr(H_2O)_6Cl_3$

c/a	a, Å	c, Å	Substance

Type $J6_1$

c/a	a, Å	c, Å	Substance
0,978	11,17	10,92	$Co(NH_3)_6 \cdot Cr(CN)_6$
0,993	10,91	10,83	$Co(NH_3)_6 \cdot Co(CN)_6$
1,01	9,64	9,70	$Fe(H_2O)_6 \cdot SiF_6$
1,01	9,68	9,77	$Mn(H_2O)_6 \cdot SiF_6$
1,01	9,79	9,87	$Mg(H_2O)_6 \cdot TiF_6$
1,01	10,76	10,87	$Co(NH_3)_5 \cdot H_2O \cdot Fe(CN)_6$
1,01	10,77	10,86	$Co(NH_3)_5 \cdot H_2O \cdot Fe(CN)_6$
1,02	10,62	10,80	$Ni(H_2O)_6 \cdot SnCl_6$
1,03	9,27	9,52	$Ni(H_2O)_6 \cdot SiF_6$
1,03	9,33	9,65	$Zn(H_2O)_6 \cdot SiF_6$
1,03	9,52	9,84	$Mg(H_2O)_6 \cdot SiF_6$
1,03	9,57	9,90	$Zn(H_2O)_6 \cdot TiF_6$
1,03	9,79	10,04	$Mg(H_2O)_6 \cdot SnF_6$
1,03	9,79	10,13	$Zn(H_2O)_6 \cdot ZrF_6$
1,03	10,62	10,91	$Co(H_2O)_6 \cdot SnCl_6$
1,03	10,70	10,97	$Fe(H_2O)_6 \cdot SnCl_6$
1,03	10,71	11,01	$Mg(H_2O)_6 \cdot SnCl_6$
1,04	9,33	9,72	$Co(H_2O)_6 \cdot SiF_6$
1,04	10,66	11,06	$Mn(H_2O)_6 \cdot SnCl_6$
1,05	9,73	10,21	$Zn(H_2O)_6 \cdot SnF_6$
1,06	(10,44)	(11,03)	$Co(NH_3)_4 \cdot (H_2O)_2 \cdot Co(CN)_6$

Type $TlSbF_6$

c/a	a, Å	c, Å	Substance
0,974	7,30	7,11	$BaGeF_6$
0,976	7,18	7,01	$BaSiF_6$
1,01	7,65	7,73	NH_4SbF_6
1,01	7,98	8,06	$CsSbF_6$
1,02	7,64	7,80	$RbSbF_6$
1,04	7,67	7,95	$TlSbF_6$

Type $Mg(H_2O)_6SO_3$

c/a	a, Å	c, Å	Substance
0,994	8,96	8,91	$MgSeO_3 \cdot 6H_2O$
1,01	8,79	9,03	$NiSO_3 \cdot 6H_2O$
1,03	8,84	9,06	$CoSO_3 \cdot 6H_2O$
1,03	8,84	9,06	$MgSO_3 \cdot 6H_2O$

Type β_1-Na_2ThF_6

c/a	a, Å	c, Å	Substance
0,612	6,13	3,75	$NaPuF_4$
0,614	6,15	3,78	$NaCeF_4$
0,618	6,54	4,04	K_2UF_6
0,619	6,18	3,83	$NaLaF_4$
0,629	5,95	3,74	Na_2UF_6
0,640	5,99	3,84	Na_2ThF_6
0,577	6,54	3,78	K_2UF_6
0,581	6,54	3,80	$KLaF_4$
0,581	6,58	3,82	K_2ThF_6

Type $K1_1$

c/a	a, Å	c, Å	Substance
0,635	10,13	6,43	$(NH_4)_2S_2O_6$
0,637	10,05	6,39	$Rb_2S_2O_6$
0,643	9,80	6,30	$K_2S_2O_6$
0,650	9,09	5,91	$Na_2S_2O_6$

c/a	a, Å	c, Å	Substance

Type K1$_2$

| 1,82 | 6,34 | 11,56 | $Cs_2S_2O_6$ |

Type K7$_1$

2,26	7,17	16,19	$K_3W_2Cl_9$
2,26	7,17	16,19	$(NH_4)_3W_2Cl_9$
2,32	7,16	16,36	$Tl_3W_2Cl_9$
2,32	7,36	17,09	$Cs_3W_2Cl_9$
2,34	7,25	16,98	$Rb_3W_2Cl_9$

Type K7$_2$

| 1,43 | 12,84 | 18,32 | $Cs_3Tl_2Cl_9$ |

Type K7$_3$

| 1,21 | 7,38 | 8,93 | $Cs_3As_2Cl_9$ |

Type S1$_3$

0,661	14,29	9,45	$(Zn, Mn)_2SiO_4$
0,662	12,46	8,24	Be_2SiO_4
0,668	13,53	9,04	$LiAlSiO_4$
0,670	13,96	9,35	Zn_2SiO_4

Type S3$_1$

| 1,00 | 9,22 | 9,22 | $Be_3Al_2(Si_6O_{18})$ |

Type S3$_2$

| 1,44 | (6,73) | (9,72) | $BaTiGe_3O_9$ |
| 1,47 | 6,61 | 9,73 | $BaTiSi_3O_9$ |

Type S3$_3$

| 0,406 | 12,70 | 5,17 | $Ca_2Na_6[Al_2Si_2O_8]_3(CO_3)_2$ |

Type S3$_4$

| 1,09 | 13,78 | 14,98 | $CaAl_2Si_4O_{12} \cdot 6H_2O$ |

Type S3$_6$

| 1,36 | 7,41 | 10,07 | $Na_2ZrSi_3O_9 \cdot 2H_2O$ |

Type S5$_7$

| 6,72 | 3,17 | 21,29 | cronstedtite |

TABLE 5-4c-2

c/a	a, Å	c, Å	Structure type	Substance
0,249	29,44	7,33		$TiBe_{12}$
(0,316)	(14,2)	4,49		$As I_3 \cdot 3S_8$
0,317	27,06	8,57		$KAlSiO_4$
0,323	9,86	3,19	$B13$	γ-NiSe
0,323	14,01	4,53	$P31c$	Cr_7C_3
0,324	20,50	6,64	$P3m$	$Pt[(NH_3)_5 Cl] Cl_3 \cdot H_2O$
0,327	13,90	4,54	$P31c$	Mn_7C_3
0,328—0,329	9,59—9,63	3,15—3,17	$B13$	NiS
0,332	15,97	5,30	$H4_{11}$	$Cd(MnO_4)_2 \cdot 6H_2O$
0,334	15,49	15,18	$H4_{11}$	$Ni(ClO_4)_2 \cdot 6H_2O$
0,334	15,62	5,21	$H4_{11}$	$Ni(MnO_4)_2 \cdot 6H_2O$
0,335	15,55	5,21	$H4_{11}$	$Co(ClO_4)_2 \cdot 6H_2O$
0,335	15,55	5,21	$H4_{11}$	$Zn(ClO_4)_2 \cdot 6H_2O$
0,336	15,61	5,25	$H4_{11}$	$Fe(ClO_4)_2 \cdot 6H_2O$
0,336	15,61	5,24	$H4_{11}$	$Zn(MnO_4)_2 \cdot 6H_2O$
0,336	15,69	5,27	$H4_{11}$	$Mg(MnO_4)_2 \cdot 6H_2O$
0,337	9,06	3,06	$P6_3/m$	$3Mg(OH,F) \cdot BO_3$
0,338	15,73	5,31	$H4_{11}$	$Mn(ClO_4)_2 \cdot 6H_2O$
0,339	15,55	5,27	$H4_{11}$	$Mg(ClO_4) \cdot 6H_2O$
0,340	15,62	5,31	$H4_{11}$	$Cd(ClO_4)_2 \cdot 6H_2O$
0,369	7,64	2,82	$P3$	AgZn
0,369	7,80	2,88		γ-Ag_3Ga
0,377	11,58	4,36	$P6_3/m$	Th_7Se_{12}
0,398	8,24	3,28	$(P6m2)$	$LiNaCO_3$
0,406	12,70	5,17	$S3_3$	$Ca_2Na_6[Al_2Si_2O_8]_3(CO_3)_2$
0,414	13,06	5,41		$KFeS_2$
0,418	12,83	5,36	$P6_3/mmc$	$(Na, Ca)_4[AlSiO_4]_3(SO_4CO_3Cl)$
0,435—0,454	15,75—16,01	6,86—7,24	$P3m$	$(Na, Ca)(Mg, Fe)_3$ $[Ba_3Al_6Si_6O_{27}](O, OH, F)_4$
0,436	28,41	12,39		$MnAl_4$
0,437	18,12	7,93		C_9Fe_{20}
0,472	16,14	7,62		$Be(Ca, Mn)SiO_4$
0,495	9,04	4,48		$(Ni, Cu)_3P$
0,497	7,88	3,92	Sl_3	$CaCl_2 \cdot 6H_2O$
0,498	7,99	3,98	Sl_3	$CaBr_2 \cdot 6H_2O$
0,504	8,53	4,30	Sl_3	$Sr I_2 \cdot 6H_2O$
0,505	8,23	4,16	Sl_3	$SrBr_2 \cdot 6H_2O$
0,506	8,4	4,26	Sl_3	$Ca I_2 \cdot 6H_2O$
0,515	7,93	4,08	Sl_3	$SrCl_2 \cdot 6H_2O$
0,517	8,92	4,61	Sl_3	$Ba I_2 \cdot 6H_2O$
0,525	6,52	3,42		PdSi
0,526	6,75	3,55	$C22$	Pt_2Ge
0,527	13,32	7,01	$R3$	$BiCl_3 \cdot 3CS(NH_2)_2$
0,528	6,48	3,42		(Pd_3B_2)
0,532	6,49	3,45	$C22$	Pd_2Si
0,534	14,69	7,85	$P\bar{3}$	CuH_2SiO_4
0,536	13,47	7,21	$P\bar{3}m1$	$(Mn, Fe)[Si_3O_7](Mn, Fe)_3(OH, Cl)_6$
0,544	11,94	6,49	$F5_3$	$Na_3(BO_2)_3$
0,545	10,67	5,82		$NaSrPO_4$
0,547	10,55	5,77		$NaCaPO_4$
0,549	10,72	5,88		$KSrPO_4$
0,551	10,62	5,85		$KCaPO_4$
0,551	10,97	6,04	$P3m$	UCl_6
0,553	7,94	4,39	UCl_3	$PrBr_3$
0,554	7,93	4,39	UCl_3	α-$NpBr_3$
0,558	7,95	4,44	UCl_3	$CeBr_3$
0,559	7,94	4,44	UCl_3	UBr_3
0,563	6,27	3,53	UCl_3	$Dy(OH)_3$
0,565	6,25	3,53	UCl_3	$Er(OH)_3$
0,565	6,27	3,54	UCl_3	$Gd(OH)_3$
0,565	6,27	3,54	UCl_3	$Sm(OH)_3$
0,566	6,24	3,53	UCl_3	$Y(OH)_3$

c/a	a, Å	c, Å	Structure type	Substance
0,566	7,97	4,51	UCl_3	$LaBr_3$
0,567	6,09	3,46	$C22$	Mn_2P
0,569	11,19	6,37		$Cu_{14}Al_5Mn$
0,573	7,40	4,24	UCl_3	$NdCl_3$
0,573	12,79	7,33	$F5_{13}$	$K_3(BO_2)_3$
0,574	7,40	4,25	UCl_3	$PuCl_3$
0,574	7,42	4,26	UCl_3	$PrCl_3$
0,575	4,37	2,52		$\gamma\text{-}LiZn_3$
0,575	5,86	3,37	$C22$	Ni_2P
0,576	7,38	4,25	UCl_3	$AmCl_3$
0,576	11,75	6,77	$G2_2$	$Nd(BrO_3)_3 \cdot 9H_2O$
0,577	6,51	3,76	$P62m$	$\beta_1\text{-}KCeF_4$
0,577	6,54	3,78	$\beta_1\text{-}K_2UF_6$	$\beta_1\text{-}K_2UF_6$
0,577	7,42	4,28	UCl_3	$NpCl_3$
0,577	15,82	9,14		$Cu_{19}Cl_4SO_4(OH)_{32} \cdot 3H_2O$
0,579	7,45	4,31	UCl_3	$CeCl_3$
0,580	8,08	4,69	UCl_3	$AcBr_3$
0,581	6,48	3,77	UCl_3	$Pr(OH)_3$
0,581	6,54	3,80	$\beta_1\text{-}K_2UF_6$	$\beta_1\text{-}KLaF_4$
0,581	6,58	3,82	$\beta_1\text{-}K_2UF_6$	$\beta_1\text{-}K_2ThF_6$
0,581	7,44	4,32	UCl_3	UCl_3
0,583	5,85	3,41	$C22$	$(NiCu)_2P$
0,583	6,43	3,75	UCl_3	$Nd(OH)_3$
0,585	7,48	4,38	UCl_3	$LaCl_3$
0,589	5,87	3,46	$C22$	Fe_2P
0,590	6,54	3,86	UCl_3	$La(OH)_3$
0,592	20,3	12,04		$Na_2Ca(CO_3)$
0,594—0,598	5,68—5,65	3,38		Ti—$Bi(20,5$—$57,1$ at. $\% Ti)$
0,596	11,91	7,07		Ni_3Sb
0,597	7,64	4,56	UCl_3	$AcCl_3$
0,598	5,50	3,29	$P6/mmm$	$\beta\text{-}BiIn_2$
0,609	16,05	9,79	$P3/m$	$Ni(H_2O)_6[Sb(OH)_6]_2$
0,612	6,13	3,75	$\beta_2\text{-}Na_2ThF_6$	$NaPuF_4$
0,612	7,38	4,52		$Al_6(Ta,Cb)_4O_{19}$
0,612	16,11	9,86	$P\bar{3}/m$	$Mg(H_2O)_6[Sb(OH)_6]_2$
0,614	6,15	3,78	$\beta_2\text{-}Na_2ThF_6$	$NaCeF_4$
0,619	9,32	5,74		$4Na_2SO_4CaSO_4$
0,618	6,54	4,04	$\beta_2\text{-}Na_2ThF_6$	$\beta_2\text{-}K_2UF_6$
0,619	6,18	3,83	$\beta_2\text{-}Na_2ThF_6$	$NaLaF_4$
0,623	12,72	7,93	$D8_{10}$	Cr_5Al_8
0,629	5,95	3,74	$\beta_2\text{-}Na_2ThF_6$	$\beta_2\text{-}Na_2UF_6$
0,630	13,46	8,47		$\alpha\text{-}Ag_{12}Te_7$
0,635	10,13	6,43	Kl_1	$(NH_4)_2S_2O_6$
0,637	10,05	6,39	Kl_1	$Rb_2S_2O_6$
0,637	21,0	13,41		$K_2Ca(CO_3)_2$
0,640	5,99	3,84	$\beta_2\text{-}Na_2ThF_6$	$\beta\text{-}Na_2ThF_6$
0,641	10,25	6,57	$O8_5$	$Na_3C_6N_9 \cdot 3H_2O$
0,643	9,80	6,30	Kl_1	$K_2S_2O_6$
0,647	4,99	3,23	$C32$	UHg_2
0,649	13,47	8,74		$Mn_3As_2O_6$
0,650	9,09	5,91	Kl_1	$Na_2S_2O_6$
0,651—0,654	7,91—7,85	5,14—5,12		$Ni_{10}Sb_{11}$
0,658	19,04	12,53		$K_5BSiW_{12}O_{40} \cdot 18H_2O$
0,660	12,63	8,33	Hl_3	Be_2GeO_4
0,661	14,29	9,45	Hl_3, Sl_3	$(Zn,Mn)_2SiO_4$
0,662	7,29	(4,83)		$Ba(ClO_4)_2 \cdot 3H_2O$
0,662	12,46	8,24	Hl_3, Sl_3	Be_2SiO_4
0,664	14,71	9,77	KUF_5	$NaUF_5$
0,665	14,23	9,47	Hl_3	Li_2WO_4
0,665	14,24	9,47	Hl_3	Li_2MoO_4
0,667	14,22	9,48	Hl_3	Zn_2GeO_4
0,668	13,53	9,04	Hl_3, Sl_3	$\alpha\text{-}LiAlSiO_4$
0,668	13,60	9,09	$P6/mmc$	$Cu_{19}SO_4Cl_4(OH)_{32} \cdot 3H_2O$

c/a	a, Å	c, Å	Structure type	Substance
0,669	13,87	9,27		Ge_3N_4
0,670	13,96	9,35	Hl_3, Sl_3	Zn_2SiO_4
0,673	13,19	8,87	Hl_3	Li_2BeF_4
0,675	13,55	9,15	$P6/mmc$	$Cu_{19}(NO_3)_2Cl_4(OH)_{32} \cdot 3H_2O$
0,679	14,43	9,80	KUF_5	$NaPuF_5$
0,683	10,23	6,98		$Pb_3(AsO_3)_3Cl$
0,684	8,73	5,97	$O8_2$	$C_3N_3(N_3)_3$
0,685	15,33	10,50	KUF_5	$KThF_5$
0,686	15,15	10,39	KUF_5	KUF_5
0,693	7,80	5,40	$H4_{18}$	$LiMnO_4 \cdot 3H_2O$
0,693	14,94	10,36	KUF_5	$KPuF_5$
0,696	6,91	4,81	$D8_8$	Mn_5Si_3
0,696	15,22	10,61	KUF_5	$RbPuF_5$
0,698	6,75	4,72	$D8_8$	Fe_5Si_3
0,704	7,73	5,43	$H4_{18}$	$LiClO_4 \cdot 3H_2O$
0,705	9,56	6,74	$H5_7$	$Ca_{8,4}Mn_{1,1}Fe_{0,5}(PO_4)_6(OH)_2$
0,706	10,47	7,39	$RbNO_3$	$RbNO_3$
0,711	6,49	4,62		$ThAl_3$
0,711	10,41	7,41	$H5_7$	$Pb_5Cl(VO_4)_3$
0,712	9,57	6,81	$H5_7$	$Ca_5Cl(PO_4)_3$
0,713	9,66	6,88	$H5_7$	$Ca_9BaCl_2(PO_4)_6$
0,713	9,66	6,88	$H5_7$	$Ca_9PbCl_2(PO_4)_6$
0,715	10,76	7,70	$RbNO_3$	$CsNO_3$
0,716	9,62	6,89	$H5_7$	$Ca_9MgCl_2(PO_4)_6$
0,717	9,61	6,89	$H5_7$	$Ca_9NiCl_2(PO_4)_6$
0,718	8,26	5,93	$D8_8$	Mg_5Hg_3
0,725	9,51	6,88		$Na_3Ca_4(SO_4)_6F_2$
0,725	9,55	6,92	$H5_7$	$Ca_{10}(F,OH)_2(SiO_4,SO_4)_3$
0,726	10,22	7,43	$H5_7$	$Pb_5Cl(AsO_4)_3$
0,729	9,35	6,81	$H5_7$	$(Ca,Mn)_5F(PO_4)_3$
0,729	9,43	6,87	$H5_7$	$(Ca,Mg)_5(OH)(PO_4)_3$
0,729	9,45	6,89	$H5_7$	$Ca_9PbO(PO_4)_6$
0,729	9,50	6,92	$H5_7$	Wilkeite
0,729	9,62	7,01	$H5_7$	$(Ca,Sr)_5(F,OH)[(P,As)O_4]_3$
0,730	9,43	6,89	$H5_7$	$Ca_9NiO(PO_4)_6$
0,730	9,45	6,89	$H5_7$	$Ca_9BaO(PO_4)_6$
0,730	9,45	6,89	$H5_7$	$Ca_9SrCO_3(PO_4)_6$
0,730	9,47	6,89	$H5_7$	$Ca_9ZnCO_3(PO_4)_6$
0,731	9,42	6,89	$H5_7$	$Ca_{10}CO_3(PO_4)_6 \cdot H_2O$
0,731	9,43	6,89	$H5_7$	$Ca_9PbCO_3(PO_4)_6$
0,731	9,45	6,90	$H5_7$	$Ca_9SrCO_3(PO_4)_6$
0,732	7,47	5,46		$LiI \cdot 3H_2O$
0,732	9,39	6,87	$H5_7$	$Ca_5(Cl,F)(PO_4)_3$
0,732	9,45	6,91	$H5_5$	$Ca_9BaCO_3(PO_4)_6$
0,733	9,44	6,92	$H5_7$	$Ca_{10}O(PO_4)_6$
0,733	9,56	7,00	$H5_7$	$Ca_{10}(OH)_2(SiO_4,SO_4)_3$
0,733	9,41	6,91	$H5_7$	$Ca_5OH(PO_4)_3$
0,734	9,0	6,62		$Cd_3(PO_4)_2$
0,734	9,56	7,00	$H5_7$	$Ca_{10}F_2(SiO_4,SO_4)_2$
0,735	9,37	6,89	$H5_7$	Staffelite I
0,735	9,38	6,89	$H5_7$	$Ca_9CdF_2(PO_4)_6$
0,735	9,38	6,89	$H5_7$	$Ca_5F(PO_4)_3$
0,735	9,63	7,03	$H5_7$	$(Ce,Ca,Na)_5(F,OH)[(Si,P)O_4]$
0,736	9,14	6,72	$H5_7$	$Ca_5Sd_5F_2(PO_4)_6$
0,736	9,35	6,88	$(H5_7)$	Osteolite
0,736	9,62	7,08	$H5_7$	$Ca_5Pb_5(OH)_2(PO_4)_6$
0,736	9,68	7,12		$Pb_3(PO_4)_2$
0,736	9,90	7,29	$H5_7$	$Pb_{10}(OH)_2(PO_4)_6$
0,736	10,04	7,38		$Pb_3(AsO_4)_2$
0,737	9,34	6,88	$H5_7$	Podolite
0,737	9,35	6,89	$H5_7$	Francolite
0,737	9,37	6,90	$H5_7$	$(Ca,K,Na,Al)_5OH(PO_4)_3$
0,737	9,53	7,02	$H5_7$	$Ca_4Na_6F_2(SO_4)_6$

c/a	a, Å	c, Å	Structure type	Substance
0,738	9,33	6,88	$H5_7$	Staffelite II
0,738	9,33	6,88	$H5_7$	$(Ca,Na,K)_5OH(PO_4,CO_3)_3$
0,739	9,76	7,21	$H5_7$	$Sr_5(OH)(PO_4)_3$
0,742	7,63	5,66		AgZn
0,743	9,87	7,33	$H5_7$	$Pb_5(Cl,F)(PO_4)_3$
0,744	9,27	6,89	$H5_7$	$Ca_9(H_2O)_2(PO_4)_6$
0,750	9,29	6,96	$H5_7$	$Ca_{10}CO_3H_2O(PO_4)_6$
0,751	10,22	7,68	$H5_7$	$Ba_5OH(PO_4)_3$
0,768	6,25	4,80		$LiHg_3$
0,779	5,43	4,23	Zn_5Ca	Zn_5La
0,781	5,40	4,22	Zn_5Ca	Zn_5Ca
0,790	11,09	8,76	$R3c$	Ag_3SbS_3
0,795	10,85	8,62		Cu_9Sb_2
0,796	4,96	3,95	Zn_5Ca	Ni_5Ca
0,796	5,17	4,12	Zn_5Ca	Cu_5La
0,796	5,33	4,25	DO_{19}	Ni_3In
0,796	(11,98)	(9,54)		B
0,798	5,66	4,52		$Mn_{11}Sn_3$
0,799	5,46	4,36	DO_{19}	Fe_3Sn
0,800	5,10	4,08	Zn_5Ca	Cu_5Ca
0,800—0,801	5,66—5,67	4,53—4,54	DO_{19}	β'-Mn-Sn (23—24,5 at.%Sn)
0,801	10,79	8,64	$R3c$	Ag_3AsS_3
0,802	5,29—5,31	4,24—4,26	DO_{19}	Ni_3Sn
0,803	4,95	3,97	Zn_5Ca	Ni_5Pr
0,804	5,14	4,13	Zn_5Ca	$Cu_{4,8}Ce_{1,2}$
0,805	5,13	4,13	DO_{19}	Co_3W
0,805	5,96	4,80	DO_{19}	$Ti_3(Sb_{0,8}Ti_{0,2})$
0,806	5,76	4,64	(Sl_{13})	K_2PtF_6
0,807	5,28	4,26	B_{35}	CoSn
0,808	4,96	4,01	Zn_5Ca	Ni_5La
0,809	6,28	5,08	DO_{19}	Mg_3Cd
(0,809)	(9,29)	(7,53)	GO_1	$CoCO_3$
0,810	6,23	5,05	DO_{19}	Cd_3Mg
0,811	4,91	3,98	Zn_5Ca	Ni_5Gd
0,811	4,92	3,99	Zn_5Ca	Ni_5Th
0,812	7,44	6,04	Sl_{13}	Cs_2PuCl_6
0,813	10,91	8,87		$CaNd_2Si_2O_8$
0,813	10,91	8,87		$KNdSO_4$
0,813	10,91	8,87		$NaNdSiO_4$
0,814	5,71	4,65	Sl_{13}	K_2MnF_6
0,815	5,72	4,66	Sl_{13}	K_2TiF_6
0,815	11,03	8,99		$KLaSiO_4$
0,816	4,95	4,04	Zn_5Ca	Co_5Th
0,816	5,86	4,79	Sl_{13}	$(NH_4)_2GeF_6$
0,816	10,81	8,82		$CaY_2Si_2O_8$
0,816	10,81	8,82		$NaYSiO_4$
0,816	11,03	8,90		$CaLa_2Si_2O_8$
0,816	11,03	8,90		$NaLaSiO_4$
0,817	5,35	4,38	DO_{19}	$Mn_{3,25}Ge$
0,818	4,96	4,06	Zn_5Ca	Co_5Ce
0,822	4,87	4,00	Zn_5Ca	Ni_5Ce
0,823	5,83	4,80	Sl_{13}	Rb_2GeF_6
0,827	5,63	4,66	Sl_{13}	K_2GeF_6
0,828	5,61	4,65	B_{35}	PtTl
0,828	5,77	4,78	Sl_6	$(NH_4)_2SiF_6$
0,834	10,05	8,38	$P6_3$	$KNa_3[AlSiO_4]_4$
0,837	10,76	9,02		Ni_5Sb_{11}
0,839	5,30	4,45	B_{35}	β-FeSn
0,846	5,24	4,45	Zn_5Ca	$Zn_{5,4}Th_{0,6}$
0,847	9,99	8,46		$NaAlSiO_4$
(0,855)	(9,98)	(8,53)	GO_1	$CaCO_3$
0,857	9,27	7,95		$NaSrPO_4$
0,866	5,58	4,83	$P6_3mc$	$BaNiO_3$

c/a	a, Å	c, Å	Structure type	Substance
0,876	6,87	6,02		ICN
0,877	6,01	5,27	$R3m$	AgCN
0,884	7,87	6,95		Mg_2Ga
0,898	14,07	12,64	$R\bar{3}m$	$BiCo(CN)_6 \cdot 6SC(NH_2)_2$
0,908	6,99	6,35	$P3,21$	$NdPO_4$
0,913	5,27	4,81	$PbSb_2O_6$	$HgSb_2O_6$
0,913	7,07	6,45	$P3,21$	$CePO_4$
0,913	7,10	6,48	$P3,21$	$LaPO_4$
0,915	5,24	4,80	$PbSb_2O_6$	$CbSb_2O_6$
0,919	5,36	4,93		$LiSb(OH)_6$
0,921	7,22	6,65		$AcPO_4 \cdot 0,5H_2O$
0,923	6,77	6,25		$2(CaSO_4) \sim 1H_2O$
0,928	6,97	6,47	$P3,21$	$BiPO_4$
0,929	9,29	8,63	$O1_3$	$(CH_2O)_3$
0,930—0,931	3,21	2,98—2,99		Hg—Sn (21,95—3,8 at. % Hg)
0,932—0,933	4,81—4,76	4,48—4,44		$Cr_{2-x}N_{1+x}$ (27,6—33,4 at. % N)
0,935	3,77	3,52	$P/6mmm$	$Ni_{0,84}Li_{2,16}N$
0,940	3,19	3,00		$Sn_{12}Hg$
0,943	4,78	4,50	$PbSb_2O_6$	$CoAs_2O_6$
0,943	5,48	5,17	$E2_3$	$LiIO_3$
0,944	10,92	10,31		$CaCO_3 \cdot CaSO_4 \cdot CaSiO_3 \cdot 15H_2O$
0,945	6,21	5,87	$F0_2$	COS ($<135°K$)
0,950	5,02	4,77		$\alpha\text{-}Al(OH)_3$
0,955	8,53	8,15	$(P6_3/m)$	$AlBO_3$
0,956	4,55	4,35	B_{35}	$\varepsilon\text{-}InNi$
0,957	11,64	11,14		$(37Mn_5SiO_7 \cdot 10Fe_3Sb_2O_8)$
0,957	19,47	18,64		$(K,Na)_5Fe_3(SO_4)_6(OH)_2 \cdot 9H_2O$
0,960	5,23	5,02	$PbSb_2O_6$	$CaSb_2O_6$
0,960	12,02	11,54		CrS
0,961	3,74	3,62	$P6/mmm$	$Co_{0,68}Li_{2,32}N$
0,964	13,89	13,39		$(C_6H_5)_3CBr$
0,966	6,83	6,60		YPO_4
0,968	5,09	4,93	$R\bar{3}m$	Po ($>348°$ K)
0,969	2,90	2,81		MoC
0,974	7,30	7,11	$BaSiF_6$	$BaGeF_6$
0,976	2,91	2,84		WC
0,976	7,18	7,01	$BaSiF_6$	$BaSiF_6$
0,978	11,17	10,92	$S6_1$	$Co(NH_3)_6Cr(CN)_6$
0,980	2,87	2,81		MoN
0,980	10,22	10,02		$(Al_2O_3 \cdot 4SiO_2 \cdot H_2O)$
0,985	5,26	5,34	$PbSb_2O_6$	$SrSb_2O_6$
0,992	7,68	7,61	$D8_{11}$	Co_2Al_5
0,993	10,91	10,83	$S6_1$	$Co(NH_3)_6 \cdot Co(CN)_6$
0,994	8,96	8,91	$Mg(H_2O)_6 \cdot SO_3$	$MgSeO_3 \cdot 6H_2O$
0,999	11,80	11,79	$S2_2$	$Al(H_2O)_6Cl_3$
0,999	11,92	11,91	$S2_2$	$Cr(H_2O)_6Cl_3$
1,00	4,28	4,28	$C32$	$PrGa_2$
1,00	4,31	4,31	$C32$	$CeGa_2$
1,00	4,32	4,32	$C32$	$CaGa_2$
1,00	9,22	9,22	$S3_1$	$Be_3Al_2(SiO_3)_6$
1,01	4,83	4,87	$PbSb_2O_6$	$CdAs_2O_6$
1,01	5,30	5,37	$PbSb_2O_6$	$PbSb_2O_6$
1,01	7,65	7,73	$TlSbF_6$	NH_4SbF_6
1,01	7,98	8,06	$TlSbF_6$	$CsSbF_6$
1,01	8,79	9,03	$Mg(H_2O)_6 \cdot SO_3$	$NiSO_3 \cdot 6H_2O$
1,01	9,64	9,70	$S6_1$	$Fe(H_2O)_6 \cdot SiF_6$
1,01	9,68	9,77	$S6_1$	$Mn(H_2O)_6 \cdot SiF_6$
1,01	9,79	9,87	$S6_1$	$Mg(H_2O)_6 \cdot TiF_6$
1,01	10,76	10,87	$S6_1$	$Co(NH_3)_5 \cdot H_2O \cdot Co(CN)_6$
1,01	10,77	10,86	$S6_1$	$Co(NH_3)_5 \cdot H_2O \cdot Fe(CN)_6$
1,02	3,00	3,06	$C32$	VB_2
1,02	3,05	3,11	$C32$	MoB_2
1,02	3,68	3,77		$Cu_{0,40}Li_{2,60}N$

c/a	a, Å	c, Å	Structure type	Substance
1,02	4,33	4,40	$C32$	$LaGa_2$
1,02	6,99	7,13	DO_{21}	Cu_3P
1,02	6,99	7,16	DO_6	SmF_3
1,02	7,07	7,23	DO_6	PrF_3
1,02—1,03	7,10—7,21	7,24—7,50	$P\bar{3}c1$	$Cu_{3+x}As_{1-x}$ (25,0—17,0 at. % As)
1,02	7,12	7,28	DO_6	CeF_3
1,02	7,13	7,29	DO_6	$(Ce,La)\,F_3$
1,02	7,14	7,29	DO_6	$(La,Ce,Pr, \ldots)\,F_3$
1,02	7,17	7,34	DO_6	LaF_3
1,02	7,64	7,80	$TlSbF_6$	$PbSbF_3$
1,02	10,62	10,80	$S6_1$	$Ni\,(H_2O)_6SnCl_6$
1,02	13,68	13,98		$Cu_2(OH)_3Cl$
1,03	4,84	4,98	$PbSb_2O_6$	$HgAs_2O_6$
1,03	7,03	7,21	DO_6	NdF_3
1,03	8,84	9,06	$Mg(H_2O)_6SO_3$	$CoSO_3 \cdot 6H_2O$
1,03	8,84	9,06	$Mg(H_2O)_6SO_3$	$MgSO_3 \cdot 6H_2O$
1,03	9,27	9,52	$S6_1$	$Ni\,(H_2O)_6SiF_6$
1,03	9,52	9,84	$S6_1$	$Mg\,(H_2O)_6SiF_6$
1,03	9,57	9,90	$S6_1$	$Zn\,(H_2O)_6TiF_6$
1,03	9,33	9,65	$S6_1$	$Zn\,(H_2O)_6SiF_6$
1,03	9,79	10,04	$S6_1$	$Mg\,(H_2O)_6SnF_6$
1,03	9,79	10,13	$S6_1$	$Zn\,(H_2O)_6ZrF_6$
1,03	10,62	10,91	$S6_1$	$Co\,(H_2O)_6SnCl_6$
1,03	10,70	10,97	$S6_1$	$Fe\,(H_2O)_6SnCl_6$
1,03	10,71	11,01	$S6_1$	$Mg\,(H_2O)_6SnCl_6$
1,04	6,32	6,56		$2Ca\,(OH)_2Ca\,(OCl)_2$
1,04	7,12	7,44		$P\,I_3$
1,04	7,67	7,95	$TlSbF_6$	$TlSbF_6$
1,04	9,33	9,72	$S6_1$	$Co\,(H_2O)_6SiF_6$
1,04	10,66	11,06	$S6_1$	$Mn\,(H_2O)_6SnCl_6$
1,05	3,09	3,24	$C32$	TaB_2
1,05	3,86	4,07	$C32$	$\beta\text{-}USi_2$
1,05	3,97	4,17	$P3m1$	$\alpha\text{-}UO_3$
1,05	4,83	5,08	$PbSb_2O_6$	$CaAs_2O_6$
1,05	6,63	6,93	$P\bar{6}\,2m$	$Al_8Si_6Mg_3Fe$
1,05	9,73	10,21	$S6_1$	$Zn\,(H_2O)_6SnF_6$
1,06	3,66	3,89	$P6/mmm$	Li_3N
1,06	10,44	11,03	$S6_1$	$Co\,(NH_3)_4\,(H_2O)_2Co\,(CN)_6$
1,07	3,03	3,23	$C32$	TiB_2
1,07	3,09	3,31	$C32$	NbB_2
1,07	7,33	7,87		$Cu_{20}Sn_6$
1,08	3,01	3,25	$C32$	AlB_2
1,08	5,30	5,75	$PbSb_2O_6$	$BaSb_2O_6$
1,08	(5,73)	(6,19)		$AlCu$
1,08	6,14	6,60	$P6_3$	BCl_3 ($\sim 80°$ K)
1,09	(3,60)	(3,92)		Cb_2O_5
1,09	5,02	5,49		$\beta\text{-}SiO_2$, β-quartz ($\sim 870°$ K)
1,09	13,78	14,98	$S3_4$	$CaAl_2Si_4O_{12} \cdot 6H_2O$ (yttroparisite)
1,10	4,01	4,41		
1,10	4,73	5,19	$(C8)$	BeF_2
1,10	4,91	5,40	$C8$	$\alpha\text{-}SiO_2$ (α-quartz)
1,10	6,19	6,79		W_2P
1,10	7,10	7,79	$(BaSiF_6)$	$\beta\text{-}KPF_6$ (243° K)
1,11	3,17	3,53	$C32$	ZrB_2
1,11	4,85	5,41	$PbSb_2O_6$	$SrAs_2O_6$
1,11	5,04	5,60	$C8$	$FePO_4$
1,12	4,90	(5,47)	$C8$	$AlPO_4$
1,12	5,04	(5,62)	$C8$	$AlAsO_4$
1,13	4,53	5,51	$D5_{13}$	In_3Pt_2
1,13	4,87	5,49	$PbSb_2O_6$	$PbAs_2O_6$
1,13	4,99	5,65	$C8$	GeO_2
1,13	5,45	6,14	$G3_2$	Na_2SO_3
1,13	8,59	9,7		$Al_5Mn_{13}O_{28} \cdot 8H_2O$

c/a	a, Å	c, Å	Structure type	Substance
1,14	4,35	4,96	$A8$	Se
1,16	3,56	4,12		Ca_3N_2
1,16	6,22	7,2		$ZnF_2 \cdot 4Zn(OH)_2$
1,16	7,25	8,43	$OH6_7$	$(NH_3C_2H_5)_2SnCl_6$
1,16	7,28	8,46	OGl_1	$SbCl_2(CH_3)_3$
1,17	8,41	9,8		$Mn_{16}O_{28} \cdot 8H_2O$
1,17	8,41	9,8		$Zn_2Mn_{14}O_{28} \cdot 8H_2O$
1,17	9,74	11,33		$MnCl_2 \cdot 2MgCl_2 \cdot 12H_2O$
1,17—1,19	11,81—12,08	13,77—14,42	$R3c$	$K_3NaFeCl_6$
1,18	4,40	5,25	$D5_{13}$	δ'-InNi
1,18	9,44	11,09		$CdCl_2 \cdot 2NiCl_2 \cdot 12H_2O$
1,18	11,49	13,58	$R3c$	CH_3CONH_2
1,20	6,15	7,37		δ-Na_2ThF_6
1,20	7,14	8,55	$OH6_7$	$(NH_3C_2H_5)_2PtCl_6$
1,20	7,39	8,92	OGl_1	$SbBr_2(CH_3)_3$
1,20	8,41	10,1		$Na_2Mn_{14}O_{28} \cdot 8H_2O$
1,20	10,80	12,91		$CuPt$
1,21	4,04	4,90	$D5_{13}$	Ni_2Al_3
1,21	4,06	4,90	$D5_{13}$	Ga_3Ni_2
1,21	4,20	5,10	$B8$	$CuSn$
1,21	4,53	5,50	$D5_{13}$	In_3Pd_2
1,21	5,41	6,57	$P6/mmc$	$(NO_2^+)(NO_3^-)$ (213° K)
1,21	5,72	6,9		$2Ca(OH)_2 \cdot Al(OH)_2Cl$
1,21	5,73	6,95		$BaTiO_3$
1,21	7,38	8,93	$K7_3$	$Cs_2As_2Cl_9$
1,22	5,45	6,66	$P6/mmc$	$(NO_2^+)(NO_3^-)$ (293° K)
1,22	11,15	13,66		$Ni_3Pb_2S_2$
1,23	4,18—4,19	5,13—5,16	$B8$	$NiIn_{1-x}$
1,23	4,23	5,18	$D5_{13}$	Ga_3Pt_2
1,23	4,23	5,21	$B8$	γ-FeSn (~44 at. % Sn)
1,23	4,28	5,25	$B8$	$CuIn_{1-x}$
1,23	5,74	7,03	$R32$	Ni_3S_2
1,23	6,31	7,79		$4Zn(OH)_2ZnCl_{2-x}OH_x$; $x=0,70$—$0,75$
1,23—1,25	8,63—8,75	10,67—10,80		$SnSb$ (45—55 at. % Sb)
1,24—1,25	4,40—4,38	5,47—5,49	$B8$	$MnSn_{1-x}$ (28,6—36,1 at. % Sn)
1,24	4,51	5,57		$ICl(II)$ (118° K)
1,24	(4,65)	(5,76)		Co_2O_3
1,24	5,73	7,1		$2Ca(OH)_2Al(OH)_2Br$
1,24	6,15	7,61		$CaCO_3 \cdot H_2O$
1,24	11,93	14,78		K_4MnCl_6
1,25	4,04	5,03	$B8$	$FeGe_{1-x}$
1,25	6,23	7,81	$R\bar{3}$	$TlIO_3$
1,26	4,05—4,15	5,12—5,21	$B8$	$CoSn_{1-x}$ (39,4—42,2 at. % Sn)
1,26—1,27	4,15—4,05	5,21—5,12	$B8$	$NiSn_{1-x}$ (39—42 at. % Sn)
1,26	4,09	5,15	$B8$	$FeSb$
(1,26)	4,09	(5,15)		Sn_3As_2
1,26	11,86	14,9	$R\bar{3}c$	K_4CdCl_6
1,26	12,48	15,71	$(R\bar{3}m)$	$(NH_4)_4CdCl_6$
1,26	12,59	15,87		$ZnCl_2 \cdot 4Zn(OH)_2$
1,265	6,47	8,18	$G0_7$	$CsBrO_3$
1,27	4,04	5,13	$C6$	$PdTe_2$
(1,27)	4,93	(6,25)		AlF_3
1,27	5,23	(6,65)		FeF_3
1,28—1,30	3,96—3,85	5,05—4,99	$B8$	$NiGe_{1-x}$
1,28	3,93	5,01	$B8$	$CoGe_{1-x}$
1,28	4,25	5,46	$B8$	$PtPb$
1,28	4,34	5,56	$B8$	$RhSn_{1-x}$
1,28	4,35	5,55	$B8$	$AuSn$
1,28	4,46—4,49	5,69—5,74	$B8$	$PdPb_{1-x}$
1,29	3,96	5,12		Ir_3Si_2
1,29	4,40—4,38	5,66—5,63	$B8$	$PdSn_{1-x}$ (36—42 at. % Sn)
1,29	5,64	7,28	$P\bar{3}m1$	$KNaSO_4$
1,30	4,02	5,21	$C6$	$PtTe_2$

c/a	a, Å	c, Å	Structure type	Substance
1,30	5,65	7,36	$P\bar{3}m1$	$BaNaPO_4$
1,30	5,67	7,34	$P\bar{3}m1$	$NaK_3(SO_4)_2$
1,30	6,20	8,09	GO_7	NH_4BrO_3
1,30	7,61	9,87	$P31c$	$NaLiSO_4$
1,30	10,29	13,35	$R3c$	P_4O_{10}
1,31	3,92	5,14	$B8$	$Ni(Sb,As)$ (10 mol. % AsNi)
1,31	3,94	5,15	$B8$	NiSb
1,31	5,39	7,06	$P\bar{3}m1$	$2Ca_2SiO_4 \cdot Ca_3(PO_4)_2$
1,31	6,16	8,09	GO_7	$TiBrO_3$
1,31	6,20	8,12	GO_7	$RbBrO_3$
(1,31)	(8,62)	11,24	$A7$	Sb
(1,31)	8,63—9,04	11,27—11,81	$A7$	Sb—Bi (16—94 at. % Bi)
(1,31)	(8,64)	11,31	$A7$	Pb_3Sb_7
(1,31)	(9,07)	11,85	$A7$	Bi
1,32	4,08	5,36	$B8$	NiBi
1,32	4,11	5,44	$B8$	PtSn
1,32	4,14	5,46	$B8$	PtTe
1,32	5,22	6,91	$P\bar{3}m1$	$Ca_2SiO_4 \cdot Ca_3(PO_4)_2$
1,32	5,76	7,61		O_2 (48° K)
1,32	7,29	9,66		$Ba(ClO_4)_2 \cdot 3H_2O$
1,32—1,34	8,58—8,54	11,34—11,43	$A7$	Sn—Sb (2—20 at. % Sn)
1,33	4,12	5,48	$B8$	CrSb
1,33	4,14	5,48	$B8$	PtSb
1,33	4,46	5,92	$A8$	Te
(1,33—1,39)	(5,4)	(7,2—7,5)		$Al_2[Si_2O_5](OH)_4 \cdot (^2/_3 - {}^1/_3)H_2O$
1,33	5,73	7,6		$2Ca(OH)_2Al(OH)_2I$
1,33	10,45	13,89	$P6/mcc$	$KCa_2(Be,Al)_2(Si_2O_5)_6$
1,34	3,48	4,67	$C6$	$Cd(OH)_2$
1,34	3,88	5,19	$B8$	CoSb
(1,34)	(4,89)	(6,55)		MoC
1,34	5,49	7,37	$P\bar{3}m1$	$SrNaPO_4$
1,34	6,06	8,15	GO_7	$TlClO_3$
1,34	6,09	6,13	GO_7	$RbClO_3$
1,35	5,39	7,27	$P\bar{3}m1$	Na_2SO_4
1,35	5,63	7,59		$K_2O \cdot 2PbO \cdot 2SiO_2$
1,36	3,08	4,19		PtO_2
1,36	3,73	5,07	$C6$	$PtSe_2$
1,36	3,76	5,10	$B8$	$NiAs \cdot NiSb$
1,36	3,96	5,37	$B8$	NiTe
1,36	5,24	7,14	$P\bar{3}m1$	$CaNaPO_4$
1,36	5,59	7,62	$P\bar{3}m1$	$CaKPO_4$
1,36	5,74	7,8		$2Ca(OH)_2 \cdot Al(OH)_2Cl \cdot 2H_2O$
1,36	6,02	8,16	GO_7	$KBrO_3$
1,36	7,41	10,07	$S3_6$	$Na_2ZrSi_3O_9 \cdot 2H_2O$
1,37	3,41	4,67	C_6	$Cd(OH)_2$—$Mn(OH)_2$
1,37	3,86	5,28	C_6	$NiTe_2$
1,37	4,08	5,59	$B8$	PdSb
1,37	4,14	5,67	$B8$	PdTe
1,37	6,26	8,61	$F5_1$	$AgFeO_2$
1,37	7,10	9,74	$G7_1$	$(Ce,La....)FCO_3$
1,37	11,88	16,32	$P\bar{3}c1$	$Cr(NH_2 \cdot CH_2 \cdot CH_2 \cdot NH_2)_3 \cdot Br_3 \cdot 3H_2O$
1,38	3,25	4,48	$C6$	$Fe(OH)_2$
1,38	3,50—3,52	4,83—4,85	$C6$	$Ca(OH)_2$—$Cd(OH)_2$ [25—75 mol.% $Ca(OH)_2$]
1,38	3,59	4,94	$C6$	$Ca(OH)_2$
1,38—1,43	3,89—3,79	5,38—5,41	$C6$	$CoTe_{2-x}$ (50,0—66 at. % Te)
1,38	3,90	5,37	$B8$	CoTe
1,38	5,72	7,88	$P\bar{3}m1$	K_2SO_4
1,39	3,61	5,02	$B8$	(AsNi—SbNi)
1,39	3,63	5,05	$B8$	AsNi—SbNi (90 mol.% AsNi)
(1,39)	4,14	(5,71)		$BiSe_3$
(1,39)	4,35	(6,05)		Bi_2Te_3
1,40	3,59	5,01	$C6$	$CdCl_2 \cdot 4Cd(OH)_2$

c/a	a, Å	c, Å	Structure type	Substance
1,40	3,62	5,05	$B8$	NiAs
1,40	3,99	5,57	$B8$	IrSn
1,40	4,13	5,79	$B8$	MnSb
1,41	3,32	4,69	$C6$	$Mn(OH)_2$
1,41	5,75	8,1		$2Ca(OH)_2Al(OH)_2Br \cdot 2H_2O$
1,42	3,55	5,03	$C6$	PtS_2
1,42	3,77	5,33		CrAs
1,42	4,31	6,13	$B8$	MnBi
1,43	3,54	5,04		$CdCl_{0,26}(OH)_{1,74}$
1,43	3,79	5,41	$C6$	$CoTe_2$
(1,43)	4,25	(6,07)		Sb_2Te_3
1,43	12,84	18,32	$K7_2$	$Cs_3Tl_2Cl_9$
1,44	4,43	6,36	$C40$	$CrSi_2$
1,44	6,73	9,72	$(S3_2)$	$BaTiGe_2O_9$
1,44	10,57	15,19		$(Na,Ca,Ce,Th)_3(Mn,Fe,Sb)$ $[(Si,P)_3(O,OH,F)_{12}]$
1,46	3,19	4,65	$C6$	$(Co,Zn)(OH)_2$
1,46	3,19	4,66	$C6$	$Zn(OH)_2$
1,46	3,20	4,66	$C6$	$Co(OH)_2$
1,46	3,62	5,29	$B8$	CoSe
1,46	3,67	5,34	$B8$	NiSe
1,46	7,43	10,59		Mg_2C_3
1,47—1,49	3,17—3,12	4,66—4,64	$C6$	$Co(OH)_2$—$Ni(OH)_2$ [30—80 mol.% $Co(OH)_2$]
1,47	3,14	4,62	$C6$	$(Ni,Zn)(OH)_2$
1,47	3,33	4,89		UHg_3
1,47	3,39	4,99		$CdF_2(2—9)Cd(OH)_2$
1,47	6,61	9,73	$S3_2$	$BaTiSi_3O_9$
1,48	3,11	4,61	$C6$	$Ni(OH)_2$
1,48	3,16	4,67	$C6$	$Co(OH)_2$—$Mg(OH)_2$
1,48	3,16	4,67	$C6$	$Co(OH)_2$—$Zn(OH)_2$ [25 mol. % $Zn(OH)_2$]
1,49	3,81	5,66	$B8$	FeTe
1,49	12,22	18,21	$R3$	Tl_2S
1,50	3,41	5,10		AuCN
1,50	5,74	8,6		$2Ca(OH)_2 \cdot Al(OH)_2NO_3 \cdot 2H_2O$
1,51	3,10	4,68	$C6$	$Ni(OH)_2$—$Mg(OH)_2$
1,51	4,55	6,89	$C6$	$PbI_2(I)$
1,51	12,41	18,72		$Ca(H_2O)_4S_2O_6$
1,51	14,13	21,34	$P3m$	$(C_6H_5)_3CBr$
1,52	3,05	4,67	$C6$	$Ni(OH)_2$—$Zn(OH)_2$ [35 mol. % $Zn(OH)_2$]
1,52	3,13	4,74	$C6$	$Mg(OH)_2$
1,52	7,1	10,8		β-Be (900° K)
1,52	14,93	22,6		$2,4,6$ $C_6H_2Br(NO_2)_3$
1,53	3,99	5,19	$B8$	CoS
1,53	3,72	5,71	$B8$	MnAs
1,53	4,23	6,47		BiTeBr
1,53	5,75	8,8		$2Ca(OH)_2 \cdot Al(OH)_2I \cdot 1,3H_2O$
1,53	6,74	10,28	Sl_3	$K_2Pt(SCN)_6$
1,54	3,10	4,77		$Zn(OH)_2$—$Mg(OH)_2$ [25 mol. % $Zn(OH)_2$]
1,54	4,25	6,54	$C6$	ZnI_2
1,54	6,78	10,47	Sl_3	$(NH_4)_2Pt(SCN)_6$
1,54	12,48	19,28		$Sr(H_2O)_4S_2O_6$
1,55—1,56	2,76—2,77	4,30	$A3$	$Cu_{1-x}Zn_{s+x}$ (79,8—85,6 at. % Zn)
1,55—1,56	3,42—3,43	5,32—5,34	$B8$	β-NiS
1,55	3,59	5,55	$C6$	$CdCl_{0,56-0,67}(OH)_{1,44-1,33}$
1,55	4,08	6,30	$D5_2$	Ac_2O_3
1,55	4,49	6,97	$C6$	CaI_2
1,55	4,49	6,97	$C6$	YbI_2
1,55	5,75	8,9		$2Ca(OH)_2 \cdot Al(OH)_2 \cdot 1/_2CO_4 \cdot nH_2O$
1,55	6,76	10,49	Sl_3	$Rb_2Pt(SCN)_6$

c/a	a, Å	c, Å	Structure type	Substance
1,56	2,82	4,39—4,38		$AuZn_7$ (86,3—88,8 at. % Zn)
1,56—1,58	2,83	4,47—4,39		$AgZn_3$ (86—70 at. % Zn)
1,56	(3,08)	(4,82)		LiMg (78° K)
1,56	3,85	6,02	$D5_2$	Nd_2O_3
1,56	3,86	6,01	$D5_2$	Pr_2O_3
1,56	3,89	6,07	$D5_2$	Ce_2O_3
1,56	3,94	6,13	$D5_2$	La_2O_3
1,56	3,98	6,21	$B8$	CrTe
1,57	2,29	3,59		α-Au—Be
1,57	2,76	4,33		Cu_3Sn (54—66 at. % Cu)
1,57	2,79	4,40		β-Li—Zn (10,5 at. % Li)
1,57	3,37	5,30	$B8$	(Fe,Co)S
1,57	5,75	9,5		$2Ca(OH)_2Al(OH)_2ClO_4$
1,57	11,40	17,91		β-$Al_{3+x}Mg_{2-x}$ (63 at. % Al)
1,58	2,29	3,61	$A3$	α-Be
1,58—1,62	2,72—2,63	4,29—4,25		OsIr [44,3—64,3 % (Os+Ru)]
1,58—1,59	2,78—2,73	4,38—4,33		Cu_3Sb (25—19 at. % Sb)
1,58	2,75	4,34		Fe_3C
1,58—1,59	2,86—2,87	4,53—4,55		$V_{2+x}C_{1-x}$ (11,8—31,5 at. % C)
1,58—1,62	2,90—2,88	4,58—4,66		γ-$Ag_{3-x}Al_{1+x}$ (43—27 at. % Al)
1,58	3,00	4,72		W_2C
1,58	3,01	4,74		Mo_2C
1,58	3,05—3,10	4,82—4,90		$AgCd_3$ (69—83 at. % Cd)
1,58	3,52	5,57	$A3$	Lu
1,58	3,54	5,60	$A3$	Er
1,58	3,56	5,63	$A3$	Ho
1,58	3,59	5,66	$A3$	Dy
1,58	3,60	5,67	$A3$	Tb
1,58	4,31	6,83		BiTeI
1,58	4,58	7,24	$D5_2$	Mg_3Sb_2
1,58	5,51	8,71		$Cu_{11}Sb_4$
1,58—1,61	5,64—5,54	8,92—8,90		ε-Mn—Zn (11,58—14,3 at. % Mn)
1,59	2,70	4,28	$A3$	Ru
1,59	2,72	4,32	$A3$	Os
1,59	2,73	4,34		Cu_2SnFe
1,59—1,63	2,89—2,87	4,58—4,66		$Ag_{5\pm x}Al_{3\mp x}$ (43—27 at. % Al)
1,59	3,09	4,90		$LiCd_3$
1,59	3,21	5,09	$A3$	Hf
1,59	3,23	5,15	$A3$	α-Zr
1,59	3,31	5,25	$A3$	Sc
1,59	3,53	5,57	$A3$	Tu
1,59	3,63	5,76	$A3$	Gd
1,59	3,65	5,81	$A3$	Y
1,59	3,69	5,86	$C6$	ZrSr
1,59	3,88	6,19	$D5_2$	Th_2N_3
1,59	4,68	7,42	$D5_2$	Mg_3Bi_2
1,60	2,54—2,55	4,07—4,09		Mn—Fe (13,4—32 at. % Mn)
1,60—1,61	2,61—2,67	4,16—4,31		Ni—N
1,60	2,64	4,21		Cu_3Ge
1,60—1,62	2,78—2,70	4,44—4,37		$Fe_{2+x}N_{1-x}$ (33,1—24,5 % N)
1,60	2,73	4,34		Os—Ru
1,60	2,74	4,39	$A3$	Tc
1,60—1,61	2,75—2,76	4,39—4,43		ε-CFeN
1,60	2,76	4,41		$MnZn_7$
1,60—1,63	2,84—2,78	4,54—4,53		Mn_2N (35,0—27,2 at. % N)
1,60	2,84	4,54—4,55		V_2N (42—27 at. % N)
1,60	2,90	4,62	$B4$	CuH
1,60	2,95	4,71	$A3$	Ti
1,60—1,61	2,97—3,00	4,76—4,84		Ag—Hg (37,0—44,8 at. % Hg)
1,60	3,10	4,94	$C6$	Ta_2C
1,60	3,12	4,99	$B4$	AlN
1,60	3,26	5,21	$B4$	ZnO
1,60	3,28	5,26	$B4$	ZnO—MnO (22 mol. % MnO)

c/a	a, Å	c, Å	Structure type	Substance
1,60	3,42	5,47	$B8$	(Ni,Fe)S
1,60	3,45	5,52	$A3$	Ti
1,60	3,46	5,53		$NaTl_3$
1,60—1,63	3,61—3,69	5,78—6,03		Cr—Se (60—53,5 at. % Se)
1,60	3,26	5,23	$B4$	ZnO—CdO (5 mol. % CdO)
1,60—1,64	4,53—4,38	7,24—7,16		$(Mg,Zn)_3Sb_2$ (12,3—64,7 at.% Zn)
1,60	4,40	7,03	$B4$	NH_4F
2,60	4,90	7,85		γ-Mg_3Ag (20—25 at. % Ag)
1,60	4,96	7,94		BiSe
1,60	5,74	9,2		$2Ca(OH)_2Al(OH)_2ClO_3$
1,61	2,54—2,53	4,09—4,08		Mn—Fe (87,4—77,5 at. % Fe)
1,61—1,62	2,71	4,37—4,38		Fe_3N (24,9—27,3at. % N)
1,61—1,63	2,76—2,73	4,44—4,45		Mn—Zn (16—50 at. % Mn)
(1,61)	(2,76—2,78)	4,44—4,48	$P321$	$Cr_{2+x}N_{1-x}$ (33,4—27,6 at. % N)
1,61	2,76	4,46	$A3$	Re
1,61	2,80	4,49	$(A3)$	$Fe_{1-x}Zn_{7+x}$ (87—92 at. % Zn)
1,61—1,63	2,98—2,95	4,79		$Ag_{3-x}In_{1+x}$, x from 0 to 1
1,61	2,97	4,78		$TiN_{0,22}$
1,61	2,99	4,81		Ag_3Sn
1,61	2,99	4,82		Ag_3Sb
1,61	3,10	4,99		(Al_2O_3)
1,61	3,54	5,70	$B4$	InN
1,61	3,65	5,88	$C6$	SnS_2
1,61	3,67	5,89	$A3$	Nd
1,61	3,80	6,12	$B8$	VTe
1,61	4,26	6,86	$C6$	CdI_2
1,61	6,64	10,66	$C14$	$BaMg_2$
1,61	10,68	17,19	$B30$	MgZn
1,62	2,51	4,07	$A3$	β-Co
1,62	2,55	4,12		$MnFe_5$ (77—88 at. % Fe)
1,62	2,61	4,22		$Cu_{6-x}As_{1+x}$
1,62	2,66	4,31		(Ni)
1,62	2,71—2,73	4,38—4,41		Fe_4N
1,62	2,93	4,75		β-Ag_3Ga ($>$ 713° K)
1,62—1,65	2,94—2,90	4,77—4,79		Au—Sn (20,8—8,0 at . % Sn)
1,62—1,65	2,94—2,91	4,76—4,80		β-$Au_{5+x}Sn_{1-x}$ (16—12 at. % Sn)
1,62—1,63	2,97—2,93	4,80—4,79		Ag—Sb (16—10 at. % Sb)
1,62—1,63	2,96—2,94	4,77—4,78		Ag_5Sn (19,7—13,3 at. % Sn)
1,62	3,00	4,86		AgHg
1,62	3,02	4,89		AgCd
1,62	3,06	4,95	$B4$	TaN
1,62—1,63	3,19—3,10	5,16—5,04	$A3$	Mg—Cd (80—42 at. % Mg)
1,62	3,20	5,18	$A3$	δ-(Mg,Ag) (3 at. % Ag)
1,62—1,63	3,20—3,21	5,19—5,21	$A3$	Mg—Zn (3—0 at. % Zn)
1,62—1,63	3,60—3,69	5,82—6,03	$B8$	$(Cr_{1-x}Se)$
1,62	3,66	5,91	$A3$	Pr
1,62	3,66	5,94	$A3$	β-Ce
1,62	3,75	6,07	$A3$	α-La
1,62	3,83	6,20	$C6$	$MnBr_2$
1,62	3,98	6,44	$B4$	MnS
1,62	4,13	6,71	$B8$	MnTe
1,62	4,14	6,71	$B4$	CdS
1,62	4,17	6,76		C_x
1,62	4,22	6,84	$C14$	$FeBe_2$
1,62	4,54	7,36	$B4$	MgTe
1,62	5,71	9,27		K_2GeF_6
1,62	5,86	9,50		Rb_2MnF_6
1,62	5,94	9,63		Rb_2GeF_6
1,62	6,23	10,12	$C14$	$CaMg_2$
1,63	2,56—2,69	4,17—4,39		Ni—H
1,63	2,60	4,19		Cu_6Si
1,63	2,60	4,23		Cu_2As
1,63	2,60	4,24		β-Cu_5Si (14,5 at. % Si)

c/a	a, Å	c, Å	Structure type	Substance
1,63	2,70	4,38	$B4$	BeO
1,63	2,73	4,43	$A3$	β-Cr
1,63	2,90—2,99	4,73—4,87		Ag—As (9,8—15,1 at. % As)
1,63—1,64	2,92	4,76—4,79		Au—In (15,7—20,2 at. % In)
1,63	3,07—3,10	4,99—5,04	$A3$	$MgCd_{2-x}$
1,63	3,13	5,09	$A3$	$Cd(Mg,Al)_2$
1,63	3,21—3,17	5,21—5,15	$A3$	Mg—Al (2—11 at. % Al)
1,63	3,19	5,18	$B4$	GaN
1,63	3,21	5,20	$A3$	Mg—Mn (1,5 at. % Mn)
1,63	3,21	5,21	$A3$	Mg
1,63	3,35	5,46		Zr_2H
1,63	3,43	5,56	$B8$	$Cr_{1-x}S$
1,63	3,58	5,83		β-Al_2S_3
1,63	3,58	5,84	$A3$	He (1,45° K, 37 atm)
(1,63)	(3,61)	(5,89)		ZrO_2
1,63	3,63	5,92	$B8$	FeSe
1,63	3,75	6,11	$A3$	Er
1,63	3,76	6,13	$A3$	H_2 (4° K)
1,63	3,76	6,13		$ZnAl_2S_4$
1,63	3,80	6,19	$C6$	$ZrSe_2$
1,63	4,13	6,73	$B4$	γ-MnSe
1,63	4,20	6,86	$B4$	Cd(S,Se)
1,63	4,24	6,92	$C14$	$MnBe_2$
1,63	4,26	6,96	$C14$	$CrBe_2$
1,63	4,31	7,03	$B4$	CdSe
1,63	4,32	7,06	$A3$	β-Sr (521° K)
1,63	4,32	7,10	$C14$	$ReBe_2$
1,63	4,40	7,14	$C14$	VBe_2
1,63	4,47	7,29		H_2O (90° K)
1,63	4,51	7,34	$C10$	H_2O (273° K)
1,63	4,53	7,36	$C10$	D_2O (272° K)
1,63	4,74	7,73	$C14$	Fe_2W
1,63	4,78	7,81	$C14$	Fe_2Ti
1,63	4,81	7,86	$C14$	Fe_2Ta
1,63	4,83	7,88	$C14$	Fe_2Cb
1,63	4,87	7,95	$C14$	Mn_2Ta
1,63	4,96×2	8,08	$C14$	Mn_2Ti
1,63	5,04	8,24	$C10$	SiO_2 (β-tridymite)
1,63—1,65	5,22—5,10	8,56—8,42		$MgZn_2$
1,63	5,11	8,32	$D0_{24}$	Ni_3Ti
1,63	5,14	8,39	$C14$	$ZrCr_2$
1,63	5,18	8,44		$MgCuAl$—$MgAl_2$
1,63	5,26	8,60	$C14$	$ZrRe_2$
1,63	5,43	8.85	$E9_2$	$4BeO \cdot NaSbO_3$
1,63	5,48	8,95	$C14$	$ThMn_2$
1,63	6,43	10,47	$C14$	$SrMg_2$
1,63	8,40	13,65		θ-Al_2O_3
1,63	9,00	14,70		Li_2SnO_3
1,64	2,66	4,34		Ni_3C
1,64	2,66	4,35		Co_3N
1,64—1,	2,91—2,92	4,77—4,84		Cd—Au (25,3—35,5 at. % Cd)
1,64	2,91—2,93	4,78—4,81		Au_3Hg (18,8—32,4 at. % Hg)
1,64	3,68	6,02		β-Ga_2S_3
1,64	3,77	6,18	$C6$	VBr_2
1,64	3,82	6,25	$B4$	ZnS
1,64	3,83	6,27	$C6$	$MgBr_2$
1,64	3,97	6,50	$A3$	β-Ca
1,64	4,04	6,62	$A3$	β-N_2 (~40° K)
1,64	4,06	6,66	$B4$	β-CuBr (664—743° K)
1,64	4,14	6,80	$C6$	GeI_2
1,64	4,17	6,83	$C6$	MnI_2
1,64	4,44	7,29	$C14$	$MoBe_2$
1,64	4,45	7,28	$C14$	WBe_2

c/a	a, Å	c, Å	Structure type	Substance
1,64	4,59	7,52	$B4$	Ag I
1,64—1,65	4,87—4,82	7,81—7,97	$C14$	$Mg(Ni,Cu)_2$
1,64	4,82	7,91		$FeTa_3$
1,64	4,87	7,97	$C14$	Mn_2Nb
1,64	5,01	7,89	$C14$	$MgCu_{1,5}Si_{0,5}$
1,64	5,04	8,24	$C14$	Mn_2Zr
1,64	5,10	8,36	$C14$	$MgCuAl—MgCu_2$
1,64—1,65	5,14—5,22	8,44—8,62	$C14$	$Mg(Zn,Cu)_2$
1,64	5,15	$(8,43)\times 2$		$Mg_3Zn_2Cu_2Al_2$
1,64	5,17	8,51	$C14$	$MgZn_2$
1,64	5,19	8,53	$C14$	$ZrOs_2$
1,64	5,29	8,66	$C14$	ZrV_2
1,64	6,26	10,25	$C14$	$CaLi_2$
1,64	6,93	11,35		β-Au—Be
1,64	7,50	12,29	$C14$	KNa_2
1,64	22,19	36,34		$K(Fe,Al,Mg,Mn)_{10}(OH)_{12}[Si_2O_5]_6$
1,65	2,55	4,20		Ni—Mo(25,0 at. % Mo)
1,65	3,56	5,88	$C6$	$TiCl_2$
1,65	3,75	6,18	$C6$	$FeBr_2$
1,65	4,07	6,74		β-Co ($\sim 61°$ K)
1,65	4,31	7,09	$B4$	β-Cu I (675—713° K)
(1,65)	4,82	(7,97)	$C36$	$MgNi_2$
(1,65)	4,87—4,82	(8,05—7,97) $\times 2$	$C36$	$Mg(Ni,Zn)_2$
(1,65)	5,10	(8,40)	$C36$	$MgCuAl$
1,65	5,20—5,22	8,60—8,62	$C14$	$Mg(Zn,Ni)_2$
1,65	5,67	9,35		K_2MnF_6
1,65	5,75	9,5		$2Ca(OH)_2 \cdot Al (OH)_2ClO_4$
1,65	5,75	9,51		K_2PdF_6
1,66—1,68	3,44—3,46	5,69—5,80	$B8$	$Fe_{0,83-0,96}S$
1,66	3,45	5,71	$B8$	CrS
1,66	3,49	5,78—5,79		Pb—Bi (20—50 at. % Bi)
1,66	3,70	6,13	$C6$	$CoBr_2$
1,66	4,11	6,82	$C6$	$Ti I_2$
1,66	4,15	6,89	$C6$	MgI_2
1,66	5,14	8,51	$C14$	$ZrRu_2$
1,66	9,94	16,51	$D2_2$	$MgZn_5$
1,66	10,02	16,61	$R32$	$K_3Cu(CN)_4$
1,67	2,91	4,72		$Ag_3Al(843°$ K$)$
1,67	3,41	5,70	$C6$	TiS_2
1,67	3,59	5,99	$B8$	VSe
1,67	3,83	6,39	$B8$	$TiTe$
1,67	4,00	6,67	$C6$	$V I_2$
1,67	4,05	6,76	$C6$	$Fe I_2$
1,67	5,75	9,6		$2Ca(OH)_2 \cdot Al(OH)_2MnO_4$
1,68	3,45	5,81	$B8$	FeS
1,68	3,97	6,66	$C6$	$CO I_2$
1,68	5,14	8,62	$H1_4$, $H2_5$	$KLi\overset{\smile}{S}O_4$
1,68	5,17	8,67	$H2_8$	$KAlSiO_4$
1,68	5,22	8,78	$H2_8$	$BaAl_2O_4$
1,68	5,45	9,13	$R3m$	KNO_3 (398° K)
1,68	6,71	11,28		Bi_2Se_3
1,69	4,72	7,98	$H3_2$	$KAl(SO_4)_2$
1,69	3,54	5,99	$C6$	$TiSe_2$
1,70	3,16	5,37	$C6$	$NiCl_2 \cdot 3Ni(OH)_2$
1,70	4,75	8,05	$H3_2$	$KCr(SO_4)_2$
1,70	9,43	16,00		$(Mn,Mg,Na)Be_2Fe_2(PO_4)_4 \cdot 6H_2O$
(1,71)	(3,23)	(5,51)	$C6$	$CoCl_2 \cdot 3Co(OH)_2$
1,71	3,28	5,60	$C6$	$Co(^1/_3Cl,^2/_3Br)_2 \cdot 3Co(OH)_2$
1,71	4,01	6,83	Ce_2O_2S	Ce_2O_2S
1,71	4,04	6,89	Ce_2O_2S	La_2O_2S
1,71	7,20	12,32	$O2_5$	$C(NH_2)_3$ I
1,72	4,24	7,27	$P6/mmm$	Cu_2Te

c/a	a, Å	c, Å	Structure type	Substance
1,72	4,84	8,33	$H3_2$	$NH_4Fe(SO_4)_2$
1,72	22,0	37,85		$K(Al,Fe,Mg,Mn)_{10}(OH)_{12}[Si_2O_5]_6$
1,73	3,36	5,81	$B8$	VS
1,73	3,41	5,91	$C6$	TaS_2
1,73	3,77	6,51	$C6$	$TiTe_2$
1,73	3,93	6,77	Ce_2O_2S	Pu_2O_2S
(1,73)	(4,23)	7,33		$TiBe_{12}$
1,73	22,5	39,1		$3MnO \cdot 4SiO_2 \cdot 4H_2O$
1,74	3,80	6,59		YC_2
1,74	4,73	8,25	$H3_2$	$NH_4Al(SO_4)_2$
1,74	5,33	9,26		$FeSn_2$
1,74	5,75	10,0		$2Ca(OH)_2 \cdot Al(OH)_2 \cdot {}^1/_2CrO_4 \cdot nH_2O$
1,74	8,25	14,34	$P3_61$	$Cu_6AlSO_4(OH)_{12}Cl \cdot 3H_2O$
1,75	3,56	6,22	$B8$	TiSe
1,75	4,20	7,42	$D0_6$	$PbThF_6$
1,75	4,28	7,48	$D0_6$	$BaUF_6$
1,75	4,79	8,39	$H3_2$	$NH_4Cr(SO_4)_2$
1,75	9,50	16,60		Na_2SnO_3
1,76	4,19	7,35	$D0_6$	$PbUF_6$
1,76	4,27	7,53	$D0_6$	AcF_3
1,76	4,29	7,54	$D0_6$	$BaThF_6$
1,76	14,33	25,15	$P\bar{3}1c$	Tl_2Cl_3
1,77	4,09	7,24	$D0_6$	PuF_3
1,77	4,11	7,27	$D0_6$	NpF_3
1,77	4,14	7,33	$D0_6$	UF_3
1,77	4,27	7,60	$D0_{18}$	Li_3P
1,77	4,71	8,33	$D0_{18}$	$\alpha\text{-}Li_3Sb$
1,77	4,99	8,82	$D0_{18}$	Na_3P
1,77	5,10	9,00	$D0_{18}$	Na_3As
1,77	5,37	9,52	$D0_{18}$	Na_3Sb
1,77	5,46	9,68	$D0_{18}$	Na_3Bi
1,77	5,79	10,24	$D0_{18}$	K_3As
1,77	6,19	10,95	$D0_{18}$	K_3Bi
1,78	4,07	7,23	$D0_6$	AmF_3
1,78	4,40	7,83	$D0_{18}$	Li_3As
1,78	4,87	8,65	$D0_{18}$	Mg_3Hg
1,78	6,04	10,71	$D0_{18}$	K_3Sb
1,78	6,04	10,72	$D0_{18}$	$K_3(Sb,Te)$
1,78	4,04	7,19	$D0_6$	$CaThF_6$
1,78	4,11	7,30	$D0_6$	$SrUF_6$
1,78	4,14	7,34	$D0_6$	$SrThF_6$
1,79	5,75	10,3		$2Ca(OH)_2 \cdot Al(OH)_2 IO_2 \cdot 29H_2O$
1,79	5,75	10,3		$2Ca(OH)_2 \cdot Al(OH)_2 \cdot {}^1/_2WO_4 \cdot nH_2O$
1,80	4,05	7,30	$D0_6$	$ThOF_2$
1,80	5,40	9,70		KNO_3 (425° K)
1,80	11,30	20,29		$K_3Rh(C_2O_4)_3 \cdot H_2O$
1,81	3,65	6,61	$C6$	$CdJ_{0,5}(OH)_{1,5}$
1,81	5,75	10,4		$2Ca(OH)_2 \cdot Al(OH)_2 \cdot {}^1/_2S_2O_3 \cdot nH_2O$
1,82	4,64	8,46	$D0_{18}$	$AuMg_3$
1,82	6,34	11,56	$K1_2$	$Cs_2S_2O_6$
1,83	3,24	5,92	$C6$	$CoBr_2 \cdot 3Co(OH)_2$
1,83	3,36	6,13	$C6$	VSe_2
1,84	3,18	5,83	$C6$	$NiBr_2 \cdot 3Ni(OH)_2$
1,84	6,80	12,48		Ni_5As_2
1,85	2,41	4,45		$CoMn_3$
1,85	3,03	5,59		CbN
1,86	2,67	4,95	$A3$	Zn
1,86	2,67	4,96	$A3$	Zn—Cd (1 at. % Cd)
1,86	2,67	4,97—4,98	$A3$	Zn—Al (0—5,0 at. % Al)
(1,86)	6,84	(12,72)		$2(CaSO_4) \sim 1H_2O$
1,87	2,67	4,97—4,99	$A3$	Hg—Zn (1,51—5,87 at. % Hg)
(1,88)	(2,94)	(5,53)	$D0_{19}$	$MgCd_3$
1,88	5,75	10,8		$2Ca(OH)_2 \cdot Al(OH)_2 \cdot {}^1/_3Fe(CN)_6$

c/a	a, Å	c, Å	Structure type	Substance
1,89—1,90	2,96—2,97	5,60—5,64		Mg—Cd (45—0 at. % Mg)
1,89—1,93	2,98—2,97	5,62—5,72		Hg—Cd (4,3—29,1 at. % Hg)
1,89	2,98	5,62	$A3$	Cd
1,89	11,18	21,04	$R3$	$Na_2B_4O_7 \cdot 5H_2O$
1,91	3,01	5,72	$C6$	Ag_2F
1,91	4,55	8,71	$(D4_1)$	B_2H_6
1,91	5,24	10,00	$S1_{12}$	$NaSbF_4(OH)_2$
1,92	4,33	8,32		B_2O_3
1,93	6,33	12,21	$F5_1$	$CsICl_2$
1,94	3,48	6,74	$A10$	Hg (177° K)
1,94	5,39	10,46	$B22$	RbSeH
1,95—1,97	5,98—5,97	11,66—11,77		$Fe_{1\mp x}S_{x=0,025}$
1,96	6,53	12,78	$S1_2$	$K_2Sn(OH)_6$
1,98—1,99	9,40	18,59—18,72		$3Mn_2O_3 \cdot MnSiO_3$
1,99	5,15	10,24	$B22$	KSeH
1,99	5,17	10,30	$B22$	RbSH
2,00	6,41	12,85	$S1_2$	$K_2Pt(OH)_6$
2,01	4,96	9,95	$B22$	KSH
2,01	8,20	16,45	$P6_3/mmc$	KU_6F_{25}
2,02	2,71	5,47		Zn—Hg (25—50 at. % Hg)
2,02	8,34	16,81	$P6_3/mmc$	KTh_6F_{25}
2,02	10,50	21,23	$P6_3/m$	$KNa_{22}(SO_4)_9(CO_3)_2Cl$
2,05	4,47	9,16	$B22$	NaSH
2,05	4,66	9,54	$B22$	NaSeH
2,07	5,91	12,21	$R3$	$Ag_2H_3IO_6$
2,07	7,46	15,45		$AgIO_3$
2,10	7,86	16,52	$R3c$	$NiO \cdot 3BaO$
(2,11)	5,06	(11,24)	$(C8)$	$AlAsO_4$
2,11	14,34	30,21	$R3m$	$(Na,Ca,Fe)_6ZrSi_6O_{18}(OH,Cl)$
2,12	6,84	14,50		$Ca_2(OH)_3Cl$
2,13	5,27	11,25		β-$LiAlSiO_4$
2,14—2,15	9,68—9,60	20,70—20,62	$R3m$	$PbSO_4 \cdot K_2SO_4$
2,16	5,61	12,14		B_4C
2,17	6,90	15,0		$CdSO_4 \cdot 3,5Cd(OH)_2$
2,19	3,18	6,97	$C6$	$Co(NO_3)_2 \cdot 3Co(OH)_2$
2,19	6,61	13,16		$NaJO_4 \cdot 3H_2O$
2,22	4,93	10,93	$C8$	$AlPO_4$
2,26	7,17	16,19	$K7_1$	$K_3W_2Cl_9$
2,26	7,17	16,19	$K7_1$	$(NH_4)_3W_2Cl_9$
2,27	7,19	16,33	Alunite	$NaFe_3(SO_4)_2(OH)_6$
2,27	7,23	16,43	Alunite	$AgFe_3(SO_4)_2(OH)_6$
2,29	4,16	9,52	$B9$	HgS
2,29	10,86	24,87	$R3$	$Na_9K_3Fe(SO_4)_6(OH)_3 \cdot 9H_2O$
2,30	5,48	12,62		Cd—Au (45,5 at. % Cd)
2,31	6,99	16,13	$R3m$	$Ca_2Al_2(PO_4)_2(OH)_4 \cdot H_2O$
2,31	6,99	16,13	$R3m$	$CaAl_3(PO_4)_2(OH)_5 \cdot H_2O$
2,32	7,16	16,36	$K7_1$	$Tl_3W_2Cl_9$
2,32	7,36	17,09	$K7_1$	$Cs_3W_2Cl_9$
2,34	6,97	16,29		$CaAl_3(SO_4)(PO_4)(OH)_6$
(2,34)	7,21	(16,83)	Alunite	$PbFe_3(SO_4)_2(OH)_6$
2,34	7,25	16,98	$K7_1$	$Rb_3W_2Cl_9$
2,36	7,17	16,93	Alunite	$Fe_3(SO_4)_2(OH)_5 \cdot 2H_2O$
2,36	7,21	17,03	Alunite	$KFe_3(SO_4)_2(OH)_6$
2,36	7,21	17,03	Alunite	$NH_4Fe_3(SO_4)_2(OH)_6$
2,37	6,98	16,55	$R\bar{3}m$	$SrAl_3(PO_4)_2(OH)_5 \cdot H_2O$
2,38	5,96	14,20	$S1_2$	$Na_2Sn(OH)_6$
2,38	7,39	17,58	$F5_{10}$	$KAg(CN)_2$
2,41	6,96	16,8	$R\bar{3}m$	$SrAl_3(SO_4)(PO_4)(OH)_6$
2,42	5,72	13,86		$(Zn,Mg,Fe^{II})(Sn,Zn)_2(Al,Fe^{III})_{12} \cdot O_{22}(OH)_2$
2,44	2,96	7,22		NiO
(2,45)	(5,58)	13,65		$Ni_3Pb_2S_2$ (shandite)
2,45	5,74	14,05	$P6_3/mmc$	$BaTiO_3$

c/a	a, Å	c, Å	Structure type	Substance
2,46	5,71	14,02		$Ba(Ti_{0,75}Pt_{0,25})O_3$
2,46	6,46	15,92	$F4_1$	$Fe_2(CO)_9$
2,48	16,26	40,28	$R\bar{3}m$	$ThSiW_{12}O_{40} \cdot 30H_2O$
2,49	16,16	40,28	$R\bar{3}m$	$ThSiW_{12}O_{40} \cdot 27H_2O$
2,50	6,96	17,41	Alunite	$KAl_3(SO_4)_2(OH)_6$
2,50	7,19	18,01		$CsCuCl_3$
2,50	15,59	38,98	$R\bar{3}m$	$Li_3HSiW_{12}O_{40} \cdot 24H_2O$
2,51	3,12	7,8		α-$Zn(OH)_2$
2,51	6,13	15,37		$Mg_6Al_2(OH)_{16}CO_3 \cdot 4H_2O$
2,51	6,21	15,60		$Mg_6Fe_2(OH)_{16}CO_3 \cdot 4H_2O$
2,51	7,05	17,66		$Ba(NO_2)_2 \cdot H_2O$
2,52	6,18	15,55		$Mg_6Cr_2(OH)_{16}CO_3 \cdot 4H_2O$
2,53	4,92	12,48	$D0_{14}$	AlF_3
2,53	5,27	13,3	$P6_222$	Mg_2Ni
2,54	7,10	18,23	$P\bar{6}c2$	$(Ce,La, \ldots)[Ca(CO_3)_2F]$
2,54	15,61	41,40		$CrHSiW_{12}O_{40} \cdot 24H_2O$
2,55	15,63	39,86	$R\bar{3}m$	$H_3PW_{12}O_{40} \cdot 24H_2O$
2,55	15,63	39,88		$FeHSiW_{12}O_{40} \cdot 24H_2O$
2,56	5,22	13,35	$D0_{12}$	FeF_3
2,59	3,10	8,0		α-$Co(OH)_2$
2,59	5,17	13,40	$D0_{12}$	VF_3
2,60	3,09	8,0		α-$(Co,Zn)(OH)_2$
2,61	4,55	11,86	$A7$	Bi
2,62	4,30	11,26		ε-Cu—Sb
2,62	4,30	11,28	$A7$	Sb
2,62	4,54	11,87		Bi_5Sn
2,62	5,07	13,30	$D0_{12}$	CoF_3
2,63	15,63	41,18	$R\bar{3}m$	$Zn_2SiW_{12}O_{40} \cdot 27H_2O$
2,64	5,14	13,58	$D5_1$	Ti_2O_3
2,64	5,30	14,01	$R6,cm$	$Mg_{1,6}Al_{1,0}Fe_{0,4}[SiAlO_5] \cdot$ $\cdot (OH)_4$ (amesite)
2,64	15,59	41,18	$R\bar{3}m$	$Li_3HSiW_{12}O_{40} \cdot 26H_2O$
2,64	15,67	41,30	$R\bar{3}m$	$Cu_2SiW_{12}O_{40} \cdot 27H_2O$
2,65	15,47	41,00	$R\bar{3}m$	$Ca_2SiW_{12}O_{40} \cdot 26H_2O$
2,65	15,61	41,40		$CrHSiW_{12}O_{40} \cdot 28H_2O$
2,65	15,63	41,44		$FeHSiW_{12}O_{40} \cdot 28H_2O$
2,66	2,52	6,69	$B12$	BN
2,67	3,08	8,2		α-$(Ni,Zn)(OH)_2$
2,67	15,59	41,43		$AlHSiW_{12}O_{40} \cdot 28H_2O$
2,68	15,26	40,93	$R\bar{3}m$	$Ba_2SiW_{12}O_{40} \cdot 24H_2O$
2,69	15,31	41,16		$Ba_3(PO_4 \cdot 12WO_3)_2 \cdot 48H_2O$
2,70	5,12	13,85	$E2_2$	$LiNbO_3$
2,71	4,96	13,42	$D5_1$	α-Ga_2O_3
2,71	5,11	13,83	$D5_1$	Rh_2O_3
2,73	4,76	13,0	$D5_1$	α-Al_2O_3
2,73	4,99—4,79	13,64—13,09	$D5_1$	$(Al,Fe)_2O_3$ (15—85 mol. % Al_2O_3)
2,74	2,30	6,27		$CdBr_2$
2,74	2,46	6,75	$A9$	C (graphite)
2,74—2,75	4,78—4,93	13,08—13,55	$D5_1$	$(Al,Cr)_2O_3$ (15—85 mol. % Cr_2O_3)
2,74	5,01—4,98	13,71—13,65	$D5_1$	$(Fe,Cr)_2O_3$ (25—75 mol. % Cr_2O_3)
2,74	5,03	13,77	$D5_1$	Fe_2O_3
2,74	5,05	13,85	$E2_2$	$NiTiO_3$
2,74	6,10	16,71	$R\bar{3}$	WCl_6
2,75	4,95	13,61	$D5_1$	Cr_2O_3
2,76—2,78	5,09—5,05	14,06—14,03		$(Ti,Fe)_2O_3$ (50—75 mol. % Fe_2O_3)
2,75	7,50	20,65	$D0_5$	BiI_3
2,76	5,05	13,93	$E2_2$	$MgTiO_3$
2,76	5,09	14,06	$E2_2$	$FeTiO_3$
2,77	5,05	13,72	$E2_2$	$CoTiO_3$
2,77	5,76	15,9		$NH_4NO_3(V)$ ($<255°$ K)
2,78	6,42	17,83		α-Al_2S_3
2,79	4,89	13,65	$D0_{12}$	RhF_3
2,79	6,39	17,82	$D0_5$	$ScCl_3$

c/a	a, Å	c, Å	Structure type	Substance
2,79	7,50	20,93	DO_5	SbI_3
2,80	3,77	10,56	$A7$	As
2,80	5,14	14,36	$E2_2$	$MnTiO_3$
2,80	5,27	14,7		$Pb(OH)_2$
2,81	2,45	6,85		CdI_2
2,81	3,67	10,29	EO_3	$Cd(OH)Cl$
2,81	5,06	14,24	DO_{12}	PdF_3
2,82	4,94	13,95	$D5_1$	V_2O_3
2,82	4,96	13,98	$E2_2$	$Fe_{1-x}Ti_{1+x}O_3$ (44,5 % FeO)
2,84	5,26	14,94	$E2_2$	$CdTiO_3$
2,86	6,13	17,54	DO_5	$TiCl_3$
2,87	6,05	17,38	DO_5	$FeCl_3$
2,87	6,42	18,40	DO_5	$FeBr_3$
2,88	6,46	18,64	DO_5	$TiBr_3$
2,89	3,43	9,92		$CdF_2 \cdot (4-6) Cd(OH)_2$
2,89	3,77	10,89	$A7$	As—Sn (70,9 at. % As)
2,89	5,10	14,72	$P6/mmm$	$CaAl_2Si_2O_8$
2,89	6,02	17,38	DO_5	VCl_3
2,90	5,99	17,38	DO_4	$CrCl_3$
2,90	6,86	19,88		CrI_3
2,91	6,29	18,30	DO_5	$CrBr_3$
2,92—2,94	6,23—6,15	18,19—18,08		$Cr(Br,Cl)_3$
2,94	6,06	17,82		$CrBrCl_2$
2,94	6,52	19,14		$CrIBr_2$
2,94	7,63	22,42	$R\bar{3}$	$K_4Ni(NO_2)_6$
(2,97)	(5,92)	(17,56)	DO_{15}	$AlCl_3$
2,97	7,19	21,38	DO_5	AsI_3
2,99	4,34	12,96	$P6_3/mmc$	Pt_2Sn_3
3,00	5,33	15,98	$E2_2$	$NaSbO_3$
3,00	6,24	18,70		$CrICl_2$
3,08	5,13	15,80		$BeCO_3 \cdot 4H_2O$
3,21	5,72	18,38	$P6_322$	$Be_4Mg_5Al_{16}O_{32}$
3,22—4,71	4,25—3,99	13,70—18,78	$C27$	$CdBr_2$—CdI_2
3,22	4,25	13,70	$C27$	CdI_2
3,22	4,76	15,30	GO_1	$ScBO_3$
3,23	4,64	14,98	GO_1	$ZnCO_3$
3,24	4,08	13,23		$BaBiO_2Br$
3,24	4,44	14,39	$R\bar{3}m$	$N_2H_6F_2$
3,24	4,60	14,91	GO_1	$MgCO_3$
3,24	4,66	15,08	GO_1	$CoCO_3$
3,24	4,78	15,49	GO_1	$InBO_3$
3,25	4,68	15,23	GO_1	$LiNO_3$
(3,26)	(2,56)	8,32	DO_{24}	Ni_3Ti
3,26	4,73	15,40	C_{36}	Ti_2Co
3,26	4,73	15,42	C_{36}	Co_2Ta
3,26	4,74	15,46	C_{36}	Co_2Nb
3,26	4,96	16,15	C_{36}	Fe_2Zr
3,26	4,96	16,15	C_{14}	Mn_2Ti
3,27	3,08	10,08	$B5$	SiC-(4H)
3,27	3,81	12,45	$B5$	ZnS (wurtzite-(4H))
3,27	4,91	16,07	GO_1	$(Cd,Mn)CO_3$ (25 mol. % $CdCO_3$)
3,28	3,85	12,6		$Cu_3Fe_2SnS_6$
3,28	4,68	15,36	$G1_1$	$FeCO_3$
3,28	4,86	15,95	GO_1	$CaSn(BO_3)_2$
3,29	4,75	15,65	GO_1	$MnCO_3$
3,29	4,82	15,86	$C36$	$MgNi_2$
3,29	4,93	16,22	GO_1	$(Cd,Mn)CO_3$ (50 mol. % $CdCO_3$)
3,29	4,93	16,22	GO_1	$(Mn,Ca)CO_3$ (75 mol.% $MnCO_3$)
3,29	4,95	16,28	GO_1	$(Cd,Mn)CO_3$ (75 mol. % $CdCO_3$)
3,29	4,96	16,33	GO_1	$CdCO_3$
3,30	6,87	22,7	$P6_322$	$Fe_{0,97}S$ (pyrrhotite)
3,31	4,84	16,02	$G1_1$	$CaMg(CO_3)_2$
3,32	5,06	16,81	GO_1	$NaNO_3$

c/a	a, Å	c, Å	Structure type	Substance
3,33	4,96	16,53		$Na_2Mg(CO_3)_2$
3,34	4,83	16,15	$G1_1$	$Ca(Mg,Fe)(CO_3)_2$
3,39	4,97	16,83	$G0_1$	$(Ca,Cd)CO_3$ (75 mol. % $CaCO_3$)
3,40	4,96	16,85	$G0_1$	$(Ca,Mn)Co_3$ (12,5 mol. % $MnCO_3$)
3,40	5,07	17,25	$G0_1$	YBO_3
3,40	5,3	18,25	$E2_2$	$KSbO_3$
3,42	4,99	17,06	$G0_1$	$CaCO_3$
3,46	5,29	18,33		$CuMg_2$
3,53	4,95	17,49		KC_{16}
3,54	6,33	22,4		$3Zn(OH)_2 \cdot 2Zn(OH)Br$
3,58	8,44	30,19	$R3c$	$Na_3Li(SO_4)_2 \cdot 6H_2O$
3,59	10,31	37,0	$R\bar{3}c$	$Ca_3(PO_4)_2$
3,63	4,95	17,99		RbC_{16}
3,67	5,39	19,79		$Sr_3(PO_4)_2$
3,67	5,8	21,3	$R3m$	$K_3MnO_4CrO_4$
3,72	6,34	23,60	$R3m$	$ZnCl_2 \cdot 4Zn(OH)_2$
3,73	4,21	15,72	$R\bar{3}m$	UO_2F_2
3,73	5,54	20,69		$PbK_2(SO_4)_2$
3,74	2,93	10,97	$P6_3/mmc$	γ'-Mo—C
3,75	4,95	18,55		CsC_{16}
3,75	5,60	21,00	$R\bar{3}m$	$Ba_3(PO_4)_2$
3,90—3,92	3,16	12,29—12,34	$C7$	$WS_{1\,90-2,0}$
3,90	3,16	12,32	$C7$	McS_2
3,91	6,07	23,74		$2(Pb,Mn)O \cdot 3Fe_2O_3$
3,92	5,89	23,07		$PbO \cdot 6Fe_2O_3$
3,93	5,87	23,08		$SrO \cdot 6Fe_2O_3$
3,94	3,29	12,97	$C7$	WSe_2
3,94	5,53	21,79		$CaF_2 \cdot 5Al_2O_3$
3,94	5,55	21,87		$CaO \cdot 6Al_2O_3$
3,94	5,57	21,93		$(Ca,Mg,Mn)O \cdot 5,8(Al,Ti)_2O_3$
3,94	5,89	23,22		$BaO \cdot 6Fe_2O_3$
3,95	5,57	21,99		$SrO \cdot 6Al_2O_3$
3,98	11,88	47,24		$PbO \cdot 2Fe_2O_3$
4,00	5,93	23,73	$D5_6$	$K_2O \cdot 11Fe_2O_3$
4,02	5,59	22,50	$D5_6$	$Na_2O \cdot 11Al_2O_3$
4,03	5,93	22,88	$D5_6$	$Rb_2O \cdot 11Fe_2O_3$
4,04	3,50	14,12	$F5_1$	$NaFHF$
4,05	3,68	14,88	$F5_1$	$CaCN_2$
4,06	5,59	22,67	$D5_6$	$(K,Na)_2O \cdot 11Al_2O_3$
4,06	5,59	22,72	$D5_6$	$K_2O \cdot 11Al_2O_3$
4,07	5,59	22,72		$BaO \cdot (5—6)Al_2O_3$
4,09	2,46	10,06	$R\bar{3}m$	C (graphite)
4,18	3,64	15,20	$F5_1$	NaN_3
4,22	3,59	15,13	$F5_1$	$NaCNO$
4,31	10,53	45,49		$Na_2Ce(Mn,Te,Fe)H_2[(Si,P)O_4]_3$
4,32	3,79	16,36	$B18$	CuS
4,32	4,95	21,38		KC_8
4,38	3,94	17,25	$B18$	$CuSe$
4,41	8,27	36,50	$R\bar{3}$	$Mn_{10}Mg_2Al_3(AsO_4)_3(OH)_{24}$
4,41	4,28	18,23	$C19$	Cs_2O
4,45	(4,10)	18,24	$(P6_3/mmc)$	$(Ce,La, . . .)Ca(CO_3)_2F$
4,50	12,83	57,7		δ_1-Zn—Fe (90 at. % Zn)
4,51	3,88	17,50	$R\bar{3}m$	$Ca(UO_2)O_2$
4,54	3,86	17,50	$C19$	$CdCl_2$
4,56	4,54	20,7		$PbI_2(11)$
4,59	3,60	16,53	$C19$	$3CdCl_2 \cdot 5Cd(OH)_2$
4,60	4,95	22,78		RbC_8
4,61	3,99	18,37	$R\bar{3}m$	$Sr(UO_2)O_2$
4,65	2,98	13,87	$P6_3/mmc$	B_5W_2
4,67	7,21	33,67	$R3m$	$PbFe_3(SO_4)_2(OH)_6$
4,69	2,98	13,98		B_5Ti_2
4,71	3,78	17,82	$C19$	$ZnCl_2$
4,75	3,69	17,51	$C19$	$MnCl_2$

c/a	a, Å	c, Å	Structure type	Substance
4,76	3,96	18,80	$C19$	$CdBr_2$
4,78	3,39	16,20	$F5_1$	$RbCrS_2$
4,78	3,93	18,77	$C19$	$ZnBr_2$
4,81	4,95	23,81		CsC_8
4,84	4,24	20,52	$P\bar{3}m1$	CdI_2 (111)
4,90	3,08	15,12	$B6$	SiC-(6H)
4,90	3,55	17,36	$C19$	$NiCl_2$
4,90	3,56	17,44	$C19$	$CoCl_2$
4,90	3,58	17,55	$C19$	$FeCl_2$
4,90	3,59	17,58	$C19$	$CdBr_{0,6}(OH)_{1,4}$
4,90	3,60	17,63	$C19$	$MgCl_2$
4,90	3,82	18,71	$B6$	ZnS (wurtzite-(6H))
4,93	3,73	18,34	$C19$	$NiBr_2$
4,96	6,15	30,5		$CaCO_3 \cdot H_2O$
4,97	3,82	18,97	$R\bar{3}m$	YOF
4,98	4,06	20,21	$R\bar{3}m$	LaOF
5,04	3,91	19,67	$C19$	NiI_2
5,20	3,28	17,02		NiCl(OH)
5,24	4,36	22,8	$(G7_1)$	$BaCO_3 \cdot 2(Ce,La, \ldots)FCO_3$
5,27	3,03	15,96	$F5_1$	$NaFeO_2$
5,34	10,77	57,51	$R\bar{3}m$	$Cu_{18}Al_2(AsO_4)_3(SO_4)_3(OH)_{27} \cdot 36H_2O$
5,41	4,74	25,63	$D8_5$	Co_7W_6
5,41	4,75	25,68	$D8_5$	Fe_7Mo_6
5,43—5,61	4,76—4,75	25,83—26,88	$D8_5$	$Fe_{7+x}W_{6-x}$
5,47	3,71	20,29	$F5_1$	$NaCrSe_2$
5,55	3,53	19,57	$F5_1$	$NaCrS_2$
5,64	3,04	17,12	$F5_1$	$CuFeO_2$
5,66	5,3	30,1		$K(Mg,Fe, \ldots)_3(OH,F)_2 \cdot [(Al,Si)_2O_5]_2$
5,74	2,98	17,10	$F5_1$	$CuCrO_2$
5,86	3,62	21,20	$F5_1$	$KCrS_2$
6,54	3,08	20,15	$P6_3mc$	SiC-(8H)
6,59	3,29	21,64	$E9_4$	Al_5C_3N
6,72	3,17	21,29	$S5_7$	$Fe_4{}^{IV}Fe_2{}^{III}(OH)_8[Fe_2{}^{III}Si_2O_{10}]$
6,82	4,10	27,99	$(G7_1)$	$CaCO_3 2(Ce,La, \ldots)FCO_3$
6,95	3,01	20,93	$R\bar{3}m$	B_5Mo_2
6,95	4,32	30,07	$C33$	Bi_2Te_2S
6,96	4,37	30,42	$C33$	Bi_2Te_3
7,03	3,44	24,2	$(F5_1)$	$K_{0,5}CrSe_2$
7,08—7,20	3,98	28,12—28,65	$R\bar{3}m$	Bi_2O_3—SrO (14—26%Sr)
7,15	4,25	30,4		Sb_2Te_3
7,51	3,33	24,98	$D7_1$	Al_4C_3
7,51	6,19	46,47		$Mg_6Cr_2(OH)_{16}CO_3 \cdot 4H_2O$
7,52	6,20	46,6		$Mg_6Fe_2(OH)_{16}CO_3 \cdot 4H_2O$
7,53	6,14	46,24		$Mg_6Al_2(OH)_{16}CO_3 \cdot 4H_2O$
7,75	3,16	24,4		$CoCl(OH) \cdot 4Co(OH)_2 \cdot 4H_2O$
7,84	3,43	26,9	$F5_1$	$RbCrSe_2$
7,85	3,89	30,53	$C12$	$CaSi_2$
7,87	3,06	24,0		$NiCl_2 \cdot (6-7)Ni(OH)_2 \cdot xH_2O$
7,89	3,14	24,8		$4Co(OH)_2 \cdot CoBr(OH)$
7,89	3,14	24,8		$4Zn(OH)_2 \cdot Zn(OH)_{0,7} \cdot (Br,Cl)_{1,3}$
7,93	3,06	24,2		$NiBr_2 \cdot 7Ni(OH)_2 \cdot xH_2O$
7,98	4,14	28,6		Bi_2Se_3
8,18	3,08	25,18	$P3m1$	SiC-(10H)
(8,37)	(5,14)	(43,00)	$S5_4$	$Al_2(OH)_4(Si_2O_5)$
8,82	4,09	36,05		Sn_3As_2
9,36	4,24	39,7	$R3m$	Bi_4TeS_2
9,67	4,11	39,75		CdBr I
9,82	3,67	36,0		$MgAl_2S_4$
9,82	3,68	36,15		$MnAl_2S_4$
9,84	3,63	35,65		$FeAl_2S_4$
12,23	3,83	46,84	$B7$	ZnS (wurtzite-(15R))

c/a	a, Å	c, Å	Structure type	Substance
12,28	3,08	37,82	B7	SiC-(15R)
13,42	4,08	54,7		Bi_2Se_3
13,66	9,72	13,28		$Ca_4(Si_6O_{15})(OH)_2 \cdot 3H_2O$
15,54	3,08	47,85	P3m1	SiC-(19H)
17,17	3,08	52,89	R3m	SiC-(21R)
22,08	3,08	68,00	R3m	SiC-(27R)
26,98	3,08	83,10	R3m	SiC-(33R)
41,88	3,08	129,0	R3m	SiC-(51R(a)),(51R(b))
61,32	3,08	188,9	R3m	SiC-(75R)
68,68	3,08	211,4	R3m	SiC-(84R)
71,14	3,08	219,1	R3m	SiC-(87R)

5-5. Method of Homologous Pairs

This is used mainly with two-phase materials; use is made of homologous pairs, which are lines of equal blackening for the two phases at various concentrations.

Phase analysis is performed by finding such lines on the pattern and deducing the composition of the mixture from the tables of homologous pairs.

Tables 5-5a to 5-5c give these pairs for certain mixtures; α and β indicate the radiation components giving rise to the lines. See [6] for a detailed discussion of the method.

5-5a. Homologous Pairs for Determining Austenite in Steels [303]

Angle of Specimen Relative to Beam: $\psi = 27°$

(hk$l$) of lines		Austenite, %	(hk$l$) of lines		Austenite, %
γ-phase	α-phase		γ-phase	α-phase	
(311) α	(220) β	5	(222) β	(220) β	62
*(220) α	(200) β	6	*(222) α	(211) α	68
*(311) α	(211) β	7	(220) β	(211) β	64
(200) α	(200) β	6	(222) α	(220) α	66
(222) α	(220) β	10	(400) β	(310) β	66
(111) α	(110) β	15	(220) β	(211) α	67
(220) α	(211) β	17	(200) β	(110) β	68
(200) α	(110) β	26	(200) α	(110) α	73
(222) α	(310) β	28	(222) β	(211) β	75
(220) α	(200) α	37	(220) β	(200) α	84
(311) β	(220) β	35	(400) β	(220) α	90
(220) β	(200) β	40	(311) β	(211) α	90
(311) α	(211) α	46	(111) β	(110) α	92
*(311) α	(220) α	43	(220) β	(211) α	95
(311) β	(211) β	51	(200) β	(110) α	94
(111) α	(110) α	59	(222) β	(211) α	96
(111) β	(110) β	59	*(222) β	(220) α	96

*These lines are too far apart, so the proportion of γ-phase is appreciably in error if they are used.

5-5b. Homologous Pairs for Quantitative Analysis of Two-Phase Brasses [9]

Line indices		β-phase, %	Line indices		β-phase, %
α-phase	β-phase		α-phase	β-phase	
(311) α	(220) β	9,7	(111) α	(110) α	59,0
(200) α	(200) β	13,0	(220) α	(220) α	60,0
(222) α	(220) β	25,0	(220) α	(211) α	63,0
(111) α	(110) β	27,0	(220) α	(110) α	72,0
(220) α	(211) β	30,0	(220) β	(210) α	73,0
(311) α	(200) α	31,0	(240) α	(123) α	77,5
(311) α	(220) α	33,0	(222) α	(211) α	78,0
(220) α	(200) α	35,5	(113) β	(211) α	82,5
(200) α	(200) α	36,5	(200) β	(110) α	91,0
(311) α	(310) α	50,0	(222) β	(112) α	94
(311) α	(211) α	53,0			

5-5c. Homologous Pairs for Analysis of Oxides on Steel [304]

Line indices		Phase, %	
Fe_3O_4	Fe_2O_3	Fe_3O_4	Fe_2O_3
(113)	(310)	90	10
(333)	(310)	80	20
(333)	(321)	70	30
(113)	(321)	60	40
(004) } or (333) }	(211) } (220) }	50	50
(044)	(101)	40	60
(113)	(211)	30	70
(224)	(321)	20	80
(044)	(112)	10	90

5-6. Method of Superposition

The ratio of the proportions by weight in a two-phase mixture is given by

$$\frac{c_1}{c_2} = \frac{I_1}{I_2}\frac{I_2'}{I_1'}\frac{\tau_1}{\tau_2}\frac{\dfrac{\mu_2}{\varrho_2}}{\dfrac{\mu_1}{\varrho_1}},$$

where I_1 and I_2 are the line intensities on the initial pattern, and I_1' and I_2' are the same for the superposition pattern recorded with the two components in turn, τ_1 and τ_2 being the exposure times.

Figure 127 gives curves for use with this method, with $j = \tau_2/\tau_1$ as abscissa and the concentration of component 1 as ordinate. The curves are for

$$\varepsilon = \frac{I_2'}{I_1'} \frac{\tau_1}{\tau_2} \frac{\dfrac{\mu_2}{\varrho_2}}{\dfrac{\mu_1}{\varrho_1}} .$$

The steps are as follows: (1) lines lying close together are selected; (2) the ratio of intensities is measured, and a superposition pattern with the same ratio is selected; (3) ε is calculated for the latter; (4) j and ε are used to find the concentration of the mixture. See [128] for details of the principles and [305] for adaptation to analysis by electron diffraction.

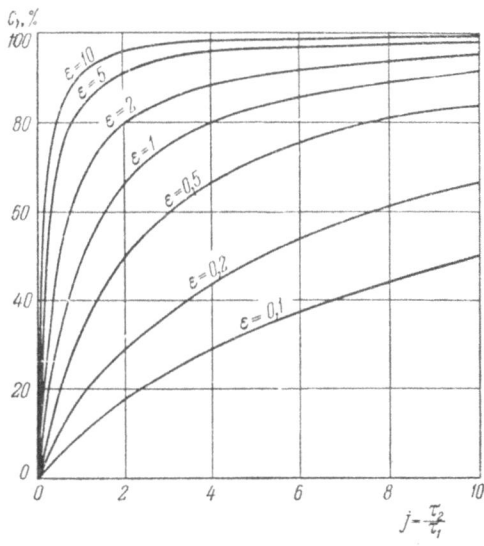

Fig. 127. Graphs for phase analysis by superposition.

5-7. Conversion of Weight Percent to Atomic Percent

Figure 128 provides means of conversion for binary systems.

The weight proportions a and b of components having atomic weights A_1 and A_2 are related to the atomic concentrations α and β by

$$\alpha + \beta = 1; \quad a + b = 1; \quad \frac{\alpha}{1-\alpha} = \frac{a}{1-a} \frac{A_2}{A_1} .$$

Scale C (right) gives A_2/A_1; scale A (left) gives a/b; and scale B (middle) gives $\alpha/(1 - \alpha)$. Given any two of these quantities, the third is found by drawing the appropriate straight line.

The nomogram also enables one to determine the composition (left part of the middle scale). The table below is used if high accuracy is needed; this table gives

$$f(x) = \lg \left[\frac{x}{100 - x} + 10 \right] .$$

Relationships used in the calculation are

$$Y = \frac{100\,X}{X + \dfrac{A_1}{A_2}(100 - X)} \ , \qquad X = \frac{100\,Y}{Y + \dfrac{A_2}{A_1}(100 - Y)} \ ,$$

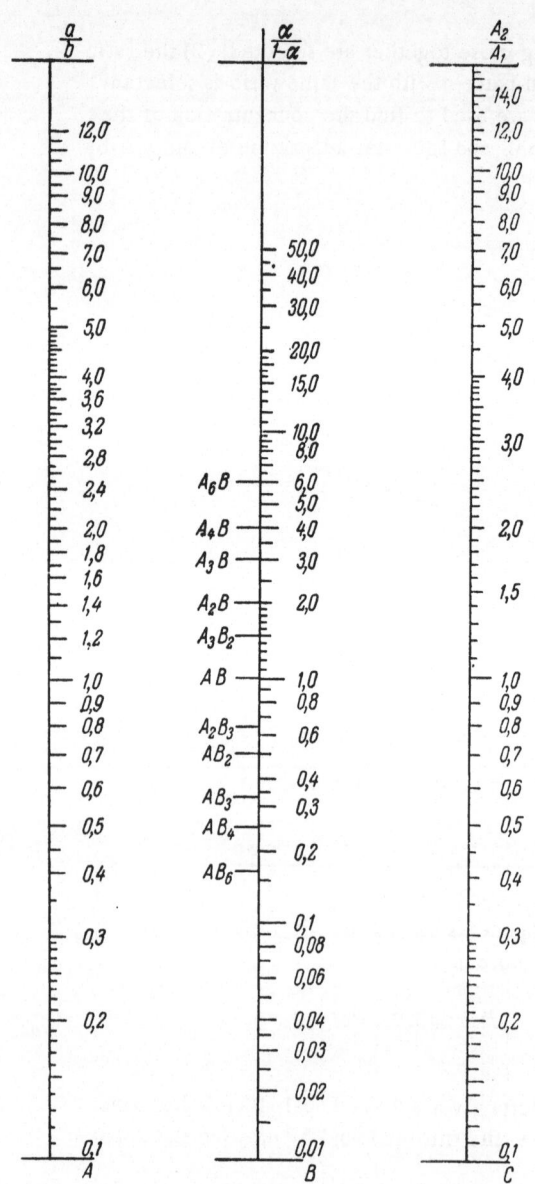

where X is the wt. % concentration for the component of atomic weight A_1 and Y is the at. % concentration for this, A_2 being the atomic weight of the second component [306]. See [129] for conversion rules for many-component systems.

The steps are as follows:

(1) The table is used to find $f(x)$ for the known concentration (weight or atomic) of the component of atomic weight A.

(2) The logarithm of the ratio of atomic weights is formed.

(3) The previous two quantities are added if the conversion is from at. % to wt. %; $f(x)$ is read back to give the wt. %.

(4) The second quantity is subtracted from the first if the conversion is from wt. % to at. %; $f(x)$ is then the desired quantity.

Examples. 20 at. % Cd in an Ag−Cd alloy is to be converted to wt. %; the table gives $f(x)$ for x = 20% as 9.3979, and the logarithm of the ratio of the atomic weights of Cd and Ag is 0.0179.

We have 9.3979 + 0.0179 = 9.4158, which the table of $f(x)$ shows to fall between 20.6 and 20.7; linear interpolation gives 20.67, so the alloy contains 20.67 wt. % Cd.

Again, 2.41 wt. % Ag in an Ag−Cd alloy is to be converted to at. %; $f(x)$ = 8.3926, the ratio of the atomic weights of Ag and Cd

Fig. 128. Nomogram for converting wt. % to at. %.

being $\overline{1}.9821$. Subtraction gives $8.3926 - \overline{1}.9821 = 8.4105$, which the table shows to correspond to x = 2.51, so the alloy contains 2.51 at. % Ag.

x	0	.01	.02	.03	.04	.05	.06	.07	.08	.09
0,0	−∞	6,0000	6,3011	6,4772	6,6022	6,6992	6,7784	6,8454	6,9034	6,9546
.1	7,0004	7,0419	7,0797	7,1145	7,1467	7,1767	7,2048	7,2312	7,2560	7,2796
.2	3019	3231	3434	3627	3812	3990	4161	4325	4484	4637
.3	4784	1927	5065	5200	5330	5456	5579	5698	5814	5928
.4	6038	6146	6251	6353	6454	6552	6648	6742	6833	6923
.5	7012	7098	7183	7266	7347	7428	7506	7584	7660	7734
.6	7808	7800	7951	8021	8090	8157	8224	8290	8355	8419
.7	8482	8544	8605	8665	8725	8783	8841	8899	8955	9011
.8	9066	9120	9174	9227	9279	9331	9382	9433	9483	9534
.9	9582	9630	9678	9725	9772	8919	9865	9910	9955	8,0000
1,0	8,0044	8,0087	8,0130	8,0173	8,0216	8,0258	8,0299	8,0340	8,0371	8,0422
.1	0462	0502	0541	0580	0619	0657	0695	0733	0770	0808
.2	0844	0881	0917	0953	0988	1024	1059	1094	1128	1162
.3	1196	1230	1263	1297	1330	1362	1395	1427	1459	1491
.4	1522	1554	1585	1616	1647	1677	1707	1738	1767	1797
.5	1826	1856	1885	1914	1943	1971	2000	2028	2056	2084
.6	2111	2139	2166	2193	2220	2247	2274	2300	2327	2353
.7	2379	2405	2431	2456	2482	2507	2532	2557	2582	2607
.8	2632	2656	2680	2705	2729	2753	2777	2800	2824	2848
.9	2871	2894	2917	2940	2963	2986	3009	3031	3054	3076
2,0	3098	3120	3142	3164	3186	3208	3229	3250	3272	3293
.1	3314	3335	3356	3377	3308	3419	3439	3460	3480	3501
.2	3521	3541	3561	3581	3601	3621	3640	3660	3680	3699
.3	3718	3738	3757	3776	3795	3814	3833	3852	3870	3889
.4	3908	3926	3945	3963	3981	3999	4018	4036	4054	4072
.5	4089	4107	4125	4142	4160	4178	4195	4212	4230	4247
.6	4264	4281	4298	4315	4332	4349	4366	4383	4399	4416
.7	4432	4449	4466	4482	4498	4514	4531	4547	4563	4579
.8	4595	4611	4627	4643	4658	4674	4690	4705	4721	4736
.9	4752	4767	4783	4798	4813	4828	4843	4858	4874	4888
3,0	4904	4918	4933	4948	4963	4978	4992	5007	5021	5036
.1	5050	5065	5079	5094	5108	5122	5136	5150	5165	5179
.2	5193	5207	5221	5235	5248	5262	5276	5290	5304	5317
.3	5331	5344	5358	5372	5385	5398	5402	5425	5438	5452
.4	5465	5478	5481	5504	5518	5531	5544	5557	5570	5583
.5	5595	5608	5621	5634	5646	5659	5672	5685	5697	5710
.6	5722	5735	5747	5760	5772	5784	5797	5809	5821	5834
.7	5846	5858	5870	5882	5894	5906	5918	5930	5942	5954
.8	5966	5978	5990	6002	6013	6025	6037	6048	6060	6072
.9	6083	6095	6107	6118	6130	6141	6152	6164	6175	6186
4,0	6198	6209	6220	6232	6243	6254	6265	6276	6288	6299
.1	6310	6321	6332	6343	6354	6365	6375	6386	6397	6408
.2	6419	6430	6440	6451	6462	6472	6483	6494	6504	6515
.3	6526	6536	6547	6557	6568	6578	6588	6599	6609	6620
.4	6630	6640	6650	6661	6671	6681	6691	6702	6712	6722
.5	6732	6742	6752	6762	6772	6782	6792	6802	6812	6822
.6	6832	6842	6852	6862	6872	6881	6891	6901	6911	6920
.7	6930	6940	6949	6959	6969	6978	6988	6998	7007	7017
.8	7026	7036	7045	7054	7064	7073	7083	7092	7102	7111
.9	7120	7130	7139	7148	7157	1677	7176	7185	7194	7203

x	0	.1	.2	.3	.4	.5	.6	.7	.8	.9
5	7212	7303	7392	7479	7565	7649	7732	7814	7894	7973
6	8050	8127	8202	8276	8349	8421	8492	8562	8631	8699
7	8766	8832	8898	8962	9026	9080	9151	9213	9274	9334
8	9393	9452	9510	9567	9624	9680	9736	9791	9845	9899
9	9952	9,0005	9,0057	9,0109	9,0160	9,0211	9,0261	9,0311	9,0360	9,0409
10	9,0458	0506	0553	0600	0647	0694	0740	0785	0831	0876
11	0920	0964	1008	1052	1095	1138	1180	1222	1264	1306
12	9,1347	9,1388	9,1429	9,1469	9,1509	9,1549	9,1589	9,1628	9,1667	9,1706
13	1744	1783	1821	1859	1896	1933	1970	2007	2044	2080
14	2116	2152	2188	2224	2259	2294	2329	2364	2398	2433
15	2467	2501	2534	2568	2602	2635	2668	2701	2734	2766
16	2798	2831	2863	2895	2926	2958	2989	3021	3052	3083
17	3114	3145	3175	3205	3236	3266	3296	3326	3356	3385
18	3415	3444	3473	3502	3531	3560	3580	3618	3646	3674
19	3703	3731	3759	3787	3815	3842	3870	3898	3925	3952
20	3979	4007	4034	4060	4087	4114	4141	4167	4193	4220
21	4246	4272	4298	4324	4350	4376	4401	4427	4453	4478
22	4503	4529	4554	4579	4604	4629	4654	4679	4703	4728
23	4752	4777	4801	4826	4850	4874	4898	4922	4946	4970
24	4994	5018	5042	5065	5089	5112	5136	5159	5182	5206
25	5229	5252	5275	5298	5321	5344	5367	5389	5412	5435
26	5457	5480	5502	5525	5547	5570	5592	5614	5636	5658
27	5680	5702	5724	5746	5768	5790	5812	5833	5855	5877
28	5898	5920	5941	5963	5984	6005	6027	6048	6069	6090
29	6111	6132	6154	6175	6196	6216	6237	6258	6279	6300
30	6320	6341	6362	6382	6403	6423	6444	6464	6484	6505
31	6525	6545	6566	6586	6606	6626	6646	6666	6686	6706
32	6726	6746	6766	6786	6806	6826	6846	6865	6885	6905
33	6924	6944	6964	6983	7003	7022	7042	7061	7081	7100
34	7119	7139	7158	7177	7197	7216	7235	7254	7273	7293
35	7312	7331	7350	7369	7388	7407	7426	7445	7463	7482
36	7501	7520	7539	7558	7576	7595	7614	7633	7651	7670
37	7689	7707	7726	7744	7763	7781	7800	7819	7837	7856
38	7874	7892	7911	7920	7948	7966	7984	8003	8021	8039
39	8057	8076	8094	8112	8130	8148	8167	8185	8203	8221
40	8230	8257	8275	8293	8311	8329	8347	8365	8383	8401
41	8419	8437	8455	8473	8491	8509	8527	8545	8563	8580
42	8598	8616	8634	8652	8680	8637	8705	8723	8741	8758
43	8776	8794	8811	8829	8847	8864	8882	8900	8917	8935
44	8953	8970	8988	9005	9023	9041	9058	9076	9093	9111
45	9129	9146	9164	9181	9199	9216	9234	9251	9269	9286
46	9304	9321	9339	9356	9374	9391	9409	9426	9443	9461
47	9478	9496	9513	9531	9548	9565	9583	9600	9618	9635
48	9652	9670	9687	9705	9722	9739	9757	9774	9792	9809
49	9826	9844	9861	9878	9896	9913	9931	9948	9965	9983
50	10,0000	10,0017	10,0035	10,0052	10,0070	10,0087	10,0104	10,0122	10,0139	10,0156
51	0174	0191	0209	0226	0243	0261	0278	0295	0313	0330
52	0348	0365	0382	0400	0417	0435	0452	0470	0487	0504
53	0522	0539	0557	0574	0592	0609	0626	0644	0661	0679
54	0696	0714	0731	0749	0766	0784	0801	0819	0836	0854
55	0872	0889	0907	0924	0942	0959	0977	0995	1012	1030
56	1047	1065	1083	1100	1118	1136	1153	1171	1189	1206
57	1224	1242	1260	1277	1295	1313	1331	1348	1366	1384
58	1402	1420	1437	1455	1473	1491	1509	1527	1545	1563
59	1581	1599	1617	1635	1653	1671	1689	1707	1725	1743

x	0	.1	.2	.3	.4	.5	.6	.7	.8	.9
60	1761	1779	1797	1815	1833	1852	1870	1888	1906	1924
61	1943	1961	1979	1998	2016	2034	2053	2071	2080	2108
62	2126	2145	2163	2182	2200	2219	2237	2256	2274	2293
63	2311	2330	2347	2367	2386	2405	2424	2442	2461	2480
64	2499	2518	2539	2555	2574	2593	2612	2631	2650	2669
65	2688	2708	2727	2746	2765	2784	2803	2823	2842	2861
66	2881	2900	2919	2939	2958	2978	2997	3017	3036	3056
67	3076	3095	3115	3135	3154	3174	3194	3214	3234	3254
68	10,3274	10,3294	10,3314	10,3334	10,3354	10,3374	10,3394	10,3414	10,3434	10,3455
69	3475	3495	3516	3536	3556	3577	3597	3618	3639	3659
70	3680	3701	3721	3742	3763	3784	3805	3826	3846	3868
71	3889	3910	3931	3952	3973	3995	4016	4037	4059	4080
72	4102	4123	4145	4167	4188	4210	4232	4254	4276	4298
73	4320	4342	4364	4386	4408	4430	4453	4475	4498	4520
74	4543	4565	4588	4611	4633	4656	4679	4702	4725	4748
75	4771	4794	4818	4841	4864	4888	4911	4935	4959	4982
76	5006	5030	5054	5078	5102	5126	5150	5174	5199	5223
77	5248	5272	5297	5322	5346	5371	5396	5421	5446	5472
78	5497	5522	5548	5573	5599	5624	5650	5676	5702	5728
79	5754	5780	5807	5833	5860	5886	5913	5940	5967	5994
80	6021	6048	6075	6103	6130	6158	6185	6213	6241	6269
81	6297	6326	6354	6383	6411	6440	6469	6498	6527	6556
82	6585	6615	6645	6674	6704	6734	6764	6795	6825	6856
83	6886	6917	6948	6979	7011	7042	7074	7105	7137	7169
84	7202	7234	7267	7299	7332	7365	7398	7432	7466	7499
85	7533	7567	7602	7636	7671	7706	7741	7776	7812	7848
86	7884	7920	7956	7993	8030	8067	8104	8142	8180	8218
87	8256	8294	8333	8372	8411	8451	8491	8531	8571	8612
88	8653	8694	8736	8778	8820	8862	8905	8948	8992	9036
89	9080	9124	9160	9215	9260	9306	9353	9400	9447	9494
90	9542	9591	9640	9680	9739	9789	9840	9891	9943	9995
91	11,0048	11,0101	11,0155	11,0210	11,0265	11,0320	11,0376	11,0433	11,0490	11,0548
92	0607	0666	0726	0787	0849	0911	0974	1038	1102	1168
93	1234	1301	1369	1438	1508	1579	1651	1742	1798	1873
94	1950	2027	2106	2186	2268	2351	2435	2521	2608	2697
95	2788	2880	2974	3070	3168	3268	3370	3474	3581	3690
96	3802	3917	4034	4154	4278	4405	4535	4669	4807	4950
97	5096	5248	5405	5568	5736	5911	6092	6282	6479	6686
98	6902	7129	7368	7621	7889	8174	8478	8804	9156	9538

x	0	.01	.02	.03	.04	.05	.06	.07	.08	.09
99,0	11,9956	12,0000	12,0045	12,0090	12,0135	12,0181	12,0228	12,0275	12,0322	12,0370
99,1	12,0418	0466	0517	0567	0618	0669	0721	0773	0820	0880
99,2	0934	0989	1045	1101	1159	1217	1275	1335	1395	1456
99,3	1518	1581	1645	1710	1776	1843	1910	1979	2040	2120
99,4	2192	2266	2340	2416	2494	2572	2653	2734	2817	2902
99,5	2988	3077	3167	3258	3352	3448	3546	3647	3749	3854
99,6	3962	4072	4186	4302	4421	4544	4670	4800	4935	5073
99,7	5216	5363	5516	5675	5839	6010	6188	6373	6566	6769
99,8	6981	7204	7440	7688	7952	8233	8533	8855	9203	9581
99,9	9996	13,0454	13,0966	13,1546	13,2216	13,3008	13,3978	13,5228	13,6989	14,0000

CHAPTER 6

PRECISION MEASUREMENT OF LATTICE CONSTANTS

Several errors of measurement must be considered here; the following classification [435] is used:

1. Random Errors. (a) Subjective errors in reading instruments and in establishing the positions of lines or intensity curves; (b) apparatus errors (unforeseen changes in adjustments, temperature, circuit parameters, positions of film and slits, movement of counter, change in line voltage, errors arising from film grain, and so on); (c) errors in the process of measurement related to design features, method of preparation of the specimen, presence of impurities, and methods of finding the lattice constants (analytic or graphical).

2. Systematic Errors. (a) Subjective errors in measuring the curvature and line shape for x-ray lines, which are related to displacement of the peak from the center of gravity, to division of the line into spots, and to overlap between adjacent lines; (b) apparatus errors: wear and drift in equipment, effects from design and method of use, compression of film, eccentricity of specimen, curvature of film, inexact focus, position of equator of film, inclination of primary beam, axial or equatorial divergence of beam, height of specimen (overlap of interference cones), errors in angular measurements, incorrect recording of pulses, absorption or refraction in the specimen, and temperature of specimen; (c) errors in the process of measurement: incorrect scales, errors in angular extrapolation functions, effects of state of specimen on the refraction correction, variation in effective wavelength, asymmetry in spectral lines, and errors in the absolute values of kX units or angstroms.

This chapter gives tables and graphs for the selection of precision methods and for the associated calculations.

See [6, 11, 12] for detailed descriptions of analytic and graphical methods for precision measurements.

The tables of Chapter 7 can be used in many instances for precision measurements.

6-1. Aspects of Precision Methods of Measuring Lattice Constants

The error of measurement of the lattice constant as a function of ϑ is given by

$$\frac{\Delta a}{a} = \pm \, \Delta \vartheta \cot \vartheta.$$

The error decreases as the angle increases; Table 1 [7] gives the error for $\Delta \vartheta = 0.001$ radian (3').

Several precision methods are discussed below.

504

TABLE 1

ϑ^0	5	10	20	30	40	50	60	70	75	80	82	84	85
$\dfrac{\Delta a}{a}$ %	1.14	0.57	0.275	0.17	0.12	0.084	0.058	0.036	0.027	0.018	0.014	0.010	0.009

6-1a. Asymmetric Recording

The film is placed asymmetrically in the camera for precision measurement of lattice constants without additional extrapolation [130]. The advantages are: (1) errors from inexact measurement of the camera diameter and from film shrinkage are eliminated; (2) effects from line displacement caused by absorption in the specimen are eliminated by the use of reflections at large angles, where such displacements are small, and by the preparation of thin specimens, which have little absorption. Errors of 5 parts in 10^6 are attainable in favorable cases [436].

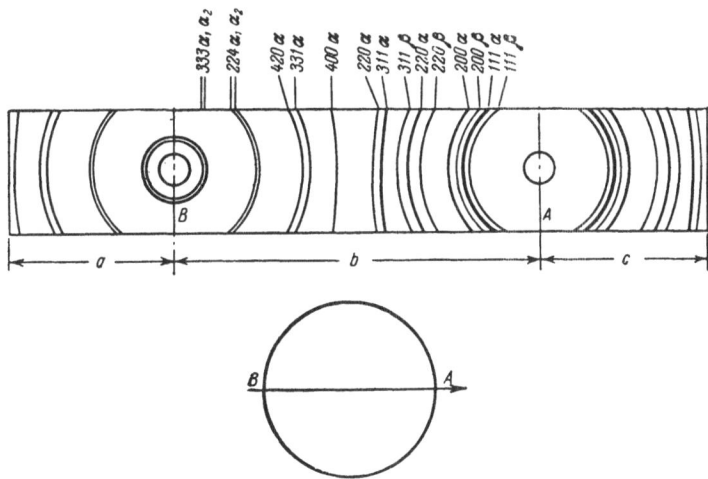

Fig. 129. X-ray pattern of aluminum recorded by the asymmetric
method with copper radiation.

E x a m p l e . An Al specimen was examined with Cu radiation, unfiltered in order to increase the number of lines at large angles. The specimen was 0.18 mm in diameter and was kept at 23.10°C during examination in a camera of diameter 57.7 mm. Figure 129 shows the form of the pattern.

The first step is to measure the positions by reference to line pairs in ranges b and c (measurements are made from point B) and a and b (from point A). The steps in the calculation are clear from Table 2.

TABLE 2

	Measured from B				
	(111)β	(200)β	(200)α	(220)β	(220)α
Distance in range b, mm	172.51	169.71	167.44	160.73	157.21
Distance in range c, mm	207.23	210.03	212.31	219.04	222.54
Sum, mm	379.74	379.74	379.75	379.77	379.75

Mean 379.75 mm

	Measured from A			
	(224)α_1	(224)α_2	(333)α_1	(333)α_2
Distance in range b, mm	120.99	120.63	108.44	107.45
Distance in range a, mm	78.33	78.72	90.91	91.89
Sum, mm	199.32	199.35	199.35	199.34

Mean 199.34 mm

TABLE 3

	Measured from A	
	(333)α_1	(333)α_2
Distance in range b, mm	108.44	107.45
Distance in range a, mm	90.91	91.89
Difference, 4φ	17.53	15.56
$90°-\vartheta = 4\varphi/1.99545$	8.7451	7.7623
$\vartheta°$	81.2549	82.2377
$\sin \vartheta$	0.988374	0.990837
a_0, Å	4.04941	4.04943
a_0 corrected for refraction, A	4.04944	4.04946
Mean a_0, Å	4.04945	

The effective length of the film is 379.75 −199.34 = 180.41 mm.

The angle corresponding to 1 mm on the film is 360°/180.41 = 1.99545°.

The next step is calculation of the lattice constant from the line pairs; Table 3 demonstrates the stages of the calculation. The geometric relations involved in the calculation have been dealt with above.

A very important point is that the ϑ of the last line should exceed 80°.

The method is applicable to materials that are not cubic if for ϑ near 80° there are two distinct lines whose indices are $(h_1k_1l_1)$ and $(h_2k_2l_2)$.

For example, for the tetragonal system

$$a_0 = \frac{\lambda}{2} \sqrt{\frac{l_1^2(h^2+k^2)-l_2^2(h_1^2+k_1^2)}{l_1^2\sin^2\vartheta_2-l_2^2\sin^2\vartheta_1}} \tag{47}$$

for two such lines; c_0 is found from the measurements on any one line:

$$c_0 = \frac{\lambda a_0 l}{\sqrt{4a_0\sin^2\vartheta-\lambda^2(h^2+k^2)}} . \tag{48}$$

The error in a_0 here decreases as $(h^2 + k^2)$ increases; the same applies to c_0 and l^2.

If a single radiation does not give two lines with $\vartheta > 80°$, use is made of two wavelengths (α_1, α_2, and β) in a single radiation or of different radiations, including the $L\alpha$ series of tungsten. Formula (47) then becomes

$$a_0 = \frac{\lambda_1 \lambda_2}{2} \sqrt{\frac{l_1^2 (h_2^2 + k_2^2) - l_2^2 (h_1^2 + k_1^2)}{\lambda_1^2 l_1^2 \sin^2 \vartheta_2 - \lambda_2^2 l_2^2 \sin^2 \vartheta_1}} \tag{49}$$

Formula (48) is unaffected.

E x a m p l e. A sample of β-Sn (tetragonal) was examined with Cu radiation at 26.57°C, which gave for the (503)α_1 and (271)α_1 lines the following: $\vartheta_{(503)} = 79.017°$, $\vartheta_{(271)} = 82.564°$.

From (47)

$$a_0 = \frac{1.54050}{2} \sqrt{\frac{3^2 (2^2 + 7^2) - 1^2 (5^2 + 0^2)}{3^2 \sin^2 82.564° - 1^2 \sin^2 79.017°}} = 5.83158 \text{ Å,}$$

and from (48) for the (271) line

$$c_0 = \frac{1.54050 \cdot 5.83158 \cdot 1}{\sqrt{4 \cdot 5.83158^2 \cdot \sin^2 82.564° - 1.54050^2 \cdot (2^2 + 7^2)}} = 3.18166 \text{ Å.}$$

Table 4 gives the a_0 and c_0 found from four line pairs by means of (47), (48), and (49).

TABLE 4

(hkl)	$\vartheta°$	a_0, Å	c_0, Å
(503) α_1 (271) α_1	79.017 82.564	5.83158	3.1816$_6$
(503) α_2 (271) α_2	79.789 83.758	5.83152	3.1809$_8$
(271) α_2 (503) α_1	83.758 79.017	5.83149	3.1814$_2$
(271) α_1 (503) α_2	82.564 79.789	5.83160	3.1812$_2$
Mean for 26.57°C		5.83155	3.1813$_2$
Values referred to 25°C		5.83140	3.1811$_6$

The error in a_0 is less than that in c_0 because here the sum of the $(h^2 + k^2)$ is 90, whereas that for l^2 is only 10.

The procedure is the same for the hexagonal system:

$$a_0 = \lambda \sqrt{\frac{l_1^2 (h_2^2 + h_2 k_2 + k_2^2) - l_2^2 (h_1^2 + h_1 k_1 + l_1^2)}{3 (l_1^2 \sin^2 \vartheta_2 - l_2^2 \sin^2 \vartheta_1)}}$$

or

$$a_0 = \lambda_1 \lambda_2 \sqrt{\frac{l_1^2(h_2^2+h_2k_2+k_2^2)-l_2^2(h_1^2+h_1k_1+l_1^2)}{3(\lambda_1^2 l_1^2 \sin^2 \vartheta_2 - \lambda_2^2 l_2^2 \sin^2 \vartheta_1)}},$$

$$\dot{c}_0 = \frac{\lambda a_0 l}{2} \sqrt{\frac{3}{3a_0^2 \sin^2 \vartheta - \lambda^2 (h^2+hk+k^2)}}.$$

6-1b. Recording at Large Distances in Divergent Beams

The film-to-specimen distance is made 400-500 mm in this method of back reflection [131], so the separation of $K\alpha_1$ from $K\alpha_2$ becomes considerable.

Focusing is provided directly by the anode without stops, so there is little increase in exposure time. A disadvantage is that the large distance tends to broaden the lines, and this hinders exact measurement of line separations.

6-1c. Use of a Standard

A single film is recorded for the specimen and a standard substance of precisely known lattice constants. The unknown constants are deduced from the distance between the two sets of lines. See [6] and also Chapters 5 and 7 for more details on the use of standards.

6-1d. Graphical Extrapolation

Here the lattice constant is deduced by the use of several lines rather than one. In general, the steps are as follows: (1) the lattice constants are deduced from several lines; (2) these quantities are plotted against ϑ or against functions of ϑ; (3) ϑ is extrapolated to 90°.

The extrapolation method is governed by the positions of the lines.

Extrapolation with Respect to Bragg Angle. The extrapolation function here is $a = f(\vartheta)$ and is represented by a curve that approaches horizontal as ϑ approaches 90°.

Extrapolation with Respect to $(\pi/2 - \vartheta) \cot \vartheta$. In this case the line is straight for large angles; not less than 3-4 points should be used for the extrapolation.

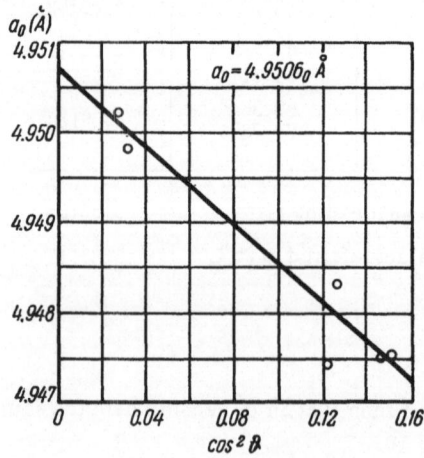

Fig. 130. Determination of lattice constant for lead by graphical extrapolation with respect to $\cos^2 \vartheta$.

Extrapolation with Respect to $\cos^2 \vartheta$. The relation of $\Delta a/a_0$ to $\cos^2 \vartheta$ should be a straight line [132, 133], so $a = f(\cos^2 \vartheta)$ is a straight line, which is more precisely extrapolated to $\vartheta = 90°$ than is $a = f(\vartheta)$. Figure 130 shows such an extrapolation for high-purity Pb at 25°C from a pattern recorded in a camera with a diameter of 114.6 mm for Cu radiation and a specimen 0.3 mm in diameter.

This method gives good results if there are several lines in the range 60-90° and if there is one line with $\vartheta > 80°$; the error in a_0 can then be ± 0.002%. If the lines do not satisfy this, one of the above methods may be used (e.g., $K\alpha_1$, $K\alpha_2$, $K\beta$, various radiations, and so on).

Extrapolation with Respect to $\frac{1}{2}[(\cos^2 \vartheta)/\sin \vartheta + (\cos^2 \vartheta)/\vartheta]$. This uses the linear relation between $\Delta a/a_0$ and $[(\cos^2 \vartheta)/\sin \vartheta + (\cos^2 \vartheta)/\vartheta]$, which applies for all angles. Moreover, it is applicable to crystals that are not cubic; here a_0, b_0, and c_0 are plotted as functions of $[(\cos^2 \vartheta)/\sin \vartheta + (\cos^2 \vartheta)/\vartheta]$ for reflections of (h00), (0k0), and (00l) types. The error in a is less than for the $\cos^2 \vartheta$ method; it may be ±0.001%.

E x a m p l e . The lattice constant of Al at 298°C may be deduced from experimental results [134] recorded with Cu radiation ($\lambda K\alpha_1 = 1.54050$ Å, $\lambda K\alpha_2 = 1.54434$ Å). Table 5 gives the results of the calculations.

TABLE 5

(hk$l$)	Radiation	$\vartheta°$	a, Å	$\frac{1}{2}\left(\dfrac{\cos^2 \vartheta}{\sin \vartheta} + \dfrac{\cos^2 \vartheta}{\vartheta}\right)$
(331)	$K\alpha_1$	55.486	4.07464	0.360
	$K\alpha_2$	55.695	4.07459	0.356
(420)	$K\alpha_1$	57.714	4.07464	0.311
	$K\alpha_2$	57.942	4.07459	0.306
(422)	$K\alpha_1$	67.763	4.07666	0.138
	$K\alpha_2$	68.107	4.07688	0.134
(333)	$K\alpha_1$	78.963	4.07778	0.032
(511)	$K\alpha_2$	79.721	4.07777	0.028

Figure 131 shows the curve constructed from Table 5, which gives $a_0 = 4.0780_8$ Å for $\vartheta \to 90°$.

The refraction correction for Al (face-centered cubic, number of atoms in the unit cell 4, atomic number 13) is $\Delta = 4.47 \cdot 10^{-6} \cdot (1.54/4.08) \cdot 4 \cdot 13 = 0.00003$ Å. So the final value is $a_0 = 4.0781$ Å.

A powder-pattern study for Cu_9Al_{14} was concerned with the effects of specimen diameter and degree of dilution with binder by the above method; Table 6 shows the agreement was good.

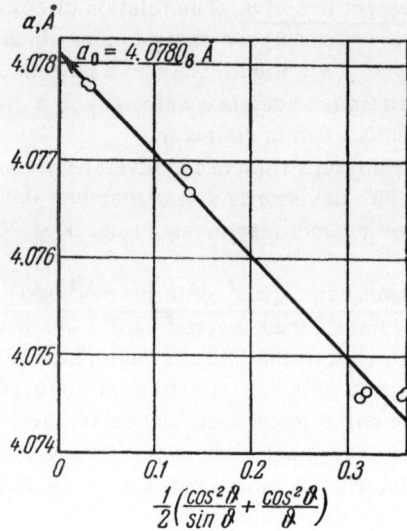

Fig. 131. Deduction of the lattice constant of aluminum at 298°C by graphical extrapolation of the quantity
$\frac{1}{2}[(\cos^2 \vartheta)/\sin \vartheta + (\cos^2 \vartheta)/\vartheta]$.

TABLE 6

Diameter, mm	Dilution with amorphous binder	Temp., °C	a_0, Å
0.59	+	15.8	8.7038
1.46	+	15.4	8.7036
0.45	−	16.2	8.7041
1.35	−	16.4	8.7039

Mean value: a_0 = 8.7039 Å.

Graphical Method of Successive Approximation. Poor results for a_0 and c_0 are obtained from asymmetric recording if a material (not cubic) gives very few (hk0) and
(00l) reflections and direct extrapolation to 90° is used. Here successive approximation
may be used [155]. An approximate value of c/a is chosen, and lines of high h and k are
used to extrapolate the a to 90°, which gives a preliminary value for a_0. The same
procedure is applied for c_0 (lines of large l), and these a_0 and c_0 are used to form a more
precise c/a; the process is then repeated several times.

The method may be illustrated by reference to the a_0 and c_0 of GeO_2 (hexagonal)
as measured with Cu radiation. Here a_0 and c_0 are given by

$$a_0 = \frac{\lambda}{2 \sin \vartheta} \; \sqrt{\frac{4}{3}(h^2+hk+k^2) + \left(\frac{a}{c}\right)^2 l^2,}$$

$$c_0 = \frac{\lambda}{2 \sin \vartheta} \; \sqrt{\frac{4}{3}\left(\frac{c}{a}\right)^2 (h^2+hk+k^2) + l^2.}$$

(50)

The relation of $(h^2 + hk + k^2)$ to l^2 is used in choosing the lines for determining a_0 and c_0; the indices are chosen to make the term in (50) containing the unknown c/a as small as possible.

Table 7 shows the sequence for the first approximation, the value assumed being $c/a = 1.14$.

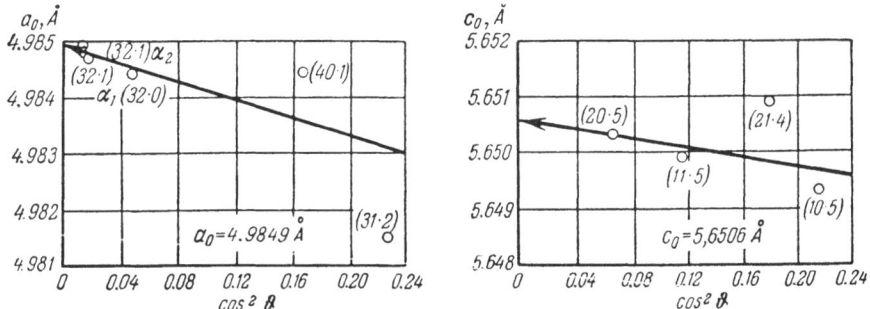

Fig. 132. Determination of lattice constants by successive approximation.

The suitability of each line for precise determination of a_0 or c_0 is considered in the extrapolation to 90°. Figure 132 illustrates the course of the extrapolation.

TABLE 7

Calculation of a_0

(hkl)	$\sin \vartheta$	$\frac{4}{3}(h^2+hk+k^2)$	$\left(\frac{a}{c}\right)^2 l^2$	a_0, Å
$(31.2)\,\alpha$	0.87912	17.333	3.076	4.9778
$(40.1)\,\alpha_1$	0.91316	21.333	0.769	4.9836
$(32.0)\,\alpha_1$	0.97747	25.333	0.000	4.9844
$(32.1)\,\alpha_1$	0.99228	25.333	0.769	4.9840
$(32.1)\,\alpha_2$	0.99427	25.333	0.769	4.9841

Calculation of c_0

(hkl)	$\sin \vartheta$	$\frac{4}{3}\left(\frac{c}{a}\right)^2(h^2+hk+k^2)$	l^2	c_0, Å
(10.5)	0.88627	1.733	25	5.6510
(21.4)	0.90674	12.133	16	5.6624
(11.5)	0.94077	5.200	25	5.6545
(20.5)	0.96711	6.933	25	5.6561

6-1e. Combined Graphical Extrapolation and Calculation

It is often found that a tetragonal or hexagonal material gives fairly many (hk0) lines with ϑ from 30 to 90° but only one (00l) line. Here a_0 is deduced by graphical extrapolation, as from $[(\cos^2 \vartheta)/\sin \vartheta + (\cos^2 \vartheta)/ \vartheta]$, while c_0 is deduced by drawing on the extrapolation graph a straight line through the point for the (00l) reflection [156]; the slope of the line for c_0 is calculated from

$$\frac{d(\Delta c_0)}{df(\vartheta)} = \frac{c_0}{a_0} \frac{d(\Delta a_0)}{df(\vartheta)} , \qquad (51)$$

i.e., from the slope of the straight line for a_0, the extrapolated value of a_0, and the value of c_0 as deduced from the (00l) reflection.

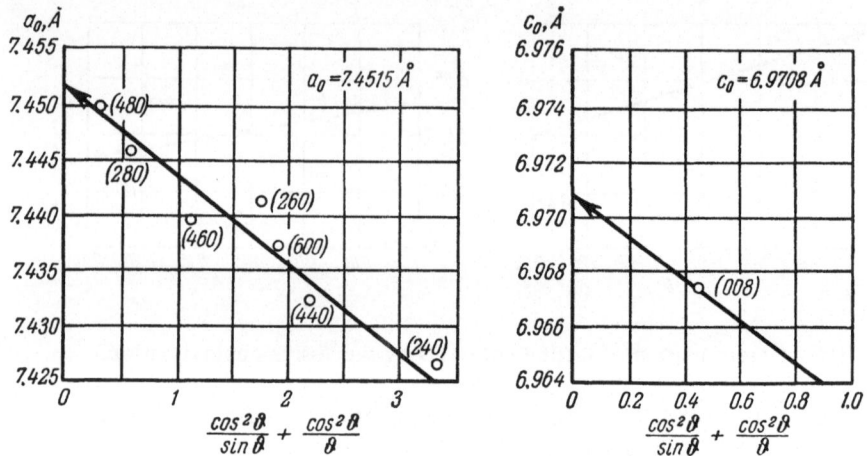

Fig. 133. Determination of lattice constants by combined graphical extrapolation and calculation.

The process is illustrated by the deduction of a_0 and c_0 for KH_2PO_4 (tetragonal); Cu radiation gave 7 (hk0) lines at angles of 30-90° but only one (00l) line.

TABLE 8

(hkl)	h^2+k^2	$\vartheta°$	$\sin \vartheta$	$\dfrac{\cos^2 \vartheta}{\sin \vartheta} + \dfrac{\cos^2 \vartheta}{\vartheta}$	a_0, Å	c_0, Å
(240) α	20	27.660	0.46422	3.336	7.4264	—
(440) α	32	35.927	0.58675	2.196	7.4322	—
(600) α	36	38.457	0.62193	1.910	7.4371	—
(260) α	40	40.937	0.65523	1.776	7.4410	—
(460) α	52	48.352	0.74724	1.122	7.4393	—
(280) α	68	58.547	0.85307	0.588	7.4456	—
(008) α_1	0	62.179	0.88441	0.446	—	6.9674
(480) α_1	80	67.640	0.92481	0.280	7.4495	—

The lattice constants are calculated from

$$a_0 = \frac{\lambda}{2} \cdot \frac{\sqrt{h^2 + k^2}}{\sin \vartheta} \quad \text{for} \quad (hk0)$$

and

$$c_0 = \frac{\lambda}{2} \cdot \frac{l}{\sin \vartheta} \quad \text{for} \quad (00l).$$

Table 8 illustrates the course of the calculation.

The extrapolated value is $a_0 = 7.4515 \pm 0.0005$ Å, so $c/a = 0.935$; the slope is $d(\Delta a_0)/d\,f(\vartheta) = -0.00803$, while (51) gives $d(\Delta c_0)/df(\vartheta) = 0.935 \cdot (-0.00803) = -0.00751$.

We draw a straight line with this slope through the point for the (008) line (Fig. 133) and by extrapolation to $\vartheta = 90°$ find that $c_0 = 6.9708 \pm 0.0010$ Å.

6-1f. Analytic Least-Squares Method

Graphical methods have the disadvantage that the placing of the line is somewhat arbitrary. Subjective errors are eliminated by the use of least squares; see [135, 136] for the theory and the principles as applied in x-ray work.

The results from the ϑ for several lines are used to compile the normal equations, which for the cubic system take the form

and

$$\left. \begin{array}{l} A_0 \sum \alpha_i^2 + D \sum \alpha_i \delta_i = \sum \alpha_i \sin^2 \vartheta_i \\[2mm] A_0 \sum \alpha_i \delta_i + D \sum \delta_i^2 = \sum \delta_i \sin^2 \vartheta_i, \end{array} \right\} \tag{52}$$

where $\alpha_i = h_i^2 + k_i^2 + l_i^2$, $\Sigma \alpha_i$ is the sum of the squares of the indices for all lines used in the calculation, $\delta_i = 10 \sin^2 2\vartheta_i$, and A_0 and D are constants. If $A_0 = \lambda^2/4a_0^2$ is known, then a_0 can be deduced.

The distinctive feature is that (52) takes no account of the doublet character of the radiation, so calculations for $K\alpha_2$ should be referred to those for $K\alpha_1$ by multiplying all $\sin^2 \vartheta$ by $(\lambda_{\alpha 1}/\lambda_{\alpha 2})^2$, which is 0.99503 for Cu radiation. Table 9 illustrates the process; it gives the results for Pb examined with Cu radiation ($\lambda_{\alpha 1} = 1.54050$ Å, $\lambda_{\alpha 2} = 1.54434$ Å).

TABLE 9

(hkl)	$\alpha = h^2 + k^2 + l^2$	$\vartheta°$	$\sin^2 \vartheta$	$\sin^2 \vartheta$ (referred to $K\alpha_1$)	$\delta = 10 \sin^2 2\vartheta$
(531) α_1	35	67.080	0.84833	0.84833	5.1
(531) α_2	35	67.421	0.85258	0.84835	5.0
(600) α_1	36	69.061	0.87230	0.87230	4.5
(600) α_2	36	69.467	0.87698	0.87263	4.3
(620) α_1	40	79.794	0.96861	0.96861	1.2
(620) α_2	40	80.601	0.97332	0.96849	1.0

$$\sum a^2 = 8242.000, \quad \sum a \sin^2 \vartheta = 199.6853, \quad \sum a\delta = 758.3,$$
$$\sum \delta \sin^2 \vartheta = 18.3767, \quad \sum \delta^2 = 92.2.$$

Then the equations of (52) take the form

$$8242.000 A_0 + 758.3 D = 199.6853$$

and

$$758.3 A_0 + 92.2 D = 18.3767.$$

The solution is

$$A_0 = 0.0242082, \quad D = 0.000213, \quad a_0 = 4.9505_2 \text{ Å},$$

or, after correction for refraction,

$$a_0 = 4.9506_6 \text{ Å}.$$

The $D\delta_i$ of (52) is very much smaller than $A_0\alpha_i$, so δ need be known only to two significant figures under normal conditions, although $\sin^2 \vartheta$ is determined to 5 or 6 significant figures.

The analytic result for a_0 agrees very closely with the 4.9506_0 Å, found by graphical extrapolation with respect to $\cos^2 \vartheta$.

Errors of measurement of the camera radius are here automatically eliminated, as in extrapolation. For example, an error of 0.2% in measuring the radius of the cassette would give the above normal equations the form

$$8242.000 \, A_0 + 761.8 \, D = 199.6012$$

and

$$761.8 A_0 + 93.2 D = 18.4512,$$

so

$$A_0 = 0.0242085, \quad D = 0.000099, \quad a_0 = 4.9505 \text{ Å},$$

which is almost without effect on a_0.

A somewhat different method is employed if the least-squares treatment is to be used for lines over a large range of angles. The proportional relation of $\Delta d/d$ to $[(\cos^2 \vartheta)/\sin \vartheta + (\cos^2 \vartheta)/\vartheta]$ is used.

The normal equations are unaltered in this case, but

$$\delta = 10 \sin^2 2\vartheta \left(\frac{1}{\sin \vartheta} + \frac{1}{\vartheta} \right).$$

The method becomes somewhat more complicated if the material is not cubic; (52) for the tetragonal and hexagonal systems becomes

$$\left. \begin{array}{l} A_0 \sum a_i^2 + C_0 \sum a_i \gamma_i + D \sum a_i \delta_i = \sum a_i \sin^2 \vartheta_i, \\ A_0 \sum a_i \gamma_i + C_0 \sum \gamma_i^2 + D \sum \gamma_i \delta_i = \sum \gamma_i \sin^2 \vartheta_i, \\ A_0 \sum a_i \delta_i + C_0 \sum \gamma_i \delta_i + D \sum \delta_i^2 = \sum \delta_i \sin^2 \vartheta_i, \end{array} \right\} \quad (53)$$

where $\alpha_i = h_i^2 + k_i^2$; $\gamma_i = l_i^2$; $\delta_i = 10 \sin^2 2\vartheta_i$; $A_0 = \lambda^2/4a_0^2$; $C_0 = \lambda^2/4c_0^2$.

Table 10 illustrates the application to the hexagonal case by reference to a_0 and c_0 for GeO_2 examined with Cu ratiation.

TABLE 10

(hkl)	$\sin^2 \vartheta$	$\sin^2 \vartheta$ (referred to $K\alpha_1$)	α	γ	δ
$(31.2)\ \alpha_2$	0.77285	0.77179	13	4	7.0
$(10.5)\ \alpha_2$	0.78547	0.78439	1	25	6.7
$(21.4)\ \alpha_1$	0.82218	0.82218	7	16	5.9
$(40.1)\ \alpha_1$	0.83386	0.83386	16	1	5.5
$(22.3)\ \alpha_1$	0.86790	0.86790	12	9	4.6
$(11.5)\ \alpha_1$	0.88505	0.88505	3	25	4.1
$(20.5)\ \alpha_1$	0.93530	0.93530	4	25	2.4
$(32.0)\ \alpha_1$	0.95545	0.95545	19	0	1.7
$(32.1)\ \alpha_1$	0.98462	0.98462	19	1	0.6
$(32.1)\ \alpha_2$	0.98857	0.98455	19	1	0.5

Then

$$\sum \alpha^2 = 1727, \quad \sum \alpha\gamma = 526, \quad \sum \alpha\delta = 357.3, \quad \sum \gamma^2 = 2231,$$
$$\sum \gamma\delta = 500.4, \quad \sum \delta^2 = 206.2, \quad \sum \alpha \sin^2 \vartheta = 102.2936,$$
$$\sum \gamma \sin^2 \vartheta = 91.9747, \quad \sum \delta \sin^2 \vartheta = 32.6681.$$

Here (53) become

$$1727 A_0 + 526 C_0 + 357.3 D = 102.2936,$$
$$526 A_0 + 2231 C_0 + 500.4 D = 91.9747,$$
$$357.3 A_0 + 500.4 C_0 + 206.2 D = 32.6681.$$

The solution is

$$A_0 = 0.0502760, \quad C_0 = 0.0293566, \quad D_0 = 0.000703,$$

so

$$a_0 = 4.9849 \text{Å}. \quad c_0 = 5.6496 \text{Å}, \quad c/a = 1.1333.$$

The error in a_0 is ± 0.0002 Å; for c_0 it is ± 0.0005 Å. The agreement with the result from successive approximation is extremely close.

The number of normal equations becomes larger for the other systems; it becomes essential to use approximate values for a, b, and c, from which are calculated assumed values of $\sin^2 \vartheta_a$, which are used in equations relating them to the true $\sin^2 \vartheta$.

For example, (53) is replaced by a set of four equations in the case of the orthorhombic system, the coefficients being

$$A_0 = \frac{\lambda^2}{4a_0^2}, \quad B_0 = \frac{\lambda^2}{4b_0^2}, \quad C_0 = \frac{\lambda^2}{4c_0^2},$$
$$\alpha = h^2, \quad \beta = k^2, \quad \gamma = l^2.$$

The lines on the pattern are indexed; then a, b, and c are deduced from a few lines, and A, B, and C are calculated for these. The normal equations for the $\sin^2 \vartheta$ are drawn up with the unknown true (experimental) values of A_0, B_0, and C_0 and with the assumed values A', B', and C'; subtraction term by term gives

$$\Delta A\alpha + \Delta B\beta + \Delta C\gamma + D\delta = \sin^2\vartheta_e - \sin^2\vartheta_a.$$

Four such equations in ΔA, ΔB, and ΔC serve to define these and hence A_0, B_0, and C_0, from which in turn a_0, b_0, and c_0 are found.

This is illustrated by the deduction for CdMg (orthorhombic) at 18°C (Table 11).

TABLE 11

(hkl)	$\sin^2\vartheta_e$ (referred to $K\alpha_1$)	$\sin^2\vartheta_a$ (referred to $K\alpha_1$)	$\Delta\sin^2\vartheta$	$\delta=10\sin^2 2\vartheta$
(331) α_1	0.74976	0.74920	0.00056	7.5
(331) α_2	0.74933	0.74920	0.00013	7.5
(332) α_1	0.81360	0.81329	0.00031	6.1
(134) α_1	0.88026	0.88020	0.00006	4.2
(430) α_1	0.89379	0.89361	0.00018	3.8
(522) α_1	0.90642	0.90629	0.00013	3.4
(504) α_1	0.93420	0.93392	0.00028	2.5
(415) α_1	0.97026	0.97025	0.00001	1.2
(514) α_1	0.99094	0.99111	0.00017	0.4

The normal equations for ΔA, ΔB, and ΔC are

$$2631\Delta A + 537\Delta B + 1370\Delta C + 431D = 1835,$$
$$537\Delta A + 423\Delta B + 255\Delta C + 277D = 1156,$$
$$1370\Delta A + 255\Delta B + 1427\Delta C + 197D = 548,$$
$$431\Delta A + 277\Delta B + 197\Delta C + 201D = 913.$$

The solution is

$$\Delta A = +0.031\cdot10^{-5}, \quad \Delta B = -2.56\cdot10^{-5}, \quad \Delta C = -0.34\cdot10^{-5}, \quad D = 8.34\cdot10^{-5}.$$

We add ΔA, ΔB, and ΔC to A', B', and C' to get A_0, B_0, and C_0, from which we have

$$a_0 = 5.0051\text{Å}, \quad b_0 = 3.2217\text{A}, \quad c_0 = 5.2700\text{Å}.$$

One of the limitations of this method is that it gives only the minimum in the expression used, not in the measured quantity. Further, lines at small angles are treated in the same way as those at large angles, although the latter give a_0 with very much less error. The calculations become very complicated if these features are incorporated. See [439] for computer programming for lattice-constant calculations.

Of course, the analytic determination may be performed by methods from the theory of probability; this also introduces much complexity, although the error is lower.

Measurement of the film-specimen distance becomes very important in back-reflection cameras; the error should not exceed 0.05 mm. Ring diameters must also be measured with great care.

Table 12 gives the errors and $\Delta d/d$ for a film-specimen distance of A = 100 mm and various angles.

TABLE 12

ϑ^0	$2l$, mm	$\dfrac{\Delta d}{d} \cdot 10^5$ for error in $\Delta(2l) = 0.05$ mm	$\dfrac{\Delta d}{d} \cdot 10^5$ for error in $\Delta A = 0.05$ mm
75.0	115.48	2.51	2.90
77.5	93.26	2.28	2.12
80.0	72.80	1.95	1.42
82.5	53.58	1.54	0.82
85.0	35.26	1.06	0.37

6-1g. Method Not Involving Use of a Standard

Back-reflection in conjunction with a standard has some disadvantages, such as the need to measure the film-specimen distance very precisely, the differences in absorption as between specimen and standard, differences in thermal expansion, etc.

The no-standard method of measuring lattice constants [137] is based on the fixed linear relation between lattice constant a and ring diameter; the steps are as follows: (1) Plots are made for $D_1/2A = f_1(a)$ and $D_2/2A = f_2(a)$, where D_1 and D_2 are the diameters of rings given by $K\alpha_1$ and $K\alpha_2$, and A is the film-specimen distance [Fig. 134a shows the

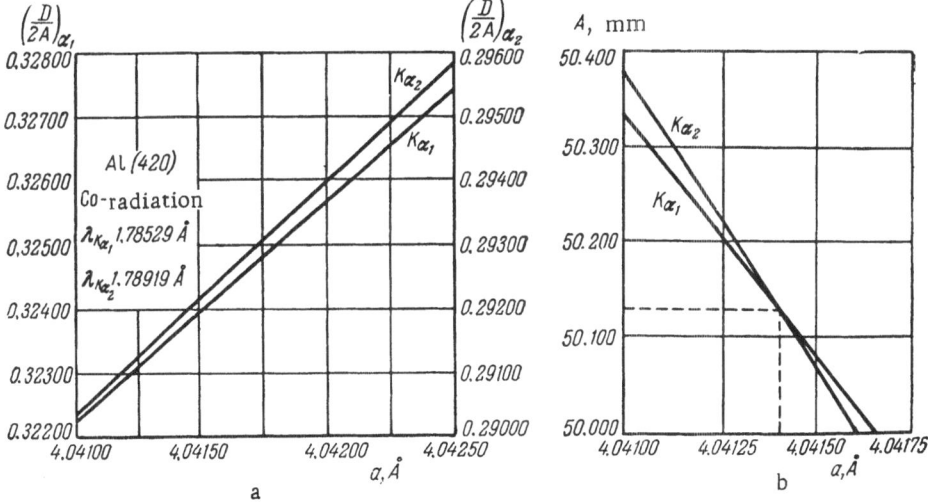

Fig. 134. Derivation of the lattice constant of aluminum by the no-standard method.

graphs for aluminum for the (420) line recorded with Co radiation]. (2) Linear plots are made for $a = f(A)$ for $K\alpha_1$ and $K\alpha_2$ from the $D_1/2A$, $D_2/2A$, and the measured D_1 and D_2 (Fig. 134b); these enable one to deduce the lattice constant with high precision from the intersection of the two lines.

The axis of abscissas in Fig. 134b can be $\Delta a = a_{\alpha_1} - a_{\alpha_2}$, which gives the point of intersection more precisely. If D_1 and D_2 are measured to ± 0.01 mm for A = 50 mm, the error in a is ± 0.00003 Å, which is much higher than when a standard is used.

Back-reflection cameras are envisaged above; see [437] for a similar method for cylindrical cameras and for the use of several radiations.

The precision method for tetragonal and hexagonal crystals is based on the linear relation between (a/c) and a^2 [455]; for the tetragonal system

$$\left(\frac{a}{c}\right)^2 = \frac{4\sin^2 \vartheta_i}{\lambda^2 l^2}\, a^2 - \frac{h^2+k^2}{l^2} \text{ (general form)}$$

$$a^2 = \frac{\lambda^2\,(h^2+k^2)}{4\sin^2 \vartheta} \text{ (for(hk0) reflections)}$$

and for the hexagonal system

$$\left(\frac{a}{c}\right)^2 = \frac{4\sin^2 \vartheta}{\lambda^2 l^2}\, a^2 - \frac{4}{3}\,\frac{h^2+hk+k^2}{3l^2} \text{ (general form)}$$

$$a^2 = \frac{\lambda^2\,(h^2+hk+k^2)}{3\sin^2 \vartheta} \text{ (for(hk0) reflections).}$$

The steps are as follows: (1) the lines are measured and indexed; (2) the equations for $(a/c)^2$ as a function of a^2 are drawn up; (3) the straight lines are plotted and their region of intersection is found; (4) a similar set of graphs is drawn on a much larger scale in the region of intersection, and the lines are numbered in order of increasing Bragg angle (the errors of experiment cause the lines to meet in a small area rather than at a point); (5) on the same sheet as the last (but above the previous lines) one constructs (with the same a^2 on the abscissa) the values of $\frac{1}{2}\,[(\cos^2 \vartheta)/\sin\vartheta + (\cos^2 \vartheta)/\vartheta]$ as ordinates, the straight line being extrapolated to meet the abscissa for the top set, the point of intersection then being transferred to the abscissa of the lower set; (6) one draws a straight line parallel to the abscissa to meet the lines corresponding to the x-ray pattern in order of increasing Bragg angle. The position for this line is found as for indexing by Hull's method (with a strip of paper or ruler); one can also use a set of graphs for $\frac{1}{2}\,[(\cos^2 \vartheta)/\sin\vartheta + (\cos^2 \vartheta)/\vartheta]$ for this purpose. The position of this line on the ordinate gives a/c.

E x a m p l e . TiAl (tetragonal) is examined; Table 13 gives the results from the x-ray pattern (Cu Kα radiation).

TABLE 13

	Indices	ϑ^0	$\frac{4\sin^2 \vartheta}{\lambda^2 l^2}$	$\frac{h^2+k^2}{l^2}$	Equation of straight line
1	202	32.76	0.12233	1	$(a/c)^2 = 0.1233\, a^2 - 1$
2	220	33.12	—	—	$a^2 = 15.9020$
3	113	39.07	0.0744	2/9	$(a/c)^2 = 0.0744\, a^2 - 0,2222$
4	131	39.73	0.6884	$\cdot 0$	$(a/c)^2 = 0.6884\, a^2 - 10$
5	222	41.62	0.1859	2	$(a/c)^2 = 0.1859\, a^2 - 2$
8	313	56.40	0.1299	10/9	$(a/c)^2 = 0.1299\, a^2 - 1,1111$
12	422	70.26	0.3733	5	$(a/c)^2 - 0.3733\, a^2 - 5$

The graphs are drawn as a^2 against $(a/c)^2$ on scales of 1 cm = 0.5 Å (a^2 being the abscissa) and 1 cm = 0.1 (a/c being the ordinate). The lines meet in the range 15.80 to 16.05, for which c/a is 0.950 to 0.970; a fresh set of graphs is drawn on a scale 10 times as large, and above this is drawn a set of graphs for a^2 against $\frac{1}{2}[(\cos^2 \vartheta)/\sin \vartheta + (\cos^2 \vartheta)/\vartheta]$. Extrapolation gives the intercept on the axis as corresponding to a = 3.997 Å ; a horizontal line is drawn on the lower set to meet the pattern lines in order of increasing Bragg angle at c/a = 1.020. Then the material has a = 3.997 Å, c = 4.077 Å, and c/a = 1.020.

6-1h. Aspects of Precision Determination by Ionization Methods

The position and width of the line are here dependent on the adjustment of the goniometer and on the position of the goniometer relative to the focus. Loss of focus and line shift have several causes, among which the most important are the flat form of the specimen, vertical divergence of the beam, and penetration of the beam into the specimen. The overall displacement of the peak is

$$\Delta\vartheta = \Delta\vartheta_f + \Delta\vartheta_p + \Delta\vartheta_s ,$$

where $\Delta\vartheta_f$ is the displacement caused by the flat form, which is

$$\Delta\vartheta_f = - \frac{b^2 \sin 2\vartheta}{24R^2} ,$$

where b is the width of the irradiated surface and R is the distance from specimen to slit or counter; $\Delta\vartheta_p$ is the displacement caused by penetration, which is given by

$$\Delta\vartheta_p = \left(\frac{\sin 2\vartheta}{4\mu R} - \frac{d_o \cos \vartheta}{R \left(\dfrac{e^{2\mu d_o}}{\sin \vartheta - 1} \right)} \right) ,$$

where μ is the absorption coefficient for the material and d_0 is the depth of the reflecting layer; $\Delta\vartheta_s$ is the displacement caused by the shift S of the plane of the specimen from the common axis of the goniometer, which is given by

$$\Delta\vartheta_s = \frac{S \cos \vartheta}{R} .$$

The displacement $\Delta\vartheta_v$ caused by vertical divergence is usually small; it is also difficult to calculate. Figure 135 illustrates the effect of each factor separately as a function of Bragg angle for aluminum.

The Bragg angle can be found precisely if the method of extrapolation with respect to $\cos^2 \vartheta$ (see above) is used in conjunction with Fig. 135.

The exact position of a peak may be ascertained by measuring the counting rate at two points (one on each side) and gradually reducing the distance between the two; use has also been made of two counters [138] and of a counter with a ring aperture [139]. The position of the counter corresponding to 0° may be established in several ways, of which the best is the knife-edge method, which enables one to set the counter to 0.001°.

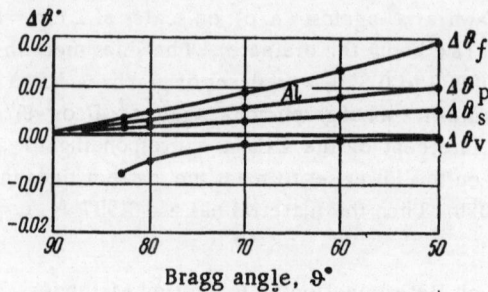

Fig. 135. Line displacement resulting from
various factors in ionization recording.

These methods enable one to measure the lattice constants of some cubic materials
with an absolute error governed by the error in the x-ray wavelength (± 0.004%) for Cu
radiation; the error in relative measurements is an order of magnitude smaller. Here the
camera is placed in a thermostat and the peak positions are determined to 0.001° in
counter setting by the above method [140].

The radiation energy can be selected by the use of a discriminator, which gives a
closer control of wavelength and thus greater accuracy in lattice constants. Correction is
made for the energy distribution at the focus of the crystal when a crystal monochromator is used [440, 441].

Lattice-constant measurement by reference to positions of diffraction peaks has some
features that tend to increase the error (difficulty in allowing for shape of the intensity
curve, for geometric conditions, and so on). The error is reduced if the center of gravity
of the intensity curve is used; the position of this is given by

$$2\vartheta_c = \frac{\sum\limits_{k=1}^{n} I_k 2\vartheta_k - \sum S_b 2\vartheta_c}{\sum\limits_{k=1}^{n} I_k - S_b} \quad ,$$

where ϑ_c is the Bragg angle corresponding to the center of gravity and I_k is the intensity
at a point corresponding to a Bragg angle ϑ_k. With the center of the counter slit placed
at $2\vartheta_k$, the counter records an average intensity over the slit width. Here k varies from
1 to n, n being some suitable number of parts; S_b is the background intensity at ϑ_c, this
ϑ being halfway between the points at which the background is measured.

It is desirable to measure with $2\vartheta_k = 5'$; if the background is then less than 1/4 of
the peak intensity of the line, it is sufficient to use a fixed counting time at each point.
Higher backgrounds require the use of a fixed number of counts (measurement is made of
the time to reach a specified count).

The counter slit must be fairly broad (about 1 mm) in precision measurements.

The following example is for tungsten [141], which was examined with Cu $K\beta$ radiation at the (321) line. The count at each point took 1 min, the background being taken as the average over 10 min. The counter slit was 1 mm wide and 8 mm high. The counter zero was displaced by 27.1'; readings were taken at 5' intervals in 2ϑ.

Table 14 gives measurements on the background on both sides of the line.

Table 15 gives the measurements on the line.

TABLE 14

k	$2\vartheta_k$	I_k	k	$2\vartheta_k$	I_k
1	109°20′	39	1	112°45′	49
2	25	40	2	50	39
3	30	33	3	55	45
4	35	40	4	113°60′	47
5	40	43	5	05	41
6	45	44	6	10	46
7	50	41	7	15	36
8	55	39	8	20	39

$$I_{b1} = 39,9 \qquad\qquad I_{b2} = 42,8$$

TABLE 15

k	$2\vartheta_k$	I_k	k	$2\vartheta_k$	I_k	k	$2\vartheta_k$	I_k
0	110°00′	44	12	111°00′	940	24	112°00′	79
1	05	48	13	05	1260	25	05	78
2	10	50	14	10	1316	26	10	68
3	15	55	15	15	1260	27	15	59
4	20	55	16	20	1972	28	20	54
5	25	70	17	25	624	29	25	57
6	30	75	18	30	396	30	30	46
7	35	94	19	35	240	31	35	44
8	40	131	20	40	180	32	40	42
9	45	210	21	45	138			
10	50	339	22	50	111			
11	55	584	23	55	97			

The quantities in the expression for the center of gravity are $S_\Sigma = \sum_k I_k = 9819$, $M_\Sigma = \sum_k I_k(2\vartheta_k - 2\vartheta_0)h = 716,150$ (origin of count $2\vartheta_0 = 110°00'$; range of count $h = 5$); $I_{b1} = 39.9$, and $I_{b2} = 42.8$. The mean background is

$$I_b = \frac{I_{b1} + I_{b2}}{2} = 41.4.$$

The background area is $S_b = I_b n = 41.4 \cdot 32 = 1366$;

$$2\vartheta_c - 2\vartheta_0 = \frac{nh}{6}, \qquad \frac{I_{b2} - I_{b1}}{I_{b2} + I_{b1}} + \frac{nh}{2} = 80.9'.$$
$$M_b = S_b(2\vartheta_c - 2\vartheta_0) = 110{,}564,$$

and the area of the curve corrected for background is

$$S_m = \sum_k I_k - S_b = 8453,$$

$$2\vartheta_c - 2\vartheta_0 = \frac{M_\Sigma - M_b}{S_m} = \frac{605\,586}{8453} = 71.6'.$$

The correction for displacement of the counter zero gives us finally $2\vartheta_c = 110°00'$ + 71.6' −27.1' = 110°44.5'; the lattice constant at 25°C is then 3.1658 Å to $\pm 20 \times 10^{-5}$ Å

Maladjustment of the goniometer could increase the error by a factor 2-2.5; relative measurements of lattice constant are much more precise than absolute ones.

See [362-364, 384] for the development of the center-of-gravity method; moments have been proposed [365] for the same purpose. Other methods of precision determination are given in [301, 366-368].

Methods have also been described recently which employ representation in reciprocal-lattice space [292, 293].

6-2. Choice of Method in Precision Determinations

6-2a. Methods of Measurement

The most commonly used methods are listed below.

No.	Method
1	Asymmetric, with calculation from latter lines
2	Asymmetric, with calculation from pairs $(h_1 k_1 l_1)$ and $(h_2 k_2 l_2)$
3	Graphical extrapolation with respect to ϑ
4	Graphical extrapolation with respect to $\cos^2 \vartheta$
5	Graphical extrapolation with respect to $[(\cos^2 \vartheta)/\sin \vartheta + (\cos^2 \vartheta)/\vartheta]$
6	Graphical extrapolation with respect to $\cot \vartheta [(\pi/2 - \vartheta)]$
7	Combination of graphical and analytic extrapolations
8	Extrapolation by successive approximation
9	Least-squares method applied to two unknowns, with $\Delta \sin^2 \vartheta = D \sin^2 2\vartheta$

No.	Method
10	Least-squares method applied to two unknowns, with $\Delta \sin^2 \vartheta = D \sin^2 2\vartheta \, [(1/\sin \vartheta) + 1/\vartheta]$
11	Least-squares method applied to three unknowns, with $\Delta \sin^2 \vartheta = D \sin^2 2\vartheta$
12	Least-squares method applied to three unknowns, with $\Delta \sin^2 \vartheta = D \sin^2 2\vartheta \, [(1/\sin \vartheta) + 1/\vartheta]$
13	Least-squares method applied to three unknowns, with use of rough values of a, b, and c, and subsequent refinement
14	Recording at large distances with a broad divergent beam
15	Use of an independent standard
16	No-standard method
17	Double graphical extrapolation

6-2b. Choice of Method

The table lists the useful methods in relation to the various structures and arrays of lines in the patterns (the methods are numbered as in Table 6-2a).

Lattice	Line distribution	Basic methods	Additional methods
Cubic	Two or more well-separated doublets with ϑ of 60-90°	3, 4, 5, 9, 6, 10, 16	1 and 14, if there is one line with $\vartheta > 75°$
	One (or two) closely placed Kα doublets with ϑ of 60-90°	5, 6, 10, 16	1 and 14, if there is one line with $\vartheta > 75°$
	Two closely placed lines of specimen and standard with $\vartheta > 70°$	15	
Tetragonal or hexagonal	One (hk0) doublet and one (00l) doublet with ϑ of 60-90°	1	
	Several (hk0) and (00l) lines with ϑ of 30-90°	5, 6, 10, 17	
	Two or more (hk$l$) and (00l) lines with $\vartheta > 75°$	2	8, 11, 12
	Several (hk0) lines with ϑ of 60-90° or one (00l) line	7, with extrapolation with respect to $\cos^2 \vartheta$	
	Several (hk0) lines with ϑ of 30-90° or one (00l) line	7, with extrapolation with respect to $(\cos^2 \vartheta)/\sin \vartheta + (\cos^2 \vartheta)/\vartheta$	
Orthorhombic	Several (hk$l$) lines with ϑ of 30-90°	13	

6-3.   Choice of Radiation for Cubic Crystals

Figure 136 is used to select from the K radiations of Cu, Ni, Co, Fe, and Cr [307]; the abscissa is the ϑ for the radiation (Kα above, Kβ below) and the ordinate is the approximate value of the lattice constant. The lines are those of constant (hk*l*).

The radiation is selected by choosing a horizontal line corresponding to the rough value of the lattice constant and observing which radiations correspond to large ϑ for the lines.

Figure 137 represents the Wulff-Bragg law

$$\frac{2\sin\vartheta}{h^2+k^2+l^2}=\frac{\lambda}{a}\ .$$

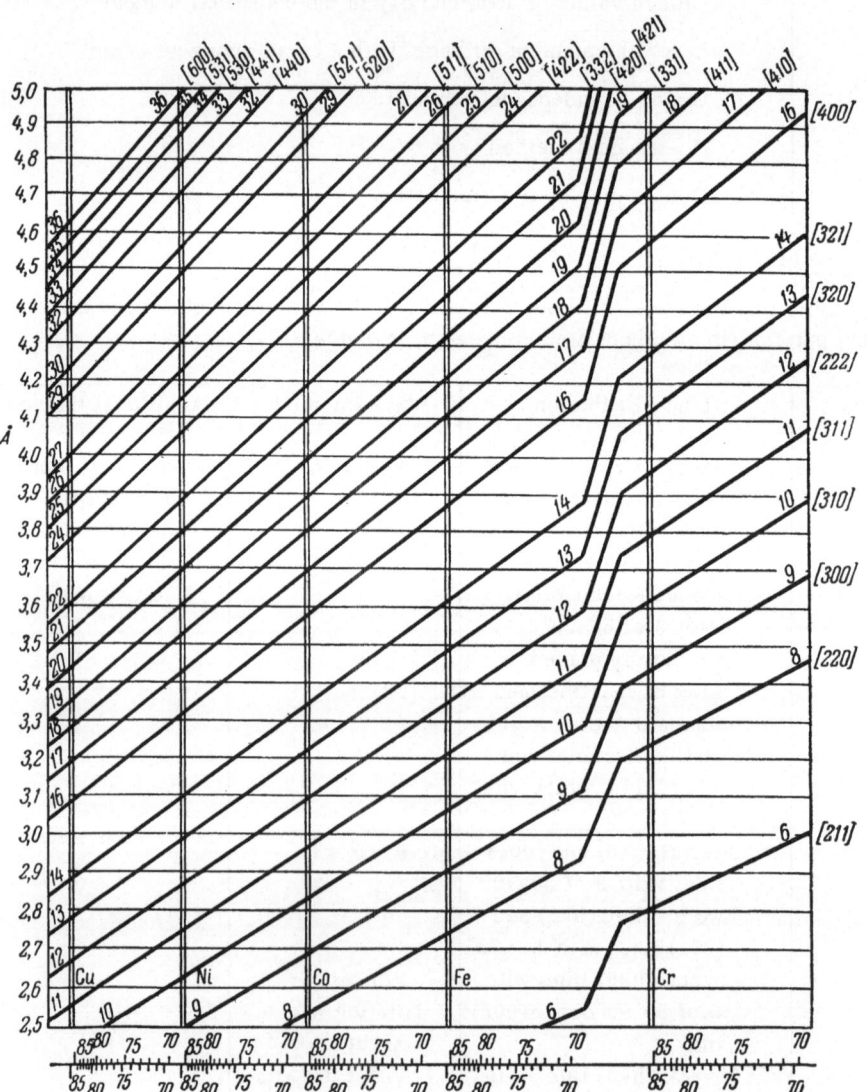

Fig. 136. Graphs for selecting radiations in precision measurements of lattice constants for cubic crystals (for the simplest cases).

The top horizontal axis gives the lattice constant; the lower one, the Bragg angle [308]. The curves running downwards to the right represent the radiations (Kα full, Kβ dot-dash); those running upwards to the right represent the reflection indices (here b denotes lines from body-centered structures and f those from face-centered ones; the others relate to primitive lattices).

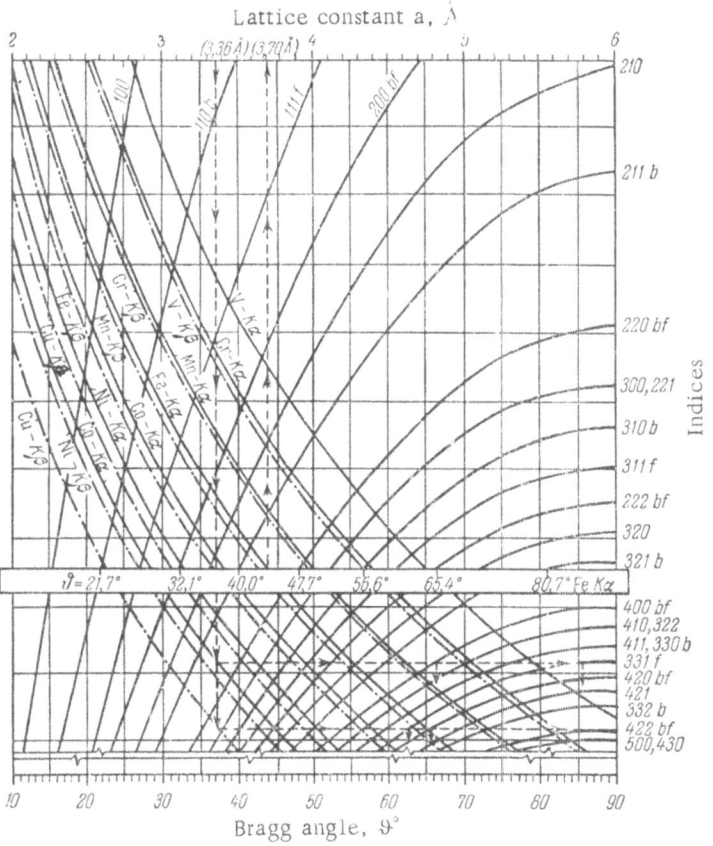

Fig. 137. Graphs for selecting radiations in precision measurements of lattice constants for cubic crystals (a of 2-6 Å).

Example. A material has a lattice constant of about 3.36 Å, and a face-centered cubic structure. From the point for 3.36 Å on the top horizontal axis in Fig. 137 a vertical straight line is drawn; we assume that the radiation from a copper anode will be used. From the point where this line meets the Cu Kα curve we draw a horizontal line to meet the (hk*l*) curves for face-centered lattices; the points of intersection give the Bragg angles of the lines. The largest angles occur for (400)α and (331)α, for which θ is 66 and 85.5°, respectively. The construction for Cu Kβ gives (331)β, (420)β, and (422)β, for which the angles are 62.5, 65.7, and 85°. Figure 137 thus enables us to select five lines at large angles.

Figure 138 is for lattice constants from 2 to 10 Å and θ from 10 to 90°; Fig. 139, for 4 to 10 Å and θ from 60 to 90°.

Figures 137-139 can also be used to index patterns from cubic materials and to deduce lattice constants. A strip of paper is marked in the scale of the graphs to represent the Bragg angles on the pattern; this strip is moved parallel to the abscissa to bring the marks into coincidence with the curves, which serves to give the indices.

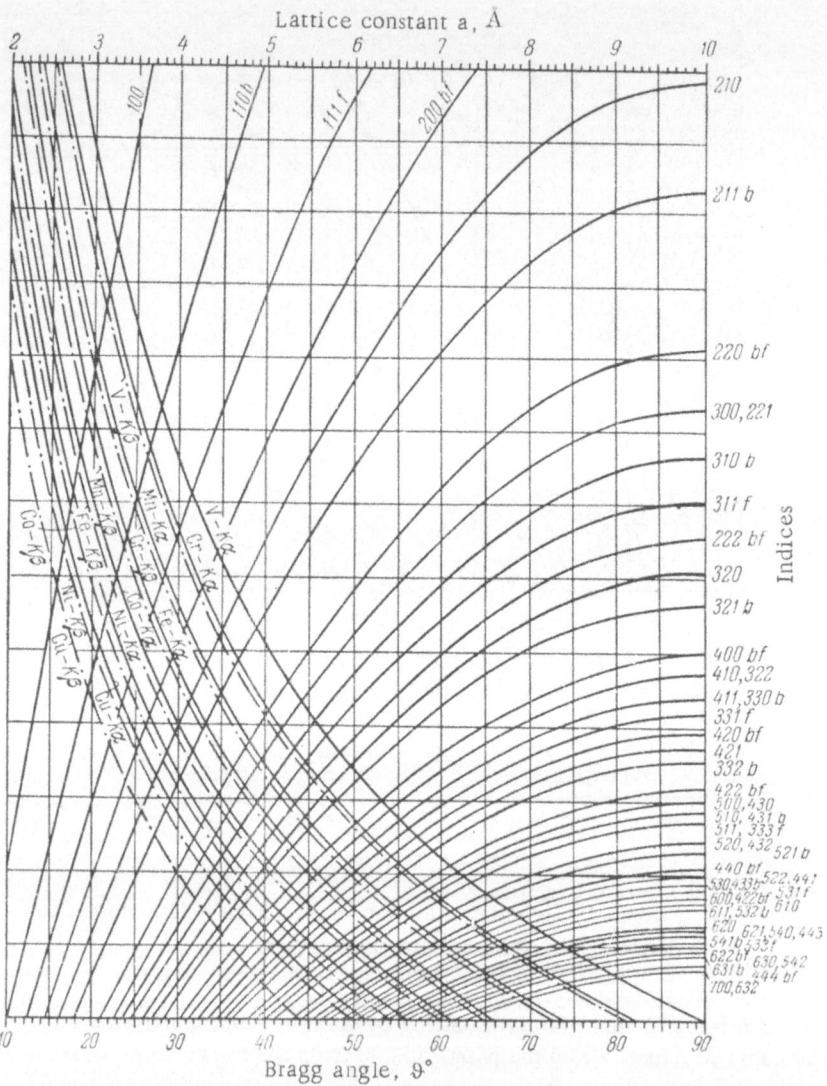

Fig. 138. Graphs for selecting radiations in precision measurements of lattice constants for cubic crystals (a of 2-10 Å).

The lattice constant is given by the abscissa of the intersection between the strip and the curve for the radiation used.

Figure 137 shows this construction for a body-centered cubic material examined with FeK radiation; the lines are (110), (200), (211), (220), (310), and (222), while a = 3.70 Å.

In general, there are two steps in the use of these curves:

(1) The lattice constant is found from the pattern without resort to precision methods.

(2) The conditions for use with precision methods are chosen by reference to the lattice constant.

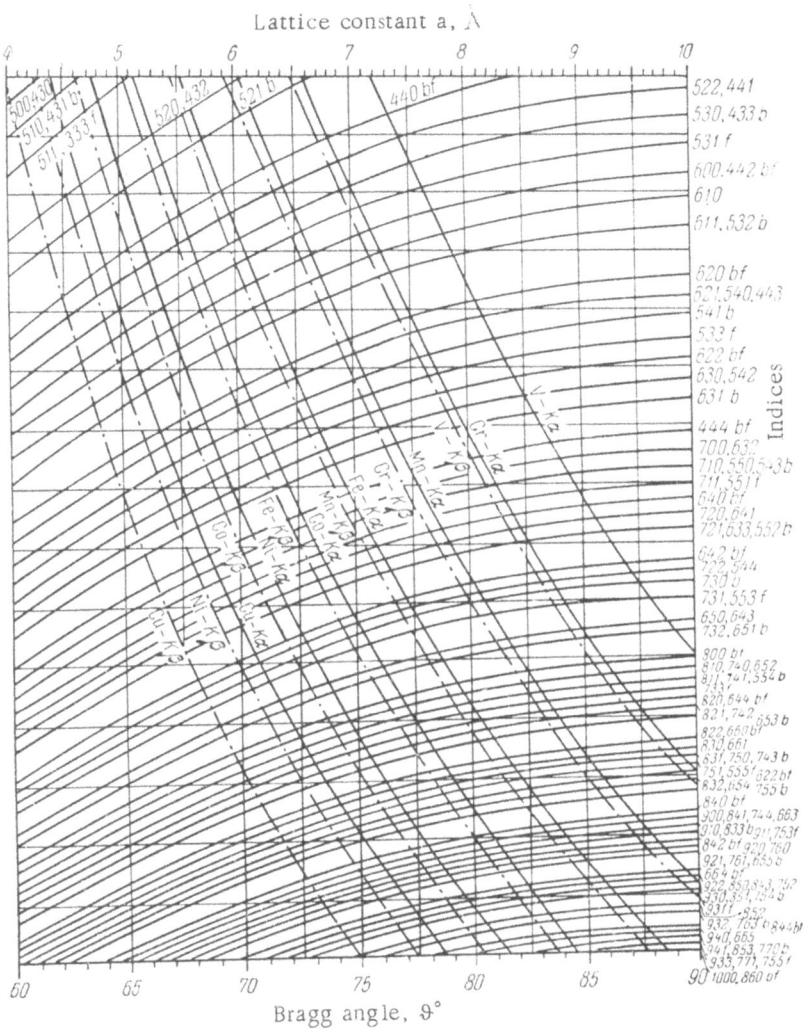

Fig. 139.   Graphs for selecting radiations in precision measurements of lattice constants for cubic crystals (a of 4-10 , large Bragg angles).

6-4. Choice of Radiation for Tetragonal Crystals

Two constants (a and c) have to be determined here, and this demands measurements on at least three lines; it is usual to use more than this in order to minimize errors, and particularly to employ lines with indices giving independent determination of one of the lattice constants, such as ($00l$).

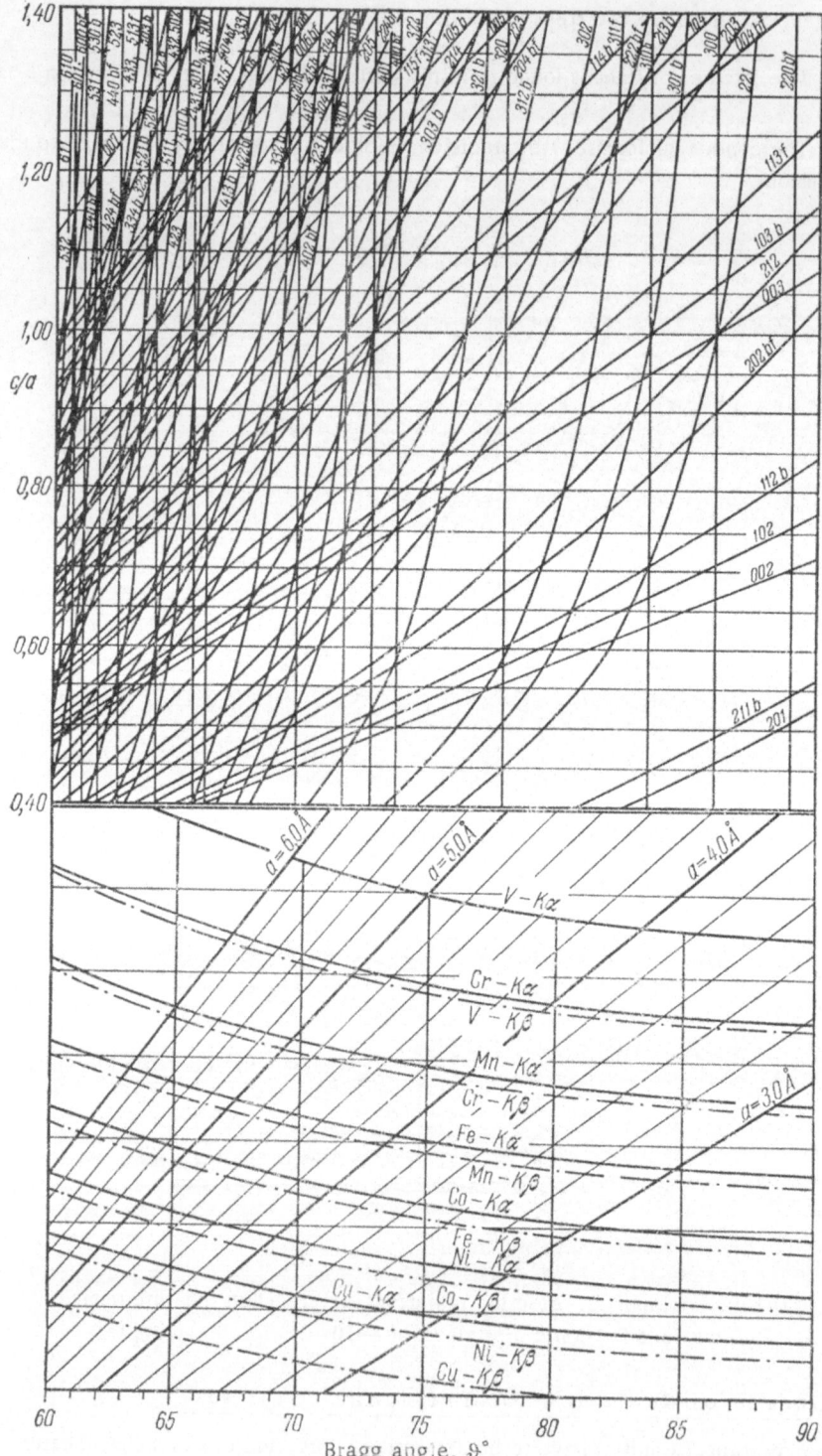

Fig. 140. Graphs for selecting radiations in precision measurements of lattice constants for tetragonal crystals.

Figure 140 is used to select radiations for use with tetragonal materials having a be-tween 3.0 and 6.0 Å and c/a from 0.4 to 1.40.

6-5. Choice of Radiation for Hexagonal and Trigonal Crystals

Idential curves can be used for these two systems if the lines from trigonal materials are indexed in the hexagonal system.

Figures 141 and 142 are used to select radiations, line indices, and angles for materials in these two systems.

Each set of curves has two parts; the lower part represents the Wulff-Bragg law in the form

$$\frac{1}{d_{hkl}} = \frac{2}{\lambda} \sin \vartheta_{hkl}$$

(curves descending to the right); the straight lines rising to the right represent the relation of d/a to d for various a.

The top part represents the Wulff-Bragg law in the form

$$\frac{d_{hkl}}{a} = \frac{1}{\sqrt{\frac{4}{3}(h^2 + hk + k^2) + \frac{l^2}{\left(\frac{c}{a}\right)^2}}} \cdot$$

The abscissas are d/a (top) and ϑ (bottom).

The full lines are for $K\alpha$ radiation; the dot-dash ones, for $K\beta$.

Figure 141 is for a from 2.0 to 3.0 Å and c/a from 1.50 to 1.80; Fig. 142, for a from 2.0 to 5.0 Å and c/a from 1.00 to 2.00.

E x a m p l e . Determine the conditions for precision examination of a hexagonal material having a = 2.50 Å and c/a = 1.62.

Figure 141 is used to select the conditions for $CoK\alpha$.

The curves are used as follows: (1) From the ends of the curve for Co $K\alpha$ on the lower part (points A and D) we draw horizontal lines to meet the straight line corresponding to a lattice constant of 2.5 Å (points B and E). (2) From B and E we draw vertical lines to meet the horizontal line for c/a = 1.62 in the upper part (points C and F). The curves representing the indices of lines lying at large angles meet the horizontal line between C and F, and from these we select (11.3) and (10.4) as being of high intensity. (3) From point L, where the curves for (11.3) and (10.4) meet, we draw a vertical line to meet the straight line in the lower part corresponding to a = 2.5 Å (point G). (4) From G we draw a horizontal line to meet the curve for Co $K\alpha$ (point H).

The abscissa of point H is ϑ = 65°, the desired Bragg angle; the corresponding construction for $CoK\beta$ gives point K (ϑ = 61.5°).

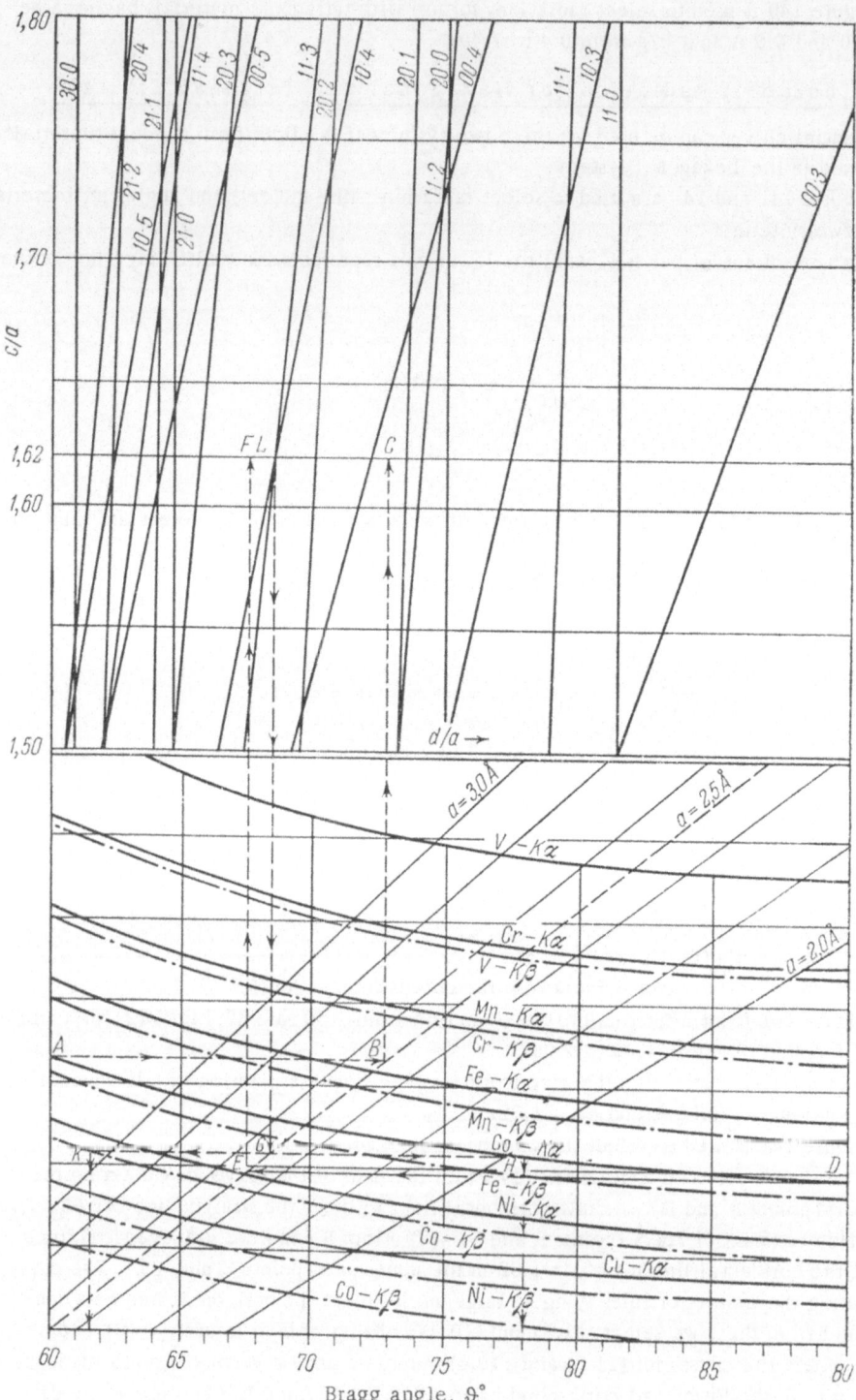

Fig. 141. Graphs for selecting radiations in precision measurements of lattice constants for hexagonal crystals (a of 2.0-3.0 Å, c/a of 1.50 to 1.80).

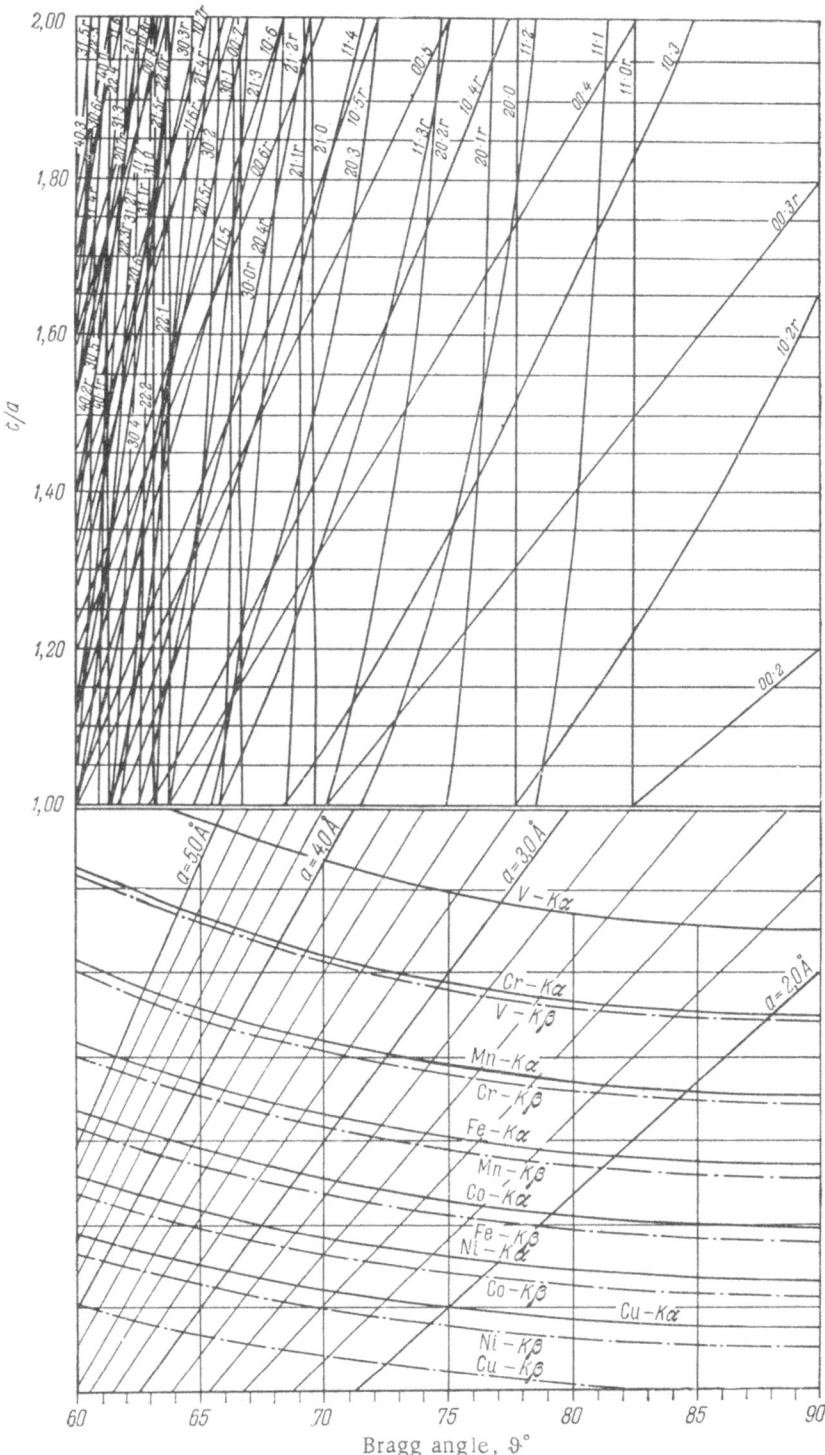

Fig. 142. Graphs for selecting radiations in precision measurements of lattice constants for hexagonal crystals (a of 2.0 to 5.0 Å, c/a of 1.00 to 2.00).

6-6. Choice of Conditions for Certain Materials

The table lists the radiations, lines, and angles best used in precision measurements on some common materials [8, 309].

In selecting a radiation, one should remember that larger Bragg angles give more precise results.

Material or phase	Radiation	Wavelength, kX	hkl	ϑ
Ferrite (and martensite)	Fe $K\beta_1$	1.753	(310)	75°39′
	Co $K\alpha_1$	1.785	(310)	81°00′
	V $K\beta_1$	2.280	(211)	77°26′
	Cr $K\alpha_1$	2.285	(211)	78°06′
Austenite	Co $K\alpha_1$	1.785	(400)	82°00′
	Cr $K\beta_1$	2.080	(311)	73°00′
	Fe $K\alpha_1$	1.932	(222)	69°00′
Nickel	Cu $K\alpha_1$	1.537	(420)	73°55′
	Mn $K\alpha_1$	2.097	(311)	80°10′
	Cr $K\beta_1$	2.080	(311)	78°43′
	V $K\alpha_1$	2.498	(220)	86°10′
Copper	Cu $K\alpha_1$	1.537	(420)	72°17′
	Co $K\alpha_1$	1.785	(400)	81°40′
	Cr $K\beta_1$	2.080	(222)	86°52′
Zinc	Cu $K\alpha_1$	1.537	(1016)	82°50′
			(1232)	69°30′
Tungsten	Ni $K\alpha_1$	1.654	(321)	78°50′
	Cu $K\alpha_1$	1.537	(400)	77°10′
Fe$_4$N (γ-phase). . .	Cr $K\alpha_1$	2.285	(311)	77°00′
Fe$_2$N (ε-phase) . . .	Cr $K\alpha_1$	2.285	(1013)	68°30′
			(1122)	80°40′
α-W$_2$C	Co $K\alpha_1$	1.785	(2132)	79°00′
WC	Co $K\alpha_1$	1.785	(211)	82°40′
VC	Cu $K\beta_1$	1.389	(531)	82°00′
TiC	Cu $K\beta_1$	1.389	(442) (600)	74°00′
Aluminum	Co $K\alpha_1$	1.785	(420)	81°06′
	Cu $K\alpha_1$	1.537	(511)	81°16′
Magnesium	Fe $K\alpha_1$	1.932	(1015)	81°00′

6-7. Tables for Determining Lattice Constants of Cubic Substances

It is convenient here to put the Wulff-Bragg law in the form $a = \sqrt{N}\, \dfrac{\lambda}{2}\, \dfrac{1}{\sin\vartheta}$, where $N = h^2 + k^2 + l^2$.

Tables 6-7a to 6-7e give $\sqrt{N}$ and $\sqrt{N}\lambda/2$ for $K\alpha_1$, $K\alpha_2$, and $K\beta$ for various elements.

The constant a is deduced from a line recorded with α_1, α_2, or β radiation by finding $\sqrt{N}\,\lambda/2$ from the tables and dividing this by the measured $\sin\vartheta$ [142].

6-7a. Use of Copper Radiation

The table gives N, $\sqrt{N}$, and $\sqrt{N}\lambda/2$ for the α_1, α_2, and β_1 lines of CuK radiation ($\lambda_{\alpha_1} = 1.54434$ Å, $\lambda_{\alpha_2} = 1.54050$ Å, $\lambda_{\beta_1} = 1.39217$ Å); the $\sqrt{N}\lambda/2$ are in angstroms. The table applies to N from 1 to 378 ($N = h^2 + k^2 + l^2$).

N	$\sqrt{N}$	$\sqrt{N}\,\lambda/2$		
		α_2	α_1	β_1
1	1,000000	0,772170	0,770250	0,696085
2	1,414214	1,092014	1,089298	0,984413
3	1,732051	1,337438	1,334112	1,205655
4	2,000000	1,544340	1,540500	1,392170
5	2,236068	1,726625	1,722331	1,556493
6	2,449490	1,891423	1,886720	1,705053
8	2,828427	2,184026	2,178596	1,968826
9	3,000000	2,316510	2,310750	2,088255
10	3,162278	2,441816	2,435745	2,201214
11	3,316625	2,560998	2,554630	2,308653
12	3,464102	2,674876	2,668225	2,411309
13	3,605551	2,784098	2,777176	2,509770
14	3,741657	2,889195	2,882011	2,604511
16	4,000000	3,088680	3,081000	2,784340
17	4,123106	3,183739	3,175822	2,870032
18	4,242641	3,276040	3,267894	2,953239
19	4,358899	3,365811	3,357442	3,034164
20	4,472136	3,453249	3,444663	3,112987
21	4,582576	3,538528	3,529729	3,189862
22	4,690416	3,621799	3,612793	3,264928
24	4,898980	3,782845	3,773439	3,410106
25	5,000000	3,860850	3,851250	3,480425
26	5,099020	3,937310	3,927520	3,549351
27	5,196152	4,012313	4,002336	3,616963
29	5,385165	4,158263	4,147923	3,748533
30	5,477226	4,229350	4,218833	3,812615
32	5,656854	4,368053	4,357192	3,937651
33	5,744563	4,435779	4,424750	3,998704
34	5,830952	4,502409	4,491291	4,058838
35	5,916080	4,568219	4,556861	4,118095
36	6,000000	4,633020	4,621500	4,176510
37	6,082763	4,696927	4,685248	4,234120
38	6,164414	4,759976	4,748140	4,290956
40	6,324555	4,883632	4,871488	4,402428
41	6,403124	4,944300	4,932006	4,457119
42	6,480741	5,004234	4,991791	4,511147
43	6,557439	5,063458	5,050867	4,564535
44	6,633250	5,121997	5,109261	4,617306
45	6,708204	5,179874	5,166994	4,669480
46	6,782330	5,237112	5,224090	4,721078
48	6,928203	5,349751	5,336448	4,822618
49	7,000000	5,405190	5,391750	4,872595
50	7,071068	5,460067	5,446490	4,922064
51	7,141428	5,514396	5,500685	4,971041
52	7,211103	5,568197	5,554352	5,019541
53	7,280110	5,621483	5,607505	5,067575
54	7,348469	5,674267	5,660158	5,115159
56	7,483315	5,778391	5,764023	5,209023
57	7,549834	5,829755	5,815260	5,255326
58	7,615773	5,880671	5,866049	5,301225
59	7,681146	5,931151	5,916403	5,346731
61	7,810250	6,030841	6,015845	5,436598
62	7,874008	6,080073	6,064955	5,480979
64	8,000000	6,177360	6,162000	5,568680
65	8,062258	6,225434	6,209954	5,612017
66	8,124038	6,273138	6,257540	5,655021
67	8,185353	6,320484	6,304768	5,697701
68	8,246211	6,367477	6,351644	5,740064
69	8,306624	6,414126	6,398177	5,782116

N	$\sqrt{N}$	$\sqrt{N}\,\lambda/2$		
		α_2	α_1	β_1
70	8,366600	6,460438	6,444374	5,823865
72	8,485281	6,552079	6,535788	5,906477
73	8,544004	6,597424	6,581019	5,947353
74	8,602325	6,642457	6,625941	5,987949
75	8,660254	6,687188	6,670561	6,028273
76	8,717798	6,731622	6,714884	6,068328
77	8,774964	6,775764	6,758916	6,108121
78	8,831761	6,819621	6,802664	6,147656
80	8,944272	6,906499	6,889326	6,225974
81	9,000000	6,949530	6,932250	6,264765
82	9,055385	6,992297	6,974910	6,303318
83	9,110434	7,034804	7,017312	6,341636
84	9,165151	7,077055	7,059458	6,379724
85	9,219545	7,119056	7,101355	6,417587
86	9,273619	7,160810	7,143005	6,455227
88	9,380832	7,243597	7,225586	6,529856
89	9,433981	7,284637	7,266524	6,566853
90	9,486833	7,325448	7,307233	6,603642
91	9,539392	7,366032	7,347717	6,640228
93	9,643651	7,446538	7,428022	6,712801
94	9,695360	7,486466	7,467851	6,748795
96	9,797959	7,565690	7,546878	6,820212
97	9,848858	7,604993	7,586083	6,855642
98	9,899495	7,644093	7,625086	6,890890
99	9,949874	7,682994	7,663890	6,925958
100	10,000000	7,721700	7,702500	6,960850
101	10,049876	7,760213	7,740917	6,995568
102	10,099505	7,798535	7,779144	7,030114
104	10,198039	7,874620	7,855040	7,098702
105	10,246951	7,912388	7,892714	7,132749
106	10,295630	7,949977	7,930209	7,166634
107	10,344080	7,987388	7,967528	7,200359
108	10,392305	8,024626	8,004673	7,233928
109	10,440307	8,061692	8,041646	7,267341
110	10,488089	8,098588	8,078451	7,300601
113	10,630146	8,208280	8,187870	7,399485
114	10,677078	8,244519	8,224019	7,432154
115	10,723805	8,280601	8,260011	7,464680
116	10,770330	8,316526	8,295847	7,497065
117	10,816654	8,352296	8,331528	7,529311
118	10,862781	8,387914	8,367057	7,561419
120	10,954451	8,458698	8,437666	7,625229
121	11,000000	8,493870	8,472750	7,656935
122	11,045361	8,528896	8,507689	7,688510
123	11,090537	8,563780	8,542486	7,719956
125	11,180340	8,633123	8,611657	7,782467
126	11,224972	8,667587	8,646035	7,813535
128	11,313709	8,736107	8,714384	7,875303
129	11,357817	8,770166	8,748359	7,906006
130	11,401754	8,804092	8,782201	7,936590
131	11,445523	8,837889	8,815914	7,967057
132	11,489125	8,871558	8,849499	7,997408
133	11,532563	8,905099	8,882957	8,027644
134	11,575837	8,938514	8,916288	8,057766
136	11,661904	9,004972	8,982582	8,117676
137	11,704700	9,038018	9,015545	8,147466
138	11,747340	9,070944	9,048389	8,177147
139	11,789826	9,103750	9,081113	8,206721

N	$\sqrt{N}$	$\sqrt{N}\,\lambda/2$		
		α_2	α_1	β_1
140	11,832160	9,136439	9,113721	8,236189
141	11,874342	9,169011	9,146212	8,265551
142	11,916375	9,201467	9,178588	8,294810
144	12,000000	9,266040	9,243000	8,353020
145	12,041595	9,298158	9,275039	8,381974
146	12,083046	9,330166	9,306966	8,410827
147	12,124356	9,362064	9,338785	8,439582
148	12,165525	9,393853	9,370496	8,468239
149	12,206556	9,425536	9,402100	8,496801
150	12,247449	9,457113	9,433598	8,525266
152	12,328828	9,519951	9,496280	8,581912
153	12,369317	9,551216	9,527466	8,610096
154	12,409674	9,582378	9,558551	8,638188
155	12,449900	9,613439	9,589535	8,666189
157	12,529964	9,675262	9,651205	8,721920
158	12,569805	9,706026	9,681892	8,749653
160	12,649111	9,767264	9,742978	8,804856
161	12,688578	9,797739	9,773377	8,832329
162	12,727922	9,828120	9,803682	8,859716
163	12,767145	9,858406	9,833893	8,887018
164	12,806249	9,888601	9,864013	8,914238
165	12,845233	9,918704	9,894041	8,941374
166	12,884099	9,948715	9,923977	8,968428
168	12,961481	10,008467	9,983581	9,022293
169	13,000000	10,038210	10,013250	9,049105
170	13,038405	10,067865	10,042831	9,075838
171	13,076697	10,097433	10,072326	9,102493
172	13,114877	10,126915	10,101734	9,129069
173	13,152946	10,156310	10,131057	9,155568
174	13,190906	10,185622	10,160295	9,181992
176	13,266499	10,243993	10,218521	9,234611
177	13,304135	10,273054	10,247510	9,260809
178	13,341664	10,302033	10,276417	9,286932
179	13,379088	10,330930	10,305243	9,312982
180	13,416408	10,359748	10,333988	9,338960
181	13,453624	10,388485	10,362654	9,364866
182	13,490738	10,417143	10,391241	9,390700
184	13,564660	10,474224	10,448179	9,442156
185	13,601471	10,502648	10,476533	9,467780
186	13,638182	10,530995	10,504810	9,493334
187	13,674794	10,559266	10,533010	9,518819
189	13,747727	10,615582	10,589187	9,569587
190	13,784049	10,643629	10,617164	9,594870
192	13,856407	10,699502	10,672897	9,645237
193	13,892444	10,727328	10,700655	9,670322
194	13,928388	10,755083	10,728341	9,695342
195	13,964240	10,782767	10,755956	9,720298
196	14,000000	10,810380	10,783500	9,745190
197	14,035669	10,837923	10,810974	9,770019
198	14,071247	10,865395	10,838378	9,794784
200	14,142136	10,920133	10,892980	9,844129
201	14,177447	10,947399	10,920179	9,868708
202	14,212670	10,974597	10,947309	9,893226
203	14,247807	11,011729	10,974373	9,917685
204	14,282857	11,028794	11,001371	9,942083
205	14,317821	11,055792	11,028302	9,966420
206	14,352700	11,082724	11,055167	9,990699
208	14,422205	11,136394	11,108703	10,039081
209	14,456832	11,163132	11,135375	10,063184

N	$\sqrt{N}$	$\sqrt{N}\lambda/2$		
		α_2	α_1	β_1
210	14,491377	11,189807	11,161983	10,087230
211	14,525839	11,216417	11,188527	10,111219
212	14,560220	11,242965	11,215009	10,135151
213	14,594520	11,269451	11,241429	10,159026
214	14,628739	11,295873	11,267786	10,182846
216	14,696939	11,348535	11,320317	10,230319
217	14,730920	11,374774	11,346491	10,253972
218	14,764823	11,400953	11,372605	10,277572
219	14,798649	11,427073	11,398659	10,301118
221	14,866069	11,479132	11,450590	10,348048
222	14,899664	11,505074	11,476466	10,371433
224	14,966630	11,556783	11,528047	10,418047
225	15,000000	11,582550	11,553750	10,441275
226	15,033296	11,608260	11,579396	10,464452
227	15,066519	11,633914	11,604986	10,487578
228	15,099669	11,659511	11,630520	10,510653
229	15,132746	11,685052	11,655998	10,533677
230	15,165751	11,710538	11,681420	10,556652
232	15,231546	11,761343	11,732098	10,602451
233	15,264338	11,786664	11,757356	10,625277
234	15,297059	11,811930	11,782560	10,648053
235	15,329710	11,837142	11,807709	10,670781
236	15,362292	11,862301	11,832805	10,693461
237	15,394804	11,887406	11,857848	10,716092
238	15,427249	11,912459	11,882839	10,738677
241	15,524175	11,987302	11,957496	10,806145
242	15,556349	12,012146	11,982278	10,828541
243	15,588457	12,036939	12,007009	10,850891
244	15,620499	12,061681	12,031689	10,873195
245	15,652476	12,086372	12,056320	10,895454
246	15,684387	12,111013	12,080899	10,917667
248	15,748016	12,160146	12,129909	10,961958
249	15,779734	12,184637	12,154340	10,984036
250	15,811388	12,209079	12,178722	11,006070
251	15,842980	12,233474	12,203055	11,028061
253	15,905974	12,282116	12,251576	11,071910
254	15,937378	12,306365	12,275765	11,093770
256	16,000000	12,354720	12,324000	11,137360
257	16,031220	12,378827	12,348047	11,159092
258	16,062378	12,402886	12,37.047	11,180780
259	16,093477	12,426900	12,396001	11,202428
260	16,124516	12,450868	12,419908	11,224034
261	16,155494	12,474788	12,443769	11,245597
262	16,186414	12,498663	12,467585	11,267120
264	16,248077	12,546278	12,515081	11,310043
265	16,278821	12,570017	12,538762	11,331443
266	16,309506	12,593711	12,562397	11,352802
267	16,340135	12,617362	12,585989	11,374123
268	16,370706	12,640968	12,609536	11,395403
269	16,401220	12,664530	12,633040	11,416643
270	16,431677	12,688048	12,656499	11,437844
272	16,492423	12,734954	12,703289	11,480128
273	16,522712	12,758343	12,726619	11,501212
274	16,552945	12,781688	12,749906	11,522257
275	16,583124	12,804991	12,773151	11,543264
276	16,613248	12,828252	12,796354	11,564233
277	16,643317	12,851470	12,819515	11,585163
278	16,673332	12,874647	12,842634	11,606056

N	$\sqrt{N}$	$\sqrt{N}\,\lambda/2$		
		α_2	α_1	β_1
280	16,733201	12,920876	12,888748	11,647730
281	16,763055	12,943928	12,911743	11,668511
282	16,792856	12,966940	12,934697	11,689255
283	16,822604	12,989910	12,957611	11,709962
285	16,881913	13,035730	13,003317	11,751267
286	16,911535	13,058580	13,026110	11,771866
288	16,970563	13,104160	13,071576	11,812954
289	17,000000	13,126890	13,094250	11,833445
290	17,029386	13,149581	13,116885	11,853900
291	17,058722	13,172233	13,139481	11,874321
292	17,088008	13,194847	13,162038	11,894706
293	17,117243	13,217422	13,184556	11,915056
294	17,146428	13,239957	13,207036	11,935371
296	17,204651	13,284915	13,251882	11,975899
297	17,233688	13,307337	13,274248	11,996112
298	17,262677	13,329721	13,296577	12,016291
299	17,291617	13,352068	13,318868	12,036435
300	17,320508	13,374377	13,341121	12,056546
301	17,349352	13,396649	13,363338	12,076624
302	17,378147	13,418884	13,385518	12,096667
304	17,435596	13,463244	13,429768	12,136657
305	17,464249	13,485369	13,451838	12,156602
306	17,492856	13,507459	13,473872	12,176515
307	17,521416	13,529512	13,495871	12,196395
308	17,549929	13,551529	13,517833	12,216242
309	17,578396	13,573510	13,539760	12,236058
310	17,606817	13,595456	13,561651	12,255841
312	17,663522	13,639242	13,605328	12,295313
313	17,691806	13,661082	13,627114	12,315001
314	17,720045	13,682887	13,648865	12,334658
315	17,748239	13,704658	13,670581	12,354283
317	17,804494	13,748096	13,713912	12,393441
318	17,832555	13,769764	13,735525	12,412974
320	17,888544	13,812997	13,778651	12,451947
321	17,916473	13,834563	13,800163	12,471388
322	17,944358	13,856095	13,821642	12,490798
323	17,972201	13,877594	13,843088	12,510180
324	18,000000	13,899060	13,864500	12,529530
325	18,027756	13,920492	13,885879	12,548851
326	18,055470	13,941892	13,907226	12,568142
328	18,110770	13,984593	13,949821	12,606635
329	18,138357	14,005895	13,971069	12,625838
330	18,165902	14,027165	13,992286	12,645012
331	18,193405	14,048402	14,013470	12,664156
332	18,220867	14,069607	14,034623	12,683272
333	18,248288	14,090781	14,055744	12,702360
334	18,275667	14,111922	14,076833	12,721418
336	18,330303	14,154110	14,118916	12,759449
337	18,357560	14,175157	14,139911	12,778422
338	18,384776	14,196172	14,160874	12,797367
339	18,411953	14,217158	14,181807	12,816284
340	18,439089	14,238111	14,202708	12,835173
341	18,466185	14,259034	14,223579	12,854034
342	18,493242	14,279927	14,244420	12,872868
344	18,547237	14,321620	14,286009	12,910453
345	18,574176	14,342421	14,306759	12,929205
346	18,601075	14,363192	14,327478	12,947929
347	18,627936	14,383933	14,348168	12,966627
349	18,681542	14,425326	14,389458	13,003941

N	$\sqrt{N}$	$\sqrt{N}\lambda/2$		
		α_2	α_1	β_1
350	18,708287	14,445978	14,410058	13,022558
352	18,761663	14,487193	14,451171	13,059712
353	18,788294	14,507757	14,471683	13,078250
354	18,814888	14,528292	14,492167	13,096761
355	18,841444	14,548798	14,512622	13,115247
356	18,867962	14,569274	14,533048	13,133705
357	18,894444	14,589723	14,553445	13,152139
358	18,920888	14,610142	14,573814	13,170546
360	18,973666	14,650896	14,614466	13,207284
361	19,000000	14,671230	14,634750	13,225615
362	19,026298	14,691537	14,655006	13,243921
363	19,052559	14,711814	14,675234	13,262201
364	19,078784	14,732065	14,695433	13,280455
365	19,104973	14,752287	14,715605	13,298685
366	19,131127	14,772482	14,735751	13,316891
369	19,209373	14,832902	14,796020	13,371356
370	19,235384	14,852986	14,816055	13,389462
371	19,261360	14,873044	14,836063	13,407544
372	19,287302	14,893076	14,856044	13,425602
373	19,313208	14,913080	14,875998	13,443634
374	19,339080	14,933057	14,895926	13,461644
376	19,390719	14,972931	14,935701	13,497589
377	19,416488	14,992830	14,955550	13,515526
378	19,442222	15,012701	14,975371	13,533439

6-7b. Use of Nickel Radiation

The table gives N and $\sqrt{N}\lambda/2$ for the α_1, α_2, β_1 lines of Ni K radiation (λ_{α_2} = 1.65783 Å, λ_{α_1} = 1.66168 Å, λ_{β_1} = 1.50008 Å); the $\sqrt{N}\lambda/2$ are in angstroms. The table applies to N from 1 to 326 ($N = h^2 + k^2 + l^2$).

N	$\sqrt{N}\lambda/2$			N	$\sqrt{N}\lambda/2$		
	α_2	α_1	β_1		α_2	α_1	β_1
1	0,830840	0,828915	0,750040	30	4,550698	4,540155	4,108139
2	1,174986	1,172263	1,060717	32	4,699941	4,689051	4,242867
3	1,439057	1,435723	1,299108	33	4,772813	4,761754	4,308652
4	1,661680	1,657830	1,500080	34	4,844588	4,833364	4,373447
5	1,857815	1,853510	1,677140	35	4,915316	4,903927	4,437297
6	2,035134	2,030419	1,837215	36	4,985040	4,973490	4,500240
8	2,349970	2,344526	2,121433	37	5,053803	5,042093	4,562316
9	2,492520	2,486745	2,250120	38	5,121642	5,109775	4,623557
10	2,627347	2,621260	2,371835	40	5,254693	5,242519	4,743669
11	2,755585	2,749200	2,487601	41	5,319972	5,307646	4,802599
12	2,878115	2,871446	2,598215	42	5,384459	5,371983	4,860815
13	2,995636	2,988695	2,704307	43	5,448183	5,435560	4,918342
14	3,108718	3,101516	2,806392	44	5,511169	5,498400	4,975203
16	3,323360	3,315660	3,000160	45	5,573444	5,560531	5,031421
17	3,425641	3,417704	3,092494	46	5,635031	5,621975	5,087019
18	3,524956	3,516789	3,182150	48	5,756228	5,742891	5,196429
19	3,621548	3,613157	3,269349	49	5,815880	5,802405	5,250280
20	3,715629	3,707021	3,354281	50	5,874926	5,861314	5,303584
21	3,807387	3,798566	3,437115	51	5,933384	5,919637	5,356357
22	3,896985	3,887956	3,518000	52	5,991273	5,977391	5,408616
24	4,070269	4,060838	3,674431	53	6,048607	6,034592	5,460374
25	4,154200	4,144575	3,750200	54	6,105402	6,091256	5,511646
26	4,236470	4,226654	3,824469	56	6,217437	6,203032	5,612786
27	4,317171	4,307168	3,897322	57	6,272704	6,258171	5,662677
29	4,474210	4,463844	4,039089	58	6,327489	6,312828	5,712134
				59	6,381803	6,367017	5,761167

N	$\sqrt{N}\,\lambda/2$			N	$\sqrt{N}\,\lambda/2$		
	a_2	a_1	β_1		a_2	a_1	β_1
61	6,489068	6,474033	5,858000	**130**	9,473033	9,451085	8,551772
62	6,542041	6,526883	5,905821	131	9,509398	9,487366	8,584600
64	6,646720	6,631320	6,000320	132	9,545625	9,523508	8,617303
65	6,698446	6,682927	6,047016	133	9,581715	9,559514	8,649884
66	6,749776	6,734137	6,093353	134	9,617668	9,595385	8,682341
67	6,800719	6,784962	6,139342	136	9,689176	9,666727	8,746894
68	6,851282	6,835408	6,184988	137	9,724733	9,702201	8,778993
69	6,901475	6,885485	6,230300	138	9,760160	9,737546	8,810975
				139	9,795459	9,772764	8,842841
70	6,951306	6,935200	6,275285	**140**	9,830632	9,807855	8,874593
72	7,049911	7,033577	6,364300	141	9,865678	9,842820	8,906231
73	7,098700	7,082253	6,408345	142	9,900601	9,877662	8,937758
74	7,147156	7,130596	6,452088	144	9,970080	9,946980	9,000480
75	7,195285	7,178614	6,495537	145	10,004639	9,981459	9,031678
76	7,243095	7,226314	6,538697	146	10,039078	10,015818	9,062768
77	7,290591	7,273699	6,581574	147	10,073400	10,050061	9,093752
78	7,337780	7,320779	6,624174	148	10,107605	10,084186	9,124630
				149	10,141695	10,118197	9,155405
80	7,431259	7,414041	6,708562	**150**	10,175671	10,152094	9,186077
81	7,477560	7,460235	6,750360	152	10,243283	10,219550	9,247114
82	7,523576	7,506144	6,791901	153	10,276923	10,253112	9,277483
83	7,569313	7,551775	6,833190	154	10,310454	10,286565	9,307752
84	7,614774	7,597131	6,874230	155	10,343875	10,319909	9,337923
85	7,659967	7,642219	6,915028	157	10,410395	10,386275	9,397974
86	7,704894	7,687042	6,955585	158	10,443497	10,419300	9,427857
88	7,793970	7,775912	7,035999				
89	7,838129	7,819968	7,075863				
90	7,882040	7,863778	7,115504	**160**	10,509387	10,485038	9,487339
91	7,925708	7,907345	7,154926	161	10,542178	10,517753	9,516941
93	8,012331	7,993767	7,233124	162	10,574867	10,550365	9,546451
94	8,055293	8,036629	7,271908	163	10,607455	10,582878	9,575869
96	8,140536	8,121675	7,348861	164	10,639944	10,615292	9,605199
97	8,182825	8,163866	7,387037	165	10,672333	10,647606	9,634439
98	8,224896	8,205840	7,425017	166	10,704625	10,679823	9,663590
99	8,266753	8,247600	7,462803	168	10,768917	10,743966	9,721629
				169	10,800920	10,775895	9,750520
100	8,308400	8,289150	7,500400	**170**	10,832828	10,807729	9,779325
101	8,349839	8,330493	7,537809	171	10,864643	10,839470	9,808046
102	8,391073	8,371631	7,575033	172	10,896364	10,871118	9,836682
104	8,472939	8,453307	7,648937	173	10,927994	10,902674	9,865236
105	8,513577	8,493851	7,685623	174	10,959532	10,934140	9,893707
106	8,554021	8,534202	7,722134	176	11,022338	10,996800	9,950405
107	8,594275	8,574363	7,758474	177	11,053608	11,027997	9,978633
108	8,634343	8,614337	7,794644	178	11,084788	11,059105	10,006782
109	8,674225	8,654127	7,830648	179	11,115881	11,090127	10,034851
110	8,713924	8,693734	7,866486	**180**	11,146888	11,121062	10,062843
113	8,831951	8,811487	7,973035	181	11,177809	11,151911	10,090756
114	8,870943	8,850390	8,008236	182	11,208645	11,182675	10,118593
115	8,909766	8,889123	8,043283	184	11,270062	11,243950	10,174038
116	8,948421	8,927688	8,078178	185	11,300646	11,274463	10,201647
117	8,986909	8,966087	8,112923	186	11,331147	11,304894	10,229182
118	9,025233	9,004322	8,147520	187	11,361566	11,335242	10,256642
				189	11,422162	11,395697	10,311345
120	9,101396	9,080309	8,216276	**190**	11,452339	11,425805	10,338588
121	9,139240	9,118065	8,250440	192	11,512457	11,485784	10,392860
122	9,176928	9,155665	8,284463	193	11,542398	11,515655	10,419889
123	9,214462	9,193112	8,318346	194	11,572262	11,545450	10,446848
125	9,289074	9,267552	8,385702	195	11,602049	11,575168	10,473739
126	9,326156	9,304548	8,419178	196	11,631760	11,604810	10,500560
128	9,399882	9,378103	8,485734	197	11,661395	11,634377	10,527313
129	9,436529	9,414665	8,518817	198	11,690955	11,663868	10,553998

N	$\sqrt{N}\,\lambda/2$			N	$\sqrt{N}\,\lambda/2$		
	α_2	α_1	β_1		α_2	α_1	β_1
200	11,749852	11,722629	10,607168	265	13,525096	13,493759	12,209767
201	11,779190	11,751898	10,633652	266	13,550590	13,519194	12,232782
202	11,808455	11,781095	10,630071	267	13,576038	13,544583	12,255755
203	11,837648	11,810221	10,686425	268	13,601437	13,569924	12,278684
204	11,866769	11,839274	10,712714	269	13,626790	13,595217	12,301571
205	11,895818	11,868257	10,738938				12,324415
206	11,924797	11,897168	10,765099	270	13,652095	13,620464	
208	11,982545	11,954782	10,817231	272	13,702565	13,670817	12,369977
209	12,011314	11,983485	10,843202	273	13,727730	13,695924	12,392695
				274	13,752849	13,720984	12,415371
210	12,040016	12,012120	10,869112	275	13,777923	13,746000	12,438006
211	12,068648	12,040686	10,894960	276	13,802951	13,770970	12,460601
212	12,097213	12,069185	10,920747	277	13,827933	13,795895	12,483153
213	12,125711	12,097617	10,946474	278	13,852871	13,820775	12,505666
214	12,154142	12,125981	10,972139				
216	12,210805	12,182513	11,023292	280	13,902613	13,870401	12,550570
217	12,239038	12,210681	11,048779	281	13,927417	13,895148	12,572962
218	12,267206	12,238783	11,074208	282	13,952176	13,919850	12,595314
219	12,295310	12,266822	11,099579	283	13,976892	13,944509	12,617626
				285	14,026194	13,993696	12,662133
221	12,351325	12,322708	11,150146	286	14,050780	14,018225	12,684328
222	12,379237	12,350555	11,175344	288	14,099823	14,067154	12,728601
224	12,434875	12,406064	11,225571	289	14,124280	14,091555	12,750680
225	12,462600	12,433725	11,250600				
226	12,490264	12,461325	11,275573	290	14,148695	14,115913	12,772721
227	12,517867	12,488864	11,300492	291	14,173069	14,140231	12,794724
228	12,545409	12,516342	11,325356	292	14,197401	14,164506	12,816690
229	12,572891	12,543760	11,350165	293	14,221690	14,188739	12,838617
				294	14,245938	14,212931	12,860507
230	12,600313	12,571118	11,374920	296	14,294312	14,261193	12,904176
232	12,654978	12,625657	11,424269	297	14,318437	14,285262	12,925955
233	12,682223	12,652839	11,448864	298	14,342523	14,309292	12,947698
234	12,709408	12,679962	11,473406	299	14,366567	14,333281	12,969404
235	12,736536	12,707027	11,497896				
236	12,763607	12,734034	11,522333	300	14,390571	14,357229	12,991074
237	12,790619	12,760984	11,546719	301	14,414536	14,381138	13,012708
238	12,817576	12,787878	11,571054	302	14,438460	14,405007	13,034305
241	12,898106	12,868222	11,643752	304	14,486191	14,452627	13,077394
242	12,924837	12,894891	11,667884	305	14,509997	14,476378	13,098885
243	12,951514	12,921506	11,691966	306	14,533764	14,500091	13,120342
244	12,978135	12,948066	11,715999	307	14,557493	14,523765	13,141763
245	13,004703	12,974572	11,739983	308	14,581183	14,547399	13,163149
246	13,031216	13,001024	11,763918	309	14,604835	14,570996	13,184500
248	13,084082	13,053767	11,811642	310	14,628448	14,594555	13,205817
249	13,110434	13,080058	11,835432	312	14,675561	14,641558	13,248348
250	13,136734	13,106297	11,859173	313	14,699060	14,665003	13,269562
251	13,162982	13,132484	11,882869	314	14,722522	14,688411	13,290743
253	13,215319	13,184700	11,930117	315	14,745947	14,711782	13,311889
254	13,241411	13,210732	11,953671	317	14,792686	14,758412	13,354083
256	13,293440	13,262640	12,000640	318	14,816000	14,781672	13,375130
257	13,319379	13,288519	12,024056	320	14,862518	14,828082	13,417124
258	13,345266	13,314346	12,047426	321	14,885722	14,851233	13,438071
259	13,371104	13,340124	12,070751	322	14,908890	14,874348	13,458986
260	13,396893	13,365853	12,094032	323	14,932023	14,897427	13,479870
261	13,422631	13,391531	12,117267	324	14,955120	14,920470	13,500720
262	13,448320	13,417161	12,140458	325	14,978181	14,943477	13,521538
264	13,499552	13,468275	12,186708	326	15,001207	14,966450	13,542325

6-7c. Use of Cobalt Radiation

The table gives N and $\sqrt{N}\lambda/2$ for the α_1, α_2, and β_1 lines of CoK radiation ($\lambda_{\alpha 1}$ = 1.78890 Å, $\lambda_{\alpha 2}$=1.79279 Å, $\lambda_{\beta 1}$= 1.62073 Å); the $\sqrt{N}\lambda/2$ are in angstroms. The table applies to N from 1 to 281 (N = $h^2 + k^2 + l^2$).

N	√N λ/2			N	√N λ/2		
	α_2	α_1	β_1		α_2	α_1	β_1
1	0,896395	0,894450	0,810365	70	7,499778	7,483505	6,780000
2	1,267694	1,264944	1,146030	72	7,606163	7,589660	6,876175
3	1,552602	1,549233	1,403594	73	7,658802	7,642284	6,923762
4	1,792790	1,788900	1,620730	74	7,711081	7,694350	6,971023
5	2,004400	2,000051	1,812031	75	7,763008	7,746164	7,017967
6	2,195711	2,190946	1,984981	76	7,814591	7,797634	7,064598
8	2,535388	2,529887	2,292058	77	7,865834	7,848767	7,110924
9	2,689185	2,683350	2,431095	78	7,916746	7,899569	7,156950
10	2,834650	2,828500	2,562599	80	8,017601	8,000204	7,248125
11	2,973006	2,966555	2,687677	81	8,067555	8,050050	7,293285
12	3,105204	3,098466	2,807187	82	8,117202	8,099589	7,338167
13	3,231998	3,224985	2,921812	83	8,166547	8,148828	7,382777
14	3,354003	3,346725	3,032108	84	8,215596	8,197769	7,427118
16	3,585580	3,577800	3,241460	85	8,264354	8,246422	7,471197
17	3,695932	3,687912	3,341221	86	8,312826	8,294789	7,515016
18	3,803082	3,794830	3,438088	88	8,408931	8,390685	7,601898
19	3,907295	3,898817	3,532299	89	8,456573	8,438224	7,644968
20	4,008800	4,000102	3,624062	90	8,503950	8,485498	7,687797
21	4,107798	4,098885	3,713559	91	8,551063	8,532509	7,730389
22	4,204465	4,195343	3,800949	93	8,644521	8,625764	7,814877
24	4,391421	4,381893	3,969962	94	8,690872	8,672015	7,856780
25	4,481975	4,472250	4,051825	96	8,782841	8,763784	7,939923
26	4,570736	4,560818	4,132067	97	8,828467	8,809311	7,981170
27	4,657805	4,647698	4,210780	98	8,873858	8,854603	8,022204
29	4,827235	4,816761	4,363949	99	8,919017	8,899665	8,063030
30	4,909758	4,899105	4,438552	100	8,963950	8,944500	8,103650
32	5,070776	5,059773	4,584116	101	9,008659	8,989112	8,144068
33	5,149398	5,138224	4,655193	102	9,053146	9,033502	8,184285
34	5,226836	5,215495	4,725199	104	9,141471	9,121636	8,264134
35	5,303145	5,291638	4,794184	105	9,185316	9,165385	8,303770
36	5,378370	5,366700	4,862190	106	9,228951	9,208926	8,343218
37	5,452558	5,440727	4,929258	107	9,272382	9,252262	8,382480
38	5,525750	5,513760	4,995425	108	9,315610	9,295397	8,421560
40	5,669299	5,656998	5,125198	109	9,358639	9,338333	8,460459
41	5,739728	5,727274	5,188868	110	9,401471	9,381071	8,499180
42	5,809304	5,796699	5,251766	113	9,528810	9,508134	8,614298
43	5,878056	5,865301	5,313919	114	9,570879	9,550112	8,652330
44	5,946012	5,933110	5,375354	115	9,612765	9,591907	8,690196
45	6,013201	6,000153	5,436094	116	9,654470	9,633522	8,727898
46	6,079647	6,066455	5,496163	117	9,695995	9,674956	8,765438
48	6,210407	6,196931	5,614373	118	9,737343	9,716214	8,802818
49	6,274765	6,261150	5,672555	120	9,819515	9,798209	8,877104
50	6,338470	6,324717	5,730146	121	9,860345	9,838950	8,914015
51	6,401540	6,387650	5,787163	122	9,901006	9,879523	8,950774
52	6,463997	6,449971	5,843625	123	9,941502	9,919931	8,987383
53	6,525854	6,511694	5,899546	125	10,022001	10,000255	9,060156
54	6,587131	6,572838	5,954942	126	10,062009	10,040176	9,096324
56	6,708006	6,693451	6,064217	128	10,141552	10,119547	9,168234
57	6,767633	6,752949	6,118121	129	10,181090	10,158999	9,203977
58	6,826741	6,811928	6,171556	130	10,220475	10,198299	9,239582
59	6,885341	6,870401	6,224532	131	10,259710	10,237448	9,275051
61	7,001069	6,985878	6,329153	132	10,298794	10,276448	9,310385
62	7,058221	7,042906	6,380820	133	10,337732	10,315301	9,345585
64	7,171160	7,155600	6,482920	134	10,376522	10,354007	9,380653
65	7,226968	7,211287	6,533372	136	10,453672	10,430990	9,450399
66	7,282347	7,266546	6,583436	137	10,492035	10,469269	9,485079
67	7,337310	7,321389	6,633124	138	10,530257	10,507408	9,519633
68	7,391862	7,375823	6,682441	139	10,568341	10,545410	9,554062
69	7,446016	7,429860	6,731397				

N	$\sqrt{N}\,\lambda/2$			N	$\sqrt{N}\,\lambda/2$		
	α_2	α_1	β_1		α_2	α_1	β_1
140	10,606289	10,583276	9,588368	210	12,989998	12,961812	11,743305
141	10,644101	10,621005	9,622551	211	13,020889	12,992637	11,771232
142	10,681779	10,658602	9,656613	212	13,051708	13,023389	11,799093
144	10,756740	10,733400	9,724380	213	13,082455	13,054068	11,826888
145	10,794026	10,770605	9,758087	214	13,113128	13,084676	11,854618
146	10,831182	10,807680	9,791678	216	13,174263	13,145677	11,909885
147	10,868212	10,844630	9,825154	217	13,204723	13,176071	11,937422
148	10,905116	10,881454	9,858516	218	13,235114	13,206396	11,964896
149	10,941896	10,918154	9,891766	219	13,265435	13,236652	11,992307
150	10,978552	10,954731	9,924904	221	13,325870	13,296955	12,046942
152	11,051500	11,027520	9,990851	222	13,355984	13,327004	12,074166
153	11,087794	11,063736	10,023662	224	13,416012	13,386902	12,128433
154	11,123970	11,099833	10,056365	225	13,445925	13,416750	12,155475
155	11,160028	11,135813	10,088963	226	13,475771	13,446532	12,182457
157	11,231797	11,207426	10,153844	227	13,505552	13,476248	12,209380
158	11,267510	11,243062	10,186130	228	13,535268	13,505899	12,236243
				229	13,564918	13,535485	12,263048
160	11,338600	11,313997	10,250397	230	13,594503	13,565006	12,289794
161	11,373978	11,349299	10,282380	232	13,653482	13,623856	12,343112
162	11,409246	11,384490	10,314263	233	13,682876	13,653187	12,369685
163	11,444405	11,419573	10,346047	234	13,712207	13,682454	12,396201
164	11,479458	11,454549	10,377736	235	13,741475	13,711659	12,422660
165	11,514403	11,489419	10,409327	236	13,770682	13,740802	12,449064
166	11,549242	11,524182	10,440823	237	13,799825	13,769882	12,475410
168	11,618607	11,593397	10,503531	238	13,828909	13,798903	12,501703
169	11,653135	11,627850	10,534745				
170	11,687561	11,662201	10,565867	241	13,915793	13,885598	12,580248
171	11,721886	11,696452	10,596898	232	13,944633	13,914376	12,606321
172	11,756110	11,730602	10,627837	243	13,973415	13,943095	12,632340
173	11,790235	11,764653	10,658687	244	14,002137	13,971755	12,658306
174	11,824262	11,798606	10,689449	245	14,030801	14,000357	12,684219
176	11,892023	11,866220	10,750706	246	14,059406	14,028900	12,710078
177	11,925760	11,899884	10,781205	248	14,116443	14,085813	12,761641
178	11,959401	11,933451	10,811618	249	14,144875	14,114183	12,787344
179	11,992948	11,966925	10,841945	250	14,173249	14,142496	12,812995
180	12,026401	12,000306	10,872187	251	14,201568	14,170753	12,838596
181	12,059761	12,033594	10,902346	253	14,258036	14,227098	12,889645
182	12,093030	12,066791	10,932422	254	14,286186	14,255188	12,915093
184	12,159293	12,132910	10,992326	256	14,342320	14,311200	12,965840
185	12,192291	12,165836	11,022156	257	14,370305	14,339125	12,991140
186	12,225198	12,198672	11,051905	258	14,398235	14,366994	13,016389
187	12,258017	12,231419	11,081574	259	14,426112	14,394811	13,041590
189	12,323394	12,296654	11,140677	260	14,453936	14,422573	13,066743
190	12,355953	12,329143	11,170111	261	14,481704	14,450282	13,091847
192	12,420814	12,393863	11,228747	262	14,509421	14,477938	13,116903
193	12,453117	12,426097	11,257950	264	14,564695	14,533092	13,166873
194	12,485337	12,458247	11,287078	265	14,592254	14,560591	13,191787
195	12,517475	12,490314	11,316131	266	14,619760	14,588038	13,216653
196	12,549530	12,522300	11,345110	267	14,647215	14,615434	13,241473
197	12,581504	12,554204	11,374015	268	14,674619	14,642778	13,266247
198	12,613395	12,586027	11,402846	269	14,701972	14,670071	13,290975
200	12,676940	12,649434	11,460292	270	14,729273	14,697313	13,315656
201	12,708593	12,681017	11,488907	272	14,783726	14,751648	13,364882
202	12,740166	12,712523	11,517450	273	14,801876	14,778740	13,389428
203	12,771663	12,743951	11,545924	274	14,837977	14,805782	13,413927
204	12,803082	12,775301	11,574327	275	14,865029	14,832775	13,438383
205	12,834423	12,806575	11,602661	276	14,892032	14,859720	13,462795
206	12,865689	12,837773	11,630926	277	14,918986	14,886615	13,487162
208	12,927992	12,899941	11,687250	278	14,945891	14,913462	13,511485
209	12,959032	12,930913	11,715311	280	14,999558	14,967012	13,560000
				281	15,026319	14,993715	13,584193

6-7d. Use of Iron Radiation

The table gives N and $\sqrt{N}\lambda/2$ for the α_1, α_2, and β_1 lines of FeK radiation ($\lambda_{\alpha 1}$ = 1.93597 Å, $\lambda_{\alpha 2}$ = 1.93991 Å, $\lambda_{\beta 1}$ = 1.75654 Å); the $\sqrt{N}\lambda/2$ are in angstroms. The table applies to N from 1 to 241 ($N = h^2 + k^2 + l^2$).

N	$\sqrt{N}\lambda/2$			N	$\sqrt{N}\lambda/2$		
	α_2	α_1	β_1		α_2	α_1	β_1
1	0,969955	0,967985	0,878270	61	7,575591	7,560205	6,859508
2	1,371724	1,368938	1,242062	62	7,637433	7,621922	6,915505
3	1,680012	1,676599	1,521208	64	7,759640	7,743880	7,026160
4	1,939910	1,935970	1,756540	65	7,820027	7,804145	7,080839
5	2,168885	2,164480	1,963871	66	7,879951	7,863947	7,135099
6	2,375895	2,371070	2,151314	67	7,939424	7,923299	7,188950
8	2,743447	2,737875	2,484123	68	7,998454	7,982209	7,242400
9	2,909865	2,903955	2,634810	69	8,057051	8,040687	7,295459
10	3,067267	3,061038	2,777334	70	8,115226	8,098743	7,348134
11	3,216977	3,210443	2,912892	72	8,230341	8,213625	7,452368
12	3,360023	3,353199	3,042417	73	8,287299	8,270468	7,503942
13	3,497222	3,490119	3,166647	74	8,343868	8,326922	7,555164
14	3,629239	3,621868	3,286185	75	8,400057	8,382996	7,606041
16	3,879820	3,871940	3,513080	76	8,455872	8,438698	7,656580
17	3,999227	3,991105	3,621200	77	8,511320	8,494034	7,706788
18	4,115171	4,106813	3,726184	78	8,566411	8,549012	7,756671
19	4,227936	4,219349	3,828290	80	8,675541	8,657921	7,855486
20	4,337771	4,328961	3,927743	81	8,729595	8,711865	7,904430
21	4,444893	4,435865	4,024739	82	8,783316	8,765477	7,953073
22	4,549492	4,540252	4,119452	83	8,836711	8,818763	8,001421
24	4,751790	4,742139	4,302627	84	8,889784	8,871729	8,049477
25	4,849775	4,839925	4,391350	85	8,942544	8,924381	8,097250
26	4,945820	4,935775	4,478316	86	8,994993	8,976724	8,144741
27	5,040034	5,029797	4,563624	88	9,098985	9,080505	8,238903
29	5,223368	5,212759	4,729629	89	9,150537	9,131952	8,285582
30	5,312663	5,301873	4,810483	90	9,201801	9,183112	8,332001
32	5,486894	5,475750	4,968245	91	9,252781	9,233988	8,378162
33	5,571968	5,560651	5,045277	93	9,353908	9,334910	8,469729
34	5,655761	5,644274	5,121150	94	9,404063	9,384963	8,515144
35	5,738331	5,726677	5,195916	96	9,503579	9,484277	8,605253
36	5,819730	5,807910	5,269620	97	9,552949	9,533547	8,649957
37	5,900006	5,888023	5,342308	98	9,602065	9,582563	8,694429
38	5,979204	5,967060	5,414020	99	9,650930	9,631329	8,738676
40	6,134534	6,122074	5,554667	100	9,699550	9,679850	8,782700
41	6,210742	6,198128	5,623672	101	9,747927	9,728129	8,826505
42	6,286027	6,273260	5,691840	102	9,796065	9,776169	8,870092
43	6,360421	6,347503	5,759202	104	9,891639	9,871549	8,956632
44	6,433954	6,420887	5,825784	105	9,939081	9,918895	8,999590
45	6,506656	6,493441	5,891614	106	9,986298	9,966015	9,042343
46	6,578555	6,565194	5,956717	107	10,033292	10,012914	9,084895
48	6,720045	6,706397	6,084833	108	10,080068	10,059595	9,127250
49	6,789685	6,775895	6,147890	109	10,126628	10,106061	9,169408
50	6,858618	6,844688	6,210307	110	10,172974	10,152313	9,211374
51	6,926864	6,912795	6,272102	113	10,310763	10,289822	9,336138
52	6,994445	6,980240	6,333295	114	10,356285	10,335251	9,377357
53	7,061379	7,047037	6,393902	115	10,401608	10,380482	9,418396
54	7,127684	7,113208	6,453940	116	10,446735	10,425519	9,459258
56	7,258479	7,243737	6,572371	117	10,491668	10,470359	9,499943
57	7,322999	7,308126	6,630793	118	10,536409	10,515009	**9,540455**
58	7,386957	7,371954	6,688705				
59	7,450366	7,435234	6,746120				

N	$\sqrt{N}\lambda/2$			N	$\sqrt{N}\lambda/2$		
	α_2	α_1	β_1		α_2	u_1	β_1
120	10,625325	10,603744	9,620966	180	13,013312	12,986882	11,783229
121	10,669505	10,647835	9,660970	181	13,049410	13,022906	11,815914
122	10,713503	10,691744	9,700809	182	13,085409	13,058832	11,848510
123	10,757322	10,735473	9,740486	184	13,157110	13,130387	11,913434
125	10,844427	10,822401	9,819357	185	13,192815	13,166020	11,945764
126	10,887718	10,865605	9,858556	186	13,228423	13,201556	11,978006
128	10,973789	10,951501	9,936491	187	13,263935	13,236995	12,010161
129	11,016571	10,994196	9,975230	189	13,334677	13,307594	12,074216
130	11,059188	11,036727	10,013818	190	13,369907	13,342753	12,106117
131	11,101642	11,079095	10,052259	192	13,440091	13,412794	12,169667
132	11,143934	11,121301	10,090554	193	13,475046	13,447677	12,201317
133	11,186067	11,163348	10,128704	194	13,509910	13,482471	12,232885
134	11,228041	11,205237	10,166710	195	13,544684	13,517175	12,264373
136	11,311522	11,288548	10,242300	196	13,579370	13,551790	12,295780
137	11,353032	11,329974	10,279887	197	13,613967	13,586317	12,327107
138	11,394391	11,371249	10,317336	198	13,648476	13,620756	12,358354
139	11,435601	11,412375	10,354650	200	13,717236	13,689376	12,420614
140	11,476663	11,453353	10,391831	201	13,751486	13,723556	12,451626
141	11,517577	11,494185	10,428878	202	13,785650	13,757651	12,482562
142	11,557348	11,534872	10,465795	203	13,819732	13,791663	12,513421
144	11,639460	11,615820	10,539240	204	13,853729	13,825591	12,544205
145	11,679805	11,656083	10,575772	205	13,887642	13,859436	12,574913
146	11,720011	11,696207	10,612177	206	13,921473	13,893198	12,605546
147	11,760080	11,736195	10,648458	208	13,988890	13,960478	12,666590
148	11,800012	11,776046	10,684616	209	14,022476	13,993997	12,697002
149	11,839810	11,815763	10,720652	210	14,055984	14,027436	12,727342
150	11,879474	11,855347	10,756567	211	14,089410	14,060794	12,757609
152	11,958408	11,934121	10,828040	212	14,122758	14,094075	12,787804
153	11,997681	11,973313	10,863600	213	14,156028	14,127276	12,817929
154	12,036825	12,012378	10,899044	214	14,189219	14,160400	12,847983
155	12,075843	12,051316	10,934374	216	14,255369	14,226416	12,907881
157	12,153501	12,128817	11,004691	217	14,288330	14,259310	12,937725
158	12,192145	12,167383	11,039683	218	14,321214	14,292127	12,967501
				219	14,354024	14,324870	12,997209
160	12,269068	12,244150	11,109335	221	14,419418	14,390132	13,056422
161	12,307350	12,282353	11,143997	222	14,452004	14,422651	13,085928
162	12,345512	12,320138	11,178552	224	14,516958	14,487473	13,144742
163	12,383556	12,358405	11,213000	225	14,549325	14,519775	13,174050
164	12,421485	12,396257	11,247344	226	14,581621	14,552005	13,203293
165	12,459298	12,433993	11,281583	227	14,613845	14,584164	13,232472
166	12,496996	12,471615	11,315718	228	14,645999	14,616253	13,261586
168	12,572053	12,546519	11,383680	229	14,678083	14,648271	13,290637
169	12,609415	12,583805	11,417510				
170	12,646666	12,620980	11,451240	230	14,710096	14,680219	13,319624
171	12,683808	12,658047	11,484871	232	14,773914	14,743908	13,377410
172	12,720841	12,695004	11,518403	233	14,805721	14,775650	13,406210
173	12,757766	12,731854	11,551838	234	14,837459	14,807324	13,434948
174	12,794585	12,768599	11,585177	235	14,869129	14,838929	13,463624
176	12,867907	12,841772	11,651568	236	14,900732	14,870468	13,492240
177	12,904412	12,878203	11,684623	237	14,932267	14,901939	13,520795
178	12,940814	12,914531	11,717583	238	14,963737	14,933346	13,549290
179	12,977113	12,950756	11,750452				
				241	15,057751	15,027169	13,634417

6-7e. Use of Chromium Radiation

The table gives N and $\sqrt{N}\lambda/2$ for the α_1, α_2, and β_1 lines of CrK radiation ($\lambda_{\alpha 1}$ = 2.28962 Å, $\lambda_{\alpha 2}$ = 2.29352 Å, $\lambda_{\beta 1}$ = 2.08479 Å); the $\sqrt{N}\lambda/2$ are in angstroms. The table applies to N from 1 to 172 ($N = h^2 + k^2 + l^2$).

N	$\sqrt{N}\lambda/2$			N	$\sqrt{N}\lambda/2$		
	α_2	α_1	β_1		α_2	α_1	β_1
1	1,146760	1,144810	1,042395	70	9,594482	9,578167	8,721302
2	1,621764	1,619006	1,474170	72	9,730581	9,714035	8,845014
3	1,986247	1,982869	1,805481	73	9,797922	9,781261	8,906227
4	2,293520	2,289620	2,084790	74	9,864802	9,848028	8,967021
5	2,564233	2,559873	2,330866	75	9,931233	9,914345	9,027405
6	2,808977	2,804201	2,553336	76	9,997222	9,980222	9,087389
8	3,243527	3,238012	2,948338	77	10,062778	10,045667	9,146979
9	3,440280	3,434430	3,127185	78	10,127910	10,110688	9,206184
10	3,626374	3,620207	3,296343	80	10,256933	10,239492	9,323464
11	3,803373	3,796905	3,457233	81	10,320840	10,303290	9,381555
12	3,972494	3,965739	3,610963	82	10,384353	10,366695	9,439288
13	4,134702	4,127671	3,758408	83	10,447481	10,429716	9,496671
14	4,290783	4,283486	3,900285	84	10,510229	10,492357	9,553708
16	4,587040	4,579240	4,169580	85	10,572605	10,554627	9,610408
17	4,728213	4,720173	4,297905	86	10,634615	10,616532	9,666774
18	4,865291	4,857018	4,422509	88	10,757563	10,739270	9,778532
19	4,998611	4,990111	4,543695	89	10,818512	10,800116	9,833935
20	5,128467	5,119746	4,661732	90	10,879121	10,860621	9,889027
21	5,255115	5,246179	4,776854	91	10,939393	10,920791	9,943815
22	5,378781	5,369635	4,889266	93	11,058953	11,040148	10,052494
24	5,617954	5,608401	5,106672	94	11,118251	11,099345	10,106395
25	5,733800	5,724050	5,211975	96	11,235907	11,216801	10,213343
26	5,847352	5,837409	5,315193	97	11,294276	11,275071	10,266400
27	5,958739	5,948607	5,416443	98	11,352345	11,333041	10,319184
29	6,175492	6,164991	5,613469	99	11,410118	11,390715	10,371699
30	6,281064	6,270383	5,709433	100	11,467600	11,448100	10,423950
32	6,487054	6,476023	5,896676	101	11,524796	11,505199	10,475940
33	6,587635	6,576433	5,988104	102	11,581708	11,562014	10,527674
34	6,686703	6,675332	6,078155	104	11,694703	11,674817	10,630385
35	6,784324	6,772788	6,166892	105	11,750794	11,730812	10,681370
36	6,880560	6,868860	6,254370	106	11,806617	11,786540	10,732113
37	6,975469	6,963608	6,340642	107	11,862177	11,842006	10,782617
38	7,069103	7,057083	6,425754	108	11,917480	11,897215	10,832887
40	7,252747	7,240414	6,592685	109	11,972526	11,952168	10,882924
41	7,342846	7,330360	6,674584	110	12,027321	12,006869	10,932732
42	7,431855	7,419217	6,755492	113	12,190226	12,169497	11,080811
43	7,519809	7,507022	6,835442	114	12,244046	12,223226	11,129733
44	7,606746	7,593811	6,914467	115	12,297631	12,276719	11,178441
45	7,692700	7,679619	6,992598	116	12,350984	12,329981	11,226938
46	7,777705	7,764479	7,069867	117	12,404106	12,383014	11,275226
48	7,944986	7,931476	7,221924	118	12,457003	12,435820	11,323309
49	8,027320	8,013670	7,296765	120	12,562126	12,540765	11,418865
50	8,108818	8,095029	7,370846	121	12,614360	12,592910	11,466345
51	8,189504	8,175578	7,444189	122	12,666378	12,644840	11,513629
52	8,269404	8,255343	7,516818	123	12,718184	12,696558	11,560720
53	8,348539	8,334343	7,588750	125	12,821167	12,799365	11,654331
54	8,426930	8,412601	7,660007	126	12,872349	12,850460	11,700855
56	8,581566	8,566974	7,800570	128	12,974109	12,952047	11,793354
57	8,657848	8,643125	7,869909	129	13,024690	13,002542	11,839332
58	8,733464	8,718613	7,938644	130	13,075075	13,052842	11,885131
59	8,808431	8,793453	8,006788	131	13,125268	13,102949	11,930756
61	8,956482	8,941252	8,141366	132	13,175269	13,152865	11,976206
62	9,029597	9,014243	8,207827	133	13,225082	13,202593	12,021486
64	9,174080	9,158480	8,339160	134	13,274707	13,252134	12,066595
65	9,245475	9,229754	8,404057	136	13,373405	13,350664	12,156310
66	9,316322	9,300480	8,468457	137	13,422482	13,399658	12,200921
67	9,386635	9,370674	8,532371	138	13,471380	13,448472	12,245368
68	9,456425	9,440345	8,595809	139	13,520101	13,497111	12,289656
69	9,525704	9,509506	8,658783				

N	$\sqrt{N}\lambda/2$			N	$\sqrt{N}\lambda/2$		
	a_2	a_1	β_1		a_2	a_1	β_1
140	13,568648	13,545575	12,333784	**160**	14,505495	14,480829	13,185370
141	13,617020	13,593865	12,377755	161	14,550754	14,526011	13,226510
142	13,665222	13,641985	12,421570	162	14,595872	14,571052	13,267522
144	13,761120	13,737720	12,508740	163	14,640851	14,615955	13,308408
145	13,808819	13,785338	12,552098	164	14,685694	14,660722	13,349170
146	13,856354	13,832792	12,595307	165	14,730399	14,705351	13,389807
147	13,903726	13,880084	12,638368	166	14,774969	14,749845	13,430320
148	13,950937	13,927215	12,681282	168	14,863708	14,838433	13,510983
149	13,997990	13,974187	12,724053	169	14,907880	14,882530	13,551135
150	14,044885	14,021002	12,766680	**170**	14,951921	14,926496	13,591168
152	14,138207	14,114166	12,851509	171	14,995833	14,970333	13,631084
153	14,184638	14,160518	12,893714	172	15,039616	15,014042	13,670882
154	14,230918	14,206719	12,935782				
155	14,277047	14,252770	12,977714				
157	14,368862	14,344428	13,061172				
158	14,414550	14,390038	13,102702				

6-8. Extrapolation Functions

The tables give values of $\cos^2\vartheta$, $1/2\,[(\cos^2\vartheta)/\sin\vartheta + (\cos^2\vartheta)/\vartheta]$, and $\vartheta\,\mathrm{tg}\,\vartheta$, which are functions used in precision methods of graphical extrapolation [12, 88]; Table 6-2a deals with the cases in which any given function is to be preferred.

6-8a. Function $\cos^2\vartheta$

$\vartheta°$	.0	.1	.2	.3	.4	.5	.6	.7	.8	.9
45	0,5000	0,4983	0,4965	0,4948	0,4930	0,4913	0,4895	0,4878	0,4860	0,4843
6	.4826	.4808	.4791	.4773	.4756	.4738	.4721	.4703	.4686	.4669
7	.4651	.4634	.4616	.4599	.4582	.4564	.4547	.4529	.4512	.4495
8	.4477	.4460	.4443	.4425	.4408	.4391	.4373	.4356	.4339	.4321
9	.4304	.4287	.4270	.4252	.4235	.4218	.4201	.4183	.4166	.4149
50	.4132	.4115	.4097	.4080	.4063	.4046	.4029	.4012	.3995	.3978
1	.3960	.3943	.3926	.3909	.3892	.3875	.3858	.3841	.3824	.3807
2	.3790	.3773	.3757	.3740	.3723	.3706	.3689	.3672	.3655	.3639
3	.3622	.3605	.3588	.3572	.3555	.3538	.3521	.3505	.3488	.3472
4	.3455	.3438	.3422	.3405	.3389	.3372	.3356	.3339	.3323	.3306
5	.3290	.3274	.3257	.3241	.3224	.3208	.3192	.3176	.3159	.3143
6	.3127	.3111	.3095	.3079	.3062	.3046	.3030	.3014	.2998	.2982
7	.2966	.2950	.2934	.2919	.2903	.2887	.2871	.2855	.2840	.2824
8	.2808	.2792	.2777	.2761	.2746	.2730	.2715	.2699	.2684	.2668
9	.2653	.2637	.2622	.2607	.2591	.2576	.2561	.2545	.2530	.2515
60	.2500	.2485	.2470	.2455	.2440	.2425	.2410	.2395	.2380	.2365
1	.2350	.2336	.2321	.2306	.2291	.2277	.2262	.2248	.2233	.2219
2	.2204	.2190	.2175	.2161	.2146	.2132	.2118	.2104	.2089	.2075
3	.2061	.2047	.2033	.2019	.2005	.1991	.1977	.1963	.1949	.1935
4	.1922	.1908	.1894	.1881	.1867	.1853	.1840	.1826	.1813	.1799
5	.1786	.1773	.1759	.1746	.1733	.1720	.1707	.1693	.1680	.1667
6	.1654	.1641	.1628	.1616	.1603	.1590	.1577	.1565	.1552	.1539
7	.1527	.1514	.1502	.1489	.1477	.1464	.1452	.1440	.1428	.1415
8	.1403	.1391	.1379	.1367	.1355	.1343	.1331	.1320	.1308	.1296
9	.1284	.1273	.1261	.1249	.1238	.1226	.1215	.1204	.1192	.1181

$\vartheta°$	.0	.1	.2	.3	.4	.5	.6	.7	.8	.9
70	0.1170	0,1159	0,1147	0,1136	0,1125	0,1114	0,1103	0,1092	0;1082	0,1071
1	.1060	.1049	.1039	.1028	.1017	.1007	.0996	.0986	.0976	.0965
2	.0955	.0945	.0934	.0924	.0914	.0904	.0894	.0884	.0874	.0865
3	.0855	.0845	.0835	.0826	.0816	.0807	.0797	.0788	.0778	.0769
4	.0760	.0751	.0741	.0732	.0723	.0714	.0705	.0696	0687	.0679
5	.0670	.0661	.0653	.0644	.0635	.0627	.0618	.0610	.0602	.0593
6	.0585	.0577	.0569	.0561	.0553	.0545	.0537	.0529	.0521	.0514
7	.0506	.0498	.0491	.0483	.0476	.0468	.0461	.0454	.0447	.0439
8	.0432	.0425	.0418	.0411	.0404	.0397	.0391	.0384	.0377	.0371
9	.0364	.0358	.0351	.0345	.0338	.0332	.0326	.0320	.0314	.0308
80	.0302	.0296	.0290	.0284	.0278	.0272	.0267	.0261	.0256	.0250
1	.0245	.0239	.0234	.0229	.0224	.0218	.0213	.0208	.0203	.0199
2	.0194	.0189	.0184	.0180	.0175	.0170	.0166	.0161	.0157	.0153
3	.0149	.0144	.0140	.0136	.0132	.0128	.0124	.0120	.0117	.0113
4	.0109	.0106	.0102	.0099	.0095	.0092	.0089	.0085	.0082	.0079
5	.0076	.0073	.0070	.0067	.0064	.0062	.0059	.0056	.0054	.0051
6	.0049	.0046	.0044	.0042	.0039	.0037	.0035	.0033	.0031	.0029
7	.0027	.0026	.0024	.0022	.0021	.0019	.0018	.0016	.0015	.0013
8	.0012	.0011	.0010	.0009	.0008	.0007	.0006	.0005	.0004	.0004
9	.0003	.0002	.0002	.0001	.0001	.0001	.0000	.0000	.0000	.0000

6-8b. Function $1/2\,[(\cos^2 \vartheta)/\sin \vartheta + (\cos^2 \vartheta)/\vartheta]$

$\vartheta°$	.0	.1	.2	.3	.4	.5	.6	.7	.8	.9
10	5,572	5.513	5,456	5,400	5,345	5,291	5,237	5,185	5,134	5,084
1	5,034	4,986	4,939	4,892	4,846	4,800	4,756	4,712	4,669	4,627
2	4,585	4,544	4,504	4,464	4,425	4,386	4,348	4,311	4,274	4,238
3	4,202	4,167	4,133	4,098	4,065	4,032	3,999	3,967	3,935	3,903
4	3,872	3,842	3,812	3,782	3,753	3,724	3,695	3,667	3,639	3,612
5	3,584	3,558	3,531	3,505	3,479	3,454	3,429	3,404	3,379	3,355
6	3,331	3,307	3,284	3,260	3,237	3,215	3,192	3,170	3,148	3,127
7	3,105	3,084	3,063	3,042	3,022	3,001	2,981	2,962	2,942	2,922
8	2,903	2,884	2,865	2,847	2,828	2,810	2,792	2,774	2,756	2,738
9	2,721	2,704	2,687	2,670	2,653	2,636	2,620	2,604	2,588	2,572
20	2,556	2,540	2,525	2,509	2,494	2,479	2,464	2,449	2,434	2,420
1	2,405	2.391	2,376	2,362	2,348	2,335	2,321	2,307	2,294	2,280
2	2,267	2,254	2,241	2,228	2,215	2,202	2,189	2,177	2,164	2,152
3	2,140	2,128	2,116	2,104	2,092	2,080	2,068	2,056	2,045	2,034
4	2,022	2,011	2,000	1,989	1,978	1,967	1,956	1,945	1,934	1,924
5	1,913	1,903	1,892	1,882	1,872	1,861	1,851	1,841	1,831	1,821
6	1,812	1,802	1,792	1,782	1,773	1,763	1,754	1,745	1,735	1,726
7	1,717	1,708	1,699	1,690	1,681	1,672	1,663	1,654	1,645	1,637
8	1,628	1,619	1,611	1,602	1,594	1,586	1,577	1,569	1,561	1,553
9	1,545	1,537	1,529	1,521	1,513	1,505	1,497	1,489	1,482	1,474
30	1,466	1,459	1,451	1,444	1,436	1,429	1,421	1,414	1,407	1,400
1	1,392	1,385	1,378	1,371	1,364	1,357	1,350	1,343	1,336	1,329
2	1,323	1,316	1,309	1,302	1,296	1,289	1,282	1,276	1,269	1,263
3	1,256	1,250	1,244	1,237	1,231	1,225	1,218	1,212	1,206	1,200
4	1,194	1,188	1,182	1,176	1,170	1,164	1,158	1,152	1,146	1,140
5	1,134	1,128	1,123	1,117	1,111	1,106	1,100	1,094	1,088	1,083
6	1,078	1,072	1,067	1,061	1,056	1,050	1,045	1,040	1,034	1,029
7	1,024	1,019	1,013	1,008	1,003	0,998	0,993	0,988	0,982	0,977
8	0,972	0,967	0,962	0,958	0,953	0,948	0,943	0,938	0,933	0,928
9	0,924	0,919	0,914	0,909	0,905	0,900	0,895	0,891	0.886	0,881

θ°	.0	.1	.2	.3	.4	.5	.6	.7	.8	.9
40	0,877	0,872	0,868	0,863	0,859	0,854	0,850	0,845	0,841	0,837
1	0,832	0,828	0,823	0,819	0,815	0,810	0,806	0,802	0,798	0,794
2	0,789	0,785	0,781	0,777	0,773	0,769	0,765	0,761	0,757	0,753
3	0,749	0,745	0,741	0,737	0,733	0,729	0,725	0,721	0,717	0,713
4	0,709	0,706	0,702	0,698	0,694	0,690	0,687	0,683	0,679	0,676
5	0,672	0,668	0,665	0,661	0,657	0,654	0,650	0,647	0,643	0,640
6	0,636	0,632	0,629	0,625	0,622	0,619	0,615	0,612	0,608	0,605
7	0,602	0,598	0,595	0,591	0,588	0,585	0,582	0,578	0,575	0,572
8	0,569	0,565	0,562	0,559	0,556	0,553	0,549	0,546	0,543	0,540
9	0,537	0,534	0,531	0,528	0,525	0,522	0,518	0,515	0,512	0,509
50	0,506	0,504	0,501	0,498	0,495	0,492	0,489	0,486	0,483	0,480
1	0,477	0,474	0,472	0,469	0,466	0,463	0,460	0,458	0,455	0,452
2	0,449	0,447	0,444	0,441	0,439	0,436	0,433	0,430	0,428	0,425
3	0,423	0,420	0,417	0,415	0,412	0,410	0,407	0,404	0,402	0,399
4	0,397	0,394	0,392	0,389	0,387	0,384	0,382	0,379	0,377	0,375
5	0,372	0,370	0,367	0,365	0,363	0,360	0,358	0,356	0,353	0,351
6	0,349	0,346	0,344	0,342	0,339	0,337	0,335	0,333	0,330	0,328
7	0,326	0,324	0,322	0,319	0,317	0,315	0,313	0,311	0,309	0,306
8	0,304	0,302	0,300	0,298	0,296	0,294	0,292	0,290	0,288	0,286
9	0,284	0,282	0,280	0,278	0,276	0,274	0,272	0,270	0,268	0,266
60	0,264	0,262	0,260	0,258	0,256	0,254	0,252	0,250	0,249	0,247
1	0,245	0,243	0,241	0,239	0,237	0,236	0,234	0,232	0,230	0,229
2	0,227	0,225	0,223	0,221	0,220	0,218	0,216	0,215	0,213	0,211
3	0,209	0,208	0,206	0,204	0,203	0,201	0,199	0,198	0,196	0,195
4	0,193	0,191	0,190	0,188	0,187	0,185	0,184	0,182	0,180	0,179
5	0,177	0,176	0,174	0,173	0,171	0,170	0,168	0,167	0,165	0,164
6	0,162	0,161	0,160	0,158	0,157	0,155	0,154	0,152	0,151	0,150
7	0,148	0,147	0,146	0,144	0,143	0,141	0,140	0,139	0,138	0,136
8	0,135	0,134	0,132	0,131	0,130	0,128	0,127	0,126	0,125	0,123
9	0,122	0,121	0,120	0,119	0,117	0,116	0,115	0,114	0,112	0,111
70	0,110	0,109	0,108	0,107	0,106	0,104	0,103	0,102	0,101	0,100
1	0,099	0,098	0,097	0,096	0,095	0,094	0,092	0,091	0,090	0,089
2	0,088	0,087	0,086	0,085	0,084	0,083	0,082	0,081	0,080	0,079
3	0,078	0,077	0,076	0,075	0,075	0,074	0,073	0,072	0,071	0,070
4	0,069	0,068	0,067	0,066	0,065	0,065	0,064	0,063	0,062	0,061
5	0,060	0,059	0,059	0,058	0,057	0,056	0,055	0,055	0,054	0,053
6	0,052	0,052	0,051	0,050	0,049	0,048	0,048	0,047	0,046	0,045
7	0,045	0,044	0,043	0,043	0,042	0,041	0,041	0,040	0,039	0,039
8	0,038	0,037	0,037	0,036	0,035	0,035	0,034	0,034	0,033	0,032
9	0,032	0,031	0,031	0,030	0,029	0,029	0,028	0,028	0,027	0,027
80	0,026	0,026	0,025	0,025	0,024	0,023	0,023	0,023	0,022	0,022
1	0,021	0,021	0,020	0,020	0,019	0,019	0,018	0,018	0,017	0,017
2	0,017	0,016	0,016	0,015	0,015	0,015	0,014	0,014	0,013	0,013
3	0,013	0,012	0,012	0,012	0,011	0,011	0,010	0,010	0,010	0,010
4	0,009	0,009	0,009	0,008	0,008	0,008	0,007	0,007	0,007	0,007
5	0,006	0,006	0,006	0,006	0,005	0,005	0,005	0,005	0,005	0,004
6	0,004	0,004	0,004	0,003	0,003	0,003	0,003	0,003	0,003	0,002
7	0,002	0,002	0,002	0,002	0,002	0,002	0,001	0,001	0,001	0,001
8	0,001	0,001	0,001	0,001	0,001	0,001	0,001	0,000	0,000	0,000

6-8c. Function ϑ tg ϑ

ϑ°	0,0	0,1	0,2	0,3	0,4	0,5	0,6	0,7	0,8	0,9
0	0,0000	0,0000	0,0000	0,0000	0,0000	0,0001	0,0001	0,0002	0,0002	0,0002
1	.0003	.0004	.0004	.0005	.0006	.0007	.0008	.0009	.0010	.0011
2	.0012	.0013	.0015	.0016	.0018	.0019	.0021	.0022	.0024	.0026
3	.0027	.0029	.0031	.0033	.0035	.0037	.0040	.0042	.0044	.0046
4	.0049	.0051	.0054	.0056	.0059	.0062	.0065	.0067	.0070	.0073
5	.0076	.0079	.0083	.0086	.0089	.0092	.0096	.0099	.0103	.0106
6	.0110	.0114	.0118	.0121	.0125	.0129	.0133	.0137	.0142	.0146
7	.0150	.0154	.0159	.0163	.0168	.0172	.0177	.0182	.0186	.0191
8	.0196	.0201	.0206	.0211	.0216	.0222	.0227	.0232	.0238	.0243
9	.0249	.0254	.0260	.0266	.0272	.0277	.0283	.0289	.0295	.0302
10	.0308	.0314	.0320	.0327	.0333	.0340	.0346	.0353	.0360	.0366
11	.0373	.0380	.0387	0394	.0401	.0408	.0416	.0423	.0430	.0438
12	.0445	.0453	.0460	.0468	.0476	.0484	.0492	.0500	.0508	.0516
13	.0524	.0532	.0540	.0549	.0557	.0566	.0574	.0583	.0592	.0600
14	.0609	.0618	.0627	.0636	.0645	.0654	.0664	.0673	.0682	.0692
15	.0701	.0711	.0721	.0731	.0740	.0750	.0760	.0770	.0780	.0791
16	.0801	.0811	.0821	.0832	.0842	.0853	.0864	.0874	.0885	.0896
17	.0907	.0918	.0929	.0940	.0952	.0963	.0974	.0986	.0997	.1009
18	.1021	.1033	.1044	.1056	.1068	.1080	.1093	.1105	.1117	.1129
19	.1142	.1154	.1167	.1180	.1192	.1205	.1218	.1231	.1244	.1257
20	.1270	.1284	.1297	.1311	.1324	.1338	.1351	.1365	.1379	.1393
21	.1407	.1421	.1435	.1449	.1464	.1478	.1493	.1507	.1522	.1537
22	.1551	.1566	.1581	.1596	.1611	.1627	.1642	.1657	.1673	.1668
23	.1704	.1720	.1735	.1751	.1767	.1783	.1800	.1816	.1832	.1848
24	.1865	.1882	.1898	.1915	.1932	.1949	.1966	.1983	.2000	.2017
25	.2035	.2052	.2070	.2087	.2105	.2123	.2141	.2159	.2177	.2195
26	.2213	.2232	.2250	.2269	.2287	.2306	.2325	.2344	.2363	.2382
27	.2401	.2420	.2440	.2459	.2479	.2499	.2518	.2538	.2558	.2578
28	.2598	.2619	.2639	.2660	.2680	.2701	.2722	.2742	.2763	.2784
29	.2806	.2827	.2848	.2870	.2891	.2913	.2935	.2957	.2979	.3001
30	.3023	.3045	.3068	.3090	.3113	.3136	.3158	.3181	.3205	.3228
31	.3251	.3274	.3298	.3321	.3345	.3369	.3393	.3417	.3441	.3466
32	.3490	.3514	.3539	.3564	.3589	.3614	.3639	.3664	.3689	.3715
33	.3740	.3766	.3792	.3818	.3844	.3870	.3896	.3923	.3949	.3976
34	.4003	.4030	.4057	.4084	.4111	.4138	.4166	.4194	.4221	.4249
35	.4277	.4306	.4334	.4362	.4391	.4420	.4448	.4477	.4506	.4536
36	.4565	.4595	.4624	.4654	.4684	.4714	.4744	.4774	.4805	.4835
37	.4866	.4897	.4928	.4959	.4991	.5022	.5054	.5086	.5117	.5149
38	.5182	.5214	.5247	.5279	.5312	.5345	.5378	.5411	.5445	.5478
39	.5512	.5546	.5580	.5614	.5649	.5683	.5718	.5753	.5788	.5823
40	.5858	.5894	.5929	.5965	.6001	.6037	.6073	.6110	.6147	.6183
41	.6220	.6258	.6295	.6333	.6370	.6408	.6446	.6484	.6523	.6562
42	.6600	.6639	.6678	.6718	.6757	.6797	.6837	.6877	.6917	.6958
43	.6998	.7039	.7080	.7122	.7163	.7205	.7247	.7289	.7331	.7373
44	.7416	.7459	.7502	.7545	.7589	.7632	.7676	.7720	.7765	.7809
45	.7854	.7899	.7944	.7990	.8035	.8081	.8127	.8173	.8220	.8267

6-9. Lattice Constants of Some Standard Substances

The table gives the lattice constants and thermal-expansion coefficients for some substances used as standards in precision lattice-constant determinations.

Several values are given for some substances; these reflect the effects of purity on the lattice constants.

Substance	Purity, %	Lattice constant, Å	Coeff., $deg^{-1} \cdot 10^6$	Source
Al	99,971	4,04958	23,29	[143]
	99,99	4,04958	—	[144]
	99,992	4,04953	—	[145]
	99,9986	4,04963	—	[146]
Ag	99,999	4,08613	18,72	[143]
		4,08610	—	[148]
Au	99,998	—	14,13	[147]
Si	99,84	5,43078	4,15	[143]
		5,43078	—	[148]
	99,9	5,43075	—	[149]
	99,97	5,43100	—	[150]
Ge	99,999	5,65758	5,92	[150]
NaCl	—	5,64009	40,49	[146]
TlCl	99,999	3,84236	54,57	[148]
TlBr	99,999	3,98584	51,2	[148]
CaF_2	99,999	5,426	—	[151]
Cs I	Traces: 0,01% Na 0,05% Rb	4,5678	48,6	[152]

6-10. Bragg Angles for Some Standard Substances

Values are given for $\sin^2 \vartheta$ and ϑ for some standard substances for various radiations and lines [11, 102, 109].

The tables can be used in the method in which a standard is employed, in measurements on macroscopic stresses, in adjusting apparatus, and so on.

6-10a. NaCl

(hkl)	K line	Cr		Fe		Co		Cu		Mo	
		$\sin^2 \vartheta$	$\vartheta°$	$\sin^2 \vartheta$	$\vartheta°$	$\sin^2 \vartheta$	$\vartheta°$	$\sin^2 \vartheta$	$\vartheta°$	$\sin^2 \vartheta$	$\vartheta°$
(111)	α_1	0,12364	20,59	0,08840	17,30	0,07548	15,95	0,05597	13,69	0,01187	6,26
	α_2	0,12406	20,62	0,08876	17,33	0,07580	15,98	0,05625	13,72	0,01201	6,29
	α	0,12378	20,60	0,08852	17,31	0,07559	15,96	0,05606	13,70	0,01191	6,27
(200)	α_1	0,16485	23,96	0,11786	20,08	0,10064	18,50	0,07463	15,85	0,01582	7,23
	α_2	0,16541	24,00	0,11834	20,12	0,10107	18,54	0,07500	15,89	0,01601	7,27
	α	0,16504	23,97	0,11802	20,09	0,10078	18,51	0,07475	15,87	0,01588	7,24
(220)	α_1	0,32970	35,04	0,2357?	29,05	0,20127	26,66	0,14926	22,73	0,03164	10,25
	α_2	0,33082	35,11	0,23668	29,11	0,20214	26,72	0,15000	22,79	0,03202	10,31
	α	0,33008	35,07	0,23604	29,07	0,20156	26,68	0,14950	22,75	0,03177	10,27
(113)	α_1	0,45334	42,32	0,32412	34,70	0,27675	31,74	0,20523	26,94	0,04351	12,04
	α_2	0,45488	42,41	0,32544	34,78	0,27795	31,82	0,20625	27,01	0,04402	12,11
	α	0,45386	42,35	0,32456	34,73	0,27715	31,77	0,20556	26,96	0,04368	12,06
(222)	α_1	0,49456	44,69	0,35358	36,49	0,30191	33,33	0,22388	28,24	0,04746	12,58
	α_2	0,49624	44,78	0,35502	36,57	0,30322	33,41	0,22500	28,32	0,04802	12,66
	α	0,49512	44,72	0,35406	36,51	0,30234	33,36	0,22424	28,26	0,04765	12,61
(400)	α_1	0,65941	54,30	0,47144	43,36	0,40254	39,38	0,29851	33,12	0,06328	14,57
	α_2	0,66165	54,43	0,47336	43,47	0,40429	39,48	0,30000	33,21	0,06403	14,66
	α	0,66016	54,34	0,47208	43,40	0,40312	39,41	0,29899	33,15	0,06354	14,60
(133)	α_1	0,78305	62,24	0,55984	48,44	0,47802	43,74	0,35448	36,54	0,07515	15,91
	α_2	0,78571	62,42	0,56212	48,57	0,48009	43,86	0,35625	36,65	0,07604	16,01
	α	0,78394	62,30	0,56060	48,48	0,47871	43,78	0,35505	36,57	0,07545	15,94
(042)	α_1	0,82426	65,22	0,58930	50,14	0,50318	45,18	0,37314	37,65	0,07910	16,33
	α_2	0,82706	65,43	0,59170	50,28	0,50536	45,31	0,37500	37,76	0,08004	16,43
	α	0,82520	65,29	0,59010	50,19	0,50390	45,22	0,37374	37,69	0,07942	16,37

6-10b. Ag (a = 4.078 Å)

(hkl)	K line	Cr		Fe		Cu		Rel. intens.
		sin² θ	θ°	sin² θ	θ°	sin² θ	θ°	
(111)	β	0,19530	26,23	0,13860	21,24	0,08700	17,20	1,8
(111)	α	0,23589	27,19	0,16869	23,62	0,10683	19,08	11,3
(200)	β	0,26040	30,02	0,18480	25,46	0,11600	19,90	1
(200)	α	0,31452	33,42	0,22492	27,66	0,14244	22,17	6,4
(220)	β	0,52080	46,19	0,36960	37,44	0,23200	28,79	0,8
(220)	α	0,62904	52,48	0,44984	40,21	0,28488	31,59	5,1
(311)	β	0,71610	57,80	0,50820	45,47	0,31900	34,40	0,9
(222)	β	0,78120	62,10	0,55440	48,13	0,34800	36,16	0,3
(311)	α	0,86493	68,44	0,61853	51,85	0,39171	38,75	6,5
(222)	α	0,94356	74,03	0,67476	54,24	0,42732	40,82	1,8
(400)	β	—	—	0,73920	59,29	0,46400	42,95	0,01
(331)	β	—	—	0,87680	69,46	0,55100	47,94	0,4
(400)	α	—	—	0,89968	71,53	0,56976	49,01	1,0
(420)	β	—	—	0,92400	72,03	0,58000	49,61	0,4
(331)	α	—	—	—	—	0,67659	55,35	4,1
(422)	β	—	—	—	—	0,69600	56,53	0,4
(420)	α	—	—	—	—	0,71220	57,51	4,1
(511) } (333)	β	—	—	—	—	0,78300	62,23	0,6
(422)	α	—	—	—	—	0,86464	68,41	6,0
(440)	β	—	—	—	—	0,92800	74,43	0,3
(511) } (333)	α	—	—	—	—	0,96147	78,68	14,4

6-10c. Au (a = 4.070 Å)

(hkl)	K line	Cr		Fe		Co		Cu		Mo	
		sin² θ	θ°	sin² θ	θ°	sin² θ	θ°	sin² θ	θ°	sin² θ	θ°
(111)	α₁	0,23638	29,09	0,16900	24,27	0,14430	22,33	0,10701	19,09	0,02268	8,66
	α₂	0,23718	29,15	0,16969	24,33	0,14492	22,38	0,10754	19,14	0,02296	8,72
	α	0,23665	29,11	0,16922	24,29	0,14451	22,34	0,10718	19,11	0,02277	8,68
(200)	α₁	0,31517	34,15	0,22533	28,34	0,19240	26,02	0,14268	22,19	0,03024	10,02
	α₂	0,31624	34,22	0,22625	28,40	0,19323	26,08	0,14338	22,25	0,03061	10,08
	α	0,31553	34,18	0,22563	28,36	0,19268	26,04	0,14291	22,21	0,03036	10,04
(220)	α₁	0,63034	52,56	0,45066	42,17	0,38480	38,34	0,28535	32,29	0,06049	14,24
	α₂	0,63249	52,69	0,45250	42,27	0,38646	38,44	0,28677	32,38	0,06122	14,33
	α	0,63106	52,60	0,45126	42,20	0,38535	38,37	0,28582	32,32	0,06073	14,27
(113)	α₁	0,86671	68,59	0,61965	51,92	0,52910	46,67	0,39236	38,78	0,08317	16,76
	α₂	0,86967	68,84	0,62218	51,07	0,53139	46,80	0,39431	38,90	0,08417	16,86
	α	0,86771	68,67	0,62049	51,97	0,52986	46,71	0,39300	38,82	0,08350	16,80
(222)	α₁	0,94550	76,50	0,67598	55,30	0,57720	49,44	0,42803	40,86	0,09073	17,53
	α₂	0,94873	76,91	0,67874	55,47	0,57970	49,59	0,43015	40,98	0,09182	17,64
	α	0,94660	76,64	0,67690	55,36	0,57803	49,49	0,42872	40,90	0,09109	17,57
(400)	α₁			0,90131	71,69	0,76960	61,32	0,57070	49,06	0,12098	20,35
	α₂			0,90499	72,05	0,77293	61,54	0,57354	49,23	0,12243	20,48
	α			0,90253	71,81	0,77070	61,39	0,57163	49,12	0,12146	20,40
(133)	α₁					0,91390	72,94	0,67771	55,41	0,14366	22,27
	α₂					0,91785	73,34	0,68107	55,62	0,14539	22,41
	α					0,91521	73,07	0,67881	55,48	0,14423	22,32
(142)	α₁					0,96200	78,76	0,71338	57,63	0,15122	22,88
	α₂					0,96616	79,40	0,71692	57,86	0,15304	23,03
	α					0,96338	78,97	0,71454	57,70	0,15182	22,93

6-10d. Polycrystalline Quartz

(hkl)	2θ°		(hkl)	2θ°	
	CuKα₁	CuKα₂		CuKα₁	CuKα₂
(100)	20,88	—	(132)	—	90,83
(101)	26,66	—	(231)	—	106,61
(110)	36,58	36,55	(134)	—	120,13
(200)	42,49	42,45	(116)	—	131,23
(112)	50,19	50,14	(143)	—	137,90
(121)	60,02	59,96	(240)	—	146,63
(302)	75,74	75,66	(134)	—	153,54

6-10e. Monocrystalline Quartz Cut to (10$\bar{1}$1)

Anode	2θ°			Order	Anode	2θ°			Order
	Kα₁	Kα₂	Kβ			Kα₁	Kα₂	Kβ	
Cu	26,64	26,70	24,04	1	Ni	28,72	28,79	25,94	1
	54,87	55,04	49,20	2		59,47	59,62	53,34	2
	87,45	87,75	77,29	3		96,14	96,44	84,62	3
Fe	33,67	33,74	33,47	1	Cr	40,07	40,14	36,20	1
	70,79	70,97	63,40	2		86,47	86,65	77,09	2
	120,64	121,04	104,04	3					
Co	31,04	31,10	28,07	1					
	64,70	64,87	58,04	2					
	106,64	107,14	93,34	3					

6-10f. Some Lines of Al, Cr, Au, Ag, and W

The table gives Bragg angles for standards often used in measurements on macro-scopic stresses (Al, Cr, Au, Ag, W) with various radiations [111].

Anode	λKα₁, Å	Al $a=4,0414$ Å		Cr $a=2,8786$ Å		Au $a=4,0700$ Å		Ag $a=4,0783$ Å		W $a=3,1577$ Å	
		(hkl)	θ°	(hkl)	θ°	(hkl)	θ°	(hkl)	θ°	(hkl)	θ°
Cu	1,537396	(333)	81,24	(213)	87,65	(333)	78,43	(333)	78,35	(004)	76,84
Co	1,785287	(024)	81,03	(013)	78,70	(024)	78,77	(024)	78,20	(222)	78,31
Fe	1,932076	(004)	72,97	—		—		—		—	
Cr	2,285033	(222)	78,32	(112)	76,46	(222)	76,51	(222)	76,04	(112)	62,41

6-11. Factors for Reducing $\sin^2 \vartheta$ to a Single Wavelength

One of the steps in a least-squares determination is that of reducing the $\sin^2 \vartheta$ for each line to a standard wavelength; the $\sin^2 \vartheta(\lambda_1)$ must be converted to $\sin^2 \vartheta(\lambda)$.

The table gives the conversion factors for the α_1, α_2, and β_1 lines of various radiations, as well as $\lambda^2/3$ and $\lambda^2/4$ [310].

E x a m p l e . The $\sin^2 \vartheta$ recorded with Fe Kβ_1 radiation are reduced to those for Fe Kα_1 by multiplying the $\sin^2 \vartheta(\lambda_\beta)$ by 0.823216.

λ	CrKα_1	MnKα_1	FeKα_1	CoKα_1	NiKα_1	CuKα_1	ZnKα_1	GaKα_1	GeKα_1	MoKα_1	AgKα_1
CrKα_1	1	$1,18676_5$	$1,39871_6$	$1,63811_7$	$1,90739_9$	$2,20901_0$	$2,54540_2$	$2,91942_8$	$3,33568_7$	$10,421_2$	$16,755_0$
α_2	$1,00340_1$	$1,19080_1$	$1,40347_3$	$1,64368_8$	$1,91388_6$	$2,21652_3$	$2,55405_9$	$2,92935_7$	$3,34502_4$	$10,456_5$	$16,812_0$
β_1	$0,829090_6$	$0,983935_6$	$1,15966_2$	$1,35814_8$	$1,58140_7$	$1,83147_0$	$2,11036_9$	$2,42047_0$	$2,76392_8$	$8,6400_8$	$13,891_4$
MnKα_1	$0,842626_9$	1	$1,17859_6$	$1,38032_2$	$1,60722_6$	$1,86137_2$	$2,14482_4$	$2,45998_8$	$2,80905_4$	$8,7811_5$	$14,118_2$
α_2	$0,845789_0$	$1,00375_3$	$1,18301_9$	$1,38550_2$	$1,61325_7$	$1,86835_7$	$2,15287_3$	$2,46922_0$	$2,81959_6$	$8,8141_0$	$14,171_2$
β_1	$0,695998_2$	$0,825986_3$	$0,973503_9$	$1,14012_7$	$1,32754_6$	$1,53746_7$	$1,77159_5$	$2,03191_7$	$2,32024_0$	$7,2531_1$	$11,661_4$
FeKα_1	$0,714941_4$	$0,848467_3$	1	$1,17111_5$	$1,36367_9$	$1,57931_3$	$1,81981_3$	$2,08722_0$	$2,38339_1$	$7,4505_2$	$11,978_8$
α_2	$0,717854_4$	$0,851924_4$	$1,00407_4$	$1,17593_0$	$1,36923_5$	$1,58574_8$	$1,82722_8$	$2,09572_4$	$2,39310_2$	$7,4808_7$	$12,027_6$
β_1	$0,588551_3$	$0,698472_0$	$0,823216_2$	$0,964116_0$	$1,12260_2$	$1,30011_6$	$1,49810_0$	$1,71823_3$	$1,96204_6$	$6,1333_8$	$9,8611_7$
CoKα_1	$0,610456_9$	$0,724468_9$	$0,853855_9$	1	$1,16438_5$	$1,34850_6$	$1,55385_8$	$1,78218_5$	$2,03507_2$	$6,3616_7$	$10,228_2$
α_2	$0,613094_2$	$0,727598_6$	$0,857544_7$	$1,00432_0$	$1,16941_5$	$1,38433_1$	$1,56057_1$	$1,78988_4$	$2,04386_4$	$6,3891_5$	$10,272_4$
β_1	$0,501077_9$	$0,594661_7$	$0,700865_7$	$0,820824_3$	$0,955755_6$	$1,10688_6$	$1,27544_5$	$1,46286_1$	$1,67043_7$	$5,2218_1$	$8,3955_5$
NiKα_1	$0,524274_1$	$0,622190_1$	$0,733310_6$	$0,858822_5$	1	$1,15812_7$	$1,33448_9$	$1,53058_0$	$1,74776_6$	$5,4635_4$	$8,7842_0$
α_2	$0,526712_0$	$0,625083_3$	$0,736720_5$	$0,862816_0$	$1,00465_0$	$1,16351_2$	$1,34069_4$	$1,53769_8$	$1,75589_3$	$5,4889_5$	$8,8250_5$
β_1	$0,429253_3$	$0,509422_7$	$0,600403_4$	$0,703167_2$	$0,818757_3$	$0,948224_9$	$1,09262_2$	$1,25317_4$	$1,43099_6$	$4,4733_1$	$7,1921_3$
CuKα_1	$0,452691_4$	$0,537238_3$	$0,633186_7$	$0,741561_6$	$0,863463_2$	1	$1,15228_2$	$1,32160_0$	$1,50913_1$	$4,7175_7$	$7,5848_4$
α_2	$0,454939_2$	$0,539905_9$	$0,636330_8$	$0,745243_3$	$0,867750_7$	$1,00496_6$	$1,15800_3$	$1,32816_2$	$1,51662_5$	$4,7409_9$	$7,6225_0$
β_1	$0,369707_0$	$0,438755_4$	$0,517115_2$	$0,605623_5$	$0,705178_9$	$0,816686_7$	$0,941053_2$	$1,07933_3$	$1,23248_8$	$3,8527_7$	$6,1944_4$
ZnKα_1	$0,392865_2$	$0,466238_6$	$0,549506_9$	$0,643159_3$	$0,749350_8$	$0,867843_3$	1	$1,14694_2$	$1,30969_0$	$4,0941_1$	$6,5824_5$
α_2	$0,394964_9$	$0,468730_5$	$0,552443_8$	$0,646998_9$	$0,753355_8$	$0,872481_7$	$1,00534_5$	$1,15307_7$	$1,31668_9$	$4,1159_9$	$6,6176_3$
β_1	$0,320007_6$	$0,379773_8$	$0,447599_8$	$0,524209_9$	$0,610382_2$	$0,706900_1$	$0,814548_0$	$0,934239_1$	$1,06680_5$	$3,3348_5$	$5,3617_2$
GaKα_1	$0,342532_8$	$0,406506_0$	$0,479106_2$	$0,561109_0$	$0,653346_9$	$0,756658_6$	$0,871883_9$	1	$1,14189_7$	$3,5695_9$	$5,7391_3$
α_2	$0,344534_7$	$0,408817_7$	$0,481906_2$	$0,564388_2$	$0,657155_2$	$0,761080_7$	$0,876979_4$	$1,00584_4$	$1,14857_1$	$3,5904_5$	$5,7726_7$
β_1	$0,278286_4$	$0,330260_6$	$0,389243_7$	$0,455865_8$	$0,530803_3$	$0,614737_6$	$0,708350_9$	$0,812437_1$	$0,927719_8$	$2,9000_7$	$4,6626_8$
GeKα_1	$0,299968_2$	$0,355991_7$	$0,419570_3$	$0,491383_0$	$0,572159_0$	$0,662632_8$	$0,763539_7$	$0,875735_5$	1	$3,1260_1$	$5,0259_6$
α_2	$0,301865_2$	$0,358243_6$	$0,422224_4$	$0,494491_4$	$0,575778_4$	$0,666824_5$	$0,768369_6$	$0,881275_1$	$1,00632_6$	$3,1457_9$	$5,0577_5$
β_1	$0,243099_5$	$0,288502_0$	$0,340027_2$	$0,398225_5$	$0,463687_9$	$0,537009_4$	$0,618786_1$	$0,709711_6$	$0,810417_8$	$2,5333_8$	$4,0731_3$
MoKα_1	$0,095958_6$	$0,11388_0$	$0,13421_9$	$0,15719_2$	$0,18303_1$	$0,21197_4$	$0,24425_3$	$0,28014_4$	$0,31989_6$	1	$1,6077_9$
α_2	$0,097120_3$	$0,11525_9$	$0,13584_4$	$0,15909_4$	$0,18524_7$	$0,21454_0$	$0,24721_0$	$0,28353_6$	$0,32376_9$	$1,0121_1$	$1,6272_5$
β_1	$0,076251_9$	$0,090493_1$	$0,10665_5$	$0,12491_0$	$0,14544_3$	$0,16844_1$	$0,19400_2$	$0,22261_2$	$0,25420_0$	$0,79463_3$	$1,2776_0$
AgKα_1	$0,059683_7$	$0,070830_6$	$0,083480_6$	$0,097769_0$	$0,11384_7$	$0,13184_7$	$0,15191_9$	$0,17424_2$	$0,19896_7$	$0,62197_3$	1
α_2	$0,060630_7$	$0,071954_4$	$0,084805_1$	$0,099320_2$	$0,11564_4$	$0,13393_4$	$0,15432_9$	$0,17700_7$	$0,20212_4$	$0,63184_2$	$1,0158_2$
β_1	$0,047119_8$	$0,055920_1$	$0,065907_2$	$0,077187_8$	$0,089876_3$	$0,10408_8$	$0,11993_9$	$0,13756_3$	$0,15708_3$	$0,49104_3$	$0,78949_1$
$\lambda^2/3$	$1,74745_3$	$1,47245_1$	$1,24932_2$	$1,06674_5$	$0,916144_5$	$0,791057_0$	$0,686513_6$	$0,598560_1$	$0,524180_4$	$0,16768_3$	$0,10429_5$
$\lambda^2/4$	$1,31059_0$	$1,10433_8$	$0,936995_0$	$0,800058_7$	$0,687108_4$	$0,593292_8$	$0,514885_2$	$0,448920_1$	$0,393135_3$	$0,12576_2$	$0,078220_9$

6-12. Correction for Refraction

The Wulff-Bragg law as corrected for refraction shift is

$$2d \left(1 - \frac{\delta}{\sin^2 \vartheta} \right) \sin \vartheta = n\lambda,$$

where δ is related to the refractive index R by $\delta = 1 - R$.

The relation for δ for cubic crystals is

$$\delta = \frac{N_0 e^2 \lambda^2}{2\pi mc^2} \varrho \frac{\sum Z}{\sum A} = 2.71 \cdot 10^{-6} \lambda^2 \varrho \frac{\sum Z}{\sum A},$$

where ρ is the density, ΣZ is the sum of the charges, and ΣA is the sum of the atomic weights for all atoms in the unit cell.

The lattice constant is corrected for refraction in precision measurements using

$$a = a_m \quad (1 + \delta),$$

where a_m is the measured constant before correction.

The table gives values of $\delta \cdot 10^6$ for various radiations and $\rho(\Sigma Z)/(\Sigma A)$ [102].

$\rho \dfrac{\sum Z}{\sum A}$ $\diagdown$ λ	V K_α	Cr K_α	Mn K_α	Fe K_α	Co K_α	Ni K_α	Cu K_α	Mo K_α	Ag $K\alpha$	W $K\alpha$
0,1	1,84	1,42	1,20	1,02	0,87	0,75	0,64	0,14	0,09	0,01
2	3,63	2,84	2,40	2,03	1,74	1,49	1,29	0,27	0,17	0,02
3	5,52	4,27	3,60	3,05	2,61	2,24	1,93	0,42	0,26	0,04
4	7,37	5,69	4,79	4,07	3,47	2,98	2,58	0,56	0,34	0,05
5	9,21	7,11	5,99	5,09	4,34	3,73	3,22	0,68	0,43	0,06
6	11,05	8,53	7,19	6,10	5,21	4,48	3,87	0,82	0,51	0,07
7	12,89	9,96	8,39	7,12	6,08	5,22	4,51	0,96	0,60	0,08
8	14,73	11,38	9,59	8,14	6,95	5,97	5,15	1,10	0,68	0,10
9	16,57	12,80	10,79	9,15	7,82	6,71	5,80	1,23	0,77	0,11
1,0	18,42	14,22	11,99	10,17	8,69	7,46	6,44	1,37	0,85	0,12
1	20,25	15,65	13,18	11,19	9,55	8,21	7,09	1,51	0,94	0,13
2	22,10	17,07	14,38	12,20	10,42	8,95	7,73	1,64	1,02	0,14
3	23,94	18,49	15,58	13,22	11,29	9,70	8,37	1,78	1,11	0,16
4	25,78	19,91	16,78	14,24	12,16	10,44	9,02	1,92	1,19	0,17
5	27,62	21,33	17,98	15,26	13,03	11,19	9,66	2,05	1,28	0,18
6	29,46	22,76	19,18	16,27	13,90	11,94	10,31	2,19	1,36	0,19
7	31,30	24,18	20,38	17,29	14,76	12,68	10,95	2,33	1,45	0,20
8	33,14	25,60	21,57	18,31	15,63	13,43	11,60	2,46	1,53	0,22
9	34,98	27,02	22,77	19,32	16,50	14,17	12,24	2,60	1,62	0,23
2,0	36,83	28,45	23,97	20,34	17,37	14,92	12,88	2,74	1,70	0,24
1	38,67	29,87	25,17	21,36	18,24	15,67	13,53	2,87	1,79	0,25
2	40,51	31,29	26,37	22,38	19,11	16,41	14,17	3,01	1,87	0,26
2	42,35	32,71	27,57	23,39	19,98	17,16	14,82	3,15	1,96	0,28
3	44,19	34,13	28,77	24,41	20,84	17,90	15,46	3,29	2,04	0,29
4	46,03	35,56	29,97	25,43	21,71	18,65	16,11	3,42	2,13	0,30
6	47,87	36,98	31,16	26,44	22,58	19,40	16,75	3,56	2,22	0,31
7	49,71	38,40	32,36	27,46	23,45	20,14	17,39	3,70	2,30	0,32
8	51,56	39,82	33,56	28,48	24,35	20,89	18,04	3,83	2,39	0,34
9	53,40	41,25	34,76	29,50	25,19	21,63	18,68	3,97	2,47	0,35
3,0	55,24	42,67	35,96	30,51	26,06	22,38	19,33	4,11	2,56	0,36
1	57,08	44,09	37,16	31,53	26,92	23,13	19,97	4,24	2,64	0,37
2	58,92	45,51	38,36	32,55	27,79	23,87	20,61	4,38	2,73	0,38
3	60,76	46,94	39,55	33,56	28,66	24,62	21,26	4,52	2,81	0,40
4	62,60	48,36	40,75	34,58	29,53	25,36	21,90	4,65	2,90	0,41

$\rho \frac{\Sigma Z}{\Sigma A}$ λ	V K_α	Cr K_α	Mn K_α	Fe K_α	Co K_α	Ni K_α	Cu K_α	Mo K_α	Ag K_α	W K_α
5	64,45	49,78	41,95	35,60	30,40	26,11	22,55	4,79	2,98	0,42
6	66,29	51,20	43,15	36,61	31,27	26,86	23,19	4,93	3,07	0,43
7	68,13	52,62	44,35	37,63	32,14	27,60	23,84	5,07	3,15	0,44
8	69,97	54,05	45,55	38,65	33,00	28,35	24,48	5,20	3,24	0,46
9	71,81	55,47	46,75	39,67	33,87	29,09	25,12	5,34	3,32	0,47
4,0	73,65	56,89	47,94	40,68	34,74	29,84	25,77	5,48	3,41	0,48
1	75,49	58,31	49,14	41,70	35,61	30,59	26,41	5,61	3,49	0,49
2	77,33	59,74	50,34	42,72	36,48	31,33	27,06	5,75	3,58	0,50
3	79,18	61,16	51,54	43,73	37,35	32,08	27,70	5,89	3,66	0,52
4	81,02	62,58	52,74	44,75	38,21	32,82	28,34	6,02	3,75	0,53
5	82,86	64,00	53,94	45,77	39,08	33,57	28,99	6,16	3,83	0,54
6	84,70	65,43	55,14	46,79	39,95	34,32	29,63	6,30	3,92	0,55
7	86,54	66,85	56,33	47,80	40,82	35,06	30,28	6,43	4,00	0,56
8	88,38	68,27	57,53	48,82	41,69	35,81	30,92	6,57	4,09	0,58
9	90,22	69,69	58,73	49,84	42,56	36,55	31,57	6,71	4,17	0,59
5,0	92,06	71,11	59,93	50,85	43,43	37,30	32,21	6,85	4,26	0,60
1	93,91	72,54	61,13	51,87	44,29	38,05	32,85	6,98	4,35	0,61
2	95,75	73,96	62,33	52,89	45,16	38,79	33,50	7,12	4,43	0,62
3	97,59	75,38	63,53	53,91	46,03	39,54	34,14	7,26	4,52	0,64
4	99,43	76,80	64,74	54,92	46,90	40,28	34,79	7,39	4,60	0,65
5	101,27	78,23	65,92	55,94	47,77	41,03	35,43	7,53	4,69	0,66
6	103,11	79,65	67,12	56,96	48,64	41,78	36,08	7,67	4,77	0,67
7	104,95	81,07	68,32	57,97	49,51	42,52	36,72	7,80	4,86	0,68
8	106,79	82,49	69,52	58,99	50,37	43,27	37,36	7,94	4,94	0,70
9	108,64	83,92	70,72	60,01	51,24	44,01	38,01	8,08	5,03	0,71
6,0	110,48	85,34	71,92	61,02	52,11	44,76	38,65	8,21	5,11	0,72
1	112,32	86,76	73,11	62,04	52,98	45,51	39,30	8,35	5,20	0,73
2	114,16	88,18	74,31	63,06	53,85	46,25	39,94	8,49	5,28	0,74
3	116,00	89,60	75,51	64,08	54,72	47,00	40,58	8,62	5,37	0,76
4	117,84	91,03	76,71	65,09	55,58	47,74	41,23	8,76	5,45	0,77
5	119,68	92,45	77,91	66,11	56,45	48,49	41,87	8,90	5,54	0,78
6	121,53	93,87	79,11	67,13	57,32	49,24	42,52	9,04	5,62	0,79
7	123,37	95,29	80,31	68,14	58,19	49,98	43,16	9,17	5,71	0,80
8	125,21	96,72	81,50	69,16	59,06	50,73	43,81	9,31	5,79	0,82
9	127,05	98,14	82,70	70,18	59,93	51,44	44,45	9,45	5,88	0,83
7,0	128,89	99,56	83,90	71,20	60,80	52,22	45,09	9,58	5,96	0,84
1	130,73	100,98	85,10	72,21	61,66	52,97	45,74	9,72	6,05	0,85
2	132,57	102,40	86,30	73,23	62,53	53,71	46,38	9,86	6,13	0,86
3	134,41	103,83	87,50	74,25	63,40	54,46	47,03	9,99	6,22	0,88
4	136,26	105,25	88,70	75,26	64,27	55,20	47,67	10,13	6,30	0,89
5	138,10	106,67	89,90	76,28	65,14	55,95	48,32	10,27	6,39	0,90
6	139,94	108,09	91,09	77,30	66,01	56,70	48,96	10,40	6,48	0,91
7	141,78	109,52	92,29	78,32	66,88	57,44	49,60	10,54	6,56	0,92
8	143,62	110,94	93,49	79,33	67,74	58,19	50,25	10,68	6,45	0,94
9	145,46	112,36	94,69	80,50	68,61	58,93	50,89	10,82	6,73	0,95
8,0	147,30	113,78	95,89	81,37	69,48	59,78	51,54	10,95	6,82	0,96
1	149,14	115,21	97,09	82,38	70,35	60,43	52,18	11,09	6,90	0,97
2	150,99	116,63	98,29	83,40	71,22	61,17	52,82	11,23	6,99	0,98
3	152,83	118,05	99,48	84,42	72,09	61,92	53,47	11,36	7,07	1,00
4	154,67	119,47	100,68	85,43	72,96	62,66	54,11	11,50	7,16	1,01
5	156,51	120,89	101,88	86,45	73,82	63,41	54,76	11,64	7,24	1,02
6	158,35	122,32	103,08	87,47	74,69	64,16	55,40	11,77	7,33	1,03
7	160,19	123,74	104,28	88,49	75,56	64,90	56,05	11,91	7,41	1,04
8	162,03	125,16	105,48	89,50	76,43	65,65	56,69	12,05	7,50	1,06
9	163,87	126,58	106,68	90,52	77,30	66,39	57,33	12,18	7,58	1,07
9,0	165,72	128,01	107,87	91,54	78,17	67,14	57,98	12,32	1,08	7,67
1	167,56	129,43	109,07	92,55	79,03	67,89	58,62	12,46	1,09	7,75
2	169,40	130,85	110,27	93,57	79,90	68,63	59,27	12,59	1,10	7,84

$\rho\dfrac{\Sigma z}{\Sigma A}$	λ	V K_α	Cr K_α	Mn K_α	Fe K_α	Co $\dot{K}_x$	Ni K_α	Cu K_α	Mo K_x	Ag K_α	W K_α
3		171,24	132,27	111,47	94,59	80,77	69,38	59,91	12,73	1,12	7,92
4		173,08	133,70	112,67	95,61	81,64	70,12	60,55	12,87	1,13	8,01
5		174,92	135,12	113,87	96,62	82,51	70,87	61,20	13,01	1,14	8,09
6		176,76	136,54	115,07	97,64	83,38	71,62	61,84	13,14	1,15	8,18
7		178,61	137,96	116,26	98,66	84,25	72,36	62,49	13,28	1,16	8,26
8		180,45	139,38	117,46	99,67	85,11	73,11	63,13	13,42	1,18	8,35
9		182,29	140,85	118,66	100,69	85,98	73,85	63,78	13,55	1,19	8,43
10,0		184,13	142,23	119,86	101,71	86,85	74,60	64,42	13,69	1,20	8,52

6-13. Linear-Expansion Coefficients for Some Metals, Alloys, and Other Materials [311]

Substance	$\alpha \times 10^6$ (linear)	Substance	$\alpha \times 10^6$ (linear)
		Metals	
Aluminum	23	Nickel	12.8
Aluminum 0° to 600°C	29	Nickel 0° to 1000°C	18
Antimony	11	Palladium	c. 11
Antimony ‖ axis	16	Palladium 0 to 1000°C	14
Antimony ⊥ axis	8	Platinum	8.9
Bismuth	13	Platinum at 800°C	11
Bismuth ‖ axis	15	Rhodium	8.4
Bismuth ⊥ axis	11	Silver	19
Cadmium	30	Silver 0 to 900°C	20.5
Cadmium ‖ axis	52	Thallium	28
Cadmium ⊥ axis	20	Thallium 0 to 290°C	34
Chromium	c.7	Tantalum	6
Chromium 0 to 900°C	11	Tantalum 0 to 400°C	7
Cobalt	c. 12	Tin	21
Cobalt 25 to 350°C	c. 18	Tin ‖ axis	30.5
Copper	16.7	Tin ⊥ axis	15.5
Copper 0 to 1000°C	20	Titanium	c. 9
Gold	14	Tungsten	4.5
Gold 0 to 500°C	15	Tungsten 600° to 1400°C	6
Iridium	6.5	Tungsten 1400° to 2200°C	7
Iridium 0 to 1750°C	9	Vanadium	c. 8
Iron (electrolytic)	11.7	Zinc (cast)	c. 30
Iron (electrolytic) 0 to 700°C	15	Zinc (crystal; ‖ axis)	c. 60
Iron (wrought)	12	Zinc (crystal; ⊥ axis)	c. 13
Lead	29	Zinc (rolled sheet; ‖ rolling)	c. 31
Lead 0 to 320°C	33	Zinc (rolled sheet; ⊥ rolling)	c. 20
Magnesium	25		
Magnesium 0° to 400°C	30		

Substance	$\alpha \times 10^6$ (linear)	Substance	$\alpha \times 10^6$ (linear)
Alloys			
Aluminum bronze, c. 92 Cu, 7 Al, 0.4 Zn	16	Nickel-steels (including Invar):	
Brass, c. 68 Cu, 32 Zn	18 to 19	10% Ni	13
Yellow brass, Muntz metal c. 60 Cu, 40 An (20° to 100°C)	19 to 23	36% Ni	0 to 1*
		43% Ni	7.9
Bronze (gunmetal, bell metal), c. 80 Cu, 20 Sn	17 to 18	58% Ni	11.5
Constantan, c. 60 Cu, 40 Ni	15 to 17	Phosphor bronze	17
Cupro-nickel (Monel metal) c. 63 Ni, 30 Cu, with about 2% each of Fe, Mn and Pb (between 500° and 600°C)	c. 14	Platinum-iridium, 90 Pt, 10 Ir	8.7
	c. 19	Speculum metal, c. 68 Cu, 32 Sn	19
Delta metal, 60 Cu, 36 Zn, 2 Sn, 2 Fe	20	Stainless steel (typical 18 Cr, 8 Ni)	10 to 11
Duralumin	23	(13% Cr, 20° to 100°C)	10
c. 95 Al, 4 Cu (20° to 500°C)	27	Steel (typical carbon steels)	c. 11
German silver (nickel silver)	18	Stellite, c. 80 Co, 20 Cr (20° to 100°C)	14
Magnalium, c. 90 Ni, 10 Mg	c. 23	20° to 600°C	16
Monel metal (see Cupro-nickel)	c. 14	Stellite No. 2, c. 55 Co, 35 Cr 10 W:	
Nichrome, c. 90 Ni, 10 Cr	13	20° to 100°C	11
87 Ni, 9 Cr, 1.5 Fe	12.5	20° to 600°C	13.5
85 Ni, 10 Cr, 3 Fe	13	Y-alloy	22
Miscellaneous Materials			
Alundum, Al_2O_3 (at 100°C)	c. 8	Glass,† soft soda, 14% Na_2O	c. 8.5
Building materials, etc.:		Glass, hard potash, 20% K_2O	c. 10
Brick	3 to 9	Glass, flint, lead, 46% PbO	8
Cement and concrete	10 to 14	Glass, borate	3 to 7
Granite	6 to 9	Glass, Pyrex	3
Limestone	4 to 9	Glass, Hysil (Chance)	3.3
Marble	3 to 15	Porcelain (insulators)	3 to 6
Masonry	4 to 7	Porcelain (refractory)	2 to 5
Portland stone	c. 3	Porcelain (laboratory)	3 to 4
Sandstone	5 to 12	Quartz ∥ to axis	7.5
Slate	6 to 12	Quartz ⊥ to axis	13.7
Fluorspar, CaF_2	19	Silica, fused,	0.4
		Silica, fused 0° to 1000°C	0.5
		Woods, typical, along grain	c. 3 to 5
		Woods, typical, across grain	c. 35 to 60

* 36% nickel-steels have small values of α only at ordinary ambient temperatures; at 500°C, α is about 9×10^{-6}.

†For most glasses, the coefficients range from 8 to 10×10^{-6} at ordinary ambient temperatures, the exceptions being those containing boron.

DETERMINATION OF MACROSCOPIC STRESSES

Chapter 7 gives tables for selecting the conditions for such measurements; the methods themselves are described in detail in [6, 10, 13, 171].

Many of the tables in Chapter 6 are also of value here.

7-1. Some Formulas for Stress Determination

1. The sum of the principal stresses is given by

$$\sigma_1 + \sigma_2 = -\frac{E}{\nu}\left(\frac{d_\perp - d_0}{d_0}\right), \tag{54}$$

where E is Young's modulus, ν is Poisson's ratio, $d_\perp$ is the mean interplanar distance for the stressed material normal to the surface, and d_0 is the same for the unstressed material.

2. The two-pattern method of deducing the stress in a given direction (with respect to the azimuth φ) employs

$$\sigma_\varphi = \frac{d_\psi - d_\perp}{d_\perp}\frac{E}{1+\nu}\frac{1}{\sin^2\psi}, \tag{55}$$

where ψ is the angle between the normal to the (hkl) plane and the normal to the surface of the specimen, d_ψ is the interplanar distance for the direction ψ, and $d_\perp$ is the same for the normal to the surface of the specimen.

3. The one-pattern method of deducing the stress in a given direction parallel to the surface employs

$$\sigma_\varphi = -\frac{E}{1+\nu}\left(\frac{d_1 - d_2}{d_1}\right)\frac{1}{\sin 2\psi_0 \sin 2\eta}, \tag{56}$$

where d_1 applies to the side of the flat film for which $\psi_1 = \psi_0 + \eta$ and d_2 to the side for which $\psi_2 = \psi_0 - \eta$, ψ_0 being the angle between the axis of the x-ray beam and the normal to the surface of the specimen; $\eta = 90° - \theta$.

It is usual to have $\psi_0 = 45°$, in which case we have

$$\sigma_\varphi = \frac{E}{1+\nu}\frac{d_1 - d_2}{d_1}\frac{1}{\sin 2\eta}.$$

4. The error of stress measurements is reduced by using five patterns for which $\psi = 0$, 30, and 45°. Extrapolation is performed by means of

$$\operatorname{cosec} \vartheta = \frac{\sigma (1+\nu) \sin^2 \psi}{E \sin \vartheta_z} + \frac{1}{\sin \vartheta_z} ,$$

where ϑ_z relates to the beam perpendicular and ϑ to the beam inclined.

5. The principal stresses are determined individually from three patterns: one with an unstressed specimen (d_0) and two with the stressed one with ψ constant and azimuths of φ (d_ψ) and $\varphi + 90°$ (d'_ψ).

The formulas are

$$\sigma_1 = K + L, \quad \sigma_2 = K - L,$$

where

$$K = \frac{1}{2} \left(\frac{d_\psi + d'_\psi - 2d_0}{d_0} \frac{E}{(1+\nu)\sin^2 \psi - 2\nu} \right), \quad L = \frac{1}{2} \frac{d_\psi - d'_\psi}{d_0} \frac{E}{(1+\nu)\sin^2 \psi} ,$$

d_ψ being the interplanar distance for incidence at an angle ψ to the normal with the plane of the specimen inclined at an angle φ, d'_ψ being the same but for ψ and $\varphi + 90°$.

6. If the stresses in the surface layer are of unknown direction and magnitude, patterns are recorded with angles of φ, $\varphi + 60°$, and $\varphi - 60°$; the relationships are

$$\sigma_1 = \frac{1}{3} [M + N], \quad \sigma_2 = \frac{1}{3} [M - N], \quad \operatorname{tg} 2\varphi = \frac{\sqrt{3} [\sigma_{\varphi-60°} - \sigma_{\varphi+60°}]}{3\sigma_\varphi - M} ,$$

where

$$M = \sigma_\varphi + \sigma_{\varphi+60°} + \sigma_{\varphi-60°}, \quad N = \sqrt{(3\sigma_\varphi - M)^2 + 3 [\sigma_{\varphi-60°} - \sigma_{\varphi+60°}]^2}.$$

See [313, 314] for new methods of stress determination.

7. The relation for back-reflection with a flat film is [463]

$$\frac{\Delta d}{d} = \cos^2 \vartheta \cos (180° - 2\vartheta) \left(\frac{l_2 - l_1}{l_1} - \frac{l_{2s} - l_{1s}}{l_{1s}} \right),$$

where l_1 and l_2 are distances from the center of the pattern to the lines corresponding to states 1 and 2 of the specimen, and l_{1s} and l_{2s} are the same for the standard. See below (text to Table 7-5) for other relationships for use with standards.

7-2. Accessory Table for Determining Stresses in Iron, Copper, Aluminum, and Alloys of These

The table serves to determine the changes in lattice constant for alloys of these metals from the line shifts.

Data are given for the (310) line for Co Kα and the (211) line for Cr Kα for iron alloys; for (400) for Co Kα for copper alloys; and for (511) and (333) for Cu Kα and for (420) for Co Kα for aluminum alloys.

The table applies to Bragg angles from 75 to 88°; it can also be used in phase analysis (determination of concentration for solid solutions) [153].

ϑ	$a_{Fe\text{-}alloy}$, kX				$a_{Cu\text{-}alloy}$, kX		$a_{Al\text{-}alloy}$, kX			
	(310) CoKα		(211) CrKα		(400) CoKα		(511), (333) CuKα		(420) CoKα	
	α_1	α_2	α_1	α_2	α_1	α_2	α_1	α_2	α_1	α_2
75°00′	2,92237	2,92875	2,89730	2,90222	3,69653	3,70461	4,13523	4,14551	4,13285	4,14188
01	2,92214	2,92852	2,89707	2,90200	3,69625	3,70433	4,13485	4,14517	4,13253	4,14155
02	2,92191	2,92830	2,89685	2,90177	3,69596	3,70403	4,13453	4,14485	4,13221	4,14123
03	2,92169	2,92807	2,89662	2,90155	3,69567	3,70374	4,13420	4,14453	4,13189	4,14091
04	2,92146	2,92784	2,89640	2,90133	3,69539	3,70346	4,13388	4,14420	4,13157	4,14059
05	2,92124	2,92761	2,89617	2,90110	3,69510	3,70317	4,13356	4,14388	4,13125	4,14027
06	2,92101	2,92739	2,89594	2,90088	3,69481	3,70289	4,13324	4,14356	4,13093	4,13995
07	2,92079	2,92716	2,89573	2,90065	3,69452	3,70261	4,13292	4,14324	4,13061	4,13963
08	2,92056	2,92694	2,89550	2,90043	3,69424	3,70232	4,13262	4,14291	4,13029	4,13931
09	2,92034	2,92671	2,89528	2,90020	3,69395	3,70203	4,13231	4,14259	4,12997	4,13899
10	2,92011	2,92648	2,89506	2,89997	3,69367	3,70174	4,13197	4,14228	4,12965	4,13867
11	2,91989	2,92626	2,89483	2,89975	3,69339	3,70145	4,13165	4,14196	4,12934	4,13835
12	2,91966	2,92603	2,89461	2,89952	3,69310	3,70117	4,13133	4,14165	4,12902	4,13803
13	2,91943	2,92581	2,89439	2,89930	3,69282	3,70089	4,13102	4,14133	4,12871	4,13771
14	2,91921	2,92558	2,89416	2,89908	3,69254	3,70060	4,13070	4,14101	4,12839	4,13740
15	2,91898	2,92535	2,89394	2,89885	3,69225	3,70032	4,13039	4,14070	4,12808	4,13708
16	2,91876	2,92513	2,89372	2,89863	3,69197	3,70004	4,13007	4,14038	4,12777	4,13677
17	2,91853	2,92490	2,89350	2,89841	3,69169	3,69975	4,12975	4,14006	4,12745	4,13645
18	2,91831	2,92468	2,89328	2,89819	3,69141	3,69947	4,12943	4,13974	4,12714	4,13613
19	2,91809	2,92446	2,89306	2,89797	3,69113	3,69919	4,12912	4,13942	4,12682	4,13582
20	2,91787	2,92424	2,89284	2,89775	3,69084	3,69891	4,12880	4,13911	4,12651	4,13550
21	2,91765	2,92402	2,89262	2,89753	3,69056	3,69863	4,12849	4,13880	4,12620	4,13518
22	2,91743	2,92380	2,89240	2,89731	3,69028	3,69834	4,12817	4,13848	4,12588	4,13487
23	2,91721	2,92358	2,89218	2,89709	3,69000	3,69806	4,12786	4,13817	4,12557	4,13456
24	2,91699	2,92336	2,89196	2,89687	3,68972	3,69778	4,12755	4,13786	4,12525	4,13424
25	2,91677	2,92314	2,89174	2,89665	3,68945	3,69750	4,12724	4,13754	4,12494	4,13393
26	2,91655	2,92292	2,89152	2,89643	3,68917	3,69722	4,12692	4,13723	4,12663	4,13362
27	2,91633	2,92270	2,89130	2,89621	3,68889	3,69695	4,12661	4,13692	4,12432	4,13331
28	2,91611	2,92248	2,89108	2,89600	3,68861	3,69667	4,12630	4,13661	4,12400	4,13300
29	2,91589	2,92226	2,89087	2,89578	3,68833	3,69639	4,12599	4,13630	4,12368	4,13269
30	2,91567	2,92203	2,89065	2,89558	3,68805	3,69611	4,12568	4,13598	4,12337	4,13238
31	2,91545	2,92181	2,89044	2,89534	3,68777	3,69583	4,12537	4,13567	4,12306	4,13207
32	2,91523	2,92159	2,89022	2,89513	3,68750	3,69555	4,12506	4,13536	4,12275	4,13176
33	2,91501	2,92137	2,89001	2,89491	3,68722	3,69527	4,12475	4,13505	4,12244	4,13145
34	2,91479	2,92115	2,88979	2,89470	3,68694	3,69500	4,12444	4,13474	4,12213	4,13114
35	2,91457	2,92093	2,88957	2,89448	3,68665	3,69472	4,12414	4,13443	4,12182	4,13083
36	2,91435	2,92071	2,88936	2,89425	3,68638	3,69444	4,12383	4,13413	4,12151	4,13052
37	2,91413	2,92050	2,88914	2,89405	3,68611	3,69416	4,12352	4,13382	4,12120	4,13021
38	2,91391	2,92028	2,88893	2,89384	3,68584	3,69388	4,12321	4,13351	4,12090	4,12991
39	2,91370	2,92006	2,88871	2,89362	3,68557	3,69361	4,12290	4,13321	4,12059	4,12960
40	2,91348	2,91985	2,88849	2,89340	3,68529	3,69334	4,12260	4,13291	4,12029	4,12929
41	2,91327	2,91963	2,88828	2,89319	3,68502	3,69307	4,12229	4,13260	4,11998	4,12898
42	2,91305	2,91942	2,88806	2,89297	3,68475	3,69280	4,12199	4,13230	4,11968	4,12868
43	2,91284	2,91920	2,88785	2,89276	3,68448	3,69253	4,12168	4,13199	4,11937	4,12837
44	2,91262	2,91898	2,88763	2,89254	3,68421	3,69226	4,12138	4,13168	4,11907	4,12807
45	2,91240	2,91876	2,88742	2,89233	3,68394	3,69199	4,12107	4,13137	4,11876	4,12776
46	2,91219	2,91855	2,88720	2,89211	3,68367	3,69171	4,12077	4,13106	4,11846	4,12746
47	2,91197	2,91833	2,88699	2,89190	3,68339	3,69144	4,12047	4,13076	4,11815	4,12715
48	2,91175	2,91812	2,88678	2,89168	3,68312	3,69117	4,12016	4,13045	4,11785	4,12685
49	2,91154	2,91790	2,88657	2,89147	3,68285	3,69090	4,11986	4,13014	4,11754	4,12654
50	2,91133	2,91769	2,88636	2,89126	3,68258	3,69063	4,11956	4,12984	4,11724	4,12624
51	2,91111	2,91747	2,88614	2,89104	3,68231	3,69035	4,11925	4,12954	4,11694	4,12594
52	2,91090	2,91726	2,88593	2,89083	3,68204	3,69008	4,11895	4,12923	4,11664	4,12563
53	2,91068	2,91704	2,88572	2,89062	3,68177	3,68981	4,11865	4,12893	4,11634	4,12534
54	2,91047	2,91683	2,88551	2,89041	3,68149	3,68954	4,11835	4,12863	4,11604	4,12503
55	2,91026	2,91661	2,88530	2,89020	3,68122	3,68927	4,11805	4,12833	4,11574	4,12473
56	2,91005	2,91640	2,88509	2,88999	3,68095	3,68900	4,11775	4,12802	4,11544	4,12443
57	2,90984	2,91619	2,88488	2,88978	3,68068	3,68873	4,11745	4,12772	4,11514	4,12413
58	2,90963	2,91598	2,88467	2,88957	3,68041	3,68846	4,11715	4,12742	4,11484	4,12383
59	2,90942	2,91577	2,88446	2,88936	3,68015	3,68819	4,11685	4,12712	4,11454	4,12353

ϑ	$a_{\text{Fe-alloy}}$, kX				$a_{\text{Cu-alloy}}$, kX		$a_{\text{Al-alloy}}$, kX			
	(310) CoKα		(211) CrKα		(400) CoKα		(511), (333) CuKα		(420) CoKα	
	α₁	α₂	α₁	α₂	α₁	α₂	α₁	α₂	α₁	α₂
76°00′	2,90921	2,91556	2,88425	2,88915	3,67988	3,68792	4,11655	4,12682	4,11424	4,12323
01	2,90900	2,91535	2,88404	2,88894	3,67961	3,68765	4,11625	4,12652	4,11396	4,12293
02	2,90879	2,91514	2,88383	2,88873	3,67935	2,68738	4,11595	4,12623	4,11366	4,12263
03	2,90858	2,91493	2,88362	2,88852	3,67908	3,68712	4,11566	4,12593	4,11336	4,12233
04	2,90837	2,91472	2,88342	2,88831	3,67882	3,68685	4,11537	4,32563	4,11306	4,12203
05	2,90816	2,91451	2,88321	2,88811	3,67855	3,68659	4,11507	4,12533	4,11276	4,12173
06	2,90795	2,91430	2,88300	2,88790	3,67829	3,68632	4,11477	4,12504	4,11247	4,12144
07	2,90774	2,91409	2,88280	2,88769	3,67802	3,68606	4,11447	4,12474	4,11217	4,12114
08	2,90753	2,91388	2,88259	2,88748	3,67776	3,68579	4,11418	4,12445	4,11188	4,12085
09	2,90732	2,91367	2,88239	2,88727	3,67749	3,68553	4,11388	4,12415	4,11158	4,12055
10	2,90711	2,91346	2,88218	2,88707	3,67723	3,68527	4,11358	4,12385	4,11128	4,12026
11	2,90690	2,91325	2,88197	2,88686	3,67698	3,68500	4,11329	4,12356	4,11098	4,11996
12	2,90669	2,91304	2,88176	2,88666	3,67672	3,68474	4,11300	4,12327	4,11069	4,11967
13	2,90648	2,91283	2,88156	2,88645	3,67644	3,68447	4,11270	4,12297	4,11040	4,11937
14	2,90628	2,91263	2,88135	2,88625	3,67617	3,68421	4,11241	4,12267	4,11010	4,11908
15	2,90607	2,91242	2,88115	2,88604	3,67591	3,68395	4,11211	4,12237	4,10981	4,11878
16	2,90587	2,91221	2,88094	2,88584	3,67565	3,68369	4,11182	4,12208	4,10952	4,11849
17	2,90566	2,91200	2,88074	2,88563	3,67539	3,68343	4,11153	4,12178	4,10922	4,11820
18	2,90546	2,91180	2,88053	2,88543	3,67513	3,68317	4,11124	4,12149	4,10893	4,11790
19	2,90525	2,91159	2,88033	2,88522	3,67487	3,68291	4,11094	4,12119	4,10864	4,11761
20	2,90504	2,91139	2,88012	2,88502	3,67461	3,68264	4,11065	4,12091	4,10835	4,11732
21	2,90483	2,91118	2,87992	2,88481	3,67435	3,68238	4,11037	4,12062	4,10806	4,11703
22	2,90463	2,91098	2,87972	2,88461	3,67409	3,68212	4,11008	4,12033	4,10777	4,11674
23	2,90442	2,91077	2,87951	2,88440	3,67383	3,68186	4,10979	4,12005	4,10748	4,11645
24	2,90422	2,91057	2,87931	2,88420	3,67358	3,68160	4,10950	4,11976	4,10719	4,11616
25	2,90402	2,91036	2,87911	2,88400	3,67332	3,68134	4,10922	4,11947	4,10690	4,11587
26	2,90381	2,91016	2,87890	2,88380	3,67306	3,68109	4,10893	4,11918	4,10661	4,11558
27	2,90360	2,90995	2,87870	2,88360	3,67281	3,68083	4,10864	4,11889	4,10632	4,11529
28	2,90340	2,90975	2,87850	2,88340	3,67255	3,68057	4,10835	4,11860	4,10603	4,11500
29	2,90320	2,90954	2,87830	2,88320	3,67229	3,68031	4,10806	4,11831	4,10574	4,11471
30	2,90300	2,90934	2,87810	2,88301	3,67204	3,68005	4,10777	4,11802	4,10546	4,11443
31	2,90279	2,90913	2,87790	2,88281	3,67178	3,67980	4,10749	4,11774	4,10517	4,11414
32	2,90259	2,90893	2,87770	2,88261	3,67152	3,67954	4,10720	4,11745	4,10489	4,11386
33	2,90239	2,90873	2,87750	2,88241	3,67127	3,67929	4,10692	4,11716	4,10460	4,11357
34	2,90219	2,90853	2,87730	2,88221	3,67101	3,67903	4,10663	4,11688	4,10432	4,11328
35	2,90199	2,90833	2,87710	2,88201	3,67076	3,67878	4,10635	4,11659	4,10403	4,11300
36	2,90179	2,90813	2,87690	2,88181	3,67050	3,67852	4,10606	4,11631	4,10375	4,11271
37	2,90159	2,90793	2,87670	2,88161	3,67024	3,67827	4,10578	4,11602	4,10346	4,11243
38	2,90139	2,90773	2,87650	2,88141	3,66998	3,67802	4,10549	4,11574	4,10318	4,11215
39	2,90119	2,90753	2,87630	2,88121	3,66973	3,67776	4,10521	4,11545	4,10289	4,11187
40	2,90099	2,90733	2,87610	2,88101	3,66948	3,67751	4,10492	4,11517	4,10261	4,11159
41	2,90079	2,90713	2,87590	2,88081	3,66923	3,67725	4,10464	4,11488	4,10232	4,11130
42	2,90059	2,90693	2,87571	2,88061	3,66898	3,67700	4,10436	4,11460	4,10204	4,11102
43	2,90039	2,90673	2,87551	2,88041	3,66873	3,67675	4,10407	4,11431	4,10176	4,11073
44	2,90019	2,90653	2,87532	2,88021	3,66848	3,67649	4,10379	4,11403	4,10148	4,11045
45	2,89999	2,90633	2,87512	2,88001	3,66823	3,67624	4,10351	4,11375	4,10120	4,11017
46	2,89979	2,90613	2,87493	2,87981	3,66798	3,67599	4,10323	4,11347	4,10092	4,10989
47	2,89959	2,90593	2,87473	2,87962	3,66773	3,67574	4,10295	4,11319	4,10064	4,10961
48	2,89939	2,90573	2,87454	2,87942	3,66748	3,67549	4,10267	4,11291	4,10036	4,10933
49	2,89919	2,90553	2,87434	2,87922	3,66723	3,67524	4,10239	4,11263	4,10008	4,10905
50	2,89900	2,90533	2,87414	2,87902	3,66698	3,67499	4,10211	4,11235	4,09980	4,10877
51	2,89880	2,90513	2,87394	2,87882	3,66673	3,67474	4,10183	4,11207	4,09952	4,10849
52	2,89860	2,90493	2,87375	2.87863	3,66648	3,67449	4,10155	4,11179	4,09924	4,10821
53	2,89841	2,90474	2,87355	2,87843	3,66622	3,67423	4,10128	4,11152	4,09886	4,10793
54	2,89821	2,90454	2,87336	2,87824	3,66598	3,67399	4,10100	4,11124	4,09869	4,10765
55	2,89802	2,90435	2,86316	2,87804	3,66573	3,67374	4,10073	4,11096	4,09841	4,10737
56	2,89782	2,90415	2,87297	2,87785	3,66549	3,67349	4,10045	4,11069	4,09814	4,10709
57	2,89763	2,90396	2,87277	2,87765	3,66524	3,67325	4,10017	4,11041	4,09786	4,10682
58	2,89743	2,90376	2,87258	2,87746	3,66500	3,67300	4,09989	4,11013	4,09758	4,10655
59	2,89724	2,90357	2,87238	2,87726	3,66476	3,67275	4,09961	4,10985	4,09730	4,10627

| ϑ | $a_{\text{Fe-alloy}}$, kX | | | | $a_{\text{Cu-alloy}}$, kX | | $a_{\text{Al-alloy}}$, kX | | | |
| | (310) CoKα | | (211) CrKα | | (400) CoKα | | (511), (333) CuKα | | (420) CoKα | |
	α_1	α_2	α_1	α_2	α_1	α_2	α_1	α_2	α_1	α_2
77°00′	2,89704	2,90337	2,87219	2,87707	3,66451	3,67251	4,09933	4,10957	4,09703	4,10599
01	2,89684	2,90317	2,87200	2,87688	3,66426	3,67227	4,09906	4,10930	4,09676	4,10571
02	2,89664	2,90298	2,87180	2,87668	3,66401	3,67202	4,09879	4,10902	4,09648	4,10544
03	2,89645	2,90278	2,87161	2,87649	3,66378	3,67177	4,09851	4,10875	4,09621	4,10516
04	2,89525	2,90259	2,87142	2,87630	3,66353	3,67152	4,09824	4,10847	4,09593	4,10489
05	2,89506	2,90239	2,87123	2,87610	3,66328	3,67128	4,09797	4,10820	4,09566	4,10461
06	2,89587	2,90219	2,87104	2,87591	3,66303	3,67104	4,09770	4,10792	4,09538	4,10434
07	2,89568	2,90200	2,87085	2,87572	3,66278	3,67079	4,09742	4,10765	4,09511	4,10407
08	2,89549	2,90181	2,87066	2,87553	3,66253	3,67054	4,09715	4,10737	4,09484	4,10380
09	2,89520	2,90162	2,87047	2,87534	3,67228	3,67029	4,09688	4,10709	4,09457	4,10352
10	2,89511	2,90143	2,87028	2,87515	3,66204	3,67004	4,09660	4,10682	4,09430	4,10325
11	2,89492	2,90124	2,87009	2,87496	3,66180	3,66980	4,09633	4,10655	4,09403	4,10298
12	2,89473	2,90105	2,86990	2,87477	3,66156	3,66956	4,09606	4,10628	4,09386	4,10271
13	2,89454	2,90086	2,86971	2,87458	3,66132	3,66933	4,09580	4,10601	4,09359	4,10244
14	2,89435	2,90077	2,86952	2,87439	3,66108	3,66908	4,09552	4,10574	4,09332	4,10217
15	2,89416	2,90048	2,86933	2,87420	3,66083	3,66884	4,09525	4,10547	4,09305	4,10190
16	2,89397	2,90029	2,86914	2,87401	3,66060	3,66860	4,09498	4,10520	4,09288	4,10163
17	2,89378	2,90010	2,86895	2,87382	3,66036	3,66836	4,09471	4,10493	4,09261	4,10136
18	2,89359	2,89991	2,86877	2,87364	3,66012	3,66812	4,09444	4,10466	4,09234	4,10109
19	2,89340	2,89972	2,86858	2,87345	3,65989	3,66788	4,09417	4,10440	4,09207	4,10082
20	2,89321	2,89953	2,86839	2,87326	3,65964	3,66764	4,09390	4,10412	4,09161	4,10055
21	2,89302	2,89934	2,86820	2,87307	3,65940	3,66740	4,09364	4,10385	4,09134	4,10028
22	2,89283	2,89915	2,86802	2,87289	3,65917	3,66716	4,09337	4,10359	4,09108	4,10002
23	2,89264	2,89896	2,85783	2,87270	3,65893	3,66692	4,09311	4,10332	4,09081	4,09975
24	2,89245	2,89877	2,86765	2,87252	3,65868	3,66668	4,09284	4,10306	4,09055	4,09949
25	2,89226	2,89858	2,86746	2,87233	3,65844	3,66644	4,09258	4,10279	4,09028	4,09922
26	2,89208	2,89839	2,86728	2,87215	3,65821	3,66621	4,09231	4,10253	4,09002	4,09895
27	2,89189	2,89820	2,86709	2,87196	3,65797	3,66596	4,09205	4,10226	4,08975	4,09868
28	2,89170	2,89801	2,86691	2,87178	3,65774	3,66573	4,09178	4,10200	4,08949	4,09842
29	2,89151	2,89782	2,86672	2,87159	3,65750	3,66549	4,09152	4,10173	4,08922	4,09816
30	2,89133	2,89764	2,86653	2,87140	3,65727	3,66526	4,09125	4,10146	4,08896	4,09789
31	2,89114	2,89745	2,86635	2,87122	3,65703	3,66502	4,09099	4,10120	4,08870	4,09763
32	2,89096	2,89727	2,86616	2,87103	3,65679	3,66477	4,09072	4,10093	4,08844	4,09736
33	2,89077	2,89708	2,86598	2,87085	3,65656	3,66455	4,09046	4,10067	4,08818	4,09710
34	2,89058	2,89690	2,86579	2,87066	3,65632	3,66431	4,09020	4,10040	4,08792	4,09683
35	2,89040	2,89671	2,86561	2,87048	3,65609	3,66408	4,08994	4,10014	4,08766	4,09657
36	2,89021	2,89652	2,86543	2,87029	3,65586	3,66384	4,08968	4,09988	4,08740	4,09630
37	2,89002	2,89634	2,86524	2,87011	3,65563	3,66361	4,08942	4,09961	4,08714	4,09604
38	2,88984	2,89615	2,86506	2,86992	3,65539	3,66338	4,08916	4,09935	4,08688	4,09578
39	2,88966	2,89597	2,86487	2,86974	3,65514	3,66315	4,08890	4,09909	4,08662	4,09552
40	2,88948	2,89579	2,86469	2,86956	3,65491	3,66292	4,08863	4,09883	4,08636	4,09526
41	2,88929	2,89560	2,86451	2,86937	3,65468	3,66267	4,08837	4,09857	4,08610	4,09500
42	2,88911	2,89542	2,86432	2,86919	3,65445	3,66243	4,08811	4,09831	4,08584	4,09474
43	2,88892	2,89523	2,86414	2,86901	3,65422	3,66219	4,08785	4,09806	4,08558	4,09448
44	2,88874	2,89505	2,86396	2,86882	3,65398	3,66197	4,08759	4,09780	4,08532	4,09422
45	2,88855	2,89481	2,86378	2,86864	3,65375	3,66174	4,08733	4,09754	4,08506	4,09397
46	2,88837	2,89468	2,86360	2,86846	3,65353	3,66151	4,08708	4,09728	4,08480	4,09371
47	2,88819	2,89450	2,86342	2,86828	3,65330	3,66123	4,08682	4,09703	4,08454	4,09345
48	2,88801	2,89432	2,86324	2,86810	3,65307	3,66103	4,08656	4,09677	4,08428	4,09320
49	2,88783	2,89414	2,86306	2,86792	3,65285	3,66082	4,08630	4,09651	4,08402	4,09294
50	2,88765	2,89396	2,86288	2,86774	3,65262	3,66060	4,08605	4,09625	4,08376	4,09268
51	2,88747	2,89378	2,86270	2,86756	3,65238	3,66037	4,08579	4,09599	4,08350	4,09243
52	2,88729	2,89360	2,86252	2,86738	3,65216	3,66014	4,08554	4,09573	4,08325	4,09217
53	2,88711	2,89342	2,86234	2,86720	3,65193	3,65990	4,08528	4,09548	4,08299	4,09192
54	2,88693	2,89324	2,86216	2,86702	3,65170	3,65968	4,08503	4,09522	4,08273	4,09166
55	2,88675	2,89306	2,86198	2,86684	3,65148	3,65945	4,08477	4,09497	4,08248	4,09140
56	2,88657	2,89288	2,86180	2,86666	3,65125	3,65923	4,08452	4,09471	4,08222	4,09115
57	2,88639	2,89270	2,86163	2,86649	3,65102	3,65900	4,08426	4,09446	4,08197	4,09089
58	2,88621	2,89252	2,86145	2,86631	3,65080	3,65877	4,08401	4,09420	4,08172	4,09064
59	2,88603	2,89234	2,86127	2,86613	3,65057	3,65855	4,08375	4,09395	4,08146	4,09038

ϑ	$a_{Fe\text{-alloy}}$, kX				$a_{Cu\text{-alloy}}$, kX		$a_{Al\text{-alloy}}$, kX			
	(310) CoKα		(211) CrKα		(400) CoKα		(511), (333) CuKα		(420) CoKα	
	a_1	a_2	a_1	a_2	a_1	a_2	a_1	a_2	a_1	a_2
78°00′	2,88585	2,89216	2,86110	2,86596	3,65035	3,65832	4,08350	4,09370	4,08121	4,09013
01	2,88567	2,89198	2,86092	2,86578	3,65012	3,65810	4,08325	4,09345	4,08096	4,08988
02	2,88549	2,89180	2,86075	2,86560	3,64990	8,65787	4,08300	4,09319	4,08071	4,08962
03	2,88531	2,89162	2,86057	2,86543	3,64967	3,65764	4,08275	4,09294	4,08046	4,08937
04	2,88513	2,89144	2,86040	2,86525	3,64945	3,68742	4,08250	4,09269	4,08021	4,08912
05	2,88495	2,89126	2,86022	2,86508	3,64922	3,65720	4,08225	4,09243	4,07996	4,08887
06	2,88477	2,89108	2,86005	2,86490	3,64900	3,65696	4,08200	4,09218	4,07971	4,08862
07	2,88460	2,89090	2,85987	2,86473	3,64878	3,65674	4,08175	4,09193	4,07946	4,08837
08	2,88442	2,89073	2,85970	2,86455	3,64856	3,65652	4,08150	4,09168	4,07921	4,08812
09	2,88425	2,89055	2,85952	2,86438	3,64833	3,65630	4,08125	4,09143	4,07896	4,08787
10	2,88408	2,89038	2,85935	2,86420	3,64811	3,65607	4,08100	4,09118	4,07871	4,08762
11	2,88390	2,89020	2,85917	2,86403	3,64789	3,65586	4,08075	4,09093	4,07846	4,08737
12	2,88373	2,89003	2,85900	2,86386	3,64766	3,65563	4,08050	4,09068	4,07821	4,08712
13	2,88355	2,88985	2,85883	2,86369	3,64744	3,65541	4,08025	4,09044	4,07797	4,08687
14	2,88337	2,88968	2,85865	2,86351	3,64722	3,65519	4,08001	4,09019	4,07772	4,08663
15	2,88320	2,88950	2,85848	2,86333	3,64699	3,65497	4,07976	4,08994	4,07747	4,08638
16	2,88302	2,88933	2,85831	2,86316	3,64677	3,65474	4,07951	4,08970	4,07723	4,08613
17	2,88285	2,88916	2,85814	2,86298	3,64655	3,65452	4,07927	4,08945	4,07698	4,08589
18	2,88268	2,88899	2,85796	2,86281	3,64634	3,65430	4,07902	4,08920	4,07674	4,08564
19	2,88251	2,88881	2,85779	2,86264	3,64612	3,65408	4,07878	4,08895	4,07649	4,08540
20	2,88234	2,88864	2,85762	2,86247	3,64590	3,65384	4,07853	4,08871	4,07624	4,08515
21	2,88216	2,88846	2,85745	2,86230	3,64568	3,65363	4,07828	4,08846	4,07600	4,08490
22	2,88199	2,88829	2,85728	2,86212	3,64546	3,65343	4,07804	4,08821	4,07575	4,08466
23	2,88181	2,88811	2,85710	2,86195	3,64524	3,65321	4,07780	4,08797	4,07550	4,08441
24	2,88164	2,88794	2,85693	2,86178	3,64503	3,65298	4,07755	4,08772	4,07526	4,08417
25	2,88147	2,88776	2,85676	2,86161	3,64481	3,65277	4,07731	4,08748	4,07501	4,08392
26	2,88130	2,88759	2,85659	2,86144	3,64458	3,65256	4,07706	4,08723	4,07477	4,08368
27	2,88113	2,88742	2,85642	2,86127	3,64437	3,65234	4,07682	4,08699	4,07453	4,08343
28	2,88096	2,88725	2,85625	2,86110	3,64415	3,65212	4,07658	4,08675	4,07429	4,08319
29	2,88079	2,88708	2,85608	2,86093	3,64394	3,65190	4,07634	4,08651	4,07405	4,08294
30	2,88062	2,88691	2,85591	2,86076	3,64373	3,65169	4,07610	4,08627	4,07381	4,08270
31	2,88045	2,88674	2,85574	2,86059	3,64352	3,65147	4,07586	4,08603	4,07357	4,08246
32	2,88028	2,88657	2,85557	2,86042	3,64330	3,65126	4,07562	4,08579	4,07333	4,08222
33	2,88011	2,88640	2,85540	2,86025	3,64308	3,65104	4,07538	4,08555	4,07309	4,08198
34	2,87994	2,88623	2,85523	2,86008	3,64286	3,65083	4,07514	4,08531	4,07285	4,08174
35	2,87977	2,88606	2,85506	2,85991	3,64265	3,65061	4,07490	4,08507	4,07261	4,08150
36	2,87960	2,88589	2,85490	2,85975	3,64244	3,65040	4,07466	4,08483	4,07237	4,08126
37	2,87943	2,88572	2,85473	2,85958	3,64223	3,65018	4,07442	4,08459	4,07213	4,08102
38	2,87926	2,88555	2,85457	2,85941	3,64201	3,64996	4,07418	4,08434	4,07189	4,08078
39	2,87909	2,88538	2,85440	2,85925	3,64179	3,64975	4,07394	4,08410	4,07165	4,08055
40	2,87893	2,87522	2,85424	2,85908	3,64158	3,64953	4,07370	4,08387	4,07142	4,08031
41	2,87876	2,88505	2,85407	2,85892	3,64137	3,64933	4,07347	4,08363	4,07118	4,08007
42	2,87860	2,88488	2,85390	2,85875	3,64116	3,64909	4,07323	4,08340	4,07095	4,07984
43	2,87843	2,88462	2,85374	2,85859	3,64095	3,64888	4,07300	4,08316	4,07072	4,07960
44	2,87826	2,88445	2,85357	2,85842	3,64074	3,64866	4,07276	4,08293	4,07048	4,07937
45	2,87809	2,88438	2,85341	2,85825	3,64051	3,64847	4,07253	4,08269	4,07025	4,07913
46	2,87793	2,88412	2,85324	2,85809	3,64031	3,64826	4,07229	4,08245	4,07001	4,07890
47	2,87776	2,88395	2,85308	2,85793	3,64009	3,64805	4,07203	4,08222	4,06978	4,07866
48	2,87760	2,88379	2,85291	2,85776	3,63989	3,64784	4,07182	4,08198	4,06954	4,07843
49	2,87743	2,88362	2,85275	2,85760	3,63968	3,64763	4,07159	4,08174	4,06930	4,07819
50	2,87726	2,88355	2,85258	2,85743	3,63947	3,64742	4,07135	4,08151	4,06907	4,07795
51	2,87710	2,88338	2,85242	2,85727	3,63927	3,64722	4,07112	4,08127	4,06883	4,07772
52	2,87693	2,88322	2,85225	2,85710	3,63906	3,64701	4,07088	4,08104	4,06860	4,07748
53	2,87677	2,88305	2,85209	2,85694	3,63886	3,64680	4,07065	4,08081	4,06836	4,07725
54	2,87660	2,88289	2,85193	2,85677	3,63865	3,64660	4,07042	4,08057	4,06813	4,07701
55	2,87644	2,88272	2,85176	2,85661	3,63844	3,64639	4,07018	4,08034	4,06790	4,07678
56	2,87627	2,88256	2,85160	2,85644	3,63823	3,64618	4,06995	4,08011	4,06767	4,07655
57	2,87611	2,88239	2,85144	2,85628	3,63803	3,64597	4,06972	4,07988	4,06744	4,07632
58	2,87595	2,88222	2,85128	2,85612	3,63782	3,64577	4,06949	4,07965	4,06721	4,07609
59	2,87579	2,88205	2,85112	2,85596	3,63761	3,64556	4,06926	4,07941	4,06698	4,07586

ϑ	$a_{Fe\text{-alloy}}$, kX				$a_{Cu\text{-alloy}}$, kX		$a_{Al\text{-alloy}}$, kX			
	(310) CoKα		(211) CrKα		(400) CoKα		(511), (333) CuKα		(420) CoKα	
	a_1	a_2	a_1	a_2	a_1	a_2	a_1	a_2	a_1	a_2
79°00′	2,87562	2,88189	2,85096	2,85580	3,63741	3,64535	4,06903	4,07919	4,06675	4,07563
01	2,87546	2,88173	2,85080	2,85564	3,63720	3,64514	4,06880	4,07896	4,06652	4,07540
02	2,87530	2,88157	2,85064	2,85548	3,63699	3,64493	4,06857	4,07873	4,06630	4,07517
03	2,87513	2,88141	2,85048	2,85532	3,63678	3,64473	4,06834	4,07850	4,06607	4,07494
04	2,87497	2,88125	2,85032	2,85516	3,63658	3,64453	4,06810	4,07827	4,06584	4,07471
05	2,87481	2,88109	2,85016	2,85500	3,63638	3,64432	4,06787	4,07804	4,06561	4,07449
06	2,87465	2,88093	2,85000	2,85484	3,63617	3,64412	4,06764	4,07781	4,06539	4,07426
07	2,87449	2,88077	2,84984	2,85468	3,63597	3,64391	4,06742	4,07758	4,06516	4,07403
08	2,87433	2,88061	2,84968	2,85452	3,63576	3,64371	4,06719	4,07736	4,06493	4,07380
09	2,87417	2,88045	2,84952	2,85432	3,63555	3,64351	4,06697	4,07714	4,06470	4,07358
10	2,87401	2,88029	2,84936	2,85420	3,63536	3,64331	4,06675	4,07691	4,06447	4,07335
11	2,87385	2,88013	2,84920	2,85404	3,63516	3,64311	4,06653	4,07668	4,06425	4,07313
12	2,87370	2,87997	2,84904	2,85388	3,63496	3,64291	4,06630	4,07645	4,06402	4,07290
13	2,87354	2,87981	2,84889	2,85372	3,63476	3,64270	4,06608	4,07622	4,06380	4,07268
14	2,87338	2,87965	2,84873	2,85356	3,63456	3,64250	4,06585	4,07599	4,06357	4,07246
15	2,87322	2,87949	2,84857	2,85341	3,63436	3,64230	4,06563	4,07577	4,06335	4,07223
16	2,87306	2,87933	2,84842	2,85325	3,63416	3,64210	4,06540	4,07554	4,06312	4,07201
17	2,87290	2,87917	2,84826	2,85309	3,63396	3,64190	4,06518	4,07532	4,06290	4,07178
18	2,87275	2,87902	2,84810	2,85293	3,63376	3,64170	4,06495	4,07510	4,06267	4,07156
19	2,87259	2,87886	2,84795	2,85278	3,63356	3,64150	4,06473	4,07488	4,06245	4,07133
20	2,87243	2,87870	2,84779	2,85262	3,63336	3,64130	4,06450	4,07465	4,06222	4,07110
21	2,87227	2,87855	2,84764	2,85247	3,63316	3,64111	4,06428	4,07443	4,06200	4,07087
22	2,87211	2,87839	2,84748	2,85231	3,63296	3,64090	4,06406	4,07420	4,06177	4,07065
23	2,87196	2,87823	2,84733	2,85216	3,63277	3,64070	4,96384	4,07398	4,06155	4,07043
24	2,87180	2,87808	2,84717	2,85200	3,63257	3,64050	4,06362	4,07379	4,06133	4,07021
25	2,87164	2,87792	2,84702	2,85185	3,63237	3,64030	4,06340	4,07357	4,06111	4,06999
26	2,87149	2,87776	2,84686	2,85169	3,63217	3,64011	4,06318	4,07335	4,06089	4,06977
27	2,87133	2,87761	2,84671	2,85154	3,63198	3,63991	4,06296	4,07312	4,06067	4,06955
28	2,87117	2,87745	2,84655	2,85138	3,63178	3,63971	4,06274	4,07288	4,06045	4,06933
29	2,87102	2,87730	2,84640	2,85123	3,63158	3,63952	4,06252	4,07266	4,06023	4,06911
30	2,87086	2,87714	2,84624	2,85107	3,63138	3,63932	4,06230	4,07243	4,06001	4,06889
31	2,87071	2,87699	2,84609	2,85092	3,63118	3,63912	4,06208	4,07221	4,05979	4,06867
32	2,87055	2,87683	2,84593	2,85076	3,63099	3,63892	4,06186	4,07200	4,05958	4,06845
33	2,87040	2,87668	2,84578	2,85061	3,63080	3,63873	4,06164	4,07178	4,05936	4,06823
34	2,87024	2,87652	2,84563	2,85046	3,63060	3,63853	4,06142	4,07156	4,05914	4,06801
35	2,87009	2,87636	2,84548	2,85030	3,63041	3,63834	4,06120	4,07134	4,05893	4,06780
36	2,86993	2,87621	2,84532	2,85015	3,63022	3,63815	4,06098	4,07112	4,05871	4,06758
37	2,86978	2,87606	2,84517	2,85000	3,63003	3,63796	4,06077	4,07091	4,05850	4,06736
38	2,86963	2,87590	2,84502	2,84985	3,62983	3,63776	4,06055	4,07070	4,05828	4,06714
39	2,86948	2,87575	2,84487	2,84970	3,62962	3,63755	4,06034	4,07048	4,05806	4,06692
40	2,86933	2,87560	2,84472	2,84955	3,62943	3,63736	4,06012	4,07027	4,05785	4,06671
41	2,86917	2,87544	2,84457	2,84940	3,62924	3,63716	4,05991	4,07005	4,05763	4,06649
42	2,86902	2,87529	2,84442	2,84925	3,62905	3,63698	4,05970	4,06984	4,05742	4,06628
43	2,86887	2,87514	2,84427	2,84910	3,62886	3,63678	4,05948	4,06962	4,05720	4,06606
44	2,86872	2,87499	2,84412	2,84895	3,62867	3,63660	4,05927	4,06940	4,05699	4,06585
45	2,86857	2,87484	2,84397	2,84880	3,62848	3,63642	4,05906	4,06918	4,05677	4,06563
46	2,86842	2,87469	2,84382	2,84865	3,62830	3,63622	4,05885	4,06897	4,05656	4,06542
47	2,86827	2,87454	2,84367	2,84850	3,62811	3,63603	4,05864	4,06875	4,05634	4,06520
48	2,86812	2,87439	2,84352	2,84835	3,62792	3,63584	4,05842	4,06853	4,05613	4,06499
49	2,86797	2,87424	2,84337	2,84820	3,62773	3,63565	4,05820	4,06832	4,05592	4,06478
50	2,86782	2,87409	2,84322	2,84805	3,62755	3,63546	4,05799	4,06811	4,05571	4,06457
51	2,86767	2,87394	2,84307	2,84790	3,62736	3,63528	4,05778	4,06790	4,05550	4,06436
52	2,86752	2,87379	2,84292	2,84775	3,62717	3,63509	4,05757	4,06768	4,05529	4,06415
53	2,86737	2,87364	2,84278	2,84760	3,62698	3,63490	4,05736	4,06747	4,05508	4,06394
54	2,86722	2,87349	2,84263	2,84746	3,62679	3,63472	4,05715	4,06726	4,05487	4,06373
55	2,86707	2,87334	2,84248	2,84731	3,62661	3,63453	4,05694	4,06705	4,05466	4,06352
56	2,86693	2,87319	2,84234	2,84716	3,62642	3,63434	4,05673	4,06684	4,05445	4,06331
57	2,86678	2,87304	2,84219	2,84702	3,62623	3,63415	4,05652	4,06663	4,05424	4,06310
58	2,86663	2,87290	2,84204	2,84687	3,62604	3,63396	4,05631	4,06642	4,05403	4,06289
59	2,86648	2,87275	2,84190	2,84672	3,62586	3,63378	4,05610	4,06621	4,05382	4,06268

| ϑ | $a_{Fe\text{-alloy}}$, kX | | | | $a_{Cu\text{-alloy}}$, kX | | $a_{Al\text{-alloy}}$, kX | | | |
| | (310) CoKα | | (211) CrKα | | (400) CoKα | | (511), (333) CuKα | | (420) CoKα | |
	α_1	α_2	α_1	α_2	α_1	α_2	α_1	α_2	α_1	α_2
80°00′	2,86634	2,87260	2,84175	2,84658	3,62568	3,63359	4,05589	4,06601	4,05361	4,06247
01	2,86619	2,87245	2,84161	2,84643	3,62549	3,63340	4,05568	4,06580	4,05340	4,06226
02	2,86605	2,87230	2,84146	2,84629	3,62530	3,63321	4,05547	4,06560	4,05320	4,06205
03	2,86590	2,87216	2,84132	2,84614	3,62511	3,63303	4,05526	4,06539	4,05299	4,06184
04	2,86575	2,87202	2,84117	2,84600	3,62492	3,63286	4,05506	4,06518	4,05279	4,06163
05	2,86561	2,87187	2,84103	2,84585	3,62474	3,63266	4,05485	4,06497	4,05258	4,06143
06	2,86546	2,87173	2,84088	2,84571	3,62455	3,63247	4,05464	4,06477	4,05238	4,06122
07	2,86532	2,87158	2,84074	2,84556	3,62436	3,63228	4,05443	4,06456	4,05217	4,06101
08	2,86517	2,87144	2,84060	2,84542	3,62418	3,63210	4,05423	4,06435	4,05197	4,06081
09	2,86503	2,87129	2,84045	2,84527	3,62400	3,63192	4,05402	4,06415	4,05176	4,06060
10	2,86488	2,87114	2,84031	2,84513	3,62381	3,63170	4,05382	4,06394	4,05155	4,06040
11	2,86474	2,87100	2,84016	2,84499	3,62363	3,63155	4,05361	4,06374	4,05134	4,06019
12	2,86459	2,87085	2,84002	2,84484	3,62345	3,63137	4,05341	4,06353	4,05114	4,05999
13	2,86445	2,87071	2,83988	2,84470	3,62327	3,63119	4,05320	4,06333	4,05093	4,05978
14	2,86430	2,87057	2,83974	2,84456	3,62309	3,63100	4,05300	4,06312	4,05073	4,05958
15	2,86416	2,87043	2,83959	2,84441	3,62291	3,63082	4,05280	4,06292	4,05053	4,05938
16	2,86402	2,87028	2,83945	2,84427	3,62273	3,63064	4,05260	4,06271	4,05033	4,05918
17	2,86388	2,87014	2,83931	2,84413	3,62255	3,63046	4,05240	4,06251	4,05013	4,05898
18	2,86373	2,86999	2,83917	2,84399	3,62237	3,63028	4,05220	4,06231	4,04993	4,05878
19	2,86359	2,86985	2,83903	2,84385	3,62219	3,63010	4,05200	4,06211	4,04973	4,05858
20	2,86345	2,86970	2,83889	2,84371	3,62201	3,62993	4,05180	4,06191	4,04953	4,05838
21	2,86331	2,86956	2,83875	2,84357	3,62183	3,62975	4,05160	4,06171	4,04933	4,05818
22	2,86316	2,86942	2,83861	2,84343	3,62165	3,62957	4,05140	4,06151	4,04913	4,05798
23	2,86302	2,86927	2,83847	2,84329	3,62148	3,62939	4,05120	4,06131	4,04893	4,05778
24	2,86288	2,86913	2,83833	2,84315	3,62130	3,62921	4,05100	4,06111	4,04873	4,05758
25	2,86274	2,86899	2,83819	2,84301	3,62112	3,62903	4,05080	4,06091	4,04853	4,05738
26	2,86260	2,86885	2,83805	2,84287	3,62094	3,62885	4,05060	4,06071	4,04833	4,05718
27	2,86246	2,86871	2,83791	2,84273	3,62076	3,62867	4,05040	4,06051	4,04813	4,05698
28	2,86232	2,86857	2,83777	2,84259	3,62058	3,62849	4,05020	4,06031	4,04793	4,05678
29	2,86218	2,86843	2,83763	2,84245	3,62040	3,62832	4,05000	4,06011	4,04773	4,05658
30	2,86204	2,86829	2,83749	2,84231	3,62023	3,62813	4,04981	4,05992	4,04754	4,05638
31	2,86190	2,86815	2,83735	2,84217	3,62004	3,62795	4,04960	4,05973	4,04734	4,05619
32	2,86176	2,86801	2,83721	2,84203	3,61986	3,62778	4,04941	4,05953	4,04715	4,05599
33	2,86162	2,86787	2,83707	2,84189	3,61969	3,62760	4,04921	4,05934	4,04695	4,05580
34	2,86148	2,86773	2,83693	2,84175	3,61951	3,62741	4,04902	4,05914	4,04676	4,05560
35	2,86135	2,86760	2,83680	2,84162	3,61934	3,62725	4,04882	4,05895	4,04657	4,05541
36	2,86121	2,86746	2,83666	2,84148	3,61917	3,62707	4,04863	4,05875	4,04637	4,05521
37	2,86107	2,86732	2,83652	2,84134	3,61899	3,62689	4,04843	4,05855	4,04618	4,05502
38	2,86093	2,86718	2,83639	2,84121	3,61882	3,62672	4,04824	4,05836	4,04598	4,05482
39	2,86080	2,86705	2,83625	2,84107	3,61864	3,62654	4,04805	4,05816	4,04578	4,05463
40	2,86066	2,86691	2,83612	2,84094	3,61847	3,62637	4,04786	4,05796	4,04559	4,05443
41	2,86053	2,86677	2,83598	2,84080	3,61830	3,65620	4,04767	4,05777	4,04540	4,05424
42	2,86039	2,86664	2,83585	2,84067	3,61812	3,62603	4,04747	4,05757	4,04520	4,05404
43	2,86026	2,86650	2,83571	2,84053	3,61795	3,62585	4,04728	4,05738	4,04501	4,05385
44	2,86012	2,86637	2,83558	2,84040	3,61778	3,62568	4,04709	4,05718	4,04481	4,05365
45	2,85998	2,86623	2,83544	2,84026	3,61762	3,62551	4,04690	4,05699	4,04462	4,05346
46	2,85985	2,86610	2,83531	2,84013	3,61745	3,62534	4,04670	4,05680	4,04443	4,05326
47	2,85972	2,86596	2,83517	2,83999	3,61728	3,62518	4,04651	4,05660	4,04424	4,05307
48	2,85958	2,86583	2,83504	2,83986	3,61711	3,62501	4,04632	4,05641	4,04405	4,05288
49	2,85915	2,86569	2,83491	2,83972	3,61694	3,62484	4,04613	4,05622	4,04386	4,05269
50	2,85931	2,86555	2,83478	2,83959	3,61677	3,62467	4,04594	4,05603	4,04367	4,05250
51	2,85918	2,86542	2,83464	2,83946	3,61660	3,62450	4,04575	4,05584	4,04348	4,05231
52	2,85905	2,86528	2,83451	2,83932	3,61644	3,62433	4,04556	4,05565	4,04329	4,05212
53	2,85891	2,86515	2,83438	2,83919	3,61627	3,62416	4,04537	4,05547	4,04310	4,05193
54	2,85878	2,86501	2,83424	2,83906	3,61610	3,62399	4,04518	4,05529	4,04291	4,05174
55	2,85865	2,86488	2,83411	2,83892	3,61593	3,62383	4,04499	4,05510	4,04272	4,05155
56	2,85852	2,86475	2,83398	2,83879	3,61575	3,62365	4,04480	4,05491	4,04254	4,05137
57	2,85838	2,86462	2,83385	2,83866	3,61559	3,62349	4,04461	4,05472	4,04235	4,05118
58	2,85825	2,86449	2,83372	2,83853	3,61542	3,62332	4,04443	4,05453	4,04216	4,05100
59	2,85812	2,86436	2,83359	2,83840	3,61525	3 62315	4,04424	4,05434	4,04198	4,05081

ϑ	$a_{Fe-alloy}$, kX				$a_{Cu-alloy}$, kX		$a_{Al-alloy}$, kX			
	(310) CoKα		(211) CrKα		(400) CoKα		(511), (333) CuKα		(420) CoKα	
	α_1	α_2	α_1	α_2	α_1	α_2	α_1	α_2	α_1	α_2
81°00′	2,85798	2,86423	2,83346	2,83827	3,61508	3,62299	4,04416	4,05415	4,04179	4,05062
01	2,85785	2,86409	2,83333	2,83814	3,61492	3,62282	4,04388	4,05396	4,04160	4,05043
02	2,85771	2,86396	2,83320	2,83801	3,61475	3,62266	4,04369	4,05378	4,04142	4,05025
03	2,85758	2,86382	2,83307	2,83788	3,61459	3,62249	4,04350	4,05360	4,04123	4,05006
04	2,85745	2,86369	2,83294	2,83775	3,61443	3,62232	4,04332	4,05341	4,04105	4,04988
05	2,85732	2,86356	2,83281	2,83762	3,61426	3,62216	4,04313	4,05323	4,04086	4,04969
06	2,85719	2,86343	2,83268	2,83749	3,61410	3,62199	4,04295	4,05304	4,04068	4,04951
07	2,85706	2,86330	2,83255	2,83736	3,61393	3,62183	4,04276	4,05286	4,04050	4,04932
08	2,85693	2,86317	2,83243	2,83724	3,61377	3,62166	4,04258	4,05267	4,04032	4,04914
09	2,85680	2,86304	2,83230	2,83711	3,61360	3,62150	4,04240	4,05249	4,04014	4,04896
10	2,85667	2,86291	2,83217	2,83698	3,61344	3,62133	4,04221	4,05230	4,03995	4,04878
11	2,85654	2,86278	2,83204	2,83685	3,61328	3,62117	4,04202	4,05212	4,03977	4,04859
12	2,85641	2,86265	2,83192	2,83673	3,61311	3,62101	4,04184	4,05193	4,03959	4,04841
13	2,85628	2,86252	2,83179	2,83660	3,61294	3,62084	4,04166	4,05175	4,03941	4,04823
14	2,85615	2,86239	2,83166	2,83647	3,61278	3,62068	4,04148	4,05157	4,03923	4,04804
15	2,85603	2,86227	2,83153	2,83635	3,61262	3,62052	4,04130	4,05149	4,03904	4,04786
16	2,85590	2,86214	2,83141	2,83622	3,61246	3,62035	4,04112	4,05121	4,03886	4,04768
17	2,85577	2,86201	2,83128	2,83609	3,61230	3,62019	4,04094	4,05103	4,03868	4,04750
18	2,85564	2,86189	2,83115	2,83597	3,61214	3,62003	4,04076	4,05085	4,03850	4,04732
19	2,85551	2,86176	2,83102	2,83584	3,61198	3,61987	4,04058	4,05067	4,03832	4,04714
20	2,85539	2,86163	2,83090	2,83571	3,61182	3,61971	4,04040	4,05049	4,03814	4,04696
21	2,85527	2,86150	2,83077	2,83559	3,61166	3,61955	4,04022	4,05031	4,03796	4,04678
22	2,85514	2,86138	2,83065	2,83546	3,61150	3,61939	4,04004	4,05013	4,03778	4,04660
23	2,85502	2,86125	2,83052	2,83534	3,61134	3,61923	4,03997	4,04995	4,03760	4,04642
24	2,85489	2,86113	2,83040	2,83521	3,61118	3,61907	4,03970	4,04977	4,03742	4,04624
25	2,85477	2,86100	2,83028	2,83508	3,61102	3,61891	4,03951	4,04960	4,03725	4,04606
26	2,85464	2,86088	2,83015	2,83496	3,61086	3,61875	4,03933	4,04942	4,03707	4,04589
27	2,85452	2,86075	2,83002	2,83483	3,61071	3,61859	4,03925	4,04925	4,03690	4,04571
28	2,85439	2,86063	2,82990	2,83471	3,61055	3,61843	4,03908	4,04907	4,03672	4,04553
29	2,85427	2,86050	2,82978	2,83459	3,61039	3,61828	4,03880	4,04889	4,03655	4,04536
30	2,85414	2,86038	2,82966	2,83447	3,61023	3,61812	4,03863	4,04871	4,03637	4,04518
31	2,85402	2,86025	2,82953	2,83434	3,61008	3,61796	4,03846	4,04853	4,03620	4,04501
32	2,85390	2,86013	2,82941	2,83422	3,60992	3,61781	4,03828	4,04836	4,03602	4,04483
33	2,85377	2,86000	2,82928	2,83409	3,60977	3,61765	4,03810	4,04819	4,03585	4,04467
34	2,85364	2,85988	2,82916	2,83397	3,60961	3,61750	4,03793	4,04801	4,03567	4,04449
35	2,85352	2,85975	2,82904	2,83384	3,60945	3,61734	4,03776	4,04784	4,03550	4,04432
36	2,85339	2,85963	2,82892	2,83372	3,60930	3,61718	4,03762	4,04766	4,03532	4,04414
37	2,85327	2,85951	2,82880	2,83360	3,60914	3,61703	4,03741	4,04749	4,03515	4,04397
38	2,85315	2,85939	2,82868	2,83348	3,60898	3,61687	4,03724	4,04733	4,03498	4,04380
39	2,85303	2,85927	2,82856	2,83336	3,60883	3,61672	4,03707	4,04716	4,03480	4,04363
40	2,85291	2,85915	2,83844	2,83324	3,60868	3,61657	4,03690	4,04700	4,03463	4,04346
41	2,85279	2,85902	2,82832	2,83312	3,60853	3,61641	4,03673	4,04681	4,03446	4,04328
42	2,85267	2,85890	2,82820	2,83300	3,60837	3,61626	4,03655	4,04664	4,03429	4,04311
43	2,85255	2,85878	2,82808	2,83288	3,60822	3,61611	4,03638	4,04647	4,03412	4,04293
44	2,85243	2,85866	2,82796	2,83276	3,60807	3,61595	4,03621	4,04630	4,03395	4,04276
45	2,85231	2,85854	2,82784	2,83264	3,60792	3,61580	4,03604	4,04613	4,03378	4,04259
46	2,85219	2,85842	2,82772	2,83252	3,60776	3,61564	4,03587	4,04596	4,03361	4,04242
47	2,85207	2,85830	2,82760	2,83240	3,60762	3,61550	4,03570	4,04581	4,03344	4,04225
48	2,85195	3,85818	2,82748	2,83228	3,60747	3,61534	4,03553	4,04562	4,03327	4,04208
49	2,85183	2,85806	2,82737	2,83217	3,60732	3,61519	4,03536	4,04545	4,03310	4,04191
50	2,85171	2,85794	2,82725	2,83205	3,60716	3,61504	4,03519	4,04528	4,03293	4,04174
51	2,85159	2,85782	2,82713	2,83193	3,60701	3,61489	4,03502	4,04511	4,03276	4,04157
52	2,85147	2,85770	2,82702	2,83181	3,60686	3,61474	4,03485	4,04494	4,03259	4,04140
53	2,85135	2,85758	2,82690	2,83170	3,60671	3,61458	4,03469	4,04477	4,03242	4,04123
54	2,85123	2,85746	2,82678	2,83158	3,60656	3,61444	4,03452	4,04460	4,03225	4,04107
55	2,85111	2,85735	2,82667	2,83146	3,60641	3,61428	4,03435	4,04443	4,03208	4,04090
56	2,85100	2,85723	2,82655	2,83135	3,60626	3,61413	4,03418	4,04427	4,03192	4,04074
57	2,85088	2,85711	2,82643	2,83123	3,60611	3,61399	4,03402	4,04410	4,03175	4,04057
58	2,85076	2,85700	2,82632	2,83111	3,60597	3,61384	4,03385	4,04393	4,03159	4,04040
59	2,85064	2,85688	2,82620	2,83100	3,60581	3,61369	4,03371	4,04377	4,03142	4,04024

ϑ	$a_{\text{Fe-alloy}}$, kX				$a_{\text{Cu-alloy}}$, kX		$a_{\text{Al-alloy}}$, kX			
	(310) CoKα		(211) CrKα		(400) CoKα		(511), (333) CuKα		(420) CoKα	
	α_1	α_2	α_1	α_2	α_1	α_2	α_1	α_2	α_1	α_2
82°00′	2,85053	2,85676	2,82608	2,83088	3,60566	3,61354	4,03352	4,04361	4,03125	4,04007
01	2,85041	2,85665	2,82597	2,83077	3,60552	3,61339	4,03336	4,04344	4,03109	4,03990
02	2,85030	2,85653	2,82585	2,83065	3,60537	3,61325	4,03320	4,04327	4,03092	4,03974
03	2,85018	2,85642	2,82574	2,83054	3,60523	3,61310	4,03303	4,04311	4,03076	4,03958
04	2,85007	2,85630	2,82562	2,83042	3,60509	3,61296	4,03287	4,04295	4,03060	4,03942
05	2,84995	2,85619	2,82551	2,83031	3,60494	3,61281	4,03270	4,04279	4,03043	4,03926
06	2,84984	2,85607	2,82539	2,83019	3,60479	3,61267	4,03254	4,04263	4,03027	4,03910
07	2,84972	2,85596	2,82528	2,83008	3,60465	3,61252	4,03239	4,04247	4,03011	4,03894
08	2,84961	2,85584	2,82516	2,82996	3,60450	3,61237	4,03223	4,04231	4,02995	4,03878
09	2,84949	2,85572	2,82505	2,82985	3,60436	3,61223	4,03208	4,04214	4,02979	4,03862
10	2,84938	2,85560	2,82494	2,82974	3,60421	3,61209	4,03191	4,04196	4,02963	4,03846
11	2,84926	2,85549	2,82482	2,82962	3,60407	3,61194	4,03175	4,04180	4,02947	4,03829
12	2,84915	2,85537	2,82471	2,82951	3,60392	3,61180	4,03159	4,04164	4,02931	4,03812
13	2,84903	2,85526	2,82460	2,82940	3,60378	3,61165	4,03143	4,04148	4,02915	4,03795
14	2,84892	2,85514	2,82448	2,82928	3,60364	3,61151	4,03127	4,04132	4,02899	4,03779
15	2,84880	2,85503	2,82437	2,82917	3,60349	3,61137	4,03111	4,04115	4,02883	4,03763
16	2,84869	2,85491	2,82426	2,82906	3,60335	3,61123	4,03095	4,04099	4,02867	4,03747
17	2,84858	2,85480	2,82415	2,82895	3,60321	3,61108	4,03079	4,04083	4,02852	4,03731
18	2,84847	2,85469	2,82404	2,82884	3,60307	3,61094	4,03063	4,04067	4,02836	4,03715
19	2,84836	2,85458	2,82393	2,82873	3,60293	3,61080	4,03047	4,04051	4,02820	4,03699
20	2,84825	2,85447	2,82382	2,82862	3,60278	3,61065	4,03030	4,04036	4,02804	4,03683
21	2,84814	2,85436	2,82371	2,82851	3,60264	3,61051	4,03014	4,04020	4,02788	4,03667
22	2,84803	2,85425	2,82360	2,82840	3,60250	3,61037	4,02998	4,04004	4,02773	4,03651
23	2,84792	2,85414	2,82349	2,82829	3,60236	3,61023	4,02982	4,03988	4,02757	4,03636
24	2,84781	2,85403	2,82338	2,82818	2,60222	3,61009	4,02967	4,03973	4,02742	4,03620
25	2,84770	2,85392	2,82327	2,82807	3,60208	3,60995	4,02951	4,03957	4,02726	4,03605
26	2,84759	2,85381	2,82316	2,82796	3,60194	3,60982	4,02936	4,03942	4,02711	4,03589
27	2,84748	2,85370	2,82305	2,82785	3,60181	3,60968	4,02920	4,03926	4,02695	4,03574
28	2,84737	2,85359	2,82294	2,82774	3,60167	3,60954	4,02904	4,03910	4,02680	4,03558
29	2,84726	2,85348	2,82284	2,82763	3,60153	3,60940	4,02888	4,03894	4,02664	4,03543
30	2,84715	2,85337	2,82273	2,82752	3,60139	3,60926	4,02873	4,03879	4,02648	4,03527
31	2,84704	2,85326	2,82262	2,82741	3,60126	3,60913	4,02857	4,03863	4,02633	4,03512
32	2,84693	2,85315	2,82251	2,82730	3,60112	3,60900	4,02842	4,03847	4,02617	4,03496
33	2,84682	2,85304	2,82241	2,82720	3,60099	3,60886	4,02827	4,03832	4,02602	4,03481
34	2,84671	2,85293	2,82230	2,82709	3,60085	3,60872	4,02812	4,03816	4,02586	4,03446
35	2,84661	2,85282	2,82219	2,82698	3,60071	3,60858	4,02797	4,03800	4,02571	4,03450
36	2,84650	2,85272	2,82209	2,82688	3,60058	3,60844	4,02782	4,03785	4,02555	4,03435
37	2,84639	2,85261	2,82198	2,82677	3,60045	3,60829	4,02767	4,03770	4,02540	4,03420
38	2,84628	2,85250	2,82187	2,82666	3,60031	3,60815	4,02752	4,03755	4,02525	4,03405
39	2,84617	2,85239	2,82177	2,82656	3,60017	3,60802	4,02737	4,03740	4,02510	4,03390
40	2,84607	2,85229	2,82166	2,82645	3,60003	3,60789	4,02721	4,03726	4,02495	4,03375
41	2,84596	2,85218	2,82156	2,82635	3,59989	3,60775	4,02706	4,03711	4,02480	4,03360
42	2,84580	2,85208	2,82145	2,82624	3,59975	3,60762	4,02691	4,03696	4,02465	4,03345
43	2,84575	2,85197	2,82135	2,82614	3,59962	3,60748	4,02675	4,03681	4,02450	4,03330
44	2,84565	2,85187	2,82124	2,82603	3,59949	3,60735	4,02660	4,03666	4,02435	4,03315
45	2,84554	2,85176	2,82114	2,82593	3,59936	3,60722	4,02645	4,03651	4,02420	4,03300
46	2,84544	2,85166	2,82103	2,82582	3,59923	3,60709	4,02631	4,03636	4,02405	4,03285
47	2,84533	2,85155	2,82093	2,82572	3,59910	3,60695	4,02616	4,03621	4,02390	4,03270
48	2,84523	2,85145	2,82082	2,82561	3,59896	3,60682	4,02601	4,03606	4,02375	4,03255
49	2,84512	2,85134	2,82072	2,82550	3,59883	3,60668	4,02586	4,03591	4,02360	4,03240
50	2,84502	2,85123	2,82062	2,82540	3,59870	3,60655	4,02572	4,03577	4,02346	4,03225
51	2,84491	2,85113	2,82051	2,82529	3,59856	3,60641	4,02557	4,03562	4,02331	4,03211
52	2,84481	2,85102	2,82041	2,82519	3,59843	3,60628	4,02543	4,03548	4,02317	4,03196
53	2,84470	2,85092	2,82030	2,82509	3,59830	3,60615	4,02528	4,03533	4,02302	4,03182
54	2,84460	2,85081	2,82020	2,82498	3,59817	3,60602	4,02513	4,03518	4,02288	4,03167
55	2,84449	2,85071	2,82010	2,82488	3,59804	3,60589	4,02499	4,03503	4,02273	4,03152
56	2,84439	2,85060	2,82000	2,82478	3,59791	3,60576	4,02484	4,03488	4,02259	4,03138
57	2,84429	2,85050	2,81990	2,82468	3,59778	3,60562	4,02470	4,03474	4,02244	4,03123
58	2,84419	2,85040	2,81980	2,82458	3,59765	3,60550	4,02455	4,03459	4,02230	4,03109
59	2,84409	2,85030	2,81970	2,82448	3,59752	3,60537	4,02441	4,03445	4,02215	4,03094

ϑ	$a_{\text{Fe-alloy}}$, kX				$a_{\text{Cu-alloy}}$, kX		$a_{\text{Al-alloy}}$, kX			
	(310) CoKα		(211) CrKα		(400) CoKα		(511), (333) CuKα		(420) CoKα	
	α_1	α_2	α_1	α_2	α_1	α_2	α_1	α_2	α_1	α_2
83°00′	2,84399	2,85020	2,81960	2,82438	3,59739	3,60525	4,02426	4,03430	4,02201	4,03080
01	2,84389	2,85010	2,81950	2,82428	3,59726	3,60512	4,02412	4,03416	4,02186	4,03065
02	2,84379	2,85000	2,81940	2,82418	3,59713	3,60500	4,02398	4,03402	4,02172	4,03050
03	2,84369	2,84990	2,81930	2,82408	3,59701	3,60486	4,02384	4,03388	4,02157	4,03036
04	2,84359	2,84980	2,81920	2,82398	3,59688	3,60473	4,02370	4,03374	4,02143	4,03022
05	2,84349	2,84970	2,81910	2,82388	3,59675	3,60461	4,02355	4,03360	4,02129	4,03008
06	2,84339	2,84960	2,81900	2,82378	3,59662	3,60448	4,02341	4,03345	4,02115	4,02994
07	2,84329	2,84950	2,81890	2,82368	3,59649	3,60436	4,02327	4,03331	4,02101	4,02980
08	2,84319	2,84940	2,81880	2,82358	3,59637	3,60424	4,02313	4,03317	4,02087	4,02966
09	2,84309	2,84930	2,81870	2,82349	3,59625	3,60411	4,02299	4,03303	4,02073	4,02952
10	2,84299	2,84920	2,81860	2,82339	3,59612	3,60398	4,02285	4,03288	4,02059	4,02938
11	2,84289	2,84910	2,81850	2,82329	3,59600	3,60386	4,02271	4,03274	4,02045	4,02924
12	2,84279	2,84900	2,81841	2,82319	3,59588	3,60374	4,02257	4,03260	4,02031	4,02910
13	2,84269	2,84890	2,81831	2,82310	3,59576	3,60361	4,02243	4,03246	4,02017	4,02896
14	2,84259	2,84880	2,81821	2,82300	3,59564	3,60348	4,02229	4,03232	4,02003	4,02882
15	2,84249	2,84870	2,81812	2,82290	3,59552	3,60335	4,02215	4,03219	4,01989	4,02868
16	2,84240	2,84860	2,81802	2,82280	3,59540	3,60323	4,02201	4,03205	4,01975	4,02854
17	2,84230	2,84851	2,81792	2,82271	3,59527	3,60311	4,02187	4,03191	4,01962	4,02840
18	2,82220	2,84841	2,81783	2,82261	3,59514	3,60298	4,02173	4,03177	4,01948	4,02826
19	2,84210	2,84831	2,81773	2,82251	3,59501	3,60286	4,02159	4,03163	4,01934	4,02812
20	2,84201	2,84822	2,81763	2,82241	3,59489	3,60274	4,02146	4,03150	4,01920	4,02798
21	2,84191	2,84812	2,81754	2,82232	3,59477	3,60262	4,02132	4,03136	4,01907	4,02785
22	2,74181	2,84802	2,81744	2,82222	3,59465	3,60249	4,02119	4,03122	4,01893	4,02771
23	2,84172	2,84793	2,81735	2,82213	3,59453	3,60237	4,02105	4,03108	4,01880	4,02758
24	2,84162	2,84783	2,81725	2,82203	3,59441	3,60225	4,02091	4,03095	4,01866	4,02744
25	2,84153	2,84774	2,81716	2,82194	3,59429	3,60213	4,02078	4,03082	4,01853	4,02731
26	2,84143	2,84764	2,81706	2,82184	3,59416	3,60201	4,02065	4,03068	4,01839	4,02717
27	2,84134	2,84755	2,81697	2,82175	3,59404	3,60189	4,02052	4,03055	4,01826	4,02704
28	2,84124	2,84745	2,81687	2,82165	3,59392	3,60178	4,02038	4,03041	4,01812	4,02690
29	2,84115	2,84736	2,81678	2,82156	3,59380	3,60166	4,02025	4,03028	4,01799	4,02677
30	2,84105	2,84726	2,81668	2,82147	3,59368	3,60154	4,02012	4,03014	4,01785	4,02663
31	2,84096	2,84717	2,81659	2,82137	3,59356	3,60142	4,01998	4,03001	4,01772	4,02650
32	2,84086	2,84707	2,81649	2,82128	3,59344	3,60130	4,01985	4,02988	4,01759	4,02636
33	2,84077	2,84698	2,81640	2,82119	3,59333	3,60118	4,01972	4,02975	4,01746	4,02623
34	2,84067	2,84688	2,81631	2,82109	3,59321	3,60106	4,01958	4,02962	4,01733	4,02610
35	2,84058	2,84679	2,81621	2,82100	3,59309	3,60094	4,01945	4,02949	4,01720	4,02597
36	2,84048	2,84669	2,81612	2,82090	3,59297	3,60082	4,01932	4,02936	4,01707	4,02584
37	2,84039	2,84670	2,81603	2,82081	3,59285	3,60070	4,01919	4,02923	4,01694	4,02571
38	2,84030	2,84661	2,81594	2,82072	3,59273	3,60058	4,01906	4,02910	4,01681	4,02558
39	2,84021	2,84642	2,81585	2,82063	3,59262	3,60046	4,01893	4,02897	4,01668	4,02545
40	2,84012	2,84633	2,81576	2,82054	3,59251	3,60034	4,01880	4,02883	4,01655	4,02532
41	2,84003	2,84623	2,81567	2,82045	3,59240	3,60023	4,01867	4,02870	4,01642	4,02519
42	2,83994	2,84614	2,81558	2,82036	3,59229	3,60012	4,01854	4,02857	4,01629	4,02506
43	2,83985	2,84605	2,81549	2,82027	3,59217	3,60001	4,01841	4,02844	4,01616	4,02493
44	2,83976	2,84596	2,81540	2,82018	3,59205	3,59988	4,01828	4,02832	4,01603	4,02480
45	2,83967	2,84587	2,81531	2,82009	3,59193	3,59977	4,01816	4,02818	4,01590	4,02467
46	2,83958	2,84578	2,81522	2,82000	3,59181	3,59966	4,01803	4,02805	4,01577	4,02454
47	2,83949	2,84569	2,81513	2,81991	3,59170	3,59955	4,01790	4,02792	4,01564	4,02441
48	2,83940	2,84560	2,81505	2,81983	3,59159	3,59943	4,01777	4,02780	4,01551	4,02428
49	2,83931	2,84551	2,81496	2,81974	3,59148	3,59931	4,01764	4,02767	4,01538	4,02415
50	2,83922	2,84542	2,81487	2,81965	3,59137	3,59920	4,01751	4,02754	4,01526	4,02402
51	2,83913	2,84533	2,81478	2,81956	3,59126	3,59909	4,01739	4,02742	4,01513	4,02390
52	2,83904	2,84524	2,81469	2,81947	3,59115	3,59898	4,01726	4,02729	4,01501	4,02378
53	2,83895	2,84515	2,81460	2,81939	3,59103	3,59887	4,01714	4,02717	4,01488	4,02365
54	2,83886	2,84506	2,81452	2,81930	3,59091	3,59876	4,01702	4,02704	4,01476	4,02353
55	2,83877	2,84497	2,81443	2,81921	3,59080	3,59864	4,01690	4,02692	4,01463	4,02340
56	2,83868	2,84489	2,81434	2,81912	3,59069	3,59853	4,01678	4,02679	4,01451	4,02328
57	2,83860	2,84480	2,81425	2,81903	3,59058	3,59842	4,01666	4,02667	4,01438	4,02315
58	2,83851	2,84471	2,81417	2,81895	3,59047	3,59831	4,01653	4,02655	4,01426	4,02303
59	2,83842	2,84462	2,81408	2,81886	3,59036	3,59820	4,01640	4,02642	4,01414	4,02291

ϑ	$a_{\text{Fe-alloy}}$, kX				$a_{\text{Cu-alloy}}$, kX		$a_{\text{Al-alloy}}$, kX			
	(310) Co$K\alpha$		(211) Cr$K\alpha$		(400) Co$K\alpha$		(511), (333) Cu$K\alpha$		(420) Co$K\alpha$	
	α_1	α_2	α_1	α_2	α_1	α_2	α_1	α_2	α_1	α_2
84°00′	2,83834	2,84454	2,81399	2,81877	3,59025	3,59809	4,01627	4,02630	4,01402	4,02279
01	2,83825	2,84445	2,81390	2,81869	3,59014	3,59798	4,01615	4,02618	4,01389	4,02266
02	2,83816	2,84436	2,81382	2,81860	3,59003	3,59787	4,01603	4,02606	4,01377	4,02254
03	2,83808	2,84428	2,81373	2,81852	3,58992	3,59776	4,01590	4,02593	4,01365	4,02241
04	2,83799	2,84419	2,81365	2,81843	3,58981	3,59765	4,01578	4,02581	4,01353	4,02229
05	2,83791	2,84411	2,81356	2,81835	3,58970	3,59754	4,01566	4,02569	4,01341	4,02217
06	2,83782	2,84402	2,81348	2,81826	3,58959	3,59744	4,01554	4,02556	4,01329	4,02205
07	2,83774	2,84394	2,81340	2,81818	3,58949	3,59733	4,01542	4,02544	4,01317	4,02193
08	2,84765	2,84385	2,81331	2,81809	3,58938	3,59722	4,01530	4,02532	4,01305	4,02184
09	2,83757	2,84377	2,81323	2,81801	3,58927	3,59711	4,01518	4,02520	4,01293	4,02169
10	2,83748	2,84368	2,81315	2,81792	3,58916	3,59700	4,01506	4,02508	4,01281	4,02157
11	2,83740	2,84360	2,81307	2,81784	3,58906	3,59689	4,01495	4,02497	4,01269	4,02145
12	2,83731	2,84351	2,81298	2,81776	3,58895	3,59679	4,01483	4,02485	4,01257	4,02133
13	2,83723	2,84343	2,81290	2,81767	3,58885	3,59669	4,01471	4,02473	4,01245	4,02121
14	2,83714	2,84334	2,81282	2,81759	3,58877	3,59658	4,01459	4,02461	4,01233	4,02110
15	2,83706	2,84326	2,81273	2,81751	3,58863	3,59647	4,01447	4,02449	4,01221	4,02098
16	2,83697	2,84317	2,81265	2,81743	3,58852	3,59636	4,01435	4,02437	4,01210	4,02086
17	2,83689	2,84309	2,81257	2,81734	3,58841	3,59625	4,01424	4,02425	4,01198	4,02074
18	2,83681	2,84301	2,81248	2,81726	3,58832	3,59615	4,01412	4,02413	4,01186	4,02063
19	2,83673	2,84293	2,81240	2,81718	3,58822	3,59605	4,01400	4,02401	4,01174	4,02051
20	2,83665	2,84285	2,81232	2,81710	3,58812	3,59595	4,01389	4,02390	4,01163	4,02040
21	2,83657	2,84276	2,81224	2,81702	3,58801	3,59585	4,01377	4,02378	4,01151	4,02028
22	2,83649	2,84268	2,81216	2,81694	3,58791	3,59575	4,01366	4,02367	4,01140	4,02017
23	2,83641	2,84260	2,81208	2,81686	3,58780	3,59564	4,01354	4,02355	4,01128	4,02005
24	2,83633	2,84252	2,81200	2,81678	3,58770	3,59553	4,01343	4,02344	4,01117	4,01994
25	2,83625	2,84244	2,81192	2,81670	3,58760	3,59543	4,01331	4,02332	4,01106	4,01982
26	2,83617	2,84236	2,81184	2,81662	3,58750	3,59533	4,01320	4,02321	4,01094	4,01971
27	2,83609	2,84228	2,81176	2,81654	3,58740	3,59523	4,01308	4,02309	4,01083	4,01959
28	2,83601	2,84220	2,81168	2,81646	3,58730	3,59513	4,01297	4,02298	4,01072	4,01948
29	2,83593	2,84212	2,81160	2,81638	3,58720	3,59503	4,01285	4,02287	4,01061	4,01936
30	2,83585	2,84204	2,81152	2,81630	3,58710	3,59493	4,01274	4,02276	4,01050	4,01925
31	2,83577	2,84196	2,81144	2,81622	3,58700	3,59483	4,01263	4,02264	4,01038	4,01914
32	2,83569	2,84188	2,81136	2,81614	3,58689	3,59473	4,01252	4,02253	4,01027	4,01903
33	2,83561	2,84180	2,81128	2,81606	3,58679	3,59463	4,01241	4,02242	4,01016	4,01892
34	2,83553	2,84172	2,81121	2,81598	3,58669	3,59452	4,01230	4,02231	4,01005	4,01881
35	2,83545	2,84165	2,81113	2,81590	3,58659	3,59442	4,01219	4,02220	4,00994	4,01870
36	2,83537	2,84157	2,81105	2,81583	3,58650	3,59432	4,01208	4,02209	4,00983	4,01859
37	2,83529	2,84149	2,81097	2,81575	3,58640	3,59422	4,01197	4,02198	4,00972	4,01848
38	2,83521	2,84141	2,81090	2,81567	3,58630	3,59413	4,01186	4,02187	4,00961	4,01837
39	2,83513	2,84134	2,81082	2,81560	3,58620	3,59403	4,01175	4,02176	4,00950	4,01826
40	2,83506	2,84126	2,81075	2,81552	3,58610	3,59393	4,01164	4,02165	4,00939	4,01815
41	2,83498	2,84118	2,81067	2,81545	3,58601	3,59383	4,01154	4,02154	4,00928	4,01804
42	2,83491	2,84111	2,81060	2,81537	3,58591	3,59374	4,01143	4,02143	4,00917	4,01793
43	2,83483	2,84103	2,81052	2,81530	3,58581	3,59365	4,01132	4,02132	4,00906	4,01782
44	2,83476	2,84096	2,81045	2,81522	3,58571	3,59355	4,01121	4,02121	4,00895	4,01771
45	2,83468	2,84088	2,81037	2,81515	3,58561	3,59346	4,01110	4,02110	4,00884	4,01760
46	2,83461	2,84081	2,81030	2,81507	3,58552	3,59336	4,01100	4,02099	4,00873	4,01749
47	2,83453	2,84073	2,81022	2,81500	3,58543	3,59326	4,01089	4,02089	4,00863	4,01738
48	2,83446	2,84066	2,81015	2,81492	3,58534	3,59318	4,01079	4,02078	4,00853	4,01727
49	2,83438	2,84058	2,81007	2,81485	3,58525	3,59308	4,01068	4,02067	4,00842	4,01717
50	2,83431	2,84050	2,81000	2,81477	3,58516	3,59298	4,01057	4,02057	4,00832	4,01707
51	2,83423	2,84042	2,80992	2,81470	3,58506	3,59288	4,01047	4,02047	4,00821	4,01696
52	2,83416	2,84035	2,80985	2,81463	3,58496	3,59279	4,01036	4,02037	4,00811	4,01686
53	2,83408	2,84028	2,80978	2,81455	3,58487	3,59270	4,01025	4,02026	4,00800	4,01675
54	2,83401	2,84021	2,80970	2,81448	3,58477	3,59260	4,01015	4,02015	4,00790	4,01665
55	2,83394	2,84014	2,80963	2,81441	3,58468	3,59251	4,01004	4,02005	4,00779	4,01654
56	2,83386	2,84006	2,80956	2,81434	3,58459	3,59242	4,00994	4,01994	4,00769	4,01644
57	2,83379	2,83998	2,80949	2,81426	3,58450	3,59233	4,00984	4,01983	4,00758	4,01634
58	2,83371	2,83991	2,80941	2,81419	3,58441	3,59223	4,00973	4,01973	4,00748	4,01624
59	2,83364	2,83983	2,80934	2,81412	3,58432	3,59214	4,00963	4,01963	4,00738	4,01614

ϑ	$a_{Fe-alloy}$, kX				$a_{Cu-alloy}$, kX		$a_{Al-alloy}$, kX			
	(310) CoKα		(211) CrKα		(400) CoKα		(511), (333) CuKα		(420) CoKα	
	α_1	α_2	α_1	α_2	α_1	α_2	α_1	α_2	α_1	α_2
85°00′	2,83357	2,83976	2,80927	2,81404	3,58423	3,59205	4,00953	4,01953	4,00728	4,01604
01	2,83350	2,83969	2,80920	2,81397	3,58414	3,59196	4,00943	4,01943	4,00718	4,01593
02	2,83343	2,83961	2,80913	2,81390	3,58404	3,59187	4,00932	4,01933	4,00708	4,01583
03	2,83336	2,83954	2,80906	2,81383	3,58395	3,59178	4,00922	4,01923	4,00698	4,01573
04	2,83329	2,83947	2,80899	2,81376	3,58386	3,59169	4,00912	4,01913	4,00688	4,01563
05	2,83322	2,83940	2,80892	2,81369	3,58377	3,59160	4,00902	4,01903	4,00678	4,01553
06	2,83315	2,83933	2,80885	2,81362	3,58368	3,59151	4,00892	4,01893	4,00668	4,01543
07	2,83308	2,83926	2,80878	2,81355	3,58359	3,59141	4,00882	4,01883	4,00658	4,01533
08	2,83301	2,83919	2,80871	2,81348	3,58350	3,59132	4,00872	4,01873	4,00648	4,01523
09	2,83294	2,83912	2,80864	2,81341	3,58341	3,59123	4,00862	4,01863	4,00638	4,01513
10	2,83287	2,83905	2,80857	2,81334	3,58332	3,59114	4,00852	4,01853	4,00628	4,01503
11	2,83280	2,83898	2,80850	2,81327	3,58323	3,59105	4,00843	4,01844	4,00618	4,01493
12	2,83273	2,83891	2,80843	2,81320	3,58314	3,59097	4,00833	4,01834	4,00608	4,01483
13	2,83266	2,83884	2,80836	2,81313	3,58306	3,59089	4,00823	4,01824	4,00598	4,01473
14	2,83259	2,83877	2,80831	2,81306	3,58297	3,59080	4,00813	4,01814	4,00588	4,01463
15	2,83252	2,83870	2,80823	2,81300	3,58289	3,59071	4,00803	4,01804	4,00579	4,01453
16	2,83245	2,83863	2,80816	2,81293	3,58280	3,59062	4,00794	4,01795	4,00570	4,01444
17	2,83238	2,83857	2,80809	2,81286	3,58271	3,59053	4,00784	4,01785	4,00560	4,01434
18	2,83231	2,83850	2,80802	2,81279	3,58263	3,59045	4,00775	4,01776	4,00550	4,01425
19	2,83224	2,83843	2,80796	2,81272	3,58254	3,59036	4,00765	4,01766	4,00541	4,01415
20	2,83218	2,83837	2,80789	2,81265	3,58245	3,59027	4,00756	4,01756	4,00531	4,01406
21	2,83211	2,83830	2,80782	2,81259	3,58237	3,59019	4,00746	4,01746	4,00522	4,01396
22	2,83205	2,83824	2,80776	2,81252	3,58229	3,59011	4,00737	4,01737	4,00512	4,01387
23	2,83198	2,83817	2,80769	2,81246	3,58220	3,59003	4,00727	4,01727	4,00503	4,01377
24	2,83191	2,83810	2,80763	2,81239	3,58212	3,58994	4,00718	4,01718	4,00493	4,01368
25	2,83185	2,83803	2,80756	2,81233	3,58204	3,58986	4,00708	4,01708	4,00484	4,01359
26	2,83178	2,83797	2,80750	2,81226	3,58195	3,58978	4,00698	4,01698	4,00474	4,01349
27	2,83172	2,83790	2,80743	2,81220	3,58186	3,58969	4,00689	4,01689	4,00465	4,01340
28	2,83165	2,83784	2,80736	2,81213	3,58178	3,58961	4,00680	4,01680	4,00455	4,01331
29	2,83158	2,83777	2,80730	2,81207	3,58170	3,58953	4,00671	4,01671	4,00446	4,01322
30	2,83152	2,83771	2,80723	2,81200	3,58162	3,58944	4,00662	4,01662	4,00437	4,01312
31	2,83145	2,83764	2,80717	2,81194	3,58153	3,58936	4,00653	4,01653	4,00428	4,01303
32	2,83139	2,83758	2,80710	2,81188	3,58146	3,58928	4,00644	4,01644	4,00419	4,01294
33	2,83132	2,83751	2,80704	2,81181	3,58137	3,58919	4,00635	4,01635	4,00410	4,01285
34	2,83126	2,83745	2,80698	2,81175	3,58129	3,58912	4,00626	4,01626	4,00401	4,01276
35	2,83120	2,83738	2,80691	2,81168	3,58121	3,58904	4,00617	4,01617	4,00392	4,01267
36	2,83113	2,83732	2,80685	2,81162	3,58113	3,58896	4,00608	4,01608	4,00383	4,01258
37	2,83107	2,83725	2,80679	2,81156	3,58106	3,58888	4,00599	4,01599	4,00374	4,01249
38	2,83100	2,83719	2,80672	2,81149	3,58098	3,58880	4,00590	4,01590	4,00365	4,01240
39	2,83094	2,83713	2,80666	2,81143	3,58090	3,58872	4,00581	4,01581	4,00356	4,01231
40	2,83088	2,83707	2,80660	2,81137	3,58082	3,58864	4,00572	4,01572	4,00347	4,01222
41	2,83082	2,83700	2,80654	2,81131	3,58074	3,58856	4,00564	4,01564	4,00338	4,01213
42	2,83076	2,83694	2,80648	2,81124	3,58067	3,58848	4,00555	4,01555	4,00329	4,01204
43	2,83070	2,83688	2,80642	2,81118	3,58058	3,58840	4,00546	4,01546	4,00320	4,01195
44	2,83063	2,83682	2,80636	2,81112	3,58050	3,58832	4,00537	4,01537	4,00312	4,01186
45	2,83057	2,83676	2,80630	2,81106	3,58042	3,58824	4,00528	4,01528	4,00303	4,01178
46	2,83051	2,83670	2,80624	2,81100	3,58035	3,58817	4,00520	4,01520	4,00298	4,01169
47	2,83045	2,83664	2,80618	2,81094	3,58027	3,58808	4,00511	4,01511	4,00286	4,01160
48	2,83039	2,83658	2,80612	2,81088	3,58020	3,58801	4,00503	4,01503	4,00278	4,01152
49	2,83033	2,83652	2,80606	2,81082	3,58012	3,58794	4,00495	4,01494	4,00269	4,01143
50	2,83027	2,83646	2,80600	2,81076	3,58004	3,58786	4,00486	4,01485	4,00261	4,01135
51	2,83021	2,83640	2,80594	2,81070	3,57997	3,58778	4,00478	4,01476	4,00253	4,01126
52	2,83015	2,83634	2,80588	2,81064	3,57989	3,58771	4,00469	4,01467	4,00244	4,01118
53	2,83009	2,83628	2,80582	2,81058	3,57982	3,58764	4,00461	4,01459	4,00236	4,01109
54	2,83003	2,83622	2,80576	2,81053	3,57974	3,58756	4,00452	4,01451	4,00227	4,01101
55	2,82997	2,83616	2,80570	2,81047	3,57967	3,58749	4,00444	4,01443	4,00219	4,01092
56	2,82991	2,83610	2,80564	2,81041	3,57959	3,58741	4,00435	4,01435	4,00210	4,01084
57	2,82985	2,83604	2,80558	2,81035	3,57952	3,58734	4,00427	4,01427	4,00202	4,01076
58	2,82979	2,83598	2,80553	2,81030	3,57944	3,58726	4,00418	4,01418	4,00194	4,01068
59	2,82973	2,83592	2,80547	2,81024	3,57937	3,58719	4,00410	4,01410	4,00186	4,01060

ϑ	$a_{\text{Fe-alloy}}$, kX				$a_{\text{Cu-alloy}}$, kX		$a_{\text{Al-alloy}}$, kX			
	(310) CoKα		(211) CrKα		(400) CoKα		(511), (333) CuKα		(420) CoKα	
	α_1	α_2	α_1	α_2	α_1	α_2	α_1	α_2	α_1	α_2
86°00′	2,82968	2,83587	2,80541	2,81018	3,57930	3,58711	4,00402	4,01402	4,00178	4,01052
01	2,82962	2,83581	2,80535	2,81012	3,57922	3,58704	4,00394	4,01394	4,00170	4,01044
02	2,82956	2,83575	2,80530	2,81007	3,57915	3,58697	4,00386	4,01386	4,00162	4,01036
03	2,82951	2,83570	2,80524	2,81001	3,57908	3,58690	4,00378	4,01378	4,00154	4,01028
04	2,82945	2,83564	2,80519	2,80996	3,57901	3,58683	4,00370	4,01370	4,00146	4,01020
05	2,82940	2,83558	2,80513	2,80990	3,57894	3,58676	4,00362	4,01362	4,00138	4,01012
06	2,82934	2,83552	2,80508	2,80985	3,57887	3,58668	4,00354	4,01354	4,00130	4,01004
07	2,82929	2,83546	2,80502	2,80979	3,57880	3,58661	4,00345	4,01346	4,00122	4,00996
08	2,82923	2,83541	2,80497	2,80974	3,57872	3,58654	4,00338	4,01338	4,00114	4,00988
09	2,82917	2,83535	2,80491	2,80968	3,57865	3,58647	4,00330	4,01330	4,00106	4,00980
10	2,82912	2,83530	2,80485	2,80962	3,57858	3,58640	4,00323	4,01322	4,00098	4,00972
11	2,82906	2,83524	2,80480	2,80957	3,57851	3,58633	4,00315	4,01315	4,00090	4,00964
12	2,82901	2,83519	2,80474	2,80951	3,57845	3,58627	4,00307	4,01307	4,00083	4,00956
13	2,82895	2,83513	2,80469	2,80946	3,57838	3,58620	4,00300	4,01299	4,00075	4,00949
14	2,82890	2,83508	2,80463	2,80940	3,57832	3,58613	4,00292	4,01291	4,00067	4,00941
15	2,82885	2,83503	2,80458	2,80935	3,57825	3,58607	4,00284	4,01283	4,00060	4,00933
16	2,82879	2,83497	2,80452	2,80929	3,57818	3,58600	4,00277	4,01276	4,00052	4,00926
17	2,82874	2,83492	2,80447	2,80924	3,57811	3,58592	4,00269	4,01268	4,00045	4,00918
18	2,82868	2,83486	2,80442	2,80918	3,57804	3,58585	4,00261	4,01260	4,00037	4,00911
19	2,82863	2,83481	2,80437	2,80913	3,57797	3,58578	4,00253	4,01252	4,00029	4,00903
20	2,82858	2,83476	2,80432	2,80908	3,57790	3,58572	4,00246	4,01245	4,00022	4,00896
21	2,82852	2,83470	2,80426	2,80902	3,57784	3,58565	4,00239	4,01238	4,00014	4,00888
22	2,82847	2,83465	2,80421	2,80897	3,57777	3,58559	4,00232	4,01230	4,00007	4,00881
23	2,82842	2,83460	2,80416	2,80892	3,57770	3,58552	4,00224	4,01223	3,99999	4,00873
24	2,82837	2,83455	2,80411	2,80887	3,57764	3,58545	4,00217	4,01216	3,99992	4,00866
25	2,82832	2,83450	2,80406	2,80882	3,57757	3,58539	4,00210	4,01209	3,99984	4,00858
26	2,82827	2,83445	2,80401	2,80877	3,57751	3,58532	4,00202	4,01201	3,99977	4,00851
27	2,82822	2,83440	2,80396	2,80872	3,57744	3,58526	4,00195	4,01194	3,99970	4,00844
28	2,82817	2,83435	2,80391	2,80867	3,57738	3,58519	4,00188	4,01187	3,99963	4,00837
29	2,82812	2,83430	2,80386	2,80862	3,57731	3,58512	4,00181	4,01180	3,99956	4,00830
30	2,82807	2,83425	2,80381	2,80857	3,57725	3,58506	4,00173	4,01172	3,99949	4,00823
31	2,82802	2,83420	2,80376	2,80852	3,57718	3,58500	4,00166	4,01165	3,99942	4,00816
32	2,82797	2,83415	2,80371	2,80847	3,57712	3,58494	4,00159	4,01158	3,99935	4,00809
33	2,82792	2,83410	2,80366	2,80842	3,57706	3,58488	4,00152	4,01151	3,99928	4,00802
34	2,82787	2,83405	2,80361	2,80837	3,57700	3,58481	4,00145	4,01144	3,99921	4,00795
35	2,82782	2,83400	2,80356	2,80832	3,57694	3,58475	4,00138	4,01137	3,99914	4,00788
36	2,82777	2,83395	2,80351	2,80827	3,57688	3,58469	4,00131	4,01130	3,99907	4,00781
37	2,82772	2,83390	2,80347	2,80823	3,57681	3,58463	4,00124	4,01123	3,99900	4,00774
38	2,82767	2,83385	2,80342	2,80818	3,57675	3,58457	4,00117	4,01116	3,99893	4,00767
39	2,82762	2,83380	2,80337	2,80813	3,57669	3,58451	4,00110	4,01109	3,99886	4,00760
40	2,82758	2,83375	2,80332	2,80808	3,57664	3,58444	4,00104	4,01102	3,99880	4,00753
41	2,82753	2,83370	2,80327	2,80803	3,57657	3,58438	4,00097	4,01095	3,99873	4,00746
42	2,82748	2,83365	2,80322	2,80798	3,57651	3,58432	4,00090	4,01088	3,99866	4,00740
43	2,82743	2,83361	2,80318	2,80794	3,57645	3,58427	4,00084	4,01082	3,99860	4,00733
44	2,82739	2,83356	2,80313	2,80789	3,57639	3,58420	4,00077	4,01075	3,99853	4,00726
45	2,82734	2,83351	2,80309	2,80785	3,57633	3,58414	4,00070	4,01069	3,99846	4,00720
46	2,82729	2,83346	2,80304	2,80780	3,57627	3,58408	4,00064	4,01062	3,99839	4,00713
47	2,82725	2,83342	2,80300	2,80776	3,57621	3,58403	4,00057	4,01056	3,99832	4,00706
48	2,82720	2,83337	2,80295	2,80771	3,57616	3,58397	4,00050	4,01049	3,99826	4,00700
49	2,82715	2,83332	2,80291	2,80767	3,57610	3,58391	4,00044	4,01043	3,99820	4,00693
50	2,82711	2,83328	2,80286	2,80762	3,57604	3,58385	4,00038	4,01036	3,99813	4,00687
51	2,82706	2,83323	2,80282	2,80758	3,57598	3,58379	4,00031	4,01030	3,99807	4,00680
52	2,82702	2,83319	2,80277	2,80753	3,57592	3,58374	4,00025	4,01023	3,99800	4,00674
53	2,82697	2,83314	2,80273	2,80749	3,57587	3,58368	4,00019	4,01017	3,99794	4,00667
54	2,82692	2,83310	2,80268	2,80744	3,57581	3,58363	4,00012	4,01011	3,99787	4,00661
55	2,82688	2,83306	2,80264	2,80739	3,57576	3,58357	4,00006	4,01005	3,99781	4,00654
56	2,82683	2,83301	2,80259	2,80735	3,57570	3,58352	4,00000	4,00998	3,99775	4,00648
57	2,82679	2,83297	2,80255	2,80730	3,57565	3,58346	3,99993	4,00992	3,99769	4,00642
58	2,82675	2,83292	2,80251	2,80726	3,57559	3,58340	3,99987	4,00985	3,99763	4,00636
59	2,82671	2,83288	2,80246	2,80722	3,57554	3,58335	3,99981	4,00979	3,99757	4,00630

ϑ	$a_{Fe\text{-}alloy}$, kX				$a_{Cu\text{-}alloy}$, kX		$a_{Al\text{-}alloy}$, kX			
	(310) CoKα		(211) CrKα		(400) CoKα		(511), (333) CuKα		(420) CoKα	
	$α_1$	$α_2$	$α_1$	$α_2$	$α_1$	$α_2$	$α_1$	$α_2$	$α_1$	$α_2$
87°00′	2,82667	2,83284	2,80242	2,80718	3,57548	3,58329	3,99975	4,00973	3,99751	4,00624
01	2,82662	2,83280	2,80237	2,80713	3,57542	3,58326	3,99969	4,00967	3,99745	4,00618
02	2,82658	2,83275	2,80233	2,80709	3,57537	3,58318	3,99963	4,00961	3,99739	4,00612
03	2,82654	2,83271	2,80229	2,80705	3,57532	3,58313	3,99957	4,00955	3,99733	4,00606
04	2,82650	2,83267	2,80224	2,80700	3,57526	3,58307	3,99951	4,00949	3,99727	4,00600
05	2,82645	2,83263	2,80220	2,80696	3,57521	3,58302	3,99945	4,00943	3,99721	4,00594
06	2,82641	2,83258	2,80216	2,80692	3,57516	3,58297	3,99939	4,00937	3,99715	4,00588
07	2,82637	2,83254	2,80212	2,80688	3,57510	3,58291	3,99933	4,00931	3,99709	4,00582
08	2,82633	2,83250	2,80208	2,80684	3,57505	3,58286	3,99927	4,00925	3,99703	4,00576
09	2,82629	2,83246	2,80204	2,80680	3,57500	3,58281	3,99921	4,00919	3,99697	4,00570
10	2,82625	2,83242	2,80200	2,80676	3,57495	3,58276	3,99915	4,00914	3,99692	4,00565
11	2,82621	2,83238	2,80196	2,80672	3,57490	3,58271	3,99910	4,00908	3,99686	4,00559
12	2,82617	2,83234	2,80192	2,80668	3,57485	3,58266	3,99904	4,00902	3,99681	4,00554
13	2,82613	2,83230	2,80188	2,80664	3,57480	3,58261	3,99899	4,00897	3,99675	4,00548
14	2,82609	2,83226	2,80184	2,80660	3,57475	3,58256	3,99893	4,00891	3,99670	4,00543
15	2,82605	2,83222	2,80180	2,80656	3,57470	3,58251	3,99888	4,00886	3,99664	4,00537
16	2,82601	2,83218	2,80176	2,80652	3,57465	3,58246	3,99882	4,00880	3,99659	4,00531
17	2,82597	2,83214	2,80172	2,80648	3,57460	3,58241	3,99877	4,00875	3,99653	4,00526
18	2,82593	2,83210	2,80169	2,80645	3,57455	3,58236	3,99871	4,00869	3,99648	4,00520
19	2,82589	2,83206	2,80165	2,80641	3,57450	3,58230	3,99865	4,00864	3,99642	4,00514
20	2,82585	2,83202	2,80161	2,80637	3,57445	3,58226	3,99860	4,00858	3,99636	4,00509
21	2,82581	2,83199	2,80157	2,80633	3,57440	3,58221	3,99854	4,00853	3,99630	4,00503
22	2,82577	2,83195	2,80154	2,80630	3,57435	3,58216	3,99849	4,00847	3,99625	4,00497
23	2,82573	2,83191	2,80150	2,80626	3,57430	3,58211	3,99844	4,00842	3,99619	4,00492
24	2,82569	2,83187	2,80146	2,80622	3,57426	3,58207	3,99838	4,00837	3,99613	4,00486
25	2,82566	2,83184	2,80143	2,80619	3,57421	3,58202	3,99833	4,00831	3,99608	4,00481
26	2,82562	2,83180	2,80139	2,80615	3,57416	3,58197	3,99828	4,00826	3,99603	4,00476
27	2,82558	2,83176	2,80135	2,80611	3,57412	3,58193	3,99822	4,00821	3,99598	4,00471
28	2,82555	2,83173	2,80131	2,80607	3,57408	3,58189	3,99817	4,00815	3,99593	4,00466
29	2,82551	2,83169	2,80128	2,80604	3,57403	3,58184	3,99812	4,00810	3,99588	4,00461
30	2,82548	2,83165	2,80124	2,80600	3,57398	3,58179	3,99807	4,00805	3,99583	4,00456
31	2,82544	2,83161	2,80120	2,80596	3,57394	3,58175	3,99802	4,00800	3,99578	4,00451
32	2,82541	2,83158	2,80117	2,80593	3,57389	3,58170	3,99797	4,00795	3,99573	4,00446
33	2,82527	2,83154	2,80113	2,80589	3,57385	3,58165	3,99792	4,00790	3,99568	4,00441
34	2,82534	2,83151	2,80110	2,80586	3,57380	3,58161	3,99787	4,00785	3,99563	4,00436
35	2,82530	2,83148	2,80106	2,80582	3,57376	3,58158	3,99782	4,00780	3,99558	4,00431
36	2,82527	2,83144	2,80103	2,80579	3,57372	3,58152	3,99778	4,00775	3,99553	4,00426
37	2,82523	2,83141	2,80099	2,80575	3,57367	3,58148	3,99773	4,00770	3,99548	4,00421
38	2,82520	2,83137	2,80096	2,80572	3,57363	3,58144	3,99768	4,00765	3,99543	4,00416
39	2,82516	2,83133	2,80093	2,80569	3,57359	3,58139	3,99763	4,00760	3,99538	4,00411
40	2,82513	2,83130	2,80090	2,80566	3,57354	3,58135	3,99759	4,00756	3,99534	4,00407
41	2,82509	2,83126	2,80086	2,80562	3,57350	3,58131	3,99754	4,00751	3,99529	4,00402
42	2,82506	2,83123	2,80083	2,80559	3,57346	3,58126	3,99749	4,00746	3,99525	4,00398
43	2,82503	2,83120	2,80080	2,80555	3,57342	3,58122	3,99745	4,00742	3,99520	4,00394
44	2,82500	2,83117	2,80076	2,80552	3,57338	3,58118	3,99740	4,00737	3,99516	4,00389
45	2,82497	2,83114	2,80073	2,80549	3,57334	3,58114	3,99736	4,00733	3,99511	4,00385
46	2,82493	2,83111	2,80070	2,80546	3,57329	3,58110	3,99731	4,00728	3,99507	4,00380
47	2,82490	2,83108	2,80067	2,80543	3,57325	3,58106	3,99726	4,00724	3,99502	4,00376
48	2,82487	2,83104	2,80064	2,80540	3,57321	3,58102	3,99722	4,00719	3,99498	4,00371
49	2,82484	2,83101	2,80061	2,80537	3,57317	3,58098	3,99717	4,00715	3,99493	4,00366
50	2,82481	2,83098	2,80058	2,80534	3,57313	3,58094	3,99713	4,00710	3,99489	4,00361
51	2,82478	2,83095	2,80055	2,80531	3,57309	3,58090	3,99709	4,00706	3,99484	4,00357
52	2,82475	2,83092	2,80052	2,80528	3,57306	3,58086	3,99704	4,00702	3,99480	4,00352
53	2,82472	2,83089	2,80049	2,80525	3,57302	3,58082	3,99700	4,00697	3,99475	4,00348
54	2,82469	2,83086	2,80046	2,80522	3,57298	3,58078	3,99696	4,00693	3,99471	4,00343
55	2,82466	2,83083	2,80043	2,80519	3,57294	3,58074	3,99691	4,00689	3,99466	4,00339
56	2,82463	2,83080	2,80040	2,80516	3,57290	3,58071	3,99687	4,00684	3,99462	4,00335
57	2,82460	2,83077	2,80037	2,80513	3,57286	3,58067	3,99683	4,00680	3,99458	4,00331
58	2,82457	2,83074	2,80035	2,80510	3,57282	3,58063	3,99678	4,00676	3,99454	4,00327
59	2,82454	2,83071	2,80032	2,80507	3,57279	3,58060	3,99674	4,00672	3,99450	4,00323
88°00′	2,82451	2,83068	2,80029	2,80504	3,57275	3,58056	3,99670	4,00668	3,99446	4,00319

7-3. Stress Producing a Line Shift of 0.1 mm for Various Materials and Conditions

The table gives the sum of the principal stresses corresponding to a line shift of 0.1 mm for various materials [13] used with standards, the diameter $2l_s$ of the ring for the standard being 50.0 or 70.0 mm. Table 7-5 gives the distance A from specimen to film; it is an extension of this table.

Under these conditions the table enables one to deduce $\sigma_1 + \sigma_2$ rapidly, the formula being

$$\sigma_1 + \sigma_2 = 10 \, \Delta l \, \Delta \, (\sigma_1 + \sigma_2),$$

where Δl is the line shift in millimeters and $\Delta(\sigma_1 + \sigma_2)$ is taken from the table.

Material	a, kX	E, kg/mm^2	ν	Rad.	Std.	$2l_s$, mm	Shift, mm	(hk$l$)	ϑ	$2l$, mm	$\Delta a \cdot 10^3$	$\Delta(\sigma_1+\sigma_2)$, kg/mm^2
Aluminum	4,0414	7 200	0,34	Cu	Au	50,0	0,1	(383)	81°14,5′	38,79	0,461	2,41
Duralumin	4,0340	7 400	0,34	Cu	Au	50,0	0,1	(333)	81°57′	35,46	0,426	2,30
Iron	2,8610	21 000	0,28	Co	Au	50,0	0,1	(310)	80°37,5′	41,04	0,350	9,18
	2,8610	21 000	0,28	Co	Ag	50,0	0,1	(310)	80°37,5′	38,85	0,370	9,70
	2,8610	21 000	0,28	Cr	Cr	50,0	0,1	(112)	78°0,5′	43,48	0,519	13,60
Copper	3,6077	12 500	0,34	Co	Au	50,0	0,1	(400)	81°46,5′	35,71	0,397	4,04
Brass	3,6880	9 000	0,35	Co	Au	50,0	0,1	(400)	75°30′	67,00	0,603	4,21
Electron	3,180*	4 500	0,30	Fe	Al	70,0	0,1	(114)	74°30′	62,20	0,197**	2,95

* $c/a = 5,166$.
** $d_{114} = 1,0025$.

7-4. Accessory Function for Calculating Macroscopic Stresses in Iron from Patterns Taken with Cobalt Radiation

For back-reflection and a flat film

$$\frac{r}{A} = \text{tg} \left(\pi - 2 \text{arc} \sin \frac{\lambda}{2d} \right) = \text{tg} \, B,$$

where r is the radius of the ring and A is the specimen-to-film distance;

$$\frac{\lambda}{2d} = \frac{\lambda \sqrt{10}}{2a} \, .$$

The table gives tg B for the (310) line as recorded with cobalt radiation [111].
The values enable one to deduce a from r (A is usually constant).
Examples. (1) r/A = 0.29461, a = 2.8520 kX; (2) r/A = 0.34134, a = 2.8614 kX.

a, kX	r/A=tg B									
	0	1	2	3	4	5	6	7	8	9
2,850	0,28388	0,28443	0,28497	0,28551	0,28605	0,28659	0,28713	0,28767	0,28821	0,28875
1	0,28929	0,28982	0,29036	0,29089	0,29142	0,29196	0,29249	0,29302	0,29344	0,29408
2	0,29461	0,29514	0,29566	0,29619	0,29672	0,29724	0,29777	0,29829	0,29881	0,29933
3	0,29985	0,30037	0,30089	0,30141	0,30193	0,30245	0,30297	0,30348	0,30400	0,30451
4	0,30503	0,30554	0,30605	0,30657	0,30708	0,30759	0,30810	0,30861	0,30911	0,30962
5	0,31013	0,31063	0,31114	0,31165	0,31215	0,31266	0,31316	0,31366	0,31416	0,31466
6	0,31516	0,31567	0,31617	0,31667	0,31716	0,31766	0,31816	0,31866	0,31915	0,31965
7	0,32014	0,31063	0,32113	0,32162	0,32211	0,32260	0,32309	0,32359	0,32407	0,32456
8	0,32505	0,32554	0,32603	0,32651	0,32700	0,32749	0,32797	0,32846	0,32891	0,32942
9	0,32991	0,33039	0,33087	0,33135	0,33183	0,33231	0,33279	0,33327	0,33375	0,33423
2,860	0,33471	0,33518	0,33566	0,33614	0,33661	0,33709	0,33756	0,33804	0,33851	0,33898
1	0,33945	0,33993	0,34040	0,34087	0,34134	0,34181	0,34228	0,34275	0,34322	0,34368
2	0,34415	0,34462	0,34508	0,34555	0,34602	0,34648	0,34694	0,34741	0,34787	0,34834
3	0,34880	0,34926	0,34972	0,35019	0,35064	0,35111	0,35156	0,35203	0,35249	0,35295
4	0,35340	0,35386	0,35432	0,35478	0,35523	0,35569	0,35614	0,35660	0,35705	0,35751
5	0,35796	0,35842	0,35887	0,35932	0,35977	0,36022	0,36068	0,36113	0,36158	0,36203
6	0,36248	0,36293	0,36338	0,36382	0,36427	0,36472	0,36516	0,36561	0,36605	0,36650
7	0,36695	0,36739	0,36784	0,36829	0,36873	6,36917	0,36961	0,37006	0,37050	0,37094
8	0,37138	0,37183	0,37227	0,37271	0,37314	0,37359	0,37403	0,37447	0,37490	0,37534
9	0,37578	0,37622	0,37666	0,37709	0,37753	0,37797	0,37840	0,37884	0,37927	0,37971
2,870	0,38014	0,38058	0,38101	0,38144	0,38188	0,38231	0,38274	0,38317	0,38361	0,38404
1	0,38447	0,38490	0,38533	0,38576	0,38619	0,38662	0,38705	0,38747	0,38790	0,38833
2	0,38876	0,38919	0,38961	0,39004	0,39047	0,39089	0,39132	0,39174	0,39217	0,39259

7-5. Values of Constants in Stress-Determination Relations

The distances between lines of standard and specimen may be measured in the deduction of macroscopic stresses; here

$$\sigma_1 + \sigma_2 = (\delta_\perp - \delta_0) C_{\perp 0}, \tag{a}$$

$$\sigma_x = (\delta_\perp - \delta_{\psi_1}) C_{\perp +}, \tag{b}$$

$$\sigma_x = (\delta_\perp - \delta_{\psi_2}) C_{\perp -}, \tag{c}$$

$$\sigma_x = (\delta_{\psi_2} - \delta_{\psi_1}) C_{+-}, \tag{d}$$

$$\sigma_x = (\delta_0 - \delta') C_{0+}, \tag{e}$$

where $\psi_1 = \psi_0 + \eta$, $\psi_2 = \psi_0 - \eta$ ($\psi_0 = 45°$) are the angles between the normal to the surface of the specimen and the beam; δ_{ψ_1} is the distance between lines of the specimen and standard on the side of the film corresponding to the angle ψ_1, δ_{ψ_2} is the same but for the side corresponding to ψ_2; the C's are constants; $\delta_\perp$, δ_0, and δ' are distances between lines of the specimen and the standard: $\delta_\perp$ is measured with the beam at right angles to the specimen; δ_0 corresponds to the state free from stress, or, in the case of uniaxial stress, to the value found by examining the specimen at the angle ψ'_0 given in the table for the $\psi'_0 - \eta$ side of the film; δ' is determined on the $\psi'_0 + \eta$ side.

The equations relate to the case in which the lines of the specimen lie within those of the standard; the sign in the right-hand side of each equation becomes minus if the reverse is the case.

The C's listed in the table have been calculated for the A's also listed there; the diameter $2l_s$ for the ring of the standard is 50.0 or 70.0 mm, as stated in Table 7-3, which is the source of Table 7-5. The equations can be used for any A if the quantity inserted in the formulas is the δ_{red} given by $\delta_{red} = \delta_{meas}[2l_s/(2l_s)_{meas}]$, where $(2l_s)_{meas}$ is the diatance between lines of the standard as measured on the film.

Material	Change, kX/mm	kg/mm³						
		C_{0+}	$C_{\perp+}$	$C_{\perp-}$	C_{+-}	$C_{\perp 0}$	ψ_0	A, mm
Aluminum	0,00461	24,1	9,4	17,6	20,4	20,9	39°1'	61,445
Duralumin	0,00426	23,0	9,1	16,0	21,1	21,7	38°18'	61,445
	0,00693	37,4	13,8	30,6	25,0	25,2	41°24'	49,13
	0,00447	24,1	9,5	17,1	21,2	21,8	38°33'	60,453
Iron	0,00350	91,8	30,4	59,2	62,5	64,8	37°16'	60,453
	0,00516	135,3	42,1	99,7	72,8	74,0	39°53'	49,13
	0,00519	136,0	42,3	100,2	73,2	74,4	39°53'	48,875
Nitrided iron	0,00242	154,3	46,4	123,6	74,4	75,0	41°23'	46,32
68/32 brass	0,00603	42,1	14,7	42,3	22,5	22,5	45°6'	60,453
Copper	0,00397	40,4	16,0	28,6	36,3	37,3	38°29'	60,453
Electron	0,00197	29,5	9,0	28,1	13,2	13,2	44°13'	51,756

7-6. Correction for Relation Between σ_x and σ_y

The σ_x given by the inclined pattern is obtained as a relation between σ_x and σ_y.
The table gives values of the correction (kg/mm²) for macroscopic stresses in various materials for various ratios of σ_x to σ_y (in kg/mm²) [111]. The table relates to expression (d) in Section 7-5.

Material	Radiation	$\sigma_x = +30$ $\sigma_y = 0$	+30 +30	+30 −30	+100 0	+100 +100	+100 −100
Duralumin	Cu	0,18	0,00	0,37	—	—	—
Iron	Co	0,08	0,02	0,15	0,91	0,20	1,62
Iron	Cr	0,08	0,02	0,15	0,93	0,21	1,65
Brass	Co	0,17	0,00	0,36	—	—	—

7-7. Correction for Film Movement

The film is rocked through a range $\pm \delta$ when coarse-grained materials are used, and in certain other cases, in order to obtain a continuous line.

The table gives values of the correction as a percent of the measured stresses for values of $n = \sigma_y/\sigma_x$ [111]. The table relates to expression (e) of Section 7-5 for iron specimens examined with Co Kα radiation.

±δ°	n, %						
	−6	−3	−1	0	+1	+3	+6
10	+ 1,1	+ 0,7	+ 0,5	+0,4	+0,2	0	−0,4
20	+ 4,9	+ 3,4	+ 2,5	+2,0	+1,5	+0,5	−0,9
30	+10,4	+ 7,3	+ 5,3	+4,3	+3,3	+1,2	−1,8
40	+17,4	+12,3	+ 8,9	+7,2	+5,5	+2,1	−3,0
45	+21,3	+15,2	+11,1	+9,1	+7,1	+3,1	−3,1

7-8. Nomogram for Determining Stresses

Figure 143 is used to deduce macroscopic stresses from patterns recorded with CoKα radiation and a silver standard; it applies to the (310)α line of Fe and the (420)α line of Ag [154].

The lattice constant is deduced from the difference in ring diameters:

$$\Delta_1 = D_{Ag\ K\alpha_1} - D_{Fe\ K\alpha_1},$$

or

$$\Delta_2 = D_{Ag\ K\alpha_1} - D_{Fe\ K\alpha_2},$$

and from $D_{Ag\ K\alpha_1}$. The nomogram applies to lattice constants between 2.8540 and 2.8670 Å.

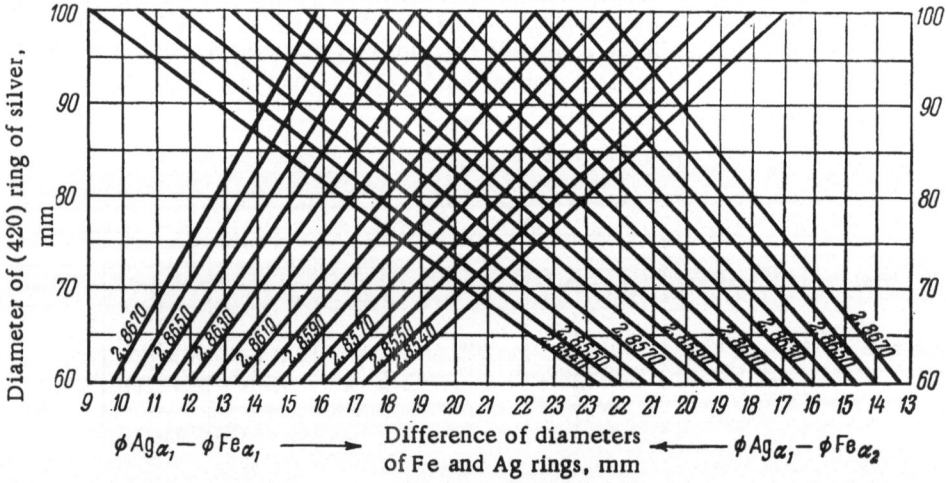

Fig. 143. Nomogram for determining macroscopic stresses in iron and steel with cobalt radiation.

The abscissa shows from left to right the difference in diameters of the Agα₁ and Feα₁ rings; from right to left it does the same for Agα₁ and Feα₂. In the first case one uses the lines running upwards to the right; in the second the lines running to the left.

For example, a difference of 27 mm in the diameter of the Agα₁ and Feα₁ rings, together with a diameter of 90 mm for the silver ring, gives a lattice constant of 2.8540 kX.

DETERMINATION OF CRYSTALLITE
AND BLOCK SIZES, OF MICROSTRESSES,
AND OF LATTICE DISTORTION

Chapter 8 gives tables and graphs for the determination of crystallite and block sizes, microstresses, and lattice distortions.

See [6-8, 10-13] for experimental techniques and for methods of calculating block sizes and lattice distortions.

8-1. Determination of Crystallite Sizes

8-1a. From Size and Number of Spots on Pattern

There is a fixed relation of grain size to spot size for crystallites over $5 \cdot 10^{-3}$ mm in size.

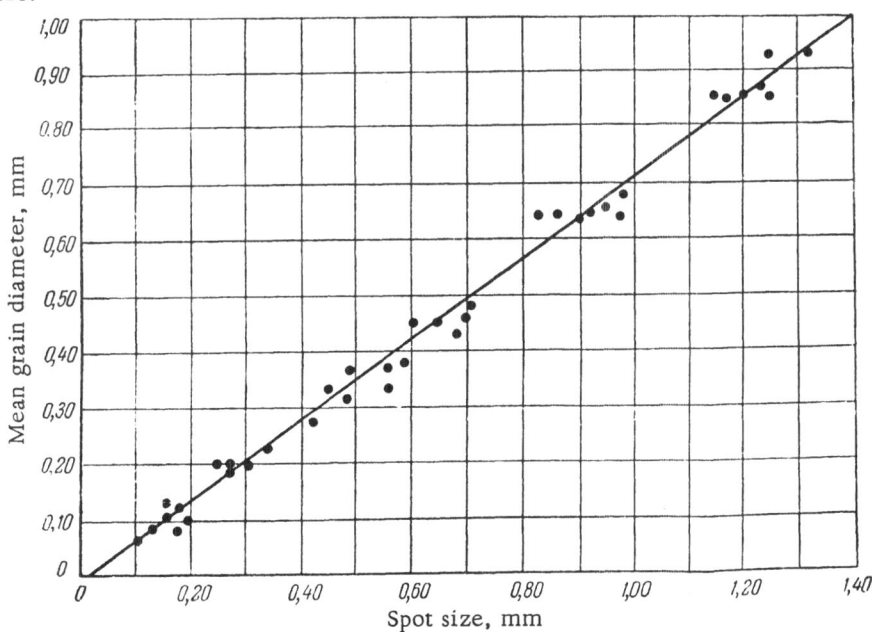

Fig. 144. Calibration curve for derivation of crystallite sizes from size of spots.

Figure 144 gives a calibration curve derived from specimens of brass, steel, carborundum, and quartz; it is applicable to other materials [315].

The number of spots on a diffraction ring, n_{hkl} , is related to block (crystallite) size by

$$D = \frac{A}{\sqrt[3]{n_{hkl}}} \ ,$$

where A is a constant.

For absolute determinations we have

$$D = \sqrt[3]{\frac{V p_{hkl} a \cos \vartheta}{\cdot 2L \, n_{hkl}}} \ ,$$

where L is the specimen-to-film distance; a is the effective diameter of the irradiated area on the specimen, V is the effective irradiated volume, which is

$$V = \frac{\pi a^2 h}{4} \ ,$$

in which h is the depth of penetration; and p_{hkl} is the multiplicity factor for the (hkl) plane. Two exposures are used to eliminate h and V; here we have

$$D = \sqrt[3]{\frac{\pi a^3 p \cos \vartheta \ln r \, (n_2 - n_1)}{8L n_{hkl}' \mu \, (1 - \sec 2\vartheta)}}$$

where n_{hkl}^* is the number of spots from the longer exposure, n_1 and n_2 are the numbers of spots from exposure times t_1 and t_2, $r = t_2 / t_1$, and μ is the linear attenuation coefficient.

8-1b. Table for Determining Block Sizes from Line Intensities

In the simplest case we have

$$\frac{I}{I_s} = \frac{\tanh nq}{nq} \ ,$$

where I is the integral line intensity; I_s is the line intensity for an ideal mosaic standard free from primary extinction; q is the reflectivity of the atomic plane, which is given by

$$q = \frac{e^2}{mc^2} \, N d \lambda \frac{1}{\sin \vartheta} \, |F| .$$

in which e is the charge on an electron, m is the mass of an electron, c is the velocity of light, $e^2/mc^2 = 2.82 \cdot 10^{-13}$ cm, N is the number of unit cells in unit volume, d is the interplanar distance for the given system of planes, λ is the wavelength, F is the structure factor; and n is the number of parallel atomic planes within a block. The block size is

$$D = d_{hkl} n.$$

The table gives I/I_s for nq from 0.1 to 3.0 [316].

nq	I/I_s	nq	I/I_s	nq	I/I_s
0,1	0,997	0,9	0,800	2,2	0,440
0,2	0,987	1,0	0,760	2,4	0,410
0,3	0,871	1,2	0,700	2,6	0,380
0,4	0,950	1,4	0,630	2,8	0,360
0,5	0,924	1,6	0,580	3,0	0,330
0,7	0,895	1,8	0,530		
0,8	0,863	2,0	0,480		

8-2. Measurement of Crystallite and Block Sizes from Line Intensities

Loss of intensity from primary extinction is most pronounced for lines of high intensity having small values of $(h + k + l)$.

The line intensity given by a coarse-grained powder may [157] be put as

$$P_{hkl} = P^0_{hkl} B_{hkl} R_{hkl}. \tag{57}$$

where P^0_{hkl} is the reflected intensity,

$$P^0_{hkl} = IQA \frac{1 - e^{-\mu V l a}}{\mu a} e^{-\mu V t}, \tag{58}$$

in which I is the mean beam strength, A is the cross section of the primary beam, μ is the linear absorption coefficient, V is the proportion of the volume taken up by the material per cubic centimeter, t is the thickness of the specimen, $a = \sec 2\vartheta$, and Q is given by

$$Q = p \frac{1 + \cos^2 2\vartheta}{\cos \vartheta \sin^2 \vartheta} (F_{hkl})^2 N^2 \lambda^3; \tag{59}$$

here p is the multiplicity factor, F_{hkl} is the structure factor, and N is the number of unit cells per cubic centimeter.

The factors B_{hkl} and R_{hkl} in (57) are related to primary extinction; the second is the particle-size factor and the first is the block-size factor, which is given by

$$B_{hkl} = \frac{\mu}{\varepsilon} \frac{e^{-\frac{2}{3} \varepsilon D} - e^{-\frac{2}{3} \varepsilon D (a+1)}}{e^{-\frac{2}{3} \varepsilon D} - e^{-\frac{2}{3} \mu D (a-1)}}, \tag{60}$$

where D is the block size (linear). Equation (60) involves the assumption that the blocks are spheres of diameter D; the D of (60) must be replaced by $D' = D(1 + \alpha)$ if the deviation from spherical is large (α is a factor to be determined by experiment). In (60), ε is the sum of μ (the linear attenuation coefficient) and μ_ε (the attenuation coefficient arising from primary extinction). The latter is given by

$$\mu_\varepsilon = \frac{3\pi}{16} \lambda | F_{hkl} | R \frac{e^2}{mc} \frac{1 + \cos^2 2\vartheta}{1 + \cos 2\vartheta}, \tag{61}$$

where e is the charge on an electron (in electrostatic units), m is the mass of an electron, c is the velocity of light, and λ is the wavelength of the radiation.

The block size is deduced from the relative line intensities on the basis that the ratio of intensities for two lines of not too large angular separation,

$$\frac{P_{hkl}}{P_{h'k'l'}} = \frac{P^0_{hkl} B_{hkl} R_{hkl}}{P^0_{h'k'l'} B_{h'k'l'} R_{h'k'l'}}, \tag{62}$$

has its two particle-size factors nearly equal, so these cancel out; (62) then becomes

$$\frac{P_{hkl}}{P_{h'k'l'}} = \frac{P^0_{hkl}}{P^0_{h'k'l'}} \frac{B_{hkl}}{B_{h'k'l'}} ,$$ (63)

where D is the unknown.

It is usually difficult to use (63) directly, so some preliminary operations are first performed. Expressions (60) and (61) are applied to the chosen line, whose indices are (hkl), to calculate B_{hkl} as a function of D; these values are applied to the chosen pair of lines, for which P^0_{hkl} is calculated in accordance with the actual conditions, and (63) is used to plot $P_{hkl}/P_{h'k'l'}$ as a function of the size of the coherent-scattering blocks. The curve enables one to deduce the block size for any ratio of intensities in the pattern.

The line intensity is governed by P_{hkl} , the energy of the rays diffracted by the specimen; for recording in transmission with a flat film, the diameter of a ring is r_{hkl} = const/cos $2\vartheta_{hkl}$. The line intensity is proportional to the energy received per unit length of arc, if the ring is evenly blackened:

$$I_{hkl} \propto \frac{P_{hkl}}{2\pi r_{hkl}} \propto P_{hkl} \cos 2\vartheta_{hkl}.$$

However, it is best to calculate block sizes not from (63) but from

$$\frac{I_{hkl}}{I_{h'k'l'}} = \frac{P^0_{hkl} \cos 2\vartheta_{hkl}}{P^0_{h'k'l'} \cos 2\vartheta_{h'k'l'}} \frac{B_{hkl}}{B_{h'k'l'}} .$$ (64)

Some examples are as follows. The block size for the hexagonal form of graphite is deduced from the intensity of the (002) line relative to the nearby (100) and (101) lines, which are taken as a single line. A specimen of powder 0.1 cm thick and of density 0.5474 was examined with copper radiation; μ = 12.37 was calculated from the tabulated μ/ρ = 5.50 and ρ = 2.25, while ϑ of the lines were deduced from a = 2.456 Å and c = 6.696 Å; μ_ε was deduced from (61).

Table 1 gives the results.

TABLE 1

Parameter	Line		
	(002)	(100)	(101)
ϑ	13°16′	21°11′	22°17′
$\lvert F \rvert$	$4f=15,24$	$f=2,72$	$\sqrt{3}f=4,55$
ε	1116,47	196,73	318,12

Table 1 and (60) together gave the B_{hkl} listed in Table 2.

TABLE 2

D, cm	B_{hkl} for		
	(002)	(100)	(101)
$1 \cdot 10^{-5}$	0,9622	0,9985	0,9976
$1 \cdot 10^{-4}$	0,9251	0,9855	0,9758
$5 \cdot 10^{-4}$	0,6772	0,9297	0,8844
$1 \cdot 10^{-3}$	0,4589	0,8652	0,7829
$2 \cdot 10^{-3}$	0,2107	0,7491	0,6135
$5 \cdot 10^{-3}$	0,0208	0,4869	0,3015
$1 \cdot 10^{-2}$	0,0003	0,2376	0,0891
$2 \cdot 10^{-2}$	0,0000	0,0575	0,0084
$5 \cdot 10^{-2}$	0,0000	0,0098	0,0000
$1 \cdot 10^{-1}$	0,0000	0,0000	0,0000

Then (58) gives the P_{hkl} , which together with the B_{hkl} are used to plot $I_{002}/(I_{100} + I_{101})$ against D from

$$\frac{I_{002}}{I_{100}+I_{101}} = \frac{P_{002}^0 \cos 2\vartheta_{002} B_{002}}{P_{100}^0 \cos 2\vartheta_{100} B_{100} + P_{101}^0 \cos 2\vartheta_{101} B_{101}} \ .$$

Table 3 gives the results.

TABLE 3

D, cm	$\dfrac{I_{002}}{I_{100}+I_{101}}$	D, cm	$\dfrac{I_{002}}{I_{100}+I_{101}}$
$1 \cdot 10^{-6}$	6,612	$1 \cdot 10^{-3}$	3,801
$1 \cdot 10^{-5}$	6,575	$2 \cdot 10^{-3}$	2,197
$1 \cdot 10^{-4}$	6,257	$5 \cdot 10^{-3}$	0,413
$5 \cdot 10^{-4}$	5,018	$1 \cdot 10^{-2}$	0,018

Figure 145 shows that the relative intensity curve is of very large slope in the range of block sizes from 10^{-4} to 10^{-2} cm; the size is measured most accurately if it falls in this range.

TABLE 4

$\mu = 93,04$, $a = 4,903$ Å, $c = 5,393$ Å

Parameter	Lines					
	(100)	(101) + (011)		(110)	(102) + (012)	
ϑ	10°26′	13°19′		18°16′	19°44′	
$\|F\|$	14,85	23,66	37,05	17,49	12,39	8,42
ε	435,31	644,28	956,26	511,01	512,53	296,15

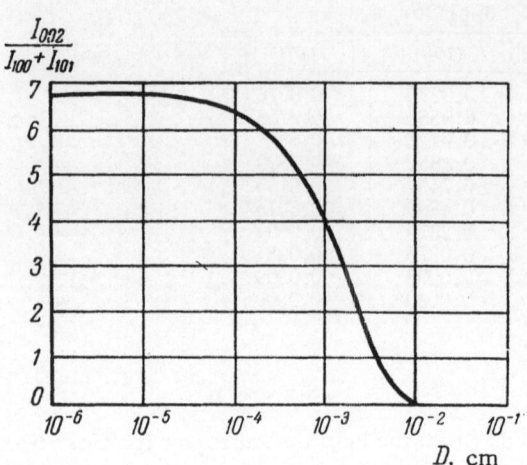

Fig. 145. Ratio of line intensities as a function of block size for hexagonal graphite.

In practice, the film can give only a restricted range of densities, and the density is related to the intensity by a fixed logarithmic formula, so it is unreasonable to apply a diagram such as Fig. 145 to lines having a large intensity ratio.

In this particular case, the (002) line should be weakened by the use of an aluminum filter, whose thickness is chosen to make the ratio of (002) to (100) + (101) close to one. The ratios given in Table 3 are then to be reduced by the factor applicable to the (002) line.

Ionization methods can accept a wide range of intensities, but here again it is desirable to attenuate one of the lines, because the lost counts increase rapidly with the intensity, which may result in serious errors in $I_{hkl}/I_{h'k'l'}$.

The method is illustrated by the application to α-quartz; the sets of lines were (100), (101), and (011) on the one hand, and (110), (102), and (012) on the other, all being recorded with copper radiation. Table 4 gives the basic parameters, which were derived as above.

Table 5 gives B.

TABLE 5

D, cm	B for					
	(100)	(101)	(011)	(110)	(102)	(012)
$1\cdot10^{-5}$	0,9977	0,9961	0,9939	0,9968	0,9968	0,9985
$1\cdot10^{-4}$	0,9767	0,9618	0,9409	0,9662	0,9684	0,9846
$5\cdot10^{-4}$	0,8909	0,7371	0,8100	0,8553	0,8518	0,9253
$1\cdot10^{-3}$	0,7897	0,6776	0,5437	0,7314	0,7257	0,8562
$2\cdot10^{-3}$	0,6236	0,4592	0,2965	0,5356	0,5271	0,7333
$5\cdot10^{-3}$	0,3151	0,1431	0,0477	0,2108	0,2030	0,4612
$1\cdot10^{-2}$	0,0944	0,0206	0,0023	0,0451	0,0421	0,2141
$2\cdot10^{-2}$	0,0089	0,0004	0,0000	0,0021	0,0019	0,0470
$5\cdot10^{-2}$	0,0000	0,0000	0,0000	0,0000	0,0000	0,0006
$1\cdot10^{-1}$	0,0000	0,0000	0,0000	0,0000	0,0000	0,0000

These values were obtained with a powder specimen 0.05 cm thick and of density 0.5037.

Table 6 gives the ratios as functions of block size.

Table 6 shows that all ratios fall sharply between $5\cdot10^{-4}$ and $1\cdot10^{-2}$ cm, which is therefore the best range.

In practice, the very strong (101) + (011) line should be attenuated (with aluminum foil) to bring the ratio close to one for block sizes roughly in the middle of the above range. Figure 146 shows curves for this, which indicate that the ratio of $I_{101} + I_{011}$ to I_{100}

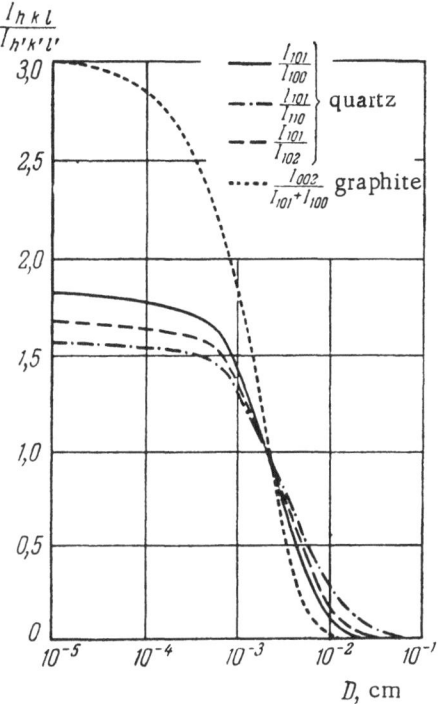

Fig. 146. Ratio of line intensities as a function of block size for quartz and graphite.

is the best in this range, not $(I_{101} + I_{011})$ to $(I_{102} + I_{012})$, as might be concluded from Table 6.

The general conditions for choice of lines are as follows. The two lines must have small ϑ not too widely different; one of the lines should have the largest possible F_{hkl}, which means that its intensity will be much dependent on block size (on account of primary extinction). The second should have a small structure factor, with an intensity varying little with block size.

Examination in reflection instead of transmission leads to somewhat different relationships [158]; B_{hkl} for a monocrystal becomes

$$B_{hkl} = \left[\frac{\mu}{\varepsilon} \frac{1 - e^{-\frac{4}{3} \varepsilon D}}{1 - e^{-\frac{4}{3} \mu D}} \right] (1 - e^{-\frac{2\mu t}{\sin \vartheta}}).$$

The spread in block orientation in a real crystal causes the secondary extinction to affect the result also; here B_n can be found from

$$B_n = \frac{\mu}{\varepsilon} \left[\frac{1 - e^{-\frac{4}{3} \varepsilon D}}{1 - e^{-\frac{4}{3} \varepsilon \mu + u(\varepsilon - \mu)}} \right] D [1 - e^{-2 [\mu + u (\varepsilon + \mu)]}] \frac{t}{\sin \vartheta}, \qquad (65)$$

where u is a factor specifying the degree of orientation. We need to know the distribution in orientation angle in order to deduce u; methods of deducing this (the mosaic angle) are given below. The relation for u is

$$u = \frac{2\Delta \vartheta}{2\Delta \vartheta'}, \qquad (66)$$

TABLE 6

D, cm	$\dfrac{I_{101} + I_{011}}{I_{100}}$	$\dfrac{I_{101} + I_{011}}{I_{110}}$	$\dfrac{I_{101} + I_{011}}{I_{102} + I_{012}}$
$1 \cdot 10^{-5}$	4,731	12,623	16,967
$1 \cdot 10^{-4}$	4,601	12,362	16,581
$5 \cdot 10^{-4}$	4,335	12,039	15,990
$1 \cdot 10^{-3}$	3,500	10,076	13,202
$2 \cdot 10^{-3}$	2,615	8,117	10,322
$5 \cdot 10^{-3}$	1,135	4,522	5,086
$1 \cdot 10^{-2}$	0,383	2,137	1,733
$2 \cdot 10^{-2}$	0,071	0,788	0,216

where $2\Delta\vartheta' = \delta$ is the mosaic angle as found by experiment, and $2\Delta\vartheta$ is the angular range of the reflection from a single block:

$$2\Delta\vartheta = \frac{4\gamma}{\sin 2\vartheta} \; ; \tag{67}$$

here γ is the relative difference between the speed of x-rays in the material and that in a vacuum, $\delta = 2.72 \cdot 10^{10} \, Z\rho\lambda^2/A$, A is the atomic weight, Z is the number of electrons in the atom, ρ is the density, and λ is the wavelength.

Table 7 gives an example for aluminum foil (0.2 cm thick) as examined with molybdenum radiation for the (111), (200), (222), and (400) lines; $\mu = 14.26$, $\rho = 2.69$, and $a = 4.04$ Å. Values of ϑ, $|F|$, and ε are given.

TABLE 7

Parameter	Line			
	(111)	(200)	(222)	(400)
ϑ	8°46′	10°7′	17°42′	20°33′
$|F|$	33,73	31,76	21,40	17,92
ε	603,04	564,53	364,69	300,40

The structure factors of Table 7 have been corrected for the thermal motion of the atoms.

Table 8 gives the block-size factor and the intensity ratios.

TABLE 8

D, cm	B_{111}	B_{222}	B_{200}	B_{400}	$\dfrac{I_{111}}{I_{222}}$	$\dfrac{I_{200}}{I_{400}}$
$1 \cdot 10^{-6}$	1,0000	1,0000	1,0000	1,0000	5,479	7,167
$1 \cdot 10^{-5}$	0,9966	0,9981	0,9971	0,9981	5,471	7,159
$5 \cdot 10^{-5}$	0,9804	0,9879	0,9825	0,9904	5,438	7,106
$1 \cdot 10^{-4}$	0,9611	0,9762	0,9644	0,9809	5,394	7,046
$5 \cdot 10^{-4}$	0,8272	0,8916	0,8405	0,9106	5,083	6,615
$1 \cdot 10^{-3}$	0,6939	0,7997	0,7101	0,8323	4,754	6,114
$2 \cdot 10^{-3}$	0,5069	0,6518	0,5271	0,7015	4,261	5,385
$5 \cdot 10^{-3}$	0,2561	0,3934	0,2722	0,4530	3,567	4,307
$1 \cdot 10^{-2}$	0,1366	0,2242	0,1459	0,2704	3,337	3,867
$5 \cdot 10^{-2}$	0,0385	0,0637	0,0412	0,0774	3,312	3,813
$5 \cdot 10^{-1}$	0,0236	0,0391	0,0252	0,0475	3,312	3,813

The ratios given in the last two columns must be multiplied by the ratio of the cosines of the ϑ if the material is polycrystalline. Figure 147 gives the relation of intensity ratio to block size; it includes, in addition to the values of Table 8, curves corrected for secondary extinction for u of 0.00184 and 0.00318 as found by experiment. Table 9 gives the final results.

TABLE 9

Growth rate, mm/min	$\delta_{max} = 2\Delta\vartheta'$	Lines	u	$\dfrac{I_{hkl}}{I_{h'k'l'}}$	D, cm	
					$u=0$	$u>0$
0,5	26,2	(111), (222)	0,00318	2,95	10^{-2}	10^{-2}
6,0	45,2	(111), (222)	0,00184	4,62	$1,25\cdot10^{-3}$	$1\cdot10^{-3}$

Table 9 shows that the secondary-extinction correction is without marked effect on the result; it can usually be neglected for block sizes of 10^{-3} to 10^{-2} cm, but it must be incorporated for sizes of the order of 10^{-4} cm.

See [295-297] for some new methods of deducing block sizes.

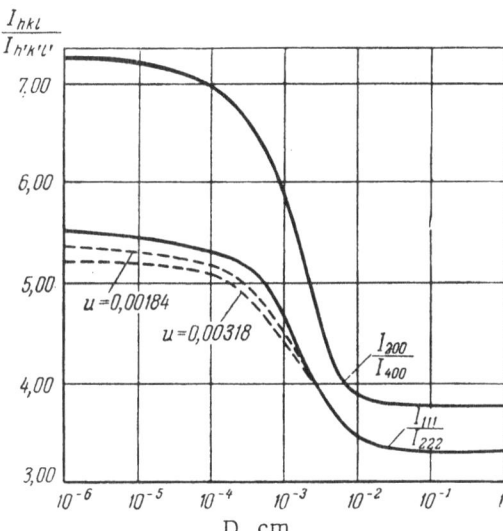

Fig. 147. Secondary-extinction correction in the deduction of block sizes from line intensities for aluminum.

The correction factor for both types of extinction, as experienced in reflection, is given by

$$g = \Phi - \frac{1}{f}x,$$

where $\Phi = lkI_0V/16\pi rP\sin\vartheta_0$, P is the energy reaching a part of the ring of length l, r is the distance from specimen to film, k is the multiplicity factor, I_0 is the incident intensity, V is the irradiated volume, $x = \mu_0/Q$, $f = (\tanh nq)/nq$ is the primary-extinction factor, μ_0 is the linear-absorption coefficient, and Q is the reflected intensity given by an ideal mosaic crystal [290, 412]. The effects may be distinguished as described in [413] for recording in transmission. The diffraction conditions indicate that the finite block size produces line displacement [456].

See [442] for a detailed analysis of methods of determining block size with allowance for both types of extinction.

8-3.   X-Ray Determination of Dislocation Densities

Dislocations are special types of structure distortion [159, 160]; the two main types are the edge (line) and screw dislocations, but there are also several special types.

The simplest case of an edge dislocation is that caused by the loss of one row or one half-plane of atoms in a small area of the crystal; this gives rise to a defect structure over a certain volume, but the structure reverts to ideal some distance away. A screw dislocation is produced by the displacement of one part of the crystal relative to the other by one interatomic distance on a certain plane.

There are several x-ray methods of deducing dislocation densities (i.e., the number of dislocation lines passing through unit area).

One of these is based on the block size as determined in one of the above ways from line intensities, line widths, or number of spots.

The minimal dislocation density is given [161] by

$$\varrho = \frac{3}{D^2}.$$

Table 1 gives densities for annealed and deformed metals as deduced from line intensities and line widths.

TABLE 1

Metal	Treatment	Width β, rad.	D, cm from intens.	D, cm from width	ρ, cm^{-2}
Aluminum	Annealed	—	$3.5 \cdot 10^{-4}$	—	$2.4 \cdot 10^7$
	Filed	$4.5 \cdot 10^{-4}$	$2.6 \cdot 10^{-4}$	—	$4.5 \cdot 10^7$
Tungsten	Annealed	10^{-3}	$4.5 \cdot 10^{-5}$	—	$1.4 \cdot 10^9$
	Filed	$6.5 \cdot 10^{-3}$	—	10^{-5}	$3 \cdot 10^{10}$
Armco iron	Annealed	—	$1.2 \cdot 10^{-4}$	—	$2 \cdot 10^8$
	Filed	$4 \cdot 10^{-3}$	—	10^{-5}	$3 \cdot 10^{10}$

Another method is based on the relation of ρ to the true width β:

$$\varrho = A\beta^2, \tag{68}$$

where A is a factor dependent on the elastic constants of the material, on the characteristics (Burgers vector) of the dislocations, and so on. Cubic metals (Al, W, Mo, Fe, and alloys of these) have A of about $2 \cdot 10^{-16}$ cm^{-2} [162, 163]. Formula (68) can [161] be put in the form $\rho = A\zeta^2$, where ζ^2 is the width of the distribution curve for the microstresses, if it is necessary to separate the broadening into parts dependent on the block size and microstresses.

Table 2 gives densities deduced from (68) for some cold-worked metals.

TABLE 2

Metal	Treatment	Width β, rad.	ρ, cm^{-2}
Aluminum	Filed	$4.5 \cdot 10^{-4}$	$4 \cdot 10^9$
Tungsten	Filed	$6.5 \cdot 10^{-3}$	$8 \cdot 10^{11}$
Molybdenum	Filed	$5.5 \cdot 10^{-3}$	$6 \cdot 10^{11}$
Armco iron	Filed	$5.5 \cdot 10^{-3}$	$6 \cdot 10^{11}$

See [434, 449 -452] for the application to phase transformations.

A special microbeam method (one in which a very small area of the surface is examined) has been used in studies on dislocation densities.

8-4. Geometric Broadening of Lines on X-Ray Patterns

The conditions for focus show that the line width produced by a cylindrical specimen in a divergent x-ray beam is

$$\Delta = x' - x'' = r \left[\cos 2\vartheta + \frac{R}{a} - \sqrt{\left(1 - \frac{R}{a} \right)^2 + \frac{2R}{a} \left(1 + \cos 2\vartheta \right)} \right] ,$$

where x' and x'' are, respectively, the minimal and maximal displacements of the rays from the position of ideal focus, r is the radius of the specimen, R is the radius of the camera, and a is the distance from the specimen to the first slit in the collimator [10]. The central angle γ_1 of the specimen giving the maximal reflecting surface is

$$\gamma_1 = \text{arctg} \frac{\sin^2 \vartheta}{\left(\dfrac{R}{a} + \cos 2\vartheta \right)}$$

A cylindrical surface gives ideal focus, in which case the width from purely geometrical causes is:

flat film: $\Delta_1 = l\,R/a$, where l is the linear size of the focus;
cylindrical: $\Delta' = \Delta_1 / \cos \psi$, where ψ is the angle between the reflected beam and the tangent to the surface at the middle of the specimen.

A flat specimen placed tangentially to the focal circle gives for unsymmetrical focusing a line width determined by the larger of the two expressions

$$\Delta_2 = \frac{L \sin \beta_2 \sin 2\vartheta}{2 \sin \varphi \sin (\psi - \beta_2)} , \Delta_1 = \frac{L \sin \beta_1 \sin 2\vartheta}{2 \sin \varphi \sin (\psi + \beta_1)} ,$$

where φ and ψ are the angles made with the plane of the specimen by the axes of the primary and reflected beams, L is the length of the specimen corresponding to the angle of divergence of the primary beam, $\beta = \beta_1 + \beta_2$, and R is the radius of the focal circle.
    For symmetrical focusing,

$$\Delta = 2r\beta^2 \frac{\sin 2\vartheta}{\sin \varphi \sin \psi} .$$

The table gives Δ (mm) for a cylindrical specimen for R = 23 mm, a = 50 mm, and r = 0.5 mm [322].

$2\vartheta°$	γ_1	$x' = x\,(\gamma_1)$	$x'' = x\,(0)$	Δ
57	39°50′	0,65	0,50	0,15
83	61°15′	0,57	0,27	0,30
111	83°45′	0,47	0,05	0,42
155	122°00′	0,34	−0,18	0,52

8-5. Correction for Heterochromatic Radiation

The line usually consists of the sum of the $K\alpha_1$ and $K\alpha_2$ components in the ordinary methods, for these are to some extent superimposed. The line width B on the pattern is the resultant of $K\alpha_1$ and $K\alpha_2$.

It is usual to isolate the stronger component ($K\alpha_1$) for detailed structural studies.

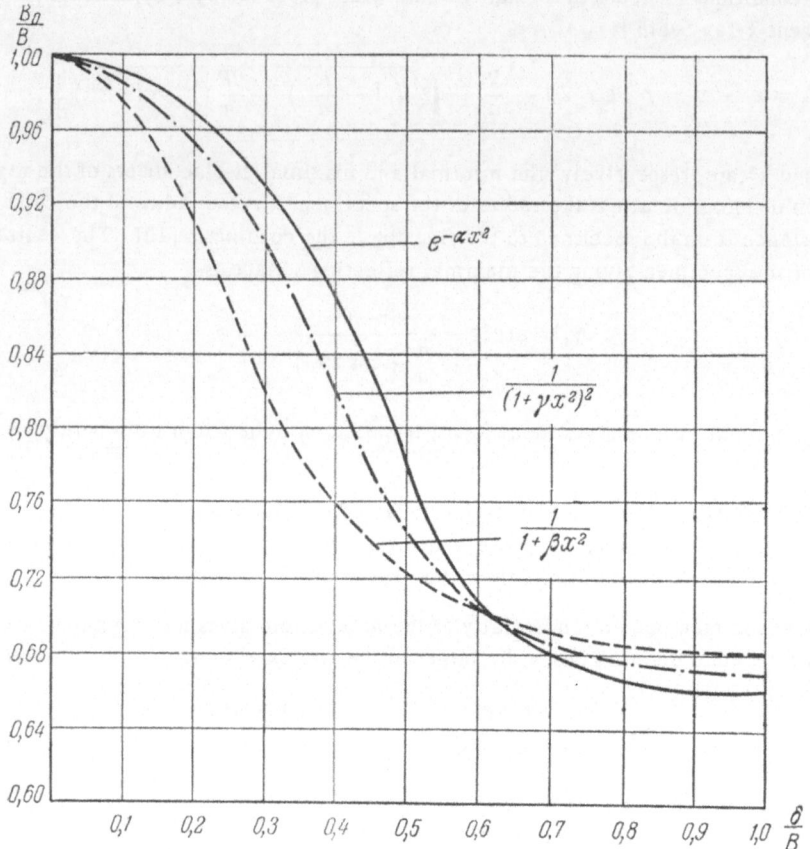

Fig. 148. Curves for the correction for heterochromatic radiation.

Figure 148 gives correction curves for the width B_0 of the $K\alpha_1$ line as a function of the resultant width B and the distance δ between the components of the $K\alpha$ doublet; B_0/B is read from δ /B, which gives B_0. The curves represent different forms of approximating functions; δ and B must be expressed in the same units (minutes of arc, radians, or millimeters). Table 2-11 gives δ for various radiations and Bragg angles.

8-6. Physical Line Broadening

The width B_0 (after correction as above) is governed by the width and height of the slits, the absorption in the specimen, divergence in the primary beam, and so on, and also by the structure of the specimen.

The broadening from the last cause is given by

$$\beta = \frac{B_0}{b_0} \int_{-\infty}^{+\infty} F(x) f(x) \, dx,$$

where β is the physical broadening (that caused by the structure), B_0 is the width corrected as above (for the α_1 component of $K\alpha$), b_0 is the width given by a standard specimen (this is governed only by the experimental conditions), $F(x)$ is an analytic function representing the intensity distribution in the physical-broadening curve, and $f(x)$ is the same but for the photometric curve of the standard.

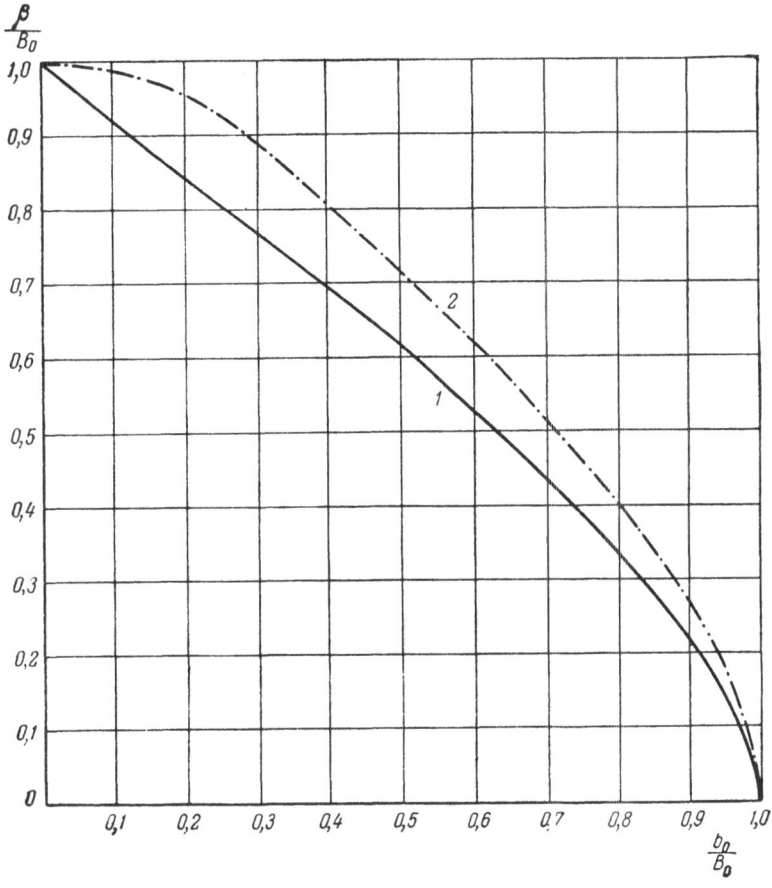

Fig. 149. Curves for physical broadening of lines on x-ray patterns.

Figure 149 shows the relation of b_0/B_0 to β/B_0 [6, 323]; curve 1 is for

$$f(x) = \frac{1}{1+ax^2}, \quad F(x) = \frac{1}{(1+\beta x^2)^2},$$

and curve 2 for

$$f(x) = \frac{1}{(1+ax^2)^2}, \quad F(x) = \frac{1}{(1+\beta x^2)^2}.$$

The standard is usually a specimen in its equilibrium state; both are examined under the same conditions.

8-7. Separation of the Effects of Block Size and of Micro-stresses

The true (physical) broadening may be put as

$$\beta = \frac{mn}{\int_{-\infty}^{+\infty} M(x)\,N(x)\,dx}\,,$$

where m and M(x) are, respectively, the part of the broadening and the intensity distribution function corresponding to the dispersion of the blocks, n and N(x) being the same for the microstresses (stresses of scale comparable with the grain size).

The system of equations for two lines then gives us

$$\frac{m_{h_1k_1l_1}}{\beta_{h_1k_1l_1}} = f_1\left(\frac{\beta_{h_2k_2l_2}}{\beta_{h_1k_1l_1}}\right) \quad \text{and} \quad \frac{n_{h_2k_2l_2}}{\beta_{h_2k_2l_2}} = f_2\left(\frac{\beta_{h_2k_2l_2}}{\beta_{h_1k_1l_1}}\right).$$

Figures 150 to 155 [6, 164, 323, 324] give the curves for some particular cases. The block size is deduced from the broadening using

$$D = \frac{K\lambda}{m\cos\vartheta}\,,$$

where K is a constant dependent on the block shape and m is as above.

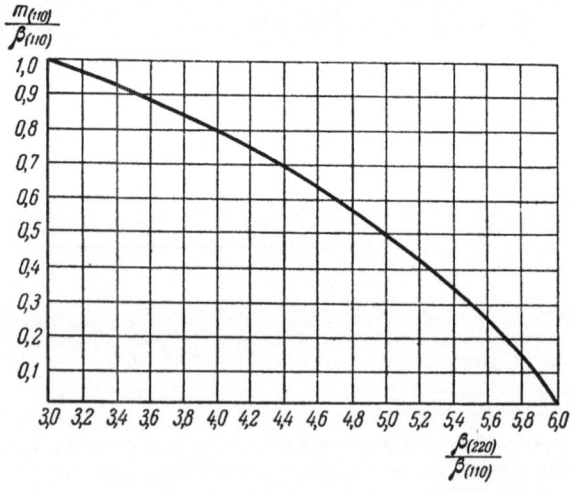

Fig. 150. Graph for the part of the broadening associated with block dispersion for the (110) line of α-iron and steel, for FeKα radiation.

Figure 156 shows that the calculated D are appreciably dependent on the relation of m to n [164].

X-ray diffraction is also affected by errors in the positions of the lattice planes and by twinning; these effects are produced by deformation and by certain other processes, and they affect the strength, width, and position of the lines.

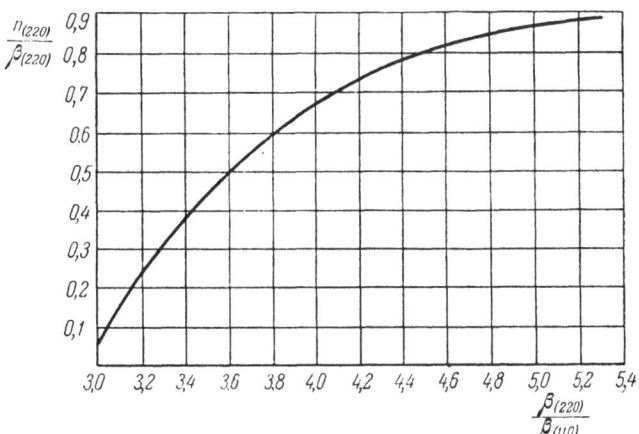

Fig. 151. Graph for the part of the broadening associated with micro-stresses for the (220) line of α-iron and steel, for FeKα radiation.

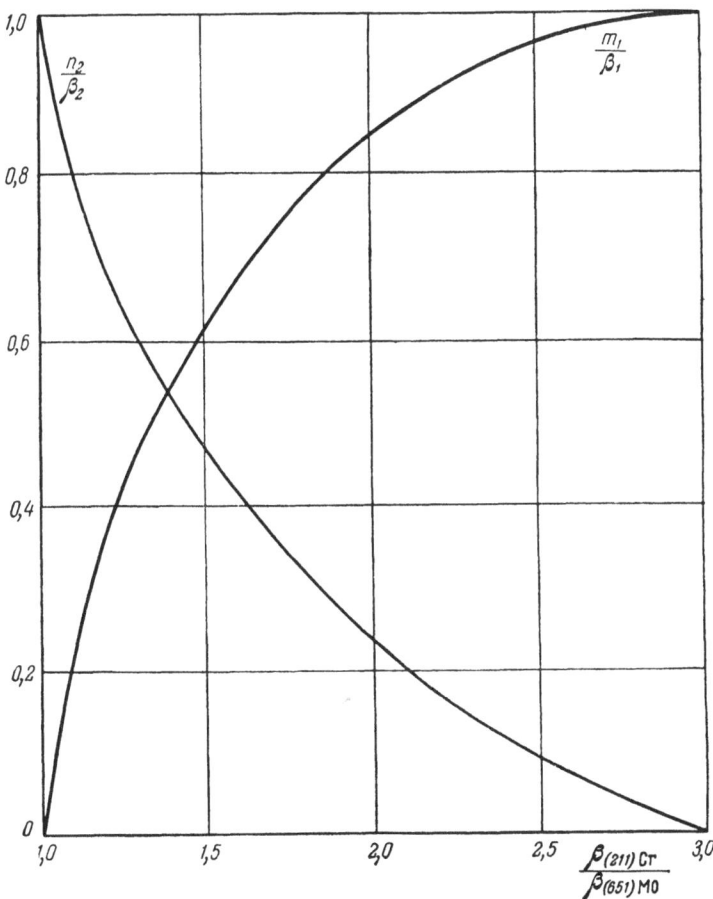

Fig. 152. Graph for the part of the broadening associated with block dispersion for the (211) line of α-iron and steel, for CrKα radiation; also graph for the part of the broadening associated with microstresses for the (651) line of α-iron and steel, for MoKα radiation.

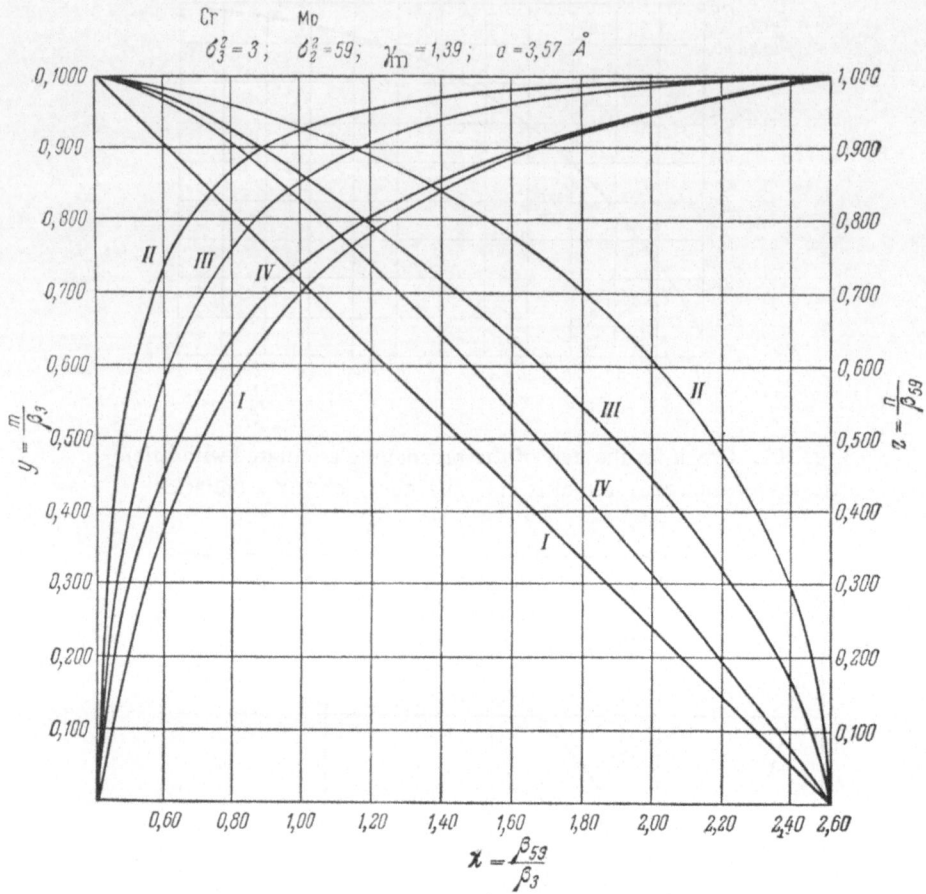

Fig. 153. Graphs for identifying contributions to the broadening for various types of relation between m, n, and β :

$$I-\beta=m+n; \qquad II-\beta=\sqrt{m^2+n^2};$$

$$III-\beta=\frac{(m+n)^3}{(m+n)^2+mn}; \qquad IV-\beta=\frac{(m+2n)^2}{m+4n}.$$

Body-centered cubic crystals are most likely to produce packing errors by slip on (211) planes along [0$\bar{1}\bar{1}$]; in orthorhombic axes A, B, and C, the C axis coincides with the [1$\bar{1}\bar{2}$] direction of the cubic lattice, B with [111], and A with [0$\bar{1}$1]. The relations between the indices are

$$H = h - k; \quad K = \frac{1}{2}(h + k + l); \quad L = -h + k + 2l.$$

The probability α of a packing error is given by

$$\frac{1}{D_{hkl}} = \frac{1}{D_0} + \frac{3}{2}\frac{\alpha \Sigma l}{h_0 ap},$$

where D_{hkl} is the block size as calculated without allowance for the error, D_0 is the true block size, L is the index of the plane in the orthorhombic axes, $h_0 = \sqrt{h^2 + k^2 + l^2}$, a is the lattice constant, and p is the multiplicity factor. This D_{hkl} is deduced by constructing a graph in coordinates of $\sin^2 \vartheta$ and $\beta^2 \cos^2 \vartheta$ from

$$\beta^2 = \frac{\lambda^2}{D^2 \cos^2 \vartheta} + 16 \left(\frac{\Delta a}{a}\right)^2 \mathrm{tg}^2 \, \vartheta.$$

It is stated [404] that $D_0 = 1.5 \cdot 10^{-5}$ cm for filings of β-brass; $\Delta a/a = 2.7 \cdot 10^{-3}$ and $\alpha = 6 \cdot 10^{-3}$. See [198-205] for other formulas and for results on certain materials. The formulas for crystals whose axes are unequal is

$$m = \frac{\lambda}{\cos \vartheta} \sqrt{\frac{\dfrac{h^2}{m_x a^4} + \dfrac{k^2}{m_y b^4} + \dfrac{l^2}{m_z c^4}}{\dfrac{h^2}{a^2} + \dfrac{k^2}{b^2} + \dfrac{l^2}{c^2}}} \, ,$$

where h, k, and l refer to the reflecting face; m_x, m_y, and m_z are the numbers of lattice constants along the x, y, and z axes corresponding to the dimensions of the crystal; and a, b, and c are the lattice constants in the x, y, and z directions.

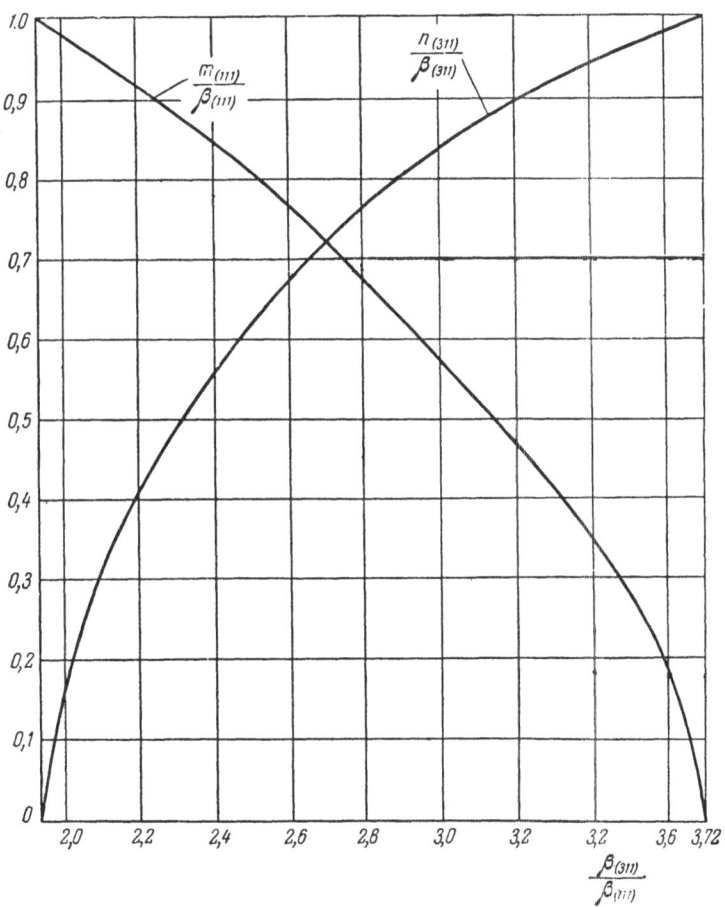

Fig. 154. Graph for the part of the broadening associated with block dispersion for line (111) of γ-iron and for the part of the broadening associated with microstresses for line (311) of γ-iron, both for Fe Kα radiation.

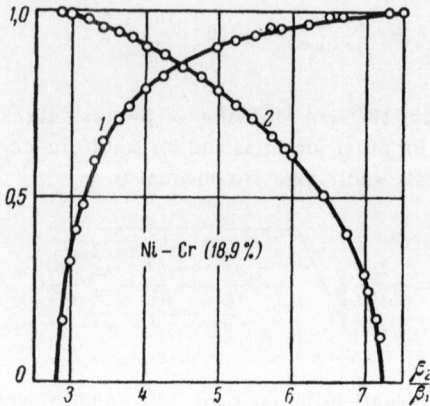

Fig. 155. Graphs for the parts of the broadening: (1) of the (331) line of nickel (and nickel alloys) from microstresses; (2) of the (111) line from block dispersion. Both curves are for CuKα radiation.

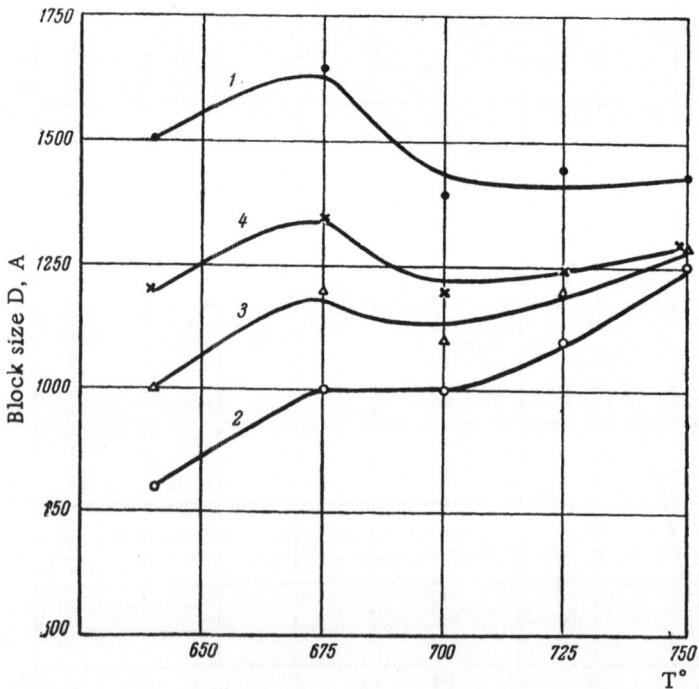

Fig. 156. Block-size changes in Fe−Ni−Ti alloys during annealing after reverse martensite transformation:

$$1 - \beta = m + n; \qquad 2 - \beta = \sqrt{m^2 + n^2};$$
$$3 - \beta = \frac{(m+n)^2}{(m+n)^2 + mn}; \qquad 4 - \beta = \frac{(m+2n)^2}{m+4n}.$$

The spread in orientation of the crystallites is given by

$$b = \frac{A}{\cos(\pi - 2\vartheta)} \, \text{tg} \, \alpha \pm S,$$

where A is the distance from specimen to film, α is the spread, and S is the size of the reflecting crystallite [302]. The plus sign corresponds to a convex crystallite; the minus, to a concave one. See [407-411] for other methods of deducing crystallite disorientation.

The microstresses are deduced from the line broadening via

$$\frac{\Delta a}{a} = \frac{n}{4R} \cot \vartheta,$$

where a is the lattice constant, n is the part of the broadening associated with these stresses, and R is the distance from specimen to detector. To convert from $\Delta a/a$ to a quantity having the dimensions of stress we have

$$\sigma \approx \frac{\Delta a}{a} E,$$

where E is the modulus of elasticity for the material.

See [164] for details of methods of calculating σ and $\Delta a/a$.

8-8. Microstress Anisotropy Factor

The broadening produced by microstresses is dependent on the indices of the line; on the assumption that the mean lattice deformation in the corresponding direction is inversely proportional to E_{hkl} (the elastic modulus), we have the following relation between the true broadening β and (hkl) for a cubic lattice:

$$\beta \cot \vartheta = A + B g_{hkl}, \quad \text{where} \quad g_{hkl} = \frac{k^2 l^2 + l^2 h^2 + h^2 k^2}{(h^2 + k^2 + l^2)^2};$$

A and B are constants. For a Gaussian distribution

$$\beta \cot \vartheta = \sqrt{A + B g_{hkl}},$$

The table gives g_{hkl} for various planes in cubic lattices [12].

(hkl)	(100)	(110)	(111)	(210)	(211)	(221)	(310)
g_{hkl}	0,000	0,250	0,333	0,160	0,250	0,296	0;090

8-9. Block-Shape Constant

The table gives values of K in the relation for line width,

$$D = \frac{K\lambda}{m \cos \vartheta}$$

for cubic materials whose blocks take the forms cube, tetrahedron, octahedron, and sphere for several reflecting planes [98].

Line	Cube	Tetra-hedron	Octa-hedron	Sphere
(100)	1,0000	1,3867	1,1006	1,0747
(110)	1,0607	0,9806	1,0376	1,0747
(111)	1,1547	1,2009	1,1438	1,0747
(210)	1,0733	1,2403	1,1075	1,0747
(211)	1,1527	1,1323	1,1061	1,0747
(221)	1,1429	1,1556	1,1185	1,0747
(310)	1,0672	1,3156	1,1138	1,0747

8-10. Determination of True Width and Shape of Lines from Fourier Series

The resultant broadened curve $h(x)$ is related to the intensity curve for the standard $f(x)$ and to the curve for the true diffraction broadening of a rectangular element of the intensity curve $F(x)$ by

$$h(x) = \int_{-\infty}^{+\infty} F(y)\, f(x-y)\, dy. \tag{69}$$

It is impracticable to solve this equation and deduce $F(x)$ in general form, so resort is made to an approximate method, which involves expanding each function in Fourier-series form and operating on the coefficients.

The expansion is performed for the limits beyond which the functions are zero, each function being represented by a sum:

$$f(x) = \sum_{-\infty}^{+\infty} f'(t) \exp\left(-2\pi i\, \frac{tx}{a}\right), \tag{70}$$

$$F(x) = \sum_{-\infty}^{+\infty} F'(t) \exp\left(-2\pi i\, \frac{tx}{a}\right), \tag{71}$$

$$h(x) = \sum_{-\infty}^{+\infty} h'(t) \exp\left(-2\pi i\, \frac{tx}{a}\right), \tag{72}$$

where a is the expansion interval $(+a/2$ to $-a/2)$, t is the summation variable, and $f'(t)$, $h'(t)$, and $F'(t)$ are complex coefficients of the series.

Substitution from (70)–(72) into (69) gives us

$$h(x) = a \sum_{-\infty}^{+\infty} f'(t)\, F'(t) \exp\left(-2\pi i\, \frac{tx}{a}\right) \tag{73}$$

or, comparing with (72), we have

$$F'(t) = \frac{1}{a}\, \frac{h'(t)}{f'(t)}. \tag{74}$$

Substitution for $F'(t)$ in (71) gives us a formula for $F(x)$:

$$F(x) = \frac{1}{a} \sum_{-\infty}^{+\infty} \frac{h'(t)}{f'(t)} \exp\left(-2\pi i \frac{tx}{a} \right). \tag{75}$$

The experimental curves are drawn up for specimen and standard; the expansion interval is taken as $2\pi = a$ for both, as both functions should be zero at the ends of this interval. The interval is divided into a suitable number of parts (usually 24, 48, or 60) and the ordinates in these parts are measured.

The process for the true (physical) broadening may be illustrated by reference to deformed manganese steel [165].

The expansion interval is chosen for the broadest line in the examination of several types of treatment. The midpoint 0 must correspond to the peak of the α_1 line. The range from −a to +a is divided into n equal parts (here 40). The Fourier coefficients are calculated in several steps:

1. The ordinates of the intensity curves are measured.

2. The values are normalized to 100 by multiplication by a constant factor (the maximum ordinate y is made 100); Table 1 gives the results from these two steps for the (222) line of the standard.

TABLE 1

№	y	y_{100}	№	y	y_{100}	№	y	y_{100}	№	y	y_{100}
0	180	100	11	2	2	−19	0	0	−9	0	0
1	151	81	12	1	1	−18	0	0	−8	0	0
2	101	56	13	0	0	−17	0	0	−7	1	1
3	114	63	14	0	0	−16	0	0	−6	3	2
4	111	62	15	0	0	−15	0	0	−5	8	4
5	88	49	16	0	0	−14	0	0	−4	15	8
6	53	29	17	0	0	−13	0	0	−3	34	19
7	27	15	18	0	0	−12	0	0	−2	74	41
8	14	8	19	0	0	−11	0	0	−1	142	79
9	7	4	20	0	0	−10	0	0			
10	3	2									

3. The Fourier coefficients are deduced for these curves; here several methods may be used. The most usual are the use of tables of sines and cosines (strips; see below), templates of various kinds, and computer methods. Templates of Lopshits' design [166] provide a simple and reliable method; these are sheets of card cut with suitable apertures.

4. The Fourier coefficients are deduced for F(x).

The coefficients for F(x) for the real (r) and imaginary (i) parts of the expansion are

$$F_r = \frac{H_r G_r + H_i G_i}{G_r^2 + G_i^2},$$

$$F_i = \frac{H_i G_r - H_r G_i}{G_r^2 + G_i^2},$$

where H is for the specimen and G is for the standard.

Table 2 illustrates stages 3 and 4 (here F_r and F_i are normalized to 1.00).

The F(x) function can be synthesized once F_r and F_i are known, and a suitable analytic representation can be chosen. It is possible to calculate β by reference to correction curves without resort to harmonic analysis if materials of a fixed type are to be examined.

TABLE 2

Coefficient	Standard		Spec. 10% deformed		F_r	F_i
	G_r	G_i	H_r	H_i		
0	12,4	0	19,5	0	1,00	0,00
1	25,1	—3,2	30,9	—2,3	0,98	0,001
2	24,1	—0,3	30,1	—4,0	0,96	0,003
4	23,3	—4,6	29,7	—4,0	0,85	0,005
8	17,3	—7,1	22,7	—4,7	0,64	0,01
12	11,8	—1,4	12,1	—2,7	0,53	0,02
16	7,2	—5,3	9,9	—0,1	0,50	0,04
20	4,4	—2,3	2,0	—2,9	0,48	0,01
24	3,5	—2,0	1,0	—1,7	0,47	0,05
28	2,5	—1,6	0,5	—1,3	0,44	0,10

The Fourier coefficients enable us to deduce the block size and lattice distortion. The F_i are usually small and are neglected. A graph is drawn up for F_r as a function of the number n. The state of the specimen can be judged to a certain extent from the shape of this curve alone; if $\left.\dfrac{dF}{dn}\right|_{n=0} = 0$, i.e., if the curve falls sharply for n small, the blocks are large and there are no microstresses. The curve becomes a straight line for n small if there are no microstresses and the blocks are small. The Fourier coefficients for the total effect, F_t, are related to those for the block dispersion F_b and those for the microstresses F_m by $F_t = F_b F_m$ for each n.

The F_b are found by taking the tangent to the F_t curve for n → 0; the ordinates of the tangent for the various n are the F_b.

Table 3 gives the F_b so found, as well as the F_m, which are found from $F_m = F_t/F_b$. All values are normalized to 1.00.

TABLE 3

n	F_b	F_m	n	F_b	F_m
0	1,00	1,00	12	0,900	0,790
1	0,995	0,995	16	0,860	0,740
2	0,993	0,990	20	0,820	0,680
4	0,978	0,973	24	0,760	0,660
8	0,950	0,980			

The block size is deduced from the intercept on the ordinate axis; the formula for cubic materials is

$$D = \frac{R}{a} \frac{2d \, \text{tg} \, \vartheta}{\sqrt{h^2 + k^2 + l^2}} \, n, \tag{76}$$

where n is here the abscissa of the point where the tangent meets the axis, R is the camera radius or distance from specimen to counter, d is the lattice constant, and a is the expansion interval. The interval is measured in the units of R.

The following example is for austenitic manganese steel, the (111) line of austenite being used. The tangent met the abscissa at n = 10.

Here R = 160 mm and a = 4.2 mm; ϑ_{111} for Fe radiation is 27°42', tg ϑ = 0.525, and a = 3.61 Å.

The results are inserted into (76) to give

$$D = \frac{160}{4.2} \frac{2 \cdot 0.0525 \cdot 3.61}{\sqrt{3}} \cdot 10 = 540 \text{ Å} = 5.4 \cdot 10^{-6} \text{ cm},$$

which agrees well with the value from line broadening.

The microstress distribution in a direction normal to a given atomic plane is found from the lattice deformation:

$$\sqrt{\overline{\Delta L^2}} = \frac{a\sqrt{-\ln F_m}}{\sqrt{2} \, \pi \, \sqrt{h^2 + k^2 + l^2}} , \tag{77}$$

where F_m is that part of the Fourier coefficient corresponding to the distortion.

The block size must be known as well as $\sqrt{\overline{\Delta L^2}}$; the size is found as described above, whereupon $\sqrt{\overline{\Delta L^2}}$ is plotted as a function of L, and $\varepsilon = \sqrt{\overline{\Delta L^2}}/L$ is plotted as a function of L. The first shows the distribution of absolute deformations in the corresponding direction, while the second does the same for the relative ones.

This may be illustrated by reference to the above steel after abrasive work-hardening; the (111) line is used, and the method described above is employed to find F_t and F_m (Table 4). See [444] regarding the effects of lattice distortion on the position of the centroid of the interference function near the (hk$l$) node in the reciprocal lattice, and also for the mean distance between (hk$l$) planes.

TABLE 4. μ = 93.04, a = 4.903 A, c = 5.393 A

F_t	F_m	L, 10^{-6}cm	$\sqrt{\overline{\Delta L^2}}$	$\varepsilon = \sqrt{\dfrac{\overline{\Delta L^2}}{L}}$
1,00	1,00		0,0564	2,1
0,94	0,99		0,1410	2,6
0,82	0,92		0,3619	3,35
0,43	0,56	4,5	0,5500	3,4
0,17	0,26		0,5640	2,6

The curve for $\sqrt{\overline{\Delta L^2}}$ vs. L usually has the form shown in Fig. 157 (for deformed tungsten). There is clearly a saturation region, which starts when L is $1-1.2 \cdot 10^{-6}$ cm. The deformation within any given block is homogeneous within these limits, so the blocks in this case are about 10^{-6} cm in size. The maximal ε is about 10^{-3}, which corresponds to the $\Delta d/d$ found from line breadth. It has also been proposed to take account of the intensity factor [443], and methods have been developed for deducing the moments of the lattice distortion and also the particle-size distribution. These are based on the fact that F(x), for the true diffraction broadening, is readily transformed to F(z), in which $z = (xd \cos \vartheta)/\lambda$ (d is the interplanar distance for the set of planes in question). The calculation given below relates to the absence of microstresses. It can be shown that a function h(n) related to the particle-size distribution is linked to the intensity F(z) by

$$h(n) = \text{const} \int F(z) \, e^{-2\pi i n z} \, dz. \tag{78}$$

F(z) is deduced by the Fourier method above, whence we can obtain either h(n) by computing the integral in (78) for x from $+\infty$ to $-\infty$ or draw a curve for h(n) by expanding F(z) as a Fourier series. In the latter case, the coefficients are the values of h(n) for the corresponding n. Analysis of h(n) shows that it can give $\overline{D}$ (the mean particle size), $\sqrt{\overline{D^2}}$ (the mean-square size), ε (the mean deviation from the mean size), and the mean-square deviations from the mean size on the side of excess (ε_+^2) and deficit (ε_-^2).

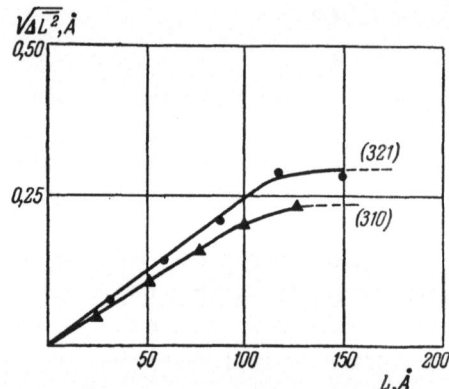

Fig. 157. Relation of $\sqrt{\overline{\Delta L^2}}$ (atomic displacement) to distance L for deformed tungsten.

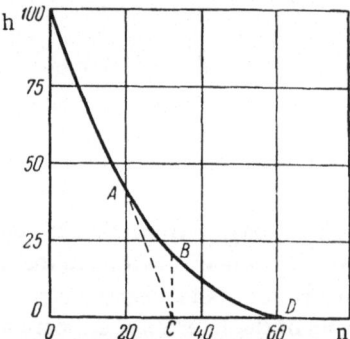

Fig. 158. Curve for finding the particle-size distribution for iron powder.

Figure 158 gives an example of a curve for use in the practical deduction of all these quantities; it is h(n) for the (110) line of iron powder. The ordinate is h(n) in a scale chosen to give h(n) = 100 for n = 0; the abscissa is n. The curve becomes a straight line for n small, and the tangent to this part meets the abscissa at C, for n = 31 (this corresponds to $\left. \dfrac{\partial h}{\partial n} \right|_{n=0}$ = 31). Point C gives $\overline{D} = n_c d = 31 \cdot 2.02 = 63$ Å [d = 2.02 Å for the (110) line]. The ordinate CB of the curve at C gives us the mean deviation from the mean size, which is 20%. The ratio of the area of ABC to the total area EOD equals the ratio of ε_-^2 to the mean-square particle size: $\varepsilon_-^2/M^2 = 0.07$. Again, the ratio of BCD to the total area under the curve is numerically equal to the ratio of ε_+^2 to the mean-square particle size: ε_+^2/M^2 = 0.185. This shows that the size distribution is skew about the mean.

With a scale of 0 to 100 for the ordinate, the area under the curve gives S = 35.2 in units of n; the mean-square particle size is then

$$V\overline{M^2} = V\sqrt{S^2 d^2} = 71 \text{ Å}.$$

That is, the mean size differs from the mean-square size by 27%. Similar calculations for other (hkl) enable one to deduce the particle size in any direction (i.e., to deduce the shape).

See [298] for the method of examining the size distribution by reference to change in the Fourier coefficients. It has been shown [457] that diffraction can displace the lines. See [463] for recent work on the use of harmonic analysis.

Less precise studies can use the experimental curve $h_1(x)$ in place of F(x), the former being broadened on account of size dispersion. This greatly facilitates the determination of the above parameters, so many different materials and treatments can be examined.

Table 8-10a gives A cos $2\pi t x$; Table 8-10b, A sin $2\pi t x$; both are for an interval of 48 parts, A being the amplitude [10].

The first column gives A from 2 to 100; the second has c for the cosine table and s for the sine table; the third has t; and columns 4 to 16 give the two functions for x from 0 to 12 (first quarter of the range). The t range from 1 to 12. The numbers in heavy type represent negative values.

The quarters are summed by writing down lines corresponding to the ordinates A for the first quarter; summation by columns gives G_r and G_i for the first quarter.

The absolute values are the same for the other quarters; Table 5 is used to determine the signs and summation parameters.

TABLE 5

Quarter	x	Summation parameter	Even or odd	cos	sin
I	0−12	x	either	+	+
II	13−24	24−x	even	+	−
II	13−24	24−x	odd	−	+
III	23−12	24+x	even	+	+
III	23−12	24+x	odd	−	−
IV	11−1	−x	either	+	−

8-10a. A cos 2πtx

0	0	0	0	0	0	0	0	0	0	0	0	0	0	0	0	0	0	0	0	0	0	0	0	0
7	7	7	8	8	8	8	9	9	9	9	10	10	10	10	11	11	11	11	12	12	12	13	13	13
14	14	15	15	16	16	17	17	18	18	19	19	20	20	21	21	22	22	23	23	24	24	25	25	26
20	21	22	22	23	24	25	25	26	27	28	28	29	30	31	31	32	33	34	35	35	36	37	38	38
26	27	28	29	30	31	32	33	34	35	36	37	38	39	40	41	42	43	44	45	46	47	48	49	50
32	33	34	35	37	38	39	40	41	43	44	45	46	47	49	50	51	52	54	55	56	57	58	60	61
37	38	40	41	42	44	45	47	48	49	51	52	54	55	57	58	59	61	62	64	65	66	68	69	71
41	43	44	46	48	49	51	52	54	56	57	59	60	62	63	65	67	68	70	71	73	75	76	78	79
45	47	48	50	52	54	55	57	59	61	62	64	66	68	69	71	73	74	76	78	80	81	83	85	87
48	50	52	54	55	57	59	61	63	65	67	68	70	72	74	76	78	80	81	83	85	87	89	91	92
50	52	54	56	58	60	62	64	66	68	70	71	73	75	77	79	81	83	85	87	89	91	93	95	97
52	54	56	58	60	61	63	65	67	69	71	73	75	77	79	81	83	85	87	89	91	93	95	97	99
52	54	56	58	60	62	64	66	68	70	72	74	76	78	80	82	84	86	88	90	92	94	96	98	100
1	1	1	1	1	1	1	1	1	1	1	1	1	1	1	1	1	1	1	1	1	1	1	1	1
c	c	c	c	c	c	c	c	c	c	c	c	c	c	c	c	c	c	c	c	c	c	c	c	c
52	54	56	58	60	62	64	66	68	70	72	74	76	78	80	82	84	86	88	90	92	94	96	98	100
0	0	0	0	0	0	0	0	0	0	0	0	0	0	0	0	0	0	0	0	0	0	0	0	0
0	1	1	1	1	2	2	2	2	3	3	3	3	4	4	4	4	5	5	5	5	6	6	6	7
1	1	2	2	3	3	4	4	5	5	6	6	7	7	8	8	9	9	10	10	11	11	12	12	13
1	2	2	3	4	5	5	6	7	8	8	9	10	11	12	12	13	14	15	15	16	17	18	18	19
1	2	3	4	5	6	7	8	9	10	11	12	13	14	15	16	17	18	19	20	21	22	23	24	25
1	2	4	5	6	7	9	10	11	12	13	15	16	17	18	19	21	22	23	24	26	27	28	29	30
1	3	4	6	7	8	10	11	13	14	16	17	18	20	21	23	24	25	27	28	30	31	33	34	35
2	3	5	6	8	9	11	13	14	16	17	19	21	22	24	25	27	29	30	32	33	35	37	38	40
2	3	5	7	9	10	12	14	16	17	19	21	22	24	26	28	29	31	33	35	36	38	40	42	43
2	4	6	7	9	11	13	15	17	18	20	22	24	26	28	30	31	33	35	37	39	41	42	44	46
2	4	6	8	10	12	14	15	17	19	21	23	25	27	29	31	33	35	37	39	41	43	44	46	48
2	4	6	8	10	12	14	16	18	20	22	24	26	28	30	32	34	36	38	40	42	44	46	48	50
2	4	6	8	10	12	14	16	18	20	22	24	26	28	30	32	34	36	38	40	42	44	46	48	50
1	1	1	1	1	1	1	1	1	1	1	1	1	1	1	1	1	1	1	1	1	1	1	1	1
c	c	c	c	c	c	c	c	c	c	c	c	c	c	c	c	c	c	c	c	c	c	c	c	c
2	4	6	8	10	12	14	16	18	20	22	24	26	28	30	32	34	36	38	40	42	44	46	48	50

52	54	56	58	60	62	64	66	68	70	72	74	76	78	80	82	84	86	88	90	92	94	96	98	100
50	52	54	56	58	60	62	64	66	68	69	71	73	75	77	79	81	83	85	87	89	91	93	95	97
45	47	48	50	52	54	55	57	59	61	62	64	66	68	69	71	73	74	76	78	80	81	83	85	87
37	38	40	41	42	44	45	47	48	49	51	52	54	55	57	58	59	61	62	64	65	66	68	69	71
26	27	28	29	30	31	32	33	34	35	36	37	38	39	40	41	42	43	44	45	46	47	48	49	50
14	14	15	15	16	16	17	17	18	18	19	19	20	20	21	21	22	22	23	23	24	24	25	25	26
0	0	0	0	0	0	0	0	0	0	0	0	0	0	0	0	0	0	0	0	0	0	0	0	0
14	14	15	15	16	16	17	17	18	18	19	19	20	20	21	21	22	22	23	23	24	24	25	25	26
26	27	28	29	30	31	32	33	34	35	36	37	38	39	40	41	42	43	44	45	46	47	48	49	50
37	38	40	41	42	44	45	47	48	49	51	52	54	55	57	58	59	61	62	64	65	66	68	69	71
45	47	48	50	52	54	55	57	59	61	62	64	66	68	69	71	73	74	76	78	80	81	83	85	87
50	52	54	56	58	60	62	64	66	68	69	71	73	75	77	79	81	83	85	87	89	91	93	95	97
52	54	56	58	60	62	64	66	68	70	72	74	76	78	80	82	84	86	88	90	92	94	96	98	100
2	2	2	2	2	2	2	2	2	2	2	2	2	2	2	2	2	2	2	2	2	2	2	2	2
c	c	c	c	c	c	c	c	c	c	c	c	c	c	c	c	c	c	c	c	c	c	c	c	c
52	54	56	58	60	62	64	66	68	70	72	74	76	78	80	82	84	86	88	90	92	94	96	98	100
2	4	6	8	10	12	14	16	18	20	22	24	26	28	30	32	34	36	38	40	42	44	46	48	50
2	4	6	8	10	12	14	16	18	20	22	24	26	28	30	32	34	36	38	40	42	44	46	48	50
2	3	5	7	9	10	12	14	16	17	19	21	22	24	26	28	29	31	33	35	36	38	40	42	43
1	3	4	6	7	8	10	11	13	14	16	17	18	20	21	23	24	25	27	28	30	31	33	34	35
1	2	3	4	5	6	7	8	9	10	11	12	13	14	15	16	17	18	19	20	21	22	23	24	25
1	1	2	2	3	3	4	4	5	5	6	6	7	7	8	8	9	9	10	10	11	11	12	12	13
0	0	0	0	0	0	0	0	0	0	0	0	0	0	0	0	0	0	0	0	0	0	0	0	0
1	1	2	2	3	3	4	4	5	5	6	6	7	7	8	8	9	9	10	10	11	11	12	12	13
1	2	3	4	5	6	7	8	9	10	11	12	13	14	15	16	17	18	19	20	21	22	23	24	25
1	3	4	6	7	8	10	11	13	14	16	17	18	20	21	23	24	25	27	28	30	31	33	34	35
2	3	5	7	9	10	12	14	16	17	19	21	22	24	26	28	29	31	33	35	36	38	40	42	43
2	4	6	8	10	12	14	16	18	20	22	24	26	28	30	32	34	36	38	40	42	44	46	48	50
2	4	6	8	10	12	14	16	18	20	22	24	26	28	30	32	34	36	38	40	42	44	46	48	50
2	2	2	2	2	2	2	2	2	2	2	2	2	2	2	2	2	2	2	2	2	2	2	2	2
c	c	c	c	c	c	c	c	c	c	c	c	c	c	c	c	c	c	c	c	c	c	c	c	c
2	4	6	8	10	12	14	16	18	20	22	24	26	28	30	32	34	36	38	40	42	44	46	48	50

0	0	0	0	0	0	0	0	0	0	0	0	0	0	0	0	0	0	0	0	0	0	0	0	0
20	21	22	22	23	24	25	25	26	27	28	28	29	30	31	31	32	33	34	35	35	36	37	38	38
37	38	40	41	42	44	45	47	48	49	51	52	54	55	57	58	59	61	62	64	65	66	68	69	71
48	50	52	54	55	57	59	61	63	65	67	68	70	72	74	76	78	80	81	83	85	87	89	91	92
52	54	56	58	60	62	64	66	68	70	72	74	76	78	80	82	84	86	88	90	92	94	96	98	100
48	50	52	54	55	57	59	61	63	65	67	68	70	72	74	76	78	80	81	83	85	87	89	91	92
37	38	40	41	42	44	45	47	48	49	51	52	54	55	57	58	59	61	62	64	65	66	68	69	71
20	21	22	22	23	24	25	25	26	27	28	28	29	30	31	31	32	33	34	35	35	36	37	38	38
0	0	0	0	0	0	0	0	0	0	0	0	0	0	0	0	0	0	0	0	0	0	0	0	0
20	21	22	22	23	24	25	25	26	27	28	28	29	30	31	31	32	33	34	35	35	36	37	38	38
37	38	40	41	42	44	45	47	48	49	51	52	54	55	57	58	59	61	62	64	65	66	68	69	71
48	50	52	54	55	57	59	61	63	65	67	68	70	72	74	76	78	80	81	83	85	87	89	91	92
52	54	56	58	60	62	64	66	68	70	72	74	76	78	80	82	84	86	88	90	92	94	96	98	100
3	3	3	3	3	3	3	3	3	3	3	3	3	3	3	3	3	3	3	3	3	3	3	3	3
c	c	c	c	c	c	c	c	c	c	c	c	c	c	c	c	c	c	c	c	c	c	c	c	c
52	54	56	58	60	62	64	66	68	70	72	74	76	78	80	82	84	86	88	90	92	94	96	98	100
0	0	0	0	0	0	0	0	0	0	0	0	0	0	0	0	0	0	0	0	0	0	0	0	0
1	2	2	3	4	5	5	6	7	8	8	9	10	11	12	12	13	14	15	15	16	17	18	18	19
1	3	4	6	7	8	10	11	13	14	16	17	18	20	21	23	24	25	27	28	30	31	33	34	35
2	4	6	7	9	11	13	15	17	18	20	22	24	26	28	30	31	33	35	37	39	41	42	44	46
2	4	6	8	10	12	14	16	18	20	22	24	26	28	30	32	34	36	38	40	42	44	46	48	50
2	4	6	7	9	11	13	15	17	18	20	22	24	26	28	30	31	33	35	37	39	41	42	44	46
1	3	4	6	7	8	10	11	13	14	16	17	18	20	21	23	24	25	27	28	30	31	33	34	35
1	2	2	3	4	5	5	6	7	8	8	9	10	11	12	12	13	14	15	15	16	17	18	18	19
0	0	0	0	0	0	0	0	0	0	0	0	0	0	0	0	0	0	0	0	0	0	0	0	0
1	2	2	3	4	5	5	6	7	8	8	9	10	11	12	12	13	14	15	15	16	17	18	18	19
1	3	4	6	7	8	10	11	13	14	16	17	18	20	21	23	24	25	27	28	30	31	33	34	35
2	4	6	7	9	11	13	15	17	18	20	22	24	26	28	30	31	33	35	37	39	41	42	44	46
2	4	6	8	10	12	14	16	18	20	22	24	26	28	30	32	34	36	38	40	42	44	46	48	50
3	3	3	3	3	3	3	3	3	3	3	3	3	3	3	3	3	3	3	3	3	3	3	3	3
c	c	c	c	c	c	c	c	c	c	c	c	c	c	c	c	c	c	c	c	c	c	c	c	c
2	4	6	8	10	12	14	16	18	20	22	24	26	28	30	32	34	36	38	40	42	44	46	48	50

52	54	56	58	60	62	64	66	68	70	72	74	76	78	80	82	84	86	88	90	92	94	96	98	100
45	47	48	50	52	54	55	57	59	61	62	64	66	68	69	71	73	74	76	78	80	81	83	85	87
26	27	28	29	30	31	32	33	34	35	36	37	38	39	40	41	42	43	44	45	46	47	48	49	50
0	0	0	0	0	0	0	0	0	0	0	0	0	0	0	0	0	0	0	0	0	0	0	0	0
26	27	28	29	30	31	32	33	34	35	36	37	38	39	40	41	42	43	44	45	46	47	48	49	50
45	47	48	50	52	54	55	57	59	61	62	64	66	68	69	71	73	74	76	78	80	81	83	85	87
52	54	56	58	60	62	64	66	68	70	72	74	76	78	80	82	84	86	88	90	92	94	96	98	100
45	47	48	50	52	54	55	57	59	61	62	64	66	68	69	71	73	74	76	78	80	81	83	85	87
26	27	28	29	30	31	32	33	34	35	36	37	38	39	40	41	42	43	44	45	46	47	48	49	50
0	0	0	0	0	0	0	0	0	0	0	0	0	0	0	0	0	0	0	0	0	0	0	0	0
26	27	28	29	30	31	32	33	34	35	36	37	38	39	40	41	42	43	44	45	46	47	48	49	50
45	47	48	50	52	54	55	57	59	61	62	64	66	68	69	71	73	74	76	78	80	81	83	85	87
52	54	56	58	60	62	64	66	68	70	72	74	76	78	80	82	84	86	88	90	92	94	96	98	100
4	4	4	4	4	4	4	4	4	4	4	4	4	4	4	4	4	4	4	4	4	4	4	4	4
c	c	c	c	c	c	c	c	c	c	c	c	c	c	c	c	c	c	c	c	c	c	c	c	c
52	54	56	58	60	62	64	66	68	70	72	74	76	78	80	82	84	86	88	90	92	94	96	98	100
2	4	6	8	10	12	14	16	18	20	22	24	26	28	30	32	34	36	38	40	42	44	46	48	50
2	3	5	7	9	10	12	14	16	17	19	21	22	24	26	28	29	31	33	35	36	38	40	42	43
1	2	3	4	5	6	7	8	9	10	11	12	13	14	15	16	17	18	19	20	21	22	23	24	25
0	0	0	0	0	0	0	0	0	0	0	0	0	0	0	0	0	0	0	0	0	0	0	0	0
1	2	3	4	5	6	7	8	9	10	11	12	13	14	15	16	17	18	19	20	21	22	23	24	25
2	3	5	7	9	10	12	14	16	17	19	21	22	24	26	28	29	31	33	35	36	38	40	42	43
2	4	6	8	10	12	14	16	18	20	22	24	26	28	30	32	34	36	38	40	42	44	46	48	50
2	3	5	7	9	10	12	14	16	17	19	21	22	24	26	28	29	31	33	35	36	38	40	42	43
1	2	3	4	5	6	7	8	9	10	11	12	13	14	15	16	17	18	19	20	21	22	23	24	25
0	0	0	0	0	0	0	0	0	0	0	0	0	0	0	0	0	0	0	0	0	0	0	0	0
1	2	3	4	5	6	7	8	9	10	11	12	13	14	15	16	17	18	19	20	21	22	23	24	25
2	3	5	7	9	10	12	14	16	17	19	21	22	24	26	28	29	31	33	35	36	38	40	42	43
2	4	6	8	10	12	14	16	18	20	22	24	26	28	30	32	34	36	38	40	42	44	46	48	50
4	4	4	4	4	4	4	4	4	4	4	4	4	4	4	4	4	4	4	4	4	4	4	4	4
c	c	c	c	c	c	c	c	c	c	c	c	c	c	c	c	c	c	c	c	c	c	c	c	c
2	4	6	8	10	12	14	16	18	20	22	24	26	28	30	32	34	36	38	40	42	44	46	48	50

0	0	0	0	0	0	0	0	0	0	0	0	0	0	0	0	0	0	0	0	0	0	0	0	0
32	33	34	35	37	38	39	40	41	43	44	45	46	47	49	50	51	52	54	55	56	57	58	60	61
50	52	54	56	58	60	62	64	66	68	70	71	73	75	77	79	81	83	85	87	89	91	93	95	97
48	50	52	54	55	57	59	61	63	65	67	68	70	72	74	76	78	80	81	83	85	87	89	91	92
26	27	28	29	30	31	32	33	34	35	36	37	38	39	40	41	42	43	44	45	46	47	48	49	50
7	7	7	8	8	8	8	9	9	9	9	10	10	10	10	11	11	11	12	12	12	13	13	13	
37	38	40	41	42	44	45	47	48	49	51	52	54	55	57	58	59	61	62	64	65	66	68	69	71
52	54	56	58	60	61	63	65	67	69	71	73	75	77	79	81	83	85	87	89	91	93	95	97	99
45	47	48	50	52	54	55	57	59	61	62	64	66	68	69	71	73	74	76	78	80	81	83	85	87
20	21	22	22	23	24	25	25	26	27	28	28	29	30	31	31	32	33	34	35	35	36	37	38	38
14	14	15	15	16	16	17	17	18	18	19	19	20	20	21	21	22	22	23	23	24	24	25	25	26
41	43	44	46	48	49	51	52	54	56	57	59	60	62	63	65	67	68	70	71	73	75	76	78	79
52	54	56	58	60	62	64	66	68	70	72	74	76	78	80	82	84	86	88	90	92	94	96	98	100
5	5	5	5	5	5	5	5	5	5	5	5	5	5	5	5	5	5	5	5	5	5	5	5	5
c	c	c	c	c	c	c	c	c	c	c	c	c	c	c	c	c	c	c	c	c	c	c	c	c
52	54	56	58	60	62	64	66	68	70	72	74	76	78	80	82	84	86	88	90	92	94	96	98	100
0	0	0	0	0	0	0	0	0	0	0	0	0	0	0	0	0	0	0	0	0	0	0	0	0
1	2	4	6	6	7	9	10	11	12	13	15	16	17	18	19	21	22	23	24	26	27	28	29	30
2	4	6	8	10	12	14	15	17	19	21	23	25	27	29	31	33	35	37	39	41	43	44	46	48
2	4	6	7	9	11	13	15	17	18	20	22	24	26	28	30	31	33	35	37	39	41	42	44	46
1	2	3	4	5	6	7	8	9	10	11	12	13	14	15	16	17	18	19	20	21	22	23	24	25
0	1	1	1	1	2	2	2	2	3	3	3	3	4	4	4	4	5	5	5	5	6	6	6	7
1	3	4	6	7	8	10	11	13	14	16	17	18	20	21	23	24	25	27	28	30	31	33	34	35
2	4	6	8	10	12	14	16	18	20	22	24	26	28	30	32	34	36	38	40	42	44	46	48	50
2	3	5	7	9	10	12	14	16	17	19	21	22	24	26	28	29	31	33	35	36	38	40	42	43
1	2	2	3	4	5	5	6	7	8	8	9	10	11	12	12	13	14	15	15	16	17	18	18	19
1	1	2	2	3	3	4	4	5	5	6	6	7	7	8	8	9	9	10	10	11	11	12	12	13
2	3	5	6	8	9	11	13	14	16	17	19	21	22	24	25	27	29	30	32	33	35	37	38	40
2	4	6	8	10	12	14	16	18	20	22	24	26	28	30	32	34	36	38	40	42	44	46	48	50
5	5	5	5	5	5	5	5	5	5	5	5	5	5	5	5	5	5	5	5	5	5	5	5	5
c	c	c	c	c	c	c	c	c	c	c	c	c	c	c	c	c	c	c	c	c	c	c	c	c
2	4	6	8	10	12	14	16	18	20	22	24	26	28	30	32	34	36	38	40	42	44	46	48	50

52	54	56	58	60	62	64	66	68	70	72	74	76	78	80	82	84	86	88	90	92	94	96	98	100
37	38	40	41	42	44	45	47	48	49	51	52	54	55	57	58	59	61	62	64	65	66	68	69	71
0	0	0	0	0	0	0	0	0	0	0	0	0	0	0	0	0	0	0	0	0	0	0	0	0
37	38	40	41	42	44	45	47	48	49	51	52	54	55	57	58	59	61	62	64	65	66	68	69	71
52	54	56	58	60	62	64	66	68	70	72	74	76	78	80	82	84	86	88	90	92	94	96	98	100
37	38	40	41	42	44	45	47	48	49	51	52	54	55	57	58	59	61	62	64	65	66	68	69	71
0	0	0	0	0	0	0	0	0	0	0	0	0	0	0	0	0	0	0	0	0	0	0	0	0
37	38	40	41	42	44	45	47	48	49	51	52	54	55	57	58	59	61	62	64	65	66	68	69	71
52	54	56	58	60	62	64	66	68	70	72	74	76	78	80	82	84	86	88	90	92	94	96	98	100
37	38	40	41	42	44	45	47	48	49	51	52	54	55	57	58	59	61	62	64	65	66	68	69	71
0	0	0	0	0	0	0	0	0	0	0	0	0	0	0	0	0	0	0	0	0	0	0	0	0
37	38	40	41	42	44	45	47	48	49	51	52	54	55	57	58	59	61	62	64	65	66	68	69	71
52	54	56	58	60	62	64	66	68	70	72	74	76	78	80	82	84	86	88	90	92	94	96	98	100
6	6	6	6	6	6	6	6	6	6	6	6	6	6	6	6	6	6	6	6	6	6	6	6	6
c	c	c	c	c	c	c	c	c	c	c	c	c	c	c	c	c	c	c	c	c	c	c	c	c
52	54	56	58	60	62	64	66	68	70	72	74	76	78	80	82	84	86	88	90	92	94	96	98	100
2	4	6	8	10	12	14	16	18	20	22	24	26	28	30	32	34	36	38	40	42	44	46	48	50
1	3	4	6	7	8	10	11	13	14	16	17	18	20	21	23	24	25	27	28	30	31	33	34	35
0	0	0	0	0	0	0	0	0	0	0	0	0	0	0	0	0	0	0	0	0	0	0	0	0
1	3	4	6	7	8	10	11	13	14	16	17	18	20	21	23	24	25	27	28	30	31	33	34	35
2	4	6	8	10	12	14	16	18	20	22	24	26	28	30	32	34	36	38	40	42	44	46	48	50
1	3	4	6	7	8	10	11	13	14	16	17	18	20	21	23	24	25	27	28	30	31	33	34	35
0	0	0	0	0	0	0	0	0	0	0	0	0	0	0	0	0	0	0	0	0	0	0	0	0
1	3	4	6	7	8	10	11	13	14	16	17	18	20	21	23	24	25	27	28	30	31	33	34	35
2	4	6	8	10	12	14	16	18	20	22	24	26	28	30	32	34	36	38	40	42	44	46	48	50
1	3	4	6	7	8	10	11	13	14	16	17	18	20	21	23	24	25	27	28	30	31	33	34	35
0	0	0	0	0	0	0	0	0	0	0	0	0	0	0	0	0	0	0	0	0	0	0	0	0
1	3	4	6	7	8	10	11	13	14	16	17	18	20	21	23	24	25	27	28	30	31	33	34	35
2	4	6	8	10	12	14	16	18	20	22	24	26	28	30	32	34	36	38	40	42	44	46	48	50
6	6	6	6	6	6	6	6	6	6	6	6	6	6	6	6	6	6	6	6	6	6	6	6	6
c	c	c	c	c	c	c	c	c	c	c	c	c	c	c	c	c	c	c	c	c	c	c	c	c
2	4	6	8	10	12	14	16	18	20	22	24	26	28	30	32	34	36	38	40	42	44	46	48	50

0	0	0	0	0	0	0	0	0	0	0	0	0	0	0	0	0	0	0	0	0	0	0	0	0
41	43	44	46	48	49	51	52	54	56	57	59	60	62	63	65	67	68	70	71	73	75	76	78	79
50	52	54	56	58	60	62	64	66	68	70	71	73	75	77	79	81	83	85	87	89	91	93	95	97
20	21	22	22	23	24	25	25	26	27	28	28	29	30	31	31	32	33	34	35	35	36	37	38	38
26	27	28	29	30	31	32	33	34	35	36	37	38	39	40	41	42	43	44	45	46	47	48	49	50
52	54	56	58	60	61	63	65	67	69	71	73	75	77	79	81	83	85	87	89	91	93	95	97	99
37	38	40	41	42	44	45	47	48	49	51	52	54	55	57	58	59	61	62	64	65	66	68	69	71
7	7	7	8	8	8	8	9	9	9	10	10	10	10	11	11	11	11	12	12	12	13	13	13	
45	47	48	50	52	54	55	57	59	61	62	64	66	68	69	71	73	74	76	78	80	81	83	85	87
48	50	52	54	55	57	59	61	63	65	67	68	70	72	74	76	78	80	81	83	85	87	89	91	92
14	14	15	15	16	16	17	17	18	18	19	19	20	20	21	21	22	22	23	23	24	24	25	25	26
32	33	34	35	37	38	39	40	41	43	44	45	46	47	49	50	51	52	54	55	56	57	58	60	61
52	54	56	58	60	62	64	66	68	70	72	74	76	78	80	82	84	86	88	90	92	94	96	98	100
7	7	7	7	7	7	7	7	7	7	7	7	7	7	7	7	7	7	7	7	7	7	7	7	7
c	c	c	c	c	c	c	c	c	c	c	c	c	c	c	c	c	c	c	c	c	c	c	c	c
52	54	56	58	60	62	64	66	68	70	72	74	76	78	80	82	84	86	88	90	92	94	96	98	100
0	0	0	0	0	0	0	0	0	0	0	0	0	0	0	0	0	0	0	0	0	0	0	0	0
2	3	5	6	8	9	11	13	14	16	17	19	21	22	24	25	27	29	30	32	33	35	37	38	40
2	4	6	8	10	12	14	15	17	19	21	23	25	27	29	31	33	35	37	39	41	43	44	46	48
1	2	2	3	4	5	5	6	7	8	8	9	10	11	12	12	13	14	15	15	16	17	18	18	19
1	2	3	4	5	6	7	8	9	10	11	12	13	14	15	16	17	18	19	20	21	22	23	24	25
2	4	6	8	10	12	14	16	18	20	22	24	26	28	30	32	34	36	38	40	42	44	46	48	50
1	3	4	6	7	8	10	11	13	14	16	17	18	20	21	23	24	25	27	28	30	31	33	34	35
0	1	1	1	1	2	2	2	3	3	3	3	4	4	4	4	5	5	5	5	6	6	6	7	
2	3	5	7	9	10	12	14	16	17	19	21	22	24	26	28	29	31	33	35	36	38	40	42	43
2	4	6	7	9	11	13	15	17	18	20	22	24	26	28	30	31	33	35	37	39	41	42	44	46
1	1	2	2	3	3	4	4	5	5	6	6	7	7	8	8	9	9	10	10	11	11	12	12	13
1	2	4	5	6	7	9	10	11	12	13	15	16	17	18	19	21	22	23	24	26	27	28	29	30
2	4	6	8	10	12	14	16	18	20	22	24	26	28	30	32	34	36	38	40	42	44	46	48	50
7	7	7	7	7	7	7	7	7	7	7	7	7	7	7	7	7	7	7	7	7	7	7	7	7
c	c	c	c	c	c	c	c	c	c	c	c	c	c	c	c	c	c	c	c	c	c	c	c	c
2	4	6	8	10	12	14	16	18	20	22	24	26	28	30	32	34	36	38	40	42	44	46	48	50

52	54	56	58	60	62	64	66	68	70	72	74	76	78	80	82	84	86	88	90	92	94	96	98	100
26	27	28	29	30	31	32	33	34	35	36	37	38	39	40	41	42	43	44	45	46	47	48	49	50
26	**27**	**28**	**29**	**30**	**31**	**32**	**33**	**34**	**35**	**36**	**37**	**38**	**39**	**40**	**41**	**42**	**43**	**44**	**45**	**46**	**47**	**48**	**49**	**50**
52	**54**	**56**	**58**	**60**	**62**	**64**	**66**	**68**	**70**	**72**	**74**	**76**	**78**	**80**	**82**	**84**	**86**	**88**	**90**	**92**	**94**	**96**	**98**	**100**
26	**27**	**28**	**29**	**30**	**31**	**32**	**33**	**34**	**35**	**36**	**37**	**38**	**39**	**40**	**41**	**42**	**43**	**44**	**45**	**46**	**47**	**48**	**49**	**50**
26	27	28	29	30	31	32	33	34	35	36	37	38	39	40	41	42	43	44	45	46	47	48	49	50
52	54	56	58	60	62	64	66	68	70	72	74	76	78	80	82	84	86	88	90	92	94	96	98	100
26	27	28	29	30	31	32	33	34	35	36	37	38	39	40	41	42	43	44	45	46	47	48	49	50
26	**27**	**28**	**29**	**30**	**31**	**32**	**33**	**34**	**35**	**36**	**37**	**38**	**39**	**40**	**41**	**42**	**43**	**44**	**45**	**46**	**47**	**48**	**49**	**50**
52	54	56	58	60	62	64	66	68	70	72	74	76	78	80	82	84	86	88	90	92	94	96	98	100
26	27	28	29	30	31	32	33	34	35	36	37	38	39	40	41	42	43	44	45	46	47	48	49	50
26	27	28	29	30	31	32	33	34	35	36	37	38	39	40	41	42	43	44	45	46	47	48	49	50
52	54	56	58	60	62	64	66	68	70	72	74	76	78	80	82	84	86	88	90	92	94	96	98	100
∞	∞	∞	∞	∞	∞	∞	∞	∞	∞	∞	∞	∞	∞	∞	∞	∞	∞	∞	∞	∞	∞	∞	∞	∞
c	c	c	c	c	c	c	c	c	c	c	c	c	c	c	c	c	c	c	c	c	c	c	c	c
52	54	56	58	60	62	64	66	68	70	72	74	76	78	80	82	84	86	88	90	92	94	96	98	100
2	4	6	8	10	12	14	16	18	20	22	24	26	28	30	32	34	36	38	40	42	44	46	48	50
1	2	3	4	5	6	7	8	9	10	11	12	13	14	15	16	17	18	19	20	21	22	23	24	25
1	2	3	4	5	6	7	8	9	10	11	12	13	14	15	16	17	18	19	20	21	22	23	24	25
2	4	6	8	10	12	14	16	18	20	22	24	26	28	30	32	34	36	38	40	42	44	46	48	50
1	**2**	**3**	**4**	**5**	**6**	**7**	**8**	**9**	**10**	**11**	**12**	**13**	**14**	**15**	**16**	**17**	**18**	**19**	**20**	**21**	**22**	**23**	**24**	**25**
1	2	3	4	5	6	7	8	9	10	11	12	13	14	15	16	17	18	19	20	21	22	23	24	25
2	4	6	8	10	12	14	16	18	20	22	24	26	28	30	32	34	36	38	40	42	44	46	48	50
1	2	3	4	5	6	7	8	9	10	11	12	13	14	15	16	17	18	19	20	21	22	23	24	25
1	2	3	4	5	6	7	8	9	10	11	12	13	14	15	16	17	18	19	20	21	22	23	24	25
2	**4**	**6**	**8**	**10**	**12**	**14**	**16**	**18**	**20**	**22**	**24**	**26**	**28**	**30**	**32**	**34**	**36**	**38**	**40**	**42**	**44**	**46**	**48**	**50**
1	**2**	**3**	**4**	**5**	**6**	**7**	**8**	**9**	**10**	**11**	**12**	**13**	**14**	**15**	**16**	**17**	**18**	**19**	**20**	**21**	**22**	**23**	**24**	**25**
1	2	3	4	5	6	7	8	9	10	11	12	13	14	15	16	17	18	19	20	21	22	23	24	25
2	4	6	8	10	12	14	16	18	20	22	24	26	28	30	32	34	36	38	40	42	44	46	48	50
∞	∞	∞	∞	∞	∞	∞	∞	∞	∞	∞	∞	∞	∞	∞	∞	∞	∞	∞	∞	∞	∞	∞	∞	∞
c	c	c	c	c	c	c	c	c	c	c	c	c	c	c	c	c	c	c	c	c	c	c	c	c
2	4	6	8	10	12	14	16	18	20	22	24	26	28	30	32	34	36	38	40	42	44	46	48	50

0	0	0	0	0	0	0	0	0	0	0	0	0	0	0	0	0	0	0	0	0	0	0	0	0
48	50	52	54	55	57	59	61	63	65	67	68	70	72	74	76	78	80	81	83	85	87	89	91	92
37	38	40	41	42	44	45	47	48	49	51	52	54	55	57	58	59	61	62	64	65	66	68	69	71
20	21	22	22	23	24	25	25	26	27	28	28	29	30	31	31	32	33	34	35	35	36	37	38	38
52	54	56	58	60	62	64	66	68	70	72	74	76	78	80	82	84	86	88	90	92	94	96	98	100
20	21	22	22	23	24	25	25	26	27	28	28	29	30	31	31	32	33	34	35	35	36	37	38	38
37	38	40	41	42	44	45	47	48	49	51	52	54	55	57	58	59	61	62	64	65	66	68	69	71
48	50	52	54	55	57	59	61	63	65	67	68	70	72	74	76	78	80	81	83	85	87	89	91	92
0	0	0	0	0	0	0	0	0	0	0	0	0	0	0	0	0	0	0	0	0	0	0	0	0
48	50	52	54	55	57	59	61	63	65	67	68	70	72	74	76	78	80	81	83	85	87	89	91	92
37	38	40	41	42	44	45	47	48	49	51	52	54	55	57	58	59	61	62	64	65	66	68	69	71
20	21	22	22	23	24	25	25	26	27	28	28	29	30	31	31	32	33	34	35	35	36	37	38	38
52	54	56	58	60	62	64	66	68	70	72	74	76	78	80	82	84	86	88	90	92	94	96	98	100
9	9	9	9	9	9	9	9	9	9	9	9	9	9	9	9	9	9	9	9	9	9	9	9	9
c	c	c	c	c	c	c	c	c	c	c	c	c	c	c	c	c	c	c	c	c	c	c	c	c
52	54	56	58	60	62	64	66	68	70	72	74	76	78	80	82	84	86	88	90	92	94	96	98	100
0	0	0	0	0	0	0	0	0	0	0	0	0	0	0	0	0	0	0	0	0	0	0	0	0
2	4	6	7	9	11	13	15	17	18	20	22	24	26	28	30	31	33	35	37	39	41	42	44	46
1	3	4	6	7	8	10	11	13	14	16	17	18	20	21	23	24	25	27	28	30	31	33	34	35
1	2	2	3	4	5	5	6	7	8	8	9	10	11	12	12	13	14	15	15	16	17	18	18	19
2	4	6	8	10	12	14	16	18	20	22	24	26	28	30	32	34	36	38	40	42	44	46	48	50
1	2	2	3	4	5	5	6	7	8	8	9	10	11	12	12	13	14	15	15	16	17	18	18	19
1	3	4	6	7	8	10	11	13	14	16	17	18	20	21	23	24	25	27	28	30	31	33	34	35
2	4	6	7	9	11	13	15	17	18	20	22	24	26	28	30	31	33	35	37	39	41	42	44	46
0	0	0	0	0	0	0	0	0	0	0	0	0	0	0	0	0	0	0	0	0	0	0	0	0
2	4	6	7	9	11	13	15	17	18	20	22	24	26	28	30	31	33	35	37	39	41	42	44	46
1	3	4	6	7	8	10	11	13	14	16	17	18	20	21	23	24	25	27	28	30	31	33	34	35
1	2	2	3	4	5	5	6	7	8	8	9	10	11	12	12	13	14	15	15	16	17	18	18	19
2	4	6	8	10	12	14	16	18	20	22	24	26	28	30	32	34	36	38	40	42	44	46	48	50
9	9	9	9	9	9	9	9	9	9	9	9	9	9	9	9	9	9	9	9	9	9	9	9	9
c	c	c	c	c	c	c	c	c	c	c	c	c	c	c	c	c	c	c	c	c	c	c	c	c
2	4	6	8	10	12	14	16	18	20	22	24	26	28	30	32	34	36	38	40	42	44	46	48	50

52	54	56	58	60	62	64	66	68	70	72	74	76	78	80	82	84	86	88	90	92	94	96	98	100
14	14	15	15	16	16	17	17	18	18	19	19	20	20	21	21	22	22	23	23	24	24	25	25	26
45	47	48	50	52	54	55	57	59	61	62	64	66	68	69	71	73	74	76	78	80	81	83	85	87
37	38	40	41	42	44	45	47	48	49	51	52	54	55	57	58	59	61	62	64	65	66	68	69	71
26	27	28	29	30	31	32	33	34	35	36	37	38	39	40	41	42	43	44	45	46	47	48	49	50
50	52	54	56	58	60	62	64	66	68	70	71	73	75	77	79	81	83	85	87	89	91	93	95	97
0	0	0	0	0	0	0	0	0	0	0	0	0	0	0	0	0	0	0	0	0	0	0	0	0
50	52	54	56	58	60	62	64	66	68	70	71	73	75	77	79	81	83	85	87	89	91	93	95	97
26	27	28	29	30	31	32	33	34	35	36	37	38	39	40	41	42	43	44	45	46	47	48	49	50
37	38	40	41	42	44	45	47	48	49	51	52	54	55	57	58	59	61	62	64	65	66	68	69	71
45	47	48	50	52	54	55	57	59	61	62	64	66	68	69	71	73	74	76	78	80	81	83	85	87
14	14	15	15	16	16	17	17	18	18	19	19	20	20	21	21	22	22	23	23	24	24	25	25	26
52	54	56	58	60	62	64	66	68	70	72	74	76	78	80	82	84	86	88	90	92	94	96	98	100
10	10	10	10	10	10	10	10	10	10	10	10	10	10	10	10	10	10	10	10	10	10	10	10	10
c	c	c	c	c	c	c	c	c	c	c	c	c	c	c	c	c	c	c	c	c	c	c	c	c
52	54	56	58	60	62	64	66	68	70	72	74	76	78	80	82	84	86	88	90	92	94	96	98	100
2	4	6	8	10	12	14	16	18	20	22	24	26	28	30	32	34	36	38	40	42	44	46	48	50
1	1	2	2	3	3	4	4	5	5	6	6	7	7	8	8	9	9	10	10	11	11	12	12	13
2	3	5	7	9	10	12	14	16	17	19	21	22	24	26	28	29	31	33	35	36	38	40	42	43
1	3	4	6	7	8	10	11	13	14	16	17	18	20	21	23	24	25	27	28	30	31	33	34	35
1	2	3	4	5	6	7	8	9	10	11	12	13	14	15	16	17	18	19	20	21	22	23	24	25
2	4	6	8	10	12	14	15	17	19	21	23	25	27	29	31	33	35	37	39	41	43	44	46	48
0	0	0	0	0	0	0	0	0	0	0	0	0	0	0	0	0	0	0	0	0	0	0	0	0
2	4	6	8	10	12	14	15	17	19	21	23	25	27	29	31	33	35	37	39	41	43	44	46	48
1	2	3	4	5	6	7	8	9	10	11	12	13	14	15	16	17	18	19	20	21	22	23	24	25
1	3	4	6	7	8	10	11	13	14	16	17	18	20	21	23	24	25	27	28	30	31	33	34	35
2	3	5	7	9	10	12	14	16	17	19	21	22	24	26	28	29	31	33	35	36	38	40	42	43
1	1	2	2	3	3	4	4	5	5	6	6	7	7	8	8	9	9	10	10	11	11	12	12	13
2	4	6	8	10	12	14	16	18	20	22	24	26	28	30	32	34	36	38	40	42	44	46	48	50
10	10	10	10	10	10	10	10	10	10	10	10	10	10	10	10	10	10	10	10	10	10	10	10	10
c	c	c	c	c	c	c	c	c	c	c	c	c	c	c	c	c	c	c	c	c	c	c	c	c
2	4	6	8	10	12	14	16	18	20	22	24	26	28	30	32	34	36	38	40	42	44	46	48	50

0	0	0	0	0	0	0	0	0	0	0	0	0	0	0	0	0	0	0	0	0	0	0	0	0
52	54	56	58	60	61	63	65	67	69	71	73	75	77	79	81	83	85	87	89	91	93	95	97	99
14	14	15	15	16	16	17	17	18	18	19	19	20	20	21	21	22	22	23	23	24	24	25	25	26
48	50	52	54	55	57	59	61	63	65	67	68	70	72	74	76	78	80	81	83	85	87	89	91	92
26	27	28	29	30	31	32	33	34	35	36	37	38	39	40	41	42	43	44	45	46	47	48	49	50
41	43	44	46	48	49	51	52	54	56	57	59	60	62	63	65	67	68	70	71	73	75	76	78	79
37	38	40	41	42	44	45	47	48	49	51	52	54	55	57	58	59	61	62	64	65	66	68	69	71
32	33	34	35	37	38	39	40	41	43	44	45	46	47	49	50	51	52	54	55	56	57	58	60	61
45	47	48	50	52	54	55	57	59	61	62	64	66	68	69	71	73	74	76	78	80	81	83	85	87
20	21	22	22	23	24	25	25	26	27	28	28	29	30	31	31	32	33	34	35	35	36	37	38	38
50	52	54	56	58	60	62	64	66	68	70	71	73	75	77	79	81	83	85	87	89	91	93	95	97
7	7	7	8	8	8	8	9	9	9	9	10	10	10	10	11	11	11	11	12	12	12	13	13	13
52	54	56	58	60	62	64	66	68	70	72	74	76	78	80	82	84	86	88	90	92	94	96	98	100
11	11	11	11	11	11	11	11	11	11	11	11	11	11	11	11	11	11	11	11	11	11	11	11	11
c	c	c	c	c	c	c	c	c	c	c	c	c	c	c	c	c	c	c	c	c	c	c	c	c
52	54	56	58	60	62	64	66	68	70	72	74	76	78	80	82	84	86	88	90	92	94	96	98	100
0	0	0	0	0	0	0	0	0	0	0	0	0	0	0	0	0	0	0	0	0	0	0	0	0
2	4	6	8	10	12	14	16	18	20	22	24	26	28	30	32	34	36	38	40	42	44	46	48	50
1	1	2	2	3	3	4	4	5	5	6	6	7	7	8	8	9	9	10	10	11	11	12	12	13
2	4	6	7	9	11	13	15	17	18	20	22	24	26	28	30	31	33	35	37	39	41	42	44	46
1	2	3	4	5	6	7	8	9	10	11	12	13	14	15	16	17	18	19	20	21	22	23	24	25
2	3	5	6	8	9	11	13	14	16	17	19	21	22	24	25	27	29	30	32	33	35	37	38	40
1	3	4	6	7	8	10	11	13	14	16	17	18	20	21	23	24	25	26	28	30	31	33	34	35
1	2	4	5	6	7	9	10	11	12	13	15	16	17	18	19	21	22	23	24	26	27	28	29	30
2	3	5	7	9	10	12	14	16	17	19	21	22	24	26	28	29	31	33	35	36	38	40	42	43
1	2	2	3	4	5	5	6	7	8	8	9	10	11	12	12	13	14	15	15	16	17	18	18	19
2	4	6	8	10	12	14	15	17	19	21	23	25	27	29	31	33	35	37	39	41	43	44	46	48
0	1	1	1	1	2	2	2	3	3	3	3	4	4	4	4	5	5	5	5	6	6	6	7	
2	4	6	8	10	12	14	16	18	20	22	24	26	28	30	32	34	36	38	40	42	44	46	48	50
11	11	11	11	11	11	11	11	11	11	11	11	11	11	11	11	11	11	11	11	11	11	11	11	11
c	c	c	c	c	c	c	c	c	c	c	c	c	c	c	c	c	c	c	c	c	c	c	c	c
2	4	6	8	10	12	14	16	18	20	22	24	26	28	30	32	34	36	38	40	42	44	46	48	50

52	54	56	58	60	62	64	66	68	70	72	74	76	78	80	82	84	86	88	90	92	94	96	98	100
o	o	o	o	o	o	o	o	o	o	o	o	o	o	o	o	o	o	o	o	o	o	o	o	o
52	54	56	58	60	62	64	66	68	70	72	74	76	78	80	82	84	86	88	90	92	94	96	98	100
o	o	o	o	o	o	o	o	o	o	o	o	o	o	o	o	o	o	o	o	o	o	o	o	o
52	54	56	58	60	62	64	66	68	70	72	74	76	78	80	82	84	86	88	90	92	94	96	98	100
o	o	o	o	o	o	o	o	o	o	o	o	o	o	o	o	o	o	o	o	o	o	o	o	o
52	54	56	58	60	62	64	66	68	70	72	74	76	78	80	82	84	86	88	90	92	94	96	98	100
o	o	o	o	o	o	o	o	o	o	o	o	o	o	o	o	o	o	o	o	o	o	o	o	o
52	54	56	58	60	62	64	66	68	70	72	74	76	78	80	82	84	86	88	90	92	94	96	98	100
o	o	o	o	o	o	o	o	o	o	o	o	o	o	o	o	o	o	o	o	o	o	o	o	o
52	54	56	58	60	62	64	66	68	70	72	74	76	78	80	82	84	86	88	90	92	94	96	98	100
12	12	12	12	12	12	12	12	12	12	12	12	12	12	12	12	12	12	12	12	12	12	12	12	12
c	c	c	c	c	c	c	c	c	c	c	c	c	c	c	c	c	c	c	c	c	c	c	c	c
52	54	56	58	60	62	64	66	68	70	72	74	76	78	80	82	84	86	88	90	92	94	96	98	100

2	4	6	8	10	12	14	16	18	20	22	24	26	28	30	32	34	36	38	40	42	44	46	48	50
o	o	o	o	o	o	o	o	o	o	o	o	o	o	o	o	o	o	o	o	o	o	o	o	o
2	4	6	8	10	12	14	16	18	20	22	24	26	28	30	32	34	36	38	40	42	44	46	48	50
o	o	o	o	o	o	o	o	o	o	o	o	o	o	o	o	o	o	o	o	o	o	o	o	o
2	4	6	8	10	12	14	16	18	20	22	24	26	28	30	32	34	36	38	40	42	44	46	48	50
o	o	o	o	o	o	o	o	o	o	o	o	o	o	o	o	o	o	o	o	o	o	o	o	o
2	4	6	8	10	12	14	16	18	20	22	24	26	28	30	32	34	36	38	40	42	44	46	48	50
o	o	o	o	o	o	o	o	o	o	o	o	o	o	o	o	o	o	o	o	o	o	o	o	o
2	4	6	8	10	12	14	16	18	20	22	24	26	28	30	32	34	36	38	40	42	44	46	48	50
o	o	o	o	o	o	o	o	o	o	o	o	o	o	o	o	o	o	o	o	o	o	o	o	o
2	4	6	8	10	12	14	16	18	20	22	24	26	28	30	32	34	36	38	40	42	44	46	48	50
12	12	12	12	12	12	12	12	12	12	12	12	12	12	12	12	12	12	12	12	12	12	12	12	12
c	c	c	c	c	c	c	c	c	c	c	c	c	c	c	c	c	c	c	c	c	c	c	c	c
2	4	6	8	10	12	14	16	18	20	22	24	26	28	30	32	34	36	38	40	42	44	46	48	50

8-10b. A sin $2\pi x$

52	54	56	58	60	62	64	66	68	70	72	74	76	78	80	82	84	86	88	90	92	94	96	98	100
52	54	56	58	60	61	63	65	67	69	71	73	75	77	79	81	83	85	87	89	91	93	95	97	99
50	52	54	56	58	60	62	64	66	68	70	71	73	75	77	79	81	83	85	87	89	91	93	95	97
48	50	52	54	55	57	59	61	63	65	67	68	70	72	74	76	78	80	81	83	85	87	89	91	92
45	47	48	50	52	54	55	57	59	61	62	64	66	68	69	71	73	74	76	78	80	81	83	85	87
41	43	44	46	48	49	51	52	54	56	57	59	60	62	63	65	67	68	70	71	73	75	76	78	79
37	38	40	41	42	44	45	47	48	49	51	52	54	55	57	58	59	61	62	64	65	66	68	69	71
32	33	34	35	37	38	39	40	41	43	44	45	46	47	49	50	51	52	54	55	56	57	58	60	61
26	27	28	29	30	31	32	33	34	35	36	37	38	39	40	41	42	43	44	45	46	47	48	49	50
20	21	22	22	23	24	25	25	26	27	28	28	29	30	31	31	32	33	34	35	35	36	37	38	38
14	14	15	15	16	16	17	17	18	18	19	19	20	20	21	21	22	22	23	23	24	24	25	25	26
7	7	7	8	8	8	8	9	9	9	9	10	10	10	10	11	11	11	11	12	12	12	13	13	13
0	0	0	0	0	0	0	0	0	0	0	0	0	0	0	0	0	0	0	0	0	0	0	0	0
$\bar{1}$	$\bar{1}$	$\bar{1}$	$\bar{1}$	$\bar{1}$	$\bar{1}$	$\bar{1}$	$\bar{1}$	$\bar{1}$	$\bar{1}$	$\bar{1}$	$\bar{1}$	$\bar{1}$	$\bar{1}$	$\bar{1}$	$\bar{1}$	$\bar{1}$	$\bar{1}$	$\bar{1}$	$\bar{1}$	$\bar{1}$	$\bar{1}$	$\bar{1}$	$\bar{1}$	$\bar{1}$
s	s	s	s	s	s	s	s	s	s	s	s	s	s	s	s	s	s	s	s	s	s	s	s	s
52	54	56	58	60	62	64	66	68	70	72	74	76	78	80	82	84	86	88	90	92	94	96	98	100
2	4	6	8	10	12	14	16	18	20	22	24	26	28	30	32	34	36	38	40	42	44	46	48	50
2	4	6	8	10	12	14	16	18	20	22	24	26	28	30	32	34	36	38	40	42	44	46	48	50
2	4	6	8	10	12	14	15	17	19	21	23	25	27	29	31	33	35	37	39	41	43	44	46	48
2	4	6	7	9	11	13	15	17	18	20	22	24	26	28	30	31	33	35	37	39	41	42	44	46
2	3	5	7	9	10	12	14	16	17	19	21	22	24	26	28	29	31	33	35	36	38	40	42	43
2	3	5	6	8	9	11	13	14	16	17	19	21	22	24	25	27	29	30	32	33	35	37	38	40
1	3	4	6	7	8	10	11	13	14	16	17	18	20	21	23	24	25	27	28	30	31	33	34	35
1	2	4	5	6	7	9	10	11	12	13	15	16	17	18	19	21	22	23	24	26	27	28	29	30
1	2	3	4	5	6	7	8	9	10	11	12	13	14	15	16	17	18	19	20	21	22	23.	24	25
1	2	3	3	4	5	5	6	7	8	8	9	10	11	12	12	13	14	15	15	16	17	18	18	19
1	1	2	2	3	3	4	4	5	5	6	6	7	7	8	8	9	9	10	10	11	11	12	12	13
0	$\bar{1}$	$\bar{1}$	$\bar{1}$	$\bar{1}$	$\bar{2}$	$\bar{2}$	$\bar{2}$	$\bar{2}$	$\bar{3}$	$\bar{3}$	$\bar{3}$	$\bar{3}$	$\bar{4}$	$\bar{4}$	$\bar{4}$	$\bar{4}$	$\bar{5}$	$\bar{5}$	$\bar{5}$	$\bar{5}$	$\bar{6}$	$\bar{6}$	$\bar{6}$	$\bar{7}$
0	0	0	0	0	0	0	0	0	0	0	0	0	0	0	0	0	0	0	0	0	0	0	0	0
$\bar{1}$	$\bar{1}$	$\bar{1}$	$\bar{1}$	$\bar{1}$	$\bar{1}$	$\bar{1}$	$\bar{1}$	$\bar{1}$	$\bar{1}$	$\bar{1}$	$\bar{1}$	$\bar{1}$	$\bar{1}$	$\bar{1}$	$\bar{1}$	$\bar{1}$	$\bar{1}$	$\bar{1}$	$\bar{1}$	$\bar{1}$	$\bar{1}$	$\bar{1}$	$\bar{1}$	$\bar{1}$
s	s	s	s	s	s	s	s	s	s	s	s	s	s	s	s	s	s	s	s	s	s	s	s	s
2	4	6	8	10	12	14	16	18	20	22	24	26	28	30	32	34	36	38	40	42	44	46	48	50

0	0	0	0	0	0	0	0	0	0	0	0	0	0	0	0	0	0	0	0	0	0	0	0	0
14	14	15	15	16	16	17	17	18	18	19	19	20	20	21	21	22	22	23	23	24	24	25	25	26
26	27	28	29	30	31	32	33	34	35	36	37	38	39	40	41	42	43	44	45	46	47	48	49	50
37	38	40	41	42	44	45	47	48	49	51	52	54	55	57	58	59	61	62	64	65	66	68	69	71
45	47	48	50	52	54	55	57	59	61	62	64	66	68	69	71	73	74	76	78	80	81	83	85	87
50	52	54	56	58	60	62	64	66	68	70	71	73	75	77	79	81	83	85	87	89	91	93	95	97
52	54	56	58	60	62	64	66	68	70	72	74	76	78	80	82	84	86	88	90	92	94	96	98	100
50	52	54	56	58	60	62	64	66	68	70	71	73	75	77	79	81	83	85	87	89	91	93	95	97
45	47	48	50	52	54	55	57	59	61	62	64	66	68	69	71	73	74	76	78	80	81	83	85	87
37	38	40	41	42	44	45	47	48	49	51	52	54	55	57	58	59	61	62	64	65	66	68	69	71
26	27	28	29	30	31	32	33	34	35	36	37	38	39	40	41	42	43	44	45	46	47	48	49	50
14	14	15	15	16	16	17	17	18	18	19	19	20	20	21	21	22	22	23	23	24	24	25	25	26
0	0	0	0	0	0	0	0	0	0	0	0	0	0	0	0	0	0	0	0	0	0	0	0	0
2	2	2	2	2	2	2	2	2	2	2	2	2	2	2	2	2	2	2	2	2	2	2	2	2
s	s	s	s	s	s	s	s	s	s	s	s	s	s	s	s	s	s	s	s	s	s	s	s	s
52	54	56	58	60	62	64	66	68	70	72	74	76	78	80	82	84	86	88	90	92	94	96	98	100
0	0	0	0	0	0	0	0	0	0	0	0	0	0	0	0	0	0	0	0	0	0	0	0	0
1	1	2	2	3	3	4	4	5	5	6	6	7	7	8	8	9	9	10	10	11	11	12	12	13
1	2	3	4	5	6	7	8	9	10	11	12	13	14	15	16	17	18	19	20	21	22	23	24	25
1	3	4	6	7	8	10	11	13	14	16	17	18	20	21	23	24	25	27	28	30	31	33	34	35
2	3	5	7	9	10	12	14	16	17	19	21	22	24	26	28	29	31	33	35	36	38	40	42	43
2	4	6	8	10	12	14	15	17	19	21	23	25	27	29	31	33	35	37	39	41	43	44	46	48
2	4	6	8	10	12	14	16	18	20	22	24	26	28	30	32	34	36	38	40	42	44	46	48	50
2	4	6	8	10	12	14	15	17	19	21	23	25	27	29	31	33	35	37	39	41	43	44	46	48
2	3	5	7	9	10	12	14	16	17	19	21	22	24	26	28	29	31	33	35	36	38	40	42	43
1	3	4	6	7	8	10	11	13	14	16	17	18	20	21	23	24	25	27	28	30	31	33	34	35
1	2	3	4	5	6	7	8	9	10	11	12	13	14	15	16	17	18	19	20	21	22	23	24	25
1	1	2	2	3	3	4	4	5	5	6	6	7	7	8	8	9	9	10	10	11	11	12	12	13
0	0	0	0	0	0	0	0	0	0	0	0	0	0	0	0	0	0	0	0	0	0	0	0	0
2	2	2	2	2	2	2	2	2	2	2	2	2	2	2	2	2	2	2	2	2	2	2	2	2
s	s	s	s	s	s	s	s	s	s	s	s	s	s	s	s	s	s	s	s	s	s	s	s	s
2	4	6	8	10	12	14	16	18	20	22	24	26	28	30	32	34	36	38	40	42	44	46	48	50

52	54	56	58	60	62	64	66	68	70	72	74	76	78	80	82	84	86	88	90	92	94	96	98	100
48	50	52	54	55	57	59	61	63	65	67	68	70	72	74	76	78	80	81	83	85	87	89	91	92
37	38	40	41	42	44	45	47	48	49	51	52	54	55	57	58	59	61	62	64	65	66	68	69	71
20	21	22	22	23	24	25	25	26	27	28	28	29	30	31	31	32	33	34	35	35	36	37	38	38
0	0	0	0	0	0	0	0	0	0	0	0	0	0	0	0	0	0	0	0	0	0	0	0	0
20	21	22	22	23	24	25	25	26	27	28	28	29	30	31	31	32	33	34	35	35	36	37	38	38
37	38	40	41	42	44	45	47	48	49	51	52	54	55	57	58	59	61	62	64	65	66	68	69	71
48	50	52	54	55	57	59	61	63	65	67	68	70	72	74	76	78	80	81	83	85	87	89	91	92
52	54	56	58	60	62	64	66	68	70	72	74	76	78	80	82	84	86	88	90	92	94	96	98	100
48	50	52	54	55	57	59	61	63	65	67	68	70	72	74	76	78	80	81	83	85	87	89	91	92
37	38	40	41	42	44	45	47	48	49	51	52	54	55	57	58	59	61	62	64	65	66	68	69	71
20	21	22	22	23	24	25	25	26	27	28	28	29	30	31	31	32	33	34	35	35	36	37	38	38
0	0	0	0	0	0	0	0	0	0	0	0	0	0	0	0	0	0	0	0	0	0	0	0	0
3	3	3	3	3	3	3	3	3	3	3	3	3	3	3	3	3	3	3	3	3	3	3	3	3
s	s	s	s	s	s	s	s	s	s	s	s	s	s	s	s	s	s	s	s	s	s	s	s	s
52	54	56	58	60	62	64	66	68	70	72	74	76	78	80	82	84	86	88	90	92	94	96	98	100
2	4	6	8	10	12	14	16	18	20	22	24	26	28	30	32	34	36	38	40	42	44	46	48	50
2	4	6	7	9	11	13	15	17	18	20	22	24	26	28	30	31	33	35	37	39	41	42	44	46
1	3	4	6	7	8	10	11	13	14	16	17	18	20	21	23	24	25	27	28	30	31	33	34	35
1	2	2	3	4	5	5	6	7	8	8	9	10	11	12	12	13	14	15	15	16	17	18	18	19
0	0	0	0	0	0	0	0	0	0	0	0	0	0	0	0	0	0	0	0	0	0	0	0	0
1	2	2	3	4	5	5	6	7	8	8	9	10	11	12	12	13	14	15	15	16	17	18	18	19
1	3	4	6	7	8	10	11	13	14	16	17	18	20	21	23	24	25	27	28	30	31	33	34	35
2	4	6	7	9	11	13	15	17	18	20	22	24	26	28	30	31	33	35	37	39	41	42	44	46
2	4	6	8	10	12	14	16	18	20	22	24	26	28	30	32	34	36	38	40	42	44	46	48	50
2	4	6	7	9	11	13	15	17	18	20	22	24	26	28	30	31	33	35	37	39	41	42	44	46
1	3	4	6	7	8	10	11	13	14	16	17	18	20	21	23	24	25	27	28	30	31	33	34	35
1	2	2	3	4	5	5	6	7	8	8	9	10	11	12	12	13	14	15	15	16	17	18	18	19
0	0	0	0	0	0	0	0	0	0	0	0	0	0	0	0	0	0	0	0	0	0	0	0	0
3	3	3	3	3	3	3	3	3	3	3	3	3	3	3	3	3	3	3	3	3	3	3	3	3
s	s	s	s	s	s	s	s	s	s	s	s	s	s	s	s	s	s	s	s	s	s	s	s	s
2	4	6	8	10	12	14	16	18	20	22	24	26	28	30	32	34	36	38	40	42	44	46	48	50

0	0	0	0	0	0	0	0	0	0	0	0	0	0	0	0	0	0	0	0	0	0	0	0	0
26	27	28	29	30	31	32	33	34	35	36	37	38	39	40	41	42	43	44	45	46	47	48	49	50
45	47	48	50	52	54	55	57	59	61	62	64	66	68	69	71	73	74	76	78	80	81	83	85	87
52	54	56	58	60	62	64	66	68	70	72	74	76	78	80	82	84	86	88	90	92	94	96	98	100
45	47	48	50	52	54	55	57	59	61	62	64	66	68	69	71	73	74	76	78	80	81	83	85	87
26	27	28	29	30	31	32	33	34	35	36	37	38	39	40	41	42	43	44	45	46	47	48	49	50
0	0	0	0	0	0	0	0	0	0	0	0	0	0	0	0	0	0	0	0	0	0	0	0	0
26	27	28	29	30	31	32	33	34	35	36	37	38	39	40	41	42	43	44	45	46	47	48	49	50
45	47	48	50	52	54	55	57	59	61	62	64	66	68	69	71	73	74	76	78	80	81	83	85	87
52	54	56	58	60	62	64	66	68	70	72	74	76	78	80	82	84	86	88	90	92	94	96	98	100
45	47	48	50	52	54	55	57	59	61	62	64	66	68	69	71	73	74	76	78	80	81	83	85	87
26	27	28	29	30	31	32	33	34	35	36	37	38	39	40	41	42	43	44	45	46	47	48	49	50
0	0	0	0	0	0	0	0	0	0	0	0	0	0	0	0	0	0	0	0	0	0	0	0	0
4	4	4	4	4	4	4	4	4	4	4	4	4	4	4	4	4	4	4	4	4	4	4	4	4
s	s	s	s	s	s	s	s	s	s	s	s	s	c	s	s	s	s	s	s	s	s	s	s	s
52	54	56	58	60	62	64	66	68	70	72	74	76	78	80	82	84	86	88	90	92	94	96	98	100
0	0	0	0	0	0	0	0	0	0	0	0	C	0	0	0	0	0	0	0	0	0	0	0	0
1	2	3	4	5	6	7	8	9	10	11	12	13	14	15	16	17	18	19	20	21	22	23	24	25
2	3	5	7	9	10	12	14	16	17	19	21	22	24	26	28	29	31	33	35	36	38	40	42	43
2	4	6	8	10	12	14	16	18	20	22	24	26	28	30	32	34	36	38	40	42	44	46	48	50
2	3	5	7	9	10	12	14	16	17	19	21	22	24	26	28	29	31	33	35	36	38	40	42	43
1	2	3	4	5	6	7	8	9	10	11	12	13	14	15	16	17	18	19	20	21	22	23	24	25
0	0	0	0	0	0	0	0	0	0	0	0	0	0	0	0	0	0	0	0	0	0	0	0	0
1	2	3	4	5	6	7	8	9	10	11	12	13	14	15	16	17	18	19	20	21	22	23	24	25
2	3	5	7	9	10	12	14	16	17	19	21	22	24	26	28	29	31	33	35	36	38	40	42	43
2	4	6	8	10	12	14	16	18	20	22	24	26	28	30	32	34	36	38	40	42	44	46	48	50
2	3	5	7	9	10	12	14	16	17	19	21	22	24	26	28	29	31	33	35	36	38	40	42	43
1	2	3	4	5	6	7	8	9	10	11	12	13	14	15	16	17	18	19	20	21	22	23	24	25
0	0	0	0	0	0	0	0	0	0	0	0	0	0	0	0	0	0	0	0	0	0	0	0	0
4	4	4	4	4	4	4	4	4	4	4	4	4	4	4	4	4	4	4	4	4	4	4	4	4
s	s	s	s	s	s	s	s	s	s	s	s	s	s	s	s	s	s	s	s	s	s	s	s	s
2	4	6	8	10	12	14	16	18	20	22	24	26	28	30	32	34	36	38	40	42	44	46	48	50

52	54	56	58	60	62	64	66	68	70	72	74	76	78	80	82	84	86	88	90	92	94	96	98	100
41	43	44	46	48	49	51	52	54	56	57	59	60	62	63	65	67	68	70	71	73	75	76	78	79
14	14	15	15	16	16	17	17	18	18	19	19	20	20	21	21	22	22	23	23	24	24	25	25	26
20	21	22	22	23	24	25	25	26	27	28	28	29	30	31	31	32	33	34	35	35	36	37	38	38
45	47	48	50	52	54	55	57	59	61	62	64	66	68	69	71	73	74	76	78	80	81	83	85	87
52	54	56	58	60	61	63	65	67	69	71	73	75	77	79	81	83	85	87	89	91	93	95	97	99
37	38	40	41	42	44	45	47	48	49	51	52	54	55	57	58	59	61	62	64	65	66	68	69	71
7	7	7	8	8	8	8	9	9	9	9	10	10	10	10	11	11	11	11	12	12	12	13	13	13
26	27	28	29	30	31	32	33	34	35	36	37	38	39	40	41	42	43	44	45	46	47	48	49	50
48	50	52	54	55	57	59	61	63	65	67	68	70	72	74	76	78	80	81	83	85	87	89	91	92
50	52	54	56	58	60	62	64	66	68	70	71	73	75	77	79	81	83	85	87	89	91	93	95	97
32	33	34	35	37	38	39	40	41	43	44	45	46	47	49	50	51	52	54	55	56	57	58	60	61
0	0	0	0	0	0	0	0	0	0	0	0	0	0	0	0	0	0	0	0	0	0	0	0	0
5	5	5	5	5	5	5	5	5	5	5	5	5	5	5	5	5	5	5	5	5	5	5	5	5
s	s	s	s	s	s	s	s	s	s	s	s	s	s	s	s	s	s	s	s	s	s	s	s	s
52	54	56	58	60	62	64	66	68	70	72	74	76	78	80	82	84	86	88	90	92	94	96	98	100
2	4	6	8	10	12	14	16	18	20	22	24	26	28	30	32	34	36	38	40	42	44	46	48	50
2	3	5	6	8	9	11	13	14	16	17	19	21	22	24	25	27	29	30	32	33	35	37	38	40
1	1	2	2	3	3	4	4	5	5	6	6	7	7	8	8	9	9	10	10	11	11	12	12	13
1	2	2	3	4	5	5	6	7	8	8	9	10	11	12	12	13	14	15	15	16	17	18	18	19
2	3	5	7	9	10	12	14	16	17	19	21	22	24	26	28	29	31	33	35	36	38	40	42	43
2	4	6	8	10	12	14	16	18	20	22	24	26	28	30	32	34	36	38	40	42	44	46	48	50
1	3	4	6	7	8	10	11	13	14	16	17	18	20	21	23	24	25	27	28	30	31	33	34	35
0	1	1	1	1	2	2	2	2	3	3	3	3	4	4	4	4	5	5	5	5	6	6	6	7
1	2	3	4	5	6	7	8	9	10	11	12	13	14	15	16	17	18	19	20	21	22	23	24	25
2	4	6	7	9	11	13	15	17	18	20	22	24	26	28	30	31	33	35	37	39	41	42	44	46
2	4	6	8	10	12	11	15	17	19	21	23	25	27	29	31	33	35	37	39	41	43	44	46	48
1	2	4	5	6	7	9	10	11	12	13	15	16	17	18	19	21	22	23	24	26	27	28	29	30
0	0	0	0	0	0	0	0	0	0	0	0	0	0	0	0	0	0	0	0	0	0	0	0	0
5	5	5	5	5	5	5	5	5	5	5	5	5	5	5	5	5	5	5	5	5	5	5	5	5
s	s	s	s	s	s	s	s	s	s	s	s	s	s	s	s	s	s	s	s	s	s	s	s	s
2	4	6	8	10	12	14	16	18	20	22	24	26	28	30	32	34	36	38	40	42	44	46	48	50

0	0	0	0	0	0	0	0	0	0	0	0	0	0	0	0	0	0	0	0	0	0	0	0	0
37	38	40	41	42	44	45	47	48	49	51	52	54	55	57	58	59	61	62	64	65	66	68	69	71
52	54	56	58	60	62	64	66	68	70	72	74	76	78	80	82	84	86	88	90	92	94	96	98	100
37	38	40	41	42	44	45	47	48	49	51	52	54	55	57	58	59	61	62	64	65	66	68	69	71
0	0	0	0	0	0	0	0	0	0	0	0	0	0	0	0	0	0	0	0	0	0	0	0	0
37	38	40	41	42	44	45	47	48	49	51	52	54	55	57	58	59	61	62	64	65	66	68	69	71
52	54	56	58	60	62	64	66	68	70	72	74	76	78	80	82	84	86	88	90	92	94	96	98	100
37	38	40	41	42	44	45	47	48	49	51	52	54	55	57	58	59	61	62	64	65	66	68	69	71
0	0	0	0	0	0	0	0	0	0	0	0	0	0	0	0	0	0	0	0	0	0	0	0	0
37	38	40	41	42	44	45	47	48	49	51	52	54	55	57	58	59	61	62	64	65	66	68	69	71
52	54	56	58	60	62	64	66	68	70	72	74	76	78	80	82	84	86	88	90	92	94	96	98	100
37	38	40	41	42	44	45	47	48	49	51	52	54	55	57	58	59	61	62	64	65	66	68	69	71
0	0	0	0	0	0	0	0	0	0	0	0	0	0	0	0	0	0	0	0	0	0	0	0	0
6	6	6	6	6	6	6	6	6	6	6	6	6	6	6	6	6	6	6	6	6	6	6	6	6
s	s	s	s	s	s	s	s	s	s	s	s	s	s	s	s	s	s	s	s	s	s	s	s	s
52	54	56	58	60	62	64	66	68	70	72	74	76	78	80	82	84	86	88	90	92	94	96	98	100
0	0	0	0	0	0	0	0	0	0	0	0	0	0	0	0	0	0	0	0	0	0	0	0	0
1	3	4	6	7	8	10	11	13	14	16	17	18	20	21	23	24	25	27	28	30	31	33	34	35
2	4	6	8	10	12	14	16	18	20	22	24	26	28	30	32	34	36	38	40	42	44	46	48	50
1	3	4	6	7	8	10	11	13	14	16	17	18	20	21	23	24	25	27	28	30	31	33	34	35
0	0	0	0	0	0	0	0	0	0	0	0	0	0	0	0	0	0	0	0	0	0	0	0	0
1	3	4	6	7	8	10	11	13	14	16	17	18	20	21	23	24	25	27	28	30	31	33	34	35
2	4	6	8	10	12	14	16	18	20	22	24	26	28	30	32	34	36	38	40	42	44	46	48	50
1	3	4	6	7	8	10	11	13	14	16	17	18	20	21	23	24	25	27	28	30	31	33	34	35
0	0	0	0	0	0	0	0	0	0	0	0	0	0	0	0	0	0	0	0	0	0	0	0	0
1	3	4	6	7	8	10	11	13	14	16	17	18	20	21	23	24	25	27	28	30	31	33	34	35
2	4	6	8	10	12	14	16	18	20	22	24	26	28	30	32	34	36	38	40	42	44	46	48	50
1	3	4	6	7	8	10	11	13	14	16	17	18	20	21	23	24	25	27	28	30	31	33	34	35
0	0	0	0	0	0	0	0	0	0	0	0	0	0	0	0	0	0	0	0	0	0	0	0	0
6	6	6	6	6	6	6	6	6	6	6	6	6	6	6	6	6	6	6	6	6	6	6	6	6
s	s	s	s	s	s	s	s	s	s	s	s	s	s	s	s	s	s	s	s	s	s	s	s	s
2	4	6	8	10	12	14	16	18	20	22	24	26	28	30	32	34	36	38	40	42	44	46	48	50

52	54	56	58	60	62	64	66	68	70	72	74	76	78	80	82	84	86	88	90	92	94	96	98	100
32	33	34	35	37	38	39	40	41	43	44	45	46	47	48	50	51	52	54	55	56	57	58	60	61
14	14	15	15	16	16	17	17	18	18	19	19	20	20	21	21	22	22	23	23	24	24	25	25	26
48	50	52	54	55	57	59	61	63	65	67	68	70	72	74	76	78	80	81	83	85	87	89	91	92
45	47	48	50	52	54	55	57	59	61	62	64	66	68	69	71	73	74	76	78	80	81	83	85	87
7	7	7	8	8	8	8	9	9	9	9	10	10	10	10	11	11	11	11	12	12	12	13	13	13
37	38	40	41	42	44	45	47	48	49	51	52	54	55	57	58	59	61	62	64	65	66	68	69	71
52	54	56	58	60	61	63	65	67	69	71	73	75	77	79	81	83	85	87	89	91	93	95	97	99
26	27	28	29	30	31	32	33	34	35	36	37	38	39	40	41	42	43	44	45	46	47	48	49	50
20	21	22	22	23	24	25	25	26	27	28	28	29	30	31	31	32	33	34	35	35	36	37	38	38
50	52	54	56	58	60	62	64	66	68	70	71	73	75	77	79	81	83	85	87	89	91	93	95	97
41	43	44	46	48	49	51	52	54	56	57	59	60	62	63	65	67	68	70	71	73	75	76	78	79
0	0	0	0	0	0	0	0	0	0	0	0	0	0	0	0	0	0	0	0	0	0	0	0	0
7	7	7	7	7	7	7	7	7	7	7	7	7	7	7	7	7	7	7	7	7	7	7	7	7
s	s	s	s	s	s	s	s	s	s	s	s	s	s	s	s	s	s	s	s	s	s	s	s	s
52	54	56	58	60	62	64	66	68	70	72	74	76	78	80	82	84	86	88	90	92	94	96	98	100
2	4	6	8	10	12	14	16	18	20	22	24	26	28	30	32	34	36	38	40	42	44	46	48	50
1	2	4	5	6	7	9	10	11	12	13	15	16	17	18	19	21	22	23	24	26	27	28	29	30
1	1	2	2	3	3	4	4	5	5	6	6	7	7	8	8	9	9	10	10	11	11	12	12	13
2	4	6	7	9	11	13	15	17	18	20	22	24	26	28	30	31	33	35	37	39	41	42	44	46
2	3	5	7	9	10	12	14	16	17	19	21	22	24	26	28	29	31	33	35	36	38	40	42	43
0	1	1	1	1	2	2	2	2	3	3	3	3	4	4	4	4	5	5	5	5	6	6	6	7
1	3	4	6	7	8	10	11	13	14	16	17	18	20	21	23	24	25	27	28	30	31	33	34	35
2	4	6	8	10	12	14	16	18	20	22	24	26	28	30	32	34	36	38	40	42	44	46	48	50
1	2	3	4	5	6	7	8	9	10	11	12	13	14	15	16	17	18	19	20	21	22	23	24	25
1	2	2	3	4	5	5	6	7	8	8	9	10	11	12	12	13	14	15	15	16	17	18	18	19
2	4	6	8	10	12	14	15	17	19	21	23	25	27	29	31	33	35	37	39	41	43	44	46	48
2	3	5	6	8	9	11	13	14	16	17	19	21	22	24	25	27	29	30	32	33	35	37	38	40
0	0	0	0	0	0	0	0	0	0	0	0	0	0	0	0	0	0	0	0	0	0	0	0	0
7	7	7	7	7	7	7	7	7	7	7	7	7	7	7	7	7	7	7	7	7	7	7	7	7
s	s	s	s	s	s	s	s	s	s	s	s	s	s	s	s	s	s	s	s	s	s	s	s	s
2	4	6	8	10	12	14	16	18	20	22	24	26	28	30	32	34	36	38	40	42	44	46	48	50

0	0	0	0	0	0	0	0	0	0	0	0	0	0	0	0	0	0	0	0	0	0	0	0	0
45	47	48	50	52	54	55	57	59	61	62	64	66	68	69	71	73	74	76	78	80	81	83	85	87
45	47	48	50	52	54	55	57	59	61	62	64	66	68	69	71	73	74	76	78	80	81	83	85	87
0	0	0	0	0	0	0	0	0	0	0	0	0	0	0	0	0	0	0	0	0	0	0	0	0
45	47	48	50	52	54	55	57	59	61	62	64	66	68	69	71	73	74	76	78	80	81	83	85	87
45	47	48	50	52	54	55	57	59	61	62	64	66	68	69	71	73	74	76	78	80	81	83	85	87
0	0	0	0	0	0	0	0	0	0	0	0	0	0	0	0	0	0	0	0	0	0	0	0	0
45	47	48	50	52	54	55	57	59	61	62	64	66	68	69	71	73	74	76	78	80	81	83	85	87
45	47	48	50	52	54	55	57	59	61	62	64	66	68	69	71	73	74	76	78	80	81	83	85	87
0	0	0	0	0	0	0	0	0	0	0	0	0	0	0	0	0	0	0	0	0	0	0	0	0
45	47	48	50	52	54	55	57	59	61	62	64	66	68	69	71	73	74	76	78	80	81	83	85	87
45	47	48	50	52	54	55	57	59	61	62	64	66	68	69	71	73	74	76	78	80	81	83	85	87
0	0	0	0	0	0	0	0	0	0	0	0	0	0	0	0	0	0	0	0	0	0	0	0	0
8	8	8	8	8	8	8	8	8	8	8	8	8	8	8	8	8	8	8	8	8	8	8	8	8
s	s	s	s	s	s	s	s	s	s	s	s	s	s	s	s	s	s	s	s	s	s	s	s	s
52	54	56	58	60	62	64	66	68	70	72	74	76	78	80	82	84	86	88	90	92	94	96	98	100
0	0	0	0	0	0	0	0	0	0	0	0	0	0	0	0	0	0	0	0	0	0	0	0	0
2	3	5	7	9	10	12	14	16	17	19	21	22	24	26	28	29	31	33	35	36	38	40	42	43
2	3	5	7	9	10	12	14	16	17	19	21	22	24	26	28	29	31	33	35	36	38	40	42	43
0	0	0	0	0	0	0	0	0	0	0	0	0	0	0	0	0	0	0	0	0	0	0	0	0
2	3	5	7	9	10	12	14	16	17	19	21	22	24	26	28	29	31	33	35	36	38	40	42	43
2	3	5	7	9	10	12	14	16	17	19	21	22	24	26	28	29	31	33	35	36	38	40	42	43
0	0	0	0	0	0	0	0	0	0	0	0	0	0	0	0	0	0	0	0	0	0	0	0	0
2	3	5	7	9	10	12	14	16	17	19	21	22	24	26	28	29	31	33	35	36	38	40	42	43
2	3	5	7	9	10	12	14	16	17	19	21	22	24	26	28	29	31	33	35	36	38	40	42	43
0	0	0	0	0	0	0	0	0	0	0	0	0	0	0	0	0	0	0	0	0	0	0	0	0
2	3	5	7	9	10	12	14	16	17	19	21	22	24	26	28	29	31	33	35	36	38	40	42	43
2	3	5	7	9	10	12	14	16	17	19	21	22	24	26	28	29	31	33	35	36	38	40	42	43
0	0	0	0	0	0	0	0	0	0	0	0	0	0	0	0	0	0	0	0	0	0	0	0	0
8	8	8	8	8	8	8	8	8	8	8	8	8	8	8	8	8	8	8	8	8	8	8	8	8
s	s	s	s	s	s	s	s	s	s	s	s	s	s	s	s	s	s	s	s	s	s	s	s	s
2	4	6	8	10	12	14	16	18	20	22	24	26	28	30	32	34	36	38	40	42	44	46	48	50

52	54	56	58	60	62	64	66	68	70	72	74	76	78	80	82	84	86	88	90	92	94	96	98	100
20	21	22	22	23	24	25	25	26	27	28	28	29	30	31	31	32	33	34	35	35	36	37	38	38
37	38	40	41	42	44	45	47	48	49	51	52	54	55	57	58	59	61	62	64	65	66	68	69	71
48	50	52	54	55	57	59	61	63	65	67	68	70	72	74	76	78	80	81	83	85	87	89	91	92
0	0	0	0	0	0	0	0	0	0	0	0	0	0	0	0	0	0	0	0	0	0	0	0	0
48	50	52	54	55	57	59	61	63	65	67	68	70	72	74	76	78	80	81	83	85	87	89	91	92
37	38	40	41	42	44	45	47	48	49	51	52	54	55	57	58	59	61	62	64	65	66	68	69	71
20	21	22	22	23	24	25	25	26	27	28	28	29	30	31	31	32	33	34	35	35	36	37	38	38
52	54	56	58	60	62	64	66	68	70	72	74	76	78	80	82	84	86	88	90	92	94	96	98	100
20	21	22	22	23	24	25	25	26	27	28	28	29	30	31	31	32	33	34	35	35	36	37	38	38
37	38	40	41	42	44	45	47	48	49	51	52	54	55	57	58	59	61	62	64	65	66	68	69	71
48	50	52	54	55	57	59	61	63	65	67	68	70	72	74	76	78	80	81	83	85	87	89	91	92
0	0	0	0	0	0	0	0	0	0	0	0	0	0	0	0	0	0	0	0	0	0	0	0	0
9	9	9	9	9	9	9	9	9	9	9	9	9	9	9	9	9	9	9	9	9	9	9	9	9
s	s	s	s	s	s	s	s	s	s	s	s	s	s	s	s	s	s	s	s	s	s	s	s	s
52	54	56	58	60	62	64	66	68	70	72	74	76	78	80	82	84	86	88	90	92	94	96	98	100
2	4	6	8	10	12	14	16	18	20	22	24	26	28	30	32	34	36	38	40	42	44	46	48	50
1	2	2	3	4	5	5	6	7	8	8	9	10	11	12	12	13	14	15	15	16	17	18	18	19
1	3	4	6	7	8	10	11	13	14	16	17	18	20	21	23	24	25	27	28	30	31	33	34	35
2	4	6	7	9	11	13	15	17	18	20	22	24	26	28	30	31	33	35	37	39	41	42	44	46
0	0	0	0	0	0	0	0	0	0	0	0	0	0	0	0	0	0	0	0	0	0	0	0	0
2	4	6	7	9	11	13	15	17	18	20	22	24	26	28	30	31	33	35	37	39	41	42	44	46
1	3	4	6	7	8	10	11	13	14	16	17	18	20	21	23	24	25	27	28	30	31	33	34	35
1	2	2	3	4	5	5	6	7	8	8	9	10	11	12	12	13	14	15	15	16	17	18	18	19
2	4	6	8	10	12	14	16	18	20	22	24	26	28	30	32	34	36	38	40	42	44	46	48	50
1	2	2	3	4	5	5	6	7	8	8	9	10	11	12	12	13	14	15	15	16	17	18	18	19
1	3	4	6	7	8	10	11	13	14	16	17	18	20	21	23	24	25	27	28	30	31	33	34	35
2	4	6	7	9	11	13	15	17	18	20	22	24	26	28	30	31	33	35	37	39	41	42	44	46
0	0	0	0	0	0	0	0	0	0	0	0	0	0	0	0	0	0	0	0	0	0	0	0	0
9	9	9	9	9	9	9	9	9	9	9	9	9	9	9	9	9	9	9	9	9	9	9	9	9
s	s	s	s	s	s	s	s	s	s	s	s	s	s	s	s	s	s	s	s	s	s	s	s	s
2	4	6	8	10	12	14	16	18	20	22	24	26	28	30	32	34	36	38	40	42	44	46	48	50

0	0	0	0	0	0	0	0	0	0	0	0	0	0	0	0	0	0	0	0	0	0	0	0	0
50	52	54	56	58	60	62	64	66	68	70	71	73	75	77	79	81	83	85	87	89	91	93	95	97
26	27	28	29	30	31	32	33	34	35	36	37	38	39	40	41	42	43	44	45	46	47	48	49	50
37	38	40	41	42	44	45	47	48	49	51	52	54	55	57	58	59	61	62	64	65	66	68	69	71
45	47	48	50	52	54	55	57	59	61	62	64	66	68	69	71	73	74	76	78	80	81	83	85	87
14	14	15	15	16	16	17	17	18	18	19	19	20	20	21	21	22	22	23	23	24	24	25	25	26
52	54	56	58	60	62	64	66	68	70	72	74	76	78	80	82	84	86	88	90	92	94	96	98	100
14	14	15	15	16	16	17	17	18	18	19	19	20	20	21	21	22	22	23	23	24	24	25	25	26
45	47	48	50	52	54	55	57	59	61	62	64	66	68	69	71	73	74	76	78	80	81	83	85	87
37	38	40	41	42	44	45	47	48	49	51	52	54	55	57	58	59	61	62	64	65	66	68	69	71
26	27	28	29	30	31	32	33	34	35	36	37	38	39	40	41	42	43	44	45	46	47	48	49	50
50	52	54	56	58	60	62	64	66	68	70	71	73	75	77	79	81	83	85	87	89	91	93	95	97
0	0	0	0	0	0	0	0	0	0	0	0	0	0	0	0	0	0	0	0	0	0	0	0	0
10	10	10	10	10	10	10	10	10	10	10	10	10	10	10	10	10	10	10	10	10	10	10	10	10
s	s	s	s	s	s	s	s	s	s	s	s	s	s	s	s	s	s	s	s	s	s	s	s	s
52	54	56	58	60	62	64	66	68	70	72	74	76	78	80	82	84	86	88	90	92	94	96	98	100
0	0	0	0	0	0	0	0	0	0	0	0	0	0	0	0	0	0	0	0	0	0	0	0	0
2	4	6	8	10	12	14	15	17	19	21	23	25	27	29	31	33	35	37	39	41	43	44	46	48
1	2	3	4	5	6	7	8	9	10	11	12	13	14	15	16	17	18	19	20	21	22	23	24	25
1	3	4	6	7	8	10	11	13	14	16	17	18	20	21	23	24	25	27	28	30	31	33	34	35
2	3	5	7	9	10	12	14	16	17	19	21	22	24	26	28	29	31	33	35	36	38	40	42	43
1	1	2	2	3	3	4	4	5	5	6	6	7	7	8	8	9	9	10	10	11	11	12	12	13
2	4	6	8	10	12	14	16	18	20	22	24	26	28	30	32	34	36	38	40	42	44	46	48	50
1	1	2	2	3	3	4	4	5	5	6	6	7	7	8	8	9	9	10	10	11	11	12	12	13
2	3	5	7	9	10	12	14	16	17	19	21	22	24	26	28	29	31	33	35	36	38	40	42	43
1	3	4	6	7	8	10	11	13	14	16	17	18	20	21	23	24	25	27	28	30	31	33	34	35
1	2	3	4	5	6	7	8	9	10	11	12	13	14	15	16	17	18	19	20	21	22	23	24	25
2	4	6	8	10	12	14	15	17	19	21	23	25	27	29	31	33	35	37	39	41	43	44	46	48
0	0	0	0	0	0	0	0	0	0	0	0	0	0	0	0	0	0	0	0	0	0	0	0	0
10	10	10	10	10	10	10	10	10	10	10	10	10	10	10	10	10	10	10	10	10	10	10	10	10
s	s	s	s	s	s	s	s	s	s	s	s	s	s	s	s	s	s	s	s	s	s	s	s	s
2	4	6	8	10	12	14	16	18	20	22	24	26	28	30	32	34	35	38	40	42	44	46	48	50

52	54	56	58	60	62	64	66	68	70	72	74	76	78	80	82	84	86	88	90	92	94	96	98	100
7	7	7	8	8	8	8	9	9	9	9	10	10	10	10	11	11	11	11	12	12	12	13	13	13
50	52	54	56	58	60	62	64	66	68	70	71	73	75	77	79	81	83	85	87	89	91	93	95	97
20	21	22	22	23	24	25	25	26	27	28	28	29	30	31	31	32	33	34	35	35	36	37	38	38
45	47	48	50	52	54	55	57	59	61	62	64	66	68	69	71	73	74	76	78	80	81	83	85	87
32	33	34	35	37	38	39	40	41	43	44	45	46	47	49	50	51	52	54	55	56	57	58	60	61
37	38	40	41	42	44	45	47	48	49	51	52	54	55	57	58	59	61	62	64	65	66	68	69	71
41	43	44	46	48	49	51	52	54	56	57	59	60	62	63	65	67	68	70	71	73	75	76	78	79
26	27	28	29	30	31	32	33	34	35	36	37	38	39	40	41	42	43	44	45	46	47	48	49	50
48	50	52	54	55	57	59	61	63	65	67	68	70	72	74	76	78	80	81	83	85	87	89	91	92
14	14	15	15	16	16	17	17	18	18	19	19	20	20	21	21	22	22	23	23	24	24	25	25	26
52	54	56	58	60	61	63	65	67	69	71	73	75	77	79	81	83	85	87	89	91	93	95	97	99
0	0	0	0	0	0	0	0	0	0	0	0	0	0	0	0	0	0	0	0	0	0	0	0	0
11	11	11	11	11	11	11	11	11	11	11	11	11	11	11	11	11	11	11	11	11	11	11	11	11
s	s	s	s	s	s	s	s	s	s	s	s	s	s	s	s	s	s	s	s	s	s	s	s	s
52	54	56	58	60	62	64	66	68	70	72	74	76	78	80	82	84	86	88	90	92	94	96	98	100
2	4	6	8	10	12	14	16	18	20	22	24	26	28	30	32	34	36	38	40	42	44	46	48	50
0	1	1	1	1	2	2	2	2	3	3	3	3	4	4	4	4	5	5	5	5	6	6	6	7
2	4	6	8	10	12	14	15	17	19	21	23	25	27	29	31	33	35	37	39	41	43	44	46	48
1	2	2	3	4	5	5	6	7	8	8	9	10	11	12	12	13	14	15	15	16	17	18	18	19
2	3	5	7	9	10	12	14	16	17	19	21	22	24	26	28	29	31	33	35	36	38	40	42	43
1	2	4	5	6	7	9	10	11	12	13	15	16	17	18	19	21	22	23	24	26	27	28	29	30
1	3	4	6	7	8	10	11	13	14	16	17	18	20	21	23	24	25	27	28	30	31	33	34	35
2	3	5	6	8	9	11	13	14	16	17	19	21	22	24	25	27	29	30	32	33	35	37	38	40
1	2	3	4	5	6	7	8	9	10	11	12	13	14	15	16	17	18	19	20	21	22	23	24	25
2	4	6	7	9	11	13	15	17	18	20	22	24	26	28	30	31	33	35	37	39	41	42	44	46
1	1	2	2	3	3	4	4	5	5	6	6	7	7	8	8	9	9	10	10	11	11	12	12	13
2	4	6	8	10	12	14	16	18	20	22	24	26	28	30	32	34	36	38	40	42	44	46	48	50
0	0	0	0	0	0	0	0	0	0	0	0	0	0	0	0	0	0	0	0	0	0	0	0	0
11	11	11	11	11	11	11	11	11	11	11	11	11	11	11	11	11	11	11	11	11	11	11	11	11
s	s	s	s	s	s	s	s	s	s	s	s	s	s	s	s	s	s	s	s	s	s	s	s	s
2	4	6	8	10	12	14	16	18	20	22	24	26	28	30	32	34	36	38	40	42	44	46	48	50

8-10c. Tables for Determining Positions of Peaks

The peaks in F(x) must be located when this function is constructed.

The table below (Booth's method) enables one to find the position of the peak from two known values, one on each side of the peak.

Example. x is the coordinate, $\rho(x)$ is the value at point x:

x	5	6	7
$\varrho(x)$	124	286	198

Let ρ_0 denote the minimal value of $\rho(x)$, and let the others be ρ_1 and ρ_2; then we have for $\Delta\rho = \rho_1 - \rho_0$ and $\Delta\rho = \rho_2 - \rho_0$:

x	5	6	7
$\Delta\varrho$	0	162	74
	ϱ_0	ϱ_1	ϱ_2

Then r = 74/124 = 0.456; the calculated r gives us Δx =1.1477 (from the table), so the precise coordinate is x = 5.0 + 1.1477 = 6.1477.

If the minimal value of $\rho(x)$ lies to the right of the peak, the resulting Δx is subtracted from the x corresponding to the minimum.

r	0	1	2	3	4	5	6	7	8	9
0,00	1,0000	1,0003	1,0005	1,0008	1,0010	1,0013	1,0015	1,0018	1,0020	1,0023
1	1,0025	1,0028	1,0030	1,0033	1,0035	1,0038	1,0041	1,0043	1,0046	1,0048
2	1,0051	1,0054	1,0056	1,0059	1,0061	1,0064	1,0066	1,0069	1,0071	1,0074
3	1,0076	1,0079	1,0081	1,0084	1,0086	1,0089	1,0092	1,0094	1,0097	1,0099
4	1,0102	1,0104	1,0107	1,0110	1,0112	1,0115	1,0118	1,0120	1,0123	1,0125
5	1,0128	1,0131	1,0133	1,0136	1,0139	1,0142	1,0144	1,0147	1,0150	1,0152
6	1,0155	1,0158	1,0160	1,0163	1,0165	1,0168	1,0171	1,0173	1,0176	1,0178
7	1,0181	1,0184	1,0186	1,0189	1,0192	1,0194	1,0197	1,0200	1,0203	1,0205
8	1,0208	1,0211	1,0214	1,0216	1,0219	1,0222	1,0225	1,0228	1,0230	1,0233
9	1,0236	1,0239	1,0241	1,0244	1,0247	1,0250	1,0252	1,0255	1,0258	1,0260
0,10	1,0263	1,0266	1,0269	1,0271	1,0274	1,0277	1,0280	1,0283	1,0285	1,0288
1	1,0291	1,0294	1,0297	1,0299	1,0302	1,0305	1,0308	1,0311	1,0313	1,0316
2	1,0319	1,0322	1,0325	1,0328	1,0331	1,0333	1,0336	1,0339	1,0342	1,0345
3	1,0348	1,0351	1,0354	1,0356	1,0359	1,0362	1,0365	1,0368	1,0370	1,0373
4	1,0376	1,0379	1,0382	1,0385	1,0388	1,0391	1,0393	1,0396	1,0399	1,0402
5	1,0405	1,0408	1,0411	1,0414	1,0417	1,0420	1,0423	1,0426	1,0429	1,0432
6	1,0435	1,0438	1,0441	1,0444	1,0447	1,0450	1,0453	1,0456	1,0459	1,0462
7	1,0465	1,0468	1,0471	1,0474	1,0477	1,0480	1,0483	1,0486	1,0489	1,0492
8	1,0495	1,0498	1,0501	1,0504	1,0507	1,0510	1,0513	1,0515	1,0519	1,0522
9	1,0525	1,0528	1,0531	1,0534	1,0537	1,0541	1,0544	1,0547	1,0550	1,0553

r	0	1	2	3	4	5	6	7	8	9
0,20	1,0556	1,0559	1,0562	1,0565	1,0568	1,0572	1,0575	1,0578	1,0581	1,0584
1	1,0587	1,0590	1,0593	1,0596	1,0599	1,0603	1,0606	1,0609	1,0612	1,0615
2	1,0618	1,0621	1,0624	1,0628	1,0631	1,0634	1,0637	1,0640	1,0644	1,0647
3	1,0650	1,0653	1,0660	1,0663	1,0666	1,0669	1,0672	1,0676	1,0676	1,0679
4	1,0682	1,0685	1,0688	1,0692	1,0695	1,0698	1,0701	1,0704	1,0708	1,0711
5	1,0714	1,0717	1,0721	1,0724	1,0727	1,0731	1,0734	1,0737	1,0740	1,0744
6	1,0747	1,0750	1,0754	1,0757	1,0760	1,0763	1,0767	1,0770	1,0773	1,0777
7	1,0780	1,0783	1,0787	1,0790	1,0794	1,0797	1,0800	1,0804	1,0807	1,0811
8	1,0814	1,0817	1,0821	1,0824	1,0828	1,0831	1,0834	1,0838	1,0841	1,0845
9	1,0848	1,0851	1,0855	1,0858	1,0862	1,0865	1,0868	1,0872	1,0875	1,0879
0,30	1,0882	1,0886	1,0889	1,0893	1,0896	1,0900	1,0903	1,0907	1,0910	1,0914
1	1,0917	1,0921	1,0924	1,0928	1,0931	1,0935	1,0938	1,0942	1,0945	1,0949
2	1,0952	1,0956	1,0959	1,0963	1,0966	1,0970	1,0974	1,0977	1,0981	1,0984
3	1,0988	1,0992	1,0995	1,0999	1,1002	1,1006	1,1010	1,1013	1,1017	1,1020
4	1,1024	1,1028	1,1031	1,1035	1,1039	1,1043	1,1046	1,1050	1,1054	1,1057
5	1,1061	1,1065	1,1068	1,1072	1,1076	1,1080	1,1083	1,1087	1,1091	1,1094
6	1,1098	1,1102	1,1105	1,1109	1,1113	1,1117	1,1120	1,1124	1,1128	1,1131
7	1,1135	1,1139	1,1143	1,1146	1,1150	1,1154	1,1158	1,1162	1,1165	1,1169
8	1,1173	1,1177	1,1181	1,1184	1,1188	1,1192	1,1196	1,1200	1,1203	1,1207
9	1,1211	1,1215	1,1219	1,1223	1,1227	1,1231	1,1234	1,1238	1,1242	1,1246
0,40	1,1250	1,1254	1,1258	1,1262	1,1266	1,1270	1,1273	1,1277	1,1281	1,1285
1	1,1289	1,1293	1,1297	1,1301	1,1305	1,1309	1,1313	1,1317	1,1321	1,1325
2	1,1329	1,1333	1,1337	1,1341	1,1345	1,1349	1,1353	1,1357	1,1361	1,1365
3	1,1369	1,1373	1,1377	1,1381	1,1385	1,1390	1,1394	1,1398	1,1402	1,1406
4	1,1410	1,1414	1,1418	1,1423	1,1427	1,1431	1,1435	1,1439	1,1444	1,1448
5	1,1452	1,1456	1,1460	1,1465	1,1469	1,1473	1,1477	1,1481	1,1486	1,1490
6	1,1494	1,1498	1,1502	1,1507	1,1511	1,1515	1,1519	1,1523	1,1528	1,1532
7	1,1536	1,1540	1,1545	1,1549	1,1553	1,1558	1,1562	1,1566	1,1570	1,1575
8	1,1579	1,1583	1,1588	1,1592	1,1597	1,1601	1,1605	1,1610	1,1614	1,1619
9	1,1623	1,1627	1,1632	1,1636	1,1641	1,1645	1,1649	1,1654	1,1658	1,1663
0,50	1,1667	1,1671	1,1676	1,1680	1,1685	1,1689	1,1693	1,1698	1,1702	1,1707
1	1,1711	1,1716	1,1720	1,1725	1,1729	1,1734	1,1739	1,1743	1,1748	1,1752
2	1,1757	1,1762	1,1766	1,1771	1,1775	1,1780	1,1785	1,1789	1,1794	1,1798
3	1,1803	1,1808	1,1812	1,1817	1,1821	1,1826	1,1831	1,1835	1,1840	1,1844
4	1,1849	1,1854	1,1859	1,1863	1,1868	1,1873	1,1878	1,1883	1,1887	1,1892
5	1,1897	1,1902	1,1906	1,1911	1,1915	1,1920	1,1925	1,1930	1,1935	1,1939
6	1,1944	1,1949	1,1954	1,1959	1,1964	1,1964	1,1973	1,1978	1,1983	1,1988
7	1,1993	1,1998	1,2003	1,2008	1,2013	1,2018	1,2022	1,2027	1,2032	1,2037
8	1,2042	1,2047	1,2052	1,2057	1,2062	1,2067	1,2072	1,2077	1,2082	1,2087
9	1,2092	1,2097	1,2102	1,2107	1,2112	1,2118	1,2123	1,2128	1,2133	1,2138
0,60	1,2143	1,2148	1,2153	1,2158	1,2163	1,2169	1,2174	1,2179	1,2184	1,2189
1	1,2194	1,2199	1,2204	1,2210	1,2215	1,2220	1,2225	1,2230	1,2236	1,2241
2	1,2246	1,2251	1,2257	1,2262	1,2267	1,2273	1,2278	1,2283	1,2288	1,2294
3	1,2299	1,2304	1,2310	1,2315	1,2321	1,2326	1,2331	1,2337	1,2342	1,2348
4	1,2353	1,2358	1,2364	1,2369	1,2375	1,2380	1,2385	1,2391	1,2396	1,2402
5	1,2407	1,2413	1,2418	1,2424	1,2429	1,2435	1,2441	1,2446	1,2452	1,2457
6	1,2463	1,2469	1,2474	1,2480	1,2485	1,2491	1,2497	1,2502	1,2508	1,2513
7	1,2519	1,2525	1,2530	1,2536	1,2542	1,2548	1,2553	1,2559	1,2565	1,2570
8	1,2576	1,2582	1,2588	1,2593	1,2599	1,2605	1,2611	1,2617	1,2622	1,2628
9	1,2634	1,2640	1,2646	1,2651	1,2657	1,2663	1,2669	1,2675	1,2680	1,2686
0,70	1,2692	1,2698	1,2704	1,2710	1,2716	1,2722	1,2728	1,2734	1,2740	1,2746
1	1,2752	1,2758	1,2764	1,2770	1,2776	1,2782	1,2788	1,2795	1,2801	1,2807
2	1,2813	1,2819	1,2825	1,2831	1,2837	1,2844	1,2850	1,2856	1,2862	1,2868
3	1,2874	1,2880	1,2887	1,2893	1,2899	1,2906	1,2912	1,2918	1,2924	1,2931
4	1,2937	1,2943	1,2950	1,2956	1,2962	1,2969	1,2975	1,2981	1,2987	1,2994
5	1,3000	1,3007	1,3013	1,3020	1,3026	1,3033	1,3039	1,3046	1,3052	1,3059
6	1,3065	1,3072	1,3078	1,3085	1,3091	1,3098	1,3104	1,3111	1,3117	1,3124
7	1,3130	1,3137	1,3143	1,3151	1,3157	1,3164	1,3170	1,3177	1,3184	1,3190
8	1,3197	1,3204	1,3211	1,3217	1,3224	1,3231	1,3238	1,3245	1,3252	1,3258
9	1,3265	1,3272	1,3279	1,3285	1,3292	1,3299	1,3306	1,3313	1,3319	1,3326

r	0	1	2	3	4	5	6	7	8	9
0,80	1,3333	1,3340	1,3347	1,3354	1,3361	1,3368	1,3375	1,3382	1,3389	1,3396
1	1,3403	1,3410	1,3417	1,3425	1,3432	1,3439	1,3446	1,3453	1,3461	1,3468
2	1,3475	1,3482	1,3489	1,3497	1,3504	1,3511	1,3518	1,3525	1,3533	1,3540
3	1,3547	1,3554	1,3562	1,3569	1,3576	1,3584	1,3591	1,3599	1,3606	1,3614
4	1,3621	1,3629	1,3636	1,3644	1,3651	1,3659	1,3666	1,3674	1,3681	1,3689
5	1,3696	1,3704	1,3711	1,3719	1,3726	1,3734	1,3742	1,3749	1,3757	1,3764
6	1,3772	1,3780	1,3788	1,3795	1,3803	1,3811	1,3819	1,3827	1,3834	1,3842
7	1,3850	1,3858	1,3866	1,3874	1,3882	1,3890	1,3897	1,3905	1,3913	1,3921
8	1,3929	1,3937	1,3945	1,3953	1,3961	1,3969	1,3977	1,3985	1,3993	1,4001
9	1,4009	1,4017	1,4025	1,4034	1,4042	1,4050	1,4058	1,4066	1,4075	1,4083
0,90	1,4091	1,4099	1,4108	1,4116	1,4124	1,4133	1,4141	1,4149	1,4157	1,4166
1	1,4174	1,4183	1,4191	1,4200	1,4208	1,4217	1,4225	1,4334	1,4242	1,4251
2	1,4259	1,4268	1,4276	1,4285	1,4294	1,4303	1,4311	1,4320	1,4329	1,4337
3	1,4346	1,4355	1,4364	1,4372	1,4381	1,4390	1,4399	1,4408	1,4416	1,4425
4	1,4434	1,4443	1,4452	1,4461	1,4470	1,4479	1,4488	1,4497	1,4506	1,4515
5	1,4524	1,4533	1,4542	1,4551	1,4560	1,4570	1,4579	1,4588	1,4597	1,4606
6	1,4615	1,4624	1,4634	1,4643	1,4653	1,4662	1,4671	1,4681	1,4690	1,4700
7	1,4709	1,4719	1,4728	1,4738	1,4747	1,4757	1,4766	1,4776	1,4785	1,4795
8	1,4804	1,4814	1,4823	1,4833	1,4843	1,4853	1,4362	1,4872	1,4882	1,4892
9	1,4901	1,4911	1,4921	1,4931	1,4941	1,4951	1,4960	1,4970	1,4980	1,4990
1,00	1,5000									

8-11. Determination of Dynamic Lattice Distortions and Characteristic Temperatures

Dynamic displacements $\sqrt{\overline{u_d^2}}$ are calculated for the cubic system on the basis of the thermal intensity factor M from

$$2M = 16\pi^2 \overline{u_d^2} \left(\frac{\sin \vartheta}{\lambda} \right)^2 . \tag{79}$$

The M for each temperature T (absolute) is given by

$$2M = \frac{12h^2}{mk\Theta} \left[\frac{\Phi(\Theta/T)}{\Theta/T} + \frac{1}{4} \right] \left(\frac{\sin \vartheta}{\lambda} \right)^2 , \tag{80}$$

where Θ is the characteristic temperature (this is dependent on the bond strengths), $\Phi(\Theta/T)$ is the Debye function (see Section 4-18 for this as a function of Θ/T), h is Planck's constant, m is the mass of an atom of the material (m = Am_H, where A is the atomic weight and m_H is the mass of a hydrogen atom), k is Boltzmann's constant, and λ is the x-ray wavelength.

It is usual to determine Θ without consideration of the Lorentz factor or changes in lattice constant by taking a pair of lines whose intensity ratio is measured at several temperatures (e.g., room temperature and liquid-nitrogen temperature):

$$\alpha_{293^\circ} = \left(\frac{I_{h_1k_1l_1}}{I_{h_2k_2l_2}} \right)_{293^\circ} , \quad \alpha_{90^\circ} = \left(\frac{I_{h_1k_1l_1}}{I_{h_2k_2l_2}} \right)_{90^\circ} .$$

A graphical method is commonly employed; a plot is made for the temperatures T_1 and T_2 of the following as a function of Θ:

$$\psi\left(\Theta, T_1, T_2\right) = \left[\frac{\Phi\left(\Theta/T_1\right)}{\Theta/T_1} - \frac{\Phi\left(\Theta/T_2\right)}{\Theta/T_2} \right]. \qquad (81)$$

For two lines of Bragg angles ϑ_1 and ϑ_2 we plot

or

$$\left.\begin{array}{c} \psi\left(\Theta\right) = \dfrac{mk}{12h^2} \dfrac{\ln \dfrac{\alpha_1}{\alpha_2}}{\dfrac{\sin^2 \vartheta_2}{\lambda^2} - \dfrac{\sin^2 \vartheta_1}{\lambda^2}} \\[4ex] \psi\left(\Theta\right) = \dfrac{mk}{12h^2} \dfrac{4a^2}{\Sigma_1 - \Sigma_2} \ln \dfrac{\alpha_1}{\alpha_2} \Theta, \end{array}\right\} \qquad (82)$$

where

$$\Sigma_1 - \Sigma_2 = \left(h_1^2 + k_1^2 + l_1^2\right) - \left(h_2^2 + k_2^2 + l_2^2\right).$$

The two graphs are plotted on the same sheet; Θ is read from the point of intersection [325]. See [405] for the deduction of Θ from the intensities of several lines.

The following example [326] is for Fe + 2%V.

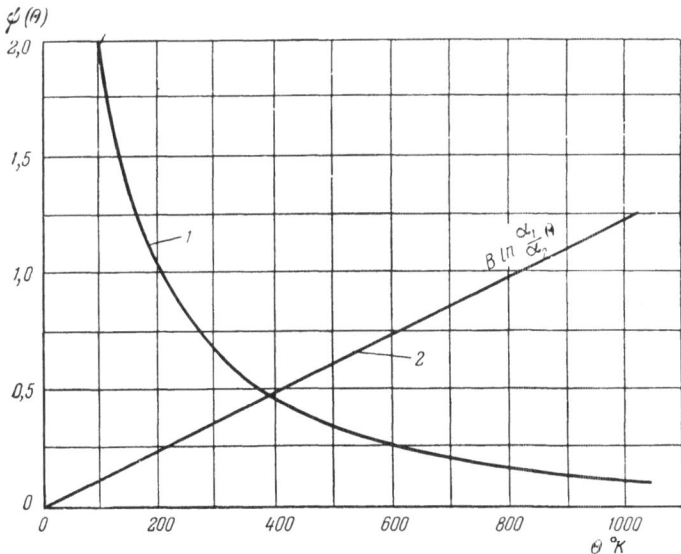

Fig. 159. Graphs for finding a characteristic temperature.

Measurements were made at 20° and −185° (liquid nitrogen) in a low-temperature camera.

The intensities were measured for the (211) and (510) lines (molybdenum radiation). The ratios were $\alpha_1 = (I_{211}/I_{510})_{293°} = 2.98$, $\alpha_2 = (I_{211}/I_{510})_{108°} = 2.24$.

From (80), $\Phi(\Theta)$ was constructed for $T_1 = 293°K$ and $T_2 = 108°K$; Fig. 159 (curve 1) shows the result.

Then $\psi(\Theta)$ is calculated from (82) on the basis of the known parameters:

$$\frac{12h^2}{mk}\frac{(h_1^2+k_1^2+l_1^2)-(h_2^2+k_2^2+l_2^2)}{4a^2} = \frac{12\cdot6,54^2\cdot10^{-54}(26-6)}{55.74\cdot1.66\cdot10^{-24}\cdot1.372\cdot10^{-16}\cdot4(2.863)^2 10^{-16}} = 246,6,$$

$$\psi(\Theta) = \frac{2.8 \ln \dfrac{2.98}{2.24}}{246.6}\,\Theta = 0.0114\,\Theta.$$

The straight line 2 in Fig. 159 corresponds to $\psi(\Theta) = 0.0114\,\Theta$; the meeting-point of lines 1 and 2 gives Θ for this Fe + 2% V at 390°K.

This, with (80), gives M, which with (79) gives the dynamic lattice distortion for any given temperature. For example, in the above case the result was $\sqrt{u_d^2}$ = 0.123 Å for 296°K and 0.080 Å for 108°K.

The characteristic temperature need not be known in order to calculate the difference of the dynamic distortions at two temperatures T_1 and T_2. The formula to use here is

$$\Delta\overline{u^2} = (\overline{u^2}_d)_{T_1} - (\overline{u^2}_d)_{T_2} = \frac{\ln \dfrac{a_1}{a_2}}{\dfrac{4}{3}\dfrac{\pi^2}{a^2}[(h_2^2+k_2^2+l_2^2)-(h_1^2+k_1^2+l_1^2)]}.$$

The following results (see table) are for Co−Al, Ni−Al, and Ti−C at 295 and 110°K, and at 473 and 295°K; Mo radiation was used with the (200) and (510) line for Co−Al and Ni−Al, and with the (400) and (842) lines for Ti−C [167].

Alloy	$\dfrac{a_{295°}}{a_{110°}}$	Δu^2, Å (295°—110°)	$\dfrac{a_{473°}}{a_{295°}}$	Δu^2, Å (473°—295°)
Co—Al (50)	1,23	0,058	1,29	0,072
Co—Al (55)	1,30	0,074	1,37	0,089
Ni—Al (45)	1,45	0,0105	1,52	0,0118
Ni—Al (50)	1,28	0,0071	1,32	0,0079
Ni—Al (60)	1,35	0,0085	1,41	0,0097
Ti—C (50)	1,09	0,00118	1,13	0,0025
Ti—C (30)	1,15	0,0029	1,23	0,0043

8-12. Determination of Static Lattice Distortions of Atomic Scale

Displacements caused by thermal motion are accompanied by ones having other causes (plastic deformation, formation of solid solutions, and so on). These also reduce line strengths, and their magnitude can be evaluated from the reduction.

The simplest method is the measurement of the ratio of line intensity to background for specimens with distorted and undistorted lattices.

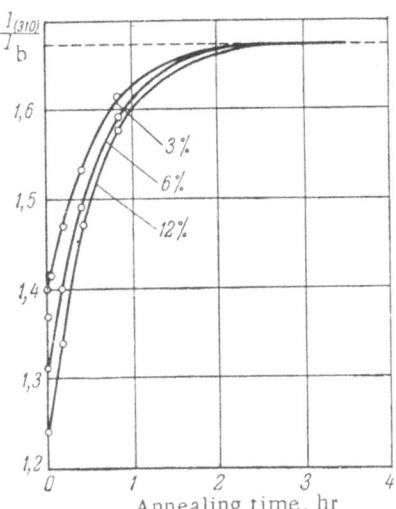

Fig. 160. Ratio of intensity of (310) line to background in annealing of deformed steel.

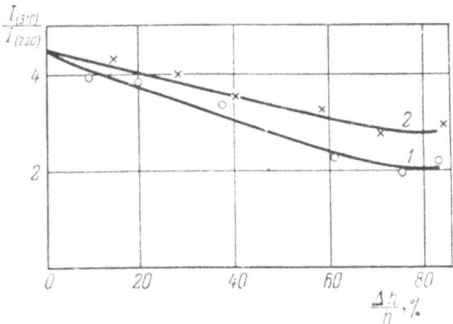

Fig. 161. Ratio of intensities of (310) and (220) lines during the plastic deformation of steel.

Figure 160 shows the ratio of the (310) line to the background at small angles for type 40 steel examined with Co radiation [168]; the graph shows the effects of annealing time (at 600°) for specimens differing in degree of plastic deformation. Increase in I_{310}/I_b corresponds to removal of distortion

Another method is to measure the ratio of intensities for two lines of the same specimen or of the specimen and a standard. Figure 161 shows curves for the (310) and (220) lines of steel, as recorded with Co radiation, for static (curve 1) and dynamic (curve 2) deformation [169]. Extinction causes considerable weakening of the earlier lines, and correction for this causes a substantial reduction in the estimates of distortions of atomic scale [405]. The displacement from the equilibrium positions in the lattice is

$$\overline{u^2}_s = \frac{3a^2 \ln\left[\dfrac{(I_1/I_2)\mathrm{def}}{(I_1/I_2)\mathrm{undef}}\right]}{4\pi^2\left[(h_2^2+k_2^2+l_2^2)-(h_1^2+k_1^2+l_1^2)\right]} \qquad (83)$$

The static and dynamic components may be distinguished if more precise values are needed; here use is made of specimens with undistorted lattices, for which the distortions are solely thermal (dynamic), and also specimens distorted by plastic deformation, formation of solid solutions, and so on (here both types of displacement are present). For undeformed materials (e.g., annealed Fe), without allowance for extinction,

$$\frac{I_{h_1k_1l_1}}{I_{h_2k_2l_2}} = a_1 = Ae^{-2(M_1-M_2)}, \qquad (84)$$

while for others (e.g., Fe-base alloys or deformed Fe)

$$\frac{I'_{h_1k_1l_1}}{I_{h_2k_2l_2}} = a_2 = Ae^{-2(M'_1-M'_2)}e^{-K}, \qquad (85)$$

where

$$K = 8\pi^2\overline{u^2}_s \;.$$

The thermal factors M_1, M_2, M'_1, and M'_2 are found from the Θ for specimen and standard (see 8-11); for example, for Fe−V alloy

$$K = \ln \frac{\alpha_{Fe-V}}{\alpha_{Fe}} - (M_1 - M_2) + (M'_1 - M'_2)$$

or

$$K = \ln \frac{\alpha_{Fe-V}}{\alpha_{Fe}} - \left\{ \frac{12h^2}{mk\Theta_{Fe-V}} \left[\frac{\Phi(\Theta_{Fe-V}/T)}{\Theta_{Fe-V}/T} + \frac{1}{4} \right] - \right.$$
$$\left. - \frac{12h^2}{mk\Theta_{Fe}} \left[\frac{\Phi(\Theta_{Fe}/T)}{\Theta_{Fe}/T} + \frac{1}{4} \right] \right\} \frac{(h_2^2 + k_2^2 + l_2^2) - (h_1^2 + k_1^2 + l_1^2)}{4a^2}. \quad (86)$$

Table 1 gives results for annealed Fe powder and for Fe + 2% V as examined with Mo radiation.

TABLE 1

Material	$\alpha = I_{211}/I_{510}$	$\alpha_{Fe-V}/\alpha_{Fe}$	K_{Fe-V}	$\sqrt{\overline{u_s^2}}$, Å
Fe	2,66	—	—	—
Fe+2%V	2,98	1,12	1,07	0,040

See [396-403, 406] for new methods of evaluating lattice distortion.

The accuracy of the result increases with $(h_2^2 + k_2^2 + l_2^2) - (h_1^2 + k_1^2 + l_1^2)$, so it is desirable to use a hard radiation giving numerous lines on the pattern. For example, molybdenum radiation is used with Fe and steels, in which case the lines may have indices whose squares sum to 6 and 62 [(211) and (732)].

If the specimen has large microstresses or very small blocks, it becomes difficult to measure the intensities of lines at large angles; resort is then made to lines of small ϑ. For example, Fe and its alloys can be examined with (211) as the first line on Mo radiation ($\Sigma = 6$, $\vartheta = 17°36'$) and with (321) as the second ($\Sigma = 14$, $\vartheta = 27°30'$), or, alternatively, (510) ($\Sigma = 26$, $\vartheta = 39°12'$). The intensities can alter greatly to either side if the specimen has a preferred orientation; for example, uniaxial stretching of chrome-nickel-molybdenum steel reduces the strength of the (211) line produced by Cr radiation by a large factor, whereas rolling of Fe produces an increase (by more than a factor 2), the difference in the preferred orientations being responsible for this. Existing methods of correcting for texture effects are complicated and inadequately developed, so it is best to ensure that preferred orientations are absent in studies on displacements.

Finally, the results can be in error on account of extinction, which may result from large block size in the standard, which is usually a specimen that has been given prolonged annealing to produce the equilibrium state.

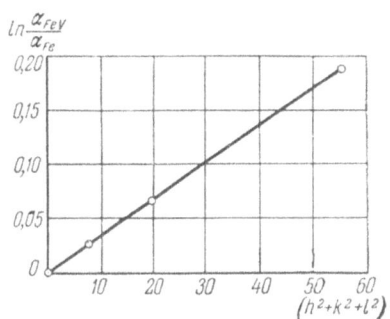

Fig. 162. Graph for the effects of extinction on line intensities for Fe−V alloy.

The above relationships show that there is a strict relation of line intensity to the sum of the squares of the indices; the natural logarithm of the ratio of the α for the distorted and undistorted lattices is directly proportional to the Σ of the lines. The intensities of lines of small Σ are much reduced by extinction, whereas those of large Σ are not much affected.

One way of detecting extinction effects is from the deviation of α_{dis}/α_s as a function of Σ from a straight line; Fig. 162 shows the line for the Fe−V alloy, which indicates that there was no extinction under the conditions used.

Extinction effects cannot be neglected in lattice-distortion measurements if the material is of large grain size; for instance, the extinction effect becomes apparent for deformed tungsten powder for block sizes exceeding $5 \cdot 10^{-5}$ cm. Here the line intensities are dependent on the block size as well as on the lattice distortion.

Three lines can be used in order to eliminate the effect of block size [170]; the unknowns B in the formula for the intensity I, which is dependent on block size D as well as on the lattice distortion, are given by

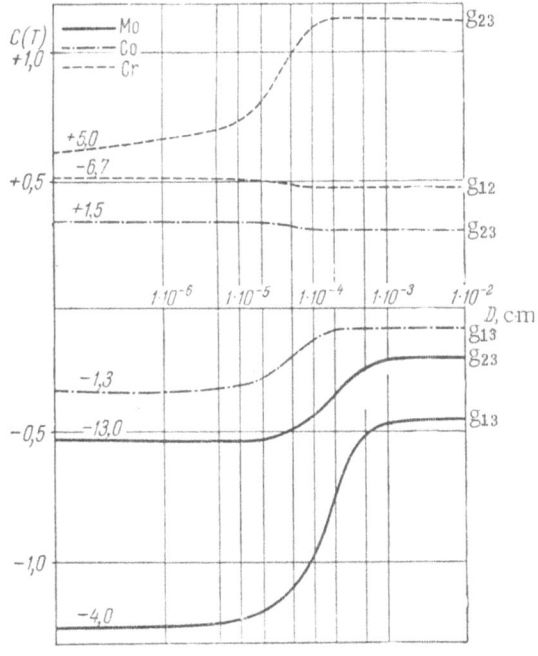

Fig. 163. Graphs for the extinction correction for line intensities for tungsten and several radiations.

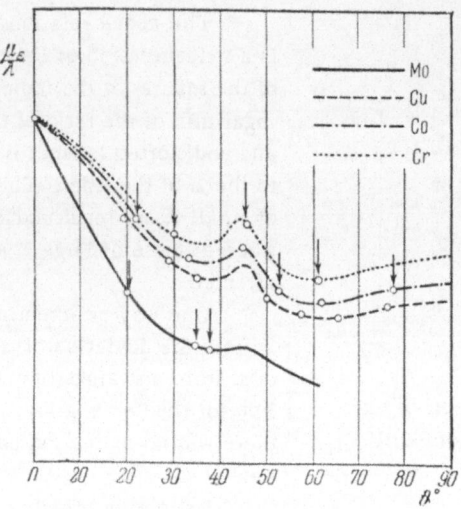

Fig. 164. Values of μ_ε/λ for tungsten for various radiations and Bragg angles.

$$\left.\begin{aligned}\frac{2B}{\lambda_2} &= g_{12} - \frac{\lg \dfrac{I_1}{I_2}}{(\sin^2 \vartheta_1 - \sin^2 \vartheta_2)\,\lg e}\,, \\[2ex] \frac{2B}{\lambda_2} &= g_{23} - \frac{\lg \dfrac{I_2}{I_3}}{(\sin^2 \vartheta_2 - \sin^2 \vartheta_3)\,\lg e}\,, \end{aligned}\right\} \tag{87}$$

where

$$g_{12} = \left(\lg \frac{p_1}{p_2} \frac{f(\vartheta_1)}{f(\vartheta_2)} \frac{|F_1|^2}{|F_2|^2} \frac{\varepsilon_2}{\varepsilon_1} + \lg \frac{1 - e^{-2\varepsilon_1 D}}{1 - e^{-2\varepsilon_2 D}} \right) \frac{1}{(\sin^2 \vartheta_1 - \sin^2 \vartheta_2)\,\lg e}\,, \tag{88}$$

p is the multiplicity factor, $f(\vartheta) = (1 + \cos^2 2\vartheta)/\cos \vartheta \sin \vartheta$, $|F|$ is the modulus of the structure factor, and ε is the effective absorption coefficient (joint effects of extinction and ordinary absorption). See Section 8-2 for the calculation of effective absorption coefficients.

In practice, (87) is used to calculate g as a function of block size; Fig. 163 gives g for tungsten for several radiations. The abscissa is block size (on a logarithmic scale), the ordinate being g. Table 2 gives the indices of lines 1, 2, and 3. The curves have been displaced vertically to facilitate use; the numbers at the left indicate the displacements. The ε of (88) is found from

$$\varepsilon = \mu + \mu_\varepsilon,$$

where μ is the linear absorption coefficient and

$$\mu_\varepsilon = \frac{3\pi}{16} \lambda |F| N \frac{e^2}{mc^2} \frac{1 + \cos^2 2\vartheta}{1 + \cos 2\vartheta}.$$

Here N is the number of unit cells in unit volume (a_W = 3.158 Å), e is the charge on an electron in esu, c is the velocity of light, and m is the mass of an electron.

Figure 164 gives values of μ_ε/λ for tungsten for various radiations and Bragg angles; the circles denote angles corresponding to diffraction lines.

The equations of (87) are solved graphically by constructing curves for g_{12} and g_{23} as functions of block size, the two curves being displaced respectively along the ordinate axis by

$$\frac{-\lg\dfrac{I_1}{I_2}}{(\sin^2\vartheta_1 - \sin^2\vartheta_2)\lg e} \quad \text{and} \quad \frac{-\lg\dfrac{I_2}{I_3}}{(\sin^2\vartheta_2 - \sin^2\vartheta_3)\lg e} \, .$$

The ordinate of the point of intersection gives $2B/\lambda$, from which the distortion on the atomic scale may be calculated; the abscissa of the point gives the block size D. A consideration of the errors shows that two lines of similar ε must be used together with one of very different ε in order to obtain B and D with the minimum error; in addition, the lines should satisfy the other conditions of selection for lattice-distortion measurements (large angles and so on).

Table 2 gives an example of the choice of lines for tungsten powder examined with various radiations.

TABLE 2

Radiation	Line	(hkl)	ϑ_{hkl}	μ	μ_ε	ε
Mo	1	013	20°46′	2013	2690	4703
	2	314	34°52′		1870	3883
	3	125	37°53′		1806	3819
Co	1	011	23°34′	4890	9574	14464
	2	022	53°04′		6886	11756
	3	222	78°15′		6975	11865
Cr	1	011	30°47′	8709	11699	20408
	2	002	46°21′		11887	20596
	3	112	62°24′		9509	18218

Table 2 indicates that primary extinction can be neglected for lines of closely similar ε; the lattice distortions are then given by the ratio of intensities of the two. A large difference in the absorption coefficients requires measurements on three lines, in which case the crystal size may be deduced as well as the distortion.

The range of block sizes providing highest accuracy is that in which the $g = f(D)$ curve rises steeply. This part of the curve moves to smaller block sizes as the x-ray wavelength is increased. For example, the range in D is $4 \cdot 10^{-5}$ to $5 \cdot 10^{-4}$ cm for tungsten powder and Mo radiation, whereas it is 10^{-5} to $5 \cdot 10^{-5}$ for Cr radiation.

CHAPTER 9

DETERMINATION OF PREFERRED
ORIENTATIONS (TEXTURES)

Chapter 9 gives graphs and nets for texture studies and for the construction of polar figures.

See [6-9, 11-13, 172, 312] for details of methods of examining preferred orientations.

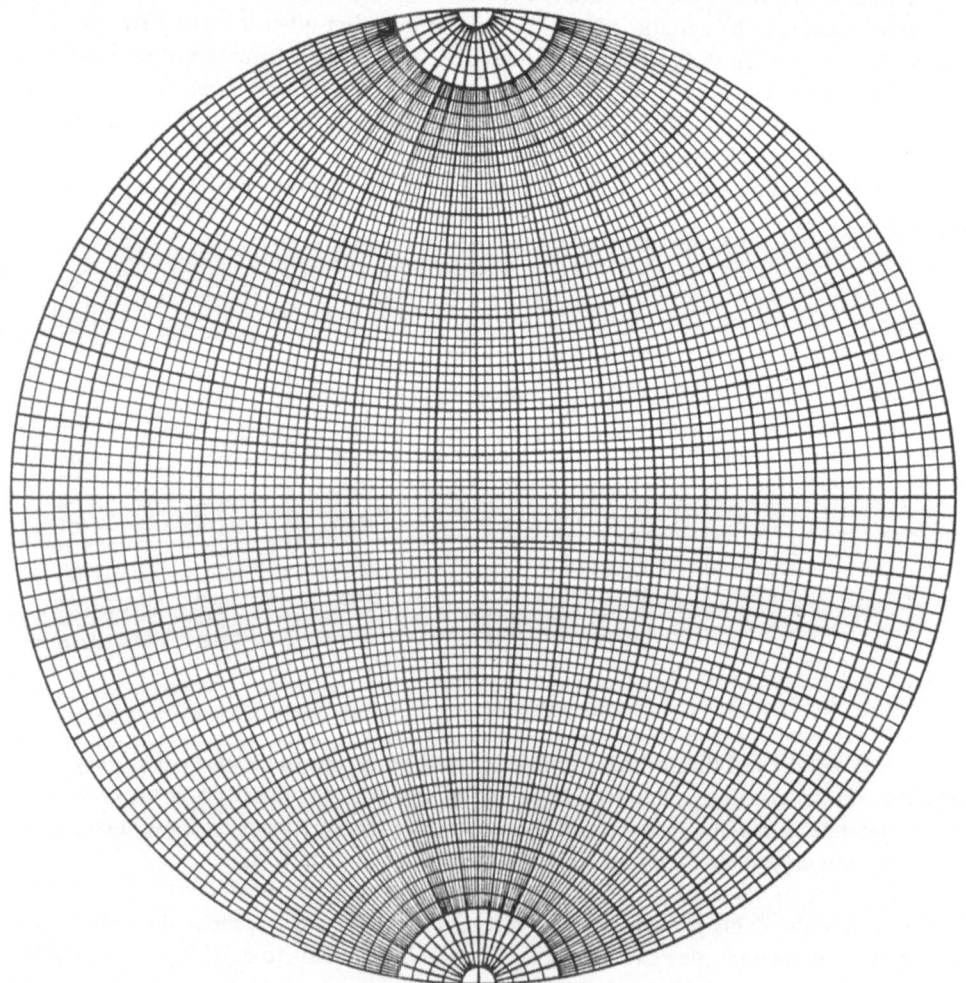

Fig. 165. Wulff net.

See [281] for cameras for examining textures in coarse-grained materials and [12] for special types of cameras.

Electron-diffraction analysis of textures is dealt with in [211, 329, 330]; new methods of examining textures, in [372, 373].

9-1. The Wulff Net

The Wulff net (Fig. 165) is used for determining the orientations of monocrystals, for constructing stereographic projections and polar figures, and so on. The meridians and parallels are at intervals of 2°; see [7, 8, 171] for constructions with the net. It is usual to use a net 20 cm in diameter.

9-2. The Polar Net

Figure 166 shows the polar net, which is used to construct polar figures.

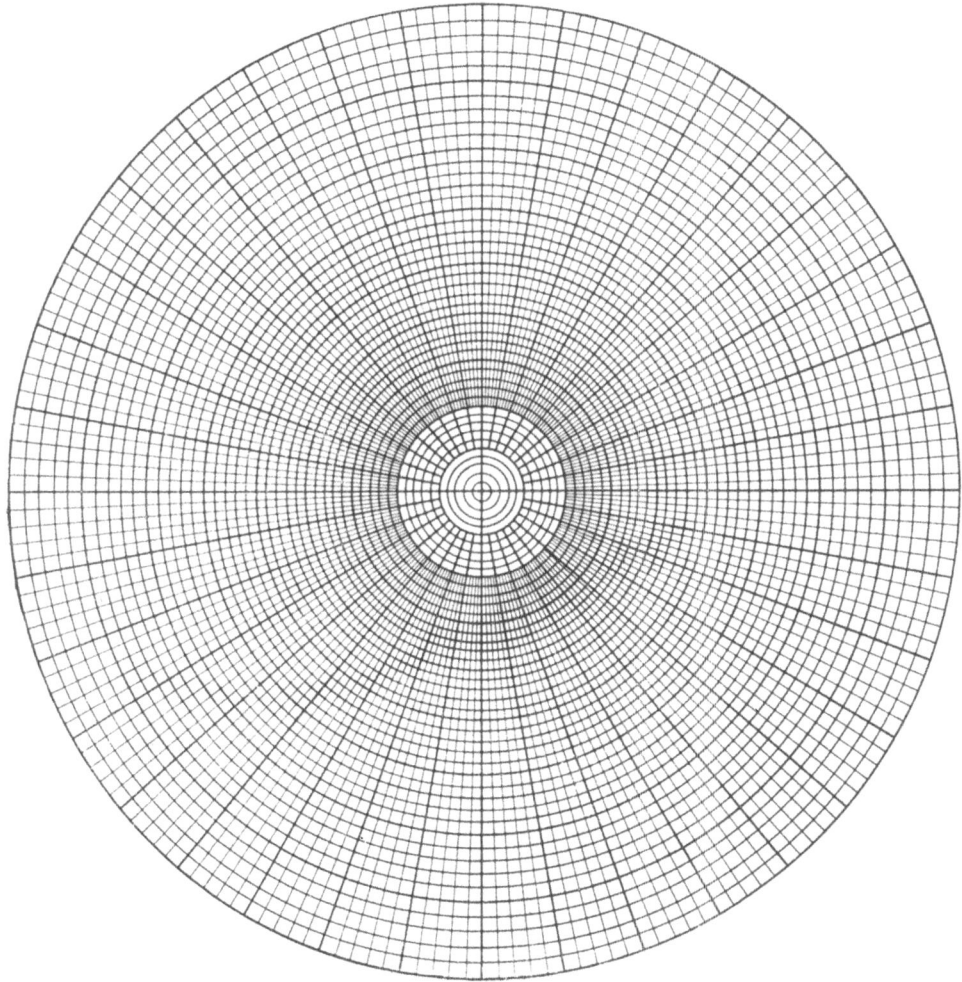

Fig. 166. Polar net.

The divisions are at intervals of 2°; see [6, 11] for descriptions of the use of the net. It is usual to use a net 20 cm in diameter.

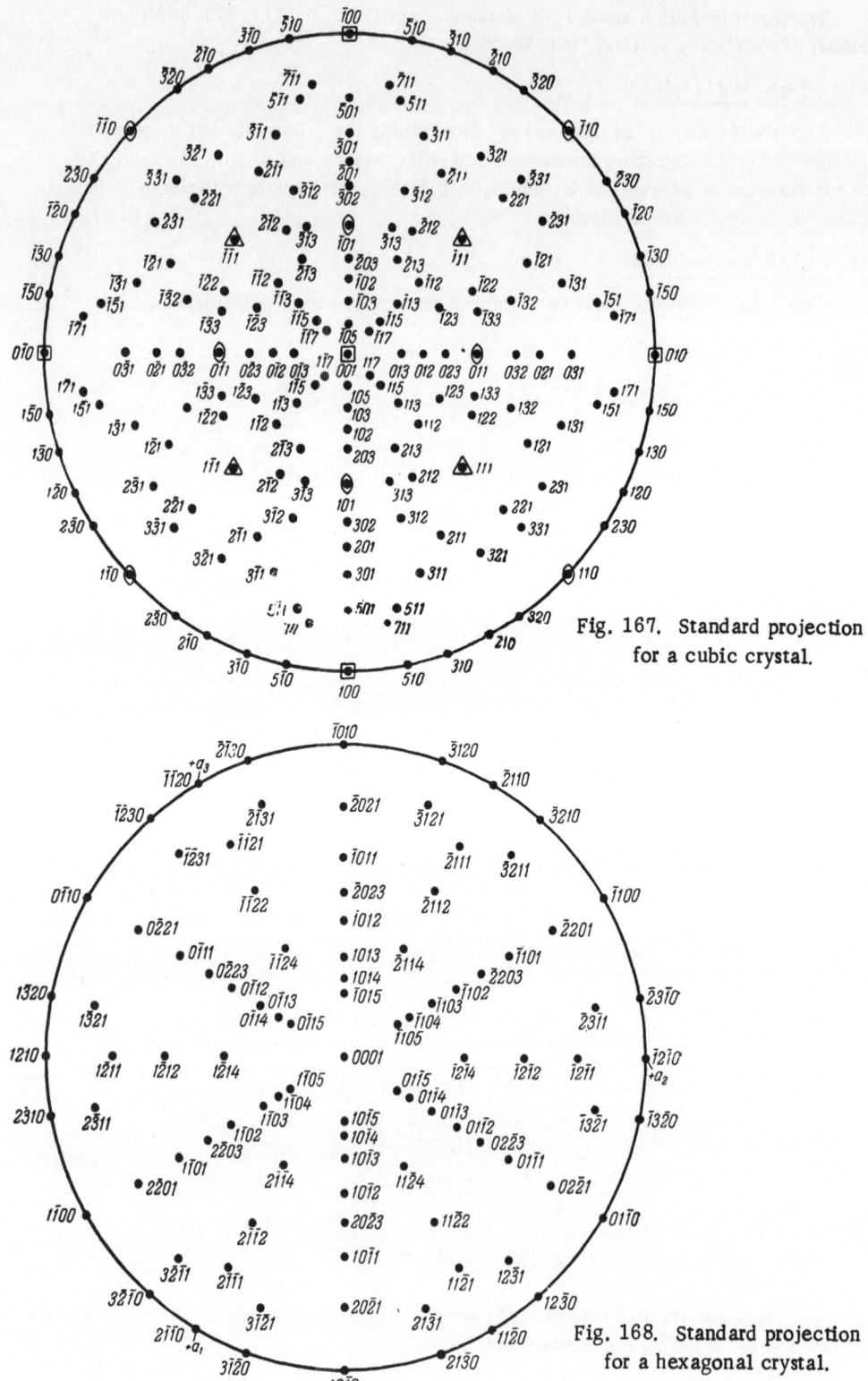

Fig. 167. Standard projection for a cubic crystal.

Fig. 168. Standard projection for a hexagonal crystal.

9-3. Standard Projections of Crystals

A standard projection is a stereographic projection of the poles of all the most important planes; all planes with low indices are shown. The points on the projection represent the emergence of the normals to the planes, whose indices are given near the points.

Figure 167 gives the standard projection for a cubic crystal whose cube faces are parallel to the projection plane. The X and Y axes of the crystal lie in that plane, so the poles of the (100) and (010) planes lie on the bounding circle. The Z axis is perpendicular to the projection plane, so the pole of (001) lies at the center of the circle.

Figure 168 gives the standard projection for a hexagonal compact crystal for c/a = 1.86 (zinc).

Here the pole of (0001) lies at the center of the circle.

A stereographic projection has the following main features [171], which are used in constructions and calculations:

1. Small circles on the sphere project as circles, but their centers do not coincide with the center of the projection; the radial displacement is dependent on the angular distance of the center on the sphere with respect to the corresponding circles.

2. A great circle on the sphere projects as a circle, which meets the bounding circle at two diametrically opposed points; a great circle lying in a plane normal to the plane of projection appears as a diameter of the bounding circle, while great circles lying in inclined sections of the sphere can be brought into coincidence with one of the meridians.

3. The angles between points on the sphere are equal to the differences of latitude on the net when the points have been brought to the same longitude by rotation. The linear distance corresponding to 1° varies by a factor of two between the center and the bounding circle.

4. The angles between points in the projection do not alter when the points rotate around the axis of the stereographic net.

9-4. Angles Between Atomic Planes

9-4a. Cubic System

The table gives the angles (in degrees) between planes $(h_1k_1l_1)$ and $(h_2k_2l_2)$ for cubic materials [12].

The angle φ for planes not listed may be found from

$$\varphi = \arccos \frac{h_1h_2 + k_1k_2 + l_1l_2}{\sqrt{(h_1^2 + k_1^2 + l_1^2)(h_2^2 + k_2^2 + l_2^2)}} \cdot \qquad (89)$$

In the table, p_1 is the multiplicity factor for $(h_1k_1l_1)$, and p_2 is that for $(h_2k_2l_2)$; the number of $(h_1k_1l_1)$ planes forming the given angle with $(h_2k_2l_2)$ is given in parentheses.

The fibrous texture in a wire is specified by the angle ρ between the axis of the wire and the normal to the (hk$l$) plane; it is given by

$$\varrho = \arccos(\cos\vartheta \cos\delta),$$

where δ is the angle between the projection of the axis on the film and a radius drawn from the center of the film to the texture peak. The table gives ρ for several crystallographic planes and directions of the axis.

$(h_1k_1l_1)$	p_1	$(h_2k_2l_2)$	p_2	Angle between planes $(h_1k_1l_1)$ and $(h_2k_2l_2)$ in degrees
(100)	6	(100)	6	00 (2), 90 (4)
		(110)	12	45 (8), 90 (4)
		(111)	8	54,73 (8)
		(210)	24	26,57 (8), 63,43 (8), 90 (8)
		(211)	24	35,27 (8), 65,90 (16)
		(221)	24	48,19 (16), 70,53 (8)
		(310)	24	18,44 (8), 71,56 (8), 90 (8)
		(311)	24	25,24 (8), 72,45 (16)
		(320)	24	33,69 (8), 56,31 (8), 90 (8)
		(321)	48	36,70 (16), 57,69 (16), 74,50 (16)
		(322)	24	43,31 (8), 60,98 (16)
		(410)	24	14,03 (8), 75,97 (8), 90 (8)
		(411)	24	19,47 (8), 76,37 (16)
		(331)	24	46,51 (16), 76,74 (8)
(110)	20	(110)	12	0 (2), 60 (2), 90 (2)
		(111)	8	35,27 (4), 90 (4)
		(210)	24	18,44 (4), 50,77 (8), 71,56 (12)
		(211)	24	30 (4), 54,73 (4), 73,22 (8), 90 (8)
		(221)	24	19,47 (4), 45 (4), 76,37 (8), 90 (8)
		(310)	24	26,57 (4), 47,87 (8), 63,43 (4), 77,08 (8)
		(311)	24	31,48 (8), 64,76 (12), 90 (4)
		(320)	24	11,31 (4), 53,96 (8), 66,91 (8), 78,69 (4), 79,11 (16)
		(321)	48	19,11 (8), 40,89 (8), 55,46 (8), 67,79 (8)
		(322)	24	30,97 (8), 46,69 (4), 80,13 (4)
		(410)	24	30,97 (4), 46,68 (8), 59,03 (4), 80,13 (8)
		(411)	24	33,55 (8), 60 (8), 70,53 (4), 90 (4)
		(331)	24	13,27 (4), 49,56 (8), 71,07 (8), 90 (8)
(111)	8	(111)	8	0 (2), 70,53 (6)
		(210)	24	39,23 (12), 75,04 (12)
		(211)	24	19,47 (6), 61,87 (12), 90 (6)
		(221)	24	15,81 (6), 54,73 (6), 78,90 (12)
		(310)	24	43,10 (12), 68,58 (12)
		(311)	24	29,50 (6), 58,52 (12), 79,98 (6)
		(320)	24	36,81 (12), 80,79 (12)
		(321)	48	22,21 (12), 51,89 (12), 72,02 (12), 90 (12)
		(322)	24	11,42 (6), 65,16 (12), 81,95 (6)
		(410)	24	45,57 (12), 65,16 (12), 81,95 (6)

Angle between planes $(h_1k_1l_1)$ and $(h_2k_2l_2)$ in degrees

$(h_1k_1l_1)$	p_1	$(h_2k_2l_2)$	p_2	Angle between planes $(h_1k_1l_1)$ and $(h_2k_2l_2)$ in degrees
		(411)	24	35,27 (6) 57,02 (12) 74,21 (6)
		(331)	24	21,99 (6) 48,53 (6) 82,39 (12)
(210)	24	(210)	24	0 (2) 36,87 (6) 53,13 (2) 66,42 (2) 78,46 (8) 90 (4)
		(211)	24	24,09 (4) 43,09 (4) 56,79 (8) 79,48 (8) 90 (4)
		(221)	24	26,57 (4) 41,81 (4) 53,40 (4) 63,43 (4) 72,65 (4) 90 (4)
		(310)	24	8,13 (2) 31,95 (2) 45 (4) 64,90 (4) 73,57 (4) 81,87 (6)
		(311)	24	19,29 (4) 47,61 (8) 66,14 (8) 82,25 (8)
		(320)	24	7,12 (2) 29,75 (4) 41,91 (4) 60,25 (4) 68,15 (4) 75,64 (4) 82,88 (2)
		(321)	48	17,02 (4) 33,21 (8) 53,30 (8) 61,44 (12) 68,99 (8) 83,13 (8) 90 (4)
		(322)	24	29,80 (4) 40,60 (4) 49,40 (4) 64,29 (4) 77,47 (4) 83,77 (4)
		(410)	24	12,53 (2) 40,60 (4) 49,40 (4) 64,29 (4) 77,47 (4) 83,77 (4)
		(411)	24	18,43 (2) 42,45 (4) 50,57 (4) 71,57 (4) 83,95 (4)
		(331)	24	22,57 (4) 44,10 (4) 59,14 (4) 72,07 (4) 84,11 (4)
(211)	24	(211)	24	0 (2) 33,56 (2) 48,19 (4) 60 (4) 70,53 (4) 80,41 (8) 90 (2)
		(221)	24	17,72 (4) 35,26 (4) 47,12 (2) 65,90 (4) 74,21 (6) 82,18 (4)
		(310)	24	25,35 (4) 49,80 (8) 58,91 (8) 75,04 (4) 82,59 (4)
		(311)	24	10,02 (2) 42,39 (2) 50,50 (6) 75,75 (6) 90 (4)
		(320)	24	25,07 (4) 37,57 (4) 55,52 (4) 63,07 (8) 83,50 (8)
		(321)	48	10,90 (4) 29,21 (4) 40,20 (8) 49,11 (4) 56,94 (4) 70,89 (12) 77,40 (4) 83,74 (4) 90
		(322)	24	8,05 (2) 26,98 (2) 53,55 (4) 60,33 (4) 72,72 (4) 78,58 (2) 84,32 (4)
		(410)	24	26,98 (4) 53,55 (4) 60,33 (4) 72,72 (4) 78,58 (4) 84,32 (2)
		(411)	24	15,80 (2) 39,67 (4) 47,66 (4) 54,73 (4) 61,24 (2) 73,22 (4) 84,48 (4)
		(331)	24	20,51 (4) 41,47 (6) 68,00 (8) 79,20 (8)
(221)	24	(221)	24	0 (2) 27,27 (4) 38,94 (4) 63,61 (2) 83,62 (8) 90 (4)
		(310)	24	32,51 (4) 42,45 (4) 58,19 (8) 76,06 (8) 83,95 (4)
		(311)	24	25,24 (4) 45,29 (6) 59,83 (6) 72,45 (8) 84,23 (2)
		(320)	24	22,41 (4) 42,30 (4) 49,67 (4) 68,30 (4) 79,34 (4) 84,70 (4)
		(321)	48	11,49 (4) 27,02 (4) 36,70 (4) 57,69 (8) 63,55 (8) 74,50 (8) 84,89 (8)
		(322)	24	14,04 (4) 27,21 (2) 49,70 (4) 66,16 (2) 71,13 (4) 75,96 (4) 90 (4)
		(410)	24	36,06 (4) 43,31 (4) 55,53 (4) 60,98 (4) 80,69 (8)
		(411)	24	30,20 (4) 45 (4) 51,06 (2) 66,87 (4) 71,68 (4) 90 (2)
		(331)	24	6,21 (2) 32,73 (6) 57,64 (6) 67,52 (4) 85,61 (8)
(310)	24	(310)	24	0 (2) 25,84 (2) 36,87 (4) 53,13 (2) 72,54 (2) 84,26 (4) 90 (4)
		(311)	24	17,55 (4) 40,29 (4) 55,10 (4) 67,58 (4) 79,01 (4) 90 (4)
		(320)	24	15,25 (2) 37,87 (2) 52,13 (2) 58,25 (2) 74,75 (6) 79,90 (4) 85,15 (4) 90 (4)
		(321)	48	21,62 (4) 32,31 (4) 40,48 (4) 47,46 (4) 53,73 (8) 59,53 (8) 65,00 (8) 75,31 (4) 90 (4)

Angle between planes $(h_1k_1l_1)$ and $(h_2k_2l_2)$ in degrees

$(h_1k_1l_1)$	p_1	$(h_2k_2l_2)$	p_2	Angle between planes $(h_1k_1l_1)$ and $(h_2k_2l_2)$ in degrees											
(311)	24	(322)	24	32,47	46,35 (4)	52,15 (4)	57,53 (4)	72,13 (4)	76,70 (4)	85,60 (6)					
		(410)	24	4,40	23,02 (2)	32,47 (4)	57,53 (2)	72,13 (2)	76,70 (4)						
		(411)	24	14,31	34,93 (4)	56,55 (4)	72,65 (4)	81,43 (4)	85,73 (4)						
		(331)	24	29,48	43,49 (4)	54,52 (4)	64,20 (4)	90 (8)							
(320)	24	(311)	24	23,09	41,18 (4)	54,17 (4)	65,28 (4)	75,47 (8)	85,20 (4)						
		(320)	24	0	35,10 (2)	50,48 (6)	62,97 (4)	84,78 (4)	85,20 (4)						
		(321)	48	14,77	36,31 (4)	49,86 (8)	61,08 (12)	71,20 (4)	80,73 (12)						
		(322)	24	18,08	36,45 (2)	48,84 (4)	59,21 (4)	68,55 (4)	85,81 (6)						
		(410)	24	18,08	36,45 (4)	59,21 (4)	68,55 (4)	77,33 (4)	85,81 (4)						
		(411)	24	5,77	31,48 (2)	44,72 (4)	55,35 (4)	64,76 (4)	81,83 (4)	90 (4)					
		(331)	24	25,95	40,46 (4)	51,50 (2)	61,04 (4)	69,77 (4)	78,02 (6)						
(321)	48	(321)	48	0	21,79 (4)	31,00 (4)	38,21 (6)	44,42 (6)	50,00 (2)	60,00 (4)	64,62 (2)	69,07 (2)	73,40 (6)	81,79 (2)	85,90 (8)
		(322)	24	13,52	24,84 (2)	32,58 (2)	44,52 (2)	49,59 (2)	63,02 (2)	71,08 (4)	78,79 (2)	82,55 (2)	86,28 (4)		
		(410)	24	24,84	32,58 (2)	44,52 (2)	49,59 (4)	54,31 (2)	63,02 (2)	67,11 (2)	71,08 (2)	82,55 (2)	86,28 (2)	86,39 (2)	
		(411)	24	19,11	35,02 (2)	40,89 (2)	46,14 (2)	50,95 (2)	55,46 (2)	67,79 (4)	71,64 (2)	75,41 (2)	79,11 (2)		
		(331)	24	11,18	30,87 (4)	42,63 (4)	52,18 (2)	60,63 (2)	68,42 (2)	75,80 (4)	82,95 (4)	90 (2)			
(322)	24	(322)	24	0	19,75 (2)	58,03 (4)	61,93 (4)	76,39 (4)	86,63 (2)						
		(410)	24	34,56	49,68 (4)	53,97 (4)	69,33 (8)	72,90 (4)	86,63 (8)						
		(411)	24	23,85	42,00 (4)	46,69 (4)	59,04 (4)	62,78 (2)	66,41 (4)	80,13 (4)					
		(331)	24	18,93	33,42 (4)	43,67 (2)	59,95 (4)	73,85 (2)	80,39 (4)	86,81 (4)					
(410)	24	(410)	24	0	19,75 (2)	28,07 (4)	61,93 (2)	76,39 (2)	86,63 (8)	90 (2)					
		(411)	24	13,63	30,96 (4)	52,78 (4)	73,39 (4)	80,13 (4)	90 (4)						
		(331)	24	33,42	43,67 (4)	52,26 (4)	59,95 (2)	67,08 (4)	86,81 (4)						
(411)	24	(411)	24	0	27,27 (2)	38,94 (4)	60,00 (2)	67,12 (4)	86,82 (8)						
		(331)	24	30,10	40,80 (4)	57,27 (4)	64,37 (6)	77,51 (4)	83,79 (2)						
(331)	24	(331)	24	0	26,52 (2)	37,86 (2)	61,73 (4)	80,91 (8)	86,98 (4)						

9-4b. Tetragonal System

The table gives the angle φ (in degrees) between $(h_1 k_1 l_1)$ and $(h_2 k_2 l_2)$ planes for tetragonal materials having c/a from 0.5 to 1.5 [102].

The φ for planes not listed may be found from

$$\varphi = \arccos\left\{ \frac{\dfrac{h_1 h_2 + k_1 k_2}{a^2} + \dfrac{l_1 l_2}{c^2}}{\sqrt{\left(\dfrac{h_1^2 + k_1^2}{a^2} + \dfrac{l_1^2}{c^2}\right)\left(\dfrac{h_2^2 + k_2^2}{a^2} + \dfrac{l_2^2}{c^2}\right)}} \right\}. \tag{90}$$

$(h_1 k_1 l_1)$	$(h_2 k_2 l_2)$		c/a									
			0,5		0,6		0,9		1,2		1,5	
(100)	(011)	(101)	90,0	63,4	90,0	59,0	90,0	48,0	90,0	39,8	90,0	33,7
	(012)	(102)	90,0	76,0	90,0	73,3	90,0	65,8	90,0	59,0	90,0	53,1
	(013)	(103)	90,0	80,5	90,0	78,7	90,0	73,3	90,0	68,2	90,0	63,4
	(014)	(104)	90,0	82,9	90,0	81,5	90,0	77,3	90,0	73,3	90,0	69,4
	(021)	(201)	90,0	45,0	90,0	39,8	90,0	29,1	90,0	22,6	90,0	18,4
	(023)	(203)	90,0	71,6	90,0	68,2	90,0	59,0	90,0	51,3	90,0	45,0
	(025)	(205)	90,0	78,7	90,0	76,5	90,0	70,2	90,0	64,4	90,0	59,0
	(110)		45,0		45,0		45,0		45,0		45,0	
	(111)		65,9		62,8		56,2		52,5		50,2	
	(112)		76,4		74,0		67,7		62,8		59,0	
	(113)		80,7		78,9		74,0		69,6		65,9	
	(120)	(210)	63,4	26,6	63,4	26,6	63,4	26,6	63,4	26,6	63,4	26,6
	(121)	(211)	70,5	48,2	69,0	44,2	66,4	36,8	65,2	33,1	64,6	31,0
	(122)	(212)	77,4	64,1	75,6	60,1	71,5	50,6	69,0	44,2	67,4	39,8
	(123)	(213)	81,0	71,8	79,5	68,6	75,6	60,1	72,7	53,4	70,5	48,2
(110)	(012)	(102)	80,1	80,1	78,3	78,3	73,1	73,1	68,7	68,7	64,9	64,9
	(013)	(103)	83,3	83,3	82,0	82,0	78,3	78,3	75,0	75,0	71,6	71,6
	(021)	(201)	60,0	60,0	57,1	57,1	51,8	51,8	49,3	49,3	47,9	47,9
	(023)	(203)	77,1	77,1	74,8	74,8	68,7	68,7	63,8	63,8	60,0	60,0
	(111)	(111)	54,7	90,0	49,7	90,0	38,2	90,0	30,5	90,0	25,2	90,0
	(112)	(112)	70,5	90,0	67,0	90,0	57,5	90,0	49,7	90,0	43,3	90,0
	(113)	(113)	76,7	90,0	74,3	90,0	67,0	90,0	60,5	90,0	54,7	90,0
(111)	(011)	(101)	24,1	56,8	27,2	65,3	33,8	85,0	37,5	81,8	39,8	72,8
	(012)	(102)	27,0	46,1	30,5	53,2	37,7	70,4	41,5	83,0	43,5	87,6
	(013)	(103)	29,3	42,4	33,1	48,9	41,3	64,4	45,8	75,8	48,1	84,5
	(021)	(201)	30,0	73,2	32,9	82,1	38,2	79,3	40,8	68,5	42,1	61,8
	(023)	(203)	25,4	49,8	28,6	57,4	35,3	75,7	39,0	80,9	41,1	81,3
	(112)	(112)	15,8	39,7	17,3	45,4	19,4	58,6	19,2	67,2	18,1	73,0
	(113)	(113)	22,0	37,4	24,5	42,8	28,9	55,3	30,0	63,8	29,5	69,6

9-4c. Hexagonal System

The table gives the angle φ (in degrees) between $(h_1 k_1 l_1)$ and $(h_2 k_2 l_2)$ planes for hexagonal materials having c/a from 1.40 to 2.00 [102].

The φ for planes not listed may be found from

$$\varphi = \arccos\left\{ \frac{h_1 k_2 + k_1 k_2 + \frac{1}{2}(h_1 h_2 + h_2 k_1) + \dfrac{3a^2}{4c^2} l_1 l_2}{\sqrt{\left(h_1^2 + k_1^2 + h_1 k_1 + \dfrac{3a^2}{4c^2} l_1^2\right)\left(h_2^2 + k_2^2 + h_2 k_2 + \dfrac{3a^2}{4c^2} l_2^2\right)}} \right\}. \tag{91}$$

$(h_1k_1l_1)$	$(h_2k_2l_2)$	c/a											
		1,40	1,50	1,55	1,60	1,65	1,70	1,75	1,80	1,85	1,90	1,95	2,00
(001)	(100)	90	90	90	90	90	90	90	90	90	90	90	90
	(101)	90	60	60,81	61,58	62,31	63,01	63,67	64,31	64,91	65,50	66,05	66,59
	(102)	90	40,89	41,82	42,73	43,61	44,46	45,30	46,10	46,89	47,65	48,39	49,11
	(103)	90	30	30,82	31,63	32,42	33,20	33,96	34,71	35,45	36,18	36,89	37,59
	(104)	90	23,41	24,11	24,79	25,47	26,14	26,80	27,46	28,10	28,74	29,38	30,00
	(105)	90	19,10	19,70	20,38	20,86	21,44	22,00	22,57	23,14	23,69	24,24	24,79
	(201)	90	73,90	74,39	74,86	75,30	75,71	76,10	76,47	76,83	77,16	77,48	77,78
	(203)	90	49,11	50,03	50,93	51,79	52,62	53,41	54,18	54,92	55,64	56,33	57,00
	(205)	90	34,72	35,60	36,46	37,31	38,14	38,95	39,74	40,48	41,27	42,01	42,73
	(207)	90	26,33	27,08	27,83	28,56	29,29	30,00	30,70	31,40	32,08	32,76	33,42
	(110)	90	90	90	90	90	90	90	90	90	90	90	90
	(112)	90	56,31	57,17	58,00	58,78	59,53	60,26	60,94	61,61	62,24	62,85	63,44
	(114)	90	36,87	37,78	38,66	39,52	40,36	41,19	41,99	42,77	43,53	44,28	45,00
	(116)	90	26,56	27,32	28,07	28,81	29,54	30,26	30,96	31,66	32,35	33,02	33,69
	(210)	90	90	90	90	90	90	90	90	90	90	90	90
	(211)	90	77,69	78,08	78,44	78,18	79,10	79,41	79,69	80,22	80,22	80,47	80,70
	(212)	90	66,42	67,10	67,75	68,36	68,94	69,49	70,00	70,51	70,99	71,44	71,88
(100)	(010)	60	60	60	60	60	60	60	60	60	60	60	60
	(110)	30	30	30	30	30	30	30	30	30	30	30	30
	(210)	19,11	19,11	19,11	19,11	19,11	19,11	19,11	19,11	19,11	19,11	19,11	19,11

9-5. Nets for Constructing Polar Figures for a Flat Film

Special nets greatly facilitate the construction in this case; the results from the pattern (the values of β, the angle between the edges of the peak and the vertical diameter of the film) are entered on the net. The numbers on the horizontal diameter correspond to the angle α between the plane of the specimen and the projection plane; a projection specified by α and β is constructed by placing over the net a piece of tracing paper bearing a circle of the same diameter, the point being marked in where the parallel corresponding to β meets the meridian corresponding to α.

The polar figure is the assembly of all such points.

The form of the net is dependent on the material, on the radiation, and on the indices of the reflecting plane.

Figures 169-174 give nets for certain materials and radiations [171, 327, 328]. These nets are best enlarged to match the Wulff and polar nets to be used.

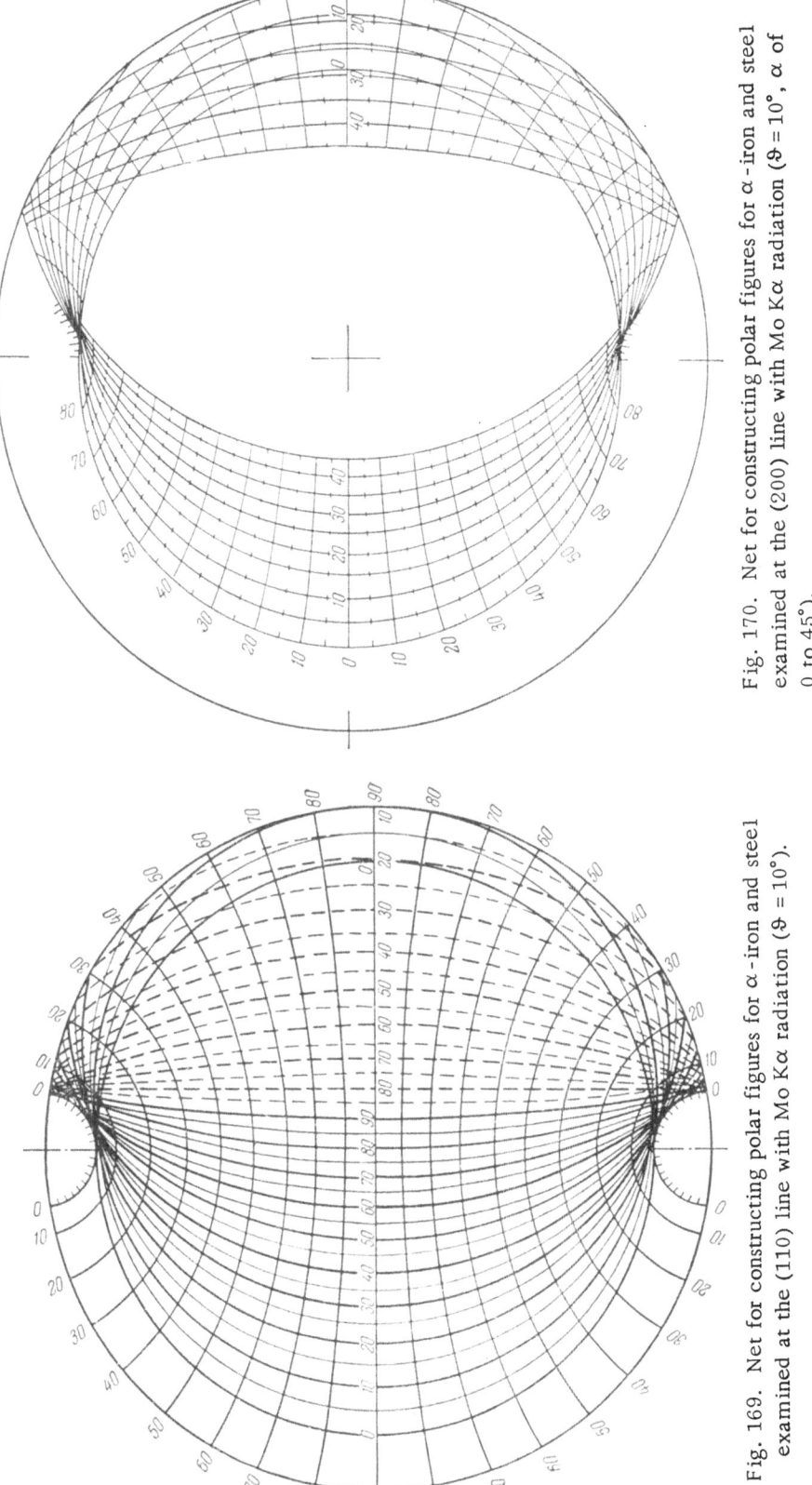

Fig. 169. Net for constructing polar figures for α - iron and steel examined at the (110) line with Mo Kα radiation ($\vartheta = 10°$).

Fig. 170. Net for constructing polar figures for α - iron and steel examined at the (200) line with Mo Kα radiation ($\vartheta = 10°$, α of 0 to 45°).

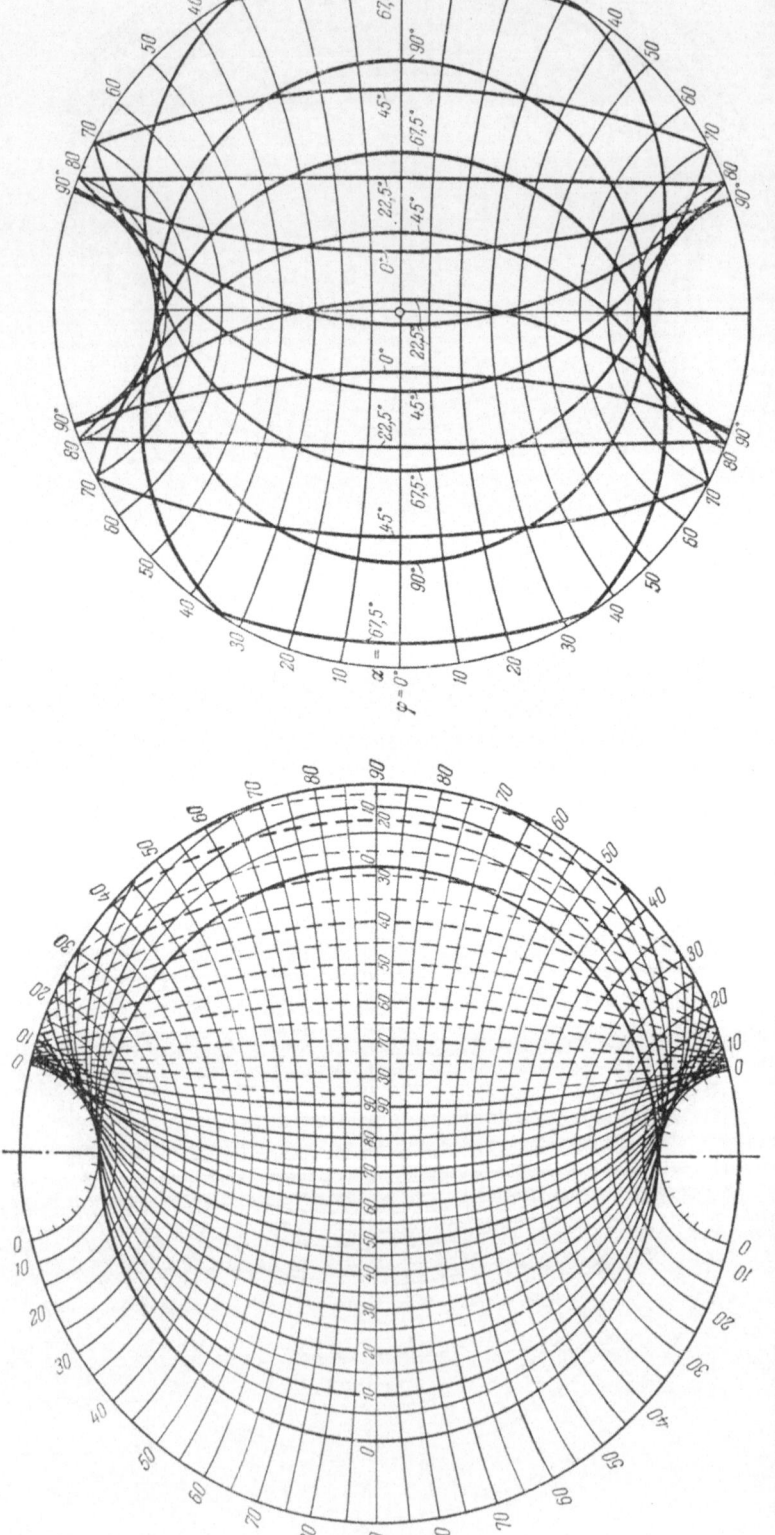

Fig. 172. Net for constructing polar figures for α-iron and steel examined at the (110) line with Fe Kα radiation.

Fig. 171. Net for constructing polar figures for α-iron and steel examined at the (200) line with Mo Kα radiation (ϑ = 10°, α of 0 to 90°).

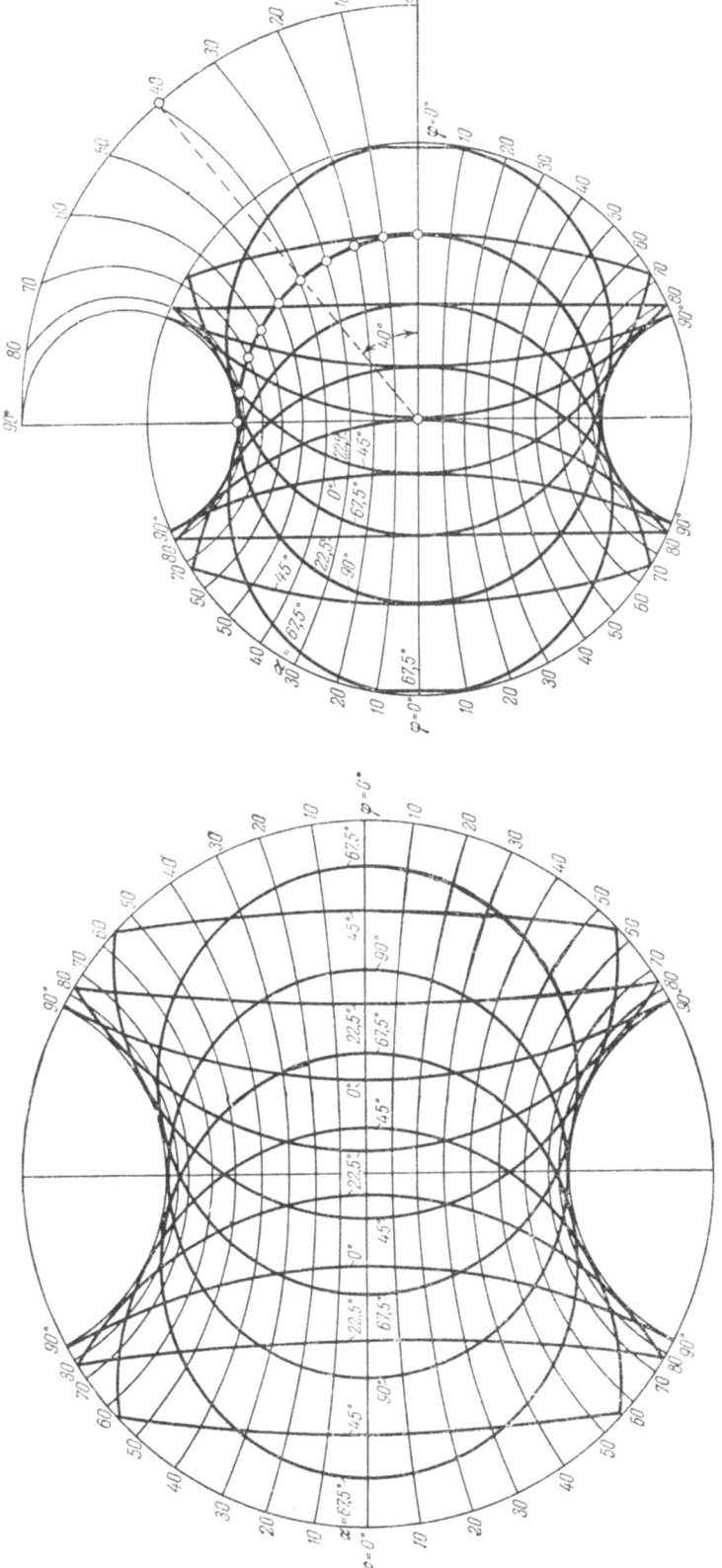

Fig. 173. Net for constructing polar figures for aluminum and its alloys examined at the (111) line with Cu Kα radiation.

Fig. 174. Net for constructing polar figures for aluminum and its alloys examined at the (200) line with Cu Kα radiation.

9-6. Net for Constructing Polar Figures When an Axial Camera Is Used

A special stop is used to isolate a single Debye cone, which is recorded on a cylindrical film with the specimen set at various angles to the primary beam.

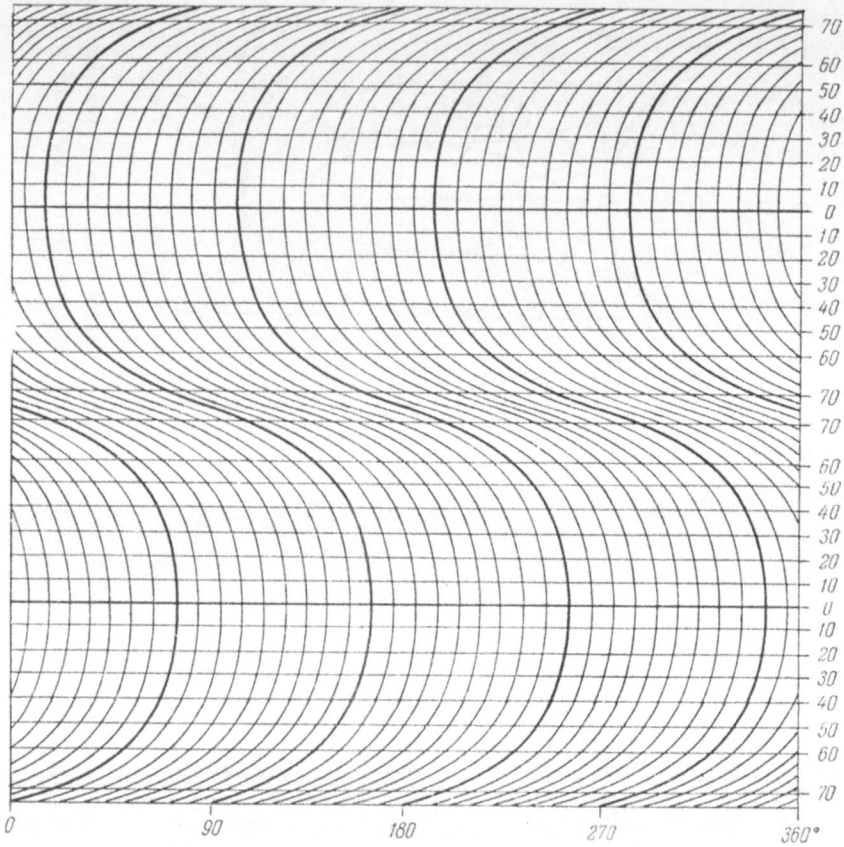

Fig. 175. Net for constructing polar figures for the (111) line of aluminum as examined with Cu Kα radiation in an axial camera.

The texture diagram (cylindrical projection) is converted to a stereographic projection or conversely by utilizing the fact that in the latter case the parallels of the Wulff net become a set of parallel lines, while the meridians are represented by curves:

$$\cos x = \cot \psi \, \mathrm{tg}\left(\psi - \frac{\omega}{v} y \right),$$

where $\psi = \pi/2 - v$, ω and v being, respectively, the angular and linear velocities of the cassette [172].

Figure 175 shows a net for constructing polar figures for the (111) line of aluminum as examined with Cu Kα radiation in an axial camera.

9-7. Correction for Absorption in Ionization Recording

Absorption corrections are very important when a texture diagram is to be analyzed quantitatively by ionization methods.

The correction for examination in transmission is

$$R = \frac{I_\alpha}{I_0} \; \frac{\cos \vartheta \left[e^{-\frac{\mu t}{\cos(\vartheta - \alpha)}} - e^{-\frac{\mu t}{\cos(\vartheta + \alpha)}} \right]}{\mu t e^{-\frac{\mu t}{\cos \vartheta} \left[\frac{\cos(\vartheta - \alpha)}{\cos(\vartheta + \alpha)} - 1 \right]}} \; ,$$

where I_α is the intensity of the diffracted rays when the specimen is rotated through an angle α clockwise, I_0 is the same for $\alpha = 0$, t is the thickness of the specimen, and μ is the linear absorption coefficient [173]. The corrected peak intensity is found by dividing the measured integral intensity by R. Figure 176 gives R for the (111) line of aluminum as examined with Cu Kα radiation ($\vartheta = 19.25°$, $\mu t = 1.0$).

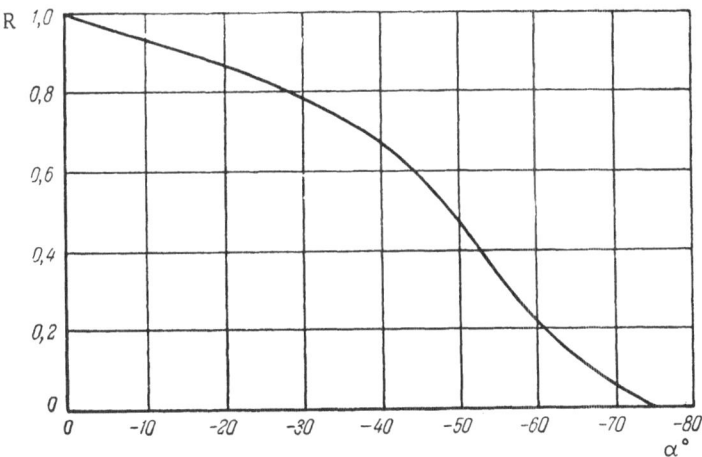

Fig. 176. Graph for the absorption correction in texture studies by ionization methods on the (111) line of aluminum (Cu Kα radiation).

The correction can be calculated as follows [88]:   The diffracted intensity in transmission ($\gamma = \vartheta \pm \alpha$, $\varphi = \pi$) is

$$I_{\pm a} = \frac{A \, (e^{Bt} - 1)}{B \cos (\vartheta \pm \alpha)} \; ,$$

where α is the angle between the normal to the plane of the sheet and the reflecting plane (positive if measurements are made clockwise, negative if counterclockwise), γ is the angle between the primary beam and the normal to the plane of the sheet, φ is the azimuthal angle along the diffraction ring, t is the thickness of the sheet, $A = \exp(-\mu t C)$, $B = -\mu[1/\cos(\vartheta \pm \alpha) - C]$, and $C = 1/\cos(\vartheta \pm \alpha)$. It is usual to employ negative values of α.

The table gives $100 I_0 / I_{-\alpha}$.

$\alpha°$	$2\vartheta°$ / $e^{-\mu t}$	10	20	30	40	50	60	70	80
5	0,01	102	103	104	105	106	107	108	109
	.02	102	103	104	105	106	107	108	109
	.03	102	103	103	104	105	107	108	109
	.04	102	102	103	104	105	107	108	109
	.05	102	102	103	104	105	106	108	109
	.06	101	102	103	104	105	106	108	109
	.07	101	102	103	104	105	106	108	109
	.08	101	102	103	104	105	106	108	109
	.09	101	102	103	104	105	106	108	109
	.10	101	102	103	104	105	106	107	109
	.20	101	102	103	104	105	106	107	109
	.30	101	102	102	103	104	105	107	108
	.40	101	102	102	103	104	105	107	108
	.50	101	101	102	103	104	105	106	108
	.60	101	101	102	103	104	105	106	108
	.70	101	101	102	103	104	105	106	107
	.80	100	101	102	103	104	105	106	107
	.90	100	101	102	103	104	105	106	107
10	0,01	107	109	111	113	116	118	121	124
	.02	106	108	110	112	115	118	121	124
	.03	106	107	109	112	114	117	120	124
	.04	105	107	109	111	114	117	120	124
	.05	105	107	109	111	113	116	120	124
	.06	104	106	108	110	113	116	119	123
	.07	104	106	108	110	113	116	119	123
	.08	104	106	108	110	112	115	119	123
	.09	104	106	108	110	112	115	118	123
	.10	104	105	107	109	112	115	118	122
	.20	103	104	106	108	110	113	116	120
	.30	102	104	105	107	110	112	115	119
	.40	101	103	105	107	109	111	114	118
	.50	101	103	104	106	108	111	113	117
	.60	101	102	104	106	108	110	113	116
	.70	101	102	104	106	107	110	112	115
	.80	100	102	104	105	107	109	112	114
	.90	100	102	103	105	107	109	111	114
15	0,01	116	120	123	127	131	136	141	145
	.02	114	117	120	124	129	134	140	146
	.03	112	115	119	123	127	133	139	146
	.04	111	114	117	121	126	131	138	146
	.05	110	113	116	120	125	130	137	145
	.06	109	112	116	120	124	130	136	145
	.07	109	112	115	119	123	129	136	144
	.08	108	111	114	118	123	128	135	143
	.09	108	111	114	118	122	128	134	143
	.10	107	110	113	117	122	127	134	142
	.20	105	107	110	114	118	123	129	137
	.30	103	106	109	112	116	120	126	133
	.40	102	105	108	111	114	118	123	130
	.50	101	104	107	110	113	117	121	127
	.60	101	103	106	109	112	115	120	125
	.70	100	103	105	108	111	114	118	123
	.80	100	102	104	107	110	113	117	121
	.90	99	102	104	107	109	112	116	120
20	0,01	131	135	141	148	155	162	170	176
	.02	125	130	135	142	150	159	169	179
	.03	122	126	132	138	146	156	167	180
	.04	119	124	130	136	144	153	165	179

α°	$e^{-\mu t}$ \ $2\theta°$	10	20	30	40	50	60	70	80
20	.05	118	122	128	134	142	151	163	178
	.06	116	121	126	132	140	150	161	177
	.07	115	120	125	131	139	148	160	175
	.08	114	119	124	130	137	147	159	174
	.09	113	118	123	129	136	145	157	173
	.10	113	117	122	128	135	144	156	172
	.20	108	112	116	121	128	136	147	161
	.30	105	109	113	118	124	131	140	153
	.40	103	107	110	115	120	127	135	146
	.50	101	105	109	113	118	124	131	141
	.60	100	104	107	111	116	121	128	136
	.70	99	103	106	110	114	119	125	132
	.80	98	102	105	109	112	117	122	129
	.90	98	101	104	107	111	115	120	126
25	0,01	152	160	169	179	190	202	214	223
	.02	142	149	158	168	181	195	212	229
	.03	136	143	152	162	174	190	209	230
	.04	132	139	147	157	170	186	205	230
	.05	129	136	144	154	166	182	202	228
	.06	127	133	141	151	163	179	199	226
	.07	125	131	139	148	160	176	197	224
	.08	123	129	137	146	158	174	194	222
	.09	121	128	135	144	156	171	192	220
	.10	120	126	134	143	154	169	190	218
	.20	112	117	124	131	141	154	172	198
	.30	107	112	118	125	134	145	160	183
	.40	104	109	114	120	128	138	151	170
	.50	101	106	111	117	124	132	144	160
	.60	100	104	109	114	120	128	138	151
	.70	98	102	107	112	117	124	132	144
	.80	97	101	105	109	115	121	128	137
	.90	95	99	103	108	112	118	124	131
30	0,01	186	198	211	227	245	263	281	297
	.02	167	178	192	208	228	252	279	309
	.03	157	168	181	197	217	243	274	312
	.04	151	161	173	189	209	235	269	311
	.05	145	155	167	183	203	229	264	309
	.06	141	151	163	178	198	224	259	307
	.07	138	147	159	174	193	219	254	304
	.08	135	144	155	170	189	214	250	300
	.09	133	141	152	167	185	211	246	297
	.10	131	139	150	164	182	207	242	293
	.20	117	125	134	145	160	181	212	260
	.30	110	117	125	134	147	165	191	232
	.40	105	111	118	127	138	153	175	209
	.50	101	107	114	121	131	144	162	190
	.60	99	104	110	117	125	136	151	174
	.70	96	101	107	113	121	130	142	160
	.80	94	99	104	110	117	124	134	148
	.90	93	97	102	107	113	120	128	138
35	0,01	242	260	281	305	332	361	391	427
	.02	208	224	245	271	303	341	388	447
	.03	190	206	226	252	285	327	380	453
	.04	178	193	213	238	271	314	372	454
	.05	170	184	203	227	260	304	364	451
	.06	163	177	195	219	251	295	357	448
	.07	158	171	188	211	243	287	349	444
	.08	153	166	183	205	236	279	342	439
	.09	149	162	178	200	230	273	336	434

α°	$e^{-\mu t}$ \ 2θ°	10	20	30	40	50	60	70	80
35	.10	146	158	174	195	224	267	329	428
	.20	125	135	147	164	188	223	277	372
	.30	114	123	133	148	167	195	240	322
	.40	107	115	124	136	152	175	212	279
	.50	102	109	117	127	141	160	190	243
	.60	98	104	112	121	132	148	171	212
	.70	94	100	107	115	125	138	156	186
	.80	91	97	103	110	119	129	143	164
	.90	89	94	100	106	113	122	132	146
40	0,01	337	366	399	437	478	526	590	701
	.02	274	300	334	376	427	493	582	736
	.03	242	267	299	341	395	468	570	746
	.04	222	245	276	317	372	447	557	748
	.05	208	229	259	298	353	430	544	745
	.06	196	217	245	284	337	415	532	740
	.07	187	207	234	271	324	401	520	733
	.08	180	199	225	261	312	389	508	725
	.09	173	192	217	252	302	377	497	717
	.10	168	186	210	244	293	367	487	708
	.20	136	149	168	194	232	293	399	612
	.30	120	131	146	167	197	246	333	520
	.40	109	119	132	149	173	212	282	436
	.50	102	111	122	136	156	186	241	362
	.60	96	104	114	126	142	166	207	298
	.70	92	99	108	118	131	149	180	243
	.80	88	95	102	111	121	135	157	198
	.90	85	91	98	105	113	124	138	161
45	0,01	520	566	618	674	741	838	1020	1564
	.02	392	436	491	559	647	774	999	1637
	.03	333	373	426	496	590	728	973	1658
	.04	296	333	384	453	548	691	948	1660
	.05	270	305	353	420	516	661	923	1653
	.06	250	283	329	394	488	634	900	1640
	.07	235	266	310	372	465	611	878	1624
	.08	222	252	294	354	445	590	857	1605
	.09	212	240	280	338	427	570	837	1586
	.10	203	230	268	325	411	553	818	1566
	.20	152	171	199	240	307	423	660	1349
	.30	128	144	165	197	248	340	538	1140
	.40	113	126	144	169	208	280	439	946
	.50	103	114	129	149	180	234	356	766
	.60	96	105	117	134	157	198	288	599
	.70	90	98	108	121	140	170	232	447
	.80	85	92	101	112	126	146	186	314
	.90	80	87	95	103	114	127	149	208
50	0,01	913	986	1056	1140	1271	1546	2412	
	.02	629	704	794	910	1080	1398	2331	
	.03	505	574	665	787	968	1300	2252	
	.04	431	495	583	705	889	1225	2181	
	.05	382	441	524	644	827	1164	2115	
	.06	345	400	480	596	777	1111	2055	
	.07	317	369	444	557	735	1065	1999	
	.08	295	343	415	524	698	1024	1946	
	.10	260	304	369	470	637	952	1849	
	.20	177	206	250	322	450	709	1470	
	.30	141	163	195	248	345	552	1185	
	.40	120	137	162	202	275	436	952	
	.50	106	120	140	170	224	345	751	
	.60	95	107	123	147	187	273	575	

α°	$e^{-\mu t}$	2θ° 10	20	30	40	50	60	70	80
50	.70	87	97	110	128	157	216	421	
	.80	81	90	100	114	134	172	292	
	.90	75	83	92	102	115	136	191	
55	0,01	1926	2000	2052	2182	2566	3946		
	.02	1179	1304	1447	1663	2104	3473		
	.03	882	1006	1165	1399	1847	3179		
	.04	717	833	991	1228	1670	2961		
	.05	610	718	871	1104	1537	2788		
	.06	534	634	780	1008	1430	2643		
	.07	477	570	710	930	1341	2518		
	.08	433	520	652	865	1265	2408		
	.09	397	478	605	810	1199	2309		
	.10	367	444	564	763	1140	2220		
	.20	220	267	345	484	769	1608		
	.30	162	195	249	349	566	1228		
	.40	130	155	195	267	429	949		
	.50	110	129	159	212	330	726		
	.60	96	111	134	172	255	543		
	.70	85	98	115	142	198	390		
	.80	77	87	100	119	155	266		
	.90	70	79	89	102	121	171		
60	0,01	5278	4960	4729	5078	7328			
	.02	2775	2906	3089	3649	5738			
	.03	1895	2100	2375	2966	4905			
	.04	1442	1656	1955	2540	4353			
	.05	1165	1373	1673	2240	3946			
	.06	978	1174	1467	2013	3628			
	.07	842	1026	1308	1834	3367			
	.08	740	912	1182	1686	3149			
	.09	660	820	1078	1563	2960			
	.10	595	745	992	1457	2796			
	.20	300	386	542	862	1802			
	.30	199	255	360	587	1285			
	.40	149	188	260	422	946			
	.50	118	147	198	312	697			
	.60	98	120	156	234	505			
	.70	84	100	126	177	354			
	.80	73	85	103	136	236			
	.90	64	74	86	105	150			

DIFFUSE SCATTERING OF X-RAYS
AND SMALL-ANGLE SCATTERING

Chapter 10 gives tables needed in the study of diffuse scattering and small-angle scattering for various materials.

See [11, 123, 174, 175] for methods of producing and working up patterns.

The tables are applicable to liquids and amorphous solids as well as to crystals.

10-1. Some Formulas for the Diffuse-Scattering Intensity

1. The measured diffuse-scattering intensity in general is

$$I_{\mathrm{m}} = (I_{\mathrm{p}} + I_{\mathrm{t}} + I_{\mathrm{c}})\, P\,(\vartheta)\, A\,(\vartheta)\, G\,(\vartheta)$$
$$= (I_{\mathrm{cs}} + I_{\mathrm{is}})\, P\,(\vartheta)\, A\,(\vartheta)\, G\,(\vartheta),$$

where I_{cs} is the coherent-scattering intensity, I_{is} is the incoherent-scattering intensity, I_{p} is the intensity dependent on the atomic positions, I_{t} is the intensity resulting from thermal motion, I_{c} is the intensity from Compton scattering, and $P(\vartheta)$, $A(\vartheta)$, and $G(\vartheta)$ are, respectively, the polarization, absorption, and geometrical factors.

The polarization factor for an unpolarized primary beam is

$$P\,(\vartheta) = \frac{1 + \cos^2 2\vartheta}{2}\,,$$

and for a polarized one (from a monochromator) it is

$$P\,(\vartheta) = \frac{1 + \cos^2 2\alpha \cos^2 2\vartheta}{1 + \cos^2 2\alpha}\,,$$

where α is the reflection angle for the monochromator.

The geometric factor for a specimen lying along the axis of a cylindrical film and for intensity measurements along the equator is

$$G\,(\vartheta) = 1;$$

for Guinier's method with a focusing monochromator it is

$$G\,(\vartheta) = \frac{1}{\cos 2\vartheta}\,,$$

and for the use of a flat film it is

$$G(\vartheta) = \cos^3 2\vartheta.$$

The absorption factor is calculated as for diffraction peaks.

2. The intensity given by a monatomic gas is

$$I(s) = Nf^2 \left[1 - \frac{\Omega}{V} \Phi(sR) \right],$$

(92)

where N is the total number of scattering atoms, Ω is the volume of a spherical atom of radius R, V is the irradiated volume, and f is the atomic scattering function.

The function $\Phi(sR)$ is

$$\Phi(sR) = \frac{3 \left[\sin(sR) - sR \cos(sR) \right]}{(sR)^3},$$

where

$$s = \frac{4\pi \sin \vartheta}{\lambda}.$$

3. The intensity given by a monatomic liquid is

$$I(s) = Nf^2 \left\{ 1 + \int_0^\infty 4\pi r^2 \left[\varrho(r) - \varrho_0(r) \right] \frac{\sin(sr)}{sr} dr \right\},$$

(93)

where $\rho(r)$ is the radial distribution function and $\rho_0(r)$ is the mean density.

The measured intensity is used to find $\rho(r)$ from

$$4\pi r^2 \varrho(r) = 4\pi r^2 \varrho_0(r) + \frac{2r}{\pi} \int_0^\infty si(s) \sin(sr) ds.$$

4. The intensity given by a polyatomic gas is

$$I(s) = N \left[\sum_p f_p^2 + 2 \sum_{p,q} f_p f_q \frac{\sin(sl_{pq})}{sl_{pq}} \right],$$

(94)

where the first summation is carried over all atoms in the molecule and the second over pairs of molecules having distances between atoms of l_{pq}.

5. The intensity given by a polyatomic liquid is

$$I = N \left[\sum_m f_m^2 + 4\pi \int_0^\infty \sum_m k_m \varrho_m(r) r^2 \frac{\sin(sr)}{sr} dr \right],$$

(95)

where m denotes the type of atom, $f_m = k_m f_e$, k_m is the effective number of electrons in the atom, and f_e is the atomic factor for an electron.

The summation is carried over structural units (e.g., molecules).

6. A cylindrical,symmetrically scattering particle gives

$$I(s) = Nf^2 \left\{ 1 + \int_0^\infty 2\pi r \left[\varrho(r) - \varrho_0(r) \right] J_0(sr)\, dr \right\},$$ (96)

where r is the distance from the axis of the cylinder and $J_0(sr)$ is the Bessel function to zero order [176].

7. The thermal diffuse scattering for a diatomic gas is allowed for by inserting a factor $\exp(-A)$, where

$$A = \frac{h}{M\nu} \operatorname{cth} \left(\frac{h\nu}{2kT} \right) \frac{\sin^2 \vartheta}{\lambda^2}.$$

Here $M = 1/m_1 + 1/m_2$, m_1 and m_2 being the masses of the atoms and ν being the frequency of the diatomic oscillator.

8. A real crystal, with allowance for the thermal motion, gives

$$I_t = Nf^2 [1 - e^{-2B \frac{\sin^2 \vartheta}{\lambda^2}}],$$ (97)

where

$$B = \frac{6h^2}{mk\Theta} \left[\frac{\Phi(x)}{x} + \frac{1}{4} \right]$$ (98)

(Θ is the characteristic temperature and $\Phi(x) = \frac{1}{x} \int_0^x \frac{\xi\, d\xi}{e^\xi - 1}$ is the Debye function).

This relation applies to noninteracting atoms; allowance for interaction during thermal vibration introduces an extra factor [206].

9. A binary alloy, with allowance for thermal and incoherent scattering, gives

$$I_t = N\,(x_A f_A + x_B f_B)^2 + [x_A f_A e^{-B_A \frac{\sin^2 \vartheta}{\lambda^2}} + x_B f_B e^{-B_B \frac{\sin^2 \vartheta}{\lambda^2}}],$$ (99)

where x_A and x_B are the atomic concentrations of the components [177].

10. The short-range order in a binary alloy gives

$$I_p = N_{x_1 x_2} (f_1 - f_2)^2\, 4\pi\varrho \int_0^\infty (\overline{g_2^{(1)}} - \overline{g_{12}^{(2)}}) \frac{\sin(sr)}{sr}\, dr,$$ (100)

where $\overline{g_2^{(1)}}$ is the mean value of the molecular distribution function for one molecule and $\overline{g_{12}^{(2)}}$ is the same but for pairs of molecules, the mean being taken over all crystallographic directions; x_1 and x_2 are the atomic concentrations of the components and ρ is the density [178].

The degree of short-range order in an alloy is [179] specified by

$$r = (\overline{g_2^{(1)}} - \overline{g_{12}^{(2)}}) = \frac{1}{2\pi^2\varrho} \int_0^\infty s\, i(s) \sin(sr)\, ds,$$

where

$$i\,(s) = \frac{I\,p(s)}{N_{x_1 x_2}\,(f_1 - f_2)} - 1.$$

11. The diffuse scattering caused by static displacements in a face-centered cubic alloy takes the form [180]

$$I_s = \overline{f}^2 H\,(x)\,(1 - e^{-2M}),$$

where $\overline{f} = m_A f_A + m_B f_B$, f_A and f_B are the atomic scattering factors for components A and B, m_A and m_B are the atomic concentrations; $M = {}^8/_3\,(\pi^2 \sin^2 \vartheta)\,\overline{u_s^2}/\lambda^2$, $\overline{u_s^2}$ is the mean static displacement from the equilibrium position, and H(x) is a modulating function, which is 1 ± 0.1.

See [181] for a general expression for the diffuse scattering from a binary solid solution.

The absorption factor for use with ionization recording takes the fixed value $A = \mu/2$, where μ is the linear absorption coefficient; the geometric factor is independent of ϑ.

The intensity is given in electron units in all the above expressions; the measured values are converted to those units by means of an amorphous standard (fused quartz, paraffin) and conversion via special relationships [177, 207, 208].

See [294] for some designs of x-ray cameras for diffuse scattering.

A distinctive feature of such measurements is that the scattering by air is usually of the same order as the diffuse scattering by the specimen, so it is usual to work with evacuated cameras [209].

See [283, 311, 414] for some new designs of camera for use at small angles with photographic and ionization recording.

Fourier transforms have been used in studies on short-range order in alloys. The equations are to be found in [177].

The diffuse scattering has two parts for a solid solution consisting of atoms differing in scattering power and displaced from their ideal positions; one is a quasi-periodic function of the Bragg angle, while the other increases with this angle. See [445, 460] for the formulas.

10-2. Values of s(ϑ) for Various Radiations

The table gives

$$s = \frac{4\pi \sin \vartheta}{\lambda},$$

which appears in all expressions for the diffuse-scattering intensity [102].

The s are for Mo Kα (λ = 0.7107Å), Cu Kα (λ = 1.5418 Å), Co Kα (λ = 1.790 Å), Fe Kα (λ = 1.9373 Å), and Cr Kα (λ = 2.291 Å) for angles between 0 and 89.5°.

ϑ°	Radiation					ϑ°	Radiation				
	Mo Kα	Cu Kα	Co Kα	FE Kα	Cr Kα		Mo Kα	Cu Kα	Co Kα	Fe Kα	Cr Kα
0	0	0	0	0	0	12,0	3,676	1,695	1,460	1,349	1,140
0,2	0,062	0,028	0,025	0,023	0,019	12,2	3,736	1,722	1,484	1,371	1,159
0,4	0,123	0,057	0,049	0,045	0,038	12,4	3,797	1,750	1,508	1,393	1,178
0,6	0,185	0,085	0,074	0,068	0,057	12,6	3,857	1,778	1,531	1,415	1,197
0,8	0,247	0,114	0,098	0,091	0,077	12,8	3,917	1,806	1,555	1,437	1,215
1,0	0,309	0,142	0,123	0,113	0,096	13,0	3,977	1,833	1,579	1,459	1,234
1,2	0,370	0,171	0,147	0,136	0,115	13,2	4,088	1,861	1,603	1,481	1,253
1,4	0,432	0,199	0,172	0,158	0,134	13,4	4,098	1,889	1,627	1,503	1,271
1,6	0,494	0,228	0,196	0,181	0,153	13,6	4,158	1,916	1,651	1,525	1,290
1,8	0,555	0,256	0,221	0,204	0,172	13,8	4,218	1,944	1,675	1,547	1,308
2,0	0,617	0,284	0,245	0,226	0,191	14,0	4,278	1,972	1,698	1,569	1,327
2,2	0,679	0,313	0,270	0,249	0,211	14,2	4,337	1,999	1,722	1,591	1,346
2,4	0,741	0,341	0,294	0,272	0,230	14,4	4,397	2,027	1,746	1,613	1,364
2,6	0,802	0,370	0,318	0,294	0,249	14,6	4,457	2,054	1,770	1,635	1,383
2,8	0,864	0,398	0,343	0,317	0,268	14,8	4,517	2,082	1,793	1,657	1,401
3,0	0,925	0,427	0,367	0,340	0,287	15,0	4,576	2,109	1,817	1,679	1,420
3,2	0,987	0,455	0,392	0,362	0,306	15,2	4,636	2,137	1,841	1,701	1,438
3,4	1,049	0,483	0,416	0,385	0,325	15,4	4,696	2,164	1,864	1,723	1,457
3,6	1,110	0,512	0,441	0,407	0,344	15,6	4,755	2,192	1,888	1,744	1,475
3,8	1,172	0,540	0,465	0,430	0,363	15,8	4,814	2,219	1,911	1,766	1,493
4,0	1,233	0,569	0,490	0,453	0,383	16,0	4,874	2,247	1,935	1,788	1,512
4,2	1,295	0,597	0,514	0,475	0,402	16,2	4,933	2,274	1,959	1,810	1,530
4,4	1,357	0,625	0,539	0,498	0,421	16,4	4,992	2,301	1,982	1,831	1,549
4,6	1,418	0,654	0,563	0,520	0,440	16,6	5,051	2,328	2,006	1,853	1,567
4,8	1,480	0,682	0,587	0,543	0,459	16,8	5,111	2,356	2,029	1,875	1,585
5,0	1,541	0,710	0,612	0,565	0,478	17,0	5,170	2,383	2,053	1,896	1,604
5,2	1,602	0,739	0,636	0,588	0,497	17,2	5,229	2,410	2,076	1,918	1,622
5,4	1,664	0,767	0,661	0,610	0,516	17,4	5,288	2,437	2,099	1,940	1,640
5,6	1,725	0,795	0,685	0,633	0,535	17,6	5,346	2,464	2,123	1,961	1,659
5,8	1,787	0,824	0,709	0,656	0,554	17,8	5,405	2,492	2,146	1,983	1,677
6,0	1,848	0,852	0,734	0,678	0,573	18,0	5,464	2,519	2,169	2,004	1,695
6,2	1,910	0,880	0,758	0,701	0,592	18,2	5,522	2,546	2,193	2,026	1,713
6,4	1,971	0,909	0,783	0,723	0,611	18,4	5,581	2,573	2,216	2,047	1,731
6,6	2,032	0,937	0,807	0,746	0,630	18,6	5,640	2,600	2,239	2,069	1,750
6,8	2,094	0,965	0,831	0,768	0,649	18,8	5,698	2,627	2,262	2,090	1,768
7,0	2,155	0,993	0,856	0,791	0,668	19,0	5,757	2,654	2,286	2,112	1,786
7,2	2,216	1,021	0,880	0,813	0,687	19,2	5,815	2,680	2,309	2,133	1,804
7,4	2,277	1,050	0,904	0,835	0,707	19,4	5,873	2,707	2,332	2,155	1,822
7,6	2,339	1,078	0,929	0,858	0,725	19,6	5,931	2,734	2,355	2,176	1,840
7,8	2,400	1,106	0,953	0,880	0,744	19,8	5,989	2,761	2,378	2.197	1,858
8,0	2,461	1,134	0,977	0,903	0,763	20,0	6,047	2,788	2,401	2,219	1,876
8,2	2,522	1,162	1,001	0,925	0,782	20,2	6,106	2,814	2,424	2,240	1,894
8,4	2,583	1,191	1,026	0,948	0,801	20,4	6,163	2,841	2,447	2,261	1,912
8,6	2,644	1,219	1,050	0,970	0,820	20,6	6,221	2,868	2,470	2,282	1,930
8,8	2,705	1,247	1,074	0,992	0,839	20,8	6,279	2,894	2,493	2,303	1,948
9,0	2,766	1,275	1,098	1,015	0,858	21,0	6,337	2;921	2,516	2,325	1,966
9,2	2,827	1,303	1,122	1,037	0,877	21,2	6,394	2,947	2,539	2,346	1,983
9,4	2,888	1,331	1,147	1,059	0,896	21,4	6,452	2,974	2,562	2,367	2,001
9,6	2,949	1,359	1,171	1,082	0,915	21,6	6,509	3,000	2,584	2,388	2,019
9,8	3,010	1,387	1,195	1,104	0,934	21,8	6,566	3,027	2,607	2,409	2,037
10,0	3,070	1,415	1,219	1,113	0,952	22,0	6,624	3,053	2,630	2,430	2,055
10,2	3,131	1,443	1,243	1,149	0,971	22,2	6,681	3,080	2,653	2,451	2,072
10,4	3,192	1,471	1,267	1,171	0,990	22,4	6,738	3,106	2,675	2,472	2,090
10,6	3,253	1,499	1,291	1,193	1,009	22,6	6,795	3,132	2,698	2,493	2,108
10,8	3,313	1,527	1,315	1,215	1,028	22,8	6,852	3,158	2,721	2,514	2,126
11,0	3,374	1,555	1,340	1,238	1,047	23,0	6,909	3,185	2,743	2,534	2,143
11,2	3,434	1,583	1,364	1,260	1,065	23,2	6,966	3,211	2,766	2,555	2,161
11,4	3,495	1,611	1,388	1,282	1,084	23,4	7,022	3,237	2,788	2,576	2,178
11,6	3,555	1,639	1,412	1,304	1,103	23,6	7,079	3,263	2,811	2,597	2,196
11,8	3,616	1,667	1,436	1,326	1,122	23,8	7,135	3,289	2,833	2,618	2,214

$\vartheta°$	Mo $K\alpha$	Cu $K\alpha$	Co $K\alpha$	Fe $K\alpha$	Cr $K\alpha$	$\vartheta°$	Mo $K\alpha$	Cu $K\alpha$	Co $K\alpha$	Fe $K\alpha$	Cr $K\alpha$
24,0	7,192	3,315	2,855	2,638	2,231	36,0	10,393	4,791	4,126	3,813	3,224
24,2	7,248	3,341	2,878	2,659	2,248	36,2	10,443	4,814	4,146	3,831	3,240
24,4	7,304	3,367	2,900	2,686	2,266	36,4	10,493	4,837	4,166	3,849	3,255
24,6	7,360	3,393	2,922	2,700	2,283	36,6	10,542	4,859	4,186	3,867	3,270
24,8	7,417	3,419	2,945	2,721	2,301	36,8	10,592	4,882	4,205	3,886	3,286
25,0	7,473	3,445	2,967	2,741	2,318	37,0	10,641	4,905	4,225	3,904	3,301
25,2	7,528	3,470	2,989	2,762	2,335	37,2	10,690	4,928	4,244	3,922	3,316
25,4	7,584	3,496	3,011	2,782	2,353	37,4	10,740	4,950	4,264	3,940	3,332
25,6	7,640	3,522	3,033	2,803	2,370	37,6	10,788	4,973	4,283	3,958	3,347
25,8	7,696	3,547	3,055	2,823	2,387	37,8	10,837	4,995	4,303	3,976	3,362
26,0	7,751	3,573	3,077	2,844	2,404	38,0	10,886	5,018	4,322	3,994	3,377
26,2	7,807	3,599	3,100	2,864	2,422	38,2	10,935	5,040	4,341	4,011	3,392
26,4	7,862	3,624	3,122	2,884	2,439	38,4	10,983	5,063	4,361	4,029	3,407
26,6	7,917	3,649	3,143	2,904	2,456	38,6	11,031	5,085	4,380	4,047	3,422
26,8	7,972	3,675	3,165	2,925	2,473	38,8	11,079	5,107	4,399	4,064	3,437
27,0	8,027	3,700	3,187	2,945	2,490	39,0	11,127	5,129	4,418	4,082	3,452
27,2	8,082	3,726	3,209	2,965	2,507	39,2	11,175	5,151	4,437	4,100	3,467
27,4	8,137	3,751	3,221	2,985	2,524	39,4	11,223	5,173	4,456	4,117	3,482
27,6	8,192	3,776	3,253	3,005	2,541	39,6	11,272	5,196	4,475	4,135	3,497
27,8	8,246	3,801	3,274	3,025	2,558	39,8	11,318	5,217	4,494	4,152	3,511
28,0	8,301	3,826	3,296	3,045	2,575	40,0	11,366	5,239	4,513	4,169	3,526
28,2	8,355	3,851	3,317	3,065	2,592	40,5	11,483	5,293	4,559	4,213	3,562
28,4	8,410	3,876	3,339	3,085	2,609	41,0	11,600	5,347	4,606	4,256	3,599
28,6	8,464	3,902	3,361	3,105	2,626	41,5	11,716	5,401	4,652	4,298	3,635
28,8	8,518	3,926	3,382	3,125	2,642	42,0	11,831	5,454	4,697	4,340	3,670
29,0	8,572	3,951	3,404	3,145	2,659	42,5	11,946	5,506	4,743	4,382	3,706
29,2	8,626	3,976	3,425	3,165	2,676	43,0	12,059	5,559	4,788	4,424	3,741
29,4	8,680	4,001	3,446	3,184	2,693	43,5	12,171	5,610	4,832	4,465	3,776
29,6	8,734	4,026	3,468	3,204	2,709	44,0	12,283	5,662	4,877	4,506	3,810
29,8	8,787	4,051	3,489	3,224	2,726	44,5	12,396	5,713	4,921	4,546	3,845
30,0	8,841	4,075	3,510	3,243	2,743	45,0	12,503	5,763	4,964	4,587	3,879
30,2	8,894	4,100	3,531	3,263	2,759	45,5	12,611	5,813	5,007	4,627	3,912
30,4	8,947	4,124	3,553	3,282	2,776	46,0	12,719	5,863	5,050	4,666	3,946
30,6	9,001	4,149	3,574	3,302	2,792	46,5	12,826	5,912	5,092	4,705	3,979
30,8	9,054	4,173	3,595	3,321	2,809	47,0	12,931	5,961	5,134	4,744	4,012
31,0	9,107	4,198	3,616	3,341	2,825	47,5	13,036	6,009	5,176	4,782	4,044
31,2	9,160	4,222	3,637	3,360	2,841	48,0	13,140	6,057	5,217	4,820	4,076
31,4	9,212	4,246	3,658	3,380	2,858	48,5	13,243	6,104	5,258	4,858	4,108
31,6	9,265	4,271	3,679	3,399	2,874	49,0	13,345	6,151	5,298	4,895	4,140
31,8	9,318	4,295	3,699	3,418	2,890	49,5	13,445	6,198	5,338	4,932	4,171
32,0	9,370	4,319	3,720	3,437	2,907	50,0	13,545	6,244	5,378	4,969	4,202
32,2	9,422	4,343	3,741	3,457	2,923	50,5	13,644	6,289	5,417	5,005	4,232
32,4	9,474	4,367	3,762	3,476	2,939	51,0	13,741	6,334	5,456	5,041	4,263
32,6	9,526	4,391	3,782	3,495	2,955	51,5	13,838	6,379	5,494	5,076	4,293
32,8	9,578	4,415	3,803	3,514	2,971	52,0	13,933	6,423	5,532	5,111	4,322
33,0	9,630	4,439	3,824	3,533	2,987	52,5	14,028	6,466	5,570	5,146	4,352
33,2	9,682	4,463	3,844	3,552	3,003	53,0	14,121	6,509	5,607	5,180	4,381
33,4	9,733	4,487	3,865	3,571	3,019	53,5	14,214	6,552	5,643	5,214	4,409
33,6	9,785	4,510	3,885	3,590	3,035	54,0	14,305	6,594	5,680	5,248	4,438
33,8	9,836	4,534	3,905	3,608	3,051	54,5	14,395	6,635	5,715	5,281	4,466
34,0	9,887	4,558	3,926	3,627	3,067	55,0	14,484	6,676	5,751	5,313	4,493
34,2	9,938	4,581	3,946	3,646	3,083	55,5	14,572	6,717	5,786	5,346	4,520
34,4	9,990	4,605	3,966	3,665	3,099	56,0	14,659	6,757	5,820	5,378	4,547
34,6	10,040	4,628	3,986	3,683	3,115	56,5	14,745	6,797	5,854	5,409	4,574
34,8	10,091	4,651	4,007	3,702	3,130	57,0	14,829	6,835	5,888	5,440	4,600
35,0	10,142	4,675	4,027	3,721	3,146	57,5	14,912	6,874	5,921	5,471	4,626
35,2	10,192	4,698	4,047	3,739	3,162						
35,4	10,243	4,721	4,067	3,758	3,177						
35,6	10,293	4,745	4,087	3,776	3,193						
35,8	10,343	4,768	4,107	3,794	3,209						

$\vartheta°$	Radiation					$\vartheta°$	Radiation				
	Mo Kα	Cu Kα	Co Kα	Fe Kα	Cr Kα		Mo Kα	Cu Kα	Co Kα	Fe Kα	Cr Kα
58,0	14,995	6,912	5,954	5,501	4,652	74,0	16,997	7,835	6,749	6,235	5,273
58,5	15,076	6,949	5,986	5,531	4,677	74,5	17,039	7,854	6,765	6,251	5,286
59,0	15,156	6,986	6,018	5,560	4,702	75,0	17,079	7,873	6,781	6,266	5,298
59,5	15,235	7,023	6,049	5,589	4,726	75,5	17,118	7,891	6,797	6,280	5,310
60,0	15,313	7,058	6,080	5,618	4,750	76,0	17,156	7,908	6,812	6,294	5,322
60,5	15,389	7,094	6,110	5,646	4,774	76,5	17,193	7,925	6,827	6,307	5,334
61,0	15,465	7,129	6,140	5,673	4,797	77,0	17,228	7,942	6,840	6,320	5,345
61,5	15,539	7,163	6,170	5,701	4,820	77,5	17,263	7,957	6,854	6,333	5,355
62,0	15,612	7,196	6,199	5,727	4,843	78,0	17,295	7,972	6,867	6,345	5,365
62,5	15,684	7,230	6,227	5,754	4,865	78,5	17,327	7,987	6,879	6,356	5,375
63,0	15,755	7,262	6,255	5,780	4,887	79,0	17,357	8,001	6,891	6,367	5,384
63,5	15,824	7,294	6,283	5,805	4,909	79,5	17,385	8,014	6,903	6,378	5,393
64,0	15,892	7,325	6,310	5,830	4,930	80,0	17,413	8,027	6,914	6,388	5,402
64,5	15,959	7,357	6,336	5,855	4,951	80,5	17,439	8,039	6,924	6,398	5,410
65,0	16,025	7,387	6,363	5,879	4,971	81,0	17,464	8,050	6,934	6,407	5,410
65,5	16,090	7,417	6,388	5,903	4,991	81,5	17,488	8,061	6,943	6,415	5,425
66,0	16,153	7,446	6,413	5,926	5,011	82,0	17,510	8,071	6,952	6,423	5,432
66,5	16,215	7,474	6,438	5,949	5,030	82,5	17,:30	8,081	6,960	6,431	5,438
67,0	16,276	7,502	6,462	5,971	5,049	83,0	17,550	8,090	6,968	6,438	5,444
67,5	16,336	7,530	6,486	5,993	5,068	83,5	17,568	8,098	6,975	6,445	5,450
68,0	16,394	7,557	6,509	6,014	5,086	84,0	17,585	8,106	6,982	6,451	5,455
68,5	16,451	7,583	6,532	6,035	5,103	84,5	17,600	8,113	6,988	6,457	5,460
69,0	16,507	7,609	6,554	6,056	5,121	85,0	17,614	8,119	6,993	6,462	5,464
69,5	16,562	7,634	6,576	6,076	5,138	85,5	17,627	8,125	6,999	6,467	5,468
70,0	16,615	7,659	6,597	6,095	5,154	86,0	17,638	8,131	7,003	6,471	5,472
70,5	16,667	7,683	6,618	6,114	5,170	86,5	17,649	8,135	7,007	6,474	5,475
71,0	16,718	7,706	6,638	6,133	5,186	87,0	17,657	8,139	7,011	6,478	5,478
71,5	16,768	7,729	6,657	6,151	5,202	87,5	17,665	8,143	7,014	6,480	5,480
72,0	16,816	7,752	6,677	6,169	5,217	88,0	17,671	8,145	7,016	6,483	5,482
72,5	16,863	7,773	6,695	6,186	5,231	88,5	17,676	8,148	7,018	6,484	5,483
73,0	16,909	7,794	6,714	6,203	5,245	89,0	17,679	8,149	7,019	6,485	5,484
73,5	16,954	7,815	6,731	6,219	5,259	89,5	17,681	8,150	7,020	6,486	5,485

10-3. Polarization Factor for Diffuse Scattering

This factor for unpolarized incident radiation is

$$P(\vartheta) = \frac{1 + \cos^2 2\vartheta}{2}.$$

For polarized radiation from a monochromator

$$P(\vartheta) = \frac{1 + \cos^2 2\alpha \cos^2 2\vartheta}{1 + \cos^2 2\alpha},$$

where α is the reflection angle for the monochromator. The table gives $P(\vartheta)$ for the first case and also for a quartz monochromator used with Mo, Cu, Co, and Cr radiations [102]. In addition, it gives values of $1/\cos^3 2\vartheta$ and $\cos 2\vartheta$, which are needed for the geometric factor.

$\vartheta°$	$\dfrac{1+\cos^2 2\vartheta}{2}$	$\dfrac{1+\cos^2 2\alpha \cos^2 2\vartheta}{1+\cos^2 2\alpha}$				$\dfrac{1}{\cos^3 2\vartheta}$	$\cos 2\vartheta$
		$\alpha = 6°43'$ Mo $K\alpha$	$\alpha = 13°24'$ Cu $K\alpha$	$\alpha = 15°37'$ Co $K\alpha$	$\alpha = 20°09'$ Cr $K\alpha$		
0	1,000	1,000	1,000	1,000	1,000	1,000	1,0000
0,2	1,000	1,000	1,000	1,000	1,000	1,000	0,9999
0,4	1,000	1,000	1,000	1,000	1,000	1,000	0,9999
0,6	1,000	1,000	1,000	1,000	1,000	1,001	0,9997
0,8	1,000	1,000	1,000	1,000	1,000	1,001	0,9996
1,0	0,999	0,999	0,999	0,999	1,000	1,002	0,9993
1,2	0,999	0,999	0,999	0,999	0,999	1,003	0,9991
1,4	0,999	0,999	0,999	0,999	0,999	1,004	0,9988
1,6	0,998	0,999	0,999	0,999	0,999	1,005	0,9984
1,8	0,998	0,998	0,998	0,998	0,999	1,006	0,9980
2,0	0,998	0,998	0,998	0,998	0,998	1,007	0,9975
2,2	0,997	0,997	0,997	0,998	0,998	1,009	0,9970
2,4	0,996	0,997	0,997	0,997	0,997	1,011	0,9964
2,6	0,996	0,996	0,996	0,997	0,997	1,012	0,9958
2,8	0,995	0,995	0,996	0,996	0,996	1,014	0,9952
3,0	0,995	0,995	0,995	0,995	0,996	1,017	0,9945
3,2	0,994	0,994	0,994	0,995	0,995	1,019	0,9937
3,4	0,993	0,993	0,994	0,994	0,995	1,021	0,9929
3,6	0,992	0,992	0,993	0,993	0,994	1,024	0,9921
3,8	0,991	0,991	0,992	0,993	0,994	1,027	0,9912
4,0	0,990	0,991	0,991	0,992	0,993	1,030	0,9902
4,2	0,980	0,990	0,991	0,991	0,992	1,033	0,9892
4,4	0,988	0,989	0,990	0,990	0,991	1,036	0,9882
4,6	0,987	0,988	0,989	0,989	0,991	1,040	0,9871
4,8	0,986	0,986	0,988	0,988	0,990	1,043	0,9860
5,0	0,985	0,985	0,987	0,987	0,989	1,047	0,9848
5,2	0,984	0,984	0,986	0,986	0,988	1,051	0,9835
5,4	0,982	0,983	0,984	0,985	0,987	1,055	0,9822
5,6	0,981	0,982	0,983	0,984	0,986	1,059	0,9809
5,8	0,980	0,980	0,982	0,983	0,985	1,064	0,9795
6,0	0,978	0,979	0,981	0,982	0,984	1,069	0,9781
6,2	0,977	0,978	0,980	0,981	0,983	1,073	0,9766
6,4	0,975	0,976	0,978	0,979	0,982	1,078	0,9751
6,6	0,974	0,975	0,977	0,978	0,981	1,084	0,9735
6,8	0,972	0,973	0,975	0,977	0,980	1,089	0,9716
7,0	0,971	0,972	0,974	0,975	0,978	1,095	0,9703
7,2	0,969	0,970	0,973	0,974	0,977	1,101	0,9685
7,4	0,967	0,968	0,971	0,972	0,976	1,107	0,9668
7,6	0,966	0,967	0,970	0,971	0,975	1,113	0,9650
7,8	0,964	0,965	0,968	0,969	0,973	1,119	0,9631
8,0	0,962	0,963	0,966	0,968	0,972	1,126	0,9612
8,2	0,960	0,961	0,965	0,966	0,971	1,133	0,9593
8,4	0,958	0,959	0,963	0,965	0,969	1,140	0,9573
8,6	0,956	0,957	0,961	0,963	0,968	1,147	0,9552
8,8	0,954	0,956	0,959	0,961	0,966	1,155	0,9531
9,0	0,952	0,954	0,958	0,960	0,965	1,162	0,9510
9,2	0,950	0,952	0,956	0,958	0,963	1,170	0,9488
9,4	0,948	0,949	0,954	0,956	0,962	1,179	0,9466
9,6	0,946	0,947	0,952	0,954	0,960	1,187	0,9443
9,8	0,944	0,945	0,950	0,952	0,959	1,196	0,9420
10,0	0,942	0,943	0,948	0,951	0,957	1,205	0,9396
10,5	0,936	0,938	0,943	0,946	0,953	1,229	0,9335
11,0	0,930	0,932	0,938	0,941	0,948	1,255	0,9271
11,5	0,924	0,926	0,932	0,936	0,944	1,282	0,9205
12,0	0,917	0,920	0,927	0,930	0,939	1,312	0,9135
12,5	0,911	0,913	0,921	0,925	0,934	1,343	0,9063
13,0	0,904	0,906	0,915	0,919	0,929	1,377	0,8987
13,5	0,897	0,900	0,909	0,913	0,924	1,414	0,8910
14,0	0,890	0,893	0,902	0,907	0,919	1,453	0,8829
14,5	0,882	0,886	0,896	0,901	0,914	1,495	0,8746

$\vartheta°$	$\dfrac{1 + \cos^2 2\vartheta}{2}$	$\dfrac{1 + \cos^2 2\alpha \cos^2 2\vartheta}{1 + \cos^2 2\alpha}$				$\dfrac{1}{\cos^2 2\vartheta}$	$\cos 2\vartheta$
		$\alpha = 6°43'$ Mo $K\alpha$	$\alpha = 13°24'$ Cu $K\alpha$	$\alpha = 15°37'$ Co $K\alpha$	$\alpha = 20°09'$ Cr $K\alpha$		
15,0	0,875	0,878	0,889	0,894	0,908	1,540	0,8660
15,5	0,867	0,871	0,882	0,888	0,902	1,588	0,8571
16,0	0,860	0,863	0,875	0,881	0,897	1,640	0,8480
16,5	0,852	0,856	0,868	0,875	0,891	1,695	0,8386
17,0	0,844	0,848	0,861	0,868	0,885	1,755	0,8290
17,5	0,836	0,840	0,854	0,861	0,879	1,819	0,8191
18,0	0,827	0,832	0,847	0,854	0,873	1,889	0,8090
18,5	0,819	0,824	0,839	0,847	0,867	1,963	0,7986
19,0	0,810	0,816	0,832	0,840	0,861	2,044	0,7880
19,5	0,802	0,807	0,824	0,833	0,854	2,131	0,7771
20,0	0,793	0,799	0,817	0,825	0,848	2,225	0,7660
20,5	0,785	0,791	0,809	0,818	0,842	2,326	0,7547
21,0	0,776	0,782	0,801	0,811	0,835	2,437	0,7431
21,5	0,767	0,774	0,794	0,804	0,829	2,556	0,7313
22,0	0,759	0,765	0,786	0,796	0,823	2,687	0,7193
22,5	0,750	0,757	0,778	0,789	0,816	2,828	0,7071
23,0	0,741	0,748	0,771	0,781	0,810	2,983	0,6946
23,5	0,733	0,740	0,763	0,774	0,803	3,152	0,6820
24,0	0,724	0,731	0,755	0,767	0,797	3,338	0,6691
24,5	0,715	0,723	0,747	0,759	0,791	3,541	0,6560
25,0	0,707	0,715	0,740	0,752	0,784	3,765	0,6427
26	0,690	0,698	0,725	0,738	0,772	4,285	0,6156
27	0,673	0,682	0,710	0,724	0,759	4,924	0,5877
28	0,656	0,666	0,695	0,710	0,747	5,719	0,5591
29	0,640	0,650	0,681	0,696	0,736	6,720	0,5299
30	0,625	0,635	0,667	0,683	0,724	8,000	0,5000
31	0,610	0,621	0,654	0,671	0,713	9,665	0,4694
32	0,596	0,607	0,642	0,659	0,703	11,871	0,4383
33	0,583	0,594	0,630	0,648	0,693	14,861	0,4067
34	0,570	0,582	0,619	0,637	0,684	19,022	0,3746
35	0,558	0,570	0,608	0,627	0,675	24,994	0,3420
36	0,548	0,560	0,599	0,618	0,667	33,887	0,3090
37	0,538	0,550	0,590	0,610	0,660	47,755	0,2756
38	0,529	0,542	0,583	0,602	0,654	70,621	0,2419
39	0,522	0,535	0,576	0,596	0,648	111,235	0,2079
40	0,515	0,528	0,570	0,590	0,643	190,840	0,1736
41	0,510	0,523	0,565	0,586	0,639	370,370	0,1391
42	0,505	0,519	0,561	0,582	0,636	877,193	0,1045
43	0,502	0,516	0,559	0,580	0,634	2941,176	0,0698
44	0,501	0,514	0,557	0,578	0,633	25000,000	0,0349
45	0,500	0,514	0,557	0,578	0,632		0,0000
46	0,501	0,514	0,557	0,578	0,633		—0,0349
47	0,502	0,516	0,559	0,580	0,634		—0,0697
48	0,505	0,519	0,561	0,582	0,636		—0,1045
49	0,510	0,523	0,565	0,586	0,639		—0,1391
50	0,515	0,528	0,570	0,590	0,643		—0,1736
51	0,522	0,535	0,576	0,596	0,648		—0,2079
52	0,529	0,542	0,583	0,602	0,654		—0,2419
53	0,538	0,550	0,590	0,610	0,660		—0,2756
54	0,548	0,560	0,599	0,618	0,667		—0,3090
55	0,558	0,570	0,608	0,627	0,675		—0,3420
56	0,570	0,582	0,619	0,637	0,684		—0,3746
57	0,583	0,594	0,630	0,648	0,693		—0,4067
58	0,596	0,507	0,642	0,659	0,703		—0,4383
59	0,610	0,621	0,654	0,671	0,713		—0,4694
60	0,625	0,635	0,667	0,683	0,724		—0,5000
61	0,640	0,650	0,681	0,696	0,736		—0,5299
62	0,656	0,666	0,695	0,710	0,747		—0,5591
63	0,673	0,682	0,710	0,724	0,759		—0,5877
64	0,690	0,698	0,725	0,738	0,772		—0,6156

$\vartheta°$	$\dfrac{1 + \cos^2 2\vartheta}{2}$	$\dfrac{1 + \cos^2 2\alpha \cos^2 2\vartheta}{1 + \cos^2 2\alpha}$				$\dfrac{1}{\cos^3 2\vartheta}$	$\cos 2\vartheta$
		$\alpha = 6°43'$ Mo $K\alpha$	$\alpha = 13°24'$ Cu $K\alpha$	$\alpha = 15°37'$ Co $K\alpha$	$\alpha = 20°09'$ Cr $K\alpha$		
65	0,707	0,715	0,740	0,752	0,784		—0,6427
66	0,724	0,731	0,755	0,767	0,797		—0,6691
67	0,741	0,748	0,771	0,781	0,810		—0,6946
68	0,759	0,765	0,786	0,796	0,823		—0,7193
69	0,776	0,782	0,801	0,811	0,835		—0,7431
70	0,793	0,799	0,817	0,825	0,848		—0,7660
71	0,810	0,816	0,832	0,840	0,861		—0,7880
72	0,827	0,832	0,847	0,854	0,873		—0,8090
73	0,844	0,848	0,861	0,868	0,885		—0,8290
74	0,860	0,863	0,875	0,881	0,897		—0,8480
75	0,875	0,878	0,889	0,894	0,908		—0,8660
76	0,890	0,893	0,902	0,907	0,919		—0,8829
77	0,904	0,906	0,915	0,919	0,929		—0,8987
78	0,917	0,920	0,927	0,930	0,939		—0,9135
79	0,930	0,932	0,938	0,941	0,948		—0,9271
80	0,942	0,943	0,948	0,951	0,957		—0,9396
81	0,952	0,955	0,958	0,960	0,965		—0,9510
82	0,962	0,963	0,966	0,968	0,972		—0,9612
83	0,971	0,972	0,974	0,975	0,978		—0,9703
84	0,978	0,979	0,981	0,982	0,984		—0,9781
85	0,985	0,985	0,987	0,987	0,989		—0,9848
86	0,990	0,991	0,991	0,992	0,993		—0,9902
87	0,995	0,995	0,995	0,995	0,996		—0,9945
88	0,998	0,998	0,998	0,998	0,998		—0,9975
89	0,999	0,999	0,999	0,999	1,000		—0,9993
90	1,000	1,000	1,000	1,000	1,000		—1,0000

10-4. Angular Intensity Factors

Values are given for various functions appearing in the relationships for diffuse scattering and small-angle scattering [102]: $s = (4\pi\sin\vartheta)/\lambda$, where r is the distance from specimen to film or counter, $J_0(sr)$ is the Bessel function of zero order, and $J_1(sr)$ is the Bessel function of the first order.

sr	$\sin sr$	$\dfrac{\sin sr}{sr}$	$\dfrac{\sin^2 sr}{sr}$	$\dfrac{\sin^3 sr}{(sr)^2}$	$3\dfrac{(\sin sr - sr \cos sr)}{(sr)^3}$	$\left[3\dfrac{(\sin sr - sr \cos sr)}{(sr)^3}\right]^2$	$J_0(sr)$	$J_1(sr)$
0,05	+0,0499	+1,000	+0,050	+0,999	+1,000	+1,000	+0,999	+0,025
0,10	+0,0998	+0,998	+0,100	+0,997	+0,999	+0,998	+0,998	+0,050
0,15	+0,1494	+0,996	+0,149	+0,993	+0,998	+0,995	+0,994	+0,075
0,20	+0,1986	+0,993	+0,197	+0,987	+0,996	+0,992	+0,990	+0,100
0,25	+0,2474	+0,990	+0,245	+0,979	+0,994	+0,988	+0,984	+0,124
0,30	+0,2955	+0,985	+0,291	+0,970	+0,991	+0,982	+0,978	+0,148
0,35	+0,3429	+0,980	+0,336	+0,960	+0,988	+0,976	+0,970	+0,172
0,40	+0,3894	+0,974	+0,379	+0,948	+0,984	+0,968	+0,960	+0,196

sr	$\sin sr$	$\dfrac{\sin sr}{sr}$	$\dfrac{\sin^2 sr}{sr}$	$\dfrac{\sin^3 sr}{(sr)^2}$	$\dfrac{3(\sin sr - sr\cos sr)}{(sr)^3}$	$\left[\dfrac{3(\sin sr - sr\cos sr)}{(sr)^3}\right]^2$	$J_0(sr)$	$J_1(sr)$
0,45	+0,4349	+0,967	+0,420	+0,934	+0,980	+0,960	+0,950	+0,219
0,50	+0,4794	+0,959	+0,460	+0,919	+0,975	+0,951	+0,939	+0,242
0,55	+0,5226	+0,950	+0,497	+0,903	+0,970	+0,941	+0,926	+0,265
0,60	+0,5646	+0,941	+0,531	+0,886	+0,964	+0,930	+0,912	+0,287
0,65	+0,6051	+0,931	+0,563	+0,867	+0,958	+0,918	+0,897	+0,308
0,70	+0,6442	+0,920	+0,593	+0,847	+0,952	+0,906	+0,881	+0,329
0,75	+0,6816	+0,909	+0,620	+0,826	+0,945	+0,893	+0,864	+0,349
0,80	+0,7173	+0,897	+0,643	+0,804	+0,937	+0,879	+0,846	+0,369
0,85	+0,7512	+0,884	+0,664	+0,781	+0,930	+0,864	+0,827	+0,388
0,90	+0,7833	+0,870	+0,682	+0,758	+0,921	+0,849	+0,808	+0,406
0,95	+0,8134	+0,856	+0,696	+0,733	+0,913	+0,833	+0,787	+0,423
1,00	+0,8414	+0,841	+0,708	+0,708	+0,904	+0,816	+0,765	+0,440
1,05	+0,8674	+0,826	+0,717	+0,682	+0,894	+0,799	+0,743	+0,456
1,10	+0,8912	+0,810	+0,722	+0,656	+0,884	+0,782	+0,720	+0,471
1,15	+0,9127	+0,794	+0,724	+0,630	+0,874	+0,764	+0,696	+0,485
1,20	+0,9320	+0,777	+0,724	+0,603	+0,863	+0,745	+0,671	+0,498
1,25	+0,9489	+0,759	+0,720	+0,576	+0,852	+0,726	+0,646	+0,511
1,30	+0,9635	+0,741	+0,714	+0,549	+0,841	+0,707	+0,620	+0,522
1,35	+0,9757	+0,723	+0,705	+0,522	+0,829	+0,688	+0,594	+0,533
1,40	+0,9854	+0,704	+0,694	+0,495	+0,817	+0,668	+0,567	+0,542
1,45	+0,9927	+0,685	+0,680	+0,469	+0,805	+0,648	+0,540	+0,550
1,50	+0,9975	+0,665	+0,663	+0,442	+0,792	+0,628	+0,512	+0,558
1,60	+0,9995	+0,625	+0,624	+0,390	+0,766	+0,587	+0,455	+0,570
1,70	+0,9916	+0,583	+0,578	+0,340	+0,739	+0,547	+0,398	+0,578
1,80	+0,9738	+0,541	+0,527	+0,293	+0,711	+0,506	+0,340	+0,582
1,90	+0,9463	+0,498	+0,471	+0,248	+0,683	+0,466	+0,282	+0,581
2,00	+0,9093	+0,455	+0,413	+0,207	+0,653	+0,427	+0,224	+0,577
2,10	+0,8632	+0,411	+0,355	+0,169	+0,623	+0,388	+0,167	+0,568
2,20	+0,8085	+0,368	+0,297	+0,135	+0,593	+0,351	+0,110	+0,556
2,30	+0,7457	+0,324	+0,242	+0,105	+0,562	+0,314	+0,056	+0,540
2,40	+0,6754	+0,281	+0,190	+0,079	+0,531	+0,282	+0,003	+0,520
2,50	+0,5984	+0,239	+0,143	+0,057	+0,499	+0,249	−0,048	+0,497
2,60	+0,5155	+0,198	+0,102	+0,039	+0,468	+0,219	−0,097	+0,471
2,70	+0,4273	+0,158	+0,068	+0,025	+0,437	+0,191	−0,142	+0,442
2,80	+0,3349	+0,120	+0,040	+0,014	+0,406	+0,165	−0,185	+0,410
2,90	+0,2392	+0,083	+0,020	+0,007	+0,376	+0,141	−0,224	+0,375
3,00	+0,1411	+0,047	+0,007	+0,002	+0,346	+0,119	−0,260	+0,339
3,10	+0,0415	+0,013	+0,001	+0,0002	+0,316	+0,100	−0,292	+0,301
3,20	−0,0583	−0,018	+0,001	+0,0003	+0,287	+0,082	−0,320	+0,261
3,30	−0,1577	−0,048	+0,008	+0,002	+0,259	+0,067	−0,344	+0,221
3,40	−0,2555	−0,075	+0,019	+0,006	+0,231	+0,054	−0,364	+0,179
3,50	−0,3507	−0,100	+0,035	+0,010	+0,205	+0,042	−0,380	+0,137
3,60	−0,4425	−0,123	+0,054	+0,015	+0,179	+0,032	−0,392	+0,096
3,70	−0,5298	−0,143	+0,076	+0,021	+0,154	+0,024	−0,399	+0,054
3,80	−0,6118	−0,161	+0,099	+0,026	+0,131	+0,017	−0,403	+0,013
3,90	−0,6877	−0,176	+0,121	+0,031	+0,108	+0,012	−0,402	−0,027
4,00	−0,7568	−0,189	+0,143	+0,036	+0,0871	+0,0076	−0,397	−0,066
4,20	−0,8715	−0,208	+0,181	+0,043	+0,0481	+0,0023	−0,377	−0,139
4,40	−0,9516	−0,216	+0,206	+0,047	+0,0141	+0,0002	−0,342	−0,203
4,60	−0,9936	−0,216	+0,215	+0,047	−0,0147	+0,0002	−0,296	−0,257
4,80	−0,9961	−0,208	+0,207	+0,043	−0,0384	+0,0015	−0,240	−0,299
5,00	−0,9589	−0,192	+0,184	+0,037	−0,0571	+0,0033	−0,178	−0,328
5,20	−0,8834	−0,170	+0,150	+0,029	−0,0708	+0,0050	−0,110	−0,343
5,40	−0,7727	−0,143	+0,111	+0,020	−0,0800	+0,0064	−0,041	−0,345
5,60	−0,6312	−0,113	+0,071	+0,013	−0,0850	+0,0072	+0,027	−0,334
5,80	−0,4646	−0,080	+0,037	+0,006	−0,0861	+0,0074	+0,92	−0,311

sr	$\sin sr$	$\dfrac{\sin sr}{sr}$	$\dfrac{\sin^2 sr}{sr}$	$\dfrac{\sin^3 sr}{(sr)^2}$	$3\,\dfrac{(\sin sr - sr\cos sr)}{(sr)^3}$	$\left[3\,\dfrac{(\sin sr - sr\cos sr)}{(sr)^3}\right]^2$	$J_0\,(sr)$	$J_1\,(sr)$
6,00	—0,2794	—0,047	+0,013	+0,002	—0,0839	+0,0070	+0,151	—0,277
6,20	—0,0830	—0,013	+0,001	+0,0002	—0,0788	+0,0062	+0,202	—0,233
6,40	+0,1165	+0,018	+0,002	+0,0003	—0,0714	+0,0051	+0,243	—0,182
6,60	+0,3115	+0,047	+0,015	+0,002	—0,0622	+0,0039	+0,274	—0,125
6,80	+0,4941	+0,073	+0,036	+0,005	—0,0517	+0,0027	+0,293	—0,065
7,00	+0,6569	+0,094	+0,062	+0,009	—0,0404	+0,0016	+0,300	—0,005
7,20	+0,7936	+0,110	+0,087	+0,012	—0,0288	+0,0008	+0,295	+0,054
7,40	+0,8987	+0,121	+0,109	+0,015	—0,0174	+0,0003	+0,279	+0,110
7,60	+0,9679	+0,127	+0,123	+0,016	—0,0064	+0,00004	+0,252	+0,159
7,80	+0,9985	+0,128	+0,128	+0,016	—0,0037	+0,00001	+0,215	+0,201
8,00	+0,9893	+0,124	+0,122	+0,015	—0,0126	+0,00016	+0,172	+0,235
8,50	+0,7984	+0,094	+0,075	+0,009	—0,0289	+0,00084	+0,042	+0,273
9,00	+0,4121	+0,046	+0,019	+0,002	—0,0354	+0,00126	—0,090	+0,245
9,50	—0,0751	—0,008	+0,001	+0,0001	—0,0329	+0,00108	—0,194	+0,161
10,00	—0,5440	—0,054	+0,030	+0,003	—0,0235	+0,00055	—0,246	+0,044
10,50	—0,8797	—0,084	+0,074	+0,007	—0,0107	+0,00011	—0,237	—0,079
11,00	—0,9999	—0,091	+0,091	+0,008	—0,0024	+0,000001	—0,171	—0,177
11,50	—0,8754	—0,076	+0,067	+0,006	—0,0127	+0,000161	—0,068	—0,228
12,00	—0,5365	—0,045	+0,024	+0,002	—0,0185	+0,000343	+0,048	—0,223
12,50	—0,0663	—0,005	+0,0004	+0,00003	—0,0193	+0,000371	+0,147	—0,166
13,00	+0,4201	+0,032	+0,014	+0,001	—0,0155	+0,000241	+0,207	—0,070
13,50	+0,8037	+0,060	+0,048	+0,004	—0,0088	+0,000078	+0,215	+0,038
14,00	+0,9906	+0,071	+0,070	+0,005	—0,0010	+0,000001	+0,171	+0,133
14,50	+0,9349	+0,064	+0,060	+0,004	—0,0060	+0,000036	+0,088	+0,193
15,00	+0,6502	+0,043	+0,028	+0,002	—0,0107	+0,000115	—0,014	+0,205
15,50	+0,2064	+0,013	+0,003	+0,0002	—0,0124	+0,000153	—0,109	+0,167
16,00	—0,2879	—0,018	+0,005	+0,0003	—0,0110	+0,000121	—	—
16,50	—0,7117	—0,043	+0,031	+0,002	—0,0073	+0,000053	—	—
17,00	—0,9614	—0,057	+0,054	+0,003	—0,0023	+0,000005	—	—
17,50	—0,9756	—0,056	+0,054	+0,003	—0,0027	+0,000007	—	—
18,00	—0,7509	—0,042	+0,031	+0,002	—0,0065	+0,000042	—	—
18,50	—0,3424	—0,019	+0,006	+0,0003	—0,0084	+0,000071	—	—
19,00	+0,1498	+0,008	+0,001	+0,0001	—0,0082	+0,000066	—	—
19,50	+0,6055	+0,031	+0,019	+0,001	—0,0060	+0,000036	—	—
20,00	+0,9129	+0,046	+0,042	+0,002	—0,0027	+0,000007	—	—
21	+0,8366	+0,040	+0,033					
22	—0,0088	—0,0004	+0,00000					
23	—0,8462	—0,037	+0,031					
24	—0,9055	—0,038	+0,034					
25	—0,1323	—0,005	+0,001					
26	+0,7625	+0,029	+0,022					
27	+0,9563	+0,035	+0,034					
28	+0,2709	+0,010	+0,003					
29	—0,6636	—0,023	+0,015					
30	—0,9880	—0,033	+0,033					
31	—0,4040	—0,013	+0,005					
32	+0,5514	+0,017	+0,010					
33	+0,9999	+0,030	+0,030					
34	+0,5290	+0,016	+0,008					
35	—0,4281	—0,012	+0,005					
36	—0,9917	—0,028	+0,027					
37	—0,6435	—0,017	+0,011					
38	+0,2963	+0,008	+0,002					
39	+0,9368	+0,025	+0,024					
40	+0,07451	+0,019	+0,014					
41	—0,1586	—0,004	+0,001					

sr	$\sin sr$	$\dfrac{\sin sr}{sr}$	$\dfrac{\sin^2 sr}{sr}$	$\dfrac{\sin^3 sr}{(sr)^2}$	$\dfrac{3(\sin sr - sr\cos sr)}{(sr)^3}$	$\left[\dfrac{3(\sin sr - sr\cos sr)}{(sr)^3}\right]^2$	$J_0\,(sr)$	$J_1\,(sr)$
42	$-0,9165$	$-0,022$	$+0,020$					
43	$-0,9317$	$-0,019$	$+0,016$					
44	$+0,0177$	$+0,0004$	$+0,00001$					
45	$+0,8509$	$+0,019$	$+0,016$					
46	$+0,9017$	$+0,020$	$+0,018$					
47	$+0,1235$	$+0,003$	$+0,0003$					
48	$-0,7682$	$-0,016$	$+0,012$					
49	$-0,9537$	$-0,019$	$+0,019$					
50	$-0,2623$	$-0,005$	$+0,001$					
51	$+0,6702$	$+0,013$	$+0,009$					
52	$+0,9866$	$+0,019$	$+0,019$					
53	$+0,3959$	$+0,007$	$+0,003$					
54	$-0,5587$	$-0,010$	$+0,006$					
55	$-0,9997$	$-0,018$	$+0,018$					
56	$-0,5215$	$-0,009$	$+0,005$					
57	$+0,4361$	$+0,008$	$+0,003$					
58	$+0,9928$	$+0,017$	$+0,017$					
59	$+0,6367$	$+0,011$	$+0,007$					
60	$-0,3048$	$-0,005$	$+0,002$					

10-5. Incoherent-Scattering Intensity

The intensity from the incoherent (Compton) scattering for Z between 1 and 14 (light elements) is given by

$$I_c = Z - \sum_i f_i^2,$$

where Z is the atomic number and f_i is the scattering function for each electron.

The f_i have been calculated by the self-consistent field method. The intensity for heavy atoms (Z from 15 to 100) is given by

$$I_c = Zs\,(b),$$

where s(b) is a function of reflection angle, wavelength, and atomic number.

The table gives I_c in electron units for neutral atoms and for some ions for Z from 1 to 94 and for $s = (4\pi \sin \vartheta)/\lambda$ from 0 to 13 [102].

Z	Element	1	2	3	4	5	6	7	8	9	10	11	12	13
1	H	0,25	0,56	0,85	0,94	0,98	0,99	1,00	1,00	1,00	1,00	1,00	1,00	1,00
2	He	0,16	0,60	1,15	1,55	1,72	1,83	1,91	1,95	1,97	1,98	1,99	2,00	2,00
3	Li⁺	0,05	0,24	0,58	0,95	1,20	1,41	0,65	1,84	1,91	1,91	1,95	1,97	2,00
3	Li	0,83	1,24	1,58	1,95	2,2	2,4	2,7	2,8	2,9	2,9	3,00	3,00	3,00
4	Be	1,35	1,95	2,22	2,5	2,8	3,0	3,3	3,5	3,7	3,8	3,8	3,9	4,0
5	B	1,50	2,4	2,9	3,3	3,5	3,7	4,0	4,3	4,5	4,5	4,6	4,7	4,7
6	C	1,49	2,7	3,6	4,2	4,5	4,6	4,9	5,1	5,2	5,3	5,4	5,5	5,5
7	N	1,35	2,9	4,2	5,1	5,3	5,5	5,7	5,9	6,0	6,1	6,1	6,2	6,3
8	O	1,38	3,0	4,7	5,7	6,0	6,3	6,5	6,6	6,7	6,8	6,9	7,0	7,1
8	O⁻²	2,40	4,7	6,7	7,8	8,2	8,4	8,5	8,6	8,7	8,8	8,9	9,0	9,1
9	F	1,26	3,2	4,9	6,1	6,7	7,0	7,3	7,4	7,5	7,6	7,8	7,9	7,9
9	F⁻	1,75	4,0	6,0	7,3	7,8	8,1	8,3	8,4	8,5	8,6	8,8	8,9	8,9
10	Ne	1,25	3,25	5,1	6,5	7,3	7,8	8,1	8,3	8,4	8,5	8,7	8,8	8,9
11	Na⁺	0,65	2,6	4,5	5,7	6,8	7,4	7,8	8,1	8,3	8,4	8,6	8,6	8,7
11	Na	1,30	3,6	5,5	6,7	7,8	8,4	8,8	9,1	9,3	9,4	9,6	9,6	9,7
12	Mg⁺²	0,45	2,0	3,2	5,0	6,1	6,9	7,5	7,8	8,2	8,3	8,5	8,6	8,7
12	Mg	1,50	3,9	5,2	7,0	8,1	8,9	9,5	9,8	10,2	10,3	10,5	10,6	10,7
13	Al	1,75	4,2	5,9	7,3	8,5	9,4	10,1	10,5	10,9	11,1	11,3	11,4	11,5
14	Si	2,25	4,8	6,4	7,8	9,0	9,9	10,7	11,3	11,7	12,0	12,2	12,3	12,5
15	P	3,3	5,25	6,7	7,81	8,8	9,5	10,1	10,6	11,1	11,5	11,9	12,1	12,4
16	S	3,5	5,47	7,0	8,0	9,1	9,9	10,6	11,2	11,6	12,1	12,5	12,8	13,1
17	Cl	3,7	5,6	7,3	8,5	9,4	10,3	11,1	11,7	12,2	12,7	13,1	13,4	13,8
18	Ar	3,9	5,9	7,6	8,8	9,8	10,7	11,5	12,2	12,7	13,2	13,7	14,1	14,4
19	K	3,9	6,1	7,8	9,1	10,2	11,2	12,0	12,7	13,3	13,8	14,3	14,7	15,1
20	Ca	4,0	6,3	8,1	9,5	10,6	11,5	12,4	13,2	13,8	14,3	14,8	15,3	15,7
21	Sc	4,2	6,4	8,2	9,7	11,0	12,0	12,8	13,7	14,3	14,9	15,4	15,9	16,3
22	Ti	4,3	6,6	8,6	9,9	11,2	12,3	13,3	14,1	14,9	15,4	15,9	16,4	16,9
23	V	4,4	6,6	8,7	10,4	11,5	12,7	13,8	14,5	15,3	15,9	16,5	17,0	17,5
24	Cr	4,6	6,9	8,8	10,6	12,0	13,1	14,0	15,0	15,9	16,4	17,0	17,6	18,1
25	Mn	4,8	6,9	9,0	10,9	12,3	13,5	14,5	15,4	16,3	17,0	17,6	18,1	18,7
26	Fe	4,9	7,2	9,3	11,3	12,6	13,8	15,0	15,9	16,8	17,6	18,1	18,7	19,2
27	Co	5,0	7,4	9,6	11,4	12,3	14,2	15,4	16,3	17,2	18,0	18,6	19,2	19,8
28	Ni	5,1	7,5	9,8	11,8	13,2	14,4	15,7	16,7	17,6	18,5	19,2	19,7	20,3
29	Cu	5,2	7,6	10,0	11,9	13,4	14,8	16,0	17,1	18,0	19,0	19,7	20,2	20,9
30	Zn	5,3	7,8	10,2	12,2	13,8	15,0	16,4	17,5	18,4	19,4	20,2	20,8	21,4
31	Ga	5,4	8,0	10,3	12,4	14,1	15,5	16,7	17,8	18,8	19,8	20,7	21,2	21,9
32	Ge	5,6	8,1	10,5	12,8	14,4	15,8	17,1	18,3	19,3	20,3	21,2	21,9	22,5
33	As	5,8	8,2	10,7	12,9	14,7	16,1	17,2	18,5	19,7	20,6	21,6	22,3	22,9
34	Se	5,9	8,3	10,8	12,9	14,8	16,2	17,6	18,9	20,0	21,1	22,2	22,9	23,5
35	Br	6,0	8,4	11,0	13,3	15,2	16,7	18,1	19,4	20,5	21,5	22,5	23,3	24,0
36	Kr	6,1	8,5	11,3	13,6	15,6	17,1	18,5	19,9	21,0	21,9	23,0	23,9	24,6
37	Rb	6,3	8,7	11,5	13,7	15,9	17,4	18,7	20,1	21,3	22,4	23,5	24,5	25,2
38	Sr	6,5	8,9	11,7	13,8	16,2	17,6	19,1	20,6	21,8	23,0	23,9	24,9	25,7
39	Y	6,6	9,2	11,9	14,0	16,5	17,9	19,5	20,9	22,1	23,3	24,1	25,4	26,1
40	Zr	6,6	9,3	12,0	14,2	16,7	18,3	19,7	21,2	22,5	23,6	24,6	25,7	26,7
41	Nb	6,7	9,4	12,1	14,5	17,0	18,6	20,2	21,4	22,8	24,1	25,1	26,2	27,2
42	Mo	6,8	9,6	12,2	14,7	17,1	18,9	20,5	21,9	23,3	24,5	25,6	26,7	27,7
43	Tc	6,9	9,8	12,4	15,0	17,4	19,2	20,9	22,1	23,6	24,8	25,9	27,0	28,1
44	Ru	7,0	9,9	12,6	15,3	17,7	19,6	21,0	22,5	23,9	25,3	26,5	27,5	28,7
45	Rh	7,2	10,0	12,8	15,6	18,0	20,0	21,5	23,0	24,4	25,8	27,1	28,1	29,3
46	Pd	7,2	10,0	12,9	15,6	18,1	20,1	21,9	23,2	24,6	26,2	27,4	28,2	29,4
47	Ag	7,2	10,2	13,1	15,8	18,2	20,4	22,2	23,5	24,9	26,5	27,7	28,6	29,8
48	Cd	7,4	10,3	13,2	15,9	18,4	20,6	22,2	23,9	25,4	26,7	28,0	29,0	30,2
49	In	7,5	10,5	13,4	16,1	18,5	20,8	22,5	24,3	25,7	27,0	28,4	29,6	30,7
50	Sn	7,7	10,7	13,6	16,3	18,6	21,1	22,9	24,6	26,1	27,5	28,9	30,2	31,3
51	Sb	7,9	10,9	13,7	16,6	18,8	21,4	23,2	25,0	26,3	27,9	29,3	30,5	31,7
52	Te	8,0	11,1	13,8	16,9	19,1	21,8	23,5	25,4	26,6	28,3	29,7	30,9	32,1
53	I	8,1	11,1	13,9	17,1	19,2	22,0	23,8	25,5	27,0	28,5	30,0	31,3	32,4
54	Xe	8,2	11,2	14,1	17,1	19,3	22,2	24,1	25,7	27,4	28,8	30,1	31,6	32,8
55	Cs	8,3	11,3	14,2	17,2	19,6	22,4	24,4	26,0	27,8	29,2	30,7	32,1	33,3
56	Ba	8,5	11,4	14,3	17,4	19,8	22,6	24,8	26,4	28,1	29,7	31,1	32,6	33,8
57	La	8,6	11,4	14,4	17,7	20,1	22,9	24,9	26,7	28,3	29,9	31,5	32,8	34,2
58	Ce	8,7	11,4	14,6	17,9	20,4	23,2	25,1	26,9	28,5	30,2	31,9	33,1	34,6

z	s Element	1	2	3	4	5	6	7	8	9	10	11	12	13
59	Pr	8,8	11,5	14,8	18,1	20,7	23,4	25,4	27,3	28,8	30,6	32,3	33,6	35,0
60	Nd	8,9	11,7	15,0	18,3	20,9	23,6	25,8	27,7	29,3	30,9	32,6	34,0	35,5
61	Pm	9,0	11,9	15,1	18,4	21,2	23,7	26,2	28,0	29,8	31,2	32,9	34,4	35,9
62	Sm	9,2	12,1	15,1	18,6	21,5	23,9	26,5	28,3	30,2	31,6	32,2	34,8	36,3
63	Eu	9,2	12,2	15,3	18,8	21,7	24,1	26,8	28,6	30,4	31,9	33,6	35,2	36,7
64	Gd	9,3	12,4	15,5	18,9	22,0	24,4	27,1	28,9	30,7	32,3	34,0	35,5	37,0
65	Tb	9,4	12,5	15,7	19,0	22,1	24,7	27,3	29,2	30,9	32,5	34,4	35,9	37,4
66	Dy	9,5	12,6	15,8	19,1	22,2	24,8	27,5	29,5	31,3	32,9	34,8	36,3	37,8
67	Ho	9,6	12,8	15,9	19,2	22,3	24,9	27,7	29,8	31,7	33,4	35,1	36,6	38,1
68	Er	9,7	12,9	16,1	19,2	22,4	25,0	27,9	30,1	32,0	33,8	35,4	37,1	38,5
69	Tu	9,8	13,1	16,3	19,4	22,6	25,1	28,1	30,3	32,3	34,1	35,7	37,3	38,9
70	Yb	9,9	13,2	16,4	19,5	22,8	25,2	28,4	30,5	32,6	34,3	35,8	37,6	39,3
71	Lu	10,1	13,4	16,5	19,7	23,0	25,3	28,5	30,7	32,9	34,6	36,1	37,9	39,7
72	Hf	10,2	13,5	16,7	19,9	23,2	25,6	28,9	31,0	33,2	35,0	36,5	38,3	40,1
73	Ta	10,3	13,7	16,9	20,1	23,4	25,8	29,2	31,4	33,4	35,4	36,9	38,7	40,5
74	W	10,4	13,8	17,1	20,4	23,6	26,1	29,5	31,8	33,7	35,8	37,4	39,1	40,9
75	Re	10,6	14,0	17,3	20,5	23,8	26,4	29,7	32,1	34,0	36,0	37,7	39,5	41,3
76	Os	10,7	14,1	17,4	20,7	24,1	26,7	29,9	32,3	34,4	36,2	38,1	39,7	41,6
77	Ir	10,8	14,2	17,6	20,8	24,3	27,0	30,0	32,5	34,7	36,3	38,5	40,0	41,8
78	Pt	10,9	14,4	17,7	21,0	24,5	27,2	30,3	32,8	34,9	36,7	38,8	40,4	42,1
79	Au	11,0	14,5	17,8	21,2	24,7	27,5	30,5	33,1	35,2	37,1	39,2	40,8	42,4
80	Hg	11,1	14,6	17,9	21,4	25,0	27,8	30,8	33,4	35,6	37,5	39,5	41,1	42,8
81	Tl	11,2	14,7	18,1	21,5	25,2	27,9	30,9	33,7	36,0	37,8	39,9	41,5	43,3
82	Pb	11,2	14,8	18,2	21,7	25,4	28,1	31,1	33,9	36,3	38,1	40,2	41,8	43,6
83	Bi	11,3	14,9	18,3	21,9	25,6	28,3	31,2	34,1	36,7	38,4	40,5	42,2	44,1
84	Po	11,4	15,0	18,5	22,1	25,8	28,5	31,3	34,4	37,0	38,7	40,8	42,5	44,4
85	At	11,5	15,0	18,6	22,3	25,9	28,6	31,3	34,6	37,2	39,0	41,1	42,8	44,7
86	Rn	11,6	15,1	18,7	22,4	26,0	28,8	31,3	34,8	37,5	39,3	41,5	43,2	45,1
87	Fr	11,7	15,3	18,9	22,6	26,1	29,0	31,4	35,1	37,7	39,7	41,8	43,5	45,4
88	Ra	11,9	15,5	19,2	22,9	26,4	29,3	31,7	35,5	38,1	40,2	42,2	43,9	46,9
89	Ac	12,0	15,7	19,4	23,1	26,7	29,5	32,0	35,9	38,5	40,6	42,6	44,3	46,4
90	Th	12,2	15,8	19,6	23,3	26,9	29,8	32,3	36,3	38,9	41,0	43,0	44,8	46,9
91	Pa	12,3	16,0	19,8	23,6	27,2	30,1	32,6	36,7	39,3	41,4	43,5	45,2	47,3
92	U	12,4	16,1	19,9	23,7	27,4	30,4	32,9	37,0	39,7	41,9	44,0	45,8	47,8
93	Np	12,6	16,3	20,1	23,9	27,7	30,8	33,3	37,4	40,1	42,3	44,9	46,3	48,4
94	Pu	12,6	16,4	20,3	24,3	28,0	31,3	33,7	37,8	40,5	42,8	45,5	46,8	48,9

10-6.   Relativistic Correction for Incoherent Scattering

A correction factor $1/B^3$ is introduced by relativistic effects:

$$I_c^{\cdot} = I_c \cdot \frac{1}{B^3}.$$

The factor is given by

$$\frac{1}{B^3} = \frac{1}{\left(1 + \dfrac{h\lambda}{8\pi^2 mc} s^3\right)^3},$$

where

$$s = \frac{4\pi \sin \vartheta}{\lambda}.$$

The table gives $1/B^3$ for Mo Kα ($\lambda = 0.7107$Å), Cu Kα ($\lambda = 1.5418$A), Co Kα ($\lambda = 1.790$Å), Fe Kα ($\lambda = 1.9373$Å), and Cr Kα ($\lambda = 2.291$Å) for s from 0 to 14.0.

s	Mo Kα	Cu Kα	Co Kα	Fe Kα	Cr Kα	s	Mo Kα	Cu Kα	Co Kα	Fe Kα	Cr Kα
0,0	1,000	1,000	1,000	1,000	1,000	5,0	0,984	0,965	0,960	0,957	0,949
.1	1,000	1,000	1,000	1,000	1,000	.2	0,982	0,963	0,957	0,953	0,945
.2	1,000	1,000	1,000	1,000	1,000	.4	0,981	0,960	0,953	0,950	0,941
.3	1,000	1,000	1,000	1,000	1,000	.6	0,980	0,957	0,950	0,946	0,937
.4	1,000	1,000	1,000	1,000	1,000	.8	0,978	0,954	0,946	0,942	0,932
.5	1,000	1,000	1,000	0,999	1,999	6,0	0,977	0,951	0,943	0,938	0,928
.6	1,000	0,999	0,999	0,999	0,999	.2	0,975	0,947	0,939	0,935	0,923
.7	1,000	0,999	0,999	0,999	0,999	.4	0,974	0,944	0,935	0,930	0,918
.8	1,000	0,999	0,999	0,999	0,999	.6	0,972	0,941	0,931	0,926	0,913
.9	0,999	0,999	0,999	0,998	0,998	.8	0,970	0,937	0,927	0,922	0,908
1,0	0,999	0,999	0,998	0,998	0,998	7,0	0,969	0,933	0,923	0,917	0,903
.1	0,999	0,998	0,998	0,998	0,997	.2	0,967	0,930	0,919	0,913	0,898
.2	0,999	0,998	0,998	0,997	0,997	.4	0,965	0,926	0,915	0,908	0,893
.3	0,999	0,998	0,997	0,997	0,996	.6	0,963	0,922	0,910	0,903	0,887
.4	0,999	0,997	0,997	0,996	0,996	.8	0,961	0,918	0,906	0,898	0,882
.5	0,999	0,997	0,996	0,996	0,995	8,0	0,959	0,914	0,901	0,894	0,876
.6	0,998	0,996	0,996	0,995	0,995	.2	0,957	0,910	0,897	0,888	0,870
.7	0,998	0,996	0,995	0,995	0,994	.4	0,955	0,906	0,892	0,883	0,865
.8	0,998	0,995	0,995	0,994	0,993	.6	0,953	0,902	0,887	0,878	0,859
.9	0,998	0,995	0,994	0,994	0,992	.8	0,951	0,898	0,882	0,873	0,853
2,0	0,997	0,994	0,993	0,993	0,992	9,0	0,949	0,893	0,877	0,868	0,847
.1	0,997	0,994	0,993	0,992	0,991	.2	0,946	0,889	0,872	0,862	0,840
.2	0,997	0,993	0,992	0,991	0,990	.4	0,944	0,884	0,867	0,857	0,834
.3	0,997	0,993	0,991	0,991	0,989	.6	0,942	0,880	0,862	0,851	0,828
.4	0,996	0,992	0,991	0,990	0,988	.8	0,940	0,875	0,857	0,846	0,822
.5	0,996	0,991	0,990	0,989	0,987	10,0	0,937	0,870	0,851	0,841	0,815
.6	0,996	0,990	0,989	0,988	0,986	.2	0,935	0,866	0,846	0,834	0,809
.7	0,995	0,990	0,988	0,987	0,985	.4	0,932	0,861	0,841	0,829	0,802
.8	0,995	0,989	0,987	0,986	0,984	.6	0,930	0,856	0,835	0,823	0,796
.9	0,995	0,988	0,986	0,985	0,982	.8	0,927	0,851	0,830	0,817	0,789
3,0	0,994	0,987	0,985	0,984	0,981	11,0	0,925	0,846	0,824	0,813	0,782
.1	0,994	0,986	0,984	0,983	0,980	.2	0,922	0,841	0,818	0,805	0,776
.2	0,993	0,986	0,983	0,982	0,979	.4	0,919	0,836	0,813	0,799	0,769
.3	0,993	0,985	0,982	0,981	0,977	.6	0,917	0,831	0,807	0,793	0,762
.4	0,992	0,984	0,981	0,980	0,976	.8	0,914	0,826	0,801	0,787	0,755
.5	0,992	0,983	0,980	0,978	0,975	12,0	0,911	0,820	0,795	0,781	0,748
.6	0,992	0,982	0,979	0,977	0,973	.2	0,908	0,815	0,790	0,775	0,741
.7	0,991	0,981	0,978	0,976	0,972	.4	0,906	0,810	0,784	0,769	0,734
.8	0,991	0,980	0,977	0,974	0,970	.6	0,903	0,804	0,778	0,763	0,727
.9	0,990	0,979	0,975	0,973	0,969	.8	0,900	0,799	0,772	0,757	0,721
4,0	0,990	0,978	0,974	0,972	0,967	13,0	0,897	0,794	0,766	0,750	0,714
.1	0,989	0,976	0,973	0,970	0,965	.2	0,894	0,788	0,760	0,744	0,706
.2	0,989	0,975	0,971	0,969	0,964	.4	0,891	0,783	0,754	0,737	0,699
.3	0,988	0,974	0,970	0,968	0,962	.6	0,888	0,777	0,748	0,731	0,692
.4	0,987	0,973	0,969	0,966	0,960	.8	0,885	0,772	0,741	0,725	0,685
.5	0,987	0,972	0,967	0,965	0,958	14,0	0,882	0,766	0,735	0,718	0,678
.6	0,986	0,971	0,966	0,963	0,957						
.7	0,986	0,969	0,964	0,961	0,955						
.8	0,985	0,968	0,963	0,960	0,953						
.9	0,984	0,967	0,961	0,958	0,951						

10-7. Values of q for Particles of Various Shapes

The small-angle scattering for a Gaussian distribution can be put as

$$I = e^{-q\,(sL)^2},$$

where 2L is the maximal dimension of a particle; q is dependent on the shape.

The table gives q for various single particles and groups [12].

Particle	q
Hollow sphere of outer radius L and inner radius cL (c < 1)	$\dfrac{1}{5}\left[c^2+\dfrac{c+1}{c^2+c+1}\right]$
Complete sphere	0.20
Infinitely thin sphere	0.167
2 spheres of radius L in contact	0.53
5 spheres of radius L in contact (centered tetrahedron)	1.28
13 spheres of radius L in contact (close packing around central sphere)	1.50
Thin disc of radius L	0.167
Long cylinder of height 2L and axial ratio 1/m	$(1+m^2)/12$
Ellipsoid of rotation, principal axis 2L, axial ratio 1/m	$(2+m^2)/15$

Some particular experimental conditions have to be considered in small-angle scattering for metals and alloys [186-195].

The results from small-angle scattering are often much distorted by double Bragg scattering; see [196, 197, 382] for correction for the latter.

Multiple scattering can be used to determine the dimensions of regions of inhomogeneity; scattering curves for specimens differing in mass are replaced by intensity measurements on the forward beam with the specimen in different positions. See [414] for details.

10-8. Scattering Functions for Systems of Homogeneous Particles

10-8a. Spherical Particles (General Values)

The scattering function for spherical particles of radius a and of uniform electron density is

$$i(s) = \Phi^2(sa) = \left[3\,\frac{\sin(sa)-sa\cos(sa)}{(sa)^3}\right]^2 = \frac{9\pi}{2}\left[\frac{J_{3/2}(sa)}{(sa)^{3/2}}\right]^2, \qquad (101)$$

where $s = (4\pi\sin\vartheta)/\lambda$, $J_{3/2}(sa)$ is the Bessel function of order $3/2$.

The table gives $\Phi^2(sa)$ for sa from 0.5 to 20.0 [175].

sa	$\Phi^2(sa)$	sa	$\Phi^2(sa)$	sa	$\Phi^2(sa)$	sa	$\Phi^2(sa)$
0,5	0,951	5,5	0,00689	10,5	0,000114	15,5	0,000153
1,0	0,816	6,0	0,00704	11,0	0,00000559	16,0	0,000121
1,5	0,628	6,5	0,00449	11,5	0,000161	16,5	0,0000528
2,0	0,427	7,0	0,00163	12,0	0,000342	17,0	0,00000515
2,5	0,249	7,5	0,000140	12,5	0,000371	17,5	0,00000727
3,0	0,119	8,0	0,000159	13,0	0,000241	18,0	0,0000423
3,5	0,0419	8,5	0,000835	13,5	0,0000777	18,5	0,0000705
4,0	0,00758	9,0	0,00126	14,0	0,00000102	19,0	0,0000664
4,5	0,00000	9,5	0,00108	14,5	0,0000358	19,5	0,0000364
5,0	0,00326	10,0	0,000554	15,0	0,000115	20,0	0,00000739

10-8b. Spherical Particles (Maxima and Minima)

The zeros of Φ^2 (sa) correspond to zeros in $J_{3/2}$(sa); the maxima, to zeros in $J_{5/2}$(sa).

The table gives the first 8 maxima and minima in Φ^2(sa) and also the intensities of the maxima. The calculations have been made for the absence of the collimator effect and for a collimator in the form of infinite slits.

No.	Perfect collimator			Infinite slits		
	sa		I_{max}	sa		I_{max}
	min	max		min	max	
0	—	0	1	—	0	1
1	4,493	5,765	0,00742	4,32	5,31	0,0165
2	7,725	9,095	0,00127	7,52	8,63	0,00383
3	10,90	12,32	0,00039	10,67	11,85	0,00145
4	14,07	15,52	0,000154	13,82	15,04	0,000696
5	17,22	18,69	0,000073	16,96	18,22	0,000282
6	20,37	21,85	0,000039			
7	23,52	25,01	0,0000228			
8	26,67	28,17	0,0000142			

10-8c. Ellipsoid of Rotation

The scattering function here is

$$ i\,(s) = \int_{0}^{\frac{\pi}{2}} \Phi^2 \left(sR\sqrt{\cos^2\vartheta + w^2 \sin^2\vartheta}\right) \cos\vartheta\, d\vartheta, \qquad (102) $$

where w is the axial ratio [182].

The tables give i(s) for various w and sR, where $R = \sqrt{(2 + w^2)/5}$ is the radius of rotation [175].

Ellipsoids with w > 1

sR	w						
	1,5	2	3	4	6	10	∞
0,0	1,000	1,000	1,000	1,000	1,000	1,000	1,000
0,5	0,920	0,920	0,921	0,921	0,922	0,922	0,920
1,0	0,713	0,718	0,726	0,730	0,735	0,737	0,738
1,5	0,461	0,477	0,505	0,522	0,537	0,546	0,551
2,0	0,242	0,274	0,327	0,359	0,389	0,407	0,418
2,5	0,0990	0,138	0,206	0,250	0,292	0,318	0,334
3,0	0,0293	0,0616	0,128	0,176	0,225	0,259	0,278
3,5	0,00638	0,0240	0,0760	0,122	0,176	0,214	0,239
4,0	0,00333	0,00789	0,0421	0,0827	0,138	0,180	0,210
4,5	0,00447	0,00294	0,0216	0,0546	0,108	0,153	0,187
5,0	0,00450	0,00270	0,0101	0,0347	0,0836	0,131	0,168

Ellipsoids with w < 1

sR	w						
	$^2/_3$	$^1/_2$	$^1/_3$	$^1/_4$	$^1/_6$	$^1/_{10}$	0
0,0	1,000	1,000	1,000	1,000	1,000	1,000	1,000
0,5	0,920	0,920	0,920	0,920	0,920	0,920	0,920
1,0	0,712	0,714	0,716	0,717	0,717	0,718	0,718
1,5	0,457	0,464	0,472	0,475	0,477	0,479	0,479
2,0	0,236	0,250	0,267	0,274	0,278	0,284	0,284
2,5	0,0931	0,112	0,136	0,148	0,154	0,162	0,166
3,0	0,0262	0,0450	0,0714	0,0846	0,0916	0,102	0,106
3,5	0,00688	0,0203	0,0428	0,0555	0,0627	0,0738	
4,0	0,00523	0,0113	0,0274	0,0385	0,0453	0,0562	
4,5	0,00525	0,00625	0,0166	0,0259	0,0320	0,0426	
5,0	0,00369	0,00307	0,00934	0,0170	0,0225	0,0328	

The tables are used by constructing graphs as in Fig. 177 with logarithmic scales [lg I against lg 2 ϑ $\approx$ lg (2 sin ϑ)]; Fig. 177a applies to oblate ellipsoids and Fig. 177b to prolate ones [123].

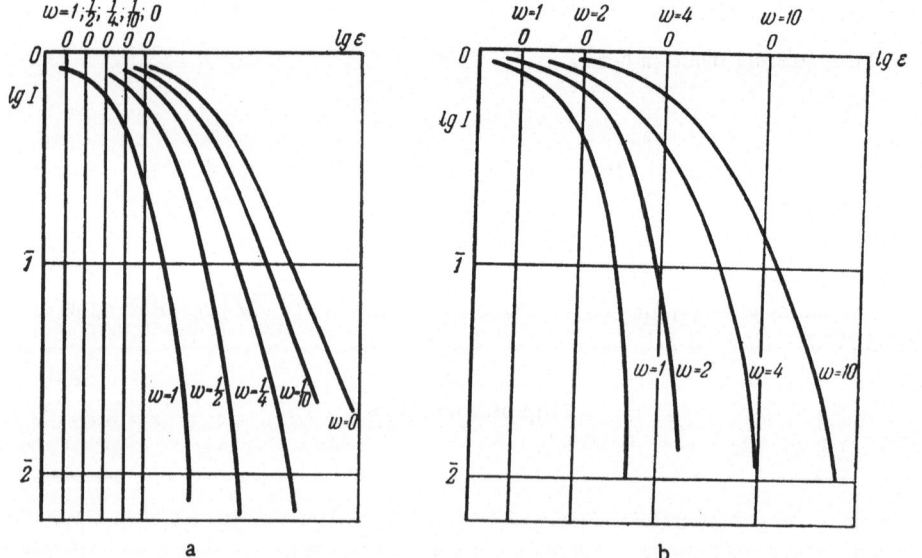

a   b

Fig. 177. Graphs for finding particle sizes: (a) oblate ellipsoids; (b) prolate ellipsoids.

The origin for each curve is shown on the abscissa; the theoretical curves are compared with the experimental one to give the best agreement (i.e., to find w). The displacement of the origin is

$$\lg sR - \lg 2 \sin \vartheta = \lg \frac{2\pi R}{\lambda},$$

so the lengths of the axes can be deduced.

10-8d. Particles in the Form of Cylinders

The scattering function for homogeneous orthogonal circular cylinders of diameter 2a and height 2H is given by

$$i(s) = \int_0^{\frac{\pi}{2}} \frac{\sin^2(sH\cos\vartheta)}{s^2H^2\cos^2\vartheta} \frac{4J_1^2(sa\sin\vartheta)}{s^2a^2\sin^2\vartheta} \sin\vartheta\,d\vartheta, \qquad (103)$$

where $J_1(sa\sin\vartheta)$ is a Bessel function of order one [183].

The table gives $i(s)$ for elongated cylinders of radius R; w equals H/a.

	w						
sR	1	2	3	4	6	8	∞
0,0	1,000	1,000	1,000	1,000	1,000	1,000	1,000
0,5	0,916	0,920	0,921	0,921	0,922	0,922	0,922
1,0	0,704	0,714	0,722	0,732	0,738	0,738	0,738
1,5	0,447	0,482	0,510	0,520	0,533	0,538	0,543
2,0	0,233	0,290	0,340	0,363	0,388	0,399	0,410
2,5	0,0820	0,168	0,235	0,270	0,302	0,314	0,333
3,0	0,0263	0,092	0,162	0,205	0,244	0,258	0,282
3,5		0,0463	0,110	0,154	0,198	0,217	0,245
4,0		0,0205	0,070	0,113	0,161	0,183	0,217
4,5				0,0816	0,132	0,155	0,193
5,0				0,0606	0,108	0,133	0,175

10-8e. Particles in the Form of Cylinders of Small Diameter

The scattering function for cylinders of very small diameter and of length 2H is

$$i(s) = \frac{Si(2sH)}{sH} - \frac{\sin^2(sH)}{s^2H^2}, \qquad (104)$$

where

$$Si(x) = \int_0^x \frac{\sin t}{t}\,dt, \qquad s = \frac{4\pi\sin\vartheta}{\lambda}.$$

The table gives $i(s)$ for s from 0.0 to 4.0 [175].

s	$i(s)$	s	$i(s)$	s	$i(s)$
0,0	1,000	1,0	0,898	2,0	0,673
0,2	0,996	1,2	0,858	2,2	0,622
0,4	0,984	1,4	0,813	2,4	0,584
0,6	0,961	1,6	0,768	3,0	0,473
0,8	0,931	1,8	0,719	3,5	0,406
				4,0	0,357

10-8f. Particles in the Form of Elliptic Cylinders

The scattering function for infinitely long cylinders of elliptic cross section is

$$F^2(r^*) = \frac{1}{2\pi} \int_0^{2\pi} \left[\frac{2J_1 \left(2\pi r^* R \left(K^2 \cos^2 \alpha + \sin^2 \alpha\right)^{1/2}\right)}{2\pi r^* R \left(K^2 \cos^2 \alpha + \sin^2 \alpha\right)^{1/2}} \right]^2 d\alpha, \qquad (105)$$

where r^* is the radial coordinate in reciprocal space [namely, $2(\sin \vartheta)/\lambda$], R is the minor axis of the ellipse, and K is the axial ratio [184].

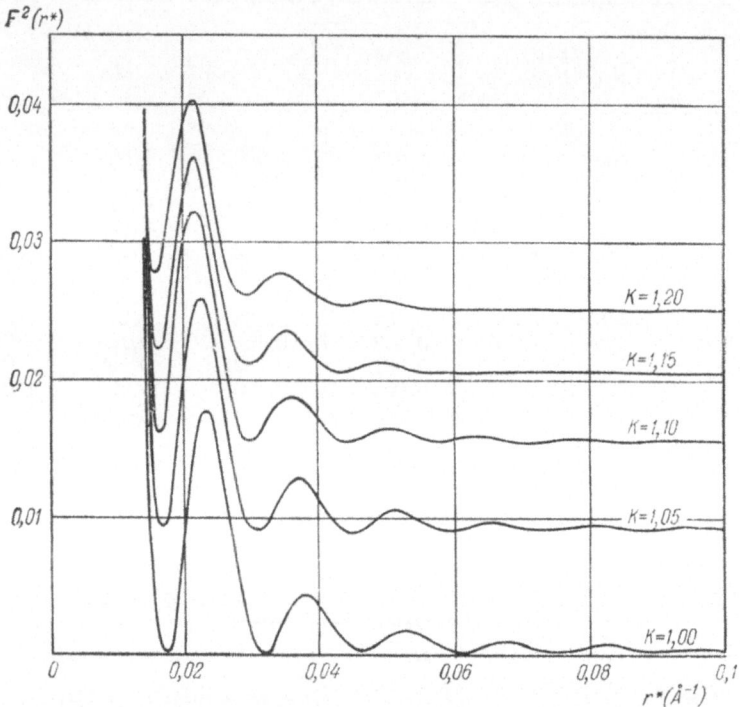

Fig. 178. Scattering function for particles in the form of elliptic cylinders.

Figure 178 shows the scattering function for R = 35 Å and various K.

10-8g. Particles in the Form of Discs

The scattering function for particles in the form of infinitely thin discs of radius R is

$$i(s) = \frac{2}{s^2 R^2} \left[1 - \frac{1}{sR} J_1(2sR) \right], \qquad (106)$$

where $s = 4\pi (\sin \vartheta)/\lambda$ and $J_1(2sR)$ is a Bessel function of order one.
Figure 179 gives i(s) for sR from 0 to 12 [175].
The broken line represents the asymptotic function i(s) = $2/s^2R^2$.

10-8h. Particles in the Form of Rectangular Prisms

The table [175] gives values of the scattering function for particles in the form of rectangular prisms having edges a, 2a, and 2wa for w from 1 to 10 and for various radii of gyration, the last being

$$R = \frac{(5 + 4w^2) a^2}{3} .$$

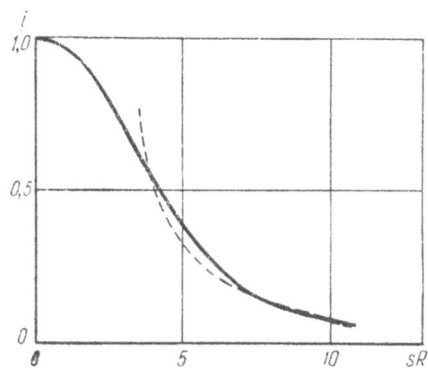

Fig. 179. Scattering function for particles in the form of discs.

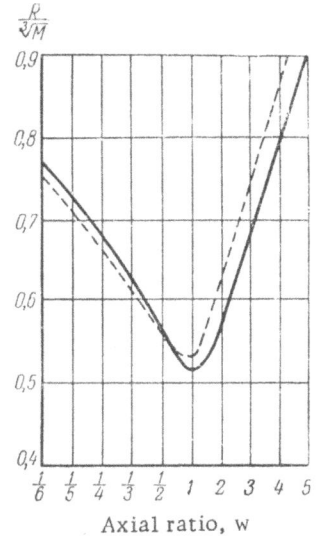

Axial ratio, w

Fig. 180. Graph for deducing radii of gyration.

sR	w						
	1	2	3	4	6	8	10
0,0	1,000	1,000	1,000	1,000	1,000	1,000	1,000
0,5	0,920	0,920	0,920	0,921	0,921	0,921	0,921
1,0	0,716	0,722	0,728	0,735	0,738	0,738	0,738
1,5	0,462	0,491	0,513	0,525	0,536	0,536	0,536
2,0	0,246	0,308	0,350	0,374	0,393	0,399	0,400
2,5	0,118	0,190	0,248	0,280	0,307	0,320	0,323
3,0	0,055	0,119	0,182	0,218	0,249	0,268	0,273
3,5	0,030	0,077	0,130	0,169	0,206	0,223	0,232
4,0		0,050	0,082	0,128	0,171	0,188	0,199
4,5		0,029	0,063	0,099	0,141	0,160	0,173
5,0				0,076	0,117	0,138	0,151

10-9. Graph for Determining Radii of Gyration of Particles

Figure 180 gives curves for determining this radius of gyration R [123]; the abscissa represents the axial ratio w (1.6 to 5.0) and the ordinate $R/\sqrt[3]{M}$, where M is the relative molecular weight. Here R is in angstroms. The full line corresponds to ellipsoids of rotation; the broken one, to cylindrical particles.

For ellipsoids of rotation

$$\frac{R}{\sqrt[3]{M}} = \frac{\sqrt{\dfrac{2+w^2}{5}}}{\sqrt[3]{\dfrac{\delta}{1,65}\dfrac{4}{3}\pi w}},$$

and for cylinders

$$\frac{R}{\sqrt[3]{M}} = \frac{\sqrt{\dfrac{1}{12}\left[6+\left(\dfrac{h}{2}\right)^2\right]}}{\sqrt[3]{\pi\delta\dfrac{h}{2}}}\sqrt[3]{1,65},$$

where h is the height and δ is the density. The graph relates to $\delta = 0.75$.

10-10. Scattering by Inhomogeneous Systems of Particles

10-10a. System of Spherical Particles

The scattering function for a system of hard particles with no interaction is

$$I(s) = I_e(s)\,\overline{N}\Phi^2(sR)\,\frac{1}{1+\dfrac{8v_0}{v_1}\varepsilon\Phi(2sr)}\,, \tag{107}$$

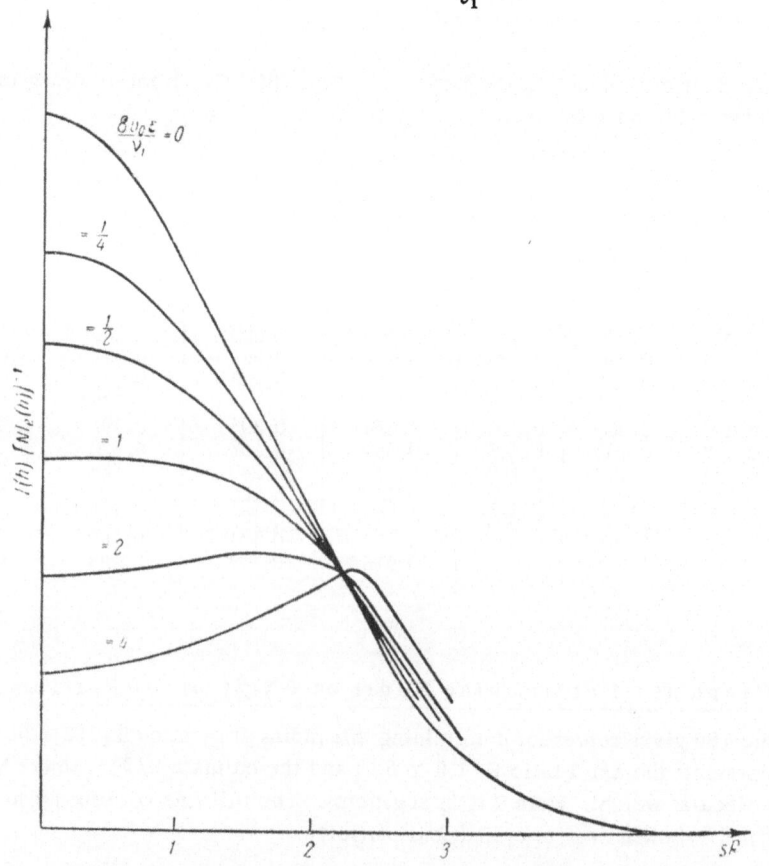

Fig. 181. Scattering function for a system of spherical particles.

where $\Phi(x) = 3\,(\sin x - x \cos x)/x^3$, $\overline{N}$ is the mean number of particles in the irradiated volume, v_1 is the mean volume of a particle ($v_1 = V_0/N_0$, in which V_0 is the total volume of the particles and N_0 is the number of particles), v_0 is the volume of each spherical particle, and ε is a constant ($\varepsilon \approx 1$).

Figure 181 gives scattering curves for $8v_0\varepsilon/v_1$ from 0 to 4 [174].

10-10b. Scattering Function for a System of Spherical Particles Separated by Gaps

The scattering function for a system of spherical particles of diameter d_0 separated each from its neighbor by a gap $\overline{d}$ is

$$I\,(s) = A\left(1 + \frac{\sin s\overline{d}}{s\overline{d}}\right)\Phi\left(\frac{sd_0}{2}\right), \tag{108}$$

where A is a constant.

Figure 182 gives I(s) in reduced form for $d_0/\overline{d}$ from 0.00 to 0.75 [175].

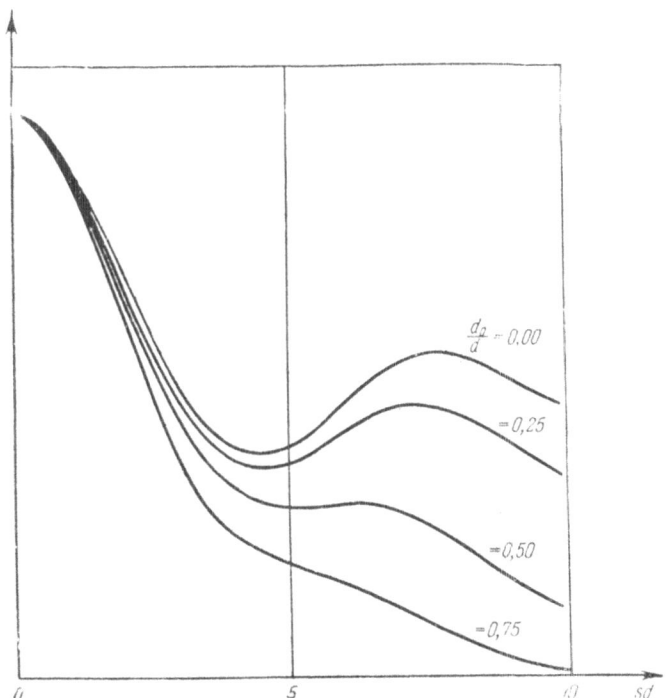

Fig. 182. Scattering function for a system of spherical particles separated by gaps.

10-10c. Scattering Function for a System of Spherical Particles of Various Radii

Figure 183 gives scattering curves for a system consisting of homogeneous spherical particles of radii R and 2R; here x is the ratio of the total mass of the smaller particles to the mass of the system and K is the ratio of the effective volume of the particles to the effective volume of the system.

The full lines correspond to K = 0; the broken lines with long dashes, to K = 0.125; and the broken lines with short dashes, to K = 0.500.

The graphs are for x from 0 to 100% [175].

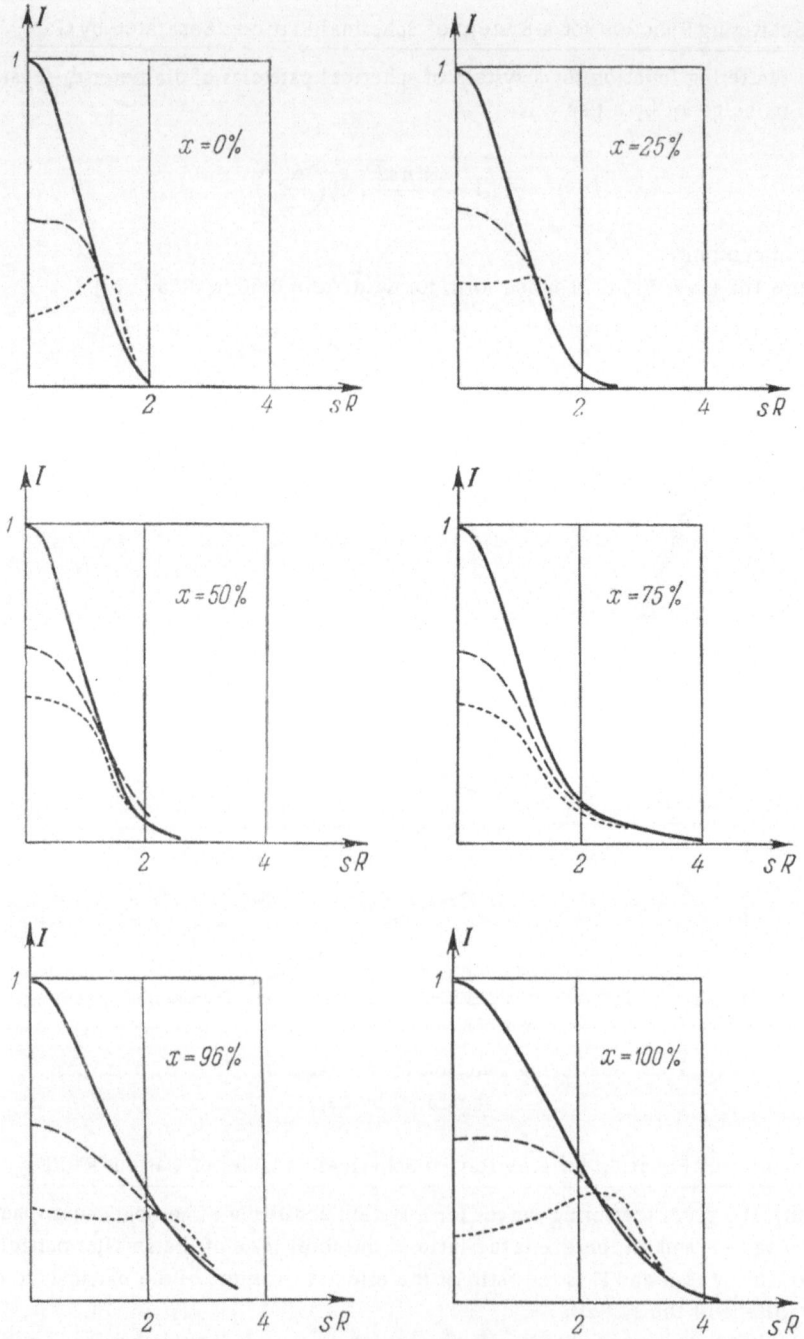

Fig. 183. Scattering function for a system of spherical particles of various radii.

10-10d. System of Particles with a Linear Structure

Figure 184 gives scattering curves for a system of particles with a linear structure [175]; the system consists of segments each of length a separated by gaps which have, for complete disorder, a size distribution given by

$$H(y) = \frac{1}{b} e^{-\frac{y}{b}},$$

where b is the mean gap between segments. The graphs relate to various values of q = (a + b)/a.

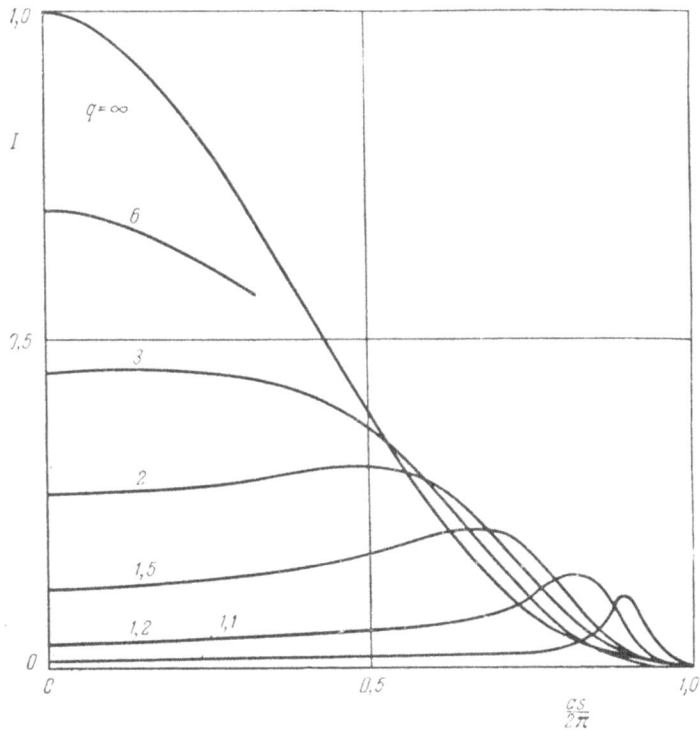

Fig. 184. Scattering curves for a system of particles with
a linear structure.

See [459] for the corresponding expressions for a system of filaments.

10-10e. Short-Range Order in Particle Disposition

An ordered array characterized by a short-range-order parameter P (packing coefficient) gives an intensity

$$I = Mn^2 \Phi^2 (sR) \left\{ 1 + P \left(5 \frac{\sin 2sR}{2sR} - 6\Phi (2sR) \right) \right\} .$$

(109)

Figure 185 gives scattering curves of I/Mn^2 vs. sR for P = 0 (complete disorder, gas scattering), P = 1/3, and P = 1/2 [175].

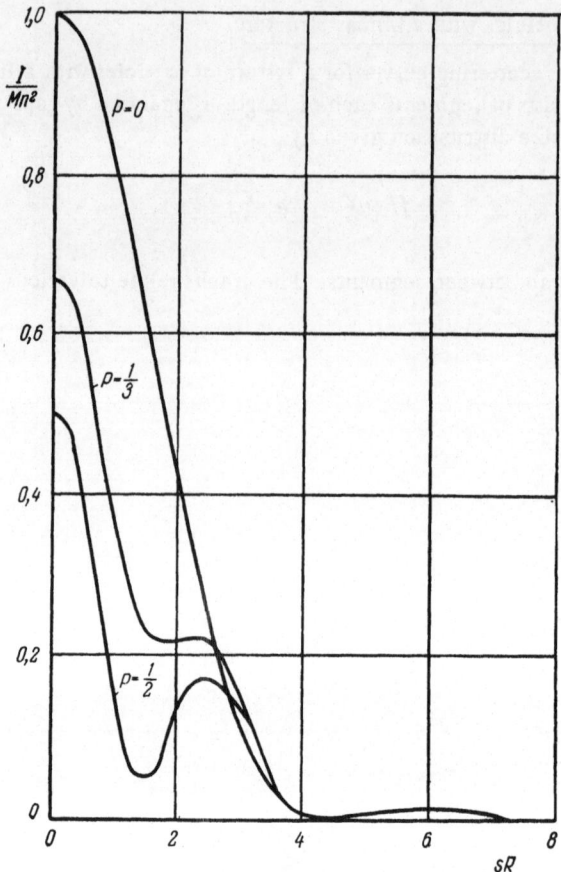

Fig. 185. Scattering function for short-range order in particle disposition.

10-11. Scattering Curves for Various Particle-Size Distributions

A Maxwellian distribution has the number of particles with radii between R and R + dR as

$$m\,(R)\,dR = \frac{2}{R_0^{n+1}\Gamma\left(\dfrac{n+1}{2}\right)}\,R^2 e^{-R^2/R_0^2}\,dR,$$

where $\Gamma\left(\dfrac{n+1}{2}\right)$ is the gamma function; R_0 and n define the mean value of R and the width of the distribution [185].

The scattering function in this case is

$$I\,(s) = k\left(1 + \frac{s^2 R_0^2}{3}\right)^{-\frac{n+4}{2}}.$$

Figures 186-189 give I as a function of $s^2 R^2$ for various n.

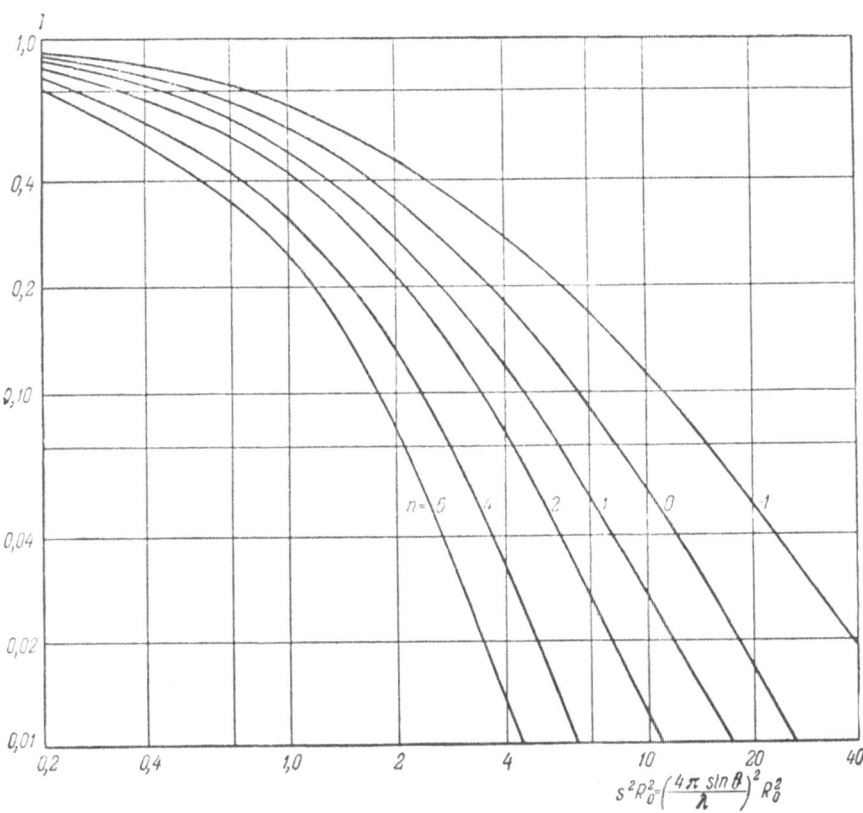

Fig. 186.   Scattering function for particles of arbitrary shape and various size distributions.

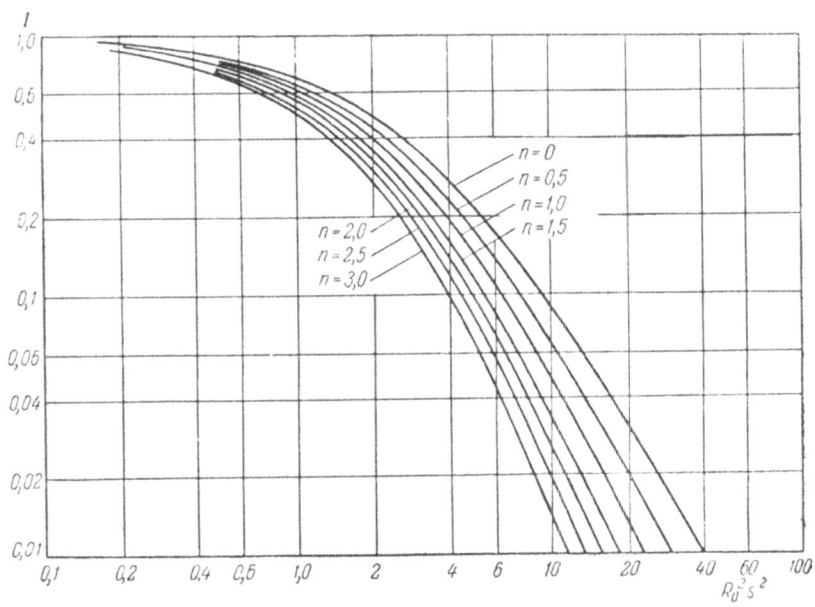

Fig. 187.   Scattering function for spherical particles having various size distributions.

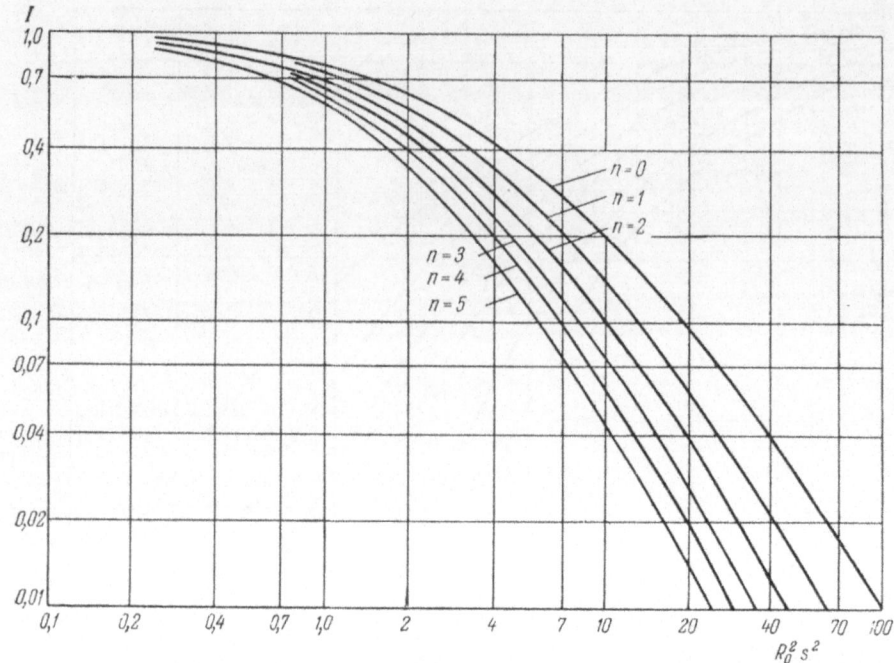

Fig. 188. Scattering function for ellipsoidal particles having various size distributions.

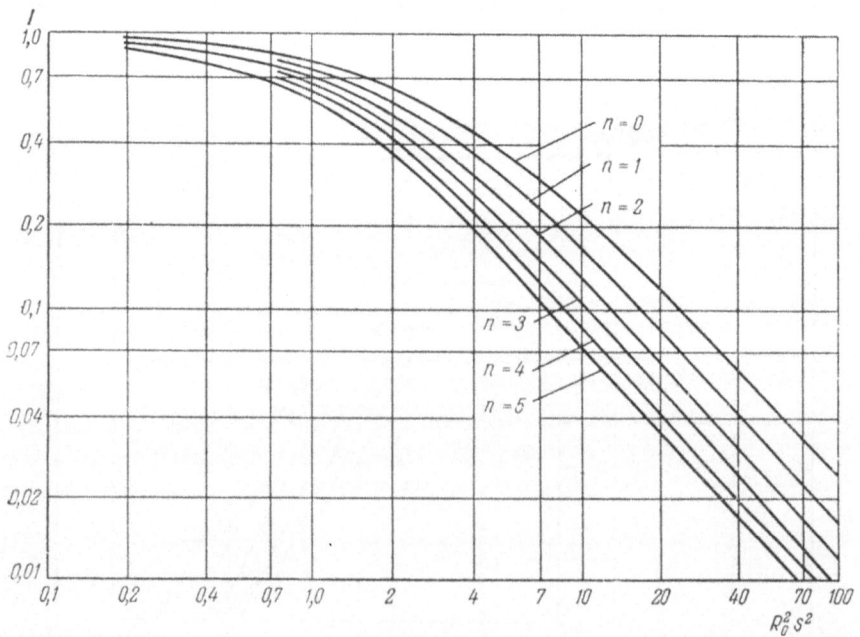

Fig. 189. Scattering function for discoidal particles having various size distributions.

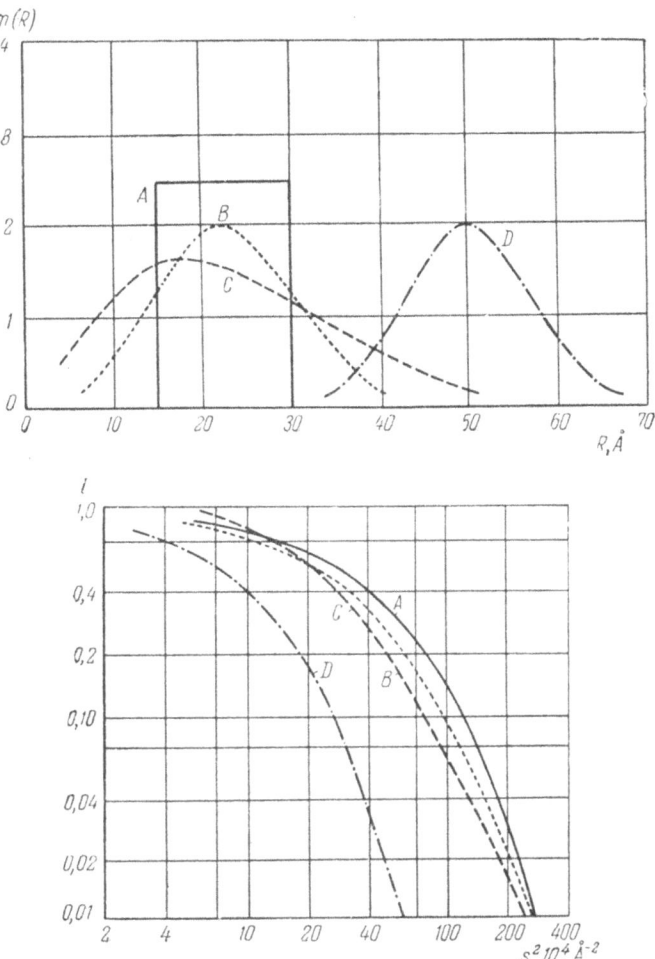

Fig. 190. Scattering functions for particles with rectangular,
Maxwellian, and Gaussian size distributions.

The mean size $\overline{R}$ is as follows:

n	0	1	2	3	4	5
$\dfrac{\overline{R}}{R_0}$	0,227	0,693	1,183	1,667	2,176	2,674

Figure 190 gives curves for various types of distribution: (A) rectangular; (C) Maxwellian; (B) and (D) Gaussian for R_0 of 22 and 50 Å.

10-12. The Collimator Effect

The correction to the scattering function can be calculated for a collimator in the form of an infinitely long slit of small width.

The tables give scattering functions for spherical particles with uniformly distributed charge.

10-12a. Effect of the Collimator on the Scattering Function

The table gives values of $P_h(sa)$ for h from 0 to 1.0 and for $0 \leq 2sa \leq 40$.

2sa \ h	0.0	0.3	0.6	0.8	0.9	1.0
1	0.94765	0.94638	0.93950	0.93109	0.92577	0.91973
2	.80454	.80007	.77653	.74877	.73173	.71289
3	.60720	.59921	.55898	.41479	.4983	.46252
4	.40145	.39143	.34486	.30018	.2775	.25618
5	.22685	.21749	.18052	.15522	.1470	.14306
6	.10554	.099597	.086171	.09247	.1030	.11770
7	.039294	.038506	.054220	.08977	.1130	.13638
8	.014626	.019105	.059126	.10887	.1327	.15134
9	.012198	.020532	.072485	.11767	.1324	.13914
10	.015636	.025551	.075443	.10396	.1080	.10670
11	.016202	.025433	.063401	.075263	.07525	.076345
12	.012575	.019550	.042758	.047896	.05303	.064510
13	.0074683	.011595	.023431	.034084	.04919	.070340
14	.0038323	.005376	.012105	.034976	.05743	.080048
15	.0027056	.002435	.009457	.042439	.06468	.080344
16	.0032057	.002010	.011626	.046415	.06188	.068804
17	.0038012	.002422	.013790	.041912	.14970	.053680
18	.0035675	.002489	.013161	.031157	.03612	.045184
19	.0025986	.002050	.009849	.020494	.02913	.046681
20	.0015764	.001587	.005746	.015180	.03066	.052919
21	.0010664	.001529	.002731	.015973	.03605	.055667
22	.0011189	.001845	.001498	.019474	.03861	.051018
23	.0013719	.002154	.001538	.021256	.03501	.042069
24	.0014346	.002101	.001881	.019098	.02731	.035357
25	.0011901	.001650	.001874	.014082	.02063	.034841
26	.00081110	.001064	.001460	.009196	.01869	.038810
27	.00054911	.0006608	.0009546	.006872	.02115	.041990
28	.00051488	.0005699	.0006470	.007384	.02438	.040464
29	.00062125	.0006827	.0005864	.009057	.02466	.034897
30	.00069560	.0007845	.0006349	.009802	.02115	.029475
31	.00063557	.0007361	.0006555	.008654	.01615	.027918
32	.00047552	.0005570	.0006288	.006223	.01292	.030285
33	.00032971	.0003684	.0006214	.003958	.01302	.033260
34	.00028259	.0002730	.0006783	.002952	.01523	.033346
35	.00032503	.0002799	.0007624	.003242	.01687	.029931
36	.00037892	.0003209	.0007928	.003990	.01605	.025574
37	.00037372	.0003284	.0007337	.004246	.01307	.023493
38	.00030297	.0002914	.0006356	.003643	.009947	.024681
39	.00021811	.0002530	.0005924	.002526	.008638	.027209
40	.00017560	.0002573	.0006545	.001554	.009475	.028144

10-12b. Effect of the Collimator on the Turning Points in the Scattering Function

The table gives the values at the maxima and minima of $P_h(sa)$ for h from 0 to 1.0. The true radius of gyration is

$$R' = kR$$

for particles having the form of ellipsoids of rotation; here R is the radius as deduced from the scattering curves and k is a factor dependent on the axial ratio w.

Figure 191 gives k(w) for w from 0.1 to 20.

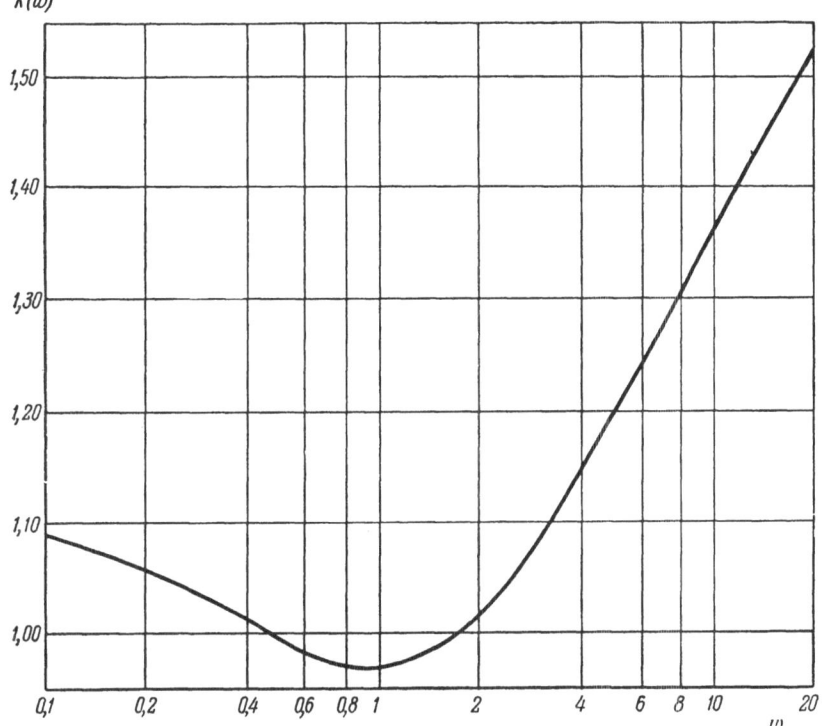

Fig. 191. Correction factor for determining the true radius of gyration for particles in the form of ellipsoids.

h	$2ha$	$P_h (sa)$	h	$2ha$	$P_h (sa)$	h	$2ha$	$P_h (sa)$
0,0	8,6506	0,011680	0,6	7,2575	0,053167	0,9	6,190	0,1359
	10,624	.016541		9,6964	.076189		8,493	.04856
	15,037	.0027045		14,882	.009421		12,71	.06508
	17,256	.0038332		17,306	.013941		15,26	.02880
	21,350	.0010339		22,400	.001416		19,28	.03863
	23,704	.0014494		24,490	.001938		21,94	.01864
	27,641	.00050296		28,819	.0005842		25,86	.02504
	30,086	.00069615		30,820	.0006535		28,58	.01257
	33,928	.00028233		32,658	.0006174		32,45	.01693
	36,436	.00038602		35,841	.0007941		35,20	.008635
				38,928	.0005921			
0,3	8,3237	.018262	0,8	6,5241	.08589	1,0	5,8843	.11738
	10,483	.026329		8,9052	.11778		8,0839	.15145
	15,736	.001972		13,398	.032989		12,076	.064458
	17,612	.002536		15,994	.046415		14,530	.081839
	20,616	.001494		20,322	.014871		18,313	.044683
	23,388	.002187		22,976	.021257		20,887	.055718
	27,808	.0005655		27,263	.006775		24,567	.034284
	30,219	.0007887		29,912	.009810		27,214	.042113
	34,380	.0002661		34,206	.002925		30,829	.027854
	36,652	.0003317		36,807	.004263		33,528	.033792
	39,411	.0002483					37,097	.023476
							39,833	.028188

10-12c. Choice of Collimator

The relation for a collimator consisting of three slits O_1, O_2, and O_3 is

$$\frac{\overline{p+r}}{v} < A, \quad w = \frac{2r}{B-A}, \quad q = r\left[1 + \frac{2pr}{v(B-A)}\right], \\ s = \frac{r}{B-A}\frac{2(p+r)+v(B-A)}{Av-(p+r)}, \tag{110}$$

where p is the width of O_1, r is the width of O_2, v is the distance O_1O_2, q is the width of O_3, w is the distance O_2O_3, A (= a/s) is the distance from specimen to film or counter, a is the width of the primary beam in the plane of the film or counter slit, and B (= b/s) is the width of a reflection at the film in the absence of the specimen.

Figure 192 gives graphs for use in selecting collimator dimensions [175].

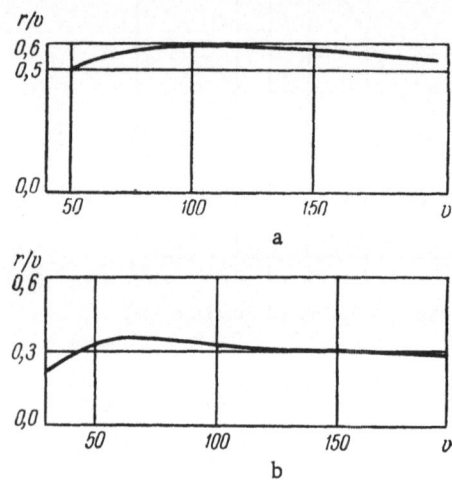

a

b

Fig. 192. Graphs for selecting collimator dimensions: (a) for ionization recording; (b) for photographic recording.

Example. The focus is 1 mm wide, so an angle of divergence of 1° in the primary beam implies that p = 20 μ; A is then 10^{-3}, which means that for Cu Kα radiation (λ = 1.54 Å) one can distinguish lines with d up to 1540 Å in the direction of the primary beam. To make such reflections accessible, parasitic scattering must be kept outside the angle corresponding to d $\approx$ 1200 Å, which implies that B = $2.5 \cdot 10^{-3}$.

The curve of Fig. 192a is used to choose the other parameters for ionization recording with a maximum s of 500 mm. The maximum on the curve corresponds to v = 100 mm, so r = 58 μ; then (110) gives q = 120 μ and w = 77 mm.

For photographic recording we assume a minimal s of 100 mm and a microphotometer slit of 20 μ; the trace of the primary beam on the film is then 5 times the width of the microphotometer slit. Figure 192b shows that the maximum corresponds to v = 70 mm, so r = 24 μ, w = 31 mm, and q = 47 mm.

A graphical method [369] or analytic methods [370, 371, 375, 376] may be used to determine the radial distribution of the intensity for small-angle scattering.

CHAPTER 11

ELECTRON DIFFRACTION

This chapter gives tables for calculations on electron-diffraction patterns.
See [210, 211] for details of methods of producing and using the patterns.
See [320, 321, 392] for the use of electron diffraction in various special problems.

11-1. Some Formulas for Electron Diffraction

1. The wavelength of an electron is

$$\lambda = \frac{1}{\sqrt{1+9.76 \cdot 10^{-7}V}} \frac{12.225}{\sqrt{V}} \mathring{A}, \tag{111}$$

where V is the accelerating potential (volts).

2. The reflections on the pattern are given by the approximate relation

$$2rd = 2L\lambda, \tag{112}$$

where 2r is the distance between symmetrically placed reflections, d is the interplanar distance, L is the distance from specimen to plate, and λ is the wavelength.

The exact relation is

$$rd = L\lambda \cdot \left(\frac{\text{tg } 2\vartheta}{2 \sin \vartheta} \right). \tag{113}$$

3. Unit-cell parameters are deduced as follows.

If all three axes of the reciprocal unit cell appear on the pattern, the constants of the reciprocal lattice are

$$a^* = \frac{2r_{h00}}{2L\lambda h}, \quad b^* = \frac{2r_{0k0}}{2L\lambda k}, \quad c^* = \frac{2r_{00l}}{2L\lambda l}. \tag{114}$$

If the third axis is lacking (falls in the dead zone), we use

$$c^* = \frac{\sqrt{4r_{hkl}^2 - 4r_{hk0}^2}}{2L\lambda l}. \tag{115}$$

For crystals with oblique cells

$$c^* = \frac{\sqrt{4r_{hkl}^2 + 4r_{h0l}^2 - 8r_{hk0}^2}}{2} \Big/ 2L\lambda l. \tag{116}$$

4. The scattered intensity produced by a unit cell is defined by the structure factor

$$\Phi_{hkl} = \sum_i f_{e,T} \, e^{2\pi i(2rH)}, \tag{117}$$

where $f_{e,T}$ is the product of the atomic and temperature factors; the latter factor, $f_T = \exp[-B(\sin^2\vartheta/\lambda^2)]$, is usually small.

5. The integral intensity of a point reflection from a mosaic monocrystalline film consisting of small blocks is

$$I_{hkl} = I_0\lambda^2 \left| \frac{\Phi_{hkl}}{\Omega} \right|^2 V' \frac{d_{hkl}}{\alpha}, \tag{118}$$

where I_0 is the intensity of the incident beam, $\Omega = a_1a_2a_3$ is the volume of the unit cell, V' is the volume of the mosaic film exposed to the beam, and α is the effective angular spread of the blocks.

6. The intensity from an ideal monocrystal is

$$I_{hkl} = I_0 S\lambda^2 \left| \frac{\Phi_{hkl}}{\Omega} \right|^2 A^2, \tag{119}$$

where S is the irradiated area and A is the thickness.

7. The dynamic theory for a mosaic monocrystal gives

$$I_{hkl} = I_0 S \cdot \frac{1}{2} Q \frac{d_{hkl}}{\alpha}, \tag{120}$$

where

$$Q = \lambda \left| \frac{\Phi}{\Omega} \right|.$$

8. The integral ring intensity given by a polycrystal is

$$I_{hkl} = I_0\lambda^2 \left| \frac{\Phi_{hkl}}{\Omega} \right|^2 V' \frac{d_{hkl}}{2} p, \tag{121}$$

where V' is the irradiated volume and p is the multiplicity factor for the (hkl) plane.

A section of length Δ in a ring has an intensity

$$I'_{hkl} = I_0\lambda^2 \left| \frac{\Phi_{hkl}}{\Omega} \right|^2 V' \frac{d^2_{hkl}\Delta p}{4\pi L\lambda}. \tag{122}$$

9. A specimen having crystallites of various sizes gives

$$I_{hkl} = I_0 \left(Q^2 V_k + \frac{1}{2} Q S_d \right) L, \tag{123}$$

where L is the Lorentz factor, V_k is the irradiated volume of the specimen containing crystallites of thickness less than A (which scatter kinematically), and S_d is the arc of the dynamically scattering crystallites (ones of thickness greater than A).

10. A textured material gives an integral intensity of

$$I_{hkl} = I_0 SQ^2 t \frac{L\lambda p}{2\pi R' \sin \varphi} \, , \tag{124}$$

where t is the thickness of the film, R' is the horizontal coordinate of the reflection on the pattern, φ is the angle of the specimen with respect to the normal to the primary beam, and p is the multiplicity factor for patterns from textures (this differs from p for polycrystals).

An arc of a ring given by a textured material has an intensity

$$I'_{hkl} = I_0 StQ^2 \frac{d_{hk0}d_{hkl}\Delta}{2\pi L\lambda\alpha} \, , \tag{125}$$

where α is the effective angular spread of the texture axis, Δ is the length of the arc, and d_{hk0} is the interplanar distance for a reflection lying on the zero layer line and on the same ellipse as the (hkl) reflection. There is no (hk0) reflection for an oblique-angled structure, so d_{hk0} is replaced by $L\lambda/R$, where R is the length of the semiminor axis of the ellipse.

See [461] for relationships used in interpreting patterns from specimens with complex textures.

11. The width Δr of a reflection is related to crystal size D by

$$\Delta r = \frac{\lambda}{D} L$$

where L is the distance from specimen to plate.

11-2. Relation of Electron Wavelength to Applied Voltage

An approximate formula for the wavelength is

$$\lambda = \frac{12.225}{\sqrt{V}} \, \text{Å}, \tag{126}$$

where V (the accelerating potential) is in volts; equation (111) is the exact formula.

The table gives values calculated from (111) [210].

V, kV	λ, Å	V, kV	λ, Å	V, kV	λ, Å
17.5	0.0921	37.5	0.0621	60.0	0.0448
20.0	0.0857	40.0	0.0599	65.0	0.0466
22.5	0.0815	42.5	0.0581	70.0	0.0447
25.0	0.0764	45.0	0.0564	75.0	0.0432
27.5	0.0728	47.5	0.0548	80.0	0.0417
30.0	0.0695	50.0	0.0534	85.0	0.0403
32.5	0.0668	52.5	0.0521	90.0	0.0391
35.0	0.0643	55.0	0.0512	100.0	0.0361

11-3. Δ Correction in Precision Interplanar Distance Measurements

The r of (113) should be replaced by $\bar{r} = r(1 - \Delta)$ in precision measurements; the table gives Δ as a function of r/L [211].

$\frac{r}{L}$	Δ	$\frac{r}{L}$	Δ
0.01	0.000037	0.06	0.001345
0.02	0.000150	0.07	0.001827
0.03	0.000337	0.08	0.002390
0.04	0.000600	0.09	0.003020
0.05	0.000935	0.10	0.003726

11-4. Universal Atomic-Scattering Function for Electrons

The universal atomic-scattering function $\Phi_e(\xi)$ is used to calculate the atomic scattering factor for electrons for elements of medium and high atomic numbers:

$$\Phi_e\,(\xi) = \frac{1}{\xi} \int_0^\infty \psi(x) \sin \xi x \, dx, \tag{127}$$

where $\psi(x)$ is the screening function for the inner electron shells.

A special correction is made if dynamic scattering must be considered [212, 374].

ξ	$\Phi(\xi)$	ξ	$\Phi(\xi)$	ξ	$\Phi(\xi)$
0	1.600	0.20	0.146	0.40	0.0510
0.02	1.330	0.22	0.126	0.42	0.0467
0.04	0.920	0.24	0.110	0.44	0.0427
0.06	0.670	0.26	0.097	0.46	0.0394
0.08	0.510	0.28	0.088	0.48	0.0366
0.10	0.400	0.30	0.079	0.50	0.0340
0.12	0.320	0.32	0.072	0.55	0.0286
0.14	0.255	0.34	0.066	0.60	0.0247
0.16	0.206	0.36	0.0605	0.65	0.0217
0.18	0.171	0.38	0.0555	0.70	0.0194

11-5. Atomic Scattering Factors for Electrons

11-5a. Scattering by Light Atoms

The atomic scattering factors for electrons and light atoms are

$$f_e\,(s) = 4\pi k \left|_0^\infty \varphi(r)\, r^2 \frac{\sin(sr)}{sr}\, dr, \tag{128}\right.$$

where $k = 2\pi m e/h^2$, $\varphi(r)$ is the radial distribution of potential in the atom, $s = 4\pi(\sin \vartheta)/\lambda$, and r is the radius.

The table gives f_e for Z from 1 to 18.

Element	z	d, Å	10.00	5.000	3.333	1.500	2.000	1.667
			\multicolumn{6}{c}{$\frac{\sin \vartheta}{\lambda}$, 10^8}
		0	0.05	0.10	0.15	0.20	0.25	0.3
H^+	1		4.000	1.000	0.444	0.250	0.160	0.110
H	1	0.221	0.210	0.190	0.160	0.130	0.106	0.084
He	2	0.154	0.150	0.144	0.135	0.123	0.110	0.097
Li	3	0.60	1.260	0.770	0.460	0.310	0.218	0.165
Be	4	1.30	1.136	0.920	0.670	0.480	0.350	0.262
B	5	1.12	1.012	0.890	0.745	0.573	0.440	0.335
C	6	0.95	0.874	0.795	0.700	0.595	0.485	0.392
N	7	0.79	0.745	0.698	0.638	0.572	0.498	0.415
O	8	0.68	0.660	0.633	0.594	0.545	0.490	0.425
F	9	0.58	0.575	0.562	0.542	0.516	0.483	0.432
Ne	10	0.53	0.528	0.514	0.496	0.470	0.440	0.404
Na	11	4.0	2.6	1.35	0.92	0.70	0.56	0.478
Mg	12	3.7	2.9	1.65	1.12	0.84	0.65	0.528
Al	13	3.45	3.0	2.00	1.35	1.01	0.75	0.583
Si	14	3.2	2.95	2.40	1.58	1.12	0.83	0.645
P	15	3.0	2.8	2.45	1.80	1.24	0.92	0.728
S	16	2.8	2.65	2.40	1.90	1.32	1.00	0.794
Cl	17	2.65	2.50	2.24	1.84	1.43	1.11	0.861
Ar	18	2.5	2.35	2.10	1.72	1.35	1.04	0.843

Element	z	d, Å	1.250	1.000	0.834	0.715	0.625	0.556	0.500	0.454
			\multicolumn{8}{c}{$\frac{\sin \vartheta}{\lambda}$, 10^8}
		0.4	0.5	0.6	0.7	0.8	0.9	1.0	1.1
H^+	1	0.062	0.040	0.028	0.020	0.015	0.012	0.010	0.008
H	1	0.055	0.036	0.027	0.020	0.015	0.012	0.010	0.008
He	2	0.075	0.058	0.045	0.036	0.029	0.023	0.019	0.015
Li	3	0.106	0.080	0.061	0.048	0.039	0.032	0.027	0.022
Be	4	0.151	0.104	0.078	0.062	0.048	0.041	0.035	0.028
B	5	0.206	0.140	0.100	0.077	0.059	0.050	0.042	0.035
C	6	0.261	0.180	0.124	0.094	0.073	0.059	0.050	0.042
N	7	0.292	0.203	0.149	0.111	0.086	0.070	0.057	0.048
O	8	0.316	0.232	0.172	0.128	0.100	0.081	0.066	0.055
F	9	0.330	0.252	0.193	0.144	0.113	0.092	0.075	0.063
Ne	10	0.325	0.262	0.204	0.160	0.125	0.103	0.085	0.070
Na	11	0.359	0.278	0.217	0.170	0.137	0.112	0.092	0.077
Mg	12	0.378	0.288	0.227	0.181	0.148	0.122	0.100	0.084
Al	13	0.400	0.300	0.236	0.190	0.157	0.129	0.107	0.091
Si	14	0.428	0.316	0.247	0.200	0.166	0.137	0.114	0.097
P	15	0.472	0.341	0.260	0.209	0.172	0.143	0.120	0.102
S	16	0.509	0.366	0.278	0.219	0.180	0.149	0.126	0.108
Cl	17	0.559	0.390	0.292	0.230	0.187	0.156	0.132	0.113
Ar	18	0.582	0.408	0.305	0.241	0.195	0.164	0.139	0.119

11-5b. Scattering by Atoms of Medium and High Atomic Numbers

The scattering factors for Z from 19 to 104 are derived by constructing curves for $\Phi(\xi)$ from Section 11-4 with the radius vector increased by a factor $Z^{1/3}$, which corresponds to

$$f_e = kZ^{1/3}\Phi_e \ (\xi), \tag{129}$$

$(\sin \vartheta/\lambda)\cdot 10^{-8}$

Element	z	1.50	1.40	1.30	1.20	1.10	1.00	0.90	0.80	0.70	0.60	0.50	0.40	0.35	0.30	0.25	0.20	0.15	0.10	0.05	0.00
K	19	—	—	—	—	0.128	0.151	0.178	0.211	0.256	0.325	0.430	0.610	—	0.930	—	1.45	—	2.56	—	4.27
Ca	20	0.18	0.20	0.23	0.27	0.31	0.37	0.44	0.53	0.65	0.82	1.07	1.45	1.72	2.06	2.52	3.13	3.85	4.57	5.08	5.4
Sc	21	0.19	0.21	0.24	0.28	0.32	0.38	0.45	0.55	0.68	0.86	1.12	1.51	1.78	2.14	2.61	3.24	3.98	4.72	5.27	5.6
Ti	22	0.20	0.22	0.25	0.29	0.34	0.40	0.47	0.57	0.71	0.89	1.16	1.57	1.85	2.21	2.70	3.35	4.12	4.88	5.46	5.8
V	23	0.21	0.23	0.26	0.30	0.35	0.41	0.49	0.60	0.74	0.93	1.20	1.62	1.91	2.29	2.79	3.45	4.24	5.03	5.65	5.9
Cr	24	0.21	0.24	0.27	0.32	0.37	0.43	0.51	0.62	0.76	0.96	1.25	1.68	1.98	2.36	2.88	3.56	4.37	5.17	5.84	6.1
Mn	25	0.22	0.25	0.29	0.33	0.38	0.45	0.53	0.64	0.79	0.99	1.29	1.73	2.04	2.43	2.97	3.66	4.49	5.34	5.93	6.2
Fe	26	0.22	0.25	0.30	0.34	0.39	0.46	0.55	0.66	0.82	1.03	1.33	1.79	2.10	2.51	3.05	3.76	4.62	5.48	6.13	6.4
Co	27	0.23	0.26	0.31	0.35	0.41	0.48	0.57	0.69	0.84	1.06	1.37	1.84	2.16	2.58	3.14	3.87	4.73	5.62	6.32	6.5
Ni	28	0.24	0.27	0.32	0.36	0.42	0.49	0.59	0.71	0.87	1.09	1.41	1.89	2.23	2.65	3.22	3.97	4.85	5.74	6.41	6.7
Cu	29	0.25	0.28	0.33	0.38	0.43	0.51	0.60	0.73	0.90	1.13	1.45	1.95	2.29	2.72	3.30	4.06	4.97	5.89	6.61	6.8
Zn	30	0.25	0.29	0.34	0.39	0.45	0.52	0.62	0.75	0.92	1.16	1.49	2.00	2.35	2.79	3.38	4.16	5.08	6.03	6.70	7.0
Ga	31	0.26	0.30	0.35	0.40	0.46	0.54	0.64	0.77	0.95	1.19	1.53	2.05	2.41	2.86	3.46	4.25	5.20	6.15	6.89	7.2
Ge	32	0.27	0.31	0.36	0.41	0.47	0.56	0.66	0.79	0.97	1.22	1.57	2.10	2.46	2.93	3.54	4.35	5.32	6.29	7.09	7.3
As	33	0.28	0.31	0.37	0.42	0.49	0.57	0.68	0.82	1.00	1.25	1.61	2.15	2.52	2.99	3.62	4.44	5.43	6.41	7.18	7.5
Se	34	0.29	0.32	0.38	0.43	0.50	0.59	0.70	0.84	1.02	1.28	1.65	2.20	2.58	3.06	3.70	4.54	5.53	6.56	7.37	7.6
Br	35	0.29	0.33	0.39	0.44	0.51	0.60	0.71	0.86	1.05	1.32	1.69	2.25	2.64	3.13	3.78	4.63	5.63	6.68	7.47	7.8
Kr	36	0.30	0.34	0.40	0.46	0.53	0.62	0.73	0.88	1.08	1.35	1.73	2.31	2.69	3.19	3.85	4.71	5.74	6.80	7.56	7.9
Rb	37	0.31	0.35	0.41	0.47	0.54	0.63	0.75	0.90	1.10	1.38	1.77	2.35	2.75	3.26	3.93	4.80	5.85	6.92	7.75	8.0
Sr	38	0.32	0.36	0.42	0.48	0.55	0.65	0.77	0.92	1.13	1.41	1.80	2.40	2.80	3.32	4.00	4.89	5.96	7.04	7.85	8.2
Y	39	0.33	0.37	0.43	0.49	0.57	0.66	0.78	0.94	1.15	1.44	1.84	2.45	2.86	3.38	4.07	4.98	6.06	7.16	8.04	8.3
Zr	40	0.33	0.38	0.44	0.50	0.58	0.68	0.80	0.96	1.17	1.47	1.88	2.50	2.91	3.45	4.15	5.06	6.16	7.28	8.14	8.5
Nb	41	0.34	0.39	0.45	0.51	0.59	0.69	0.82	0.98	1.20	1.50	1.92	2.54	2.97	3.51	4.22	5.15	6.27	7.40	8.23	8.6
Mo	42	0.35	0.40	0.46	0.52	0.60	0.71	0.84	1.00	1.22	1.53	1.95	2.59	3.02	3.57	4.29	5.24	6.36	7.52	8.42	8.7
Tc	43	0.36	0.41	0.47	0.53	0.62	0.72	0.85	1.02	1.25	1.56	1.99	2.64	3.08	3.63	4.36	5.31	6.47	7.63	8.52	8.9
Ru	44	0.37	0.42	0.48	0.55	0.63	0.74	0.87	1.04	1.27	1.58	2.03	2.68	3.13	3.69	4.43	5.40	6.56	7.75	8.62	9.0
Rh	45	0.38	0.43	0.49	0.56	0.64	0.75	0.89	1.06	1.30	1.61	2.06	2.73	3.18	3.75	4.50	5.48	6.66	7.85	8.81	9.1
Pd	46	0.39	0.44	0.50	0.57	0.66	0.77	0.90	1.08	1.32	1.64	2.10	2.77	3.23	3.81	4.57	5.56	6.75	7.97	8.90	9.3
Ag	47	0.40	0.45	0.51	0.58	0.67	0.78	0.92	1.10	1.34	1.67	2.13	2.82	3.28	3.87	4.64	5.64	6.85	8.07	9.00	9.4

$(\sin \vartheta / \lambda) \cdot 10^{-8}$

Element	Z	0,00	0,05	0,10	0,15	0,20	0,25	0,30	0,35	0,40	0,50	0,60	0,70	0,80	0,90	1,00	1,10	1,20	1,30	1,40	1,50
Cd	48	9,5	9,19	8,19	6,95	5,72	4,71	3,93	3,34	2,86	2,17	1,71	1,37	1,12	0,94	0,79	0,68	0,59	0,52	0,46	0,40
In	49	9,6	9,29	8,31	7,03	5,80	4,78	3,99	3,39	2,91	2,20	1,73	1,39	1,14	0,95	0,81	0,69	0,60	0,53	0,46	0,41
Sn	50	9,8	9,38	8,40	7,13	5,88	4,84	4,05	3,44	2,95	2,24	1,76	1,41	1,16	0,97	0,82	0,71	0,61	0,54	0,47	0,42
Sb	51	9,9	9,48	8,50	7,22	5,95	4,91	4,10	3,49	3,00	2,27	1,79	1,44	1,18	0,99	0,84	0,72	0,62	0,55	0,48	0,43
Te	52	10,0	9,57	8,62	7,31	6,03	4,97	4,16	3,54	3,04	2,31	1,81	1,46	1,20	1,00	0,85	0,73	0,63	0,55	0,49	0,44
I	53	10,1	9,77	8,71	7,39	6,11	5,04	4,22	3,59	3,08	2,34	1,84	1,48	1,22	1,02	0,87	0,74	0,64	0,56	0,50	0,44
Xe	54	10,2	9,86	8,81	7,49	6,19	5,10	4,27	3,64	3,13	2,38	1,87	1,51	1,24	1,04	0,88	0,76	0,66	0,57	0,51	0,45
Cs	55	10,4	9,96	8,93	7,57	6,26	5,17	4,33	3,68	3,17	2,41	1,90	1,53	1,26	1,05	0,89	0,77	0,67	0,58	0,52	0,46
Ba	56	10,5	10,05	9,02	7,66	6,34	5,23	4,39	3,73	3,21	2,45	1,93	1,55	1,28	1,07	0,91	0,78	0,68	0,59	0,52	0,47
La	57	10,6	10,15	9,12	7,75	6,40	5,30	4,44	3,78	3,26	2,48	1,95	1,57	1,30	1,09	0,92	0,79	0,69	0,60	0,53	0,47
Ce	58	10,7	10,24	9,21	7,84	6,49	5,36	4,50	3,83	3,30	2,51	1,98	1,60	1,32	1,10	0,94	0,80	0,70	0,61	0,54	0,48
Pr	59	10,8	10,44	9,31	7,92	6,56	5,42	4,55	3,88	3,34	2,55	2,01	1,62	1,33	1,12	0,95	0,82	0,71	0,62	0,55	0,49
Nd	60	10,9	10,53	9,41	8,01	6,63	5,48	4,60	3,93	3,38	2,58	2,03	1,64	1,35	1,13	0,96	0,83	0,72	0,63	0,56	0,50
Pm	61	11,0	10,63	9,53	8,10	6,70	5,55	4,66	3,97	3,43	2,61	2,06	1,66	1,37	1,15	0,98	0,84	0,73	0,64	0,57	0,50
Sm	62	11,1	10,72	9,62	8,17	6,77	5,61	4,71	4,02	3,47	2,65	2,09	1,69	1,39	1,17	0,99	0,85	0,74	0,65	0,57	0,51
Eu	63	11,2	10,82	9,72	8,25	6,85	5,67	4,77	4,07	3,51	2,68	2,11	1,71	1,41	1,18	1,00	0,86	0,75	0,66	0,58	0,52
Gd	64	11,4	10,92	9,79	8,34	6,91	5,73	4,82	4,11	3,55	2,71	2,14	1,73	1,43	1,20	1,02	0,88	0,76	0,67	0,59	0,53
Tb	65	11,5	11,01	9,88	8,42	6,98	5,79	4,87	4,16	3,59	2,74	2,17	1,75	1,45	1,21	1,03	0,89	0,77	0,68	0,60	0,53
Dy	66	11,6	11,11	9,98	8,50	7,05	5,85	4,92	4,20	3,63	2,78	2,19	1,77	1,47	1,23	1,05	0,90	0,78	0,69	0,61	0,54
Ho	67	11,7	11,20	10,08	8,58	7,12	5,91	4,98	4,25	3,67	2,81	2,22	1,80	1,48	1,25	1,06	0,91	0,79	0,70	0,61	0,55
Er	68	11,8	11,30	10,17	8,66	7,19	5,97	5,03	4,30	3,71	2,84	2,25	1,82	1,50	1,26	1,07	0,92	0,80	0,70	0,62	0,56
Tu	69	11,9	11,49	10,27	8,74	7,26	6,03	5,08	4,34	3,75	2,87	2,27	1,84	1,52	1,28	1,09	0,94	0,81	0,71	0,63	0,56
Yb	70	12,0	11,59	10,36	8,82	7,33	6,09	5,13	4,39	3,79	2,91	2,30	1,86	1,54	1,29	1,10	0,95	0,82	0,72	0,64	0,57
Lu	71	12,1	11,68	10,44	8,90	7,40	6,15	5,18	4,43	3,83	2,94	2,32	1,88	1,56	1,31	1,11	0,96	0,83	0,73	0,65	0,58
Hf	72	12,2	11,78	10,53	8,98	7,46	6,20	5,23	4,48	3,87	2,97	2,35	1,90	1,58	1,32	1,13	0,97	0,84	0,74	0,66	0,58
Ta	73	12,3	11,87	10,63	9,05	7,53	6,26	5,28	4,52	3,91	3,00	2,38	1,93	1,59	1,34	1,14	0,98	0,85	0,75	0,66	0,59
W	74	12,4	11,97	10,72	9,13	7,59	6,32	5,33	4,56	3,95	3,03	2,40	1,95	1,61	1,35	1,15	0,99	0,86	0,76	0,67	0,60
Re	75	12,5	12,06	10,79	9,21	7,66	6,38	5,38	4,61	3,99	3,06	2,43	1,97	1,63	1,37	1,17	1,01	0,87	0,77	0,68	0,61
Os	76	12,6	12,16	10,89	9,29	7,72	6,43	5,43	4,65	4,03	3,09	2,45	1,99	1,65	1,38	1,18	1,02	0,89	0,78	0,69	0,61

$(\sin \vartheta/\lambda)\cdot 10^{-8}$

Element	Z	0,00	0,05	0,10	0,15	0,20	0,25	0,30	0,35	0,40	0,50	0,60	0,70	0,80	0,90	1,00	1,10	1,20	1,30	1,40	1,50
Ir	77	12,7	12,26	10,96	9,36	7,79	6,49	5,48	4,70	4,07	3,12	2,48	2,01	1,66	1,40	1,19	1,03	0,90	0,79	0,70	0,62
Pt	78	12,8	12,35	11,06	9,44	7,86	6,55	5,53	4,74	4,11	3,16	2,50	2,03	1,68	1,42	1,21	1,04	0,91	0,80	0,70	0,63
Au	79	12,9	12,45	11,13	9,51	7,92	6,60	5,58	4,78	4,14	3,19	2,53	2,05	1,70	1,43	1,22	1,05	0,92	0,80	0,71	0,64
Hg	80	13,0	12,54	11,23	9,58	7,98	6,66	5,63	4,83	4,18	3,22	2,55	2,07	1,72	1,45	1,23	1,06	0,93	0,81	0,72	0,64
Tl	81	13,1	12,64	11,32	9,66	8,05	6,71	5,68	4,87	4,22	3,25	2,58	2,10	1,74	1,46	1,25	1,07	0,94	0,82	0,73	0,65
Pb	82	13,2	12,69	11,39	9,74	8,11	6,77	5,72	4,91	4,26	3,28	2,60	2,12	1,75	1,48	1,26	1,09	0,95	0,83	0,74	0,66
Bi	83	13,2	12,75	11,49	9,81	8,18	6,82	5,77	4,95	4,30	3,31	2,63	2,14	1,77	1,49	1,27	1,10	0,96	0,84	0,74	0,66
Po	84	13,3	12,83	11,56	9,87	8,24	6,88	5,82	4,99	4,33	3,34	2,65	2,16	1,79	1,51	1,28	1,11	0,97	0,85	0,75	0,67
At	85	13,4	12,93	11,66	9,95	8,30	6,93	5,87	5,04	4,37	3,37	2,68	2,18	1,81	1,52	1,30	1,12	0,98	0,86	0,76	0,68
Rn	86	13,5	13,02	11,73	10,02	8,36	6,98	5,92	5,08	4,41	3,40	2,70	2,20	1,82	1,54	1,31	1,13	0,99	0,87	0,77	0,69
Fr	87	13,6	13,12	11,80	10,10	8,42	7,04	5,96	5,12	4,44	3,43	2,73	2,22	1,84	1,55	1,32	1,14	1,00	0,88	0,78	0,69
Ra	88	13,7	13,22	11,90	10,16	8,49	7,09	6,01	5,16	4,48	3,46	2,75	2,24	1,86	1,56	1,34	1,15	1,01	0,88	0,78	0,70
Ac	89	13,8	13,31	11,97	10,24	8,55	7,14	6,06	5,20	4,52	3,49	2,78	2,27	1,87	1,58	1,35	1,16	1,02	0,89	0,79	0,71
Th	90	13,9	13,41	12,04	10,30	8,61	7,20	6,10	5,24	4,55	3,52	2,80	2,29	1,89	1,59	1,36	1,18	1,03	0,90	0,80	0,71
Pa	91	14,0	13,50	12,14	10,37	8,67	7,25	6,15	5,28	4,59	3,55	2,82	2,31	1,91	1,61	1,37	1,19	1,04	0,91	0,81	0,72
U	92	14,1	13,60	12,21	10,45	8,73	7,31	6,19	5,33	4,63	3,58	2,85	2,33	1,93	1,62	1,39	1,20	1,04	0,92	0,82	0,73
Np	93	14,2	13,69	12,28	10,51	8,79	7,35	6,24	5,37	4,66	3,61	2,87	2,35	1,94	1,64	1,40	1,21	1,05	0,93	0,82	0,73
Pu	94	14,3	13,77	12,38	10,59	8,85	7,41	6,28	5,41	4,70	3,63	2,90	2,37	1,96	1,65	1,41	1,22	1,06	0,94	0,83	0,74
Am	95	14,4	13,83	12,45	10,69	8,91	7,46	6,33	5,45	4,74	3,66	2,92	2,39	1,98	1,67	1,43	1,23	1,07	0,95	0,84	0,75
Cm	96	14,4	13,90	12,52	10,71	8,97	7,51	6,38	5,49	4,77	3,69	2,94	2,41	1,99	1,68	1,44	1,24	1,08	0,95	0,85	0,76
Bk	97	14,5	13,98	12,59	10,79	9,03	7,56	6,42	5,53	4,81	3,72	2,97	2,43	2,01	1,70	1,45	1,25	1,09	0,96	0,85	0,76
Cf	98	14,6	14,08	12,69	10,85	9,09	7,61	6,47	5,57	4,84	3,75	2,99	2,45	2,03	1,71	1,46	1,26	1,10	0,97	0,86	0,77
Es	99	14,7	14,17	12,76	10,92	9,14	7,67	6,51	5,61	4,88	3,78	3,01	2,47	2,04	1,73	1,48	1,28	1,11	0,98	0,87	0,78
Fm	100	14,8	14,27	12,83	10,99	9,20	7,72	6,56	5,65	4,91	3,81	3,04	2,49	2,06	1,74	1,49	1,29	1,12	0,99	0,88	0,78
Mv	101	14,9	14,37	12,90	11,05	9,26	7,77	6,60	5,69	4,95	3,84	3,06	2,51	2,08	1,75	1,50	1,30	1,13	1,00	0,88	0,79
No	102	15,0	14,46	12,96	11,12	9,33	7,82	6,64	5,73	4,98	3,87	3,09	2,53	2,10	1,77	1,51	1,31	1,14	1,01	0,89	0,80
	103	15,1	14,56	13,05	11,18	9,37	7,86	6,69	5,76	5,02	3,89	3,11	2,54	2,11	1,78	1,53	1,32	1,15	1,01	0,90	0,80
	104	15,2	14,66	13,12	11,25	9,43	7,91	6,73	5,80	5,05	3,92	3,13	2,56	2,13	1,80	1,54	1,33	1,16	1,02	0,91	0,81

where k = $(me^2/2h^2) \cdot 10^{-14} = 2.393 \cdot 10^{-8}$ cm, and Z is the atomic number, $\Phi_e(\xi)$ being the universal function of Section 11-4.

The atomic factors for these atoms (apart from K) have been calculated with allowance for electron exchange in the statistical model of the atom [213].

11-5c. Scattering by Ions

The table gives values of the scattering factor for ions:

$$f_e \approx \frac{Z - f_p}{\left(\dfrac{\sin \vartheta}{\lambda} \right)^2} ,$$

(130)

where $f_p = \int \varrho(r)\, e^{i(sr)}\, dV$, dV being a volume element [211]. The f_p have been calculated for the ions, not for atoms.

Ion	z	$\dfrac{\sin \vartheta}{\lambda}\ 10^{-8}$				
		0	0,1	0,2	0,3	0,4
Li$^+$	3	$+\infty$	1,04*			
Be^{+2}	4	$+\infty$	2,00*			
B^{+3}	5	$+\infty$	3,01	0,77	0,35*	
N^{+5}	7	$+\infty$	5,00	1,25	0,57	0,30*
N^{+3}	7	$+\infty$	3,30	1,00	0,52*	
O^{-2}	8	$-\infty$	0	0,50	0,42*	
F$^-$	9	$-\infty$	0,30	0,48	0,40*	
Na$^+$	11	$+\infty$	1,55*			
Mg^{+2}	12	$+\infty$	2,25	0,85		
Al^{+3}	13	$+\infty$	3,30	1,02*		
Si^{+4}	14	$+\infty$	4,25	2,21	0,64*	
P^{+5}	15	$+\infty$	5,20	1,44	0,73*	
P^{-3}	15	$-\infty$	1,70	1,24	0,73*	
S^{+6}	16	$+\infty$	6,15	1,65	0,81*	
S^{-2}	16	$-\infty$	1,70	1,32	0,79*	
Cl$^-$	17	$-\infty$	1,80	1,38	0,85*	
K$^+$	19	$+\infty$	2,60	1,46*		
Ca^{+2}	20	$+\infty$	3,20	1,52*		

* The f_e for ions coincide with the f_e for atoms for subsequent values of $(\sin \vartheta)/\lambda$.

11-6. Symmetry of Point Patterns

Electron-diffraction point patterns have only 6 symmetry classes, in contrast to the 11 Laue classes in x-ray analysis.

Figure 193 compares the symmetry classes for electron-diffraction patterns with the 11 Laue classes [211]; the second column gives diffraction classes that can appear in electron-diffraction patterns of the corresponding class.

Symmetry class of pattern	Possible diffraction symmetries	Highest symmetry of pattern for the diffraction class
C_2	All classes	C_i
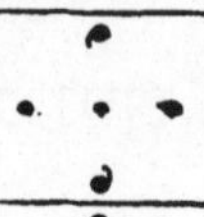 C_4	C_{4h}	C_{4h}
 C_6	C_{6h}, C_{3i}, T_h	C_{6h}, C_{3i}, T_h
 C_{2v}	All classes except C_i, C_{3i}	C_{2h}, D_{2h}
 C_{4v}	D_{4h}, O_h	D_{4h}, O_h
 C_{6v}	D_{3d}, D_{6h}, O_h	D_{3d}, D_{6h}, O_h

Fig. 193. Symmetry of point electron-diffraction patterns.

The third column gives the diffraction class for which the symmetry of the pattern is the highest (the crystal is examined in the direction of highest symmetry).

CHAPTER 12

NEUTRON DIFFRACTION

This chapter deals with the properties and diffraction characteristics of neutrons. See [174, 214] for details of practical methods and of calculations on intensities. Current designs of diffractometers and accessory equipment are dealt with in [299, 300, 332-360].

12-1. Some Formulas for Neutron Diffraction

Properties of the Neutron

Mass = 1.008982 ± 0.000007 atomic unit
= 1.67470 ± 0.00004 · 10^{-24} g
Lifetime ($t_{1/2}$) = 12.8 ± 2.5 min
Spin = $^1/_2$
Magnetic moment = −1.91319 ± 0.00006 nuclear magnetons
= −0.96623 ± 0.00004 · 10^{-23} erg/gauss

Scattering of Neutrons by Free Nuclei

1. The wave function for large distances from the scattering nucleus is

$$u\,(r,\,\vartheta_s,\,\varphi) = A_0 \left[e^{ikr} + \frac{a\,(\vartheta_s,\,\varphi)}{r}\,e^{ikr} \right],$$ (131)

where ϑ_s is the angle between the incident and reflected waves, φ is the azimuthal angle, A_0 is related to the neutron flux, r is the distance from the center of the nucleus, and a is the scattering amplitude.

2. The number of neutrons entering a solid angle dΩ in a direction (ϑ_s, φ) is

$$\frac{d\sigma_r\,(\vartheta_s,\,\varphi)}{d\Omega} = |\,a\,(\vartheta_s,\,\varphi)|^2.$$ (132)

3. The total scattering cross section (the proportion of neutrons scattered in all directions for one nucleus per square centimeter) is

$$\sigma_r = \int |\,a\,(\vartheta_s,\,\varphi)|^2 \sin \vartheta_s\,d\vartheta_s\,d\varphi.$$ (133)

4. The scattering amplitude for a bound nucleus is

$$b = \frac{M + M_a}{M_a}\, a \approx \frac{1 + A}{A}\, a, \tag{134}$$

where A is the mass number, M_a is the mass of the atom, and M is the rest mass of the neutron.

5. The structure factor for a unit cell of a nonmagnetic crystal is

$$F_N = \sum_i b_i e^{-w_i} e^{i \varkappa r_i}, \tag{135}$$

where b_i is the scattering amplitude for bound nucleus i, $w_i = 8\pi^2 \overline{u_{i\chi}^2} (\sin^2 \vartheta)/\lambda^2$, and $\overline{u_{i\chi}^2}$ is the mean-square thermal displacement of atom i in the direction χ.

The scattering amplitude is of the form

$$A_N = \frac{A_{0N}}{r_D}\, F_N, \tag{136}$$

where A_{0N} is the incident amplitude and r_D is the distance from specimen to detector.

6. The amplitude of the neutron wave from a small perfect crystal is

$$A_N = \frac{A_{0N}}{r_D}\, F_N \sum_L e^{i \varkappa A_L}, \tag{137}$$

where $\mathbf{A}_L$ is the vector joining the origin to the L-th unit cell. The scattered intensity given by a small crystal in the form of a parallelepiped is

$$I_N = \frac{I_{0N}}{r_D^2}\, |F_N|^2 \frac{\sin^2\left(\frac{1}{2} n_a \varkappa a\right)}{\sin^2\left(\frac{1}{2} \varkappa a\right)} \frac{\sin^2\left(\frac{1}{2} n_b \varkappa b\right)}{\sin^2\left(\frac{1}{2} \varkappa b\right)} \frac{\sin^2\left(\frac{1}{2} n_c \varkappa c\right)}{\sin^2\left(\frac{1}{2} \varkappa c\right)}, \tag{138}$$

where $n_a \mathbf{a}$, $n_b \mathbf{b}$, and $n_c \mathbf{c}$ are the dimensions along the x, y, and z axes; $\varkappa \mathbf{a}/2\pi = h$, $\varkappa \mathbf{b}/2\pi = k$, $\varkappa \mathbf{c}/2\pi = l$ and (hkl) are the indices of the reflecting plane.

The structure factor for the diffraction peaks is

$$F_N(hkl) = \sum_j b_j e^{-w_j} e^{2\pi i (hx_j + ky_j + lz_j)}, \tag{139}$$

where $x_j = \mathbf{r}_j \mathbf{a}$, $y_j = \mathbf{r}_j \mathbf{b}$, $z_j = \mathbf{r}_j \mathbf{c}$.

The integral intensity of a reflection is

$$R_{sN} = \frac{N^2 \lambda^3}{\sin 2\vartheta}\, (F_N)^2\, \Delta v,$$

where Δv is the volume of the crystal.

7. The integral intensity of a reflection from a large perfect crystal is

$$R_{pN} = \frac{N\lambda^2 |F_N|}{\sin 2\vartheta}\, \text{th}\left(\frac{N\lambda |F_N| t_s}{\sin \vartheta}\right), \tag{140}$$

where t_s is the thickness of the specimen; for t_s large

$$R_{pN} = \frac{N\lambda^2 \, |F_N|}{\sin 2\vartheta} \, .$$

8. The integral intensity of a reflection from an ideal mosaic crystal is

$$\frac{1}{\eta} R_{LN} = \int_{-\infty}^{+\infty} \frac{u\, d\left(\dfrac{\Delta}{r}\right)}{(1+u) + \sqrt{1+2u}\; \mathrm{ctg}\left(\dfrac{\mu_N t_s}{\sin\vartheta}\sqrt{1+2u}\right)} \, , \qquad (141)$$

where $u = Q_N w(\Delta)/\mu_N$ is the disorientation angle of the blocks, η is the standard deviation in this angle, t_s is the thickness of the crystal, $\mu_N = \sum\limits_{j} N_j(\sigma_a + \sigma_{ic})$, N_j is the
number of atoms of type j per cubic centimeter, σ_a is the absorption cross section, and
σ_{ic} is the incoherent-scattering cross section. Figure 194 gives curves of R_{LN}/η for various
$\mu_N t_s/\sin\vartheta$.

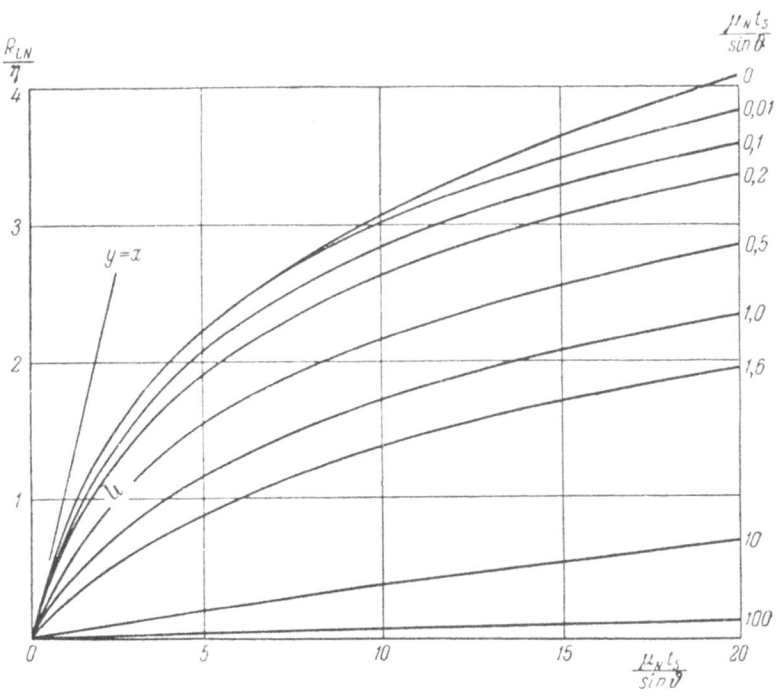

Fig. 194. Integral-intensity curves for reflections from an ideal
mosaic crystal.

9. The detector records a fraction $H/2\pi r_D \sin 2\vartheta$ of the total intensity in the case
of a polycrystal or powder; here H is the length of the path of the neutrons in the detector.

A powder consisting of randomly oriented particles gives

$$p_N = I_{0N} \frac{H Q_N v \varrho'}{8\pi r_D \varrho \sin \vartheta} ,$$

where $Q_N = N^2 \lambda^3 |F_N|^2 / \sin 2\vartheta$, p_N is the number of neutrons recorded per second by the detector, I_{0N} is the intensity of the primary beam (n/cm^2-sec), v is the volume of the specimen, ρ' is the density of the powder, and ρ is the density of the crystalline material.

The scattering cross section for a polycrystal is

$$\sigma_{pc} = \frac{N\lambda^3}{4j} \sum \frac{|F_N|^2}{\sin \vartheta} , \tag{142}$$

where j is the number of atoms in the cell; the summation is for all d > λ / 2.

10. Thermal inelastic scattering covers several distinct types; the coherent elastic scattering causes an intensity change of

$$w_\tau = \frac{6\pi^2 \hbar^2 \tau^2}{M_a k \Theta} \left[\Phi\left(\frac{\Theta}{T}\right) + \frac{1}{4} \right] , \tag{143}$$

where τ is the reciprocal-lattice vector, M_a is the mass of the atom, and $\Phi(\Theta/T)$ is the Debye function (see Section 6-11). The incoherent elastic scattering has a cross section

$$\sigma_{pie} = \frac{\sigma_i}{4k^2 y} (1 - e^{-4k^2 y}),$$

where

$$y = \frac{3\hbar^2}{M_a k \Theta} \left[\Phi\left(\frac{\Theta}{T}\right) + \frac{1}{4} \right] .$$

11. The total scattering cross section for a polycrystalline material giving coherent and incoherent scattering is

$$\sigma_t = \sigma_{ic'} \frac{M}{M_a} \sum_{n=0,2,3,4} \left(\frac{k\Theta}{E}\right)^{1/2 - n/2} \left[a_n \frac{T}{\Theta} - b_n + c_n \left(\frac{T}{\Theta}\right)^{-1} \right] +$$

$$+ \sigma_{ic} \frac{M}{M_a} \sum_{n=0,2,3,4} \left(\frac{k\Theta}{E}\right)^{1/2 - n/2} \left[d_n \left(\frac{T}{\Theta}\right) + e_n \left(\frac{T}{\Theta}\right) + f_n^2 + g_n \left(\frac{T}{\Theta}\right)^{-1} \right] . \tag{144}$$

This applies for neutron energies E < kΘ at sufficiently high temperatures (T $\gg$ Θ /2π).

Here σ_{ic} is the incoherent-scattering cross section (equal to the difference of the total and coherent cross sections), M_a is the mass of the atom (nucleus), M is the rest mass of the neutron, Θ is the characteristic temperature, and E is the neutron energy.

The table gives the coefficients of (144) as functions of n:

n	a_n	b_n	c_n	d_n	e_n	f_n	g_n
0	1,2	0,4286	0,0556	1,3221	2,1141	0,7410	—0,1044
2	5	1,5	0,1786	—8,0590	3,8562	1,660	0,1877
3	—12	0	—0,333	0	0	0	0
4	5,25	0,825	0,0825	—41,042	—5,439	—1,262	0,2666
5	0	0	0	96	0	5,333	0

Figure 195 may be used to find σ_t; this applies to iron without allowance for the magnetic scattering, but it can be applied to other materials. It is assumed that $\sigma_t \sim 1/M_a^\Theta$.

12. The refractive index for neutrons is

$$n = 1 - \frac{2\pi N_A \bar{b}}{k^2},$$

where N_A is the number of atoms per cubic centimeter, $\bar{b}$ is the mean scattering amplitude, and k is the wave number.

The angle of critical reflection is

$$\vartheta_c = \frac{2}{k} (\pi N_A \bar{b})^{1/2},$$

where ϑ_c for neutrons with $\lambda = 2.0$ Å is about 0.003 radian.

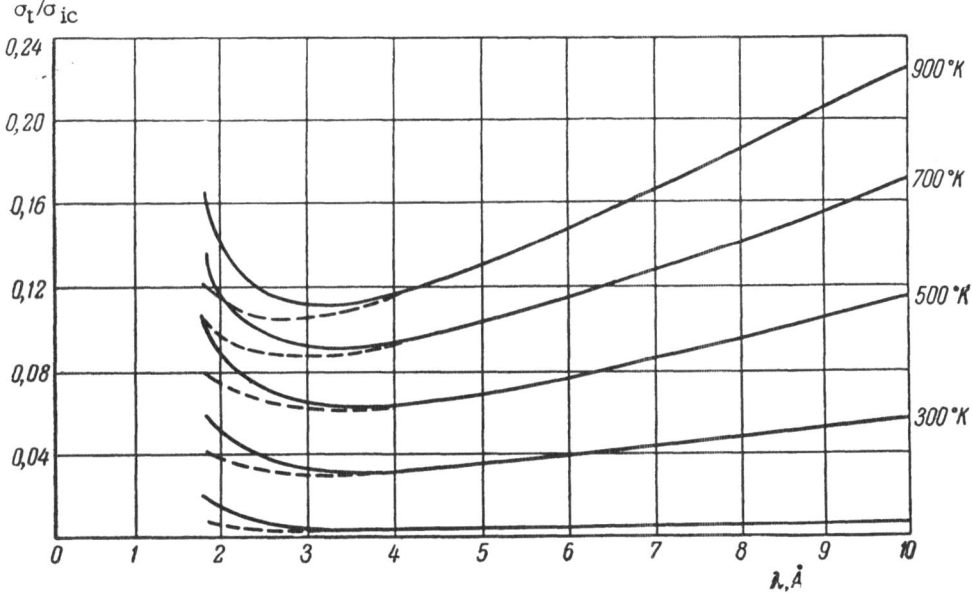

Fig. 195. Ratio of the total cross section to the incoherent-scattering cross section for neutrons in iron.

13. The neutron flux from a reactor having a Maxwellian velocity distribution is

$$d\varphi = \frac{2\varphi v^3}{v_0^4} e^{-\left(\frac{v}{v_0}\right)^2} dv, \tag{145}$$

where v is the neutron velocity, φ is the integral flux $(cm^{-2}\text{-}sec^{-1})$, v_0 is the velocity at the maximum in the Maxwellian distribution, $v_0 = (2kT/M)^{1/2} = 1.284 \cdot 10^4 T^{1/2}$, and T is the absolute temperature.

The integral flux recorded by the detector is $\varphi_t = \varphi A_p/4\pi r_p^2$, in which φ is the integral flux from the reactor, A_p is the cross section of the beam, and r_p is the distance from neutron source to detector.

14. Fermi (time-of-flight) monochromatization gives the resolving power of the monochromator as

$$\frac{v}{\Delta v} = \frac{\pi}{2\Delta\vartheta_B},$$

where Δv is the total width of the velocity spectrum and $\Delta\vartheta_B$ is the angular width of the beam in a plane normal to the monochromator.

The resolving power of a mechanical modulator with an oblique slit is

$$\Delta v = \Delta\vartheta_M \frac{v^2}{\omega_M r_M},$$

where $\Delta\vartheta_M$ is the angular width of the beam in the direction of motion of the monochromator, ω_M is the angular velocity of the monochromator, and r_M is the radius of the rotating part.

15. The total number of neutrons entering the counter per second from a flat polycrystalline specimen is

$$P_N = P_{0N} \frac{LQ_N t_p \varrho'}{4\pi r_{DQ} \sin 2\vartheta} e^{-\mu_t t_p \sec \vartheta}, \tag{146}$$

where t_p is the thickness and μ_t is the attenuation coefficient, which includes losses from absorption, incoherent scattering, and secondary extinction. The exponential attenuation factor is measured by reference to the intensities of the incident and transmitted beams.

For cylindrical specimens

$$P_N = P_{0N} \frac{LQ_N v \varrho'}{8\pi r_{DQ} \sin \vartheta} A(\vartheta, \mu_t r_s), \tag{147}$$

where $A(\vartheta, \mu_t r_s)$ is the absorption factor and r_s is the radius.

Inelastic scattering raises the background for polycrystals; it is defined by

$$\frac{d\sigma_i}{d\omega} = \sum (1 - e^{-2\omega\tau}) b^2, \tag{148}$$

where the summation is carried over all atoms.

16. An approximation used in small-angle scattering is

$$\varrho_p = 2k\delta R_p,$$

where δ is the difference of the refractive indices for powder and binder, k is the wave number, and R_p is the radius of a particle. The diffraction approximation is used for $\rho_p \ll 1$, the differential scattering cross section being

$$\frac{d\sigma}{d\Omega} = 4R_p^6 k^4 \delta^2 \left(\frac{\sin x}{x^3} - \frac{\cos x}{x^2} \right)^2, \tag{149}$$

where $x = R_p k \sin \vartheta_s$, and ϑ_s is the scattering angle. For $\rho_p \gg 1$ the scattering approximation is

$$\langle \vartheta_s^2 \rangle = 4N_p \pi R_p^2 \delta^2 \left[\ln \frac{2}{|\delta|} + 1 \right].$$

17. The scattering cross section for an atom of a ferromagnetic material is

$$\frac{d\sigma}{d\Omega} = b_n^2 + 2b_n Dq\lambda + D^2 q^2 = b_n^2 + b_m^2 + D^2 q^2, \tag{150}$$

where b_n is the coherent-scattering amplitude, $\boldsymbol{\lambda}$ is a unit vector in the direction of polarization, $\mathbf{q} = \mathbf{e}\,(\mathbf{e}_{\boldsymbol{\chi}_m}) - \boldsymbol{\chi}_m$ is a unit vector parallel to $(\mathbf{k} - \mathbf{k'})$, $\boldsymbol{\chi}_m$ is a unit vector in the direction of magnetization, and D is the magnetic-scattering coefficient:

$$D = \left(\frac{e^2}{mc^2}\right) \gamma S f_m;$$

here $e^2/mc^2 = 2.818 \cdot 10^{-13}$ cm is the radius of the electron, γ is the magnetic moment of the neutron in Bohr magnetons, S is the spin quantum number, f_m is the magnetic form factor; and b_m is the magnetic-scattering amplitude for unchanged spin.

Random orientation in the magnetic moments of the atoms gives

$$\frac{d\sigma}{d\Omega} = b_n^2 + \frac{2}{3} D^2$$

for unpolarized neutrons.

For paramagnetic materials

$$\frac{d\sigma}{d\Omega} = b_n^2 + \frac{2}{3} S\,(S+1) \left(\frac{e^2\gamma}{mc^2}\right)^2 f_m^2.$$

The scattering cross section as corrected for the orbital moments is

$$\frac{d\sigma}{d\Omega} = \frac{2}{3} \left(\frac{e^2\gamma}{mc^2}\right) g^2 J\,(J+1),$$

where J is the inner quantum number and g is the Landé factor,

$$g = 1 + \frac{J\,(J+1) + S\,(S+1) - L\,(L+1)}{2J\,(J+1)},$$

L being the additional quantum number.

18. Polarization in transmission alters the scattering cross section from σ to σ_+ $= \sigma + p$ if the spin is parallel to $\boldsymbol{\chi}_m$ and to $\sigma_- = \sigma - p$ if the spin is antiparallel.

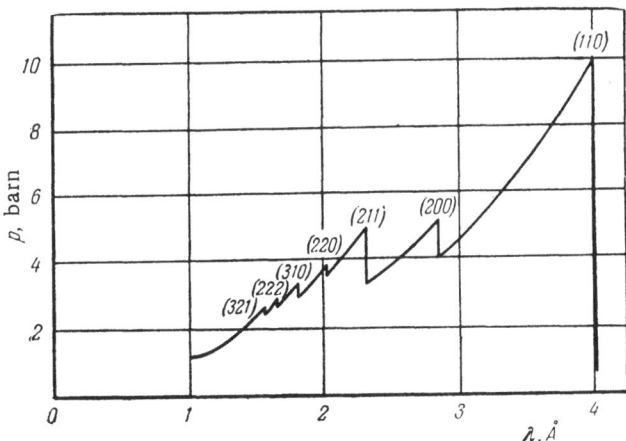

Fig. 196. Neutron polarization by reflection from polycrystalline iron.

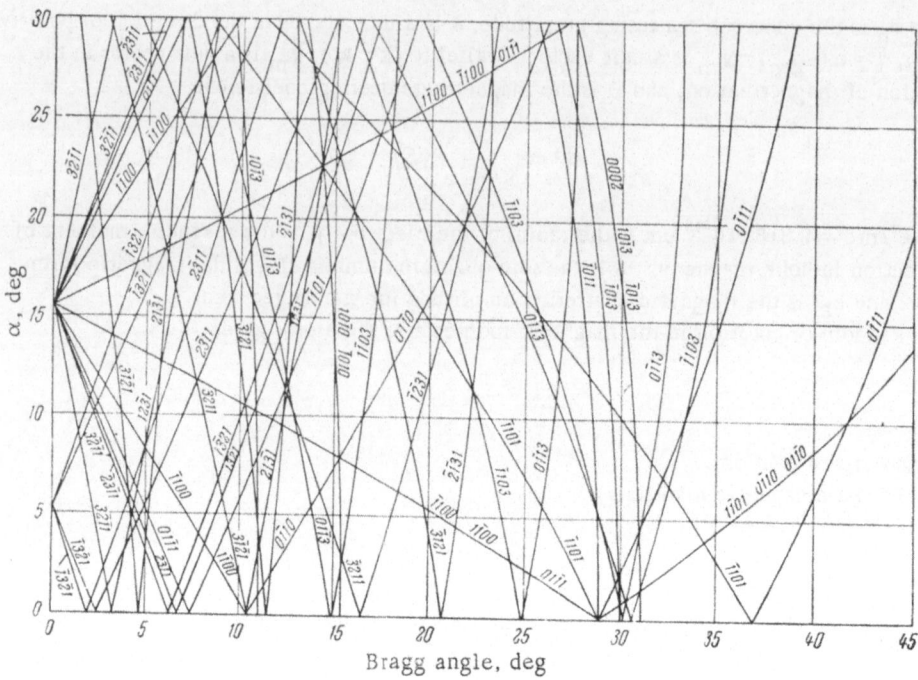

Fig. 197. Planes of competing extinction for neutrons reflected from the (1011) plane of a beryllium monochromator.

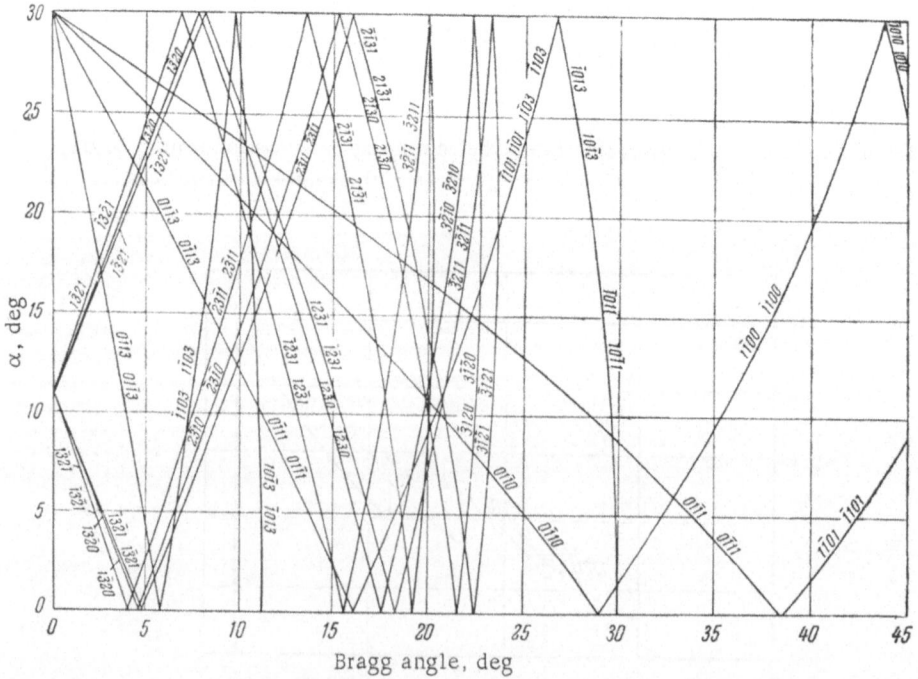

Fig. 198. Planes of competing extinction for neutrons reflected from the (0002) plane of a beryllium monochromator.

The p for a body-centered cubic material and for a beam incident normal to χ_m is

$$p = b_m \frac{e^2}{mc^2} \gamma S \frac{\lambda}{2a_0} \sum \frac{M_g}{l_p} \left(1 + \frac{l_p^2 \lambda^2}{4a_0^2}\right) f_m \left(\frac{l_p}{2a_0}\right),$$

where a_0 is the lattice constant and M_g is the number of (hkl) corresponding to one value of $l_p^2 = (h^2 + k^2 + l^2)$, the summation being for all $d < \lambda/2$. Figure 196 gives p for poly-crystalline iron.

19. The refractive index of a magnetic material is

$$n^2 = 1 - \frac{4\pi N_A b}{k^2} \pm \frac{|\boldsymbol{\mu}||\boldsymbol{B}|}{E},$$

where $\boldsymbol{\mu}$ is the magnetic moment of the neutron, $\boldsymbol{B}$ is the magnetic induction, and E is the neutron energy in ergs. The minus sign corresponds to $\boldsymbol{\mu}$ and $\boldsymbol{B}$ parallel; the plus, to antiparallel.

A beryllium monocrystal may be used as monochromator, in which case some of the beam is lost in the monocrystal on account of reflection by other planes. The geometry of the beam implies that the crystal must be turned with respect to the Y axis to bring the axis of the zone vertical. See [448, 462] for the theory of competing extinction.

Figures 197 and 198 relate the angle α of the crystal setting to the Bragg angle for competing extinction [448]; the numbers are the indices of the planes that give the effect.

A smoother reflection spectrum is provided by a crystal whose lattice is simpler than that of Be (e.g., NaCl), but the reflectivity of the monochromator is then usually much lower.

Example. Determine α for reflection from the (0002) plane of a beryllium monochromator: $\alpha = 0$ for $[0\bar{1}0]$ vertical, and $\alpha = 19°6.4'$ for $[2\bar{1}0]$ vertical.

See [214-217] for the general features of neutron diffraction, [218, 219] for the theory of neutron diffraction in crystals, [220-223] for the theory of magnetic scattering, [224-229] for the theory of inelastic scattering, and [230-234] for some experimental results. Recent data are surveyed in [177, 235].

12-2. Characteristics of Neutrons of Various Energies

The wavelength (uncorrected for relativistic changes) is

$$\lambda = \frac{h}{Mv}, \tag{151}$$

where λ is in centimeters, h is Planck's constant (erg/sec), M is the rest mass of the neutron in grams, and v is the speed of the neutron (cm/sec). The numerical value of λ is

$$\lambda = \frac{(3.95603 \pm 0.00005) \cdot 10^{-3}}{v} \text{ cm} \tag{152}$$

or

$$\lambda = \frac{(2.86005 \pm 0.00004) \cdot 10^{-9}}{E^{1/2}} \text{ cm}, \tag{153}$$

where E is the energy (eV).

The velocity spectrum is

$$dn = n \left(\frac{2}{\pi}\right)^{1/2} \left(\frac{M}{kT}\right)^{3/2} v^2 e^{-\frac{Mv^2}{kT}} dv, \tag{154}$$

where n is the number of neutrons per cubic centimeter with velocity v and dn is the number with velocities between v and v + dv. The most probable velocity at a given temperature is that corresponding to an energy kT.

The table gives energies in electron volts, velocities in centimeters per second, temperatures T in °K for which v is the most probable velocity, and also wavelengths[174].

E, eV	v, cm/sec	T, °K	λ, cm
10^{-3}	$4,374 \cdot 10^4$	11,61	$9,044 \cdot 10^{-8}$
10^{-2}	$1,383 \cdot 10^5$	116,1	$2,860 \cdot 10^{-8}$
0,0253	$2,200 \cdot 10^5$	293,6	$1,798 \cdot 10^{-8}$
10^{-1}	$4,374 \cdot 10^5$	1161	$9,044 \cdot 10^{-9}$
1	$1,383 \cdot 10^6$	$1,161 \cdot 10^4$	$2,860 \cdot 10^{-9}$
10^2	$1,383 \cdot 10^7$	$1,161 \cdot 10^6$	$2,860 \cdot 10^{-10}$
10^6	$1,38 \cdot 10^9$	$1,161 \cdot 10^{10}$	$2,860 \cdot 10^{-12}$

12-3. Properties of X-Rays and Neutrons

The table [214] compares properties for the wavelengths and energies used in diffraction work.

Property	X-Rays	Neutrons
Wavelength	Characteristic line spectra, e.g., λ for Cu Kα is 1.542 Å	Range of wavelengths, e.g., 1.1 ± 0.05 Å , isolated from Maxwellian spectrum by crystal monochromator. Minor component (intensity less than 1%) of wavelength λ/2
Energy for λ =1 Å,	10^{18}h	10^{13}h, which is of the order of the quanta for lattice vibrations
General nature of scattering by atoms	Electron scattering, with factor dependent on $(\sin \vartheta)/\lambda$ and polarization factor dependent on angle. The scattering amplitude increases monotonically with atomic number and can be deduced from the known electron configuration. No differences between isotopes; scattering amplitude always positive	Nuclear isotropic scattering, form factor not dependent on angle; no regular relation of scattering amplitude to atomic number. The amplitude is governed by the nuclear structure and can be determined only by experiment; it varies from one isotope to another and is also dependent on the nuclear spin, so isotope and spin incoherence can occur. The scattering amplitude is positive for most elements, but it is negative for H, Li, Ti, V, Mn, and ^{62}Ni

Property	X-Rays	Neutrons
Magnetic scattering	None	Additional scattering from atoms with magnetic moments: 1. Diffuse scattering by paramagnetics. 2. Coherent diffraction peaks for ferromagnetics and antiferromagnetics. Scattering amplitude decreases as $(\sin \vartheta)/\lambda$ increases; it can be calculated from the magnetic moment, and it depends on the spin quantum number of the ion, e.g., Fe^{2+} and Fe^{3+}
Absorption coefficient	Very large, with true absorption much greater than scattering; μ is about 10^2 to 10^3 and increases with atomic number	True absorption usually very small and less than scattering ($\mu \approx 0.1$), though there are marked exceptions (high absorption in B, Cd, and rare earths). The absorption also varies from one isotope to another
Thermal effects	Coherent scattering decreases exponentially	
Inelastic scattering	Broad diffuse peaks	
	Slight dependence on wavelength	Marked dependence on wavelength
Reflection from monocrystal	Ideal reflection from crystal restricted by primary extinction. A mosaic crystal gives an integral intensity of Qv	
	Secondary extinction in thick crystals is of secondary importance	Secondary extinction dominant in thick crystals (criterion for a "thin" crystal)
Usual methods of detection	Film, Geiger counter	BF_3 proportional counter; photographic methods may be used indirectly, but these are insensitive. Scintillation techniques under development
Absolute intensity measurement	Difficult; the interpretation is dependent on the accuracy with which atomic-scattering curves are known	Simple, especially for diffraction by polycrystals.

12-4. Neutron Scattering by Isotopes

The table [174] gives the basic parameters, with the atomic factors for x-rays for comparison, for the cases

$$\frac{\sin \vartheta}{\lambda} = 0. \text{ and } 0.5 \text{ Å}^{-1}.$$

The atomic factor for x-rays is dependent on ϑ, whereas the coherent-scattering amplitude is the same for all angles for neutrons.

Notation. b is the coherent-scattering amplitude for neutrons; $S = 4\pi b_r^2$ is the same for the element; b_r is the amplitude for the bound state (which corresponds to the atomic factor for x-rays); $\sigma = S + s$ is the total scattering cross section; s is the incoherent-scattering cross section; f_X is the atomic-scattering function for x-rays.

Element	z	A	Isotope	Nuclear spin	Neutrons			X-Rays $f_x \cdot 10^{12}$, cm	
					$b \cdot 10^{12}$, cm	$S \cdot 10^{24}$, cm	$\sigma \cdot 10^{21}$, cm	$\dfrac{\sin\vartheta}{\lambda}=0$	$\dfrac{\sin\vartheta}{\lambda}=0{,}5\ \text{Å}^{-1}$
H	1	—	H¹	1/2	—0,38	1,8	81	0,28	0,02
			H²	1	0,65	5,4	7,4	0,28	0,02
Li	3	6,94	—	—	—0,18	0,4	—	0,84	0,28
			Li⁶	1	0,7	6	—	0,84	0,28
			Li⁷	3/2	—0,25	0,8	2	0,84	0,28
Be	4	—	Be⁹	3/2	0,78	7,7	6,5	1,13	0,39
C	6	—	C¹²	0	0,66	5,4	5,2	1,69	0,48
			C¹³	—	0,60	4,5	5,5	1,69	0,48
N	7	—	N¹⁴	1	0,94	11,1	20	1,97	0,53
O	8	—	O¹⁶	0	0,58	4,2	4,2	2,25	0,62
F	9	—	F¹⁹	1/2	0,55	3,8	3,5	2,53	0,75
Na	11	—	Na²³	3/2	0,35	1,5	3,5	3,09	1,14
Mg	12	24,3	—	—	0,54	3,6	3,7	3,38	1,35
Al	13	—	Al²⁷	5/2	0,35	1,5	1,5	3,65	1,55
Si	14	28,06	—	—	0,40	2,0	—	3,95	1,72
P	15	31,03	—	—	0,50	3,1	—	4,23	1,83
S	16	—	S³²	0	0,31	1,2	1,2	4,5	1,9
Cl	17	35,5	—	—	0,99	12,2	15	4,8	2,0
K	19	39,1	—	—	0,35	1,5	2	5,3	2,2
Ca	20	40,1	—	—	0,49	3,0	3,5	5,6	2,4
			Ca⁴⁰	0	0,49	3,0	3,2	5,6	2,4
			Ca⁴⁴	0	0,18	0,4	—	5,6	2,4
Sc	21	—		—	1,02	13,1	—	—	—
Ti	22	47,9	—	—	—0,38	1,8	6	6,2	2,7
V	23	—	V⁵¹	7/2	—0,05	0,03	5,1	6,5	2,8
Cr	24	52,0	—	—	0,35	1,56	3,8	6,8	3,0
Mn	25	—	Mn⁵⁵	5/2	—0,37	1,7	2,2	7,0	3,1
Fe	26	55,8	—	—	0,96	11,4	11,7	7,3	3,3
			Fe⁵⁴	0	0,42	2,2	2,5	7,3	3,3
			Fe⁵⁶	0	0,01	12,8	13	7,3	3,3
			Fe⁵⁷	0	0,23	0,64	2	7,3	3,3
Co	27	—	Co⁵⁹	7/2	0,28	1,0	5	7,6	3,4
Ni	28	58,7	—	—	1,03	13,4	18,0	7,9	3,6
			Ni⁵⁸	0	1,44	25,9	27,0	7,9	3,6
			Ni⁶⁰	0	0,30	1,1	1	7,9	3,6
			Ni⁶²	0	—0,87	9,5	9	7,9	3,6
Cu	29	63,6	—	—	0,76	7,3	7,8	8,2	3,8
Zn	30	65,4	—	—	0,59	4,3	4,2	3,5	3,9
Ge	32	72,6	—	—	0,84	8,8	8,5	9,0	4,2
As	33	—	As⁷⁵	3/2	0,63	5,0	7	9,3	4,4
Se	34	79,0	—	—	0,89	10,0	10	9,6	4,5
Br	35	79,9	—	—	0,67	5,7	6,0	9,8	4,7
Rb	37	85,5	—	—	0,55	3,8	5,5	10,4	5,0
Sr	38	87,6	—	—	0,57	4,1	9,5	10,7	5,2
Zr	40	91,2	—	—	0,62	4,9	7	11,3	5,5
Nb	41	—·	Nb⁹³	9/2	0,69	6,0	6,2	11,5	5,7
Mo	42	95,9	—	—	0,66	5,5	7,4	11,8	5,9
Pd	46	106,7	—	—	0,63	5,0	4,8	12,9	6,5
Ag	47	107,9	—	—	0,61	4,6	7	13,3	6,7
			Ag¹⁰⁷	1/2	0,83	8,7	10	13,3	6,7
			Ag¹⁰⁹	1/2	0,43	2,3	6	13,3	6,7
Su	50	118,7	—	—	0,61	4,6	4,9	14,1	7,2
Sb	51	121,8	—	—	0,54	3,7	4,2	14,4	7,4
Te	52	127,5	—	—	0,51	3,3	3,9	14,7	7,6
I	53	—	I¹²⁷	5/2	0,52	3,4	3,8	15,0	7,7
Cs	55	—	Cs¹³³	7/2	0,49	3,0	7	15,5	8,1
Ba	56	137,4	—	—	0,53	3,5	—	15,8	8,3
La	57	138,9	—	—	0,83	8,7	9,3	16,1	8,4
Ce	58	140,25	—	—	0,46	2,7	2,7	16,3	8,6

Element	Z	A	Isotope	Nuclear spin	Neutrons			X-Rays $f_x \cdot 10^{12}$, cm	
					$b \cdot 10^{12}$, cm	$S \cdot 10^{24}$, cm	$\sigma \cdot 10^{24}$, cm	$\frac{\sin \vartheta}{\lambda} = 0$	$\frac{\sin \vartheta}{\lambda} = 0{,}5$ Å^{-1}
			Ce^{140}	—	0,47	2,8	2,8	16,3	8,6
			Ce^{142}	—	0,45	2,6	2,6	16,3	8,6
Pr	59	140,9		—	0,44	2,4	4	16,6	8,8
Nd	60	144,3		—	0,72	6,5	24	16,9	9,0
			Nd^{142}	—	0,77	7,5	—	16,9	9,0
			Nd^{144}	—	0,28	1,0	—	16,9	9,0
			Nd^{146}	—	0,87	9,5	—	16,9	9,0
Sm	62	150,43		—	—	—	—	17,5	9,3
			Sm^{152}	—	−0,5	3	—	17,5	9,3
			Sm^{154}	—	0,8	8	—	17,5	9,3
Er	68	167,64		—	0,79	7,8	15	19,2	10,4
Ta	73		Ta^{181}	$7/2$	0,70	6,1	7,0	20,5	11,3
W	74	183,9		—	0,47	2,7	6,8	20,8	11,4
Pt	78	195,2		—	0,95	11,2	11,2	22,0	12,1
Au	79	—	Au^{197}	$3/2$	0,76	7,3	9	22,2	12,3
Hg	80	200,6		—	1,31	21,5	26,5	22,5	12,5
Tl	81	204,4		—	0,75	7,1	—	22,8	12,7
Pb	82	207,2		—	0,96	11,5	11,6	23,1	12,9
Bi	83	—	Bi^{209}	$9/2$	0,89	10,1	10·	23,3	13,1
Th	90	—	Th^{232}	0	1,01	12,8	12,8	25,2	14,4
U	92	238,1		—	0,85	9,0	—	25,8	14,7

12-5. Neutron Absorption

The table [214] gives effective (true) absorption coefficients σ_a, mass attenuation coefficients μ/ρ, and linear attenuation coefficients μ for neutrons ($\lambda = 1.08$ Å) and for Cu Kα x-rays ($\lambda = 1.54$ Å).

Element	Symbol	Z	$\sigma_a \cdot 10^{24}$, cm^2		μ/ρ, cm^{-1}		μ, cm^{-1}	
			neutrons	x-rays	neutrons	x-rays	neutrons	x-rays
Hydrogen	H	1	0,19	—	0,11	—	—	—
Lithium	Li	3	40	—	3,5	—	1,87	—
Beryllium	Be	4	0,005	20	0,0003	1,3	0,00054	2,3
Boron	B	5	430	55	24	3,1	60	7,7
Carbon	C	6	0,003	109	0,00015	5,5	0,0005	19,2
Nitrogen	N	7	1,1	196	0,048	8,5	—	—
Oxygen	O	8	<0,0005	336	<0,00002	12,7	—	—
Fluorine	F	9	<0,01	547	<0,0003	17,5	—	—
Neon	Ne	10	0,2	819	0,006	24,6	—	—
Sodium	Na	11	0,28	1 170	0,007	30,9	0,007	30
Magnesium	Mg	12	0,04	1 630	0,001	40,6	0,0017	70
Aluminum	Al	13	0,13	2 170	0,003	48,7	0,008	131
Silicon	Si	14	0,06	2 810	0,002	60,3	0,004	145
Phosphorus	P	15	0,09	3 730	0,002	73,0	0,004	146
Sulphur	S	16	0,28	4 820	0,0055	91,3	0,011	182
Chlorine	Cl	17	19,5	5 040	0,33	103	—	—
Argon	Ar	18	0,4	7 420	0,0060	113	—	—
Potassium	K	19	1,2	9 230	0,018	143	0,016	125
Calcium	Ca	20	0,25	11 380	0,0037	172	0,0057	266
Scandium	Sc	21	7,1	13 800	0,09	185	0,28	555
Titanium	Ti	22	3,5	16 160	0,044	204	0,20	920
Vanadium	V	23	2,8	19 100	0,033	227	0,20	1360

Element	Symbol	Z	$\sigma_a \cdot 10^{24}$, cm²		μ/ρ, cm⁻¹		μ, cm⁻¹	
			neutrons	x-rays	neutrons	x-rays	neutrons	x-rays
Chromium	Cr	24	1,8	22 200	0,021	259	0,15	1840
Manganese	Mn	25	7,6	25 700	0,083	284	0,60	2040
Iron	Fe	26	1,4	29 800	0,015	324	0,12	2570
Cobalt	Co	27	21	34 300	0,21	354	1,87	3160
Nickel	Ni	28	2,7	4 750	0,028	49,2	0,25	438
Copper	Cu	29	2,2	5 620	0,021	52,7	0,19	470
Zinc	Zn	30	0,6	6 350	0,0055	59,0	0,039	420
Gallium	Ga	31	1,8	7 310	0,015	63,3	0,089	374
Germanium	Ge	32	1,3	8 300	0,011	69,4	0,058	374
Arsenic	As	33	2,5	9 450	0,020	76,5	0,12	436
Selenium	Se	34	7,4	10 700	0,056	82,8	0,27	398
Bromine	Br	35	3,8	12 020	0,029	92,6	—	—
Krypton	Kr	36	0,027	13 600	0,0002	100	—	—
Rubidium	Rb	37	0,42	15 400	0,0029	109	0,0044	167
Strontium	Sr	38	0,7	17 100	0,0048	119	0,012	303
Yttrium	Y	39	0,83	19 000	0,0056	129	0,031	710
Zirconium	Zr	40	0,10	21 200	0,0006	143	0,0041	915
Niobium	Nb	41	0,63	23 400	0,0041	153	0,034	1290
Molybdenum	Mo	42	1,4	26 000	0,009	164	0,08	1670
Ruthenium	Ru	44	1,5	30 900	0,009	185	0,10	2250
Rhodium	Rh	45	90	33 600	0,53	198	6,6	2480
Palladium	Pd	46	4,0	36 400	0,023	207	0,28	2530
Silver	Ag	47	36	39 600	0,20	223	2,0	2340
Cadmium	Cd	48	2100	43 400	11,2	234	97	2020
Indium	In	49	115	47 600	0,6	252	4,4	1840
Tin	Sn	50	0,35	51 600	0,002	265	0,011	1540
Antimony	Sb	51	3,2	56 900	0,016	284	0,10	1910
Tellurium	Te	52	2,7	60 600	0,013	289	0,081	1800
Iodine	I	53	3,7	65 600	0,018	314	0,09	1540
Xenon	Xe	54	18	71 000	0,083	330	—	—
Cesium	Cs	55	17	75 900	0,077	347	0,15	660
Barium	Ba	56	0,6	81 300	0,0027	359	0,010	1260
Lanthanum	La	57	5,3	86 500	0,023	378	0,14	2340
Cerium	Ce	58	0,48	94 100	0,0021	407	0,015	2810
Praseodymium	Pr	59	6,7	97 800	0,029	422	0,19	2740
Neodymium	Nd	60	26	104 000	0,11	437	0,75	3020
Samarium	Sm	62	6350	115 800	25	467	190	3670
Europium	Eu	63	2520	115 600	10	461	—	—
Gadolinium	Gd	64	22000	122 000	84	470	—	—
Terbium	Tb	65	26	114 000	0,09	435	—	—
Dysprosium	Dy	66	535	124 000	2,0	462	—	—
Holmium	Ho	67	40	34 500	0,15	128	—	—
Erbium	Er	68	100	36 900	0,36	133	—	—
Thulium	Tu	69	71	38 600	0,25	139	—	—
Ytterbium	Yb	70	22	41 000	0,076	144	—	—
Lutecium	Lu	71	65	43 600	0,22	151	—	—
Hafnium	Hf	72	61	46 100	0,20	157	2,7	2080
Tantalum	Ta	73	13	49 000	0,044	164	0,73	2720
Tungsten	W	74	11	51 800	0,036	171	0,70	3300
Rhenium	Re	75	50	—	0,16	—	3,3	—
Osmium	Os	76	9	58 500	0,028	186	0,63	4190
Iridium	Ir	77	260	61 800	0,80	194	18	4360
Platinum	Pt	78	5	66 000	0,02	205	0,36	4400
Gold	Au	79	57	69 500	0,17	214	3,3	4140
Mercury	Hg	80	210	73 600	0,63	223	8,4	3020
Thallium	Tl	81	2,0	77 600	0,006	231	0,07	2720
Lead	Pb	82	0,1	82 200	0,0003	241	0,003	2720
Bismuth	Bi	83	<0,01	86 800	<0,00003	253	<0,0003	2870
Radon	Rn	86	—	102 000	—	278	—	—
Radium	Ra	88	—	113 400	—	304	—	—
Thorium	Th	90	—	125 000	—	327	—	3690
Uranium	U	92	2,1	138 000	0,005	352	—	6600

12-6. Nuclear and Magnetic Scattering Amplitudes for Neutrons

The table [174] gives the nuclear and magnetic scattering amplitudes σ and p for the forward direction for (sin ϑ)/λ of 0 and 0.25 Å^{-1}.

These σ and p apply to the neutral atoms and ions.

Atom or ion	σ, 10^{-12}cm	Effective spin quantum number S	p, 10^{-12} cm	
			$\dfrac{\sin\vartheta}{\lambda}=0$	$\dfrac{\sin\vartheta}{\lambda}=0,25\text{Å}^{-1}$
Cr^{++}	0,35	2	1,08	0,45
Mn^{++}	—0,37	$^5/_2$	1,35	0,57
Fe (metal)	0,96	1,11	0,60	0,35
Fe^{++}	0,96	2	1,08	0,45
Fe^{+++}	0,96	$^5/_2$	1,35	0,57
Co (metal)	0,28	0,87	0,47	0,27
Co^{++}	0,28	2,2	1,21	0,51
Ni (metal)	1,03	0,3	0,16	0,10
Ni^{++}	1,03	1,0	0,54	0,23

12-7. Form Factors for Magnetic Scattering for Atoms and Ions

Figure 199 gives the form factor f_m as a function of (sin ϑ)/λ for some transition elements; curves 1 and 2 represent f_m for Fe (as the metal) for different values of the scattering function, and curve 3 is f_m for Mn^{2+}.

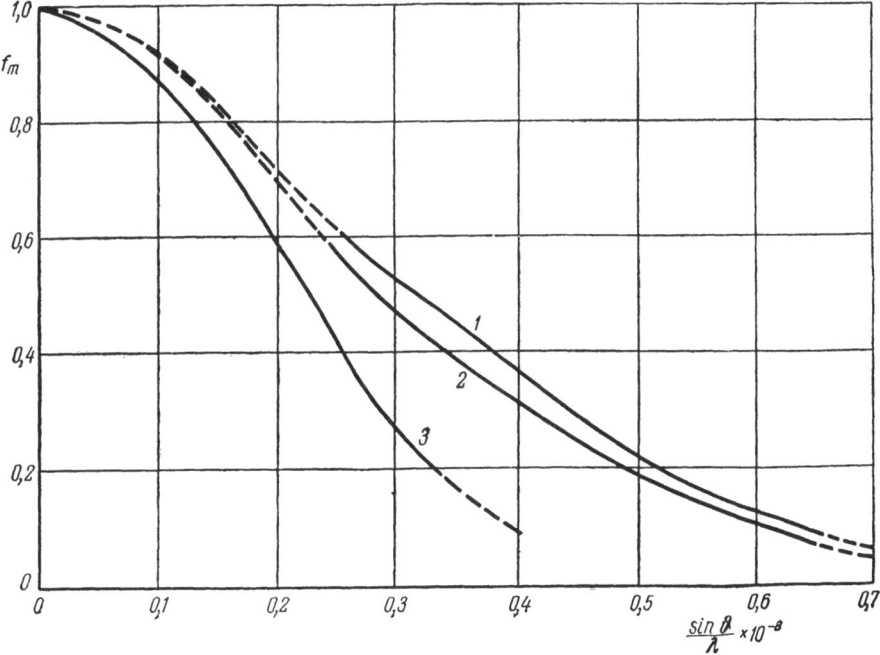

Fig. 199. Form factors for neutron scattering: (1) and (2) atoms of iron; (3) Mn^{2+}.

Figure 200 gives similar curves for Nd^{3+}, Pr^{3+}, and Er^{3+} [174].

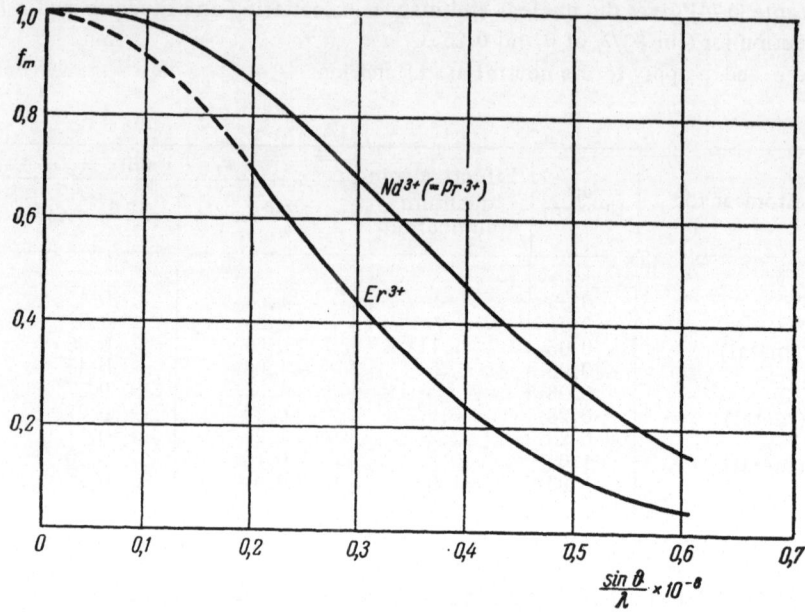

Fig. 200. Form factors for neutron scattering by paramagnetic ions of the rare earths.

12-8. Effective Neutron-Scattering Cross Sections for Metals and Alloys

The table gives experimental and theoretical values of the effective cross sections for the free and bound states [214].

The theoretical values σ_t for alloys have been found from

$$\sigma_{AB} = 4\pi \left[\frac{1}{2} \bar{b}_A + \frac{1}{2} \bar{b}_B \right]$$

which is correct for equal atomic concentrations of the components.

The table gives σ_{AB} for scattering by elements A and B with the same phase and with opposed phases.

Substance	σ_e, barn			σ_t, barn	
	σ_f	σ_b	S	same phase	opposed phases
Copper	8,2	8,5	6,6	—	—
Manganese	1,80	1,87	1,9	—	—
Nickel	17,4	18,0	13,9	—	—
Cu−Ni alloy	12,4	12,8	10,1	9,9	0,35
Mn−Ni alloy	13,6	14,1	5,5	9,5	5,4

LITERATURE CITED*

1. M. A. Blokhin, The physics of x-rays, Moscow, Gostekhizdat, (1957).
2. Internationale Tabellen zur Bestimmung von Kristallestrukturen, Berlin, Borntraeger (1935).
3. Y. Cauchois and H. Hulubei, Longuers d'onde de emissions X et des discontinuités d'absorption, Paris (1947).
4. G. W. Grodstein, X-ray attenuation coefficients, New York (1957).
5. Laboratory Metallography, Moscow, Metallurgizdat (1957).
6. N. N. Kachanov and L. L. Mirkin, X-ray analysis of polycrystals: a practical handbook, Moscow, Mashgiz (1960).
7. G. S. Zhdanov and Ya. S. Umanskii, X-radiography of metals, Part I, Moscow, Metallurgizdat, (1938); Part II, (1941).
8. Ya. S. Umanskii, A. K. Trapeznikov, and A. I. Kitaigorodskii, X-radiography, Moscow, Mashgiz (1951).
9. M. V. Mal'tsev, X-radiography of metals, Moscow, Metallurgizdat (1952).
10. B. Ya. Pines, Microfocus x-ray tubes and applied x-ray structural analysis, Moscow, Gostekhizdat (1955).
11. H. P. Klug and L. E. Alexander, X-ray diffraction procedures, New York (1954).
12. X-ray diffraction by polycrystalline materials, London (1955).
13. R. Glocker, Materialprüfung mit Röntgenstrahlen, Berlin (1958).
14. A. K. Trapeznikov, X-ray flaw detection, Moscow, Mashgiz (1948).
15. M. M. Umanskii, Pribory i Tekhn. Eksperim. 3, No. 3 (1959).
16. A. Weyerer, Z. angew. Phys. 8, No. 3, 135 (1956).
17. H. Lipson, J. Sci. Instr. 19, 63 (1942).
18. I. V. Isaichev, Zh. Tekhn. Fiz. 8, No. 12, 1180 (1938).
19. R. Berthold and H. Bohm, Metallwirtschaft 11, 567 (1932).
20. A. J. Buergers, B. Buerger, and D. Chesley, Am. Mineral. 28, 285 (1943).
21. D. Chesley, Rev. Sci. Instr. 17, 558 (1946).
22. A. Connell, Rev. Sci. Instr. 18, 367 (1947).
23. O. Dorn and A. Glochler, Rev. Sci. Instr. 7, 389 (1936).
24. A. Edwards, G. Speiser, and K. Johnston, Rev. Sci. Instr. 20, 343 (1949).
25. A. Ellwood, J. Inst. Metals 66, 87 (1940).
26. G. Goldschmidt, J. Sci. Instr. 27, 177 (1950).
27. G. Gordon, J. Appl. Phys. 20, 908 (1949).
28. N. P. Goss, Metal Progr. 28, 163 (1935).

*The individual methods of analysis and the derivations of some formulas are given in detail in the literature cited.

29.   W. Hume-Rothery and P. W. Reynolds, Proc. Roy. Soc. A167, 25 (1938).

30.   A. H. Jay, Proc. Phys. Soc. 45, 635 (1933).

31.   K. Kubo and D. Akabori, J. Phys. and Coll. Chem. 54, 1121 (1950).

32.   A. Owen, J. Sci. Instrum. 20, 190 (1943).

33.   A. Owen, J. Sci. Instrum. 26, 114 (1949).

34.   F. Schossberger, Z. Krist. 98, 259 (1938).

35.   M. Straumanis, A. Jevins, and K. Karlsons, Z. anorg. allgem. Chem. 238, 175 (1938).

36.   M. Straumanis, J. Appl. Phys. 20, 726 (1949).

37.   A. Vand, J. Appl. Phys. 20, 726 (1949).

38.   D. Taylor, Rev. Sci. Instr. 2, 751 (1931).

39.   A. Wilson, Proc. Phys. Soc. 53, 235 (1941).

40.   E. Z. Kaminskii and T. I. Stelletskaya, Coll.: Problems of metal science and of the physics of metals, p. 240 (1951).

41.   A. Westgren and G. Phragmen, Z. phys. Chem. A102, 1 (1922).

42.   W. Kohn, Z. Phys. 50, 123 (1928).

43.   O. Ruff and F. Ebert, Z. Krist. 180, 19 (1929).

44.   H. Brekken and L. Harang, Z. Krist. 75, 538 (1930).

45.   A. Göt and A. Hergenroter, Phys. Rev. 40, 643 (1932).

46.   Wangsgard, Trans. ASM 30, 1303 (1942).

47.   J. Birks and A. Friedman, Rev. Sci. Instr. 18, 576 (1947).

48.   A. van Valkenberg and G. McMurdie, J. Res. Nat. Bur. Stand. 38, 415 (1947).

49.   B. Heal and N. Mykura, Metal Treat. 17, 129 (1950).

50.   L. S. Zevin and D. M. Kheiker, Zavodsk. Lab. No. 5, 636 (1958).

51.   I. V. Isaichev, Zh. Tekhn. Fiz. 8, No. 12, 1177 (1938).

52.   G. Campbell and H. Hildebrand, J. Chem. Phys. 11, 334 (1943).

53.   G. Clifton, Rev. Sci. Instr. 21, 339 (1950).

54.   A. Hengstenberg and M. Mark, Z. Krist. 69, 271 (1928).

55.   W. Hume-Rothery and A. Strawbridge, J. Sci. Instr. 24, 89 (1947).

56.   N. Kaufman and F. Fankuchen, Rev. Sci. Instr. 20, 733 (1949).

57.   A. Keeling, C. Fraser, and R. Pepinsky, Rev. Sci. Instr. 24, 89 (1947).

58.   A. Keesom and B. DeSmedt, Proc. Acad. Sci. Amsterdam 25, 118 (1923).

59.   K. Lonsdale and R. Smith, J. Sci. Instr. 18, 133 (1941).

60.   A. McKeehan and D. Gioffi, Phys. Rev. 19, 444 (1922).

61.   A. McFarlan, Rev. Sci. Instr. 7, 82 (1936).

62.   A. Owen and D. Williams, J. Sci. Instr. 31, 49 (1954).

63.   A. Post, K. Schwarz, and F. Fankuchen, Rev. Sci. Instr. 22, 218 (1951).

64.   A. Ubellohde and G. Woodward, Proc. Roy. Soc. A185, 448 (1946).

65.   A. Wallwork and G. Harding, J. Sci. Instr. 31, 163 (1954).

66.   A. Wood, Rev. Sci. Instr. 24, 325 (1953).

67.   B. Tombs, J. Sci. Instr. 19, 364 (1952).

68.   G. Pohland, Z. phys. Chem. B26, 238 (1934).

69.   A. Barnes and K. Hampton, Rev. Sci. Instr. 6, 342 (1935).

70.   D. Abrahams, A. Collin, G. Gipsaub, and A. Reed, Rev. Sci. Instr. 21, 396 (1950).

71.   M. Wolf, Z. Phys. 53, 72 (1929).

72.   R. Mehl and C. S. Barrett, Trans. ASME 89, 575 (1930).

73.   F. Veer and F. Cletzer, Z. Elektrochem. 41, 850 (1935).

74.   W. H. Barnes and W. F. Hampton, Rev. Sci. Instr. 6, 342 (1935).

75.   G. A. Gol'der, Coll.: X-ray methods of examination in the chemical industry, p. 139 (1953).

76. A. Trewlis and W. Davey, J. Sci. Instr. 32, 79 (1955).

77. G. K. Williamson and A. Moore, J. Sci. Instr. 33, 107 (1956).

78. G. Jan, Proc. Phys. Soc. 45, 635 (1933).

79. A. Wilson, Proc. Roy. Soc. A143, 465 (1934).

80. G. Stoner and A. Wilson, Proc. Phys. Soc. 53, 657 (1941).

81. G. Jan, Proc. Phys. Soc. 53, 400 (1941).

82. A. Edwards, G. Speiser, and A. Johnston, J. Appl. Phys. 22, 424 (1951).

83. A. S. Belikov and Ya. S. Umanskii, Proceedings of the Ministry of Higher Education (Ministerstvo vysshego obrazovaniya), Metallurgy section, No. 1 (1958).

84. S. S. Kvitka and M. M. Umanskii, Izvest. Akad. Nauk SSSR, Ser. Fiz., 15, No. 2, 271 (1951).

85. D. M. Kheiker, Zavodsk.Lab. 24, No. 9, 1077 (1958).

86. Yu. A. Bagaryatskii and E. V. Koıontsova, Zavodsk. Lab. 15, No. 9, 1062 (1949).

87. G. B. Bokii and M. A. Porai-Koshits, A practical course of x-ray structural analysis, Vol. 1, Moscow, Izd. Moskov. gosudarst. univ. (1952).

88. International tables for x-ray crystallography, Vol. I, 1952, II, 1959.

89. G. S. Zhdanov and V. A. Pospelov, Zh. Eksperim. i Teoret. Fiz. 15, 709 (1945).

90. M. Cernohorsky, Acta Acad. Sci. Ceskosl. 30, No. 4 (1958).

91. I. I. Kozhina, E. V. Stroganov, and S. S. Tolkachev, Handbook on laboratory work in structural crystallography, Leningrad, Izd. Leningrad. gosudarst. univ. (1958).

92. N. F. M. Henry, H. Lipson, and W. A. Wooster, The interpretation of x-ray diffraction photographs, London (1951).

93. L. Zsoldos, Acta Cryst. 11, 835 (1958).

94. V. I. Mikheev, X-ray tables for the identification of minerals, Moscow, Gosgeolizdat (1957).

95. A. I. Kitaigorodskii, X-ray structural analysis, Gostekhizdat (1950).

96. G. S. Zhdanov, Principles of x-ray structural analysis, Moscow, Gostekhizdat (1940).

97. B. Ya. Pines, Lectures on structure analysis, Khar'kov, Izd. Khar'kov. gosudarst. univ. (1957).

98. R. James, The optical principles of the diffraction of x-rays [Russian translation], Moscow, Izd. inostr. lit. (1950).

99. W. H. Zachariasen, Theory of x-ray diffraction in crystals, New York (1945).

100. D. R. Hartree, The calculation of atomic structures, New York (1957).

101. E. Hellner, Z. Krist. 106, 2 (1954).

102. K. Sagel, Tabellen zur Röntgenstrukturanalyse, Berlin (1958).

103. R. McWeeny, Acta Cryst. 4, 513 (1951).

104. J. Berghuis, Acta Cryst. 8, 478 (1955).

105. C. S. Abrahams, Acta Cryst. 8, 661 (1955).

106. M. Qurashi, Acta Cryst. 7, 310 (1954).

107. L. H. Thomas and K. Umeda, J. Chem. Phys. 26, 293 (1957).

108. A. Vand, Acta Cryst. 1, 290 (1948).

109. A. I. Kitaigorodskii, Handbook on x-ray structural analysis, Moscow, Gostekhizdat (1948).

110. M. J. Buerger, X-ray crystallography [Russian translation], Moscow, Izd. inostr. lit. (1958).

111. Landolt–Börnstein, Tabellen Zahlwerte und Funktionen, Berlin (1950-1959).

112. B. F. Ormont, Structures of inorganic materials, Moscow (1950).

113. G. B. Bokii, Introduction to chemical crystallography, Moscow, Izd. Moskov. gosudarst. univ. (1954).

114. D. M. Kheiker and L. S. Zevin, Kristallografiya 1, 739 (1956).
115. L. I. Mirkin, Zavodsk. Lab. 24, 569 (1958).
116. N. N. Kachanov and L. I. Mirkin, Tekhnol. avtomobilestroeniya, No. 5, 76 (1957).
117. A. McCreery, J. Am. Ceram. Soc. 32, 4 (1949).
118. G. Mauson, J. Appl. Phys. 26, 1254 (1954).
119. H. Klug, Anal. Chem. 25, 704 (1953).
120. R. Black, Anal. Chem. 25, 743 (1953).
121. A. Leroux, Anal. Chem. 30, 886 (1958).
122. ASTM Diffraction data card file, 1957.
123. A. I. Kitaigorodskii, X-ray structural analysis of cryptocrystalline and amorphous bodies, Moscow, Gostekhizdat (1952).
124. V. I. Mikheev, X-ray tables for the identification of minerals, Moscow, Gosgeolizdat (1957).
125. L. K. Frevel, Ind. Eng. Chem. 14, 687 (1942).
126. L. K. Frevel, H. W. Rinn, and H. C. Anderson, Ind. Eng. Chem. 18, 83 (1946).
127. L. K. Frevel and H. W. Rinn, Anal. Chem. 25, 1697 (1953).
128. L. S. Palatnik, Izvest. Akad. Nauk SSSR, Ser. Fiz. 15, 134 (1951).
129. V. Ya. Anosov and S. A. Pogodin, Principles of physicochemical analysis, Moscow (1947).
130. M. E. Straumanis and A. Jevins, Die Präzisionbestimmung von Gitterkonstanten, Berlin (1940).
131. A. Z. Zhmudskii, Zavodsk. Lab. No. 9 (1949); No. 6 (1952).
132. B. E. Warren, J. Appl. Phys. 16, 614 (1945).
133. A. Taylor and G. Sinclair, Proc. Phys. Soc. 57, 108 (1945).
134. H. Lipson and A. Wilson, Proc. Phys. Soc. 53, 245 (1941).
135. M. Cohen, Rev. Sci. Instr. 6, 68 (1935).
136. A. Hess, Acta Cryst. 4, 109 (1951).
137. B. M. Rovinskii and E. P. Kostyukova, Kristallografiya 3, No. 3 (1958).
138. G. Berthold and A. Gerold, Z. Metallkunde 46, 9, 599 (1955).
139. M. Tournarie, J. Phys. et Rad. 15, 1, 11A (1954).
140. A. Smakula and B. Kalnais, Phys. Rev. 99, 6, 1736 (1955).
141. D. M. Kheiker, Dissertation, Moscow (1958).
142. W. Parrish, M. G. Ekstein, and B. W. Irwin, Data for x-ray analysis, Vol. II (1953).
143. E. R. Jette and F. Foote, J. Chem. Phys. 3, 605 (1935).
144. H. van Bergen, Ann. der Phys. 39, 553 (1941).
145. A. J. C. Wilson, Proc. Phys. Soc. 53, 235 (1941).
146. A. Jevins and M. Straumanis, Z. Phys. B34, 402 (1936).
147. H. Weyerer, Z. angew. Phys. 8, 297 (1956).
148. M. Straumanis, J. Appl. Phys. 20, 726 (1949).
149. H. Lipson and L. Rogers, Phil. Mag. 35, 544 (1944).
150. M. Straumanis and E. Aka, J. Appl. Phys. 23, 330 (1952).
151. H. Swanson and E. Tatge, Nat. Bur. Stand. Diss. 1, 69 (1953).
152. T. Rymer and P. Hambling, Acta Cryst. 4, 565 (1951).
153. A. Z. Zhmudskii, Tables of lattice constants for iron, copper, aluminum, and alloys of these, Kiev (1953).
154. M. Ya. Fuks, Izvest. Akad. Nauk SSSR, Ser. Fiz. 17, 357 (1953).
155. H. Lipson and A. J. C. Wilson, J. Sci. Instr. 18, 144 (1941).
156. A. Taylor and R. W. Floyd, Acta Cryst. 3, 285 (1950).

157. A. Kochanovska, Czech. Phys. J. 3, 1 (1953); 4, 1 (1954).

158. D. E Ovsienko and E. I. Sosnina, Fiz. Metal. i Metalloved. 3, 516 (1956).

159. A. Cottrell, Dislocations and plastic flow in crystals [Russian translation], Izd. inostr. lit. (1956).

160. W. Read, Dislocations in crystals [Russian translation], Izd. inostr. lit. (1957).

161. G. K. Williamson and R. Smallman, Phil. Mag. 1, 34 (1956).

162. L. I. Mirkin and Ya. S. Umanskii, Izv. vyssh. uchebn. zavedenii, Fizika No. 3 (1960).

163. L. I. Mirkin and Ya. S. Umanskii, Fiz. Metal. i Metalloved. 9, No. 6 (1960).

164. Ya. M. Golovchiner, Zavodsk. Lab. 26, 431 (1960).

165. O. V. Bogorodskii, Dissertation, Moscow (1956).

166. A. M. Lopshits, Templates for harmonic analysis, Moscow (1947).

167. S. M. Nikolaeva and Ya. S. Umanskii, Izvest. Akad. Nauk SSSR, Ser. Fiz., 20, No. 6, 631 (1956).

168. Yu. S. Terminasov and A. P. Feklistov, Izvest. Akad. Nauk SSSR, Ser. Fiz., 20, 695 (1956).

169. A. A. Smirnova and Yu.S. Terminasov, Izvest. Akad. Nauk SSSR, Ser. Fiz., 20, 679 (1956).

170. A. Kochanovska, Czech. Phys. J. 4, No. 3, 290 (1954).

171. C. S. Barrett, The structure of metals [Russian translation], Moscow (1948).

172. G. S. Zhdanov, Study of crystal orientations in metals and alloys by means of polar figures, Moscow (1934).

173. B. D. Cullity, Elements of x-ray diffraction, Reading, Mass. (1956).

174. Handbuch der Physik, Vol. 32, Berlin (1957).

175. A. Guinier and G. Fournet, Small-angle scattering of x-rays, New York (1955).

176. G. Oster and D. P. Riley, Acta Cryst. 5, 272 (1952).

177. Modern physical test methods in metallurgy, Moscow, Metallurgizdat (1958).

178. A. Münster and K. Sagel, Z. Phys. Chem. 12, 145 (1957).

179. A. Münster, Statistische Thermodynamik, Berlin (1956).

180. C. R. Houska and B. L. Averbach, J. Appl. Phys. 30, 1532 (1959).

181. B. Borie, Acta Cryst. 10, 89 (1957).

182. G. Porod, Acta Phys. Austriaca 2, 255 (1948).

183. G. Fournet, Bull. Soc. Franc. Miner. Crist. 74, 39 (1951).

184. R. D. B. Fraser and T. P. McRal, Acta Cryst. 12, 171 (1959).

185. R. Hosemann, Ergebn. exact. Naturwiss. 24, 142 (1951).

186. J. Blin, The scattering of x-rays by crystals [Russian translation], Moscow (1959).

187. R. E. Smallmann and K. H. Westmacott, J. Appl. Phys. 30, 603 (1959).

188. H. H. Atkinson, J. Appl. Phys. 30, 637 (1959).

189. H. H. Atkinson, R. E. Smallman, and K. H. Westmacott, J. Appl. Phys. 30, 646 (1959).

190. K. Thomas and A. Franks, J. Appl. Phys. 30, 649 (1959).

191. V. M. Kalikhman and Ya. S. Umanskii, Izvest. vyssh. shkoly, Fizika (in press).

192. S. M. Astrakhantsev and Ya. S. Umanskii, Izvest. vyssh. shkoly, Tsvetnaya Metallurgiya, No. 6 (1959).

193. R. H. Neyhaber, W. G. Brammer, and W. W. Beckman, J. Appl. Phys. 30, 656(1959).

194. H. Fricke and W. Gerold, J. Appl. Phys. 30, 661 (1959).

195. J. C. Grosshrentz and F. R. Rollins, J. Appl. Phys. 30, 668 (1959).

196. A. Guinier and A. Guyon, J. Appl. Phys. 30, 622 (1959).

197. A. Guinier, J. Appl. Phys. 30, 601 (1959).

198. O. J. Guentert and B. E. Warren, J. Appl. Phys. 29, 40 (1958).

199. C. R. Houska and B. L. Averbach, Acta Cryst. 11, 139 (1958).

200. R. Smallman and K. H. Westmacott, Phil. Mag. 2, 669 (1957).

201. R. Hirsh and H. Otte, Acta Cryst. 10, 447 (1957).

202. A. Spreadborough, Phil. Mag. 3, 1167 (1958).

203. A. Wagner, Acta Metall. 5, 477 (1957).

204. A. Wagner, Acta Metall. 5, 427 (1957).

205. H. Otte, Acta Metall. 5, No. 11 (1957).

206. F. H. Herbstein and B. L. Averbach, Acta Cryst. 8, 843 (1955).

207. N. Norman and B. E. Warren, J. Appl. Phys. 22, 483 (1951).

208. C. B. Walker, J. Appl. Phys. 23, 118 (1952).

209. A. S. Kagan, V. A. Somenkov, and Ya. S. Umanskii, Kristallografiya, No. 3 (1960).

210. Z. G. Pinsker, Electron diffraction, Moscow, Gostekhizdat (1949).

211. B. K. Vainshtein, Structural electron diffraction, Moscow (1956).

212. Z. G. Pinsker and G. G. Dvoryankina, Kristallografiya 3, No. 4 (1958).

213. J. Ebers and B. K. Vainshtein, Kristallografiya 4, 611 (1959).

214. G. Bacon, Neutron diffraction [Russian translation], Moscow, Izd. inostr. lit. (1957).

215. D. Hughes, Neutron optics [Russian translation], Moscow, Izd. inostr. lit. (1956).

216. D. J. Hughes, Pile neutron research, Cambridge, Mass. (1954).

217. D. Hughes, Effective neutron cross-sections [Russian translation], Moscow, Izd.
      inostr. lit. (1959).

218. M. L. Goldberger and F. Seitz, Phys. Rev. 71, 294 (1947).

219. G. E. Bacon and R. O. Lowde, Acta Cryst. 1, 303 (1948).

220. O. Halpern and M. H. Johnson, Phys. Rev. 55, 898 (1939).

221. O. Halpern and T. Holstein, Phys. Rev. 59, 960 (1941).

222. O. Halpern, M. Hammermesh, and M. H. Johnson, Phys. Rev. 59, 981 (1941).

223. L. van Hove, Phys. Rev. 95, 1974 (1954).

224. R. Weinstock, Phys. Rev. 65, 1 (1944).

225. J. M. Cassels, Progr. Nucl. Phys. 1, 185 (1950).

226. G. L. Squires, Proc. Roy. Soc.. A212, 192 (1952).

227. G. Placzek and L. van Hove, Phys. Rev. 93, 1207 (1954).

228. L. van Hove, Phy . Rev. 95, 249 (1954).

229. G. Placzek and L. van Hove, Nuovo Cim. 1, 233 (1955).

230. E. Fermi and L. Marshall, Phys. Rev. 71, 666 (1947).

231. E. O. Wollan and C. G. Shull, Phys. Rev. 73, 830 (1948).

232. C. G. Shull et al., Phy . Rev. 73, 842 (1948).

233. C. G. Shull et al., Phys. Rev. 83, 333 (1951).

234. C. G. Shull et al., Phys. Rev. 84, 912 (1951).

235. S. S. Sidhu, L. Heaton, and M. H. Muellar, J. Appl. Phys. 30, 1923 (1959).

236. A. Rose, Tables et abaques, Paris (1956).

237. N. V. Belov, Dokl. Akad. Nauk SSSR 59, 487 (1948).

238. X-ray apparatus and equipment, Moscow, Standartgiz (1960).

239. F. N. Kharadzha, A general course in x-ray technique, Moscow, Gosenergoizdat (1956).

240. A. Paskin, Acta Cryst. 10, 667 (1957).

241. R. B. Roof, J. Nucl. Mater. 2, No. 1, 39 (1960).

242. V. K. Latyshev and A. K. Felinger, Prob. Metalloved. i Fiz. Metal. No. 6, 453
      (1959).

243. L. Fankuchen, Anal. Chem. No. 6 (1960).

244. I. B. Borovskii, The physical principles of x-ray spectral analysis, Moscow (1956).

245. H. Granicher, Z. Krist. 110, 432 (1958).
246. K. Lonsdale, Proc. Roy. Soc. A247, 424 (1958).
247. H. Smithells, Metal Reference Book, New York (1950).
248. O. Klein and U. Nischina, Zs. Phys. 52, 853 (1929).
249. R. Böklen and S. Geiling, Zs. Metallk. 40, 157 (1949).
250. M. M. Umanskii, Dissertation, Moscow (1947).
251. E. E. Vainshtein and M. M. Kakhana, Tables for x-ray spectroscopy, Moscow (1953).
252. The URS-70K1 apparatus for x-ray structural and spectral analysis, Leningrad, Zavod Burevestnik (1955).
253. The URS-55 apparatus for x-ray structural analysis, Leningrad, Zavod Burevestnik (1958).
254. The URS-50I apparatus for x-ray structural analysis, Leningrad, Zavod Burevestnik (1959).
255. The URS-60 apparatus for x-ray structural analysis, Leningrad, Zavod Burevestnik (1959).
256. P. Müller, Arch. Eisenhüttw. 15, 402 (1958).
257. K. A. Aglintsev, Dosimetry of ionizing radiations, Moscow (1959).
258. N. N. Kachanov and L. I. Mirkin, Tekhnol. Avtomobilestroeniya, No. 5, 72 (1957).
259. V. A. Alekseev, I. E. Konstantinov, and D. M. Kheiker, Zavodsk. Lab. 26, 501 (1960).
260. D. M. Kheiker, Coll., Proceedings of the All-Union Scientific Research Institute of the Asbestos Cement Industry, Moscow (1959).
261. D. M. Kheiker and L. S. Zevin, X-ray diffractometry (in press).
262. The Burevestnik-1 attachment: technical description, Moscow (1960).
263. T. C. Furnas, V Intern. Congr. IUC, Cambridge (1960).
264. P. J. Brown and J. B. Forsyth, V Intern. Congr. IUC, Cambridge (1960).
265. J. Shimura, V Intern. Congr. IUS, Cambridge (1960).
266. J. Adam, V Intern. Congr. IUC, Cambridge (1960).
267. D. M. Kheiker and L. S. Zevin, Kristallografiya, No. 6 (1956).
268. Yu. K. Ioffe, Kristallografiya 1, No. 2 (1956).
269. Tables for laboratory work in x-radiography, Moscow Steel Institute (1959).
270. L. Azaroff and R. Buerger, Powder method in x-ray crystallography, New York (1958).
271. F. Ebert, Z. Krist. No. 1 (1958).
272. L. I. Gurarii, Current x-ray apparatus, Moscow (1957).
273. W. P. Davey, Study of crystal structure and its applications, New York (1934).
274. C. W. Bunn, Chemical crystallography, London (1945).
275. M. A. Porai-Koshits, A practical course of x-ray structural analysis, Vol. 2, Moscow (1960).
276. N. V. Belov, Tr. Inst. Kristallogr. Akad. Nauk SSSR, No. 9, 277 (1954).
277. L. Fankuchen, Z. Krist. 90, 284 (1935).
278. H. T. Evans and M. G. Ekstein, Acta Cryst. 5, 540 (1952).
279. L. E. Copeland and R. H. Bragg, Anal. Chem. 30, 196 (1958).
280. E. I. Bodneva and A. A. Katsnel'son, Zavodsk. Lab. 26, 1014 (1960).
281. M. N. Borodkina, Zavodsk. Lab. 26, 491 (1960).
282. V. P. Ovcharov, F. I. Chuprinin, and P. P. Petrosyan, Zavodsk. Lab. 26, 496 (1960).
283. L. B. Nepomnyashchii, V. I. Sushin, and T. V. Treskunova, Zavodsk. Lab. 26, 498 (1960).
284. O. S. Osipov and D. M. Kheiker, Zavodsk. Lab. 26, 363 (1960).
285. V. R. Golik, Zavodsk. Lab. 26, 364 (1960).
286. V. A. Landa, Zavodsk. Lab. 26, 71 (1960).

287. A. S. Kagan and Ya. S. Umanskii, Zavodsk. Lab. 26, 108 (1960).

288. N. N. Serebrennikov, R. P. Kreptsis, and P. V. Gel'd, Zavodsk. Lab. 26, 109 (1960).

289. A. G. Khachaturyan, Kristallografiya 5, 354 (1960).

290. V. M. Kardonskii, Kristallografiya 5, 359 (1960).

291. W. A. Wooster, Kristallografiya 5, 375 (1960).

292. L. G. Popov, Papers from the Fedorov Scientific Meeting, 5 (1952).

293. P. M. Wolff, Acta Cryst. 11, 664 (1958).

294. A. S. Kagan, V. A. Somenkov, and Ya. S. Umanskii, Kristallografiya 5, 468 (1960).

295. L. V. Tikhonov, Kristallografiya 5, 194 (1960).

296. L. V. Tikhonov, Dokl. Akad. Nauk SSSR 122, 389 (1958).

297. Z. W. Wilchinsky, Acta Cryst. 4, No. 1 (1951).

298. Ya. S. Umanskii, X-radiography of metals, Moscow (1960).

299. R. P. Ozerov et al., Kristallografiya 5, 317 (1960).

300. V. S. Kagan, B. G. Lazarev, G. S. Zhdanov, and R. P. Ozerov, Kristallografiya 5,
      320 (1960).

301. M. M. Umanskii, V. V. Zubenko, and Z. K. Zolina, Kristallografiya 5, 51 (1960).

302. E. P. Kostyukova, Kristallografiya 4, 826 (1959).

303. L. S. Palatnik, Uch. Zap. Khar'kovsk. Gos. Univ. 115 (1958).

304. A. I. Alekseeva, Zavodsk. Lab. No. 3 (1952).

305. L. S. Palatnik and B. G. Boiko, Fiz. Metal. i Metalloved. 8, 318 (1959).

306. M. Hansen and A. Anderko, Structure of binary alloys, New York (1958).

307. C. R. Smith, J. Appl. Phys. No. 5 (1946).

308. A. Carapella, J. Appl. Phys. No. 3 (1939).

309. V. I. Iveronova, Zh. Tekhn. Fiz. 4, 459 (1934).

310. A. Wilson, Acta Cryst. 4, No. 1 (1951).

311. G. Kaye and T. Laby, Tables of physical and chemical constants, 12th ed., John
      Wiley, New York (1960).

312. V. I. Arkharov, Crystallography of the quenching of steel, Moscow (1948).

313. H. Stroppe, Freiberger Forschung B, 136 (1960).

314. H. Taylor, Brit. J. Appl. Phys. No. 3 (1960).

315. A. Taylor, Introduction to x-ray metallography, New York (1945).

316. Ya. S. Umanskii and A. I. L'vovskaya, Zh. Tekhn. Fiz. No. 3 (1955).

317. G. S. Zhdanov, Izvest. Akad. Nauk SSSR, Ser. Fiz., 17, 156 (1953).

318. V. I. Arkharov, Izvest Akad. Nauk SSSR, Ser. Fiz., 17, 145 (1953).

319. G. S. Zhdanov, Coll.: X-ray methods of examination in the chemical industry,
      Moscow (1953), p. 5.

320. V. L. Karpov, Coll.: X-ray methods of examination in the chemical industry,
      Moscow. (1953), p. 22.

321. N. A. Shishakov, Coll.: X-ray methods of examination in the chemical industry,
      Moscow (1953), p. 37.

322. G. V. Kurdyumov, Zs. Phys. 43, 921 (1927).

323. L. I. Lysak, Problems of the physics of metals and of metal science, No. 3, Kiev
      (1953).

324. M. P. Arbuzov, Fiz. Metal. i Metalloved. 8, 110 (1959).

325. V. A. Il'ina, V. K. Kritskaya, and G. V. Kurdyumov, Problems of metal science and
      of the physics of metals, Moscow (1951), p. 222.

326. L. I. Mirkin, Izvest. Vyssh. Uch. Zaved., Chem. Metallurg., No. 11, 93 (1959).

327. F. Wever and W. E. Schmid, Mitt. Kais.-Wilhelm Inst. Eisenforsch. 11 (1929).

328. F. Wever, Trans. AIME 93, 51 (1931).

329. Yu. I. Sozin, Fiz. Metal. i Metalloved. 8, 240 (1959).

330. Yu. I. Sozin, Fiz. Metal. i Metalloved. 9, 892 (1960).

331. H. Jagodzinsky and K. Wohlleben, Zs. Elektrochem. 64, 212 (1960).

332. W. I. Sturm. Phys. Rev. 71, 757 (1947).

333. R. Sawyer, et al., Phys. Rev. 72, 109 (1947).

334. Yu. G. Abov, Papers from the meeting of the Academy of Sciences of the USSR on the Peaceful Use of Atomic Energy, Moscow (1955).

335. R. Lowde, Nature 167, 243 (1951).

336. B. G. Lyashchenko et al., Kristallografiya 2, 65 (1957).

337. B. G. Lyashchenko et al., Kristallografiya 3, 148 (1958).

338. A. W. McReynolds and T. Riste, Phys. Rev. 95, 1161 (1954).

339. V. I. Goman'kov, S. N. Kasatkin, S. V. Kiselev, A. A. Loshmanov, and R. P. Ozerov, Pribory i Tekhn. Eksperim. (1960).

340. L. B. Borst and V. L. Soilor, Rev. Sci. Instr. 24, 141 (1953).

341. E. O. Wollan and W. C. Kohler, Phys. Rev. 100, 545 (1955).

342. Yu. Ya. Konakhovich and I. S. Panasyuk, Pribory i Tekhn. Eksperim., No. 3, 26 (1959).

343. G. E. Bacon and R. F. Dyer, Rev. Sci. Instr. 36, 419 (1959).

344. G. E. Bacon and R. F. Dyer, Rev. Sci. Instr. 32, 256 (1955).

345. V. N. Bykov, et al., Kristallografiya 2, 634 (1957).

346. I. I. Yamzin, Kristallografiya 4, 423 (1959).

347. R. Pepinsky, et al., Rev. Sci. Instr. 31, 699 (1954).

348. S. Pasternack, et al., Phys. Rev. 81, 326 (1951).

349. R. J. Weiss, et al., Phys. Rev. 81, 863 (1951).

350. R. A. Alikhanov, Zh. Eksperim. i Teoret. Fiz. 36, 1690 (1959).

351. D. F. Litvin, Kristallografiya 4, 663 (1959).

352. R. Maddin and W. R. Asher, Rev. Sci. Instr. 27, 881 (1950).

353. V. N. Bykov and V. A. Levdik, Pribory i Tekhn. Eksperim., No. 6, 113 (1959).

354. Yu. G. Abov and B. A. Averkin, Pribory i Tekhn. Eksperim. (1960).

355. R. A. Alikhanov, Zh. Eksperim. i Teoret. Fiz. 38, 806 (1960).

356. R. Lowde, Rev. Mod. Phys. 30, 69 (1958).

357. B. W. Brockhouse, Bull. Am. Phys. Soc. 3, 233 (1958).

358. Yu. G. Abov and D. F. Litvin, Pribory i Tekhn. Eksperim. No. 3, 3(1960).

359. N. P. Glazkov, Pribory i Tekhn. Eksperim., No. 3, 16 (1960).

360. Yu. G. Abov et al., Pribory i Tekhn. Eksperim., No. 4, 51 (1960).

361. J. Cermak, Czech. J. Phys. 12, 87 (1959).

362. E. R. Pike, Acta Cryst. 10, 215 (1960).

363. E. R. Pike and A. J. C. Wilson, Brit. J. Appl. Phys. 10, 57 (1959).

364. J. Laddell, W. Parrish, and J. Taylor, Acta Cryst. 12, 561 (1959).

365. A. Fingerland, Czech. J. Phys. 10, 233 (1960).

366. A. Franks, V Intern. Congress IUC, Cambridge (1960).

367. W. Parrish and J. Laddell, V Intern. Congress IUC (1960).

368. M. Cernohorsky, Czech. J. Phys. 10, 225 (1960).

369. V. Synecek and M. Simerska, Czech. J. Phys. 10, 240 (1960).

370. O. Kratky, G. Porod, and L. Cahovec, Zs. Elektrochem. 55, 53 (1951).

371. R. E. Franclin, Acta Cryst. 3, 158 (1950).

372. V. Synecek, V Intern. Congress IUC, Cambridge (1960).

373. J. H. Hardwich, V Intern. Congress IUC, Cambridge (1960).

374. J. M. Cowley and S. Kuwabara, V. Intern. Congress IUC (1960).

375. P. W. Schmidt and R. Hight, Acta Cryst. 13, 480 (1960).

376. V. Synecek, Acta Cryst. 13, 378 (1960).

377. E. J. W. Whittaker, Acta Cryst. 6, 222 (1953).

378. L. V. Azaroff, Acta Cryst. 8, 701 (1955).

379. W. L. Bond, Acta Cryst. 12, 375 (1959).

380. L. V. Azaroff, Acta Cryst. 9, 315 (1956).

381. H. A. Levy and R. D. Ellison, Acta Cryst. 13, 270 (1960).

382. B. E. Warren, Acta Cryst. 12, 837 (1959).

383. L. Cavalca and M. Nardelli, Acta Cryst. 12, 701 (1959).

384. J. Laddell, et al., Acta Cryst. 12, 567 (1959).

385. G. Albrecht, Rev. Sci. Instr. 10, 221 (1939).

386. D. E. Henschaw, Acta Cryst. 11, 302 (1958).

387. D. Rogers and R. H. Moffett, Acta Cryst. 9, 1037 (1956).

388. D. K. Smith, Acta Cryst. 12, 479 (1959).

389. L. I. Mirkin, Izvest. Vyssh. Uch. Zaved., Fiz., No. 5 (1960).

390. M A. Blokhin, Methods of x-ray spectral examination, Moscow (1959).

391. S. S. Tolkachev, Tables of interplanar distances, Leningrad (1955).

392. P. D. Dankov, D. V. Ignatov, and N. A. Shishakov, The electron-diffraction examination of oxide and hydroxide films on metals, Moscow (1953).

393. X-ray methods of identification and the crystal structure of minerals and clays, Moscow (1955).

394. Tabulated data on the structures of borides and silicides, ed. by A. E. Koval'skii, Moscow (1958).

395. National Bureau of Standards, Circular 539, Suppl. 1-6, New York (1950-1956).

396. M. A. Krivoglaz, Zh. Eksperim. i Teoret. Fiz. 34, 204 (1958).

397. B. N. Brockhouse and A. T. Stewart, Rev. Mod. Phys. 30, 236 (1958).

398. M A. Krivoglaz and E. A. Tikhonova, Ukr. Fiz. Zh. 5, 174 (1960).

399. M. A. Krivoglaz and E. A. Tikhonova, Ukr. Fiz. Zh. 3, 297 (1958).

400. M. A. Krivoglaz, Fiz. Metal. i Metalloved. 7, 650 (1959).

401. M. A. Krivoglaz, Fiz. Metal. i Metalloved. 8, 648 (1959).

402. M. A. Krivoglaz, Fiz. Metal. i Metalloved. 8, 514 (1959).

403. M. A. Krivoglaz, Fiz. Metal. i Metalloved. 10, 169 (1960).

404. V. I. Iveronova, I. I. Popova, and G. P. Revkevich, Kristallografiya 5, 530 (1960).

405. D. Batsur', V. I. Iveronova, and G. P. Revkevich, Izvest. Akad. Nauk SSSR, Ser. Fiz., 23, 591 (1959).

406. M. A. Krivoglaz, Kristallografiya 4, 813 (1960).

407. B. M. Rovinskii, Zh. Tekhn. Fiz. 16, No. 11 (1946).

408. B. M. Rovinskii and L. M. Rybakova, Izvest. Akad. Nauk SSSR, Ser. Fiz. 15, No. 1 (1951).

409. O. B. Hirsch and J. N. Kellar, Acta Cryst. 5, No. 2 (1952).

410. Ya. M. Golovchiner, R. A. Landa, and L. M. Khalin, Problems of metal science and of the physics of metals, No. 5 (1958).

411. V. I. Iveronova, Dissertation for doctorate, Moscow (1948).

412. D. Batsur', Dissertation, Moscow (1959).

413. V. Vand, J. Appl. Phys., 26, 1191 (1955).

414. G. M. Plavnik and B. M. Rovinskii, Fiz. Tverd. Tela 2, 1099 (1960).

415. S. L. Nudel'man, Zh. Tekhn. Fiz. 5, 773 (1952).

416. J. Fortey and E. Kohen, J. Sci. Instr. 31, 11 (1954).

417. S. S. Kvitka, Kristallografiya 1, 485 (1958).

418. R. C. Evans, P. B. Hirsch, and J. N. Kellar, Acta Cryst. 1, 124 (1948).

419. J. Fankuchen, Phys. Rev. 53, 210 (1938).

420. D. W. Berreman, Rev. Sci. Instr. 26, 1048 (1955).

421. B. Ya. Pines, Coll.: 70th birthday of A. F. Ioffe (1950), p. 448.

422. D. W. Berreman, J. W. M. DuMond, and P. E. Marmier, Rev. Sci. Instr. 25, 1219 (1954).

423. B. E. Warren, J. Appl. Phys. 25, 814 (1954).

424. E. A. W. Mueller, Arch. f. Techn. Mess. 74-11, 74-12 (1951).

425. A. Francs, Proc. Phys. Soc. 68, 1054 (1955).

426. J. Hildebrand, Fortschr. der Phys. 4, 1 (1956).

427. J. A. Lely and T. W. van Russell, Philips Cechn. Rundschau 13, 66 (1951).

428. S. S. Kvitka, E. V. Kolontsova, and M. M. Umanskii, Izvest. Akad. Nauk SSSR, Ser. Fiz., 16, 372 (1952).

429. H. Lipson, J. B. Nelson, and D. P. Riley, J. Sci. Instr. 22, 184 (1945).

430. R. Renninger, Acta Cryst. 7, 677 (1954).

431. M. M. Umanskii, Apparatus for x-ray structure work, Fizmatgiz (1960).

432. S. L. Nudel'man, Pribory i Tekhn. Eksperim. 4, 83 (1957).

433. S. L. Nudel'man, Pribory i Tekhn. Eksperim. 6, 125 (1959).

434. L. I. Mirkin, Izvest. Vyssh. Uch. Zaved., Mashinostroenie, No. 6, 158 (1959).

435. H. Weyerer, Acta Cryst. 13, 821 (1960).

436. M. E. Straumanis, Acta Cryst. 13, 818 (1960).

437. M. Cernohorsky, Acta Cryst. 13, 823 (1960).

438. M. Wilkins, Acta Cryst. 13, 826 (1960).

439. M. H. Mueller, Acta Cryst. 13, 828 (1960).

440. H. Barth, Acta Cryst. 13, 830 (1960).

441. J. Cermak, Acta Cryst. 13, 832 (1960).

442. O. N. Shivrin, Kristallografiya 5, 797 (1960).

443. D. H. Kobe, Acta Cryst. 13, 767 (1960).

444. R. Asimov, Acta Cryst. 13, 510 (1960).

445. V. I. Iveronova and A. A. Katsnel'son, Kristallografiya 5, 795 (1960).

446. H. Fischmeister and A. Niggli, Acta Cryst. 13, 508 (1960).

447. S. L. Nudel'man, Kristallografiya 5, 819 (1960).

448. R. R. Spenser and J. R. Smith, Nucl. Sci. and Engng. 8, 393 (1960).

449. L. I. Mirkin, Fiz. Metal. i Metalloved. 9, 459 (1960).

450. L. I. Mirkin, Izvest. Vyssh. Uch. Zaved., Chernaya Metallurg., No. 11, 93 (1959).

451. L. I Mirkin, Fiz. Metal. i Metalloved. 10, 312 (1960).

452. L. I. Mirkin, Izvest. Vyssh. Uch. Zaved., Chernaya Metallurg., No. 12 (1960).

453. R. E. Watson and A. J. Freeman, Acta Cryst. 14, 27 (1961).

454. A. J. Freeman and R. E. Watson, Acta Cryst. 14, 231 (1961).

455. E. J. Myers and F. C. Davies, Acta Cryst. 14, 194 (1961).

456. A. V. Kuznetsov and Yu. S. Terminasov, Kristallografiya 6, 111 (1961).

457. J. Laddell, Acta Cryst. 14, 47 (1961).

458. L. I. Mirkin, Fiz. Metal. i Metalloved. 11, No. 3 (1961).

459. V. Luzatti and H. Benoit, Acta Cryst. 14, 297 (1961).

460. V. I. Iveronova and A. A. Katsnel'son, Fiz. Metal. i Metalloved. 11, 40 (1961).

461. M. A. Rumsh and T. M. Zimkina, Kristallografiya 6, 56 (1961).

462. A. O. Konnor and J. Sosnowski, Acta Cryst. 14, 292 (1961).

463. D. M. Vasil'ev and B. I. Smirnov, Uspekhi Fiz. Nauk 73, 503 (1961).

464. M. V. Schwarz and O. Summa, Praktische Auswertungshilfsmittel für Feinstrukturun-
       tersuchungen, München (1932).

SUBJECT INDEX